STP 169C

Significance of Tests and Properties of Concrete and Concrete-Making Materials

Paul Klieger and Joseph F. Lamond,
editors

ASTM Publication Code Number (PCN)
04-169030-07

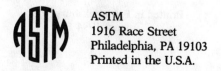

ASTM
1916 Race Street
Philadelphia, PA 19103
Printed in the U.S.A.

Library of Congress Cataloging-in-Publication Data

Significance of tests and properties of concrete and concrete-making materials / Paul Klieger and Joseph F. Lamond, editors. —4th ed.
  (STP ; 169c)
  "ASTM publication code number (PCN) 04-169030-07."
  Includes bibliographical references and index.
  ISBN 0-8031-2053-2
  1. Concrete—Testing. I. Klieger, Paul. II. Lamond, Joseph F., 1933–   . III. Series: ASTM special technical publicaltion; 169c.
TA440.S556 1994
620.1'36'0287—dc20
                                                                94-16746
                                                                    CIP

Copyright © 1994 AMERICAN SOCIETY FOR TESTING AND MATERIALS, Philadelphia, PA. All rights reserved. This material may not be reproduced or copied, in whole or in part, in any printed, mechanical, electronic, film, or other distribution and storage media, without the written consent of the publisher.

Photocopy Rights

Authorization to photocopy items for internal or personal use, or the internal or personel use of specific clients, is granted by the AMERICAN SOCIETY FOR TESTING AND MATERIALS for users registered with the Copyright Clearance Center (CCC) Transactional Reporting Service, provided that the base fee of $2.50 per copy, plus $0.50 per page is paid directly to CCC, 222 Rosewood Dr., Danvers, MA 01923; Phone: (508) 750-8400; Fax: (508) 750-4744. For those organizations that have been granted a photocopy license by CCC, a separate system of payment has been arranged. The fee code for users of the Transactional Reporting Service is 0-8031-2053-2/94 $2.50 + .50.

Peer Review Policy

Each paper published in this volume was evaluated by three peer reviewers. The authors addressed all of the reviewers' comments to the satisfaction of both the technical editor(s) and the ASTM Committee on Publications.

The quality of the papers in this publication reflects not only the obvious efforts of the authors and the technical editor(s), but also the work of these peer reviewers. The ASTM Committee on Publications acknowledges with appreciation their dedication and contribution to time and effort on behalf of ASTM.

Printed in Fredericksburg, VA
August 1994

Foreword

This publication is a revision and expansion of *Significance of Tests and Properties of Concrete and Concrete-Making Materials (STP 169B)* published in 1978. That publication in turn replaced editions of *Report on Significance of Tests of Concrete and Concrete Aggregates* published in 1935, 1943, 1956, and 1966. The present publication includes a number of new materials and test methods which have been developed, or materials which have increased in importance since the 1978 edition. A most useful addition is the inclusion of two new chapters on cement prepared by authors who are members of ASTM Committee C1 on Cement. Previous editions did not contain chapters specifically devoted to cement.

As in the previous publications, chapters have been authored by individuals selected on the basis of their knowledge of their subject areas, and in most cases because of their participation in the development of pertinent specifications and test methods by ASTM Committee C9, and in some cases by ASTM Committee C1. Authors developed their chapters in conformance with general guidelines only. Each chapter has been reviewed and, where necessary, coordinated with chapters where overlap of subject matter might occur.

This latest editon, has been developed under the direction of the Executive Committee of ASTM Committee C9 on Concrete and Aggregates by coeditors Paul Klieger, Consultant on Concrete and Concrete Materials, and Joseph F. Lamond, Consulting Engineer, both members of ASTM Committee C9.

Contents

CHAPTER 1—Introduction—PAUL KLIEGER AND JOSEPH F. LAMOND ... 1

PART I
GENERAL

CHAPTER 2—The Nature of Concrete—RICHARD A. HELMUTH ... 5

CHAPTER 3—Techniques, Procedures, and Practices of Sampling of Concrete and Concrete-Making Materials—
EDWARD A. ABDUN-NUR AND TOY S. POOLE ... 15

CHAPTER 4—Statistical Considerations in Sampling and Testing—
GARLAND W. STEELE ... 23

CHAPTER 5—Variability of Concrete-Making Materials—
ANTHONY E. FIORATO ... 31

CHAPTER 6—The Role of Cement and Concrete Testing Laboratories—JAMES H. PIELERT ... 38

CHAPTER 7—Research Needs—W. L. DOLCH ... 42

PART II
FRESHLY MIXED CONCRETE

CHAPTER 8—Factors Influencing Concrete Workability—
JOHN M. SCANLON ... 49

CHAPTER 9—Air Content, Temperature, Unit Weight, and Yield—
LAWRENCE R. ROBERTS ... 65

CHAPTER 10—Making and Curing Concrete Specimens—
JOSEPH F. LAMOND ... 71

CHAPTER 11—Time of Setting—VANCE H. DODSON ... 77

CHAPTER 12—Bleeding—STEVEN H. KOSMATKA ... 88

CHAPTER 13—Cement and Water Content of Fresh Concrete—
DEBORAH J. LAWRENCE ... 112

PART III
HARDENED CONCRETE

CHAPTER 14—Concrete Strength Testing—PEGGY M. CARRASQUILLO ... 123

CHAPTER 15—Prediction of Potential Concrete Strength at Later Ages—NICHOLAS J. CARINO ... 140

CHAPTER 16—Freezing and Thawing—HOWARD NEWLON, JR., AND TERRY M. MITCHELL ... 153

CHAPTER 17—Corrosion of Reinforcing Steel—WILLIAM F. PERENCHIO ... 164

CHAPTER 18—Embedded Metals and Materials Other Than Reinforcing Steel—BERNARD ERLIN ... 173

CHAPTER 19—Abrasion Resistance—TONY C. LIU ... 182

CHAPTER 20—Elastic Properties and Creep—ROBERT E. PHILLEO ... 192

CHAPTER 21—Bond with Reinforcing Steel—LEROY A. LUTZ ... 202

CHAPTER 22—Petrographic Examination—BERNARD ERLIN ... 210

CHAPTER 23—Volume Change—P. KUMAR MEHTA ... 219

CHAPTER 24—Thermal Properties—JOHN M. SCANLON AND JAMES E. MCDONALD ... 229

CHAPTER 25—Pore Structure and Permeability—NATALIYA HEARN,
R. DOUGLAS HOOTON, AND RONALD H. MILLS   240
CHAPTER 26—Chemical Resistance of Concrete—G. W. DEPUY   263
CHAPTER 27—Resistance to Fire and High Temperature—PETER SMITH   282
CHAPTER 28—Air Content and Unit Weight of Hardened Concrete—
KENNETH C. HOVER   296
CHAPTER 29—Analyses for Cement and Other Materials in Hardened
Concrete—WILLIAM G. HIME   315
CHAPTER 30—Nondestructive Tests—V. MOHAN MALHOTRA   320

PART IV
CONCRETE AGGREGATES

CHAPTER 31—Petrographic Evaluation of Concrete Aggregated—
RICHARD C. MIELENZ   341
CHAPTER 32—Alkali-Silica Reactions in Concrete—DAVID STARK   365
CHAPTER 33—Alkali-Carbonate Rock Reaction—MICHAEL A. OZOL   372
CHAPTER 34—Degradation Resistance, Strength, and Related
Properties—RICHARD C. MEININGER   388
CHAPTER 35—Grading, Shape, and Surface Properties—
JOSEPH E. GALLOWAY, JR.   401
CHAPTER 36—Soundness, Deleterious Substances, and Coatings—
STEPHEN W. FORSTER   411
CHAPTER 37—Unit Weight, Specific Gravity, Absorption, and Surface
Moisture—ROBERT LANDGREN   421
CHAPTER 38—The Pore System of Coarse Aggregates—
DOUGLAS WINSLOW   429
CHAPTER 39—Thermal Properties of Aggregated—D. STEPHEN LANE   438

PART V
OTHER CONCRETE MAKING MATERIALS

CHAPTER 40—Hydraulic Cements—Physical Properties—LESLIE STRUBLE
AND PETER HAWKINS   449
CHAPTER 41—Hydraulic Cement—Chemical Properties—
SHARON M. DEHAYES   462
CHAPTER 42—Mixing and Curing Water for Concrete—JAMES S. PIERCE   473
CHAPTER 43—Curing and Curing Materials—EPHRAIM SENBETTA   478
CHAPTER 44—Air-Entraining Admixtures—PAUL KLIEGER   484
CHAPTER 45—Chemical Admixtures—BRYANT MATHER   491
CHAPTER 46—Mineral Admixtures—CRAIG J. CAIN   500

PART VI
SPECIALIZED CONCRETES

CHAPTER 47—Ready Mixed Concrete—RICHARD D. GAYNOR   511
CHAPTER 48—Lightweight Concrete and Aggregates—THOMAS A. HOLM   522
CHAPTER 49—Cellular Concrete—LEO A. LEGATSKI   533
CHAPTER 50—Concrete for Radiation Shielding—DOUGLAS E. VOLKMAN   540
CHAPTER 51—Fiber-Reinforced Concrete—COLIN D. JOHNSTON   547
CHAPTER 52—Preplaced Aggregate Concrete—RAYMOND E. DAVIS, JR.   562
CHAPTER 53—Roller-Compacted Concrete (RCC)—KENNETH L. SAUCIER   567
CHAPTER 54—Polymer-Modified Concrete and Mortar—
LOU A. KUHLMANN AND MICHAEL O'BRIEN   577
CHAPTER 55—Shotcrete—I. LEON GLASSGOLD   589
CHAPTER 56—Organic Materials for Bonding, Patching, and Sealing
Concrete—RAYMOND J. SCHUTZ   599
CHAPTER 57—Packaged, Dry, Cementitious Mixtures—OWEN BROWN   604

INDEXES
Subject Index   611

Introduction

Paul Klieger[1] and
Joseph F. Lamond[2]

ASTM STP 169B, "Significance of Tests and Properties of Concrete and Concrete-Making Materials," was published in 1978. ASTM Committee C9 on Concrete and Concrete Aggregates has once again decided the time was appropriate to update and revise this useful publication to reflect changes in the technology of concrete and concrete-making materials that have taken place since that time. As Robert E. Philleo said in the Introduction to the 1978 edition ". . . it embarks on a job that will never be finished." New materials have appeared on the scene, along with a greater appreciation of the capabilities of concrete as a basic construction material. Committee C9 and its subcommittees have made significant changes in many of its specifications and test methods to reflect these changes. New specifications and testing techniques have been developed to provide for informed use of new materials and new uses for concrete.

As noted in the Introduction to *ASTM STP 169B*, the 1978 edition, Committee C9 first published "Report on Significance of Tests of Concrete and Concrete Aggregates," *ASTM STP 22*, in 1935. Subsequent up-dated reports were published in 1943. *ASTM STP 169* was published in 1956, followed by *ASTM STP 169A* in 1966, and the 1978 edition, *ASTM STP 169B*.

Following this brief Introduction is a revised chapter on the Nature of Concrete which serves as a general informational chapter on the physical structure of concrete, chemical aspects of importance, and glimpses into the future for this basic construction material.

In Part I, the chapters are directed to the necessity for providing a quality product, an objective faced by all producers of products. A new chapter deals with variability of concrete-making materials and the impact of such variability on concrete quality. Revised chapters deal with statistical considerations in sampling and testing, techniques for sampling, and the role of testing laboratories in this search for quality. An updated chapter on Research Needs is also included in Part I.

Part II deals with freshly-mixed concrete, with subjects such as concrete workability, air content and yield, setting time, bleeding, cement and water contents, and making and curing test specimens. These are basic chapters which were present in *ASTM STP 169B* but have been revised and updated to reflect changes which have taken place since the last edition.

Part III concerns itself with hardened concretes. Most of these chapters were present in the last edition and have been revised and updated. A new chapter on prediction of concrete strength from early age test results has been added to reflect the need for reducing the time for acceptance-rejection to ensure quality and yet recognize the need to achieve quality within a shortened time frame.

Part IV deals with concrete aggregates. Since the last edition, there have been significant achievements in the development of a better understanding of the chemical reactions between cements and aggregates, both siliceous and carbonate aggregates. Two chapters deal with these reaction problems.

Part V includes two new chapters on Cement—Physical Properties and Chemical Properties. This is the first edition to include chapters on what is obviously a most important component in concrete. These have been provided by authors who are members of ASTM Committee C1 on Cement and are welcome and useful additions to this publication.

Part VI on Specialized Concretes contains four new chapters on fiber reinforced concrete, roller compacted concrete, polymer modified concrete, and shotcrete. These are indicative of the special needs and developments which have taken place since the last edition. The subcommittee structure of Committee C9 has been modified to accommodate these needs.

The editors, along with ASTM Committee C9 on Concrete and Concrete Aggregates, believe this new edition will serve the concrete industry well. Authors were selected by the editors and chapters were reviewed in accordance with ASTM's peer review procedures. Technical subcommittee chairmen having jurisdiction over the subjects for pertinent chapters participated informally in the review process. The editors appreciate the help and guidance of these people and the cooperation of ASTM Committee C1 on Cement in providing authors for the two chapters on cement.

[1] Consultant, Northbrook, IL 60065-2275.
[2] Consulting engineer, Springfield, VA 22151.

PART I
General

The Nature of Concrete
Richard A. Helmuth[1]

PREFACE

T. C. Powers authored the first version of this chapter, which was published in *ASTM STP 169A* in 1966. His chapter was reprinted without revision in *ASTM STP 169B* in 1978. Readers are advised that his version is still worthwhile reading. The present chapter condenses some of that work and includes more recent material.

INTRODUCTION

For thousands of years, mankind has explored the versatility of materials that can be molded or cast while in a plastic state and then harden into strong and durable products [1]. As with ceramics and gypsum plasters, lime mortars and pozzolanic concretes provided engineers with economical materials for production of diverse utilitarian and aesthetically pleasing structures. Modern concretes preserve these ancient virtures while greatly extending the range of technically achievable goals.

Concrete-Making Materials—Definitions

Concrete is defined in ASTM Terminology Relating to Concrete and Concrete Aggregates (C 125) as a composite material that consists essentially of a binding medium within which are embedded particles or fragments of aggregate; in hydraulic-cement concrete, the binder is formed from a mixture of hydraulic cement and water. Hydraulic-cement concretes are those most widely used in the United States and world wide. Hydraulic cement is defined in ASTM Terminology Related to Hydraulic Cement (C 219) as a cement that sets and hardens by chemical interaction with water and that is capable of doing so under water. Portland cement is the most important hydraulic cement. It is produced by pulverizing portland cement clinker, consisting essentially of hydraulic calcium silicates, usually by intergrinding with small amounts of one or more forms of calcium sulfate in order to control reaction rates.

Aggregate is defined in ASTM C 125 as granular material, such as sand, gravel, crushed stone, or iron blast-furnace slag, used with a cementing medium to form hydraulic-cement concrete or mortar. Detailed descriptions of these and other materials for making concrete and their effects on concrete properties are given in other chapters in this work.

Typical hydraulic-cement concretes have volume fractions of aggregate that range approximately from 0.7 to 0.8. The remaining volume is occupied initially by a matrix of fresh cement paste consisting of water, cement, and admixtures, that also encloses air voids. While the aggregates occupy most of the volume, they are relatively inert and intended to be stable. It is the cement paste matrix that undergoes the remarkable transformation from nearly fluid paste to rock-hard solid, transforms plastic concrete into an apparent monolith, and controls many important engineering properties of hardened concretes.

Scope

Hydraulic-cement concretes may be properly designed to provide properties required for widely varying applications at low life-cycle cost. If not properly designed or produced, or if exposed to service conditions not understood or unanticipated, premature failures may result. Successful use depends on understanding the nature of concrete.

The scope of this examination of the materials science of concrete is mainly confined to concretes made with portland cements, with or without mineral and chemical admixtures. The focus is mainly on how we understand concrete performance in ordinary construction practice. That understanding is based on knowledge of its constituents, and their physical and chemical interactions in different environments.

FRESHLY MIXED CEMENT PASTE AND CONCRETE

Water in Concrete

The properties of fresh cement paste (FCP) and concrete depend on the structure and properties of ordinary water, which are unusual for a substance of such low molecular weight. Each molecule has a permanent dipole moment. Strong forces of attraction between these highly polar molecules result in unusually high melting and boiling temperatures, heats of fusion and vaporization, viscosity, and surface tension [2].

---

[1] Materials research consultant, Construction Technology Laboratories, Skokie, IL 60077–1030.

In addition to dipole interactions, hydrogen bonding between water molecules and thermal agitation affect the structure of water and aqueous solutions. Hydrogen bonding causes formation of clusters of molecules, the degree of association depending on the temperature. Thermal agitation, including translational, rotational, and vibrational motions, tends to disrupt the structure.

In the liquid state, the molecules are easily oriented in an electric field so that water has a large dielectic constant (78.6 at 25°C). This orientation, as well as molecular, polarization means that the electric field strength and the forces between charged particles, such as ions in solution, are reduced to 1/78.6 relative to that in vacuum (or air). Because of its high dielectric constant, water is an excellent solvent of salts; the energy of separation of two ions in solution is an inverse function of the dielectric constant of the solvent. Ions in solution are not separate entities but have water molecules attached to them by ion-dipole bonds.

A few minutes after mixing begins, about half of the cement alkalies are dissolved so that the concentration of the alkali and hydroxyl ions may commonly be 0.1 to 0.4 mol/L, depending mainly on the water/cement ratio and the cement alkali content [3]. At 0.3 mol/L, each ion would be separated from like ions, on the average, by about 1.7 nm, or about five water molecules.

Interparticle Forces

Atoms in solids near the surface are distorted and shifted relative to their positions in the interior because of the unsatisfied atomic bonds at the surface. These distortions of the surface produce net positive or negative surface charge, and elastic excess surface free energy. Silicon in the surface of quartz in aqueous solutions attract hydroxyl ions, reduce pH, and produce surfaces with excess negative charge [4]. Particles with surface charges of the same sign repel each other in suspensions and tend to remain dispersed. Particles of opposite sign attract each other and flocculate [5].

In addition to these coulombic and polar forces, which can be attractive as well as repulsive, there are forces between solids, atoms, and molecules that are always attractive. These van der Waals, or dispersion, forces exist because even neutral bodies constitute systems of oscillating charges that induce polarization and oscillating dipole interactions [5]. The combined action of the different forces cause sorption of water molecules and ions from solution, which can neutralize surface charge and establish separation distances of minimum potential energy between solid particles [6]. The mechanical properties of both fresh (FCP) and hardened (HCP) cement pastes and concretes depend on these forces.

Fresh Cement Paste Structure

Modern portland cements have mass median particle sizes that are about 12 to 15 μm (diameter of an equivalent sphere), almost all particles being smaller than 45 μm, and very little of the cement being finer than 0.5 μm. During grinding, calcium sulfates grind faster and usually become much finer than the clinker. After mixing with water, the solid surfaces are covered by diffuse layers of adsorbed ions, oriented-dipole water molecules, and solvated ions, forming film thicknesses at least several times the size of the water molecule (0.3 nm). These films both separate and weakly bind the solid particles into the floc structure.

In FCP and concretes made with high doses of water-reducing admixtures, cement particles may become almost completely dispersed (deflocculated) because large organic molecules are adsorbed on their surfaces, displacing water films, and greatly reducing attractive forces between cement particles. Mineral admixtures that contain small percentages of ultrafine (submicron) particles may also produce dispersion of cement particles by adsorption of the ultrafine particles on the surfaces of the larger particles. This specific kind of fine-particle effect is responsible for the improved flow of many portland cement/fly ash mixtures [7,8].

The average thickness of films of water separating dispersed particles in the paste depends on the water-cement ratio (w/c) and the cement fineness. A first approximation to the average thickness of these films is given by the hydraulic radius: the volume of water divided by the specific surface area, if it assumed that the films are thin compared with the particle sizes. If that were so, the calculated thickness is 1.2 μm for cement of specific surface of 430 m^2/kg, mixed at 0.5 w/c [9]. Since the assumption is not valid for the finer-size fractions (Fig. 1) and much of the fine fraction in portland cement is composed of calcium sulfates and other phases that dissolve within minutes after mixing begins, the average film thickness for the larger particles in that paste is probably about 2 μm. For focculated particles, the films are much thinner between adjacent particles, so that much of the water is

FIG. 1—Idealized model of the structure of fresh paste with cement particles (up to 30 μm diameter) uniformly dispersed in water.

forced into relatively large cavities or capillary-like channels.

CEMENT HYDRATION AND STRUCTURE FORMATION

Early Hydration Reactions

Chemical reactions between portland cements, water, and admixtures are usually well controlled and follow a definite sequence [10]. Initial partial dissolution of alkali sulfates, calcium sulfates, aluminates, and silicates rapidly increase concentrations of ions in solution to supersaturation levels (except for alkali hydroxides), and produce calcium aluminate, sulfoaluminate, and silicate hydrates on the surfaces of the cement and other particles. Within minutes, these reactions are slowed by coatings of hydrates on the reaction surfaces that produce a "dormant" period of slow reaction as ettringite continues to form and calcium supersaturation persists. Eventually, calcium hydroxide begins to crystallize from solution, reducing the calcium concentration and accelerating the dissolution and reaction of the silicates. Setting results from the growth of acicular crystallites of calcium silicate hydrates that bridge the water-filled void spaces. Reaction of the aluminates to form ettringite continues until the readily soluble calcium sulfates are depleted. The concentration of sulfate in solution then decreases, causing dissolution of some ettringite, and a period of acceleration of hydration of the aluminate and ferrite phases.

During the reactions prior to setting, two and sometimes three kinds of volume changes occur. Sedimentation causes subsidance of the floc structure and collection of bleeding water on the top surface, if evaporation is not excessive. If the surface becomes partly dried, capillary tension in the water can cause plastic shrinkage and cracks. Chemical shrinkage is the volume change that results from formation of hydrates that have less solid volume than the volume of water and solids reacted. While the paste is plastic, the entire volume of paste undergoes chemical shrinkage. After setting, the external dimensions remain essentially fixed and additional water must be imbibed to keep the pores saturated with water. If sufficient water is not imbibed, the paste becomes self-desiccated.

If the early reactions of C_3A and C_4AF are not well controlled by sufficiently rapid dissolution of calcium sulfates, thin platy crystals of calcium aluminoferrite monosulfate hydrate (AFm) form in the solution instead of dense coatings of ettringite, the AFt or trisulfate phase, resulting in premature stiffening, or in extreme cases, "flash set" [11]. If, on the other hand, the aluminate reactions are too slow to consume all of the rapidly dissolving calcium sulfates, crystallization of fine needles of gypsum from solution will cause early stiffening referred to, in the extreme, as "false set." If there is a slight imbalance in reaction rate control, rapid slump loss of concrete may result from adsorption of water-reducing admixtures from solution by high surface area reaction products. The balance of these reaction rates may be upset in some (incompatible) cement/admixture combinations so that premature stiffening occurs, especially in low water/cement ratio concrete.

Hardening Reactions and Microstructure

Portland cements continue to react with water at diminishing rates following setting. After 24 h at room temperature, 30 to 40% of the cement is usually hydrated, forming coatings of increasing density and thickness around each particle (Fig. 2). Larger clinker particles hydrate by partial dissolution and partly by in situ reaction so that a pseudomorph of inner products is formed within the boundaries of the original grain. Depth of reaction increases with time, but at decreasing rates so that the larger particles may have unhydrated cores even after years of moist curing. The dissolved portion forms outer products in the water-filled space near the grains. The calcium silicates produce crystalline calcium hydroxide and nearly amorphous calcium silicate hydrate (C-S-H gel) that engulf crystalline phases formed by the early reactions. Capillary pores remaining in mature HCP increase in size with w/c and have diameters ranging from 10 µm to 10 nm [12].

Powers defined the hydration products of portland cements as "cement gel," recognizing that they contained both C-S-H gel and crystalline products, and essential micropores [13]. Early research showed that typical cement gels had minimum porosities of about 30%, and specific surface areas of about 200 m^2/g, as calculated by BET theory from water-vapor adsorption data obtained after first drying to remove all of the evaporable water. These studies also showed that at 0.38 w/c all of the capillary pore space was just filled by maximun density gel when all the cement was hydrated. Mixtures made with water/cement ratios less than 0.38 cannot be completely hydrated; the amount of cement that can hydrate is proportionately less because hydration virtually stops when the capillary space is filled with gel of minimum porosity. Saturated HCP made at water/cement ratios above 0.38 have remaining capillary pore space (by definition), when completely hydrated, equal to the excess above 0.38. Partially hydrated mixtures have proportionately less gel and more capillary space. Cements of different compositions behave similarly, with similar values for the constants.

Later work has shown that drying and rewetting alter the microstructure and that different adsorbates measure different surface areas. The sheet-like crystallites are imperfectly stacked and separated by interlayer adsorbed water at relative humidities down to 11% relative humidity (RH). Before drying or aging, cement gels have specific surface areas of C-S-H monomolecular sheets, about 700 m^2/g, as measured by small-angle X-ray scattering [14]. Because of the large internal surface area, the distances between solid surfaces of the pores in the gel approach the size of water molecules; most of the gel water is close to the surfaces. In such systems, it is not certain how much of the volume change of chemical shrinkage should be attributed to the reaction itself, and how much to the possible change of density of water in pores as it is adsorbed on newly created surfaces. If it is assumed [15] that the adsorbed solution has the same density as that

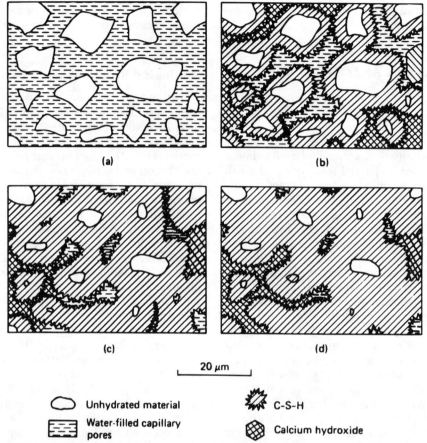

FIG. 2—Representation of microstructural development in portland cement pastes; fine particles omitted for clarity; (a) initial mix, (b) 7 days, (c) 28 days, and (d) 90 days [12]. Reprinted with permission by Academic Press.

in large pores, the apparent specific volumes of the nonevaporable (hydrate) water and solids were found to be 0.74 and 0.398 cm³/g, respectively, and minimum porosity of the gel to be 30%. The amount of chemical shrinkage is expressed in terms of the change in the apparent specific volume of the reacted water, from 0.99 to 0.74 cm³/g, and the amount, w_n, of that nonevaporable water: 0.25 w_n.

Hydration of each unit volume of cement produces about 2.2 volumes of gel. This value does not depend on the assumption concerning specific volumes. Although chemical shrinkage slightly reduces the space filling by solid hydrates, cement gel is an even more effective filler of the capillary space than the solid hydrates because of the 30% porosity of the gel.

Effects of Drying

Loss of moisture in drying atmospheres partially empties the largest capillaries at exposed surfaces. Adsorbed water remains on capillary walls as concave menisci form and progress into smaller interconnected pores. Menisci curvature and capillary tension in the remaining water are increased as the relative humidity is decreased down to about 45% RH, below which sorption effects prevail. Reductions in relative humidity slow the hydration rate; at 80% RH, hydration is insignificant. Drying causes shrinkage of hardened cements and major alterations of the gel microstructure. Shrinkage and stabilization of HCP by drying are complex and partially irreversible processes involving capillary, sorption, and dehydration effects.

Capillary tension in the pore water increases as the relative humidity decreases below that of the pore solution. For dilute solutions, tensions increase to about 97 MPa (14 000 psi) at 50% RH. At lower relative humidities, 40 to 45% RH, the tension exceeds the cohesive strength of water in capillaries and the meniscii can no longer exist [13]. Above 45% RH, capillary tensile stress in the water must be balanced by compressive stresses in the solid structure, in which stress concentrations can produce irreversible effects. When the pores are nearly water-filled, the average stress is that produced by tension in the cross-sectional area that is pore water; the resulting strain in the solid structure is the beginning of the drying shrinkage. As the capillaries empty the water-filled cross-section is reduced, but the tension increases and causes local collapse of less dense regions of the outer product, and enlargement of large pores. Desorption causes shrinkage by permitting solids to come together, and by increasing solid surface tension. Well-crystallized AFm and AFt hydrated phases also dehydrate, decreasing lattice spacings, so that elastic restraint of the shrinking C-S-H gel is reduced.

HCP cured for six months before drying at 47% RH for the first time shows both reversible and irreversible water loss and shrinkage [16]. Increased drying time causes increased water loss, shrinkage, and greatly reduced internal surface area. Rewetting causes sorption and swelling that only partly reverses the first water loss and shrinkage (Fig. 3). The irreversible component of shrinkage increases with water-cement ratio (0.4 to 0.6) from 0.2 to 0.4%, whereas the reversible component (after stabilization by drying) is only 0.2% and nearly independent of porosity. Even without drying, long-term aging in moist conditions causes age stabilization at water-cement ratios above 0.4 so that even the irreversible shrinkage tends to become porosity independent at about 0.2%. The irreversible shrinkage volume is only about half of the volume of the irreversibly lost water, if we assume its specific volume to be 0.99, which suggests that some pores emptied during drying become closed off and are not accessible during rewetting [17], or have reduced capacity. Below 11% RH, loss of interlayer water is accompanied by large partially irreversible shrinkage and water loss effects [18].

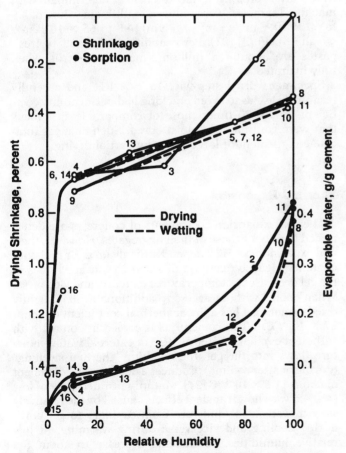

FIG. 3—Drying shrinkage, rewetting swelling, and sorption of 0.6 w/c hardened portland cement paste. Numbers indicate sequence of measurements [16]. Reprinted with permission by the Portland Cement Association.

CEMENT PASTE STRUCTURE-PROPERTY RELATIONSHIPS

Rheology of Fresh Cement Paste

When cements are mixed with sufficient water and intensity, dry agglomerates of fine particles are first dispersed and then tend to form a floc structure, which is continously broken down by mixing if the early reactions are well controlled. When mixing is stopped, the floc structure reforms until it becomes an essentially continuous floc. Cement pastes in this condition are actually weak solids with measureable shear stress yield values that depend on water/cement ratio, cement fineness, and other factors. Typical values for portland cements without admixtures range roughly from 10 to 100 Pa (0.0014 to 0.014 psi) for w/c from 0.6 to 0.35 [9]. Prior to yield, the pastes are elastic and shear strains can reach about 20 deg, indicating a rubber-like elasticity.[2] They are also plastic solids with typical values for plastic viscosity that range roughly from 10 to 100 mPa · s (centipoise). High water/cement ratio pastes seem to be liquid and may be poured easily because their yield stress values are so low. At low water/cement ratios, they are obviously plastic and can be deformed by moderate forces. Standard test pastes made at normal consistency have w/c about 0.25 and yield stress values of about 2000 Pa (0.29 psi) [9].

At ordinary temperatures, portland cement hydration reactions cause progressive stiffening and setting during the first few hours. Yield stress values increase to 2×10^4 Pa (2.9 psi) at initial set and 1×10^5 Pa (14 psi) at final set [9].

Elasticity and Creep

Hardened cement pastes are not perfectly elastic, but are viscoelastic solids. Internal friction, creep, and stress relaxation are useful in dissipating vibrational energy, and preventing excessive stress concentrations in concrete. They are a result of redistribution of moisture, viscous flow of gel, and dissolution of solids under stress and recrystallization in pores. These processes, and slow growth of cracks, are thermally activated processes in which random thermal motions provide sufficient energy, in addition to the applied stress, at sites of adsorbed water molecules, or solid-solid bonds, to exceed the bond energy. Short-term loading tests of water-saturated HCP show that creep and creep recovery versus time curves are bimodal and consist of a component with retardation times ranging from 0.2 to 2 s and a slower component that ranged over weeks. The short-time component was identified with redistribution of water in capillary pores [19]. Diffusion of strongly adsorbed and hydrate water, recrystallization, and other irreversible changes are believed to contribute to the slower processes. Long-term creep of HCP can be several times the elastic deformation.

---

[2] Unpublished work done at Construction Technology Laboratories for the Portland Cement Association under Project HR 7190.

Elastic moduli can be measured precisely by dynamic methods and are found to vary with porosity, ϵ, according to

$$E = E_0(1 - \epsilon)^3$$

in which E_0 is the modulus at zero porosity [20]. If the capillary porosity is used, E_0 is the modulus of the cement gel, about 34 GPa (5×10^6 psi) for Young's modulus of water-saturated pastes. If the total porosity (including that of the gel) is used, E_0 is an average modulus for the solids, about 76 GPa (11×10^6 psi). Equations of the same form apply for the shear and bulk moduli. Drying significantly reduces Poisson's ratio, from about 0.3 to 0.18, and the bulk modulus; stresses are carried by at least some of the water in pores.

Elastic moduli of saturated pastes increase moderately as the temperature is decreased to 0°C. At temperatures in the freezing range down to about −60°C, ice formation in capillaries increases the moduli as ice contents increase. At still lower temperatures, the moduli increase more rapidly; internal friction reaches a peak at about −90°C as the gel water viscosity increases as it approaches its glass transition temperature at −119°C [21]. The gel water (adsorbed solution) does not freeze to ice, but becomes a glassy solid.

Compressive Strength

The fraction, X, of the available space that is filled by cement gel at any stage of hydration is called the gel/space ratio. It can be calculated from the w/c, the fraction of the cement that has hydrated, and the volume of gel produced. For fully hydrated cement pastes, it is the same as $(1 - \epsilon)$ for the capillary porosity. Compressive strength, f_c, at different w/c and ages can be simply expressed as

$$f_c = f_{cg}X^n$$

in which f_{cg} is the intrinsic strength of the gel (at $X = 1$), and n has a value of about 3. Use of this equation indicated intrinsic strengths of cement gels ranging from 90 to 130 MPa (13 000 to 18 500 psi) in mortars made with five different cements [13]. However, mortars probably do not provide accurate measures of intrinsic strengths of pastes because of transition zones at aggregate surfaces. Tests of cement pastes yielded higher strengths at gel/space ratios calculated to be equal to those of mortars made with the same cement.

More recent testing of pastes made with both normally ground and controlled-particle-size-distribution (CPSD) portland cements has shown that intrinsic strengths of the gel do not depend on cement particle size distributions over the range investigated, although rates of strength development do. However, paste strengths at several ages, defined different straight lines for each w/c when plotted against X^3, and indicated intrinsic strengths of 134 and 97 MPa (19 400 and 14 000 psi) at 0.36 and 0.54 w_0/c, respectively [22]. This result indicates that the intrinsic strength of the gel formed at w/c above about 0.38 decreases with increasing w/c, in contrast with Powers' mortar data at different w/c.

If we consider fresh cement pastes to have strengths equal to their yield stress values, typically 10 to 100 Pa, and ultimately harden to compressive strengths of 10 to 100 MPa, the increase is about one million times.

Permeability

HCP and concretes are porous and permeable to water, dissolved material, and gases. When water-saturated, flow is proportional to hydraulic pressure differences, if corrections are made for osmotic effects; concentration gradients cause osmotic flow to higher concentrations, and diffusion of ions to lower concentrations. When partly dried, relative humidity and moisture gradients cause flow because of capillary tension, and diffusion along surfaces, and in the vapor phase. The same changes of microstructure that cause great changes of elastic moduli and strength of cement pastes during hardening cause reductions of permeability.

Permeability coefficients of fresh portland cement pastes of 0.5 and 0.7 w/c, calculated from bleeding data, range from 5.7 to 20×10^{-5} m³/(s · m² · MPa/m) [6.1 to 22×10^{-4} in.³/(s · in.² · psi/in.)], respectively. These coefficients for hardened pastes of the same water-cement ratios after prolonged moist curing, determined with machined samples, were reduced to ultimate values of 4.5 to 60×10^{-12} m³/(s · m² · MPa/m) [4.8 to 65×10^{-11} in.³/(s · in.² · psi/in.)] [23]. Permeability coefficients of fresh pastes are about ten million times as great as when fully hydrated.

Specimens dried step-wise to 79% RH and carefully resaturated so as to avoid cracking had permeability coefficients about 70 times those of comparable specimens that were continuously moist cured. Such changes indicate enlargement of large pores by partial drying.

Thermal Expansion

Thermal expansion coefficients of concretes are determined mainly by those of their aggregates. However, thermal expansions of HCP depend strongly on their moisture contents because retention of water by surface forces in the gel decreases as temperatures increase, and vice versa. When cooled without access to additional water, slightly dried cement gel has a linear thermal coefficient of about 27×10^{-6}/°C. When cement gel is cooled in contact with sufficient capillary water in HCP or external water, moisture flows into the gel; the resulting "thermal swelling" (during or after cooling) produces a net thermal coefficient of about 11.6×10^{-6}/°C [24]. Mature saturated HCP of low (< 0.45) w/c show transient effects caused by the relatively slow movement of moisture from capillary to gel pores during cooling and vice versa during warming. At low relative humidities, coefficients decrease to about the same value as for saturated pastes. Such differences between thermal expansion coefficients of pastes and aggregates may cause excessive local stresses in concretes unless relieved by creep.

CONCRETE AGGREGATES

The major constituents of ordinary concretes are crushed rocks or gravels and sands used as coarse and fine aggregates. Materials used in concrete usually need to be processed to be of proper size gradation and relatively free of deleterious substances.

Specific Gravity and Porosity

It is useful to classify aggregates by specific gravity and porosity into lightweight, ordinary, and heavy-weight materials (ASTM C 125). Lightweight aggregates are used to reduce dead loads and stresses, especially in tall structures, and to provide thermal insulation. Heavy-weight aggregates are used mainly for radiation shielding. Ordinary aggregates, such as sandstone, quartz, granite, limestone, or dolomite, have specific gravities that range from about 2.2 to 3.0. Densities of ordinary concretes range from about 2.24 to 2.4 Mg/m^3 (140 to 150 lb/ft^3).

Porosity reduces weight, elastic moduli, and strength of aggregates and concretes, although the effect on strength may be significant only in high-strength concrete. Porosity increases permeability to fluids and ions in pore solutions, especially if the pores are open (interconnected) rather than closed. Freezing of water in pores in aggregate particles in moist concrete can cause surface pop-outs or D-cracking in concrete pavements [12].

Strength of Aggregate Particles

Strength test results of individual samples of rock from any one source show wide variations that are caused by planes of weakness, and their different orientations, in some of the samples. Such weaknesses in the rock samples may not be significant once the rock has been crushed to the sizes used in concrete so that the higher, or at least average, strengths may be more significant. Ten different common types of rock tested at the Bureau of Public Roads had average compressive strengths that ranged from 117 MPa (16 900 psi) for marble to 324 MPa (47 000 psi) for felsite [25]. A good average value for concrete aggregates is about 200 MPa (30 000 psi), but many excellent aggregates range in strength down to 80 MPa (12 000 psi) [26]. These values are generally above strengths of ordinary concretes.

Permeability

Measurements of coefficients of permeability to water of selected small (25-mm (1-in.) diameter) pieces of rock, free of visible imperfections, yield values several orders of magnitude smaller than for larger specimens, which probably contained flaws [23]. Values for the small specimens ranged from 3.5×10^{-13} $m^3/(s \cdot m^2 \cdot MPa/m)$ [3.8×10^{-12} $in.^3/(s \cdot in.^2 \cdot psi/in.)$] for a dense trap rock, to 2.2×10^{-9} $m^3/(s \cdot m^2 \cdot MPa/m)$ [2.4×10^{-8} $in.^3/(s \cdot in.^2 \cdot psi/in.)$] for a granite. These values are equal to those measured for mature hardened portland cement pastes made at water cement ratios of 0.38 and 0.71, respectively, despite the low (less than 1%) porosities of these rocks.

CONCRETE PROPORTIONING, STRUCTURE, AND PROPERTIES

Proportioning and Consistency

Two basically different kinds of concrete mixtures must be distinguished. Nonplastic mixtures made with relatively small amounts of water show considerable bulking as water is added, and after compaction have sufficient strength to support their own weight. The concrete block industry is based on such nonplastic but cohesive mixtures. Void space in such mixtures is relatively high and filled mostly by air. The strength of the cohesive mixture results from capillary tension under meniscii bounding the water films on and between the solid particles, and solid surface forces. Strength and bulking increase to a maximum as water is added, and then decrease as the void space becomes nearly water filled and capillary tension is diminished. With sufficient water, the mixture is wetted so that surface meniscii and capillary tension disappear, void contents reach a minimum, and limited plastic deformation becomes possible. The remaining cohesive force is a result of interparticle attraction between closely spaced fine particles. This minimum void space contains about 12% air when such mixtures are compacted by ordinary means, and the cement content is not below a certain limit. The water content at minimum voids content is called the "basic water content" [6].

The consistency of cement paste at its basic water content is nearly the same as the normal consistency as defined in ASTM standards. Normal consistency pastes, and mortar or concrete mixtures made with different aggregates at their basic water contents, have slump values of about 42 mm (1.7 in.) in the standard test. Such concretes are much stiffer than the plastic mixtures commonly used in American practice that usually contain chemical admixtures and higher water contents. Further additions of water increase void volume, reduce interparticle forces, and increase the capacity for plastic deformation.

The main effect of adding increments of aggregate to paste is to reduce the amounts of voids and cement per unit volume. The total effect is not just that of volume displacement, because the cement paste void space is dilated by the added aggregate surfaces, as described in the next section. Also, when aggregate is introduced, plastic strains in the paste during compaction are necessarily greater and the mixture is stiffer than the paste. If such additions are begun using cement paste of the standard normal consistency, and if the same compacting force or energy is applied to the mixtures as to the paste, that consistency can be maintained constant if increments of water are added with each increment of aggregate. The ratio of the volume of water plus air to the total volume of solids (voids ratio) decreases with added aggregate, but not as much as without the added water, until a minimum voids content is reached, and then increases.

Consistency of concrete depends on consistency of cement paste as well as on dispersion of aggregate by sufficient paste volume for each particular aggregate. Although concrete yield stress values can be calculated from slump values, there is as yet no valid method of calculation of concrete slumps from paste yield stress values for concretes made with different aggregates and proportions. For fixed proportions, the stiffer the paste, the stiffer the concrete, and such changes can be calculated from paste data for ordinary concretes.[3] In mixtures that are relatively rich in cement and paste volume, adding increments of aggregate does not greatly increase water requirements for flow. In leaner mixtures, particularly those with aggregate contents above those at minimum voids ratios, but below those very lean mixtures that require excess amounts of entrapped air, the water requirement is proportional to the volume fraction of aggregate in the total solids [6]. This range comprises much of the concrete made for ordinary use.

Structure

The fine structure of fresh and HCP were described in earlier sections. However, the HCP matrix in mortars and concretes differ in pore structures from those of neat HCP usually used for study [27–29]. Differences are caused by (1) entrapped air in normally compacted mixtures, (2) deficiencies of large cement particles and higher porosity at aggregate surfaces (Fig. 4) because particles cannot pack as in bulk paste, and (3) sedimentation of cement particles prior to setting that releases bleeding water from the paste under larger aggregate particles, creating voids, especially at high water/cement ratios. Mineral admixtures such as fly ash and condensed silica fume reduce the latter two defects and greatly reduce permeability [8].

The coarse structure of concrete is determined by the volume and particle size grading of the total aggregate, the volumes of cement, admixtures, water, and air in the concrete mixture, mixing and placement or sampling procedures, and curing conditions. Over a considerable range of proportions of adequately mixed acceptable materials, concretes form cohesive, plastic, workable mixtures. Such mixing results in cement and admixture particles being dispersed in water and aggregate particles being desegregated and distributed in a three-dimensional, apparently random, array in a matrix of paste and air voids. However, systematic local deviations are forced by flat bounding surfaces of forms where larger particles cannot pack together as they can within the mass, so that smaller particles fill more of the space adjacent to forms. Similar effects occur within the mass at surfaces of particles that are much larger than their neighbors. During hardening of the paste matrix, concrete becomes more nearly homogeneous as the paste properties approach those of the aggregates.

In order for concrete to possess plasticity, the aggregate must be dispersed by a sufficient volume of cement paste to permit deformation under shear stress. For any aggregate size gradation, the minimum voids ratio indicates the volume required to fill the voids in compacted (dry-rodded) aggregate. If the concrete is plastic, it must contain an excess volume of paste and air above that minimum to disperse the aggregate, that is, to provide some separation between particles that would be in contact in the absence of paste. Fine aggregate disperses coarse aggregate but also reduces average paste film thicknesses. For concretes made with nearly the same voids ratios (about 0.20), at 75 to 100 mm (3 to 4 in.) slump, and different aggregate finenesses, Powers calculated the minimum separation distances between aggregate particles from the excess paste volumes, by two different methods with dissimilar results [6]. The average values by the two different methods ranged from 26 μm to 121 μm for lean to rich mixtures, respectively, the latter having the highest percentage of fine aggregate (43%) and being close to the minimum voids ratio. Such results indicate that many concretes, especially very lean mixtures, suffer from poor workability because of particle interference to flow by the larger ($\geq$ 30 μm) cement particles. This indication has been confirmed by recent research. Although cement pastes made with cements of 30-μm maximum particle size were stiffer than those made with ordinary cements, improved flow was obtained using such CPSD cements in standard mortars and ordinary (not lean) concretes [30]. Particle interference by large particles is also one of the reasons that some fly ashes increase water requirements of concretes [8].

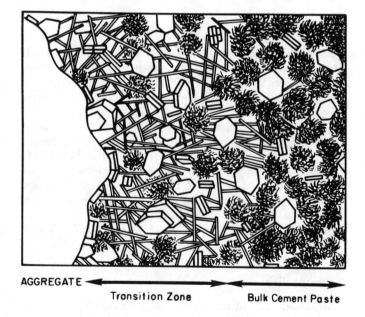

FIG. 4—Representation of transition zone at paste/aggregate interface in concrete, showing more coarsely crystalline and porous microstructure than in interzonal mass [27]. Reprinted with permission by P.K. Mehta.

---

[3] As yet unpublished research at Construction Technology Laboratories for the Portland Cement Association (Project HR 7190) and W. R. Grace & Co., (Project CR 7895) on test methods for early stiffening and slump loss.

Hardened Concrete Properties

Other chapters in this publication give comprehensive treatments of many properties of hardened concretes. Some details are noted here to relate previous sections to specific properties of the HCP/aggregate composite.

In short-term loading tests for compressive strength, stress/strain relationships for aggregates and HCP can be sensibly linear up to near the compressive strength of HCP, while those for concrete exhibit curvature with increasing strain and pseudo-plasticity at stresses above about 0.4 times strength; microcracking at paste-aggregate interfaces develops progressively with increasing strain [26].

Creep of many concretes, except possibly those loaded at early ages, is proportional to stress/strength ratios up to 0.3 to 0.6; microcracking also begins in about this same range, depending on the heterogeneity of the mixture. Mortars, for example, exhibit proportionality up to perhaps 0.85. In concretes, stress/strength ratios near this value produce failure in time [26].

Long-term durability of concrete depends strongly on exposure and service conditions, concrete properties, especially porosity and pore structure, and structural design. Exposure to acidic or neutral waters causes leaching of calcium hydroxide from HCP, which increases its porosity and permeability, and can eventually soften even the much less soluble C-S-H. Sulfates in fresh waters can penetrate into concrete to cause sulfate attack by reaction with aluminate phases in HCP to produce ettringite, which can be destructively expansive if there is insufficient space to accommodate the volume expansion of that highly hydrated reaction product. Calcium and sodium chloride solutions react to form Friedel's salt and other complex salts, some of which are also expansive reactions under some conditions. Seawater causes leaching and contains sodium, chloride, magnesium, and sulfate ions in amounts sufficient to cause significant reactions, but the main effect is that of erosion or loss of constituents [1].

Moisture and freezing temperatures can cause damage by ice formation in large pores in the HCP and some aggregates. Ice formation at frozen surfaces is propagated through capillaries large enough to freeze. Because of ice/water interfacial tension, smaller capillaries require lower temperatures to be penetrated by the growing tips of ice crystals. These crystals in frozen capillaries, and also those in entrained air voids, grow by osmotic accretion of ice formed by diffusion of water from the gel pores. If entrained air voids are closely spaced, so that their void spacing factors are less than about 0.20 mm (0.008 in.), diffusion of moisture from both capillary and gel pores to ice in the air voids dries the HCP and prevents excessive expansions, otherwise caused by ice formation in capillary pores [31].

Approximate values for many properties of hardened concretes can be calculated from relatively simple equations for composites if values for properties of their aggregates and HCP (and water contents) are known, or can be estimated with reasonable accuracy. Density and heat capacity are simple examples. In recent years, our understanding of the nature of concrete has been formalized and exploited through mathematical modeling, despite differences between structures of bulk HCP and HCP in mortar and concrete. Predictive models are now widely used for engineering design for creep, shrinkage, and temperature effects [32]. Modeling the development of microstructure and properties of concrete has been used to predict hydration rates, total porosity reductions, pore volumes of different sizes, heat evolution, permeability and diffusion coefficients, and compressive strength of concretes of diverse compositions at different temperatures and moisture contents [33,34]. Results of variations of mix proportions and other variables in tests for alkali-silica reactivity have also been modeled with considerable success [35]. Many other examples could be cited. Such studies continue to advance our understanding and ability to design and to test concretes and concrete-making materials, for performance in specific applications.

CONCLUDING DISCUSSION

It must be recognized that the apparent successes of mathematical modeling have been achieved in a relatively narrow range of compositions and properties of portland and blended cements, for which decades of research and field experience have provided a large base of knowledge and understanding. This base has been used to write mainly prescription-based specifications for the materials. Such successes must also be contrasted with the limitations of standard test methods to guarantee performance or to yield results concerning performance problems that are not misleading. ASTM tests that have been subject to such criticism include ASTM Test Method for Autoclave Expansion of Portland Cement (C 151) [36,37], ASTM Test Method for Potential Alkali Reactivity of Cement-Aggregate Combinations (Mortar-Bar Method) (C 227) [8,38], ASTM Method for Sampling and Testing Fly Ash or Natural Pozzolans for Use as a Mineral Admixture in Portland Cement Concrete (C 311) [8], and ASTM Test Method for Effectiveness of Mineral Admixtures or Ground Blast-Furnace Slag in Preventing Excessive Expansion of Concrete Due to the Alkali-Silica Reaction (C 441) [8].

These limitations should be overcome if we are to continue to move towards total reliance on performance testing, without the prescription-based specifications on which our knowledge base has been built, especially for long-term durability. Advocates of performance testing argue persuasively that their course will permit innovation, more extensive utilization of waste materials, and advanced technology. Performance test-based specifications will open the way for formulations of cements and concrete mixtures ranging well outside even the present base, which does not yet have altogether satisfactory test methods. Research on improved test methods, and the use of those results together with mathematical modeling to predict long-term performance, should therefore be given high priority.

REFERENCES

[1] Lea, F. M., *The Chemistry of Cement and Concrete*, 3rd ed., Chemical Publishing Co., New York, 1971, pp. 1–10, and 625.

[2] "Water" in *Van Nostrand's Scientific Encyclopedia*, 5th ed., Van Nostrand Reinhold, New York, 1976, pp. 2311–2312.
[3] Gartner, E. M., Tang, F. J., and Weiss, S. J., *Journal*, American Ceramic Society, Vol. 68, No. 12, Dec. 1985, pp. 667–673.
[4] Weyl, W. A. in *Structure and Properties of Solid Surfaces*, R. Gomer and C. S. Smith, Eds., University of Chicago Press, Chicago, 1953, pp. 147–180.
[5] Adamson, A. W., *Physical Chemistry of Surfaces*, 2nd ed., Interscience Publishers, New York, 1960, pp. 209–245 and 317–379.
[6] Powers, T. C., *The Properties of Fresh Concrete*, Wiley, New York, 1968, pp. 97–110, 121–125, 156, and 392–436.
[7] Helmuth, R. in *Fly Ash, Silica Fume, Slag, and Natural Pozzolans in Concrete*, SP-91, V. M. Malhotra, Ed., American Concrete Institute, Detroit, MI, 1986, pp. 723–740.
[8] Helmuth, R. A., *Fly Ash in Cement and Concrete*, Portland Cement Association, Skokie, IL, 1987, pp. 14, 75, 76–79, 112–117, and 156–163.
[9] Helmuth, R. A. in *Principal Reports*, Seventh International Congress on the Chemistry of Cement, Paris, 1980, Paris Editions Septima, Vol. III, 1980, pp. VI-0/1–30.
[10] Seligmann, P. and Greening, N. R., *Proceedings*, Fifth International Symposium on the Chemistry of Cement, Cement Association of Japan, Tokyo, Vol. II, 1969, pp. 170–199.
[11] Taylor, H. F. W., *Cement Chemistry*, Academic Press, London, 1990, pp. 233–234.
[12] Mindess, S. and Young, J. F., *Concrete*, Prentice-Hall, Englewood Cliffs, NJ, 1981, pp. 95–99 and 566–567.
[13] Powers, T. C., *Proceedings*, Fourth International Symposium on the Chemistry of Cement, Monograph 43, National Bureau of Standards, U.S. Department of Commerce, Washington, DC, Vol, II, 1962, pp. 577–608.
[14] Winslow, D. N. and Diamond. S., *Journal*, American Ceramic Society, Vol. 57, No. 5, May 1974, pp. 193–197.
[15] Copeland, L. E. and Verbeck, G. J., "Structure and Properties of Hardened Cement Paste," *Proceedings*, Sixth International Congress on the Chemistry of Cement, English preprints, Moscow, 1974, Paper II-5.
[16] Helmuth, R. A. and Turk, D. H., *Journal*, Portland Cement Association, Vol. 9, May 1967, pp. 8–21.
[17] Parrott, L. J., Hansen, W., and R. L. Berger, *Cement and Concrete Research*, Vol. 10, No. 5, Sept. 1980, p. 647.
[18] Ramachandran, V. S., Feldman, R. F., and Beaudoin, J. J., *Concrete Science*, Heyden, Philadelphia, 1981, pp. 77–88.
[19] Sellevold, E. J., "Anelastic Behavior of Hardened Portland Cement Paste," Technical Report No. 113, Department of Civil Engineering, Stanford University, Stanford, CA, Aug. 1969.
[20] Helmuth, R. A. and Turk, D. H., *Symposium on Structure of Portland Cement Paste and Concrete*, Special Report 90, Highway Research Board, Washington, DC, 1966, pp. 135–144.
[21] Helmuth, R. A., "Investigation of the Low-Temperature Dynamic-Mechanical Response of Hardened Cement Paste," Technical Report No. 154, Department of Civil Engineering, Stanford University, Stanford, CA, Jan. 1972.
[22] Helmuth, R. A., "Phase I: Energy Conservation Potential of Portland Cement Particle Size Distribution Control," Construction Technology Laboratories Final Report, CR7523-4330, to U.S. Department of Energy, Washington, DC, March 1979.
[23] Powers, T. C., Copeland, L. E., Hayes, J. C., and Mann, H. M., *Journal*, American Concrete Institute *Proceedings*, Vol. 51, Nov. 1954, pp. 285–298.
[24] Helmuth, R. A., *Proceedings*, Highway Research Board, Vol. 40, 1961, pp. 315–336.
[25] Woolf, D. O. in *Concrete and Concrete-Making Materials*, ASTM STP 169A, American Society for Testing and Materials, Philadelphia, 1966, pp. 462–475.
[26] Neville, A. M., *Properties of Concrete*, Longman Scientific and Technical, Harlow, Essex, UK, (Wiley, New York), 3rd ed., 1987, pp. 131, 364, and 401–402.
[27] Mehta, P. K., *Concrete Structure, Properties and Materials*, Prentice-Hall, Englewood Cliffs, NJ, 1986, p. 38.
[28] Winslow, D. and Ding Liu, *Cement and Concrete Research*, Vol. 20, No. 2, March 1990, pp. 227–235.
[29] Mindness, S. in *Materials Science of Concrete I*, J. P. Skalny, Ed., The American Ceramic Society, Westerville, OH, 1989, pp. 163–180.
[30] Helmuth, R. A., Whiting, D. A., and Gartner, E. M., "Energy Conservation Potential of Portland Cement Particle Size Distribution Control Phase II," Final Report to U.S. Department of Energy, Washington, DC.
[31] Helmuth, R. A., *Proceedings*, Fourth International Symposium on the Chemistry of Cement, Monograph 43, National Bureau of Standards, U.S. Department of Commerce, Washington, DC, Vol. II, 1962, pp. 829–833.
[32] *ACI Manual of Concrete Practice*, ACI Committee 209, American Concrete Institute, Detroit, MI, 1988, pp. 209R1–92.
[33] Parrott, L. J. in *Research on the Manufacture and Use of Cements*, G. Frohnsdorff, Ed. Engineering Foundation, New York, 1986, pp. 43–74.
[34] Parrott, L. J. in *Materials Science of Concrete I*, J. P. Skalny, Ed., The American Ceramic Society, Westerville, OH, 1989, pp. 181–196.
[35] Hobbs, D. W., *Magazine of Concrete Research*, Vol. 33, 1981, pp. 208–220.
[36] Mehta, P. K., "History and Status of Performance Tests for Evaluation of Soundness of Cements," *Cement Standards—Evolutions and Trends*, ASTM STP 663, P. K. Mehta, Ed., American Society for Testing and Materials, Philadelphia, 1978, pp. 35–60.
[37] Helmuth, R. and West, P., "Reappraisal of the Autoclave Expansion Test," ASTM Subcommittee C01.31 Task Group Report, American Society for Testing and Materials, Philadelphia, Nov. 1987.
[38] Stark, D. C., *Cement, Concrete, and Aggregates*, Vol. 2, No. 2, 1980, pp. 92–94.

Techniques, Procedures, and Practices of Sampling of Concrete and Concrete-Making Materials

Edward A. Abdun-Nur[1] and Toy S. Poole[2]

PREFACE

The subject of sampling practices was covered in all three previous editions of ASTM Special Technical Publication, *ASTM STP 169*, *ASTM STP 169A*, and *ASTM STP 169B*. The chapter in *ASTM STP 169* was authored by C. E. Proudley. The chapter in *STP 169A* and *169B* was authored by E. A. Abdun-Nur. Ed had planned to write the chapter for the current edition, but passed away shortly before the effort was initiated. The current chapter is largely Ed's work from the *ASTM STP 169B*, with references updated, and some additional information added on selected topics.

INTRODUCTION

Millions upon millions of dollars change hands daily in various segments of the concrete industry, based on evidence obtained from samples. Yet, in many instances, the sample is obtained in an indifferent manner by someone who is ignorant of the basic principles of sampling, the use to which the sample is to be put, the tests to be performed on the sample, and the final decisions to be based on or derived from the test results—all very important factors that should be considered in the sampling process. There appears to be no adequate appreciation of the importance of sampling—those who must make important decisions and set policy based on test results from samples—despite the cost of the large volume of samples and testing in the concrete field and the larger economic significance of conclusions that are derived from these.

On the other hand, other industries confronted with similar problems have gone to great lengths to develop reliable and efficient methods and procedures. The coal, fertilizer, ore, and abrasives industries [1] that have sampling problems similar to those found in aggregates, for example, have developed criteria and methods that increase the probabilities of proper sampling. Basic principles and approaches that can be helpful in providing sampling details for the various segments of the concrete industry have been developed by the American Society for Testing and Materials (ASTM) Committee E11 on Statistical Methods and others [2–13]. Other sampling procedures for concrete and concrete ingredients that range from indifferent to very good may be found in various ASTM standards (see Table 1 and other handbooks and manuals [14–16].

WHAT IS SAMPLING

The Sample

A sample is a small portion of a larger universe of a material (such as a lot, shipment, stockpile, batch, carload, truckload, or continuous production stream) about which information is desired. The characteristics of the samples are presented as evidence of the properties of the larger universe from which it is taken. A series of such samples can, if properly taken, provide a pattern of the variations in properties of that universe. The measurements from many samples from the universe constitute a statistical population.

Sampling

Sampling is the process of obtaining samples from the larger universe. When the universe is perfectly homogeneous, sampling becomes the simple physical act of taking a sample from the unit about which information is desired. In this case, any sample can truly represent the larger homogeneous whole.

Unfortunately, nature, in general, and the concrete field, in particular, rarely if ever, present us with a homogeneous universe of any material or process. If the rare occasion of perfect homogeneity should occur, one would not know it ahead of time and would have to treat it as being variable. Each lot, truckload, batch, or stockpile, not only varies in some measure from similar units intended to be identically the same, but frequently varies within each such unit. The more heterogeneous the universe, the more work it takes to develop a reliable estimate of its characteristics. Therefore, it pays to save sampling and testing funds where little variability occurs and expend them where more heterogeneity is prevalent.

Sampling is thus much more than the physical act of taking a sample of the larger unit as evidence of the properties of the latter. First of all, a sampling plan must be formulated to reflect the variability characteristics of the

---

[1] Deceased.
[2] Chemist, U.S. Army Corps of Engineers, Waterways Experiment Station, Vicksburg, MS 39180-6199.

universe. After that, individual samples taken in accordance with such a plan must be obtained in such a manner that each sample is truly representative of the unit from which it comes.

This broadened concept of sampling is, in effect, an acceptance of the fact that sampling is a complex business and the needed skill lies as much or more in the one exercising overall control of the work as in the one actually taking the sample. A major reorientation of ideas on the subject is needed.

The term "probability sampling" refers to the process of selecting samples in such a way that resultant descriptive properties of the larger universe of the material is a unbiased estimate of its true properties. Establishing proper probability sampling procedures is part of the sampling plan.

Sampling Plan

A sampling plan should include information on the number, size, and location or timing (if from a production stream), and the extent of compositing, if applicable. Also, procedures for reduction of the gross sample to a lab sample should be included.

If a lot of material is variable, but that variability occurs more or less randomly in space throughout it, then samples should be taken randomly in space. This is often impossible because of the physical dimensions of the lot of material. A reasonable alternative is to take samples during loading or unloading of containers at random time intervals. This procedure relies on the assumption that the spatial variation is accurately reflected as temporal variation during loading or unloading. Automatic sampling at regular intervals is a common practice, but may hide some variation, particularly if variation also occurs at regular time intervals.

If it is known or suspected that a material varies in some predictable way, for example, if gradients over space or time or both are suspected to exist, then an appropriate sampling plan would include the construction of one or more transects through the material that run parallel to the gradient, then take samples at random intervals along that transect at random distances from the transect (at 90° to the transect),

If one has no knowledge of such predictable variation, then one should take samples completely at random. This practice will avoid bias, but it may not use the data to maximum advantage in describing variation if nonrandom variation does exist.

The amount of detail needed in describing the nature of a lot of material is a strong determinant of the number of samples needed. If it is necessary to describe all of the variation that exists in a lot of a material, a large number of samples may be required. Depending on the size of the lot and processing involved, quite a complex pattern of variation may exist and a very large number of samples may be required to completely map it. This may be necessary for quality control purposes during production, but much of this variation may be insignificant to the user for the production of reasonably uniform concrete. If it is desirable only to estimate whether or not a lot of material meets specification requirements, a much smaller number may be adequate. Determination of specification compliance requires that a mean and standard deviation be estimated and, from these parameters, the fraction of the material in a lot that meets the requirements can be calculated. For this purpose, it is necessary to have reasonable estimates of the mean and the standard deviation.

It is commonly believed that the estimate of the mean will be improved (the estimate will be closer to the true mean) and the estimated standard deviation will become smaller with increasing sample numbers. The latter interpretation is incorrect. The correct interpretation is that estimates of both the mean and the standard deviation of a property both tend to improve with increasing numbers of samples. In other words, the standard deviation can be as easily underestimated as overestimated with small sample numbers. However, the effect of increasing sample numbers tends to be a case of diminishing returns. There comes a point when the effect of adding one or two additional samples is small, and noticeable improvement in estimates is realized only with large increases in sample numbers. In common practice, sample sizes in excess of ten tend to show this effect.

The development of a sampling plan requires the mastery of the sophisticated fundamentals of probability sampling, an intimate knowledge of the product being sampled, and a high degree of skill, experience, background, and creativeness. In some cases, the help of a statistician will prove advantageous in developing the theoretical aspects and procedures for the practical person to follow. Useful references to existing ASTM standards are found in the Table 1. Setting up sampling plans without the use of probability techniques will, in most instances, introduce subconscious bias.

Taking of Sample

The actual physical manipulation needed to take a sample representing a larger unit is simpler than formulating the sampling plan, and the *Annual Book of ASTM Standards* provides guidelines under the various designations

TABLE 1—ASTM Sampling Standards for Concrete and Concrete-Making Materials.

Practice for Sampling Freshly Mixed Concrete (C 172)
Method for Sampling and Testing of Hydraulic Cement (C 183)
Method for Sampling and Testing Fly Ash for Use as an Admixture in Portland Cement Concrete (C 311)
Method for Reducing Field Samples of Aggregate to Testing Size (C 702)
Practice for Examination and Sampling of Hardened Concrete in Construction (C 823)
Practice for Sampling Aggregate (D 75)
Methods for Sampling and Testing Calcium Chloride for Roads and Structural Applications (D 345)
Practice for Random Sampling of Construction Materials (D 3665)
Practice for Probability Sampling of Materials (E 105)
Practice for Choice of Sample Size to Estimate the Average Quality of a Lot or Process (E 122)
Practice for Acceptance of Evidence Based on the Results of Probability Sampling (E 141)

as shown in Table 1. The procedures can be mastered easily, and if the instructions are followed carefully, one can expect a reasonably reliable sample to result. Other guidelines are found in various handbooks and manuals [14-16]. In general, these procedures were developed many years ago, and they are becoming improved materially in the continuing process of reexamination by the issuing body.

The relationship of a sampling plan to the actual taking of a sample may best be illustrated by an example in which the universe is the total concrete of any given class on a project. The concrete reaches the project or is made at the site in a series of batches. Although all batches are intended to be exact duplicates, this is not so in practice. Therefore, for proper evidence of the variations of the concrete properties in the universe, a plan is needed that designates which of the batches to sample in order to develop the same pattern of variation found in the universe; this is known as random or probability sampling. The second step is to sample each batch that has been tagged for sampling, in conformance with the sampling plan, in such a manner as to reflect the variation pattern in the concrete as it comes out of the mixer. This is best done by taking several subsamples at different stages of discharge from each batch and compositing them for a total sample. In case, on the other hand, it is desired to study the efficiency of mixing of the mixer, then the subsamples are tested separately to develop a variation pattern of the mixing operation to determine if such a pattern falls within the tolerances applicable to the situation.

This example shows strikingly that sampling in the concrete field is not a simple matter and must not be delegated to the untrained or careless, or relegated to a laborer for convenience.

SIGNIFICANCE OF VARIOUS FACETS OF SAMPLING

Economic Significance

Adequate sampling plans and procedures have more far reaching effects than appear on the surface. In the first place, sampling involves the costs of sampling, packaging, and shipping, which represents an investment. But samples are usually taken for the purpose of making specimens for testing, which normally costs far more than the sampling and involves time, sophisticated equipment, and skilled personnel, all of which add to the sample cost. A common consequence of this cost is a reduction in number of samples taken and tested. However, the greatest economic significance based on the test results from the samples is manifested in the decisions and conclusions that run into millions of dollars daily in the concrete industry alone. In many cases, such decisions are not only economic factors, but also involve safety and life.

Thus, it is seen that sampling is the most important link in the whole chain of construction events leading to decisions, exchange of funds between parties, contractual relationships, and even safety. Under such circumstances, can one sufficiently stress the importance of sampling plans, methods, and procedures?

Random or probability sampling permits one to attach measures of confidence to the estimates of population parameters. Biased or subjective sampling precludes the use of such confidence limits. This affects the reliability of decisions based on test results.

Why, then, have sampling concepts been somewhat neglected in the concrete field? Many factors have caused this situation to develop.

1. In the concrete field, samples are heavy, bulky, and dirty, and therefore, in many cases, the engineer passes the buck by sending someone at the lower echelon to take care of sampling, feeling that he would be wasting his expensive time to do so himself.
2. In the past, when testing was not used to reach conclusions by statistical or probability means, probability sampling was not nearly as important as it is now, when statistics has come into the picture. Yet, sampling procedures have not been revised substantially.
3. Automation, which makes "quality control" possible and places a premium on homogeneity, has only recently come into being in the concrete field. "Perfect" homogeneity was not a practical aim or achievement as long as most operations were conducted by hand, and the influence of personal differences between individuals was a paramount factor.
4. With some construction agencies, decisions have been made to a large extent by whether the engineer felt the job to be satisfactory or not and not so much by test results. The latter were something to stick in the file for legal protection, and when an occasional test turned out to be low, everyone got into a huff and blamed the sample or the contractor and made the latter the goat. Since the engineer in those days was very close to the job and observed everything that was taking place, his judgment was very effective. An exception to this practice is government construction, where test results often do govern decisions on whether or not to use a material.

No one ever dreamed that occasional low test results are part of everyday patterns of variation and are to be expected. If they fall within the proper variation pattern, they should not be a source of concern. Fortunately, these low test results showed up only occasionally since the natural tendency for bias in sampling is towards the better side rather than the poorer side. This problem has been discussed in detail by Abdun-Nur [17].

Sampling and Inspection

In most instances, sampling (actual taking of the sample), which becomes one of the duties of the inspector, is frequently and unfortunately passed on to a laborer furnished by the contractor. In many instances, it has been observed that the inspector handles sampling as a matter of convenience, neglecting the fact that sampling is a most important duty, to be performed in exact accord with specified techniques.

As an example, on one project, written instructions, the specifications, and the formal inspector-indoctrination meetings all stressed the obtaining of three sets of cylinders from three distinct batches of concrete to represent the variations in a given volume. Field observations indicated that several inspectors were making the three sets of cylinders, all at one time, from one single batch. This of course yielded erroneous records, inasmuch as the test results were being analyzed as though the tests had come from three distinct batches rather than from one batch. The record thus indicated a higher degree of homogeneity than actually existed on the project. Inasmuch as variability was tied to strength requirements, this permitted the contractor to get by with lower strengths than desired and required for the integrity of the design by the engineer. When an inspector was asked why he was not following instructions, the answer was that, "It saves making three messes and gets the job done all in one mess." In general, when the effect of such disregard of instruction is explained, inspectors are more careful.

Thus, it is not only important that the inspector be adequately trained, for example by ACI certification, and qualified to do his job properly and to pay special attention to his sampling plan and procedures, but it is also just as important to have surveillance to assure that such procedures are being carried out properly.

Inspection can, of course, be either internal—by the producer or contractor, to control his product—or external—by the owner, his engineer, or his representative—to ascertain compliance with specifications. At the present time, there is need for both internal and external inspections to assure proper construction. A detailed discussion of this subject is outside the scope of this paper and has been treated by Abdun-Nur [18].

Protection of Samples

Sampling plans, methods, and procedures are not enough to assure proper samples reaching the laboratory for the preparation of specimens for testing. Wills [19] has shown the undesirable effects on concrete properties due to the use of bags contaminated with sugar or flour, or bags treated with chemical preservatives.

Every engineer has seen concrete cylinder specimens unprotected, permitted to dry out in the hot sun or to freeze on the job. Occasionally, carelessness has resulted in heavy equipment running over specimens or causing vibration of specimens such that they were damaged beyond usefulness; yet, they were sent to the laboratory for testing with the expectation that usable results would be obtained. Proper protection, packaging, and care of samples in the field and in shipping are essential if test results obtained from specimens prepared from these samples are to mean much and if reliable conclusions are to be drawn from such results. These are serious hazards; yet data obtained in such a manner are being used every day as a guide for the uninitiated to follow.

Compositing Samples

The validity of composited samples depends on the purpose and end use of the test results from the specimens prepared therefrom. If average properties of a unit of a material are to be obtained, then taking subsamples from various portions of the unit and compositing them is in order. This practice may be useful for description of a unit of material when it is believed that subsequent handling of the material will result in substantial mixing. But if the sampling is for the study of the variations within a unit or variation among batches of concrete, then the subsamples should be kept separate, and individual specimens should be made and tested separately. The question of compositing should be decided upon in the development of the sampling plan and passed down to the sampler in the form of instructions to be followed explicitly; otherwise, the analyzed results may be unknowingly very misleading.

Compositing practices can be the cause of substantial disagreements in acceptance testing. Producers of some materials commonly perform analysis on composited samples. Acceptance testing is often performed on grab samples. Grab samples will typically reflect much greater variation in properties than will composite samples, depending on the inherent variation in the material and the extent of compositing.

Sampling and Quality Control

The control of any product is only possible through proper sampling. This is true whether the product be aggregates, various packaged ingredients that enter into the concrete as constituents at various times, or the concrete itself in the plastic stage. Sampling permits the determination of whether the process is producing what is needed and when adjustments and their extent may be required to bring the process within tolerances and up to the desired quality. To attempt to control a product by occasional sampling of the finished material is unrealistic, ineffective, and too late—by then it has already been produced and, in most cases, cannot be modified. Yet in the concrete field, occasional sampling of this latter type is almost standard routine. It is naive to think that this is quality control.

To obtain adequate quality control of concrete, every step of the procedure must be controlled as well as every ingredient that enters into it. Such control has to be developed by the producer or contractor (who is a producer in the broad sense of the word) and be built into his processes or operations. It cannot be imposed by the purchaser or owner from outside.

The most the owner or purchaser can do is to sample and test to satisfy himself that the product complies with the specification requirements. Incentives and or penalties built into the specifications by the owner or purchaser are of tremendous help in getting the producer to develop proper quality control in his operations.

Modern automated computer-controlled batch and mixing plants are now available that will record, control, and correct weights and slumps—even reject a batch that is outside of tolerances [20].

SAMPLING CONCRETE AND ITS CONSTITUENTS

General

Table 1 gives a list of the various ASTM standards whose major purpose is to describe the sampling of concrete and its constituents. The list of references at the end of this chapter gives papers, manuals, handbooks, and bibliographies that treat approaches to sampling plans, methods, and procedures in various fields that can be adapted to the needs of the concrete field. In these references, the reader can also find references to other items in case he desires to go deeper into the subject. Steele (Chapter 4) treats the statistical facets of sampling concrete and its ingredients.

The purpose of sampling will determine to a large extent the desirable sampling plan and procedures. Is the purpose of sampling to investigate the quality of a deposit or a product, to determine the homogeneity for a quality control operation, or to determine whether the material meets specification requirements?

In general, where an operation is functioning under proper and effective quality control procedures, a variable sampling plan can be most effective and economical [21]. In such a plan, a basic full sampling plan and inspection requirement is first developed. When results indicate control and acceptable homogeneity for a long period of time, the plan can be reduced to less frequent sampling, but the basic plan is reinstated immediately upon the appearance of any signs that control has been relaxed.

Aggregates

The workhorse of sampling methods used in aggregate work has been issued by ASTM Committee D4 under ASTM D 75. It treats various procedures for sampling aggregates at various locations such as quarries, pits, railroad cars, bins, and stockpiles. It is quite definite in its instructions and cautions about homogeneity or variability of the material in the locations being sampled, but not about the tools. It requires the taking of a sufficient number of samples to reflect the variability of the material being sampled but does not indicate how to develop a sampling plan for such a series of samples or under what conditions compositing is valid. ASTM E 105, E 122, E 141, and D 3665 provide guidelines for developing such sampling plans.

Various methods for sampling, whether natural deposits, finished aggregates, other concrete-making materials, or concrete itself, can be found elsewhere [14–16]. The references on coal sampling, bulk sampling, and sampling in general [1–13,21–23] provide additional information that is applicable, even though not standardized specifically for the aggregate field.

Admixtures (Chemical), Curing, Patching, Bonding, and Sealing Materials

These materials reach the job or the concrete products plant in packaged form (cans, buckets, barrels, bags, and so on) or are delivered in some instances in bulk solution form. All of such materials have had some factory processing, and the processor invariably claims very close quality control.

In practice, most large users test the source for approval and subsequently rely on the processor's certification that the material is the same as that submitted for original approval. This is, in many ways, a similar procedure to that used for cement for practically all jobs with a few exceptions.

The acceptance of such a procedure is based on the false assumption that, because it is a processed material, there are no significant variations in the final product. This assumption is made for convenience rather than because it represents a true picture—it eliminates the need for continuous testing that is costly, takes time, and can delay work.

This works fine as long as everything turns out smoothly, but if some problem develops in the concrete, it becomes impossible to determine with any certainty what the significant factors might be. There are not only the normal variations in the materials that are not on record, but it is always possible that some human error somewhere along the line of processing, warehousing, and delivery may have been made and gone undetected. Whether this calculated risk is worth taking is up to the responsible engineer to determine in each case. But if proper original or continuous sampling is desired, the procedures and plans are much simpler than in the case of bulk aggregates.

The proper sampling for these materials becomes a problem similar to that of sampling any manufactured packaged article or item. Many plans have been worked out and proven satisfactory and usually can be adapted to the concrete field [1–14,21–23]. Usually, this adaptation consists of a two-step procedure. The first step is the random sampling of a lot or shipment that results in the selection of a number of the packages out of the whole to be used as evidence of the variation pattern of the lot. The second step is to take individual samples from each package that has been set aside for sampling.

The first step is based on probabilities and statistical principles, while the second step requires manipulative care to ensure that the sample is representative of the container. When the package is relatively small and the product is in the form of a liquid of some type, the second step is easy, as proper shaking or stirring will permit the obtaining of a representative sample without much trouble. But when one gets into the larger packages, such as a 50-gal drum, a tank car, or tank truck, there is always the possibility that there is stratification, and the sample has to be taken in such a manner that it is a composite of the various strata.

Compositing assumes that in the unloading, transferring, and handling, various strata will be intermixed. If this is not the case, then the sampling has to be related to the handling and dispensing procedures in order to develop the variation patterns. This is likely to get rather complicated and can be avoided if the storage container from which the material is dispensed is kept continuously stirred to maintain the homogeneity of its contents, refer to ASTM Specification for Liquid Membrane-Forming Compounds for Curing Concrete (C 309).

Detailed guidelines for sampling these classes of materials, which have proven satisfactory in general industrial applications and which in many cases can be adapted to the concrete field, may be found in the industrial product control literature [2,21,23].

Admixtures (Mineral)

Sampling of mineral admixtures (almost totally coal fly ash in the United States) is the subject of many of the same considerations as hydraulic cements, however, practices do vary. Whereas cement is a manufactured product, coal fly ash is a waste by-product of electricity generation, over which the fly ash broker has incomplete control at best, and no control at worst. The purpose of sampling then is to determine which material in the waste stream is suitable for sale as a concrete admixture and which must be diverted to other uses or to the waste heap. Factors that influence fly ash properties include coal source, grinding prior to combustion, rate of combustion, air-to-fuel ratio, collection systems, and chemical additions. These things vary considerably among sources and, consequently, quality control sampling plans also vary considerably. It is generally true that a limited amount of testing is done on a daily or hourly basis, usually limited to loss on ignition, fraction larger than 45 μm, and density. Full testing is commonly done on composite samples representing a month or more of production.

Acceptance testing is commonly done on grab samples (see ASTM C 311). If producer quality control sampling procedures include some degree of compositing, differences in uniformity estimates, which are specification requirements for mineral admixtures (see ASTM Specification for Fly Ash and Raw or Calcined Natural Pozzolan for Use as a Mineral Admixture in Portland Cement Concrete (C 618), are inevitable. Guidance needs to be more specific on compositing.

Fresh Concrete

ASTM Specification for Ready-Mixed Concrete (C 94) discusses sampling at some length. ASTM C 172 on sampling fresh concrete outlines in detail the manipulations in taking a sample from a batch but does not describe sampling plans that permit the selection of the batches to be sampled. ASTM D 3665 provides guidance for developing such sampling plans. In general, the same problems that are encountered in bulk aggregate sampling from truckloads are found also in sampling batches of concrete; it is fortunate that the most difficult problem of sampling aggregates from a stockpile does not exist for concrete. However, there are practical problems that are unique to concrete. For example, some concrete is usually discharged into forms before concrete is sampled from a batch. This may leave little time to conduct tests without delaying placing the remaining part of the batch. Also, if the batch is found to be out of compliance, then removal of placed concrete could be difficult. Sampling procedures, compositing, details of manipulations, and sampling plans should depend on the purpose for which the sampling is being done.

Hardened Concrete

ASTM Test Method for Obtaining and Testing Drilled Cores and Sawed Beams of Concrete (C 42) governs the taking of cores or the sawing of beams from hardened concrete, but no mention is made of how to decide at what point or points such specimens should be secured. ASTM C 823 describes this as well as many other aspects of sampling of hardened concrete in considerable detail. Other sources include Abdun-Nur [24] and the American Concrete Institute (ACI) Committee 214 on Evaluation of Results of Tests Used to Determine the Strength of Concrete. Standardization of this sampling process is difficult because it involves a lot of judgment that is often difficult to standardize. These are important as safety decisions involving potential life hazard, and economic decisions involving large sums are frequently made on the basis of testing of specimens from hardened concrete.

A recent example of such problems is a case where a large number of cores were taken and tested by several laboratories. These tests indicated a certain level of strength and a variation pattern—all with the exception of one laboratory, where three cores indicated consistently a 50% higher strength than those tested by other laboratories—a most improbable situation if sampling had been properly conducted.

Hydraulic Cement

Cement is generally sampled and tested extensively during manufacture and then spot-checked by the user. During manufacture, samples are typically taken from the production stream at regular intervals, usually about an hour, and then composited every 24 h of production, with full testing done on the 24-h composite. Selected tests may be run on the hourly samples, particularly on properties that run close to specification requirements. Thus, mill test data represent average daily properties and are not particularly sensitive to variations that might occur over shorter time intervals.

Acceptance testing is governed by ASTM C 183. It allows either grab or compositing of grab samples, but gives no firm guidance on this practice. It is common practice to test cement for acceptance on grab samples. Since grab samples will reflect the maximum variation in a material (but mean properties will be unaffected), there is the reasonable probability that the differences in production sampling and acceptance sampling may result in considerable disagreements. Currently, there are no uniformity requirements in ASTM specifications for hydraulic cements, which are extremely sensitive to the type of sampling, so such disagreements are a serious problem only when they involve a specification compliance dispute.

NEEDED IMPROVEMENTS AND THEIR BENEFITS

ASTM Standards

In many cases, the existing ASTM standards contain the manipulative instruction for taking samples of the various concrete ingredients and other materials used in the curing, protection, and repair of concrete, and also for the sampling of hardened concrete. In many cases, modifications or revisions and broadening of scope of these procedures are desirable. More detailed guidance on compositing practices is probably needed.

Recognition of Variability

Essentially, the big problem in sampling in the field of concrete technology is the dissemination of the ideas regarding variability. Since it must become commonly accepted that there is no such thing as a "representative batch" or a "typical batch," a sampling plan that reflects the variations in any given situation or universe is needed. Such a plan would provide the proper information for control of concrete and its ingredients as a basis for determining specification conformance and permit making realistic decisions when problems arise.

Probability Sampling

Probability sampling is our greatest need at the present time. To bring it about requires training of many people, from the engineer down to the inspector and sampler. The importance of proper sampling and its effect on testing and making the final most important decisions needs to be stressed repeatedly. The basic fundamentals for probability sampling are available. Just as the continued stirring of a liquid or solution keeps it so homogeneous that one can take a sample at any time and at any place in the container, probability sampling achieves the equivalent of homogenization mathematically and permits a calculation of its reliability.

It has been said that the probability testing of aggregates is of no importance since the testing of the plastic concrete is the final criterion. This may be true for determining compliance with specifications in the few cases where the latter are based on final performance only. But, for the concrete producer who is trying to control his product, it is essential that he use the best probability sampling techniques to determine the pattern of variation of the aggregates or any other ingredients that go into his concrete. This enables him to make the proper allowances in the production of his concrete to meet a given specification requirement.

Inasmuch as the probability approach to sampling provides the required degree of reliability from the smallest number of samples, in most cases, the reduced quantity of sampling will result in an overall decrease of the total sampling cost. Added to this is the fact that the cost of testing specimens prepared from unreliable samples will be eliminated, and the total saving becomes very appreciable. Of course, the main advantage is that the resulting data are more reliable and, therefore, the decisions based on such information become more realistic. This permits designing to closer tolerances or lower factors of safety, and in the long run would result in safer and more economical engineering structures.

Notwithstanding all that has been stressed about the need and usefulness of probability or random sampling, it must be stressed that because of the widespread availability of computers and because engineers and statisticians are so comfortable with large amounts of data, one must keep in mind that a balance must be kept between the practical and theoretical; and frequently judgment must be used in lieu of complicated theoretical considerations.

Trained Personnel

Such an approach requires sophisticated and properly trained persons at various levels, such as ACI certified field technicians. But once plans and procedures have been set, the actual sampling becomes a question of following instructions and being sure that there is adequate supervision by the more sophisticatedly trained personnel of the lower grade personnel who actually take samples. Thus, the cost of supervision and higher grade personnel might increase, but this will be compensated for by the lower cost of the less-skilled samplers.

SUMMARY

Sampling is probably the most important step in the testing sequence. On this information, millions of dollars change hands daily—yet, it is the most neglected activity in this area of engineering.

The one thing needed above all else is the realization that variations in composition and properties of concrete and concrete-making materials are basic and can only be reduced but never completely eliminated. The degree of homogeneity that can be attained in practice depends on the cost of reducing variability as compared with the benefits derived therefrom [17].

Probability sampling and sampling plans are essential to the determination of the true pattern of the existing variations, and fundamental to the problem of controlling any of the products going into concrete and of the concrete itself.

Control must be made an integral part of every step in production. Control cannot be imposed from outside by the job owner or the purchaser of over-the-counter products; all the owner or purchaser can do is determine whether the material as-produced does or does not meet specifications. Incentives built into the specifications can motivate the producer to develop proper control [25–28].

It is necessary to sample not only at all the steps of production to guide production procedures, but sampling should also be done on the finished product as close as possible to the point of use to ascertain compliance with specifications. Handling of the materials after they have been sampled needs much attention; otherwise, the samples will not reflect the materials as they are incorporated into the structure.

More detailed instructions and standards can provide better guidance. Better qualified and trained personnel, under proper supervision, are a necessity to achieve maximum reliability with minimum sampling as a basis for final important economic and safety decisions. This would improve the present sampling situation immensely.

REFERENCES

[1] *Bulk Sampling, ASTM STP 242*, American Society for Testing and Materials, Philadelphia, 1958.

[2] "Usefulness and Limitations of Samples," *Proceedings*, American Society for Testing and Materials, Philadelphia, Vol. 48, 1948, pp. 857–895.

[3] Tanner, L. and Deming, E., "Some Problems in the Sampling of Bulk Materials," *Proceedings*, American Society for Testing and Materials, Philadelphia, Vol. 49, 1949, pp. 1181–1188.

[4] Slonim, M. J., *Sampling in a Nutshell*, Simon and Schuster, New York, 1960.

[5] Montgomery, D. A. *Introduction to Statistical Quality Control*, Wiley, New York, 1985.

[6] Tanner, L., "Probability Sampling Methods for Wool," *ASTM Materials Research and Standards*, March 1961, pp. 172–175.

[7] Cochran, W. G., *Sampling Techniques*, 2nd ed., Wiley, New York, 1963.

[8] Bicking, C. A., "Bibliography on Sampling of Raw Materials and Products in Bulk," *Technical Association of the Pulp and Paper Industry*, Vol. 47, No. 5, May 1964.

[9] Bicking, C. A., "The Sampling of Bulk Materials," *ASTM Materials Research and Standards*, March 1967, pp. 95–116.

[10] Duncan, A. J., "Contributions of ASTM to the Statistical Aspects of the Sampling of Bulk Materials," *ASTM Materials Research and Standards*, Nov. 1967, pp. 477–485.

[11] Visman, J., "A General Sampling Theory," *ASTM Materials Research and Standards*, Nov. 1969, pp. 8–13, 51–56, 62, 64, and 66.

[12] *Sampling, Standards, and Homogeneity, ASTM STP 540*, American Society for Testing and Materials, Philadelphia, 1973.

[13] "Acceptance Sampling," *ASTM Standardization News*, Sept. 1975.

[14] *Handbook for Concrete and Cement*, Waterways Experiment Station, U.S. Army Corps of Engineers, Vicksburg, MS, 1949 and subsequent revisions.

[15] *ACI Manual of Concrete Inspection*, 6th ed., American Concrete Institute, Detroit, MI, 1981.

[16] *Concrete Manual*, 8th ed., U.S. Bureau of Reclamation, Washington, DC, 1975.

[17] Abdun-Nur, E. A., "How Good is Good Enough," *Proceedings*, American Concrete Institute, Vol. 59, No. 1, Jan. 1962, pp. 31–46 and 1219–1244.

[18] Abdun-Nur, E. A., "Inspection and Quality Control," *Proceedings*, 6th Annual Concrete Conference, Utah State University, Logan, UT, March 1964.

[19] Wills, M. H., Jr., "Contamination of Aggregate Samples," National Sand and Gravel Association, July 1964.

[20] Abdun-Nur, E. A., "Accelerated, Early, and Immediate Evaluation of Concrete Quality," *Proceedings*, International Symposium on Accelerated Strength Testing of Concrete, American Concrete Institute Convention, Mexico City, Mexico, 24–29 Oct. 1976.

[21] Sampling Procedures and Tables for Inspection by Variables for Percent Defective, MIL-STD-414, Superintendent of Documents, Washington, DC, 11 June 1957.

[22] *Single-Level Continuous Sampling Procedures and Tables for Inspection by Attributes, Handbook H-107*, Superintendent of Documents, Washington, DC, 30 April 1959.

[23] Sampling Procedures and Tables for Inspection by Attributes, MIL-STD-105D, Superintendent of Documents, Washington, DC, 29 April 1963.

[24] Abdun-Nur, E. A., "Sampling of Concrete in Service," Special Report 106, Highway Research Board, Washington, DC, 1970, pp. 13–17.

[25] Abdun-Nur, E. A., "Designing Specifications—A Challenge," *Proceedings, Journal*, Construction Division, Separate No. 4315, American Society of Civil Engineers, New York, May 1965.

[26] Abdun-Nur, E. A., "Product Control and Incentives," *Proceedings, Journal*, Construction Division, Separate No. 4900, American Society of Civil Engineers, New York, Sept. 1966.

[27] Abdun-Nur, E. A., "Adapting Statistical Methods to Concrete Production," *Proceedings*, National Conference on Statistical Quality Control Methodology in Highway and Airfield Construction, Nov. 1966.

[28] Abdun-Nur, E. A., "What is the Quality Assurance System?" *Proceedings*, Transportation Research Board, 55th Annual Meeting, Washington, DC, 19–23 Jan. 1976.

Statistical Considerations in Sampling and Testing

Garland W. Steele[1]

PREFACE

The application of statistical considerations to the sampling and testing of concrete and concrete-making materials has been addressed by chapters in each of the three previous editions of *ASTM STP 169*. The first edition, published in 1956, contained the chapter entitled "Size and Number of Samples and Statistical Considerations in Sampling," by W. A. Cordon. The second edition, published in 1966, contained the chapter entitled "Evaluation of Data," by J. F. McLaughlin and S. J. Hanna. The second edition also contained the chapter by W. A. Cordon that first appeared in the 1956 edition. The third edition, published in 1978, contained the chapter entitled "Statistical Considerations in Sampling and Testing," by H. T. Arni. The contributions of each of these authors to the practical application of statistical probability in the field of concrete is hereby acknowledged.

INTRODUCTION

The purpose of this chapter is to provide suggestions regarding the use of statistical applications that are practical, valuable, and appropriate when used in the sampling, testing, and evaluation of concrete and concrete materials. If needed, detailed texts on statistical methods and procedures are available from many sources. A few are listed in the references.

GENERAL CONSIDERATIONS [1]

Statistical Parameters

If the characteristics of concrete or of a material used in the concrete are to be determined with a known probability of being correct, a plan is required. Such plans are commonly called acceptance plans or other similar names that appeal to the designer of the plan.

When a plan is designed to obtain the desired information through inspection by attributes, each item or group of items will usually be classified only as satisfactory or unsatisfactory so that the relevant parameter is percent satisfactory or percent unsatisfactory. The number of each is recorded for use in decisions concerning the use or other disposition of the item(s). Attributes inspection is occasionally used on precast concrete items.

The majority of plans designed for concrete and concrete materials use inspection by variables. Inspection by variables is done by measuring the selected characteristic of a material or product and recording the value. The values are used to calculate fundamental statistical parameters needed to describe the characteristic. Such descriptions are used in decisions concerning the use or other disposition of the material or product.

The fundamental statistics derived from variables inspection that are most useful in making decisions concerning concrete and concrete materials are the mean and the standard deviation. The mean (arithmetic), when derived from data gathered by a properly designed plan, conveys, in one figure, the average value (central tendency) of the measured characteristic of the material or product. It also indicates that approximately one half the characteristic will have greater values and approximately one half will have smaller values when measurements are made. Although the mean is a very important statistic, it does not show how far or in what way the greater or lesser values may be distributed from the mean. This information is conveyed by the standard deviation, a powerful statistic that measures dispersion. The standard deviation indicates, in one figure, how far above or below the mean other values will be and how many values will likely be found at any distance from the mean.

Other statistics occasionally derived from variables inspection of concrete and concrete-making materials include skewness and kurtosis. Skewness indicates whether the distribution of values tend to be grouped unequally above or below the mean (nonsymmetrical) rather than equally occurring on each side (symmetrical). Kurtosis indicates whether the frequency of occurrence of values at any distance from the mean is greater or less than the frequency expected from a normal distribution.

Other statistics concerning central tendency, that is, geometric mean, mode, etc., are seldom calculated for concrete and concrete materials characteristics. Likewise, statistics concerning dispersion, such as the average deviation, are usually not calculated. However, two statistics concerning dispersion are frequently calculated. These are the range and the coefficient of variation. The range conveys the difference between the largest and smallest value of the measured characteristic. It is not uncommon to see

[1] President, Steele Engineering, Inc., Tornado, WV 25202.

the range used in decisions concerning the use of materials or products. The coefficient of variation is the ratio of the standard deviation to the mean of a group of values, usually expressed as a percentage for construction materials. This statistic is calculated extensively for measured characteristics of concrete and concrete materials.

Generally, as previously noted, the most useful statistic concerning the central tendency of data collected during the measurement of concrete and concrete materials characteristics is the arithmetic mean. The most useful statistic concerning the dispersion of said data is less straightforward. The simplest of the three statistics that are commonly encountered is the range. The range is also the least powerful and it's use is normally limited to specific applications (for example, control charts). The range is also used occasionally to estimate the standard deviation. However, since many pocket calculators used in construction activities have keys for direct calculation of the standard deviation, it is preferable not to use the range for this purpose. The standard deviation is the preferred statistic, particularly when it is constant over the range of values collected for the measured characteristic. When the standard deviation is not reasonably constant for the range of values encountered, the coefficient of variation should be considered. If the coefficient of variation is constant, it would normally be used. If neither statistic is constant over the range of values obtained, either statistic may be used. However, the standard deviation is preferred in such cases unless standard practice dictates otherwise.

When the behavior of one variable is to be compared to the behavior of another variable, a correlation coefficient may be determined. A good, or significant, correlation implies(1)[2] that as one of the variables being correlated increases, the other tends to increase also. Alternatively, the second variable may tend to decrease as the first increases (negative correlation). All that is indicated by a good correlation is that the trend exhibited by the data when one variable is plotted against the other is sufficiently well defined to permit rejection of the hypothesis that there is no relationship between the two variables, in other words, that the relationship is completely random.

Regression Lines

The standard method of using paired data as a source of developing means for predicting one variable from another is by calculating a linear regression line using the method of least squares [4–8]. This method is very useful in situations that involve actual calibration of a measuring device and in which the plotted points approximate the calculated regression line very closely with small scatter (2).

When two separate measurements, each obtained from different test methods, are statistically related to each other and one of the test methods is used to obtain measurements that are then used to predict measurements of the other type, there is almost always a large variation in the data. While both sets of measurements are affected by changes in the property of interest, each is actually measuring two different quantities and each is affected by different sets of influences extraneous to the property of interest. This is usually accounted for by recognizing that many measurements are needed to derive a reliable line. It should also be recognized that if the regression line represents the current measurements in the same way that it represented the measurements from which it was derived, then new measurements are subject to the same variations as the original measurements, and the derived measurement has a wide uncertainty associated with it (3).

If regression lines showing the relationship between data from two measurement systems are to be calculated and used for purposes other than illustrating the relationship existing in the data, several points should be considered.

1. Derivation of a linear regression line involves determination of two parameters, the slope of the line, b_1, and the Y intercept of the line, b_0. Due to the departures of the data points from the line, each of these parameters has an estimated variance and corresponding standard error (the square root of the variance): s_{b1}^2 for the slope and s_{b0}^2 for the intercept. In addition, there is an estimated variance, s_y^2, and corresponding standard deviation, s_y, calculated from the sum of the squares of the deviations of the measured Ys above and below the fitted line. The three measures of variation (variances or their corresponding square roots) should always be given whenever a regression line is reported. The standard errors of the slope and intercept indicate the significance of the relationship (4). In addition, the number of pairs of data from which the regression was calculated and the range of spread of both X and Y values should be reported along with the other parameters.
2. A confidence interval for the fitted line should be shown on the graph of the line (5) (Fig. 1).
3. There are three kinds of confidence intervals that can be calculated for a fitted regression line: the line as a whole, a point on the line, or a future value of Y corresponding to a given value of X (6).
4. When appropriate confidence limits are correctly calculated and reported, proper use requires a practical understanding of the confidence interval given (7).

It should be reiterated when using regression lines that new measurements are probably subject to the same degree of variation that characterized the original data.

In general, then, if any system of measurements is valid, "the system must have validity within itself." The use of a regression equation to convert measurements obtained in one system to measurements that might have been obtained by some other system introduces an additional degree of uncertainty into the process. Further, the use of relationships other than straight linear equations will, in appropriate situations, provide more reliable data (8).

---

[2] Numbers in parentheses refer to the endnotes attached to the end of this chapter. Endnotes 10 through 13 are adapted from Ref 1.

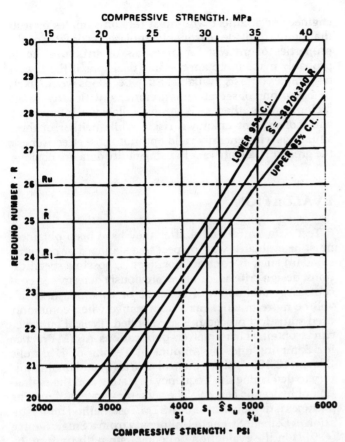

FIG. 1—Regression curve and confidence limits for compressive strength versus rebound numbers [10].

SAMPLING

The goal of sampling concrete and concrete materials is to obtain samples from which an unbiased estimate of the characteristic of interest of a lot of material or product can be obtained. Therefore, when a sample is used to evaluate a characteristic of a lot (quantity) of material, all material in the lot should have an equal chance of being included in the sample. Alternatively, if the probability of inclusion is not the same for all parts of the lot, this fact should be considered when evaluating data derived from the sample.

Given the preceding goal, the simplistic answer may seem to be a requirement that all material in the lot should have an equal chance of being included in any sample obtained from the lot. However, in many situations this may not be practical. Consider, for example, a 100-yd³ (76-m³) concrete placement.

1. Using an appropriate plan, the fresh concrete could be sampled during the placement process in a manner that would assure that all concrete had an equal chance of being included in the sampling operation.
2. Using an appropriate plan for hardened concrete sampling, it is likely that in most situations the concrete in different segments of the placement would have an unequal chance of being included in the sampling operation (or even no chance). Final geometry, critical stresses, reinforcement, and lack of accessibility all contribute to the situation.

It is, therefore, very important that the circumstances in each sampling situation be recognized and receive competent consideration.

The term "representative sample" has been used in many different ways in the construction industry. The most useful definition would paraphrase the goal of sampling previously noted; that is, a representative sample is a sample that is obtained from a lot or quantity of material, using procedures that will allow an unbiased estimate of a particular characteristic of the material to be derived from the sample. If the stated definition is to be operant, the sample must be selected randomly. ASTM Practice for Random Sampling of Construction Materials (D 3665) contains a random number table that can assist the sampler in obtaining random samples.

The definition should also assist in a better understanding of composite samples. If, for example, a particular characteristic of several randomly selected 2000-lb (900-kg) units of aggregate is to be used to estimate that characteristic in 10 000 tons (9000 metric tons) of material, the characteristic can be determined from testing each of the 2000-lb (900-kg) units in total. Alternatively, five (or some other suitably determined number) randomly selected subsamples may be selected from each of the 2000-lb (900-kg) units for compositing. The five randomly selected subsamples from each of the 2000-lb (900-kg) units may then be combined to form a single sample obtained from each 2000-lb unit randomly selected for sampling. Results from the single composited samples may then be used to estimate the characteristic of the 10 000-ton (9000-metric ton) lot of aggregate. Obviously, this alternative provides a possible advantage in that much smaller quantities of aggregate are handled in the final sample. However, a possible disadvantage to be considered is that variations in the characteristic of interest that exist in the 2000-lb (900-kg) unit of aggregate will remain unknown. A second alternative for compositing is available that provides the information needed for estimating the variations within the 2000-lb (900-kg) units of aggregate for the characteristic of interest. The second alternative requires that each subsample be individually retained and used for testing. The data derived from each subsample in a 2000-lb (900-kg) unit may then be averaged and used in the same manner as the data derived from the composited sample in the first alternative.

Similar types of examples could be presented for cement, other concrete materials, and concrete, either fresh or hardened.

The number of samples needed to provide the information necessary for estimating a particular characteristic of a lot or quantity of material with a given confidence in the results can be easily determined. ASTM Practice for Choice of Sample Size to Estimate an Average of Quality for a Lot or Process (E 122) contains equations for calculating sample size. Note that "sample size" as used in ASTM E 122 is equivalent to the "number of samples" as commonly used in the construction industry.

A useful concept when sampling concrete and concrete materials is "stratified random sampling." This concept can be easily implemented by dividing the lot or quantity of material into a number of sublots or subquantities. The most convenient number is usually equal to the number of samples to be obtained. Each sample is then randomly obtained from one sublot or subquantity. The purpose of stratified random sampling is to prevent the possibility that several samples would be obtained from one segment of the lot or quantity of material being sampled.

An adequate plan for evaluating specific characteristics of a material (or process) will provide sampling direction. This should include the type, frequency, method, and location of sampling. Further, the quantity and handling of samples subsequent to selection should be clearly detailed [17].

Testing

The reason for testing concrete and concrete materials is to produce data from which unbiased estimates of certain characteristics of the material can be derived. The reliability of these estimates improves as the number of test results increases. Also, although depending somewhat on the purpose for which the estimates will be used, the reliability tends to increase as the quantity of material undergoing test increases. To illustrate, consider the 2000-lb (900-kg) unit of aggregate example set forth in the previous section on sampling. Without resorting to mathematical proof (which could be done), it is intuitively seen that the performance of a test on the entire 2000-lb (900-kg) unit will likely yield a better estimate of the characteristic being determined than would the performance of a test on a 10-lb (4½-kg) subsample obtained from the 2000-lb (900-kg) unit. Likewise, estimates based on ten 10-lb (4½-kg) subsamples obtained from the 2000-lb (900-kg) unit will be more reliable than the estimate derived from one 10-lb (4½-kg) subsample. Then, as the number of subsamples is increased, the reliability of estimates based thereon will approach that of the 2000-lb (900-kg) unit. Finally, when the number of subsamples obtained from the 2000-lb (900-kg) unit equals 200, the reliability of the estimate will be equal. However, as noted in the original example, the information concerning the characteristic that can be derived from the 200 subsamples will be far greater than that which can be derived from treating the 2000-lb (900-kg) unit as a single sample.

It is also intuitively evident from the previous example that testing additional samples or samples of greater size increases the cost of testing. It is necessary, therefore, to establish the reliability required in each case commensurate with the resources that are to be made available and with how much information it is necessary to obtain. These decisions should be based on a determination of the consequences of using defective material.

Similar examples could be presented for cement, other concrete materials, and concrete, either fresh or hardened.

Although it is seldom articulated, it may be better to make no tests than to make tests with poor samples that do not portray the actual properties of the materials. An engineer or architect who must rely on samples or tests that do not provide reasonably unbiased estimates of the properties of interest for materials or structures could probably make more appropriate decisions if there were no samples or tests available. Such decisions would likely be based on conservative assumptions, with large safety factors, rather than reliance on fallacious information with consequent unknown risks. While neither of these situations will generally yield optimum cost effectiveness, the greater risks inherent in using faulty data are obvious.

EVALUATION

Test data, to be useful after they have been obtained, must be evaluated with respect to a standard or potential standard such as contract documents, possible specifications, design criteria, or other previously determined concepts. Any evaluation of concrete or concrete materials will be more practical and informative when commonly used statistical procedures are applied. Typical guidance can be obtained in numerous publications. ASTM Practice for Sampling and the Amount of Testing of Hydraulic Cement (C 183) utilizes control charts (9) and quality history to determine the frequency of testing and the evaluation of hydraulic cement. The versatile concept of moving averages is demonstrated by ASTM Test Method for Evaluation of Cement Strength Uniformity from a Single Source (C 917) in the evaluation of cement strength uniformity. ACI 214, Recommended Practice for Evaluation of Strength Test Results of Concrete (10), provides detailed procedures for the evaluation of concrete strength tests. The principles contained in the three example standards noted earlier can be extended to many other concrete and concrete materials tests. However, for a few nonparametric (11) tests this is not the case. Examples are ASTM Test Method for Scaling Resistance of Concrete Surfaces Exposed to Deicing Chemicals (C 672) that uses an ordinal scale of measurement and ASTM Test Method for Organic Impurities in Fine Aggregates for Concrete (C 40) that uses a nominal or classification scale. The median can be used to indicate central tendency for such tests. Reference 24 should be consulted for other procedures applicable to these types of test data.

PRECISION AND BIAS STATEMENTS

One of the most important facts concerning a test method that is used to determine acceptance and rejection of materials and construction in a buyer-seller relationship is the information contained in the precision and bias statement.

Numerical limits, based on standard test data, which are included in specifications to govern acceptance decisions, should be fully compatible with the precision and bias information contained in the test standard used for generating the data [25]. Test standards that contain no information concerning precision are of limited value in making acceptance or rejection decisions.

Standards (12) best suited for the determination of precision and bias of tests for concrete and concrete materials are ASTM Practice for Preparing Precision and Bias Statements for Test Methods for Construction Materials (C 670) and ASTM Practice for Conducting an Interlaboratory Test Program to Determine the Precision of Test Methods for Construction Materials (C 802). The indices of precision used in these standards are, in order of preference, the "difference two-sigma limit (d2S), and the difference two-sigma limit in percent (d2S%). The indices are determined by multiplying the factor, $2(2)^{1/2}$, by the standard deviation of test results or the coefficient of variation of test results, respectively. When one of these indices is provided in a test standard, the user of the standard can presume that only about one time in twenty will the difference between two test results (13), obtained under approximately the same conditions noted in the precision information, exceed the value indicated by the index. If more than two results are to be compared, different multipliers are required to determine the index. ASTM C 670 contains a table of multipliers for determining indices for use in comparing up to ten test results. The following items are very important in the maintenance of valid precision information in standard test methods:

1. Requirements for the designated number of test results (one or more) that constitute a valid test should be stated in a test standard.
2. Criteria used in deriving the precision information should be outlined in the standard.
3. When a standard test is revised, other than editorially, the precision information should be reviewed to determine whether a new precision index should be derived.

Although other indices based on different criteria can be derived and used, those noted earlier have been recommended as the most appropriate for test standards developed by ASTM Committees C1, C9, D4, and D18.

An economical procedure for the detection and reduction of variations in test methods before initiating a complete interlaboratory study is provided by ASTM Practice for Conducting A Ruggedness or Screening Program (C 1067).

Procedures for determination of precision based on values from other related tests are presented in ASTM Practice for Calculating Precision Limits Where Values are Calculated from Other Test Methods (D 4460).

An additional standard now under development by Committees C1, C9, D4, and D18 entitled "Practice for Inclusion of Precision Statement Variation in Specification Limits" will provide guidance concerning the appropriate interfacing of precision and bias values in a test method with specifications containing limits that are based on data obtained by use of the test method.

ENDNOTES

(1) For example, many in-place tests on concrete tend to show increases in the measurements obtained as concrete strength increases. A good correlation (or high correlation coefficient) between the results of any test that exhibits such behavior and the results of standard strength tests does not in itself constitute evidence that the relationship is sufficiently close to permit the use of one type of test as a means of predicting what will happen in another type of test. In fact, the correlation coefficient has limited use in the field of analyzing engineering data. For further discussion of this point, see Refs 2 and 3. Reference 3 contains quotations from a number of statistical authorities on the use of the correlation coefficient.

(2) An example is the calibration of proving rings that are used for calibrating testing machines. This use is discussed by Hockersmith and Ku [9].

(3) Even when the existence of this uncertainty is recognized, its size and affect on the derived measurements may be underestimated. This point is illustrated in the case concerning prediction of compressive strength measurements from penetration and rebound tests in Refs *10* and *11*.

(4) In general, if the slope is not at least twice as large as its standard error, this is insufficient evidence to conclude that the true slope is other than zero (that is, the relationship is random).

(5) Calculation of the quantities needed to plot the confidence interval is described elsewhere [4]. It should be noted that the upper and lower confidence limits are represented by two branches of the hyperbola that are closest together at the point where $X = \bar{X}$, the average of all the X values used in calculating the relationship. Thus uncertainty of the calculated Y increases as the X value departs from $\bar{X}$ in either direction. Occasionally, confidence intervals have been plotted by multiplying the S_Y by the t value for the number of points used and drawing parallel straight lines above and below the regression line. This is not correct and gives an optimistic picture of the uncertainty of estimated Y values.

(6) Reference 4 describes how to calculate these three confidence intervals. It is important to note that only the confidence interval for the location of the line as a whole, is appropriate to use if the calculated line is to be used repeatedly for predicting future values of Y from future observed values of X.

(7) Figure 1 (taken from Ref *10*) illustrates the point. The regression line shown is based on 16 plotted points relating strengths of 28-day cylinders to the average of 20 Swiss hammer rebound numbers obtained on slabs made from the same batches of concrete as the cylinders. The figure also shows the hyperbolic curves representing the upper and lower 95% confidence limits for location of the line referred to earlier.

For a hypothetical rebound number of 25, this figure indicates a calculated average compression strength of 31.2 MPa (4530 psi), with the 95% confidence interval extending from 30.0 to 40.5 MPa (4350 to 4710 psi). Since the rebound numbers themselves have a distribution with a characteristic scatter, illustrated in this case by a standard deviation of 0.50 for averages of 20 rebound numbers, the 95% confidence interval for the average rebound measurement is from 24 to 26. These figures combined with the confidence interval for the line give an approximately

90% confidence interval for the predicted compressive strength of 27.6 to 35.1 MPa (4010 to 5070 psi).

Unfortunately, this is not the limit of the final uncertainty of the predicted result. A 95% confidence limit is often interpreted as meaning that 95% of future results will be within the limits given. What it actually means, however, is that if the experiment is repeated a large number of times, each with the same materials and conditions and with the same number of determinations, and each time the 95% confidence interval is calculated, then 95% of the intervals so calculated will include the true average. This does not mean that any particular one of the intervals will actually contain the true average in its exact center. For any given determination of the line and its confidence interval, the circumstance that the calculated line coincides exactly with the true line is highly unlikely.

(8) One example that illustrates the successful use of nonlinear equations for determining the relationship between certain variables in concrete is contained in ASTM Method for Developing Early-Age Compression Test Values and Projecting Later-Age Strengths (C 918). This standard involves the application of the maturity concept. The concept goes back a long way, but detailed work was done on its application to evaluation of concrete strengths in the 1970s [13–16].

The standard involves establishing a relationship between the strength and the logarithm of a quantity called maturity of the concrete that is defined as the product of temperature at which the curing is taking place and the time in hours (degree-hours).

(9) Monitoring Production-Continuous Evaluation (adapted from reference 18)

One of the most effective means of maintaining the quality of a manufactured product is by continuously monitoring the quality by means of regularly performed tests throughout the process of production. The best tool for doing this is by means of control charts.

The control chart became a well-established technique in production quality control during the World War II era. The control chart is a combination of both graphical and analytical procedures. The basis of the theory arises from the fact that the variation of a process may be divided into two general categories. One portion of the variation can be described as random or chance variation of the process and the other as the variation due to assignable causes. A process that is operating with only chance variation should result in some distribution of the measured characteristics, and one should be able to predict a range within which a certain percentage of the data should fall. If some assignable cause (such as an increased water-cement ratio) results in a change in the distribution, then the values of the measured characteristic could fall outside the predicted range.

Three types of control charts that are frequently used are control charts for averages (or moving averages), control charts for standard deviations (or moving standard deviations), and control charts for ranges (or moving ranges). Also, control charts for other measures, such as percent defective, may be useful. Detailed treatment of this subject and tables of control chart constants for determining upper and lower control limits are presented in tests on statistical quality control [12,17,19]. Table 27 in Ref 20 contains the control chart constants for averages, standard deviations, and ranges.

(10) Evaluation of Strength Tests

One of the earliest and most widely used applications of statistics in the concrete field has been in the area of evaluation of strength tests both of mortar cubes for the testing of cement strength and, more extensively, for the analysis of strengths of concrete specimens, usually, in the United States, in the form of 6 by 12-in. cylinders. The chief pioneer in this effort was Walker who published his study in 1944 [21]. In 1946, largely at the instigation of Walker, the American Concrete Institute (ACI) began work on statistical evaluation of compression tests that eventually resulted in the publication of ACI Standard Recommended Practice for Evaluation of Strength Test Results of Concrete (ACI 214-77). First published as a standard in 1957, this document has undergone a number of revisions.

In 1971 a symposium was conducted at the ACI Fall Convention on the subject "Realism in the Application of ACI Standard 214-65." This symposium presented valuable information on the meaning and use of ACI 214 and resulted in a symposium volume that included seven papers presented at the symposium, a reprint of ACI 214-65, and reprints of two earlier papers dealing with evaluation of concrete strengths [22].

(11) Nonparametric Tests

There are some test methods that do not provide numbers for which the customary processes of calculating means, standard deviation, (D2S) limits, and other so-called parametric statistics are applicable. Such tests measure on a nominal or classification scale, or on an ordinal or ranking scale [23]. Test methods of the latter type sometimes cause problems because of the fact that numbers are assigned to the different levels of quality of performance in the method, and then the numbers are treated as though they represented measurement on an interval scale, which is the type of measurement scale appropriate to most concrete or concrete materials test methods. Because of the numbers derived, there is a temptation to average results of several specimens and even to calculate standard deviations to indicate scatter. Such calculations are inappropriate when the magnitudes of the numbers indicate only order or rank and not measurements of quantities. When lengths are measured, for instance, the difference between an object that measures 5 cm and one that measures 6 cm is a length of 1 cm. The same difference applies to two objects that measure 9 and 10 cm, respectively. The increment of one between scaling ratings of one and two, however, is not necessarily the same increment as that between three and four. Adding ranking numbers of this type and dividing by the number of measurements may have little significance. Central tendency and scatter can be indicated by giving the median and the range.

A test that provides measurement on a nominal or classification scale is one in which results merely fall into different categories without any judgment being made that one category is higher or lower than another, that is, ASTM

C 40. In one procedure, a solution from the test sample is compared to a reference solution and judged to be lighter, darker, or the same. In another procedure, five color standards may be used. The latter may be treated as an ordinal scale if one end of the scales is judged to be better than the other end and the stages in-between represent progression from one level to another.

(12) As a result of concern about problems connected with precision statements and how to develop and use them, a joint task group of ASTM Committee C1 on Cement, C9, D4 on Road and Paving Materials, and D18 on Soil and Rock for Engineering Purposes developed two practices: ASTM Practice for Preparing Precision and Bias Statements for Test Methods for Construction Materials (C 670) and ASTM Practice for Conducting an Interlaboratory Test Program to Determine the Precision of Test Methods for Construction Materials (C 802). ASTM C 670 gives direction and a recommended form for writing precision statements when the necessary estimates (usually standard deviations) for precision and or bias are in hand. ASTM C 802 describes a recommended method for conducting an interlaboratory study and analyzing the results in order to obtain the necessary estimates. Both of these standards appear in several volumes of the *Annual Books of ASTM Standards* and should be studied and followed closely by any task group that is charged with writing a precision and bias statement for construction materials.

(13) If two results differ by more than the (D2S) limit, a number of interpretations are possible. Which interpretation is most appropriate depends on various circumstances connected with the situation, and in most cases a degree of judgment is involved.

The limit in a precision statement is to provide a criterion for judging when something is wrong with the results. Thus the failure of a pair of results to meet the (D2S) criterion causes concern that the conditions surrounding the two tests may not be the same as those existing when the precision index was derived or that the samples used in the two tests are not unbiased samples from the same type material. The appropriate action to take depends on how serious the consequences of failure are. In most cases, a single isolated failure to meet the criterion is not cause for alarm, but an indication that the process under consideration should be watched to see if the failure persists. If appropriate, the tests can be repeated, and usually, the procedures of the laboratory(s) involved should be examined to make sure that the test is performed in accordance with the standard from which the precision statement was developed.

Failure to meet a multilaboratory precision limit may entail more serious consequences than those connected with failure to meet a single-operator criterion. The latter is sometimes used to check the results and procedures of a single operator in a laboratory, and failure to meet the criterion leads to reexamination of the materials and procedures. If the test is being used to determine compliance with a specification, the single-operator (D2S) limit should be used to check whether or not the results obtained are a valid test for the purpose. The former may occur in situations where there is a dispute about acceptance of materials. In these cases, both laboratories should obtain two results by the same operator who was used in the multilaboratory tests, and use the single-operator difference as a check on proper performance of the test method within the laboratories.

Also note that conditions, materials, apparatus, operators, etc., change with time. In many cases, the subcommittee responsible for the test method can obtain proficiency sample data from the Cement and Concrete Reference Laboratory or the AASHTO Materials Reference Laboratory from which appropriate revisions to update a precision statement can be drafted as shown in the ASTM C 670 appendix.

REFERENCES

[1] Arni, H. T., "Statistical Considerations in Sampling and Testing," *Significance of Tests and Properties of Concrete and Concrete-Making Materials, ASTM STP 169B*, American Society for Testing and Materials, Philadelphia, 1978.

[2] Arni, H. T., "The Correlation Coefficient in Analysis of Engineering Data-Its Significance of Limitations," *Public Roads*, Vol. 36, No. 8, June 1971, pp. 167–174.

[3] Arni, H. T., "The Significance of the Correlation Coefficient for Analyzing Engineering Data," *Materials Research and Standards*, Vol. 11, No. 5, May 1971, pp. 16–19.

[4] Natrella, N. G., *Experimental Statistics*, Handbook 91, National Bureau of Standards, Gaithersburg, MD, 1963.

[5] Dixon, W. J. and Massey, F., Jr., *Introduction to Statistical Analysis*, McGraw-Hill, New York, 1957.

[6] Daniel, C. and Wood, F. S., *Fitting Equations to Data*, Wiley, New York, 1971.

[7] Acton, F. S., *The Analysis of Straight Line Data*, Wiley, New York, 1959.

[8] Mandell, J., *The Statistical Analysis of Experimental Data*, Wiley, New York, 1964.

[9] Hockersmith, T. E. and Ku, H. H., "Uncertainty Associated with Proving Ring Calibration," Reprint Number 12.3-2-64, Instrument Society of America, 1964, (reprinted in *Precision Measurement and Calibration*, National Bureau of Standards, Special Technical Publication 300, Vol. 1, 1969.)

[10] Arni, H. T., "Impact and Penetration Tests of Portland Cement Concrete," Report No. FHWA-RD-73-5, Federal Highway Administration, Washington, DC, Feb. 1972.

[11] Arni, H. T., "Impact and Penetration Tests of Portland Cement Concrete," *Highway Research Record No. 378*, Highway Research Board, Washington, DC, 1972, pp. 55–67.

[12] Burr, I. W., *Engineering Statistics and Quality Control*, McGraw-Hill, New York, 1953.

[13] Hudson, S. B. and Steele, G. W., "Prediction of Potential Strength of Concrete from the Results of Early Tests," *Highway Research Record No. 370*, Highway Research Board, Washington, DC, 1971, pp. 25–28.

[14] Hudson, S. B. and Steele, G. W., "Developments in the Prediction of Potential Strength of Concrete from Results of Early Tests," *Transportation Research Record No. 558*, Transportation Research Board, Washington, DC, 1975, pp. 1–12.

[15] Lew, H. S. and Reichard, T. W., "Prediction of Strength of Concrete from Maturity," *Proceedings*, ACI Symposium Volume on Accelerated Strength Testing, to be published.

[16] Hudson, S. B., Bowery, F. J., and Higgins, F. T., "Research Study to Refine Methods and Procedures for Implementing the Method of Early Prediction of Potential Strength of Portland Cement Concrete," West Virginia Department of Highways Research Project 47, Final Report, Woodward-Clyde Consultants, Rockville, MD, 1976.

[17] Bennett, C. A. and Franklin, N. L., *Statistical Analysis in Chemistry and the Chemical Industry*, Wiley, New York, 1954.
[18] Mclaughlin, J. F. and Hanna, S. J., "Evaluation of Data," *Significance of Tests and Properties of Concrete and Concrete-Making Materials, ASTM STP 169A*, American Society for Testing and Materials, Philadelphia, 1966, p. 36.
[19] Duncan, A. J., *Quality Control and Industrial Statistics*, rev. ed., Richard D. Irwin, Inc., Homewood, IL, 1959.
[20] *Manual on Presentation of Data and Control Chart Analysis, ASTM STP 15D*, American Society for Testing and Materials, Philadelphia, 1976.
[21] Walker, S., "Application of Theory of Probability to Design of Concrete for Strength," *Concrete*, Vol. 52, No. 5, Part 1, May 1944, pp. 3–5.
[22] "Realism in the Application of ACI Standard 164–65," *SP-37*, American Concrete Institute, Detroit, MI, 1973.
[23] Siegel, S. in *Nonparametric Statistics for the Behavioral Sciences*, McGraw-Hill, New York, 1956, pp. 16–28.
[24] Siegel, S. in *Nonparametric Statistics for the Behavioral Sciences*, McGraw-Hill, New York, 1956, pp. 26–30.
[25] Philleo, R. E., "Establishing Specification Limits for Materials," *ASTM Cement, Concrete, and Aggregates Journal*, Vol. 1, No. 2, 1979, pp. 83–87.

Variability of Concrete-Making Materials

Anthony E. Fiorato[1]

PREFACE

The subject of variability of concrete-making materials was not covered in previous editions of *ASTM STP 169*. It has been derived from the work of the joint ASTM C01/C09 Task Group on Variability of Concrete-Making Materials. The Task Group is providing the impetus for development of guidelines on determining uniformity of concrete-making materials.

INTRODUCTION

The goal of the concrete supplier is to provide a material that consistently meets requirements set out by the buyer, whether these are defined in the form of prescriptive or performance specifications. The question is, "How do we define and assure uniformity of concrete?" To answer this question, it is necessary to consider those factors that affect concrete properties and performance.

The steps to obtaining concrete performance are conceptually illustrated in Fig. 1. The process starts with a mix design and specification developed for the particular application. It is followed by selection and acquisition of constituent materials and processing of those materials in accordance with the specifications. Presumably, if the design, selection, and implementation steps are properly conducted, the concrete properties and performance will meet job requirements. However, it is naive to assume that the steps to obtaining properties and performance can be achieved without accommodating variations. But what level of variation can be accepted without detrimental impact on performance? This chapter will address the variability of concrete-making (constituent) materials and their effects on performance.

With improvements in concrete technology, concrete has become more versatile, but also more complex in that the number of mix constituents has increased. It is rare to encounter concrete that consists only of cement, fine aggregate, coarse aggregate, and water. Today, most mixes also contain chemical admixtures or mineral admixtures (supplementary cementitious materials) or both. To minimize the variability of concrete, it is necessary to control the uniformity of constituent materials. Uniformity of properties may be as important to the concrete supplier as the individual properties themselves. Within relatively broad limits, the supplier can adapt concrete mixes to accommodate individual properties of constituent materials. However, once that is done, it is essential to maintain uniformity to assure consistent concrete properties and performance. For example, once a mixture has been developed for a specific cement, water reducer, retarder, air-entraining admixture, aggregates, and batch water, unanticipated changes in critical properties of individual components can cause problems with fresh or hardened concrete performance.

An ASTM C01/C09 Uniformity Task Group is working to identify important properties of constituent materials, how much variation is acceptable, and how uniformity from a single source of these materials can be controlled.

PROPERTIES OF CONSTITUENT MATERIALS THAT AFFECT CONCRETE PERFORMANCE

In 1988, members of ASTM Committee C1 on Cement and Committee C9 on Concrete and Concrete Aggregates were surveyed to obtain their impressions on the relative importance of concrete-making materials. Members were asked to rank major constituent materials in their order of importance relative to variability, and also in the context of field practices for three categories of construction (residential, low-rise commercial, and high tech/high strength). In addition, each constituent material was rated relative to its own properties and attributes.

Twenty-eight members responded to the survey. Since no attempt was made to scientifically select the sample population, no claim can be made for statistical significance. However, the respondents are among the world's most knowledgeable and experienced individuals in concrete materials technology. Therefore, the survey can be considered a valid representation of industry experience and perceptions regarding those materials' characteristics that affect concrete performance.

The survey provides guidance on specific materials' properties and performance attributes that impact concrete properties and performance. This information is valuable in identifying properties that must be controlled to achieve uniformity of performance.

Figure 2 is a summary of responses (27 for this part) to a question that required the respondent to rank ten

---
[1] Vice president, Research and Technical Services, Portland Cement Association, Skokie, IL 60077.

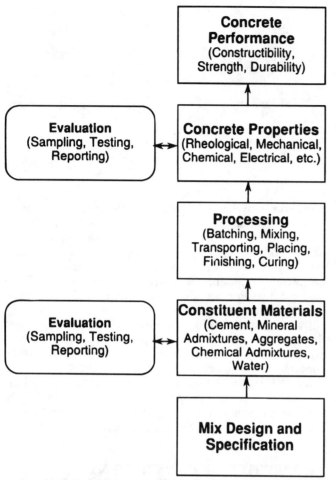

FIG. 1—Uniformity of concrete is a function of the entire design and construction process.

major constituent materials in order of importance from 1 being most important to 10 being least important. In the context of the question, "important" relates to what impact variations in the constituent material would have on concrete performance. No distinction was made as to what performance aspect—constructibility, strength, or durability—might be affected, but it is likely that strength was the most commonly considered attribute. Variations in cement are identified as the most important by a significant margin. Variations in batch water are identified as least important. The relatively "unimportant" rankings given to slag and silica fume may be related to a belief that these materials have little variability, or to the fact that they are less frequently used than the other constituents.

Figure 3 shows results when constituent materials were rated within different construction types (residential, low-rise commercial, and high tech/high strength). The intent of the question was to determine the overall importance of potential variability in the constituent material for selected types of construction. Answers were to reflect whether the variability of the constituent material can be considered to produce few or numerous field problems. In this part of the survey, the materials were rated (not ranked) on a scale of 1 (important) to 10 (not important). Not surprisingly, the overall importance (lower rating numbers) increased from residential to commercial to high tech. Cement was considered the most important for all construction categories. For residential and low-rise commercial construction, silica fume was considered least important (not likely to be used), while for high-tech/high-strength concrete, batch water was considered least important.

While the results in Figs. 2 and 3 provide a rather general picture of perceptions about the relative importance of constituent materials, another valuable part of the survey is summarized in Table 1. For this part, major constituent materials were evaluated independently of each other to

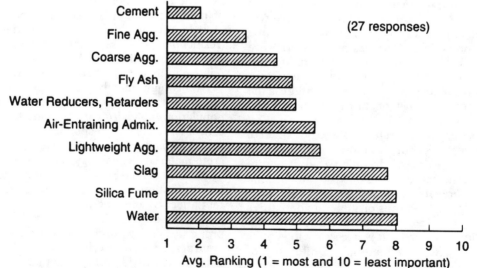

FIG. 2—A 1988 survey of ASTM C1 and C9 committee members revealed their perceptions about the impact of variability of constituent materials on concrete performance.

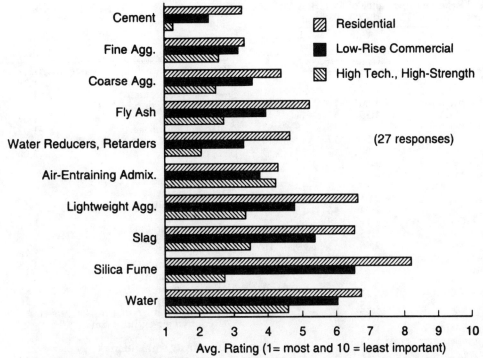

FIG. 3—ASTM C1/C9 1988 Survey of perceptions about the impact of variability of constituent materials on concrete performance based on field practice for different construction types.

identify those characteristics that are important to performance. Respondents were asked to rate each material property or performance attribute on a scale of 1 to 3 with 1 being most important and 3 being least important. Performance attributes reflect the behavior of the constituent material when incorporated in paste, mortar, or concrete.

Although further quantification is needed, data such as listed in Table 1 serve as the basis for uniformity standards. Before such standards can be developed, specific characteristics to be controlled must be identified and then quantified with respect to their impact on concrete performance variations. Table 1 provides a comprehensive list of properties and attributes for the major constituents in concrete. It also provides an indication of their perceived level of importance relative to defining potential variations in concrete performance. This identifies critical characteristics that should be considered in uniformity standards for constituent materials. Such was the case for the only existing uniformity standard for concrete-making materials, that is, ASTM Test Method for Evaluation of Cement Strength Uniformity from a Single Source (C 917).

EVALUATION OF UNIFORMITY

An Example: ASTM C 917

Development of ASTM C 917 took place over a number of years, starting in the 1960s, and culminating in its first edition in 1979 [1]. It is worth reviewing the development of ASTM C 917 because it is representative of efforts that will be needed to implement uniformity standards for properties or attributes of other concrete-making materials. Therefore, the following discussion is presented not to focus on cement strength issues, but to illustrate the process of developing a uniformity standard.

The fact that cement strength was selected as the first attribute to be standardized is not too surprising given the earlier discussion of Figs. 2 and 3 and Table 1 from the 1988 ASTM survey. In fact, the initial impetus for development of ASTM C 917 can be traced to work by Walker [2] and Walker and Bloem [3]. The key point is that a specific attribute of a constituent material for concrete was identified as important to the uniformity of concrete. This led to establishment of a joint committee of the Portland Cement Association and the National Ready Mixed Concrete Association to address strength uniformity [4]. The joint committee planned a program to develop data on uniformity of cement strengths from individual cement plants [5].

The joint committee selected 7- and 28-day strengths of mortar cubes that conform to ASTM Test Method for Compressive Strength of Hydraulic Cement Mortars (Using 2-in. or 50-mm Cube Specimens) (C 109) as the reference for cement strength. After a pilot program in 1975, a one-year voluntary sampling and testing program was initiated in 1976. Forty-six cement companies, representing over 100 plants in the United States and Canada, participated. Testing was conducted on grab samples representing 25 ton (23 Mg) lots of cement at the rate of 30 samples per calendar quarter (preferably ten per month and not more than one per day). Mortar cubes were prepared in each plant's laboratory with ten duplicate sets of cubes prepared each quarter to evaluate within-laboratory

TABLE 1—Perceived Relative Importance of Materials Characteristics to Concrete Quality (ASTM C01/C09 1988 Survey).

	Average	Number of Responses[a]		
		1	2	3
CEMENT UNIFORMITY				
Material property				
Sulfate form and content	1.36	19	8	1
Fineness	1.39	17	11	0
C3A, C3S, C2S	1.43	18	8	2
Alkalies	1.50	17	8	3
Solubility of alkalies	1.68	12	13	3
Chemical composition	1.78	9	15	3
Microscopically determined composition	1.93	8	14	6
Heat of hydration	2.04	6	15	7
Air content	2.04	6	15	7
Loss on ignition (L.O.I.)	2.11	8	9	11
SiO_2, Al_2O_3, MgO	2.15	3	17	7
Specific gravity	2.46	3	9	16
Performance attribute (in paste, mortar, concrete)				
Strength	1.14	25	2	1
Strength gain from 7–28 days	1.21	22	6	0
Setting time	1.25	22	5	1
Early age strength gain	1.25	21	7	0
Slump loss with admixtures	1.43	17	10	1
Slump loss with temperature	1.71	12	12	4
Drying shrinkage	1.72	10	12	3
Sulfate expansion	1.79	7	20	1
Volume changes	1.80	8	14	3
Bleeding characteristics	1.86	10	12	6
Air entrainment dosage	1.88	8	13	5
Finishing characteristics	1.96	8	13	7
Strength gain beyond 28 days	2.07	8	10	10
Autoclave expansion	2.07	5	15	7
Expansion in moist storage	2.12	5	12	8
FINE AGGREGATE UNIFORMITY				
Material property				
Grading including fineness modulus	1.18	23	5	0
Deleterious particles	1.43	18	8	2
Particles −200 sieve (amount and type)	1.50	15	12	1
Moisture content	1.68	16	5	7
Particle shape	1.82	8	17	3
Absorption	2.07	3	20	5
Specific gravity	2.18	4	15	9
Attrition (grinding during mixing)	2.29	4	12	12
Performance attribute (in paste, mortar, concrete)				
Water requirement	1.21	22	6	0
Air entrainment	1.36	20	6	2
Concrete strength	1.57	14	12	2
Water reducer effectiveness	1.96	6	17	5
COARSE AGGREGATE UNIFORMITY				
Material property				
Grading	1.18	23	5	0
Deleterious particles (amount and type)	1.46	17	9	2
Particle shape	1.68	10	17	1
Particles −200 sieve (amount and type)	1.75	11	13	4
Moisture content	1.82	14	5	9
Absorption	2.04	6	15	7
Attrition (grinding during mixing)	2.07	7	12	9
Specific gravity	2.14	5	14	9
Temperature	2.29	3	14	11
Chemical composition	2.32	3	13	12

TABLE 1—continued.

	Average	Number of Responses[a]		
		1	2	3
Performance attribute (in paste, mortar, concrete)				
Concrete strength	1.39	17	11	0
Water requirement	1.43	17	10	1
Freeze-thaw durability (D-cracking)	1.63	13	11	3
Drying shrinkage	1.82	7	19	2
Thermal vol. changes (cracking, etc.)	2.00	5	18	5
FLY ASH UNIFORMITY				
Material property				
Loss on ignition (L.O.I)	1.11	25	1	1
Fineness	1.37	17	10	0
Variations in CaO	1.65	13	9	4
Variations in SO_3	1.76	11	9	5
Alkalies	1.78	10	13	4
Variations in SiO_2	2.15	7	8	11
Variations in Al_2O_3	2.27	4	11	11
Variations in Fe_2O_3	2.48	2	9	14
Specific gravity	2.26	3	13	10
Performance attribute (in paste, mortar, concrete)				
Required air-entrainment dosage	1.26	21	5	1
Reactivity with different cements	1.54	14	7	3
Time of set	1.58	13	11	2
Reactivity at different temperatures	1.71	12	7	5
Response to admixtures	1.81	9	13	4
Pozzolanic activity index	1.81	10	12	5
Shrinkage	2.00	3	18	3
WATER REDUCERS, HRWR, RETARDERS UNIFORMITY				
Material property				
Sensitivity to cement composition	1.30	19	8	0
Sensitivity to time of addition	1.48	15	11	1
Compatibility with other admixtures	1.48	16	9	2
Percent solids	1.59	15	8	4
Composition and concentration	1.59	15	8	4
Sensitivity to temperature	1.62	15	6	5
Variations in chlorides	1.78	9	15	3
Temperature stability (freezing, etc.)	1.81	11	10	6
Variation in alkalies (HRWR)	1.85	8	14	4
Stability in storage	1.93	8	13	6
Performance attribute (in paste, mortar, concrete)				
Time of set	1.11	24	3	0
Rapid stiffening	1.15	23	4	0
Early-age strength	1.41	17	9	1
Later-age strength	1.85	8	15	4
Finishing characteristics	1.93	10	9	8
AIR-ENTRAINING ADMIXTURE UNIFORMITY				
Material property				
Percent solids (specific gravity)	1.54	14	7	3
Composition (infrared spectra)	1.91	7	11	5
pH	2.00	7	10	7
pH in deionized water	2.17	3	13	7
Performance attribute (in paste, mortar, concrete)				
Stability of air with fly ash	1.11	24	3	0
Air void system characteristics	1.27	20	5	1
Sensitivity to cement composition	1.37	19	6	2
Sensitivity to temperature	1.44	16	10	1

TABLE 1—continued.

	Average	Number of Responses[a]		
		1	2	3
Generation of air voids	1.50	16	7	3
Compatibility with other admixtures	1.63	14	9	4
Sensitivity to aggregate grading	1.78	10	13	4
Sensitivity to mix water composition	2.15	7	9	11
LIGHTWEIGHT AGGREGATE UNIFORMITY				
Material property				
Unit weight	1.23	20	6	0
Absorption	1.27	19	7	0
Grading	1.38	16	10	0
Moisture content	1.38	19	4	3
Specific gravity	1.60	14	7	4
Particle shape	1.77	8	16	2
Attrition (grinding during mixing)	1.81	8	15	3
Performance attribute (in paste, mortar, concrete)				
Concrete strength	1.31	20	4	2
Shrinkage and volume changes	1.58	12	13	1
Air entrainment	1.85	10	10	6
Absorption of admixtures	1.96	8	10	7
SLAG UNIFORMITY				
Material property				
Fineness	1.24	19	6	0
Glass content	1.48	14	10	1
Variation in chemical composition	1.74	10	9	4
Specific gravity	2.12	5	12	8
Performance attribute (in paste, mortar, concrete)				
Activity index	1.25	19	4	1
Temperature	1.63	13	7	4
Required air-entrainment dosage	1.83	7	14	3
Required water reducer, HRWR dosage	2.00	7	10	7
Shrinkage	2.00	5	14	5
SILICA FUME UNIFORMITY				
Material property				
Composition	1.40	16	8	1
Percent solids	1.68	11	7	4
Stability in storage	1.73	11	6	5
Specific gravity	2.23	6	5	11
Performance attribute (in paste, mortar, concrete)				
Air entrainment and air void system	1.44	16	7	2
MIX WATER UNIFORMITY				
Material property				
Chloride content	1.52	16	5	4
Organics content	1.62	14	8	4
Alkali content	1.83	9	10	5
Sulfate content	1.92	8	11	6
Hardness	2.25	4	10	10
pH	2.29	3	11	10
Solids content	2.38	2	11	11
Performance attribute (in paste, mortar, concrete)				
Air entrainment	1.46	14	9	1
Time of set (Cl,Na$_2$CO$_3$)	1.72	10	12	3
Temperature	1.80	10	10	5
Durability (ASR, sulfate resist., etc.)	2.00	7	10	7

[a] 1 = important to 3 = unimportant.

test error. Data were submitted quarterly for statistical analysis.

Data from the program that includes information on standard deviations, coefficients of variation, 7- and 28-day average strengths, and ratios of 28-day to 7-day strengths are published in the Appendix to ASTM C 917. They are arranged in terms of cumulative percentage of plants falling below the value indicated for the statistic of interest, and provide a reference point for comparing strength uniformity results from a particular source.

In addition to comprehensive data that quantified potential strength uniformity, the PCA/NRMCA program provided extensive information on sampling and testing procedures, and correction factors for testing variations. Compilation of this information into a draft recommended practice document greatly facilitated the ASTM development process for the new standard. Thus, the first ASTM standard for uniformity of a concrete-making material, ASTM C 917, was approved in 1979, approximately two years after completion of the test program.

Even with approval of a document such as ASTM C 917, there is a continuing development and educational process that must take place to foster appropriate use. A uniformity standard should provide a communication tool between manufacturer and customer that will improve overall concrete quality and performance [1,6,7]. Education of both those providing and those using uniformity data is an important step once a standard is introduced [8,9]. A final test of the standard is whether it is used. Widespread adoption of ASTM C 917 has been slow [10,11]. However, with the increasing trend toward total quality management, it is anticipated that its use will grow.

The early work by Walker recognized that concrete strength uniformity was not solely a function of cement strength uniformity; and included discussion of such factors as sampling and testing variations, temperature effects, and age effects [2]. Batching, mixing, transporting, placing, and curing also have important implications. Walker also noted that strength was not the only concrete performance attribute of importance; he recognized constructibility and durability as other critical attributes [2].

Given that today's concretes are more sophisticated, it is appropriate to consider standardization of uniformity provisions for other constituent material properties and attributes that affect concrete constructibility, strength, and durability. The following section discusses a protocol (guide) for future standards that is being developed by the ASTM C01/C09 Task Group [12].

Guidelines for Future Standards

Figure 4 illustrates the process for developing a uniformity standard. First, the critical property or attribute of the particular constituent material must be identified. Potential properties and attributes are listed in Table 1. It would be prohibitive and unnecessary to develop uniformity requirements for each specific property or attribute in Table 1, so those that are most critical must be selected. This is an appropriate responsibility for ASTM committees that govern standards for materials listed in Table 1.

Once the attribute has been identified for evaluation, the test method must be selected. In most cases, an appro-

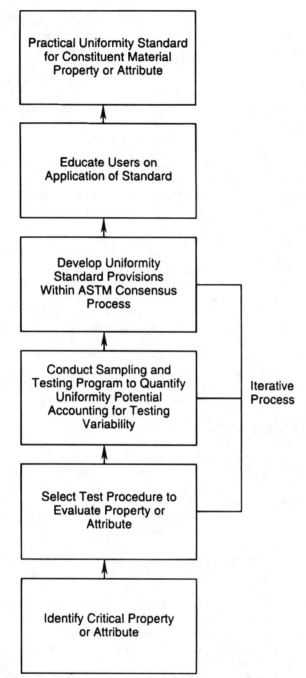

FIG. 4—Process for developing uniformity standards for concrete-making materials.

priate ASTM method exists, particularly for material properties. However, for some attributes, standard procedures may need to be developed or existing procedures modified.

After test methods are selected, a comprehensive sampling and testing program to quantify uniformity potential and testing variability for the particular property should be conducted. This will provide data for specific standard's provisions on sampling frequency and protocol, testing criteria, statistical corrections for testing variations, and reporting requirements. The development of comprehensive test data is considered an essential part of the process because it is necessary to define the effect of inherent variability of testing on the property or attribute being measured.

Testing errors can be significant, particularly for between-laboratory tests, and must be accommodated. Generally, single-laboratory testing is preferred for uniformity work. For example, as pointed out by Gaynor [10], the precision statement in ASTM C 109 implies that duplicate batches of mortar made in a single laboratory should give strengths that do not differ by more than 10.7% one time in 20. However, single batches mixed by two different laboratories should not differ by more than 20.7%.

The ASTM C01/C09 Task Group Guide should speed the standards development process because the test program data for any property or attribute can be "plugged in" to a standard format. The Task Group Guide addresses the following components for determining the uniformity of properties of a material from a single source:

1. Sampling

(a) Sampling is performed by trained personnel.
(b) Grab samples are taken at a frequency defined by maximum lot sizes.
(c) Sampling protocols are defined by existing ASTM standards.
(d) A minimum rate of sampling of ten per month or two per week is required.

2. Evaluation Procedure

(a) Samples are tested in accordance with standard ASTM procedures.
(b) Variations from single source are corrected for variations inherent in the test procedures.
(c) Within- and between-laboratory testing variations are considered as required.
(d) Single-laboratory test variations are established by duplicate testing if no history is established.
(e) Between-laboratory variations are quantified by sample exchange or standard reference samples.

3. Statistical calculations

(a) Equations are defined for average and total standard deviations of the measured values.
(b) Equations are defined for testing standard deviation and coefficient of variation.
(c) Equations are defined for standard deviation corrected for testing variations.

4. Report requirements

(a) Reports identify materials tested.
(b) Reports cover a minimum of three and a maximum of 12 months.
(c) Reports include duplicate test results.
(d) Reports include specific statistical results for time period covered.

This format follows that of ASTM C 917 and provides a "generic" approach to accommodating any property or attribute.

As discussed here, even after the development of the consensus standard, it will be necessary to educate users. This will be a continuing process. In addition, uniformity of concrete-making materials is necessary, but not sufficient, for obtaining uniform concrete performance. Field practices must also be addressed.

SUMMARY

This chapter has addressed uniformity of concrete-making materials. Those properties and attributes of constituent materials that are considered to affect concrete uniformity are identified and a protocol for developing uniformity standards is discussed. With the growing sophistication of concrete mixtures, and the importance of total quality to concrete users, it is essential that the industry be prepared to provide uniform product performance. This will lead to increasing reliance on uniformity standards by suppliers of constituent materials and concrete producers.

REFERENCES

[1] Peters, D. J., "Evaluating Cement Variability—The First Step," NRMCA Publication 161, National Ready Mixed Concrete Association, Silver Spring, MD, 1 April 1980.

[2] Walker, S., "Uniformity of Concrete Strength as Affected by Cement," NRMCA Publication No. 77, National Ready Mixed Concrete Association, Silver Spring, MD, March 1958.

[3] Walker, S. and Bloem, D. L., "Variations in Portland Cement," *Proceedings*, American Society for Testing and Materials, Vol. 58, 1958; also reprinted as NRMCA Publication No. 76.

[4] Davis, R. E., "Uniformity of Portland Cements: Facts and Fantasies," *Modern Concrete*, April 1975, p. 61.

[5] Klieger, P., "The PCA/NRMCA Strength Uniformity Study," Research and Engineering Standards Committee Meeting, National Ready Mixed Concrete Association, Houston, TX, 2 Feb. 1976.

[6] Davis, R. E., "How Uniform is a Portland Cement From a Single Source," *Concrete Construction*, Feb. 1977, p. 87.

[7] Davis, R. E., "The Cement Producer's Role in Providing Information for an ASTM C 917 Method for Evaluation of Cement Strength From a Single Source," Research and Engineering Standards Committee Meeting, National Ready Mixed Concrete Association, St. Louis, MO, 22 July 1980.

[8] Gaynor, R. D., "Use of Cement Strength Uniformity Information," National Ready Mixed Concrete Association, Silver Spring, MD, Feb. 1978.

[9] Gaynor, R. D., "Has There Been a Change in Strength Level? (Or How to Look at C 917 Data When It Comes In)," NRMCA Seminar on Cement Strength Uniformity, St. Louis, MO, 22 July 1980.

[10] Gaynor, R. D., "Cement Strength Variability and Trends in Cement Specifications in the U.S.," European Ready Mixed Concrete Organization (ERMCO) Congress, Rome, May 1986.

[11] *Uniformity of Cement Strength, ASTM STP 961*, E. Farkas and P. Klieger, Eds., American Society for Testing and Materials, Philadelphia, 1987.

[12] "Guide for Determining Uniformity of Ingredients of Concrete From a Single Source," ASTM C01/C09 Task Group on Variability of Concrete-Making Materials, Draft No. 3, American Society for Testing and Materials, Philadelphia, July 1992.

The Role of Cement and Concrete Testing Laboratories

James H. Pielert[1]

PREFACE

The contributions of J. J. Waddell and J. R. Dise, who prepared the chapters on laboratory testing in *ASTM STP 169A*, and *ASTM STP 169B*, respectively, are acknowledged.

INTRODUCTION

Field inspection (including sampling) and laboratory testing of concrete and concrete-making materials are key activities in the construction process. The results of these activities are important to building officials, contractors, concrete producers, owners, and architectural and engineering firms in determining whether the qualities of the materials in the construction comply with contract documents. The role of field inspection and laboratory testing in promoting the quality of construction, and protecting the public safety is receiving added consideration by the building community, both nationally and internationally.

CONCERNS IN TESTING

Testing of concrete and its component materials consists of sampling, performing specified tests, and reporting results of tests.

Procurement of specimens through sampling is often regarded as the most important step in the testing process, and it is recognized that poor specimens and faulty sampling techniques will defeat the purposes for which tests are made. A laboratory can not produce satisfactory information if the samples it receives are carelessly taken or are altered by mistreatment in initial storage, curing, or shipment and do not represent the material under consideration. Sampling procedures are described in standards for concrete and concrete-making materials prepared by the American Society for Testing and Materials (ASTM). Other recognized national standards such as those prepared by the American Association of State Highway and Transportation Officials (AASHTO) for highways and bridge construction may vary slightly from ASTM standards. Adequate instruction of sampling personnel is essential, and supervisors at all levels must be well acquainted with sampling plans and procedures, and fully prepared to ensure that the plans and procedures are faithfully followed.

Concrete specifications enumerated in construction contracts are often based on the results of test methods developed by ASTM. For example, the Standard Building Code of the American Concrete Institute (ACI 318) that is frequently referenced specifies that tests of material and concrete be made in accordance with ASTM standards. No matter who the originator of the chosen standards may be, it is imperative that every effort be made to avoid the use of obsolete, rejected, or modified testing techniques. Employment of an unsatisfactory procedure is a waste of time and effort, and is potentially dangerous because it can lead to completely erroneous conclusions about the characteristics of the concrete in a structure.

In studying quality assurance for highway construction materials, it has been found that variance in quality can be divided into material or process variance, sampling variance, and testing variance. In a study conducted by the Bureau of Public Roads in the late 1960s, it was found that 50% or more of the overall variance could be attributed to two of these factors; sampling and testing [1]. These two processes must be constantly monitored if optimum results are to be obtained.

On completion of testing, a laboratory customarily submits a written report to its client. Reports should be complete and factual, citing the name of the laboratory, report identification and date issued, name of the client and project identification, sample identification, identification of the standard test method(s) and known deviations, test results, and other information required by the appropriate standard. This may include comments about the appearance or behavior of specimens that might in any way have affected the results obtained.

When a final report is available, it can be determined by the laboratory or another representative of the client if either a material complies or fails to comply with specification requirements [2]. If there is a question about the sampling or testing used to generate the report, then the reliability of the entire report may be questioned.

TRENDS IN PROMOTING QUALITY OF TESTING

The key to competent testing is use of an effective and comprehensive system by the laboratory involving both

---

[1] Manager, Cement and Concrete Reference Laboratory, National Institute of Standards and Technology, Gaithersburg, MD 20899.

quality assurance and quality control activities. ASTM Committee E36 defines quality assurance as "the activity of providing the evidence needed to establish confidence that laboratory data are of the requisite accuracy;" and quality control as "the process through which a laboratory measures its performance, compares its performance with standards and acts on any differences." There is increasing emphasis, both nationally and internationally, on the need for a laboratory to establish and maintain a quality system. [3] ISO Guide 25 [4] defines a quality system as "the organizational structure, responsibilities, procedures, processes and resources for implementing quality management." Such a quality system must be tailored to the unique characteristics and capabilities of each laboratory.

Standards are moving toward the requirement that a laboratory's policies, procedures, and practices that represent the quality system be documented in a quality manual. ISO Guide 25 defines a quality manual as "a document stating the quality policy, quality system, and quality practices of an organization." Such a manual provides the staff with an understanding of the laboratory's quality policies and operating procedures and the extent of their duties and responsibilities. Assessment of a laboratory by an evaluating authority is based on the existence of a comprehensive written quality manual and documentation that demonstrates that the laboratory operates in the way stated in the manual.

A quality manual is needed since it serves as the basic reference to a laboratory's quality system. Typical topics to be covered in the quality manual are described in ISO Guide 49 [5]. These topics include:

1. Table of contents
2. Quality policy
3. Terminology
4. Description of the laboratory structure
5. Staff
6. Equipment testing, calibration, and maintenance
7. Environment
8. Test methods and procedures
9. Updating and control of documents affecting quality
10. Handling of items to be tested
11. Verification of results
12. Test reports
13. Diagnostic and corrective actions
14. Records
15. Responses to outside complaints and comments
16. Subcontracting
17. Cooperation between laboratories

ASTM Committee E36 on Laboratory and Inspection Agency Evaluation and Accreditation has been active in preparing standards related to testing laboratory quality [6]. In some technical areas, the work of the committee has lead to similar international standards, while in other areas the international work has preceded the ASTM standards.

The relevant standards prepared by Committee E36 include:

1. ASTM Practice for Use in the Evaluation of Testing and Inspection Agencies as Used in Construction (E 329)—Standard deals with laboratories testing and inspecting construction materials and references other ASTM standards for specific materials.
2. ASTM Guide for General Criteria Used for Evaluating Laboratory Competence (E 548)—Criteria are provided for evaluating the technical competence of a testing laboratory inspection agency, or other organization involved in testing, measuring, inspecting, or calibrating activities.
3. ASTM Guide for Laboratory Accreditation Systems (E 994)—Guide identifies important features of systems that accredit testing laboratories or other organizations involved in testing, measuring, inspecting, and calibrating activities. ASTM E 994 was the basis of ISO Guide 54 [7] and ISO Guide 55 [8].
4. ASTM Terminology Relating to Laboratory Accreditation (E 1187)—This standard provides definitions of terms under the jurisdiction of ASTM Committee E36.
5. ASTM Guide for Categorizing Fields of Testing for Laboratory Accreditation Purposes (E 1224)—This guide provides a uniform means of categorizing field of testing in terms of products, services, or test methods.
6. ASTM Guide for Development and Operation of Laboratory Proficiency Testing Programs (E 1301)—Guidance is provided on the development and operation of proficiency sample programs including management structure, specimen preparation and distribution, and the analysis and reporting of results.
7. ASTM Guide for the Selection, Training, and Evaluation of Assessors for Laboratory Accreditation Systems (E 1322)—Guidance is provided to administrators of laboratory accrediting bodies in dealing with assessors who assess the performance of laboratories. This includes the selection, training, and evaluation of assessor performance.
8. ASTM Guide for Evaluating Laboratory Measurement Practices and the Statistical Analysis of the Resulting Data (E 1323)—Guidance is provided for assessors to evaluate measurement practices of laboratories and protocols for statistically analyzing the resulting data.

ASTM Committees C9 on Concrete and Concrete Aggregates and Committee C1 on Cement have prepared documents for evaluating concrete and cement testing laboratories. In 1987 Committee C9 published ASTM Practice for Laboratories Testing Concrete and Concrete Aggregates for Use in Construction and Criteria for Laboratory Evaluation (C 1077). This standard, which is referenced in ASTM E 329, is utilized for the evaluation and accreditation of laboratories testing concrete and aggregates. ASTM C 1077 establishes minimum requirements for the testing laboratory's personnel, equipment, and quality system. The standard lists ten mandatory ASTM test methods for concrete and concrete aggregates that the laboratory must be capable of performing. It also gives a list of optional ASTM concrete and concrete aggregate test methods that a laboratory may request to be evaluated. A laboratory complying with ASTM C 1077 is required to establish and maintain a quality system that includes procedures for personnel evaluation and training, participation in a proficiency sample program, procedures for record keeping, procedures for equipment calibration and maintenance inventory of test equipment, procedures for handling technical complaints, and procedures for assuring the quality of external technical services. ASTM C 1077 also requires that testing services of the laboratory be under the full-time technical direction of a professional engineer, and that the laboratory be inspected by an evaluating authority at intervals of approximately 24 months.

ASTM Practice for Evaluation Criteria of Hydraulic Cement Testing Laboratories (C 1222) was published by Committee C1 in 1993. It identifies minimum training and experience requirements of personnel, and equipment requirements for cement testing laboratories. ASTM C 1222 does not identify test methods that a laboratory must be able to perform, but requires that it have the capability of performing all laboratory testing associated with its intended functions according to standard chemical or physical requirements listed in ASTM Specification for Masonry Cement (C 91); ASTM Specification for Portland Cement (C 150); ASTM Specification for Blended Hydraulic Cements (C 595); or ASTM Specification for Expansive Hydraulic Cement (C 845). The scope of a testing laboratory may be either chemical testing or physical testing, or both. A laboratory complying with ASTM C 1222 is required to establish and maintain a quality system similar to ASTM C 1077. The manager of the laboratory should be a chemist, materials analyst, or an engineer with supervisory experience in the testing of hydraulic cement; or a person with equivalent science-oriented education or experience. A periodic assessment by an evaluating agency is also required.

EVALUATING AUTHORITIES

ASTM C 1077 defines an evaluating authority as "an independent entity, apart from the organization being evaluated, that can provide an unbiased evaluation of that organization." The standard lists the Cement and Concrete Reference Laboratory (CCRL) of the National Institute of Standards and Technology (NIST) that provides laboratory inspection and proficiency samples services; and accrediting bodies including the American Association for Laboratory Accreditation (A2LA), AASHTO Accreditation Program (AAP), Concrete Materials Engineering Council (CMEC), and the National Voluntary Laboratory Accreditation Program (NVLAP).

CCRL along with the AASHTO Materials Reference Laboratory (AMRL) comprise the Construction Materials Reference Laboratories (CMRL) at NIST [9]. CCRL and AMRL are NIST Research Associate Programs that are managed by NIST under Memoranda of Agreement between the sponsoring organizations and NIST. ASTM is the sponsor of CCRL and AASHTO is the sponsor of AMRL. ASTM provides programmatic and technical oversight through a Joint C1/C9 Subcommittee on the CCRL, while AASHTO provides oversight to AMRL through its Subcommittee on Materials.

The CMRL promotes the quality of testing through assessment of the performance of construction materials testing laboratories; support to national standards committees in the preparation of test methods; and through use of its programs by accrediting bodies, governmental agencies, and other organizations involved in quality assessment. The primary functions of the CMRL are the inspection of testing laboratories and the distribution of proficiency test samples. Inspectors visit laboratories to evaluate equipment and procedures according to the requirements of the test methods and provide a report of findings. Concrete materials included in the CCRL laboratory inspection program are cements, concrete, reinforcing steel, aggregates, and pozzolans. Samples of concrete materials routinely distributed to laboratories by CCRL include portland cement, blended cement, masonry cement, portland-cement concrete, and fly ash; while AMRL distributes aggregate samples. Data from these programs are provided to standards committees of ASTM and AASHTO for assessing the adequacy of test methods, determining the impact of revisions to standards, and developing precision statements. Over 1000 different laboratories in the United States and 15 other countries currently participate in the CMRL laboratory inspection and proficiency sample programs. Utilization of the CMRL programs is voluntary and laboratories are not rated, certified, or accredited.

A2LA was formed in 1978 as a nonprofit, scientific, membership organization dedicated to the formal recognition of testing organizations that have been shown to be competent [3]. A2LA has granted accreditation in the following fields of testing: biology, chemistry, construction materials, geotechnical, electrical, mechanical, nondestructive testing, and thermal. Cement, concrete, and aggregates are included in the construction materials field of testing. A2LA requires laboratories to participate in the applicable proficiency sample programs of CCRL and AMRL, and uses peer assessors for on-site assessment of the laboratory.

AAP was started by AASHTO in 1988 for construction materials testing laboratories [10]. AAP certifies the competency of testing laboratories in carrying out specific tests on soils, asphalt cements, cut-back asphalts, emulsified asphalts, bituminous mixtures, bituminous concrete aggregates, and portland-cement concrete and aggregates. The concrete laboratory inspection program of CCRL and proficiency sample programs of CCRL and AMRL are used to evaluate the performance of laboratories that test portland-cement concrete and aggregates.

CMEC was incorporated in Florida in 1983 for the purpose of improving concrete quality and testing and initiated a laboratory accreditation program for concrete testing laboratories in 1984 [11]. The program follows the guidelines established by ASTM C 1077 and includes an inspection of a laboratory's human resources, laboratory and field equipment, quality systems, and testing procedures. Verification and calibration of testing machines is included as part of the program. CMEC operates its own proficiency sample program and uses its own assessors for on-site visits.

NVLAP, which is administered by NIST, accredits public and private testing laboratories based on evaluation of their quality system, equipment, test procedures, and technical qualifications and competence for conducting specific test methods. NVLAP accreditation is based on conformance to the U.S. Code of Federal Regulations, Title 15 Part 7—The National Voluntary Laboratory Accreditation Program Procedures, and the standard test methods for which the laboratory is seeking accreditation. The NVLAP assessor reviews the laboratory's quality system, equipment and facilities, test procedures, and competence to conduct specific methods in a given field of testing [12]. NVLAP accreditation in the construction testing services is available for selected methods of test for concrete, aggregates, cement, admixtures, geotextiles, soil and rock, bituminous materials, and steel materials. Participation in the CCRL and AMRL proficiency sample programs may be used by laboratories to meet the NVLAP proficiency testing requirement for cement, concrete, soil, and bituminous material. NVLAP does upon request provide accreditation to concrete testing laboratories for ASTM E 329 and ASTM C 1077. NVLAP uses peer assessors for onsite assessment of the laboratory.

TECHNICIAN COMPETENCY

The competency of laboratory technicians conducting the tests plays a vital role in assuring the quality of testing. As a means of ensuring this competency, technician training and certification programs are offered by organizations including the American Concrete Institute (ACI), National Institute for Certification in Engineering Technologies, National Ready Mixed Concrete Association, and the Portland Cement Association. ASTM C 1077 references the programs of these organizations as a means of demonstrating the competency of a technician in performing tests on concrete materials.

ACI operates certification programs for field and laboratory testing technicians [13]. Both written and performance tests on a specified number of ASTM test methods are given to applicants seeking certification. The written examination is a closed-book multiple-choice test, and the performance test requires actual demonstration of the method without use of notes. Recertification is required every five years and requires successful completion of a written examination.

CONCLUSIONS

High-quality field inspection and laboratory testing services are important to achieving safe, efficient, and cost-effective concrete structures. Mechanisms are being developed, standardized, and implemented that will promote the quality of these services. The importance of concrete sampling and accurate testing will increase as structural design and modeling by computers require that material properties be determined more precisely. Additionally, the development of high-performance concretes with enhanced mechanical, volume stability, durability, and placement properties require more accurate test results.

REFERENCES

[1] McMahon, T. F. and Halstead, W. J. "Quality Assurance in Highway Construction Part 1—Introduction and Concepts," *Public Roads*, Vol. 35, No. 6, Feb. 1969, p. 1929.

[2] Waddell, J. J., "Quality of Testing," *Significance of Tests and Properties of Concrete and Concrete-Making Materials, ASTM STP 169A*, American Society for Testing and Materials, Philadelphia, 1975, p. 32.

[3] Locke, J. W., "Quality Assurance in the Construction Materials Laboratory," *Proceedings*, Workshop on Evaluation of Cement and Concrete Laboratory Performance, NIST Special Publication 788, National Institute of Standards and Technology, Gaithersburg, MD, July 1991, p. 105.

[4] *General Requirements for the Competence of Calibration and Testing Laboratories*, ISO Guide 25, International Organization for Standardization, Geneva, Switzerland, 1990.

[5] *Guidelines for Development of a Quality Manual for a Testing Laboratory*, ISO Guide 49, International Organization for Standardization, Geneva, Switzerland, 1986.

[6] Locke, J. W., "Development of ASTM Standards for Laboratory Accreditation and Other Accreditation Activities in the United States," *Proceedings*, Symposium on Test Quality for Construction, Materials and Structures, RILEM, Paris, Oct. 1990, pp. 266–276.

[7] *Testing Laboratory Accreditation Systems, General Recommendations for the Acceptance of Accreditation Bodies*, ISO Guide 54, International Organization for Standardization, Geneva, Switzerland, 1988.

[8] *Testing Laboratory Accreditation Systems—General Recommendations for Operation*, ISO Guide 55, International Organization for Standardization, Geneva, Switzerland, 1988.

[9] Pielert, J. H., "Construction Materials Reference Laboratories at NIST—Promoting Quality in Laboratory Testing," *ASTM Standardization News*, Dec. 1989, pp. 40–44.

[10] Pielert, J. H., "Activities of the Construction Materials Reference Laboratories Related to Laboratory Accreditation," *Accreditation Practices for Inspections, Tests, and Laboratories, ASTM STP 1057*, American Society for Testing and Materials, Philadelphia, 1988, pp. 30–36.

[11] Roebuck, J. P., "Accreditation of Testing Laboratories in Florida," *Proceedings*, Conference on Accreditation of Construction Materials Testing Laboratories, NBS Special Publication 736, National Bureau of Standards, Nov. 1987, pp. 35–38.

[12] Gladhill, R. L., "Impact of Accreditation on Laboratory Quality Assurance," *Proceedings*, Symposium on Test Quality for Construction, Materials and Structures, RILEM, Paris, Oct. 1990, pp. 283–289.

[13] *ACI Certification Guide*, American Concrete Institute, Detroit, MI.

Research Needs

W. L. Dolch[1]

PREFACE

This chapter is the fourth in a line of similar titles that appeared in the predecessors of this volume, *ASTM STPs 169*, *ASTM STP 169A*, and *ASTM STP 169B*. The other titles were all "Needed Research." This title is "Research Needs," and the proposal is to treat it as an unfinished sentence, and the second word as a verb rather than as a noun. The aim is to fill in the rest of the sentence and discuss what research in concrete needs in order to prosper in the future. As such, this is essentially subjective, an opinion.

INTRODUCTION

Two of the previous chapters, those by Goldbeck (in *ASTM STP 169*) and Walker (in *ASTM STP 169B*), were mostly lists of things important and unknown about concrete—specific topics that needed investigating and projects that needed doing. These lists make interesting reading, because they imply the question of what has been accomplished. Opinions will differ, but it is disappointing to see how many of these questions remain unanswered, now many years later. Although we have greatly increased our knowledge, we still do not properly understand the mechanisms of shrinkage, or freezing-and-thawing, or alkali-aggregate reactions.

The real problem with such lists is that the possibilities are so numerous that they have expanded beyond any usefulness. This expansion is the result of the variety and complexity of concrete and of the research done on it. If topics of needed research are to include all aspects of the composition and behavior of concrete, then they include the cement, the paste and its components, the aggregate, the admixtures, the interfaces, the loads, the surroundings, and so on. This number is ever increasing, as a glance at the new chapters in this volume shows. There has been a great increase in the number of substances used in concrete and in the ways in which concrete is used. These changes have resulted in such an expansion of the possibilities for topics to be included in lists of needed research that almost anything is fair game. But one person's interest is another's indifference. Besides, are not such lists a little presumptuous? Any competent researcher knows what topics are important; he does not need someone else to tell him.

The chapter by Bates in *ASTM STP 169A* was different. It was a plea for a pause, for a look at what we already knew, and for efforts to coordinate, consolidate, and disseminate this knowledge. To a reasonable extent, this has been accomplished. One can think of the abstracting services, the statements of research in progress, the bibliographies, the reviews, the texts and monographs, the symposia and conferences, and recent advances such as the American Concrete Institute's (ACI) CD ROM program.

In spite of an enormous amount of effort, and much progress, it may be that we have not come as far as we might. And it is disconcerting to reflect on how few of the really significant advances in cement and concrete have been the result of deliberate attack from research. Two of the more important advances—control of set by gypsum and the use of air entrainment—were pure chance. So perhaps there is room for another point of view and for some changes in attitude. Some suggestions follow, in no special order.

FIRST NEED

There has been, all along but increasing in recent years, a persistent tendency to emphasize the utility and immediate practicality of concrete research. Even in publicly funded research, the emphasis has been on the applied rather than the basic. So, we have all these requirements for statements of anticipated usefulness, and benefit-to-cost ratios, and implementation sections. This is not to deny the need for limited, practical, and short-term investigations, but the frequent result is a house built on sand. We end up with knowledge, but with no understanding that places that knowledge in the proper context and defines its limitations. This pragmatism then becomes the basis for design. Usually, the result is merely waste, but sometimes it is tragedy.

Proverbs enjoins us, ". . . with all thy getting get understanding." The aim of research surely should be understanding, not merely knowledge. And the motive that drives this kind of research is curiosity, not practicality or profit. But when the money gets tight, the first thing that goes is "basic" research, on the grounds that we can not afford it. The understanding should come first, then

---

[1] Professor of Engineering Materials, School of Civil Engineering, Purdue University, West Lafayette, IN 47907.

the development. It is no accident that the "R" is first in R&D.

The plain fact is that much of the concrete research nowadays is narrow, restricted, parochial, and defined in terms of the researcher's needs rather than the understanding of the subject. Look at all the published papers that are about "how we shot the bear" in this place or that state or the other region. Most of these papers contribute nothing to the understanding of the subject and are merely excuses for a trip to a nice conference.

Much of today's research work is the result of an agenda that has been concocted by some ad hoc committee, whose wishes must be followed, because they control the purse strings. The history of such efforts is disappointing. They start off with great fanfare and end with a padded report tucked away out of sight. The basic problem is that the results are determined before the work begins. All that is left is to fill in the blanks. The researcher follows the predetermined agenda, because he must, in order to get the funding. But isn't this going about it the wrong way? A research agenda should be the result of the interests of the researcher or team, not those of some committee. And there must be room to vary from the plan, and to explore side alleys, and sniff around rat holes. This is the way the all-important curiosity gets satisfied.

The other great impeller of doubtful research is the "publish or perish" imperative of the universities. This need is very much in the mind of every junior professor, and he realizes that, when it comes down to it, the quantity of his publications is more important than their quality, official protestations to the contrary. Combine this with the steadily increasing financial crisis of the universities, and you have a ready-made scene for ad hoc research and more trivia for the already bulging journals.

Having said all this, it is recognized, of course, that research is inherently an inefficient process. Benjamin Franklin compared research with a baby; you may not know if it will be important until long after. So a lot that will never amount to anything has to be tolerated, in order to be sure of getting the smaller amount that is good. But even so, the tail is wagging the dog nowadays.

So the upshot of this first need is that all concerned in the research process put first the things that are really first. The researchers, their administrators, the funding agencies, the committees, the reviewers and editors, should keep in mind that you can't grow a tree from the branches first. Everyone must do a better and more critical job of selecting, planning, evaluating, and rewarding.

SECOND NEED

Another need is for a change in the training of the people who will do the research on concrete. In the past, those who have done this kind of research have, for the most part, been civil engineers, or chemists who wandered into the field, or the occasional materials scientist. Recently, larger team efforts have recruited narrow specialists who bring exotic expertise to the problems. But there is a need for students trained in the wider spectrum of disciplines of concrete research, from pure mechanics on one end to pure chemistry on the other.

Other materials fields, such as soils and metals and polymers, have developed a specialized training for their research students. We have geotechnical engineers and metallurgical engineers and polymer chemists, and so on. Part of the reason for this specialized training has been an economic impetus. Other fields have financed their research with much greater generosity than has been the case with concrete, and the progress has been commensurately great. Considering that concrete is the most widely used of all materials, the support for its research from the industry, or from the society, has been small.

Whatever the reason, the fact is that other materials fields have developed an appropriate curriculum for aspiring students, and have become recognized branches or specialties within the scientific and engineering community. Concrete has not become so recognized, in spite of its extraordinary complexity and the high degree of scientific sophistication that is needed now for its investigation.

Only a few universities have any program at all in concrete, and even there it is sometimes regarded as an outrigger on the main effort. The Europeans and the Japanese have done a better job. It is not coincidental that much of the quality research in concrete has passed abroad from the United States.

Needed is the development by educators of a curriculum that includes the requisite variety of basic science and engineering along with modern materials science and specialized courses such as fracture mechanics and surface chemistry. Only then will the universities produce the kind of researcher who is properly equipped, from the start, to do productive work in concrete.

THIRD NEED

A third need is for workers in the concrete field to remember that the word "research" derives from the Latin word for circle. The circular nature of the effort is important. Many of the researchers today are so anxious to get to the electron microscope that they do not even see the library as they rush past. The vitally important step of a measured, even leisurely, reappraisal of what has already been done and learned is even more important today, when the literature on concrete has grown so extensive that we are on the point of developing subspecialties, and the day of the concrete generalist is passing.

An eminent concrete researcher, L. E. Copeland, once proposed that a moratorium of five years be declared on electron microscopy of cement and concrete, so we could look at everything that had already been done. This is not to pick on electron microscopy; the point is applicable to most aspects of the concrete research scene.

Part of the overall reappraisal effort is the review and synthesis advocated by Bates and mentioned earlier. This volume is a portion of that effort. A really good review, with all its necessary selection and judgment, is more important than most of the new results that hurry into print.

FOURTH NEED

The fourth need is to found our measurements on fundamental parameters. Even our everyday acceptance test methods must be greatly improved. The need is to replace empiricism with science. We must measure properties that have real meaning in terms of what we are trying to find out.

It will, of course, be argued that most of the standard tests are the result of custom and experience. All they are supposed to do is come up with a number that the same experience interprets as being satisfactory or not. But these methods keep on being used as research tools. All kinds of influences on this parameter or that are determined by using tests that themselves have little meaning for understanding the material and how it behaves. These methods, in a certain sense, get in the way of the development of better measurements. So we keep on talking of the "flexural strength" of concrete as if it means something.

The three most important properties of concrete are its workability, strength, and durability.

Workability

The basic problem with the measurement of workability is its subjective definition. It is defined as the "ease" of something, and ease cannot be measured quantitatively. So we have all these substitutes. It is remarkable that slump is still considered by many to be a measure of workability. Kelly ball and VB are useful, but strictly empirical. Good tries at rationality were made with remolding number and compacting factor, but they are cumbersome and never caught on.

What, then, is a rational measure of workability? There is none, if we stick to the subjective definition. But fresh concrete is a viscoelastic composite system made up of a more-or-less dispersed particulate phase in a liquid medium. The things that ought to be measured are the descriptors of such a system. A good start has been made on the viscoelastic properties. The parameters of a Bingham body have been described and determined for concrete. It is these, the yield value and the plastic viscosity rather than slump or compacting factor, that we should investigate, and then determine how they are influenced by water content, grading, fines, admixtures, etc.

Refined techniques exist for measuring more fundamental properties of other materials that are granular plastic systems. The literature on soil parameters and their determination is extensive. Just to take the simplest expression, the cohesion and internal friction angle are properties of soils related to their failure. And the placing of concrete, especially fairly stiff concrete, is analogous to the shear failure of soils. Perhaps the measurement of these parameters for concrete, and their change as time goes by and setting occurs, would be of value. Hardly anything has been done along such lines.

Strength

Strength is often said to be the most important property of concrete. Yet none of the conventional tests determines a parameter that is directly related to the mechanism of failure. Most design procedures make use of the unconfined compressive strength value. But unconfined concrete fails by a tensile strain criterion. The compressive strength of concrete is really just an indirect determination of the water-cement ratio and of the maturity of the paste. The various indirect tension tests have the virtue of acknowledging the tensile failure mode, but they make no provision for the critical question, that is, the resistance of the material to cracking.

The correct approach is surely the route of fracture mechanics. Cracks are, after all, the fourth ingredient of concrete. It is full of them, even before loading. So the stability of cracks under load, and the conditions for their propagation, are the main aspects of the failure of the material. Fortunately, after a slow start, the study of the fracture mechanics of concrete is becoming popular and has made much progress. We are perhaps within sight of the development of fracture parameters that can be used in design procedures, especially with the new, high-strength concretes. These developments are one of the brighter spots of modern concrete research.

Durability

Durability is even more important than strength. Most of the distress of concrete is due to failures of durability, rather than to insufficient strength. One can always get concrete strong enough, if he is willing to pay for it. Durability is the aspect of concrete about which the least has been discovered, considering the great effort that has been put into its investigation over many years.

Cracking

Concrete continues to crack, from thermal and shrinkage stresses, about as much as it ever did. For most structures it is accepted as inevitable and is controlled with joints or held together with reinforcement or prestressing. Shrinkage compensating cement was one try at crackless concrete. But there has been little attempt to study the fundamentals of either drying shrinkage or thermal volume change, except for pragmatic control of mass concrete construction.

Shrinkage mechanisms have been postulated, but the significant properties of the paste have been little studied, especially the surface energy concepts that are probably important to the larger amount of the volume change. All aspects of the development of structure in the paste are pertinent, and many of them have been studied extensively. But these results have been mostly descriptive, based on simplified models or visual observation; it is, after all, an extremely complicated system. For an understanding of the shrinkage mechanism, more needs to be known about surface energy of the hydrates, particle size and habit, pore size distributions, and so on.

Thermal stress cracking has also been little studied. A search of the entire publication of *Cement and Concrete Research* shows, for example, only four papers for which the key words are both "thermal" and "stress." Except for temperature records in the field, thermal properties have been measured mostly on small laboratory specimens.

Concrete is frequently considered homogeneous, from the thermal point of view. Better models of the dynamics of thermal change are required, and the appropriate measurements still have to be made. The inclusion of fibers and mineral admixtures increases the complication.

Alkali-Aggregate Reaction

The renaissance of interest in the alkali-aggregate reaction, principally the alkali-silica reaction (ASR), that has occurred in recent years has resulted in it becoming one of the most popular and studied aspects of the behavior of concrete. A great deal has been discovered about susceptible aggregates, field appearance, and the influence of alkali content, admixtures, etc. Several new test methods have been devised for the prediction of difficulty. This is all to the good. But can it be said that we understand the ASR much better than we used to?

Some fundamental work has been done. One thinks of the studies of the composition of the fluid phase of concrete, by the use of the expulsion technique. After all, that is what is reacting with the aggregate components. And some details have been learned that help in the sticky problem of reaction versus expansion. The expansive properties of gels have been studied, and we know something about the circumstances that promote large expansions.

However, there are other important matters that need attention. When a chemist studies a chemical reaction, he studies its stoichiometry, its kinetics, and its thermodynamics. Granted, the study of a chemical reaction that takes place inside such a complicated system as concrete is more difficult than most. But little has been done on any aspect except the stoichiometry, and not much on that. This is a problem that would profit greatly from being broken down for study into its smaller parts. What has been done, however, is mostly pot walloping and the measurement of a lot of expansions that are dubious because doing even that properly involves all sorts of difficulties.

The alkali-carbonate reaction would seem to be in an even worse condition. The fundamental reasons for the expansion and cracking are not as well understood as those for the ASR. Again, most of the work has been pragmatic in an attempt to identify aggregates that will give trouble. This is a worthy objective, but it is no substitute for understanding what goes on.

Freezing and Thawing

In cold regions the most important durability problem is freezing-and-thawing failure. During the fifty years this problem has been studied, the test methods have changed little. They are entirely pragmatic. The concrete sample is exposed to an accelerated temperature regime. Deterioration is measured as a change in either dynamic modulus of elasticity or length of the specimen.

The two methods of dynamic measurement—pulse velocity and resonant frequency—do not agree. The question of which, if either, is a reliable determinator of internal cracking is unresolved. Presumably expansion is a better measure, because the only place the expansion can come from is the cracks that have been formed. By its very definition, durability factor is an inadequate measure of what has happened. All this is to say nothing of the general disagreement among laboratories that have tried to coordinate freezing-and-thawing test results.

For difficulty arising in the paste, the proposed mechanisms for expansion have been long established. It would seem that the hydraulic pressure hypothesis, and its attendant emphasis on the degree of saturation, has been proved, although not everyone agrees. The gel water diffusion (osmotic) hypothesis has yet to be demonstrated for concrete, although its presumed importance in paste samples was established long ago.

The importance of air entrainment as a control measure has been known for a long time. Its exact mechanism is another matter, although the general idea seems understood.

For trouble arising in the aggregate component (D-cracking), the mechanisms proposed long ago remain unchanged. This kind of difficulty has been greatly increasing in importance, and many pavements are in serious trouble with no real solution in sight. A test method based on the pore size distribution of the aggregate has been proposed, but remains to be confirmed and refined.

Some fundamental work has been done. The calorimetric studies of the freezing of water in paste are important. Freezing and thawing is perhaps the most complicated of the durability problems. The details involved are the permeability and pore structure of both the paste and the aggregate, the thermal regime, thermodynamic considerations, and the purely mechanical properties of strain development within a porous, partially saturated system of various components. Hardly any aspect of these fundamentals has been properly studied.

CONCLUSION

The conclusion of these considerations is that research in concrete needs people who are better trained, concepts and measurements based on fundamentals, and a devotion to basic research.

PART II
Freshly Mixed Concrete

Factors Influencing Concrete Workability

John M. Scanlon[1]

PREFACE

As the current author of this chapter, I feel honored to have been chosen to follow in the footsteps of the original authors, I. L. Tyler and Fred Hubbard, who prepared chapters on "Uniformity, Segregation and Bleeding," and "Workability and Plasticity," respectively, which were published in the original *ASTM STP 169*. C. A. Vollick combined, revised and updated these two chapters into "Uniformity and Workability" for *ASTM STP 169A*, and D. T. Smith revised and updated this chapter for *ASTM STP 169B*. The chapter has now been revised, updated, and the title changed to "Factors Influencing Concrete Workability." Mr. Tyler was with the Portland Cement Association, Mr. Hubbard with the National Slag Association, Mr. Vollick with Sika Chemical Corporation, and Mr. Smith with Marquette Cement Manufacturing.

INTRODUCTION

Workability of concrete is generally determined in the eyes of the beholder. Each of the parties having a responsibility for the completed concrete structure views workability as the avenue through which the concrete can be controlled to better assure that his or her responsibilities are fulfilled. These parties may include the owner, engineer, contractor, concrete materials suppliers, testing agency, governmental inspectors, and others. Many projects have been put in jeopardy because of conflicts of interest pertaining to the workability of the freshly mixed concrete. A concrete construction project can not be successful without all of the responsible parties establishing reasonable guidelines pertaining to the workability of the concrete that needs to be maintained for each portion of the work.

It is important to note that by using today's technology, concrete mixtures can be proportioned with practically any workability and still have the capabilities to develop essentially any hardened properties demanded by the structure. Such technology may include the use of specialty products such as nonlocal aggregates and admixtures, both mineral and chemical.

[1] Senior consultant, Wiss, Janney, Elstner Associates, Inc., Northbrook, IL 60062-2095.

TERMINOLOGY

The American Concrete Institute (ACI) defines workability and related terms in ACI 116R [1], but the actual day-to-day interpretation of degree of workability demanded for a concrete construction project needs to be established by the engineer and contractor, either before, or during, the preconstruction conference. Hopefully, the workability requirements are clearly defined in the project specification, but, if not, this is a subject for the prebid conference. Without a true understanding between these two groups of how workability is going to be controlled, the chances of their having major confrontations during construction are assured.

In ACI 116R, there are three primary terms that have to be understood. These terms are rheology, consistency, and, of course, workability itself. Rheology is defined as "the science dealing with the flow of materials, including studies of deformation of hardened concrete, the handling and placing of freshly mixed concrete, and the behavior of slurries, pastes, and the like." Related terminology, consistency, may be more relative to the needs of the concrete industry than workability, because it relates to the ability to establish a workability level and to reasonably maintain this level. Consistency is defined as "the relative mobility or ability of freshly mixed concrete or mortar to flow; the usual measurements are slump for concrete, flow for mortar or grouts, and penetration resistance for neat cement paste." Most problems in mixing, transporting, placing, consolidating, and the hardening of concrete relate to the variability from batch to batch. Once an acceptable level of workability has been established, the concrete quality relies on control of variability by maintaining uniform consistency.

If rheology is the science dealing with the flow of materials and their affects on workability, then just what is workability? ACI 116R defines workability as "that property of freshly mixed concrete or mortar which determines the ease and homogeneity with which it can be mixed, placed, consolidated, and finished." Consequently, if the science of rheology is used to produce concrete having the workability necessary to be beneficial to mixing, transporting, placing, and finishing concrete, and to maintaining that consistency, then the finished product should most likely comply with the requirements for the structure. More information on behavior of fresh concrete during vibration and rheology on consolidation of fresh concrete can be found in ACI 309.1R [2].

UNIFORMITY OF CONCRETE

Concrete mixture proportions should always be developed in such a way that the finished hardened concrete will attain the required physical properties and be able to withstand exposure to the anticipated environmental conditions. Equally important is that the freshly mixed concrete should possess the workability and other characteristics that permits it to be mixed, transported, placed, consolidated, and finished without undue panic and hardship under whatever conditions that may prevail. After mixtures have been developed and the necessary characteristics that affect workability established, it is absolutely necessary that the quantities of ingredients, hence, the workability be maintained relatively uniform.

Nonuniformity may be evident in freshly mixed and hardened concrete. There are two distinct measurements of uniformity, that is, within-batch variations and batch-to-batch variations. Within batch variations refer to variations relating to concrete discharged from the front, middle, and rear of a mixer. These variations may be attributed to excessive wear of the mixer blades, inadequate or excessive mixing time and speed, improper loading sequence, or possibly overloading of the mixer.

Batch-to-batch variations in concrete may be attributed to variations in aggregate gradings, inaccurate weighing or volumetric dispensing equipment, and, of course, all of the variations related to within-batch variations. The concrete mixture proportions may result in segregation of the mixture due to a relatively high water-cement ratio, poor mixture proportions, or in separation of the mixture due to the use of improper materials-handling equipment. Many concrete mixtures have been accused of segregating, when it was actually separating due to improper loading and unloading of wet batch hoppers, such as buckets, belts, pumps, and nonagitating equipment.

The uniformity of concrete production and delivery should always be evaluated, because the better the uniformity during production, transportation, and placing, the greater the opportunity to obtain desired hardened concrete properties. Experience has shown that variations affecting the properties of freshly mixed concrete will also affect the properties of hardened concrete.

CONTROL OF CONCRETE PRODUCTION

Production of uniform concrete can be obtained only through systematic control of all operations from selection and production of materials through batching, mixing, transporting, conveying, placing, consolidation, finishing, and curing. All materials must have uniform properties. Control must be maintained at the batching and mixing plants. Aggregates should be tested for grading, densities, and moisture content, and mixtures must be adjusted to correct for changes in these properties. Mixed concrete should be tested for consistency, air content, temperature, and unit weight. Concrete specimens must be fabricated for evaluating hardened concrete properties.

Routine instructions for measuring, mixing, and placing concrete are given by the American Concrete Institute (ACI) *Manual of Concrete Inspection* [3]. Additional information on practices that lead to better uniformity are found in the ACI 304 [4]. Some variation must be accepted, but consistent concrete of satisfactory quality can be obtained if proper control is maintained.

Concrete uniformity generally has been measured in terms of compressive strength, slump, unit weight, air content, and content of coarse aggregate and cement. Uniformity tests have been used to establish required mixing time, mixing speed, mixer capacity, and to verify efficient batching procedures. Tests by a number of investigators have been considered in the preparation of project specifications and ASTM Specification for Ready-Mixed Concrete (C 94).

ASTM C 94 also contains tolerances of test results that are requirements of uniformity of a single batch of concrete.

Methods of Measuring Uniformity

Dunagan Test

Dunagan [5] proposed a method for measuring the proportions of cement, water, sand, and coarse aggregate in fresh concrete by a series of wash separations and weighing in air and water. This method has been used by Slater [6], Hollister [7], Cook [8], and others to study the effects of different rates of rotation of truck mixers, effect of time of haul, and effect of mixing time on uniformity of concrete. The Dunagan test has limited usefulness because of sampling errors and difficulties in distinguishing between cement and very fine sand.

U.S. Bureau of Reclamation Test of Mixer Performance

The U.S. Bureau of Reclamation's mixer performance test [9] is used to evaluate the ability of a mixer to mix concrete that will be within prescribed limits of uniformity. The uniformity of freshly mixed concrete is evaluated by comparing variations in quantity of coarse aggregate and unit weight of air-free mortar of two samples, one taken from each of the first and last portions of the batch.

Large variations in this test may indicate that the batching procedure is incorrect or mixer blades are worn. Additional mixing time may be required if the unit weight of air-free mortar varies more than 24 kg/m³ (1.5 lb/ft³).

Air-Free Unit Weight Test

A study designed to establish test methods and limits for variations in truck-mixed concrete was reported by Bloem et al. [10]. Variations in slump, air content, percent of coarse aggregate, air-free unit weight of mortar, water content by oven drying, and compressive strength of concrete obtained after approximately $1/6$, $1/2$, and $5/8$ of discharge from a truck mixer were determined. They concluded that the air-free unit weight test was an improvement over the unit weight test because the number of variables was reduced and excessive changes in this property reflected changes in water or in proportions of cement and sand. According to their data, a difference in air-free unit weight of mortar of 17.6 kg/m³ (1.1 lb/ft³) corresponds to a change in water of about 9.9 L/m³ (2 gal/yd³) when the proportions of sand to cement were maintained constant and the water alone was varied. They suggested that a variation of more than 16.0 kg/m³ (1 lb/ft³) in this test indicates real differences in the proportions of the mortar ingredients, and differences of more than 32.0 kg/m³ (2 lb/ft³) should be considered evidence of unsatisfactory uniformity.

U.S. Army Corp of Engineers Method of Test for Concrete Mixer Performance (CRD-C55-91) [11]

This method of evaluation tests three samples of concrete for water content, unit weight of air-free mortar, coarse aggregate content, air content, slump, and seven-day compressive strength. Tests are usually performed on samples of concrete representing each of the three thirds of the batch, but not the very first and very last portions of the batch to be discharged. The tests are performed also to determine the feasibility of altering the mixing time. The standard guide specifications that are used to prepare project specifications include limits for these various tests.

Centrifuge Test

This test, also known as the Willis-Hime method, is described in detail elsewhere [12]. It provides a basis for within-batch comparisons of cement content of concrete and employs a liquid with a density greater than sand but less than cement. This liquid is used to separate the components of a carefully prepared mortar sample extracted from the concrete. This test is one of several mixer performance tests specified in ASTM C 94. A variability of 7% or more in the cement content of two samples from the mixer is considered evidence of incomplete mixing.

The centrifuge test was used in the study by Bloem et al. [10]. They concluded that the test is quite involved and, in most cases, the information gained is not commensurate with the time and labor required. Data cited from previous tests indicated that the test was highly reproducible provided extreme care was used and corrections were made for sample errors in coarse aggregate content of the small test portions.

Unit Weight

Unit weight of the concrete is influenced by the density and quantity of coarse aggregate, air content, proportions of sand to cement, and water content. Consequently, variations in unit weight are difficult to evaluate as to cause or significance. The test is more definitive when the weight and solid volume of coarse aggregate and volume of air are eliminated as in the unit weight of air-free mortar test.

The unit weight test is recommended as a job control measure for lightweight aggregate concrete in conjunction with air determinations and slump. If the slump and air content are maintained constant, a change in unit weight indicates a change in weight of aggregate. If the weight of aggregate per cubic meter of lightweight concrete changes, it may be the result of change in moisture content, grading, or density of the aggregate. Additional tests, including unit weight, moisture content, and grading of the aggregate are required in order to determine the cause of the variation.

Air Content

Air content has an important influence on workability, compressive strength, and durability of concrete. The strength of a concrete decreases uniformly with an increase in air content of the fresh concrete provided the water-cement ratio and fine aggregate content are held constant. Air entrainment also increases slump, each 1% of additional air being approximately equivalent to 2.5 cm (1 in.) in slump. When the slump is maintained constant by reduction in water content, strength reduction due to air entrainment is not so great and the strength of lean mixes especially may be slightly increased. Strength reductions of 16 to 20% at 28 days have been reported for concretes containing 307 to 363 kg of cement/m³ (517 to 611 lb/yd³) when 5% entrained air was added to the concrete and the slump of the concrete was not maintained at a constant level by lowering the water-cement ratio.

Too little air will detrimentally affect workability and durability. Sudden loss of workability may indicate a rapid reduction in air content. Resistance of concrete to freezing and thawing is increased significantly by incorporating the correct amount of air in the concrete. Consequently, it is important that the concrete contain a uniform quantity of air. Within-batch variations should not exceed 1%.

Variation in air content obtained from a given dosage of air-entraining admixture may result from changes in concentration of the admixtures, brand or type of cement, sand grading, mineral or chemical admixture, temperature of the mix, slump, or length of mixing. Air content is generally maintained at the correct level by increasing or decreasing the dosage of the air-entraining admixture used. When air-entraining cement is used, control of air content is more difficult because any change in the dosage of air-entraining additive must be accompanied by a corresponding change in cement content. For this reason, many engineers prefer to use regular cement and add the air-entraining admixture during the batching sequence.

Air content of normal weight concrete may be computed by comparing the actual unit weight of concrete with the theoretical weight based on the density of the materials used, as outlined in ASTM Test Method for Unit Weight, Yield, and Air Content (Gravimetric) of Concrete (C 138). Results obtained by this method are influenced by variations in mixture proportions, density of ingredients, and changes in moisture content of aggregates.

Several methods have been developed to determine the air content of fresh concrete directly. The principally accepted methods include the pressure method, ASTM Test Method for Air Content of Freshly Mixed Concrete by the Pressure Method (C 231), and the volumetric method, ASTM Test Method for Air Content of Freshly Mixed Concrete by the Volumetric Method (C 173).

The pressure meter method, ASTM C 231, consists of a special pressure-tight container and accessories designed to hold a precalibrated volume of concrete. The container is filled with concrete, the lid is securely fastened, and a predetermined pressure is applied by a hand pump. The apparatus is so calibrated that the percentage of air is read directly when the pressure is released. This method is used more than any other and is considered satisfactory for all types of concrete and mortar except those made with highly porous lightweight aggregate. This apparatus must be calibrated periodically to guard against changes caused by rough usage, and, if the elevation of the place at which the apparatus is used changes by more than 183 m (600 ft) in elevation, it should be recalibrated. An aggregate correction factor should be determined with the materials used and subtracted from the apparent reading to determine the actual air content. The aggregate correction factor varies only slightly for the same type of aggregate and need only be checked when there is a definite change in materials.

Due to the advent of new air-entraining admixtures and their ability to entrain much smaller air bubbles, it is recommended that the results of the pressure meter tests, be periodically verified by the unit weight test.

The volumetric method, ASTM C 173, consists of removing air from a concrete sample by mixing it with water in a special container. The volume of air is determined from the difference in volume of the sample containing entrained air and the volume of the sample after it has been agitated to permit the air to escape. This method is recommended particularly for lightweight concrete, but it may be applied to other types of concrete as well.

The non-ASTM standard air indicator is a miniature device that uses the volumetric principle. A small sample of carefully selected mortar is obtained from the concrete and placed in a brass cup measuring 1.9 cm ($3/4$ in.) in diameter by 1.3 cm ($1/2$ in.) high and compacted with a wire or knife blade. The glass tube that comes with the device is filled to the top line with isopropyl alcohol, the brass tube is inserted in the tube, and the liquid level is adjusted to the top line. The finger is placed over the stem to prevent alcohol from escaping, and the indicator is rolled several times until all mortar has been removed from the cup. With the indicator in a vertical position, the finger is carefully removed from the stem, and the number of graduations from the top to the new liquid level gives an indication of the air content in the mortar sample. A correction factor, based on the mixture proportion, must be applied to convert to percent air in concrete. Meticulous care must be used in the selection of the mortar sample, method of inserting the stopper, agitation of the sample, and removal of the finger from the tube. The test can only provide an indication of relative air content and cannot be considered as reliable as the pressure meter or the volumetric method and should not be used to accept or reject concrete.

Slump

The slump test, ASTM Test Method for Slump of Hydraulic Cement Concrete (C 143), is essentially an indication of the consistency of an individual batch of concrete. Large within-batch variations in slump indicate incomplete mixing and nonuniform distribution of water or other ingredients throughout the batch. Batch-to-batch variations may result from batching errors, uncorrected changes in moisture content, grading of the aggregate, or variations in temperature. In reasonably uniform concrete, the slump measurement should not vary more than about 2.5 cm (1 in.) within a batch.

Cement Content of Fresh Concrete

Constant neutralization by 3 N hydrochloric acid (HCl) for a fixed period of time (1 h in this procedure) has been employed by the California Department of Transportation [13]. Accuracy of this procedure was reported to be within about 14 kg of cement/m^3 (24 lb/yd^3). Field tests indicate the procedure is most useful for evaluating the performance of a concrete mixer. Relative cement contents of various portions of a batch can be determined in about 1 h. The report states that "adjustments to aggregate and/or cement feed can be made as needed to improve mixing uniformity."

This test procedure was apparently used on concrete that did not contain calcareous aggregates; at least, there was no discussion of the reaction of HCl on the aggregates, and this is a phenomenon that should have a profound effect on the end results of the test.

The authors of the report state in their conclusions that the test was not proven to be of sufficient accuracy for routine control of cement content during normal concrete production and, at present, the most applicable use of the procedure is evaluating mixer efficiency. For the latter purpose, they found the titration method to be more effective than the "wash-out-method" and the "heavy method" that were also tried in this program.

The procedure requires the development and use of calibration curves when determining the relative variation of cement content within a batch. In the event of a change in source of aggregate, cement, or working solution of acid, a new calibration curve must be established.

Rapid Analysis for Determining Water Content of Freshly Mixed Concrete [14]

These methods of tests can be used in either the laboratory or in the field. ASTM Test Methods for Determining Water Content of Freshly Mixed Concrete (C 1079) contains two chemical procedures for determining the free-water content of a sample of freshly mixed concrete. The choice of which procedure to use is at the discretion of the user. The environment in which these test methods are used may have some bearing on the choice of procedure.

A given mass of freshly mixed concrete is intermixed with a chloride solution of a given strength and volume. The chloride ion concentration of the intermixed solution is directly related to the water content of the concrete sample and is determined by volumetric titration or coulometric reference technique. A blank test is required. The blank test intermixes a given volume of distilled water with a concrete sample of given mass and determines the chloride ion concentration of the intermixed blank solution by volumetric titration or coulometric reference technique. Procedure A uses the volumetric titration and Procedure B uses the coulometric reference technique.

Rapid Analysis for Determining the Cement Content of Freshly Mixed Concrete [14]

These test methods, ASTM Test Methods for Determining Cement Content of Freshly Mixed Concrete C 1078, also cover two procedures applicable to all freshly mixed portland-cement concrete for which calibration has been attained in advance, except those containing certain aggregates or admixtures, that, when washed over a 150 μm (No. 100) sieve, yield significant and varying amounts of calcium ions in solution.

A given mass of freshly mixed concrete is washed with a given volume of water over a nest of sieves. The water is agitated so that the cement and other fine particles washed from the concrete (those particles passing the finest sieve) are uniformly suspended. A constant-volume representative of the cement suspension is obtained and diluted with a known volume of nitric acid and water. The diluted sample is agitated, without heat, to dissolve some of the calcium compounds in the cement. The calcium-ion concentration of the resulting solution is determined by manual volumetric titration in Procedure A or instrumental fluorometric determination in Procedure B. This calcium-ion concentration is correlated to the cement content of the specimen by a previously developed calibration curve.

The previously discussed rapid analysis test methods can be used to determine the water and cement contents in a batch of concrete and also the variability of their contents between batches of nominally identical concrete.

The water-cement ratio of a concrete sample can be estimated when these test methods are used to determine the water and cement contents.

Compressive Strengths

Measured concrete strength is used widely as a criterion of concrete quality. Other factors such as durability, abrasion resistance, thermal properties, dimensional stability, placeability, and compactability may be more critical, but strength tests are easily made and variations in strength are assumed to be indicative of variations in other properties.

Compressive strength is a control test that is useful in determining the degree of uniformity of concrete. ACI Committee 214 [15] has developed ACI 214-77, Standard Recommended Practice for Evaluation of Compressive Test Results of Field Concrete. This evaluation is statistical in nature and is a method for control of strength based on the coefficient of variation. ASTM C 94 lists requirements for uniformity of concrete, and these are expressed as the maximum permissible difference in results of samples taken from two locations in a concrete batch. The permissible difference for the average compressive strength at seven days for each sample (not less than three cylinders), based on the average strength of all comparative specimens, is 7.5%. In a well-controlled laboratory, compressive strength of cylinders fabricated from the same concrete sample may vary 3 to 5%. Variations in excess of this amount must be attributed to variations within the batch of concrete.

WORKABILITY

Workability is an everyday concern in concrete construction, and it is a factor easily appreciated in practice. Workability means different things to different people and for different placing conditions. Various nonstandard methods have been developed for its measurement. None of these tests evaluate all characteristics that are involved in this property.

Glanville [16] defined workability as "that property of the concrete which determines the amount of useful internal work necessary to produce full compaction." Powers [17] defined it as "that property of the plastic concrete mixture which determines the ease with which it can be placed and the degree to which it resists segregation." Both relate to the physical characteristics of the concrete alone, being independent of the methods of placing and compacting.

Workability of concrete is defined in ASTM Terminology Relating to Concrete and Concrete Aggregates (C 125) as "that property determining the effort required to manipulate a freshly mixed quantity of concrete with minimum loss of homogeneity."

In actual practice, workability is related directly to the type of construction and methods of placing, mixing, and transporting. Concrete that can be placed readily without segregation or separation in a mass dam could be entirely unworkable in a thin structural member. Workable concrete compacted by means of high-frequency vibrators would be unworkable if vibrators could not be used and hand tamping and spading were required. Concrete having suitable workability for a pavement might be unsuitable for use in a thin, heavily reinforced section.

Properties involved in workability include finishing characteristics, consistency or fluidity, pumpability, mobility, segregation, and bleeding. None of the test methods proposed or in use today simultaneously measures all of these properties. Consequently, measurement of workability is determined to a large extent by judgement, based on experience.

Workability is dependent upon the physical and chemical properties of the individual components and the proportions of each in the concrete. The degree of workability required for proper placement and consolidation of concrete is governed by the type of mixing equipment, size and type of placing equipment, method of compaction, and type of concrete.

Factors Affecting Workability

Some of the factors that affect the workability [16] of concrete are quantity of cement, characteristics of cement, consistency, grading of fine aggregate, shape of sand grains, grading and shape of coarse aggregate, proportion of fine to coarse aggregate, percentage of air entrained, type and quantity of pozzolan, quantity of water, mixture and ambient temperatures, amount and characteristics of admixtures used, and time in transit.

Cement

Very lean mixes tend to produce harsh concrete having poor workability. Rich mixes are more workable than lean mixes, but concrete containing a very high proportion of cement may be sticky and difficult to finish. An increase in the fineness of cement increases the cohesiveness of the concrete mix as well as the rate at which the cement hydrates and the early strength development. Differences in bleeding tendency, accumulation of laitance, and other properties of concrete made with cements having the same fineness and chemical analysis have been observed.

Consistency

Consistency (according to ASTM C 125) and plasticity are terms often used to indicate workability. Consistency generally denotes the wetness of the concrete that is commonly measured by the slump test. It must not be assumed that the wetter the mix the more workable the concrete. If a mix is too wet, segregation may occur with resulting honeycomb or sand streaking on the exposed surface; finishing properties will be impaired because of the accumulation of laitance on the surface. If a mix is too dry, it may be difficult to place and compact, and separation may occur because of the tendency for larger particles to roll towards the outer edge of the heap formed when it is deposited. It is agreed generally that concrete should have the driest consistency that is practicable for placement with available consolidation equipment. The consistency necessary for full compaction varies with the type of structure, type and size of aggregate, and type of compaction equipment available.

Sand

Concrete containing fine sand requires more water for the same consistency, as measured by the slump test, than an equivalent amount of coarse sand. Very coarse sand can have an undesirable effect on finishing quality. Neither very fine nor very coarse sand is desirable but both have been used satisfactorily. Rounded river sand gives greater workability than crushed sand composed of sharply angular pieces with rough surfaces. Angular sand particles have an interlocking effect and less freedom of movement in the freshly mixed concrete than smooth rounded particles. Natural sand may give satisfactory results with a coarser grading than would be permitted with crushed manufactured sand. In addition, concrete must contain 2 to 3% more sand by absolute volume of total aggregate and 6 to 9 kg more water/m^3 (10 to 15 lb/yd^3) when crushed sand is used.

Coarse Aggregate

The particle size distribution of coarse aggregate influences water requirements and workability of concrete. Coarse aggregates meeting standard grading requirements, such as ASTM Specification for Concrete Aggregates (C 33) should be used. After the grading is established, it should be maintained within rather close tolerances to avoid sudden changes in workability and other concrete properties. Segregation is reduced and uniformity improved by separating the aggregate into several size fractions and recombining these fractions when concrete is manufactured. This recombined grading should then be beneficial in establishing the fine aggregate (sand) content.

Breakage, separation, and contamination of aggregate can occur during handling and stockpiling. Introduction into the mixer of a large quantity of undersize material that may have accumulated will result in a sudden change in workability resulting in a demand for additional water. ACI Committee 304 has suggested rescreening at the batch plant as the method most likely to eliminate undesirable undersize and promote uniformity.

Production of workable concrete with sharp, angular, or crushed aggregates generally requires more sand than similar concrete made with rounded aggregates. The water content may be increased 9 to 15 kg/m^3 (15 to 25 lb/yd^3). If the water-cement ratio is held constant, more cement is required. Flat or elongated particles that are defined as particles having a ratio of width to thickness or length to width, respectively, greater than 3:1, are detrimental to concrete workability. More sand, cement, and water are required when the coarse aggregate contains flat and elongated particles.

The maximum size of aggregate that can be used to produce workable concrete is limited by practical considerations including type and size of structure, amount and spacing of reinforcing bars, method of placing, and availability of materials. Generally, aggregate should not be larger than three fourths of the maximum clear spacing between reinforcing bars nor larger than one fifth of the wall thickness or narrowest dimension between sides of forms. ACI Recommended Practice for Selecting Proportions for Concrete (ACI 211.1) provides recommendations on the maximum sizes of aggregate for various types of construction. The U.S. Bureau of Reclamation [9] experience in pumping concrete indicates that concrete containing 6.4 cm (2$^1/_2$ in.) maximum size aggregate can readily be pumped through a 20.3 cm (8 in.) pipe, but aggregate larger than 6.4 cm (2$^1/_2$ in.) may cause difficulty. Concrete containing aggregate graded to 2.5 cm (1 in.) maximum size can be placed by pneumatic equipment.

Air Entrainment

Entrained air increases the paste volume, acts as a lubricant, and improves the workability of concrete. It reduces bleeding and segregation during handling and placing of concrete and increases cohesiveness or "fattiness" of the concrete. Improvement in workability resulting from air entrainment is more pronounced in lean mixes that are harsh and unworkable because of poor aggregate grading or type of aggregate used.

Finely Divided Material

Addition of finely divided material, including inert or cementitious materials or pozzolans, generally improves the workability of the concrete. Improvement is more noticeable in lean mixes than in rich mixes. These materials have been used to improve the grading of sands deficient in fines. Cementitious and pozzolanic materials are usually substituted volumetrically for 10 to 70% of the cement. Workability will be improved if these materials are added as a replacement for part of the sand, instead of substituted for part of the cement.

Chemical Admixtures

Water-reducing admixtures, when added to concrete, permit a reduction in mixing water with no loss in slump, or, if the water content is held constant, produce an increase in slump. Set-retarding admixtures reduce the early rate of hardening and permit concrete to be handled and vibrated a longer period of time.

It has been reported that there is a decrease in the frequency of plugged pump lines when water-reducing retarders are used in the concrete. In addition, less power may be required to pump the concrete.

The use of high-range water-reducing (HRWR) admixtures have greatly increased the placeability capabilities for high-strength, low-water-cement ratio concretes. HRWR admixtures have made it possible to use condensed silica fume as a mineral admixture (normally 3.5 to 15% by weight of cement); such mixtures have provided highly workable concrete capable of attaining compressive strengths in the range of approximately 69 to 138 MPa (10 000 to 20 000 psi).

Mixture Proportions

Workability can be controlled by proper proportioning of the constituent materials. As the proportion of mortar, including sand, cement, water, and air, is increased, the grading and angularity of the coarse aggregate become less important. There should be sufficient mortar to fill the voids in the coarse aggregate plus a sufficient amount to permit the concrete to be placed readily in forms and vibrated around reinforcement. An excess of mortar increases workability, but excess workability is inefficient. Too much mortar can result in a sticky mixture. It should not be more than is required for consolidation by available equipment. The quantity of mortar required to produce the desired workability with a given coarse aggregate can be determined more effectively by laboratory tests. ACI Recommended Practice for Selecting Proportions for Concrete provides a basis for estimating the proportions of coarse aggregate to be used in trial mixes.

Methods of Measuring Workability

Slump Test

The slump test (Fig. 1) is the most commonly used method of measuring the consistency of concrete. It is not suitable for very wet or very dry concrete. It does not measure all factors contributing to workability, nor is it always representative of the placeability of the concrete. However, it is used conveniently as a control test and gives an indication of the uniformity of concrete consistency from batch to batch. Repeated batches of the same mix, brought to the same slump, will have the same water content and water-cement ratio provided weights of aggregate, cement, and admixtures are uniform and aggregate gradings are within acceptable tolerances.

Additional information on the mobility of the concrete can be obtained if, after removing the slump cone and measuring the slump, the concrete is tapped on the side with the tamping rod. Two concretes with the same slump may behave differently, that is, one may fall apart after tapping and be harsh with a minimum of fines, and the other may be very cohesive with surplus workability. The first concrete may have sufficient workability for placement in pavements or mass concrete, but the other concrete may be required for more difficult placement conditions.

The slump test should be performed in strict accordance with the requirements of ASTM C 143. Tests are usually made at the point of placement and should be made whenever specimens are molded for strength testing. Slump tests may be made at the mixing plant in order to check the uniformity of batching operations.

Popovics [18] has presented data indicating that the relationship between consistency values, as measured by the slump test and the water content of concrete, is parabolic, that is, the percentage change in water content required to increase the slump 2.5 cm (1 in.) may vary from 2.0% when the initial slump is 12.7 cm (5 in.) to approximately 4.5% when the initial slump is 5.1 cm (2 in.). An average change in water content of 3% generally is considered necessary for a 2.5 cm (1 in.) change slump.

As the temperature of the concrete increases, the slump decreases. Concrete placed at a slump of 10 cm (4 in.) at 21°C (70°F) may only have a 7.6 cm (3 in.) slump when placed at 32°C (90°F), or the same concrete may have a slump of 14 cm (5.5 in.) when placed at 10°C (50°F).

Air-entrainment and water-reducing admixtures will increase the slump of concrete if all other conditions remain the same. Each 1% increase or decrease in air content will produce approximately the same influence as a change in water content of 3%.

The slump generally is reported to the nearest 0.6 cm (1/4 in.). Slump reported by different operators on the same batch of concrete may vary by as much as 1.3 cm (1/2 in.). The most unsatisfactory form of slump is the shear slump, that is, a falling away or shearing off of a portion of the concrete from the mass. If this condition exists, the concrete probably lacks the necessary plasticity for the slump test.

Recently, a number of innovative manufacturers have been developing data to verify the acceptability of slump cones made of plastic, and it appears that such equipment may be permissible for use soon.

K-Slump Tester Nonstandard Test

The K-slump tester (Fig. 2) is reported to measure slump directly in 1 min after the tester is inserted in the concrete [19]. The apparatus is comprised of the following four principal parts.

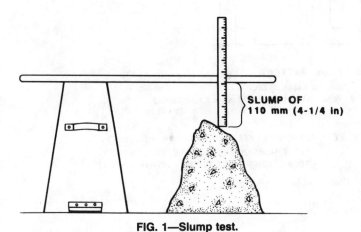

FIG. 1—Slump test.

56 TESTS AND PROPERTIES OF CONCRETE

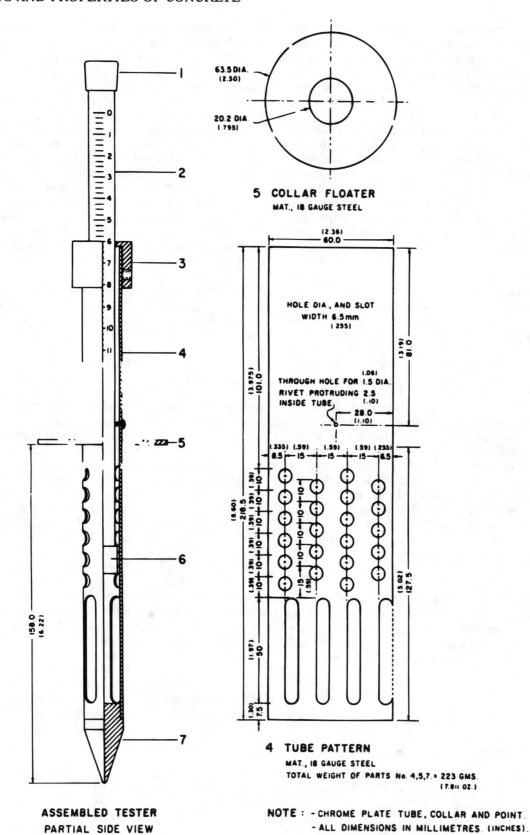

FIG. 2—K-slump tester.

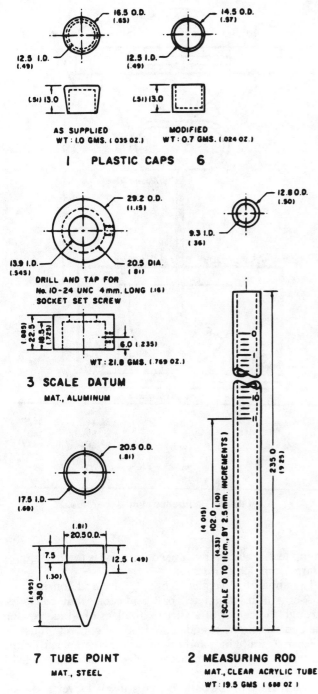

FIG. 2—Continued.

parts; the upper part serves as a handle and the lower one is for testing. The disk also serves to prevent the tester from sinking into the concrete beyond the preselected level.
3. A hollow plastic rod, 1.3 cm (0.51 in.) in diameter and 25 cm (9.84 in.) long, that contains a graduated scale in centimetres. This rod can move freely inside the tube and can be used to measure the height of mortar that flows into the tube. The rod is plugged at each end with a plastic cap to prevent concrete or any other material from entering.
4. An aluminum cap, 3 cm (1.18 in.) in diameter and 2.25 cm (0.89 in.) long, that has a small hole and a screw that can be used to set and adjust the reference zero of the apparatus. In the upper part of the tube, there is also a small pin that is used to support the measuring rod at the beginning of the test.

The K-slump tester is reported to measure an index that is related to workability after the device is removed from the concrete. The first reading is taken after the tester has been in place (the disk resting on the concrete surface) for 60 s. This reading, in centimetres, is referred to as the K-slump. The device is removed from the concrete and the measuring rod is again lowered to rest on the surface of the concrete remaining in the tube; this reading, in centimeters, represents the workability of the mix.

Studies have been made and reported on 420 concrete batches by five laboratories. Statistical determinations and equations are reported elsewhere [19].

Placeability

An article by Angles [20] described a placeability apparatus that would simulate the operation of placing concrete under typical conditions. The equipment is comprised of three units:

1. A rectangular, heavy-gage container of suitable size with two vertical channels at the midpoints on the inside of the longer sides.
2. A screen, secured by removable clamps, made of round horizontal bars with openings between them of a size appropriate to the maximum size of aggregate in the mix under test.
3. A plate to act temporarily as a shutter during the placing of the concrete in Compartment B of the container.

The test is performed by placing a known weight, for example 25 kg (55 lb) of concrete into Compartment B with a standardized scoop. The shutter is removed and the vibrating table, on which the test apparatus rests, is switched on. The time required for the concrete to pass through the screen and attain the same level in both compartments is measured with a stop watch. Placeability is the time expressed in seconds required from turning on the switch to leveling of the sample along the full length of the box.

The author also suggests that internal vibration could be used by fixing diameter, frequency, amplitude, and location of the vibrator. Other data and observations that could be obtained, according to Angles, are bleeding, segregation, and bulk density.

1. A chrome-plated steel tube with external and internal diameters of 1.9 and 1.6 cm (0.63 and 0.75 in.). The tube is 25 cm (9.84 in.) long, and its lower part is used to make the test. The length of this part is 15.5 cm (6.10 in.) that includes the solid cone that facilitates inserting the tube into the concrete. Two types of openings are provided in this part: four rectangular openings are distributed uniformly in the lower part and 22 circular openings are evenly distributed above the rectangular openings.
2. A disk floater, 6 cm (2.36 in.) in diameter and 0.24 cm (0.09 in.) in thickness, that divides the tube into two

Remolding Test

The remolding test apparatus (Fig. 3) was developed by Powers [17] to measure "the relative effort required to change a mass of concrete from one definite shape to another by means of jigging." The equipment consists of a metal cylinder mounted inside a larger cylinder and a suspended plate that fits inside the smaller cylinder. A slump cone is placed inside the smaller cylinder so that the bottom rests on the base. It is filled with concrete, the slump cone is removed, and the plate is placed on top of the concrete. The flow table on which the apparatus is mounted is then operated. The number of 0.6-cm (¹/₄-in.) drops, required to mold the concrete to a cylindrical form, is a measure of the workability of the concrete. This method has not found widespread use and no ASTM standard has been written about it.

The remolding test and slump test were used by Cordon [21] in an extensive series of tests on air-entrained concrete. He found that workability, as measured by the remolding test, was increased when air content was increased, and slump and percent sand were maintained constant. When the percent sand was reduced as the air content was increased and slump was constant at 10.2 cm (4 in.), the remolding effort was also constant after approximately 42 jigs. It might be concluded that the remolding test is more sensitive to changes in air content than the slump test.

Ball Penetration Test

The Kelly ball test (see Fig. 4) [22] was developed principally as a convenient method of measuring and controlling consistency in the field. The ball test can be performed on concrete in the forms, and it is claimed that tests can be performed faster and precision is greater than with the

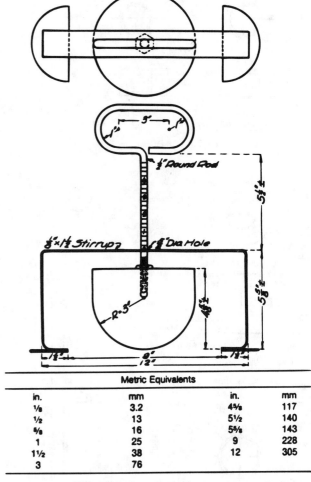

in.	mm	in.	mm
¹/₈	3.2	4⁵/₈	117
¹/₂	13	5¹/₂	140
⁵/₈	16	5⁵/₈	143
1	25	9	228
1¹/₂	38	12	305
3	76		

FIG. 4—Ball penetration apparatus.

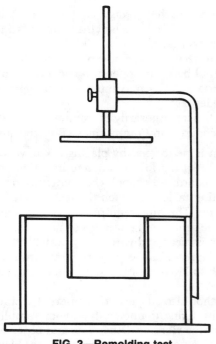

FIG. 3—Remolding test.

slump test. One disadvantage of this test is that it requires a large sample of concrete.

The apparatus weighs 13.6 kg (30 lb) and consists of a 15.2-cm (6-in.) diameter ball and stem that can slide through the center of a stirrup, the legs of which rest on the concrete to be tested. The depth of concrete must be at least 20 cm (8 in.), and the minimum distance from the center of the ball to the nearest edge of the concrete is 23 cm (9 in.).

The surface of the concrete is struck off level, avoiding excess working. The ball is lowered gradually onto the surface of the concrete, released, and the depth of penetration read immediately on the stem to the nearest 0.6 cm (¹/₄ in.). The ratio of slump to the penetration of the ball is between 1.5 and 2 and is fairly constant for a given mix but varies according to the mix. The method has been adopted as the ASTM Test Method for Ball Penetration in Fresh Portland Cement Concrete (C 360).

The ball penetration test and the slump test, along with their results, were compared by Howard and Leavitt [23]. Twenty tests were made using each method. The ball penetration averaged 1.32 with a standard deviation of 0.45, and the slump averaged 2.5 with a standard deviation of 0.81. They concluded that the slump test required approximately 10 min, and the ball penetration only required 10 s when used on paving concrete.

Thaulow Concrete Tester

The Thaulow concrete tester (Fig. 5) [24] consists of a 10-L container with a graduation mark at 5 L that is equipped with a slump cone and fitted with a stainless steel handle, and a drop table actuated by a crank that drops the table through 1 cm (0.394 in.) four times for each revolution.

The slump cone is fastened in the container, filled with concrete in the usual manner, and lifted off. The handle is allowed to fall freely from the vertical position alternately on each side of the container until the concrete is remolded and the entire periphery is at the 5-L mark. The number of blows required is an indication of the workability.

If the concrete is very dry, the container is fastened to the drop table. The slump cone is fastened, filled in the usual manner, and further compacted by 15 drops of the table. The cone is removed, and the number of revolutions of the crank handle necessary to bring the concrete to the 5-L mark is a measure of the consistency.

ACI has prepared the Recommended Practice for Selecting Proportions for No-Slump Concrete (ACI 211). Differences in consistency of very dry mixes cannot be measured with the slump cone, but the Thaulow equipment is considered to have merit for this application. Concrete with a slump of 2.5 to 5.1 cm (1 to 2 in.) requires 14 to 28 drops, and concrete with a slump of 7.6 to 10.2 cm (3 to 4 in.) requires less than 7 drops.

Vebe Apparatus

The Vebe consistometer (Fig. 6) [25] includes a vibrating table, a sheet metal pan, slump cone, and plastic plate attached to a graduated, free-moving rod that serves as a reference end point. The cone is placed in the pan, filled with concrete, and removed. The plastic disk is brought

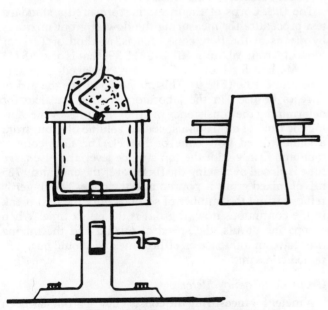

FIG. 5—Thaulow concrete tester.

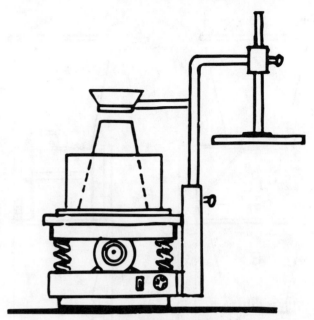

FIG. 6—Vebe apparatus.

into position on top of the concrete, and the vibrating table is set in motion. The number of seconds required to remold the cone of truncated concrete to the shape of the cylinder is the measure of consistency and is reported as the number of Vebe seconds or degrees. This method is very suitable for very dry concrete, but the variation is too vigorous for concrete with a slump greater than about 5.1 cm (2 in.). For example, 0 to 3 s is required for concrete with a slump of 7.6 to 10.2 cm (3 to 4 in.), and 10 to 52 s may be required for concrete with less than zero slump.

Compacting Factor

The compacting factor test (see Fig. 7) was developed in Great Britain [16]. The apparatus consists of two conical hoppers fitted with strong doors at their base and a 15 by 30-cm (6 by 12-in.) cylinder. The top hopper is filled with the concrete to be tested and struck off without compacting it. The door at the bottom of the hopper is opened, and the concrete drops by gravity into the somewhat smaller hopper below. The door of the second hopper is opened, and the concrete is allowed to fall into a cylinder that is struck off and weighed. The ratio of the weight of concrete in the cylinder mold to the weight of concrete from the same batch fully compacted in the mold is the compacting factor. The sensitivity of this test is considered good for medium-consistency concrete, but less than some other tests for very dry concrete. An average compacting factor of 0.75 corresponds to a slump of 0 to 2.5 cm (0 to 1 in.), and 0.90 is approximately $7^{1}/_{2}$ to 10-cm (3 to 4-in.) slump.

Wigmore Consistometer

The Wigmore consistometer (Fig. 8) is described by Orchard [26].

This apparatus consists of a galvanized container and hand-operated compaction table. A 5.1-cm (2-in.) diameter ball that is fastened to a sliding stem is mounted in the lid of the container. The container is filled with con-

60 TESTS AND PROPERTIES OF CONCRETE

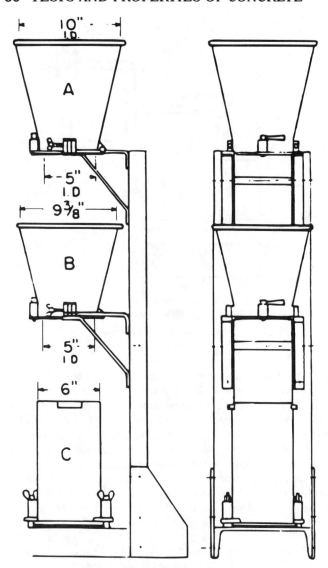

FIG. 7—Compacting factor apparatus.

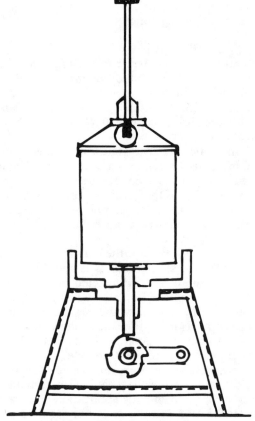

FIG. 8—Wigmore consistometer.

crete that is compacted on the table by 8 drops. The container is again filled with concrete and leveled off; then the lid and the ball are placed in position with the ball resting on the surface of the concrete. The apparatus is placed on the table and the concrete is compacted by turning the handle attached to the cam at the rate of about 1 rps. The table drops 0.6 cm ($7/_{32}$ in.) four times per revolution of the cam, and the number of drops required to lower the ball and stem 19.7 cm (7 $3/_4$ in.) into the concrete is considered a measure of the consistency of the concrete.

The number of drops required may vary from 20 for very wet concrete, 15-cm (6-in.) slump, to 200 for very stiff concrete.

It is claimed that the Wigmore consistometer is an improvement over the slump test because work is actually done on the concrete in a way that resembles field conditions. Variations in results may be expected if the ball comes in contact with large aggregate.

Flow Cone

Contraction joints in dams, cleavage planes in rock foundations, cavities behind tunnel linings, voids in preplaced aggregate, and openings around post-tensioned cables may be filled with grout pumped under pressure. Grouts consist of cement and water or combinations of sand, cement, water, finely divided filler, and admixtures. Grouts must be very fluid to penetrate small cavities. The slump test and other described methods of measuring concrete consistency are unsuitable.

The U.S. Corps of Engineers has prepared a standard test procedure for measuring the flow of grout mixtures by means of the flow cone, that is, Method of Test for Flow of Grout Mixtures (CRD-C611-89) that is also ASTM Test Method for Flow of Grout for Preplaced-Aggregate Concrete C 939, (Fig. 9). This method outlines the procedure to be used in the laboratory and in the field for determining the consistency of grout mixtures by measuring the time of efflux of a specified volume of grout from a standardized flow cone or funnel. The flow cone is mounted firmly with the top surface level, the discharge tube is closed by placing the finger over the end, and 1725 mL of mixed grout is poured into the cone. The finger is removed, and the number of seconds until the first break in the continuous flow of grout is the efflux time. When comparing grouts, the speed of mixing and the mixing time have an influence on efflux time and should be maintained constant.

Grout Consistency Meter

A meter for measuring the consistency of grout has been developed at the University of California and is described

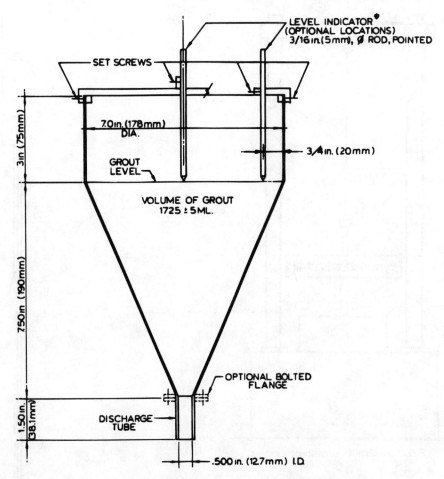

NOTE—Other means of indicating grout level may be used as long as accurate indication of grout level on volume is obtained.

FIG. 9—Flow cone.

elsewhere [27]. The grout consistency meter (Fig. 10) is essentially a torque meter. The sample of grout is placed in a metal pan mounted on a platform that can be rotated at a constant speed of 60 rpm. Suspended from a music wire is a 7.3-kg (16-lb) paddle assembly to which a torque is applied as the sample of grout is rotated. The angle of twist or consistency factor is read by an index pointer attached to a cross strut.

Otto Graf Viscosimeter

The Otto Graf viscosimeter (see Fig. 11) has been used in Europe to measure fluidity of grouts. It consists of a brass tube, approximately 90 cm (35 $^1/_2$ in.) long and 6 cm (2 $^7/_{16}$ in.) inside diameter, mounted on a base in an upright position, and an immersion body weighing 5 kg (11.1 lb) that is 28.2 cm (11 $^3/_{32}$ in.) long. The immersion body has protruding cams on the sides so that grout can pass between them when the immersion body is placed in the tube containing the grout. In practice, the tube is filled to a prescribed level with grout. The immersion body is then placed in the tube and released. The number of seconds required for the body to assume its final position is a measure of the fluidity of the grout.

This equipment was specified as a control measure for grouting operations on a post-tensioned bridge. Consistency was also measured with the flow cone. Tests were made immediately after mixing and again 30 min after mixing. The tests indicate that the flow cone provided the same information as the viscosimeter and was as effective for control purposes. The ratio of viscosimeter reading to flow cone reading was approximately 1:5.

Two-Point Workability Test

Tattersall [28] discusses the principles of measurements of the workability of fresh concrete and a proposed simple two-point test. In this paper, the author points up that an understanding of workability of fresh concrete is important: (*a*) to make possible the design of mixes for particular purposes; (*b*) to provide a method of control in the manufacture of the mix; and (*c*) to contribute to the efficient use of manufacturing processes such as vibration, pumping extrusion, and the like.

Tattersall contends that in spite of all the efforts (papers that have been written and the many proposed tests) over the past 30 years, there is no test that is fully satisfactory and the property of workability cannot even be defined except in the most general terms. Each of the test methods is capable of classifying, as identical, concretes that can be shown to be dissimilar.

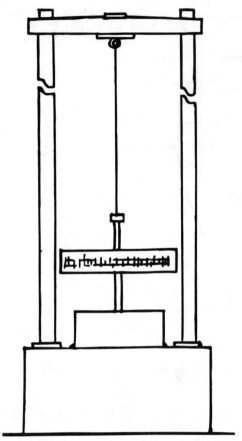

FIG. 10—Grout consistency meter.

Tattersall suggests the following summary of proposed terminology as an effort toward standardization.

I. Qualitative
   A. Workability
      1. Flowability
      2. Compactability
      3. Stability
      4. Finishability
      5. Pumpability
      6. Extrudability

II. Quantitative empirical
   A. Slump
   B. Compacting factor
   C. V-B time

III. Quantitative fundamental
   A. Viscosity
   B. Yield value
   C. Mobility

The objection to the several workability tests are that they were almost without exception single-point tests, whereby only one measurement was made at one specific rate of shear or set of shearing conditions. Tattersall states that such a procedure is valid only for a simple Newtonian liquid whose flow properties are completely defined by the constant ratio of stress to shear rate, and that ordinary observation shows that fresh concrete is not a Newtonian liquid and it consequently follows that any test based explicitly or implicitly on the assumption that it is, will be inadequate.

Tattersall also states that there is evidence to indicate that, in practice, it may be sufficient to treat the material as conforming with the Bingham model and that it approximately obeys the following relationship: the stress at a rate of shear—the yield value is equal to the product of the plastic viscosity and the rate of shear. The yield value and plastic viscosity are constants, and it follows that measurements at two shear rates are required to determine them. The balance of this paper by Tattersall [29] discusses test procedures, results, and conclusions, and further modifications to allow application of vibration.

Tattersall [28,30] further explains the rationale of a two-point workability test and the relationships between slump, compacting factor, Vebe time, and the two-point test.

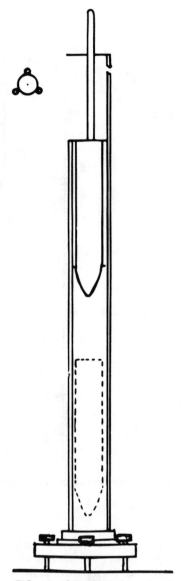

FIG. 11—Otto Graf viscosimeter.

The test procedure constitutes the measurement of power required at three separate speeds to operate a 18.9-L (20-qt) food mixer when empty; then repeat the power measurement at all three speeds when the bowl contains 21 kg (46.5 lb) of concrete. Values for yield and plastic viscosity are obtained by plotting $(P - P_E)/w$ against w, where w is speed, P is power under load, and P_E is power when the bowl is empty.

The two-point test appears to yield more information concerning the performance of a concrete mixture and appears to have good potential for determining and consequently controlling concrete uniformity.

CONCLUSIONS

Concrete knowledge and technology have advanced slowly during the years, but additional knowledge will be required if concrete is to maintain the position it has established as the universal building material. New products and technologies are being developed in all phases of industry at a rapid rate, and improvements in concrete production, control, delivery, and placing techniques must also be developed. At the same time, more widespread use of available knowledge for controlling uniformity, quality, and workability will improve the competitive position of concrete.

New methods of mixing, placing, consolidating, and finishing concrete may permit the use of much less water and improve concrete quality. Eventually, it may be possible to place and consolidate 0 to 5-cm (0 to 2-in.) slump concrete in walls, floors, and so on, as readily as 8 to 10-cm (3 to 4-in.) slump is placed today. New tests and methods of quality control and measurement of workability must be developed concurrently with methods of mixing and placing concrete. The slump test is an old friend and has served its purpose well. We do not like to see old friends pass away, but change is inevitable. The slump test and other tests used to measure uniformity may be replaced by more efficient test methods in the future.

Meters have been developed to measure moisture content of sand and coarse aggregate in the bins at concrete plants and promote better control of mixing water. Van Alstine [31] attributed the uniformity obtained at Denver Reservoir No. 22 Dam, where sand bins were filled 17 to 20 times each day with sand of widely varying moisture content, to the use of an electrical resistance moisture meter. More general use of devices of this type will improve quality control of concrete throughout the industry.

Methods of measuring the consistency of concrete while it is being mixed in the mixer drum should be developed so that corrections can be made immediately. The slump test and other tests used to indicate consistency can be made only after the concrete is discharged, and corrections can be applied only to subsequent batches. Polatty [32] reported that such a device, called the "plastograph," gave a better indication of workability than the slump test and indicated a moisture change in time to make a correction in the same batch. The equipment was used at Allatoona Dam with mixes containing 15.2-cm (6-in.) maximum aggregate and three bags of cement per cubic meter. When used with 7.6 or 3.8-cm (3 or 1½ in.) aggregate, the equipment was less efficient.

The development of improved testing procedures for freshly mixed concrete remains a requirement of the concrete industry.

REFERENCES

[1] Cement and Concrete Terminology (ACI 116R-90), *ACI Manual of Concrete Practice 1991*, Part 1, American Concrete Institute, Detroit, MI, 1991.

[2] Behavior of Fresh Concrete During Vibration (ACI 309.1R), *ACI Manual of Concrete Practice*, Part 2, American Concrete Institute, Detroit, MI, 1993.

[3] *ACI Manual of Concrete Inspection*, SP-2, 7th ed., (ACI 311.1R-81) American Concrete Institute, Detroit, MI, 1981.

[4] Guide for Measuring, Mixing, Transporting, and Placing Concrete (ACI 304R-89), *ACI Manual of Concrete Practice 1991*, Part 2, American Concrete Institute, Detroit, MI, 1993.

[5] Dunagan, W. M., "A Method of Determining the Constituents of Fresh Concrete," *Journal*, American Concrete Institute; *Proceedings*, Vol. 26, 1930.

[6] Slater, W. A., "Tests of Concrete Conveyed from a Central Mixing Plant," *Proceedings*, American Society for Testing and Materials, Vol. 31, Part II, 1931.

[7] Hollister, S. C., "Tests of Concrete From a Transit Mixer," *Proceedings*, American Concrete Institute, Vol. 28, 1952, pp. 405–417.

[8] Cook, G. C., "Effect of Time of Haul on Strength and Consistency of Ready-Mixed Concrete," *Journal*, American Concrete Institute; *Proceedings*, Vol. 39, 1952, pp. 413–428.

[9] "Variability of Constituents in Concrete (A Test of Mixer Performance), (Test 26)," *Concrete Manual*, 8th ed., U.S. Bureau of Reclamation, Denver, CO, 1975.

[10] Bloem, D. L., Gaynor, R. D., and Wilson, J. R., "Testing Uniformity of Large Batches of Concrete," *Proceedings*, American Society for Testing and Materials, Vol. 61, 1961.

[11] *U.S. Army Corps of Engineers Handbook for Concrete and Cement*, Updated Quarterly, Waterways Experiment Station, Vicksburg, MS.

[12] "Proposed Tentative Method of Test for Cement Content of Freshly Mixed Concrete," *Bulletin*, American Society for Testing and Materials, No. 239, July 1959, p. 48.

[13] Report No. CA-DOT-TL-5149-1-76-65, California Department of Transportation, Sacramento, CA, Sept. 1976.

[14] CERL Technical Report M-212, U.S. Army Construction Engineering Research Laboratory, Champaign, IL, May 1977.

[15] Recommended Practice for Evaluation of Strength Test Results of Concrete (ACI 214-77), *ACI Manual of Concrete Practice*, Part 2, American Concrete Institute, Detroit, MI, 1973.

[16] Granville, W. R., Collins, A. R., and Mathews, D. D., "The Grading of Aggregate and Workability of Concrete," Road Research Technical Paper No. 5, Her Majesty's Stationary Office, London, 1947.

[17] Powers, T. C., "Studies of Workability of Concrete," *Journal*, American Concrete Institute, Vol. 28, Feb. 1932, p. 419.

[18] Popovics, S., "Relations Between the Change of Water Content and the Consistence of Fresh Concrete," *Magazine of Concrete Research*, London, July 1962.

[19] Nasser, K. W., *Journal*, American Concrete Institute, Oct. 1976.

[20] Angles, J., *Concrete*, Dec. 1974.

[21] Cordon, W. A., "Entrained Air—A Factor in the Design of Concrete Mixes," *Journal*, American Concrete Institute; *Proceedings*, Vol. 51, May 1955, p. 881.
[22] Kelley, J. W. and Polivka, M., "Ball Test for Field Control of Concrete Consistency," *Journal*, American Concrete Institute; *Proceedings*, Vol. 51, May 1955, p. 881.
[23] Howard, E. L. and Leavitt, G., "Kelly Ball versus Slump Cone," *Journal*, American Concrete Institute; *Proceedings*, Vol. 48, 1952, pp. 353–354.
[24] Standard Practice for Selecting Proportions for No-Slump Concrete (ACI 211, 3-75), *ACI Manual of Concrete Practice*, Part 1, American Concrete Institute, Detroit, MI, 1993.
[25] Bahrner, V., "Report on Consistency Tests on Concrete Made by Means of the Vebe Consistometer," Report No. 1, Joint Research Group on Vibration of Concrete, Svenska Cementforeningen, March 1940.
[26] Orchard, D. F., *Concrete Technology*, Vol. 2, Wiley, New York, 1962.
[27] Davis, R. E., Jansen, E. C., and Neelands, W. T., "Restoration of Barker Dam," *Journal*, American Concrete Institute; *Proceedings*, Vol. 44, April 1948.
[28] Tattersall, G. H., "The Rationale of a Two-Point Workability Test," *Magazine of Concrete Research*, Vol. 25, No. 84, Sept. 1973.
[29] Tattersall, G. H., "Principles of Measurement of the Workability of Fresh Concrete and a Proposed Simple Two-Point Test," *Fresh Concrete*, Rilem Seminar Proceedings, Vol. 1, 1973.
[30] Tattersall, G. H., "The Relationships Between the British Standard Tests for Workability and the Two-Point Test," *Magazine of Concrete Research*, Vol. 28, No. 96, Sept. 1976.
[31] Van Alstine, C. B., "Water Control by Use of a Moisture Meter," *Journal*, American Concrete Institute; *Proceedings*, Nov. 1955, pp. 341–347.
[32] Polatty, J. M., "New Type of Consistency Meter Tested at Allatoona Dam," *Journal*, American Concrete Institute; *Proceedings*, Oct. 1949, pp. 129–136.

Air Content, Temperature, Unit Weight, and Yield

Lawrence R. Roberts[1]

PREFACE

This chapter continues the tradition of previous editions of *ASTM STP 169*. Due to the fundamental nature of these topics, much of what has appeared before in *ASTM STP 169B*, authored by F. F. Bartel [1], is equally applicable today and will be borrowed freely as appropriate. The author would also like to acknowledge the support and encouragement of his company, and the many colleagues who have provided useful insight.

INTRODUCTION

Measurement of air content, temperature, and unit weight of freshly-mixed concrete are the backbone of field quality control of concrete construction. Without these, proportioning of mixtures, control of yield, and assurance of placeability, strength development, and durability of concrete exposed to freezing conditions are not possible. The tests involved are relatively straightforward, use simple equipment, and are thoroughly described in the applicable ASTM test methods. Despite this, job problems, ranging from simple disputes over yield to complete performance failures, frequently can be attributed directly to faulty or absent application of these methods. The purpose of this chapter is to help the interested concrete technician understand some of the key issues controlling the correct application and interpretation of the methods, and hopefully to encourage more careful and frequent application.

OVERVIEW OF SIGNIFICANCE AND USE

Temperature measurement of fresh concrete is vital to ensure adherence to maximum temperature specifications usually imposed to control thermal gradients and resultant possible thermal cracking as the cement hydrates and then cools to ambient conditions, and to minimum temperature specifications imposed in cold weather to ensure adequate setting and strength performance.

Tests for air content and unit weight are made of fresh concrete to provide a control of these properties in the hardened concrete, and to determine the volume of concrete being produced from a given batch. The yield data so obtained are then available for the calculation of unit cement and aggregate contents that is essential in mixture development and may be required in some specifications. Tests for unit weight are also made to control concrete weight per se of both lightweight and heavyweight concretes.

While air content is most commonly determined to ensure the presence of air entrainment for freeze-thaw durability, knowledge of air content of nonair-entrained concrete is also important, due to the strong negative impact that unexpected increases in air content can have on compressive and flexural strengths.

Air contents of air-entrained concretes can vary for a large number of reasons. Among these are changes in air entraining admixture type or dosage, changes in cement alkali content [2], fine aggregate grading, slump, concrete temperature, mixing intensity and duration, and many others. While a discussion of the impact of each of these is beyond the scope of this work and is well-covered elsewhere [3–5], suffice it to say that the possibility of a change due to any one factor is great enough to warrant close control of the air content. Further, nominally nonair-entrained concretes can also develop significant air contents, due to errors in batching or contamination of materials. Thus, checking air content is also a necessary safety precaution for these concretes.

The interrelationship between air content and unit weight should be obvious: an increase in volume of air results in a lower unit weight, all other material contents being unchanged. The calculation of material quantities per unit volume from the unit weight depends not only on the correct determination of the concrete unit weight, but also on the accurate knowledge of the weights actually batched. If the unit weight test is to be used to estimate air content directly from proportions, then the specific gravities of the aggregates and cement (usually taken as 3.15 when portland cement is used) must be accurately known. Slight errors can result in significant error in estimating the air content. For this reason, the gravimetric determination of air is much more useful to monitor variation in air content of mixtures, the reference air content of which is determined by other methods. Also, when air contents vary in concretes being controlled to a fixed slump, the unit weight changes will be less than predicted by air content changes alone, due to the reduction of required water for equal slump in many concrete mixtures

---

[1] Director of Technology Planning and Transfer, Grace Construction Products, Cambridge, MA 02140.

as air content increases, and the gravimetric method can be used only if all the water weights are accurately known.

It must be emphasized that all methods to be discussed here measure the total air in the concrete, subject to the limitations of each method. While the adjectives "entrapped" and "entrained" are sometimes applied in technical discussion, and unfortunately sometimes in contract documents, to distinguish between large and small air voids, respectively, these methods cannot in any way make such a distinction. Furthermore, the fresh concrete air contents specified by the American Concrete Institute (ACI) and others for concrete durability under freeze-thaw are uniformly the total measured air contents, yet are referred to as entrained air. While some conclusions may be made for research purposes about the expected amount of small air voids by comparing the air contents of similar air-entrained and nonair-entrained concretes, there is no clear dividing line even by microscopic examination, and such distinctions are purely arbitrary. In some cases, contract language for "entrained air content" has been interpreted to mean that the specified level should be in addition to the base air content of nonair-entrained concrete. This is an inappropriate application of these fresh concrete methods, since no such distinction is possible.

Sampling

These methods are regularly applied under a variety of conditions—in the laboratory during mixture proportion development, at the plant for control of production, at the job site discharge point—to ensure compliance with specifications, and, at the point of placement, to best estimate the resulting hardened concrete properties. Due to the limitations of these various locations, a clear understanding of proper sampling procedures and acceptable deviations from test specifications is needed.

For example, all the methods to be discussed require that concrete be sampled according to ASTM Practice for Sampling Freshly Mixed Concrete (C 172), which requires that the concrete come from two or more portions of the load taken at regularly spaced intervals during discharge of the middle of the batch. This is clearly not possible if a truck is being sampled prior to discharge for compliance with specifications, which is perhaps the most common application of these methods. Accordingly, samples must be taken from the initial discharge, but enough concrete must be allowed to discharge, usually about 10%, to obtain concrete representative of the load, and all the other directions of care described in ASTM C 172 must be adhered to.

One source of error in sampling during application of these methods deserves special mention. Frequently, one or more of the air content methods is applied to a load of concrete arriving at a job site at the same time the slump test, ASTM Test Method for Slump of Hydraulic Cement Concrete (C 143), is run. The air is found to be within specifications, but the slump is too low, and the specifications allow addition of retempering water. This water is added, the slump re-run and found to be within specifications, but the air content of the retempered concrete is not determined. The addition of water and further mixing alter the air content, and subsequent determinations of the hardened air contents do not agree with the results recorded for the fresh concrete properties. Questions about the accuracy of the air content measurements then arise, solely because the concrete tested in the fresh state was altered after testing but before being allowed to harden.

TEST METHODS

Temperature

Concrete temperature measurement is determined in accordance with ASTM Test Methods for Temperature of Freshly-Mixed Portland-Cement Concrete (C 1064). This test method describes the types and precision (0.5°C) required of the temperature measuring devices to be used. Although liquid in glass thermometers may be used, the conditions of field concrete testing in many cases make metal dial thermometers more practical, and these are most frequently employed. Yearly calibration against 0.2°C precision liquid in glass thermometers is required by the test method, using two temperatures at least 15°C apart; but due to the ease with which some metal dial thermometers can lose calibration, it is recommended that a single temperature comparison against a reference thermometer be performed daily as an equipment check, with a full recalibration being run if deviation is noted.

Another key provision of ASTM C 1064 relates to sample size. Clearly, if the sample is small enough to gain or lose significant heat to its surroundings during the time of testing, the result will not be representative of the mass of the concrete. Accordingly, ASTM C 1064 calls for the sample to be large enough for a minimum of 75 mm of concrete to surround the temperature-measuring device.

Air Content

The three tests for air content of fresh concrete, using pressure, volumetric, and gravimetric methods, each have their own advantages and limitations. We will discuss each in order, with focus on the proper selection of method.

Pressure Air Measurement

The pressure method for determining air content of fresh concrete is based on Boyle's law, which states that the volume of a gas is inversely related to the pressure. By applying pressure to a known volume of concrete containing air voids, the voids are compressed, the concrete is reduced in volume, and the volume change can be measured and related to the initial volume. Knowledge of the pressure difference allows calculation of the total volume of air. This principle was first applied by Klein and Walker in 1946 [6], while Menzel [7] refined the apparatus and proposed a standard test procedure. This method has the advantage that no knowledge of specific gravities or batch weights are required to obtain the required answer.

Two types of meters are defined in ASTM Test Method for Air Content of Freshly Mixed Concrete by the Pressure Method (C 231). The Type A meter relies on direct measurement of the volume change, by means of a column of water above the known volume of concrete. A calibrated sight tube allows measurement of the volume reduction directly as pressure is applied. This method is very straightforward, but has the drawback that recalibration is necessary if barometric pressure or elevation changes exceeding 183 m occur. A change of 183 m is approximately equal to a 2% change in barometric pressure at sea level.

In the Type B air meter, a known volume of air at an established higher pressure is allowed to equilibrate with the known volume of concrete in a sealed container. The drop in pressure measured in the high-pressure air chamber can be related to the amount of air within the concrete. This method does not have the ambient air pressure recalibration requirement of the Type A meter, but the complexity of valves and seals make the apparatus prone to leakage; operators should be prepared with tools and replacement parts for repair and be sensitive to instability in dial readings that may signify leakage and therefore incorrect results. A common mistake is to close the petcocks prior to pressurizing the high-pressure chamber. If this is done, it is possible to not observe a leak in the needle valve, which normally would be noticed due to air bubbles escaping the open petcock. This would result in incorrectly low air content readings.

This method requires complete consolidation of the concrete in the bowl; any large air voids due to lack of consolidation will be measured as air content of the concrete. To ensure proper consolidation, rodding is required above a slump of 75 mm, internal vibration below 25 mm, and either in between. Vibration should cease when all the coarse aggregate is submerged and the surface takes on a smooth, glistening appearance. Extreme care must be taken to avoid removal of the intentionally entrained air by over-vibration. Concretes containing aggregates larger than 50 mm must be screened using the 37.5 mm sieve prior to testing, since representative sampling becomes difficult with larger aggregate.

Strike-off of the concrete is possible with either a bar or a strike-off plate. In the case where only a pressure air test is being run, the precise volume of concrete is less critical, and the bar is acceptable. When the air meter base is being used for determination of unit weight, a strike-off plate must be used as described later.

The pressure method is limited to use with concretes containing relatively dense aggregates. Air in the interconnected porosity within the aggregate particles will be compressed just as air within the cement paste, thus indicating a higher than true air content. This is corrected for by application of the required aggregate correction factor, the lack of which application is a common flaw in observed field testing procedures. While this correction factor will compensate for the air in the aggregate, it is not appropriate to apply the pressure test to concrete containing lightweight aggregates where the aggregate correction factor exceeds about 0.5%, because with aggregate of this high porosity, relatively minor changes in aggregate porosity will lead to significant variation in measured air content.

Due to the use of water, the concrete must be discarded at the end of the test.

In recent years, a degree of controversy has arisen regarding the ability of the pressure meter to measure air content when the air voids are very small. This will occur when the air contents are high and may be to some extent dependent on the type of air-entraining agent used. Several authors have reported significantly higher air contents in hardened concrete, measured according to ASTM Practice for Microscopical Determination of Air-Void Content and Parameters of the Air-Void System in Hardened Concrete (C 457), than was obtained using the pressure method [8–12], as shown in Fig. 1. Meilenz et al. discussed the theoretical possibility of air in the smaller, higher pressure air voids dissolving, then coming out of solution in lower pressure, larger voids, leading to a net increase in air volume [13]. Hover [14] calculated the effect of the incompressibility of very small air voids. Other authors have failed to find such an increase [15].

Closer examination of the data indicate the problem may largely be a manifestation of the sampling problem mentioned earlier. While Hover's calculations [14] show that significant underestimation could occur if the very small air voids comprised a significant portion of the air volume, this is not the case, even in concretes of high air content [15]. Thus, this underestimation is unlikely to exceed around 1% air content, not the 3 to 6% reported by some.

Ozyldirim [16] followed up on field reports of higher air in the hardened concrete by a thorough test program in which all types of fresh concrete air content tests were compared with hardened air results by ASTM C 457. In no case did the differences display the extreme variation previously reported. Only when the concrete was retempered and the hardened air contents compared to the fresh concrete results prior to retempering were significant differences seen. Reexamination of the data included in reports of such underestimation problems [10] shows that the pressure test corresponds well with the gravimetric method results, Fig. 2. This would seem to preclude a specific problem with the pressure test, although some change in the actual concrete from fresh to hardened state could be invoked.

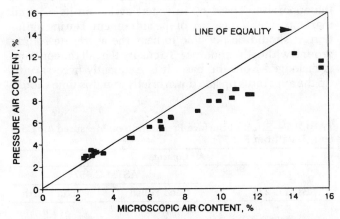

FIG. 1—Pressure versus microscopic air content (data from Ref 10).

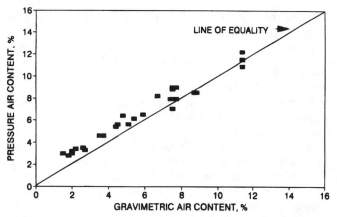

FIG. 2—Pressure versus gravimetric air content (data from Ref 10).

Further consideration makes this unlikely. The air differences reported are up to 6%. Accepting that the fresh concrete air content measurements are accurate, a net growth of the concrete would be required to cause this measured difference. This would result in up to an 18-mm increase in height of a typical 300-mm cylinder. Such growth is not reported. The answer may well lie in the difficulties associated with the ASTM C 457 test, which are covered elsewhere [3]. Some preliminary data showing this effect have been presented [17], as seen in Table 1. Poor sample surface preparation tends to erode the void edges, effectively making them seem larger, and thus contributing higher measured air contents.

Volumetric Method

The ASTM Test Method for Air Content of Freshly Mixed Concrete by the Volumetric Method (C 173) measures air content by washing the air out of the concrete through agitation of the concrete with an excess of water, primarily by a rolling action. A known volume of concrete is covered by water and the water level adjusted to a zero mark in a calibrated clear neck on the described apparatus. The apparatus is sealed; then the concrete and water are agitated by rocking and rolling until the air in the original concrete is displaced and rises to the top of the neck of the apparatus. The drop in water level from its original mark provides a measure of the air content. During agitation, it should be possible to hear the aggregate rolling around within the chamber. To ensure that all the concrete is dislodged from the base, it is frequently necessary to tip the apparatus upside down briefly, but this time should be minimized, as aggregate particles can lodge in the neck of the meter. If this occurs, they usually may be dislodged by sloshing the apparatus from side to side. The agitation must be repeated until there is no further drop in the water level. Typically, although not defined in ASTM C 173, "no further drop" is considered to be when there is less than 0.1% change in the measured air content.

This method has the advantage that air trapped in the aggregates has no impact on the test results, and thus is the method of choice for lightweight aggregate concretes. However, it has two significant disadvantages. First, the effort involved in agitating the filled apparatus is significant and can lead to severe operator fatigue over a work day. Second, it is absolutely vital for accurate measurements that the agitation indeed be repeated until no significant change in air content occurs. These two factors work in opposition to each other in practice, especially in high cement factor, sticky concretes having high air contents. In these situations, many repetitions of the agitation becomes necessary, which may take a long period of time. Times in excess of 30 min have been reported.

The operator should be alert to any free water on or around the apparatus, indicative of a failure to properly seal the unit. Such a leak will result in a higher air content being measured than actually is present in the concrete. Fine particles adhering to gaskets and the cap frequently cause such leaks.

A further constraint is that with high air content concretes it may be very difficult to dispel residual bubbles of air in the neck of the apparatus. When needed, increments of isopropyl alcohol may be added using the small cup provided, the volume of which is equal to 1% air content as measured in the calibrated neck. Care must then be exercised to note the volumes added and add to the air content finally read as described in the calculation section of ASTM C 173.

If the air content is expected to be high or very stable, a few cup-fulls of isopropyl alcohol may be added with the initial water filling. Experience will tell how far up the neck to fill with water before "topping off" with the alcohol. Care must be taken *not* to count these initial cups of alcohol as one must do with the cups of alcohol added subsequent to agitation.

Recent reports of problems with volumetric measurement of high air content concretes have paralleled the problems attributed to the pressure method as described earlier, and may be related to the wider application of air-entrained high cement factor concrete in recent years. In most cases, the air contents measured by pressure and volumetric methods agree with each other reasonably well, and with the gravimetric method, leading to the conclusion that the problem is not in these test methods. Figure 3, based on data from one paper in which a lack of correspondence with hardened air contents is shown [10], confirms the good cross-correlation of fresh concrete methods. This is in line with previous experience [18,19].

However, in some instances [20], it does appear that with extremely sticky concretes it is so difficult to wash out all the air that at the very least the volumetric method takes an impractically long time to complete. Therefore, care should be exercised in use of this method when such

TABLE 1—Effect of Surface Preparation on Measured Air Content (Data from Ref 17).

	Air Content, %	
	ASTM C 457	
ASTM C 231	Good Polish	Bad Polish
8.0	7.7	11.0
8.8	8.4	11.2
7.5	7.23	8.69

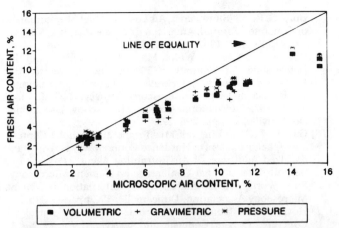

FIG. 3—*Fresh versus microscopic air content (data from Ref 10).*

concretes are encountered, with special attention paid to continuation of agitation until no air increases are registered and to the checking of the bowl after completion to ensure that all the concrete was loosened from the base. If such attention results in unacceptably long test times, alternate methods should be employed.

Gravimetric Method

In the gravimetric procedure, the unit weight of concrete is determined in accordance with ASTM Test Method for Unit Weight, Yield, and Air Content (Gravimetric) of Concrete (C 138). The concrete is placed in a container of known volumetric capacity and weighed. Using the known volume, the unit weight of the concrete is calculated and, from this, the air content and yield can be determined provided the requisite information is available. This dependence on extraneous information is a major limitation of this method when used for other than unit weight per se.

Aggregate size determines the minimum capacity of the measures used, since as the aggregate increases in size, edge effects tend to bias the sampling toward lesser aggregate contents. The minima given in ASTM C 138 are reproduced in Table 2, from which it is important to note the very high volumes required when aggregate size exceeds 25 mm. Safety considerations require that means be provided for movement of the buckets on and off the scales without undue strain on the individuals conducting the test.

TABLE 2—*Minimum Capacity of Measures for Use in ASTM C 138.*

Nominal Maximum Size of Coarse Aggregate, mm (in.)	Minimum Measure Capacity, dm^3 (ft^3)
25.0 (1)	6 (0.2)
37.5 (1 ½)	11 (0.4)
50 (2)	14 (0.5)
75 (3)	28 (1.0)
114 (4)	71 (2.5)
152 (6)	99 (3.5)

As with the other methods discussed here, the results of this test are completely dependent on proper filling and consolidation of the measure. Rodding may be used above 25-mm slump; vibration is required for lower slumps. Vibration may not be used if the slump exceeds 75 mm. ASTM C 138 requires that measures smaller than 11 dm^3 be consolidated by rodding, due to the possibility of excessive loss of entrained air, while ASTM C 231 permits vibration in this size. Thus, the smallest measures cannot be used in accordance with ASTM C 138 for concretes of less than 25-mm slump.

The top surface of the concrete must be obtained with a strike-off plate. This is best accomplished by covering about two thirds of the surface with the plate, pressing down, and withdrawing with a sawing motion. Then replace the plate on the smoothed two thirds, press down, and advance it over the unsmoothed portion with a sawing motion. Striking off with a bar is specifically not permitted, as it leads to a less precise filling. Do not overfill the measure, then push the coarse aggregate down into the measure with the strike-off plate, as this results in mortar being squeezed out, giving a nonrepresentative sample.

Subsequent calculation of volume of concrete produced per batch, the relative yield, actual cement content per unit volume, and air content are outlined in ASTM C 138. Each requires precise knowledge of batch weights and material properties not readily available in the field. When the gravimetric procedure for determining air content of fresh concrete is used, the unit weight of concrete as determined is compared with the theoretical unit weight of air-free concrete. This is calculated from the weight and specific gravity of each ingredient of the concrete mixture, and requires knowledge of the saturated, surface dry specific gravities and their moisture contents as batched. This can lead to serious errors. For example, when there is a 2% error in the moisture content of the fine aggregate or an error of 0.04 in the aggregate specific gravities, an approximate 1% error in the air content computed according to ASTM C 138 will result. Thus, ASTM C 138 is not appropriate for determining air contents of lightweight aggregate concrete, since aggregate-specific gravities and moisture contents are both subject to wide variation.

Further, when air contents change, required water contents will be reduced in most concretes; the amount of reduction is greatest in the lower cement content mixtures [21], due to the air substituting for the water needed to fill in the voids between the fine aggregate particles. Accordingly, air-content increases measured by ASTM C 138 would be reduced by approximately one-third unless completely accurate water-content information were available.

Since no water comes in contact with concrete testing according to ASTM C 138, it may be used in other testing.

SUMMARY AND FUTURE TRENDS

Tests run in accordance with these methods will yield accurate and useful results providing the operator follows the methods carefully and care is taken to avoid procedural pitfalls and use of methods inappropriate in certain situations as described earlier. The specification writer should take cognizance of these limitations and not ask more of the methods or the operators than they can provide.

The methods described here are all based on mechanical measurement techniques. While digital thermometers and scales can be used to make the work more convenient, no fundamental difference in these test methods is needed or possible. The growing emphasis on high quality, durable concrete, coupled with advances in digital electronics are poised to change this situation. A complex, but potentially valuable test has been developed [22], based on buoyancy of released air voids, that shows promise to allow determination not only of air content but also of air-void size in fresh concrete. Methods based on the use of fiber optics to measure reflectance off air voids in fresh concrete have been proposed, as well as vacuum methods wherein the displacement of the surface of the concrete under vacuum is directly measured by use of a digital strain gage.

It is expected that the effort to improve the convenience and rapidity of test methods will yield significant advances in the next decade. For now, the methods discussed here provide excellent quality control of fresh concrete.

REFERENCES

[1] Bartel, F. F., "Air Content and Unit Weight," *Significance of Tests and Properties of Concrete and Concrete-Making Materials, ASTM STP 169B*, American Society for Testing and Materials, Philadelphia, 1978, pp. 122–131.

[2] Greening, N. A., "Some Causes for Variation in Required Amount of Air-Entraining Agent in Portland Cement Mortars," *Journal*, PCA Research and Development Laboratories, Vol. 9., No. 2, May 1967, pp. 22–36.

[3] Hover, K. C., "Air Content and Unit Weight," in this publication.

[4] Kleiger, P., "Air Entraining Admixtures," in this publication.

[5] Kosmatka, S. H. and Panarese, W. C., *Design and Control of Concrete Mixtures, Thirteenth Edition*, Portland Cement Association, Skokie, IL, 1988.

[6] Klein, W. H. and Walker, S., "A Method for Direct Measurement of Entrained Air in Concrete," *Journal*, American Concrete Institute, Vol. 42, p. 657.

[7] Menzel, C. A., "Procedures for Determining the Air Content of Freshly-Mixed Concrete by the Rolling and Pressure Methods," *Proceedings*, American Society for Testing and Materials, Vol. 47, 1947, p. 833.

[8] Burg, G. R. U., "Slump Loss, Air Loss, and Field Performance of Concrete," *Journal*, American Concrete Institute, Vol. 80, No. 4, July-Aug. 1983, pp. 332–339.

[9] Gay, F. T., "A Factor Which May Affect Differences in the Determined Air Content of Plastic and Hardened Air-Entrained Concrete," *Proceedings*, Fourth International Conference on Cement Microscopy, Las Vegas, 28 March–1 April 1982, International Cement Microscopy Association, Duncanville, TX, pp. 296–292.

[10] Gay, F. T., "The Effect of Mix Temperature on Air Content and Spacing Factors of Hardened Concrete Mixes with Standardized Additions of Air-Entraining Agent," *Proceedings*, Seventh International Conference on Cement Microscopy, Fort Worth, 25–28 March 1985, International Cement Microscopy Association, Duncanville, TX, pp. 305–315.

[11] Hover, K. C., "Some Recent Problems with Air-Entrained Concrete," *Cement, Concrete, and Aggregates*, Vol. 11, No. 1, Summer 1989, pp. 67–72.

[12] Khayat, K. H. and Nasser, K. W., "Comparison of Air Contents in Fresh and Hardened Concretes Using Different Airmeters," *Cement, Concrete, and Aggregates*, Vol. 13, No. 1, Summer 1991, pp. 16–17.

[13] Meilenz, R. C., Wolkodorff, V. E., Backstrom, J. E., and Flack, H. L., "Origin, Evolution, and Effects of the Air Void System in Concrete, Part 1—Entrained Air in Unhardened Concrete," *Proceedings*, American Concrete Institute, Vol. 30, No. 1, July 1955, pp. 95–122.

[14] Hover, K. C., "Analytical Investigation of the Influence of Bubble Size on the Determination of Air Content in Fresh Concrete," *Cement, Concrete, and Aggregates*, Vol. 10, No. 1, Summer 1988, pp. 29–34.

[15] Roberts, L. R. and Scheiner, P., "Microprocessor-Based Linear Traverse Apparatus for Air-Void Distribution Analysis," *Proceedings*, Third International Conference on Cement Microscopy, Houston, 16–19 March 1981, International Cement Microscopy Association, Duncanville, TX, pp. 211–227.

[16] Ozyldirim, C., "Comparison of the Air Contents of Freshly Mixed and Hardened Concretes," *Cement, Concrete, and Aggregates*, Vol. 13, No. 1, Summer 1991, pp. 11–17.

[17] Roberts, L. R. and Gaynor, R. D, discussion of paper by Ozyldirim, C., "Comparison of the Air Contents of Freshly Mixed and Hardened Concretes," *Cement, Concrete, and Aggregates*, Vol. 13, No. 1, Summer 1991, pp. 16–17.

[18] *Proceedings*, Symposium on Measurement of Entrained Air in Concrete, American Society for Testing and Materials, Vol. 47, 1947, p. 832.

[19] Britton, R. E., "Report of Investigation of Different Methods for Determining the Amount of Air Entrained in Fresh Concrete," Pennsylvania Slag Association, 11 April 1949.

[20] Gaynor, R. D., personal communication, 1992.

[21] Klieger, P., "Effect of Entrained Air on Concretes Made with So-Called 'Sand Gravel' Aggregates," *Journal*, American Concrete Institute, Oct. 1948; *Proceedings*, Vol. 42, p. 629.

[22] Jensen, B. J., U.S. Patent No. 4,967,588, 6 Nov. 1990.

Making and Curing Concrete Specimens

Joseph F. Lamond[1]

PREFACE

This chapter covers the importance of properly making and curing concrete test specimens in both the field and in the laboratory. This chapter was previously authored in *ASTM STP 169B* [1] by R. F. Adams and in *ASTM STP 169A* [2] by T. B. Kennedy. In *ASTM STP 169* [3], the information was in a section of the chapter, Static and Fatigue Strength, authored by Clyde E. Kesler and Chester P. Siess.

INTRODUCTION

The 1914 Committee Report [4] is the basis for the presently accepted procedures for testing concrete cylinders and beams to determine the compressive or flexural strength. Two ASTM standards have been developed for making and curing test specimens. One is ASTM Practice for Making and Curing Concrete Test Specimens in the Field (C 31) and was originally published in 1920 and updated periodically. The other is ASTM Practice for Making and Curing Concrete Test Specimens in the Laboratory (C 192) and was originally published in 1944 and updated periodically. These two standards have had considerable use in concrete research and concrete construction. Specimens have to be made and cured properly since departures from the standard procedures to make and cure specimens will affect the test results.

APPLICATIONS

The strength of concrete is one of its most important and useful properties and one of the most easily determined. The strength indicated by the specimens is affected by many variables encountered in making and curing test specimens. These include size of the aggregate, size and shape of the test specimen, consolidation of the concrete, type of mold, capping procedure, curing, temperature, and moisture content at time of setting. The effect that any of these variables has on the apparent strength of the specimen will often vary depending on the particular circumstances. Among the many who have written about factors that influence the strength of concrete were Price [5], Sparkes [6], *ACI Manual of Concrete Inspection* [7], Walker and Bloem [8], and Richardson [9].

The strength of concrete, in compression, tension, and shear, or a combination of these, has in most cases a direct influence on the load-carrying capacity of both plain and reinforced structures. In most structural applications, concrete is used primarily to resist compressive stresses. In those cases where strength in tension or in shear is of primary importance, correlation to the compressive strength is often used. However, strength may not necessarily be the most critical factor in the overall performance of the concrete. For example, the need for acceptable durability may impose lower water-cement ratios than required to meet the strength requirements. In such cases, the required compressive strength may be in excess of structural requirements. Compressive strength tests in accordance with ASTM Test Method for Compressive Strength of Cylindrical Concrete Specimens (C 39) are often used as an indication of the suitability of other properties. Pavements are often designed for tensile stresses in the concrete slab. Traffic loads have been found to induce critical tensile stresses in a longitudinal direction at the top of the slab near the transverse joint, and in the transverse direction near the longitudinal edges. Accordingly, a flexural strength test either ASTM Test Method for Flexural Strength of Concrete (Using Simple Beam with Third-Point Loading) (C 78) or ASTM Test Method for Flexural Strength of Concrete (Using Simple Beam with Center-Point Loading) (C 293) is used for most pavement concrete.

It is important to keep in mind that the test specimens indicate the potential rather than the actual strength of the concrete in the structure. To be meaningful, conclusions on strength must be derived from a pattern of tests from which the characteristics of the concrete can be estimated with reasonable accuracy. Insufficient tests will result in unreliable conclusions. Statistical procedures provide tools of considerable value in evaluating results of strength tests. ACI Recommended Practice 214 [10] discusses variations that occur in the strength of concrete, and presents statistical procedures that are useful in interpretation of these variations.

Testing Personnel

There is increasing emphasis and a requirement in many building codes, political jurisdictions, project specifications, and ASTM Specification for Ready-Mixed Concrete

---

[1] Consulting engineer, Springfield, VA 22151.

(C 94) that personnel who make concrete test specimens be certified. Over 20 000 technicians have been certified under the American Concrete Institute's program for "Concrete Field Testing Technicians, Grade I."

Adams [1] discussed the importance that specimens be made by trained personnel and that the details of the various practices be followed precisely. Only in this way meaningful and reproducible test results, which are not open to question, can be obtained. The significant differences of trained and untrained personnel were reported by Wagner [11]. Wagner's study showed that strength tests made from the same concrete made by trained personnel were higher and more uniform than those made by untrained personnel.

In addition to proper training, engineers and technicians responsible for supervising or making and curing test specimens or both, must be thoroughly familiar with the test procedures. Personnel should have the required tools and equipment and review the procedure periodically for changes. Also, supervisors should review periodically the making and curing of test specimens and the procedure to determine that it is being done correctly. When strength test results fail to meet a specification requirement and rejection of the concrete is considered, the tests are almost always questioned, particularly if it can be shown that the person making the test had not complied with all the details of the test procedure.

SAMPLING

For evaluation of the test results by statistical procedures to be valid, the data must be derived from samples obtained by means of a random sampling plan. Chapter 3 of this publication on sampling contains information on a random sampling plan.

Richardson [9] pointed out that various specifications require different numbers of replicate cylinders to be tested at differing time intervals, usually 7, 14, and 28 days. It is important to make sure that the concrete for a set of cylinders comes from a single truck. A set of cylinders that does not come from the same truck will cause a considerable amount of consternation should the 14-day measured strength be lower than the 7-day strength.

MOLDS

Specifications

The molds for casting concrete specimens must have the following properties: (1) made of a nonabsorbent material, (2) nonreactive with the concrete (aluminum and magnesium are examples of reactive materials), (3) hold their dimensions and shape, and (4) be watertight. Compressive strength test specimens are cast and hardened in an upright position, with the length equal to twice the diameter. The molds for these specimens are vertical. Flexural strength specimens are rectangular beams cast and hardened with the long axis horizontal. Molds for casting vertical specimens are covered in ASTM Specification for Molds for Forming Concrete Test Cylinders Vertically (C 470). Details for field-cast beam molds are covered in ASTM C 31. Details of molds for laboratory-cast flexural strength and freeze-thaw beams are covered in ASTM C 192. Creep specimens are cylindrical specimens cast horizontally. They are covered in ASTM C 192 and ASTM Test Method for Creep of Concrete in Compression (C 512). The requirements for molds may be covered in more than one standard; therefore, all the requirements must be considered for type of specimen being molded.

Types

There are two types of molds, the reusable type and the single-use type. The reusable molds are those designed to be used more than a single time. Most reusable molds are made from castings with separate base plate and side walls. The assembled mold must be watertight; therefore, suitable sealants must be used where necessary to prevent leakage through the joints. The single-use mold may be made of any material that passes the test requirements of ASTM C 470. Samples of single-use molds should be selected at random from each shipment in such numbers as are necessary to satisfy the purchaser that the molds comply with the specification.

Richardson [9] reported on a number of studies and observations on the rigidity, water absorption, and expansion of molds made with various kinds of materials. Reusable steel molds that were difficult to seal had a tendency to leak at the joints. In high-strength concrete, compressive strength test results may be lower with plastic molds than reusable steel molds. Paraffined cardboard molds may suffer from absorption and elongation problems. All molds can suffer from being out of round. This was noted even in some molds made with light gage metal sidewalls. The shape tends to be oval at the unsupported tops for plastic and cardboard molds. The side walls of molds should be of a sufficient stiffness to prevent the mold from becoming out of round. This can be attained with sufficient wall thickness alone or in combination with a stiffened top. The molds that result in out-of-shape cylinders have an effect on the proper capping and the specimens not meeting dimensional tolerances.

MAKING AND CURING TEST SPECIMENS IN THE FIELD (ASTM C 31)

This practice is a definitive procedure for performing specific operations to produce a test specimen. These specimens are used in various test procedures to determine a property of concrete. Specimens are fabricated using field concrete that is consolidated in cylindrical or prismatic molds by either rodding or vibration. The concrete slump measured by ASTM Test Method for Slump of Hydraulic Cement Concrete (C 143) is used to select the consolidation procedure; if the slump is greater than 3 in. (75 mm), the concrete is rodded; when between 1 and 3 in. (25 to 275 mm), the concrete may be rodded or vibrated; and when less than 1 in. (25 mm), it is vibrated. Concretes of such low water contents that they cannot be consolidated properly by rodding or vibration, or requiring other sizes and shapes of specimens to represent the product or structure, are not covered by this practice.

Uses

ASTM C 31 states that specimens may be used to develop information for a number of purposes. The specimens may be used to check the adequacy of mixture proportions for strength. A concrete mixture can be proportioned from field experience (statistical data) or from trial mixtures. When proportions are checked for strength, the specimens are cured by the standard curing procedure. Specimens may be used for a comparison of laboratory, field, or in-place tests. Test results of specimens made in the field by this standard would be compared to those made in the laboratory made in accordance with ASTM C 192 or various nondestructive test methods discussed in ACI 228 [12]. The differences in the strength results between the various test methods and curing procedures can be compared. The specimens may also be used to determine the time of forms and shoring removal or the time the structure may be put in service. When used for this purpose, the specimen curing must be in the same manner as the structure. Specimens are most widely used for determination of compliance with strength specifications. These specimens must be cured by the standard curing procedure.

Samples

The samples used to fabricate field test specimens are obtained in accordance with ASTM Practice for Sampling Freshly Mixed Concrete (C 172), unless an alternative procedure has been approved. The size of the sample for field test specimens is a minimum of 1 ft^3 (28 L). The sample must be representative of the nature and condition of the concrete being sampled. The sample is collected by taking two or more portions at regularly spaced intervals during discharge of a stationary mixer, truck mixer, or agitator. The elapsed time between first and final portions should not exceed 15 min. The portions are combined and remixed to ensure uniformity and transported to where the test specimens are to be made. Molding of the specimens must begin within 15 min after fabricating the composite sample. The time should be kept as short as possible and the sample should be protected from contamination, wind, sun, and other sources of evaporation.

Concrete is a hardened mass of heterogeneous materials and its properties are influenced by a large number of variables related to differences in types and amounts of ingredients, differences in mixing, transporting, placing, and curing. Because of these many variables, methods of checking the quality of the concrete must be employed. Strength test specimens can only measure the potential strength of concrete in the structure because of different size and curing conditions between the specimen and the structure. Therefore, multiple test results based on a random pattern should be used as a basis for judging quality rather than placing reliance on only a few tests to check uniformity and other characteristics of concrete. For this purpose, statistical methods given in ACI Recommended Practice 214 [10] should be used.

Test Data on Specimens

Measure the slump by ASTM C 143. The result is used to select the method of consolidation of the specimens and record the consistency of concrete in the specimens. Measure the air content by either ASTM Test Method for Air Content of Freshly Mixed Concrete by the Volumetric Method (C 173) or ASTM Test Method for Air Content of Freshly Mixed Concrete by the Pressure Method (C 231) and record the air content of the concrete in the specimens. Chapter 9 in this publication contains information on the effects of various air contents on the properties of concrete. Measure the temperature by ASTM Test Methods for Temperature of Freshly-Mixed Portland Cement Concrete (C 1064) and record the temperature of the concrete in the specimens. The concrete temperature may help explain some unusual strength results.

Specimen Sizes

It is generally accepted that the diameter of the cylinder should be at least three times the nominal size of the coarse aggregate. For compressive strength specimens, a 6 by 12-in. (150 by 300-mm) cylinder is the standard for aggregate smaller than 2 in. (50 mm). For flexural strength specimens, the standard beam is 6 by 6 in. (150 by 150 mm) in cross section with a length of 20 in. (500 mm). Test specimens for ASTM Test Method for Electrical Indication of Concrete's Ability to Resist Chloride Ion Penetration (C 1202) have a diameter of 4 in. (102 mm).

If the aggregate is too large for the size of mold available, the oversize aggregate is either removed by wet screening as described in ASTM C 172 or a larger specimen mold is used so that the diameter of the cylinder or the smaller cross-sectional dimension of the beam is at least three times the nominal size of the coarse aggregate in the concrete. Attention must be called to the fact that the size of the cylinder itself affects the observed compressive strength; for example, the strength of a cylinder 36 by 72 in. (920 by 1830 mm) may be only 82% of the standard 6 by 12-in. (150 by 300-mm) cylinder. A cylinder smaller in size than the standard 6 by 12-in. (150 by 300-mm) cylinder will yield a somewhat greater compressive strength [5,13]. The difference in the strength of 4 by 8-in. (100 by 200-mm) and 6 by 12-in. (150 by 300-mm) cylinders increases with an increase in the strength level of the concrete [14].

Making Specimens

Where the specimens are to be molded is important. It should be as near as practicable to where the specimens are to be stored. The molds should be placed on a firm and level surface that is free of vibrations and other disturbances. Select each scoopful, trowelful, or shovelful of concrete to be representative of the batch. Concrete should be placed in the mold to the required depth keeping the coarse aggregate from segregating as it slides from the scoop, trowel, or shovel. If the specimen is rodded, carefully count the strokes for each layer. The rod should not strike the bottom of the mold when rodding the first layer. If the specimen is to be vibrated, determine the best vibrator and the best uniform time of vibration for the particular concrete. Carefully follow the procedure in the practice for rodding or vibrating specimens. The procedure is not expected to produce optimum consolidation but is used in order to permit reproducibility of results with different technicians. To obtain optimum consolidation, a different procedure would be required for each mixture, depending mainly on its workability. Close any holes left by rodding or vibration after each layer is consolidated by tapping the outside of the mold.

The exposed concrete surface should be finished to produce a flat even surface that is level with the rim or edge of the mold and that has no depressions or projections larger than $1/8$ in. (3 mm). It is the author's experience that finishing the exposed surface of a cylinder is one of the most violated requirements. This causes poor capping and dimensional tolerances in cylinders.

Mark the specimens to positively identify them and the concrete they represent. The specimens should be moved, if necessary, to curing storage with a minimum amount of handling and immediately after finishing. Cover the top of the specimens with a glass or metal plate, or a sheet of plastic, or seal them in a plastic bag, or seal the top of the specimens with a plastic cap. Caps may leave depressions in the concrete surface greater than $1/8$ in. (3 mm) in depth making capping for testing difficult. Wet fabric may be used over the specimens to help retard evaporation, but the fabric must not be in contact with the surface of the concrete or cardboard molds.

Curing Specimens

The practice permits two methods for curing the specimens. These two curing methods are not interchangeable. If the specimens are for checking adequacy of mixture proportions for strength, or the basis for acceptance or quality control, the curing method with initial and standard curing must be used. If curing is for determining form removal time or when a structure may be put in service, the field curing method must be used.

Curing is the exposure of the specimens to standard conditions of moisture and temperature prior to testing. Control of these conditions is very important since variations can dramatically affect the concrete properties and test results.

The initial curing period is in a moist environment with the temperature between 60 to 80°F (16 to 27°C) for up to 48 h. It may be necessary to create an environment during the initial curing period to provide satisfactory moisture and temperature. Insufficient moisture during the initial curing can lower measured strength. One study [15] showed that even at proper temperatures, air curing can lower the strength 8% at one day, 11% at three days, and 18% at seven days. Early-age results may be lower when stored near 60°F (16°C) and higher when stored near 80°F (27°C).

During the setting and initial hardening, the concrete can be damaged by harsh treatment. For traffic-induced vibrations, Harsh and Darwin [16] reported that wet mixtures exhibited as much as a 5% loss in strength through segregation as opposed to a 4% gain in strength in dry mixtures due to improved consolidation.

Transporting

The specimens that are transported to the laboratory before 48 h should remain in the molds, then be demolded and placed in standard curing in the laboratory. If specimens are not transported within 48 h, the molds should be removed within 16 to 32 h and standard curing used until the specimens are transported. Transportation time should not exceed 4 h. Specimens should be transported in such a manner that moisture loss is prevented, exposed to freezing temperatures, and protected from jarring. Richardson [9] indicated that rolling and bumping around in the back of a pickup truck can result in a 7% loss of strength, dropping cylinders from waist level can lower strength at least 5%. Cylinders and beams should be cushioned during transport and handled gently at all times.

Standard Curing

Standard curing is at a temperature of 73.4 ± 3°F (23 ± 1.7°C) and a moist condition with free water maintained on the surface of the specimens. Moist rooms and water tanks are used usually for creating the moist environment. A moist room is a "walk-in" storage facility with controlled temperature and relative humidity, commonly called a fog room when the prescribed relative humidity is achieved by atomization of water. All fog rooms should be equipped with a recording thermometer. Water storage tanks constructed of noncorroding materials are also permitted. Automatic control of water temperature and recording thermometer with its sensing element are required in the storage water. The water should be clean and saturated with lime. Continuous running water, or demineralized water may affect results due to excessive leaching of calcium hydroxide from the concrete specimens and should not be used in storage tanks. ASTM Specification for Moist Cabinets, Moist Rooms, and Water Storage Tanks Used in the Testing of Hydraulic Cements and Concretes (C 511) covers the requirements for fog rooms and water storage tanks used for standard curing. Price [5] stated that water-cured specimens with a water-cement ratio of 0.55 were about 10% stronger at 28 days than those cured in a fog room at 100% relative humidity.

Field Curing

Field curing is curing the specimens as nearly as possible in the same manner and at the temperatures as the concrete they represent. These specimens will reflect the influence of ambient conditions on the properties of the concrete. They give little indication of whether a deficiency is due to the quality of the concrete as delivered or improper handling and curing. ACI Standard Practice 308 [17] gives procedures for checking the adequacy of curing.

MAKING AND CURING TEST SPECIMENS IN THE LABORATORY (ASTM C 192)

This practice, like ASTM C 31, is also a definitive procedure for performing specific operations that does not produce a test result. The procedures are for a wide variety of purposes such as: (1) mixture proportioning trials for job concrete, (2) evaluation of different mixtures and materials, or (3) providing specimens for research purposes.

Equipment

The equipment needed in the laboratory includes the following: molds; rods or vibrators; mallet; small tools; sampling pan, mixing pan, or concrete mixer; air-content apparatus; slump cone; thermometer; and scales. Hand-mixing is permitted but not for air entrained or no-slump concrete. Machine mixing is preferred, especially when a laboratory regularly mixes concrete. Scales for weighing batches of materials and concrete should be checked for accuracy prior to use and be within acceptable tolerances. A sampling pan is required to receive the entire batch discharged from the concrete mixer. Conformance of the molds to the applicable requirements should be verified prior to mixing the concrete. Different tests may require different molds. The dimensions of the molds also vary for different tests and according to aggregate size; for example, where correlation or comparison with field-made cylinders are performed, in which case the cylinders shall be 6 by 12 in. (150 by 300 mm). Some test methods require specimens that are other than cylindrical or prismatic in shape. They should be molded following the general procedures in this practice. A program to determine the number of batches of concrete, number of specimens for all the various tests, and various test ages needs to be performed prior to laboratory mixing of the concrete.

Materials Conditioning and Testing

Before mixing the concrete, all the materials must be at room temperature in the range of 68 to 86°F (20 to 30°C) unless the design is being performed at other than room temperature. Store the cement as required and check it for fineness. Determine the specific gravity and absorption of the coarse aggregate by ASTM Test Method for Specific Gravity and Absorption of Coarse Aggregate (C 127) and fine aggregate by ASTM Test Method for Specific Gravity and Absorption of Fine Aggregate (C 128). The moisture content of the aggregates must be known before batching the concrete. The weights of the cement, aggregates, admixtures and water must be known accurately prior to batching. Proportioning concrete mixtures are covered in ACI Standard Practices 211.1 [18] and 211.2 [19].

Mixing and Testing

Mix the concrete in a mixer that will provide a uniform, homogeneous mixture in the mixing times required. The size of the batch has to be about 10% in excess of the quantity required for molding the specimens. Machine-mixed concrete should be mixed for 3 min after all the ingredients are in the mixer followed by a 3-min rest and 2 min of final mixing. To eliminate segregation, deposit concrete onto a clean damp mixing pan and remix by shovel or trowel. Perform the slump and temperature tests in accordance with ASTM C 143 and C 1064, respectively. When required, perform the air-content test in accordance with ASTM C 231 or C 173 and the yield test by ASTM Test Method for Unit Weight, Yield, and Air Content (Gravimetric) of Concrete (C 138). Discard concrete used for determination of the air-content test. Make the specimens following the procedures in ASTM C 192, taking the precautions as previously discussed in the section on ASTM C 31. Care should be taken that specimens are cast and stored in accordance with the applicable test methods.

Consolidation

This practice permits specimens be consolidated by rodding and internal and external vibration. The selection of the method is similar to ASTM C 31 using the slump of the concrete as guidance for the method to be used. However, a particular method of consolidation may be required by the test method or specification for which the specimens are being made. When vibration is permitted or required, either internal or external vibration may be used. When using external vibration, care has to be taken that the mold is rigidly attached to the vibrating unit. Concrete with low water contents such as roller-compacted concrete are not covered by this practice. Specimens consolidated with low water contents may require a surcharge weight on the specimen as they are consolidated with external vibration. Additional information on consolidating low-water-content specimens is in Chapter 53 of this publication.

Finishing and Curing

Finish the specimens as required. Cover and cure the specimens for 16 to 32 h prior to the removal of the molds. Immediately subject the specimens to standard curing conditions after removal of the molds in an moist room or water tanks in meeting the requirements of ASTM C 511. Flexural test specimens must be stored in lime water at standard curing temperature for a minimum period of 20 h prior to testing.

Evaluation

A precision statement in the practice is based on data from the concrete proficiency sample program of the Cement and Concrete Reference Laboratory. Each laboratory should, as part of its quality system, analyze their data against the values in Table 3 of the practice.

CONCLUSIONS

The making and curing of concrete test specimens are covered by two ASTM standards; ASTM C 31 for field use, and ASTM C 192 for laboratory use. The field standard is used to make specimens to comply with specification requirements for concrete used on construction projects. The failure to meet specification requirements has resulted in many investigations to determine the adequacy of in-place concrete, as-delivered concrete, and the making, curing, and testing of concrete specimens. The laboratory standard is used to develop mixture proportions for field concrete and research studies. The failure of the laboratory-developed mixtures to perform in the field has caused considerable problems. Research studies on test specimens made in the laboratory have been questioned on many occasions when one researcher's results are different from those of another researcher on the same type study.

Concrete test specimens should be made, cured, and tested to obtain accurate and representative results. ASTM C 31 or C 192 and the standards they reference should be followed.

REFERENCES

[1] Adams, R. F., "Making and Curing Test Specimens," *Significance of Tests and Properties of Concrete and Concrete-Making Materials, ASTM STP 169B*, American Society for Testing and Materials, Philadelphia, 1978.

[2] Kennedy, T. B., "Making and Curing Test Specimens," *Significance of Tests and Properties of Concrete and Concrete-Making Materials, ASTM STP 169A*, American Society for Testing and Materials, Philadelphia, 1966.

[3] Kesler, C. E. and Siess, C. P., "Static and Fatigue Strength," *Significance of Tests and Properties of Concrete and Concrete Aggregates, ASTM STP 169*, American Society for Testing and Materials, Philadelphia, 1956.

[4] "Report of Committee on Specifications and Methods of Test for Concrete Materials," *Proceedings*, National Association of Cement Users (now American Concrete Institute), Detroit, MI, Vol. 19, 1914.

[5] Price, W. H., "Factors Influencing Concrete Strength," *Proceedings*, American Concrete Institute, Detroit, MI, Vol. 47, 1951, pp. 417–432.

[6] Sparkes, F. N., "The Control of Concrete Quality: A Review of the Present Position," *Proceedings*, Cement and Concrete Association, London, 1954.

[7] *ACI Manual of Concrete Inspection, SP-2*, 7th ed, American Concrete Institute, Detroit, MI, 1981.

[8] Walker, S. and Bloem, D. L., "Studies of Flexural Strength of Concrete, Part 2, Effects of Curing and Moisture Distribution," *Proceedings*, Highway Research Board, Washington, DC, Vol. 36, 1957.

[9] Richardson, D. N., "Review of Variables that Influence Measured Concrete Compressive Strength," *Journal of Materials in Civil Engineering*, May 1991.

[10] "Recommended Practice for Evaluation of Strength Test Results of Concrete," ACI Committee Report 214, American Concrete Institute, Detroit, MI, 1989

[11] Wagner, W. K., "Effect of Sampling and Job Curing Procedures on Compressive Strength of Concrete," *Materials Research and Standards*, Vol. 3, No. 8, Aug. 1963, p. 629.

[12] "In-Place Methods for Determination of Strength of Concrete," ACI Committee 228, American Concrete Institute, Detroit, MI, 1989.

[13] Gonnerman, H. F., "Effect of Size and Shape of Test Specimens on the Compressive Strength of Concrete," *Proceedings*, American Society for Testing and Materials, Vol. 25, Part 2, 1925, pp. 237–250.

[14] Malhotra, V. M., "Are 4 × 8 Inch Concrete Cylinders as Good as 6 × 12 Cylinders for Quality Control of Concrete," *Journal*, American Concrete Institute, Jan. 1976.

[15] Bloem, D. L., "Effect of Curing Conditions on the Compressive Strength of Concrete Test Specimens," National Ready Mixed Concrete Association, Publication No. 53, Silver Spring, MD, 1969.

[16] Harsh, S. and Darwin, D., "Traffic-Induced Vibrations and Bridge Deck Repairs," *Concrete International*, Vol. 8, No. 5, May 1986, pp. 36–42.

[17] "Standard Practice for Curing Concrete," ACI Committee 308, American Concrete Institute, Detroit, MI, 1986.

[18] "Standard Practice for Selecting Proportions for Normal, Heavyweight, and Mass Concrete," ACI Committee 211.1, American Concrete Institute, Detroit, MI, 1989.

[19] "Standard Practice for Selecting Proportions for Structural Lightweight Concrete," ACI Committee 211.2, American Concrete Institute, Detroit, MI, 1989.

Time of Setting
Vance H. Dodson[1]

PREFACE

Although the original version of *ASTM STP 169*, published in 1956, had a chapter written by Scripture entitled "Setting Time," there was no ASTM standard covering the subject [*1*]. A method was, however, proposed in 1957 for determining the setting time of concrete under ASTM C 403-57T that was reviewed and discussed by Kelly in *ASTM STP 169A* in 1966 [*2*]. In 1978, Sprouse and Peppler continued and updated the work of Kelly in *ASTM STP 169B* [*3*]. This chapter is an attempt to merge the historical facts and background information provided by the earlier authors with today's thinking on the subject of time of setting.

INTRODUCTION

As soon as water comes in contact with portland cement in a concrete mixture, a chemical reaction (called hydration) between the two begins. While the exact nature of the chemical interactions that take place is not clearly understood, they are exothermic and serve to chemically tie-up a portion of the free water in the cementitious system. Because of the uncertainties surrounding the nature of the hydration products and their early rates of formation, it can only be approximated by stoichiometry that 45.4 kg (100 lb) of a Type I cement will combine with about 0.9 kg (2 lb) of the concrete mix water during the first few minutes of mixing and make that amount of water unavailable for workability purposes. These hydration reactions continue to proceed, but at different rates, as the concrete matures. For example, it can be estimated, from the amounts of the four phases in the cement that have hydrated during the first five hours of mixing, that 45.4 kg (100 lb) of a typical Type I cement will have combined with about 4 to 6 kg (9 to 12 lb) of water [*4*]. This removal of free water, accompanied by the build-up of hydration products, serves to stiffen the concrete and increase its rigidity.

HISTORY

The gradual stiffening, or increase in rigidity, of fresh concrete is commonly referred to as "setting" and is important because the degree to which it has occurred determines whether the concrete can be conveniently placed, compacted, finished, and subjected to loading. The question of what happens during the setting phenomenon led to another question, that is, what criterion should be used to designate when the concrete had undergone set. Scripture was of the opinion that as long as concrete maintained any significant degree of workability it had not set, and as soon as the concrete attained any appreciable strength it had set. Therefore, the time of setting was the time elapsed between the beginning of mixing and the attainment of conditions between these two extremes [*1*]. The time of setting of concrete is also important because it is during this period in the life of the concrete that it is most vulnerable to environmental influences.

The progress up to 1956 in developing methods for measuring the time of setting of concrete has been summarized by Scripture [*1*]. Although a number of criteria had been suggested for defining time of setting, such as loss of workability before placing, degree of bond between fresh concrete and concrete in place, time of finishing for floors, ability to strip forms, and early strength gain, it was concluded that all of these were purely subjective and that none was representative of the time of setting of concrete.

Scripture also described a number of less subjective methods that had been proposed prior to 1956 for measuring, as a function of time, changes in certain properties of concrete, the results of which might be relatable to the setting of concrete and hence the time of setting [*1*].

Electrical Measurements

The electrical resistance of plain concrete had been found to increase with time until it reached a plateau, and the time at which this occurred seemed to correlate with the time of setting of the cement itself [*5,6*]. However, when an admixture containing an electrolyte, such as calcium chloride, was present, the electrical resistance of the concrete remained at a low and fairly steady level. Therefore, the method was considered unsuitable because one of the purposes of a suitable test method was to demonstrate the effects of chemical admixtures on the time of setting of concrete.

Consistency Measurements

The use of the Vebe apparatus, modified Gillmore and Vicat needles, rod penetration, and the Kelly ball had been

---

[1] Technical consultant, Construction Products Division, W. R. Grace, Cambridge, MA 02140.

suggested, but they were found to be generally unsatisfactory due to the interference of large pieces of aggregate. As long as there was any penetration with these apparatus, the concrete was not considered as having set, but the time when penetration ceased to occur was not easily determined and not reproducible. Another approach was to determine the time when it was just possible to finish the surface of the concrete by steel troweling, but whether it measured the point of setting or not, it could only be of interest to those concerned with floor finishing.

Sound Velocity and Frequency Measurements

Attempts had been made to follow the setting of concrete by measuring the ultrasonic pulse velocity of concrete or by measuring fundamental frequencies of concrete confined in a metal tube [7]. However, these methods were not developed further because of the high cost of the apparatus and the delicate technique required.

Bleeding Characteristics

This method had been suggested but no data were available on its application to the determination of time of setting. Because of a wide variation in bleeding characteristics of different concrete mixtures, the proposed method was not considered promising.

Heat of Hydration

When the temperature of concrete was plotted as a function of time, a curve was obtained that exhibited a sharp change in slope where the time of setting of the cement was expected. The applicability of the method in determining the time of setting of concrete was doubted on the basis that it measured a chemical property of cement and that the shape of the curve could be changed by altering the gypsum content of the cement without actually changing the time of setting of the cement or concrete.

Changes in Volume

When the volume change of fresh concrete (measured in a dilatometer) was plotted versus time, a sharp change in the slope of the curve was noted at a time corresponding to when setting of the cement was expected. The method was considered to be open to the same objections as those stated in connection with heat of hydration determinations.

Stiffness and Strength Determinations

One method consisted of casting concrete specimens in flexible containers and determining deflection. Some work had also been done on both compressive and flexural strength measurements, but sufficient viable data were not available to demonstrate the suitability of any of these methods for measuring time of setting. Limited test results obtained by a tension test method indicated that a fairly sharp change of the tensile strength versus time curve appeared at a time that might be considered as the time of setting.

Compressibility Changes

The pressure-type air meter had been employed to determine changes in compressibility of concrete containing entrapped or purposely entrained air, or both. Curves in which the apparent percentage of air was plotted versus time after mixing showed definite points of inflection when the concrete became sufficiently rigid to resist deformation by the air pressure used in this type of meter. However, the point of inflection in the curve became less definitive as the amount of air in the freshly mixed concrete decreased.

Summary

After reviewing the historical methods and the available data, Scripture concluded that

1. it probably would not be possible to define the time of setting of concrete in terms that would be applicable to all concrete mixtures for all applications since the meaning of time of setting depended to a considerable extent on the needs and interests of the particular individual, and
2. any definition would probably have to be in terms of a maximum change in some physical property (yet to be determined) of the concrete and a suitable test method had to be established to determine that particular property.

PENETRATION RESISTANCE METHOD

Of all the test methods proposed prior to 1956, the concept of relating the time of setting of concrete to its change in consistency, as measured by the resistance to penetration of various probing devices proved to be the precursor of today's accepted method. The problem created by the presence of large pieces of aggregate was effectively erased by screening the concrete through a No. 8 (2.36 mm) sieve prior to testing. When the sieved mortar was subjected to penetration tests, the results were reasonably reproducible. However, there was doubt whether the results really represented the time of setting of concrete because of the observed discrepancy between the times of setting of neat cement pastes and mortar pats.

Soon after Scripture's publication, Tuthill and Cordon described a method for determining the setting characteristics of concrete based on the change in resistance to penetration of mortar sieved from that concrete using Proctor needles of various bearing surface areas [8]. These authors claimed that the setting characteristics of concrete mortar, as determined by their method, were not equal to those of the corresponding concrete but were of similar character and reliably indicative of what may be expected of the concrete.

In the Tuthill-Cordon test procedure, the mortar that had been screened through a No. 4 (4.75 mm) sieve from the concrete was; (1) placed in a container at least 15.2 cm (6 in.) deep and having a diameter large enough to provide an exposed mortar surface area to permit a minimum of ten undisturbed readings of penetration resistance; (2) vibrated in its container; (3) covered; and (4) placed in a room maintained at the desired temperature of test. Bleeding water was poured off before making a penetration test. Penetration resistance readings were taken at such intervals as necessary to define the setting characteristics by means of penetration needles of appropriate bearing surface areas. The first reading was taken with a needle having a 6.45 cm^2 (1 in.^{2}) bearing area, and subsequent readings were taken with that needle until the penetration resistance reached or approached the maximum capacity of the loading spring. Smaller needles were used in subsequent penetrations, the size being dictated by the penetration resistance of the mortar and the maximum capacity of the loading spring. In making penetration resistance readings, the pressure on the needle was applied steadily and gradually until the needle had penetrated the mortar to a depth of 2.54 cm (1 in.).

Tuthill and Cordon compared the characteristics of the concrete with those of the mortar sieved from the same concrete and found that when the concrete could no longer be made mobile by vibration, the mortar had a penetration resistance of approximately 3.5 MPa (500 psi). This condition of the concrete is defined today as its vibration limit, that is, the age at which fresh concrete has set sufficiently to prevent its becoming mobile when subjected to vibration [9]. The elapsed time, after initial contact of cement and water, required for the mortar sieved from the concrete to reach a penetration resistance of 3.5 MPa (500 psi) has also come to be known as time of initial setting [10]. It is also considered to be the point at which fresh concrete will not become monolithic with previously placed concrete. The same authors also found that when the concrete was completely set and had a compressive strength of 0.7 MPa (100 psi), as determined on a 15.2-by-30.5 cm (6-by-12 in.) cylinder, the penetration resistance of the mortar was 27.6 MPa (4000 psi). The elapsed time, after initial contact of cement and water, required for the mortar sieved from the concrete to reach the penetration resistance of 27.6 MPa (4000 psi) has come to be known as time of final setting [10].

A test method based on the procedure developed by Tuthill and Cordon was proposed by Subcommittee III-n, ASTM Committee C9 on Setting Time of Concrete.[2] In 1957, ASTM Tentative Method of Test for Rate of Hardening of Mortars Sieved from Concrete Mixtures by Proctor Penetration Resistance Needles (C 403-57T) was issued. The method has undergone many revisions since its first adoption, but the principles set forth by the originators have been basically retained and is presently entitled ASTM Test Method for Time of Setting of Concrete Mixtures by Penetration Resistance (C 403). The users of ASTM C 403 are not in complete agreement that the 3.5

MPa (500 psi) penetration resistance of the sieved mortar really represents the vibration limit of the concrete and that 27.6 MPa (4000 psi) penetration resistance of the mortar is truly indicative that the concrete is completely set. However, these arbitrarily defined points serve as convenient reference points for determining the relative rates of setting of mortars from concretes fabricated from different materials, stored under different environmental conditions, and made with different proportions. Some of the differences between the concrete stiffening time test specified by the British Standards Institute, BS 5075, and ASTM C 403 are listed in Table 1 [11].

During the test procedure, the force necessary to overcome the resistance to penetration is measured. The force is divided by the bearing area of the needle and the penetration resistance of the mortar is expressed in units of stress. Since it is very unlikely that during the test, the measured penetration resistance will coincide with the 3.5 MPa (500 psi) and the 27.6 MPa (4000 psi) values, the times of setting must be obtained by interpolation. Until 1990, this was done by plotting the penetration resistance versus the elapsed time on plain graph paper and by drawing a best-fitting smooth curve through the data points. The times of initial and final setting were determined by the intersection of the curve with the lines corresponding to 3.5 and 27.6 MPa (500 and 4000 psi).

BOND PULLOUT PIN TEST

During the year that ASTM C 403-57T was published, a second method for measuring the setting of concrete, employing a bond pullout pin technique, was proposed

---

[2] Subcommittee III-n is currently (as of 1992) Section 5, Subcommittee C.09.23, Chemical Admixtures, ASTM Committee C9 on Concrete and Concrete Aggregates.

TABLE 1—A Brief Comparison of BS 5075, Part 1, with ASTM C 403.

Item	BS 5075, Part 1	ASTM C 403
Sieve size used to separate mortar from concrete	5.00 mm	4.75 mm (No. 4)
Test specimen container	75 mm dia 50 to 100 mm deep	152 mm dia, minimum 152 mm deep, minimum
Storage temperature of test specimen	20 ± 2°C	20 to 25°C
Bleed water	not removed	removed before each penetration reading
Penetration needle bearing surface area	30 mm^2	varies from 645 to 16 mm^2
Key penetration resistance values	0.5 N/mm^2 (72 psi) (limit for placing and compaction of concrete)	...
	3.5 N/mm^2 (500 psi) (guide to time available for avoidance of cold joints)	3.5 MPa (500 psi) (vibration limit and avoidance of cold joints— initial set)
	...	27.6 MPa (4000 psi) (final set)

by Kelly and Bryant [12]. In this test procedure, the whole concrete was utilized rather than the mortar portion, and the method was based on the measurement of the bond strength between concrete and stainless steel pins, 9.5 mm (⅜ in.) in diameter. The pins were embedded vertically to a depth of 12.7 cm (5 in.) by vibrating the concrete around them in a beam mold. The pins were held firmly in a vertical position by a special jig during specimen fabrication. Individual pins were subsequently pulled out of the concrete at increasing time intervals by using a spring scale, and the load to pull out the pin was measured. The bond strength (between pin and concrete) was calculated by dividing the load read from the spring scale by the surface area of the pin available for bonding. The bond strength was plotted versus elapsed time, and time of setting would be the time to reach a certain level of bond strength.

A favorable feature of the bond pullout method is that the measurements are made on the whole concrete rather than on the sieved mortar. Therefore, the results could be interpreted without concern as to the possible effects of removal of coarse aggregate or of manipulation and exposure of the sample during the process of sieving the concrete to obtain the mortar for penetration tests. On the other hand, modification and refinement of the method was deemed necessary. For example, the relationship between the embedded surface area of the pin and combined weight of the specimen and its mold were such that the maximum load was limited unless the specimen (and its mold) were physically restrained from being lifted during the later stages of setting.

A proposed tentative test method for measuring setting time of concrete based on pin pullout was drafted for consideration by ASTM Subcommittee III-n in 1959. The test procedure was essentially the same as that described by Kelly and Bryant [12]. Bond strength versus elapsed time curves obtained by the pin pullout method for concretes made from three different brands of Type I cement are shown in Fig. 1. The concretes were similar in all respects except for differences in brand of cement. The shapes of the three curves are similar to those of the curves of penetration resistance versus elapsed time as determined by ASTM C 403 and shown in Figs. 5 through 10. Although the method has been under consideration by ASTM for almost 30 years, little action has been taken to make it a viable alternative method to ASTM C 403. One reason for this is the lack of correlation between pin pullout strength, penetration resistance, and the criteria set forth by Scripture for initial and final times of setting as determined by penetration resistance. Another reason is the wide acceptance of the ASTM C 403 method and the preponderance of test data that has been accumulated using the method.

CURRENT PRACTICES

Treatment of Time of Setting Data

The penetration resistance and elapsed time data in Table 2 will be used to illustrate the manner for determining times of setting. Figure 2 is a plot on a linear scale of the penetration resistance versus the elapsed time values listed in Table 2. The curve was drawn by hand using a

TABLE 2—Penetration Resistance Data.

Penetration Resistance (PR), psi	Elapsed Time (t), min	Log (PR)	Log (t)
280	184	2.447	2.265
500	198	2.699	2.297
1200	224	3.079	2.350
1920	252	3.283	2.401
2560	263	3.408	2.420
3120	270	3.494	2.431
4500	294	3.653	2.468

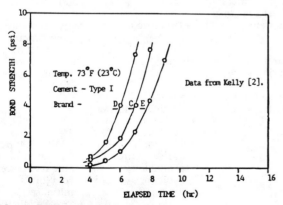

FIG. 1—Effect of brand of cement on setting behavior, as determined by pullout pin method.

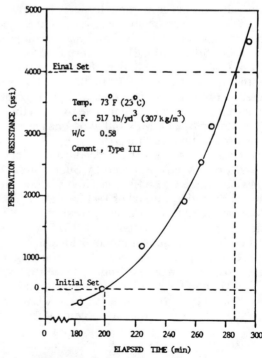

FIG. 2—Linear plot of penetration resistance versus elapsed time and hand fit curve used to determine times of setting of a Type III cement, from Table 2.

flexible drawing curve and drawn so as to achieve a visual best-fit curve to the data. Horizontal lines are drawn at penetration resistance values of 3.5 MPa (500 psi) and 27.6 MPa (4000 psi). The intersections of the horizontal lines with the curve define the times of initial and final setting, which in this case are 200 min (3 h and 20 min) and 286 min (4 h and 46 min), respectively.

Figure 3 is a log-log plot of the penetration resistance versus elapsed time values. The plot shows that for this particular data there is approximately a straight line relationship between the logarithms of the two variables. The straight line is obtained by linear regression analysis, and in this case the correlation coefficient is 0.979. In cases where the correlation coefficient is less than 0.98, the manual method previously described should be used. If log-log paper is not available, a plot of the logarithm values on linear graph paper can be used as illustrated in Fig. 3. The presentation and analysis of time of setting data using the log-log relationship was accepted by ASTM in 1990.

A semi-logarithmic plot of the data in Table 2, wherein the logarithm of the penetration resistance is plotted versus elapsed time, is shown in Fig. 4. This method of data presentation and analysis was suggested in 1960 [13]. Those that preferred the semi-logarithmic scale to the linear scale maintained that a best-fit straight line could be drawn more accurately than a best-fit curve through the data points. Critics of this manner of plotting the data claimed that the data points seldom formed a perfectly

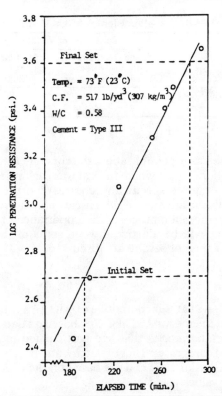

FIG. 4—Semi-logarithmic plot of penetration resistance (log) versus elapsed time and best straight line (determined by linear regression analysis) of a Type III cement, from Table 2.

straight line and that the position of the line was a matter of judgement on the part of the operator; while with a linear plot, a smooth curve could usually be fitted closely to all the data points. It must be kept in mind that the semi-log approach was suggested before calculators were readily available to determine the best-fit straight line and its correlation coefficient. For the semi-log plot in Fig. 4, the straight line has a correlation coefficient of 0.988. A comparison of the times of setting derived from the three methods of data plotting (Figs. 2, 3, and 4) is summarized in Table 3. The differences in the results are not significant in light of the precision of the method.

Information on Concrete Mixtures

A number of factors are known to influence the time of setting of concrete (and of the mortar sieved from it). Therefore, time of setting data should include as much

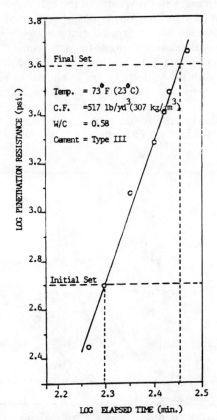

FIG. 3—Log-log plot of penetration resistance versus elapsed time and best straight line (determined by linear regression analysis) of a Type III cement, from Table 2.

TABLE 3—Time of Setting of Concrete from Data in Table 2.

Method of Plotting	Time of Setting, min (h:min)	
	Initial	Final
Linear (Fig. 2)	200 (3:20)	286 (4:46)
Log-log (Fig. 3)	198 (3:18)	283 (4:43)
Semi-log (Fig. 4)	194 (3:14)	284 (4:44)
Average	197 (3:17)	284 (4:44)
Coefficient of variation, %	1.6	0.6

TABLE 4—Precision of ASTM C 403.

Time of Setting	Coefficient of Variation, %	
	Single Operator, Multiday, Three Individual Results	Multilaboratory Average of Three Tests
Initial	7.1	5.2
Final	4.7	4.5

information as is possible about the concrete mixture and the conditions under which the test was performed. ASTM C 403, for example, specifically requires reporting of information, all, or anyone, of which can play an important role in any decision process. The importance of many of these items will be illustrated when some of the factors that affect time of setting are discussed.

Precision

In 1973, ASTM Subcommittee C09.03.14 approved precision statements for ASTM C 403, based on data received from five laboratories that had participated in a round-robin test program. Those statements have remained unchanged as of 1992 and are based on the coefficient of variation values listed in Table 4.

FACTORS THAT INFLUENCE TIME OF SETTING

Type of Cement

Penetration resistance data obtained using ASTM C 403 for a Type I, a Type II, and a Type III cement are shown in Fig. 5. The difference in the positions of the three curves is the result of differences in the three types of cement, such as fineness, particle size distribution, and chemical composition, that is, C_3A, C_3S, SO_3, and soluble alkali contents and the nature of the calcium sulfate in the cement (gypsum, hemihydrate, natural anhydrite). Other than the presence of the different types of cement and only minor differences in slump and air contents, the three concretes were identical, that is, same cement content, water-cement ratio, aggregate content, and temperature.

Brand of Cement

The setting behavior of mortars screened from concretes made from three different brands of a Type I cement is illustrated in Fig. 6. The three concretes were identical in all respects except for the three different cement brands and minor differences in their slumps and air contents. The differences previously cited between types of cement can also exist between brands of a given type, and these affect the setting behavior as measured by ASTM C 403.

Fineness of Cement

The Type I cement used to show the effect of cement type (Fig. 5) had a fineness of 3220 cm^2/g, as measured by ASTM Test Method for Fineness of Portland Cement by Air Permeability Apparatus (C 204). Portions of that cement were ground in a laboratory steel ball mill to two different, higher fineness values, at 23°C (73°F) to minimize any changes in the nature of the calcium sulfate due to the heat of grinding. The times of setting of concretes made from each of the ground cements were compared with those of the concrete made from the cement in the "as received" condition. The setting behavior of the mortars screened from each of concretes is shown in Fig. 7. The three concretes were essentially the same, differing only in the fineness of the cement and having only minor differences in slump and air content. The increase in cement fineness resulted in a reduction in times of setting, both

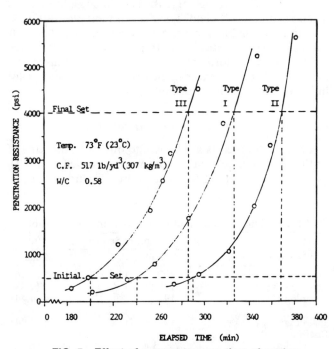

FIG. 5—Effect of cement type on time of setting.

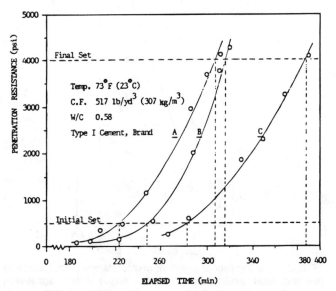

FIG. 6—Effect of cement brand on time of setting.

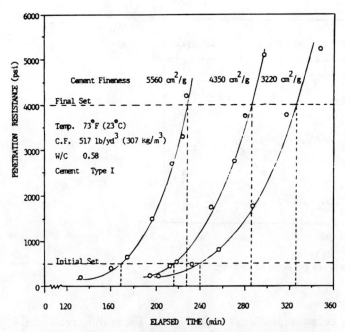

FIG. 7—Effect of cement fineness on time of setting.

cal in all respects except for differences in the masses of cement and aggregates per cubic yard (cubic metre) and their slumps and air contents. However, the mass ratio of coarse to fine aggregate was maintained at the same value of 55:45. The data show that an increase in C.F. shortens the initial and final times of setting.

Water-Cement Ratio

The influence of water-cement ratio (w/c) on the setting of mortar sieved from concrete is illustrated in Fig. 9. When the w/c is reduced, the initial and final times of setting are also reduced [14]. Reducing the w/c increases the concentration of the cement in the paste (and mortar); and the rate of setting of the sieved mortar, as well as the concrete from which it was screened, increases. The data in Fig. 9 are for mortars screened from concretes of essentially the same character except for differences in w/c, slump, air content, and amount of fine aggregate. The amount of fine aggregate in the lower w/c concrete was increased over the amount in the higher w/c concrete to compensate for the lower amount of water and to maintain the proper yield.

initial and final, and this is probably the principal reason why the concrete made from a Type III cement exhibited shorter times of setting than those of the concretes containing the Types I and II cements (Fig. 5). If the hydration of the cement particles that takes place initially and during the process of setting is a surface phenomenon, then an increase in the cement's fineness, and hence its surface area, should lead to an acceleration of that process.

Cement Content

The data in Fig. 8 illustrate the influence of the cement factor (C.F.), mass of cement per cubic yard of concrete, on the setting of sieved mortar. Both concretes were identi-

Temperature

The setting process of cement paste, mortar, and concrete is the result of chemical reactions between the cement and the water. Therefore, the laws of chemical kinetics should be in effect, and it is not surprising to find that those chemical reactions, and hence the setting process, proceed faster as the temperature is increased. The setting behavior of concretes whose raw materials were pre-conditioned (for 48 h), mixed, screened, and whose mortars were then stored and tested, at four different temperatures, is shown in Fig. 10. It is evident that as the temperature is increased from 10 to 32°C (50 to 90°F) the initial and final times of setting are progressively

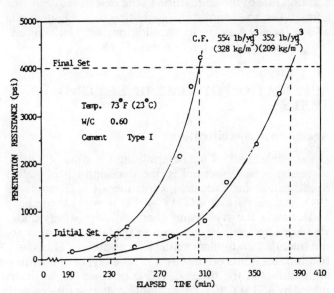

FIG. 8—Effect of cement factor (C.F.) on time of setting.

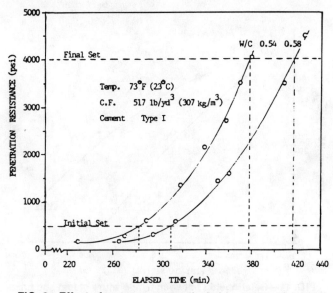

FIG. 9—Effect of water-cement ratio (w/c) on time of setting.

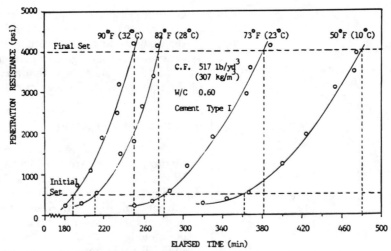

FIG. 10—Effect of temperature on time of setting.

reduced. The four concretes were identical in all respects except for their slumps, air contents, and temperatures at which they were mixed and tested. A plot of initial and final times of setting derived from Fig. 10 versus the temperature of test is shown in Fig. 11. The relationship appears to approach linearity. Although the test specimens were fabricated, stored, and tested at the four arbitrarily selected temperatures, the temperature of the test specimens in all probability exceeded those temperatures while the test was in progress because of the heat generated by the hydrating cement.

Set-Retarding and Set-Accelerating Admixtures

The influence of a set-retarding admixture on the time of setting of mortars screened from concrete mixtures whose raw materials were pre-conditioned at 28°C (82°F) for 48 h, then mixed, sieved, and tested, at the same temperature, is illustrated in Fig. 12. The addition rate of the retarder was sufficient to delay the times of setting of the concrete to a degree that they were essentially equivalent to those obtained at 23°C (73°F) without a retarder. The admixture used in this case was based on a salt of a hydroxylated carboxylic acid and added at the rate of 0.06% by mass of cement.

Also shown in Fig. 12 is the effect of a set-accelerating admixture on the setting of concrete at 10°C (50°F). The amount of set acceleration produced by the admixture fell somewhat short of reducing the time of setting of the 10°C (50°F) concrete to those of the plain concrete at 23°C (73°F). In this instance, the set-accelerating admixture was based on calcium chloride and added at the rate of 2.0% by mass of cement.

The effect of set-retarding and set-accelerating admixtures on the times of setting depends on a number of factors, such as concrete temperature, ambient temperature, nature of the cement, nature of the admixture, and addition rate of the admixture. In the case of set-retarding admixtures, the time of their addition to the concrete is also a factor. A delay in its addition, after mixing has begun, increases its effectiveness [15,16].

APPLICATION OF TIME OF SETTING TESTS

Acceptance Specifications

A suitable method for determining the time of setting of concrete was essential in the development of ASTM Specification for Chemical Admixtures for Concrete (C 494). The purpose of ASTM C 494 is to provide methods and criteria for evaluating chemical admixtures under standardized testing conditions in the laboratory that are not intended to simulate actual job conditions. This specification sets limits on the effect of seven types of chemical admixtures on the time of setting of concrete, as determined by ASTM C 403. For example, concrete treated with a set-retarding admixture must reach initial set at least 1

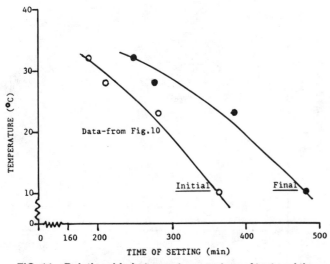

FIG. 11—Relationship between temperature of test and times of setting.

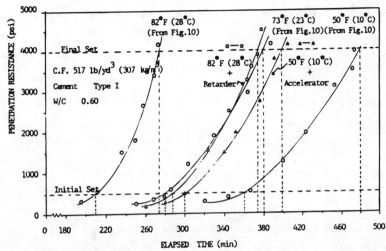

FIG. 12—Effect of set-retarding and set-accelerating admixtures on time of setting.

h later but no more than 3.5 h later than the reference concrete and must reach final set not more than 3.5 h later than the reference concrete. The temperature at which the ASTM C 403 tests are conducted is also well defined, that is, the temperature of each of the ingredients of the concrete mixtures, just prior to mixing, and the temperature at which the time of setting specimens are stored during the test period shall be 23 ± 1.7°C (73 ± 3°F). The basic approach of ASTM C 403 was used to develop a test method for measuring the time of setting of shotcrete, which was adopted in 1989 as ASTM Test Method for Time of Setting of Shotcrete Mixtures by Penetration Resistance (C 1117).

Field Concreting

Job conditions often require a greater degree of and closer control of retardation or acceleration than is required by ASTM C 494. This may result from (1) unusual environmental conditions at the time of concrete placement; (2) structural considerations, such as avoiding cracking of the concrete deck when steel bridge girders deflect during placement; (3) avoidance of cold joints in mass concrete; and (4) in special types of concreting operations, such as slip form construction or in the manufacture of steam-cured, precast concrete units. The results of time of setting tests conducted in the laboratory under simulated job conditions and utilizing the concreting materials to be used in the work can be valuable aids in planning concreting operations and determining whether special measures will be required for set control. If further control is required, the type and dosage of admixture can be tentatively selected, subject to verification of time of setting tests made on the job during actual construction.

Published works are devoid of precise correlations between time of setting test results and slab finishing operations. In one instance, where a 152-mm (6-in.) thick floor slab was being placed, samples of the concrete were taken as it was being discharged into the form and sieved. The mortars were tested according to ASTM C 403 and, during the test period, were exposed to the same environmental conditions as the concrete in the slab [17]. The test specimens exhibited initial set some 60 min after the finishers had completed their work. This discrepancy in time between initial set and finishing was attributed to the difference in the mass of the slab and of the test samples. Differences in exposure conditions between the time of setting specimen and the in-place concrete can lead to misapplication of time of setting results. This is particularly true in flat work where the time of setting of the in-place concrete can be influenced by wind velocity, nature of the subgrade and the radiant heat of the sun [18]. Others have reported that tests performed on samples of concrete at the batch plant do not reflect the influence of the increased temperature occurring within massive blocks in a dam [19].

In the precast, steam-cured concrete industry, the concrete is batched and immediately transferred to and placed in forms or beds. The forms and their contents are completely enclosed and allowed to remain undisturbed until the concrete has attained initial set. This step in the process is referred to as the "preset," or presteaming, period. Steam is introduced into the enclosure and the remainder of the curing process follows. If the preset period is ended before initial set has occurred, the early as well as the later strengths of the concrete will be reduced. Excessive time in the preset period, that is, beyond initial set, costs the producer money in terms of re-use time of the casting forms or beds, or both. The most advantageous time to end the preset period is not easily determined, even when a screened sample of the concrete is stored next to the precast member within the enclosure during the measurement of its penetration resistance. This is another example of the influence of mass, that is, of the precast member versus that of the test sample, on the time of setting [20].

SOME OBSERVATIONS RELATED TO TIME OF SETTING

It is evident from the data (Figs. 8 and 9) that both initial and final times of setting, as measured by ASTM C

TABLE 5—Concretes Illustrating the Relationship Between Time of Setting and C.F. ÷ w/c.

Concrete Components and Properties	Concrete Number[a]							
	1	2	3	4	5	6	7	8
Cement, kg/m^3 (lb/yd^3)	303(511)	242(408)	181(306)	241(406)	302(509)	243(408)	182(307)	237(400)
Coarse Aggregate, kg/m^3 (lb/yd^3)	1070(1803)	1098(1851)	1125(1897)	1090(1838)	1065(1796)	1099(1853)	1128(1902)	1074(1811)
Fine Aggregate, kg/m^3 (lb/yd^3)	852(1437)	872(1471)	896(1511)	866(1461)	849(1432)	874(1473)	899(1516)	878(1480)
Water, kg/m^3 (lb/yd^3)	173(292)	173(291)	173(291)	161(271)	176(297)	173(292)	173(292)	165(279)
w/c	0.57	0.71	0.96	0.67	0.58	0.71	0.95	0.70
C.F. ÷ w/c kg/m^3	532	341	189	360	521	342	192	339
Air, %	1.8	1.9	2.2	2.0	2.0	2.0	1.8	1.8
Slump, mm (in.)	114(4 ½)	102(4)	76(3)	63(2 ½)	76(3)	102(4)	108(4 ¼)	63(2 ½)
Time of setting, min								
Initial	230	272	303	262	297	324	342	323
Final	306	358	401	345	395	427	460	428

[a] Concrete numbers—1 through 4 are made from a Type I cement; 5 through 8 are made from a Type I/II cement.

403, are directly related to the water-cement ratio (w/c) and inversely related to the cement content (C.F.) of the concrete. Eight concretes are described in Table 5 and their times of initial and final setting are plotted as a function of their C.F.s ÷ w/cs in Fig. 13. Plotting C.F. ÷ w/c on the abscissa is preferred rather than w/c ÷ C.F., because the former yields a quotient greater than unity and has units of kg/m^3 (lb/yd^3). The eight concretes differ in most aspects except that the ratio, by mass, of coarse to fine aggregate was maintained at a value of 56:44 and the temperature of the concretes and their environment was 23°C (73°F). The plots are linear and have correlation coefficients ranging from 0.994 to 0.997. It is not surprising that the straight lines in Fig. 13 assigned to the concretes, made from two different cements, have different slopes and occupy different positions with respect to the X-Y axes. This manner of presentation has been used to account for the influence of fly ash on the time of setting of portland cement-pozzolan concretes [14].

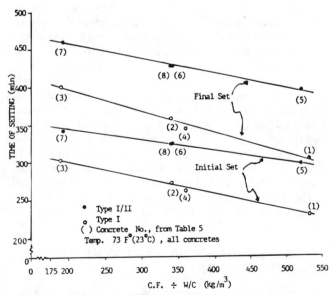

FIG. 13—Combined effect of cement content (C.F.) and water-cement ratio (w/c) on time of setting.

From the relative shapes of the penetration resistance versus elapsed time curves illustrated in this work, it is evident that as the time of initial setting of concrete increases, its time of final setting also increases, but not proportionally. When the 25 different sets of initial and final times of setting that are reported in the various figures and tables are analyzed, the ratio of final to initial time of setting is found to average 1.33, with a standard deviation of 0.03 and a coefficient of variation of 2.3%. When the precision statement for ASTM C 403 is considered, these numbers are quite acceptable. It is important to keep in mind that the 25 sets of data were derived from testing concretes that; (1) were made from a wide variety of cements; (2) had various amounts of cement; (3) had a broad range of water-cement ratios; (4) contained and did not contain admixtures; and (5) were mixed and tested at various temperatures. It is also of interest to note that the ratio of times of setting derived from the example cited in ASTM C 403-90 (p. 217) is 1.35. The average value for the ratio of final to initial time of setting, reported here, is in no way meant to be an alternative to completing the test beyond initial time of setting, but can serve as a guide in determining the credibility of the results of a given test.

The relationship between the time of setting of a given concrete and its compressive strength is, at best, relegated to generalities. For example, concretes whose times of setting have been

1. reduced by the addition of set accelerating admixtures or an increase in temperature, etc., exhibit reductions in 28-day compressive strengths and increases in three-day compressive strengths, both of which are essentially proportional to the degree of set acceleration; and
2. delayed by the addition of set-retarding admixtures, or a decrease in temperature, etc., display increases in 28-day compressive strengths that are almost proportional to the degree of set retardation [21].

FUTURE CONSIDERATIONS

With respect to ASTM C 403 as it applies to the measurement of the time of setting under carefully controlled laboratory conditions, an acceptable device is needed that will carry out the resistance to penetration measurements in the absence of an attending technician. The apparatus should be designed to minimize the test time of the technician, be self-recording, and accept a large number of test specimens over a short period of time.

The impact-echo method has been shown to be a promising technique for nondestructively monitoring the development of mechanical properties in concrete from initial setting to ages of several days [22]. In this method, the test object is subjected to point impact and the surface displacement adjacent to the impact point is monitored. From the measured displacement waveform and the thickness of the object, the P-wave velocity is determined. A strong correlation has been found between the time of initial setting of mortars sieved from concrete, as determined by ASTM C 403, and the onset of P-wave velocity development in the concrete.

The author is of the opinion that within the near future it will be possible to predict the time of setting of concrete with the same precision as that assigned to the current ASTM C 403 method. This will come about as the industry's research learns more about the variables that influence this property of concrete. However, a better understanding of the chemistry of portland cement, the reaction kinetics involved in the cement hydration processes, and the mechanisms through which chemical admixtures alter those processes and their influence on the hydration kinetics will be needed.

Because concrete will be placed in ever-escalating hostile environments in the future, the need for measuring its time of setting in the field will increase. One instrument that can be used for this purpose has been available for some 20 years, that is, the concrete initial set penetrometer.[3] It is lightweight, easily transported to the job site, simple to use, equipped with a dial that has dual scales of 0 to 50 kg/cm^2 and 0 to 700 psi and the resistance to penetration readings are locked into place until released by push button.

REFERENCES

[1] Scripture, E. W., Jr., "Setting Time," *Significance of Tests and Properties of Concrete and Concrete Aggregates, ASTM STP 169*, American Society for Testing and Materials, Philadelphia, 1956, pp. 53–60.

[2] Kelly, T. M., "Setting Time," *Significance of Tests and Properties of Concrete and Concrete Aggregates, ASTM STP 169A*, American Society for Testing and Materials, Philadelphia, 1966, pp. 102–115.

[3] Sprouse, J. H. and Peppler, R. B., "Setting Time," *Significance of Tests and Properties of Concrete and Concrete-Making Materials, ASTM STP 169B*, American Society for Testing and Materials, Philadelphia, 1978, pp. 105–121.

[4] Brunauer, S. and Copeland, L. E., "Chemistry of Concrete," *Scientific American*, Vol. 210, No. 4, 1964, pp. 80–92.

[5] Calleja, J., "New Techniques in the Study of Setting and Hardening of Hydraulic Materials," *Journal*, American Concrete Institute, Vol. 48, 1952, p. 525.

[6] Calleja, J., "Effect of Current Frequency on Measurement of Electrical Resistance of Cement Pastes," *Journal*, American Concrete Institute, Vol. 49, 1952, p. 329.

[7] Whitehurst, E. A., "Use of the Soniscope for Measuring Setting Time of Concrete," *Proceedings*, American Society for Testing and Materials, Vol. 51, 1951, p. 1166.

[8] Tuthill, L. H. and Cordon, W. A., "Properties and Uses of Initially Retarded Concrete," *Proceedings, Journal*, American Concrete Institute, Vol. 52, Part 2, 1955, p. 273; *Proceedings*, Vol. 52, 1956, p. 1187.

[9] *Cement and Concrete Terminology, Publication SP-19*, American Concrete Institute, Detroit, 1990, p. 65.

[10] ASTM Test Method for Time of Setting of Concrete Mixtures by Penetration Resistance (C 403), *Annual Book of ASTM Standards*, Vol. 04.01, American Society for Testing and Materials, Philadelphia, 1990, p. 214.

[11] Concrete Admixtures, Part 1 (B S 5075), British Standards Institute, London, 1982, pp. 1–8.

[12] Kelly, T. M. and Bryant, D. E., "Measuring the Rate of Hardening of Concrete by Bond Pullout Pins," *Proceedings*, American Society for Testing and Materials, Vol. 57, 1957, p. 112.

[13] Polivka, M. and Klein, A., "Effect of Water-Reducing Admixtures and Set-Retarding Admixtures as Influenced by Portland Cement Composition," *Effect of Water-Reducing and Set-Retarding Admixtures on Properties of Concrete, ASTM STP 266*, American Society for Testing and Materials, Philadelphia, 1960, p. 124.

[14] Dodson, V. H., "The Effect of Fly Ash on the Setting Time of Concrete—Chemical or Physical," *Proceedings, Symposium N*, Materials Research Society, University Park, PA, 1988, pp. 166–171.

[15] Bruere, G. M., "Importance of Mixing Sequence When Using Set-Retarding Agents with Portland Cements," *Nature*, Vol. 199, 1963, p. 32.

[16] Dodson, V. H. and Farkas, E., "Delayed Addition of Set-Retarding Admixtures to Portland Cement Concrete," *Proceedings*, American Society for Testing and Materials, Vol. 64, 1964, p. 816.

[17] Dodson, V. H., *Concrete Admixtures*, Van Nostrand Reinhold, New York, 1990, pp. 36–37.

[18] Schutz, R. J., "Setting Time of Concrete Controlled by the Use of Admixtures," *Journal*, American Concrete Institute; *Proceedings*, Vol. 55, 1959, p. 769.

[19] Wallace, G. B. and Ore, E. L., "Structural and Lean Mass Concrete as Affected by Water-Reducing, Set-Retarding Agents," *Effect of Water-Reducing and Set-Retarding Admixtures on Properties of Concrete, ASTM STP 266*, American Society for Testing and Materials, Philadelphia, 1959, p. 38.

[20] Dodson, V. H., *Concrete Admixtures*, Van Nostrand Reinhold, New York, 1990, pp. 122–123.

[21] Dodson, V. H., *Concrete Admixtures*, Van Nostrand Reinhold, New York, 1990, pp. 96–111.

[22] Pessiki, S. P. and Carino, N. J., "Measurement of the Setting Time and Strength of Concrete by the Impact-Echo Method," NBSIR 87-3575, U.S. Department of Commerce, National Bureau of Standards, Washington, DC, 1987, pp. 1–110.

---

[3] Available from vendors of concrete testing equipment.

Bleeding

Steven H. Kosmatka[1]

PREFACE

The subject of bleeding was briefly addressed in the first edition of *ASTM STP 169*, published in 1956. Ivan L. Tyler, manager of the Field Research Section of the Portland Cement Association, concisely described the general significance of tests for bleeding in his article on Uniformity, Segregation, and Bleeding in the Freshly Mixed Concrete section of *ASTM STP 169*. *ASTM STP 169A* and *ASTM STP 169B* did not address bleeding. This chapter elaborates on fundamental concepts presented by Mr. Tyler. The effects of concrete ingredients on bleeding as well as the significance of bleeding with modern concretes are presented. The chapter also reviews the standard ASTM test methods on bleeding and provides data on the bleeding characteristics of a variety of cement pastes, mortars, and concretes.

INTRODUCTION

Bleed water is the clear water that can gradually accumulate at the surface of freshly placed concrete, mortar, grout, or paste (Fig. 1). Bleed water is caused by sedimentation or settlement of solid particles (cement and any aggregate) and the simultaneous upward migration of water. This upward migration of water and its accumulation at the surface is called bleeding, also referred to as water gain, weeping, and sweating in some countries. A small amount of bleeding is normal and expected on freshly placed concrete. It does not necessarily have an adverse effect on the quality of the plastic or hardened concrete. However, excessive bleeding can lead to some performance problems with plastic or hardened concrete. With proper mix proportioning, mixture ingredients, placing equipment, and proper construction practices, bleeding can be controlled to a desirable level.

SIGNIFICANCE

Bleeding is not necessarily a harmful property nor is excessive bleeding desirable. Because most concrete ingredients today provide concrete with a normal and acceptable level of bleeding, bleeding is usually not a concern and bleeding tests are rarely performed. However, there are situations in which bleeding properties of concrete should be reviewed prior to construction. In some instances lean concretes placed in very deep forms have accumulated large amounts of bleed water at the surface. This not only creates a placing problem but also reduces the strength and durability of the concrete near the surface. Excessive bleeding also delays finishing as finishing should not proceed with observable bleed water present. On the other hand, lack of bleed water on concrete flat work can sometimes lead to plastic shrinkage cracking or a dry surface that is difficult to finish.

The first reported case of bleeding in North America is in 1902 during the construction of the stadium at Harvard University [1,2]. During placement, up to ⅔ m of bleed water would develop. Up to 150 mm of concrete was removed from the top of each lift prior to the sequential placements in order to remove the less durable and weaker concrete. Even with the high degree of bleeding, this structure survived the elements for 90 years and will be serviceable for many years to come (Fig. 2). Structures where severe exposures exist and where porous concrete was not adequately removed have not performed as well as structures where the bleeding-damaged concrete was properly removed.

During the construction of massive structures, such as deep foundations, tall walls, or dams, bleeding became of concern in early concrete projects. To study and help control bleeding, a variety of bleeding tests were developed. These tests will be discussed later under the section on Test Methods. By understanding the process of bleeding, Powers [3] and others provided means to control bleeding and today bleeding is rarely a problem.

Bleeding can occur at any time during the transportation, handling, and placing of concrete, as well as, shortly after placement. Most of the discussion of this chapter will focus on concrete after it is placed in a form and will no longer be agitated. This chapter will address the effects of concrete ingredients and placing practices on bleeding and the effect of bleeding on various concrete properties. Much of the discussion pertains to cement paste or mortar as bleeding in concrete is a direct function of the paste or mortar bleeding properties.

FUNDAMENTALS OF BLEEDING

A fresh concrete mixture is merely a mass of concrete ingredients that are temporarily suspended due to the agi-

---

[1] Manager, Research and Development, Portland Cement Association, Skokie, IL 60077-1083.

FIG. 1—Bleed water on the surface of a freshly placed concrete slab.

FIG. 2—Harvard University stadium after 90 years of service (photo courtesy of Tim Morse).

tation and mixing of the material. Once the agitation stops, the excess water rises through the plastic mass to the surface or, more appropriately, the solid ingredients settle. Although the actual volume of the total ingredients does not change, the height of the hardened concrete is less than the original plastic height as the bleed water will come to the surface and evaporate away.

The accumulation of water at the surface of a concrete mixture can occur slowly by uniform seepage over the entire surface or at localized channels carrying water to the surface. Uniform seepage is referred to as normal bleeding. Localized channels of water coming up through the concrete sometimes carrying fine particles with it are termed channel bleeding and usually occurs only in concrete mixes with very low cement contents, high water contents, or concretes with very high bleeding properties.

As bleeding proceeds, the water layer at the surface maintains the original height of the concrete sample in a vessel, assuming that there is no pronounced temperature change or evaporation. The surface subsides as the solids settle through the liquid (Fig. 3). Figure 4 illustrates a typical bleeding curve relating subsidence of the surface to time. The initial subsidence occurs at a constant rate, followed by a decreasing bleeding rate (Fig. 4).

In the interval between the beginning of subsidence and setting there are three primary zones describing the nature of the bleeding process (Fig. 5). These are the zones of (1) clear water at the sample surface, (2) constant water

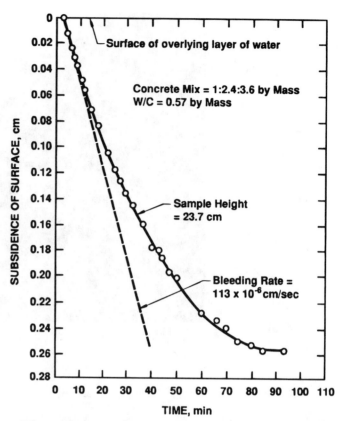

FIG. 4—Typical bleeding curve for concrete illustrating surface subsidence with respect to time [3].

FIG. 3—Demonstration of bleeding or settlement of cement particles in cement paste with water-cement ratios by weight of 0.3, 0.7, and 2.2 *(left to right)*. All cylinders contain 250 mL of paste and were photographed 1 h after the paste was mixed and placed in the cylinders. Observe the accumulation of bleed water for pastes with the higher water-cement ratios.

content or density, and (3) compression. Figure 5 is a simplified version of a five-zone analysis of the bleeding process of paste presented by Powers [3]. In the zone of constant water content, the water-to-cement ratio and density are essentially constant, even though some water is moving through the zone. The compression zone is a transition zone where the paste is being densified. The water-to-cement ratio is also being reduced and the solid particles represent a lesser volume than when originally placed. The paste densifies until the paste stabilizes and stops settling or bleeding. At equilibrium, the paste achieves a stable volume and the degree of consolidation in this fresh state dictates the hardened properties of the paste, such as strength and durability.

Bleeding is often analyzed in terms of bleeding rate and bleeding capacity. Bleeding rate is the rate at which the bleed water moves through the plastic concrete, mortar, or paste. Bleeding rate can be expressed in terms of cubic centimetres of bleed water per second per square centimetre of sample surface, centimetres per second, millimetres per second, or other applicable units. The bleeding rate of pastes with different cements and water-to-cement ratios is shown in Fig. 6. The rate of bleeding is controlled by the permeability of the plastic paste. As the solid particles settle, the flow of water is controlled by the permeable space or capillaries between particles.

The bleeding rate of concrete is less than that of paste alone with the same water-cement ratio. However, this would be expected as the aggregate in the concrete

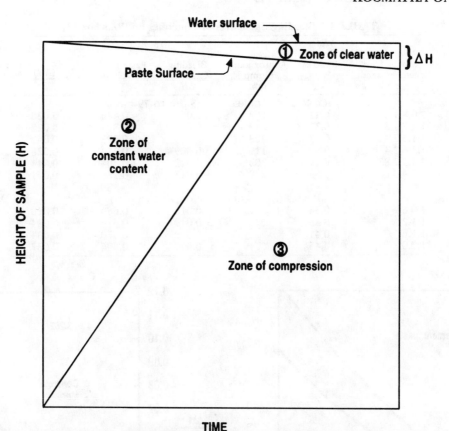

FIG. 5—Illustration of the process of bleeding in cement paste.

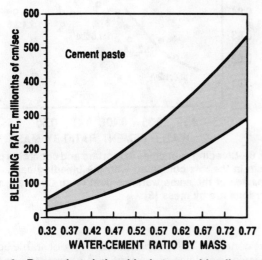

FIG. 6—Range in relationship between bleeding rate and water-cement ratio of pastes made with normal portland cement and water. The range is attributed to different cements having different chemical composition and fineness [4].

replaces some of the volume of the paste. The velocity of the water through the paste within the concrete is greater than it is for paste alone. This is due to the greater unit weight of the concrete. Consequently, the hydraulic force induced by the aggregate in the concrete disrupts the paste structure more so than in paste alone. Because of this, channeled bleeding develops in concrete sooner than it does in paste alone and it develops at lower water-cement ratios for concrete than for paste. The rate of bleeding in a concrete mixture is controlled by many variables that will be discussed later under the section on Effects of Ingredients on Bleeding.

Bleeding capacity is the quantity of bleed water that a particular concrete, mortar, grout, or paste mixture can release to the surface with respect to a certain depth. It is usually expressed in terms of the settlement or change in height of the paste- or mortar-solids surface per unit of original sample height (or, in other words, the ratio or percentage of the total decrease in sample height to the initial sample height). Bleeding capacity can also be expressed as a percent of the mix water. Figure 7 illustrates the bleeding capacity for pastes with a range of cements and water-cement ratios. Table 1 shows bleeding capacities for concrete and mortar. Figure 8 demonstrates these data in comparison to paste. Bleeding capacity is directly related to the water and paste content of concrete mixtures. Higher water contents especially increase bleeding capacity. As expected, bleeding capacity is closely related to bleeding rates (Fig. 9).

DURATION OF BLEEDING

The length of time that the concrete bleeds depends upon the depth of the concrete section as well as the setting properties of the cementitious materials. A thin slab of

TABLE 1—Bleeding Capacities of Concretes and Mortars [3].

Mix by Mass, cement: sand: gravel	Water-Cement Ratio by Mass, w/c	Approximate Slump, mm	Bleeding Capacity, ΔH	Paste per Unit Volume, p	$\dfrac{(\Delta H)}{p}$
Concrete (Aggregate: 75 μm to 19 mm)					
1:0.8:1.2	0.31	102	0.009	0.446	0.020
1:1.2:1.8	0.38	213	0.011	0.374	0.028
1:1.6:2.4	0.43	203	0.013	0.323	0.042
1:1.6:2.4	0.40	119	0.009	0.314	0.028
1:1.9:2.85	0.49	203	0.012	0.304	0.041
1:2.4:3.60	0.53	229	0.013	0.266	0.041
Mortar (Aggregate: 75 μm to 4.75 mm)					
1:0.8	0.34	...	0.018	0.688	0.028
1:1.2	0.38	...	0.019	0.614	0.034
1:1.6	0.41	...	0.019	0.570	0.034
1:2.0	0.45	...	0.019	0.506	0.037

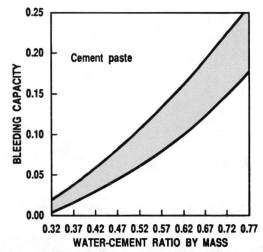

FIG. 7—Range in relationship between bleeding capacity (total settlement per unit of original paste height) and water-cement ratio of pastes made with normal portland cement and water. The range is attributed to different cements having different chemical composition and fineness [4].

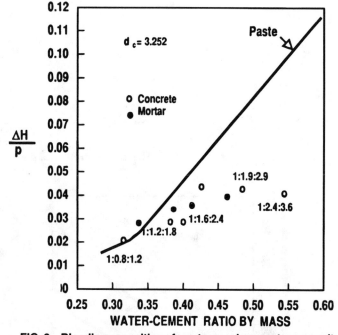

FIG. 8—Bleeding capacities of mortars and concretes per unit of paste in the mix compared with the bleeding capacities of neat pastes of the same water-cement ratio. Concrete mixture proportions are by mass [3].

concrete will settle or bleed for a shorter period of time than a deep section of concrete. Likewise, a concrete that sets up quickly will bleed much less than a concrete that takes many hours to set up. Most bleeding occurs during the dormant period, when cementing materials have little to no reaction. The dormant period is commonly around an hour. However, chemical and mineral admixtures as well as different compositions and fineness of cements can greatly affect the dormant period. Figure 10 illustrates the increase in bleeding with increased paste height and dormant period.

EFFECTS OF BLEEDING ON PLASTIC CONCRETE

Volume Change

Combining cement, water, and aggregates in a mixer creates a disbursed and suspended state of particles in plastic concrete. This suspended state is not stable because the heavier particles of cement and aggregate are forced downward through the lighter water by gravity. The downward movement of the solid particles continues until settlement ceases when the particles are in contact with one another and densify. As the surface of the disbursed solids is replaced by water, the volume of solid matter decreases. Although the total volume of materials is relatively constant, the volume after bleeding will be less than that of the original plastic mixture.

The total amount of settlement is proportional to the depth of the freshly placed concrete. Settlement can occur even though bleed water is not observed at the surface. This is because on many occasions, such as warm windy

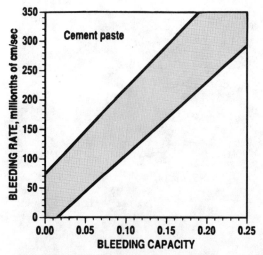

FIG. 9—Relationship between bleeding rate and bleeding capacity for cement paste using a variety of cements. Approximately 100 data points were used to develop the range [4].

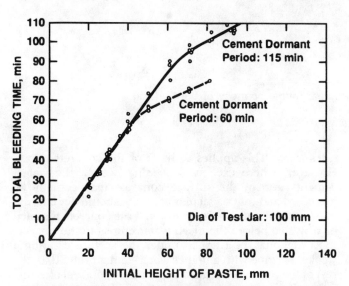

FIG. 10—Bleeding time versus height of paste sample for cements with different dormant periods [4].

days, the rate of evaporation is sufficient to remove the bleed water as it comes to the surface. The amount of volume reduction is clearly demonstrated in Fig. 4 in terms of settlement of the surface.

The small amount of settlement or volume reduction is not of concern for most general construction practices or applications. However, applications where concrete is being placed under an item that it must support, such as a machine base, should have little to no bleeding to prevent the formation of a void between the concrete surface and the object the concrete is to support.

Postbleeding Expansion

Following the bleeding period, expansion occurs within the paste. This postbleeding expansion is caused by a combination of physical and chemical reactions occurring during the first stages of setting. In effect, the gel coating on cement grains may, in their disruption during hydration, be exerting enough pressure to cause the sample to increase in volume. Most of this expansion occurs within the first day. Typical one-day expansions range from 0.05% to 1% for portland cement pastes at a water-cement ratio of 0.38 by weight [5,6]. Expansion beyond the first day can be expected to be less than half that which occurs within the first day and is not likely to add more than 0.05% expansion. Table 2 illustrates the postbleeding expansion of a cement paste. The bleeding period ended at 1 h and 12 min and the expansion began at 1 h and 30 min. It must be realized that the amount of expected expansion for concrete would be much less than that for paste alone or even nondetectable. It must also be realized that the samples in Table 2 were submerged continuously during the test. Otherwise, autogenous shrinkage[2] may oppose the expansion as the cement hydrates.

Plastic Shrinkage

Plastic shrinkage, sometimes called setting shrinkage in older literature, is shrinkage that occurs before the concrete has hardened. This shrinkage results from a loss of free water in the mixture. The water loss and resulting shrinkage is caused by a combination of loss of free water from the concrete due to bleeding and surface evaporation as well as consumption of the water by the cement during hydration (autogenous shrinkage). The amount of bleeding and surface evaporation predominantly control the amount of plastic shrinkage. If the rate of evaporation at the sample surface exceeds the bleeding rate, plastic shrinkage occurs. Autogenous shrinkage, due to water consumption during cement hydration, is very small and usually does not contribute significantly to total plastic shrinkage. The evaporation and removal of the bleeding water from the concrete creates tensile stresses near the surface. These tensile stresses can pull the concrete away from the form as well as form plastic shrinkage cracks in the concrete that resemble parallel tears (Fig. 11).

The disappearance of the water from the surface of the concrete indicates when the rate of evaporation has

TABLE 2—Post-Bleeding Expansion of Cement Paste[a] [5].

Age Interval, h	Observed Rate of Expansion, %/h	Total Expansion, %
0 to 1½	0	0
1½ to 2½	0.10	0.10
2½ to 3½	0.04	0.14
3½ to 4½	0.04	0.18
4½ to 5½	0.02	0.20
5½ to 6½	0.02	0.22
6½ to 23½	0.01	0.38

[a] Type I cement was used with a water-cement ratio of 0.38 by mass. The age of the paste at the end of the bleeding period was 1 h 12 min. The age at the beginning of the expansion was 1 h 30 min. Expansion is expressed as the percent of the depth of the sample.

[2] Autogenous shrinkage is also referred to as chemical shrinkage or shrinkage upon absorption of water due to cement hydration.

94 TESTS AND PROPERTIES OF CONCRETE

FIG. 11—Plastic shrinkage cracks in concrete (photo courtesy of the Portland Cement Association).

exceeded the bleeding rate. Shortly after this time, plastic shrinkage and cracking occur. The time required to obtain this condition is controlled by the air temperature, relative humidity, wind velocity, concrete temperature, and bleeding characteristics of the concrete (Fig. 12). At this stage, the concrete has obtained a small amount of rigidity yet is unable to accommodate the rapid volume change induced by plastic shrinkage. Consequently, tensile stresses develop and plastic shrinkage cracks form.

The best way to prevent plastic shrinkage cracking is to prevent surface evaporation. Plastic shrinkage cracks can penetrate from one fourth to full depth of a concrete slab. Although plastic shrinkage cracks may be unsightly, they often do not hinder performance of nonreinforced concrete. Plastic shrinkage cracks can also be reduced by the use of concrete with higher bleeding characteristics as well as the use of fibers, evaporation retarders, shades, windbreaks, and plastic sheets or wet burlap covering the slab. Concretes with high cementing materials contents (around 500 kg/m^3) or that use silica fume and low water-to-cementitious material ratios aggravate plastic shrinkage crack development.

Water-Cement Ratio

The water-cement ratio of a concrete mixture before bleeding is higher than after bleeding. As illustrated in Figs. 3 and 5, a densification of particles occurs. As the solid material compresses, some of the water leaves the concrete and rises to the surface at which time it evaporates away. This applies to most of the concrete depth. However, where excessive bleeding occurs, the water-cement ratio of the surface concrete may actually be increased and that particular concrete should be removed in extreme situations. In addition, if the concrete is sealed or troweled before the bleed water comes to the surface, the water may be trapped under the surface creating a weakened zone with a higher water-cement ratio than the rest of the concrete. In most applications, normal bleeding is beneficial in reducing the actual water-cement ratio of most of the concrete mixture in place.

Thixotropic Mixtures

Thixotropic mixtures of concrete, mortar, or grout have low bleeding properties. They exhibit a cohesive nature and, because of the ingredients in the mixture, they have little to no bleeding. This is illustrated in the left cylinder in Fig. 3 that contains a paste with thixotropic properties. Thixotropic mixtures are important in applications where the volume of the fresh concrete or mortar must equal the volume of the hardened concrete or mortar. An example would be in the use of supporting grout under machine base plates [8].

Placing and Finishing

Normal bleeding usually does not interfere with the placing and finishing of concrete mixtures. However, excessive bleeding of low cement content mixtures may

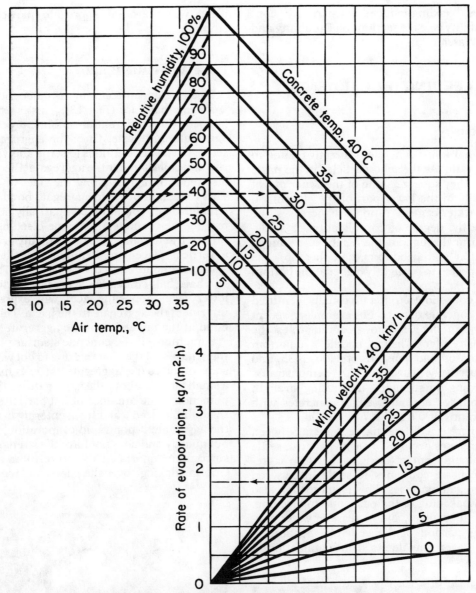

Instructions for using this chart:
1. Enter with air temperature, move *up* to relative humidity.
2. Move *right* to concrete temperature.
3. Move *down* to wind velocity.
4. Move *left*; read approximate rate of evaporation.

FIG. 12—*Nomograph demonstrating the effect of concrete and air temperatures, relative humidity, and wind velocity on the rate of evaporation of surface moisture from concrete. Evaporation rates exceeding 1 kg/m²/h are prone to induce plastic shrinkage cracks in concrete [adapted from Ref 7].*

cause undesirable early segregation during transportation and placement. Some minor bleed water is desirable to help keep the surface paste moist and help provide lubrication during the finishing of concrete; however, bleeding should not be so excessive as to interfere with the finishing operation. *Concrete must never be finished with visible bleed water on the surface as such practices promote dusting, scaling, blisters and other surface defects* [9]. Usually bleed water is allowed to evaporate away before finishing commences.

Concrete placed on a subbase of low permeability, such as clay, plastic sheeting, metal-deck forms, bituminous concrete, or vapor barriers, bleeds noticeably more than concrete placed on a granular base. Therefore, special care and planning should be used where differences in bleeding rates are caused by isolated vapor barrier locations.

If excessive bleeding occurs between lifts in a wall placement, the water and about 100 mm of surface concrete should be removed prior to placing the next lift. This practice, although not usually needed for concrete with normal bleeding, removes bleeding-induced low-strength concrete of poor durability (see the sections on Durability and Removing Bleed Water).

EFFECT OF BLEEDING ON HARDENED CONCRETE

Strength and Density

The strength of hardened concrete is directly related to the water-cement ratio. As the solid particles in the paste or concrete settle, they squeeze some of the water out of the paste, especially in the lower part of the placement. This lowers the water-cement ratio and increases the strength. Because the degree of consolidation or settlement is not uniform throughout the height of a sample (more consolidation at the bottom than the top), the strength can be expected to be slightly higher at the bottom than at the top.

The differential consolidation effect is usually identified by an increase in concrete density. For example, a recent caisson placement in the Chicago area illustrates the consolidation/density effect. The top, middle, and bottom of a 12-m-deep caisson had densities of 2400, 2435, and 2441 kg/m^3, respectively. Hoshino [10] demonstrated the effect of bleeding on strength—illustrating strength increase with depth for normal-strength concretes with water-to-cement ratios of 0.6 and 0.7 by weight.

With regular concrete, a differential settlement between the paste and aggregates occurs. The aggregate can settle only until point-to-point contact (bridging) between aggregates occurs. The paste continues to settle in between the stabilized aggregate particles. This phenomenon also contributes to a weakening of the paste due to bleeding, reduction of cement particles, and increase in water content in the upper portion of the concrete. Separation between paste and aggregate due to the accumulation of bleed water around aggregate particles also reduces strength.

Paste-Aggregate Bond

Bleed water can accumulate under and alongside coarse aggregate particles (Fig. 13). This is especially prone to happen when differential settlement occurs between the aggregate and paste. Once the aggregate can no longer settle, the paste continues to settle allowing bleed water to rise and collect under the aggregate. Bleed-water channels also tend to migrate along the sides of coarse aggregate. This reduction of paste-aggregate bond reduces concrete strength. This condition can partially be reduced by revibrating the concrete after some bleeding has occurred. A broken fracture face of a horizontally oriented crack from a vertical placement demonstrates the reduced paste-aggregate bond caused by bleeding. The upper portion will have half-embedded aggregate particles, whereas the lower face (opposing face) will expose the socket or imprint of the aggregate in the paste. This reveals the poor bond at the underside of the aggregate particle.

Sometimes, if the concrete's surface mortar sets faster than the rest of the concrete due to hot weather conditions, some of the coarse aggregate particles might settle leaving a small air void above the aggregate particle. Another phenomenon is the presence of flat particles near the surface that inhibit bleed water from entering mortar that is above the aggregate. Upon rapid evaporation, this mortar dries out quickly and does not have the strength or durability of surrounding mortar. This can result in a condition called "mortar flaking" over the coarse aggregate particles.

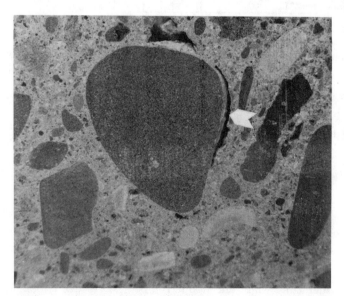

FIG. 13—Cross-section view of concrete illustrating bleed water accumulation along a coarse aggregate particle.

FIG. 14—Imprint in concrete illustrating the collection of bleed water voids under a smooth steel bar held firmly in a horizontal position during and after placement [2].

Paste-Steel Bond

Bleed water is prone to collect under reinforcing steel and other embedded items (Fig. 14). This is because not only does bleed water have an affinity to collect under large objects in concrete, but as the concrete settles, the concrete pulls away from the steel leaving an air void and an easily accessible location for water to collect.

A minor collection of bleed water or settlement under reinforcement is not detrimental to strength development in the bar because most of the stress is applied to the bar deformations. However, excessive water collection and void development can reduce bar embedment strength, paste-steel bond, and possibly promote corrosion of the steel at void locations—especially in the presence of moisture, carbonation, or chlorides. Corrosion can occur because the steel is not in contact with the corrosion resistive paste. Welch and Patten [11] demonstrated the effect of bleeding or settlement on the bond stress of reinforcing steel. For plain and deformed bars, bond stress was reduced with settlement. As expected, top bars developed less bond than bottom bars because more bleed water accumulated under top bars than bottom bars (Fig. 15). Some differential bond stress can also be attributed to differential strength development between the top and bottom of the specimen.

Durability

Concrete mixtures are designed to be durable in the environment in which they are exposed. A normal, small amount of bleeding does not reduce durability; however, excessive bleeding can have a serious effect if special precautions are not observed (Fig. 16). Durability and concrete's resistance to aggressive chemicals, chlorides, acids, and sulfates are directly related to the permeability and water-cement ratio of the concrete. An increase in water-cement ratio and permeability caused by excessive bleeding would reduce freeze-thaw, deicer-scaling, and sulfate resistance, and allow aggressive materials to enter the concrete.

The relationship between bleeding and durability is demonstrated in a study of plasticized, flowing concrete experiencing excessive bleeding. Bleed channels extending 12 mm below the surface and areas of high water-cement ratio were observed by petrographic analysis. The weakened surface layer and high water-cement ratio, induced by bleeding, contributed to poor deicer-scaling resistance [12]. Weakened surface layers can sometimes delaminate upon exposure to impact or abrasion. Dusting of a concrete surface can also develop.

Corrosion of reinforcement is more likely when bleed water collects under reinforcement steel. Also, if the upper few centimetres of a deep concrete section become significantly porous due to excessive bleeding, carbonation, chlorides, air, and moisture would more readily reach the steel, thereby inducing corrosion.

Bleeding channels can carry lightweight materials to the surface that can reduce abrasion resistance as well as allow popouts to form. A chemical analysis of bleed water indicates that channeled or uniform bleeding can leach alkalies up from the concrete to near the surface [13]. Upon evaporation of the water, an accumulation of alkalies can develop in the upper few millimetres of the surface. If alkali-aggregate-reactive particles are in this zone, the increased concentration of alkalies can aggravate alkali-aggregate reactivity, possibly resulting in popouts or surface cracking. A similar accumulation of alkalies can develop if the surface is sealed too early during finishing.

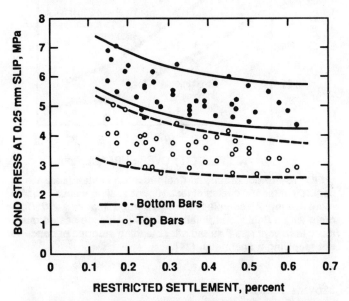

FIG. 15—The effect of concrete settlement on the bond strength of horizontally embedded deformed bars using ASTM Test Method for Comparing Concretes on the Basis of the Bond Developed with Reinforcing Steels (C 234). Bar diameter was 19 mm. Concrete slump was 50 to 125 mm. The top and bottom bars were 75 and 230 mm above the sample base [11].

Scaling

The relationship between bleeding and scaling depend upon the placing, finishing, and curing practices. Figure 17 illustrates the relationship between scaling and subsidence. In this particular example, increased degrees of subsidence were formed by increasing the rate of evaporation over the concrete. Contrary to common belief, this figure illustrates that increased bleeding or subsidence can actually improve scale resistance for this nonair-entrained concrete. It must be realized, however, that this concrete was finished after the bleed water had evaporated away from the surface and after maximum bleeding had occurred [14].

The scale resistance of a concrete surface can be jeopardized whenever the plastic material near the surface has its water content changed or its ability to transmit bleed water is reduced. Usually, this is achieved by reducing the water-cement ratio or increasing the cement content of the surface material. This can be done by finishing the slab prior to the accumulation of bleed water, or by sprinkling dry cement onto the slab to take up excess water to facilitate finishing.

FIG. 16—Deterioration at top of lifts in a dam where the porous-nondurable concrete was not adequately removed prior to sequential lift placement. The poor quality concrete at the lift surface was caused by excessive bleeding and deteriorated due to frost action.

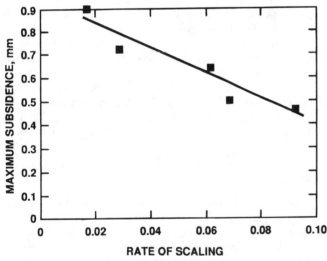

FIG. 17—The relationship between scaling and subsidence. The rate of scaling = numerical scale rating divided by the number of cycles. Higher scaling values indicate more scaling. Concrete was wood floated after maximum subsidence [14].

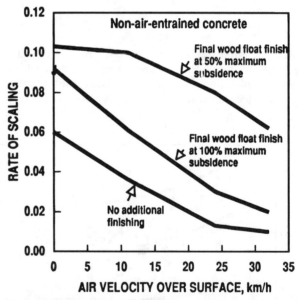

FIG. 18—Effect of time of final finish on scale resistance of concrete after 100 cycles of test. Higher scaling values indicate more scaling. All samples received a wood float strikeoff. Concrete with a final wood finish at 50% subsidence scaled much more than concrete finished after bleeding stopped or concrete not receiving a final finish [14].

Steel troweling early seals the surface much more than wood floating. The water rises to the surface and hits the surface stratum of more impermeable cement paste. It cannot penetrate through and merely accumulates under the surface. This accumulation of water either creates a weakened zone of paste or, in some cases, actually creates a water void beneath the surface. Upon freezing in the presence of water, this weakened zone or water-filled void can scale off the surface. Figure 18 illustrates that final finishing prior to the completion of bleeding significantly increases scaling.

Mortar Flaking

Mortar flaking resembles scaling or a flat popout. It is identified by a loss of mortar over flat coarse aggregate particles at the surface. Large flat coarse aggregate particles block the migration of bleed water to the mortar over the aggregate. If rapid evaporation occurs, as with an

FIG. 19—Mortar flaking over coarse aggregate (photo courtesy of the Portland Cement Association).

unprotected surface on a hot windy day, the mortar dries out; shrinks slightly more than the surrounding mortar; and, due to a lack of water for hydration, does not develop adequate strength for frost resistance. Upon repeated freezing in a wet condition, the surface mortar deteriorates and exposes the underlying coarse aggregate, usually with a flat surface parallel to the concrete surface (Fig. 19). Further deterioration to the surrounding mortar usually does not occur, as it received the necessary water for proper strength gain. This condition should not be confused with popouts caused by aggregate that swells excessively upon water saturation or freezing.

Surface Delamination

Surface delamination here refers to the separation of a large area of surface mortar from the base concrete (Fig. 20). The ¼- to 1-cm-thick surface delamination can occur in sizes ranging from 10 to 100 cm in diameter. The remaining exposed surface resembles a scaled surface with coarse aggregate exposed. The cause of surface delamination is the accumulation of bleed water under the surface creating a void or weakened zone (Fig. 21). Upon freezing of water in the void or weakened zone, the surface delaminates in sheet-like form. In some instances, delamination can occur with interior slabs not exposed to freezing simply because of the large void under the finished surface.

The consolidation of a surface by floating and troweling too early squeezes the water out of the top surface layer reducing the water-cement ratio. The addition of cement to facilitate finishing also reduces the water-cement ratio. These two practices both reduce the settling rate of the surface and make it more impermeable. This allows planes of weaknesses to develop and bleed water to accumulate under the surface forming a void. Finishing operations should be delayed as long as the setting time will permit, and the sprinkling of cement on to the surface should be avoided to minimize the risk of developing a plane of weakness or void beneath the surface. Early steel troweling is especially prone to trap bleed water beneath the surface. Delaminated areas, still intact, can be located by chaining, hammering, or by electro-mechanical sounding procedures all outlined in ASTM Practice for Measuring Delaminations in Concrete Bridge Decks by Sounding (D 4580). Delaminated areas produce a hollow sound upon impact and can often be lifted off the base concrete with a knife or screwdriver.

Blisters

Blisters are small bubbles of water that form under the surface during finishing (Fig. 22). They usually occur during or shortly after steel troweling, but before bleeding has stopped. If punctured while the concrete is plastic, water will usually squirt out. Spaced a few centimetres or decimetres apart, blisters are usually 1 to 10 cm in diameter, ¼ to 1 cm thick, and visibly rise above the surface. They can form by the accumulation of water under the

FIG. 20—Delaminated surface caused by early finishing that trapped bleed water under the surface (photo courtesy of The Aberdeen Group).

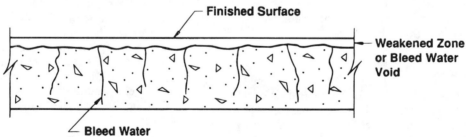

FIG. 21—Illustration of weakened zone or bleed water void under a finished surface.

surface at particular locations—often at the top end of a bleed-water channel (Fig. 23). Formation of blisters is usually an indication that the surface was finished or closed-up too early. They are more apt to occur on interior steel-troweled floors. Blisters can also form due to an excess of air in the concrete. An excess of fines or a lack of adequate vibration can also trap air under the finished surface [9].

Surface Appearance

Uniform bleeding on flat work should not affect the color of the surface; however, concretes placed adjacent to one another that have different bleeding rates or different bleeding properties can induce a color change in the surface primarily because of a potential change in water-cement ratio. A concrete that bleeds enough to increase the water-cement ratio at the surface will create a lighter-

FIG. 22—Blisters (photo courtesy of the National Ready Mixed Concrete Association).

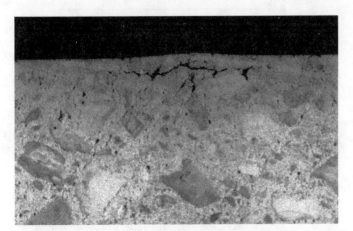

FIG. 23—Cross section of a blister, illustrating the bleed water void under the surface (photo courtesy of the Portland Cement Association).

colored surface. Consequently, a concrete placement with nonuniform bleeding can possibly result in blotchy-colored areas of light and gray. In wall placements, sand streaks can form as the bleed water collects and rises along the form face. As the bleed water moves upward along the form in long bleed-water channels, it washes away some of the paste leaving behind a somewhat sandy appearance (Fig. 24).

EFFECTS OF INGREDIENTS ON BLEEDING

The individual ingredients and the amount of each ingredient in concrete have a major effect on the bleeding characteristics of concrete. This section will briefly review the effects of some of these common ingredients and their proportions.

Water Content and Water-Cement Ratio

The water content and water-cement ratio predominantly control the bleeding of concrete (Fig. 25). Any increase in the amount of water or water-cementitious material ratio results in more available water for bleeding. Both bleeding capacity and bleeding rate are increased with increased water content. A one-fifth increase in water content of a normal concrete mixture can increase bleeding rate more than two and one-half times [3].

Cement

The type, content, chemistry, and fineness of cement can all influence bleeding properties. As the fineness of the cement increases, the amount of bleeding decreases (Fig. 26). Increases in cement content, as it relates to the reduction of water-cement ratio, also reduces bleeding (Fig. 27).

FIG. 24—Sand streaks along a wall caused by excessive bleed water rising along the form (photo courtesy of The Aberdeen Group).

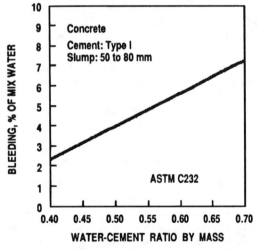

FIG. 25—Relationship between water-cement ratio and bleeding of concrete. Bleeding is expressed as a percent of mix water [15].

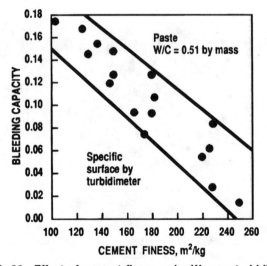

FIG. 26—Effect of cement fineness by Wagner turbidimeter on bleeding capacity of paste. Note that Wagner values are a little more than half of Blaine values [4].

Because the chemical and physical properties of cement are interdependent on one another in how they affect bleeding, it is difficult to isolate the effect of a particular property or chemical compound. Only reactions that occur during the mixing period or bleeding period will affect the bleeding rate. Figure 10 illustrates the effect of cements with different dormant periods on bleeding.

The amount of SO_3 that can be leached from cement in a short time has been found to correlate rather well to bleeding. An increase in SO_3 reduces bleeding [4]. The correlation between water-soluble alkalies in the cement and bleeding is not good. The trend is that an increase in alkalies reduces bleeding; however, other factors such as

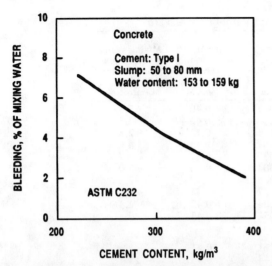

FIG. 27—Relationship between cement content and bleeding of concrete. The increased cement content reflects a decreased water-cement ratio. Bleeding is expressed here as a percent of mix water [15].

the precipitation of gypsum and SO_3 content probably have an overshadowing effect.

The correlation between bleeding and heat liberation demonstrates that bleeding can be reduced by a higher degree of initial chemical reactivity occurring with the cement shortly after mixing. Tricalcium aluminate has a high degree of reactivity and heat-producing capacity and is considered to influence bleeding properties.

Supplementary Cementing Materials

Fly ash, slag, silica fume, rice husk ash, and natural pozzolans can reduce bleeding by their inherent properties and by increasing the amount of cementitious materials in a mixture. However, each material and different source all have varying effects.

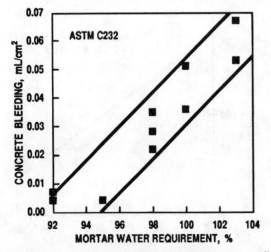

FIG. 28—Comparison of bleeding concrete versus fly ash mortar water requirement of ASTM C 618 [16].

Fly ash usually reduces bleeding. Table 3 compares the bleeding of concretes containing ten different fly ashes with respect to their water reduction and performance in comparison to two controls. Most of the fly ashes reduce bleeding compared to the concretes with cement only. Class C fly ashes reduce bleeding much more than Class F ashes in this study. The ability of fly ashes to reduce bleeding appears to be in their ability to reduce the water demand in the concrete to achieve a particular slump. This is not always the case as can be observed with Fly Ash H that increased water demand and yet still reduced bleeding. Fly Ash J, which has no effect on water, also demonstrated a reduction in bleeding. However, a direct correlation between concrete bleeding and the water requirement of the ASTM Specification for Fly Ash and Raw or Calcined Natural Pozzolan for Use as a Mineral Admixture in Portland Cement Concrete (C 618) mortar test exists (Fig. 28). The retardation effect of the fly ashes did not correlate with bleeding [16].

Ground granulated blast furnace slags usually give concrete a smaller bleeding capacity compared to mixes with portland cement only. The effect of slag on bleeding is primarily due to the fineness and calcium sulfate content of the slag [17].

Natural pozzolans, such as calcined clay or ground diatomite, usually reduce bleeding. The primary influence is related to the pozzolan's fineness and effect on water demand. In a study on calcined kaolin clay in mass gravel concrete, pozzolan dosages at 30 and 50% by volume of cementitious material resulted in bleeding capacities of 9.4 and 6.9%, respectively. The control mix with 120 kg/m^3 of portland cement had 11.9% bleeding [18].

Silica fume can greatly reduce bleeding, primarily due to its extreme fineness. Compared to a control with 20 mL of bleeding in samples 40 cm high, one study demonstrated that concretes with silica fume at dosages of 3, 7, and 13 kg/m^3 had bleeding capacities of 12, 8, and 3 mL, respectively [19]. Silica-fume concretes with very low water-to-cementing materials ratios essentially have no bleed water available to rise to the surface. Consequently, such concrete mixtures are prone to plastic shrinkage cracking if proper precautions are not taken to reduce or eliminate surface evaporation while the concrete is in the plastic state.

Rice husk ash (also called rice hull ash) reduces bleeding proportionately with the amount of ash in the paste. The fineness of the material is primarily responsible for the reduction of bleeding [20].

Aggregate

Ordinary variations in aggregate grading have little effect on the bleeding of concrete. This assumes that there is no appreciable change in the minus 75-μm material. Table 4 demonstrates that the specific surface area of the sand, or the fineness, has little effect on the bleeding rate of mortar at four different ranges of water-cement ratio. However, aggregates that contain a high amount of silt, clay, or other material passing the 75-μm sieve can have a significant effect in reducing the bleeding of concrete. This is not surprising as the aggregate represents only a

TABLE 3—Bleeding of Concretes With and Without Fly Ash [16].[a]

Fly Ash Identification	Class of Fly Ash (ASTM C 618)	Bleeding, % of mix water	Bleeding, mL/cm² of surface	Water-Cementing Material Ratio by mass, W/CM	Change in Mixing Water Requirement Compared to 307 kg/m³ Control, %
A	C	0.22	0.007	0.40	−7.0
B	F	1.11	0.036	0.42	−2.3
C	F	1.61	0.053	0.42	−2.3
D	F	1.88	0.067	0.45	+4.6
E	F	1.18	0.035	0.41	−4.7
F	C	0.13	0.004	0.40	−7.0
G	C	0.89	0.028	0.42	−2.3
H	F	0.58	0.022	0.44	+2.3
I	C	0.12	0.004	0.42	−2.3
J	F	1.48	0.051	0.43	0
Average of Class C		0.34	0.011	0.41	...
Class F		1.31	0.044	0.43	...
Control Mixtures	Cement Content, kg/m³				
1	307	1.75	0.059	0.43	...
2	282	2.42	0.080	0.48	...

[a] Concretes had a slump of 75 to 100 mm and air content between 6 and 7%. Test mixtures contained 75% cement and 25% fly ash by mass of cementitious material, based on a 307 kg/m³ cementitious material content. Control mixtures contained no fly ash. Bleeding tested as per ASTM C 232.

small amount of the surface area within a concrete mixture. For example, consider a concrete mixture with proportions of one part cement to six parts aggregate (coarse plus fine) by mass. The total surface area of the aggregate is only 5% of the total surface area of the concrete mixture. The surface area of coarse aggregate is essentially negligible [3]. Similar results have been found by other researchers, including a negligible reduction in bleeding with reduced particle size [10]. Aggregates that increase water demand, such as crushed rock, tend to increase bleeding due to the higher water content in the mix.

TABLE 4—Effect of the Specific Surface Area of Sand on the Bleeding Rate of Mortars [3].

Reference Number	Water-Cement Ratio by Mass	Bleeding Rate, 10⁻⁶ cm/s	Specific Surface Area of Sand, cm²/cm³
47	0.384	30	86
51	0.393	36	99
55	0.402	37	113
59	0.393	32	126
48	0.431	43	86
52	0.443	50	99
56	0.452	47	113
60	0.439	48	126
49	0.477	52	86
53	0.490	63	99
57	0.508	60	113
61	0.490	60	126
50	0.646	114	86
54	0.668	121	99
58	0.693	113	113
62	0.668	112	126

Chemical Admixtures

Today the most predominantly used admixtures are air-entraining admixtures and water reducers. In laboratory and field applications, it has been observed that entrained air reduces bleeding. The relationship between air content and bleeding rate for paste is illustrated in Fig. 29. Bruere found that not only do the air-entraining agents reduce bleeding by inducing an air-void system in the concrete, but also the air-entraining admixture itself can slightly reduce bleeding rates [21].

Because water reducers reduce the water content of a concrete mixture, it is expected that they likewise reduce the bleeding. The effect of high-range-water reducers or

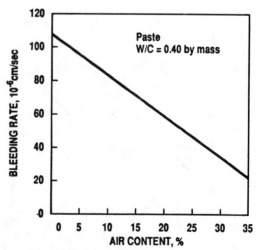

FIG. 29—The effect of entrained air on bleeding rate of paste [3].

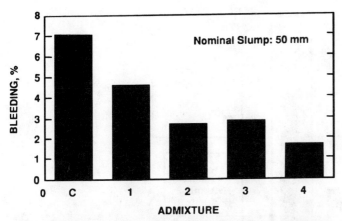

FIG. 30—Effect of high range water reducers on concrete bleeding when used to reduce water content. Mix C is the control and Mixes 1 to 4 contain HRWR admixtures inducing water reductions of 13, 10, 10, and 17%, respectively. ASTM C 232 bleeding expressed in percent of mix water. Nominal cement content is 223 kg/m^3 and control water content is 157 kg/m^3 15].

plasticizers have a similar effect to that of normal-range water reducers. A reduction in water content through use of a water reducer results in reduced bleeding. When high-range water reducers are used to reduce the free water content in concrete significantly, they likewise dramatically decrease the bleeding of concretes as illustrated in Fig. 30. This data compares concretes of equal slump.

Table 5 illustrates the effect of plasticizers on increasing slump without influencing the water-cement ratio and also compares a high-slump versus low-slump control. At a constant water-cement ratio, the concretes with high-range water reducers have slightly more bleeding than the low-slump control, but significantly less bleeding than the high-slump control. At equivalent water-cement ratios, the flowing concretes with the plasticizers bleed more than the control, but significantly less than the control of the higher water-cement ratio and high slump [22].

Figure 31 illustrates the effect of plasticizers on concrete bleeding with two different cements to make flowing concrete. All the concretes have the same nominal water and cement contents. The admixture was added to increase slump to between 175 and 225 mm. These mixes had more bleeding than the flowing concrete in Table 5 because of a higher initial water content. Figure 31 demonstrates that increased fluidity induced by the plasticizer increased bleeding and it demonstrates the effect of different cement and admixture combinations on bleeding. Excessive bleeding in plasticized, flowing concrete can reduce surface durability and deicer-scaling resistance [12]. Excessive bleeding can be avoided by optimal material selection and mixture proportioning.

Calcium chloride is commonly used as an accelerator in nonreinforced concrete. Sodium chloride and calcium chloride both can reduce bleeding significantly. The influence of accelerating salt such as calcium chloride may be largely due to the effect of its acceleration of setting. The effect of calcium chloride on bleeding time and bleeding capacity are illustrated in Table 6. Calcium chloride tends to make cement paste more susceptible to channeling. Little is known about how the different chemical compounds and chemical admixtures affect bleeding as bleeding is not a required test in ASTM admixture specifications.

Special bleed-reducing admixtures based on cellulose derivatives, water absorbing resins, or various other chemical formulations can significantly reduce or eliminate bleeding. These admixtures make concrete mixtures very cohesive and thixotropic and therefore are very effective at preventing segregation. These admixtures are rarely used for normal construction but are very effective for special applications.

EFFECTS OF PLACEMENT CONDITIONS ON BLEEDING

Placement Size and Height

The depth of a concrete placement directly affects the amount of bleeding. As shown in Fig. 32, the amount of bleeding increases with sample depth. This phenomenon has been observed especially in placements of low-cement-content concrete mixtures in deep dam sections, where many centimetres of water can accumulate over a lift just a few metres deep.

The shape of vertical forms can produce sufficient strain in the plastic concrete to induce cracks or zones of weakness due to bleeding and settlement. Deep and narrow placements in which the walls are not parallel are more prone to settlement strain. These situations can create stresses within the plastic concrete upon settlement and bleeding. For example, assume that the settlement along a vertical face is constant. The settlement along an inward incline would be less than that along the vertical face. This would create some possible movement from one side of the form to the other, inducing shear strain and possibly

TABLE 5—Bleeding of Concretes With and Without Plasticizers (ASTM C 232 Bleeding Test) [22].

Mix Identification	Water-Cement Ratio by Mass	Water Content, kg/m^3	Slump, mm	Bleeding, % by mass of mix water	Bleeding, mL/cm^2 of surface
Control 1	0.47	143	75	1.09	0.031
Control 2	0.58	171	215	3.27	0.143
Melamine sulfonate	0.47	143	215	1.59	0.060
Naphthalene sulfonate	0.47	144	225	1.50	0.059

106 TESTS AND PROPERTIES OF CONCRETE

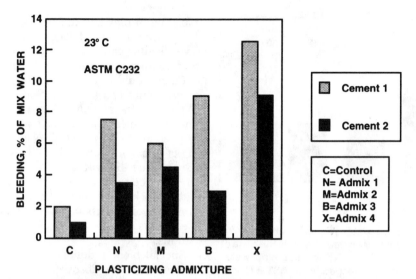

FIG. 31—Effect of plasticizers on bleeding of flowing concrete. The nominal cement, water, and air contents for all mixes was 323 kg/m³, 161 kg/m³, and 6 ± 1%, respectively. The initial slump was 75 to 125 mm and, after the admixture addition, the slump was 175 to 225 mm. The increased fluidity increased bleeding. Mix C is the control and Mixes N, M, B, and X contain different plasticizers [12].

TABLE 6—Effect of Calcium Chloride on Bleeding Rate and Bleeding Capacity [4].[a]

ASTM C 150 Cement Type	Admixture Addition	Bleeding Time, min	Bleeding Rate, 10^{-6} cm/s	Bleeding Capacity, ΔH
I	none	71	163	0.085
	CaCl$_2$	38	146	0.040
II	none	71	139	0.072
	CaCl$_2$	46	103	0.039
III	none	46	129	0.049
	CaCl$_2$	26	89	0.020

[a] Amount of CaCl$_2$ was 1% of the cement mass and about 2.1% of water. The water-cement ratio was 0.466 by mass.

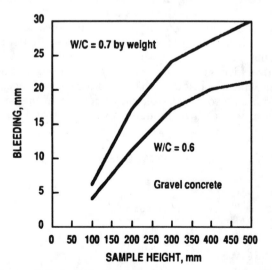

FIG. 32—Relationship between the amount of bleeding and sample height for concrete [10].

faults to form depending on the angle of the slope, distance between the wall faces, the depth of the placement, rate of filling, and bleeding capacity of the concrete. The opposite situation occurs when the concrete particles settle away from an outward incline where a layer of water can develop and collect along the sloped surface. Such conditions can create localized bleed channels along the form resulting in sand streaks.

Settlement can also form zones of weakness in forms that have areas of significantly reduced cross section. For example, if a T section is filled in one placement, faulting can develop up from the corners of the narrow form. Obviously, the way to avoid this problem is to fill the narrow portion first and then place the rest of the concrete after the concrete in the narrow section has settled. The greater the bleeding capacity of the concrete the greater the tendency is to form such faults and arches. These faults can be eliminated by revibration.

Impermeable Subbases

Concrete placed on subbases of low permeability such as plastic sheet vapor barriers, bituminous, or clay appea

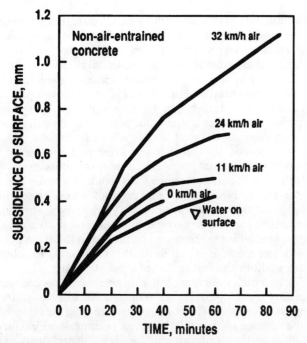

FIG. 33—Effect of wind velocity on subsidence [14].

to bleed noticeably more than concretes placed on granular subbases. All bleed water must rise to the surface when concrete is placed on an impermeable subbase. When concrete is placed on a granular material, some of the bleed water can flow out the bottom of the slab into the granular material. Bleeding can be reduced by placing 50 to 75 mm of sand on the impermeable surface. This practice also helps prevent slab curling. Construction in which only parts of a placement are on a vapor barrier will need special care in finishing practices due to nonuniform bleeding in order to provide a uniformly durable surface.

Weather Conditions

Weather conditions can have a significant effect on bleeding. Figure 33 illustrates how the increase in wind velocity and, consequently, rate of evaporation significantly increase the amount of surface subsidence. The wind velocity and increased evaporation greatly increase the capillary force at the surface pulling the bleed water out of the concrete.

The bleeding rate increases with an increase in temperature, however, the bleeding capacity tends to be nearly constant. In a test on 25 different cements, going from 23.5 to 32°C resulted in a 20% increase in the rate of bleeding for cement paste. This is primarily due to the decrease in water viscosity with an increase in temperature. In general, there was little change in bleeding capacity with the change in temperature for most of the cements studied [4].

Klieger illustrated that changes in temperature for concrete did not affect the subsidence of the surface [14]. At a wind velocity of zero for nonair-entrained concrete, the subsidence at temperatures of 10, 23, and 32°C, was 0.502, 0.413, and 0.454 mm, respectively. The average subsidence was 0.456 mm for this concrete. Whiting found bleeding of concrete at 23 and 32°C to be 2 and 3.2% by mass of mix water for normal concrete. The addition of plasticizers in some cases reduced bleeding at higher temperatures [12].

Consolidation and Revibration

Surface or internal vibration should not significantly affect the amount of bleeding. However, some studies indicate that a 2- to 3-s vibration period can slightly increase the bleeding capacity; however, the use of an internal vibrator in concretes from 20 s to 10 min tends to reduce bleeding capacity. This is partly due to the reduced volume of the matrix as well as the decreasing degree of dispersion of the aggregate and the expulsion of some of the entrapped air. Normal field vibration would not be expected to greatly affect bleeding [23].

After concrete has completed its settlement, it can bleed and settle again upon internal revibration of the concrete. This is illustrated in Table 7 where consistent degrees of additional bleeding were obtained for revibration intervals up to 4 h. Also note that the compressive strength was also increased. This would be expected as removal of some of the bleed water would have resulted in a lower water-cement ratio [23].

Remixing

Just as revibration allows bleeding to occur after concrete is revibrated, remixing of the cementitious system after it has settled has a similar effect with little change in bleeding characteristics. Table 8 illustrates that particular cement pastes remixed within the first hour bleed about the same amount as before remixing with only slightly reduced bleeding rates and capacities compared to the first settlement. This demonstrates that concrete in a ready mixed truck for some period of time can, after remixing, resume its normal bleeding characteristics.

CONTROLLING BLEEDING

Usually, little attention need be given to controlling bleeding of concretes made of normal ingredients at normal proportions. However, this section will provide some guidance as to adjusting the bleeding to solve particular problems.

TABLE 7—Effect of Revibration on Bleeding and Strength of Concrete [24].[a]

Interval Before Revibration	Bleeding, % of mix water	Compressive Strength, MPa
0	2.9	28
1 h	3.5	32
2 h	3.4	39
3 h	3.2	31
4 h	2.8	29

[a] The cement was an ASTM C 150 Type I. The aggregate was natural sand and crushed trap rock with a maximum size of 25 mm. The cement content was 307 kg/m³. The slump of the concrete was 75 mm. Compressive strength was estimated by impact hammer.

TABLE 8—Bleeding Tests on Remixed Pastes [5].[a]

Rest Period,[b] min	Bleeding Rate, 10^{-6} cm/s	Bleeding Capacity, ΔH	Duration of Bleeding, min	Time Between Initial Mix and End of Bleeding, min
0	194	0.122	55	55
15	189	0.113	55	70
30	196	0.125	57	87
45	192	0.106	52	97
60	185	0.103	48	108
90	172	0.090	47	137
120	167	0.075	45	165

[a] The water-cement ratio was 0.469 by mass. The schedule for the initial mixing was: 2 min mix, 2 min wait, followed by 2 min mix. The final remixing was ½ min continuously. The depth of the paste was 36 mm. The temperature was 23.5°C. The results for remixed pastes are averages of two or three tests.

[b] This is the period after the initial mixing, at the end of which the final remixing was done.

Increasing Bleeding

The easiest way to increase the bleeding is to increase the amount of water in the concrete mixture as well as reduce the amount of the fines in the sand and amount of cementing materials. Bleeding may need to be increased to help prevent plastic shrinkage cracking or improve finishability in dry weather.

Reducing Bleeding

Bleeding may need to be reduced for a variety of reasons including facilitating finishing operations, minimizing the formation of weak concrete at the top of lifts, reducing sand streaking in wall forms, or to stabilize the hardened volume with respect to the plastic volume of the concrete. The most important ways of reducing bleeding in concrete are as follows:

1. Reduction of the water content, water-cementitious material ratio, and slump.
2. Increase the amount of cement resulting in a reduced water-cement ratio.
3. Increase the fineness of the cementitious materials.
4. Increase the amount of fines in the sand.
5. Use supplementary cementing materials such as fly ash, slag, or silica fume.
6. Use blended hydraulic cements.
7. Use chemical admixtures that reduce water-cement ratio or by other means are capable of reducing the bleeding of concrete.
8. Use air-entrained concrete.

Removing Bleed Water

Once the observation of bleed water has become a noticeable problem, there is little time or opportunity to change the concrete mixture, especially once the concrete is in the forms. Therefore, certain techniques can be used to remove excess bleed water if normal evaporation is not satisfactory to remove the bleed water from the surface. If the concrete is placed within an enclosure, the temperature of the air can be increased and large fans can be used to evaporate away some of the water. However, care must be taken when using this technique to not remove so much bleed water that the evaporation exceeds bleeding resulting in potential plastic shrinkage crack development.

Vacuum dewatering can be accomplished by special equipment that uses a filter mat placed on the surface of the concrete. As the mat settles with the surface, bleed water is forced up through the filter and the water is removed by a vacuum pump. Because this method consolidates a slab of concrete more than would normally occur due to normal bleeding, the end product is a concrete with a higher strength, lower permeability, and a more abrasion resistant surface. For vertical applications such as wall forms, special form liners can be placed in the form prior to concrete placement that uniformly drain away the bleed water without the formation of sand streaking.

For normal concrete slabs experiencing occasional excessive bleeding that is not removed due to evaporation, a squeegee or garden hose can be used to drag the water off of the surface so that finishing can commence. Finishing should continue only after it has been ascertained that bleeding has nearly or completely stopped. If a finisher waits too long for the bleed water to evaporate away, the concrete can harden and will not be able to be finished. For precast concrete elements experiencing settlement or bleeding problems, a centrifuge can be used to compact the concrete and remove the bleeding by centrifugal force.

TEST METHODS

Three standard ASTM test methods can be used to analyze the bleeding properties of concretes, mortars, paste, and grouts. These are ASTM Test Method for Bleeding of Concrete (C 232), ASTM Test Method for Bleeding of Cement Pastes and Mortars (C 243), and ASTM Test Method for Expansion and Bleeding of Freshly Mixed Grouts for Preplaced-Aggregate Concrete in the Laboratory (C 940). Another related ASTM standard is ASTM Test Method for Water Retentivity of Grout Mixtures for Preplaced-Aggregate Concrete in the Laboratory (C 941).

ASTM C 243 is based on a test method published in 1949 [25]. The ASTM C 243 test apparatus is shown in Fig. 34. The sample container is filled with mortar or paste. The collecting ring is then placed to penetrate the sample. The burette and funnel are filled with carbon tetrachloride or 1,1,1-trichloroethane, which are liquids that are denser than water. As the sample settles and bleed water rises through the carbon tetrachloride or 1,1,1-trichloroethane to the top of the burette, bleeding measurements are taken at regular intervals until settlement stops. Bleeding rate is reported as cubic centimetres of bleed water per square centimetre of sample surface per second. Bleeding capacity is reported as cubic centimetres of water per cubic centimetres of paste or mortar.

ASTM C 232 has two procedures—one with vibration and one without. Method A allows the concrete to be undisturbed in a container with a volume of about 0.014 m^3 and the accumulation of bleed water is removed from

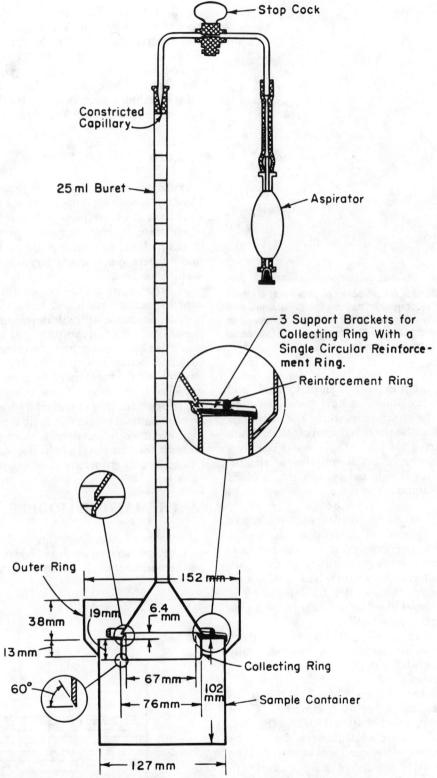

FIG. 34—ASTM C 243 test apparatus to determine the bleeding properties of paste and mortar.

FIG. 35—ASTM C 232 test for bleeding of concrete; Method A without vibration. The container has an inside diameter of about 255 mm and height of about 280 mm. The container is filled to a height of about 255 mm. The container is covered to prevent evaporation of the bleed water.

the surface (Fig. 35). Method B uses a similar sample and apparatus, along with a clamped-on cover, with vibration to help consolidate the concrete mixture. The sample is temporarily tilted when the water is drawn off to facilitate water collection. During the first 40 min, water readings are taken every 10 min and every 30 min thereafter until bleeding stops. Bleeding is reported as millilitres of bleed water per square centimetre of surface or as a percent of the mix water.

ASTM C 940 is used to test the bleeding of grout, where 800 mL of grout are placed into a 1000-mL graduated cylinder and the cylinder is covered to prevent evaporation. The upper surfaces of the grout and bleed water are recorded at 15-min intervals for the first hour, and hourly thereafter until bleeding stops. The final bleeding is recorded as the amount of bleed water decanted as a percentage of the original sample volume.

ASTM C 941 is used to determine the water retentivity of grout under vacuum atmosphere. The test essentially determines the relative ability of grout to bleed, or retain its water, under pressure.

Several nonstandard test methods have also been used for analyzing the bleeding properties of concrete. One of the first was the Powers' Float Method (Fig. 36). The float method was used to measure settlement of bleeding by measuring the subsidence of the small area at the top of the sample. The float consisted of a disk of lucite or bakelite to which a straight glass fiber was mounted like the mast on a sailboat. The disk had a diameter of about 13 mm and thickness of 3 mm. Movement of the float was measured with a micrometer microscope. Because of the thixotropic structure of the cement paste, the float remains in a relatively fixed position at the top of the paste during settlement. A layer of water is placed on the surface immediately after the float is installed to prevent the development of capillary forces from evaporation. The water depth is about 6 mm deep. The float is used both for paste, mortar, and concrete samples. However, for concrete, the sample diameter was at least 500 mm. The float method was used in most of the early major findings on the bleeding properties of paste and concrete. This test method was used in most bleeding research at the Portland Cement Association in the 1930s to 1950s.

SPECIAL APPLICATIONS

Under some situations, it is desirable to eliminate the bleeding capacity as much as possible. These are usually situations in which the surface level of the plastic volume of concrete, mortar, or grout must be maintained. This would include, for example, the construction of superflat floors where nonuniform bleeding would interfere with surface tolerance.

In other applications, grout for preplaced-aggregate concrete must have a low level of bleeding in order to provide adequate strength development within the entire mass. Grout used in grouting post-tensioning ducts must minimize bleed water from forming voids to prevent corrosion of tendons within the duct. Grout placed under base plates for machinery or column applications must not allow bleed water to form voids. As the grout provides the support for these elements, bleeding must essentially be eliminated. Certain admixtures can be added to grout to prevent such bleeding and actually induce a small amount of expansion to help eliminate bleed-water voids. Often, priority grouts are used for grouting base plates [8].

MATHEMATICAL MODELS

Ever since Powers' work in the 1930s on bleeding, several researchers, including Powers, developed several mathematical models for the bleeding rate and bleeding

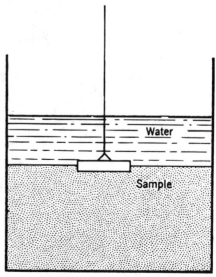

FIG. 36—Powers float test for measuring bleeding.

capacity for paste, mortar, and concrete. Many of these relationships, pertaining to the permeability of the paste, were based on Darcy's law and Poisseuille's law. Stoke's law was also used in empirical relationships. Based on Poisseuille's law of capillary flow, Powers and Steinour developed equations relating to the initial constant bleeding rate of paste. Unfortunately, the bleeding rate diminishes with time. In addition, the early equations to determine bleeding capacity assumed that the bleeding process was not influenced by hydration of the cement.

As opposed to Powers' consideration of bleeding as merely sedimentation, Tan et al. [26] used a self-weight consolidation model to express the relationship between the different characteristics of the bleeding process. By using the consolidation model, Tan et al. were able to obtain very good agreement with the bleeding process, even beyond the period of initial constant bleeding rate.

CONCLUSIONS

Bleeding is a fundamental property of concrete. By understanding its influences on plastic and hardened concrete properties and by understanding the effects of ingredients and ingredient proportions on bleeding, bleeding can be economically controlled. With proper control, bleeding should not hinder concrete construction or adversely influence concrete strength or durability.

REFERENCES

[1] Johnson, L. J., *Lecture Notes on Materials of Engineering*, Harvard Engineering School, Harvard University, Cambridge, MA, 1926.

[2] Mardulier, F. J., *The Bleeding of Cement: Its Significance in Concrete*, Paper No. 67, American Society for Testing and Materials, Philadelphia, 1967.

[3] Powers, T. C., *The Bleeding of Portland Cement Paste, Mortar and Concrete*, RX002, Portland Cement Association, Skokie, IL, July 1939.

[4] Steinour, H. H., *Further Studies of the Bleeding of Portland Cement Paste*, RX004, Portland Cement Association, Skokie, IL, Dec. 1945.

[5] Steinour, H. H., *Tests of Special Cements for Basic Research Program*, Major Series 290, Portland Cement Association, Skokie, IL, April 1941.

[6] Brownyard, T. L. and Dannis, M. L., *Expansion of Pastes, Mortars and Concretes During the First 24 Hours*, Junior Series 354, Portland Cement Association, Skokie, IL, 1941.

[7] Menzel, C. A., "Causes and Prevention of Crack Development in Plastic Concrete," *Proceedings*, Portland Cement Association, Skokie, IL, 1954, pp. 130–136.

[8] Kosmatka, S. H., *Cementitious Grouts and Grouting*, EB111, Portland Cement Association, Skokie, IL, 1990.

[9] Panarese, W. C., *Cement Mason's Guide*, PA122, Portland Cement Association, Skokie, IL, 1990.

[10] Hoshino, M., "Relationship Between Bleeding, Coarse Aggregate, and Specimen Height of Concrete," *ACI Materials Journal*, March-April 1989, pp. 185–190.

[11] Welch, G. B. and Patten, B. J. F., "Bond Strength of Reinforcement Effected by Concrete Sedimentation," *Journal*, American Concrete Institute, Vol. 62, No. 2, Feb. 1965, pp. 251–263.

[12] Whiting, D. and Dziedzic, W., *Effects of Conventional and High-Range Water Reducers on Concrete Properties*, RD107, Portland Cement Association, Skokie, IL, 1992.

[13] Kobayashi, K. and Uno, Y., "Chemical Composition of Bleeding Water on Fresh Concrete," *Seisan Kenkyu*, Tokyo, Vol. 41, No. 2, 1989, pp. 128–130.

[14] Klieger, P., *The Effect of Atmospheric Conditions During the Bleeding Period and Time of Finishing on the Scale Resistance of Concrete*, RX072, Portland Cement Association, Skokie, IL, 1956.

[15] Whiting, D., *Effects of High-Range Water Reducers on Some Properties of Fresh and Hardened Concretes*, RD061, Portland Cement Association, Skokie, IL, 1979.

[16] Gebler, S. H. and Klieger, P., *Effect of Fly Ash on Some of the Physical Properties of Concrete*, RD089, Portland Cement Association, Skokie, IL, 1986.

[17] Blondiau, L., "Bleeding of Blast Furnace Slag Cements and Portland Cements," *Silicates Industriels*, Vol. 21, Belgian Ceramic Society, Mons, Belgium, 1956, pp. 339–348.

[18] Saad, M. N. A., Pacelli de Andrade, W., and Paulon, V. A., "Properties of Mass Concrete Containing an Active Pozzolan Made from Clay," *Concrete International*, July 1982, pp. 59–65.

[19] Scheissl, P. and Schmidt, R., "Bleeding of Concrete," *RILEM Colloquium of Properties of Fresh Concrete*, Paper No. 4, The European Cement Association, Brussels, 1990.

[20] Hwang, C. L. and Wu, D. S., "Properties of Cement Paste Containing Rice Husk Ash," *Fly Ash, Silica Fume, Slag, and Natural Pozzolans in Concrete*, SP-114, American Concrete Institute, Detroit, MI, 1989, pp. 733–762.

[21] Bruere, G. M., "Mechanism by Which Air-entraining Agents Affect Viscosities and Bleeding Properties of Cement Paste," *Australian Journal of Applied Science*, Vol. 9, Commonwealth Scientific and Industrial Research Organization, Melburn, 1958, pp. 349–359.

[22] Gebler, S. H., *The Effects of High-Range Water Reducers on the Properties of Freshly Mixed and Hardened Flowing Concrete*, RD081, Portland Cement Association, Skokie, IL, 1982.

[23] Powers, T. C., *The Properties of Fresh Concrete*, Wiley, New York, 1968.

[24] Vollick, C. A., "Effects of Revibrating Concrete," *Proceedings*, American Concrete Institute, Vol. 54, 1958, pp. 721–732.

[25] Valore, R. C., Jr., Bowling, J. E., and Blaine, R. L., "The Direct and Continuous Measurement of Bleeding in Portland Cement Mixtures," *Proceedings*, American Society for Testing and Materials, Vol. 49, 1949, p. 891.

[26] Tan, T. S., Wee, T. H., Tan, S. A., Tam, C. T., and Lee, S. L., "A Consolidation Model for Bleeding of Cement Paste," *Advances in Cement Research*, Vol. 1, No. 1, Thomas Telford Services Ltd., London, Oct. 1987, pp. 18–26.

Cement and Water Content of Fresh Concrete

Deborah J. Lawrence[1]

PREFACE

The subject of determining the cement or water content in fresh concrete was not covered in previous editions of *ASTM STP 169*. It has been derived from ASTM Test Method for Determining Cement Content of Freshly Mixed Concrete (C 1078), ASTM Test Method for Determining the Water Content of Freshly Mixed Concrete (C 1079), and other similar tests studied to determine the amount of these materials in fresh concrete. The ASTM standards were first approved in 1987.

INTRODUCTION

Technological advances in the construction field have greatly increased production levels and therefore decreased construction times. Now more than ever, there is a need to have real-time quality control/quality assurance techniques available to the field. Poor or questionable quality materials need to be identified as early as possible to avoid early failure or costly rework later due to those materials being built into the construction project.

There are several procedures that have been proposed and used to determine the cement and water content of fresh concrete. The only ones currently approved as ASTM standard test methods are ASTM Test Methods for Determining Cement Content of Freshly Mixed Concrete (C 1078) and ASTM Test Methods for Determining Water Content of Freshly Mixed Concrete (C 1079). This chapter will address some of the questions that may arise when performing ASTM C 1078 and C 1079, as well as explain the basic reasons behind how these procedures work.

The two procedures in ASTM C 1078 and C 1079 represent the second and third generation (Concrete Quality Monitor) of a method originally proposed by R. T. Kelly and J. W. Vail of the Greater London Council. All three generations use the same basic water and cement test principles but vary the equipment and analytical techniques. With minor modifications, that is, sample size and sieve/sample washing procedure, ASTM C 1078 and C 1079 have been shown to work for roller compacted concrete and soil cement [1]. These test methods can also be used to determine variability of cement content in a batch of concrete, the variability of cement content between batches of nominally identical concrete and in mixer performance testing.

This chapter will also include brief summaries of other methods that have been used to determine cement and water contents, some of which have found favor outside of the United States.

CEMENT CONTENT ANALYSIS PROCEDURES

ASTM C 1078

A given mass of freshly mixed concrete is washed with a given volume (10 gal or 37.9 L) of water over a nest of sieves. The water is agitated in a manner so that the cement and other fine particles washed from the concrete are uniformly suspended. If the water is not agitated sufficiently, particles will settle out and a representative sample can not be obtained. An insufficient agitation can be detected by the failure to get a straight-line cement calibration curve or failure to get as high a calcium content reading as would be expected when running calibration check samples. Insufficient agitation usually results in lower calcium readings and, therefore, lower cement content calculations than should be obtained. This can be quickly determined by running known cement content concrete samples and plotting their test calcium readings onto a calibration chart.

The representative sample, 125 mL for Procedure A and 30 mL for Procedure B, is then mixed with dilute HNO_3. The HNO_3 puts the calcium from the cement in solution and, in Procedure B, serves to clean all of the cement solution out of the syringe. The sample and HNO_3 are further diluted with water and uniformly mixed/suspended using a magnetic stirrer. A sample of this dilute mixture is then titrated, manually in Procedure A and by a calcium analyzer in Procedure B, to determine the calcium content of the sample. The calcium content is linearly related to the cement content in the sample. By mixing known cement content concrete samples with the same materials and proportions as the concrete to be tested, determining the calcium content of the samples, and plotting the calcium content against the cement content, a linear calibration curve can be developed. The calibration curve will allow the calcium content values determined by this method to be converted to the cement content of concrete test samples.

[1] Champaign, IL 61820.

The calibration curve should be developed using the same materials as the constituents of the concrete to be tested. When any of the sources of the materials comprising the concrete change, the calibration curve should be reconstructed to allow for the changed materials.

There are some conditions that affect the results of this test method. Varying calcium contents of the raw materials comprising the concrete may cause variation in the test results. The existence of variation can be determined by running known cement content samples and plotting the calcium content against the known cement content. If the calcium contents of the materials are varying, a straight-line calibration curve will not be formed.

Test results on some concrete samples containing certain chemical admixtures and silica fume showed increasing calcium contents as testing time increased. Because of this problem, a calibration curve cannot be developed and the test method cannot be used.

The calibration curve should be a straight line. Besides reasons such as varying calcium amounts in the concrete materials and certain admixtures that test with rising calcium contents, there are some normal operating procedures that can affect the accuracy of the test results from this test method. These normal procedures that should be checked are: (a) verify that all of the fines are being washed from the aggregate into the 10 gal (37.9 L) of dilution water—some cement may still be on the aggregate; (b) verify that the sieves are not being overloaded and therefore all of the fines are not being rinsed into the 10 gal (37.9 L), that is, the water containing the fines may overflow the sieve nest and some of the cement may be lost; (c) in Procedure B, the 100-μL sample may not be consistently 100 μL, therefore, the user should visually look at the 100-μL sample to verify that it looks consistently like a complete 100-μL sample; that is, the sample may have an air bubble, void at the tip or include relatively large particles; and (d) ensure that the water has been sufficiently agitated in the 10-gal (37.9-L) cement solution when the test sample is taken, as mentioned earlier.

ASTM C 1078 has also been found to allow the determination of the cement contents of soil-cement mixtures and roller-compacted concrete (RCC) with minor modifications. For soil cement, the procedure may be followed by decreasing the sample size to 1 kg. For RCC, the sample size needs to be increased to 4 kg (due to the larger aggregate and lower cement content); care must be taken when washing all of the fines from the aggregate, and 200 g of Calgon should be added to the 10 gal (37.9 L) of water to aid in the aggregate washing process.

Some skill in chemistry is beneficial in performing Procedure A, but is not necessary. A few of the solutions, once mixed have a shelf life of two months, but this should not impact the decision on which procedure to use as new mixtures can be quickly prepared.

OTHER CEMENT CONTENT PROCEDURES

There are several methods of determining cement content of fresh concrete that have not yet been accepted as ASTM methods but have application to field use. The methods can be broken into four main categories: (1) chemical determination, (2) separation, (3) nuclear related, and (4) other, such as density and electrical means.

Chemical

Constant Neutralization Method

The Constant Neutralization Method [2] is another test relying on chemical means for the determination of cement content. This test takes the approach that the concentration of hydroxide ions in the concrete liberated during the hydration process is directly proportional to the amount of cement in the concrete. This test consists of maintaining a concrete sample neutral using phenolphthalein as the indicator and hydrochloric acid as the titrant. At 1-min intervals, the sample is brought back to a neutral state for a total of 60 min. The sieved and diluted sample is weighed at the beginning and end of the test period. The difference of masses determines the hydrochloric acid used that is then related to cement content via a preestablished calibration curve.

The calibration curve is formed by testing known cement content concrete samples, recording their results, and plotting these on a graph. As in ASTM C 1078, when any of the raw material sources, types, or proportions change, new calibration charts must be made.

One problem, the limiting factor for full-fledged field use of this test, is the time required to perform it. The test requires continual attention (every minute) for 60 min. Another area of possible concern is the necessity that the neutralization of the solution be done carefully, as excess acid can attack the aggregate, especially towards the end of the testing when the hydration process has slowed considerably.

The test has been used successfully in the field, on a few projects, by the California Department of Transportation for mixer efficiency testing. The method has been found to be accurate within 23.5 lb (10.7 kg) of cement per cubic yard of concrete.

Absorption Method

A limited amount of laboratory testing has been done on mortar passing a No. 40 (425-μm) mesh sieve using the Absorption Method [3]. This test relies on the amount of $KMnO_4$ absorbed by a cement-sand mixture. $KMnO_4$ is added to the sieved sample, shaken for 30 min, then the $KMnO_4$ not absorbed is determined by titrating the sample with oxalic acid. The quantity of $KMnO_4$ absorbed is found to increase as cement content increases for both plastic and hardened mortars.

$KMnO_4$ was selected as the absorption solution because it is chemically nonreactive with both cement and sand, and the amount of $KMnO_4$ absorbed can easily be determined by titrating with a standard oxalic acid solution. Problems existing in the use of this method relate to variations in the physical and chemical characteristics of the sand and cement, that is, the state of water saturation of the aggregate. Another factor in the accuracy of the method is temperature, as it relates to absorption characteristics.

Because of the limited laboratory testing with this method, an accuracy statement is not available. Due to the prevalence of variations possible in aggregate absorption, etc., affecting accuracy, combined with a 30-min wait for absorption to occur, this test is not likely to succeed as a rapid cement determination for field use. The test does have simplicity on its side, which is needed in the field.

Separation

Rapid Analysis Machine

Of the separation methods, the Rapid Analysis Machine (RAM) is the most likely to be approved for field use in the United States. The RAM (Fig. 1) has been used extensively in England, where it was developed. The RAM [4–7], developed by the Cement and Concrete Association, London, determines the cement content of fresh concrete by a wet analysis process using a series of automatically controlled devices. The RAM takes 5 min to test a sample.

An 8-kg sample is placed into the elutriation column. The RAM then turns on a pump that pushes water up and out over the top of the elutriator tube. The flow of water up the tube is regulated so that particles greater than 0.01 in. (0.254 mm) are left in the cone, while the finer particles are carried by the water. The water carrying the fines is then separated into three sampling channels (approximately 10% of slurry) and waste. The sample is then sieved by a 150-μm sieve before entering the conditioning vessel. In the conditioning vessel, a flocculent is added to aid the settling process of the fines. The excess water is then drained off to a predetermined level, and the conditioning vessel is weighed.

The mass of the conditioning vessel is related to the cement content by means of a calibration chart. The calibration chart is formed by testing a sample of concrete with a known cement content, and a sample with no cement. Again, as in many of the methods, the test relies

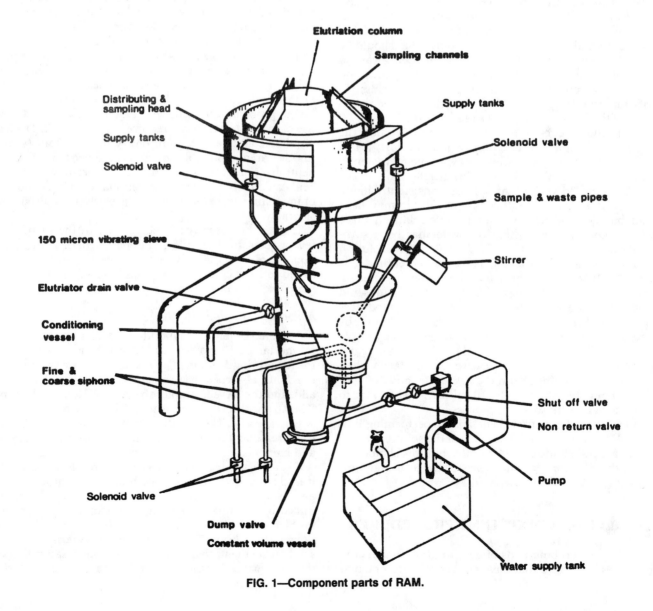

FIG. 1—Component parts of RAM.

on consistent volumes of silt and other fines from sample to sample.

Laboratory testing has defined several problems with this method and solutions in some cases. Among the problems is that as the concrete age increases before testing, the cement content determined by the RAM decreases [6,7]. Laboratory evaluations determined that the reduction of cement content with time is predictable and consistent though, and can therefore be corrected for by formula or by addition of retarder. Air-entrained concrete caused underestimates of cement content (the Waterways Experiment Station (WES) found −25.9% of the known cement content) [5]. Small quantities of tributylphosphate mixed with the concrete prior to addition to the elutriator have been reported to detrain air in the subsequent flocculation process [4,6].

The RAM must be properly maintained. Testing with the RAM has shown that careful checks on timing sequences must be performed periodically. WES [5] experienced several mechanical problems with their RAM, which were caught when timing of the sequenced steps was performed. Dhir et al. [6] also noted the impact of the timing sequence, the cleanliness of the 150-μm sieve, 40-mm coarse aggregate on repeatability in sampling, high silt content concretes, and low workability concretes.

Under laboratory conditions, the RAM was found to vary within ±10.5% of the known cement content. In the field, the variation would probably be larger due to variations in silt content and sampling errors.

By adding retarder to the concrete, up to 48 h can elapse between sampling of the concrete and testing without impacting the accuracy of the determination. Equally, the RAM has been used as a mobile unit in England; thus, providing on-site or off-site testing capabilities. Further research should be done on the mobility capabilities of the RAM though, in light of WES's [5] mechanical problems. There is also a supply and availability problem in the United States limiting the ease of use of the RAM. The supplier of the RAM in the United States does not stock or supply replacement parts. The Ram method is rapid and relatively simple to perform.

Flotation Method

The Flotation Method [8,9] is being developed by E. Nagele and H. K. Hilsdorf in Germany. The flotation method relies on solid particles being suspended in water. Then, surface-active substances are added as "collectors." They react selectively with a particular component of the mixture and make it hydrophobic. If air is blown into the suspension, the hydrophobic particles attach themselves to the air bubbles and rise to the surface. The nonhydrophobic particles remain in suspension. At the surface of the suspension a froth containing the hydrophobic components is formed and can be removed easily. Thus, a particular component may be separated from the other solid particles. If a collector can be found that makes either the cement or the aggregate particles hydrophobic, then cement and aggregate may be separated by flotation.

The froth is removed, filtered, and dried, leaving the cement as the dried product. Nagele has been able to identify collectors that selectively float cement and fly ash separately; surfactants with SO_4^{-2} and SO_3^{-2} end groups. Some problems were encountered identifying the correct collectors for different cement types for optimization of this method. Nagele et al. found that by adding K_2SiF_6 as a modifier, the amount of quartz powder floated can be greatly decreased. Despite these problems, Nagele and Hilsdorf found they could determine cement content of an unknown concrete to about ±8.5% [9]. The results obtained from the flotation method indicate the method is not affected by the temperature, age of the fresh concrete mix, type of cement, or chemical admixtures. The test, including 5 min to dry filtered samples, takes about 15 min.

A prototype field unit has been built. This equipment still needs optimization such as in air distribution in the concrete pulp and the stirring system. This method appears to be promising, but there still needs to be large-scale field testing of the system. In the laboratory, several cements, admixtures, aggregates, and fly ashes have been tested with success. There still needs to be work done on absorption properties of cement and aggregates with respect to collectors and modifiers.

Willis and Hime Separation Method

The Willis and Hime Separation Method [10] relies on heavy liquid centrifugal separation. In this method, a 5 lb (2.27 kg) concrete sample is sieved over a No. 30 (600-μm) sieve into a 3000-mL beaker three-fourths full of water containing a flocculating agent. The beaker is allowed to stand for 5 min to allow the solids to settle. Excess water is then siphoned off. The settled material is then dried by heating it in a cast iron skillet. The dried material is weighed and two 25-g samples are put into centrifuge tubes. The centrifuge tubes are then filled to the 50-mL mark with an acetylene tetrabromide/carbon tetrachloride mixture and centrifuged for specified times with intermittent stirring. The volume of the cement layer is then determined, and the equivalent volume of cement is found by determining the volume of the cement layer that would be obtained if all dried minus No. 30 (600-μm) sieve material had been tested instead of the 25 g. The cement content of the plastic concrete is then read from a calibration curve using the equivalent cement volume as entry to the curve. The calibration curve is developed by running known cement content concrete samples through the test procedure and plotting results.

The test requires about 75 min to complete. Roughly, 140 tests were performed by Hime and Willis using Type I and Type III cement achieving errors of 0.14 sack per cubic yard to 0.22 sack per cubic yard, respectively. During this testing, several potential sources of error were pinpointed. The main source of error appears to be associated with specific surface areas for the type of cement being used. Cement with higher specific surfaces exhibited higher readings than cements with lower specific surfaces. This problem should be avoided through use of calibration curves for specific cements. The efficiency of the heavy media mixture used is very sensitive to specific gravity and must be carefully adjusted to ensure all the fine aggregate floats in the centrifuging operation. The centrifugal forces also must be monitored because they impact the measured volume of the cement layer's compaction.

Several testing areas still need to be investigated for this method, including the effects of fly ash in the concrete. Because of the sources of error, further testing required, and time required to perform this test, this method needs further development.

Dunagan Buoyancy Method

The Dunagan Buoyancy Method [11] was presented in 1930, but various versions of it, that is, Part 2 of B.S. 1881 developed by Kirkam [12], are still being proposed/used. This method relies on separation of the concrete by component size. The volume of the cement is the difference between the volume of the samples and the volume of the aggregates. This approach requires that the specific gravities of the coarse aggregate, the fine aggregates, and the cement be known as well as the weights of the ingredients passing the selected sieves.

An 8- to 10-lb (3.6- to 4.5-kg) sample of plastic concrete is weighed in air and in water. The sample is then washed over a No. 4 (4.75-mm) and No. 100 (150-μm) sieve to separate the components by size. The materials retained on the sieves are then weighed in water. The weight of the material on the No. 100 (150-μm) sieve is corrected for silt, and the quantity of cement is determined by subtracting the submerged weights of the aggregates retained on the sieves from the submerged weight of the concrete sample. The submerged weight of cement is then converted to weight in air.

The method takes about 10 to 20 min to perform. The test was found to determine cement content to within 7% of the cement present [4]. There are several major problems with this test under today's concrete designs. Although a calibration to determine raw materials other than cement that pass the No. 100 (150-μm) sieve is performed, variations of the amounts (as by adding fly ash) could easily go undetected. Because particles passing the No. 100 (150-μm) sieve appear as cement, any process or addition to the concrete affecting the fines passing the No. 100 (150-μm) sieve will cause error in determining cement content.

Nuclear Related

Nuclear Cement Gage

The nuclear cement gage measures cement content by means of backscatter and absorption of low-energy gamma rays [13]. The cement-content gage probe is immersed in a concrete sample and six 20-s readings are taken. The probe contains a 14-m Ci low-energy gamma-ray source (Americium-241) and a radiation detector. The detector is shielded on the direct line from the sources, so that the only path by which the gamma rays can reach the detector is through the sample. The gamma rays are both scattered and absorbed by the concrete sample [14]. The absorption coefficient has dependence on the atomic number of the elements to the fourth power, therefore, the amount absorbed depends strongly on the chemical composition of the samples. Calcium is generally among the highest atomic number elements present and is therefore a major contributor to the absorption amount. The cement content is related by a calibration curve made from known cement content concrete samples to the nuclear cement content gage readings.

The nuclear gage relies on absorption by the elements composing the concrete. This fact poses the limitations of the gage. The gage has been found to determine cement content of siliceous aggregate mixes to within ±22 lb/yd³ (10 kg) and ±31 lb/yd³ (14 kg) for calcareous aggregate mixes. The major limitations of the gage are: (a) the necessity for recalibration when the aggregate source is changed or when the ratio of coarse to fine aggregates is changed; (b) its reduced accuracy for calcareous and certain siliceous aggregate concretes [14]; and (c) the presence of higher atomic elements, such as iron, that also absorb the gamma rays. The accuracy is also affected by variations in temperature but, these problems are resolved by using a polymer-impregnated concrete standard.

The nuclear cement content gage test takes about 10 min and looks promising for rapid on-site analysis of fresh concrete. There are several field tests with this method being currently performed. The major problems lie in making sure the calibration curves are accurate and the concrete does not contain considerably varying amounts of high atomic number elements. Care in terms of safety and maintenance of equipment must also be carefully monitored.

Thermal Neutron Activation

Other neutron activation methods are briefly presented by Malhorta [15]. One method is using thermal neutron activation with a nuclear reactor to determine calcium and therefore cement content, but this method is not suitable for concretes made with calcareous aggregate. Concrete samples are activated for 1 s. This produces short-lived calcium isotope, calcium-49 (^{49}Ca). A 400-s count is then taken, after a 720-s decay, using a lithium-drifted germanium detector. This test can be used in the field by using isotopic neutron sources for activation of large samples.

Because of the nonsuitability of this method to calcareous aggregate concretes and the lack of field tests, much work is needed before this test could be fully evaluated for usefulness. This test would have similar drawbacks as the nuclear cement content gage in terms of equipment safety and maintenance required.

Other Methods

Hydrometer Analysis Method

The Hydrometer Analysis Method [16] relies on measurements of the density of a cement-water suspension. A sample of concrete is washed over a No. 100 (150-μm) sieve. All of the fines passing the No. 100 (150-μm) sieve are collected in a container and water is added to produce a consistent testing volume. The water is well stirred for 1 min, then hydrometer readings on the solution are taken every 10 s for 1½ to 2 min. The stirring and hydrometer reading process is repeated until four series of readings are obtained. The readings are then plotted against time and extrapolated to zero time. The average value of the tests is then used to determine the cement content by referencing a previously developed calibration chart. The calibration chart is made by performing the test on known concrete mixes.

The test takes 15 to 30 min to perform and is reported to determine cement content to within 4% of the total weight of the cement. Problems with the test are that hydrometer readings are temperature dependent, and must be taken into account, and the test relies on evaluation of everything that passes the No. 100 (150-μm) sieve. In terms of field use, inspectors would have to continually plot data. It has also been reported that the accuracy of the results of repeated tests decreased for quantities of cement exceeding 40 g/L. This brings in factors relating to high fines content concrete not being accurately/reliably tested.

Concrete Consistency Method

The Concrete Consistency Method proposed by Popovics [17] utilizes the relationship of information produced from two different consistency tests or repeated measurements from the same consistency test. The relationship was formulated by Popovics and only applies to the methods he used, the materials composing the concrete he tested, and with the same testing equipment. His statistical relationship is good but would be difficult to use accurately in the field. The test requires large amounts of back-up laboratory work, consistency in material types, sources, proportions, and consistency in performing the consistency tests, and is therefore impractical for field use.

Conductimetric Method

A Conductimetric Method of determining cement content was investigated by Chadda [3]. This method involves a sieved sample (since his tests only involved sand-cement mixtures) diluted with distilled water and shaken for 30 min. The conductivity of the solution is then measured using a Dionic Water Tester. The test then requires a calibration chart relating cement concentration to mixture conductivity.

The conductimetric test is very simple but because of all the factors that influence conductivity such as varying admixture amounts, aggregate fines variations, presence of chlorides, etc., the method would probably not produce accurate enough results for field quality control.

WATER CONTENT ANALYSIS PROCEDURES

ASTM C 1079

A given mass of freshly mixed concrete is intermixed with a chloride solution of a given strength and volume. The chloride solution mixes with the free water in the concrete. The chloride ion concentration of the intermixed solution is directly related to the water content of the concrete sample and is determined by volumetric titration or coulometric reference technique. Since some concretes contain chlorides, a blank test is required. The blank test intermixes a given volume of distilled water with a concrete sample of given mass and determines the chloride ion concentration of the intermixed blank solution by volumetric titration or coulometric reference technique.

This method is based on the principle that the free water in concrete will intermix with a known chloride content solution and therefore dilute the chloride solution. The amount of dilution is directly related to the water in the concrete. A blank test is required because some concrete contains chloride. The blank test facilitates the determination of chlorides in the concrete and, therefore, the existing concrete chlorides can be allowed for when determining the water content. If the blank test indicates no chlorides are present in the concrete, the blank test may be waived in subsequent testing. If the source of any of the concrete materials changes, the blank test should be run. In both the blank test and the chloride solution test, the jars should be turned end over end or agitated so that the water in the concrete thoroughly mixes with the added solution. This agitation should not be too fast as not all of the concrete water would mix uniformly with the added solution. Sufficient time of mixing should also be allowed to produce a uniform thoroughly mixed solution.

In Procedure A, after the salt solution or water for the blank have been mixed with the concrete, the jars should be allowed to settle so that a clear 25-mL sample can be obtained for titration. In Procedure B, the 100-μL sample obtained from the centrifuge tubes should also be clear. Because 100-μL is a small sample, the sample should be free of particles or air bubbles to maintain accuracy of the test results. Air entrainers and some other admixtures in concrete may cause a layer of foam to form in the centrifuge tubes. The layer of foam should be removed from the tube with a swab or other device and the tube recentrifuged. This procedure should be repeated until a clear 100-μL sample can be obtained.

In Procedure B, a chloride meter is used to determine the chloride concentration of the test solutions. The silver electrodes of the meter need to be periodically cleaned to assure the chloride meter will calibrate and titrate properly. The electrodes may appear black, the chloride meter will not complete the condition cycle or the chloride standard will not test at its known concentration when the electrodes need cleaning.

Some of the conditions that can affect the accuracy of the test are: (a) not thoroughly mixed concrete and salt solution samples, (b) test samples that are not clear from the intermixed water or chloride (salt) and concrete solution, and (c) 100-μL samples that are not clear in Procedure B.

OTHER WATER CONTENT PROCEDURES

There are several other methods for quickly determining the water content of fresh concrete. The microwave oven method is probably the easiest of the methods to use. The methods presented here are broken into four categories: (1) chemical and other, (2) separation, (3) electrical, and (4) nuclear related.

Chemical and Other Methods

Concrete Consistency Method

The Concrete Consistency Method depends on performing repeated consistency tests on a sample to determine the water content. One particular method is to measure the slump of a concrete sample (S1), add a known amount of water (W1), determine the slump of this mixture (S2), then, via formula, relate these pieces of information to the original water content (WO). An example formula developed by Popovics [17] is

$$WO = \frac{W1}{\frac{S2^{0.1}}{S1} - 1}$$

Problems with this method are similar to the problems related to cement determination by this method. Specific relationships between consistency, water content, and cement content are not generally applicable. Different raw materials would require different relationships to be developed. Outside of the substantial laboratory back-up work required, this method could drastically be impacted by changes in the raw materials, that is, plasticizers with time.

Thermal Conductivity Methods

Thermal Conductivity Methods [18] rely on the ability of water as a thermal conductor. The basic principle involves a probe being inserted into the mixture, the probe is heated (via electric current or otherwise), and the rate at which the temperature of the probe increases is measured. The higher the water content of the surrounding material, the faster the heat is conducted away and the slower the temperature rise.

The test may be calibrated in terms of evaporable water content but each material tested must have its own calibration relationship. Proper thermal contact between the probe and porous material must also be monitored. This method is simple, but several field variables could impact on the test, that is, admixtures. The accuracy of this method was not stated. The method is probably not as accurate as other methods mentioned here but would certainly be quick and easy.

Separation Methods

Microwave Oven

The microwave oven is probably the easiest of the separation methods to use. This test simply requires a known mass of concrete to be placed in a microwave oven on defrost cycle for 1 h. The sample is then weighed again at the end of the hour. The difference between the before and after weights represents the water loss. If the effective water content of the sample is desired, the aggregate absorption must be known.

The main sources of error are evaporation of water occurring between sampling time and entrance into the microwave oven and admixtures that are driven off at the same time as the water. The admixtures can be corrected for, roughly 70% of the water reducing admixtures and 15% of the air-entraining agents are removed during the hour test. The average error of the test found by North Dakota investigators, was -0.003 gal (-0.011 L) per sack [19]. The only drawback of the test is the hour time lag from the beginning to the end of the test.

Mr. Hime et al. [20] were able to modify the microwave oven procedure to shorten the water-loss determination time to about 1 min. Their modification involves the use of an ashing block assembly. The assembly consists of two heating inserts that fit into an insulating ceramic support block. This unit within the microwave oven absorbs heat and allows the samples to achieve a higher temperature more rapidly. However, there is a sample size limitation imposed by the relatively small ashing block assembly.

Dunagan Buoyancy Method

The Dunagan Buoyancy Method [11] determines the water content of the concrete sample by subtracting the total calculated weight of the solids in air (cement plus aggregates and fines) from the weight of the sample in air. The problem with this method is that all of the variations in the other constituents of the concrete affect the water content determination. The accuracy can be improved by further drying the aggregates and weighing them or by frequently monitoring aggregate-specific gravities.

This method is not very accurate, the error in determining water content was found to be 25% of the water present. Further details of the method are given in the previous section on cement determination.

Flotation Method

Water content may also be determined using the Flotation Method [9]. Again, this is an indirect determination of the water. The water is determined by subtracting the cement, aggregates, and other fines determined from the original concrete sample weight after drying. This method has been found to be fairly accurate but again additives would be counted as water and the method is not a direct determination. Further details of the method are given in the previous section on cement determination.

Electrical Methods

Capacitance Methods

Some work has been done with Capacitance Methods [18] to determine water contents of soil mixtures and concrete mixtures. Since the dielectric constant of water is much higher than that of the dry constituents of concrete, this procedure has merit as being a rapid water test. This method uses the fact that the dielectric constant of water is about 80 and is relatively insensitive to changes in temperature from 15 to 35°C, and to changes in frequency from 10^5 to 3×10^8 Hz. Researchers have found some success with this method via use of gypsum blocks in soils and concrete between insulated electrodes.

Electrical Resistance

Direct measurement of electrical resistance [18] can give estimations of water contents of concrete. Testing has shown that the presence of sodium chloride in water influences capacitance. This leads to possible problems with the admixtures in concrete that affect capacitance. More work is needed in this area to determine relative impacts that the admixtures and other similar substances have on this method, as the method is simple for field use.

Some error is inherent in a direct measurement due to variable contact resistance between electrodes and material. These errors may be lessened by embedding the electrodes in a material such as gypsum, then embedding the gypsum block into the concrete. Another version of this method to lessen the impact on accuracy of the resistance variation is to use four electrodes instead of two. This method exhibits hysteresis effects [18], but, the evaporable water content of the concrete can be calibrated in terms of the electrical resistance of the gypsum block.

More work is needed in this area to determine what other factors impact this method as the test is very straight-forward. It is known that variations in the concentration of soluble salts will impact the resistance; more information is required in this area to define the impact. The question also remains as to how much impact admixtures have on the test and whether the impact can be calculated out of the measurements.

Nuclear Related Methods

Microwave–Absorption Method

Absorption of electromagnetic radiation (Microwave-Absorption Method) by a material is another method for determining water content. This absorption is a function of the dielectric constant and is therefore closely related to the capacitance methods described before.

Measurements are made by determining the attenuation of the radiation as it passes from a transmitting horn antenna through the specimen to a receiving horn antenna. There is a linear relationship between evaporable water content on a volume basis and absorption of microwaves, but it varies with the material. Therefore, calibration is needed for the particular material being tested [18].

Because of the possible variation caused by aggregates, if the test was performed on sieved material, it would probably be more accurate. More testing is needed to see if sieving would help and the impact of admixtures on the accuracy of the method.

Neutron-Scattering Methods

Neutron-Scattering Methods rely on the principle that if a source of high-energy neutrons and a detector of either low-energy or thermal neutrons are placed in a hole in a material, or are placed side by side on a surface of a material, the number of neutrons that are reduced to thermal energies and diffuse back to the detector will be largely a function of hydrogen nuclei in the vicinity [18].

Scattering caused by other nuclei in the concrete also impact the amount absorbed, therefore necessitating a calibration curve for the specific materials to be tested. The volume of material that influences the thermal neutron flux near subsurface gages depends on the water content and is reported to vary in radius from 4 to 18 in. (102 to 457 mm) and its thickness from 4 to 24 in. (102 to 610 mm) [18]. These large spheres of influence have caused difficulties and left most of the experimental efforts directed to laboratory methods. The method does not allow differentiation between evaporable and nonevaporable water.

With the right calibration correlations this method should be fieldable. More experimentation is needed. Nuclear moisture readings were taken on the Willow Creek Dam on roller compacted concrete (RCC) [21] with apparent success. The method was very quick and provided very good reference information. The nuclear readings had a smaller coefficient of variation than provided by the Concrete Quality Monitor on the RCC.

REFERENCES

[1] Lawrence, D. J., "Operations Guide and Modification Analysis for use of the CE Concrete Quality Monitor on Roller—Compacted Concrete and Soil Cement," USA-CERL Technical Report M-85/06 (Revised), July 1985.

[2] Woodstrom, J. H. and Neal, B. F., "Cement Content of Fresh Concrete," Report No. FHWA-CA-TL-5149-76-55, California Department of Transportation, Sacramento, CA, 1976.

[3] Chadda, L. R., "The Rapid Determination of Cement Content in Concrete and Mortar," *Indian Concrete Journal*, 15 Aug. 1955, pp. 258–260.

[4] Head, W. J. and Phillippi, H. M., "State of Technology for Quality Assurance of Plastic Concrete: Phase I—Feasibility Study," Report W. Va. DOH 59, West Virginia Department of Highways, June 1980, pp. 1–88.

[5] Tom, J. G., "Investigation of the Rapid Analysis Machine (RAM) for Determining the Cement Content of Fresh Concrete," Technical Report SL-82-4, Army Corps of Engineers, Waterways Experiment Station, Vicksburg, MS Oct. 1981, p. 36.

[6] Dhir, R. K., Munday, J. G. L., and Ho, N. Y., "Analysis of Fresh Concrete: Determination of Cement Content by the Rapid Analysis Machine," *Magazine of Concrete Research*, Vol. 34, No. 119, June 1982, pp. 59–73.

[7] Sangha, C. M. and Walden, P. J., "Fresh Concrete Analysis Off-Site," *Concrete*, Aug. 1980, pp. 27–28.

[8] Nagele, E. and Hilsdorf, H. K., "A New Method for Cement Content Determination of Fresh Concrete," *Cement and Concrete Research*, Vol. 10, No. 1, 1980, pp. 23–34.

[9] Nagele, E. and Hilsdorf, H. K., "The Analysis of Fresh Concrete Using the Flotation Method," Institut für Massivbau und Baustofftechnologie, University of Karlsruhe, Germany, July 1983, p. 66.

[10] Hime, W. G. and Willis, R. A., "A Method for the Determination of Cement Content of Plastic Concrete," *Research Department Bulletin 61*, Portland Cement Association, Skokie, IL, Oct. 1955.

[11] Dunagan, W. M., "A Proposed System for the Analysis and Field Control of Fresh Concrete," *Bulletin 113*, Iowa Engineering Experiment Station, Iowa State College, Ames, IO, 1933.

[12] Williamson, G. R., "Rapid Analysis of Fresh Concrete—A State-of-the-Art Paper," paper sent to members of RILEM 33-AC Committee, RILEM, Paris, 1984, p. 49.

[13] Mitchell, T. M., "FCP Annual Progress Report—Year Ending September 30, 1982," Project No. 6F, *Develop More Significant and Rapid Test Procedures for Quality Assurance*, p. 23.

[14] Mitchell, T. M., "Measurement of Cement Content by Using Nuclear Backscatter-and-absorption Gauge," Research

Record No. 692, Transportation Research Board, Washington, DC, 1978, pp. 34–40.
[15] Malhorta, V. M., *Testing Hardened Concrete: Nondestructive Methods*, The Iowa State University Press and American Concrete Institute, Detroit, MI, 1976, p. 188.
[16] Murdock, L. J., "The Determination of the Proportions of Concrete," *Cement and Lime Manufacture*, Vol. 21, No. 5, Sept. 1948, pp. 91–96.
[17] Popovics, S., "Determination of the Composition of Fresh Concrete by Non Nuclear Means," *Rapid Testing of Fresh Concrete, Conference Proceedings M-128*, Construction Engineering Research Laboratory, 5–7 May 1975.
[18] Monfore, G. E., "A Review of Methods for Measuring Water Content of Highway Components in Place," *Highway Research Record Number 342*, Environmental Effects on Concrete, Highway Research Board, Washington, DC, 1970, pp. 17–26.
[19] Peterson, R. T. and Leftwich, D., "Determination of Water Content of Plastic Concrete using a Microwave Oven," Item (4)-77B, North Dakota State Highway Department, Bismark, ND, 1978.
[20] Hime, W. G., "Instantaneous Determination of Water-Cement Ratio in Fresh Concrete," NCHRP 10-25A, 31 May 1989.
[21] Schrader, E., *Concrete Report Willow Creek Dam*, Corps of Engineers Walla Walla District Report, Walla Walla, WA, July 1983, p. 95.

PART III
Hardened Concrete

Concrete Strength Testing

Peggy M. Carrasquillo[1]

PREFACE

In the preparation of this chapter, the contents of the corresponding chapters in previous editions of this publication were drawn upon heavily. The author acknowledges the work of the following authors: Clyde E. Kesler and Chester P. Seiss in a chapter entitled, "Static and Fatigue Strength," published in *ASTM STP 169* [1]; C. E. Kesler in a chapter entitled, "Strength," published in *ASTM STP 169A* [2]; and K. N. Derucher in a chapter entitled, "Strength," published in *ASTM STP 169B* [3]. The work of each of these authors was a complete treatment of the topic of concrete strength testing at the time of its publication. The current chapter will review the topics as addressed by the previous authors, and introduce any new technology that has been developed. This is not intended to be a detailed analysis of the state-of-the-art in the area of concrete strength; a complete book would be required for that. Rather, general guidelines are given for better understanding the behavior of concrete under load.

INTRODUCTION

The most common concrete property tested is its strength. There are three main reasons for this. First, the strength of the concrete gives a direct indication of its capacity to resist loads in structural applications, whether they be tensile, compressive, shear, or any combination of these. Second, strength tests are relatively easy to conduct. Finally, correlations can be developed relating concrete strength to certain other concrete properties that involve much more complicated tests. It should be noted that in many instances it is necessary to know in more detail the suitability of a given concrete for a certain application, such as when durability aspects are of concern. When this is the case, the specific property should be tested according to applicable standards, rather than relying on strength correlations.

Strength tests are used for three main purposes. These are: (1) for research, (2) for quality control and quality assurance, and (3) in determining in-place concrete strength. In research, strength tests are used to determine the effects of various materials or proportions on the strength properties of concrete. Also, strength tests are used as controls when other properties of concrete are being studied, such as durability problems or structural design applications. In construction, strength tests are conducted on samples of concrete taken at job sites, either to determine the adequacy of the mix proportions developed for the particular job or to check for changes in strength that could indicate problems in quality control for the concrete supplier or changes in ambient conditions. When appropriately cured, these specimens can also be used to determine when the concrete in the structure is ready for application of construction loads, or for form removal. In cases where the quality of the concrete in-place is in doubt, samples may be cut from existing concrete placements and tested for strength.

In all instances of concrete strength testing, it is desirable that the certainty of the strength determination be as high as possible while remaining sensitive to the properties of the concrete being tested. To this end, standard testing procedures have been developed, and are still being developed and refined, that control those variables that may inadvertently affect strength test results. Following these procedures when testing concrete specimens will produce strength results that are both reproducible and meaningful.

NATURE OF CONCRETE STRENGTH

Unfortunately, concrete strength is not an absolute property. Results obtained from any given concrete will depend on specimen geometry and size, preparation, and loading method. This section will discuss the nature of concrete strength and those factors that influence its inherent strength. Factors involved in the testing of concrete specimens will be discussed in later sections.

Concrete can be considered as a two-phase composite, consisting of mortar and coarse aggregate, and its behavior will be the result of the interaction of the various properties of the two phases, and their interfacial region. Failure of concrete occurs as a result of the development of a network of microcracking through its structure to the point when it can resist no further load. The coarse aggregate particles act as inclusions within the mortar and as such can both initiate and arrest crack growth.

Various factors affect the strength of the mortar. Probably the most important factor in determining the strength of the concrete mortar is its density, which is in turn highly

---

[1] Consulting engineer, Carrasquillo Associates, Marble Falls, TX 78654.

dependent on the water-cement ratio. As the water-cement ratio is increased, the density of the paste decreases as does its strength. Also, the cement type and content will affect the mortar strength, as well as its strength gain characteristics with time. The use of pozzolanic admixtures such as fly ash will affect both the mortar strength and strength gain, since these materials have secondary as well as possibly some primary cementitious properties. Similarly, chemical admixtures can be used for water reduction, retardation, or acceleration, among many other potential uses. The entrainment of air will decrease the strength due to the increase in voids in the mortar. Retempering of the mix with water can cause a decrease in the mortar strength due to uneven dispersion of the retempering water, which leads to pockets of mortar having a high water-cement ratio. Finally, curing temperature and moisture conditions will have a marked effect on the mortar strength. High early curing temperatures, while yielding higher early strengths, will cause reduced strengths at later ages due to the formation of a less dense mortar structure. Further, hydration within the mortar will continue only as long as free moisture is available for reaction with the unhydrated cement. Thus, if the concrete is allowed to dry, hydration will cease. It has been reported that in poorly cured structural members, concrete may never reach the same strength as obtained from standard cured cylinders at 28 days [4–6].

The strength of coarse aggregates is determined mainly by their mineralogy. Beyond this, however, a smaller-sized or crushed aggregate may have strength advantages in that internal weak planes may be less likely to exist.

Factors involving both the mortar and coarse aggregate will affect the strength of the mortar-coarse aggregate interface. The bond between the mortar and coarse aggregate particles will be stronger for smaller-sized aggregates, which have a higher curvature. Also, a rough, angular surface texture such as exists in crushed aggregates will increase the bond strength [7]. Coatings, such as clay, on the aggregate surface will reduce the interfacial strength. When concretes bleed, the bleed water is often trapped beneath coarse aggregate particles as it rises, thus weakening the interfacial zone. It has been reported that the use of pozzolans such as fly ash in concrete increases the strength of the mortar-aggregate interface [8].

Other factors that affect the strength of the concrete include mixture proportions, degree of mixing of the concrete, mold material in which specimens are cast, and consolidation of the specimens. For a given set of materials, there is an optimum mixture proportion for achieving a given concrete strength, and once proportions have been selected, the materials must be adequately mixed to achieve a homogeneous mixture. The concrete must then be placed in the forms or molds and thoroughly compacted; poorly compacted concrete will produce lower strengths, with honeycombed concrete being an extreme example of this. Excessive bleeding and segregation of the concrete mixture can lead to reduced strengths, especially in the top portion of the concrete placement which will have a higher water-cement ratio due to the rise of the bleed water. The use of cardboard or plastic molds in preparing specimens has been reported to cause reduced concrete strengths as opposed to the use of rigid steel molds [9,10]. Also, concrete that has been damaged due to the application of loads or through deleterious reactions may also produce lowered capacity for resisting load.

OBTAINING TEST SPECIMENS

Concrete strength tests are conducted on both molded specimens and specimens cut from existing structures. Although the test procedures are predominantly the same for similar specimens regardless of their origin, the significance of the information obtained can be quite different depending on specimen preparation and handling prior to testing. Following are brief descriptions of current standard procedures for the treatment of test specimens up to the time of testing.

Molded Specimens

Current standard test methods call for specimens in the shape of cylinders or beams. Preparation and storage of these specimens is dictated in ASTM Test Method for Making and Curing Concrete Test Specimens in the Field (C 31) for specimens cast in the field, ASTM Test Method for Making and Curing Concrete Test Specimens in the Laboratory (C 192) for specimens cast under laboratory conditions, and ASTM Test Method for Compressive Strength of Concrete Cylinders Cast in Place in Cylindrical Molds (C 873) for push-out cylinders cast in concrete slabs. The first two of these standards are discussed in detail in Chapter 10 of this publication. ASTM C 873 is a standard test method involving push-out cylinders that are cast in molds placed within the structural slab, receiving curing identical to that of the concrete in the structure until the time of test. Since specimens made according to ASTM C 31 and C 192 are very similar in their preparation, which is dissimilar to that of specimens prepared according to ASTM C 873, the former will be referred to as molded specimens and the latter as push-out cylinders.

ASTM C 192 specifies strict control in specimen preparation and is generally used in research and mix proportioning studies. This procedure is considered to be the "ideal" condition for specimen preparation, and will yield the most consistent results due to the high degree of control involved.

Specimens prepared according to ASTM C 31 can be used for a variety of reasons. Those given moist curing can be used to check the adequacy of the mix design as supplied to the job, monitor the producer's quality control, and indicate changes in materials and other conditions. These test results verify the potential strength of the concrete mixture as supplied [11,12]. Specimens given curing to parallel that of the structural component are tested to indicate the in-place strength of the concrete, prior to form removal or application of construction loads. Even though efforts are made to give equivalent curing to both the concrete in the structure and the molded specimens, questions can be raised regarding potential differences in consolidation and early-age temperatures, as well as size effects, and their effect on strength test results.

Push-out cylinders, ASTM C 873, are used to determine the concrete strength in-place, similar to those specimens given field curing under ASTM C 31. However, it is very likely that the curing of push-out cylinders is more like that of the structure than for molded specimens. Seemingly the only difference between the concrete in the push-out specimen and that in the structural member may be due to the mold. It has been reported that molded specimens exhibit a "wall effect" that results in a smaller proportion of coarse aggregate in the molded specimen relative to the mass concrete [10,13]. This effect is present to some degree in all molded specimens, but is more pronounced in specimens having higher surface-to-volume ratios. Further, the fact that the push-out cylinder is surrounded by an impermeable material rather than concrete is a condition different from that of the mass concrete.

Specimens Cut from Existing Structures

The procedure for cutting strength test specimens from existing hardened concrete is specified in ASTM Test Method for Obtaining and Testing Drilled Cores and Sawed Beams of Concrete (C 42). The drilled cores obtained are tested either in compression or indirect tension. Sawed beams are used in flexural strength testing. In addition to placing dimensional requirements on specimens, this standard places further restrictions on sampling to ensure that the specimens are comprised of intact, sound concrete, as free of flaws as the particular construction will allow.

Generally, drilled cores or sawed beams are obtained when doubt exists as to the quality of the concrete as placed. This can be due to low strength test results during construction or signs of distress in the structure. Also, cut specimens are useful if strength information is required for older structures, or if service loads are to be increased above original design levels. All other factors being equal, the strength of these specimens is most likely to be representative of the strength of the concrete in the structure. However, cutting these specimens is both time consuming and costly, and the drilling or sawing processes themselves may introduce variables affecting strength test results.

Although the concrete in drilled or sawed specimens is more likely to be representative of that in the structure than molded specimens, one must be aware of various factors that are likely to affect the strength of the concrete samples thus obtained. Excess voidage in the concrete sample, due to poor consolidation, will cause strength reductions. Further, it is possible, and even likely, that the drilling or sawing of specimens causes damage that may affect strength test results, and that this factor may become more pronounced as the ratio of cut surface to specimen volume increases [14], or when the sawed surface of beams is subjected to tension during flexural tests, as stated in ASTM Test Method for Compressive Strength of Concrete Using Portions of Beams Broken in Flexure (C 116). The resultant strength reductions have been reported to be greater in higher strength concretes [15].

Besides the drilling or sawing processes themselves, the location and orientation of the cut specimens will affect the strength test results. Specimens cut from the tops of concrete placements give lower strength results than those cut from the bottom of the concrete placement due to bleeding and segregation of the concrete [10,13,14,16], and curing effects [17]. Bloem [17] reported that cores taken from the top of an 8 in. (203 mm) slab tested approximately 5% weaker than those from the bottom when the slab was subjected to good curing; the difference increased to 15%, however, when the slab was poorly cured. Studies on high-strength concrete [18] indicated no effect of core height on strength test results, which can be attributed to the fact that high-strength concretes generally exhibit very little bleed. Cores drilled horizontally have been shown to yield lower strength test results than those drilled vertically [10,13,14,16]. This is attributed to the alignment of flaws parallel to the loading direction in cores drilled horizontally, due to water gain under coarse aggregate particles when the fresh concrete bleeds.

The loading to which the concrete member has been subjected may also affect the cut specimen's strength. Cores taken from highly stressed regions of concrete, where cracking is likely to have occurred, give lower test results than those from unstressed regions [14,19].

The inclusion of reinforcing steel in the specimen is undesirable, but at times unavoidable. Its effect on compressive strength is variable. However, its occurrence in the tensile region of flexural beams or in splitting tensile strength specimens is not allowed due to its pronounced effect on the strength test results.

COMPRESSIVE STRENGTH TEST PROCEDURES

Compressive strength testing of concrete cylinders prepared according to ASTM C 31 or C 192 is specified in ASTM Test Method for Compressive Strength of Cylindrical Concrete Specimens (C 39). The compressive strength test procedures for cores and push-out cylinders are ASTM C 42 and C 873, respectively, both of which refer to ASTM C 39 for testing. All of these test procedures specify tolerances on specimen geometry, size, and end conditions, the loading apparatus and rate of load application, and specimen moisture condition at the time of testing. In addition, if the tolerances on specimen end conditions are not met by the concrete specimens, the test methods call for grinding or sawing of the ends to meet the requirements, or for capping of the ends according to ASTM Method for Capping Cylindrical Concrete Specimens (C 617). All of these factors will influence the strength test results obtained and will be discussed in following sections.

An additional compressive strength test involves the beam ends resulting from flexural strength tests. The test procedure is specified in ASTM C 116. Again, aspects such as specimen geometry, load application, and moisture condition are stipulated. This test procedure is not commonly used. However, it has the advantage of allowing both compressive strength and flexural strength to be determined from one specimen. The results obtained from compressive strength tests on beam ends are not interchangeable with those obtained using cylindrical specimens; correlations must be developed and utilized, as will be discussed.

Ideally, the test result obtained from compressive strength tests would be a direct indication of the concrete's ability to withstand a uniaxial compressive force. However, the stress state in compressive strength test specimens is much more complex than uniaxial compression. Friction between the bearing faces of the testing machine and the test specimen restrains the specimen laterally, thereby inducing lateral compression in the specimen ends [10,13]. The magnitude of this effect decreases with distance away from the specimen ends. Generally, at distances away from the specimen ends equal to the specimen diameter, the effects are negligible. Thus, for cylindrical or prismatic specimens having an aspect ratio of two, cross sections at midheight should be free of the effect of the end restraint. Details of this effect will be discussed in the following section.

FACTORS AFFECTING COMPRESSIVE STRENGTH

The purpose of standardizing test procedures is to provide guidelines for controlling test variables that could potentially affect strength results. In order to better understand the significance of the requirements of the test specifications, the effect of various factors on compressive strength test results will be discussed.

Effect of Specimen End Conditions

ASTM C 39 states that the ends of cylindrical specimens must depart from perpendicularity with the specimen axis by no more than 0.5° (approximately $1/8$ in. in 12 in. or 3 mm in 300 mm), and that the ends must be plane to within 0.002 in. (0.050 mm). The bearing surfaces of beam ends must also be plane to within 0.002 in. (0.050 mm), as specified by ASTM C 116. If these tolerances are not met by the uncapped specimen, it must be capped according to ASTM C 617. ASTM C 31 and C 192 on specimen preparation allow for depressions or projections on the finished surface of cylinders and beams of up to $1/8$ in. (3.2 mm). Formed surfaces of beams are to be smooth and plane such that the maximum variation from the nominal cross section shall not exceed $1/8$ in. (3.2 mm) for cross-sectional dimensions of 6 in. (152 mm) or more, or $1/16$ in. (1.6 mm) for smaller dimensions. ASTM C 42 and C 873, for cores and push-out cylinders, respectively, allow projections of up to 0.2 in. (5 mm) from the specimen ends and call for perpendicularity of the ends to the specimen axis of within 5° prior to capping. It is obvious that the restrictions on end conditions of cores and push-out cylinders prior to capping are much less strict than for molded cylinders.

The purpose of specifying end condition requirements of planeness and perpendicularity are to achieve a uniform transfer of load to the test specimen. Surface irregularities will lead to local concentrations of stress, as will nonperpendicular ends, even in specimens that are capped to meet the specified requirements [11]. The effect of cylinder end conditions prior to capping on strength test results has been reported by several authors [20–24]. In general, specimen ends that do not meet the specified requirements prior to capping cause lower strength test results, and the degree of the strength reduction increases for higher strength concretes.

ASTM C 617 covers procedures for capping the ends of cylindrical concrete specimens. Freshly molded cylinders may be capped with a neat cement paste that is allowed to harden with the concrete. It is important to keep the caps moist, since they are susceptible to drying shrinkage and possible cracking. Hardened cylinders and drilled cores may be capped with either high-strength gypsum plaster or sulfur mortar, 2-in. (50 mm) cubes of which yield a minimum of 5000 psi (34.5 MPa) in 2 h. The caps shall be plane to within 0.002 in. (0.05 mm), shall not depart from perpendicularity with the specimen axis by more than 0.5°, and shall not be off-center with respect to the specimen axis by more than $1/16$ in. (2 mm). The caps must be at least as strong as the concrete. Caps on hardened concrete specimens should be approximately $1/8$ in. (3 mm) thick, but no more than $5/16$ in. (8 mm) thick, and well bonded to the specimen end.

The use of sulfur mortar is the most common capping technique. Sulfur capping compounds must be kept molten at a temperature of approximately 265°F (130°C). These compounds have a range of fluidity, above and below which they become viscous and difficult to pour [11]. Further, volatilization of the sulfur occurs upon heating; thus, ASTM C 617 restricts the use of any material that has been used five times. Retrieval of capping material from specimen ends introduces oil and other contaminants that can reduce its fluidity and strength.

The requirements on capping materials and procedures are meant to eliminate possible detrimental effects of the caps on strength test results, since, depending on the capping method used, either lateral compression or tension may be introduced into the specimen end due to differences in lateral deformation of the capping material and the concrete under load [25]. Many capping procedures have been proposed and much research conducted to study the effect of different capping materials and methods on compressive strength test results [21–24,26,27]; the result of these studies has been the development of ASTM C 617 as it currently exists. In general, the research indicates that strength test results are highly dependent on properties of specimen capping. The caps must be strong and intact. Weak caps or those that are allowed to expand laterally under load cause reductions in apparent specimen strength; chips removed from the cap edges also result in reduced strength. These effects are more pronounced as the level of concrete strength being tested increases. Similar capping compounds from different manufacturers may have different strength gain properties, and must be tested for compliance with the strength requirements in ASTM C 617. Specimens tested with thick caps generally produce lower strength test results when compared to specimens capped with thinner caps of the same material.

Currently, the use of an additional capping method is being widely investigated [9,20,28–31]; indeed, a task group within ASTM Subcommittee C09.03.01 is examining its use as a proposed standard capping method. This proposed system consists of an elastomeric pad confined within a metal retaining ring. The purpose of the pad insert is to conform under load to the specimen surface, thereby distributing the applied load uniformly. The use of the retaining ring is essential in restricting lateral flow of the pad that would otherwise induce lateral tension in the specimen ends, thereby reducing the apparent compressive strength. Satisfactory results have been obtained comparing this system to sulfur mortar caps on concretes having strengths of up to 10 000 psi (69 MPa). However, such factors as pad durometer, retaining ring diameter, and allowable number of pad reuses are still being studied.

Effect of Specimen Size

The standard field molded cylinder size is 6 in. (152 mm) in diameter with a length-to-diameter ratio of two. However, smaller sizes are acceptable under current standards, and there is increasing interest in their use. Smaller specimens require less material to make and are much easier to handle: a standard concrete cylinder weighs 30 lb (13.6 kg), compared to 9 lb (4 kg) for a 4 by 8-in. (102 by 203-mm) cylinder. The size of drilled cores is often determined by the dimensions of the concrete member and the coarse aggregate size. In addition to these factors, the use of concretes having increasingly higher strengths in construction requires that testing machines have higher load capacities; smaller diameter specimens fail at lower ultimate loads. In cases of mass concrete placements, such as dams, the use of very large aggregates requires the use of larger diameter specimens in order to maintain a minimum diameter-to-aggregate ratio of 3 to 1, or wet-sieving to remove larger aggregate sizes.

It is commonly accepted that as specimen size increases, both the concrete's apparent strength and the variation in test results decrease [5,10,13,32–37]. The magnitude of the size effect decreases as the specimens get progressively larger. This effect is shown in Fig. 1. The reasoning behind the size effect is that the strength of a concrete specimen will be governed by the weakest part of that specimen, and that the probability of the occurrence of a critical flaw increases as specimen size increases. Further, it has been reported that the strength difference due to specimen size increases as the concrete strength increases [33,37]. Drilled cores will follow the same trend of increasing strength with decreasing specimen size for larger diameter specimens. For small diameter cores, however, the ratio of cut surface to specimen volume becomes significant, and it is possible that coring damage will cause strength reduction for decreasing diameters below 4 in. (102 mm) [14].

Although the testing of smaller specimens is more convenient, precision of strength determination should not be sacrificed. Equal precision can be obtained with smaller

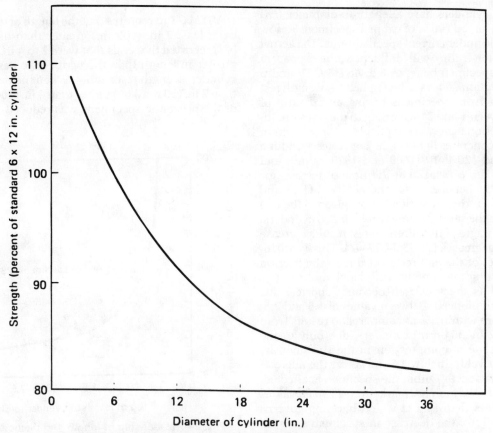

FIG. 1—Effect of cylinder size on compressive strength of concrete [5]. (From W. H. Price, *Journal of the American Concrete Institute*, Vol. 47, No. 6, 1951, pp. 417–432).

specimens if the number of specimens tested is increased. Correlations between specimens of diameters other than 6 in. (152 mm) and the standard cylinders can be made.

Effect of Diameter-to-Aggregate Ratio

Current specifications for molded specimens and push-out cylinders require that the minimum specimen dimension be at least three times the maximum nominal aggregate size. Gonnerman [34] reported that test results were satisfactory for specimens when the diameter-to-aggregate ratio was only 2; however, he reported difficulty in compacting the specimens so that they were homogeneous, and thus recommended the minimum ratio of 3 to 1. For molded specimens, larger-sized aggregates may be removed by hand picking or by wet sieving so that smaller specimen dimensions may be used. However, it has been reported that the practice of removing larger aggregate sizes from concrete will result in higher compressive strengths [38]. For drilled cores, the preferable condition is that the core diameter is at least three times the maximum nominal aggregate size used in the concrete placement. This condition may be relaxed as long as the core diameter is at least twice the maximum size of coarse aggregate in the core. The accuracy of the strength test results has been found to decrease as the diameter-to-aggregate ratio decreases [14].

Effect of Aspect Ratio

Standard test cylinders have a length-to-diameter (l:d) ratio of 2. However, l:d ratios of capped specimens as low as 1 are allowable under current specifications. The actual length-to-diameter ratio will influence the apparent strength of the specimen being tested. ASTM C 39 and C 42 provide correction factors to be applied to strength test results obtained from specimens having an l:d ratio of some value between 1 and 2. According to these test methods, these correction factors are applicable to specimens of normal-weight concrete, lightweight concretes weighing between 100 and 120 lb/ft^3 (1600 and 1920 kg/m^3), and concrete that is dry or soaked at the time of testing, for concrete strengths between 2000 and 6000 psi (13.8 and 41.4 MPa). Further, the correction factors given in the two test methods are the same for cores and molded cylinders.

It is commonly agreed that strength test results increase as the l:d ratio decreases [10,13,34,39–41]. This is attributed to the effect of the end restraint due to the friction between the testing machine bearing block and the test specimen ends. As the specimen becomes shorter, this effect is more pronounced. Between values of 1.5 and 2.5, l:d corrections are within 5% as compared to results from 6 by 12-in. (152 by 304-mm) cylinders, due only to the introduction of lateral compression in the specimen by end restraint. At values below 1.5, however, the angle at which failure may occur within the specimen is affected, and corrections for l:d increase markedly [41]. At l:d values above 2.5, but less than that at which buckling governs the specimen failure, the decrease in specimen strength with increasing l:d is due to the increasing volume of concrete that is free of the end effect, and so the correction factor approaches a limit for large l:d ratios. A typical relationship between strength correction factor and l:d ratio is shown in Fig. 2.

It has been reported that higher strength concretes require less correction for l:d ratio than lower strength concretes [10,13,40,41], and also that lightweight concretes are less affected by l:d ratio than are normal-weight concretes [40]. Further, cores have been shown to require greater corrections than molded cylinders of equal diameter, especially at low l:d values. For specimens having an l:d ratio of less than equality, strength correction factors have been shown to be on the order of 0.60 for length-to-diameter ratios of 0.50 [34,39,40].

In testing beam ends according to ASTM C 116, no correction for height-to-depth (h:d) ratio is given. However, examination of Gonnerman's data indicates that the effect of h:d in prismatic specimens is similar to that of l:d in cylindrical specimens. Beams having an h:d of two yielded strengths of between 70 and 90% of those obtained from cubes (h:d equal to 1) of equal cross section. It should be noted that the effect of h:d on strength of prismatic specimens is not as pronounced as for cylindrical specimens. If this test method is to be used, correlation for a particular specimen geometry and concrete mixture should be developed relative to the standard specimen used.

Effect of Length of Broken Beam Ends

ASTM C 116 requires that the length of the test specimen be at least 2 in. (50 mm) greater than its width. Mather [42] reported that ends that were 2 to 4 in. (50 to 100 mm) greater in length than the width gave best correlation to cylinder strength test results. When the overhang was less than 2 in. (50 mm) or greater than 4 in. (100 mm), strength of the prismatic specimens was reduced.

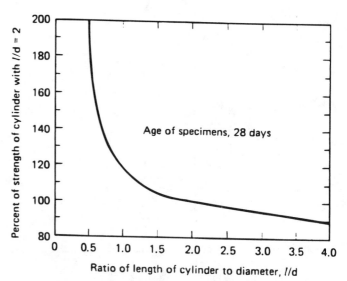

FIG. 2—Relationship of length and diameter of specimen to compressive strength [5]. (From W. H. Price, *Journal of the American Concrete Institute*, Vol. 47, No. 6, 1951, pp. 417–432).

Effect of Specimen Moisture Condition and Temperature

The moisture condition of the specimen at the time of testing can have a significant influence on strength test results. In general, specimens yield lower compressive strengths when tested in a moist condition than they would if tested in a dry state [6,10,13,14,16,17,43]. It has been reported that reductions in the range of 5 to 20% may occur. In fully dry specimens, this has been attributed to their decreased volume that results in increased strength of secondary bonds within the paste structure. In drying specimens, the specimen's outer surface attempts to shrink when drying, thereby inducing lateral compression in the specimen interior, which increases its apparent compressive strength [43]. A wetting or fully saturated condition will have the opposite effects on compressive strength.

When choosing a moisture condition for testing, the purpose of the test must be considered, as well as the effect that the moisture condition will have on the test results. For example, in core testing, if results are to be compared to standard cylinder test results, then a moist condition may be preferred; however, if determination of the in situ strength of the concrete is desired, then the specimen should be tested as closely as possible in the moisture condition that exists in the structure.

The temperature of the test specimens at the time of testing has also been reported to affect the test results. Specimens tested at elevated temperatures produce lower strengths than those tested at lower temperatures [13]. However, for ambient temperatures normally encountered, the effect is negligible.

Effect of Loading Direction versus Casting Direction

Molded concrete cylinders are tested parallel to their casting direction. Beams and drilled cores, however, may be tested either parallel or perpendicular to the casting direction, depending on the circumstances involved. In general, specimens tested in the same direction as cast will yield higher strengths than those tested perpendicular to it. This is attributed to the occurrence of flaws aligned perpendicular to the casting direction due to water gain under coarse aggregate particles, as shown in Fig. 3. Cores tested parallel to the casting direction have been reported to produce higher strengths than those tested perpendicular to the casting direction by approximately 8% [13]. Modified cubes tested perpendicular to the casting direction are reported to lose roughly 12% when compared to strengths from specimens tested parallel to the casting direction [44].

Effect of Testing Machine Properties

ASTM C 39 gives required features for the loading apparatus to be used in compressive strength testing. Among these are the capacity for smooth and continuous load application; accurate load measurement; and two bearing blocks, one being spherically seated and one being solid, both of which must satisfy further requirements of surface planeness and hardness, minimum and maximum diameters, etc. Failure to meet these requirements has been shown to reduce strength test results [12,23].

Small deflections in the bearing surfaces of the machine can lead to strength reductions. The spherically seated

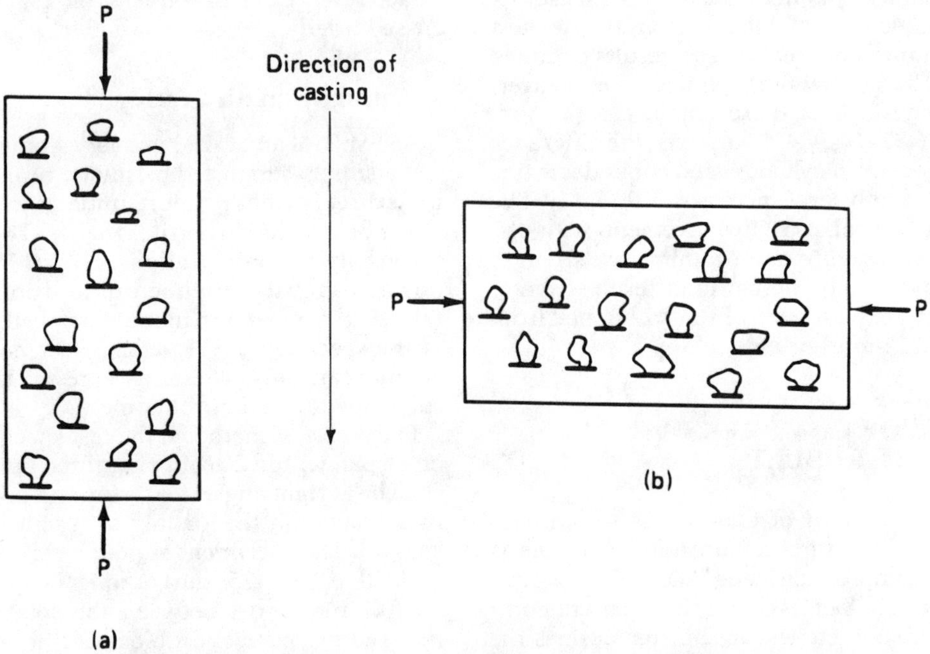

FIG. 3—Planes of weakness due to bleeding: (a) axis of specimen vertical and (b) axis of specimen horizontal [10]. (Mindess/Young, *Concrete*, © 1981, p. 439 represented by permission of Prentice-Hall, Inc., Englewood Cliffs, NJ).

bearing block must be free to rotate so that the load is transferred uniformly to the specimen end; strength reductions of up to 20% have been observed with no spherical seating [23]. Placing the specimen off-center with respect to the loading axis by only 1/2 in. (13 mm) can cause strength reductions of 10%. Also, distortion of the testing machine frame under load leads to nonuniform loading of the specimen, and possible strength reduction.

Effect of Loading Rate

The loading rates required by ASTM C 39 and C 116 are, for screw-type machines, approximately 0.05 in. (1.3 mm)/min running idle, or for hydraulically operated machines, between 20 and 50 psi/s (0.14 and 0.34 MPa/s). For cylindrical specimens, the first half of the load may be applied at a more rapid rate. The loading rate may not be adjusted during specimen failure.

The apparent strength of concrete specimens increases as the rate of load application increases [5,10,12,13,36,45,46]. Abrams reported that higher strength concretes were more affected by loading rate, and that the ultimate strength was unaffected when up to 88% of the ultimate was applied at a rapid rate [46]. The dependence of ultimate strength on loading rate is thought to be related to mechanisms of creep and microcracking [10,13,36]. This would appear to be in agreement with the observation that when subjected to a sustained load of approximately 75% of its ultimate capacity obtained using ASTM C 39, concrete will eventually fail with no further load application.

Effect of Specimen Geometry

Currently, compressive strength of concrete can be determined through testing of either cylinders or beams. Test results are significantly affected by specimen geometry, however, and thus results obtained from the two test procedures are not interchangeable. Test results obtained from 6 by 12-in. (152 by 304-mm) cylinders are, on average, between 75 and 85% of those obtained from 6-in. (152-mm) cubes [10,12–14,34,47]. However, the difference in strengths obtained from cylinders and cubes decreases as the concrete strength level increases [10,13,47]. The increased strength of cubes is in agreement with the expected effect of end conditions on compressive strength test results and aspect ratio. Strength test results from 6-in. (152-mm) cubes are within 6% of those obtained from 6 by 6-in. (152 by 152-mm) cylinders [34].

SIGNIFICANCE OF COMPRESSIVE STRENGTH TEST RESULTS

The compressive strength of concrete is of primary importance since in structural applications, concrete is often used to resist compression loads, such as in columns. It is obvious that many factors affect the apparent compressive strength of concrete test specimens. Besides the variables that have been discussed, there are also differences between the conditions that exist in concrete within the structure and those within the test specimen. These include restraint, loading conditions, and long-term effects such as creep and shrinkage. However, an understanding of what factors may affect the strength test results facilitates the decisions as to what testing conditions are to be specified. In addition, when member dimensions or concrete mix characteristics dictate specimen geometry, knowledge as to the effects of these variables allow meaningful interpretation and application of the strength test results obtained.

TENSILE STRENGTH TEST PROCEDURES

There are currently no standardized test procedures for determining directly the tensile strength of concrete. This is due to the difficulty involved in inducing pure axial tension within a specimen without introducing localized stress zones. However, knowledge of the tensile capacity of concrete is necessary in that it is the tensile strength of concrete that will determine its resistance to cracking. Therefore, several test procedures have been developed to indicate indirectly the tensile capacity of concrete. These include three standard test procedures: ASTM Test Method for Flexural Strength of Concrete (Using Simple Beam with Third-Point Loading) (C 78), ASTM Test Method for Flexural Strength of Concrete (Using Simple Beam with Center-Point Loading) (C 293), and ASTM Test Method for Splitting Tensile Strength of Cylindrical Concrete Specimens (C 496). An additional indirect tension test has been proposed that is known as the double-punch test. This test is similar in principal to the splitting tension test, and will be discussed briefly. Further, a direct tension test for concrete is being examined within ASTM Subcommittee C09.03.01. This proposed test method is an adaptation of ASTM Test Method for Direct Tensile Strength of Intact Rock Core Specimens (D 2936) and will be discussed briefly.

Flexural Strength Testing

ASTM C 78 and C 293 are very similar in that they both test a simply supported prismatic beam specimen loaded in flexure. For both, failure initiates in the extreme fibers of the beams due to tensile stresses. They differ, however, in the loading configuration. ASTM C 78 specifies application of load to the specimen at third points along the span, resulting in maximum moment applied to the middle third of the specimen. ASTM C 293 specifies load application at midspan only, which as a result is the only section subjected to maximum moment.

For both test methods, the test specimen is to be tested on a span within 2% of being three times its depth. Sides shall be at right angles to the top and bottom, and surfaces in contact with the loading or reaction blocks should be smooth. The specimens should be tested on their sides as molded. If 1-in. (25-mm) or longer gaps in excess of 0.004 in. (0.1 mm) exist between the specimen surfaces and the loading or reaction blocks at no load, the specimen surfaces should be ground or capped, or, if the specimen surfaces are within 0.015 in. (0.38 mm) of being plane, leather shims may be used. The loading apparatus must apply load to the specimen perpendicular to its face without eccentricity, and reactions must be parallel to the direction of load application. Further requirements are specified for the testing apparatus to ensure that sufficient flexibility of the supports and loading blocks exists to maintain uniform distribution of load over the specimen width as tested. The first half of the load may be applied rapidly; afterwards, the loading rate should be between 125 and 175 psi/min (861 and 1207 kPa/min) in the extreme fiber until failure.

Splitting Tensile Strength

ASTM C 496 gives requirements for the testing apparatus, test specimen geometry, and load application for determining splitting tensile strength of cylindrical test specimens. The testing machine should meet the requirements of ASTM C 39. The specimen is placed on its side and subjected to a diametral compressive force along its length. If either the upper or lower bearing block of the testing machine is shorter than the cylinder, a bearing bar or plate shall be used that is at least as thick as the distance from the edge of the machine bearing block to the end of the cylinder, at least 2 in. (51 mm) wide, and within 0.001 in. (0.025 mm) of planeness. Bearing strips at least as long as the cylinder and $1/8$ in. (3.2 mm) thick and 1 in. (25 mm) wide shall be placed between the specimen and the loading faces. The load should be applied at a rate of between 100 and 200 psi/min (689 and 1380 kPa/min) splitting tensile stress.

Failure of these specimens occurs along a vertical plane containing the specimen axis and the applied load. The load configuration of this test method induces tensile stresses along the failure plane over approximately two-thirds to three-fourths of the specimen diameter. The regions of the specimen in the vicinity of the loading strips are subjected to large compressive forces. However, tensile rather than compressive failure occurs because the stress state at the loading strips is triaxial compression, allowing the concrete to resist higher compressive stresses [48].

Double-Punch Test

First reported in 1970, the double-punch test is similar to the split-cylinder test in that it induces tension on planes containing the specimen axis through the application of a compressive force to the cylinder [49]. However, in the double-punch test, the load is applied to the cylinder ends by means of cylindrical bearing blocks having a diameter smaller than that of the concrete specimen. This allows failure to occur on any plane containing the cylinder axis, as opposed to predetermining the plane of failure as in the split cylinder test [50,51]. To the author's knowledge, the double punch test is not being widely used; nor is it a standard ASTM test method.

Direct Tension Test

The proposed test method for direct tensile strength of concrete or mortar specimens is an adaptation of an existing standard test method for rock cores, ASTM D 2936. In it, cylindrical test specimens are loaded in axial tension through the use of metal caps adhered to the specimen ends. The direct tensile strength is determined based on the applied axial load and the specimen cross-sectional area. Specimen tolerances are prescribed to facilitate the application of a direct tensile load. To the author's knowledge, this test procedure is not widely used for concrete specimens.

FACTORS AFFECTING FLEXURAL STRENGTH

Derivations of the formulas included in ASTM C 78 and C 293 include several assumptions that do not hold true for the testing of concrete beams to failure [52]. One such assumption is that the concrete behaves linearly throughout the test, which is obviously not true, especially at stresses approaching failure. The equations apply to long, thin beams, whereas the actual test specimens are short and deep. The failure stress calculated using the two test methods is slightly higher than the actual extreme fiber stress due to the simplifying assumption that the stress distribution over the depth of the beam is linear. It is likely, however, that compared to the variability inherent to concrete strength, these effects are not significant. Various other factors have been found to affect flexural strength test results obtained using either third-point or center-point loading. These will be discussed.

Effect of Specimen Dimensions

ASTM C 78 and C 293 call for the test specimen to have a span of three times its as-tested depth. While the standard beam dimensions are 6 in. by 6 in. by 20 in. (152 mm by 152 mm by 507 mm), tested on an 18-in. (456-mm) span, other beam dimensions may be used depending on maximum coarse aggregate size. As in compressive strength specimens, the minimum specimen dimension must be at least three times the nominal maximum coarse aggregate size, and the ratio of width-to-depth as-molded must not exceed 1.5. For specimens prepared in the laboratory, larger coarse aggregate sizes may be removed by hand-picking or wet-sieving.

For a constant beam width and test span, the apparent flexural strength of specimens tested in third-point loading decreases as the depth of the beam increases; however, test results seem to be independent of beam width for a given depth and span [53]. For a constant beam cross section, the effect of test span is unclear. Kellerman [54] reported that strength decreased for both center-point and third-point loading as test span increased. Reagel and Willis [53], on the other hand, reported no effect of test span on third-point loading.

Effect of Specimen Size

It is commonly agreed that as the size of the test specimen increases, the flexural strength test results decrease [53–56]. This has been found to be true for both center-point and third-point loading, and is due to the size effect as was discussed for compressive strength test specimens. Uniformity of test results increases with specimen size [56].

Effect of Coarse Aggregate Size

The flexural strength of concrete beams is higher for smaller coarse aggregate sizes. This was found to be true even when the use of the smaller coarse aggregate resulted in a mix with a higher water-cement ratio [54]. This effect should be considered when removing larger aggregate sizes from concrete by hand-picking or wet-sieving. Uniformity of test results is higher with smaller coarse aggregate size.

Effect of Loading Rate

As with compression test specimens, the apparent flexural strength of specimens increases with loading rate [55]. A linear relationship between the flexural strength and the logarithm of the rate of increase of stress has been established.

Effect of Moisture Condition and Temperature

Flexural strength test results are very sensitive to the specimen moisture condition at testing. When specimens are tested in a drying condition, the apparent flexural strength is lower than when tested in a saturated condition [10,13,57,58]. Losses of up to 33% have been reported [57]. When the surface of the specimen is allowed to dry rapidly, it attempts to shrink, but this shrinkage is restrained by the specimen core; if the induced tensile stresses are greater than the tensile strength of the concrete, cracks will develop in the outer surface of the specimen. In flexural strength testing, it is the stress in the extreme fibers of the specimen that causes failure. When cracks exist due to drying, they act both as stress concentrators and they reduce the effective cross section of the test specimen. In addition, the tensile stresses in the specimen skin due to drying act as a preload condition; that is, the tensile stresses due to drying and to applied test load are cumulative, thereby resulting in a lower observed load at failure. The drying condition discussed occurs very rapidly under normal ambient conditions if sufficient care is not taken to keep the specimen surfaces moist. Due to the significant effect of moisture condition on the test results obtained, if flexural strength specimens are used to indicate the strength of the concrete in-place, they should be cured under conditions similar to the concrete structure, but tested in a saturated condition.

As the temperature of the specimen at the time of testing increases, the apparent flexural strength decreases. It is worth noting that the effect of drying on flexural strength test results is opposite to that for compressive strength test results.

Effect of Center-Point versus Third-Point Loading

As was discussed previously, the main difference between ASTM C 78 and C 293 is the location of load application. In center-point loading, the load is applied at the specimen midspan. In third-point loading, the load is applied at third-points along the test span. For the latter case, the middle third of the beam span is subjected to maximum moment, and thus maximum extreme fiber stress. In center-point loading, however, only the cross section at midspan is subjected to maximum moment and extreme fiber stress. As a result, for a given beam size, test results obtained from center-point loading are higher than those obtained from third-point loading [54–56]; differences of 15% are not unusual. Uniformity of test results is also generally higher for third-point loading.

It is sometimes observed that fracture of test specimens tested in center-point loading occurs at a location other than midspan. Since the moment distribution along the beam span is linear from zero at the support to its maximum at midspan, fracture at a location other than midspan corresponds to a lower extreme fiber stress than exists at midspan. Thus, when the failure stress is calculated at the point of fracture, the flexural strength of the specimen is lower than when calculated at midspan [54,56]. However, ASTM C 293 does not require making note of the location of fracture. ASTM C 78, on the other hand, requires that if fracture occurs outside of the maximum moment region by a distance of within 5% of the span length, a formula is to be used to calculate the flexural strength that takes into account the location of the fracture plane. If fracture occurs outside of the maximum moment region by a distance of more than 5% of the span length, the test results are to be discarded.

FACTORS AFFECTING SPLITTING TENSILE STRENGTH

In theory, the application of a line load perpendicular to the axis of a cylinder and in a diametral plane produces a uniform tensile stress over that plane. However, the actual case of testing a concrete cylinder departs from the theoretical case in several ways [48]. First, the theory applies to a homogeneous material, which concrete is not. Secondly, concrete is not linear elastic as is assumed in the analysis. Third, the load is not applied in a line, but rather is distributed, which results in large compressive stresses in the surface of the specimen under the loading strips. Still, the splitting tensile strength test is reasonably easy to conduct, and results are good for comparative values, even if absolute tensile strength values are not obtained. Several factors found to affect splitting tensile strength test results will be discussed.

Effect of Specimen Dimensions and Size

The length of a cylinder for a given diameter does not seem to affect test results, other than possibly producing more uniformity of results for longer specimens [48]. However, cylinders having a diameter of 4 in. (102 mm) were observed to yield splitting tensile strengths that were roughly 10% higher than those obtained from cylinders having 6-in. (152-mm) diameters [48,59]. Uniformity of test results increases with specimen diameter as well.

Effect of Bearing Strips

ASTM C 496 requires the use of bearing strips made of $1/8$-in. (3.2-mm) thick plywood, 1-in. (25-mm) wide, and at least as long as the cylinder. The purpose of these strips is to conform to the specimen surface and transfer the load uniformly from the loading apparatus. Increasing the thickness of the bearing strips may cause strength reductions [48]. Steel bearing strips have been shown to cause significant strength reductions, probably due to their inability to conform to the specimen surface.

Effect of Specimen Moisture Condition

It is not expected that drying of the specimen surface will affect the apparent splitting tensile strength as significantly as it does the flexural strength, since the specimen surface contained within the failure plane is subjected to high triaxial compressive stresses. Indeed, it is possible that the effect is more similar to that in compressive strength cylinders, where the restrained shrinkage of the outer surface induces compression in the specimen interior.

Effect of Loading Rate

As with strength testing of compressive strength specimens and flexural beams, higher splitting tensile strengths are obtained when the specimens are loaded at a more rapid rate.

FACTORS AFFECTING THE DOUBLE-PUNCH TEST

In the double-punch test, vertical diametral planes are subjected to tension much as in the splitting tension test. Results of this test are lower than those obtained from the splitting tension test, however, since there are many possible planes of failure rather than one, and the probability of a weaker plane is higher [50,51]. It has been reported that results of the double-punch test are dependent on specimen size, higher results being obtained as the specimen size decreased, and that the results obtained from 3 by 3-in. (76 by 76-mm) double-punch specimens are within 10% of those obtained from the split cylinder test [50]. It is possible that this test procedure would be more sensitive to moisture condition than the splitting tension test since the failure plane could be subjected to tensile stress at the specimen surface.

RELATIONSHIP OF FLEXURAL TO SPLITTING TENSILE STRENGTH

Since in the flexural strength tests, fracture is determined by the strength of a portion of the tension surface of the beam, and in the splitting tension test failure can be initiated anywhere in a failure plane, it is expected that results from splitting tension tests would be lower than those obtained from flexural tests. This has been shown to be the case, with an average ratio of splitting tensile strength to flexural beam strength, center-point loading, of 0.65 [59].

SIGNIFICANCE OF TENSILE STRENGTH TEST RESULTS

There is currently no standard test method for determining the tensile strength of concrete. However, an indirect indication is obtained through the use of three standardized test methods. Results from each test method are particular to that method, and cannot be used interchangeably. The flexural strength test is more similar to the loading encountered in pavements than is the splitting tension test. However, it is likely that the former is more sensitive to moisture conditions, and thus more care must be taken to keep the sample surface from drying. The splitting tension test subjects a plane to tensile forces, and thus more closely resembles a direct tension test. When choice of a concrete tension test is being made, the loading of the concrete in the structure should be considered, as well as the expected use of the test results when obtained. If the result is to be used in design calculations, the proper test must be performed depending on the development of the design formulas. Even so, the results obtained are not true tensile strengths, and may be more or less sensitive to factors that affect tensile strength. In addition, similar to compressive strength test limitations, conditions in the structure are much more complex than can be modeled in strength tests on small samples.

STRENGTH RELATIONSHIPS

Due to the convenience of performing one strength test, most often the compressive strength of concrete cylinders, strength relationships have been developed so that other strength properties may be inferred from results of one test type. The relationship of compressive to tensile strength has been found to be influenced by many factors, including concrete strength level, coarse aggregate properties, testing age, curing, air entrainment or voids, and the type of tension test performed [13]. General relationships between the compressive strength of concrete cylinders and the various tension testing methods will be discussed as found in the literature [5,6,10,13,36,47,60].

The ratio of direct tension to compressive strength of concrete ranges between 0.07 and 0.11. All of the indirect tension tests overestimate the tensile capacity of the concrete. The ratio of splitting tensile strength to compressive strength falls between the values of 0.08 and 0.14. Flexural strength test results obtained by third-point loading are higher yet than splitting tensile strength, and the ratio of these results to compressive strength has been found to lie between 0.11 and 0.23. The ratios for center-point loading flexural strength tests to compressive strength would be higher still.

SHEAR AND TORSIONAL STRENGTH OF CONCRETE

The case of pure shear in concrete structures is seldom if ever encountered. Still, shear is an important factor in the failure of concrete, and so will be discussed briefly.

In tests performed to determine directly the shear strength of concrete, a wide range of values have been reported due to the difficulty of producing pure shear. In applying torsion to a hollow concrete beam, failure occurs in tension due to the large tensile stresses developed on other planes within the concrete specimen, rather than due to shear stresses. However, when combined stress tests are conducted and the results plotted to determine a Mohr failure envelope, as will be discussed in the next section, the value of concrete shear strength has been determined to be roughly 20% of the uniaxial compressive strength [10].

CONCRETE UNDER COMBINED STATES OF STRESS

In structures, concrete members are most often subjected to combinations of compression, tension, and shear stresses. Thus, it is important to predict the behavior of concrete to different stress states. ASTM Test Method for Determining the Mechanical Properties of Hardened Concrete Under Triaxial Loads (C 801) provides a standardized method for determining such behavior. According to the test method, right cylindrical test specimens are subjected to triaxial stress states such that two of the three principal stresses are equal. The test method specifies specimen and platen surface conditions, loading method, and data reduction techniques, among other requirements.

In general, the most common use of data obtained from application of combined stresses is in the construction of a Mohr failure envelope. The use of ASTM C 801 allows only the portion of the envelope to be developed for concrete under compressive stresses. However, the shear strength of the concrete can be estimated from this procedure by drawing a tangent to the various Mohr circles obtained. A more complete description of the behavior of concrete under the action of combined stresses would include the application of tensile stresses, as shown in Fig. 4.

In general, the state of biaxial stress is assumed to be the same as triaxial stress, with emphasis placed on the minimum and maximum applied stresses and the effects of the intermediate stress value being neglected. It is well established that the compressive strength capacity of concrete subjected to biaxial compression is higher than its uniaxial compressive strength, and that the uniaxial tensile strength is roughly unaffected by the application of biaxial tensile stresses [10,13], as shown by the failure envelope in Fig. 5. In this figure, stress combinations within the envelope correspond to no failure. In research reported to date, it is apparent that end conditions had a significant effect on test results, resulting in a ratio of biaxial to uniaxial compressive strength being reported in the range of 1.10 to 3.50 [62]. Later research conducted using brush platens seems to have removed the effect of end conditions to a great degree. As determined with these end conditions, an average increase in compressive strength capacity with biaxial compressive loading is 30%, and occurs at a principal stress ratio of approximately 0.5 [62].

Higher strength concretes are seeing increasing use in construction. In tests performed on high-strength concretes subjected to biaxial compression, it was reported that the increase in compressive capacity under a biaxial stress state depended on the type of coarse aggregate used in the concrete [63]. Like normal-strength concrete, average strength increases of roughly 30% were observed over the uniaxial compressive strength, ranging from a minimum of 26% for trap rock coarse aggregate to a maximum

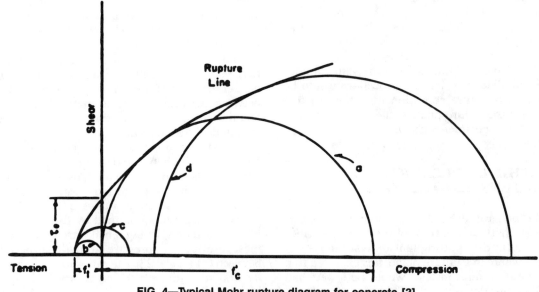

FIG. 4—Typical Mohr rupture diagram for concrete [2].

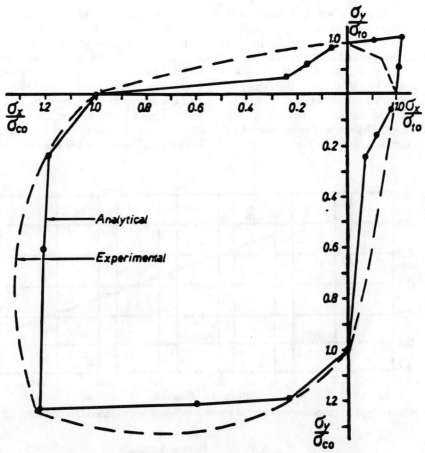

FIG. 5—Failure envelope for normal-strength concrete [61].

of 34% for granite coarse aggregate. In addition, it was noted that the proportional limit for high-strength concrete as well as its discontinuity point were at higher stress-strength ratios than for normal-strength concrete, and increased with increasing biaxial stress ratio. The discontinuity point was defined as the point at which unstable progressive crack growth commenced. The failure modes of the specimens indicated that fracture was determined by a limiting tensile strain.

Calixto [64] performed biaxial tension-compression tests on high-strength concrete specimens. In general, the application of lateral tension caused a significant decrease in the compression capacity of the concrete, whereas the application of lateral compression had a much less pronounced effect on the tensile capacity. Further, it was suggested that high-strength concrete can be subjected to a higher percentage of its ultimate strength than can normal-strength concrete without eventual failure due to unstable crack growth.

The increased capacity of concrete under a biaxial compressive stress state is significant in that the lateral restraint provided in structures through such means as steel spirals will enhance the performance of the concrete. However, lateral tension causes a significant decrease in concrete's compressive strength capacity. The higher stress ratios required to initiate progressive, unstable crack growth in higher strength concretes would seem to provide an additional safety margin in these concretes when subjected to the same maximum stress ratios as apply to normal-strength concrete.

FATIGUE STRENGTH OF CONCRETE

In numerous structural applications, such as bridge decks, concrete members are subjected to repeated applications of load at a level below the ultimate strength capacity of the concrete. Concrete has been shown to exhibit fatigue behavior; that is, when subjected to cyclic loading of a given level but below its ultimate capacity, it will eventually fail. Thus, for structural design purposes, it is desirable to ascertain the response of concrete to fatigue loading. However, a standardized test method is not available.

Concrete is not thought to exhibit a fatigue limit; that is, a maximum ratio of applied stress to ultimate strength below which the concrete can withstand an infinite number of loading cycles without failure [10,13]. Therefore, in the case of concrete, the fatigue limit is referred to as the stress ratio at which failure occurs only after a given number of cycles, usually 10^7. In normal-strength concrete, the fatigue limit thus defined is roughly 55% of its ultimate strength.

The fatigue strength of concrete is sensitive to several factors [65], most associated with the details of the applied loading. The number of cycles of loading to failure depends on the level of stress applied to the specimen; as the stress ratio is increased, the cycles to failure decrease, as shown in the S-N diagram in Fig. 6. Also, the range of maximum to minimum applied stress has an effect. As the difference in the two stresses is increased, the number of cycles to failure decreases. This is best represented in a modified Goodman diagram, shown in Fig. 7.

The sequence of loading cycles is significant; Miners rule, that the effects are cumulative, does not strictly apply. Rather, the fatigue life of a specimen is different if subjected first to high stress ratios followed by low stress ratios, than if subjected to the reverse sequence. Cycling below the fatigue limit increases both the fatigue strength and the static strength by 5 to 15% [10,13,66]. When the frequency of load application is very slow, the fatigue life is shortened as compared to a more rapid frequency of load application. This is probably due to mechanisms of

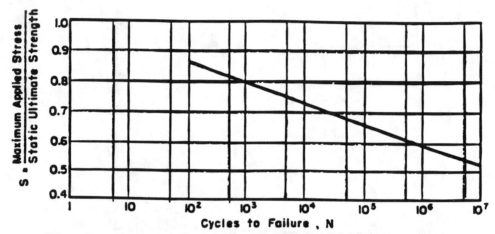

FIG. 6—Typical fatigue curve for concrete subjected to repeated flexural load [2].

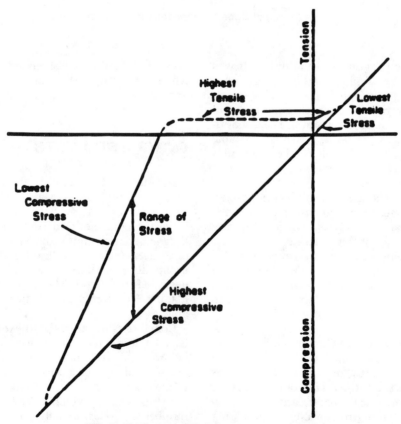

FIG. 7—Typical modified Goodman diagram [2].

creep and crack propagation, and the effect disappears at moderate loading rates. Rest periods of up to 5 min during fatigue testing also increase the fatigue life of the specimen; increasing the length of the rest period beyond 5 min has no additional beneficial effect.

Specimens subjected to a stress gradient also exhibit enhanced fatigue life over those subjected to a uniform stress distribution, possibly because the lower stressed areas of the specimen inhibit crack growth. Finally, specimens tested in a moist condition exhibit premature failure when compared to those tested in a dry condition.

In biaxial compression fatigue tests on high-strength concrete, Nelson [65] reported that a biaxial compressive stress state decreased the concrete fatigue strength for maximum stress levels below 76% of the ultimate strength; the fatigue limit decreased with increasing biaxial stress ratio. Similar to normal strength concrete, the fatigue life of high-strength concrete increases with decreasing maximum stress ratio and stress range, and wet specimens fail very rapidly compared to specimens tested dry. Cycling of the specimens below the fatigue limit resulted in increases in the specimen uniaxial static strength on the order of 40%. However, whereas normal-strength concrete is reported to have a fatigue limit of approximately 55% of its ultimate strength, the corresponding value for high-strength concrete was found to be in the range of only 47 to 52%.

Structures such as bridge decks and pavements undergo numerous load cycles under normal service conditions. Thus, it is important to study the response of concrete to this type of loading. Currently, there is little published information in this area, especially for biaxial stress states and high-strength concrete. Further, since factors such as the magnitude of load applied, the stress state, its frequency, duration of rest periods, and concrete moisture condition are largely unknown under the actual service conditions of the structure, information obtained from laboratory studies can only give basic information about a very complex phenomenon.

SUMMARY

Throughout the discussion of concrete strength testing contained within this chapter, the limitations of the test results obtained from small companion specimens in predicting any property of the concrete in the structure has been stressed. However, these tests are the best we have, and with thorough knowledge of their limitations, results obtained from correctly performed test procedures will supply useful information. It is essential that the purpose of the test be defined, and that those factors that will influence the test results be identified and controlled. The conditions that exist in a concrete structure are very complex. The current practice is to design a structure with specified concrete strengths, and to prepare and test companion specimens for strength to determine the concrete's adequacy for use in the structure, following ASTM C 31 and C 39. While not exact, it has been proven to work. However, due to the pressures of economics, new technology is constantly being introduced that could affect the adequacy of this process. Thus, the knowledge of concrete behavior and the development of improved test methods must keep pace with new technology.

REFERENCES

[1] Kesler, C. E. and Seiss, C. P., "Static and Fatigue Strength," *Significance of Tests and Properties of Concrete and Concrete Aggregates, ASTM STP 169*, American Society for Testing Materials, Philadelphia, 1956, pp. 81–93.

[2] Kesler, C. E., "Strength," *Significance of Tests and Properties of Concrete and Concrete-Making Materials, ASTM STP 169A*, American Society for Testing and Materials, Philadelphia, 1966, pp. 144–159.

[3] Derucher, K. N., "Strength," *Significance of Tests and Properties of Concrete and Concrete-Making Materials, ASTM STP 169B*, American Society for Testing and Materials, Philadelphia, 1978, pp. 146–161.

[4] Bloem, D. L., "Concrete Strength in Structures," *Proceedings*, Vol. 65, American Concrete Institute, 1968, pp. 176–187.

[5] Price, W. H., "Factors Influencing Concrete Strength," *Proceedings*, Vol. 47, American Concrete Institute, 1951, pp. 417–432.

[6] Gonnerman, H. F. and Shuman, E. C., "Compression, Flexure and Tension Tests of Plain Concrete," *Proceedings*, Vol. 28, Part II, American Society for Testing and Materials, 1928, pp. 527–564.

[7] Blick, R. L., "Some Factors Influencing High-Strength Concrete," *Modern Concrete*, Vol. 36, No. 12, April 1973, pp. 38–41.

[8] Mehta, P. K., "Pozzolanic and Cementitious By-Products as Mineral Admixtures for Concrete—A Critical Review," *SP-79*, Vol. I, American Concrete Institute, Detroit, MI, 1983, pp. 1–46.

[9] Carrasquillo, P. M. and Carrasquillo, R. L., "Evaluation of the Use of Current Concrete Practice in the Production of High-Strength Concrete," *Materials Journal*, Vol. 85, No. 1, American Concrete Institute, 1988, pp. 49–54.

[10] Mindess, S. and Young, J. F., *Concrete*, Prentice–Hall, Inc., Englewood Cliffs, NJ, 1981.

[11] *Concrete Testing–Alternative Methods Available*, Concrete Construction, Addison, IL, April 1973, pp. 14–23.

[12] Jones, R. and Wright, P. J. F., "Some Problems Involved in Destructive and Non-destructive Testing of Concrete," *Mix Design and Quality Control of Concrete, Proceedings*, Cement and Concrete Association, London, May 1954, pp. 441–464.

[13] Neville, A. M., *Properties of Concrete*, 3rd ed., Pitman Publishing Ltd., London, 1981.

[14] Bungey, J. H., *The Testing of Concrete in Structures*, 2nd ed., Chapman and Hall, New York, 1989, pp. 94–110.

[15] Malhotra, V. M., "Concrete Strength Requirements—Cores Versus In-Situ Evaluation," *Proceedings*, Vol. 74, No. 4, American Concrete Institute, 1977, pp. 163–172.

[16] Suprenant, B. A., "An Introduction to Concrete Core Testing," *Civil Engineering for Practicing and Design Engineers*, Vol. 4, No. 8, Pergamon Press, New York, 1985, pp. 607–615.

[17] Bloem, D. L., "Concrete Strength Measurement—Cores Versus Cylinders," *Proceedings*, Vol. 65, American Society for Testing and Materials, 1965, pp. 668–696.

[18] Yuan, R. L., Ragab, M., Hill, R. E., and Cook, J. E., "Evaluation of Core Strength in High Strength Concrete," *Concrete International*, Vol. 13, No. 5, May 1991, pp. 30–34.

[19] Szypula, A. and Grossman, J. S., "Cylinder Vs. Core Strength," *Concrete International*, Vol. 12, No. 2, 1990, pp. 55–61.

[20] Richardson, D. N., "Effects of Testing Variables on the Comparison of Neoprene Pad and Sulfur Mortar-Capped Concrete Test Cylinders," *Materials Journal*, Vol. 87, No. 5, American Concrete Institute, 1990, pp. 489–495.

[21] Werner, G., "The Effect of Type of Capping Material on the Compressive Strength of Concrete Cylinders," *Proceedings*, Vol. 58, American Society for Testing and Materials, 1958, pp. 1166–1186.

[22] Troxell, G. E., "The Effect of Capping Methods and End Conditions Before Capping Upon the Compressive Strength of Concrete Test Cylinders," *Proceedings*, Vol. 41, American Society for Testing and Materials, 1942, pp. 1038–1044.

[23] Gonnerman, H. F., "Effect of End Condition of Cylinder in Compression Tests of Concrete," *Proceedings*, Vol. 24, Part II, American Society for Testing and Materials, 1924, pp. 1036–1065.

[24] Vidal, E. N., "Discussion," *Proceedings*, Vol. 41, American Society for Testing and Materials, 1942, pp. 1045–1047.

[25] Murray, W. M., "Discussion," *Proceedings*, Vol. 41, American Society for Testing and Materials, 1942, pp. 1047–1048.

[26] Kennedy, T. B., "A Limited Investigation of Capping Materials for Concrete Test Specimens," *Proceedings*, Vol. 41, American Concrete Institute, 1944, pp. 117–126.

[27] McGuire, D. D., "Testing Concrete Cylinders Using Confined Sand Cushion," *Proceedings*, Vol. 30, Part I, American Society for Testing and Materials, 1930, pp. 515–517.

[28] Carrasquillo, P. M. and Carrasquillo, R. L., "Effect of Using Unbonded Capping Systems on the Compressive Strength of Concrete Cylinders," *Materials Journal*, Vol. 85, No. 3, American Concrete Institute, 1988, pp. 141–147.

[29] Bromham, S. B. and Meadley, M. D., "Use of Rubber Pads for Capping Concrete Cylinders," *Proceedings*, National Conference Publication No. 83/12, The Institution of Engineers, Australia, Perth, Oct. 1983, pp. 131–132.

[30] Oxyildirim, C., "Neoprene Pads for Capping Concrete Cylinders," *Cement, Concrete, and Aggregates*, Vol. 7, No. 1, 1985, pp. 25–28.

[31] Grygiel, J. S. and Amsler, D. E., "Capping Concrete Cylinders With Neoprene Pads," Research Report 46, Engineering Research and Development Bureau, New York State Department of Transportation, April 1977.

[32] Nasser, K. W. and Al-Manaseer, A. A., "It's Time for a Change from 6 × 12- to 3 × 6-in. Cylinders," *Materials Journal*, Vol. 84, No. 3, American Concrete Institute, 1987, pp. 213–216.

[33] Malhotra, V. M., "Are 4 × 8 Inch Concrete Cylinders as Good as 6 × 12 Inch Cylinders for Quality Control of Concrete?" *Proceedings*, Vol. 73, American Concrete Institute, 1976, pp. 33–36.

[34] Gonnerman, H. F., "Effect of Size and Shape of Test Specimen on Compressive Strength of Concrete," *Proceedings*, Vol. 25, Part II, American Society for Testing and Materials, 1925, pp. 237–255.

[35] Peterman, M. B. and Carrasquillo, R. L., "Production of High Strength Concrete," Research Report 315-1F, Center for Transportation Research, University of Texas at Austin, Oct. 1983.

[36] Mirza, S. A., Hatzinikolas, M., and MacGregor, J. G., "Statistical Descriptions of Strength of Concrete," *Journal*, Structural Division, American Society of Civil Engineers, Vol. 105, No. ST6, June 1979.

[37] Janak, K. J., "Comparative Compressive Strength of 4 × 8-in. Versus 6 × 12-in. Concrete Cylinders Along With the Investigation of Concrete Compressive Strength at 56-Days," Report 3-1-4-116, Materials and Tests Division, Texas State Department of Highways and Public Transportation, March 1985.

[38] McMillan, F. R., "Suggested Procedure for Testing Concrete in Which the Aggregate is More Than One-Fourth the Diameter of the Cylinders," *Proceedings*, Vol. 30, Part I, American Society for Testing and Materials, 1930, pp. 521–535.

[39] Chung, H.-W., "On Testing of Very Short Concrete Specimens," *Cement, Concrete, and Aggregates*, Vol. 11, No. 1, 1989, pp. 40–44.

[40] Kesler, C. E., "Effect of Length to Diameter Ratio on Compressive Strength—An ASTM Cooperative Investigation," *Proceedings*, Vol. 59, American Society for Testing and Materials, 1959, pp. 1216–1229.

[41] Tucker, J., Jr., "Effect of Length on the Strength of Compression Test Specimens," *Proceedings*, Vol. 45, American Society for Testing and Materials, 1945, pp. 976–984.

[42] Mather, B., "Effect of Type of Test Specimen on Apparent Compressive Strength of Concrete," *Proceedings*, Vol. 45, American Society for Testing and Materials, 1945, pp. 802–812.

[43] Popovics, S., "Effect of Curing Method and Final Moisture Condition on Compressive Strength of Concrete," *Proceedings*, Vol. 83, American Concrete Institute, 1986, pp. 650–657.

[44] Butterfield, E. E., "Discussion," *Proceedings*, Vol. 25, Part II, American Society for Testing and Materials, 1925, pp. 251–253.

[45] Jones, P. G. and Richart, F. E., "The Effect of Testing Speed on Strength and Elastic Properties of Concrete," *Proceedings*, Vol. 36, Part II, American Society for Testing and Materials, 1937, pp. 380–392.

[46] Abrams, D. A., "Effect of Rate of Application of Load on the Compressive Strength of Concrete," *Proceedings*, Vol. 17, Part II, American Society for Testing and Materials, 1917, pp. 364–377.

[47] Kesler, C. E., "Statistical Relation Between Cylinder, Modified Cube, and Beam Strength of Plain Concrete," *Proceedings*, Vol. 54, American Society for Testing and Materials, 1954, pp. 1178–1187.

[48] Wright, P. J. F., "Comments on an Indirect Tensile Test on Concrete Cylinders," *Magazine of Concrete Research*, Vol. 7, No. 20, Cement and Concrete Association, London, July 1955, pp. 87–96.

[49] Chen, W. F., "Double-Punch Test for Tensile Strength of Concrete," *Proceedings*, Vol. 67, No. 12, American Concrete Institute, 1970, pp. 993–995.

[50] Marti, P., "Size Effect in Double-Punch Tests on Concrete Cylinders," *Materials Journal*, Vol. 86, No. 6, American Concrete Institute, 1989, pp. 597–601.

[51] Chen, A. C. T. and Chen, W. F., "Nonlinear Analysis of Concrete Splitting Tests," *Computers & Structures*, Vol. 6, No. 6, Pergamon Press, New York, 1976, pp. 451–457.

[52] Goldbeck, A. T., "Apparatus for Flexural Tests of Concrete Beams," *Proceedings*, Vol. 30, Part I, American Society for Testing and Materials, 1930, pp. 591–597.

[53] Reagel, F. V. and Willis, T. F., "The Effect of the Dimensions of Test Specimens on the Flexural Strength of Concrete," *Public Roads*, Vol. 12, No. 2, April 1931, pp. 37–46.

[54] Kellermann, W. F., "Effect of Size of Specimen, Size of Aggregate and Method of Loading Upon the Uniformity of Flexural Strength Results," *Public Roads*, Vol. 13, No. 11, Jan. 1933, pp. 177–184.

[55] Wright, P. J. F. and Garwood, F., "The Effect of the Method of Test on the Flexural Strength of Concrete," *Magazine of Concrete Research*, No. 11, Cement and Concrete Association, London, Oct. 1952, pp. 67–76.

[56] Carrasquillo, P. M. and Carrasquillo, R. L., "Improved Concrete Quality Control Procedures Using Third Point Loading," Research Report 1119-1F, Center for Transportation Research, University of Texas at Austin, Nov. 1987.

[57] Nielsen, K. E. C., "Effect of Various Factors on the Flexural Strength of Concrete Test Beams," *Magazine of Concrete Research*, No. 15, Cement and Concrete Association, London, March 1954, pp. 105–114.

[58] Johnston, C. D. and Sidwell, E. H., "Influence of Drying on Strength of Concrete Specimens," *Proceedings*, Vol. 66, American Concrete Institute, 1969, pp. 748–755.

[59] Melis, L. M., Meyer, A. H., and Fowler, D. W., "An Evaluation of Tensile Strength Testing," Research Report 432-1F, Center for Transportation Research, University of Texas at Austin, Nov. 1985.

[60] Johnson, A. N., "Concrete in Tension," *Proceedings*, Vol. 26, Part II, American Society for Testing and Materials, 1926, pp. 441–450.

[61] Carino, N. J., "The Behavior of a Model of Plain Concrete Subjected to Compression-Tension and Tension-Tension Biaxial Stresses," Ph.D. thesis, Cornell University, Ithaca, NY, Aug. 1974.

[62] Herrin, J. C., "Behavior of a High Strength Concrete Model Subjected to Biaxial Compression," Masters thesis, The University of Texas at Austin, Dec. 1982.

[63] Chen, R. L., Fowler, D. W., and Carrasquillo, R. L., "Behavior of High Strength Concrete in Biaxial Compression," Research Report AF-3, Defense Technical Information Service, U.S. Air Force, Dec. 1984.

[64] Calixto, J. M. F., "Microcracking of High Strength Concrete Subjected to Biaxial Tension-Compression Stresses," Masters thesis, The University of Texas at Austin, May 1987.

[65] Nelson, E. L., Fowler, D. W., and Carrasquillo, R. L., "Fatigue of High Strength Concrete Subjected to Biaxial-Cycle Compression," Research Report AF-4, Defense Technical Information Service, U.S. Air Force, May 1986.

[66] Nordby, G. M., "Fatigue of Concrete—A Review of Research," *Proceedings*, Vol. 55, American Concrete Institute, 1959, pp. 191–219.

Prediction of Potential Concrete Strength at Later Ages

Nicholas J. Carino[1]

PREFACE

This chapter deals with methods for estimating the potential, later-age strength of a concrete mixture based upon the compressive strength measured on cylindrical specimens at early ages. The results of the cooperative research program leading to the development of ASTM Test Method for Making, Accelerated Curing, and Testing Concrete Compression Test Specimens (C 684) are reviewed. This review is based on Chapter 13 of *ASTM STP 169B*, which was written by M. H. Wills, Jr., and was titled "Accelerated Strength Tests." The earlier text has been augmented by the addition of information on the high-temperature and pressure accelerated test method that was added to ASTM C 684 in 1989. In addition, the chapter has been expanded by including the basis of ASTM Test Method for Developing Early-Age Compression Test Values and Projecting Later-Age Strengths (C 918).

INTRODUCTION

Rapid construction practices throughout the concrete industry have brought increasing pressure on specifying agencies to assess the quality of concrete at an earlier age than the traditional 28 days after placement. Currently, the later age is still specified for compression tests to determine the quality of concrete as delivered to the job site. During the 28-day period between the preparation and testing of the specimens, a multistory building could rise four or more floors. Many consider this situation to be too precarious for construction to proceed on a sound technical basis and with adequate assurance of safety. Furthermore, expensive and costly delays are encountered when 28-day test results are low, because a field investigation may be necessary to verify the load-carrying capacity of the structure. Further delay is certain if concrete must be reinforced or replaced. An earlier assessment of concrete quality is, therefore, absolutely essential for overall construction economy and safety.

Recognizing the need for alternative test methods for assessing the quality of concrete at early ages, ASTM Committee C9 on Concrete and Concrete Aggregates formed a subcommittee on Accelerated Strength Testing in 1964 (In 1992, the subcommittee number was changed from C09.02.09 to C09.63.). Since that time the committee has developed and maintains two methods for estimating the later-age strength of concrete specimens based upon early-age tests. One of these methods, ASTM C 684, is based on testing cylinders whose strength development has been accelerated by elevated curing temperatures. Depending on the specific procedure that is used, the accelerated strength is measured at ages ranging between 5 and 48 h. A previously established relationship between accelerated strength and standard-cured strength is used to estimate the later-age strength of future accelerated test specimens. The other method, ASTM C 918, uses the maturity method to estimate later-age strength based on the early-age strength of specimens whose temperature history has been measured. This chapter reviews the background of these two techniques and provides supporting information to the standard test methods to assist persons who are contemplating specifying these procedures in contract documents.

ACCELERATED CURING METHODS

The first assignment for ASTM Subcommittee C09.02.09 was to canvass the needs of the concrete industry and to study the suitability of standardizing several accelerated testing procedures being developed by King [1,2], Akroyd [3], Smith and Chojnacki [4], Malhotra and Zoldners [5], and Smith and Tiede [6]. A positive response to this canvass led to a cooperative test program among nine laboratories to evaluate various procedures (Table 1) involving the use of hot or boiling water or autogenous curing in an insulated container to accelerate the strength development of concrete [7].

Experimental Program

Each of the accelerated testing procedures was conducted using ASTM Type I (or Type II meeting Type I specifications) and Type III portland cement. Cement contents were 265, 325, and 385 kg/m³ (450, 550, and 650 lb/yd³). Sufficient air-entraining admixture and mixing water were used to produce air contents from 5.0 to 6.0% and slumps from 50 to 75 mm (2 to 3 in.). Concretes were mixed without a retarder or with a normal dosage of a

---

[1] Research civil engineer, Building and Fire Research Laboratory, National Institute of Standards and Technology, Gaithersburg, MD 20899.

water-reducing retarder. Fine and coarse aggregates were graded to a No. 57 size, that is, 25.0 to 4.75 mm (1 in. to No. 4). Each laboratory used materials available in their locality without interchange between laboratories. In all cases, compressive strengths of the accelerated specimens were measured at one- or two-day ages and were compared with strengths of standard-cured specimens of 28- and 364-day ages. Two replicate specimens were tested at each age, and their averages were used in subsequent analyses. All specimens were prepared, cured, and tested according to applicable ASTM standards. All materials conformed to appropriate ASTM specifications.

The nine participating laboratories were asked to conduct Procedures A, B, and C listed in Table 1. Additionally, they had the option of conducting one or more of Procedures D through G (Table 1). Despite the plans of the subcommittee, it was necessary to curtail some aspects of the experimental program. Procedure C, involving the measurement of final setting time followed by boiling, had to be abandoned because it required too many overtime man-hours to conduct. Consequently, Procedures E, F, and G were also abandoned because they also required measurement of setting time and overtime man-hours. Only four of the nine participating laboratories conducted Procedure D (Autogenous Curing), which was optional. However, all laboratories conducted Procedure A (warm water) and Procedure B (modified boiling).

Accelerated Curing Apparatus

The accelerated procedures involving the use of either hot or boiling water were conducted in thermostatically controlled tanks (Fig. 1). The tops of the cylinders were covered by about 100 mm (4 in.) of water. Water volume and heater capacity were sized to prevent an appreciable reduction in the desired water temperature when the specimens were immersed. Specimens were not placed in a tank already containing those being cured.

The autogenous curing container for Procedure D was similar to that used by Smith and Tiede to cure a single 152 by 305-mm (6 by 12-in.) cylinder [6]. It was made by using a plastic garbage can and polyurethane foam insulation (Fig. 2). The polyurethane foam retained the heat of hydration of the cement that, in turn, accelerated the strength development of the concrete. Once curing was initiated, the container was not opened until the end of the specified curing period.

TABLE 1—Accelerated Testing Procedures.

Procedure	Description
A—Warm water	cylinders placed immediately in water at 35°C (95°F)
B—Modified boiling	standard curing for 23 h followed by boiling for 3.5 h
C—Final set and boiling	at final setting (ASTM C 403),[a] cylinders boiled for 15 h
D—Autogenous curing	at 1-h age, place in insulated container for 46 h
E—Initial set and hot water, 55°C (130°F)	at initial setting (ASTM C 403), place in hot water for 15 h
F—Initial set and hot water, 75°C (175°F)	at initial setting, place in hot water for 15 h
G—Initial set and hot water, 90°C (195°F)	at initial setting, place in hot water for 15 h

[a] ASTM Test Method for Time of Setting of Concrete Mixtures by Penetration Resistance (C 403).

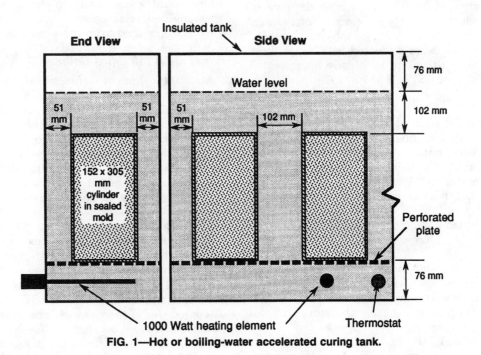

FIG. 1—Hot or boiling-water accelerated curing tank.

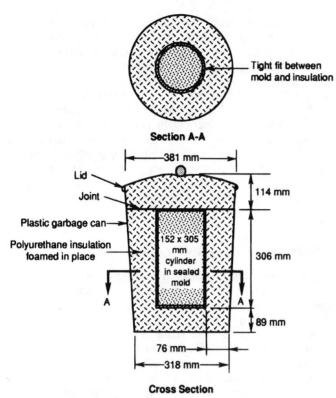

FIG. 2—Autogenous curing container.

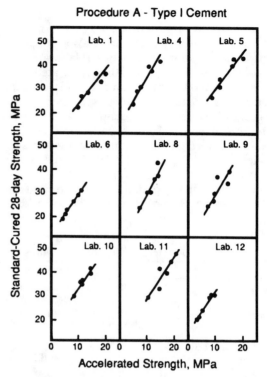

FIG. 3—Relationships between accelerated and standard-cured strengths for Procedure A using Type I cement.

Results

The results of the cooperative testing program were summarized by Mills [7], and only the main conclusions are highlighted here. The objective was to establish the nature of the relationships between accelerated and standard-cured strengths, and to examine whether these relationships were affected by factors such as the type of accelerated procedure, cement type, cement content, presence of set retarder, and laboratory. Only the results of Procedures A, B, and D are discussed here, since these were the procedures that were eventually incorporated into the first version of ASTM C 684 in 1974.

Procedure A (Warm Water)

Cylinders were cast in steel molds with tight closing lids. Immediately after casting, the cylinder molds were immersed in the water bath at a temperature of 33 to 37°C (92 to 98°F), where they remained for a period of 24 h ± 15 min. Sulfur mortar caps were applied to the cylinders and aged at least 1 h prior to measuring compressive strength at an age of 26 h ± 15 min. The warm-water procedure accelerated concrete strengths from 1.1 to 1.6 times those achieved after one day of standard moist curing.

Figure 3 shows the relationships between accelerated and 28-day strengths for concretes made with Type I cement. Within each laboratory, there were good correlations; however, it appeared that each laboratory obtained significantly different results.

Although not shown, the same general trends were observed with concretes made with Type III cement, except that the relationships were at a higher strength level. Since the test procedures were carefully controlled, the results emphasized the impact of materials, particularly the type of cement, on the relationships between accelerated and standard-cured strength. Based upon the good correlations, however, it was concluded that Procedure A could be used with a high degree of confidence in assessing concrete quality when the tests are made on mixtures produced from the same materials.

Procedure B (Modified Boiling)

After the cylinders were cast, the sealed molds were placed in a standard moist room. At 23 h ± 15 min from the time of casting, the cylinders, including molds and covers, were immersed in boiling water. Reduction in water temperature was limited to 3°C (5°F), and the temperature was required to recover to the boiling point in no more than 15 min. After boiling for 3.5 h ± 5 min, the cylinders were removed from the molds and allowed to cool for about 45 min. They were capped with sulfur mortar, the sulfur was allowed to age for 1 h, and strengths were measured at an age of 28.5 h ± 15 min. The modified boiling procedure accelerated concrete strength between 1.1 to 2.1 times that measured after one day of standard moist curing.

Figure 4 shows the results for concretes cured by Procedure B and made with Type III cement. The trends discussed for Procedure A are reemphasized. Most importantly, the laboratories also obtained what seemed to be significantly different results. Again, this is attributed to each laboratory using locally available materials. How-

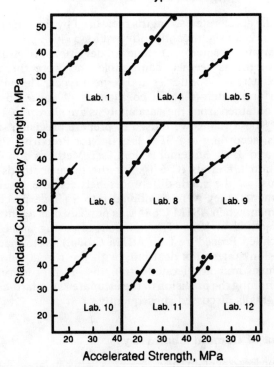

FIG. 4—*Relationships between accelerated and standard-cured strengths for Procedure B using Type III cement.*

ever, within a given laboratory, good correlations between accelerated strength and 28- or 364-day values were obtained [7]. Therefore, Procedure B was also considered to have equal merit in assessing concrete quality.

Procedure D (Autogenous Curing)

Cylinders were molded in light-gage steel molds. One hour after the start of mixing, the molded cylinders were sealed in plastic bags and placed inside the autogenous curing chambers (Fig. 2). They remained there for 46 h, after which they were removed from the molds and allowed to cool for 45 min. Sulfur mortar caps were applied and aged at least 1 h before strength was measured at an age of 49 h ± 15 min.

Only four laboratories performed this procedure, but their data were sufficient to justify including autogenous curing in the subsequent ASTM standard. The two-day strengths after autogenous curing ranged from 1.4 to 2.5 times those obtained after an equal length of standard moist curing, which were the highest level of strength acceleration.

High concrete temperatures at early ages are detrimental to the early hydration reactions of cement and to the developing paste microstructure. With the autogenous curing procedure, the heat of hydration of the cement causes the acceleration but at a low rate of temperature increase, which is beneficial to early hydration reactions. This, plus the fact that higher curing temperatures are attained than in Procedure A, may explain the high values of accelerated strength for Procedure D.

Significance of Test Procedures

The subcommittee was convinced that laboratories, test procedures, and cement types had significant effects on the correlation between accelerated and standard-cured strength, and the data were analyzed from that viewpoint [7]. It was assumed that the correlation for each set of conditions could be represented by a straight-line relationship as follows

$$\hat{S}_{28} = B_0 + B_1 S_a \qquad (1)$$

where

$\hat{S}_{28}$ = the standard-cured 28-day strength,
S_a = the accelerated strength for a particular procedure,
B_0 = the intercept, and
B_1 = the slope.

For each laboratory, cement type, and procedure, the best fit values of B_0 and B_1 were obtained by least-squares fitting. The resulting values of B_0, B_1, correlation coefficients (r), and residual standard deviations are summarized in Tables 2, 3, and 4 for Procedures A, B, and D, respectively. The residual standard deviation is a measure of the error between the data and the best-fit straight line and is computed as follows

$$s_e = \sqrt{\frac{\sum_1^n (S_{ai} - \hat{S}_{28})^2}{n - 2}} \qquad (2)$$

where

S_{ai} = the average accelerated strength of the ith specimens,
$\hat{S}_{28}$ = the estimated 28-day strength corresponding to S_{ai}, and
n = number of pairs of strength values used in the regression analysis.

TABLE 2—*Linear Regression Analysis Results for Procedure A.*

Laboratory Number	Cement Type	B_0,[a] MPa	B_1	r[b]	RSD,[c] MPa[d]
1	I	13.54	1.120	0.910	2.10
	III	15.65	0.960	0.975	0.86
4	I	19.10	1.515	0.955	1.96
	III	18.96	1.095	0.925	2.76
5	I	15.27	1.375	0.960	1.86
	III	17.86	0.905	0.980	1.07
6	I	14.17	1.525	0.985	0.86
	III	14.31	1.290	0.940	1.93
8	I	9.10	1.985	0.905	2.90
	III	15.44	1.285	0.935	2.31
9	I	16.31	1.320	0.765	3.69
	III	23.48	0.540	0.870	1.59
10	I	19.13	1.440	0.915	1.62
	III	24.20	0.770	0.815	1.28
11	I	14.34	1.475	0.905	2.83
	III	13.76	1.090	0.725	3.62
12	I	17.65	1.440	0.915	1.96
	III	15.65	0.580	0.825	2.07

[a] See Eq 1 for meaning of B_0 and B_1.
[b] r = correlation coefficient.
[c] RSD = residual standard deviation.
[d] 1 MPa ≈ 145 psi.

TABLE 3—Linear Regression Analysis Results for Procedure B.

Laboratory Number	Cement Type	B_0,[a] MPa	B_1	r[b]	RSD,[c] MPa[d]
1	I	12.82	1.060	0.970	1.34
	III	16.55	0.910	0.985	0.66
4	I	17.79	1.290	0.965	1.55
	III	13.96	1.145	0.955	2.17
5	I	9.86	1.225	0.975	1.21
	III	14.96	0.840	0.930	1.28
6	I	11.58	1.280	0.970	0.79
	III	13.13	1.050	0.965	1.07
8	I	15.62	1.280	0.810	4.03
	III	14.86	1.220	0.950	2.03
9	I	16.89	1.015	0.865	1.58
	III	19.55	0.780	0.960	1.24
10	I	19.48	1.195	0.950	1.24
	III	17.79	1.000	0.975	0.90
11	I	9.52	1.515	0.975	1.86
	III	13.41	1.095	0.800	3.48
12	I	18.13	1.020	0.860	2.48
	III	18.27	1.143	0.830	3.07

[a] See Eq 1 for meaning of B_0 and B_1.
[b] r = correlation coefficient.
[c] RSD = residual standard deviation.
[d] 1 MPa ≈ 145 psi.

TABLE 4—Linear Regression Analysis Results for Procedure D.

Laboratory Number	Cement Type	B_0,[a] MPa	B_1	r[b]	RSD,[c] MPa[d]
4	I	11.45	1.000	0.990	0.96
	III	3.79	1.230	1.000	0.41
5	I	22.03	1.315	0.780	4.07
	III	22.65	0.875	0.960	1.48
8	I	7.17	1.400	0.950	1.72
	III	9.96	1.125	0.970	1.93
12	I	16.86	0.930	0.900	2.10
	III	9.82	0.810	0.930	1.45

[a] See Eq 1 for meaning of B_0 and B_1.
[b] r = correlation coefficient.
[c] RSD = residual standard deviation.
[d] 1 MPa ≈ 145 psi.

For example, the regression equation for Laboratory 6 using Procedure A and Type I cement was

$$\hat{S}_{28} = 14.17 + 1.525\, S_a \text{ (MPa)} \qquad (3)$$

Note that the correlation coefficient had a high value of 0.985 and residual standard deviation had a correspondingly low value of 0.86 MPa (125 psi). On the other hand, Laboratory 8 had a much lower correlation coefficient and a higher residual standard deviation. Therefore, the data for Laboratory 8 did not fit its straight-line relationship as well as the data for Laboratory 6.

After studying the residual standard deviations, it was concluded that all three procedures were equal in correlating accelerated and later-age strengths. Correlation coefficients were quite high in most cases, as would be expected by the close fit of the data to the linear relationships shown in Figs. 3 and 4 for each laboratory. Further, it was found that accelerated compressive strength correlated with 364-day strength as well as with 28-day strength. Subsequently, the subcommittee embarked upon the preparation of a tentative test method that included Procedures A, B, and D.

Test Precision

While accelerated strengths measured by the nine laboratories were significantly different for a given procedure, the variances (squares of standard deviations) in strength measurements seemed compatible and were therefore pooled across laboratories and cement types. The within-batch and batch-to-batch coefficients of variation that were obtained from the data analysis are shown in Table 5 [7]. These values were used to prepare precision statements, according to ASTM Practice for Preparing Precision and Bias Statements for Test Methods for Construction Materials (C 670), for the three methods [7]. Because of the small differences between the resulting statements, they were combined into a single precision statement when ASTM C 684 was published as a tentative test method in 1971 (The autogenous curing procedure was called Procedure C in ASTM C 684.). Note that the precision statements deal only with the repeatability of the measured accelerated strengths. A procedure for determining the precision of the estimated standard-cured strengths is discussed subsequently.

Effect of Cement Chemistry

In order to explain the differences between laboratories, a program was conducted to determine the chemical or physical properties, or both, governing the one-day accelerated strength of concrete. Similar concrete batches were made using eight Type I cements mixed with the same source of sand and gravel. An attempt was made to hold the slump constant at 75 to 100 mm (3 to 4 in.). Four 152 by 305-mm (6 by 12-in.) cylinders were molded from each batch. Two cylinders were cured according to ASTM C 684, Procedure A, to obtain one-day accelerated strengths and two cylinders were moist-cured at 23°C (73°F) in 100% relative humidity to obtain 28-day strengths.

After all cylinders were tested, the physical and chemical properties of each cement along with the corresponding concrete strengths were analyzed to search for correlations. A strong correlation was found between the sodium alkali (Na_2O) content and the one-day accelerated strengths (Fig. 5).

Several multiple correlations were also examined. For two independent variables, only combinations involving Na_2O produced useful correlations with one-day accelerated strength; moreover, only loss on ignition (LOI) coupled with the Na_2O produced a better multiple correlation than Na_2O alone.

Therefore, the variation between the accelerated strengths for the different Type I cements was attributable mainly to variations in alkali content. Undoubtedly, this

TABLE 5—Precision of Accelerated Strength Tests [7].

Procedure	Within-Batch Coefficient of Variation, %	Batch-to-Batch Coefficient of Variation, %
Warm water	2.9	8.2
Modified boiling	3.0	8.5
Autogenous curing	3.6	8.5

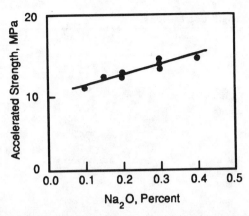

FIG. 5—Relationship between alkali content and accelerated strength using Procedure A.

caused the principal variation between laboratories in the cooperative test program since each used a different Type I cement.

High Temperature and Pressure Method

In 1978, Nasser reported on a different accelerated strength procedure that could produce test results within 5 h [8]. Acceleration of strength development is achieved by a combination of elevated temperature and pressure. A schematic of the apparatus is shown in Fig. 6. Special molds with heating wires and insulation are used to prepare 76 by 152-mm (3 by 6-in.) cylinders. After three molds are filled, they are stacked in a compression testing frame, a compressive stress of 10.3 ± 0.2 MPa (1500 ± 25 psi) is applied, and the electrical heaters are turned on. The heaters raise and maintain the concrete temperature at 149°C ± 3°C (300 ± 5°F). After three hours, the heaters are turned off, the axial stress is maintained, and the specimens are allowed to cool for 2 h. The hardened cylinders are extruded from the molds and tested for compressive strength. Usually, capping materials are not needed because the metal end caps result in sufficiently smooth ends.

In 1980, the subcommittee was requested to modify the existing version of ASTM C 684 to permit the high temperature and pressure (HTP) procedure as an alternative to the other methods. Data were provided to the subcommittee to demonstrate that the method resulted in correlations similar to the other methods [9]. The concrete mixtures were made with Type I, III, and IV cements; water-cement ratios (w/c) between 0.45 to 0.90; with and without fly ash, air-entraining agent or water reducer; and with normal-weight and lightweight aggregates. The accelerated strengths were compared with the strengths of standard-cured 152 by 305-mm (6 by 12-in.) cylinders. The accelerated strengths were between 22 and 90% of the 28-day strengths.

A task force of the subcommittee compared the residual standard deviations of correlations obtained with the HTP procedure with those obtained by others using the standard procedures (Table 6). The relationships for the HTP method were expressed as power functions as follows

$$\hat{S}_{28} = B_0 S_a^{B_1} \qquad (4)$$

Based on the comparisons, the task force recommended that the HTP method should be added to ASTM C 684, because the 5-h accelerated strength correlated reasonably well with standard-cured strength. Subsequently, a member of the subcommittee reported on a comparative study of the HTP method and the modified-boiling method (Procedure B).[2] In this study, mixtures were made with fly ash from different sources, with w/c values of 0.4, 0.5, and 0.7, and with an air-entraining agent. The accelerated strengths were correlated with standard-cured, 28-day strengths (Fig. 7). Power functions were fitted to the data. The residual standard deviations were 1.21 MPa (175 psi) and 1.77 MPa (257 psi) for Procedure B and the HTP method, respectively. Surprisingly, the best-fit equations were very similar. This study provided further evidence of the suitability of the HTP method, and steps were taken to incorporate this procedure into ASTM C 684.

During the balloting process, concern was expressed that the hydration reactions due to the high temperature and pressure would not be representative of those due to normal curing. Thus it was felt that the accelerated strength from the HTP method might not represent the

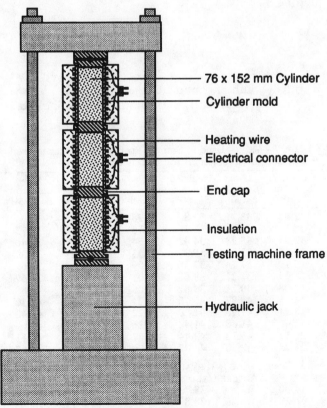

FIG. 6—Schematic of apparatus for high temperature and pressure accelerated curing method.

---

[2] V. M. Malhotra, private communication, Canada Centre for Mineral and Energy Technology, Ottawa, Ont., Canada, 11 Jan. 1985.

TABLE 6—Comparison of Accelerated Strength Correlations.

Reference	Procedure	Cement Type	Regression Equation	Number of Points	RSD,[a] MPa[b]
Bickley [10]	autogenous curing	I	linear	43	2.35
		IV		147	2.46
		I and IV		68	2.08
Lapinas [11]	modified boiling	I	linear	312	1.82
Roadway and Lenz [12]	modified boiling	I	linear	219	3.66
		I		36	2.34
		I		76	2.51
		V		61	1.58
Bisaillon [13]	autogenous curing	I	linear	not given	1.48
Bisaillon et al. [14]	autogenous curing	I	linear	213	2.30
Malhotra [15]	modified boiling	I	linear	336	2.40
		I		40	2.64
		I		265	3.54
Nasser and Beaton [9]	high temperature and pressure	I	power function	171	3.59
		III		99	3.36
		IV		65	4.03

[a] RSD = residual standard deviation.
[b] 1 MPa ≈ 145 psi.

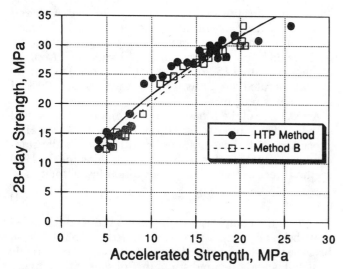

FIG. 7—Comparison of relationships between accelerated and standard 28-day strengths.

potential strength under standard curing. Proponents of the method noted that the data obtained by using a variety of materials showed good correlations. Eventually, the HTP method was adopted in 1989 as alternative Procedure D.

Prediction of Later-Age Strength

To predict the potential later-age strength from a measured early-age accelerated strength, the laboratory must first conduct enough tests to establish the regression equation and its residual standard deviation. To account for the uncertainty in the regression line, confidence bands for the line are established [16,17]. Then, for a new accelerated strength, the confidence interval for the average later-age strength can be estimated.

To illustrate the procedure, consider the 12 pairs of accelerated and 28-day strengths given in the first two columns of Table 7. Each number is the average strength of two cylinders. Using ordinary least-squares regression analysis, the best-fit regression equation for the data (Fig. 8) is

$$\hat{S}_{28} = 19.51 + 1.19 S_a \text{ (MPa)} \quad (5)$$

The residual standard deviation of the line, s_e, is 1.24 MPa (180 psi).

The 95% confidence band for the line [16,18] is constructed by calculating $\hat{S}_{28}$ for selected values of S_a and plotting $\hat{S}_{28} \pm W_i$,

$$W_i = s_e \sqrt{2F} \sqrt{\frac{1}{n} + \frac{(S_{ai} - \overline{S}_a)^2}{S_{aa}}} \quad (6)$$

where

W_i = half-width of confidence band at S_{ai},
s_e = residual standard deviation for best-fit line (Eq 2),

TABLE 7—Values Used in Sample Problem to Illustrate Prediction of Confidence Interval for 28-Day Strength.

Accelerated Strength, S_a, MPa[a]	28-day Strength, S_{28}, MPa	Predicted Strength, $\hat{S}_{28}$, MPa	W_i, MPa	Lower Confidence Band, MPa	Upper Confidence Band, MPa
12.06	33.71	33.86	1.78	32.07	35.64
12.15	34.33	33.96	1.76	32.20	35.72
12.96	35.23	34.92	1.54	33.38	36.46
13.85	35.05	35.99	1.33	34.66	37.31
15.19	37.74	37.58	1.09	36.49	38.67
16.09	37.21	38.65	1.03	37.62	39.67
17.08	40.71	39.82	1.06	38.76	40.88
18.15	40.97	41.10	1.21	39.89	42.30
18.24	41.96	41.20	1.22	39.98	42.42
18.42	41.60	41.42	1.26	40.16	42.67
20.12	45.73	43.44	1.67	41.77	45.11
21.28	42.50	44.82	2.00	42.82	46.82
CONFIDENCE INTERVAL FOR ESTIMATED STRENGTH AT ACCELERATED STRENGTH OF 17.00 MPa					
17.00		39.73	1.05	38.68	40.78
16.29		38.89	1.02	37.87	
17.71		40.57	1.13		41.70

[a] 1 MPa ≈ 145 psi.

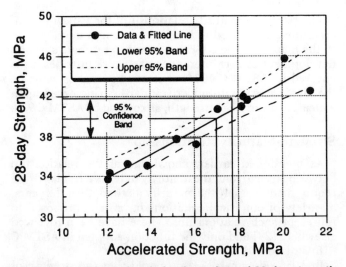

FIG. 8—Confidence bands for the estimated 28-day strength based on a measured accelerated strength and the previously established strength relationship.

F = value from F-distribution for 2 and n degrees of freedom and significance level 0.05,
n = number of data points used to establish regression line,
S_{ai} = selected value of accelerated strength,
$\overline{S}_a$ = grand average value of accelerated strength for all data points used to establish the regression line,
$S_{aa} = \Sigma (S_a)^2 - (\Sigma S_a)^2/n$, and
S_a = values of average accelerated strength for each point used to establish the regression line.

The third column in Table 7 lists the predicted average 28-day strengths for the accelerated strengths in Column 1. The value of W_i at each value S_a is listed in the fourth Column of Table 7. Finally, columns five and six list the values of the lower and upper 95% confidence bands that are shown in Fig. 8. Note that the width of the confidence band is narrowest when S_{ai} equals $\overline{S}_a$, because the second term under the square root sign in Eq 6 equals zero.

Suppose that the average accelerated strength of two cylinders made from similar concrete is 17 MPa (2460 psi). From the regression equation, the estimated average 28-day strength is 39.68 MPa (5750 psi). If the accelerated strength were known without error, the 95% confidence interval for the average 28-day strength would be 38.68 to 40.78 MPa (5610 to 5910 psi) (see the bottom of Table 7). However, the accelerated strength has an uncertainty that is described by the within-batch standard deviation. Assume that the compressive strengths measured by the specific accelerated test method has a single-laboratory, within-batch coefficient of variation of 3.0%. Therefore, the standard deviation, s, at an average strength of 17 MPa (2460 psi) is 0.51 MPa (74 psi). The 95% confidence interval for the average accelerated strength of the two cylinders is

$$17 \pm z_{0.025} \frac{s}{\sqrt{2}} = 17 \pm 1.96 \times 0.51 \times 0.707$$
$$= 17 \pm 0.71 \text{ MPa}$$

Where $z_{0.025}$ is the value from the standard normal distribution corresponding to 2.5% of the area under the curve. Thus the 95% confidence interval[3] for the average accelerated strength is 16.29 to 17.71 MPa (2360 to 2570 psi). Projecting the limits of this interval to the lower and upper confidence bands of the regression line results in 37.87 to 41.70 MPa (5490 to 6050 psi) for the approximate 95% confidence interval for the average 28-day strength. Each different measurement of accelerated strength produces a new confidence interval for the average 28-day strength. The use of a personal computer is recommended for implementing the preceding calculations for routine use.

In summary, to predict the average 28-day strength based upon accelerated strength test results, a procedure is used that accounts for the uncertainty in the regression line and in the measured accelerated strength. It is insufficient to simply use the regression equation to convert the accelerated strength to an equivalent 28-day strength. Additional information on the procedure presented in the example may be found in the references by Moore and Taylor [20] and in Miller [18]. Finally, it is emphasized that a particular regression equation is valid only for a specific test method and combination of materials. Therefore, each laboratory must conduct enough tests with a given set of materials and a certain procedure to establish the regression line and its confidence bands before predictions of later-age strengths are possible.

LATER-AGE STRENGTH PREDICTIONS USING THE MATURITY METHOD

Overview

The other method developed by the subcommittee on Accelerated Strength Testing to estimate the later-age strength from early-age tests is based upon the maturity method. Unlike the accelerated strength methods, there are no special curing requirements for this procedure, but it does require measuring the temperature history of the test specimens. The method is an outgrowth of research performed in the late 1960s and early 1970s by Hudson and Steele [21,22] at the West Virginia Department of Highways, and it was adopted as ASTM C 918 in 1980. The basic principles of the maturity method are discussed by Malhotra in Chapter 30 of this publication and a more comprehensive review is also available [23].

The motivation for the development of ASTM C 918 is discussed first. Concrete mixtures cured at a standard

---

[3] The 95% confidence interval is often interpreted to mean that there is a 95% probability that the true mean falls within the interval. However, the correct interpretation is as follows: If 100 repeated samples are taken from the same population and the 95% confidence intervals for the mean are computed in each case, 95 of the intervals would include the true mean. The 95% confidence band for the regression line has a similar interpretation: If 100 groups of data are taken from the same population and the 95% confidence bands are computed for the regression equations, 95 of those bands would include the true regression line for the population. See Mendenhall and Sincich [19] for further explanations on the proper interpretation of confidence intervals.

temperature gain strength predictably, that is, there are fixed relationships between the strengths at early ages and at a later age. Attempts have been made, therefore, to predict the later-age strength, such as at 28 days, by multiplying the early-age strength by an empirical factor. However, such predictions have not been found to be reliable. As has been explained [21,22], there are two pitfalls to practical application of this simple approach: (1) on a construction project, it is not possible to perform the early-age tests at precisely the specified age, and (2) the early-age temperatures cannot be controlled as accurately as needed. The procedure in ASTM C 918 overcomes these limitations by requiring measurement of actual temperature histories of the early-age specimens and by using a "maturity index," rather than age, to relate to the level of strength development.

To use ASTM C 918, the testing laboratory first establishes a relationship between the strength of concrete cylinders and the maturity index. Subsequently, cylinders made from similar concrete are tested at early ages. The maturity index from the time of molding each set of cylinders until the time of testing is recorded. The early-age strengths and the corresponding values of maturity index are then used to estimate the later-age strength based upon the previously established strength-maturity index relationship.

The following provides a brief review of the terminology associated with the maturity method.

Maturity

In concrete technology, the term "maturity" refers to the extent of the development of those properties that are dependent on cement hydration and pozzolanic reactions in concrete. At any age, the maturity depends on the previous curing history, that is, the temperature history and the availability of water to sustain the chemical reactions. In the early 1950s, the idea was developed that the combined effects of time and temperature could be accounted for by using a "maturity function" to convert the temperature history to a "maturity index" that would be indicative of strength development. In practice, the maturity index is obtained by using electronic instruments that monitor the concrete temperature and automatically compute the maturity index as a function of the age. The maturity index can be displayed on paper tape, a digital display, or a computer terminal. The reader is urged to read the appropriate section of Chapter 30 and ASTM Practice for Estimating Concrete Strength by the Maturity Method (C 1074) for background information on the maturity functions used to compute a maturity index. In this chapter, the "equivalent age at 23°C" is used as the maturity index. While this index is not as well known in the United States as the "temperature-time factor" (expressed in degree-hours or degree-days), the equivalent age is a more meaningful quantity and is expected to be used more widely in the future. The equivalent age is the curing age at a specified temperature that would result in the same maturity as has occurred at the actual age under the actual temperature history.

The "strength-maturity relationship" for a specific concrete mixture is used to relate the strength development to the maturity index. It can be used to estimate strength development under different temperature histories, as is covered by ASTM C 1074, or it can be used to estimate the later-age strength under standard curing based upon a measured early-age strength, as covered by ASTM C 918.

Strength-Maturity Relationship

As mentioned earlier, knowledge of the relationship between strength and the maturity index is a fundamental requirement to apply the maturity method. Over the years, a variety of empirical equations have been proposed to represent such relationships. These have been reviewed elsewhere [24] and only the function adopted in ASTM C 918 is discussed here.

In 1956, Plowman [23] proposed that the strength of concrete could be related to the maturity index by the following semi-logarithmic function

$$S_M = a + b \log M \qquad (7)$$

where

S_M = compressive strength at M,
M = maturity index, and
a, b = regression constants.

According to Eq 7, strength is a straight line function of the logarithm of the maturity index. This is generally a reasonable approximation for strength development between one and 28 days under standard room temperature curing. The parameter a represents the strength at a maturity index of 1 (the logarithm of 1 equals zero).[4] The parameter b, which is the slope of the line, represents increase in strength for a tenfold increase in the maturity index. For example, if the equivalent age at 23°C goes from one day to ten days, the strength increase equals b. The values of a and b depend on the materials and mixture proportions of the specific concrete [21–23].

Application

To use ASTM C 918 to predict the potential later-age strength, for example, at 28 days, it is first necessary to establish the strength-maturity relationship, that is, the "prediction equation," for the concrete mixture. This is done by preparing standard cylindrical specimens and subjecting them to standard curing procedures. A maturity meter is used to monitor the maturity index of the specimens.[5] Pairs of cylinders are tested at regular intervals (for

---

[4] Note that the value of a depends on the units used for the maturity index. If the maturity index is expressed as the temperature-factor (degree-hours or degree-days), the value of a will be a negative number which has no physical significance. If the maturity index is expressed in terms of equivalent age at 23°C in days, the value of a is the strength at one day of curing at 23°C.

[5] In the original version of ASTM C 918, the measurement of the ambient temperature history was considered adequate. However, revisions in 1993 call for monitoring the actual concrete temperature. The availability of relatively inexpensive maturity meters justifies this more precise approach.

example, at 1, 3, 7, 14 and 28 days), and the corresponding maturity indexes are recorded along with the average strengths. The values of a and b are obtained by least-squares fitting of Eq 7 to the data.

To estimate the later-age strength of a similar concrete mixture based on the result of subsequent early-age tests, the maturity index of the early-age test specimens is monitored from the time of molding until the time of testing. This is conveniently done by casting a "dummy" cylinder into which the sensor of the maturity meter is embedded. The instrumented cylinder is exposed to the same environment as the cylinders that will be tested for strength. The values of the early-age strength, S_m, and maturity index, m, are used along with the previously established strength-maturity relationship to estimate the potential later-age strength.

During a project, the concrete-making materials will usually be the same from batch to batch, and the main purpose of cylinder strength tests is to assure that the maximum water-cement ratio is not exceeded. However, the value of a in Eq 7 depends on the water-cement ratio since it represents the strength gained at a particular maturity index. A change in a shifts the straight line along the strength axis. Thus it is not appropriate to assume that the value of a obtained in the laboratory testing program is applicable to the cylinders being evaluated by early-age testing. However, it is reasonable to assume that the value of b is applicable, because large changes in water-cement ratio are needed to change the value of b [22]. The purpose of the early-age results is to establish the appropriate value of a as follows

$$a = S_m - b \log m \tag{8}$$

where

m = maturity index at time of early-age test, and
S_m = average early-age strength measured at maturity index, m.

By substituting Eq 8 into Eq 7, one obtains the following prediction equation for estimating the strength at a later age

$$S_M = S_m + b (\log M - \log m) \tag{9}$$

where

M = maturity index at the later age when strength is to be estimated, and
S_M = estimated strength at maturity index M.

Example

The following example illustrates the application of ASTM C 918. Assume that a set of cylinders are molded in the laboratory and that one of them is instrumented with a maturity meter that computes the equivalent age at 23°C. Table 8 gives the average strengths (two cylinders) that are obtained at different values of the maturity index.

These strengths have been plotted as a function of the logarithm of the equivalent age (Fig. 9). The best-fit equation is

$$S_M = 10.25 + 13.29 \log M, \text{ MPa} \tag{11}$$

Further assume that cylinders fabricated in the field

TABLE 8—Equivalent Age and Compressive Strength Values Used in Illustrative Example.

Age, days	Equivalent Age at 23°C, days	Average Strength, MPa (psi)
1.0	0.91	9.44 (1370)
3.0	3.10	17.09 (2480)
7.0	7.25	21.77 (3160)
14.0	14.5	25.56 (3710)
28.0	27.3	29.28 (4250)

were cured at the job site under standard conditions (curing for checking adequacy of mixtures) as specified in ASTM Practice for Making and Curing Concrete Test Specimens in the Field (C 31). After 24 h, two cylinders were removed from their molds, capped, and tested for compressive strength. The average strength of two cylinders was 10.00 MPa (1450 psi), and the corresponding equivalent age was 1.13 days based on temperature measurement of a dummy cylinder. According to Eq 9, the predicted 28-day, standard-cured strength would be

$$\begin{aligned}
S_{28} &= 10.00 + 13.29 (\log 28 - \log 1.13) \\
&= 10.00 + 13.29(1.4472 - 0.0531) \\
&= 10.00 + 18.49 \\
&= 28.49 \text{ MPa (4130 psi)}
\end{aligned}$$

Interpretation of Results

The intent of early-age tests is to provide an early indication of the potential strength of the concrete sample. It is unrealistic to expect that the traditional standard tests at later ages, such as at 28 days will be eliminated in the near future. Thus the 28-day strengths of standard-cured specimens will probably continue to be measured. By keeping records of predicted and measured 28-day strengths of companion specimens from the same batch, one can continually correct and improve the slope, b, of

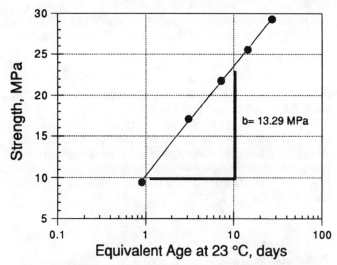

FIG. 9—Example of a strength versus maturity index relationship; the slope is used in the prediction equation to estimate later-age strength based upon early-age strength.

the prediction equation by using the following relationship [21]

$$b = \frac{\Sigma(S_{28} - S_m)}{\Sigma(\log M_{28} - \log m)} \quad (11)$$

where

S_{28} = measured standard-cured 28-day strength,
M_{28} = maturity index corresponding to standard curing for 28-days, and
S_m = measured early-age strength at maturity index m.

In addition, the differences between the predicted and measured 28-day strengths can be used to calculate a confidence limit for future predictions, and thereby establish an acceptance criterion for future early-age results. First, the average difference, $\bar{d}$, between the measured and estimated 28-day strengths is computed

$$\bar{d} = \frac{\sum_{i=1}^{n}(S_{28} - S_M)}{n} \quad (12)$$

$$\bar{d} = \frac{\sum_{i=1}^{n} d_i}{n}$$

where

S_M = the predicted 28-day strength,
S_{28} = measured standard-cured 28-day strength,
d_i = the difference between the ith pair of strength values, and
n = number of pairs of strength values.

The value of $\bar{d}$ is the "bias" of the prediction equation, and it should be close to zero if the value of b is updated as new data are accumulated. The standard deviation for the difference between the measured and predicted strengths is calculated as follows

$$s_d = \sqrt{\frac{\sum_{i=1}^{n}(d_i - \bar{d})^2}{(n-1)}} \quad (13)$$

The upper 95% confidence limit for the average difference between the measured and predicted 28-day strength is

$$K = \bar{d} + t_{0.95,n-1}\frac{s_d}{\sqrt{n}} \quad (14)$$

where $t_{0.95,n-1}$ = value from the t-distribution at the 95% point for $n - 1$ degrees of freedom.

From such past data, it is possible to establish an acceptance criterion for the concrete based on the predicted strength of a future early-age test. If the design strength is f_{28}, the concrete represented by the early-age results would be considered acceptable if the following condition were satisfied

$$S_M \geq f_{28} + K \quad (15)$$

Precautions

The prediction equation given by Eq 9 assumes that concrete strength is a linear function of the logarithm of the maturity index. Before implementing ASTM C 918, the user should verify that this assumption is valid. As was stated, past experience shows that the linear equation is usually adequate for strengths up to 28 days. However, there is no reason why the procedure cannot be applied to estimate strength at ages later than 28 days. If the laboratory results reveal that the linear approximation is not applicable, the basic principle of ASTM C 918 can still be applied provided an appropriate equation is used to represent the actual strength-maturity relationship. Examples of equations that may be applicable are given in Ref 24.

SUMMARY

This chapter has discussed the basis of two procedures to estimate the potential later-age strength of specimens tested at early ages. The results of these early-age tests can provide timely information on the concrete production process for quality control. These procedures attempt to overcome the inherent deficiencies of the traditional practice, which relies on strength tests at ages of 28 days or later, to judge the adequacy of concrete batches.

One of the techniques (ASTM C 684) involves subjecting the early-age test specimens to specific curing conditions that accelerate strength development. Four alternative accelerated curing methods are specified that permit strength determinations at ages ranging from about 5 to 48 h after preparing the specimens. Prior to using these accelerated strengths to predict the later-age strength, laboratory testing is required to establish the relationship between the accelerated strength and the standard-cured strength. Typically, the standard-cured strength is measured at 28-days, but other ages are possible, such as 56 or 91 days, depending on the job specifications.

The cooperative testing program leading to the initial development of ASTM C 684 was reviewed. Analysis of the data from that program resulted in the following conclusions [7]:

1. Significantly different relationships were obtained by the nine laboratories due to the use of different local materials.
2. Different types of cements resulted in different relationships within the same laboratory.
3. The omission or addition of a normal amount of chemical retarder in the concrete did not affect the relationships.
4. Batch-to-batch variations due to mixing on different days did not have significant effects on the relationships.
5. Correlation of accelerated strength with 28- and 364-day standard strength ranged from good to excellent.
6. For each cement type, accelerated strength correlated with 364-day strength as well as with 28-day strength.

Subsequent tests showed that significantly different strength relationships were obtained by the laboratories because of variations in alkali content of the cements. The cooperative study showed that accelerated strength testing by a single laboratory, using the same procedure and materials, results in an assessment of concrete quality at ages less than two days that is as reliable as that obtained after 28 days of standard moist curing [7]. Appropriate statistical techniques, as discussed in this chapter, should be used to establish a confidence interval of the estimated potential, later-age strength.

The other method (ASTM C 918) is based on the principle of the maturity method, in which the early-age curing history is converted to a maturity index indicative of the level of strength development. A previously established relationship is also required to estimate the potential, later-age strength based on the measured early-age strength. In this case, the relationship is between strength and the maturity index. Unlike the accelerated strength testing methods, ASTM C 918 does not require a precisely controlled curing procedure. The specimens are cured according to the standard-curing procedures in ASTM C 31. The only additional requirement is that the temperature history of the early-age specimens must be monitored to determine the maturity index at the time of testing. Early-age strengths can be measured at any age 24 h after molding the specimens.

The basis of ASTM C 918 is research performed at the West Virginia Department of Highways [21,22]. In a statistically designed testing program, it was shown that the potential, later-age strength could be estimated with sufficient precision and accuracy for quality control purposes. This method also requires use of statistical principles for a reliable lower-bound estimate of the potential strength.

In summary, techniques are available to overcome the deficiencies of quality control programs based on the traditional 28-day strengths. The potential benefits arising from early-age testing should be considered carefully prior to setting up the quality control program to monitor the concrete production process on major construction projects.

REFERENCES

[1] King, J. W. H., "Accelerated Testing of Concrete." *Proceedings*, 50th Anniversary Conference of the Institution of Structural Engineers, London, Oct. 1958, pp. 376–381, 386–387.

[2] King, J. W. H., "Accelerated Test for Strength of Concrete," *Journal of Applied Chemistry* (London), Vol. 10, June 1960, pp. 256–262.

[3] Akroyd, T. N. W., "The Accelerated Curing of Concrete Test Cubes," *Journal*, Institution of Civil Engineers, London, Vol. 19, Paper No. 6441, May 1961, pp. 1–22.

[4] Smith, P. and Chojnacki, B., "Accelerated Strength Testing of Concrete Cylinders," *Proceedings*, American Society for Testing and Materials, Philadelphia, Vol. 63, 1963, pp. 1079–1101.

[5] Malhotra, V. M. and Zoldners, N. G., "Some Field Experience in the Use of an Accelerated Method of Estimating 28-day Strength of Concrete," Internal Report MP1 68-42, Minerals Processing Division Department of Energy, Mines and Resources, Ottawa, Ont., Canada, Aug. 1968.

[6] Smith, P. and Tiede, H., "Earlier Determination of Concrete Strength Potential," *Highway Research Record*, No. 210, 1967, pp. 29–61.

[7] Wills, M. H., Jr., "Early Assessment of Concrete Quality by Accelerating Compressive Strength Development with Heat (Results of ASTM's Cooperative Test Program)," *Journal of Testing and Evaluation*, Vol. 3, No. 4, July 1975, pp. 251–262.

[8] Nasser, K. W., "A New Method and Apparatus for Accelerated Strength Testing of Concrete," *Accelerated Strength Testing, SP-56*, V. M. Malhotra, Ed., American Concrete Institute, Detroit, MI, 1978, pp. 249–258.

[9] Nasser, K. W. and Beaton, R. J., "The K-5 Accelerated Strength Tester," *Journal*, American Concrete Institute, Vol. 77, No. 3, May–June 1980, pp. 179–188.

[10] Bickley, J. A., "Accelerated Concrete Strength Testing at the CN Tower," *Accelerated Strength Testing, SP-56*, V. M. Malhotra, Ed., American Concrete Institute, Detroit, MI, 1978, pp. 29–38.

[11] Lapinas, R. A., "Accelerated Concrete Strength Testing by Modified Boiling Method: Concrete Producer's View," *Accelerated Strength Testing, SP-56*, V. M. Malhotra, Ed., American Concrete Institute, Detroit, MI, 1978, pp. 75–93.

[12] Roadway, L. E. and Lenz, K. A., "Use of Modified Boiling Method in Manitoba and Alberta, Canada," *Accelerated Strength Testing, SP-56*, V. M. Malhotra, Ed., American Concrete Institute, Detroit, MI, 1978, pp. 129–146.

[13] Bisaillon, A., "Accelerated Strength Test Results form Expanded Polystyrene Molds with Emphasis on Initial Concrete Temperature," *Accelerated Strength Testing, SP-56*, V. M. Malhotra, Ed., American Concrete Institute, Detroit, MI, 1978, pp. 201–228.

[14] Bisaillon, A., Frechétte, and Keyser, J. H., "Field Evaluation of Expanded Polystyrene Molds for Self-Cured, Accelerated Strength Testing of Concrete," *Transportation Research Record*, No. 558, 1975, pp. 50–60.

[15] Malhotra, V. M., "Canadian Experience in the Use of the Modified Boiling Method," *Transportation Research Record*, No. 558, 1975, pp. 13–18.

[16] Natrella, M. G., *Experimental Statistics*, NBS Handbook 91, National Bureau of Standards, Washington, DC, Aug. 1963.

[17] Draper, N. R. and Smith, H., *Applied Regression Analysis*, 2nd ed., Wiley, New York, 1981.

[18] Miller, R. G., *Simultaneous Statistical Inference*, 2nd ed., Springer-Verlag, New York, 1981.

[19] Mendenhall, W. and Sincich, T., *Statistics for Engineering and the Sciences*, 3rd ed., Dellen Publishing Co., San Francisco, CA, 1992.

[20] Moore, J. K. and Taylor, M. A., "Statistical Properties of Techniques for Predicting Concrete Strength and Examples of Their Use," *Accelerated Strength Testing, SP-56*, V. M. Malhotra, Ed., American Concrete Institute, Detroit, MI, 1978, pp. 259–283.

[21] Hudson, S. B. and Steele, G. W., "Prediction of Potential Strength of Concrete from the Results of Early Tests," *Highway Research Record*, No. 370, 1971, pp. 25–36.

[22] Hudson, S. B. and Steele, G. W., "Developments in the Prediction of Potential Strength of Concrete from Results of Early Tests," *Transportation Research Record*, No. 558, 1975, pp. 1–12.

[23] Plowman, J. M., "Maturity and Strength of Concrete," *Magazine of Concrete Research*, Vol. 8, No. 22, March 1956, pp. 13–22.

[24] Carino, N. J., "The Maturity Method," *Handbook on Nondestructive Testing of Concrete*, V. M. Malhotra and N. J. Carino, Eds., CRC Press, Boca Raton, FL, 1991, pp. 101–146.

BIBLIOGRAPHY

American Concrete Institute, Detroit, MI, ACI Committee 214, "Use of Accelerated Strength Testing," *Manual of Concrete Practice*, ACI 214.1R-81(1986), 1986.

Brockenbrough, T. W. and Larason, R. R., "Early Strength Test for Quality Control of Concrete," *Transportation Research Record*, No. 558, 1975, pp. 61–68.

Chin, F. K., "Strength Tests at Early Ages and at High Setting Temperatures," *Transportation Research Record*, No. 558, 1975, pp. 69–76.

Falcão Bauer, L. A. and Olivan, L. I., "Use of Accelerated Tests for Concrete Made with Slag Cement," *Accelerated Strength Testing, ACI SP-56*, V. M. Malhotra, Ed., American Concrete Institute, Detroit, MI, 1978, pp. 117–128.

Ferrer, M. M., "Quality Control of Concrete by Means of Short-Termed Tests at La Angostina Hydroelectric Project, State of Chiapas, Mexico," *Accelerated Strength Testing, ACI SP-56*, V. M. Malhotra, Ed., American Concrete Institute, Detroit, MI, 1978, pp. 51–73.

Kalyanasundaram, P. and Kurien, V. J., "Accelerated Testing for Prediction of 28-day Strength of Concrete," *Transportation Research Record*, No. 558, 1975, pp. 77–86.

Malhotra, V. M., Zoldners, N. G., and Lapinas, R., "Accelerated 28-Day Test for Concrete," *Canadian Pit and Quarry*, March 1966, pp. 51–54.

Orchard, D. F., Jones, R., and Al-Rawi, R. K., "The Effect of Cement Properties and the Thermal Compatibility of Aggregates on the Strength of Accelerated Cured Concrete," *Journal of Testing and Evaluation*, Vol. 2, No. 2, March 1974, pp. 95–101.

Ramakrishnan, V. and Dietz, J., "Accelerated Methods of Estimating the Strength of Concrete," *Transportation Research Record*, No. 558, 1975, pp. 29–44.

Ramakrishnan, V., "Accelerated Strength Testing—Annotated Bibliography," *Accelerated Strength Testing, ACI SP-56*, V. M. Malhotra, Ed., American Concrete Institute, Detroit, MI, 1978, pp. 285–312.

Sanchez, R. and Flores-Castro, L., "Experience in the Use of the Accelerated Testing Procedure for the Control of Concrete During Construction of the Tunnel 'Emissor Central' in Mexico City," *Accelerated Strength Testing, ACI SP-56*, V. M. Malhotra, Ed., American Concrete Institute, Detroit, MI, 1978, pp. 15–28.

16

Freezing and Thawing

Howard Newlon, Jr.,[1]
and Terry M. Mitchell[2]

PREFACE

The subject of freeze-thaw durability of concrete was covered in all three previous editions of the ASTM Special Technical Publication on significance of tests and properties; the chapters on this subject in *ASTM STP 169*, *ASTM STP 169A*, and *ASTM STP 169B*, were authored by C. H. Scholer and T. C. Powers, H. T. Arni, and H. Newlon, Jr., respectively. Because little has changed in this subject area since the publication of *ASTM STP 169B*, this chapter includes only minor changes, that is, an update of the reference list and a few revisions in the text, from the version that appeared there.

INTRODUCTION

The purpose of this paper is to discuss the significance of currently standardized testing procedures for evaluating the resistance of concrete to weathering under service conditions. The published literature on the subject is extensive and only a somewhat superficial treatment can be given in a brief paper. The reader should consult the similar papers in earlier editions of this special technical publication by Scholer [1,2], Powers [3], Arni [4], and Newlon [5]. A monograph by Woods [6] presents a comprehensive picture of durability, and one by Cordon [7] discusses freezing and thawing in detail. Perhaps the most comprehensive treatment of the influence of aggregates on resistance to freezing and thawing is that of Larson et al. [8] An extensive annotated bibliography on concrete durability published by the Highway Research Board (HRB) in 1957 [9] with a supplement in 1966 [10] contains 534 citations. A more current annotated bibliography, with some 600 citations, was published by the Strategic Highway Research Program in late 1992 [11]. The report of American Concrete Institute (ACI) Committee 201 on Durability of Concrete [12] contains valuable recommendations for production of concrete to provide resistance to the various destructive processes encountered in field exposures. ACI symposia on durability of concrete, in 1975, 1987, and 1991, resulted in 17, 111, and 70 papers, respectively, on various theoretical and operational aspects of producing durable concrete [13–15].

In the absence of contact with aggressive fluids or incorporation of aggregates susceptible to detrimental expansion by reaction with alkalies in cement, the resistance of concrete to weathering is determined by its ability to withstand the effects of freezing and thawing in the presence of moisture.[3]

The testing of concrete for resistance to freezing and thawing had its genesis in similar tests on building stones reported as early as 1837 by Vicat [16]. It was only natural that the advent of "artificial stone" in the form of portland cement concrete would result in corresponding evaluations. Few published tests from the nineteenth century survive, but, when freezing equipment became commercially available for food processing in the 1930s, the reporting of such testing increased. With the recognition that tests conducted in a variety of freezers designed for other purposes naturally gave variable results, several agencies such as the U.S. Bureau of Reclamation, the Corps of Engineers, the Bureau of Public Roads (now the Federal Highway Administration), the Portland Cement Association, and the National Sand and Gravel Association (now the National Aggregates Association) constructed specialized equipment for testing concrete by freezing and thawing. Descriptions of some of this equipment will be found in reports by the agencies [17–22]. These efforts marked the beginning of a systematic quest for standardization and an understanding of factors influencing the resistance of concrete to freezing and thawing.

There are currently three standard test methods and one recommended practice under the jurisdiction of ASTM Committee C9 on Concrete and Concrete Aggregates intended to aid in evaluating the resistance of concrete to freezing and thawing. In addition, several petrographic procedures are standardized that provide invaluable information for predicting the resistance of concrete to freezing and thawing and for interpreting the results of exposure in either the laboratory or under field conditions. Two older freezing and thawing methods have been discontinued. The presently standardized methods as well as those discontinued have evolved over more than 60 years and reflect the inevitable compromise between the need for rapid assessment of resistance to weathering and the

---

[1] Research consultant, Charlottesville, VA 22901.
[2] Materials research engineer, Federal Highway Administration, McLean, VA 22101–2296.

[3] Attack by aggressive fluids and cement-aggregate reactions are treated elsewhere in this publication.

difficulty of translating the results from accelerated laboratory testing to the varied conditions encountered in field exposures.

Of the currently standardized methods, ASTM Test Method for Resistance of Concrete to Rapid Freezing and Thawing (C 666) covers the exposure of specimens to cyclic freezing and thawing. This test, the most widely used of those available, is a consolidation of two earlier methods (ASTM C 290 and C 291) and provides for two procedures: Procedure A in which both freezing and thawing occurs with the specimens surrounded by water, and Procedure B in which the specimens freeze in air and thaw in water. ASTM Test Method for Critical Dilation of Concrete Specimens Subjected to Freezing (C 671) employs a single cycle of cooling through the freezing point with specimens that are continuously wet. This test was developed to accommodate certain theoretical criticisms of the cyclic method. ASTM Practice for Evaluation of Frost Resistance of Coarse Aggregates in Air-Entrained Concrete by Critical Dilation Procedures (C 682) describes the use of ASTM C 671 to evaluate the influence of coarse aggregates on the resistance of concrete to freezing and thawing. ASTM Test Method for Scaling Resistance of Concrete Surfaces Exposed to Deicing Chemicals (C 672) provides a procedure for evaluating the effect of deicing chemicals on concrete and the effectiveness of modifications of the concrete or of surface coatings in mitigating the detrimental influence of such chemicals. Two of the initially approved standards were discontinued in 1971. These methods, ASTM Test for Resistance of Concrete Specimens to Slow Freezing and Thawing in Water or Brine (C 292) and ASTM Test for Resistance of Concrete Specimens to Slow Freezing in Air and Thawing in Water (C 310), both of which provided for a single cycle every 24 h, were dropped because of lack of use.

While much is still to be learned, an extensive body of knowledge has been developed that permits evaluations of concrete to aid in minimizing premature deterioration from environmental factors.

While an extensive discussion of the theory and historical development of freezing and thawing tests is beyond the scope of this summary, a brief treatment is necessary. The reader should consult the cited references for additional information on these subjects.

Historical Evolution

As early as 1837, Vicat, in his famous *Treatise on Calcareous Mortars and Cements*, reported the results from experiments by Brard [23] to distinguish the building stones that were injured by frost from those that were not [16]. These tests were conducted "by substituting for the expansive force of the congealing water, that of an easily crystallizable salt, the sulfate of soda." In 1856, Joseph Henry reported testing, by 50 cycles of freezing and thawing, samples of marble used in the extension of the U.S. Capitol [24].

During the two decades before and those after the turn of the twentieth century, various tests of stone, concrete, brick, and other porous materials were reported. In 1928, Grun [25] in Europe and Scholer [26] in the United States reported results from accelerated freezing and thawing of concrete in the laboratory. Following Scholer's initial paper, accelerated laboratory testing greatly increased. In 1936 [27], 1944 [28], and again in 1959 [29], the Highway Research Board (HRB) Committee on Durability of Concrete reported results of cooperative freezing and thawing tests designed to identify factors that influence resistance of concrete to freezing and thawing. The first two cooperative test series were conducted before any methods were standardized. The 1936 series concentrated on the influence of cement using mortar prisms incorporating ten commercial cements. The 1944 series of tests used concrete specimens. Both of these series emphasized the necessity for carefully regulating the methods of making and curing the specimens, the air content of the specimens, the degree of saturation of the aggregates at the time of mixing the concrete, and the degree of saturation of the concrete at the time of freezing.

In 1951, ASTM Committee C9 formed Subcommittee III-0 on Resistance to Weathering (subsequently C09.03.15 and currently C09.67) with the major responsibility for proposing methods for evaluating the resistance to freezing and thawing of aggregates in concrete and standardizing the methods of tests. Drawing heavily on the results of the HRB cooperative test series and the experience of several laboratories that had developed specialized equipment for conducting freezing and thawing tests, ASTM approved four tentative test methods in 1952 and 1953 that later became standards. These methods were largely representative of the methods and procedures then in use by the membership of Subcommittee III-0. The general characteristics of the methods are reflected in their titles: ASTM Test for Resistance of Concrete Specimens to Rapid Freezing and Thawing in Water (C 290), ASTM Test for Resistance of Concrete Specimens to Rapid Freezing in Air and Thawing in Water (C 291), ASTM Test for Resistance of Concrete to Slow Freezing and Thawing in Water or Brine (C 292), and ASTM Test for Resistance of Concrete Specimens to Slow Freezing in Air and Thawing in Water (C 310). The two slow tests were adopted to cover tests usually conducted in conventional freezers with manual transfer of specimens between the freezing chamber and thawing tank. These methods were dropped in 1971, because neither was in general use nor required by any other ASTM specification. In 1971, the two rapid tests (ASTM C 290 and C 291) were combined as two procedures (A and B) in a single test (ASTM C 666).

When ASTM C 666 was originally adopted, deterioration of specimens was evaluated only by the resonant frequency method, that is, ASTM Test Method for Fundamental Transverse, Longitudinal, and Torsional Frequencies of Concrete Specimens (C 215). In 1984, length change was incorporated as an optional, additional means for that evaluation (length change measurements are made with the apparatus described in ASTM Specification for Apparatus for Use in Measurement of Length Change of Hardened Cement Paste (C 490)).

In the third cooperative test series, initiated in 1954 and reported in 1959 [29] soon after standardization of the four methods, 13 laboratories participated using three concrete formulations representing a concrete with good-quality coarse aggregate and adequate air entrainment, one with good-quality coarse aggregate and deficient air entrainment, and one with poor-quality coarse aggregate and adequate air entrainment.

Arni [4] summarized the general conclusions from the 1944 and 1954 test series as follows:

"1. Methods involving freezing in water were more severe, that is, produced failure in fewer cycles, than were those involving freezing in air.
2. Rapid freezing was more severe than slow freezing when done in air but not when done in water.
3. Rapid freezing and thawing in water (ASTM Test C 290) appeared to do the best job of detecting a difference between concretes both of which had high durability.
4. Only the slow freezing in air and thawing in water method (ASTM Test C 310) was able to discriminate adequately between concretes of low durability.
5. In general, the four methods tended to rate different concretes in the same order of durability when there was a significant difference."

One of the difficulties with the four methods that was emphasized especially by the HRB cooperative programs was poor repeatability and reproducibility of results within and between laboratories. Good reproducibility generally was obtained only for concretes that were very high or very low in durability. For concretes in the middle range, a wide spread in results was obtained. While the variabilities are large in the middle range, they are amenable to establishing a precision statement that has been incorporated in ASTM C 666 as discussed later.

As noted earlier, Subcomittee III-0 was formed initially in 1951 with the primary mission of developing freezing and thawing tests to be applied to the evaluation of coarse aggregates. In 1961, the subcommittee reported that standardization of such tests was not warranted because of the high levels of variability associated with the existing methods. However, in an unpublished report, the subcommittee did outline procedures for using ASTM C 290 if evaluations of coarse aggregates were required [30].

Lack of a standardized test method for evaluating resistance to freezing and thawing of coarse aggregates in concrete coupled with theoretically based criticisms of the cyclic methods, particularly by Powers [31], as discussed later, led to the development of methods designed to determine the length of time required for an aggregate to become critically saturated in concrete. Critical saturation was defined when specimens exposed to continuous soaking and subjected to a cycle of cooling through the freezing point exhibited dilation greater than a specified value. This method was first used by the California Department of Highways [32] and later was refined and extensively evaluated by Larson and his co-workers [33]. The literature survey prepared by Larson et al., as part of the research, is a particularly valuable reference on all aspects of freezing and thawing studies related to aggregates [8]. As an outgrowth of this work, ASTM C 671 was approved in 1971 along with ASTM C 682, which provides guidelines for applying ASTM Test C 671 to the evaluation of coarse aggregates.

More recently, agencies studying "D-cracking" of concrete have generally found a modification of the older procedure (ASTM C 666) to be useful in identifying coarse aggregates susceptible to this type of deterioration [34]. According to Stark [35], for example, aggregates susceptible to causing distress in concrete pavements can be identified by freezing and thawing of concrete in water at a rate of two cycles per day. Equipment, specimen preparation, and procedures vary considerably, and no standardized failure criterion is available. Agencies desiring to use these procedures for accepting or rejecting aggregates develop failure criteria by relating the freeze-thaw test results with field performance.

Concern with surface mortar deterioration or deicer scaling, particularly on highway and bridge deck pavements, led in 1971 to standardization of ASTM C 672, which combined features of methods that had been developed and used by various agencies for a number of years. The method uses blocks fabricated to permit ponding of water on surfaces that are subjected to freezing and thawing in the presence of various deicing agents.

Of the methods, ASTM C 666 continues to be the most widely used. Specialized equipment is commercially available for conducting the tests under controlled conditions, but the essential elements of the method are those that have been used for more than a century.

Theoretical Considerations

Cyclic freezing and thawing tests were developed on a pragmatic rather than a theoretical basis. It was assumed that the destruction resulted from the 9% volume expansion accompanying the conversion of water to ice, and that this process was reproduced artificially in the laboratory environment. Powers and his co-workers, from their comprehensive study of the structure of cement paste in 1945 [36], advanced the hypothesis that the destructive stress is produced by the flow of displaced water away from the region of freezing, the pressure being due to the viscous resistance of such flow through the permeable structure of the concrete. According to this theory, when the flow path exceeds a critical length, the pressure exceeds the strength of the paste. Such flow would occur when the water content exceeds the critical saturation point. This concept is called the hydraulic pressure theory. The theory was amplified in 1949 [37] to explain the beneficial influence of entrained air. Since the resistance to flow at a given rate is proportional to the length of the flow path, the air bubbles were conceived as spaces into which the excess water produced by freezing could be expelled without generating destructive pressures. Powers calculated a critical dimension of the order of 0.25 mm (0.01 in.), a value that was approximately the same as that suggested by Mielenz and his co-workers from experimental studies [38]. Powers' initial studies suggested that the hypothesis advanced by Taber [39] to explain frost heaving of soils did not apply to mature concrete. By this hypothesis, the stress is produced not by hydraulic pressure, but by the segregation of ice into layers that enlarge as unfrozen water is drawn toward the region of freezing rather than forced away from it.

Studies by Verbeck and Landgren [40], as well as those of Powers [41], make clear that the paste and aggregate should be considered separately when explaining the resistance of concrete to freezing and thawing. This is because the paste not only may become critically saturated by moisture from external sources but also must withstand pressure generated by water expelled from the aggregate particles during freezing.

The temperature at which water freezes in various pores within the paste decreases with the size of the pore so that, even if the concrete is at a uniform temperature throughout, the water will be at various stages of conversion to ice. As the water freezes, the solution in the pore becomes more concentrated. The existence of solutions of various concentrations in the pores of the paste causes unfrozen water to move to the site of freezing in order to lower the concentration made higher at the freezing site than the more dilute solution of the unfrozen water. This flow generates stress somewhat like osmotic pressures, hence, the designation "osmotic pressure hypothesis."

Powers, in a subsequent summary of his and other research [42], concludes that all three of the theories, with some modification, are required to account for the behavior of concrete subjected to freezing and thawing.

When freezing and thawing takes place in the presence of deicing chemicals, localized failures of the exposed surfaces occur that is called "scaling," or surface mortar deterioration. Except where the concentration of deicing chemicals is high enough to cause chemical attack, scaling results from freezing and thawing. Various theories have been advanced to explain the increased severity of the damage as compared with freezing and thawing in water. Browne and Cady [43] have recently summarized these theories and their own experiments. Although definitive answers have not been obtained, the mechanism is probably most influenced by concentration gradients. In addition to the lowering of the freezing temperature that accompanies increased deicer concentration, flow of water from areas of lower to those of higher concentrations generate stresses such as were described earlier. Verbeck and Klieger [44], in work confirmed by others, found that deterioration was greater for intermediate concentrations of deicing chemical (3 to 4%) than for lower or higher concentrations (to 16%). Such findings can be explained by generation of osmotic-like pressures accompanying the movement of water between areas of varying concentrations.

RAPID FREEZING AND THAWING TESTS

In 1971, ASTM C 290 and C 291 were combined as Procedures A and B in ASTM C 666. The method is designated as "rapid" because it permits alternately lowering the temperature of specimens from 4.4 to −17.8°C (40 to 0°F) and raising it from −17.8 to 4.4°C (0 to 40°F) in not less than 2 h nor more than 5 h. Thus, a minimum of four and a maximum of twelve complete cycles may be achieved during a 24-h period. The conventionally accepted term of testing is 300 cycles, which can be obtained in 25 to 63 days. For Procedure A, both freezing and thawing occur with the specimens surrounded by water, while in Procedure B, the specimens freeze in air and thaw in water. For Procedure A, the thawing portion is not less than 25% of the total cycle time, while for Procedure B, not less than 20% of the time is used for thawing. The time required for the temperature at the center of any single specimen to be reduced from 2.8 to −16.1°C (37 to 3°F) shall be not less than one half of the length of the cooling period, and the time required for the temperature at the center of any single specimen to be raised from −16.1 to 2.8°C (3 to 37°F) shall be not less than one half of the length of the heating period. For Procedure A, each specimen is surrounded by approximately 3 mm (⅛ in.) of water during the freezing and thawing cycles, while in Procedure B, the specimen is surrounded completely by air during the freezing phase of the cycle and by water during the thawing phase. The requirements for Procedure A are met by confining the specimen and surrounding water in a suitable container. The specimens are normally prisms not less than 76 mm (3 in.) nor more than 127 mm (5 in.) in width and depth, and between 279 and 406 mm (11 and 16 in.) long.

During the early years of testing by freezing and thawing, laboratories constructed specialized equipment. Currently, several manufacturers produce off-the-shelf or custom-built freeze-thaw equipment that meets the requirements of ASTM C 666. Capacities typically range from 18 to 80 specimens, although custom-built units have been designed for up to 120. Some of the equipment is designed for Procedure A only, while the larger units typically can be used for both A and B. Some aspects of these machines have been subject to criticism, but they do meet the needs for rapid testing within practical limits.

Deterioration of specimens is determined by the resonant frequency method, ASTM C 215. The fundamental transverse frequencies are determined at intervals not exceeding 36 cycles of exposure and are used to calculate the relative dynamic modulus of elasticity

$$P_c = \left(\frac{n_1^2}{n^2}\right) \times 100$$

where

P_c = relative dynamic modulus of elasticity, after c cycles of freezing and thawing, %;
n = fundamental transverse frequency at 0 cycles of freezing and thawing; and
n_1 = fundamental transverse frequency after c cycles of freezing and thawing.

The fundamental transverse frequency is determined with the specimens at a temperature of 5.0 ± 1.6°C (41 ± 3°F). Calculation of P_c assumes that the weight and dimensions of the specimens remain constant throughout the test. While this assumption is not true in many cases because of disintegration, the test is usually used to make comparisons between the relative dynamic moduli of specimens and P_c is adequate for the purpose.

The durability factor is calculated as

$$DF = PN/M$$

where

DF = durability factor of the specimen;
P = relative dynamic modulus of elasticity at N cycles, %;
N = number of cycles at which P reaches the specified minimum value for discontinuing the test or the specified number of cycles at which the exposure is to be terminated, whichever is less; and
M = specified number of cycles at which the exposure is to be terminated.

Because of the danger of damage to specimen containers and other parts of the equipment, testing usually is terminated when the relative dynamic modulus of elasticity falls below 50%.

The scope of ASTM C 666 states that "both procedures are intended for use in determining the effects of variations in the properties of concrete on the resistance of the concrete to the freezing and thawing cycles specified in the particular procedure. Neither procedure is intended to provide a quantitative measure of the length of service that may be expected from a specific type of concrete."

Procedure A is currently required in three ASTM specifications: namely, ASTM Specification for Air-Entraining Admixtures for Concrete (C 260), ASTM Specification for Chemical Admixtures for Concrete (C 494), and ASTM Specification for Chemical Admixtures for Use in Producing Flowing Concrete (C 1017). In these admixture specifications, the performance requirement is stated in terms of a "relative durability factor" calculated as

$$DF \text{ (or } DF_1) = PN/300$$

and

$$RDF = (DF/DF_1) \times 100$$

where

DF = durability factor of the concrete containing the admixture under test;
DF_1 = durability factor of the concrete containing a reference admixture (or in the case of ASTM C 494, only an approved air-entraining admixture);
P = relative dynamic modulus of elasticity in percent of the dynamic modulus of elasticity at zero cycles (values of P will be 60 or greater since the test is to be terminated when P falls below 60%);
N = number of cycles at which P reaches 60%, or 300 if P does not reach 60% prior to the end of the test (300 cycles); and
RDF = relative durability factor.

ASTM C 260, C 494, and C 1017 require that the relative durability factor of the concrete containing the admixture under test be at least 80 when compared with the reference concrete. The value of 80 is not intended to permit poorer performance than the reference concrete, but rather to assure the same level of performance, with appropriate recognition of the variability of the test method. The value of 80 was established before levels of precision were established for ASTM C 666, but is consistent with the now-established precision values.

One criticism of rapid freezing and thawing tests has been variability of the results both within and between laboratories. In response to the requirement by ASTM for precision statements in all test methods, ASTM Subcommittee C09.03.15 (now C09.67) reviewed the data from major published studies. Statistical analyses of these data, as described by Arni [45], showed that the variability was primarily a function of the level of durability factor of the concrete for the ranges of N and P normally used. In 1976, a precision statement was added to ASTM C 666 for expected within-batch precision for both Procedures A and B. Values for standard deviation (1S) and acceptable range (D2S), as defined by ASTM Practice for Preparing Precision and Bias Statements for Test Methods for Construction Materials (C 670), are given in Tables 2 and 3 of ASTM C 666 for ranges of average durability factor in ten increments and numbers of specimens averaged from two through six. Values for six specimens (as required by ASTM C 260, C 494, and C 1017) are given in Table 1.

These values confirm the long-recognized facts that the variability is less for very good or very poor concrete than for concretes of intermediate durability and that results from Procedure A are somewhat less variable than those obtained from Procedure B. An example of earlier work showing the variability at different durability factors is given in Fig. 1. This figure, from Cordon [7], shows durability factors of concretes containing a variety of aggregates, and with varying cement contents, water-cement ratios, and air contents. For very low and very high air contents, the measured durability factors showed relatively small variations. At intermediate values of air content, a "transition zone" reflected much greater variability of performance. One might conclude that, for intermediate levels of performance as indicated by durability factors, an undeterminable portion of the variability reflects the real variability of concretes with such characteristics and, therefore, uniform precisions would not be expected for the entire range of durability factors.

While ASTM C 666 is specified in the ASTM Specification for Lightweight Aggregates for Structural Concrete (C 330), no minimum durability factor values are given. The specification states that "in the absence of a proven

TABLE 1—Within-Laboratory Precisions for Averages of Six Beams Tested in Accordance With ASTM C 666.

Range of Average Durability Factor	Procedure A		Procedure B	
	Standard Deviation (1S)	Acceptable Range (D2S)	Standard Deviation (1S)	Acceptable Range (D2S)
0 to 5	0.3	0.9	0.4	1.2
5 to 10	0.6	1.8	1.7	4.7
10 to 20	2.4	6.8	3.3	7.4
20 to 30	3.4	9.7	4.3	12.2
30 to 50	5.2	14.7	6.3	17.8
50 to 70	6.4	18.1	8.2	23.2
70 to 80	4.7	13.4	7.0	19.7
80 to 90	2.3	6.5	3.6	6.8
90 to 95	0.9	2.4	1.6	4.5
Above 95	0.4	1.3	0.8	2.3

158 TESTS AND PROPERTIES OF CONCRETE

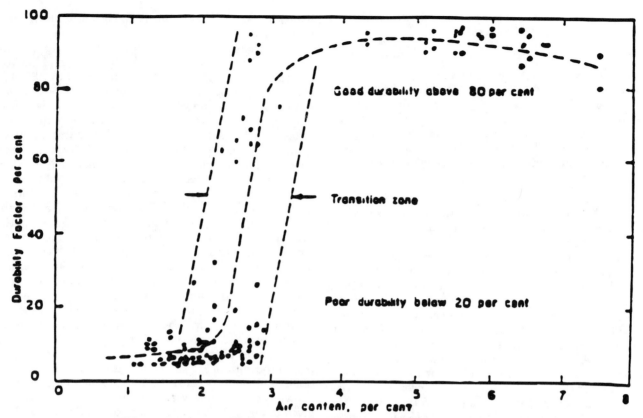

FIG. 1—Durability factors for concretes containing various aggregates, cement contents, and water-cement ratios, as a function of air content (after Cordon [7]).

freezing and thawing test satisfactory to the purchaser." ASTM C 666 is similarly listed as a method of sampling and testing in ASTM Specification for Concrete Aggregates (C 33), but again no minimum durability factors or other requirements are stated.

As noted, the primary measure of deterioration is the relative dynamic modulus calculated from determinations of resonant transverse frequency. Length change is noted in ASTM C 666 as an acceptable additional way of assessing deterioration. Weight loss is also sometimes used for such assessments, while determinations of reductions in tensile or compressive strength are used, but only infrequently, because such testing is destructive.

The question of which measure of deterioration is best to use is complicated by the facts that the different available tests measure different things, and the manner and extent to which the properties measured are related to freezing-and-thawing damage, especially under natural conditions, are matters of disagreement. Thus, the particular measure used often depends on the philosophy of the laboratory using it and on the particular purpose for which the tests are being made. Weight loss measures loss of material or sloughing from the surface of the specimens. Length change is based on the fact that internal damage is accompanied by expansion rather than contraction during cooling or by a permanent dilation after a freezing and thawing cycle, as discussed by Powers [31]. Either weight loss due to sloughing or reduction in resonant frequency may occur without accompanying length change, and, in Procedure A, weight loss often occurs without a reduction in resonant frequency or expansion. Resonant frequency and expansion reflect internal disruptions that are caused by unsound aggregates or deficient air-void characteristics, while weight loss reflects primarily surface mortar deterioration.

Users of ASTM C 666 have not reached a consensus on limiting values for the method's measurement quantities. A 1987 survey of State and Canadian provincial highway agencies indicated seven used ASTM C 666 to qualify coarse aggregates for construction on agency projects [46]. The seven had "seven different specification limiting values, based on the fact that each had a unique way of testing regarding aggregate grading, moisture conditioning, cement contents, air contents, coarse aggregate contents, curing methods and time, cycle length, number of cycles, method of measuring deterioration, failure criteria, and number of specimens constituting a test." Further, three agencies used durability factor, three used length change or another measure of expansion, and one used both for specification limits. A standard practice, which would include research-supported limiting values, is clearly desirable.

One important potential influence on ASTM C 666 results is the effect of the container used to hold the specimen during testing with Procedure A. Rigid containers have the potential to damage specimens and are not permitted. A note explains further: "Experience has indicated that ice or water pressure, during freezing tests, particularly in equipment that uses air rather than a liquid as the heat transfer medium, can cause excessive damage to rigid metal containers, and possibly to the specimens therein. Results of tests during which bulging or other distortion of containers occurs should be interpreted with caution." This situation is particularly noticeable when the specimens are exposed with the long dimension vertical. Ice forms quickly on the open top rather than uniformly along the long dimension of the container. Pressures from continued conversion of water to ice between the specimen and container cannot be relieved, with consequent bulging of the container. Various adjustments have been made to mitigate this problem, including exposure of the specimens in containers with the long dimensions horizontal to provide a larger open area; inclusion of flexible windows, or corners, in containers; or inclusion of a rubber ball in the bottom of containers to absorb the pressures. All of these have been reported to overcome the problem by some and not by others. That the degree of restraint offered by the container influences the number of cycles required to reach a specified level of relative dynamic modulus is clear, as shown by Cook [47], who found that the number of cycles necessary to reach a given level of durability factor was increased dramatically when the specimens were tested in rubber containers as compared with containers made of steel. One agency has minimized the container influence by using cylindrical specimens and surrounding them with rubber boots [46].

Another important influence on the results is the degree of saturation of the concrete and the aggregates both at the time of mixing and throughout the course of the testing. The specifications citing the use of ASTM C 666 require that testing begin after 14 days of moist curing. (A note in ASTM C 666 itself says that the tests should be started when the specimens are 14 days old "unless some other age is specified.") The mixing procedures referenced in ASTM Test Method for Making and Curing Concrete Test Specimens in the Laboratory (C 192) require that coarse aggregates be immersed for 24 h prior to mixing. Thus, the comparatively high degree of internal saturation and the early age at which testing begins result in a relatively severe test when compared with field exposures in which a period of drying normally occurs before exposure to freezing and thawing. Even a brief period of drying greatly improves resistance to freezing and thawing, since it is difficult to resaturate concrete that has undergone some drying. Strategic Highway Research Program researchers have recently confirmed the effect of drying. They found that Procedure B results approach those from Procedure A if the specimens are wrapped in terrycloth during the test; this keeps the outer surface of the concrete moist, that is, at or near saturation, rather than allowing it to dry out during the freezing-in-air portion of the test cycle.

Still another influence on freezing and thawing tests is the use of salt water (typically 2% sodium chloride in water) as the freezing-and-thawing medium. Some agencies have used salt water since that is the medium encountered in many field situations where water ponds on structures regularly treated with deicers. Use of salt water in Procedure A increases the severity of the test beyond that attained with the use of fresh water.

Philleo has taken ASTM C 666 to task, recommending that it be "modified or replaced by a more realistic standard for judging the acceptability of concrete for field applications" [48]. He points out that most concrete undergoes both drying and curing longer than 14 days before encountering its first freezing. Nevertheless, accommodating the many possible variations of saturation that might be encountered is impractical so that the most consistently reproducible condition is that of continued moist curing.

Because the testing is initiated at a fixed age, considerable variation of strength at the time of exposure to freezing and thawing may be encountered with cements of different strength gain characteristics for concrete with sound aggregates and satisfactory air-void characteristics. There is not a great body of data on the influence on resistance to freezing and thawing of strength at the time that exposure begins. Buck et al. [49], reported tests indicating that, because of its relationship with the amount of freezable water, a given level of maturity (strength) was necessary to provide an acceptable degree of frost resistance as indicated by a durability factor of 50 for concrete containing satisfactory aggregates and entrained air. They cited earlier work by Klieger [50], who reported similar findings in his studies of salt scaling. Consideration of strength at the time of initial freezing is particularly important in testing concrete made with blended cements and pozzolanic admixtures that gain strength more slowly than concrete without such admixtures. This influence is minimized when the evaluations are made by comparing concretes made with similar materials as required in ASTM C 260 and C 494.

Visual examination of specimens during ASTM C 666 cycling may also give warnings of the likelihood of popout problems in a concrete. Popouts are shallow, usually conical spalls of the concrete through aggregate particles due to internal pressure and can be attributed to defects in the aggregate. The small size of the specimens in ASTM C 666 has been criticized because a large piece of popout-producing aggregate in the center of a relatively small specimen could cause it to fail; in the field, the popout material would presumably only cause superficial surface defects [51].

Cyclic freezing and thawing methods were developed and applied only for laboratory-mixed concrete until 1975, when they were extended to cores or prisms cut from hardened concrete. Experience with testing of specimens from hardened concrete is limited; the results of a 1987 survey of State and Canadian provincial highway agencies gave no indication that any were using anything other than specimens cast to size specifically for freeze-thaw testing [46].

DILATION METHODS

In 1955, Powers published a critical review of existing cyclic methods for freezing and thawing tests [31]. He was particularly critical of what he considered unrealistically high freezing rates of from 6 to 60°C/h (10 to 100°F/h) as compared with natural cooling rates that seldom exceed 3°C/h (5°F/h). He also noted the significance of moisture conditioning of aggregates and concretes in the test methods that generally provided for only saturated concrete as compared with natural exposures where some seasonal drying is possible.

He stated that durability was not a measurable property but that expansion that occurred during a slow cooling cycle when the concrete or its aggregates became critically saturated was measurable and would provide an indication of potential resistance to damage by freezing and thawing. He proposed that specimens be prepared and conditioned so as to simulate field conditions and then be subjected to periodic slow-rate freezing and storage in water at low temperature between freezing exposures. Concrete subject to frost damage should reach some critical saturation level, after which it would expand on freezing. The length of time required to become critically saturated would be compared with the field exposures to be encountered. If the period during which freezing would be expected was less than the time to reach critical saturation, then no damage would be expected. As opposed to a single durability factor, the time to critical saturation could be more readily interpreted for various field exposures.

The California Division of Highways was first to report, in 1961, a practical application of Power's proposed method [32]. They developed specimen preparation and conditioning methods, testing and measuring techniques, and performance criteria. The method was used to evaluate several aggregates for a major highway construction project. The aggregates judged acceptable by the California procedure would have been rejected by other conventionally accepted criteria. The concrete is now more than 30 years old; sections have been overlaid because they are structurally inadequate, but the freeze-thaw performance of all of the concrete has been reported as continuing to be satisfactory.

The procedures proposed by Powers and the method developed in California were extensively studied and refined by Larson and others [33]. In 1971, ASTM C 671 was standardized. This method provides for cylindrical specimens 75 mm (3 in.) in diameter and 150 mm (6 in.) long that are stored in water at 1.7 ± 0.9°C (35 ± 2°F). At two-week intervals, the specimens are cooled in water-saturated kerosene at a rate of 2.8 ± 0.5°C/h (5 ± 1°F/h). During the cooling cycle, the specimen is placed in a strain frame to permit measurement of length change. A typical plot of length change versus temperature is shown in Fig. 2. Prior to critical saturation, the length change will proceed along the dashed curve without dilation. Critical dilation is defined as a sharp increase (by a factor of 2 or more) between dilations on successive cycles. Highly frost-resistant concrete may never exhibit critical dilation. ASTM C 682 was also standardized in 1971. This practice is based largely on the work of Larson and his coworkers. Procedures essentially in accordance with ASTM C 671 have been used by Buck [52], and he found that a specimen that is frost resistant will not show increasing dilation with continuously decreasing temperature; it will show some limited initial dilation as all moist specimens do, but dilation will not continue throughout the cooling. In

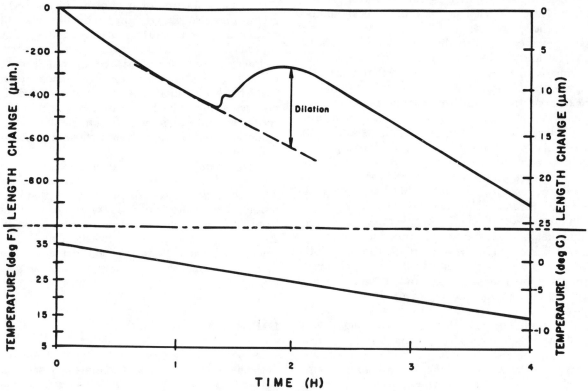

FIG. 2—Typical length change and temperature charts.

addition to ASTM C 671 criterion for critical dilation in terms of increase from one cycle to another, he suggested that a criterion for critical dilation applicable to results of a single test be as follows:

(a) If the dilation is 0.005% (= 50 millionths) or less, the specimen may be regarded as frost resistant, that is, the dilation is not critical.
(b) If the dilation is 0.020% (= 200 millionths) or more, the specimen may be regarded as not frost resistant, that is, critical dilation has been exceeded.
(c) If the dilation is in the range between 0.005 and 0.020%, an additional cycle or more should be run.

ASTM C 671 and C 682 have not been used extensively. The 1987 survey of State and Canadian provincial highway agencies indicated only one was using ASTM C 671 to any extent [46]. Although the apparatus is comparatively inexpensive and larger numbers of specimens can be processed than with ASTM C 666 equipment, significant storage capacity is required. The procedures of ASTM C 682 are extensive and complex, largely because of requirements designed to bracket a broad range of potential exposure conditions. It is suggested that aggregates and concrete be "maintained or brought to the moisture condition representative of that which might be expected in the field." However, it is noted that "aggregate moisture states other than dry or saturated are very difficult to maintain during preparation of specimens. Reproducibility of overall test results is likely to be affected adversely by variability in aggregate moisture." While the complexity of the evaluation procedure limits its general applicability, the procedures may be justified where large projects or economic consequences of detailed aggregate evaluation are warranted. ASTM C 671 currently says it "is suitable for ranking concretes according to their resistance to freezing and thawing for defined curing and conditioning procedures." Both ASTM C 671 and C 682 warn that "the significance of the results in terms of potential field performance will depend upon the degree to which field conditions can be expected to correlate with those employed in the laboratory."

SCALING RESISTANCE

In the early 1960s, it became apparent that the increasing use of deicing chemicals as part of a "bare pavement" policy adopted for the nation's highways was being reflected in widespread surface scaling of pavements and bridge decks. It has long been known that dense, high quality concrete, with adequate entrained air and with adequate curing and a period of drying before the first application of deicing agents, is essential in preventing damage [12]. Widespread scaling demonstrated that all of these requirements were not being met consistently and brought forth a plethora of remedial or preventive products including admixtures, surface treatments, and curing agents. In 1971, ASTM C 672 was standardized. The test is based on the experience of a number of agencies who used blocks that were fabricated to permit ponding of water on one surface and that could be exposed to freezing and thawing in the presence of deicing agents. The test "covers determination of the resistance to scaling of a horizontal concrete surface subject to freezing and thawing cycles in the presence of deicing chemicals. It is intended for use in evaluating the surface resistance qualitatively by visual examination. This test method can be used to evaluate the effect of mixture proportioning, surface treatment, curing, or other variables on resistance to scaling. This test method is not intended to be used in determining the durability of aggregates or other ingredients of concrete." The report of ACI Committee 201 says, on the other hand, the "use of ASTM C 672 will demonstrate the acceptability or failure of a given concrete mixture" [12].

The specimens must have a surface area of at least 0.046 m^2 (72 $in.^2$) and be at least 75 mm (3 in.) deep. The specimens are placed in a freezing space after moist curing for 14 days and air storage for 14 days. Provisions are included for applications of protective coatings if desired at the age of 21 days. The method calls for covering the surface with approximately 6 mm (¼ in.) of a solution of calcium chloride and water having a concentration such that each 100 mL of solution contains 4 g of anhydrous calcium chloride. Modifications of the deicer and application procedures, including freezing of water and addition of the solid deicer, are allowed where there is need to evaluate the specific effect.

The specimens are cycled through a freezing environment for 16 to 18 h, followed by laboratory air for 6 to 8 h. While the method describes laboratory procedures, it has been used for outdoor exposures as well.

The specimens are rated visually according to a scale from 0 (no scaling) through 5 (severe scaling) after 5, 10, 15, 25, and every 25 cycles thereafter. Some investigators have weighed the detritus, but this is not required by the method.

As noted in the method, the ratings are ranks, and as such may not be subjected appropriately to analyses based on the calculation of averages and standard deviations or other techniques that assume continuous distributions. If groups of similar specimens are to be reported or compared with other groups, such nonparametric quantities as the median and range may be used.

Experience with ASTM C 672 generally has been satisfactory for evaluating the variables for which it was developed.

OTHER WEATHERING PROCESSES

Other weathering processes that have been suspected of causing deterioration include heating and cooling and wetting and drying. As noted, chemical attack and alkali-aggregate reactions are treated elsewhere in this publication. Although cases have been reported where deterioration was attributed to aggregate with an abnormally low coefficient of thermal expansion [53] or where the freezing and thawing resistance was influenced by aggregates with different coefficients of thermal expansion [54, 55], "thermal incompatibility" is now generally believed to have at most a minor effect on concrete durability within the normal temperature range. Elevated temperatures such as are encountered in certain parts of nuclear construction are beyond the scope of this paper.

While wetting and drying induce variations in moisture content that influence the resistance to freezing and thawing, aggregate cracking from excessive drying shrinkage, and an increased concentration of dissolved salts, all of which reduce resistance to weathering, the writers are not aware of cases where alternate wetting and drying per se have caused deterioration.

SUMMARY

The resistance of concrete to weathering in the absence of chemical attack or detrimental cement-aggregate reactions depends on its ability to resist freezing and thawing. Dry concrete will withstand freezing and thawing indefinitely, whereas highly saturated concrete exposed to particularly severe conditions such as hydraulic head or very low temperatures may be severely damaged in a few cycles. This damage is more likely to occur when the air content of the concrete is at the lower end of the recommended range. Research and experience have shown that resistance to freezing and thawing requires a low water/cement (w/c) ratio, an adequate volume of entrained air with the proper void distribution and characteristics, and exposures that reduce the opportunity for critical saturation. While adequate air entrainment with proper air-void parameters protects the paste, it may not overcome the effect of aggregate that is susceptible to damage by freezing and thawing. Direct translation of results from laboratory freezing and thawing tests is difficult at best because of the variety of exposures encountered, but the currently approved standards are very useful when properly conducted and interpreted and have undoubtedly resulted in a significant improvement in the resistance of concrete to weathering from natural forces.

REFERENCES

[1] Scholer, C. H., "Durability of Concrete," *Report on Significance of Tests of Concrete and Concrete Aggregates, ASTM STP 22A*, American Society for Testing and Materials, Philadelphia, 1943.
[2] Scholer, C. H., "Hardened Concrete, Resistance to Weathering—General Aspects," *Significance of Tests and Properties of Concrete and Concrete Aggregates, ASTM STP 169*, American Society for Testing and Materials, Philadelphia, 1955.
[3] Powers, T. C., "Resistance to Weathering—Freezing and Thawing," *Significance of Tests and Properties of Concrete and Concrete Aggregates, ASTM STP 169*, American Society for Testing and Materials, Philadelphia, 1955.
[4] Arni, H. T., "Resistance to Weathering," *Significance of Tests and Properties of Concrete and Concrete-Making Materials, ASTM STP 169A*, American Society for Testing and Materials, Philadelphia, 1966.
[5] Newlon, H., Jr., "Resistance to Weathering," *Significance of Tests and Properties of Concrete and Concrete-Making Materials," ASTM STP 169B*, American Society for Testing and Materials, Philadelphia, 1978.
[6] Woods, H., "Durability of Concrete Construction," *ACI Monograph No. 4*, American Concrete Institute, Detroit, MI, 1968.
[7] Cordon, W. A., "Freezing and Thawing of Concrete—Mechanisms and Control," *ACI Monograph No. 3*, American Concrete Institute, Detroit, MI, 1966.
[8] Larson, T., Cady, P., Franzen, M., and Reed, J., "A Critical Review of Literature Treating Methods of Identifying Aggregates Subject to Destructive Volume Change When Frozen in Concrete and a Proposed Program of Research," *Special Report 80*, Highway Research Board, Washington, DC, 1964.
[9] "Durability of Concrete: Physical Aspects," *Bibliography 20*, NAS-NRC Publication 493, Highway Research Board, Washington, DC, 1957.
[10] "Durability of Concrete: Physical Aspects: Supplement to Bibliography No. 20," *Bibliography 38*, NAS-NRC Publication 1333, Highway Research Board, Washington, DC, 1966.
[11] Snyder, M. B. and Janssen, D. J., "Freeze-Thaw Resistance of Concrete—An Annotated Bibliography," Report SHRP-C/UFR-92-617, Strategic Highway Research Program, Washington, DC, 1992; available from Transportation Research Board, Washington, DC.
[12] "Guide to Durable Concrete," *Manual of Concrete Practice*, ACI 201.2R-92, Part 1, American Concrete Institute, Detroit, MI, 1993.
[13] "Durability of Concrete," *Publication SP-47*, American Concrete Institute, Detroit, MI, 1975.
[14] "Concrete Durability—Katherine and Bryant Mather International Conference," *Publication SP-100*, American Concrete Institute, Detroit, MI, 1987.
[15] "Durability of Concrete," *Publication SP-126*, American Concrete Institute, Detroit, MI, 1991.
[16] Vicat, L. J., *Treatise on Calcareous Mortars and Cements*, (translated from the French by J. T. Smith), J. Weale, London, 1837.
[17] "Symposium on Freezing-and-Thawing Tests of Concrete," *Proceedings*, American Society for Testing and Materials, Vol. 46, 1946.
[18] Wuerpel, C. E. and Cook, H. K., "Automatic Accelerated Freezing-and-Thawing Apparatus for Concrete," *Proceedings*, American Society for Testing and Materials, Vol. 45, 1945.
[19] Walker, S. and Bloem, D. L., "Performance of Automatic Freezing-and-Thawing Apparatus for Testing Concrete," *Proceedings*, American Society for Testing and Materials, Vol. 51, 1951.
[20] Arni, H. T., Foster, B. E., and Clevenger, R. A., "Automatic Equipment and Comparative Test Results for the Four ASTM Freezing-and-Thawing Methods for Concrete," *Proceedings*, American Society for Testing and Materials, Vol. 56, 1956.
[21] Cordon, W. A., "Automatic Freezing-and-Thawing Equipment for a Small Laboratory," *Bulletin No. 259*, NAS-NRC Publication 768, Highway Research Board, Washington, DC, 1960.
[22] Cook, H. K., "Automatic Equipment for Rapid Freezing-and-Thawing of Concrete in Water," *Bulletin No. 259*, NAS-NRC Publication 768, Highway Research Board, Washington, DC, 1960.
[23] Hericart de Thury, "On the Method Proposed by Mr. Brard for the Immediate Detection of Stones Unable to Resist the Action of Frost," *Annales de Chimie et de Physique*, Vol. 38, 1828, pp. 160–192.

[24] Henry, J., "On the Mode of Building Materials and an Account of the Marble Used in the Extension of the U.S. Capitol," *American Journal of Science*, Vol. 72, 1856.

[25] Grun, R., "Investigations of Concrete in Freezing Chambers," *Zement*, Vol. 17, 1928.

[26] Scholer, C. H., "Some Accelerated Freezing and Thawing Tests," *Proceedings*, American Society for Testing and Materials, Vol. 28, 1928.

[27] Mattimore, H. S., "Durability Tests of Certain Portland Cements," *Proceedings*, Highway Research Board, Vol. 16, 1936.

[28] Withey, M. O., "Progress Report, Committee on Durability of Concrete," *Proceedings*, Highway Research Board, Vol. 24, 1944.

[29] Foster, B. E., "Report on Cooperative Freezing and Thawing Tests of Concrete," *Special Report No. 47*, Highway Research Board, Washington, DC, 1959.

[30] ASTM Committee C9, minutes of December 1961 Meeting.

[31] Powers, T. C., "Basic Considerations Pertaining to Freezing-and-Thawing Tests," *Proceedings*, American Society for Testing and Materials, Vol. 55, 1955.

[32] Tremper, B. and Spellman, D. L., "Tests for Freeze-Thaw Durability of Concrete Aggregates," *Bulletin 305*, Highway Research Board, Washington, DC, 1961.

[33] Larson, T. D. and Cady, P. D., "Identification of Frost-Susceptible Particles in Concrete Aggregates," NCHRP Report No. 66, Highway Research Board, Washington, DC, 1969.

[34] Schwartz, D. R., "D-Cracking of Concrete Pavements," *NCHRP Synthesis No. 134*, Transportation Research Board, Washington, DC, 1987.

[35] Stark, D., "Characteristics and Utilization of Coarse Aggregates Associated with D-Cracking," *Living with Marginal Aggregates, ASTM STP 597*, American Society for Testing and Materials, Philadelphia, 1976, p. 45.

[36] Powers, T. C., "A Working Hypothesis for Further Studies of Frost Resistance of Concrete," *Journal*, American Concrete Institute, Detroit, MI, Feb. 1945.

[37] Powers, T. C., "The Air-Requirement of Frost Resistant Concrete," *Proceedings*, Highway Research Board, Vol. 29, 1949.

[38] Mielenz, R. C., Wolkodoff, V. E., Burrows, R. W., Backstrom, J. L., and Flack, H. E., "Origin, Evolution and Effects of the Air Void System in Concrete," *Journal*, American Concrete Institute, Detroit, MI, July 1958, Aug. 1958, Sept. 1958, and Oct. 1958.

[39] Taber, S., "The Mechanics of Frost Heaving," *The Journal of Geology*, Vol. 38, No. 4, 1930.

[40] Verbeck, G. and Landgren, R., "Influence of Physical Characteristics of Aggregate on the Frost Resistance of Concrete," *Proceedings*, American Society for Testing and Materials, Vol. 60, 1960.

[41] Powers, T. C., "The Mechanism of Frost Action in Concrete," Stanton Walker Lecture No. 3, National Sand and Gravel Association, National Ready-Mix Concrete Association, Silver Spring MD, 1965.

[42] Powers, T. C., "Freezing Effects in Concrete," *Durability of Concrete, Publication SP-47*, American Concrete Institute, Detroit, MI, 1975.

[43] Browne, F. P. and Cady, P. D., "Deicer Scaling Mechanisms in Concrete," *Durability of Concrete, Publication SP-47*, American Concrete Institute, Detroit, MI, 1975.

[44] Verbeck, G. J. and Klieger, P., "Studies of 'Salt' Scaling of Concrete," *Bulletin 150*, Highway Research Board, Washington, DC, 1957.

[45] Arni, H. T., "Precision Statements Without an Interlaboratory Test Program," *Cement, Concrete, and Aggregates*, Vol. 1, No. 2, 1979.

[46] Vogler, R. H. and Grove, G. H., "Freeze-Thaw Testing of Coarse Aggregate in Concrete: Procedures Used by Michigan Department of Transportation and Other Agencies," *Cement, Concrete, and Aggregates*, Vol. 11, No. 1, 1989.

[47] Cook, H. K., "Effects of Fluid Circulation and Specimen Containers on the Severity of Freezing and Thawing Tests," informal presentation, Session 13, Annual Meeting of Highway Research Board, Washington, DC, Jan. 1963.

[48] Philleo, R. E., "Freezing and Thawing Resistance of High-Strength Concrete," *NCHRP Synthesis No. 129*, Transportation Research Board, Washington, DC, 1986.

[49] Buck, A. D., Mather, B., and Thornton, H. T., Jr., "Investigation of Concrete in Eisenhower and Snell Locks St. Lawrence Seaway," Technical Report 6-784, Waterways Experiment Station, Vicksburg, MS, July 1967.

[50] Klieger, P., "Curing Requirements for Scale Resistance of Concrete," *Bulletin 150*, Highway Research Board, Washington, DC, 1957.

[51] Sturrup, V., Hooton, R., Mukherjee, P., and Carmichael, T., "Evaluation and Prediction of Concrete Durability—Ontario Hydro's Experience," *Concrete Durability—Katherine and Bryant Mather International Conference, Publication SP-100*, American Concrete Institute, Detroit, MI, 1987.

[52] Buck, A. D., "Investigation of Frost Resistance of Mortar and Concrete," Technical Report C-76-4, U.S. Army Engineer Waterways Experiment Station, Vicksburg, MS, 1976.

[53] Pearson, J. C., "A Concrete Failure Attributed to Aggregate of Low Thermal Coefficient," *Proceedings*, American Concrete Institute, Vol. 48, 1952.

[54] Callan, E. J., "Thermal Expansion of Aggregates and Concrete Durability," *Proceedings*, American Concrete Institute, Vol. 48, 1952.

[55] Higginson, E. C. and Kretsinger, D. G., "Prediction of Concrete Durability from Thermal Tests of Aggregate," *Proceedings*, American Society for Testing and Materials, Vol. 53, 1953.

17

Corrosion of Reinforcing Steel

William F. Perenchio[1]

PREFACE

The first volume, *ASTM STP 169*, did not contain a separate chapter on corrosion. When it was published in 1956, corrosion of reinforcing steel had not yet become a widespread problem. Heavy use of deicing salts in urban and industrial areas of the north soon changed that. By the time *ASTM STP 169A* was published with a corrosion chapter in 1966, the author, Bailey Tremper, a consultant in California, included 35 references, 22 of which had been published since *ASTM STP 169*. Clearly, the problem had grown. By 1978, then-current research on corrosion was widespread. Philip D. Cady, professor of civil engineering at Pennsylvania State University and an active researcher himself, rewrote the corrosion chapter. His list of references included over 40 written since *ASTM STP 169A* was published. For this publication I have concentrated on presenting the present state of the art, fully aware of all of the solid background covered by my predecessors.

INTRODUCTION

Many metals, including reinforcing steel, are normally well protected against corrosion by embedment in high-quality concrete. However, many of the bridges, parking garages, and marine structures in this country have been severely damaged by corrosion of reinforcement and the resultant increased volume. These and other structures have suffered damage due to the penetration of chloride in sufficient amounts to the depth of the reinforcing steel. In the case of bridges and parking structures, the source of the chloride is usually deicing salts. In marine structures, it is seawater. In others, it may be chemical admixtures added to the concrete that contain chlorides for set acceleration.

Cracks are not necessary for chloride penetration into the concrete. Salts are capable of penetrating solid concrete along with the water in which they are dissolved. Variations in the amount of chloride, or moisture, or oxygen at different points along a reinforcing bar, or between different bars, can cause voltage differentials to develop that greatly increase the rate of corrosion. The increased volume due to the development of corrosion products can cause spalling of the concrete cover, and sufficient reduction in the cross section of the bar to render a structure incapable of safely supporting its design loads.

Such chloride-induced corrosion is a serious problem. Durable repair procedures are available; however, all are expensive. Significant improvements in methods of protecting embedded metals from corrosion at the design stage have been made in recent years. This chapter will discuss the mechanisms of corrosion, typical damage caused by it, investigative techniques for evaluating its extent, means of preventing it in new construction, and remedial measures.

MECHANISMS OF CORROSION

When a metal corrodes, it returns to its natural state, which is usually the oxide or the hydroxide. Metals tend to do this because, in their metallic state, they are at a higher energy level. Materials tend to seek lower energy levels; hence, the tendency to corrode. Few metals (except for the noblest ones: platinum, gold, silver) are ever found in nature in the metallic (elemental) state.

Because of this tendency to return to the state from which they were refined, one might expect that metals would be unusable, quickly disappearing while in service. However, a film of tightly adhering corrosion products usually forms, as in concrete, that protects the remaining metal from the corrosive conditions [1]. If these products become soluble in adjacent liquid, rapid corrosion can be expected. This is the apparent reason that chloride is effective in instigating rapid corrosion of steel in concrete.

Any chemical action can be regarded as electrochemical, since it involves transfer or displacement of electrons. However, the term is sometimes applied only to cases of corrosion where anodes and cathodes are some finite distance apart, making the flow of electrical current over measurable distances a part of the process [2]. Corrosion, of course, can take place on reinforcing steel that is exposed only to air and rain. This is, therefore, sometimes termed atmospheric corrosion [3,4]. An electrolyte is not necessary; however, even a thin film of rain water can act as one, carrying ions and completing the electrical circuit. In this case, the anodes and cathodes are contained in the same bar, separated by only small distances.

Concrete provides protection for embedded metals, if it is high quality and if the depth of cover is about 2 in. (50 mm) or more. However, if either the quality or the depth of cover, or both, are low enough, corrosion of steel can occur when:

---

[1] Senior consultant, Wiss, Janney, Elstner Associates, Inc., Northbrook, IL 60062.

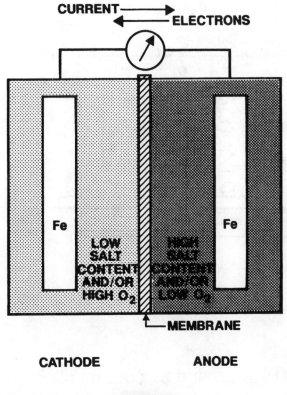

FIG. 1—Electolytic cell.

Primary reactions:
$O_2 + 2H_2O + 4e^- \rightarrow 4OH^-$ $Fe \rightarrow Fe^{++} + 2e^-$

1. the concrete cover becomes carbonated, dropping the pH from about 13 to about 9; or
2. chloride ions have penetrated into the concrete such that the concrete adjacent to the steel contains at least 1.3 lb of chloride ions/yd³ (0.77 kg of chloride ions/m³).

In the case of reinforcing steel, if the corrosion develops because the cover concrete has become carbonated, the rate of corrosion in relatively slow. If, however, the corrosion is instigated by the presence of chloride ions, the rate can be many times as fast as atmospheric corrosion [2] and it causes much greater damage. This type of corrosion involves voltage differences and transfer of electrons and electrical current, in addition to various chemical reactions.

Chloride-induced corrosion is a very common cause of concrete deterioration in all types of structures along the sea coast and in northern states where salt is used to remove ice from roads, streets, parking structures, balconies on high-rise buildings, etc. The damage occurs at a more rapid rate in warmer climates because it is due primarily to chemical reactions. Most chemical reactions double in rate for every 18°F (10°C) increase in temperature [3].

A simplified electrolytic cell is shown in Fig. 1. In the figure, a voltage difference has been set up by a low salt content surrounding the iron electrode on the left side and a high salt content on the right. The difference in voltage could be increased by bubbling oxygen (or air) into the left side. It is this difference in voltage that drives the cell and causes the corrosion to proceed rapidly.

The primary reactions that take place at each electrode are shown below the cell. Corrosion, or the conversion of metallic iron into ferrous ions in solution, occurs at the anode, leaving two electrons behind for every iron ion released. These remain within the electrode and, if no suitable cathode is available, will eventually stop the corrosion reaction. This is called anodic polarization. If a cathode is available and is connected to the anode by a material, such as metal or carbon, that is capable of transferring electrons (carrying electrical current[2]), these extra electrons can be shed by the reaction shown. Oxygen and water combine with the electrons to produce hydroxyl ions. The electrical circuit is completed through the solutions surrounding the electrodes and the semi-permeable membrane by the movement of ions such as OH^-, Cl^-, Na^+, K^+, and Ca^{++}.

The common (or now, old-fashioned) flashlight battery is an example of an electrolytic cell that produces a usable product, light. (This is called the Leclanche' cell in physical chemistry textbooks.) The case of the battery is made of zinc metal and the center pole is made of carbon. Between them is a mixture of manganese oxide, pulverized carbon, ammonium chloride, and water. It is of paste consistency rather than a solution, but still acts as an electrolyte. The battery operates by the zinc going into solution, leaving electrons behind, which travel to the carbon cathode in the center of the cell. This flow of electrons excites the filament in the flashlight bulb, producing light.

A more detailed and complete description of electrolytic cells is given by Uhlig [2].

Extending this battery concept to a more practical problem, that of corrosion of reinforcing steel in concrete, consider the schematic diagram in Fig. 2. This figure depicts the typical conditions found in a bridge deck or a floor in a parking structure, in areas where deicing salts

---

[2] Although rubbing two dissimilar metals together was found to attract bits of paper or pith, and sparks had been made to jump gaps, for over a hundred years previously, in 1729, Stephan Grey found he could transfer this "force" along a damp thread several hundred feet long and attract a feather at the other end. Because of this capacity for flow, electricity was thought to be some kind of fluid, or two different fluids. Shortly thereafter, Charles Dufay named these two "vitreous" and "resinous" electricity for the opposite charges produced by rubbing glass or resin.
In 1747, Benjamin Franklin, well known for his electrical experiments, suggested a "one-fluid system," calling it positive or negative depending on the direction of flow. Because he had no way of knowing which direction the current was flowing, he arbitrarily decided it was from positive to negative. One hundred years later, J. J. Thompson demonstrated that electrons flow from negative to positive, but Franklin's convention was so well established by then that we still refer to "current" as flowing in the direction opposite to electron flow. (Taken from Maxey Brooks, "Why is a Cathode Called a Cathode," *Materials Performance*, National Association of Corrosion Engineers, Vol. 30, No. 6, Houston, TX, June 1991.)

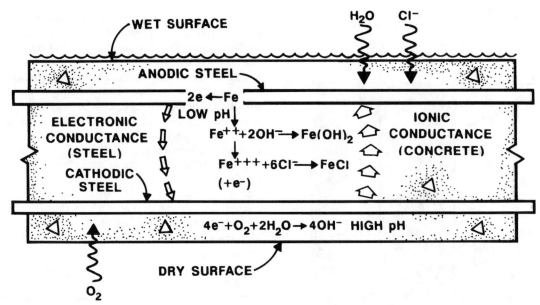

FIG. 2—Electrolytic corrosion inside concrete.

are used. The salts typically consist essentially of sodium chloride.

The top layer of steel is termed the anodic steel because this is where most of the corrosion occurs. It becomes anodic because it is much closer to the source of water and chloride ions; therefore, the surrounding concrete becomes inundated with more of each than does the concrete near the bottom steel, which becomes cathodic. Therefore, the iron that makes up the majority of the steel composition goes into solution, leaving electrons behind. The electrons find their way to the bottom mat by metallic continuity brought about by bent bars, bar chairs, etc. The electrons are then shed into the solution in the pores of the portland cement paste by combining with water and oxygen to produce hydroxyl ions, as indicated by the reaction shown, the same as that in Fig. 1. Ionic conductance within the concrete completes the circuit.

The voltage difference between the top and bottom mats of steel are due in part to differences in concentration of water, chloride ion, and oxygen. As previously stated, the top steel has more water and chloride in the cement pores surrounding it, while the bottom steel has more oxygen because there is less water in the concrete at that depth and below. Oxygen (and other gases) can diffuse into dry concrete much easier than wet concrete because, with wet concrete, it must diffuse through pore water rather than pore air, a much slower process.

The preceding discussion is presented as a common example of macrocell corrosion. However, it is not meant to imply that separations of anode and cathode are always so great. Microcells can be set up along a given bar, separated only by tiny fractions of an inch. Bits of mill scale can act as cathodes that drive an adjacent anode in the steel [2]. Local chemical or physical differences in or on the metal can also cause small voltage differences. Such localized cells can result in serious pit corrosion.

DAMAGE CAUSED BY CORROSION

Damage to concrete, in the form of cracking, delamination, and spalling, can occur due to atmospheric corrosion; that is, the type brought on by carbonation of the cover concrete. However, this is slow to develop and, depending on the external environment, may never cause serious deterioration. The author was asked to visit Tiger Stadium at Louisiana State University to inspect the football stadium, parts of which were 70, 50, 40, and as few as 10 years old. The oldest portions were badly stained by reinforcing steel corrosion products, but very little deterioration was observed. The oldest concrete was also of the poorest quality. Therefore, carbonation progressed rapidly, allowing atmospheric corrosion to take place. But the same high porosity that allowed the rapid carbonation also provided space for the corrosion products, and disruptive pressures did not soon develop. Before carbonation, the steel was passivated by the high pH of the concrete. After carbonation and the resultant drop in pH, the passivation was lost.

Chloride-induced corrosion, by comparison, is usually much more destructive. It does not depend on carbonation and can develop even in good quality concrete. All that is necessary is a reasonable supply of water (75% humidity appears to be sufficient), and at least the threshold amount of chloride ions at the anode. The cathode must have good access to oxygen. The voltages that develop cause corrosion to advance so quickly that cracking, delamination, and spalling occurs in the cover concrete. The deterioration, of course, occurs because the solid volume of the corrosion products is many times the solid volume of the original metal, in the case of iron. This increase in volume is largely due to the production of solids from reaction of the metals with gases or liquids or dissolved solids.

When steel corrodes, it produces many types of corrosion products (collectively called rust), depending primarily on oxygen, chloride, and water availability. Most of the products are amorphous (no crystalline structure determinable by X-ray diffractometry) ferrous (Fe^{++}) and ferric (Fe^{+++}) oxides, hydroxides, chlorides and hydrates, and complexes of these.

The only usual crystalline substance detected is magnetite (Fe_3O_4), but exposure to air and slow drying produces more crystalline oxides. Occasionally, someone new to X-ray analysis interprets the pattern as simply magnetite, ignoring the characteristic "amorphous hump" in the pattern that indicates that the rust is almost entirely amorphous. Crystalline corrosion products such as magnetite occupy two to three times the space that the original iron (or steel) did, while the amorphous products are generally more voluminous and variable, depending on specific conditions.[3]

Some examples of damage due to chloride-induced corrosion are shown in Figs. 3 and 4. Figure 3 shows a spall over the top reinforcing mat in a parking structure. Such spalls commonly start to occur after 5 to 15 years of service, depending on concrete cover and quality. Figure 4 shows cracking, leaching and spalling on the underside of a floor slab in a 27-year-old underground parking structure that had been exposed to deicing salt. When it was built, the technology of corrosion of metals embedded in concrete was in its infancy; therefore, no effective anti-corrosion measures had been taken.

The very large spall near the center of the photo was caused by a 4-in. (100-mm) zinc alloy electrical conduit, which corroded galvanically due to its proximity to the steel. (Galvanic corrosion will be discussed in the next Chapter.) This may not appear to follow the description that attended Fig. 2, and indeed it does not. However, the cross section of this slab did follow that description very well when the structure was much younger. The top surface spalls were "repaired" with open-graded asphalt,

FIG. 4—Severe corrosion of reinforcing steel and an embedded electrical conduit on the underside of a parking garage floor.

which allowed the entry of large quantities of water and salt into the top of the slab, which then found its way to the bottom through cracks. After the electrical conduit corroded away, local anodes and cathodes became established within the bottom mat of steel. These microcells developed rapidly because the distance that ions must travel through the concrete to complete the cell is small; therefore, the total current resistance is small.

THE PECULIAR EFFECT OF CHLORIDE ION

Thus far, we have discussed only one ion as an instigator of corrosion, the chloride ion. Corrosion specialists in general agree that any ion of the halogen family would also be destructive. However, chloride ions are the only ones of this group that are normally found in large quantities in concrete.

The peculiar action of the chloride ion is not entirely understood. Some believe that, when the chloride ion concentration becomes large enough, ferrous chloride, or a ferrous chloride complex, is formed on the steel surface, replacing the ferric oxide[4] film that was stabilized by the high pH of the cement paste. Being more mobile (soluble) than the oxide, the chloride salt or complex moves away from the steel, exposing fresh iron to the now-corrosive local environment, and instigating the electrochemical cell. Certainly, the presence of the chloride, and the water that carried it in, increases the current-carrying capacity of the concrete, normally a very poor conductor.

Until the late 1970s, the most commonly used procedure for testing concrete for chloride ion permeability was that described in AASHTO T259, "Resistance of Concrete to Chloride Ion Penetration." In this method, concrete prisms

FIG. 3—Concrete spall over a reinforcing bar in a parking garage floor.

[3] William G. Hime, personal communication.

[4] This stabilized layer causes "passivation" of the steel, but only in an uncarbonated portland-cement paste. According to some investigators, this passivation is due to the formation of a very thin gamma Fe_2O_3 film that prevents further corrosion.

are continuously ponded with 3% sodium chloride solution for 90 days. At the end of this time, the chloride content at several depths is determined. The obvious disadvantage to this test is the time required.

In 1977, the Federal Highway Administration entered into a contract with Construction Technology Laboratories to develop both a rapid field and laboratory test procedure for determining chloride permeability of concrete. The results were to correlate well with the 90-day ponding test. The rapid test method is now described by AASHTO as the rapid chloride permeability test [6]. It was also published in the *1992 Annual Book of ASTM Standards* as ASTM Test Method for Electrical Indication of Concrete's Ability to Resist Chloride Ion Penetration (C 1202). It involves forcing chloride ions to move through a small sample of concrete by applying direct current voltage.

Subsequent work has shown that the rapid method has an excellent correlation with chloride penetration as tested by AASHTO T259, for a wide range of concrete quality and types. However, some concretes that have altered electrical properties can give false readings. Concrete containing silica fume, for example, is rated at lower permeability than it should be [7–9], presumably because the silica fume is such an active pozzolan. This ties up the normally free sodium, potassium, and calcium ions within insoluble calcium silicate hydrate structures, reducing the capacity of the concrete for conduction of electricity. This is not to say that silica fume concrete has high chloride permeability, but it is not as low as the rapid test indicates.

Because of such situations, it is best to interpret the test results carefully. The test is an indirect method for determining chloride permeability, done under artificial conditions of high saturation and applied voltage. Its speed, however, makes the method attractive to those who need such information.

PRECAUTIONARY STEPS AGAINST CORROSION

Concrete water/cement ratio is especially important in determining the rate of ingress of water, and therefore dissolved salts, into concrete. This ratio above all else in plain concrete is the most important factor [10]. Some recently introduced admixtures and additives are very effective in further reducing the permeability of concrete. These will be discussed in more detail in a later section.

The effect of aggregates on the permeability of concrete is generally to increase it. Because the aggregates occupy approximately three-quarters of the volume of concrete, and their permeabilities can be as much as 1000 times greater than that of a high quality portland cement paste [11], they have a profound effect on the protection concrete offers to reinforcing steel against corrosion. Recent work [10] has shown the extremely important effects of cover and low water/cement ratios in resisting the ingress of the chloride ion into concrete, as shown in Fig. 5. The reason that 1 in. (25 mm) of cover was relatively ineffective in resisting chloride intrusion is considered to be due to the fact that $3/4$ in. (19 mm) maximum size aggregate was used. Because of its intrinsically higher permeability and the high permeability of the paste adjacent to some aggre-

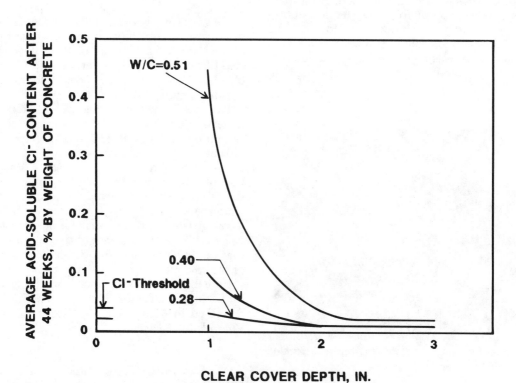

FIG. 5—Chloride content profiles at 44 weeks for different w/c ratio concretes.

gates, one large aggregate particle can easily "short-circuit" the distance between the concrete surface and the steel. This explains why 2 in. (50 mm) of cover is so much more effective than one, and three is, comparatively, little better than two.

Other methods of decreasing the permeability of the concrete cover are available, some of which have only recently been evaluated. A synthetic latex admixture was first used in a concrete overlay on a bridge deck in Michigan in the early 1960s. Since that time, because the latex proved to be so effective in resisting moisture and chloride penetration, a great deal of work has been done in the laboratory [7] and field to develop the technology of the physical properties of latex-modified concrete.

Latex reduces the permeability of concrete in two ways. The physical presence of a lattice of latex (or rubber) within the concrete presents a more tortuous route to water attempting to penetrate the concrete. And the reduction in mix water, made possible by the solids portion of the latex that imparts workability of its own to the concrete, results in lower permeability due to a decreased water/cement ratio paste.

Another method of greatly reducing the amount of water required to make concrete workable is the addition of a high-range water reducer (HRWR) (ASTM Specification for Chemical Admixtures for Concrete C 494, Types F & G). These materials became available in this country in the mid 1970s, having been developed for use in concrete almost simultaneously in Germany and Japan. They are extremely effective in reducing water requirements. For the first time, commercially prepared concrete could be supplied with water/cement ratios as low as 0.30 by weight, with good workability. All of the physical properties of concrete with equal proportions, with or without the presence of a HRWR, are essentially identical.

Soon after the HRWRs were introduced, another type of material, solid rather than liquid, appeared. This was silica fume, an extremely fine form of silica ("finer than tobacco smoke") and therefore extremely active as a pozzolan, a byproduct of the manufacture of silicon metal and ferro-silicon alloys. Because of its fineness, 50 to 100 times finer than portland cement, used as an additive by itself it would cause such a great increase in concrete water requirement that the material would be of very low quality. However, in combination with a HRWR, concrete can be produced with permeabilities as low [7,8] as that of latex-modified concrete, at a fraction of the cost.

Another additive available as an anti-corrosion agent is calcium nitrite. This material acts in some manner, as yet not clearly defined, to reduce the rate of chloride-induced corrosion [12]. However, it is likely that the nitrite may act as a reducing agent, removing the oxygen needed at cathodic sites by reacting with it to form nitrate.

A recent paper [13] suggested using a combination of this material and silica fume. The philosophy was that the silica fume would slow the entry of water and the chloride ion, and the calcium nitrite would slow the corrosion rate after the chloride concentration reached the threshold level at the steel.

A different approach to the prevention of penetration of water and dissolved salts into concrete is to apply a surface sealer after the concrete has hardened. A 1981 study [13] investigated 22 materials used for this purpose. A screening test, amounting to nothing more than immersion of 4-in. (100-mm) concrete cubes in salt water for three weeks, was used to determine which of these appeared to have merit. Five were outstanding. These were tested further by subjecting them to wetting and drying, and either freezing and thawing cycles or extremely severe exposure to ultraviolet light. This testing eliminated two materials, leaving a silane, an epoxy, and a methyl methacrylate as the best performers. The silane had the added advantages that it could not be seen, it penetrated about ⅛ in. (3 mm) into the concrete, and it did not restrict the outward movement of water vapor. Silanes also are long-lasting [14]. The use of such effective sealers as those identified by this study is actually more effective against corrosion than producing a very impermeable concrete [10]. It is also much less expensive.

Another approach to preventing corrosion is to coat the steel. Zinc has been used for this purpose, and is somewhat effective; however, it must be used on all the steel within the structure rather than only that which is expected to become anodic [10,15].

The usual product that results from corrosion of zinc is zinc oxide, which occupies only 50% more space than the original metal. However, if large amounts of chloride are present in concrete containing the zinc, zinc hydroxy chloride can also form. This occupies about $3\frac{1}{2}$ times the volume of the metal and, although this is less than the iron products, it is still capable of producing large expansive forces.

A far more effective material to use for surface protection of the steel is fusion-bonded epoxy [10,12]. Early work showed that, although the epoxy typically has small holes or holidays in the finished coating, and the coating can be damaged by rough handling or bending of the bars, it can do a very good job of protecting the steel, particularly if all the steel in the structure is coated and the number of holes, or holidays, are minimized. Although high corrosion densities may occur at breaks in the coating, the rate of metal loss is moderate. Later work showed that no perceptible corrosion had occurred in the pinhole areas nearly a year after unprotected bars in control specimens had started to corrode, under identical conditions.

PRESTRESSED CONCRETE

Generally, the same factors are involved in the corrosion of prestressing steel as in mild reinforcing steel. However, the same percentage loss of metal can cause catastrophic failures in prestressed structures while causing nothing more than spalls and delaminations in normally reinforced structures. Also, prestressing steel is subject to stress corrosion cracking, a phenomenon that does not occur in mild steel. This is sometimes brought about, or at least is associated with, the presence of certain ions. Two of these that can be present in concrete are the nitrate and bisulfide ions (NO_3^- and HS^-) [17]. Chloride ions, in this case, often appear to be innocuous [16,17], except at high temperatures.

Cases have been reported [16] where chloride levels have been high in prestressed concrete members but no corrosion is evident. However, cases have also been reported where corrosion of prestressing steel has been instigated by chloride ions.

Recent research [10,18] has shed some light on this subject. In one study [10], the corrosion threshold of chloride for prestressing steel was found to be approximately six times that for mild steel, or 1.2% by weight of the cement. Part or all of the reason for the greater resistance to corrosion may be that stearate compounds are used during the manufacturing process. The study also used commercially available epoxy-coated strand. The coating was approximately 60 mil (1.5 mm) thick, as compared to 7 or 8 mil (0.3 mm) for mild steel. Fine sand is applied to the surface of the epoxy before it sets, to assist in stress transfer.

The later study [18] of pretensioned and post-tensioned prestressing systems evaluated grouts, ducts, and anchorage systems. Highlights of the results were that polyethylene ducts were more effective in shielding the strand from corrosion than steel ducts were, but the plastic tended to wear through at bends during stressing of the steel, exposing the steel to the surrounding concrete. Traditional bare and galvanized steel ducts deteriorated badly under chloride exposure, allowing chloride to enter the grout. Joints in duct material were a serious problem, but it was overcome by the use of shrink-fit tubing. Providing an excellent grout was largely unsuccessful, because of the minimal grout cover possible within the duct at the inside surfaces of bends.

The anchorages proved to be vulnerable behind the traditional dry-pack mortar. These should be coated or covered with epoxy. The epoxy coating on the strand was never breached, even within the wedge grips.

ASSESSING THE SEVERITY OF CORROSION IN EXISTING STRUCTURES

The simplest technique for assessing the present condition of a structure deteriorated by corrosion is a visual survey [19]. Visual survey data can be used to produce maps that show locations of cracks, spalls, and other features of deterioration. This is usually followed by a delamination survey, done by striking the concrete surface with a metallic object and listening for hollow sounds. The visual and delamination survey results can be used to select areas for the in-depth studies described later, usually those areas that are typical of the worst, moderate, and best conditions.

Half-cell potential surveys are very useful in determining which areas are actively corroding [20]. The action of an electrochemical cell produces differences in electrical potential of the steel. By measuring these potentials with a half-cell and a voltmeter on a grid pattern, diagrams that resemble contour maps can be constructed, with lines connecting points of equal potential, or voltage. Closely spaced lines are typically observed near areas of high corrosion activity. Recent work [10] has shown that active corrosion is indicated wherever the half-cell potential is more negative than −0.23 V. Figure 6 shows the relationship of half-cell potential to actual corrosion current that was determined in this study. The test method is described in ASTM Test Method for Half-Cell Potentials of Uncoated Reinforcing Steel in Concrete (C 876).

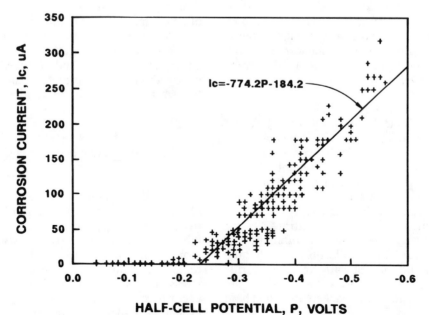

FIG. 6—Relationship between corrosion current and half-cell potential.

A magnetic rebar locator may be used to assess the amount of concrete cover over reinforcing steel. This instrument is useful in determining the potential for future corrosion in various parts of a structure. It can also be used to locate a near-surface bar for grounding the half-cell voltmeter.

CONCRETE CORES

For detailed petrographic examination, chloride ion determination, or compressive strength tests, concrete cores removed from the structure can be useful. The petrographic, or microscopical, examination, ASTM Practice for Petrographic Examination of Hardened Concrete (C 856), can yield information on the quality of the cement paste, degree of curing, stability of the aggregates, air content, and damage caused by chemical attack, freezing and thawing, corrosion, etc.

CHLORIDE SAMPLES

When concrete cores are not removed, powder samples can be taken for chloride analyses. This is usually done with a rotary hammer. Samples are taken at various depths, down to and slightly beyond the depth of the steel. Analysis can be done on an acid-soluble or water-soluble basis. Historically, the acid-soluble technique (AASHTO T260) has been used, with full awareness of the fact that all of this chloride is not available to support corrosion, due to chemical combination with the cement or because it is tightly held within aggregate particles. The results of water-soluble tests are greatly affected by the degree of grinding of the sample and by the length and temperature of leaching [21]. The currently accepted test for water-soluble chloride is described in ASTM Water-Soluble Chloride in Mortar and Concrete (C 1218).

REPAIRS TO DETERIORATED STRUCTURES

Bridges and parking structures in areas of deicer use, along with marine structures of all sorts, are constantly exposed to corrosive conditions and, therefore, are the usual structures needing repair and rehabilitation. Typically, these repairs consist of removal of spalled and delaminated concrete (sufficiently below the top reinforcing steel so that it will be encapsulated within the repair concrete), replacement of the concrete, and some means for restricting the entry of water and salt solutions. In most instances, long-term durability of such repairs has not been evaluated. In extreme cases, the structure must be demolished. In others, cathodic protection may be a viable alternate.

After the deteriorated concrete has been removed and replaced, some form of waterproofing must be applied. If it is not, new delaminations will soon form in the original concrete surrounding the repairs. This is called the "ring anode effect." Starving the concrete for water is the most effective way to reduce the corrosion rate, outside of cathodic protection. Overlays of specialty concretes, such as latex-modified, low-water cement ratio, or silica fume-containing, can be used, but they are expensive. Difficulties in achieving bond with the substrate are sometimes encountered. Thin proprietary membranes of polyurethane alone, or combined with an epoxy-sand wearing course, are also used with good success. However, they are expensive and must be maintained.

A simpler and less expensive method is the application of a penetrating sealer such as a silane or a siloxane. Silanes penetrate the surface more readily [22] and therefore are expected to last longer. These materials polymerize within the concrete and also combine with the siliceous portions of the cement and aggregates, becoming chemically bound adjuncts. Because they are inside the concrete, ultraviolet light, a major disintegrator of organic molecules, cannot reach them. Silanes are now available as water emulsions or as 100% silane, rather than the more traditional alcohol solutions, doing away with volatile organic compound concerns.

CATHODIC PROTECTION

Cathodic protection is truly the only known method of completely preventing corrosion of reinforcing steel embedded in chloride-contaminated concrete. A metal is cathodically protected when sufficient current is applied to polarize the cathodes to the open circuit potential of the anodes [23].

This can be accomplished easily and safely, given the proper conditions, with a sacrificial anode system. The technique has been used for many years on concrete piles by the state of Florida, and as part of the overall corrosion protection systems for offshore drilling platforms. An anode of zinc or magnesium is connected to the reinforcing cage through a lead wire, then immersed in the sea. Corrosion of the anode supplies electrons to the steel, preventing its corrosion. A similar system was used in a research project a decade ago [24] to protect a deteriorating interstate highway bridge deck. It was considered a success by the researchers, but the technique has not gained popular acceptance among transportation agencies.

Systems that employ impressed current, while not extremely popular, have met with somewhat greater success. The first instance of the use of such a system was in 1974, on a bridge deck in California [25]. Since that time, hundreds of bridge decks have been so treated, but serious difficulties remain [26]. Controlling current densities and maintaining uniform potentials continue to be basic problems, although several viable anode systems have been developed. Impressed current systems have not been put to use commercially on prestressed structures due to the possibility of hydrogen generation [27], which can give rise to hydrogen embrittlement. A general review of cathodic protection was presented recently by two British authors [28].

REFERENCES

[1] Berk, N. S., Pfeifer, D. W., and Weil, T. G., "Protection Against Chloride-induced Corrosion," *Concrete International*, Dec. 1988, pp. 44–55.
[2] Uhlig, H. H., *The Corrosion Handbook*, Wiley, New York, 1948.
[3] Schweitzer, P. A., *Corrosion and Corrosion Protection Handbook*, Marcel Dekker, Inc., New York, 1983, pp. 31 and 502.
[4] Mansfield, F., *Corrosion Mechanisms*, Marcel Dekker, Inc., New York, 1987, p. 211.
[5] Clear, K. C., "Evaluation of Portland Cement Concrete for Permanent Bridge Deck Repair," Report No. FHWA-RD-74-5, Federal Highway Administration, Feb. 1974.
[6] "Rapid Determination of the Chloride Permeability of Concrete," AASHTO T277, American Association of State Highway and Transportation Officials.
[7] Whiting, D., "Rapid Determination of the Chloride Permeability of Concrete," Final Report No. HWA/RD-81/119, Federal Highway Administration, Washington, DC, Aug. 1981.
[8] Whiting, D., "Permeability of Selected Concretes," *Permeability of Concrete*, SP-108, American Concrete Institute, Detroit, MI, 1988, pp. 195–222.
[9] Pfeifer, D. W., "Corrosion Protection for Concrete Structures: The Past and the Future," *Civil Engineering Practice*, Vol. 6, No. 1, Boston Society of Civil Engineers, 1991.
[10] Pfeifer, D. W., Landgren, J. R., and Zoob, A., "Protective Systems for New Prestressed and Substructure Concrete," Federal Highway Administration, Report No. FHWA/RD-86/193, Washington, DC, April 1987.
[11] Neville, A. M., *Properties of Concrete*, Wiley, New York, 1963.
[12] Virmani, Y. P., Clear, K. C., and Pasko, T. J., "Time-to-corrosion of Reinforcing Steel in Concrete Slabs. Volume 5: Calcium Nitrite Admixture or Epoxy-Coated Reinforcing Bars as Corrosion Protection System," Report No. FHWA/RD-83/012, Federal Highway Administration, Washington, DC, 1983.
[13] Pfeifer, D. W. and Scali, M. J., "Concrete Sealers for Protection of Bridge Structures," NCHRP Report No. 244, Transportation Research Board, Washington, DC, 1981.
[14] Perenchio, W. F., "Durability of Concrete Treated with Silanes," *Concrete International*, Nov. 1988, pp. 34–40.
[15] Stark, D. and Perenchio, W., "The Performance of Galvanized Reinforcement in Concrete Bridge Decks," Final Report, Project No. 2E-206, Construction Technology Laboratories, Skokie, IL, July 1974–Oct. 1975.
[16] Moore, D. G., Klodt, D. T., and Hensen, R. J., "Protection of Steel in Prestressed Concrete Bridges," Report 90, National Cooperative Highway Research Program, Washington, DC, 1970.
[17] Verbeck, G. J., "Mechanisms of Corrosion of Steel in Concrete," *Corrosion of Metals in Concrete*, SP-49, American Concrete Institute, Detroit, MI, 1975, pp. 21–38.
[18] Perenchio, W. F., Fraczek, J., and Pfeifer, D. W., "Corrosion Protection of Prestressing Systems in Concrete Bridges," NCHRP Report 313, Transportation Research Board, Washington, DC, Feb. 1989.
[19] Perenchio, W. F., "The Condition Survey," *Concrete International*, Vol. 11, No. 1, Jan. 1989, pp. 59–62.
[20] Escalante, E., Whitenton, E., and Qiu, F., "Measuring the Rate of Corrosion of Reinforcing Steel in Concrete—Final Report," NBSIR 86-3456, National Bureau of Standards, Washington, DC, Oct. 1986.
[21] Hope, B. B., Page, J. A., and Poland, J. S., "Determination of the Chloride Content of Concrete," *Cement and Concrete Research*, Vol. 15, No. 5, Sept. 1985, pp. 863–870.
[22] Kottke, E., "Evaluation of Sealers for Concrete Bridge Elements," BRI/Reports/287PD010, Alberta Transportation and Utilities, Edmontons, Alberta, Canada, Aug. 1987.
[23] Mears, R. B. and Brown, R. H., "A Theory of Cathodic Protection," *Transactions*, Electrochemical Society, 1938.
[24] Whiting, D. and Stark, D., "Galvanic Cathodic Protection for Reinforced Concrete Bridge Decks—Field Evaluation," NCHRP Report No. 234, Transportation Research Board, June 1981.
[25] Stratfull, R. F., "Experimental Cathodic Protection of a Bridge Deck," *Transportation Research Record No. 500*, Transportation Research Board, Washington, DC, 1974, pp. 36–44.
[26] Schell, H. G. and Manning, D. G., "Research Direction in Cathodic Protection for Highway Bridges," *Material Performance*, National Association of Corrosion Engineers, Houston, TX, Oct. 1989, pp. 11–15.
[27] Hope, B. B. and Poland, J. S., "Cathodic Protection and Hydrogen Generation," *ACI Materials Journal*, Vol. 87, No. 5, Sept.-Oct. 1990, pp. 469–472.
[28] Wyatt, B. S. and Irvine, D. J., "A Review of Cathodic Protection of Reinforced Concrete," *Materials Performance*, National Association of Corrosion Engineers, Houston, TX, Dec. 1987, pp. 12–21.

Embedded Metals and Materials Other Than Reinforcing Steel

Bernard Erlin[1]

PREFACE

The initial work on this chapter was done by Hubert Woods, a concrete consultant who is now deceased. The chapter was published 27 years ago in *ASTM STP 169A*. A current search of the literature by the John Crerar Library of the University of Chicago for the effects of concrete on embedded nonferrous metals revealed little new information than reported previously in this chapter. What has been located can be found herein. The basic principals of chemistry and physics do not change; hence the response of materials to their environment is the same now as it was when this chapter was originally written. And, in spite of lessons learned, or because of a different environmental exposure or some peculiarity thereof, adverse performance may occur. Let caution be a guiding light in material usages, and let a value of the principals that make embedded materials stable a code of use.

INTRODUCTION

Metals other than conventional reinforcing steel are sometimes used in conjunction with concrete. Emphasis will be given here to the possible degradable aspects relative to their use, and conditions that may render them unserviceable.

The materials to be described include metals, and other inorganic and organic substances. Among the metals are aluminum, lead, copper and copper alloys, zinc, special alloys of iron, monel metal, stellite, silver, and tin. Among the inorganic materials are glass, asbestos, and concrete; and among the organic materials are a variety of plastics and wood and similar cellulosic materials. Fiber-reinforced concrete is gaining in use, so information about some fibers has been included.

A general condition necessary for the chemical degradation of any material in concrete is exposure to moisture, and a short discussion on that aspect precedes the discussions of material behavior.

GENERAL CONDITION

Moisture is usually necessary for the chemical degradation of any material. It can be in the form of water vapor, liquid water, or solutions of which water is a component. Any concrete is never truly dry, because it contains air and that air will hold moisture in vapor form. For example, because of the chemical nature of the cement paste, and the fact that most concrete contains some unhydrated cement, air in concrete will contain sufficient moisture to maintain a relative humidity of about 80%, the vapor pressure for the paste at equilibrium.

Furthermore, although a concrete may be relatively dry, carbonation of the portland-cement paste will generate moisture due to the release of hydrate water when carbonation products are formed, such as the following

$$CaSiO_2 \cdot xH_2O + CO_3 \rightarrow CaCO_3 + SiO_2 + xH_2O$$

Concrete is wet when mixed and for some time after setting, depending principally on the quality of the concrete and the distance through which free water must move to an external surface where it can evaporate. The rate of drying also depends on external humidity, but even at low external humidity, a long drying time may be required to lower the internal humidity to a point where corrosion is unlikely. Thus, concrete that will eventually be dry may be internally wet long enough to promote serious corrosion of susceptible metals and, of course, concrete exposed continuously or frequently to a damp environment may remain wet enough to support any corrosion that may occur. The free moisture in wet concrete provides an aqueous medium that facilitates transport of soluble chemical substances, such as oxygen, calcium hydroxide, alkalies, and chlorides toward metals, and of any soluble corrosion products away from the metal. It also increases the electrical conductivity of the concrete, thus aiding any tendency for electrochemical corrosion.

The relatively long time required for concrete to dry to various internal humidity levels is not commonly appreciated. The rates of drying of 15- by 90- by 90-cm (6- by 36- by 36-in.) slabs of normal-weight concrete are shown in Table 1 [1].

Table 1 shows that at an environmental relative humidity of 50%, 36 days was required to lower the mid-depth relative humidity to 90%, and 240 days was required to reach 75%. Additional tests showed that when the mid-depth relative humidity reached 90%, the relative humidity at a point only 1.9 cm (¾ in.) from an exposed face was 87%. The moisture condition of concrete required to support active electrochemical corrosion of aluminum and other susceptible metals is not known with any accuracy, but it seems almost certain that such corrosion could

---

[1] Petrohrapher, The Erlin Company, Latrobe, PA 15650.

TABLE 1—Rates of Drying 15- by 90- by 90-cm (6- by 36- by 36-in.) Slabs of Normal-Weight Concrete.[a]

Environmental Relative Humidity (RH)	Drying Time to Reach Various Relative Humidities in the Concrete Slab at Middepth, days		
	90% RH	75% RH	50% RH
10	18	80	620
35	30	110	840
50	36	240	...
75	36	...	...

[a] From Ref 1.

proceed, other conditions being favorable, at 90% relative humidity and probably also at 60% relative humidity, though at a much lower rate.

Concrete made with lightweight aggregates dries more slowly than concrete made with normal-weight aggregates. Thicker sections, such as beams and columns, will dry more slowly than thin sections, such as walls, or supported floor slabs.

ALUMINUM

Aluminum has been given widespread attention because it has been the cause of numerous problems in the past. These prompted laboratory studies designed to describe conditions necessary for corrosion, and methods for circumventing the corrosion.

Aluminum reacts in fresh concrete principally with alkali hydroxides derived from the cement. One reaction product is hydrogen gas, and for this reason aluminum powder sometimes is used to form extremely lightweight cellular concrete. When in smaller amounts, the gas provides a slight expansion of the grout, and this expansion has been used, for example, for bedding machinery base plates. Aluminum in rod, sheet, or pipe form will react much less vigorously than will the powdered metal, because of the lesser surface area exposed to the alkalies, but the reactions will continue until the metal is dissolved.

Tests carried out by Jones and Tarleton [2] indicate that the corrosion of embedded aluminum can crack concrete under unfavorable circumstances. However, it has been shown that the situation can be worse if the concrete contains calcium chloride, and much worse if it also contains steel, such as reinforcing steel that is connected (coupled) to the aluminum.

Wright [3] has described a case of corrosion of sufficient severity to cause collapse of aluminum conduit in reinforced concrete containing calcium chloride. An instance of extensive concrete spalling over aluminum conduit in Washington Stadium has also been published [4]. Corrosion of aluminum balusters embedded in concrete have been reported [5]. Several dozen additional cases of concrete cracking due to corrosion of embedded aluminum conduit and posts, and window frames in contact with concrete have occurred. In every case, chloride was present. An example of concrete spalled over corroded aluminum conduit is shown in Fig. 1.

The laboratory investigations by Monfore and Ost [6] show that calcium chloride concentration, alkali content

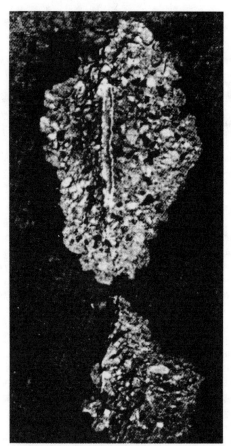

FIG. 1—Concrete spalled by corroding aluminum conduit (from Ref 5).

of cement, electrical coupling of steel and aluminum, and the ratio of steel area to aluminum area are all interrelated in the corrosion of the aluminum and the subsequent cracking of concrete. In their studies, 15-cm (6-in.) concrete cubes were prepared using cements of high and low alkali contents in which pieces of nominally 1.25-cm (½-in.) aluminum conduit was embedded 1.25 cm (½ in.) from one face. C-shaped sheets of mild steel were also embedded in the cubes. The steel and aluminum were externally connected in some tests and not connected in others. Various amounts of calcium chloride were used. The cubes were removed from their molds at 24 h, coated with a curing compound, and then stored at 23°C (73°F) and 50% relative humidity for 28 days and observed regularly for cracks. After 28 days, the aluminum pieces were removed, cleaned, and weighed. The principal results of the studies are given in Table 2.

Several important findings were noted:

1. All cubes that cracked contained calcium chloride.
2. Cubes containing 1% flake calcium chloride ($CaCl_2 \cdot 2H_2O$) by weight of cement did not crack, but metal losses in these cubes were as high as, or higher than, those in other cubes that cracked. (Exposure to a damper environment might have caused cracking to occur.)
3. With no calcium chloride and no electrical coupling, the corrosion was greatest with the high alkali cement.

TABLE 2—Corrosion of 6063 Aluminum Conduit Embedded in 15-cm (6-in.) Concrete Cubes Stored at 50% Relative Humidity for 28 Days.[a]

Cement	Cement Alkalies as Na_2O, %	$CaCl_2 \cdot 2H_2O$, % by weight of cement	Ratio of Steel Area to Aluminum Area	Electrodes	Days to Cracking	Loss in Surface Thickness, mils[b]
C	0.24	0	28	uncoupled	no crack	0.16
		2	14	coupled	5	0.92
		2	28	coupled	4	1.2
		4	7	coupled	3	1.6
		4	14	coupled	3	2.3
		4	28	coupled	3	2.4
		0	28	uncoupled	no crack	0.09
		2	14	uncoupled	no crack	0.07
		2	28	uncoupled	no crack	0.10
		4	7	uncoupled	no crack	0.07
		4	14	uncoupled	no crack	0.07
		4	28	uncoupled	no crack	0.04
		0	0	...	no crack	0.09
		4	0	...	no crack	0.04
D	0.89	0	28	coupled	no crack	0.12
		1	3.5	coupled	no crack	0.54
		1	7	coupled	no crack	0.77
		1	14	coupled	no crack	1.0
		1	28	coupled	no crack	0.85
		2	3.5	coupled	3	1.0
		2	7	coupled	3	1.4
		2	14	coupled	4	1.4
		2	28	coupled	4	1.6
		4	3.5	coupled	2	1.5
		4	7	coupled	2	1.7
		4	14	coupled	2	2.2
		4	28	coupled	7	3.3
		0	28	uncoupled	no crack	0.33
		2	14	uncoupled	no crack	0.06
		2	28	uncoupled	no crack	0.06
		4	7	uncoupled	no crack	0.07
		4	14	uncoupled	no crack	0.09
		4	28	uncoupled	no crack	0.08
		0	0	...	no crack	0.17
		4	0	...	no crack	0.05

NOTE—Conversion factor: 1 mil = 2.54 × 10⁻⁵ m.
[a] From Ref 5.
[b] Calculated from weight losses.

4. With a 2 or 4% calcium chloride addition, and with metals electrically coupled, corrosion was slightly greater when the higher alkali cement was used.
5. With calcium chloride present, and with metals electrically coupled, corrosion generally increased with increasing ratio of steel area to aluminum area, and invariably increased as the amount of calcium chloride increased.
6. Cubes that cracked did so within seven days.

In the case of coupled metals, considerable galvanic currents were detected in the circuit connecting the aluminum and steel.

Some measured currents are shown in Fig. 2 as a function of time and calcium chloride content. Current flow progressively increased with increasing calcium chloride contents at all periods (up to end of test at 28 days). When the total electrical flow during 28 days (in ampere hours per unit area of aluminum) was plotted against the amount of corrosion measured as loss in thickness, straight curves of varying shapes were obtained with the slopes depending on the amount of calcium chloride. The greater slopes occurred for the higher amounts of calcium chloride.

Tests somewhat similar to those reported by Monfore and Ost were done on 31-cm (12-in.) concrete cubes by Wright and Jenks [7]. With electrically coupled steel and aluminum (area ratio 10:1), cracks did not occur when no calcium chloride was used. However cracks did occur at various ages; after 61 days with 1.1% flake calcium chloride by weight of cement, to eight days with 5.7% flake calcium chloride by weight of cement.

McGeary [8] also found that (a) for aluminum conduit electrically coupled to steel, cracks did not occur in the encasing concrete when chlorides were not present; (b) the cause of corrosion of the aluminum was due unquestionably to galvanic cells, activated when chlorides were present in the concrete; (c) the corrosion of the aluminum was of the cubic type, which is also referred to as intergranular corrosion; and (d) there was an enrichment of chloride at the aluminum conduit surface.

Tests [6] were made to determine the effectiveness of several different coatings, applied to the aluminum before embedment in 1.25-cm (½-in.) concrete cubes, made with cement having an alkali content of 0.89% and containing 4% calcium chloride by weight of cement. The results are

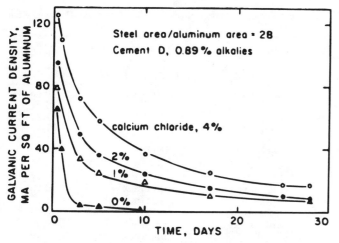

FIG. 2—Effect of calcium chloride on galvanic current (from Ref 5).

shown in Table 3. The data show that a silicone coating was ineffective; that Lacquer B prevented cracking within 28 days but permitted some corrosion; and that Lacquer C and Bituminous Coatings A and D were all effective in preventing both corrosion and cracking.

Other tests [8] indicated that resistance to galvanic corrosion was provided by the following coatings: certain bitumens, epoxies, fluidized-bed plastics, certain metallic pigmented coatings, and alkyd and phenolic materials.

McGeary [8] found the following to be effective repair procedures for concrete cracked because of corroded aluminum: (a) remove all loose concrete; (b) remove all corrosion products adjacent to the embedded aluminum surface; (c) patch with epoxy or epoxy polysulfide systems, or both; and (d) engage an experienced patching specialist to complete the repairs.

The results of these various investigations and field observations show that reinforced concrete is likely to crack and spall from corrosion of non-isolated aluminum embedded therein if calcium chloride is present. Also, insulating coatings for aluminum to be used in concrete are commercially practical.

In view of chemical similarities between calcium chloride and sodium chloride, it seems evident that the latter would also facilitate corrosion of aluminum. Sodium chloride is the principal constituent of sea salt, and it therefore seems prudent not to use aluminum in concrete exposed in or near seawater. The role of the chloride is that of an electrolyte, a necessary part of the galvanic cell. Obviously, other salts that hydrolyze and provide a strong electrolyte can be damaging.

Based upon laboratory studies and field experiences, it is wise not to use calcium chloride in concrete in which aluminum will be used, or where the concrete is exposed to chlorides.

LEAD

Lead has a high resistance to certain chemical actions, but in contact with damp concrete it is attacked by the calcium hydroxide in the concrete and becomes converted to lead oxide or to a mixture of lead oxides. If the dampness persists, the attack will continue. A lead pipe, for example, may be destroyed in a few years. If the lead is electrically coupled to reinforcing steel in the concrete, galvanic cell action may accelerate the attack [9a], in which case the rate of corrosion may be several millimetres per year.

Lead partially embedded in concrete, and thus partially exposed to the air, is susceptible to corrosion because of the differential electrical potential that results. The air-exposed portion, in the presence of water, will become cathodic with respect to the embedded portion, which will become anodic, and a galvanic cell will form. Corrosion and deterioration of the embedded lead will result.

A protective coating or covering always should be used when lead pipe or cable sheaths are to be embedded in concrete. Bituminous coatings have been used successfully. Synthetic plastic and other organic coatings or sleeves, which are themselves unaffected by damp concrete, are also appropriate. Lead in contact to concrete should be protected by suitable coatings, or otherwise isolated from contact to the concrete [9b]. Also, lead and lead salts are very strong retarders for portland cement.

There appears little, if any, likelihood of concrete itself being damaged by corrosion of lead because of the softness of the metal and its capability for absorbing stress.

COPPER AND COPPER ALLOYS

Copper will not corrode in concrete unless soluble chlorides are present. Copper pipes are used successfully in concrete except under unusual circumstances where ammonia is present [10]. Very small amounts of ammonia, and possibly nitrates, can cause stress corrosion cracking. Brass wall ties have reportedly failed by stress corrosion and manganese bronze bolts have sheared below their ultimate strength. However, because such phenomena can occur under circumstances unrelated to concrete, it is not clear what role, if any, the concrete plays. Very little systematic work has been reported on the behavior of copper and its alloys in contact with concrete, probably because these metals have given satisfactory service under such conditions.

TABLE 3—Effect of Protective Coatings on Corrosion of Aluminum Conduit Embedded for 28 Days in 15-cm (6 in.) Concrete Cubes Containing 4% Calcium Chloride and Steel Electrically Coupled to the Aluminum.[a]

Protective Coating Material	Thickness, mil	Days to Cracking	Loss in Surface Thickness, mil[b]
None	...	3	2.5
Silicone	...	2	2.1
Lacquer B	1	no crack	0.47
Lacquer C	2	no crack	nil
Bitumen A	5	no crack	nil
Bitumen D	15	no crack	nil

NOTE—Conversion factors: 1 mil = 2.54 × 10^{-5} m; 1 in. = 2.5 cm.
[a] From Ref 5.
[b] Calculated from weight losses.

Copper, brass, red brass, bronzes, aluminum-bronze, and copper-silicon alloys embedded in concrete have good resistance to corrosion [11,12].

When copper is connected or adjacent to steel reinforcement, and an electrolyte such as chloride is present, corrosion of the steel due to galvanic action is likely to occur. Thus, under such circumstances, it is desirable to insulate the copper with suitable coatings.

ZINC

Zinc reacts chemically with alkaline materials, but normally in concrete the reaction is superficial and may be beneficial to bond zinc (galvanized) to steel. The primary chemical reaction with calcium hydroxide is

$$Zn + Ca(OH)_2 \rightarrow CAZnO_2 + H_2 \qquad (1)$$

When zinc is used in concrete, it is generally as a coating for steel. The reaction is self-limiting and products of reaction are not voluminous; consequently, damaging stresses are not created.

Although good concrete normally provides a nearly ideal environment for protecting embedded steel from corrosion, embedded galvanized steel reinforcement has been used in concrete exposed in marine and other environments where chlorides are present (for example, bridge decks and columns, and docks). Galvanized reinforcing steel has not always been successful in stopping corrosion of the steel and thus conflicting reports result [13–15]. Stark [16] investigated embedded galvanized reinforcement in seawater-exposed concretes that were from 7 to 23 years old. The concrete cores examined were from mean tide and above high tide zones. Chloride concentrations to depths of 10.1 cm (4 in.) were always exceedingly high at about the level of the galvanized bars (from 1.9 to 6.4 kg/m^3 (3.2 to 10.7 lb/yd^3). Concrete cover over the galvanized bars was from 5.7 to 13.3 cm (2¼ to 5¼ in.). The amount of galvanize coating remaining was from 92 to 100%. A summary of the data is given in Table 4. Unpublished data by Hime and Erlin for galvanized bars threaded through brick hollow cores packed with mortar of 12-year-old masonry facades containing exceedingly high chloride contents revealed that the galvanize coating was completely corroded. Galvanized wall ties of brick masonry facade construction can undergo similar corrosion when chlorides are present.

Galvanized corrugated steel sheets often are used as permanent bottom forms for concrete roof or floor construction. Both satisfactory and unsatisfactory performances have been reported. Figure 3 shows one instance of multiple perforation and corrosion of such sheets under a roof slab [10]. Most of the corroded spots were dry to the touch, but some were moist and had acidic pH values of 2.7 to 4.8. Analysis of the corrosion protuberances indicated the presence of iron, zinc, and chloride. Further investigation of this and other similar cases showed that in each instance calcium chloride was used in the concrete. This abetted the corrosion by chemical action on the zinc and by increasing the electrical conductivity of the concrete. The chemical reaction is probably

$$Zn + CaCl_2 + 2H_2O \rightarrow CaZnO_2 + 2HCl + H_2 \qquad (2)$$

This chemical action produces hydrochloric acid, which explains the observed acidity of the corrosion product. It is probable that this reaction takes place only after local depletion of calcium hydroxide, by the reaction expressed in Eq 1. Other reported zinc corrosion products are zinc oxide and zinc hydroxy-chlorides.

Admixtures containing major amounts of chloride should not be used in concrete that will be exposed to moisture and contains, or is in contact with, galvanized steel. It is also advisable to keep chloride-containing solutions from permeating the concrete.

Zinc galvanize can corrode when in contact with relatively fresh concrete. The corrosion phenomenon results in pitting of the surface due to reactions of the zinc and alkalies in the cement paste. Thus, zinc-coated steel forms, for example, in precast or cast-in-place concrete, may cause concrete surface disfiguration due to contact of the zinc.

Passivation of the zinc by use of chromate dips has been reported to be effective in protecting galvanized products. The dips are solutions of sodium or potassium dichromate, acidified with sulfuric acid.

Concrete in which galvanized reinforcing steel is located close to galvanized forms has a tendency to stick to the forms. A chromate treatment, such as previously described, has been used as a method for avoiding that problem.

TABLE 4—Galvanize Coating Remaining on Reinforcing Bars After Indicated Years of Concrete Exposure to Seawater.

Years of Age	Galvanize Coating Remaining, %
7	98
8	100
10	95
10	96
10	99
12	92
23	98

FIG. 3—Corrosion and perforation of galvanized steel from under concrete slab (from Ref 11).

OTHER METALS

The following metals have been reported to have good resistance to corrosion in concrete: stainless steels, chrome-nickel steels, chromium-aluminum-silicon steels, cast silicon-iron, alloyed cast iron, nickel, chrome-nickel, iron-chrome-nickel alloys, monel metal, stellite, silver, and tin [12]. The resistance of some of these metals to corrosion may be affected seriously by the presence of "corrosion promoters" such as soluble chlorides. Monel metal and Type 316 stainless steel are well-known for their resistance to sodium chloride and other constituents of seawater and should work well in concrete. Special circumstances might justify the use of these more costly metals.

Nickel and cadmium-coated steel will not corrode in chloride-free concrete if the coatings are continuous [17]. However, the corrosion resistance of these materials becomes questionable if chlorides are present in the concrete or in solutions that permeate the concrete [18].

GLASS

Glass sometimes is embedded in mortar or concrete as artificial aggregate [19,20] used for decorative or aesthetic purposes, as reinforcing as a substitute for steel, as wall blocks or tile, and as frameless windows or lights. Some glasses are reactive with alkalies in portland cement paste. The resulting expansion may cause severe damage to the glass or the concrete or both [21].

Whenever glass is to be used in concrete, it is mandatory that the glass be tested to ensure that it will be chemically stable in concrete and not cause expansions due to alkali-silica reactivity. For example, waste bottle glass used as aggregate in a decorative concrete facing of outside-exposed precast concrete panels caused sufficient expansion to warp the panels. The warping was due to the expansion of the facing, which was bonded to the nonexpansive concrete backing. Prisms cut from the panels and tested using accelerated methods in the laboratory had expansions of 0.2% after three months of test. The glass aggregate removed from the concrete, tested using the procedures in the ASTM Test Method for Potential Reactivity of Aggregates (Chemical Method) (C 289) was found to be deleteriously reactive. Other reports [21] have also indicated the deleterious behavior of reactive glass.

Before glass is used in portland-cement concrete, it should be tested using the methods provided in the Appendix of ASTM Specification for Concrete Aggregates (C 33), and specifically those given in Paragraph X1.1.2 of ASTM C 289 and Paragraph X1.1.3 of the ASTM Test Method for Potential Alkali Reactivity of Cement-Aggregate Combinations (Mortar-Bar Method) (C 227).

Nondeleteriously reactive glass for use in concrete is manufactured and is thus available for use in concrete.

WOOD

Current trends in the use of new or unusual materials in concrete are due to the emphasis on conservation of energy and the utilization of wastes and by-products. Among the materials prepared for use in concrete is wood (including bamboo, fibers, bark, jute, cotton, and rice stalks and hulls).

Problems incidental to the use of natural-cellulosic materials have included adverse effects of sugars on concrete setting and degradation of the fibers due to the high alkalinity of concrete. Further, high differential thermal coefficients of expansion of some of these materials and unaccommodative volume changes can cause cracks to develop. These factors are perhaps foremost in precluding the use of many cellulosic materials.

Additional problems incidental to the use of natural-cellulosic materials include swelling with moisture absorption and subsequent shrinkage, and chemical degradation due to contact with calcium hydroxide. Prior treatment of the cellulosic materials such as impregnation or coating techniques, or carbonation of the portland-cement paste [22], are possible ways for improving their utilization.

Sawdust, wood pulp, and wood fibers have been incorporated in mortars and concretes, and timbers have been embedded in or placed in intimate contact with concrete in composite constructions. The use of fresh untreated sawdust, wood chips, or fibers in concrete commonly results in very slow setting and abnormally low strength because of interference with normal setting and hardening processes by carbohydrates, tannins, and possibly other substances in the wood. The amount of such substances differs with wood species and from time-to-time and place-of-origin within a single species. Softwoods generally give less trouble in this respect than hardwoods.

Many admixtures and wood treatments have been proposed or used to circumvent the influence of wood constituents on setting and hardening. Addition of hydrated lime to the mixture, in an amount equivalent to one third to one half of the cement by volume, has been found effective in overcoming this action [23]. The treatment is usually effective with mixed softwoods, except when a high proportion of larch or Douglas fir is present.

Five percent calcium chloride by weight of the cement is sometimes added as well as hydrated lime. With woods of high tannin or carbohydrate content, the addition of lime with or without calcium chloride is not effective. Other treatments that have been suggested include soaking in sodium silicate solution, moistening the wood with 1% sulfuric acid for 4 to 14 h, then neutralizing with "milk of lime"; treating with 37% aluminum chloride solution or 50% zinc-chloride solution in a rotary barrel with beater. A treatment found by Parker [23] to be effective with all woods tried consists of the following consecutive steps: (a) boiling sawdust in water, (b) draining and washing with water, (c) reboiling with a 2% water solution of ferrous sulfate, and (d) draining and rewashing.

Concrete made with wood aggregate has considerably greater volume change on wetting and drying, or simply on change in external humidity, than concrete made with mineral aggregates. If the element is restrained, drying may lead to cracking. If drying is not uniform, the element may warp. The pretreatments mentioned previously have only a small influence on these volume changes. Various methods have been employed to lessen the changes in volume consequent upon changes in moisture. Some of these methods involve encasement of the wood particles or of the finished product in a material of low permeability to moisture, but the details of such treatments and the results achieved have not been revealed in general.

Timbers embedded in concrete sometimes have been observed to deteriorate. The harm is done by calcium hydroxide, which causes dissolution of lipins, and decomposition, chiefly of pentosans, to a smaller extent of lignin, and least of all of cellulose. The most suitable wood for embedment is said to be pine or fir, preferably of a type with high resinous content [24].

PLASTICS

The use of plastics in concrete and concrete construction has increased significantly. Plastic products are now being used as pipes, conduit shields, sheaths, chairs, waterstops, and joint fillers. Their compatibility with concrete is thus important.

The principal chemicals in concrete that could conceivably attack plastics are calcium hydroxide, sodium hydroxide, and potassium hydroxide, which create a minimum pH of 12.4. The following plastic groups have excellent resistance to all three of these alkalies at 24°C (75°F) [25]: polyethylene, styrene copolymer rubber-resin blends poly(vinyl chlorides), Types I and II, and polytetrafluoroethylene.

Another source, Ref 26, provides information on the resistance of plastics to strong alkalies, see Table 5.

FIBERS

In a broad sense, fibers used for reinforcing concrete are small versions of conventional steel reinforcement, and they provide a similar service. For fibers to be useful and effective, they must enhance the physical attributes of concrete, and be durable. Among the desirable characteristics that fibers can impart to concrete are flexural strength, resistance to fatigue and impact, and increased fracture resistance. Applications of fiber-reinforced concrete are shown in Table 6. A report on fiber-reinforced concrete was prepared by the American Concrete Institute [27], but little is given regarding corrosion characteristics of embedded materials.

Among the materials that have been used as fibers in concrete are steel, glass, polyethylene, polypropylene, nylon, asbestos, and carbon. Organic fibers are considered to rely upon mechanical interlock, and steel and glass upon chemical interactions, to develop adhesion to the cement matrix. Steel fibers have also been produced as crimped or deformed fibers so that mechanical interlock is present in addition to chemical bond.

STEEL

Good durability of steel-fiber-reinforced concrete has been reported [28,29]. However, steel fibers are subject to the same type of corrosion as reinforcing steel, and thus the durability of concrete made with steel fibers and used in environments chemically aggressive to the steel may be poor. The deterioration of the steel is enhanced particularly by chlorides. However, it has been reported that outside-exposed, steel-reinforced concrete used for pavements, bridge decks, and other similar usages, where chloride deicing agents are employed, can suffer reductions in strength [30,31]. If cracks are present, rusting will be initiated at locations where steel is exposed in the cracks. Obviously, chlorides should not be a component of admixtures used in steel fiber-reinforced concrete, and the concrete should have as low a permeability as workability and water/cement (w/c) ratio will permit.

Studies of steel-fiber-reinforced beams in a simulated seawater environment for eight years have revealed that rusting of fibers to depths of 0.34 cm (1/16 in.) has occurred [32]. True electro-chemical corrosion cannot develop on a large scale, however, due to lack of electrical continuity among the steel fibers.

TABLE 5—Resistance of Plastics to Strong Alkalies [26].

Class of Material	Resistance
Polyethylene	excellent
Polymethyl methacrylate	poor
Polypropylene	excellent to good
Polystyrene	excellent
Polystyrene acrylonitrile	excellent to good
Polytetrafluoroethylene	excellent
Polytrifluorochloroethylene	excellent
Polyvinyl chloride and polyvinyl chloride vinyl acetate (rigid)	excellent
Polyvinyl chloride and polyvinyl chloride vinyl acetate (plasticized)	fair to good
Saran (monofilament grade)	fair to good
Epoxy (unfilled)	excellent
Melamine (formaldehyde)	poor
Phenol (formaldehyde)	poor
Polyester styrene-alkyd	poor
Urea (formaldehyde)	poor

TABLE 6—Some Applications of Fibers in Concrete.

Type of Use	Type of Fiber
Cast-in-place and precast bridge deck units; overlays; structural bridge deck elements; pavement; dolosse; boats; poles; tunnel linings; rock slope stabilization; highway, street, and airfield pavements; sluices; industrial floors	steel and polypropylene
Maintenance and repairs to dams, slabs, pavements, bridges, culverts, etc.	steel and polypropylene
Industrial floors, slab overlays, pile caps, pavements	steel, glass, and polypropylene
Miscellaneous small precast items (burial vaults, steps, garden units, etc.)	glass and polypropylene
Pipes, sheets, boards, fence posts, panels, piles, building panels	asbestos, glass cellulose, and polypropylene

Steel fibers and steel filings are used in proprietary floor toppings to enhance wear characteristics, desirable in warehouses. The volume percentage of metal used is significantly greater than for steel fiber-reinforced concrete. Deterioration of surfaces has resulted because of rusting of the steel. In one such case, chloride equivalent to a purposeful calcium chloride addition (of 1% weight of cement) was present in the concrete to which a topping shake had been applied. This situation points out the desirability of avoiding the use of chloride, either as a component of the original concrete mix or as an application to concrete surfaces.

GLASS FIBERS

Glass fibers can be sensitive to alkalies in portland cement paste (see the earlier discussion on Glass). Early work using glass fibers reflected the adverse effects of alkali-silica reactivity [33].

Glass fibers formulated with zirconia were thought to be resistant to alkali degradation [34,35]. However, the loss of fracture resistance caused by chemical changes within glass-fiber reinforced concrete (which occurs prior to five-year exposures in wet environments) was as dramatic as previously experienced. The embrittlement associated with that exposure is thought to result because of reformation of cement hydration products in and tightly around the fibers, so that fiber encasement is enhanced and failures are due to fracture rather than pull-out of fibers [36]. That explanation does not entirely satisfy all aspects of the situation, however, and further research is needed [37,38]. New glass fibers coated with chemically inert coating have been used in an attempt to overcome the strength-loss phenomenon.

ORGANIC MATERIALS

Polymer fibers, such as nylon and polypropylene, improve impact strength of concrete, but not tensile or flexural strengths, because they have a low modulus of elasticity. The polypropylene fibers are in common use. They increase impact strength and minimize the concrete potential for early cracking such as due to plastic shrinkage.

ASBESTOS

The use of asbestos in portland cement-based products is a thing of the past because of its carcinogenic effect. However, asbestos has been used in conjunction with portland cement since about 1900. Asbestos minerals are naturally occurring and include a variety of different materials that fall into two groups: amphibole and chrysotile. The asbestos minerals may differ in composition but have the common composition of magnesium silicate, and alkalies, magnesium, and iron that may substitute freely for each other.

Although the durability of the embedded asbestos fibers is not considered a problem, some corrosion of the fibers may occur after prolonged periods because of chemical reactions with calcium hydroxide in the portland-cement paste. The chemical reactions have been found to be topochemical and occur on fiber surfaces and on cleavage planes. Corrosion products are probably magnesium hydroxide, magnesium carbonate, magnesium silicate hydrates, and low lime calcium silicate hydrates [39]. The alteration of the asbestos fibers does not adversely affect properties of the concrete because the alteration of the fibers is restricted to only its surface.

CONCRETE

The current emphasis on recycling of materials, and the razing of old concrete structures and pavements, has resulted in the use of old concrete as aggregates for new concrete. For such use, the old concrete is crushed and graded, and used either alone or blended with other aggregates [40–42]. Crushed concrete has had principal use as a base for concrete made with conventional aggregates.

The workability of concrete made with recycled concrete aggregates is about that of concrete made with conventional aggregates. However, its compressive strength is about 75%, and its modulus of elasticity is about 60%, of that for conventional concrete [43].

Buck reported that concrete made with recycled aggregate had resistance to cyclic freezing [44]. Aggregates for use in concrete must possess certain necessary physical and chemical characteristics, such as those required in ASTM C 33. Because the concrete has seen prior service in a given exposure and environment for lengthy periods does not necessarily mean that as concrete aggregate in other service and exposures it will perform similarly. For example, concrete made using chemically unstable aggregate and used in an elevated structure where the exposure has been dry, when used in a moist environment will probably respond differently. Concrete made with a chloride addition, if used in reinforced nonchloride-containing concrete, could promote corrosion of reinforcing steel. These are but two examples that demonstrate the need for establishing new specifications and requirements to ensure that concrete made with recycled concrete will provide adequate service.

REFERENCES

[1] Abrams, M. S. and Orals, D. L., "Concrete Drying Methods and Their Effect on Fire Resistance," *Research and Development Bulletin 181*, Portland Cement Association, Skokie, IL, 1965.

[2] Jones, F. E. and Tarleton, R. D., "Effect of Embedding Aluminum and Aluminum Alloys in Building Materials, Research Paper 36, Great Britain Department of Scientific and Industrial Research, Building Research Station, London, 1963.

[3] Wright, T. E., "An Unusual Case of Corrosion of Aluminum Conduit in Concrete," *Engineering Journal*, Canada, Vol. 38, No. 10, Oct. 1955, pp. 1357–1362.

[4] "Spalled Concrete Traced to Conduit," *Engineering News-Record*, Vol. 172, No. 11, March 1964, pp. 28-29.

[5] Copenhagen, W. J. and Costello, J. A., "Corrosion of Aluminum Alloy Balusters in a Reinforced Concrete Bridge," *Materials Protection and Performance*, Vol. 9, No. 9, Sept. 1970, pp. 31-34.

[6] Monfore, G. E. and Ost, B., "Corrosion of Aluminum Conduit in Concrete," *Journal*, Portland Cement Association, Research and Development Laboratories, Vol. 7, No. 1, Jan. 1965, pp. 10-22.

[7] Wright, T. E. and Jenks, I. H., "Galvanic Corrosion in Concrete Containing Calcium Chloride," *Proceedings*, American Society of Corrosion Engineers, Journal Structural Division, Vol. 89, *ST 5*, Oct. 1963, pp. 117-132.

[8] McGeary, F. L. "Performance of Aluminum in Concrete Containing Chlorides," *Journal*, American Concrete Institute; *Proceedings*. Vol. 63, Feb. 1966, pp. 247-264.

[9a] "Corrosion of Lead by Cement," *Concrete Construction Engineering*, Vol. 44, No. 11, 1949, p. 348; Dodero "Accelerating Reactions of the Corrosion of Lead in Cement," *Metaux & Corrosion*, Vol. 25, No. 282, 1949, pp. 50-56.

[9b] Biczok, I., "Concrete Corrosion and Concrete Protection," *Akadimiai Kiado*, Budapest, 1964.

[10] Mange, C. E., "Corrosion of Galvanized Steel in Contact with Concrete Containing Calcium Chloride," Preprint, National Association of Corrosion Engineers, 13th Annual Conference, St. Louis, MO, 1957.

[11] Halstead, P. E., "Corrosion of Metals in Buildings. The Corrosion of Metals in Contact with Concrete," Reprint No. 38, Great Britain Cement and Concrete Association; reprinted from *Chemistry and Industry*, 24 Aug. 1957, pp. 1132-1137.

[12] Rabald, E., *Corrosion Guide*, Elsevier, New York, 1951.

[13] Stark, D. and Perenchio, W., "The Performance of Galvanized Reinforcement in Concrete Bridge Decks," Portland Cement Association, Skokie, IL, Oct. 1975.

[14] Griffin, D. F., "Effectiveness of Zinc Coatings on Reinforcing Steel in Concrete Exposed to a Marine Environment," Technical Note N-1032, U.S. Naval Civil Engineering Laboratory, Port Hueneme, July 1969, June 1970, and June 1971.

[15] "Use of Galvanized Rebars in Bridge Decks," *Notice*, No. 5, 140.10, Federal Highway Administration, Washington, DC, 9 July 1976.

[16] Stark, D., "Galvanized Reinforcement in Concrete Containing Chlorides," *Construction Technology Laboratories*, Project No. ZE-247, Portland Cement Association, Skokie, IL, April 1978.

[17] Freedman, S., "Corrosion of Nonferrous Metals in Contact with Concrete," *Modern Concrete*, Vol. 36, Feb. 1970.

[18] "Guide to Durable Concrete," *Journal*, ACI Committee 201, American Concrete Institute, Dec. 1977, pp. 591-592.

[19] "Waste Materials in Concrete," *Concrete Construction*, Vol. 16, No. 9, Sept. 1971, pp. 372-376.

[20] Johnson, C. D., "Waste Glass as Coarse Aggregate for Concrete," *Journal of Testing and Evaluation*, Vol. 2, No. 5, 1974, pp. 344-350.

[21] Waters, E. H., "Attack on Glass Wall Tiles by Portland Cement," Report S-41, Australia Commonwealth Scientific and Industrial Research Organization, Division of Building Research, 1956.

[22] "Fiber Reinforced Concrete," Centre de Recherches de Pont-a-Morrison, French Patent No. 1,369,415, 14 Aug. 1964.

[23] Parker, T. W., "Sawdust-Cement and Other Sawdust Building Products," *Chemistry and Industry*, 1947, pp. 593-596.

[24] Dominik, W. and Hans, M., "Action of Cement on Wood," *Przemsyl Cheniczny*, Vol. 22, 1938, pp. 74-82.

[25] Seymour, R. B. and Steiner, R. H., *Plastics for Corrosion-Resistant Applications*, Reinhold, New York, 1955.

[26] *Handbook of Chemistry and Physics*, 41st ed., Chemical Rubber Company, Cleveland, 1959-1960.

[27] "State-of-the-Art Report on Fiber Reinforced Concrete," *Journal*, Title No. 70-65, ACI Committee 544, American Concrete Institute, Detroit, MI, Nov. 1973, pp. 729-744.

[28] Shroff, J. K., "The Effect of a Corrosive Environment on the Properties of Steel Fiber Reinforced Portland Cement," Masters thesis, Clarkson College of Technology, Potsdam, New York, Sept. 1966.

[29] "A Status Report on Fiber Reinforced Concretes," *Concrete Construction*, Jan. 1976, pp. 13-16.

[30] Bateson, G. B. and Obszarski, J. M., "Strength of Steel Fiber Reinforced Concrete Exposed to a Salt Water Environment," *Materials Performance*, July 1977, Technical Note, p. 48; Paper No. 27, NACE Annual Conference, Houston, March 1976.

[31] "Corrosion Behavior of Cracked Fibrous Concrete," Materials Systems and Science Division of the Construction Engineering Research Laboratory, Project OK1, Task 03, Work Unit 006, 1975 (unpublished).

[32] Lankard, D., private communications, formerly Battelle Memorial Institute, now with Materials Laboratory, Inc., 1977.

[33] Marek, R. et al., "Promising Replacements for Conventional Aggregates for Highway Use," HRB NCHRP, Final Report, Project 4-10, University of Illinois, Urbana, IL, Jan. 1971.

[34] Ironman, R., "Stronger Market Seen for Glass-Fiber Concrete," *Concrete Products*, Jan. 1976, pp. 42-44.

[35] Tallentire, A. G., "Glass Fibre Cement Applications," *Precast Concrete*, Feb. 1977, pp. 95-97.

[36] Grimer, F. J. and Ali, M. A., "The Strengths of Cements Reinforced with Glass Fibers," *Magazine of Concrete Research*, Vol. 21, No. 66, March 1969, pp. 23-30.

[37] Majundar, A. J., "Properties of Fiber Cement Composites," *Proceedings*, Symposium on Fiber Reinforced Cement and Concrete, RILEM, Paris, 1975, pp. 279-313.

[38] Cohen, E. B. and Diamond, S., "Validity of Flexural Strength Reduction as an Indication of Alkali Attack on Glass in Fibre Reinforced Cement Composites," *Proceedings*, Symposium on Fiber Reinforced Cement and Concrete, RILEM, Paris, 1975, pp. 315-325.

[39] Opoczky, L. and Pentek, L., "Investigation on the 'Corrosion' of Asbestos Fibers in Asbestos Cement Sheets Weathered for Long Times," *Proceedings*, Symposium on Fiber Reinforced Cement and Concrete, RILEM, Paris, 1975, pp. 269-276.

[40] "Rubble Recycling Saves Time, Energy, and the Environment," *Rock Products*, Vol. 80, No. 5, May 1977, pp. 107-108.

[41] "Reusing Concrete," *Engineering News-Record*, Vol. 186, No. 22, June 1971, p. 15.

[42] "Recycled Slab is New Runway Base," *Highway and Heavy Construction*, Vol. 120, No. 7, July 1977, pp. 30-33.

[43] Frondistou-Yamas, S., "Waste Concrete as Aggregate for New Concrete," *Journal*, American Concrete Institute, Vol. 74, No. 8, Aug. 1977, pp. 373-376.

[44] Buck, A. D., "Recycled Concrete," *Highway Research Record No. 430*, 1973, p. 108.

Abrasion Resistance

Tony C. Liu[1]

PREFACE

The subject of abrasion resistance was covered in all three previous editions of *ASTM STP 169*, *ASTM STP 169A*, and *ASTM STP 169B* authored by H. L. Kennedy and M. E. Prior, M. E. Prior, and R. O. Lane, respectively. The general presentation of this paper is similar to its predecessors; however, additional information on the abrasion resistance of several newly available concrete materials, such as fiber-reinforced concrete, silica-fume concrete, and polymer concrete, is included to reflect the current state of the art. This edition also includes discussions of two new abrasion test methods that have been adopted by ASTM since the publication of *ASTM STP 169B* in 1978.

DEFINITIONS AND TYPES OF ABRASION

ASTM Terminology Relating to Erosion and Wear (G 40) defines abrasion as "wear due to hard particles or hard protuberances forced against and moving along a solid surface." Abrasion resistance, according to American Concrete Institute (ACI) Committee on Terminology and Notation (ACI 116), is the "ability of a surface to resist being worn away by rubbing and friction" [*1*].

Wear of concrete surface can be classified as follows [*2*]:

1. Wear on concrete floors, due to foot traffic, light trucking, and skidding, scraping, or sliding of objects on the surface (attrition).
2. Wear on concrete floors and road surfaces due to forklift, heavy trucking, and automobiles, with and without chains (attrition plus scraping plus percussion).
3. Wear on hydraulic structures such as stilling basins, spillways, bridge abutments, and tunnels due to the action of abrasive materials carried by water flowing at low velocities (attrition plus scraping).
4. Wear on concrete dams, spillways, tunnels, and other water-carrying systems where high velocities and negative pressure are present. This is generally known as cavitation erosion (cavitation).

The first type of wear is essentially a rubbing action and is usually greatly increased by the introduction of foreign particles, such as sand, metal scraps, or similar materials. Normal wear without the benefit of such abrasive materials would be negligible on a good concrete surface. The second type of wear is caused by a rubbing action similar to that found in the first type, plus an impact-cutting type of wear. This latter type is brought about by the use of chains on automobile and truck tires or metal vehicle wheels. As the wheel revolves, it brings the metal into contact with the concrete surface with considerable impact, a process that tends to cut the surface of the concrete. Wear on concrete floors from forklift trucks is highly abrasive and subject to impact because hard rubber wheels pick up grit and nails from pallets, which cause forklift truck wheels to impact the floor. The third type of wear is primarily grinding and cutting actions. The action of the abrasive particles carried by the flowing water, of course, is controlled largely by the velocity of the water, the angle of contact, the type of abrasive material, and the general surrounding conditions. The fourth type of wear, cavitation, is caused by the abrupt change in direction and velocity of a liquid to such a degree that the pressure at some point is reduced to the vapor pressure of the liquid. The vapor pockets so created, upon entering areas of high pressure, collapse with a great impact, which eventually causes pits or holes in the concrete surface. Discussion of cavitation resistance of concrete is beyond the scope of this paper; therefore, it will not be covered. An excellent source of information on cavitation erosion can be found in Refs *3* and *4*.

FACTORS AFFECTING ABRASION RESISTANCE

Factors that may affect the resistance of concrete to abrasive action should be considered in the design and construction of concrete surfaces that are to withstand abrasion due to rubbing, scouring, sliding, impact, scraping, attrition, percussion, gouging, or cutting from mechanical or hydraulic forces. Frequently, the failure of concrete to resist abrasion can be traced to cumulative effects such as soft aggregate, inadequate compressive strength, improper curing or finishing, or over-manipulation of the plastic concrete surface. The following discussion of significant factors relative to concrete resistance to abrasion illustrates the importance of the proper selection,

---

[1] Chief, Materials Engineering Section, Geotechnical and Materials Branch, Engineering Division, Directorate of Civil Works, Headquarters, U.S. Army Corps of Engineers, Washington, DC 20314.

composition, and application of concrete based on the specific type of service condition.

Quality of Aggregates

Studies by Liu [5] and Laplante et al. [6] indicated that the abrasion resistance of concrete is strongly influenced by the hardness of its coarse aggregate. The abrasion resistance can be increased appreciably by the use of maximum amount of dense, hard coarse aggregates such as traprock, chert, granite, or metallic aggregate. For example, the abrasion loss of concrete containing limestone aggregate as approximately twice as much as that of the concrete containing chert [5].

Abrasion tests carried out by Liu [5] indicated that no correlation existed between abrasion resistance of concrete and the quality of the coarse aggregate as determined by ASTM Test Method for Resistance to Degradation of Small-Size Coarse Aggregate by Abrasion and Impact in the Los Angeles Machine (C 131). As can be seen from Fig. 1, the Los Angeles abrasion losses are approximately equal for soft aggregate, such as limestone, and relatively hard aggregate, such as chert. However, the abrasion losses of the concrete containing these aggregates vary widely; the abrasion loss of the limestone concrete is much more than that containing chert. A similar finding was reported by Smith [7].

Schuman and Tucker [8] pointed out that the shape of aggregate particles regulates the water requirements for placing and finishing and has a direct influence on the abrasion resistance of concrete. Angular- to subangular-shaped aggregate is known to improve bond, usually resulting in increased abrasion resistance.

For concrete subject to light-to-medium abrasion, good quality aggregates meeting the requirements of ASTM Specification for Concrete Aggregates (C 33) are generally acceptable. Heavy duty floors and slabs exposed to more severe abrasive action demand well-grade hard mineral aggregates [9].

Compressive Strength

Witte and Backstrom [10], as researchers before and after them, considered the compressive strength as one of the most important factors responsible for the abrasion resistance of concrete. For the same aggregate and finishing procedure, the abrasion resistance of concrete increases with an increase in compressive strength. For example, as shown in Fig. 2, the average abrasion resistance (the reciprocal of abrasion-erosion loss) for concrete containing limestone increases approximately 44% as the compressive strength increases from 20.7 MPa (3000 psi) to 62.1 MPa (9000 psi) [5].

Mixture Proportioning

Abrasion test results by many researchers clearly indicated that, for a given aggregate, the abrasion resistance of concrete increases with a decrease in water-cement ratio [5]. A maximum water-cement ratio of 0.45 has been specified by the U.S. Army Corps of Engineers for concrete subject to abrasion in hydraulic structures.

The effects of various admixtures, including air-entraining and water-reducing admixtures and retarders, on the abrasion resistance of concrete could not be established

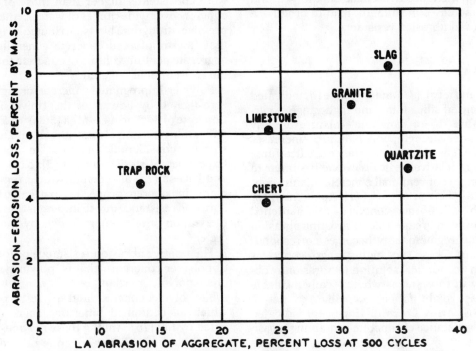

FIG. 1—Relationship between resistance of aggregate to abrasion and concrete abrasion loss.

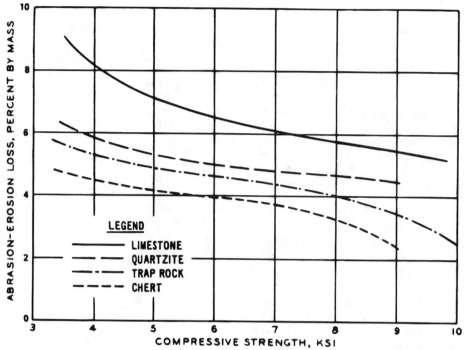

FIG. 2—Relationship between abrasion resistance and compressive strength.

conclusively on the basis of the literature review. The general trend in most studies reflects an increase in abrasion loss proportional to increase in air content for any given water-cement ratio (w/c). As the compressive strength is adversely affected by increased air contents, so will the resistance of concrete to abrasion. Water-reducing admixtures, by virtue of their positive effect on the w/c and consequently on increases in strength, tend to produce a concrete of improved abrasion resistance.

Concrete Types

Studies by Holland, et al. [11] and Laplante [6] indicated that adding condensed silica fume and high-range water-reducing admixture to a concrete mixture greatly increases compressive strength, which, in turn, increases abrasion resistance. The abrasion resistance of silica-fume concrete containing relatively soft limestone is similar to that of a high-strength conventional concrete mixture containing a very hard chert aggregate [11]. Apparently, for the high-strength silica-fume concrete, the hardened cement paste assumes a greater role in resisting abrasion and as such the aggregate quality becomes correspondingly less important. These very high-strength concretes appear to offer an economical solution to abrasion problems, particularly in those areas where locally available aggregate otherwise might not be acceptable.

Tests by the U.S. Army Corps of Engineers indicated that the abrasion resistance of concrete containing fly ash is comparable to that of the concrete without fly ash, as long as the concretes contain the same type of aggregate and have the comparable compressive strength at the time of tests.

While the addition of steel fibers in the concrete mixture would be expected to increase the impact resistance of concrete, fiber-reinforced concrete is less resistant to abrasion than conventional concrete of the same aggregate type and water-cement ratio [12]. The abrasion losses of fiber-reinforced concrete as determined by ASTM C 1138 were consistently higher than those of the conventional concrete over wide ranges of water-cement ratio and compressive strength. This is attributed primarily to the fact that fiber-reinforced concrete generally has less coarse aggregate per unit volume of concrete than does comparable conventional concrete.

Polymer-impregnated concrete for highway bridges was investigated by Fowler of the University of Texas [13]. Concrete of 34.5- to 44.8-MPa (5000- to 6500-psi) compressive strength, non-air-entrained, treated with polymer showed considerable increase in abrasion resistance. Experimental studies by Liu [5], Krukar and Cook [14], and Dikeou [15] also demonstrated remarkable improvements in concrete properties, particularly in strength and abrasion resistance due to impregnation and in-place polymerization of monomers in overlays and in mature concrete.

The relative abrasion resistance of four different types of polymer concretes (that is, polymer-impregnated concrete, polymer portland cement concrete, methyl methacrylate polymer concrete, and vinyl ester polymer concrete), which all contained limestone aggregates, was investigated by Liu [16]. Among these polymer concretes tested, the vinyl ester polymer concrete ranked first in abrasion resistance, followed by methyl methacrylate polymer concrete, polymer-impregnated concrete, and polymer portland cement concrete.

Finishing Procedures

Abrasion resistance of concrete is affected by the finishing procedures used. Tests by Kettle and Sadegzadeh [17] indicated that the abrasion resistance of concrete subjected to power finishing is significantly higher than that subjected to hand finishing. This is attributed to surface compaction and to reduction of the w/c of the surface matrix.

Finishing techniques, including wood float, magnesium float, steel trowel, and hard-steel trowel finishes, are also compared with abrasion resistance of concrete in a study by Fentress [18]. The wood float tends to tear the surface and to displace the aggregate. The magnesium float, despite ease of finishing, causes a rough-textured surface and a lowering in abrasion resistance. Both the steel trowel and hard-steel trowel produce smooth surfaces, closing any existing imperfections and providing excellent resistance to abrasion.

The key to durable, cleanable concrete slabs can be achieved by proper finishing procedures and in a reduction in the w/c [19]. Combining these demands in one application, Ytterberg demonstrated that a deferred topping finish from which surface water is removed by a vibratory absorption process yields highest abrasion resistance [19]. The technique requires that after the water of workability has been removed, the surface is blade-floated, and upon stiffening, troweled.

Abrasion tests carried out by Kettle and Sadegzadeh [17], the U.S. Bureau of Reclamation [20], and the U.S. Army Corps of Engineers [16] confirmed that vacuum-treated concrete surfaces are considerably more resistant to abrasion than surfaces with either non-treated lean or richer concrete. The improvement in abrasion resistance of vacuum-treated concrete is due principally to the reduction of water content in the concrete mixture. This treatment is most effective on concrete with a high w/c.

Concrete wearing surfaces can be improved by the use of dust coats or "dry shakes" consisting of dry cement or cement-aggregate mixtures sprinkled on the surface just prior to initial set. Pockets of bleed water should be removed prior to the application of the dry shake. After the dry shake is applied to the concrete, it should be given enough time to absorb water prior to floating. Once the dry shake has absorbed water, it is power-floated and then steel troweled into the surface without the use of additional water to produce a hard, dense topping layer.

Curing

Efficient curing increases the abrasion resistance. A correlation of curing time and abrasion resistance reported by Sawyer [21] involved a series of tests comprising a wide range of cement contents, w/c, and incremental curing. From this study, it is apparent that marked improvement in abrasion resistance can be expected with extended curing time, especially for surfaces composed of leaner concrete.

A laboratory program conducted by the California Division of Highways [22] considered the effect on abrasion of such variables as slump, finishing, curing, and surface treatments using linseed oil. Test data indicated, among these variables, that the greatest abrasion losses encountered were associated with less-than-adequate curing procedures.

For concrete floor and slab construction using Type I portland cement, at least five days of curing at concrete temperatures of 21°C (70°F) or higher and a minimum of seven days at temperatures of 10 to 21°C (50 to 70°F) are required to assure adequate abrasion resistance [9].

Surface Treatment

Certain chemicals (magnesium and zinc fluorosilicates, sodium silicate, gums, and waxes) serve to prolong the life of older floors and are considered an emergency measure for treatment of deficiencies in relatively pervious and soft surfaces that wear and dust rapidly [23]. Treatment with fluorosilicates densifies and hardens surfaces and improves abrasion resistance, but the improvement by the two types of fluorosilicates is not of equal magnitude [24]. While the magnesium fluorosilicate is more resistant to rubbing action simulated best by a revolving-disk-type abrasion machine, the zinc fluorosilicate shows greater resistance to impact-type wear as imposed by the dressing-wheel-type machine.

For poorly cured or porous concrete surfaces, beneficial effects from treatment with linseed oil were demonstrated by an increase in abrasion resistance of as much as 30% [25].

Paints and coatings used to seal concrete surfaces and to protect the concrete from the attack of the environment or chemicals [26] possess only limited abrasion resistance and any test simulation to evaluate their effectiveness is difficult. However, some coatings, among them vinyl and heavy rubber, if properly bonded remain fairly resilient and effective in protecting concrete from abrasive actions. Several types of surface coatings including polyurethane, epoxy-resin mortar, furan-resin mortar, acrylic mortar, and iron-aggregate toppings have exhibited good abrasion resistance in laboratory tests [16]. However, problems in field application of surface coatings have been reported. These have been due primarily to improper surface preparation or thermal incompatibility between coatings and concrete. More recently, formulations have been developed that have coefficients of thermal expansion more similar to that of the concrete substrate [3].

ABRASION TEST METHODS

The techniques and test methods that have been employed for over a century to evaluate the abrasion resistance of concrete have attempted with varied success to reproduce the typical forces detrimental to concrete surfaces. Preceding any standardization of laboratory methods or of testing machines, reliance was placed initially on the use of rattlers that provide abrasive action based on the tumbling of steel balls impacting a test surface [27]. Numerous modified versions of the rattler principle continued to be the dominant test procedure along with the revolving metal disk pressed against a small concrete specimen [28]. The drill-press-type abrasion machine [29] in modified form still is used by highway departments and enjoys some popularity because of its simple design.

Currently, there are four standard ASTM test methods for evaluating the resistance of concrete subjected to various types of abrasive actions. The background, significance, and applicabilities of these ASTM test methods are presented here.

ASTM C 418

ASTM Test Method for Abrasion Resistance of Concrete by Sandblasting (C 418) dates back to 1958 and is based on the principle of producing abrasion by sandblasting. This procedure simulates the action of waterborne abrasives and abrasives under traffic on concrete surface. It performs a cutting action that tends to abrade more severely the less resistant components of the concrete. Adjustments in the pressure used and the type of abrasive permit a variation in the severity of abrasion that may be used to simulate other types of abrasion.

This test method was modified by ASTM Subcommittee C09.03.13 on Methods of Testing Concrete for Resistance to Abrasion from the original Ruemelin blast cabinet equipped with an injector-type blast gun with high-velocity air jet (Fig. 3). The control over such variables as gradation of the silica sand, air pressure, rate of feed of the abrasive, distance of the nozzle from the surface, and the area of the shielded surface is critical. The 1958 standard specified carborundum or silica sand. This procedure was revised in 1964 calling for the explicit use of silica sand that allows for closer control of the abrasive action. The 1976 revision specifies the use of oil-base modeling clay of known specific gravity for the filling of voids in the abraded concrete surface. This permits accurate measurement of abrasion by volume displacement and replaces the old weight-loss determination. Rushing [30] compared the 1958 and 1964 procedures of ASTM C 418 in a laboratory study and found the later revision permits a more accurate determination of abrasion loss.

ASTM C 779

For almost two decades, ASTM C 418 remained the only specified standard for testing concrete resistance to abrasion. From 1964 through 1971, ASTM Subcommittee C09.03.13 initiated a number of major test programs conducted by independent laboratories. The results of these comparative programs established such parameters as severity and reproducibility, leading to the selection of three abrasion machines and their inclusion in one ASTM standard. ASTM Test Method for Abrasion Resistance of Horizontal Concrete Surfaces (C 779) covers three testing procedures: (a) the revolving-disk machine, (b) the dressing-wheel machine, and (c) the ball-bearing machine. All three machines are portable and adaptable for laboratory and in-place field abrasion testing.

The revolving-disk machine (Fig. 4) introduces frictional forces by rubbing and grinding. Sliding and scuffing is accomplished by rotating steel disks in conjunction with abrasive grit. The revolving-disk machine was redesigned by Master Builders Company, Cleveland, Ohio, and was essentially patterned after a machine developed by the National Institute of Standards and Technology. Among the three standard methods, the revolving-disk machine provides the most reproducible results. A supply of No. 60 silicon carbide abrasive is fed to the disks at a rate of 4 to 6 g/min. A test period of 30 min generally produces significant wear on most concrete surfaces, but it is recommended to extend the test period to 60 min, if information on the longtime abrasion resistance is desired. Davis and Troxell [31] reported that the depth of wear for periods of 30 to 60 min to be about the same for ordinary slabs as for heavy-duty concrete surfaces. A plane surface resulted in either case regardless of the variation in the quality of the hardened cement paste or the toughness of

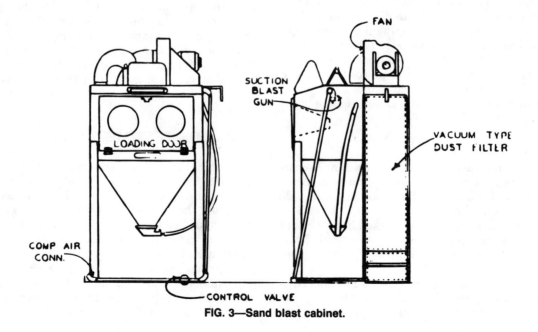

FIG. 3—Sand blast cabinet.

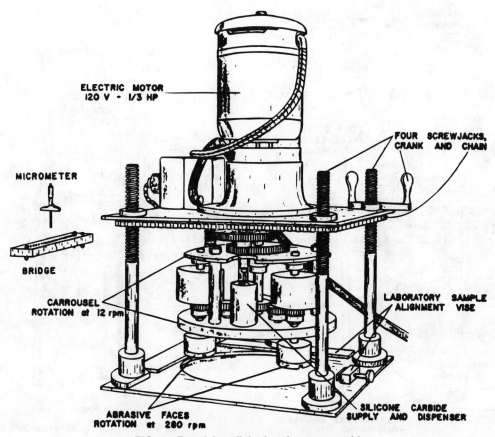

FIG. 4—Revolving-disk abrasion test machine.

the aggregate that became exposed as the depth of wear increased. The abrasive mode of this procedure best simulates wear by light-to-moderate foot traffic and light-to-medium tire-wheeled traffic or moving of light steel racks, etc.

In contrast to the disks, the dressing-wheel machine (Fig. 5) imparts high concentrated compressive forces with high impact stresses. The dressing-wheel machine is similar to the revolving-disk machine in appearance except for three sets of seven dressing wheels mounted on horizontal shafts that take the place of the three rotating steel disks. However, no abrasive material is employed with this machine. Initial and intermediate measurements of the test path are taken with a depth micrometer. Tracking of the dressing wheel normally leaves a grooved path with the test surface being irregular and fairly rough as harder aggregate particles stand out from the softer aggregate particles and mortar that are abraded more quickly. On ordinary-finished concrete slabs, the dressing wheels produce a depth of wear more than double that obtained with the revolving-disk machine for the same test duration. Yet, approximately equal depths of wear are obtained from both machines for the same test period when hard troweled finished floors are tested. The coefficient of variation established by the Berkeley study [31] for slabs abraded with the dressing-wheel machine is several times as great as that for the revolving-disk machine. These results are essentially concurrent with a parallel study conducted in the Portland Cement Association laboratories [32]. Abrasion of concrete induced by the dressing-wheel machine closely simulates the rolling, pounding, and cutting action of steel wheels or the effect of studded tires.

Repeated dynamic loading through strong impact, compressive forces, and high-speed rolling constitutes the abrasive action of the ball-bearing machine (Fig. 6). The ball-bearing machine operates on the principle of a series of eight ball bearings rotating under load at a speed of 1000 rpm on a wet concrete test surface. Water is used to flush out loose particles from the test path, bringing the ball bearing in contact with sand and stone particles still bonded to the concrete surface, thus providing impact as well as sliding friction. During the test, abrasion readings are taken every 50 s with a dial micrometer mounted directly to the supporting shaft allowing readings "on the fly." Readings are continued for a total of 1200 s or until a maximum depth of 3.0 mm (0.12 in.) is reached. In its abrasive severity, the ball-bearing machine exceeds both the revolving disk and the dressing wheel, also producing the highest coefficient of variation among the three procedures (Table 1). The effects of the ball-bearing machine on concrete surfaces indicate abrasive action becomes progressively more severe as the test continues due to soft and hard spots causing the core barrel to bounce to high angular speed. This test procedure has merit in that the ball bearings are similar to rolling wheels and are more typical of actual loadings by steel-wheel traffic to which a concrete floor may be exposed.

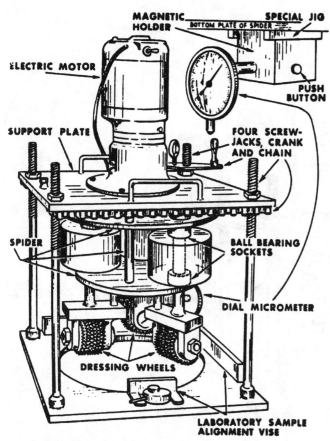

FIG. 5—Dressing-wheel abrasion test machine.

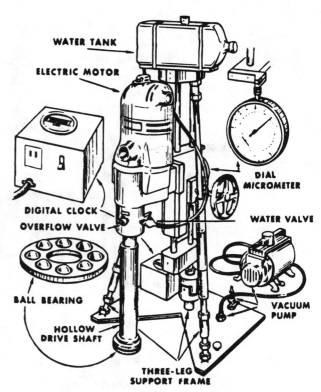

FIG. 6—Ball-bearing abrasion test machine.

Experience with testing of abrasion by ASTM C 779 demonstrates relatively good reproducibility over a fairly wide range of concrete surface types and conditions when the abrasive action is moderate, for instance, through attrition. More severe abrasion due to percussion, impact, cutting, or rolling creates erratic action and is less conducive to reproducibility. In general, loss in test accuracy may develop due to the following factors: (1) rapid wear of dressing wheels, revolving disks, and steel balls and failure to replace them regularly as specified; (2) failure to remove dust and loose-abraded material regularly from the test surface; (3) improper selection of a representative test area or specimen; (4) differences in age levels of concrete at which test results are compared; (5) arbitrary increase in the length of test or flow of sand; and (6) insufficient number of wear readings. Excluding these variables, the within-laboratory precision shown in Table 1 indicates the lowest coefficient of variation is obtained with the revolving-disk method, increasing in order with dressing wheels and ball bearings.

ASTM C 944

ASTM Test Method for Abrasion Resistance of Concrete or Mortar Surfaces by the Rotating-Cutter Method (C 944) gives an indication of the relative abrasion resistance of mortar and concrete based on testing of cored or fabricated specimens. This test method has been successfully

TABLE 1—Within-Laboratory Precision for Single Operator, ASTM C 779.

Procedure	Coefficient of Variation, percent of mean[a]	Acceptable Range of Two Results, percent of mean[a]
A. Revolving disk	5.51	15.6
B. Dressing wheel	11.69	33.1
C. Ball bearing	17.74	50.2

[a] These numbers represent respectively the 1S% and D2S% limits as described in ASTM Practice for Preparing Precision and Bias Statements for Test Methods for Construction Materials (C 670), in the section on Alternative Form of the Precision Statement.

used in the quality control of highway and concrete bridges subject to traffic. This method is primarily intended for use on the top ends of 152-mm (6-in.) diameter concrete cores, mortar specimens, or other samples of concrete of insufficient test area to permit the conduct of tests by ASTM C 418 and C 779. This test method produces a much more rapid abrasive effect than the other test methods. The test apparatus is a fairly simple piece of equipment consisting of a rotating cutter and a drill press or similar device with a chuck capable of holding and rotating the abrading cutter at a speed of 200 rpm. Figure 7 shows a rotating-cutter drill press. The difficulty in maintaining a constant load on the abrading cutter when using the lever, gear, and spring system of a drill press has been eliminated by placing a constant load of 98 N (22 lbf) directly upon the spindle that turns the cutter. General practice is to clean the surface occasionally during the

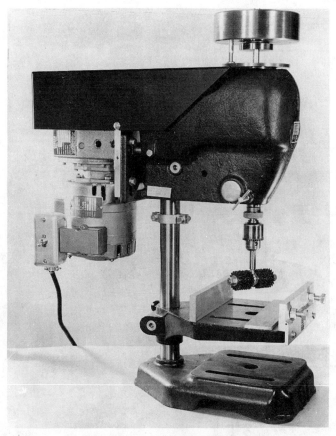

FIG. 7—Rotating-cutter drill press.

test by blowing the dust off the test specimen. The reproducibility of test results has been poor, with the single-operator coefficient of variation of more than 20%.

ASTM C 1138

ASTM Test Method for Abrasion Resistance of Concrete (Underwater Method) (C 1138) was originally developed by Liu [16] in 1980 for evaluating the resistance of concrete surfaces subjected to the abrasive action of waterborne particles on hydraulic structures such as stilling basins and outlet works. ASTM adopted this test procedure with some minor modifications in 1989. The apparatus consists of essentially a drill press, an agitation paddle, a cylindrical steel container that houses a disk-shaped concrete specimen, and 70 steel grinding balls of various sizes. The detailed cross-sectional view of the test apparatus is shown in Fig. 8. The water in the container is circulated by the immersed agitation paddle that is powered by the drill press rotating at 1200 rpm. The circulating water, in turn, moves the abrasive charges (steel grinding balls) on the surface on the concrete specimen, producing the abrasion effects. Testing, totaling 24 h, generally produces significant abrasion in most concrete surfaces. The standard test consists of six 12-h test periods for a total of 72 h. Additional testing time may be required for concrete that is highly resistant to abrasion. This test method can duplicate well the abrasive action of waterborne particles in the stilling basins. This method is not, however, intended to provide a quantitative measurement of length of service that may be expected from a specific concrete.

The abrasion loss, as determined by the original Corps of Engineers' test method [16], was expressed as a percentage of the original mass of the specimen. ASTM Subcommittee C09.03.13 revised the method by calculating the volume loss, or the average depth of wear, at the end of the test time increment to eliminate differences in results resulting from the allowable variations in specimen size. The precision of this test procedure with the revised method of calculating the abrasion loss is being determined.

APPLICATION OF TEST METHODS

As discussed in the previous section, ASTM C 418, C 779, C 944, and C 1138 offer six distinct procedures that simulate various degrees of severity and types of abrasive or erosive forces. Table 2 serves as a general guide for possible applications of the six standard procedures for various categories of abrasion. This tabulation attempts to correlate the severity of in-place wear with the abrasive action particular to each of the test methods. However, it is necessary to bear in mind that there is overlapping in the type of abrasion imparted by these test applications and that a specific service condition may be reproduced by more than one procedure.

TABLE 2—Application of Test Procedures.

Type of Abrasion	ASTM C 418	ASTM C 779 A	B	C	ASTM C 944	ASTM C 1138
Foot traffic or light-to-medium tire-wheeled traffic, etc.		X			X	
Forklift, heavy tire-wheeled traffic, automobile with chains, heavy steel-wheeled traffic, or studded tires, etc.			X	X	X	
Abrasive erosion of waterborne particles on hydraulic structures	X					X

CONCLUSION

Concrete materials are susceptible to deterioration due to the abrasive action of environmental and man-made factors. In addition, intrinsic conditions related to properties of concrete and construction techniques contribute to the reduction in abrasion resistance. Proper selection of concrete-making materials, mixture proportion, curing, and finishing procedures are the basic requirements applying to any concrete subjected to abrasion from vehicular traffic or water-borne particles. Effects of variables on reduced abrasion resistance of concrete have proved to be cumulative, which may explain why only certain portions of a concrete surface may fail in resisting abrasion while the remainder gives satisfactory performance.

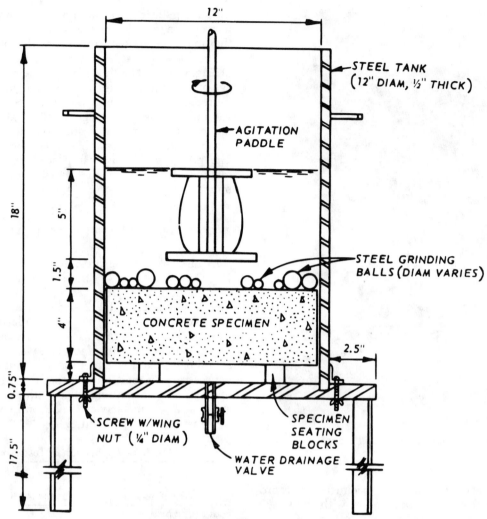

FIG. 8—Test apparatus, ASTM C 1138.

There are four ASTM test methods, that is, ASTM C 418, C 779, C 944, and C 1138, for evaluating the resistance of concrete subjected to various types of abrasive actions. These test methods serve to (1) evaluate, predict, or accept the quality of concrete surfaces; (2) evaluate specific effects of variables such as concrete-making materials, curing, finishing procedures, surface hardeners, or coating materials; (3) compare various types of concrete surfaces under simulated abrasion conditions; and (4) verify products or systems to meet specifications.

REFERENCES

[1] "Cement and Concrete Terminology (ACI 116R-85)," ACI Manual of Concrete Practice, Part 1, ACI Committee 116, American Concrete Institute, Detroit, MI, 1990.
[2] "Guide to Durable Concrete (ACI 201.2R-77)," ACI Manual of Concrete Practice, Part 1, ACI Committee 201, American Concrete Institute, Detroit, MI, 1990.
[3] "Erosion of Concrete in Hydraulic Structures (ACI 210R-87)," ACI Manual of Concrete Practice, Part 1, ACI Committee 210, American Concrete Institute, Detroit, MI, 1990.
[4] "Erosion of Concrete by Cavitation and Solids in Flowing Water," Laboratory Report No. C-342, U.S. Bureau of Reclamation, Washington, DC, 1947.
[5] Liu, T. C., "Abrasion Resistance of Concrete," *ACI Journal, Proceedings*, Vol. 78, No. 5, Sept-Oct. 1981, pp. 341–350.
[6] Laplante, P. C., Aitcin, P. C., and Vezina, D., "Abrasion Resistance of Concrete," *Journal of Materials in Civil Engineering*, Vol. 3, No. 1, Feb. 1991, pp. 19–28.
[7] Smith, F. L., "The Effect of Aggregate Quality on Resistance of Concrete to Abrasion," *Cement and Concrete, ASTM STP 205*, American Society for Testing and Materials, Philadelphia, 1956.
[8] Schuman, L. and Tucker, J., Jr., "A Portable Apparatus for Determining the Relative Wear Resistance of Concrete Floors," Research Paper No. RP 1252, National Bureau of Standards, Washington, DC, 1939.
[9] "Guide for Concrete Floor and Slab Construction (ACI 302.1R-89)," ACI Manual of Concrete Practice, Part 2, ACI Committee 302, American Concrete Institute, Detroit, MI, 1990.
[10] Witte, L. P. and Backstrom, J. E., "Some Properties Affecting the Abrasion Resistance of Air-Entrained Concrete," *Proceedings*, American Society for Testing and Materials, Vol. 51, 1951, p. 1141.
[11] Holland T. C., Krysa, A, Luther, M. D., and Liu, T. C., "Use of Silica-Fume Concrete to Repair Abrasion-Erosion Dam-

age in the Kinzua Dam Stilling Basin," *Proceedings*, Second International Conference on Fly Ash, Silica Fume, Slag, and Natural Pozzolans in Concrete, Madrid, Spain, 1986.

[12] Liu, T. C. and McDonald, J. E., "Abrasion-Erosion Resistance of Fiber-Reinforced Concrete," *ASTM Cement and Aggregate*, Vol. 3, No. 2, Winter 1981.

[13] Fowler, D. W., Houston, J. T., and Paul, D. R., "Polymer Impregnated Concrete for Highway Application," Research Report No. 1141, Center for Highway Research, University of Texas, Austin, TX, Feb. 1973.

[14] Krukar, M. and Cook, J. C., "The Effect of Studded Tires on Different Pavement and Surface Texture," Washington State University, Pullman, WA, Feb. 1973.

[15] Dikeou, J. T., "Radiation Polymerization of Monomers," U.S. Bureau of Reclamation, Washington, DC, Jan. 1970.

[16] Liu, T. C., "Abrasion-Erosion Resistance of Concrete," Technical Report C-78-4, Report 3, U.S. Army Corps of Engineers, Waterways Experiment Station, Vicksburg, MS, July 1980.

[17] Kettle, R. and Sadegzadeh, M., "The Influence of Construction Procedures on Abrasion Resistance," *Proceedings*, Katharine and Bryant Mather International Conference on Concrete Durability, *ACI SP-100*, Vol. 2, American Concrete Institute, Detroit, MI, 1987.

[18] Fentress, Blake, "Slab Construction Practices Compared by Wear Tests," *Journal*, American Concrete Institute, Detroit, MI, July 1973.

[19] Ytterberg, R. F., "Wear Resistance of Industrial Floors of Portland Cement Concrete," *Journal*, American Society of Civil Engineers, Civil Engineering, Jan. 1971.

[20] "Erosion Resistance Tests of Concrete and Protective Coatings," Concrete Laboratory Report No. C 445, U.S. Bureau of Reclamation, Washington, DC, Feb. 1952.

[21] Sawyer, J. L., "Wear Test on Concrete Using the German Standard Method of Test and Machine," *Proceedings*, American Society for Testing and Materials, Vol. 57, 1957, pp. 1143–1153.

[22] Spellman, D. L. and Ames, W. H., "Factors Affecting Durability of Concrete Surfaces," Materials and Research Department, California Division of Highways, Sacramento, CA, Report No. M&R 250908-1, BRPD-3-7, 1967.

[23] "Surface Treatments for Concrete Floors," Concrete Information ST-37, Portland Cement Association, p. 2.

[24] "The Effect of Various Surface Treatments Using Magnesium and Zinc Fluorosilicate Crystals on Abrasion Resistance of Concrete Surfaces," Concrete Laboratory Report No. C 819, U.S. Bureau of Reclamation, Feb. 1956, Washington, DC,

[25] Mayberry, D., "Linseed Oil Emulsion for Bridge Decks," U.S. Department of Agriculture, Agriculture Research Service, Peoria, IL, Sept. 1971.

[26] "A Guide to the Use of Waterproofing, Dampproofing, Protective, and Decorative Barrier Systems for Concrete (ACI 515.1R-79, Revised 1985)," ACI Manual of Concrete Practice, Part 5, ACI Committee 515, American Concrete Institute, Detroit, MI, 1990.

[27] Scofield, H. H., "Significance of Talbot Jones Rattler as Test for Concrete in Road Slabs," *Proceedings*, Highway Research Board, Washington, DC, 1925, p. 127.

[28] Guttman, A., "Abrasion Tests on Concrete," *Chemical Abstract*, Vol. 30, 1936, p. 5750.

[29] Haris, D. H., "Apparatus for Testing the Hardness of Materials," *Chemical Abstract*, Vol. 38, 1944, p. 1338.

[30] Rushing, H. B., "Concrete Wear Study," Report PB 183410, Louisiana Department of Highways, Baton Rouge, LA, June 1968.

[31] Davis, R. E. and Troxell, G. E., "Methods of Testing Concrete for Resistance to Abrasion—Cooperative Testing Program," University of California, Berkeley, CA, Oct. 1964.

[32] Klieger, P. and Brinkerhoff, C. H., "Cooperative Study on Methods of Testing for Abrasion of Concrete Floor Surfaces," Research and Development Laboratories, Portland Cement Association, Skokie, IL, March 1970.

Elastic Properties and Creep

Robert E. Philleo[1]

PREFACE

A chapter on elastic properties of concrete first appeared in *ASTM STP 169*, 1956, authored by L. W. Teller of the U.S. Bureau of Public Roads (now FHWA). Robert E. Philleo authored the chapters appearing in *ASTM STP 169A*, 1966, and *ASTM STP 169B*, 1978, under the title "Elastic Properties and Creep" Mr. Philleo died in May 1990, shortly before this revision, *ASTM STP 169C*, was begun. The co-editors of this revision have reviewed the chapter in *ASTM STP 169B* and found it to be as pertinent and informative today as it was in 1978. It is being reprinted as it appeared in *ASTM STP 169B* as a memorial to Mr. Philleo.

It is appropriate to note that subsequent to *ASTM STP 169B*, the Society, in 1981, adopted ASTM C 801, Standard Practice for Determining the Mechanical Properties of Hardened Concrete Under Triaxial Loads, which describes procedures for determining elastic properties and creep under triaxial loading conditions.

INTRODUCTION

When a load is applied to a body, the body is deformed. For a particular body loaded in a particular environment, the amount of the deformation depends upon the magnitude of the load, the rate at which it is applied, and the elapsed time after the load application that the observation is made. Different materials vary widely in their response to load. This response is known as rheological behavior. While instantaneous effects and time-dependent effects are not entirely separable, it is common to consider them separately as elastic properties (instantaneous) and creep (time-dependent).

A knowledge of the rheological properties of concrete is necessary to compute deflections of structures, to compute loss of prestress in prestressed structures, to compute stresses from observed strains, and, when working stress design procedures are employed, to proportion sections and to determine the quantity of steel required in reinforced concrete members.

Although a vast amount of work has been done on the rheology of materials, much of it is not applicable to concrete. Because of the peculiar "gel" structure of cement paste, concrete behaves quite differently under applied load than does a crystalline material.

ELASTIC PROPERTIES

A body returns to its original dimensions after enduring stress is elastic. A quantitative measure of elasticity is the ratio of stress to corresponding strain. Robert Hooke in 1678 discovered that for many materials this ratio is constant over a fairly wide range of stress. This ratio is termed the modulus of elasticity, and it has become one of the most commonly used parameters to describe material properties even though many materials do not exhibit a linear stress-strain relationship. Two additional terms are used to describe limits of elastic behavior: (1) proportional limit and (2) elastic limit. The proportional limit is defined in ASTM Definitions of Terms Relating to Methods of Mechanical Testing (E 6) as "the greatest stress which a material is capable of sustaining without any deviation from proportionality of stress to strain (Hooke's law)." The elastic limit is "the greatest stress which a material is capable of sustaining without any permanent strain remaining upon complete release of the stress."

Concrete has neither a definite proportional limit nor elastic limit. Therefore, the manner in which its modulus of elasticity is defined is somewhat arbitrary. Various forms of the modulus which have been used are illustrated on the stress-strain curve in Fig. 1. They are defined in ASTM E 6 as follows:

1. Initial Tangent Modulus—The slope of the stress-strain curve at the origin.
2. Tangent Modulus—The slope of the stress-strain curve at any specified stress or strain.
3. Secant Modulus—The slope of the secant drawn from the origin to any specified point on the stress-strain curve.
4. Chord Modulus—The slope of the chord drawn between any two specified points on the stress-strain curve.

Modulus of elasticity may be measured in tension, compression, or shear. The modulus in tension is usually equal to the modulus in compression and is frequently referred to as Young's modulus of elasticity. The shear modulus, also called the modulus of rigidity or torsional modulus, is the ratio of shear stress to shear stain. Shear stress is

[1] Deceased.

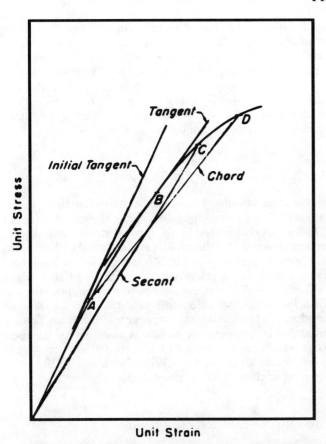

FIG. 1—Various forms of static modulus of elasticity.

defined in ASTM E 6 as "the stress or component of stress acting tangential to a plane," and shear strain is defined as "the tangent of the angular change between two lines originally perpendicular to each other."

When stress is applied in a given direction, there are changes in dimension in directions perpendicular to the direction of the applied stress, as well as in the direction of the stress. The magnitudes of the lateral strains are different for different materials. Thus, two parameters are required to describe the elastic behavior of a material. The parameters may take many forms, but the two most commonly used are Young's modulus of elasticity and Poisson's ratio. Poisson's ratio is defined in ASTM E 6 as "the absolute value of the ratio of transverse strain to the corresponding axial strain resulting from uniformly distributed axial stress below the proportional limit of the material." The transverse strains are opposite in direction to the axial strains. For a material obeying Hooke's law, Poisson's ratio is constant below the proportional limit. It can be shown that the following relationship exists among Young's modulus of elasticity, modulus of elasticity in shear, and Poisson's ratio

$$\mu = \frac{E}{2G} - 1$$

where

μ = Poisson's ratio,

E = Young's modulus of elasticity, and
G = modulus of elasticity in shear.

Thus, if any two of these quantities are determined, the third can be calculated.

It can be shown that the natural frequency of vibration of an elastic body is proportional to the square root of either Young's modulus or the shear modulus, depending on the mode of vibration. Likewise, the velocity with which a compressional shock wave travels through an elastic body is proportional to the square root of Young's modulus. Since these phenomena and their application to concrete are discussed in Chapter 23 of this publication, only the measurements of modulus of elasticity by "static" test methods will be described here.

Modulus of Elasticity in Compression

Since structural concrete is designed principally for compressive stresses, by far the greatest amount of work on the elastic properties of concrete has been done on concrete in compression. The only ASTM test method for static modulus of elasticity of concrete, ASTM Test Method for Static Modulus of Elasticity and Poisson's Ratio of Concrete in Compression (C 469), is a compressive test method. It stipulates a chord modulus between two points on the stress strain curve defined as follows: the lower point corresponds to a strain of 50 millionths and the upper point corresponds to a stress equal to 40% of the strength of concrete at the time of loading. The lower point is near the origin but far enough removed from the origin to be free of possible irregularities in strain readings caused by seating of the testing machine platens and strain measuring devices. The upper point is taken near the upper end of the working stress range assumed in design. Thus, the determined modulus is approximately the average modulus of elasticity in compression throughout the working stress range.

The 150 by 300-mm (6 by 12-in.) cylinder is the specimen size most commonly used for the determination of the modulus of elasticity in compression. In order to compensate for the effect of eccentric loading or nonuniform response by the specimen, strains should be measured along the axis of the specimen or along two or more gage lines uniformly spaced around the periphery of the cylinder. The selection of the gage length is important. It must be large in comparison with the maximum aggregate size so that local strain discontinuities do not unduly influence the results, and it must be large enough to span an adequate sample of the material. It must not, however, encroach on the ends of the specimen. Because of restraint where the specimen is in contact with the steel platens of the testing machine, strains near the ends may differ somewhat from strains elsewhere in the specimen. ASTM C 469 specifies that the gage length shall be not less than three times the maximum size of aggregate nor more than two thirds of the height of the specimen. Half the specimen height is said to be the preferred gage length. A convenient device for measuring the strains is a compressometer,

FIG. 2—Compressometer.

such as the one pictured in Fig. 2. The lower yoke is rigidly attached to the specimen, whereas the upper yoke is free to rotate as the specimen shortens. The pivot rod and dial gage are arranged so that twice the average shortening of the specimen is read on the dial. This type of device was used in the first comprehensive investigation of modulus of elasticity by Walker [1], and it is cited in ASTM C 469 as an acceptable device. Electrical strain gages have also been used successfully. These include both gages embedded along the axis of the specimens and those bonded to the surface. There are also available compressometers in which the strain, instead of being observed on a dial gage, is indicated and recorded by an electrical device such as a linear differential transformer.

Because the test is intended to measure only time-dependent strains, it is important that the specimen be loaded expeditiously and without interruption. For this purpose, an automatic stress-strain recorder is helpful but not essential. Since it is desired that only those length changes due to load be measured, temperature and moisture conditions should be controlled during the test.

A large numer of results have been reported in the literature [2–4]. The range of results has been from 7000 to 20 000 MPa (1 000 000 to 3 000 000 psi) for structural lightweight concrete and from 14 000 to 35 000 MPa (2 000 000 to 5 000 000 psi) for normal weight concrete. A simple relationship between modulus of elasticity and other easily measured properties of concrete, such as strength and unit weight, would be useful. While no theoretical relationship exists, an approximate equation adopted by the American Concrete Institute (ACI) Building Code [5], which is discussed later, has practical value.

Although the standard method of test is not concerned with the behavior of concrete at stresses above 40% of the strength, the shape of the stress-strain curve at high stresses is of significance in determining the ultimate load-carrying capacity of a concrete member. In most testing machines, concrete cylinders, except those with very low strength, fail suddenly shortly after the maximum load has been attained. It has been demonstrated [6–10,11] that such failures are related to the properties of the testing machine rather than the properties of the concrete. By using a suitably stiff testing machine or by artifically stiffening a testing machine by surrounding the concrete specimen with steel springs, it is possible to obtain stress-strain curves covering a strain range several times as great as that required to attain maximum stress. Figure 3 illustrates such stress-strain curves.

Modulus of Elasticity in Tension

A limited amount of work has been done on the determination of Young's modulus in tension of concrete [12–14]. The test is complicated by the limited stress range of concrete in tension and the problems associated with gripping the tension test specimens. Recent work has been concerned with the development of specimen shapes which will ensure uniform stress distribution throughout the section in which measurements are made. Since Young's modulus in tension does not appear to differ from Young's modulus in compression at low stresses and since there is no strain range beyond maximum tensile stress to be investigated, there is relatively little stimulus for the development of data on the tensile modulus.

Modulus of Elasticity in Flexure

Since a principal use of reinforced concrete is in flexural members, several investigators have determined Young's modulus on specimens loaded as beams. An obvious approach is to measure deflections caused by known loads and to calculate the modulus of elasticity from well-known beam deflection formulas. Unfortunately, the depth-to-span ratios of concrete beams normally used for such tests are so large that shear deflection comprises a significant part of the total deflection. In applying shear corrections, certain other corrections must be made to take care of discontinuities in the shear deflection curves at load points. The corrections most commonly used are those of Seewald [15]. For center-point loading, he gives the following deflection formula

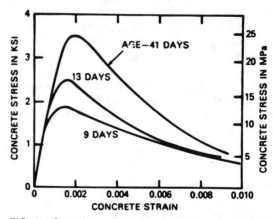

FIG. 3—Complete stress-strain (taken from Ref 11).

$$\delta = \frac{Pl^3}{48EI}\left[1 + (2.4 + 1.5\mu)\left(\frac{h}{l}\right)^2 - 0.84\left(\frac{h}{l}\right)^3\right]$$

where

- δ = maximum deflection,
- P = applied central load,
- l = distance between supports,
- E = modulus of elasticity,
- I = moment of inertia of the section with respect to the centroidal section,
- μ = Poisson's ratio, and
- h = depth of the beam.

The portion of the expression outside the brackets is the simple beam formula.

In some tests, strain gages have been placed on the tensile and compressive faces to determine strain as a function of applied load. The load may be converted to fiber stress by standard beam formulas. For such tests, the beam is usually loaded at two symmetrical points so that there is a constant stress condition between the two load points. Such tests have the advantage of indicating the position of the neutral axis, but if the stress-strain curve is nonlinear the computed stresses and, therefore, the computed modulus of elasticity are in error.

In a comprehensive investigation of the compression side of a concrete beam, Hognestad et al. [11] utilized specimens of the types shown in Fig. 4. The central prismatic portion of the specimen is loaded both concentrically and eccentrically. The eccentric load is continuously adjusted so that the strain on one surface remains zero. Thus, the section simulates that portion of a beam between the neutral axis and the extreme compressive fiber. The complete stress-strain curve in flexure can be determined from the observed loads and strains if the following two assumptions are made:

1. The distribution of strain across the section is linear.
2. All fibers follow the same stress-strain curve.

The first has been amply verified in these and other tests. The second depends on the speed of testing. If the specimen is loaded expeditiously, time-dependent effects are held to a minimum. In these tests, the stress-strain curves in flexure agreed very well with stress-strain curves obtained on companion cylinders concentrically loaded in compression both below and above the maximum stress.

Modulus of Elasticity in Shear

The shear modulus, or modulus of rigidity, is most often determined dynamically or by calculation from Young's modulus and Poisson's ratio. When a direct static determination is desired, a torsion test is the common procedure. When a given torque is applied to a body, the angle of twist in a given length is inversely proportional to the shear modulus as indicated by the following equation

$$G = \frac{L}{\phi I}$$

where

- G = modulus of elasticity in shear,
- L = torque,
- ϕ = angle of twist per unit length, and
- I = polar moment of inertia of the cross section.

A method for testing concrete cylinders has been described by Andersen [16]. He used a level bar with micrometer adjustment for measuring differences in angular change of two radial arms placed a fixed distance apart on a horizontally positioned specimen.

Poisson's Ratio

Static determinations of Poisson's ratio are made by adding a third yoke and second dial gage to a compressometer so that a magnified transverse strain may be measured, as well as a magnified axial strain or by mounting strain gages on the surface of a specimen perpendicular to the direction of loading. The same considerations apply to gage length for lateral strain measurement as for longitudinal strain measurement except that for lateral strains there is no upper limit since end restraint is not a factor. Procedures for determination of Poisson's ratio are included in ASTM C 469. Poisson's ratio is also commonly computed from results of Young's modulus and shear modulus determined dynamically. The static value at stresses below 40% of the ultimate strength is essentially constant; for most concretes the values fall between 0.15 and 0.20. The dynamic values are usually in the vicinity of 0.25. At high stresses or under conditions of rapidly alternating loads, a different picture emerges. Probst [17] has shown a systematic increase in the value of Poisson's ratio with stress repetition, and Brandtzaeg [18] has shown a marked increase at very high stresses. When the value is below 0.50, there is a decrease in volume of the body as a compressive load is applied. Brandtzaeg's work indicates that above about 80% of the strength there is an increase in volume as additional compressive loads are applied.

CREEP

Creep is defined in ASTM E 6 as "the time-dependent increase in strain in a solid resulting from force." All mate-

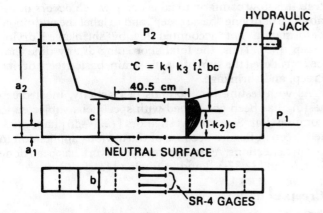

FIG. 4—Test specimens for stress distribution in a beam (taken from Ref 11).

rials undergo creep under some conditions of loading. In metals and other crystalline materials, creep has been attributed to slip in crystals. While slip of this nature undoubtedly occurs in aggregate particles and within crystalline particles that are part of hydrated paste, there is ample evidence that these are only secondary factors in the creep of concrete. Crystalline slip is normally detectable only above some threshold level of stress. Creep of concrete is observed at all stresses. Furthermore, creep of concrete is approximately a linear function of stress up to 35 to 40% of its strength. This behavior is not associated with crystalline slip. Finally, the order of magnitude of concrete creep is much greater than that of crystalline materials except for metals in the final stage of yielding prior to failure. Thus, concrete creep is considered to be an isolated rheological phenomenon associated with the gel structure of cement paste. This unique aspect of the problem has the advantage that creep may be measured without extremely sensitive equipment but the disadvantage that little of the research that has been done on other materials is applicable to concrete. A review of knowledge on creep is to be found in Ref 19.

The particular aspect of the gel structure of concrete that causes its unusual behavior is the accessibility of its large internal surface to water. In fact, Mullen and Dolch [20] found no creep at all in oven-dried pastes. The movement of water into and out of the gel in response to changes in ambient humidity produces the well-known shrinking and swelling behavior of concrete. A principal view among investigators [21–25] is that creep is closely related to shrinkage. In creep, gel water movement is caused by changes in applied pressure instead of differential hygrometric conditions between the concrete and its environment. This concept is supported by the similar manner in which creep and shrinkage curves are affected by such factors as water-cement ratio, mix proportions, properties of aggregate, compaction, curing conditions, and degree of hydration.

Another explanation of the effect of gel water [26,27] is delayed elasticity. If a load is suddenly imposed on a body consisting of a solid elastic skeleton with its voids filled with a viscous fluid, the load will be carried initially by the fluid and will gradually be transferred to the skeleton as the fluid flows under load. This is the behavior exhibited by the rheological model known as a Kelvin body which consists of a spring and dashpot in parallel. The concept of delayed elasticity has been chiefly responsible for the widespread attempts to reproduce the rheological behavior of concrete by means of rheological models.

Creep of concrete has been attributed by some [28–31] to viscous flow of the cement paste. The reduction in strain rate with time has been attributed by these investigators both to the increasing viscosity of the paste and to the gradual transfer of load from the cement paste to the aggregate. This concept is supported by the concept that creep strain is proportional to the applied stress over a wide range of stress. A convincing argument against it is the fact that the volume of concrete does not remain constant while it creeps. In fact, Poisson's ratio for creep has usually been found to be less than for elastic stress. In at least one investigation [32], Poisson's ratio was found to be zero. Another argument against the concept is the partial recovery of creep when the load is removed.

The fact that creep is associated primarily with the cement paste phase of concrete produces a serious difficulty in the interpretation and application of creep data. Unless the work is restricted to very mature concrete, the specimens do not maintain constant physical properties throughout the test. Creep measurements must necessarily be made over a considerable period of time and, during that time, the cement paste continues to hydrate. Frequently, information is desired at early ages when the cement is hydrating relatively rapidly.

Measurement of Creep

The use to be made of creep measurements usually determines the age at which creep tests are begun and the stress level to which specimens are loaded. A test procedure has been standardized in ASTM Test Method for Creep of Concrete in Compression (C 512). The method stipulates loading moist-cured specimens at an age of 28 days to a stress not exceeding 40% of the strength of the concrete at the time of loading, although provision is made for other storage conditions or other ages of loading. The stress is restricted to the range throughout which creep has been found to be proportional to stress. Limitations on gage lengths similar to those in the test for modulus of elasticity apply. The age of loading for a standard test is necessarily arbitrary. The method is intended to compare the creep potential of various concretes. Testing at a single age of loading is satisfactory for this purpose. It is required in the test method that the stress remain constant throughout the one-year duration of the test within close tolerances. The load may be applied by a controlled hydraulic system or by springs, provided in the latter case the load is measured and adjusted frequently. It is also required that there be companion unloaded specimens. Length changes of these specimens are measured and subtracted from the length changes of the loaded specimens to determine creep due to load. This correction is intended to eliminate the effects of shrinkage and other autogenous volume change. While this correction is qualitatively correct and yields usable results, most modern theories deny the independence of shrinkage and creep and thus indicate that the two effects are not additive as assumed in the test. It is now common to label creep which occurs in the absence of drying "basic creep" and to label the additional deformation not accounted for by shrinkage "drying creep" [33]. Thus, the total shortening at any time may be considered the sum of elastic strain, basic creep, drying creep, and shrinkage.

As with testing for modulus of elasticity, most creep testing has been concerned with specimens subjected to compression. Several investigators [12,34–36] have studied creep in flexure because of the obvious application to beam deflections. A limited amount of work has been done in tension [32,37,38] and in torsion [16,39,40].

Creep Equations

Creep under constant stress always proceeds in the same direction at a rate that is a decreasing function of time.

It is, therefore, common to plot creep test results on a semilogarithmic graph in which the linear axis represents creep strain and the logarithmic axis represents time. Many sets of data show an approximately straight line over a considerable period of time. This has led to the development of numerous logarithmic equations for creep. Such an equation is cited in ASTM C 512. In that test method, the results are reported by listing the strains at specified ages up to a year. In addition, it is suggested that the report contain the parameters of the following equation

$$\epsilon = \frac{1}{E} + F(K) \log_e(t + 1)$$

where

ϵ = total strain per unit stress;
E = instantaneous elastic modulus;
$F(K)$ = creep rate, calculated as the slope of a straight line representing the creep curve on the semilog plot; and
t = time after loading, in days.

While it is not intended that a theoretical logarithmic law should be inferred from the equation, the slope of the least squares line is a convenient parameter for comparing the creep characteristics of different concretes.

Most of the other equations which have been used to describe creep have been based on the assumption that there is a limiting value of creep. Typical of these is the Lorman [24] equation

$$c = \frac{mt}{n + t} \sigma$$

where

c = creep strain after time, t, for a sustained stress, σ;
m = ultimate creep strain per unit of stress; and
n = the time at which half the ultimate creep is attained.

McHenry [41] added coefficients to take into account the effects of hydration during the loading period with the following triple exponential equation

$$\epsilon_c = \alpha(1 - e^{-rt}) + \beta e^{-pk}(1 - e^{-mt})$$

where

ϵ_c = specific creep,
k = the age at time of loading,
t = time after loading, and
$\alpha, \beta, r, p,$ and m = empirical constants.

A summary of creep equations is given in Ref 42.

Rheological Models

Many investigators have used mechanical models as an aid in setting up mathematical equations. The elements normally used are the ideal spring, in which force is proportional to strain, and the ideal dashpot, in which force is proportional to the rate of strain. They are usually grouped together in pairs. A spring and dashpot in parallel form a Kelvin unit, while a spring and dashpot in series form a Maxwell unit. By selecting an appropriate group of elements, it is possible to produce a system empirically that can reproduce any given set of creep data. The models are frequently simplified somewhat if some of the spring or dashpot constants are made time-dependent or stress-dependent. While the model is an aid in visualizing creep behavior and in writing equations for that behavior, it should not be considered as representing the structure of the real material. Typical of models that have been proposed is that of Freudenthal and Roll [43], shown in Fig. 5.

Principle of Superposition

Of considerable practical value is a knowledge of the response of concrete to changing loads. While many of the investigations have included complete unloading as a portion of the study, there has been only a limited study of other forms of variable loading. McHenry [41] proposed a theory of superposition whereby the effect of each change in load is assumed to be added algebraically to the effects of all previous loads. The material is assumed to have a perfect memory. The effect of each increment of load lasts forever. Figure 6 shows how the theory applied to creep recovery. Three sets of specimens are required to produce such a figure. Two are loaded at an age of 28 days. One of these is unloaded at 90 days. A third set is loaded at 90 days to the same stress level as the original sets. The principle of superposition is valid if the strain in the last set equals the difference in strains of the first two sets. The principle is consistent with the preponderance of observations that only partial recovery occurs when a specimen is unloaded. While there is not unanimous agreement [44] on the principle of superposit-

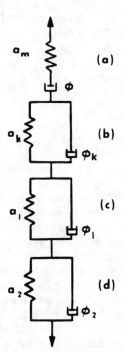

FIG. 5—Typical rheological model for representing creep (taken from Ref 43).

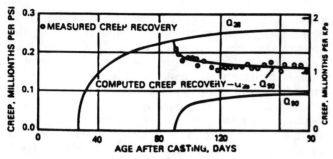

FIG. 6—Principle of superposition (taken from Ref 41).

ion, it is used regularly in computing stresses in mass concrete from measured strains.

Effect of Paste Content

Creep is influenced by mixture proportions, type of aggregate, conditions of storage, and age at loading. Most of the creep studies which have been conducted have been for the purpose of determining the effect of one or more of these variables on creep. One aspect of the effect of mixture proportions is of continual interest in the field of mass concrete. That is the relationship between creep and paste content. Mass concrete commonly contains 150-mm (6-in.) aggregate. Fabrication and testing of concrete specimens containing such large aggregate is expensive. Work at the Waterways Experiment Station of the Corps of Engineers [45] and at the University of California [46] has demonstrated that for paste contents normally used in concrete the creep of sealed specimens, having identical water-cement ratios and air contents in the mortar phase but different amounts and sizes of a given coarse aggregate, is proportional to the paste content. This finding has made it possible to develop significant data for mass concrete from small specimens. A general treatment of the effect of aggregate properties and paste contents covering the entire range of paste contents from 0 to 1 has been given by Neville [47].

Effect of Specimen Size

It has been demonstrated [45,48] that creep of sealed specimens is independent of specimen size. This observation plus the observation concerning mass concrete in the preceding paragraph indicate that the techniques and specimens of ASTM C 512 are applicable to all types of concrete sealed to prevent loss of moisture. For unsealed specimens exposed to a drying atmosphere, it is evident that there must be a size effect associated with the moisture gradients within the specimen. The creep of a structure may be only a fraction of that in a test specimen. Ross [49] investigated the effect of specimen size on shrinkage and found at each age an excellent correlation between shrinkage and the ratio of exposed surface to volume of specimen independent of the shape of the specimen. Hansen and Mattock [50], in an investigation of both size and shape of specimens, found that shrinkage and creep were dependent only on the ratio of surface to volume. Information of this sort may make it possible to apply correction factors to the data obtained from ASTM C 512 to determine the creep in any size and shape of structure.

Creep Prediction Equations

A series of creep prediction equations has been published by ACI Committee 209, Creep and Shrinkage in Concrete [51]. The equations take into account the effects of type and quantity of cement, consistency of concrete, air content, age at loading, relative humidity, and size of member.

SIGNIFICANCE OF PROPERTIES

Deflection of Flexural Members

Concrete flexural members undergo deflection upon application of load and continue to deflect with the passage of time. It is not uncommon for a reinforced concrete flexural member eventually to reach a deflection three times as great as its initial deflection. While a precise prediction of these deflections is possible only if the elastic and creep properties are known, the required precision is usually not great. Relatively little creep testing is directed to predicting deflections of specific structures.

Loss of Prestress

In contrast to the lack of precision needed for deflection measurements, an accurate knowledge of the early-age rheological properties of concrete is valuable to the prestressed concrete industry. After the prestress is applied, there is a loss of prestress resulting from creep of the concrete, shrinkage of the concrete, and relaxation of the steel. Since the initial pretress is limited by the strength of the steel and the load-carrying capacity of the member is limited by the residual prestress, a knowledge of the factors governing loss of prestress has important economic implications.

Structural Design

When reinforced concrete is designed by working stress theory, perfect bond between the concrete and steel is assumed under design load conditions. Therefore, the load carried by the steel is a function of the ratio of the modulus of elasticity of steel to that of concrete. Design is impossible unless these quantities are known or assumed. For many types of structures, such as arches, tunnels, tanks, and flat slabs, a knowledge of Poisson's ratio as well as modulus of elasticity is needed.

As discussed previously, a knowledge of the complete stress-strain curve for concrete has assisted in the refinement of strength design theory.

Stress Calculations

A prerequisite for the calculation of stresses from measured strains is a complete knowledge of a material's rheological behavior. When stresses result from nonuniform temperature or humidity as well as from applied load, the thermal coefficient of expansion and shrinkage data must also be available. Perhaps the most extensive use of creep data has been in the stress analysis of mass concrete structures [52]. These structures are unreinforced and contain complicated stress distributions because of temperature conditions resulting from the heat of hydration of cement. The procedure requires the installation of a large number of strain meters within the structure during construction. It is necessary to run creep tests on sealed specimens in the laboratory at enough ages of loading so that by interpolation and extrapolation a complete knowledge of rheological behavior is available. Such data can be represented pictorially by a creep surface such as that shown in Fig. 7. To convert strain data to stress, it is assumed that creep is a linear function of stress and that the principle of superposition applies. At the location of each strain meter, stress calculations must be made continually from the time the concrete hardens to the end of the period for which results are desired. The time is divided into small intervals. If during any interval the change in strain is different from that which would be expected from the creep produced by the sum of all the stresses present at the point at the start of the interval, the change in strain must be accounted for by adding or subtracting an amount of stress sufficient to produce the desired strain increment at the appropriate age of loading. By this technique, it has been possible to obtain an accurate assessment of the safety of dams and to evaluate the effectiveness of various methods of temperature control. A discussion of the accuracy of the method has been given by Pirtz and Carlson [53].

A method developed by Houk et al. [54] measures the strain capacity in slow loading directly by laboratory flexural testing of relatively small beams. The technique takes into account the effects of mixture proportions and aggregate properties, and it has been used to establish construction controls required to reduce cracking in mass concrete structures [55].

SPECIFICATIONS

Limits on rheological properties are almost never included in specifications. In some structures in which designers wished to minimize deflections, contractors have been restricted to those aggregates among the economically available materials which have been demonstrated to produce concrete having the lowest creep.

It has been pointed out that a knowledge of modulus of elasticity is required for some aspects of structural design, and where building codes impose limits on deflections a knowledge of creep behavior is also necessary. Specifications for structural concrete are primarily concerned with strength. It has not been considered feasible to complicate the specifications by additional rheological requirements. The problem has been approached in the ACI Building Code [5] by assuming that the modulus of elasticity is related to strength and unit weight as follows

$$E = W^{1.5}\ 43\ f_c'$$

where

E = modulus of elasticity, in Pa;
W = unit weight, in kg/m^3; and
f_c' = specified comprehensive strength, in Pa.

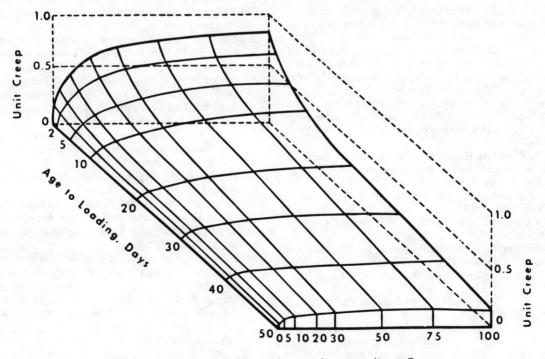

FIG. 7—Typical creep surface.

The ratio of long-time deflection to immediate deflection is assumed to vary from 1.8, when the amount of compression reinforcement equals the tension requirement, to 3.0, when there is no compression reinforcement. Neither of these estimates can claim a high degree of precision; both are probably adequate for the intended purpose. However, as structural design increases in sophistication and as applications of prestressed concrete increase, the interest in rheological properties is increasing. This interest is evident in the steadily increasing amount of creep testing being performed.

REFERENCES

[1] Walker, S., "Modulus of Elasticity of Concrete," *Proceedings*, American Society for Testing and Materials, Vol. 21, Part 2, 1919, p. 510.

[2] Teller, L. W., "Digest of Tests in the United States for the Determination of the Modulus of Elasticity of Portland Cement Mortar and Concrete," *Proceedings*, American Society for Testing and Materials, Vol. 30, Part 1, 1930, p. 635.

[3] "Bibliographies on Modulus of Elasticity, Poisson's Ratio, and Volume Changes of Concrete," *Proceedings*, American Society for Testing and Materials, Vol. 28, Part 1, 1928, p. 377.

[4] Philleo, R. E., "Comparison of Results of Three Methods for Determining Young's Modulus of Elasticity of Concrete," *Journal*, American Concrete Institute, Jan. 1955; *Proceedings*, Vol. 51, pp. 461–469.

[5] *Building Code Requirements for Reinforced Concrete* (ACI 318-77), American Concrete Institute, Detroit, MI, Oct. 1977.

[6] Whitney, C. S., discussion of a paper by V. P. Jensen, "The Plasticity Ratio of Concrete and Its Effect on the Ultimate Strength of Beams," *Journal*, American Concrete Institute, Nov. 1943, Supplement, *Proceedings*, Vol. 39, pp. 584–2 to 584–6.

[7] Saliger, R., "Bruchzustand und Sicherheit im Eisenbetonbalken," *Beton und Eisen*, Berlin, Vol. 35, Nos. 19 and 20, Oct. 1936, pp. 317–320 and 339–346.

[8] Kiendl, O. G. and Maldari, T. A., "A Comparison of Physical Properties of Concrete Made of Three Varieties of Coarse Aggregate," B. S. thesis, University of Wisconsin, Madison, WI, 1938.

[9] Ramaley, D. and McHenry, D., "Stress-Strain Curves for Concrete Strained Beyond the Ultimate Load," Laboratory Report No. SP-12, U.S. Bureau of Reclamation, Denver, CO, March 1947.

[10] Blanks, R. F. and McHenry, D., "Plastic Flow of Concrete Relieves High-Load Stress Concentrations," *Civil Engineering*, Vol. 19, No. 5, May 1949, pp. 320–322.

[11] Hognestad, E., Hanson, N. W., and McHenry, D., "Concrete Stress Distribution in Ultimate Strength Design," *Journal*, American Concrete Institute, Dec. 1955; *Proceedings*, Vol. 52, pp. 455–479.

[12] Davis, R. E., Davis, H. E., and Brown, E. H., "Plastic Flow and Volume Changes in Concrete," *Proceedings*, American Society for Testing and Materials, Vol. 37, Part 2, 1937, p. 317.

[13] Johnson, A. N., "Tests of Concrete in Tension," *Public Roads*, Vol. 7, No. 4, June 1926.

[14] Johnson, J. W., "Relationship Between Strength and Elasticity of Concrete in Tension and in Compression," *Bulletin No. 90*, Engineering Experiment Station, Ames, IA, 1928.

[15] Seewald, F., "Abhandlungen," Aerodynamischen Inst. an der Technischen Hochschule, Aachen, Vol. 7, 1927, p. 3.

[16] Andersen, P., "Experiments with Concrete in Torsion," *Transactions*, American Society of Civil Engineers, Vol. 100, 1935, p. 949.

[17] Probst, E., "The Influence of Rapidly Alternating Loading on Concrete and Reinforced Concrete," *The Structural Engineer*, British, Vol. 9, No. 12, Dec. 1931.

[18] Brandtzaeg, A. "The Failure of Plain and Spirally Reinforced Concrete in Compression," *Bulletin No. 190*, University of Illinois Engineering Experiment Station, Urbana, IL, April 1929.

[19] *Symposium on Creep of Concrete, ACI SP-9*, American Concrete Institute, Detroit, MI, 1964.

[20] Mullen, W. G. and Dolch, W. L., "Creep of Portland-Cement Paste," *Proceedings*. Vol. 64, 1964, pp. 1146–1171.

[21] Lynam, C. G., *Growth and Movement in Portland Cement Concrete*, Oxford University Press, London, 1934.

[22] Lea, F. M. and Lee, C. R., "Shrinkage and Creep in Concrete," *Proceedings*, Symposium on Shrinkage and Cracking of Cementive Materials, The Society of Chemical Industry, London, 1947, pp. 17–22.

[23] Seed, H. B., "Creep and Shrinkage in Reinforced Concrete Structures," *The Reinforced Concrete Review* (London), Vol. 1, No. 8, Jan. 1948, pp. 253–267.

[24] Lorman, W. R., "Theory of Concrete Creep," *Proceedings*, American Society for Testing and Materials, Vol. 40, 1940, pp. 1082–1102.

[25] Hansen, T. C., "Creep of Concrete," *Proceedings*, Swedish Cement and Concrete Research Institute, Royal Institute of Technology, Stockholm, 1958, p. 48.

[26] Freyssinet, E., "The Deformation of Concrete," *Magazine of Concrete Research*, Vol. 3, No. 8, Dec. 1951, pp. 49–56.

[27] Torroja, E. and Paez, A., "Set Concrete and Reinforced Concrete," *Building Materials, Their Elasticity and Plasticity*, M. Reiner, Ed., North-Holland Publishing Co., Amsterdam, 1954, pp. 290–360.

[28] Freudenthal, A. M., *The Inelastic Behavior of Engineering Materials and Structures*, Wiley, New York, 1950, p. 587.

[29] Reiner, M., *Deformation, Strain and Flow*, H. K. Lewis and Co., Ltd., London, 1960.

[30] Thomas, F. G., "Conception of Creep of Unreinforced Concrete and an Estimation of the Limiting Values," *Structural Engineering* (London), Vol. 11, 1933, p. 69.

[31] Glanville, W. H. and Thomas, F. G., "Further Investigations on the Creep or Flow of Concrete Under Load," *Building Research Technical Paper 21*, Department of Scientific and Industrial Research, London, 1939.

[32] Ross, A. D., "Experiments on the Creep of Concrete Under Two-Dimensional Stressing," *Magazine of Concrete Research*, Vol. 6, No. 16, June 1954, pp. 3–10.

[33] Neville, A. M., *Properties of Concrete*, 2nd ed., Wiley, New York, 1973.

[34] Chang, T. S. and Kesler, C. E., "Correlation of Sonic Properties of Concrete with Creep and Relaxation," *Proceedings*, American Society for Testing and Materials, Vol. 56, 1956, pp. 1257–1272.

[35] Vaishnav, R. N. and Kesler, C. E., "Correlation of Creep of Concrete with its Dynamic Properties," *T. & A. M. Report No. 603*, University of Illinois, Urbana, IL, Sept. 1961.

[36] Chang, T. S., "Prediction of the Rheological Behavior of Concrete from Its Sonic Properties," Ph.D. thesis, Theoretical and Applied Mechanics, Report No. 522, University of Illinois, Urbana, IL, 1955, p. 207.

[37] McMillan, F. R., "Shrinkage and Time Effects in Reinforced Concrete," *Bulletin No. 3*, University of Minnesota, Minneapolis, MN, 1915, p. 41.

[38] Wajda, R. L. and Holloway, L., "Creep Behavior of Concrete in Tension," *Engineering*, London, Vol. 198, No. 5130, 14 Aug. 1964.

[39] Duke, C. M. and Davis, H. E., "Some Properties of Concrete Under Sustained Combined Stresses," *Proceedings*, American Society for Testing and Materials, Vol. 44, 1944, pp. 888–896.

[40] Glucklich, J. and Ishai, O., "Creep Mechanism in Cement Mortar," *Journal*, American Concrete Institute, July 1962; *Proceedings*, Vol. 59, pp. 923–946.

[41] McHenry, D., "A New Aspect of Creep in Concrete and Its Application to Design," *Proceedings*, American Society for Testing and Materials, Vol. 43, pp. 1069–1084.

[42] "The Creep of Structural Concrete," Report of a Working Party of Materials Technology Divisional Committee, *Concrete Society Technical Paper No. 101*, Jan. 1973.

[43] Freudenthal, A. M. and Roll, F., "Creep and Creep Recovery of Concrete Under High Stress," *Journal*, American Concrete Institute, June 1958; *Proceedings*, Vol. 54, pp. 1111–1142.

[44] Davies, R. D., "Some Experiments on Applicability of the Principle of Superposition to the Strain of Concrete Subjected to Changes of Stress with Particular Reference to Prestressed Concrete," *Magazine of Concrete Research*, Vol. 9, No. 27, 1957, pp. 161–172.

[45] "Creep of Mass Concrete," Miscellaneous Paper No. 6-132 Report 3, U.S. Army Engineer Waterways Experiment Station, Vicksburg, MS, Jan. 1958, p. 22.

[46] Polivka, M., Pirtz, D., and Adams, R. F., "Studies of Creep in Mass Concrete," *Symposium on Mass Concrete, ACI SP-6*, American Concrete Institute, Detroit, MI, 1963, pp. 257–283.

[47] Neville, A. M., "Creep of Concrete as a Function of its Cement Paste Content," *Magazine of Concrete Research*, Vol. 16, No. 46, March 1964, pp. 21–30.

[48] Karapetrin, K. S., "Influence of Size upon Creep and Shrinkage of Concrete Test Specimens," I. Tekhiniceskie Nauki, Armenian Academy of Sciences (Yerevan), Akad. Nauk Armianskoi SSR, Fiziko—Matematicheskie, Estestvennye, Vol. 9, No. 1, 1956, pp. 87–100 (in Russian).

[49] Ross, A. D., "Shape, Size, and Shrinkage," *Concrete and Constructional Engineering*, Vol. 39, No. 8, Aug. 1944.

[50] Hansen, T. C. and Mattock, A. H., "The Influence of Size and Shape of Member on the Shrinkage and Creep of Concrete," American Concrete Institute, Feb. 1966; *Proceedings*, Vol. 63, pp. 267–290.

[51] "Prediction of Creep, Shrinkage, and Temperature Effects in Concrete Structures," *Special Publication 27*, ACI Committee 209, American Concrete Institute, Detroit, MI, 1971.

[52] Raphael, J. M., "The Development of Stresses in Shasta Dam," *Transactions*, American Society of Civil Engineers, Vol. 118, 1953, pp. 289–321.

[53] Pirtz, D. and Carlson, R. W., "Tests of Strain Meters and Stress Meters under Simulated Field Conditions," *Symposium on Mass Concrete, ACI SP-6*, American Concrete Institute, Detroit, MI, 1963, pp. 287–299.

[54] Houk, I. E., Jr., Paxton, J. A., and Houghton, D. L., "Prediction of Thermal Stress and Strain Capacity of Concrete by Tests on Small Beams," *Journal*, American Concrete Institute, March 1970; *Proceedings*, Vol. 67, pp. 253–261.

[55] Houghton, D. L., "Concrete Strain Capacity Tests, Their Economic Implications," *Economical Construction of Concrete Dams, Proceedings*, Engineering Foundation Conference, American Society of Civil Engineers, New York, 1972, pp. 75–99.

Bond with Reinforcing Steel

LeRoy A. Lutz[1]

PREFACE

A review of the significance of various tests for the bond between reinforcing steel and concrete is presented. The effect of the more important bond strength parameters such as bar diameter, concrete cover, and spacing of bars is appraised, and the appropriate provisions of the standards for reinforced concrete construction are discussed. The effect of epoxy coating the reinforcing bar is introduced. A brief discussion of the significance of bond and bond tests for prestressed concrete is also included. David Watstein of the National Bureau of Standards was the author of this section prior to 1979.

INTRODUCTION

Early experimenters and designers of reinforced concrete recognized that slip of the reinforcement had to be prevented in order to develop adequate bond and shear strengths and maintain the integrity of reinforced concrete members.

The first comprehensive series of bond tests was conducted by Duff Abrams [1] who employed both pullouts and beam specimens and measured the slip with care. This investigation laid the foundation for many subsequent studies that finally culminated in the adoption of ASTM Specification for Minimum Requirements for the Deformations of Deformed Steel Bars for Concrete Reinforcement (A 305–56T). These requirements are now incorporated into ASTM A 615, A 616, and A 617. This standard defined the minimum requirements for deformed reinforcing bars as we know them today.

TESTS FOR BOND OF REINFORCEMENT IN BEAMS

Bond stress in a beam may be defined either as the nominal "flexural bond" [2] value calculated from the expression $u = V/\Sigma ojd$, or it may be defined as the average bond stress from a point of maximum tension to the end of the bar. The latter has been termed "anchorage" or "development bond," and it is generally considered to be more meaningful than the "flexural" bond value.

Bond strength is determined using either beam or pullout specimens. Pullout tests are generally satisfactory for measuring the relative bond values of bars with different deformations. However, experience has shown that bond values derived from pullout tests cannot be applied, in general, to design of reinforced concrete beams and that beam tests are necessary to develop design criteria.

The difference between bond values obtained from pullout tests and beams are due primarily to the different stress states developed in the two types of specimens. In pullout specimens, the concrete is placed in compression, while in the beam specimens, both the concrete and the reinforcement are in tension. In beams, transverse cracks develop as the steel stresses increase, and each new transverse crack tends to initiate a new longitudinal or splitting crack. Transverse cracks are absent entirely in pullout specimens and the cracking is confined to longitudinal splitting. The bonding in the portion of the beam between the support and the nearest transverse crack is probably the only bonding that the pullout specimen properly models according to work done by Mains [3].

The Bond Stress Committee of the American Concrete Institute in its report of February 1945 [4] described a test procedure providing a uniform basis for comparison of bond values of different reinforcing bars. This test procedure included length of embedment and casting position of bar (top or bottom) as parameters. This procedure was extensively used in bond studies [5,6] that culminated in the development of ASTM A 305 covering the geometric requirements for deformed bars. The test procedure was later adopted by the American Concrete Institute as ACI 208-58. This standard did not include any minimum performance criteria, since its primary purpose was to establish relative bond values for different bars based on tests of one bar size at one embedment length in a concrete with a single concrete strength.

To study the effect of the various parameters on the bond values of deformed bars in simply supported beams subjected to concentrated loads, a beam specimen and a test procedure were developed [7] which represent a considerable departure from the ACI 208-58. This procedure, described in the report of ACI Committee 408 [8] in 1964, was intended to provide greater flexibility in the design of the recommended test specimen and test procedure in order to permit the use of bars of different diameters, more than one strength of concrete, and the longer

---

[1] Vice president, Computerized Structural Design, Inc., Milwaukee, WI 53217.

lengths of embedment needed to develop the newer deformed bars with 414 MPa (60 ksi) yield strength and greater.

This test beam illustrated in Fig. 1 was designed to permit the measurement of the average value of bond stress and the slip at both the loaded and free ends of the portion of the bar between the supports and the load points. The beam was provided with T-shaped ends in order to shift the loaded areas to points where they would not contribute to the restraint of longitudinal splitting. A strip of metal embedded in the concrete directly opposite each load point assured formation of a crack at that plane, and slip measurements were made at each load point plane. The test evaluated development of a bar from a support location.

This bond test specimen was also intended to provide a research tool yielding data that could be readily translated into design criteria. This test beam still has drawbacks in that the beam size, stirrup reinforcement, and test bar placement are all fixed. Although this standardized beam is fine for providing comparison of test results, the beam test still is not truly representative of the real-life beam.

As research work on bond and development continued, various test specimens were developed in an effort to simulate conditions found in beams. The beam specimen shown in Fig. 2 was designed for investigating bond in the region of bar cutoffs and points of inflection [9,10]. This arrangement permitted the calculation of the maximum steel stress and the average bond stress over an arbitrary development length, L. The bar size, bar cover, and beam width were varied.

A stub-cantilever beam shown in Fig. 3 was used in several investigations [11–13]. This beam permits easy

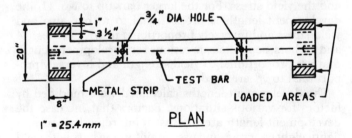
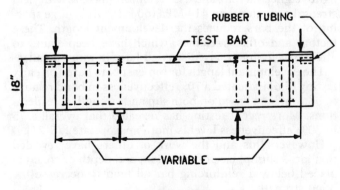

FIG. 1—Bond test beam recommended by ACI Committee 408 in 1964 [8].

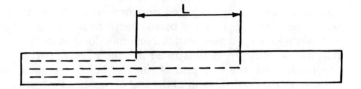

PLAN

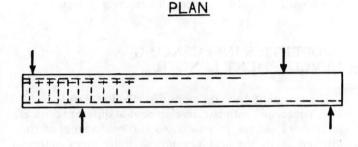

ELEVATION

FIG. 2—University of Texas Beam [9].

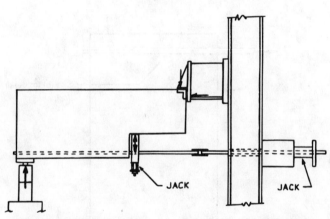

FIG. 3—Stub-cantilever beam used at West Virginia University [11].

variation of shear and moment as well as dowel forces in evaluating development to a support. However, the benefit in bond obtained from confinement of the bar by the support reaction is present with this test specimen.

A pair of bond tests can be conducted for the L lengths with the beam specimen illustrated in Fig. 4. Alternatively, the specimen can be designed so that bond tests can be

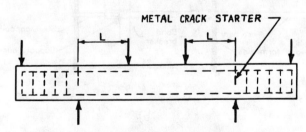

FIG. 4—Symmetrical beam for development tests.

conducted on the overhanging portions on each end as was done with the standard beam specimen of Fig. 2.

A beam-end specimen with a bar embedded in only a portion of its depth is illustrated in Fig. 5 [14,15]. This specimen, with the free end of the bar being recessed, eliminates the confinement at the end of the development length that is present in the specimens in Figs. 3 and 4. This specimen is constructed so that it can be used for two separate bond tests merely by turning it over.

PROPERTIES INFLUENCING DEVELOPMENT LENGTH

The two bond investigations described in Refs 7 and 9 were concerned with bar development at different critical sections of beams, as was noted earlier. In spite of the differences in the test specimens, both investigations showed that the ultimate bond stress is proportional to $\sqrt{f_c'}$, where f_c' is the 28-day compressive strength of concrete and is inversely proportional to the bar diameter when other factors were constant. This relationship was recognized by the American Concrete Institute in the formulation of the allowable bond stresses in ACI 318-63 for bars with sizes and deformations conforming to ASTM A 305.

An upper limit value functioned as the ultimate bond stress for the smaller reinforcing bars in ACI 318-63. The $\sqrt{f_c'}$ was the sole parameter used in an expression for the largest tension bars conforming to ASTM A 408 (Nos. 14S and 18S). Numbers 14 and 18 bars are now covered in ASTM A 615.

The position of the bar in the beam was also a factor influencing the bond stress; the bond stress for top-cast bars was limited to 70% of the bond stress used for bottom and vertically cast bars. A top-cast bar is horizontal reinforcement with more than ~ 305 mm (12 in.) of concrete placed below it. An ultimate bond stress for reinforcing bars in compression was also introduced in ACI 318-63.

Two basic changes were reflected by the bond provisions that were in ACI 318-71 [16]. Anchorage, or development bond, was considered of such importance relative to flexural bond that all bond provisions were presented in terms of development length rather than bond stress. Secondly, the development lengths were made about 20% longer than those calculated using the provisions of ACI 318-63; this was done due to the observation that closely spaced bars, which are quite typical, required this longer embedment length. It was noted by Ferguson [17] that the intent was to have these code-imposed development lengths be about 25% higher than test values to ensure that bar yielding would occur prior to bond failure.

The development length expressions were based on the bond stress expressions of ACI 318-63 and consequently involve the same variables. As a result, the development length of small bars is proportional to the bar diameter and the yield stress. For the larger bars up to No. 11, the development length is proportional to to bar area and yield stress and inversely proportional to $\sqrt{f_c'}$. For Nos. 14 and 18 bars, separate expressions for development length that are proportional to yield strength and inversely proportional to $\sqrt{f_c'}$ are used.

The development lengths calculated were modified by factors for various situations. Factors that increase the development length are top-cast reinforcing bars, use of lightweight concrete, and use of reinforcing bars with yield strengths greater than 414 MPa (60 ksi), while large spacing of the bars reduces the development length. These factors and other variables, which have been found to affect development, are discussed later in more detail.

The development length for top-cast bars based on early testing [5,6] indicated a 70% effectiveness in bond relative to bottom bars based on both slip and strength considerations. More recent testing has revealed that overall a 75 to 80% effectiveness level is more appropriate [18–21].

However, this and the work of others have revealed that large slump, lack of vibration, and depth of concrete placed below a reinforcing bar all tend to decrease the bond strength.

The development length of reinforcing bars has been found not to be directly proportional to the yield strength. The bond effectiveness decreases as the yield strength increases in a bar [18,19].

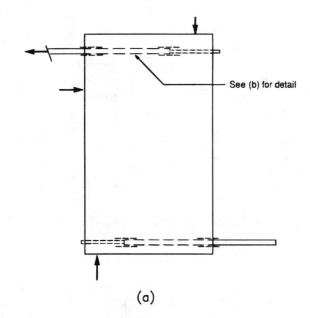

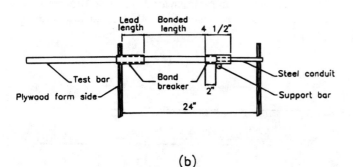

FIG. 5—Beam-end specimen used at University of Kansas [14,15].

The development length for small bars as well as the larger bars has been found to be inversely proportional to $\sqrt{f_c'}$ [18,19]. It is not known whether this parameter is valid for concretes with f_c' in excess of 55 to 69 MPa (8000 to 10 000 psi). Review of the data from Ref 18 for low strength concretes might suggest caution in using $\sqrt{f_c'}$ as a parameter for concretes with f_c' less than 17.2 MPa (2500 psi).

The most significant impact of Refs 18 and 19 plus others [22,23] is the realization that confinement of the reinforcement by concrete and transverse ties or stirrups has an important effect on the development length. Furthermore, these references revealed that the development expressions in ACI 318-71 [16] could be unconservative in many situations based on laboratory tests.

The concrete confinement can be evaluated by a parameter, C, that was developed in Ref 19 and based on stress analysis around a single bar by Tepfers [24]. The C parameter represents the smaller of the cover on the reinforcing bar, or half of the clear spacing between bars. The splitting failure that controls the bond strength will either occur along the row of bars (when close spacing controls) or out to the surface as illustrated in Fig. 6.

If transverse reinforcement crosses the splitting crack, additional confinement can be realized from this reinforcement. The benefit is proportional to the area of transverse reinforcement, the yield strength of this reinforcement, and inversely proportional to the spacing of these ties or stirrups.

There appears to be a limit to both the effectiveness of transverse steel and the effectiveness of the overall confinement in reducing the development length. When this point is reached, a maximum ultimate bond stress is reached and pullout occurs by failure along the bonded surface rather than a splitting away of the concrete around the bars.

Confinement factors were introduced into new provisions for bars with hooks in ACI 318-83 [25]. Confinement factors were included for straight bar tension development length determination in ACI 318-89 [26].

BOND OF LAPPED SPLICES

One of the areas in reinforced concrete design in which bond plays a critical role is in tensile lapped splices. Longitudinal splitting is produced along the splice length with increase in bar stress and is the basic cause for failure of lapped splices as occurs with bar development.

Tests of splices [27–30] have been conducted typically in test setups that place the spliced bars in a constant moment region as illustrated in Fig. 7. However, a number of investigations have examined the splice in a varying moment region as produced by a single concentrated load on a simple beam span [31–33].

Early research work [27–29] on lap splices has emphasized the following factors that influence the lap splice length required:

1. The increase of spacing between lapped splices decreases the splice length required.
2. An increase in concrete cover over the splice decreases the required splice length.
3. Confinement of the splice by some form of lateral reinforcement decreases the required splice length.
4. The length of splice required increases as the tensile force requirement increases.

The code provisions [25] reflected some of these points by increasing the required splice length as the percentage of spliced bars is increased and the level of stress in the bars is increased. However, the means to quantify the effect of confinement on lap splice length came with the Orangun, Jirsa, Breen work [18,19]. This work indicated that lap splice length can be considered the same as the development length when confinement effects of the concrete and transverse reinforcement are addressed as indicated earlier for development length.

Thus, if confinement effects of concrete and transverse reinforcement are addressed for development length, as done in ACI 318-89 [26], the lap splice length required for safety can truly be considered equal to the calculated development length. The modifying factors for classes of splices in ACI 318-89 are thus for the purpose of improving splice behavior at service loads and to discourage the splicing of a majority of the reinforcement at one location and in regions of relatively high stress.

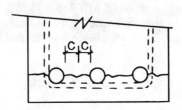

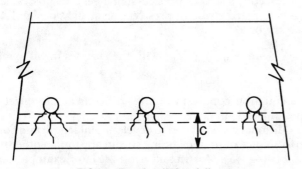

FIG. 6—Bond splitting failures.

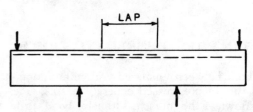

FIG. 7—Symmetrical beam for lap splice tests.

EFFECTS OF EPOXY COATING ON BOND OF BARS

Use of epoxy coating to seal the bar surface in an attempt to eliminate attack from chlorides began in the 1970s. Fusion-bonding of epoxy with a thickness ranging from 5 to 12 mils, as specified by the ASTM Specification for Epoxy-Coated Reinforcing Steel Bars (A 775), has been found to be an effective and economical means of controlling bar deterioration from corrosion. As of 1991, approximately 5% of all bars produced are epoxy coated.

Mathey and Clifton [34] were the first to investigate the bond of coated bars. Using pull-out specimens, they found that epoxy-coated bars with 1 to 11 mil thickness performed satisfactorily as compared to bars with a 25-mil coating of epoxy or of polyvinylchloride. Extensive testing of the anchorage length of epoxy-coated bars did not begin until usage became established in the early 1980s.

Johnson and Zia [35] tested primarily beam-end specimens and concluded that the embedment lengths for epoxy-coated bars were about 15% longer than values for corresponding uncoated bars. Treese and Jirsa [36] tested spliced bars of two diameters using beam specimens (Fig. 7). Concrete confinement, concrete strength, coating thickness, and casting position were varied in the test program. Their main conclusion was that the bond strength is dependent on the mode of failure (pullout or splitting) for epoxy-coated bars. They recommended that the basic development length be 15% longer for highly confined epoxy-coated bars, and for bars with lower confinement, 50% longer than the length of corresponding uncoated bars. For epoxy-coated bars with lower confinement, the top-cast bar length increase factor is about 15% rather than the typical 30%.

More recent work [15,37] using beam tests for splices and beam-end specimens for development length has increased the experimental data base significantly. A 35% increase, rather than 50% was found to be more appropriate for epoxy-coated bars with lower confinement. Confinement from concrete and transverse reinforcement was found to significantly improve the bond strength of epoxy-coated bars.

This work [15,37] and earlier work [38] also address the effects of the thickness of epoxy-coating and the deformation pattern of the bars. Thicker coatings appear to cause a greater reduction than thinner coatings for No. 5 bars and smaller. The magnitude of the reduction in bond strength from epoxy-coating is less for bars of any size that have larger rib bearing areas. Research in progress in 1991 may lead to suggestions for revising the criteria for bar deformations in ASTM A 615, A 616, and A 617 in an effort to help reduce development and lap-splice lengths.

BOND IN PRESTRESSED CONCRETE

The bond between tensioned tendons and concrete plays a dual role in prestressed concrete beams. One type of bond known as the "prestress transfer bond" is developed at the ends of concrete members prestressed by the pretensioning method. In this case, tension in the steel is transferred to the surrounding concrete primarily by friction that is made possible by the normal force produced by the Poisson effect, or lateral swelling of the steel in the transfer zone.

The second type of bond stresses in prestressed concrete is brought about by flexural action. When a prestressed beam is subjected to flexure, the tendons function to some extent as ordinary reinforcing bars and bond stresses develop in accordance with shears and moments in the beam. However, prior to cracking of concrete, the stresses in the prestressing steel change quite slowly and the bond stresses are low. As cracking occurs, the steel stresses at the cracked sections increase rapidly and bond stresses increase accordingly. These bond stresses which are associated with flexural cracking are commonly referred to as "flexural bond stresses."

In a study of the nature of bond in pretensioned prestressed concrete, Janney [39] investigated the variation of the prestress transfer bond with various parameters, as well as the interaction between prestress transfer bond and flexural bond. Considerable variation in the anchorage length and the shape of bond stress distribution curves were noted for wires of different diameters and different surface conditions of the wire ranging from rusted to lubricated. Tests of beams prestressed by pretensioning indicated that as cracks occurred the flexural bond stresses increased sharply and the bond stress "wave" progressed from the cracked midsection toward the ends of the beam. As the diameter is reduced and radial pressure is relieved by the sharp increase in tension in the tendon after cracking, the flexural bond capability is reduced considerably near the ends of the beam. As a result, a bond failure may be expected when the flexural bond stress wave reaches the anchorage zone, even prior to reaching the free end of the tendon. Janney [39] demonstrated this phenomenon by strain measurements in the pretensioned beams.

An expression for development length of three-wire and seven-wire pretensioning strand based primarily on the work of Hanson and Kaar [40] appeared initially in ACI 318-63 [2] and remains in ACI 318-89 [26]. The length is proportional to strand diameter, d_b, and composed of two parts, a transfer bond length and a flexural bond length, respectively, as follows

$$\ell_d = \frac{f_{se}}{3}d_b + (f_{ps} - f_{se})d_b$$

where

f_{se} = effective stress in prestressing steel after losses, ksi;
f_{ps} = calculated stress in prestressing steel at design load, ksi; or

$\ell_d = \frac{1}{7}\left[\frac{f_{se}}{3}d_b + (f_{ps} - f_{se})d_b\right]$ with the stresses in MPa and ℓ_d and d_b in mm.

The equation considers failure to occur at the onset of general bond slippage.

Additional work related to development length and bond performance of pretensioning strand appears in other references [41–45]. Martin and Scott [42], for example, suggested a set of equations that would limit f_{ps} if the full development length could not be achieved in a particular situation.

Pretensioning strand with epoxy coating was developed in the mid 1980s. The epoxy coating is considerably thicker than used for bar and is covered with grit made of crushed glass. The bond integrity of this coated product is a function of the density of the grit and is superior to that of the uncoated strand [46–49].

These experimental results which include tests on uncoated strand plus tests by others have been compared to the ACI expression (given earlier) and other expressions for transfer and development length [49]. Preliminary indications are that the ACI expression may be unconservative for uncoated strand. A new expression for development length was proposed that includes the concrete strength in terms of $\sqrt{f_c'}$. Additional research is being conducted to further examine the appropriateness of the ACI expression. Little is still known as to how the size of member, position of the strand, and spacing of the strand may influence the bond integrity.

INFLUENCE OF BOND IN CONTROL OF CRACKING AND DEFLECTIONS

The efficient and economical use of high-yield-strength reinforcement is limited by the necessity of avoiding excessively wide cracks that may be objectionable from the standpoint of appearance or lead to corrosion of reinforcement. In recent years, considerable research has been carried out with new types of high-strength deformed reinforcing bars to study the mechanism of cracking and to determine the effect of such parameters as diameter of bars, ratio of reinforcement, and cross-sectional dimensions of beams on the width and spacing of cracks. Clark [50] investigated these variables in a comprehensive series of tests of beams and slabs, and presented a formula relating the average width of cracks to the computed stress in the reinforcement, the diameter of the bars, and the ratio of reinforcement. Chi and Kirstein [51] made use of Clark's data in addition to their own to develop a simplified formula for the width of cracks, taking into account the distance from the bar surface to the nearest beam surface as a parameter. Similar conclusions were reached by Broms [52,53].

The earlier studies by Watstein and Parsons [54], and Watstein and Seese [55] have demonstrated, by means of tension tests of axially reinforced cylindrical specimens, the direct relationship between the bond strength and both the spacing and the width of tensile cracks. A tension specimen devised to measure the bonding efficiency of a bar [55] gave values that correlated well with the spacing and width of cracks observed in an independent set of long axially reinforced tension specimens.

A statistical study conducted by Gergely and Lutz [56] of beam crack width data from several investigations revealed that the bar stress, concrete cover, and area of concrete surrounding a reinforcing bar were the primary variables affecting crack width. The crack width expression recommended by Gergely and Lutz formed the basis of the parameter used in ACI 318-71 to control the distribution of reinforcement (and crack width) in beams and one-way slabs.

The bonding efficiency of reinforcing bars also affects the rigidity of flexural members because the quality of bond affects the contribution of the concrete between tensile cracks that, in turn, improves the stiffness of the beam. In other words, reduction of the slip between the concrete and the reinforcement along the entire length of the beam does increase the beam stiffness. Some work using epoxy-coated bars [35,57] has indicated that beam deflection and crack width is approximately 10% greater than when using uncoated bars.

The resistance of the concrete between tensile cracks to flexure was considered at length in the 1957 RILEM Symposium on Bond and Crack Formation in Reinforced Concrete [58]. The papers that discuss the various parameters affecting the distribution of strain between cracks and their effect on the flexural stiffness are by Soretz [59], Baker, Ashdown, and Wildt [60], and Murashev [61].

REFERENCES

[1] Abrams, D., "Tests of Bond Between Concrete and Steel," *Bulletin No. 71*, University of Illinois, Urbana, IL, Dec. 1913.

[2] "ACI Standard Building Code Requirements for Reinforced Concrete (ACI 318-63)," *Journal*, American Concrete Institute, Vol. 60, June 1963.

[3] Mains, R. M., "Measurement of the Distribution of Tensile and Bond Stresses Along Reinforcing Bars," *Journal*, American Concrete Institute, Vol. 48, Nov. 1951, p. 225.

[4] "Report of ACI Committee 208, Proposed Test Procedure to Determine Relative Bond Value of Reinforcing Bars," *Journal*, American Concrete Institute, Vol. 41, Feb. 1945, p. 273.

[5] Clark, A. P., "Comparative Bond Efficiency of Deformed Concrete Reinforcing Bars, RP 1755," *Journal of Research*, National Bureau of Standards, Vol. 37, Dec. 1946.

[6] Clark, A. P., "Bond of Concrete Reinforcing Bars, RP 2050," *Journal of Research*, National Bureau of Standards, Vol. 43, Dec. 1949.

[7] Mathey, R. G. and Watstein, D., "Investigation of Bond in Beam and Pullout Specimens with High-Yield-Strength Deformed Bars," *Journal*, American Concrete Institute, Vol. 57, March 1961, p. 1071.

[8] "Report of ACI Committee 408, A Guide for Determination of Bond Strength in Beam Specimens," *Journal*, American Concrete Institute, Vol. 61, Feb. 1964, p. 129.

[9] Ferguson, P. M. and Thompson, J. N., "Development Length of High Strength Reinforcing Bars in Bond," *Journal*, American Concrete Institute, Vol. 59, July 1962, p. 887.

[10] Ferguson, P. M. and Thompson, J. N., "Development Length of Large High-Strength Reinforcing Bars," *Journal*, American Concrete Institute; *Proceedings*, Vol. 62, No. 1, Jan. 1965, pp. 71–94.

[11] Wilhelm, W. J., Kemp, E. L., and Lee, Y. T., "Influence of Deformation Height and Spacing on the Bond Characteristics of Steel Reinforcing Bars," Civil Engineering Studies Report No. 2013, Department of Civil Engineering, West Virginia University, Morgantown, WV, 1971, No. PB-200635, National Technical Information Service Accession.

[12] Gergely, P., "Splitting Cracks Along the Main Reinforcement in Concrete Members," report to the Bureau of Public Roads, U.S. Department of Transportation, Department of Struc-

tural Engineering, Cornell University, Ithaca, NY, April 1969.

[13] Mirza, M. S. and Hsu, C. T., "Investigation of Bond in Reinforced Concrete Models—Eccentric Pullout Test," *Proceedings*, International Conference on Shear, Torsion, and Bond in Reinforced and Prestressed Concrete, Coimbatore, India, Jan. 1969.

[14] Choi, O. C., Hadje-Ghaffari, H., Darwin, D., and McCabe, S. L., "Bond of Epoxy-Coated Reinforcement to Concrete: Bar Parameters," SL Report 90-1, University of Kansas Center for Research, Lawrence, KS, Jan. 1990.

[15] Hadje-Ghaffari, H., Darwin, D., and McCabe, S. L., "Effects of Epoxy-Coating on the Bond of Reinforcing Steel to Concrete," SM Report No. 28, The University of Kansas Center for Research, Inc., Lawrence, KS, July 1991.

[16] "Building Code Requirements for Reinforced Concrete (ACI 318-71)," ACI Committee 318, American Concrete Institute, Detroit, 1971 (plus cumulative supplement issued annually).

[17] Ferguson, P. M., *Reinforced Concrete Fundamentals*, 3rd ed., Wiley, New York, 1973, p. 172.

[18] Orangun, C. O., Jirsa, J. O., and Breen, J. E., "The Strength of Anchored Bars: A Re-evaluation of Test Data on Development Length and Splices," preliminary review copy, Research Report 154-3F, Center for Highway Research, The University of Texas at Austin, Austin, TX, Jan. 1975.

[19] Orangun, C. O., Jirsa, J. O., and Breen, J. E., "A Re-evaluation of Test Data on Development Length and Splices," *Journal*, American Concrete Institute; *Proceedings*, Vol. 74, No. 3, March 1977, pp. 114–122.

[20] Luke, J. J., Hamad, B. S., Jirsa, J. O., and Breen, J. E., "The Influence of Casting Position on Development and Splice Length on Reinforcing Bars," Research Report No. 242-1, Center for Transportation Research, Bureau of Engineering Research, The University of Texas at Austin, Austin, TX, June 1981.

[21] Brettmann, B. B., Darwin, D., and Donahey, R. C., "Effect of Superplasticizers on Concrete—Steel Bond Strength," SL Report 84-1, University of Kansas Center for Research, Lawrence, KS, April 1984.

[22] Kemp, E. L. and Wilhelm, W. J., "An Investigation of the Parameters Influencing Bond Cracking," *Proceedings*, Interaction Between Steel and Concrete Symposium, American Concrete Institute Convention, San Diego, CA, 17 March 1977.

[23] Ferguson, P. M., "Small Bar Spacing or Cover—A Bond Problem for the Designer," *Journal*, American Concrete Institute; *Proceedings*, Vol. 74, No. 9, Sept. 1977, pp. 435–439.

[24] Tepfers, R., "A Theory of Bond Applied to Overlapped Tensile Reinforcement Splices for Deformed Bars," *Publication 73:2*, Division of Concrete Structures, Chalmers University of Technology, Goteborg, 1973.

[25] "Building Code Requirements for Reinforced Concrete (ACI 318-83)," ACI Committee 318, American Concrete Institute, Detroit, MI, 1983.

[26] "Building Code Requirements for Reinforced Concrete (ACI 318-89)," ACI Committee 318, American Concrete Institute, Detroit, MI, 1989.

[27] Chamberlin, S. J., "Spacing of Spliced Bars in Beams," *Journal*, American Concrete Institute, Vol. 54, Feb. 1958, p. 689.

[28] Chinn, J., Ferguson, P. M., and Thompson, J. N., "Lapped Splices in Reinforced Concrete Beams," *Journal*, American Concrete Institute, Vol. 52, Oct. 1955, p. 201.

[29] Ferguson, P. M. and Breen, J. E., "Lapped Splices for High Strength Reinforcing Bars," *Journal*, American Concrete Institute; *Proceedings*, Vol. 62, Sept. 1965.

[30] Thompson, M. A., Jirsa, J. O., Breen, J. E., and Meinheit, D. F., "The Behavior of Multiple Lap Splices in Wide Sections," Research Report 154-1, Center for Highway Research, The University of Texas at Austin, Austin, TX, Jan. 1975.

[31] Ferguson, P. M. and Briceno, A., "Tensile Lap Splices—Part 1: Retaining Wall Type, Varying Moment Zone," Research Report 113-2, Center for Highway Research, The University of Texas at Austin, Austin, TX, July 1969.

[32] Ferguson, P. M. and Krishnaswamy, C. N., "Tensile Lap Splices—Part 2: Design Recommendations for Retaining Wall Splices and Large Bar Splices," Research Report 113-3, Center for Highway Research, The University of Texas at Austin, Austin, TX, April 1971.

[33] Zekany, A. J., Neumann, S., Jirsa, J. O., and Breen, J. E., "The Influence of Shear on Lapped Splices in Reinforced Concrete," Research Report 242-2, Center for Transportation Research, The University of Texas at Austin, Austin, TX, July 1981.

[34] Mathey, R. G. and Clifton, J. R., "Bond of Coated Reinforcing Bars in Concrete," *Journal*, Structural Division, American Society of Civil Engineers, Vol. 102, ST1, Jan. 1976, pp. 215–229.

[35] Johnston, D. W. and Zia, P., "Bond Characteristics of Epoxy Coated Reinforcing Bars," Report No. FHWA-NC-82-002, Federal Highway Administration, Washington, DC, 1982.

[36] Treece, R. A. and Jirsa, J. O., "Bond Strength of Epoxy-Coated Reinforcing Bars," *Materials Journal*, American Concrete Institute, Vol. 86, No. 2, March–April 1989, pp. 167–174.

[37] Hester, C. S., Sulamizavaregh, S., Darwin, D., and McCabe, S. L., "Bond of Epoxy-Coated Reinforcement to Concrete: Splices," SL Report 91-1, University of Kansas Center for Research, Inc., Lawrence, KS, May 1991.

[38] Choi, O. C., Hadje-Jhaffari, H., Darwin, D., and McCabe, S. L., "Bond of Epoxy-Coated Reinforcement: Bar Parameters," *Materials Journal*, American Concrete Institute, Vol. 88, No. 2, March–April 1991, pp. 207–217.

[39] Janney, J. R., "Nature of Bond in Pretensioned Prestressed Concrete," *Journal*, American Concrete Institute, Vol. 50, May 1954, p. 717.

[40] Hanson, N. W. and Kaar, P. H., "Flexural Bond Tests of Pretensioned Prestressed Beams," *Journal*, American Concrete Institute, *Proceedings*, Vol. 55, No. 7, Jan 1959, pp. 783–803; also, *Development Department Bulletin D28*, Portland Cement Association, Skokie, IL, Jan. 1959.

[41] Kaar, P. H., LaFraugh, R. W., and Mass, M. A., "Influence of Concrete Strength on Strand Transfer Length," *Journal*, Prestressed Concrete Institute, Vol. 8, No. 5, Oct. 1963, pp. 47–67; also, *Development Department Bulletin D71*, Portland Cement Association, Skokie, IL, Oct. 1963.

[42] Martin, L. D. and Scott, N. L., "Development of Prestressing Strand in Pretensioned Members," *Journal*, American Concrete Institute, Vol. 73, Aug. 1976, p. 453.

[43] Anderson, A. R. and Anderson, R. G., "An Assurance Criterion for Flexural Bond in Pretensioned Hollow Core Units," *Journal*, American Concrete Institute, Vol. 73, Aug. 1976, p. 457.

[44] Salmons, J. R. and McCrate, T. E., "Bond Characteristics of Untensioned Prestressing Strand," *Journal*, Prestressed Concrete Institute, Vol. 22, No. 1, Jan.–Feb. 1977.

[45] Zia, P. and Mostafa, P., "Development Length of Prestressing Strand," *Journal*, Prestressed Concrete Institute, Vol. 22, No. 5, Sept.–Oct. 1977, pp. 54–66.

[46] Cousins, T. E., Johnston, D. W., and Zia, P., "Bond of Epoxy Coated Prestressing Strand," FHWA/NC/87-005, Center for Transportation Engineering Studies, Department of Civil Engineering, North Carolina State University, Raleigh, NC, Dec. 1986.

[47] Cousins, T. E., Johnston, D. W., and Zia, P., "Transfer Length of Epoxy Coated Prestressing Strand," *Materials Journal*, American Concrete Institute, Vol. 87, No. 3, May–June 1990, pp. 193–203.

[48] Cousins, T. E., Johnston, D. W., and Zia, P., "Development Length of Epoxy Coated Prestressing Strand," *Materials Journal*, American Concrete Institute, Vol. 87, No. 4, July–Aug. 1990, pp. 309–318.

[49] Cousins, T. E., Johnston, D. W., and Zia, P., "Transfer and Development Length of Epoxy Coated and Uncoated Prestressing Strand," *Journal*, Prestressed Concrete Institute, Vol. 35, No. 4, July–Aug. 1990, pp. 92–103.

[50] Clark, A. P., "Cracking in Reinforced Concrete Flexural Members," *Journal*, American Concrete Institute, Vol. 52, April 1956, p. 851.

[51] Chi, M. and Kirstein, A. F., "Flexural Cracks in Reinforced Concrete Beams," *Journal*, American Concrete Institute, Vol. 54, No. 4, April 1958, p. 865.

[52] Broms, B., "Crack Width and Crack Spacing in Reinforced Concrete Members," *Journal*, American Concrete Institute, Vol. 62, No. 10, Oct. 1965.

[53] Broms, B. and Lutz, L. A., "Effects of Arrangement of Reinforcement on Crack Width and Spacing of Reinforced Concrete Members," *Journal*, American Concrete Institute, Vol. 62, No. 11, Nov. 1965.

[54] Watstein, D. and Parsons, D. E., "Width and Spacing of Tensile Cracks in Axially Reinforced Concrete Cylinders, RP 1545," *Journal of Research*, National Bureau of Standards, Vol. 31, July 1943.

[55] Watstein, D. and Seese, N. A., Jr., "Effect of Type of Bar on Width of Cracks in Reinforced Concrete Subjected to Tension," *Journal*, American Concrete Institute, Vol. 41, Feb. 1945, p. 293.

[56] Gergely, P. and Lutz, L. A., "Maximum Crack Width in Reinforced Concrete Flexural Members," *Causes, Mechanism, and Control of Cracking in Concrete*, SP-20, American Concrete Institute, Detroit, MI, 1968, pp. 1–17.

[57] Kobayashi, K. and Takewaka, K., "Experimental Studies on Epoxy Coated Reinforcing Steel for Corrosion Protection," *The International Journal of Cement and Lightweight Concrete*, Vol. 6, No. 2, May 1984, pp. 99–116.

[58] *Proceedings*, Symposium on Bond and Crack Formation in Reinforced Concrete, RILEM, Stockholm, 1957.

[59] Soretz, S., "Factors Affecting the Stiffness of Reinforced Concrete Beams," *Proceedings*, RILEM Symposium on Bond and Crack Formation, Stockholm, 1957.

[60] Baker, A. L. L., Ashdown, A. J., and Wildt, R. H., "The Deformation of Cracked Rectangular Reinforced Concrete Beams," *Proceedings*, RILEM Symposium on Bond and Crack Formation, Stockholm, 1957.

[61] Murashev, V. I., "Calculation of Crack Width in, and Flexural Rigidity of, Reinforced Concrete Members in Flexure," *Proceedings*, RILEM Symposium on Bond and Crack Formation, Stockholm, 1957.

Petrographic Examination

Bernard Erlin[1]

PREFACE

Twenty-seven years have past since the publication of *ASTM STP 169A*, where this chapter was first published. Kay Mather, who wrote the previous version of this chapter in *ASTM STP 169A and 169B*, in her closure quoted from St. Paul, "Things which are seen"—concrete and mortar—"were not made of things which do appear." St. Paul is not with us—neither is Kay who died in 1991. The bulk of this paper is still her "quote" on petrography, but with modest changes and additions. The next version may be entirely different. But the emphasis will remain the same.

INTRODUCTION

Twenty-seven years have past since the inception of *ASTM STP 169A* and 15 years have past since the updated *ASTM STP 169B* was published. At that time, there was little reference material available on petrography of concrete. Today, there are literature references to petrographic examinations available where the theme is on petrography as a tool used to provide information about concrete instead of petrography that is incidental to the paper. Both approaches, however, are needed because doing petrography for the sake of solely providing information can be an exercise of futility, while doing petrography for the purpose of providing links connecting "activity" within the concrete to its behavior is indeed a fulfillment of the science.

Significant advances in petrography and petrographic methods have been published about petrographic examinations related to concrete. Once considered by many to principally revolve around light optical microscopes, the science has greatly expanded to include new types of instrumentation and techniques from light optical microscopy to specimen preparation, "wet" chemical analyses to infrared spectroscopy, X-ray diffractometry and spectroscopy, scanning electron microscopy with attendant elemental analysis, plus differential thermal analysis and other analytical tools.

In the history of petrography of concrete as we know it today, at least from published papers, was the work of Johnson in 1915 [1] who described and related microscopical observations of the composition and texture of deteriorated concrete to its performance. He further applied what he saw toward a philosophy of what makes inferior concrete inferior—and said that "even with the very best of materials, only concrete of inferior strength is commonly produced." There were trailblazers before Johnson, but his work was, perhaps, the first widespread enough so that it truly reached out to engineers. His work was published in a six-part series, from January through March in 1915 [1].

Subsequent and more recent informative documents include Refs 2 through 15. The International Cement Microscopy Association (ICMA) was founded in 1978. That organization, whose main emphasis is on cement, provides publications of interest to concrete petrographers.

In June of 1989, the first symposium specifically directed toward presenting information on petrography of concrete and concrete aggregates was sponsored by ASTM. The papers presented at that symposium resulted in *ASTM STP 1061* [16].

The closing remarks of Kay Mather who authored the initial versions on petrographic examination of concrete in *ASTM STP 169A* and *ASTM STP 169B* foretold the potential advances in the science of petrography. Today, the imagination extends even further, and it is gratifying that her analytical visions are in use and being extended.

PETROGRAPHY

Within the realm of petrography, which includes the use of a broad variety of analytical and physical methods, is the scientific description of the composition and texture of materials, including the systematic classification of rocks. It also includes almost anything that can be said about a concrete, from its mineralogy to its strength and volume sensitivity.

Petrographic analysis of concrete—a man-made rock—is its examination by analytical techniques that will identify procedures and the sequence of its production, its composition and internal structure, and allow its classification as to type, original and existing conditions, and future serviceability.

The physical and chemical properties of concrete, especially immediately and shortly after it is made, is a physical wonder. Within a very short time measurable in terms of days, it becomes hard and strong and usually endures

---

[1] Petrographer, The Erlin Company, Latrobe, PA 15650.

for long periods. Its strength originates within itself by complex chemical reactions. It is recreated "rock" akin to the rock conglomerate, which mother nature has made.

Like rocks and minerals, concrete is a "mirror" with a memory. Petrographic examinations allow us to interpret the concretes' past as it really was—to identify, beyond all of the obscurities essential facts about its manufacture and performance. Its makeup and past performance, in light of research and practical experiences, allows projections of its future serviceability.

Petrography is used, frequently, to assist in forensic evaluations, where it is vital for supplying factual information. That information can relate to mix proportions that include coarse and fine aggregates, portland cement and other cementitious materials, water-cement ratios, air contents, mixing, placing, bleeding, cement hydration, finishing, curing, cracking, scaling, spalling, low strength, excessive wear, blistering, delamination, and other features. To be effective, the petrographer should have a good understanding of all concrete-making materials, concrete manufacture, and the influence of environmental exposure on its stability and performance.

This chapter provides insight into aspects of petrographic examinations by giving information about its usefulness in evaluating hardened concrete, noting problems inherent in its applications, outlining what it involves, describing the kinds of information that it can produce, and showing how this information can be applied.

The questions that materials testing and evaluation tries to answer are: (a) Does the concrete conform to specification requirements; (b) How will it behave in use? (c) Why did it behave the way it did? and (d) What can be anticipated in the future? The most useful method for developing practical information from which to answer these questions is to study the concrete in the field and in the laboratory.

Testers of materials try to compress time and to anticipate and reproduce the environment that the material will experience. Generally, they use standardized procedures not directly related to the specific environment or that do not determine the particular properties relevant to performance in the specific instance. Thus, the testing of construction materials amounts to obtaining various kinds of information about certain samples in specified conditions and extrapolating to the conditions of use insofar as they can be predicted.

Petrographic examination of hardened concrete is included among the subjects in this volume because it helps to improve the extrapolation from test results to performance. It offers direct observational information on what is being tested and what is in the concrete structure, giving another way of appraising the relationship between samples being tested and materials in service by judging how similar the two are.

COMMUNICATIONS

A petrographic examination of concrete ordinarily begins with communication between the person who requests the examination and the petrographer. Unless the two succeed in producing a clear mutually understood statement of the problem, they cannot expect a clear useful answer to be obtained economically. In nearly all cases, whoever asks for an examination knows that the concrete is unusual as reflected by its performance; the more clearly he defines the features prompting his interest, the more he directs the petrographer toward the important aspects. The person may not be familiar with the techniques that the petrographer may use or with his approach; conversely, the petrographer may not realize the individual's responsibility for a decision or action, may not have all the background information that could help him better understand the concrete and the problem, and may not realize which petrographic findings are useful and relevant.

The petrographer should not expect petrographic results to be taken on faith; the rationality of the techniques producing them should be demonstrable. Both should remember that the essentials of petrographic examination of concrete are practiced anytime anyone looks intelligently at concrete either in a structure or as a specimen and tries to relate and project what he sees to the past or future performance of the concrete. On this basis, useful "petrographic information" can come from inspectors, engineers, chemists, physicists—anyone concerned with the production and laboratory examination or use of concrete. No one should hesitate to examine concrete with all available means. Anyone, from novice to expert, is entitled to question data and conclusions and verify them in as many ways possible.

METHODS—STANDARDIZATION AND DESCRIPTION

The recommended practice for petrographic examination of aggregates, ASTM Guide for Petrographic Examination of Aggregates for Concrete (C 295) was published in 1952. ASTM Practice for Examination and Sampling of Hardened Concrete in Constructions (C 823) was adopted in 1975. It gives guidance for both the field examination of concrete in constructions and the petrographic examination of the samples taken. Steps to be taken before examination and preliminary investigations are outlined and include: (1) the desirability of assembling reports and legal documents concerning the construction, and (2) the usefulness of interviews with contractors and others connected with the construction and with the owners, occupants, and users of the constructions. Procedures for detailed investigations of the concrete in place in constructions are described. Sampling hardened concrete is discussed along with the preparation of appropriate sampling plans and selection of the number and size of samples. Information needed to accompany samples is described. ASTM C 823 was prepared to be useful not only to petrographers but also to engineers and others who have reason to examine constructions.

ASTM Practice for Petrographic Examination of Hardened Concrete (C 856) was adopted in 1977. It turned out to be a more complicated document than was expected, because there may be different purposes for examining concrete specimens including: constructions, specimens after periods of field exposure, specimens in simulated service, specimens from laboratory tests, and specimens after initial laboratory preparation and curing. What may be recognized as concrete features produced in field exposures, simulated service, laboratory tests, and laboratory curing may be features interesting and revealing in themselves in laboratory experiments. The laboratory observations may also make similar or related features of concrete from constructions more comprehensible. ASTM C 856 (1) serves those who buy a petrographic examination and want more understanding of what they are getting and why, (2) serves petrographers approaching the examination of concrete, and (3) reminds concrete petrographers of things they may have forgotten or neglected.

In 1976, Ref 17 was published. It provides help in (1) identifying alite and belite residues in paste, and (2) examining paste by X-ray diffraction, differential thermal analysis, and infrared absorption spectroscopy. Pastes of three cements, that is, a Type I, a white cement, and a cement containing no tricalcium aluminate (C_3A), were examined at different ages up to one year and a number of pure phases were synthesized and examined by these methods. Another publication for help in petrographic examination of hardened concrete is Ref 18. Other publications of interest are in volumes of the Fifth International Symposium on the Chemistry of Cement [19] and those of the Sixth International Symposium on the Chemistry of Cement [20]. A stimulating publication originating in 1971 [21] covers a wide field of cement and concrete topics and has the advantage of being truly international because the editorial board and the contributing authors come from centers in various parts of the world where cement and concrete research is carried out.

Before ASTM C 856 was published, petrographic techniques for the examination of hardened concrete had been described and discussed but were not standardized. In 1966, Kay Mather wrote:

> At present I know of no laboratory where petrographic examinations of concrete are made that is equipped to use all the methods that have yielded useful information and no one person who has digested the available approaches and developed the ability to choose the particular combination of techniques best suited to each problem encountered. Concretes are more complex than most rocks used as aggregates; their constituents are less well known; concretes change through time more rapidly than most aggregates. All of these circumstances combine to make each petrographic examination of concrete unique and thus to make the methods harder to generalize and standardize. Each examination presents some new facet for the petrographer who is willing to learn.

Today, in 1994, sophisticated equipment is available in many laboratories that is bringing about a breakthrough by orders of magnitude in our ability to decipher and to interpret the composition and performance aspects of concrete in constructions. These methods can bring with them the danger of losing touch with the primary purposes of an examination, that is, due to great ranges in scale such as constructions with dimensions in hundreds of metres (kilometres in the case of pavements); core samples with dimensions in hundreds of millimetres that can be examined with great advantage using low-power stereomicroscopes; thin sections with scales ordinarily about 800 mm^2 by 15- to 30- μm thick examined using petrographic microscopes; X-ray diffraction samples that may be milligrams of material hand-picked under the stereomicroscope or a few grams of material concentrated by hand-picking paste or aggregate from carefully broken concrete; scanning electron microscope specimens that may be 200 mm^3 or much smaller and thinner; single crystals a few angstroms cubed that are examined using microprobes; and the nanometer-sized material examined using transmission electron microscopes. Few cases will be found in which all of these steps in scale will be used. The preliminary important steps are (1) looking at the construction itself and (2) observing the samples from it with the naked eye and the low-power stereomicroscope. Then the petrographer can form one or more working hypotheses for best making the transition from the macroscale to the microscale so that knowledge gained from each unit of effort, preferably using more than one method, provides independent and complimentary verification. Care is needed to ensure that proper sampling is done at all levels of scale. For example, field samples should include cracks, specimens for detailed laboratory analysis should be taken from deposits in cracks and voids in several samples, and paste concentrates should be made from several samples so that a sensible sampling plan is carried out from the macro to the microscale.

"The straw that broke the camel's back" theory of concrete deterioration is archaic. Unless the concrete was hit by a truck or a large object from outer space, it is highly probable that evidence of more than one deteriorative mechanism (or what appears to be such evidence) will be present in any sample of deteriorated concrete. Sorting out the major cause or causes from the minor causes or secondary effects of deterioration requires the exercise of an experienced petrographer's best judgment.

The history of the concrete construction and its subsequent behavior may be of great importance, and for this kind of information the petrographer should consult engineers, inspectors, and contractor's personnel who have been present during construction and those who have later inspected or occupied the construction.

The newer analytical techniques offer an opportunity to understand internal chemical reactions in much more details so that it is possible to characterize the hydration and reaction products more clearly. Consequently, the chemical reactions of normal hydration or abnormal deterioration, or both, in concrete can be evaluated with much more clarity and certainty. The relative roles of chemical attack and physical attack in producing deterioration can then be better understood. The goal of relating better-established and more-familiar techniques in petrographic examination of hardened concrete to the more intimate and detailed insights that are made possible by the newer techniques remains to be achieved completely. While progress has been considerable, the task is still far from complete.

PURPOSE AND APPROACH

Purpose

A petrographic examination of concrete attempts to answer three general objective questions: (1) What is the concrete composition? (2) How is it put together? and (3) What chemical or physical features are present now that were not present when it was made? The first question refers to the recognizable individual constituents present on the scale at which they are seen. The second question refers to structural fabric, that is, the articulation or packing in space of the component elements making up any sort of external form [22] or heterogeneous solid body. The third question refers to those forces, chemical or physical, that have altered the original concrete. These questions may be answered on any useful scale by choosing the technique or techniques of appropriate resolving power. The resolving power needed differs depending on the specific questions to be answered.

Approach

Step one, in any case, is to define the problem in order to find and ask the right questions—those that need to be answered to solve the problem that caused the examination to be requested. These questions should be answered as long as they are in the context, limited as it may be, of money, time, instrumentation, and the state of the art. The best petrographic examination is the one that answers the questions with maximum economy in minimum time, with a demonstration clear to all concerned that the right questions were answered with all necessary and no superfluous detail. The approach to this ideal varies depending on the problem, the skill with which the questions are asked, and the skill of the petrographer. One measure of the petrographer's skill is knowing when to stop, either because the problem is adequately solved, or, in some cases, because it has been shown to be unsolvable under the circumstances. The petrographers skill includes not only deftness in obtaining data, but also in putting the data together. Knowing when to stop the petrographic examination is not easy because of a synergism of phenomena sometimes encountered, or when two phenomena occur concurrently. If the petrographer has too little skill and understanding about concrete and the concrete-making materials, he may stop short or dwell too long on the examination. If there is a choice, the latter is preferred.

FABRIC AND COMPOSITION

Fabric is the packing and relationship of components and is the textural and compositional heterogeneity obvious on a weathered concrete surface or a broken or sawed surface of concrete. Fabric includes all of the structural elements, ranging in scale from gross to atomic, and comprises both structure and texture as those terms are used in rock description. The fabric is reflected on the magnitude and scale of the concrete lift or batch, the structural crack, the coarse aggregate, the sand grains, the air voids, the microcracks, the residual (unhydrated) cement, the calcium hydroxide crystals, the hydrous calcium aluminates, the almost amorphous crystalline hydrous calcium silicates in the hydrated cement paste, and the atomic structure of any crystal forming a part of any of the structural components. The closest naturally occurring analogue among rocks to the fabric of concrete is greywacke conglomerate with an abundant matrix. The closest naturally occurring analogue to hardened cement paste is silty clay.

Fabric and composition together define, characterize, and form the basis for descriptive classification of solid multicomponent substances. Composition and fabric are so closely interrelated in concrete that they cannot be clearly separated.

One important aspect to be derived from petrographic examinations is the determination of whether problems were created by the physical structure of the concrete or because of its chemical composition. For example, problems arising from the physical structure or fabric include inadequacies in mixing or consolidation and inadequacy of the air-void system to provide frost resistance. Problems arising from the chemical composition include those resulting from errors in batching, reactions between cement and aggregate or cement and admixtures, reactions between a contaminant and cement paste, and reactions between cement paste and solutions from external sources.

The petrographic examinations can be used to evaluate whether (1) construction practices were suitable for producing concrete capable of giving satisfactory service in the particular environment and exposure; (2) the concrete-making materials were susceptible to chemical reactions that have deleterious consequences; and (3) there was a failure to modify the environment, for instance, by improving the drainage so as to increase the ability of the concrete to survive the chemical or physical character of the environment. Usually, several causes have interacted, but one is probably the originator. If the originating cause can be identified before the petrographic studies, the most direct and appropriate techniques for the studies can be used.

Investigating composition and fabric provides a specific, unique definition of what is being examined. The standard tests do not always supply information that permits discrimination between one piece of concrete and another, but direct observation on the relevant scale does. For example, there are n possible concretes all having 50-mm (2-in.) slumps, with air contents of 5%, with 31-MPa (4500-psi) compressive strengths at 28 days, but the No. 2 cylinder in the set of three broken on Day A in Laboratory B is unique and different, perceptibly, from Nos. 1 and 3 and from all the members of the other possible sets, and its top is different from its bottom as cast. A petrographic study of this unique cylinder and a "normal" cylinder will define their differences. The salient lesson from this study of composition and fabric is the individuality and uniqueness of this cylinder, or of each structure or part of a structure, or of each specimen and each thin section. This individual combination of fabric and composition reflects a part of the history of the concrete and can be used to understand past performance and to forecast future performance of the concrete. What is investigated at any time is a particular concrete, not concrete in general. Each concrete and each part of a concrete is unique in terms of composition, fabric, history, and exposure.

COMPARISONS

To say that each structure and specimen is unique does not mean that comparisons are useless or impossible. They are essential, and concretes can be grouped rationally and compared usefully within classes and between classes, if the basis for the grouping is objective. Each comparison may leave out some characteristics of the things compared, so that any evaluations should consider the accidentally or deliberately omitted factors that may prove to be important.

Paste, mortar, or concrete of known proportions, materials, age, and curing history offer the logical basis for comparison and extrapolation; laboratory specimens made to be examined or salvaged after having been tested for strength provide a good source of such comparative material. Specimens exposed to laboratory air outside the moist room or curing tank for more than a few hours are much less suitable, because pastes of specimens that are cracked or that have slender cross sections sometimes carbonate very rapidly. Specimens exposed to simulated weathering tests, or wetting and drying cycles, or prolonged drying are not to be considered as representative of normally cured or naturally weathered concretes, but nevertheless, may be instructive. Natural weathering differs from part to part of a structure, as well as from climate to climate, elevation to elevation, and subgrade to subgrade.

However, based on field and laboratory data, there are expectations of what will happen to concrete after it is made, handled, tested, or exposed. With an understanding of what can happen, an experienced petrographer can separate the expected features from the artifacts.

INTERPRETATION OF OBSERVATIONS

Normal Concrete

"Normal" constituents and fabrics are here defined as those in serviceable concrete of the class and age in the region. "Serviceable" is used here instead of "undeteriorated," because it is possible to tell whether concrete in a structure is serving as it was intended. However, criteria that distinguish inevitable chemical and physical changes from deterioration in concrete 20 or 50 years old have not been well established.

Very valuable information that can be obtained by petrographic methods comes from the examination of normal concrete. By comparing the range of constituents and fabrics in the normal satisfactory concrete, differences from the normal can be identified and specifically defined. Unless it can be demonstrated that the constituents, or the proportions of constituents or the fabric, depart from those found in serviceable concrete of the age and class in the region, there is no logical basis for assuming any connection between constituents, or proportions or fabric, and service behavior. When petrographic studies reveal that a concrete has a peculiar service record and some unique feature or features not shared by others of comparable class, age, and provenance, the unique feature(s) can be evaluated in light of the peculiar service record, and a cause-and-effect relationship established.

Class of Concrete

Proper comparisons of different kinds of concretes is necessary because changes in cement content, water/cement ratio (w/c), and maximum size of aggregate can entail major changes in specified properties so that no close comparison will be significant. If, for example, the criteria for paving concrete are applied to mass concrete, all mass concrete appears very inferior, which it is not for the purpose it is intended to serve. The class of concrete is important because it implies relative homogeneity in mixture proportions, particularly in w/c, cement content, and maximum size of aggregate. It is possible by petrographic methods to identify concretes that differ in cement content and w/c ratio regardless of age, and it is possible to identify concretes fairly homogeneous in age having different cement contents. The ability to evaluate cement contents and w/c can be in the range of ± 30 kg/m^3 (50 lb/yd^3), and about ± 0.05 in w/c.

AGE OF CONCRETE

The age of the concrete may be important for judging the significance of observations. For example, calcium sulfoaluminate (ettringite) found in many voids as far as 130 mm (5 in.) from the outer surfaces of a concrete pavement of unknown age and of high flexural and compressive strength may be of importance relative to projected service. That observation is of particular importance when in other 15-year-old, similarly exposed, field concretes from the region calcium sulfoaluminate is not abundant. If the concrete of unknown age is in fact five or seven years old and it differs conspicuously, the difference probably justifies some concern about its future; if it is 15 years old, it is peculiar, but the peculiarity is probably of less practical importance.

SOURCES OF CONCRETE

Restriction of an investigation to one region assists in rational comparison from several points of view. The aggregates economically available in an area are determined by the regional geology and consequently show some homogeneity of composition resulting from similarity of origin and history. In a particular region, cements and aggregates economically available are used in making concrete exposed to the climate characteristic of the region—the prevailing temperature range and temperature frequency distribution and the characteristic amount and sequence of precipitation.

The extent of a region may vary from a few square miles to many thousands, depending on variations in:

(a) regional geology—as it determines quantity and uniformity of aggregates;

(b) topography—a region of low relief and generally uniform slope, such as the Great Plains, or the Atlantic or Gulf Coastal Plain, that has widespread comparable range and distribution of temperature and precipitation. In a region of high relief and broken slopes, temperature varies considerably with altitude, and precipitation varies with orientation to prevailing winds, making important differences in exposure over short distances; and

(c) patterns of distribution of aggregates and cement from different sources—in some areas, only one type of natural sand and gravel is available, no manufactured aggregate is produced, and synthetic aggregate sources are not common. Metropolitan marketing areas served by water transportation usually have available a selection of natural coarse and fine aggregates, manufactured coarse aggregates, and synthetic aggregates.

All variations between these extremes can be found.

There currently are 119 portland-cement plants in the United States [23] of which three produce white portland cement and eight are used only for grinding clinker. Twelve states have no producing plants, and the largest number of plants, ten, are in Pennsylvania. Foreign imports of portland cement amount to about 9% of the U.S. production, and serve costal and Great Lakes ports.

An additional influence that may appear is a prevailing engineering opinion within an organization placing concrete in different regional areas on what is desirable in proportions of components or methods of placing or consolidation. The existence of satisfactory structures built in many different ways underlines the need to define "normal" concrete in objective and regionally restricted terms.

NORMAL AND UNUSUAL CONCRETE

Although petrographic examinations of normal concrete is important, usually the concrete that a petrographer is asked to examine has behaved in an unexpected way. Before and during the early stages of the petrographic examination, the petrographer should seek background information about the history and behavior of the concrete. He can compare what he finds to the manufacturing processes used for making the concrete, and by secondary features present, evaluate interactions within the concrete and between the concrete and its environment.

Among the questions he should answer are: (1) What process or processes could produce the described results? (2) What observable traces could the process or processes leave in the concrete? and (3) Would such traces be unique and specific evidence of what is supposed to have happened?

RECONSTRUCTION OF HISTORY OF FIELD CONCRETE

To progress from consideration of simple petrographic examination to the petrographic examination of concrete that has aged and perhaps deteriorated in service introduces two important new unknowns—time and the precise environment of the structure. The effects of the passage of relatively short periods of time on the constituents present in several cement pastes of known w/c stored under laboratory conditions have been investigated. However, anomalies remain in the results, even though compositions of pastes and nature of the environments were known and controlled far more thoroughly than the composition and environment of any field concrete. Today, where there is extensive use of fly ash, silica fume, and slag, there are even more anomalies.

COMPOSITION

If the changes in composition of cement paste with time in laboratory conditions were known for a representative number of cement compositions and water cement ratios, effects of both cement-aggregate interactions and of environmental influences would be easier to recognize and could be interpreted more usefully.

Environment

Why do exposed vertical walls of chert-gravel concrete in the vicinity of St. Louis, Missouri, generally have fewer popouts than similar walls in the vicinity of Memphis, Tennessee? The winters are colder in St. Louis, but the mean annual rainfall is lower; in Memphis, a larger proportion of the higher mean annual rainfall occurs in winter. The difference probably is that the chert gravel in the Memphis walls is more likely to be critically saturated when it freezes. The Weather Bureau's climatological data for the location are a valuable source of information that can assist in interpreting the data from many petrographic examinations of hardened concrete. In Mississippi, several highway pavements and associated structures were affected by sulfate attack and by combined sulfate and acid attack [24]. The petrographic examinations can reveal the cause of the problem. Available information on the ground-water environment can tie the cause and effect together.

The examination of samples of field concrete after extended service involves an increase in analytical complexity, a decrease in available information, and, to a certain extent, a decrease in the confidence that may be placed in the answer, as compared to examinations of new concrete or laboratory test specimens of concrete.

Sometimes the petrographer must admit that he cannot find an answer to the questions posed—neither may the chemist, the physicist, nor the engineer. Sometimes the petrographer can recover evidence not accessible by other approaches. There may be difficulty in reconstructing the history of field concretes. Yet, the results of a petrographic study can be used to eliminate certain factors or direct attention to others.

Deteriorated field concrete submitted to a laboratory or to a petrographer may have performed abnormally because of more than one cause. Simple cases can sometimes be explained on the spot to the satisfaction of those concerned. Usually, the cases are more complex. The field concrete examined by a petrographer is concrete that has worried some responsible person enough to make the effort and expense of sampling and testing appear justified. There is thus a built-in bias in the sampling process. Normally concrete that a petrographer sees as part of his assigned duties is controversial concrete sent in by organizations with alert conscientious concrete technologists or concrete that has become the subject of controversy under other situations. In practice, this generally means that he usually sees only the poor concrete produced under conditions where a degree of control was intended, unless samples representing all conditions are included. Concrete produced where there was little intent for control to be exercised, that is, the worst concrete, rarely is sampled and sent to a petrographer; good concrete is rarely controversial.

Furthermore, the older the concrete the less information is likely to be available about materials, proportions, conditions of placing, and the characteristics that undeteriorated comparable concrete would have.

Although it can be deduced from concrete that the w/c was high or low, that cement content was high or low or medium, and general quality of workmanship, the alkali content of the cement can usually not be reconstructed. Aggregate sources, particularly of natural sand and gravel, can be located from their composition—the constituents present and their size distribution are usually diagnostic of the region and sometimes of the particular source.

Finally, deteriorated field concrete usually shows superimposed traces of several processes, with at least one in an advanced stage. The most advanced process may conceal evidence of others that were more important in effect.

Laboratory test exposures are simple compared to natural exposures because some factors are excluded and those that are retained and often regulated or "accelerated" by altering some factor so as to remove it from the range possible in nature. Consequently, a laboratory procedure often results in symptoms different from symptoms encountered in a field example of the process that the test is intended to simulate.

Samples of field concrete, when examined using light microscopy, are frequently found to contain secondary calcium carbonate near their outer surfaces, along old cracks, and sometimes in the interior. Such calcium carbonate, when examined by optical methods, is generally found to be calcite rarely aragonite, and almost never vaterite, in the form—birefringent spherulitic calcite with interstitial water. Vaterite, however, was found by optical methods to be common on mortar bars that had been tested according to ASTM Test Method for Potential Alkali Reactivity of Cement-Aggregate Combinations (Mortar-Bar Method) (C 227) and had been found on concrete specimens tested for resistance to freezing and thawing according to ASTM Test Method for Resistance of Concrete to Rapid Freezing and Thawing (C 666). The use of X-ray diffraction to examine cement-paste concentrates from field concrete has revealed that vaterite, not recognized by optical methods, is frequently a major constituent of the secondary calcium carbonate [25], especially on samples from seawater exposures or from other wet environments. Vaterite is known to persist for several months in laboratory specimens stored in room conditions. The sequence from poorly crystallized vaterite, calcite, and aragonite to well-crystallized calcite in the carbonation of pastes and mortars has been clarified [26], and vaterite is now known as a natural mineral [27].

Accelerated freezing and thawing in water according to ASTM C 666, Procedure A, produces a characteristic loss of surface skin and loss of mortar, which is not like the condition of specimens exposed on the mean-tide rack at Eastport, Maine [28]. Field concrete that is not air-entrained and is deteriorated by natural freezing and thawing develops sets of subparallel cracks normal to the placing direction of the concrete or deteriorated regions parallel to the nearest free surface. These phenomena may not be reproduced in accelerated freezing and thawing in water.

Field concrete that has deteriorated because of alkali-silica reactions usually has much more advanced and conspicuous internal symptoms of this reaction than are found in mortar bars of expansive combinations examined after they are tested according to ASTM C 227. On the other hand, some field concrete regarded as undeteriorated has shown a range of evidence of alkali-silica reaction.

Alkali-carbonate and alkali-silica reactions exist together in varying degrees of development in some concretes, and inconspicuous degrees of reaction may be the only recognizable peculiarities in cases of unsatisfactory service with possibly expensive consequences. Alkali-carbonate and alkali-silica reactions are described in other papers in this publication.

CONCLUSION

The kinds of information about hardened concrete that petrographic examinations reveal can be used to identify, for example; unsound aspects of portland cement that have lead to bulk concrete expansion [29]; improprieties of aggregates that can cause low concrete strength [30]; stray naturally occurring organic chemicals in aggregates that can cause variable concrete air contents [8]; the causes of concrete failure to set; rapid slump loss; poor proportioning and batching; factors resulting in large concrete shrinkage; early and late concrete instability due to a variety of causes; conformance to specifications; malperformance due to chemical reactions of aggregate; surface distress due to improper finishing, curing, and inadequate air entrainment; and poor performance due to composition, manufacture, and exposure to aggressive environments for which it is not designed.

Before the petrographic trail was blazed, during its early adventuresome period, and even today where new and exciting methods and techniques are constantly developed, we still have some ignorance about the factors that lead to good or poor concrete performance. However, there is no need to view our analytical level of understanding as insignificant.

It makes no difference if the concrete is a 10-ton cherry pie 14 ft (4.2 m) in diameter and 2 ft (0.6 m) deep constructed in Charlevoix, Michigan, to commemorate the birthday of George Washington, or a dam, nuclear reactor facility, bridge, foundation, wall, pavement, sidewalk, electrical insulator, or bank safe liner anywhere in the world; the basic constituents are similar at any age. Petrographic methods are ideal for evaluating how they have been put together and changes that have occurred, and provide an understanding of performance.

Petrography of concrete is flourishing. Like in other sciences, the development of new methods and techniques of examination, more sophisticated equipment, more people and laboratories available to complete work, and better understanding of the physical and chemical make-up of concrete, now make possible better interpretations for the causes of concrete performance and malperformance.

Along with improvements in analytical methods and equipment comes a new problem—the development of petrographers who are able to obtain the right analytical data, but also have the background to interpret the data. There is a need for hands-on appreciation of mixtures of portland cements; blended cements; mineral and chemical admixtures; water; varieties of coarse and fine aggregates; methods of mixing, placing, and consolidation; the physics and chemistry of what happens to and within plastic and hardened concrete; and the effects of time and environmental exposure on performance. Too frequently, the analyst becomes a specialist who operates in a narrow walkway of self- or organization-imposed semi-isolationism.

In this day of specialization, the techniques, and particularly the analytical equipment, are very expensive. The cost of getting petrographic information dramatically increases as the size and scale of what is analyzed progressively decreases. However, there is still a lot of information obtainable at moderate and justifiable costs.

Kay Mather included in her closure to this chapter in *ASTM STP 169B* a quote from St. Paul, "Things which are seen"—concrete and mortar—"were not made of things which do appear." St. Paul is not with us—neither is Kay who died in 1991. The bulk of this paper is still her "quote" on petrography, but with modest changes and additions. The next version may be entirely different. But the emphasis will remain the same.

REFERENCES

[1] Johnson, N. C., "Application of the Microscope to the Study of Concrete," *Engineering Record*, 23 Jan., 6 Feb., 13 Feb., 27 Feb., 6 March, and 13 March 1915.

[2] Brown, L. S., "Petrography of Cement and Concrete," *Journal*, Portland Cement Association, Research and Development Laboratories, Vol. 1, No. 3, Sept. 1959, pp. 23–24.

[3] Mielenz, R. C., "Petrography Applied to Portland-Cement Concrete," *Reviews in Engineering Geology, Significance of Tests and Properties of Concrete and Concrete-Making Materials*, Geological Society of America, Boulder, CO, Vol. 1, 1962, pp. 1–38.

[4] Mielenz, R. C., "Diagnosing Concrete Failures," *Stanton Walker Lecture Series on the Materials Sciences*, University of Maryland, College Park, MD, Nov. 1964.

[5] Mielenz, R. C., "Why Do Some Concretes Fail," *Concrete Construction*, Sept. 1972.

[6] Mather, K., "Petrographic Examination," *Significance of Tests and Properties of Concrete and Concrete-Making Materials, ASTM STP 169A*, American Society for Testing and Materials, Philadelphia, 1966, pp. 125–143.

[7] Dreizler, I., "The Microscopy of Concrete," *Zement-Kalk-Gips*, No. 5, 1966, pp. 216–222.

[8] Erlin, B., "Methods Used in Petrographic Studies of Concrete," *Analytical Techniques for Hydraulic Cement and Concrete, ASTM STP 395*, American Society for Testing and Materials, Philadelphia, 1966, pp. 3–17.

[9] Erlin, B., "Analytical Techniques," *Observations of the Performance of Concrete in Service*, Special Report 106, Highway Research Board, Washington, DC, 1970, pp. 29–37.

[10] Idorn, G. M., *Durability of Concrete Structures in Denmark*, Technical University of Denmark, Copenhagen, Jan. 1967.

[11] Klemm, W. A., Skalny, J., Hawkins, P., and Copeland, L. E., "Cement Research: Boon or Boondoggle," *Rock Products, Cement International*, Vol. 80, No. 4, 1977, pp. 156–170.

[12] Porvar, T. O. and Hammersley, G. P., "Practical Concrete Petrography," *Journal*, Concrete Society, England, Aug. 1978, Vol. 12, No. 8, 1978, pp. 27–31.

[13] Bennitt, J. G., "Petrography: New Man on the Construction Team," *Concrete Construction*, 1981, pp. 585–587.

[14] Jensen, A. D., Eriksen, M., Chatterji, S., Thaulow, N., and Brandt, I., "Petrographic Analysis of Concrete," Danish Building Export Council, Ltd., Denmark, 1984.

[15] Pitts, J., "The Role of Petrography in the Investigation of Concrete and Its Constituents," *Journal*, Concrete Society, July 1987, pp. 5–7.

[16] *Petrography Applied to Concrete and Concrete Aggregates, ASTM STP 1061*, B. Erlin and D. Stark, Eds., American Society for Testing and Materials, Philadelphia, 1990.

[17] "Evaluation of Methods of Identifying Phases of Cement Paste," *Transportation Research Circular No. 176*, W. L. Dolch, Ed., Transportation Research Board, National Research Council, Washington, DC, June 1976.

[18] "Guide to Compounds of Interest in Cement and Concrete Research," Special Report No. 127, Highway Research Board, National Research Council, Washington, DC, 1972.

[19] *Proceedings*, Fifth International Symposium on the Chemistry of Cement, Tokyo, 1968, Vols. I–IV, 1969.

[20] *Proceedings*, Sixth International Symposium on the Chemistry of Cement, 1974; Moscow Stroyizdat 1976; in part, available in English preprints.

[21] *Cement and Concrete Research, an International Journal* (bimonthly), D. M. Roy, Ed., Pergamon Press, New York, 1971.

[22] *Structural Petrology*, Memoir 6, Geological Society of America, Boulder, CO, 1938, p. 12 (translation of B. Sander, *Gefügekunde der Gesteine*).

[23] Private communication, Portland Cement Association, Skokie, IL, 1993.

[24] Lossing, F., "Sulfate Attack on Concrete Pavements in Mississippi," Symposium on Effects of Aggressive Fluids on Concrete, *Highway Research Record No. 113*, Highway Research Board, National Academy of Sciences, Washington, DC, 1966, pp. 88–102.

[25] Mather, K. and Mielenz, R. C., "Cooperative Examination of Cores from the McPherson Test Road," *Proceedings*, Highway Research Board, Washington, DC, 1960, pp. 205–216.

[26] Cole, W. F. and Kroone, B., "Carbonate Minerals in Hydrated Portland Cement," *Nature*, Vol. 184, 1959, p. B.A.57.

[27] McConnell, J. D. C., "Vaterite from Ballycraigy, Larne, Northern Ireland," *Mineralogical Magazine*, Vol. 32, No. 250, Sept. 1960, pp. 535–544.
[28] Kennedy, T. B. and Mather, K., "Correlation Between Laboratory Accelerated Freezing and Thawing and Weathering at Treat Island, Maine," *Proceedings*, American Concrete Institute, Vol. 50, 1953, pp. 141–172.
[29] Erlin, B., "The Magic of Investigative Petrography: The Practical Basis for Resolving Concrete Problems," *Petrography Applied to Concrete and Concrete Aggregates, ASTM STP 1061*, B. Erlin and D. Stark, Eds, American Society for Testing and Materials, Philadelphia, 1990, pp. 171–181.
[30] Davis, R. E., Mielenz, R. C., and Polivka, M., "Importance of Petrographic Analysis and Special Tests Not Usually Required in Judging Quality of Concrete Sand," *Journal of Materials*, Vol. 2, No. 3, Sept. 1967, pp. 461–486.

Volume Change

P. Kumar Mehta[1]

PREFACE

In *ASTM STP 169* and *ASTM STP 169A*, the chapter on volume changes in hardened concrete was authored by George W. Washa [1,2]. In *ASTM STP 169B*, the chapter was authored by James L. Sawyer [3]. In the interest of consistency, the essential format as well as some of the fundamental aspects of the earlier publications are retained in this paper.

INTRODUCTION

Concrete is subject to several types of volume changes during its service life. In engineering practice, the volume change is generally expressed as "strain," which is defined as the change in length per unit length. For instance, the instantaneous strain on loading that is fully reversible and also is proportional to the applied stress is called "elastic strain." On sustained loading, especially when the applied stress level is 50% or more of the ultimate strength of concrete, the material shows nonlinear stress-strain behavior. The gradual increase in strain with time under a constant load, which is not completely reversible on unloading, is termed as "creep strain." The strains resulting from concrete's response to applied stress are discussed elsewhere in this publication and therefore are not covered by this chapter.

There are other types of strain that occur as a result of concrete's response to environmental effects, such as wetting, drying, heating, cooling, freezing, and thawing. Environmental effects also include attack on concrete from atmospheric CO_2 (which results in shrinkage) and attack from exposure to sulfate-bearing water (which may result in expansion under certain conditions). At least three other phenomena are associated with expansion of concrete, namely, corrosion of embedded steel, hydration of free CaO and MgO present in cement, and interaction between reactive aggregates and alkalies that are normally contributed by the hydration of cement.

Many of the volume changes mentioned in the preceding paragraph are covered individually by relevant chapters of this publication. Those that are not covered elsewhere include carbonation shrinkage, drying shrinkage, and expansion due to hydration of free CaO and MgO in hardened concrete; these three types of volume changes will be discussed in this chapter. Also discussed here is the autogenous shrinkage, associated with cement hydration, which can occur in sealed and unloaded specimens. The last section of this chapter contains a brief description of volume change in expansive cements that are used in the production of shrinkage-compensating concrete.

From the standpoint of their relative significance, it is believed that under conditions of restraint, the volume changes involving shrinkage are relatively more deleterious. It should be noted that structural elements are almost always under some restraint from connecting members, foundations, subgrade friction, steel reinforcement, etc. Shrinkage of hardened concrete, when restrained, produces tensile stress. Since the tensile strain capacity of concrete is very low, it usually cracks. Cracking usually is not serious enough to jeopardize the structural integrity of an element, but is undesirable for a variety of reasons. First, cracks look bad and are the most frequent subject of complaints about concrete, particularly architectural concrete. More importantly, large external cracks, interlinking with internal voids and microcracks that are always present in concrete, make it possible for water and harmful chemicals and gases to penetrate with relative ease into the interior of the concrete mass. This phenomenon is probably the cause of numerous concrete durability problems with field structures. Unrestrained, expansive volume changes also cause cracking and are therefore harmful.

AUTOGENOUS VOLUME CHANGES

Autogenous volume changes are associated with cement hydration alone and do not include environmental effects due to variation in moisture and temperature. The autogenous volume change with ordinary portland cement concrete, is usually small, that is, less than 0.010%, or 100 microstrain expansion or shrinkage. The magnitude is dependent on the overall effect of two opposing phenomena: (*a*) the increase in the disjoining pressure in poorly crystalline C-S-H[2] and ettringite due to water adsorption, and (*b*) the reduction in the disjoining pressure due to

---

[1] Roy W. Carlson Distinguished Professor, Department of Civil Engineering, University of California, Berkeley, CA 94720.

[2] Cement chemistry abbreviations are used: C = CaO, S = SiO_2, A = Al_2O_3, F = Fe_2O_3, $\bar{S}$ = SO_3, and H = H_2O.

removal of adsorbed water by desiccation. According to Washa [2], in most instances the initial expansions obtained during the first few months do not exceed 0.003%, while the ultimate shrinkage obtained after several years usually does not exceed 0.010%. Further, according to the author, the autogenous volume changes are influenced by the composition and fineness of cement, the amount of mixing water, the concrete mixture proportions, and the curing conditions. The magnitude of autogenous volume change reportedly increases as the fineness of cement and the cement content in concrete for a given consistency are increased. Ultimate contraction appears to be somewhat greater for high C-S-H forming cements (namely, ASTM Type IV cement) than for normal portland cement (Type I).

With ordinary portland-cement concrete, since the magnitude of autogenous shrinkage when compared to other types of shrinkage is very small, it is ignored for practical purposes except in the case of dams [4]. This is because in the interior of mass concrete there is little likelihood of occurrence of any other type of shrinkage. The development of high-strength systems with very low water-cement ratio has brought the phenomenon of autogenous shrinkage to the attention of researchers again. Recently, using cement pastes made with a 0.23 or 0.30 water-cement ratio, a high early strength portland cement (430 m²/kg, Blaine) and a superplasticizer, Tazawa and Miyazawa [5] reported that the autogenous shrinkage of sealed specimens at an age of 70 days was of the order of 1000 microstrain; this autogenous shrinkage value increased to almost double when the cement was replaced with 10 or 20% condensed silica fume by weight. The authors believe that due to capillary tension the adsorbed water moves from the exterior parts of the sealed specimen into the interior parts where it is needed for hydration of unhydrated cement and silica fume particles. This phenomenon that caused the autogenous shrinkage is also believed to be responsible for the observation that flexural strengths of sealed beams of cement paste were much lower than those obtained for specimens cured under water when the water-cement ratio was less than 0.4. It appears that, due to the desiccation phenomenon, the tensile stress generated near the surface of the specimen must have caused some microcracking that resulted in the reduction of flexural strength.

CARBONATION SHRINKAGE

Carbonation shrinkage occurs as a result of chemical interaction between atmospheric carbon dioxide (CO_2) and hydration products of cement. Since it takes place concurrently with drying shrinkage, most reported data do not distinguish between the two and designate both as drying shrinkage. This is convenient when the carbonation shrinkage is low, such as with low-permeability concrete kept in a continually wet or dry condition. However, with permeable concrete, the magnitude of carbonation shrinkage may approach the magnitude of drying shrinkage in a CO_2-rich environment at 50 to 65% relative humidity [6].

All of the constituents of hydrated portland cement, namely C-S-H, CH, ettringite, and monosulfate hydrate, are subject to carbonation eventually. The rate of carbonation is dependent on several factors such as porosity of concrete, size of the member, relative humidity, temperature, CO_2 concentration, time of exposure, method of curing, and the sequence of drying and carbonation. Carbonation proceeds slowly and usually produces little shrinkage at relative humidities below 25% or near saturation. Concrete that has been subjected earlier to carbonation shrinkage will still shrink or swell with changes in relative humidity; however, the magnitude of these volume changes is smaller than before carbonation.

Carbonation reactions, such as the one shown here, tend to release moisture

$$Ca(OH)_2 + CO_2 \rightarrow CaCO_3 + H_2O$$

The porosity of the cement paste and the specimen size are controlling factors in carbonation because they determine the rates at which CO_2 can diffuse into the interior and the moisture released by CO_2 can diffuse to the exterior of concrete. Less dense concrete products such as block made with lightweight aggregate are more susceptible to carbonation than dense concrete products. While autoclaved blocks are relatively free from carbonation, concrete blocks cured in steam at atmospheric pressure show maximum carbonation shrinkage near 50% relative humidity [6]. Also, it is reported that precarbonation improves the volume stability of blocks cured in steam at atmospheric pressure.

The possibility of improving the volume stability of concrete block cured in steam at atmospheric pressure and subsequently subjected to drying and carbonation by hot-flue gases was investigated [7]. Under favorable conditions a 30% reduction in the shrinkage of the finished product was obtained. Conditions for an effective precarbonation treatment included: relative humidity between 15 and 35%, a minimum CO_2 content of 1.5%, temperature between 65 and 100°C, and an exposure period of 24 h.

Ying-yu and Qui-dong [8] investigated the mechanism of carbonation of cement mortars and the dependence of carbonation on pore size. Since the gaseous-phase carbonation process is a chemical reaction controlled by diffusion, the authors suggest "prediction of the carbonation coefficient" by the equation

$$a_1(2C_1/kP)^{1/2}$$

where C_1 is the partial pressure of CO_2, k is a constant, and P is porosity. The carbonating sample was divided into three areas—the carbonated area, the carbonating area, and the uncarbonated area—and a set of differential equations were used to describe the diffusion process in each area. The authors found that pores with radii above 32 nm increased the carbonation coefficient greatly. Regarding the mechanism of carbonation, it is concluded that there are some active sites on the pore walls. When carbonation occurs, $CaCO_3$ first nucleates on these sites. Hydration products near the pore dissolve continuously to furnish Ca^{++} ions to the liquid phase from which the ions migrate to the $CaCO_3$ nucleating site, where the crystal growth takes place.

Factors influencing the depth of carbonation in different concrete types were discussed in several recent studies. Schubert [9] reported the carbonation behavior of mortars and concretes containing different types of cements and fly ashes. Effects of various parameters on the depth of carbonation, the carbonation rate, and compressive strength were investigated for laboratory storage conditions (20°C, 65% relative humidity) for up to 10 years. The author found an inverse linear relationship, ($1/\sqrt{fc}$), between the carbonation rate and the 28-day compressive strength. For specimens with similar compressive strengths, no significant differences in the carbonation behavior were found between concretes and mortars, with or without fly ash, as long as cements with normal CaO content were used, the cement content was not too low (that is, <240 kg/m^3), and the fly ash content was not too high (that is, <30% by weight of blended cement). Cements with unusually low lime contents (that is, 70% slag cement) showed increases in carbonation rates. Carbonation depths of mortar specimens moist-cured for 28 days were much less than the seven-day moist-cured specimens. From a similar study, Ohga and Nagataki [10] confirmed that concrete with fly ash is more affected by the initial curing period in water than concrete without fly ash. The authors claim that carbonation depths of concrete both with and without fly ash can be predicted by an equation obtained from an accelerated test. The equation is expressed by the 28-day compressive strength of concrete. Kokubu and Nagataki [11] reported similar results from a 20-year-old field test. According to the authors, these test results have been incorporated into the Japanese standard specification for concrete cover. For fly ash concretes with seven days of standard moist curing, Ho and Lewis [12] found that carbonation depth depended mainly on the water/cement, not water/binder ratio regardless of the mix constituents used in their study. Also, the results from a survey of existing structures in Australia showed that with comparable 28-day strengths, binder content, and water/binder ratio, portland-cement, fly ash concretes showed lower resistance to carbonation than neat portland-cement concretes.

Dhir et al. [13] studied the effect of 28-day concrete strength on depth of carbonation in 20 years, using an accelerated test method. Both ordinary concrete and a superplasticized concrete showed similar trends with regard to the effect of concrete strength on carbonation; namely, with increasing strength, the depth of carbonation was reduced proportionately. However, for a given strength level, it was found that the superplasticized concrete always had a lower depth of carbonation than the corresponding normal concrete. For instance, with 40 MPa strength level, the normal concrete showed 15 mm depth of carbonation against 10 mm for the superplasticized concrete. Interestingly, 60- and 70-MPa superplasticized concretes showed negligible carbonation in 20 years in the accelerated test. Malhotra et al. [14] reported that a superplasticized, high-volume fly ash concrete containing 56% fly ash by weight of the cementitious material, even with 30 MPa compressive strength at 28-days showed only 3-mm carbonation in one year and 7-mm in five years in the laboratory air (50 ± 5% relative humidity).

Oye and Justness [15] investigated the carbonation resistance of latex-modified cement mortars and reported that, contrary to a general improvement in other properties, the carbonation resistance of latex modified mortars showed no significant improvement when compared to unmodified mortars. In fact, some of the epoxy-modified systems performed very poorly in the carbonation resistance test. It seems that CO_2 diffusion in the polymer-modified cement mortar occurred through channels in the cement paste that are found between the three-dimensional network of the polymer phase. Note that latex-modified concrete is air-cured for 24 h instead of moist-cured.

Sakuta et al. [16] discussed methods to reduce the rate of carbonation in concrete. It was found that the addition of amino alcohol and glycol ether derivatives was effective in controlling carbonation. Amino alcohol derivatives that are used for desulfurization and deoxidation during petroleum refining also possess the ability to absorb atmospheric CO_2 and slow down the rate of surface carbonation of concrete. The use of a water-insoluble glycol ether derivative, which is an efficient anti-foaming agent, reduced the size of air bubbles in concrete and made it less permeable. This slowed down the progress of carbonation considerably. The authors proposed that the carbonation of concrete can be effectively controlled by the use of low water-cement ratios, the two derivatives discussed by them and by thorough curing of concrete after placement. Aimin and Chandra [17] investigated the effect of acrylic polymer additions on the rate of carbonation of portland-cement paste. Increasing the polymer content reduced the porosity and water absorption of the cement paste. However, this increased the rate of carbonation that was maximum at 10 to 15% polymer loading and decreased thereafter.

Discussing the Italian experience with cements containing natural pozzolans, Massazza and Oberti [18] reported that the carbonation depths with pozzolanic cements were similar to those of portland cements provided that concretes having similar strengths were compared. The carbonation behavior of blended cements was inferior to portland cement only when the moist-curing period was too short. With prolonged moist curing, there was usually a reversal of the behavior. The existence of a correlation between strength and depth of carbonation did not seem surprising to the authors since both strength and permeability depend on the porosity of concrete. While carbonation affected the porous and permeable concretes negatively (that is, high carbonation depths are recorded), it affected the dense and strong concretes favorably since their porosity and permeability are greatly reduced.

Paillere et al. [19] investigated the effect of freezing and thawing (F-T) cycles on air-entrained concrete containing fly ash or slag. With increased F-T cycles, carbonation depths were increased. Whereas the presence of entrained air improved the resistance to freezing and thawing cycles, it did not help against the penetration of CO_2; in fact, it enhanced the carbonation depths. Portland cements containing up to 29% Class F fly ash or blastfurnace slag showed similar behavior to carbonation as neat portland cements; at increasing levels of cement replacement, the carbonation of concrete was enhanced.

Bijen and van Selst [20] compared the carbonation resistance of normal portland-cement concrete with portland blast-furnace slag cement concrete, when a part of the cement was replaced by fly ash. Due to the lower alkalinity of the portland blast-furnace slag cement paste, the fly ash does not show significant pozzolanic activity with this cement. As expected, therefore, the carbonation rate of concrete made with portland blast-furnace slag cement containing the fly ash was found to be higher than the corresponding concrete made with normal portland cement and the fly ash. Bijen et al. [21] reported earlier that substantial differences in carbonation rates between portland cement concrete and portland blast-furnace slag cement concrete that were observed in the laboratory curing at 20°C and 65% relative humidity, were not observed in outdoor exposure tests.

From field tests and actual structures where the concrete contained different levels of ground blast-furnace slag as cement replacement material, Osborne [22] concluded that the two main factors influencing the depth of carbonation were the level of cement replaced by slag and the environmental conditions around the concrete. Carbonation depths were greater for higher slag contents when the concrete was in a sheltered or dry climate. As the cement content was increased from 350 to 450 kg/m³ and slag replacement level was reduced from 70 to 40%, both the carbonation depth and the permeability of concrete were substantially reduced.

Sakai et al. [23] compared the carbonation characteristics of a Type K expansive cement concrete with ordinary portland cement concrete from a 22-year-old building in Nigata Prefecture, Japan. From the core samples, the compressive strengths and the carbonation depths of the two concretes, both made with 0.57 water-to-cement ratio, were found to be similar and of the order of about 58 Mpa and 12 mm, respectively. Mineralogical analysis of the carbonated material confirmed that C-S-H was decomposed to calcite and silica gel, whereas the Aft (ettringite) phase in the expansive cement concrete decomposed to calcite, gypsum, and Al(OH)$_3$ gel. The authors observed that the decomposition rate of AFt by carbonation was somewhat slower than that of C-S-H.

DRYING SHRINKAGE

When plain, normal-weight concrete is dried from a saturated condition to a state of equilibrium with air at 50% relative humidity, the shrinkage associated with moisture loss is in the range of 0.04 to 0.08% (400 to 800 microstrain). The source of drying shrinkage in concrete is the adsorbed water and the water held in small capillary pores of the hydrated cement paste [24]. It has been suggested that the adsorbed water causes a disjoining pressure when confined to narrow spaces between two solid surfaces. The removal of the adsorbed water reduces the disjoining pressure and brings about the shrinkage of hydrated cement paste on exposure to drying conditions. In regard to capillary water, it has been suggested that water minisicus in small capillaries (5 to 50 nm) exerts hydrostatic tension, and removal of this water tends to induce a compressive stress on the walls of the capillary pores, thus contributing to the overall contraction of the system.

Among the more important factors influencing the drying shrinkage of concrete are the content of cement paste and its quality (that is, water-cement ratio and degree of hydration), the elastic modulus of aggregate, the characteristics and amounts of admixtures used, the time and the relative humidity of exposure, the size and shape of concrete mass, and the amount and distribution of reinforcing steel. Drying shrinkage of concrete and the factors influencing it are discussed in numerous reports, such as those authored by Ytterberg [25] and Meininger [26]. A brief discussion of these factors follows.

Concrete is made up of two constituents: cement paste and aggregate; the former shrinks and the latter is shrinkage-restraining. The effectiveness of an aggregate to restrain the shrinkage of cement paste is related to the elastic modulus of the aggregate. From the data obtained by an experimental study with concrete mixtures of water-cement ratios of 0.35 or 0.50, Powers [27] developed the following expression showing how the drying shrinkage of concrete (S_c) is related to the drying shrinkage of cement paste (S_p), the volume fraction of the cement paste ($1 - g$), and a constant (n) that is dependent on the elastic modulus of aggregate

$$S_c = Sp(1 - g)^n$$

Thus, for a given aggregate and cement paste the drying shrinkage may be doubled when the aggregate volume fraction is decreased from 80 to 50%. Similarly, long-term studies on the effect of aggregate type on drying shrinkage [28] showed that with a given concrete mixture the 23-year shrinkage was more than twice for the low elastic modulus aggregates when compared with the high elastic modulus aggregates. Concrete made with dirty sand or with unwashed aggregates containing silt and clay shrinks significantly more than concrete made with clean aggregates. To the extent that the water requirement for a given consistency of concrete is influenced by the maximum size, grading, shape, and surface texture of aggregate, these factors also affect drying shrinkage. The influence of the water content of concrete on drying shrinkage is discussed next.

The quality of the cement paste is primarily a function of the water content for a given cement content in concrete (water-cement ratio), and the degree of hydration that in turn depends on the fineness and chemical composition of cement as well as the curing time, temperature, and humidity conditions. Since unhydrated cement particles can offer restraint to shrinkage of hydrated cement paste in the same manner as aggregate particles, the factors leading to a greater degree of hydration generally contribute to a higher drying shrinkage. Thus, finer cements usually show greater shrinkage values. Portland cements with high-C_3A and high-alkali contents tend to give high drying shrinkage, which can be controlled by using optimum gypsum content via ettringite formation and the associated shrinkage-compensating effect (see the last section of this chapter that deals with testing methods). Admixtures that increase the water requirement of concrete for a given consistency (namely, diatomaceous earth or calcined clays) generally increase drying shrinkage but many water-reducing admixtures that reduce the water content do not reduce drying shrinkage. Accelerating admixtures such as calcium chloride and triethanolamine tend to increase drying shrinkage.

Atmospheric diffusion of the adsorbed water present within cement paste and the water held by capillary tension takes place over a long period of time and is accelerated by high environmental temperatures and low relative humidity. For a wide range of concrete mixtures only 20 to 25% of the 20-year drying shrinkage was realized in two weeks, 50 to 60% in three months, and about 75% in one year. Almost twice as much drying shrinkage is obtained at 45% relative humidity as compared to 80% relative humidity exposure.

The size and shape of the concrete mass have a considerable effect on the rate and total amount of drying shrinkage. The rate and ultimate shrinkage of a large mass of concrete are smaller than the values for small-size concrete elements, although the shrinkage process continues over a longer period for the large mass. Under given drying conditions, a 6-in. (152-mm) thick concrete member may reach equilibrium in one year, while a 12-in. (304 mm) thick member may require four years. Also, the ultimate shrinkage of the larger member may not exceed two-thirds of that obtained for the smaller one. It seems that slower drying conditions yield lower ultimate shrinkage values. Also, in massive concrete members differential drying conditions produce larger shrinkage at and near the surface. This gives rise to tensile stresses at and near the surface while compressive stresses are developed in the interior. Consequently, surface cracking may occur if the tensile stresses exceed the tensile strength of the material and the stress relaxation provided by creep.

Shrinkage of reinforced unrestrained structures produces tension in concrete and compression in steel. By increasing the amount of reinforcement, the shrinkage can be correspondingly reduced but it will increase the tensile stresses in concrete such that concrete is likely to crack when excessive reinforcement is used. Reinforced concrete elements with normal amounts of reinforcing steel may show drying shrinkages of the order of 0.02 to 0.03%. An increase in steel reinforcement will cause cracks at a closer spacing and also reduce the crack widths.

According to Washa [2], drying shrinkage values for structural lightweight concrete may vary between 0.04 to 0.15% and are likely to be more pronounced with concretes containing aggregates that have high rates of absorption and require high cement contents to obtain the specified strength. Moist-cured, cellular (foamed) products made with neat cement and weighing between 10 and 20 lb/ft^3 (0.22 to 0.44 kg/m^3) may have drying shrinkage as high as 0.3 to 0.6%. Autoclaved concrete products containing fine siliceous additives may weigh 40 lb/ft^3 (0.88 kg/m^3) and have drying shrinkage in the range of 0.02 to 0.1%.

In regard to the effect of wetting and drying cycles on drying shrinkage, after a few cycles the shrinkage becomes completely reversible although the original length before the first drying is never reached. According to Washa [2], wetting and drying cycles combined with alternations of high and low temperatures caused residual expansion that increased as the number of cycles were increased. One hundred and twenty cycles consisting of 9 h of oven drying at 82°C, followed by 48 h immersion in water at 21°C and then 15 h of air storage at 21°C, caused expansion of 0.1 to 0.25% for concrete mixtures made with different cements, water-cement ratios, and methods of placement.

Condensed silica fume is being increasingly used as a supplementary cementing material for concrete elements, and there is some controversy in regard to its influence on the drying shrinkage. Published reports show that the drying shrinkage can be higher, similar, or lower depending on the water-cement ratio and the period of initial curing, as discussed next.

Pistilli et al. [29] reported the drying shrinkage values for high water-cement ratio concretes. At 64 weeks, the concrete containing 237 kg/m^3 cement, a ratio of 0.7 between water to cement + silica fume, and 10% silica fume by weight of cement showed some increase in drying shrinkage, whereas a concrete with 297 kg/m^3 cement and 0.6 ratio between water to cement + silica fume, showed the same drying shrinkage as concrete without silica fume addition. Maage [30] reported that concrete specimens containing silica fume exhibited higher drying shrinkage than the control concrete when they were exposed to drying immediately after demolding. Similarly, Wolsiefer [31] measured the drying shrinkage strains according to ASTM Test Method for Length Change of Hardened Hydraulic Cement Mortar and Concrete (C 157) of high-strength, superplasticized, silica fume concretes initially moist-cured for 1 and 14 days; it was reported that the drying shrinkage was higher for specimens moist-cured for one day.

Carette and Malhotra [32] reported the test results for 420 days of drying according to ASTM C 157, using 28-day moist-cured concrete specimens containing 15 and 30% silica fume (w/c + SF = 0.4). The control concrete and the concrete with 15% silica fume showed similar values of drying shrinkage strain, whereas concrete with 30% silica fume showed slightly lower values.

The drying shrinkage strains of a normal-strength concrete (w/c = 0.57) and several high-strength silica fume concretes (w/c + SF) = 0.22, 0.25, 0.28 containing 10% silica fume by weight of cement were investigated by Tachibana et al. [33]. At one year, the drying shrinkage strains of the high-strength concretes were in the range of 540 to 610 × 10^{-6}, whereas that of the normal-strength concrete was about 50% higher. De Larrad [34] believes that the low drying shrinkage in high-strength concretes is due to the very low water content that is responsible for an increase in the autogenous shrinkage and a corresponding reduction in the drying shrinkage. Incorporation of silica fume leads to reduction of pore sizes in the cement paste, thus increasing the surface tension in small capillary pores and therefore the autogenous shrinkage. According to Sellevold [35], the autogenous (sealed) shrinkage due to self-desiccation in high-strength concretes can be as much as one half of the total shrinkage measured in a 50% relative humidity environment. Tazawa and Yonekura [36] confirmed that, at the same water-cement ratio, the drying shrinkage of concrete with silica fume was lower than that of concrete without silica fume. However, at the same compressive strength, the values of drying shrinkage per unit volume of cement paste were approximately the same for both concrete types with standard curing, whereas higher values for the silica fume concrete were obtained with autoclave curing (180°C, 5 h).

EXPANSION DUE TO HYDRATION OF FREE CaO AND MgO

It has been often reported in the published literature that hydration of crystalline MgO (periclase) or CaO, when present in significant amounts in a portland cement, can cause expansion and cracking. Laboratory tests on early portland cements showed that the cement pastes made with low-MgO cements containing 3% or more free (crystalline) CaO gave considerable expansion, which caused cracking of the unrestrained paste [37]. The phenomena is virtually unknown in modern portland cements due to better manufacturing controls as a result of which the content of uncombined or crystalline CaO seldom exceeds 1%. Expansive additives for cement containing a large amount of crystalline CaO are being used in Japan for the purpose of obtaining controlled expansions under restraint in shrinkage-compensating concretes, which are described in the next section of this chapter.

In regard to crystalline MgO or periclase, some early portland cements, which were made at much lower temperatures than used today, contained large amounts of this compound. The expansion and cracking of cement pastes and concretes containing these cements was attributed to periclase. It is now accepted that the periclase formed in modern portland-cement clinker at a kiln temperature of 1400 to 1500°C is either inert to moisture or too slow to hydrate under ambient temperature conditions. Only under autoclaving conditions, in accelerated laboratory tests, such as ASTM Test Method for Autoclave Expansion of Portland Cement (C 151), periclase is known to hydrate and cause considerable expansion in unrestrained pastes (see the section on test methods and specification for additional discussion). Since portland cements meeting ASTM Specification for Portland Cement (C 150) are required to comply with the maximum limits of expansion in the autoclave test, it is one of the chemical requirements of the ASTM C 150 that the total MgO content of portland cement shall not exceed 6%. Note that approximately 2 to 3% MgO goes into the solid solution of portland-cement clinker compounds (C_3S, C_2S, C_3A, and F_{ss}[3]), and the remainder, if any, show up in the form of periclase.

VOLUME CHANGES IN EXPANSIVE CEMENTS

ASTM Specification for Expansive Hydraulic Cement (C 845) covers hydraulic cements that expand during the early hardening period after setting. Large expansion occurring in an unrestrained cement paste can cause cracking. However, if the expansion is adequately restrained, its magnitude will be reduced but a self-stress will develop. When the magnitude of expansion is small such that the prestress developed in concrete is on the order of 25 to 100 psi (0.2 to 0.7 MPa), which is generally enough to offset the effect of tensile stress due to subsequent drying shrinkage, the cement is known as shrinkage-compensating. During the last 20 years, the cements of this type have been commercially used for making crack-free industrial floors, airport taxiways, water-storage tanks, and post-tensioned concrete members for parking garages.

Formation of ettringite and hydration of crystalline CaO are the two phenomena known in concrete technology that are capable of causing disruptive expansion. Both the phenomena have been harnessed to produce shrinkage-stress-compensating concretes. The expansive additive in U.S. practice is a modified portland-cement clinker that contains significant amounts of anhydrous calcium aluminosulfate ($4CaO \cdot 3A\ell_2O_3 \cdot SO_3$), and calcium sulfate in addition to the cementitious compounds of portland cement, $3CaO \cdot SiO_2$, $2CaO \cdot SiO_2$, and $4CaO \cdot A\ell_2O_3 \cdot Fe_2O_3$. The cement produced by blending this additive with normal portland cement is called Type K expansive cement.

In Fig. 1, a graphical representation is given of the concept showing how a Type K shrinkage-stress-compensating cement, when compared to normal portland cement, works to reduce the risk of drying shrinkage cracking [38]. Immediately following the start of hydration of Type K cement, it is observed that large amounts of ettringite start forming. After initial set, the concrete will bond to the reinforcing steel and any expansion of concrete associated with the ettringite formation will be restrained by the steel. Under these conditions, the steel will go into tension and concrete into compression. At the end of the moist-curing period when the element is exposed to environmental drying conditions, it will shrink in the same manner as normal portland cement concrete. However, the shrinkage will first relieve the precompression in concrete before tensile stresses have a chance to develop. Thus by preventing the development of large tensile stresses, the expansive cements can become instrumental in reducing the risk of cracking in concrete from drying shrinkage. Due to their ability to produce reinforced concrete members that suffer little or no dimensional change during their service life, the shrinkage-compensating cements are sometimes called, non-shrinking cements. This, however, is misleading because concretes made with expansive cements do show almost the same amount of drying shrinkage as normal portland cement concrete (Fig. 1a). With the former the overall dimensional change is negligible because of expansion during moist curing, which precedes the shrinkage on drying.

As with normal portland cement concrete, the rate and magnitude of drying shrinkage with Type K cement concrete is influenced by the aggregate content and type, and water content. With increasing water-cement ratio, the expansion during this moist-curing period is reduced proportionately; therefore, the residual dimensional change

---
[3] F_{ss} stands for the calcium aluminoferrite solid solution phase.

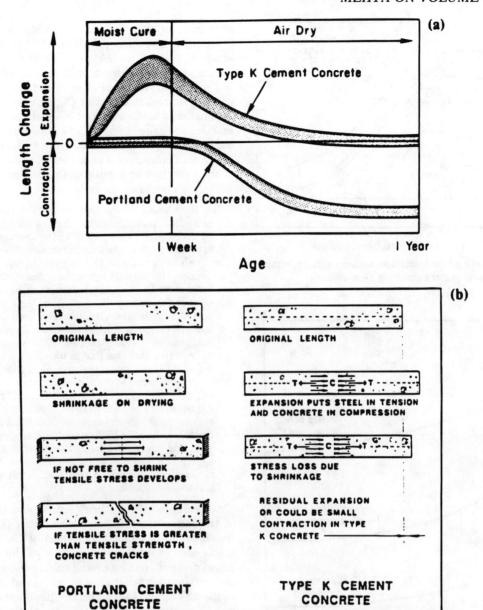

FIG. 1—(a) Comparison of length change characteristics between portland cement and Type K cement concretes; and (b) illustration showing why Type K cement concrete is resistant to cracking from drying shrinkage (Ref 38).

after the drying shrinkage changes from positive to negative at high water-cement ratios. This effect is shown by test data from Polivka and Willson [39] in Fig. 2. With a water-cement ratio of 0.53 or less, the magnitude of initial expansion was large enough to ensure a residual expansion after two months of drying shrinkage. Since the magnitude of expansion and degree of pre-compression is reduced considerably with water-cement ratios above 0.6, it is recommended that water-cement ratios lower than 0.6 should be used with Type K expansive cements, even when it is not needed from the standpoint of structural strength.

To develop adequate expansion and pre-compression in concrete, Kesler [40] found that good moist-curing conditions are absolutely essential (Fig. 3).

TEST METHODS OF DETERMINING VOLUME CHANGES

For evaluation of volume changes discussed in this chapter, ASTM offers several methods are listed here:

1. ASTM Test Method for Autoclave Expansion of Portland Cement (C 151)
2. ASTM Test Method for Length Change of Hardened Hydraulic Cement Mortar and Concrete (C 157)
3. ASTM Test Method for Drying Shrinkage of Mortar Containing Portland Cement (C 596)
4. ASTM Test Method for Restrained Expansion of Expansive Cement Mortar (C 806)

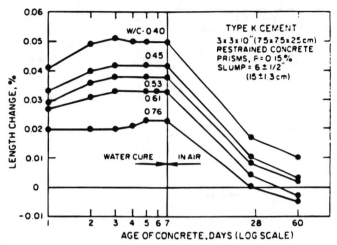

FIG. 2—Effect of w/c on restrained expansion and shrinkage of a Type K expansive cement concrete (Ref 39).

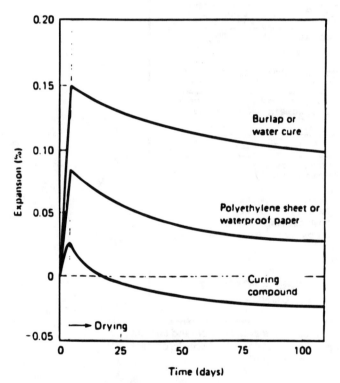

FIG. 3—Effect of curing conditions on restrained expansion of expansive cement concrete (Ref 40).

5. ASTM Test Method for Restrained Expansion of Shrinkage-Compensating Concrete (C 878)

A brief description of these methods including their scopes and significance follows.

Portland Cement

ASTM C 151 provides an index of potential delayed expansion caused by the hydration of CaO or MgO, or both. The method covers determination of the autoclave expansion of portland cement by means of a test on neat cement paste specimens, 25.4 by 25.4 by 285.8 mm, which are moist cured for 24 ± ½ h and then exposed to the action of steam under a pressure of 2 ± 0.07 MPa for 3 h. The rates at which pressure is increased in the beginning of the test and released at the end are specified. Linear expansion is measured by a micrometer comparator over an effective gage length of 254 mm. ASTM C 150 for portland cement permits a maximum expansion of 0.80%.

After a comprehensive review of the history and status of the autoclave test method, this author [38] came to the following conclusion:

> The proponents of the test claim that the test is able to provide a quantitative measure of the expansion caused by delayed hydration of free CaO and crystalline MgO. The critics of the test say that the measured expansion has no significance because: (1) exaggerated expansion values are obtained by the destroying the cohesive forces that would be present in normally hardened cement pastes; (2) the test conditions force the crystalline MgO present to hydrate and expand, whereas in normally cured commercial portland cements, within permissible chemical limits (i.e., maximum 6% total MgO), the MgO present is either inert or hydrates at a rate that is too slow to be of any consequence; (3) the magnitudes of expansion in *neat cement paste bars* due to hydration of the free CaO normally present may be high (that is high enough not to meet the requirements of the ASTM C 150 specification limit of maximum 0.80% expansion), but in corresponding concretes it will not be high enough to cause significant expansion and cracking (due to the restraining effect of the aggregate); and (4) no correlation has ever been shown between the autoclave test specification limit and the soundness (cracking potential) of concrete.

Accordingly, the author recommended a reevaluation of the autoclave test for determining the soundness of cement due to delayed hydration of CaO and MgO [38]. It was also pointed out that most of the countries of the world prefer to use Le Chatelier's method, which does not distort the hydration characteristics of cement as much as the ASTM C 151 test. The Le Chatelier apparatus is a 30-mm, longitudinally split, cylindrical mold with indicator needles. At the end of a 24-h normal moist-curing period, the mold containing the cement paste is exposed to boiling water at atmospheric pressure for a period of 3 h. The specification for cement soundness, when the Le Chatelier's test is used, permits a maximum distance of 10 mm between the indicator points. Recently, the European Committee for Standardization has approved the use of the Le Chatelier method for testing cement soundness (ENV 197). However, after 15 years of deliberations on the question of the validity of the autoclave test and the associated specification limit [41], ASTM has yet to take any action on the status of ASTM C 151.

Portland Cement Mortar and Concrete

ASTM C 157 provides a method for potential volumetric expansion or contraction of mortar or concrete due to various causes other than applied stress or temperature change. The method is particularly useful for comparative evaluation of potential expansion or shrinkage in different hydraulic-cement mortar or concrete mixtures. The test utilizes 25.4 by 25.4 by 280-mm mortar prisms, or 76.2 by 76.2 by 280-mm concrete prisms with 25.4-mm maximum size aggregate (254-mm gage length in both cases). The specimens are kept in molds for 23.5 ± 0.5 h, or longer if necessary to prevent damage, then demolded and moist-cured for 28 days. After curing, they are stored at 23.0 ± 1.7°C in water or in air at 50 ± 4% relative humidity. The specimens in air storage have a clearance of at least 25 mm on all sides. Comparator readings after air storage are taken at 4, 7, 14, and 28-day ages, and after 8, 16, 32, and 64 weeks unless otherwise specified. Similarly, water storage readings are taken at the age of 8, 16, 32, and 64 weeks.

It may be pointed out that ASTM C 157 is intended for use under a standard laboratory environment. Much greater variability and, in some cases, higher drying shrinkage values result when specimens are cast in the field under temperature and humidity conditions that are different from the laboratory.

ASTM C 596 covers determination of the effect of portland cement on the drying shrinkage of a graded standard sand mortar subjected to stated conditions of temperature, relative humidity, and rate of evaporation in the environment. In regard to significance and use of the method, it is stated that the drying shrinkage of mortar as determined by this method has a linear relationship to the drying shrinkage of concrete made with the same cement and exposed to same drying conditions. Since drying shrinkage of concrete is greatly influenced by the aggregate content, aggregate stiffness, and water content, many researchers question the validity of extrapolating the data on mortar shrinkage to concrete shrinkage. For instance, Swayze [42] found it inconsistent to rely on the behavior of either neat pastes or rich mortars to predict the ultimate shrinkage of concrete, especially when the tests are concluded at early ages.

Expansive Cement Mortar and Concrete

ASTM C 806 covers the determination of length changes of expansive cement mortar while under restraint due to the development of internal forces resulting from hydration of the cement. The test specimen is a 50 by 50 by 260-mm prism having a 250-mm gage length, which is restrained by using a standard restraining cage. The mix components of the mortar and the mixing procedure are specified. The specimens are removed from the molds at the age of 6 h and exposed to curing in lime-saturated water at 23.0 ± 1.7°C until an age of seven days. At the end of the water-curing period, the length change measurement is taken and percent expansion is calculated. ASTM Specification for Expansive Hydraulic Cement (C 845) requires the seven-day retrained expansion to range between 0.04 and 0.10%.

Since the potential for expansion, under conditions of controlled restraint, of concrete made with shrinkage-compensating cement cannot always be satisfactorily predicted from test mortars made in accordance with ASTM C 806, a need has been recognized for a test method in which concrete specimens are tested. ASTM C 878 can be adopted readily to studies of expansion involving differences in degree of restraint, cement composition, cement content, mixture proportions, or environmental treatments that vary from the standard procedures prescribed by this test method. The procedure calls for a standard restraining cage for a 76 by 76 by 254-mm concrete prism that is demolded at the age of 6 h, then water-cured until the age of seven-days when the length change is measured.

REFERENCES

[1] Washa, G. W., "Volume Changes and Creep," *Report on Significance of Tests of Concrete and Concrete Aggregates, ASTM STP 169*, American Society for Testing and Materials, Philadelphia, 1956.

[2] Washa, G. W., "Volume Changes," *Significance of Tests and Properties of Concrete and Concrete-Making Materials, ASTM STP 169A*, American Society for Testing and Materials, Philadelphia, 1966, pp. 189–201.

[3] Sawyer, J. L., "Volume Change," *Significance of Tests and Properties of Concrete and Concrete-Making Materials, ASTM STP 169B*, American Society for Testing and Materials, Philadelphia, 1978, pp. 228–241.

[4] Houk, J. E., Borge, O. E., and Houghton, D. L. "Studies in Autogenous Volume Changes in Concrete for Dwarshak Dam," *Journal*, American Concrete Institute, Detroit, MI, 1969, pp. 560–568.

[5] Tazawa, E. and Miyazawa, S., "Autogeneous Shrinkage of Cement Paste with Condensed Silica Fume," *Proceedings*, Fourth International Conference on Fly Ash, Silica Fume, Clay, and Natural Pozzolas, Supplementary Papers, CANMET, Ottawa, Canada, 1992, pp. 875–894.

[6] Verbeck, G., "Carbonation of Hydrated Portland Cement," *Cement and Concrete, ASTM STP 205*, 1958, pp. 17–36.

[7] Toennier, H. T. and Shiedler, J. J., "Plant Drying and Carbonation of Concrete Block, NCMA-PCA Cooperative Program," *Journal*, American Concrete Institute, Vol. 60, No. 5, 1963, p. 617.

[8] Ying-yu, L. and Qui-dong, W., "The Mechanics of Carbonation of Mortars and the Dependence of Carbonation on Pore Structure," *SP-100*, American Concrete Institute, Detroit, MI, 1987, pp. 1915–1943.

[9] Schubert, P., "Carbonation Behavior of Mortars and Concretes Made with Fly Ash," *SP-100*, American Concrete Institute, Detroit, MI, 1987, pp. 1945–1962.

[10] Ohga, H. and Nagataki, S., "Prediction of Carbonation Depth of Concrete with Fly Ash," *SP-114*, American Concrete Institute, Detroit, MI, 1989, pp. 275–294.

[11] Kokubu, M. and Nagataki, S., "Carbonation of Concrete with Fly Ash and Corrosion of Reinforcements in 20-year Tests," *SP-114*, American Concrete Institute, Detroit, MI, 1989, pp. 315–330.

[12] Ho, D. W. and Lewis, R. K., "Carbonation of Concrete and its Prediction," *Cement and Concrete Research*, Vol. 17, No. 3, 1987, pp. 489–504.

[13] Dhir, R., Tham, K., and Dransfield, J., "Durability of Concrete with a Superplasticizing Admixture," *SP-100*, American Concrete Institute, Detroit, MI, 1987, pp. 741–764.

[14] Malhotra, V. M., Carrette, G. G., Bilodeau, A., and Sivasundaram, V., "Some Aspects of Durability of High Volume ASTM Class F Fly Ash Concrete," *SP-126*, American Concrete Institute, Detroit, MI, 1991, pp. 65–82.

[15] Oye, B. A. and Justness, H., "Carbonation Resistance of Polymer Cement Mortar," *SP-126*, American Concrete Institute, Detroit, MI, 1991, pp. 1031–1046.

[16] Sakutu, M., Urano, T., Izumi, I., Sugiyama, M., and Tanaka, K., "Measures to Restrain Rate of Carbonation in Concrete," *SP-100*, American Concrete Institute, Detroit, MI, 1987, pp. 1963–2977.

[17] Aimin, X. and Chandra S., "Influence of Polymer Additions on the Rate of Carbonation of Portland Cement Paste," *Journal of Cement Composites*, Vol. 10, No. 1, 1988, pp. 49–52.

[18] Massazza, F. and Oberti, G., "Durability of Pozzolanic Cements and Italian Experience in Mass Concrete," *SP-126*, American Concrete Institute, Detroit, MI, 1991, pp. 1259–1283.

[19] Paillere, A. M., Raverdy, M., and Grimaldi, G., "Carbonation of Concrete with Low-Calcium Fly Ash and Granulated Blast-Furnace Clay," *SP-91*, American Concrete Institute, Detroit, MI, 1986, pp. 541–562.

[20] Bijen, J. and van Selst, R., "Effects of Fly Ash on Carbonation of Concrete with Blast-Furnace Clay Cement," *SP-126*, American Concrete Institute, Detroit, MI, 1991, pp. 1001–1030.

[21] Bijen, J., van der Wegen, G., and van Selst, R., "Powder Coal Fly Ash as a Filler in Portland Blast-Furnace Slag Cement Concrete," INTRON Report 88105, Maastricht, University of Delft, The Netherlands, 1988.

[22] Osborne, G. J., "Carbonation and Permeability of Blast-Furnace Clay Cement Concretes from Field Structures," *SP-114*, American Concrete Institute, Detroit, MI, 1989, pp. 1209–1238.

[23] Sakai, E., Kosuge, K., Terumura, S., and Nakagawa, K., "Carbonation of Expansive Cement Concrete and Change of Hydration Products," *SP-126*, American Concrete Institute, Detroit, MI, 1991, pp. 989–999.

[24] Mehta, P. K. and Monteiro, P. J. M., *Concrete: Structure, Materials, and Properties*, Prentice Hall, 1993, p. 36.

[25] Ytterberg, R. F., "Shrinkage and Curling of Slabs on Grade," *Concrete International*, Vol. 9, Nos. 4, 5, and 6, 1987.

[26] Meininger, R. C., "Drying Shrinkage of Concrete," NRMC Report Rk_3, National Ready-Mix Concrete Association, Silver Spring, MD, 1966.

[27] Powers, T. C., Review of Material Construction, No. 545, Paris, 1961, pp. 79–85.

[28] Troxell, G. E., Raphael, J. M., and Davis, R. E., "Influence of Aggregate Type on Long Term Drying Shrinkage and Creep of Concrete," *Proceedings*, American Society for Testing and Materials, Vol. 58, 1958, pp. 1101–1120.

[29] Pistilli, M. F. Wintersteen, R., and Cechner, R., "The Uniformity and Influence of Silica Fume on the Properties of Portland Cement Concrete," *ASTM Journal of Cement, Concrete, and Aggregates*, Vol. 6, No. 21, 1984, pp. 105–119.

[30] Maage, M., "Modified Portland Cement," Sintef. Report No. STF 64 A 83063, Norwegian Institute of Technology, Trondheim, Norway, 1983.

[31] Wolsiefer, J., "Ultra-high Strength, Field Placeable Concrete with Silica Fume Admixtures," *Concrete International*, Vol. 6, No. 4, 1984, pp. 25–31.

[32] Carrette, G. and Malhotra, V. M., "Mechanical Properties, Durability, and Drying Shrinkage of Portland Cement Concrete Containing Silica Fume," *Concrete International*, 1983, pp. 3–13.

[33] Tachibana, D., Imai, Y., Kawai, T., and Inada, Y., "High Strength Concrete Incorporating Several Admixtures," *SP-121*, American Concrete Institute, Detroit, MI, 1990, pp. 309–330.

[34] De Larrad, F., "Creep and Shrinkage of High-Strength Field Concretes," *SP-121*, American Concrete Institute, Detroit, MI, 1990, pp. 577–598.

[35] Sellevold, E., "Shrinkage of Concrete: Effect of Binder Composition and Aggregate Volume Fraction," *Nordic Concrete Research*, No. 11, 1992, pp. 139–152.

[36] Tazawa, E. and Yonekura, A., "Drying Shrinkage and Creep of Silica Fume Concrete," *SP-91*, American Concrete Institute, Detroit, MI, 1986, pp. 903–921.

[37] Lea, F. M., *The Chemistry of Cement and Concrete*, Chemical Publishing Company, New York, 1971.

[38] Williams, J. V., "Recommendations for Use of Shrinkage—Compensating Concrete in Sanitary Structures," *Concrete International*, Vol. 3, No. 4, 1981, pp. 57–61.

[39] Polivka, M. and Willson, C., "Properties of Shrinkage—Compensating Concrete," *SP-38*, American Concrete Institute, Detroit, MI, 1973, pp. 227–237.

[40] Kesler, C. E., "Control of Expansive Concretes During Construction," *Proceedings*, American Society of Civil Engineers, *Journal*, Construction Division, Vol. 102, No. C 01, 1976, pp. 41–49.

[41] Mehta, P. K., "History and Status of Performance Tests for Evaluation of Soundness of Cements," *Cement Standards—Evolution and Trends, ASTM STP 663*, American Society for Testing and Materials, Philadelphia, 1978, pp. 35–60.

[42] Swayze, M. A., "Volume Changes in Concrete," *Materials Research and Standards*, Vol. 1, No. 9, 1961, p. 703.

24

Thermal Properties

John M. Scanlon[1]
and James E. McDonald[2]

PREFACE

The authors feel very honored to have been chosen to review and update this chapter on Thermal Properties, originally written in 1965 by L. J. Mitchell, a supervisory engineer with the Bureau of Reclamation, Denver, Colorado, and later updated by J. A. Rhodes, who retired in 1980 as chief of the Concrete Branch, Office Chief of Engineers. At the time of authoring this chapter, their total experience dealing with thermal properties of mass concrete approached 60 years.

INTRODUCTION

The thermal properties of hydraulic cement concrete, whether the concrete is massive or in thin sections, are the properties that are most ignored and the least understood by the general concrete engineering and construction industry. The thermal characteristics of concrete covered in this chapter are conductivity, expansion (or contraction), diffusivity, specific heat, and heat of hydration of cementitious materials. This last property is included because of its influence on the thermal and physical behavior of concrete.

One of the earliest and most comprehensive studies of the fundamental thermal properties of concrete was carried out by the U.S. Bureau of Reclamation in conjunction with the design and construction of Boulder (Hoover) Dam between 1930 and 1935. The investigations were stimulated by the realization that in massive structures heat generated by hydration of cement could be responsible for volumetric changes that would affect the integrity of a structure, and that this heat, under ordinary circumstances, could take as much as a century or more to dissipate. Results are given in the Boulder Canyon Reports [1].

As with most construction materials, there is a direct relationship between a change in temperature of concrete and its change in length or volume. This fact has been long recognized for such structures as highways, bridges, walls, and buildings. Only at very high and very low temperatures do the expansion characteristics vary from those under normal conditions. The favorable thermal insulation characteristic of normal-weight concrete, and the even more favorable properties of lightweight concretes have been used effectively in building construction and for other applications in which resistance to steady-state heat flow is needed. The unique ability of concrete to damp out annual and daily ambient temperature variations (unsteady state) and to store or release heat over significant time periods has been utilized by designers.

THERMAL CONDUCTIVITY

Definition and Units

Thermal conductivity, a measure of the ability of a material to conduct heat, is defined as the ratio of the rate of heat flow to the temperature gradient. In normal metric use, it can be considered to be the number of kilocalories passing between opposite faces of a 1-m cube per unit of time when the temperature difference is 1°C. An alternate set of dimensions is joules per second · square meter · degree Celsius per metre. By dimensional manipulation and substitution, the approved SI (Systéme International) units are obtained, watts per meter · Kelvin, thus keeping the unit of time at 1 s. In U.S. customary and British units, conductivity is frequently expressed in Btu per hour · square foot · degree Fahrenheit per inch. Values of conductivity in these units may be converted to the SI units by multiplying by 0.1441314. Typically, engineering disciplines frequently express temperature gradients, areas, and time in units of measure most useful to them, and users of conductivity values are cautioned to assure that compatibility and consistency exist, both within and between systems of units [2].

Parameters and Values

Three principal conditions (water content, density, and temperature) significantly influence the thermal conductivity of a specific concrete. The mineralogical character of the aggregates largely determines the thermal conductivity for normal-weight concrete, while with lightweight concretes air voids and moisture content mask the effect of aggregate type. Other factors of slight, or negligible, importance in their effect on conductivity are cement type and content, entrained air, water/cement (w/c), and age.

---

[1] Senior consultant, Wiss, Janney, Elstner Associates, Inc., Northbrook, IL 60062-2095.
[2] Research civil engineer, U.S. Army Engineer Waterways Experiment Station, Vicksburg, MS 39180.

Neat cement pastes, with w/c ranging from 0.3 to 0.6 and ages from three days to one year [3], exhibit a fairly constant thermal conductivity value of 1.2 W/m · K (8.0 Btu · in./h · ft² · °F) at normal air temperatures and in a moist condition. Thermal conductivity measured under reduced moisture conditions has little meaning because the specimens suffer extensive cracking because of drying.

The amount of free water in concrete, regardless of density, has a major factor influence on the thermal conductivity. While water is a relatively poor conductor of heat as compared to rock, its thermal conductivity as shown in Table 1 is many times that of air, which it replaces in concrete [3]. Thermal conductivity of concrete varies directly with moisture content [4,5]. The effect of moisture on thermal conductivity values from oven-dry to a moist condition (not necessarily saturated) is given in Table 2.

For heavyweight, normal-weight, and structural lightweight concretes, the mineralogical characteristics of the aggregate markedly affect the conductivity of the concrete, as shown in Tables 3 and 4. Insulating lightweight concretes, those with densities less than 960 kg/m³ (60 lb/ft³), may have been aerated (foamed) or may contain a very lightweight porous aggregate. Thermal conductivities [6,7] given in Table 5 are for air-dry or low moisture contents.

Over a temperature range from room temperature to $-157°C$ ($-250°F$), the thermal conductivity of oven-dry, normal-weight and lightweight concrete is essentially con-

TABLE 1—Thermal Conductivity of Water.

Water Temperature		Conductivity	
°C	°F	W/m · K	Btu · in./h · ft² · °F
20	68	0.59	4.1
0	32	0.56	3.9
−18	0	2.3	16.0
−59	−75	2.6	18.0
−101	−150	3.3	23.0
−157	−250	5.2	36.0

TABLE 2—Typical Variations in Thermal Conductivity With Moisture at Normal Temperatures.

Moisture Condition	Conductivity	
	W/m · K	Btu · in./h · ft² · °F
LIMESTONE CONCRETE		
Moist	2.2	15.0
50% relative humidity	1.7	11.0
Dry	1.4	10.0
SANDSTONE CONCRETE		
Moist	2.9	20.0
50% relative humidity	2.2	15.0
Dry	1.4	10.0
QUARTZ GRAVEL CONCRETE		
Moist	3.3	23.0
50% relative humidity	2.7	19.0
Dry	2.3	16.0
EXPANDED SHALE CONCRETE		
Moist	0.85	5.9
50% relative humidity	0.79	5.5
Dry	0.62	4.3

TABLE 3—Effect of Aggregate Type on Conductivity of Dry Concrete at Normal Temperatures.

Aggregate Type	Dry Density		Conductivity	
	lb/ft³	kg/m³	Btu · in./h · ft² · °F	W/m · K
Hematite	179	2870	18	2.6
Marble	143	2290	12	1.7
Sandstone	120	1920	10	1.4
Limestone	126	2020	10	1.4
Dolerite	136	2180	8.6	1.2
Barite	180	2880	8.5	1.2
Expanded shale	89	1430	4.3	0.62
Expanded slag	103	1650	3.2	0.46
Expanded slag	60	960	1.5	0.22

TABLE 4—Effect of Aggregate Type on Conductivity of Moist Concrete at Normal Temperatures.

Aggregate Type	Moist Density		Conductivity	
	lb/ft³	kg/m³	Btu · in./h · ft² · °F	W/m · K
Hematite	190	3040	28	4.1
Quartzite	150	2400	28	4.1
Quartzite	152	2440	24	3.5
Dolomite	156	2500	23	3.3
Quartzite	...	...	23	3.3
Limestone	153	2450	22	3.2
Quartzite	147	2350	21	3.1
Sandstone	133	2130	20	2.9
Sandstone	150	2400	20	2.9
Granite	151	2420	18	2.6
Limestone	151	2420	18	2.6
Marble	152	2440	15	2.2
Limestone	152	2440	15	2.2
Basalt	157	2520	14	2.0
Rhyolite	146	2340	14	2.0
Barite	190	3040	14	2.0
Dolerite	147	2350	14	2.0
Basalt	158	2350	13	1.9
Expanded shale	99	1590	5.9	0.85

TABLE 5—Thermal Conductivity of Insulating Concrete.

Density		Thermal Conductivity	
kg/m³	lb/ft³	W/m · K	Btu · in./h · ft² · °F
AERATED			
320	20	0.07	0.5
480	30	0.11	0.75
640	40	0.14	1.0
800	50	0.20	1.4
960	60	0.26	1.8
VERMICULITE			
400	25	0.10	0.72
EXPANDED CLAY			
825	52	0.17	1.2

stant [3,8]. For moist normal-weight concrete, the conductivity at $-157°C$ ($-250°F$) has been found to be about 50% greater than at normal temperatures [3]. At elevated temperatures, up to about 750°C (1380°F), conductivities of cement pastes, mortars, and concrete decrease in a consistently uniform manner. This decrease has been attributed to disruption of the intercrystalline bonds in the aggregate caused by excessive thermal expansion [9]. Conductivity values at about 400°C (750°F) are given in Table 6 [5]. Above temperatures of 400°C (70°F), gradual

TABLE 6—Conductivity Values.

	W/m · K	Btu · in./h · ft² · °F
Cement paste	0.56	3.9
1:3 mortar	0.75	5.2
Sandstone concrete	1.6	10.9
Ilmenite concrete	1.2	8.2

disintegration of the fully hydrated cement paste occurs, resulting in further decreases in conductivity [9].

TEST METHODS

Values for the thermal conductivity of concrete are usually calculated from the diffusivity and specific heat because they are easier to measure. The Corps of Engineers [10] Method for Calculation of Thermal Conductivity of Concrete (CRD-C 44) is suitable for calculating the thermal conductivity of concrete from results of tests for diffusivity and specific heat. However, the conductivity can be determined directly with any of the steady-state or transient (nonsteady-state) test methods described in the following.

The test method for thermal conductivity developed for the Boulder Canyon Project [1] used 200-mm (8-in.) diameter cylinders subjected to steady heat flow conditions. Use of water as the heating and cooling mediums limited the results to a specific portion of the temperature range between the freezing and boiling points of water and to saturated or near-saturated concrete.

The ASTM Test Method for Steady-State Heat Flux Measurements and Thermal Transmission Properties by Means of the Guarded-Hot-Plate Apparatus (C 177) covers the achievement and measurement of steady-state heat flux through flat-slab specimens. This test method is applicable to the measurement of a wide variety of specimens, ranging from opaque solids to porous or transparent materials, and a wide range of environmental conditions.

The ASTM Test Method for Steady-State Heat Flux Measurements and Thermal Transmission Properties by Means of the Heat Flow Meter Apparatus (C 518) is a comparative method of measurement since specimens of known thermal transmission properties must be used to calibrate the apparatus. The method has been used at ambient conditions of 10 to 40°C (50 to 104°F) with flat slab thicknesses up to approximately 250 mm (10 in.).

The Corps of Engineers [10] Method of Test for Thermal Conductivity of Lightweight Insulating Concrete (CRD-C 45) measures conductivity directly under steady-state conditions in a manner similar to that of ASTM C 177. Nominal specimen thickness is 25 mm (1 in.), and temperature differential through the oven-dry specimen is from 32 to 60°C (90 to 140°F).

For experimental work, Campbell-Allen and Thorne [5] developed a hollow-cylinder, steady-state system for measuring conductivity at temperatures up to 200°C (392°F). The specimen, at a selected moisture content, was coated with mercury on all surfaces to retain moisture and electrically heated internally and cooled externally. Its acceptability or modifications are not known.

Lentz and Monfore, with the Portland Cement Association [8], developed a transient hot-wire method for determining thermal conductivity of concrete or rock. A thermocouple cast along the axis of a prism (or in the center of a split and lapped specimen) measures temperature response to a measured alternating current input. There are no moisture or density restrictions, and measurements are completed within a few minutes time.

SPECIFIC HEAT

Definition and Units

The rigorous definition of specific heat is the ratio of the amount of heat required to raise a unit weight of the material 1° to the amount of heat required to raise the same weight of water 1°. In those systems of units in which the heat capacity of water is 1.0 (either Btu/lb · °F or cal/g · °C), the specific heat (and heat capacity) values are the same. In SI units, specific heat is expressed in joules/kilogram · K, which is obtained from either the foot-pound or CGS values by multiplying by 4.1868×10^3.

Parameters and Values

The mineralogical differences among such aggregates as generally used have little effect on specific heat of the concrete, with values at normal temperatures ranging between 0.22 and 0.24 cal/g · °C (Btu/lb · °F). Increased water content, according to the Boulder Canyon studies [1], tended to increase the specific heat of the concrete. The actual values found were 0.22 cal/g · °C at 4% of mixing water by weight of concrete and 0.24 cal/g · °C at 7% mixing water by weight and higher.

Specific heat varies directly with concrete temperatures, as indicated in Table 7. These values represent mass concrete mixtures and several aggregate types [11]. Specific heats of most rock types tend to increase up to at least 400°C (750°F), and hydrated cement pastes up to 1000°C (1830°F). Investigations by Harmathy and Allen [12] indicate the specific heat of expanded aggregate lightweight concrete differs little from that of normal-weight concrete at ordinary temperatures and also increases up to at least 600°C (1110°F).

Testing Methods

Some type of calorimeter apparatus is employed in all procedures to measure the specific heat (heat capacity) of hardened concrete, aggregates, cement, and cement pastes. Among the most suitable of the procedures is the

TABLE 7—Typical Specific Heats of Concrete.

Temperature, °C	Specific Heat	
	J/kg · K	cal/g · °C
10	917	0.219
38	971	0.232
66	1038	0.248

Corps of Engineers [10] Method of Test for Specific Heat of Aggregates, Concrete, and Other Materials (CRD-C 124). In this procedure, approximately 1 kg (2 lb) of the material, with no particles larger than 25 mm (1 in.) in size, is tested. When more precise values for specific heat are desired and the specimen may be pulverized or ground to pass a 840-μm (No. 20) sieve, the Corps of Engineers [10] Method of Test for Mean Specific Heat of Hydraulic Cements, Cement Pastes, and Other Materials (CRD-C 242) should be used. This test method and ASTM Test Method for Specific Heat of Liquids and Solids (D 2766) are applicable to materials, such as cement and cement pastes, which would react with water. These tests require smaller samples (up to 100 g weight) and provide for selection of nonreactive cooling media whose physical or chemical properties differ from those of water.

The Bureau of Reclamation uses a procedure described as Procedure No. 4907, Specific Heat of Aggregates, Concrete, and Other Materials, that can be found in the Bureau's Concrete Manual [13].

THERMAL DIFFUSIVITY

Definition and Units

The diffusivity property is described as a measure of the facility with which temperature changes take place within a mass of material. Thermal diffusivity is defined numerically as thermal conductivity divided by the product of specific heat and density, or $\alpha = k/(c \cdot p)$. Thus, diffusivity results from the consolidation of three other properties that appear in differential equations that define heat flow and heat storage under unsteady-state conditions.

When calculating diffusivity from its parts, one must take care to assure that the dimensional units of the three constituents are compatible. These are shown in Table 8.

In approved SI base units, diffusivity values are usually small. The following formulas are used to convert British to SI units.

$$\text{ft}^2/\text{h} \times 2.58064 \times 10^{-5} = \text{m}^2/\text{s}$$

or

$$\text{ft}^2/\text{h} \times 9.29030 \times 10^{-2} = \text{m}^2/\text{h}$$

Parameters and Values

Those variables and conditions that influence thermal conductivity, specific heat, and density also affect thermal diffusivity. Both thermal conductivity and density are sensitive to moisture content of the concrete, and derived thermal diffusivity values should be based on conductivities and densities that correspond to the condition of the concrete in service. Within the same system of units, higher diffusivity values are associated with concrete that heats or cools most easily.

Thermal diffusivity of concrete is determined largely by the mineralogical characteristics of the coarse aggregate. Since specific heat varies directly with temperature, diffusivity values for a specific concrete will decrease as the concrete temperature increases.

Typical diffusivity values in Table 9, taken from Refs 1, 11, and 12, show the general range that can be expected for normal-weight and structural lightweight concrete. The Bureau of Reclamation [1] reported diffusivities for neat cement pastes ranging from 0.0012 to 0.0016 m²/h (0.013 to 0.017 ft²/h) at and somewhat above, normal room temperatures.

Tests Methods

The Corps of Engineers [10] Method of Test for Thermal Diffusivity of Concrete (CRD-C 36) outlines a procedure for determining the thermal diffusivity of partially saturated concrete. This method is directly applicable to a 152- by 305-mm (6- by 12-in.) cylinder; however, specimens of other sizes and shapes may be accommodated. A test specimen is heated in boiling water and then transferred to a bath of running cold water. The thermal diffusivity of the concrete is determined from the relationship between time and the temperature differential between the interior and the surface of the specimen as it cools.

The Corps of Engineers [10] Method of Test for Thermal Diffusivity of Mass Concrete (CRD-C 37) determines diffusivity with large moist concrete cube specimens. The 0.23-m³ (8-ft³) test specimen is heated by hot air until the specimen is in thermal equilibrium at 38°C (100°F); then the surfaces are cooled by a water spray. The cooling procedure creates a moisture gradient from the surface of the specimen inward, and the concrete mass is considered to be "moist" rather than either "dry" or "saturated." While the results of both tests are applicable only to moderate temperature ranges, they are quite reliable.

Another standard procedure (USBR Procedure 4909), for determining thermal diffusivity can be found in Ref 13.

One practice, not necessarily always preferable, is to calculate thermal diffusivity from conductivity, specific heat, and density as determined from laboratory tests. Test methods for conductivity are essentially limited to an

TABLE 8—Diffusivity Calculations.

British system units	
Conductivity	Btu/h · ft² · °F per ft
Specific heat	Btu/lb · °F
Density	lb/ft³
Thus α	ft²/h
SI units	
Conductivity	W/m · K
Specific heat	J/kg · K
Density	kg/m³
Thus α	m²/s, where W = J/s

TABLE 9—Typical Thermal Diffusivity Values.

Type of Aggregate in Concrete	Thermal Diffusivity			
	m²/h	ft²/h	ft²/day[a]	m²/day
Quartz	0.0079	0.085	2.04	0.190
Quartzite	0.0061	0.065	1.56	0.146
Limestone	0.0055	0.059	1.42	0.132
Basalt	0.0025	0.027	0.65	0.060
Expanded shale	0.0015	0.016	0.38	0.036

[a] Convenient when computing heat flow in large structures.

oven-dry condition, and diffusivity values are applicable to dry environments only.

THERMAL EXPANSION

Definition and Units

As with most construction materials, concrete has a positive coefficient of thermal expansion, which can be defined as the change in linear dimension per unit length divided by the temperature change. While a general value of 10 millionths/°C (5.5 millionths/°F) has been widely used, change in length is a complex process reflecting principally materials, moisture, and temperature individually and together. The actual thermal expansion is the net result of two actions occurring simultaneously. The first is a normal expansion typical of anhydrous solids. Second, there is a hygrothermal expansion or contraction associated with the movement of internal moisture from capillaries or from gel pores.

Parameters and Values

Since aggregate comprises from 80 to 85% of concrete, its thermal properties greatly influence the behavior of the concrete. Thermal expansion can vary widely among aggregates because of differences in mineralogical content. Quartzites and other siliceous aggregates exhibit high thermal expansion properties, and concrete containing such aggregates frequently show values up to 13×10^{-6}/°C at normal temperatures. Some limestone aggregate concretes exhibit expansion values of 5.6×10^{-6}/°C for comparable conditions, and other natural rock types usually have values between these two.

Cement paste occupies only about 15 to 20% of the concrete volume, but its expansion coefficient ranges from 9 to 22×10^{-6}/°C, which may be several times that of the aggregate itself. In addition, most of the capillary water and essentially all of the absorbed water are contained with the gel pore system, which makes the paste sensitive to water movements caused by temperature changes. Powers [14] showed one aspect of the significance of the moisture content of paste specimens, in which the expansion coefficients for oven-dry and saturated conditions were half the maximum expansion coefficient of 11×10^{-6}/°C that occurred at 75% relative humidity. A similar relationship was found for concrete, except that the dry and saturated condition values were from 65 to 80% of the maximum value, which occurred at about 60% relative humidity. Generally, the coefficient of expansion increases with decrease in the w/c.

Some sources report a slight increase in the coefficient of thermal expansion with age up to three months but a general decreasing trend thereafter. For concretes used in Ilha Solteira Dam in Brazil, the thermal expansion coefficient was found to increase significantly with age when the aggregate was all quartzite and less when only the fine aggregate was quartzite (Table 10).

For a given concrete mixture, the magnitude of thermal expansion or contraction at temperatures from freezing

TABLE 10—Effect of Age on Thermal Expansion[a] (Units Are Millionths/K) (Millionths/°F).

Approximate Age, days	Quartzite Concrete[b]	Basalt Concrete[c]	Units are millionths/K (°F) Quartzite Mortar[d]
4	12.9(7.14)	10.2(5.64)	12.2(6.75)
9	12.9(7.18)	10.2(5.67)	12.8(7.12)
17	13.2(7.34)	10.2(5.65)	14.0(7.75)
30	13.4(7.42)	10.7(5.96)	15.0(8.33)
48	13.5(7.51)	10.8(5.98)	15.6(8.69)
62	13.7(7.62)	10.6(5.89)	16.0(8.90)

[a] Results are from University of California (Berkeley) tests with Ilha Solteria (Brazil) materials.
[b] All natural quartzite aggregate.
[c] Basalt coarse aggregate and quartzite sand.
[d] Quartzite natural sand.

to about 65°C is the same for each unit temperature change, so the coefficient of thermal expansion is a constant figure. Below the freezing point of water, the length changes are smaller per unit of temperature, so the thermal expansion coefficient decreases. The slowing rate of contraction at low temperatures results in continued contraction of the concrete (and of the ice already formed) and an opposite expansion resulting from the formation of additional ice. Berwanger and Sarker [15] found the coefficient decreased from 7.7 to 5.1 millionths/°C for temperatures above and below 0°C, respectively, and Monfore and Lentz [16] showed a similar typical decrease from 9.4 to 6.8 millionths/°C. There is some evidence that the coefficient continues to decrease with further decreases in temperatures.

As stated previously, the expansion of concrete exposed to increasing temperature is the net result of the inherent thermal expansion property of the aggregate and a complex hygrothermal volume change of the cement paste. Up to about 100°C, the paste has achieved its natural expansion, and at higher temperatures starts to shrink [17], continuing to do so up to about 500 or 600°C. At this level, only the original dry ingredients remain. For calcareous aggregate concrete, Philleo [17] reported thermal expansion coefficient of 8.5 millionths/°C (4.7 millionths/°F) below 260°C (500°F), and 22.5 millionths/°C (12.5 millionths/°F) above 425°C (800°F). For expanded shale aggregate concrete, the values were 5.0 millionths/°C (2.8 millionths/°F) and 8.8 millionths/°C (4.9 millionths/°F) for similar temperature levels.

Over a number of years, the Bureau of Reclamation and Corps of Engineers have conducted thermal expansion coefficient tests of specimens composed of concrete mixtures that have moderate to low cement factors and large-size aggregates. The aggregates, from the actual job sources, are frequently complex in terms of type and mineralogical content; and the expansion coefficients do not lend themselves to unique groupings by rock type. Some typical results are listed in Table 11. The values therein are applicable for an average temperature of about 38°C.

Test Methods

The significant effect that moisture and temperature have on the behavior of concrete and the wide variety of

TABLE 11—Thermal Expansion Coefficients for Mass Concrete.

Dam Name	Aggregate Type	Coefficient of Thermal Expansion	
		millionths/K	millionths/°F
Hoover	limestone and granite	9.5	5.3
Hungry Horse	sandstone	11.2	6.2
Grand Coulee	basalt	7.9	4.4
Table Rock	limestone and chert	7.6	4.2
Greers Ferry	quartz	12.1	6.7
Dworshak	granite-gneiss	9.9	5.5
Libby	quartzite and argillite	11.0	6.1
Jupia (Brazil)	quartzite	13.6	7.5

environmental conditions to which concrete is exposed have precluded development of a standard test method for general use. Physical measurement of specimen length change with the necessary accuracy is not a major obstacle, but conditioning of the specimen and control of the conditions during the test may be difficult. Many test procedures, some quite sophisticated, have been developed in conjunction with research studies [8,17].

The Corps of Engineers [10] Test Method for Coefficient of Linear Thermal Expansion for Concrete (CRD-C 39) determines the thermal coefficient of concrete by measuring length changes of a specimen alternately immersed in 5 and 40°C (40 and 140°F) water baths. A length comparator similar to that described in ASTM Specification for Apparatus for Use in Measurement of Length Change of Hardened Cement Paste, Mortar, and Concrete (C 490) is used to measure length changes. When laboratory molded specimens are used, strain meters embedded along the longitudinal axes of 152- by 405-mm (6- by 16-in.) cylinders are used in lieu of the length comparator. Because the thermal coefficient of concrete varies with moisture condition, being a minimum when saturated or oven dry and a maximum at about 70% saturated, it is important to select the relevant moisture condition for the tests to be made.

The Bureau of Reclamation has a procedure (No. 4910) that can be found in Ref 13 for determining the linear thermal expansion for concrete.

HEAT OF HYDRATION

Definition and Units

When water is added to cement, the reaction is exothermic, and a considerable amount of heat is generated over an extended period of time. The heat liberated up to a specific time or age is measured in calories per unit weight of cement (cal/g) or kilojoules per kilogram (kJ/kg). When the amount of this heat and the heat capacity of the paste, mortar, or concrete are known, the resulting temperature rise can be calculated, assuming no heat loss to the surroundings.

Parameters and Values

The hydration reactions of portland cement have been studied and reported by many authors for many years. All the compounds present in cement are anhydrous and, when in contact with water, are all attacked at various rates and at varying times. There is general agreement that the tricalcium aluminate (C_3A) is the largest single compound contributor to the evolved heat, followed by tricalcium silicate (C_3S) and tetracalcium aluminoferrite (C_4AF) with about equal contributions, and finally dicalcium silicate (C_2S). Approximate contributions of the four principal calculated compounds in 0.40 w/c cement pastes cured at 21°C are given in Table 12. Although the data apparently indicates progressive increases in the total heats generated by each compound, values at intermediate ages for both C_3A and C_4AF deviate from this pattern by generating a higher proportion of heat after one year. Possible causes and significance of the regression characteristics are discussed in Ref 18. When water is first brought into contact with portland cement, there is a very rapid and very brief heat evolution with the evolution of heat reaching possibly a rate of 1 cal/g/min. There then occurs a 1- or 2-h delay, followed by a gradually increasing hydration rate to about 6 or 8 h, and a slow decline in rate thereafter. Relative proportions of the four major compounds, other minor compounds, alkalies, and gypsum content cause variations in both early- and later-age hydration rates. The nature and interrelationships of the chemical reactions are complex, variable, and sometimes inconsistent or controversial (the reader is referred to Lea [6] or other comparable texts for a comprehensive presentation on the chemistry of cements).

As with most chemical reactions, the rate of cement hydration increases with temperature. The accelerating effect of high curing temperatures is limited to early ages, and subsequently the hydration process slows to a rate less than that corresponding to normal placement and curing temperatures. Verbeck [19] suggests this slowing is the result of a dense hydration product created by high temperatures at early ages surrounding the cement grains, which thereby substantially retards subsequent hydration. This degradation in hydration is generally true for temperatures above 24°C (75°F) and becomes significant at 52°C (125°F) and above. At and below 0°C (32°F), a sharp drop in rate of hydration occurs, and between −10 and −15°C hydration ceases.

The w/c in pastes has a significant influence on the amount of heat generated at ages of three days and greater. The differences over the range 0.4 to 0.8 w/c are greatest at intermediate ages but continue to be evident at up to six years. According to Verbeck [19], the hydration product steadily increases in volume, filling the capillary void space in the paste. If the capillary void space is small (low w/c), the available space will become filled completely with hydration products and hydration of the remaining

TABLE 12—Typical Compound Contribution to Heat of Hydration [18].

Compound	Heat Evolved in cal/g		
	3 Day	1 Year	13 Year
C_3S	58	117	122
C_2S	12	54	59
C_3A	212	279	324
C_4AF	69	90	102

cement will cease. Thus, a high w/c will tend to result in more complete hydration and in more heat developed. Data reported by Verbeck [19] show that an increase in w/c from 0.4 to 0.8 had only a slight effect on heat liberation for Type IV (low-heat) cements but produced an increase of 10.8 cal/g, or 14%, for Type III (high-early) at three days of age. The increases in heat of hydration caused by an increase in w/c from 0.4 to 0.8 ranged from 11.4 cal/g for Type II (moderate-heat) cement to 15.9 cal/g for Type III at one year of age.

Prior to about 1935, the fineness of a cement was a major factor in the rate and amount of heat developed during hydration. The finer cements presented much more surface area to be wetted and, therefore, caused more rapid and complete hydration. In recent years, especially after World War II, all cements are so finely ground that the moderate variation in fineness of different cements is no longer an important factor in cement hydration. However, high-early strength cements are finer than other types to the degree that their extra fineness works with their more active chemical compounds to produce earlier strength and more rapid release of heat.

Pozzolans are defined in ASTM Specification for Fly Ash and Raw or Calcined Natural Pozzolan for Use as a Mineral Admixture in Portland Cement Concrete (C 618) as siliceous or siliceous and aluminous materials that in themselves possess little or no cementitious value but will, in finely divided form and in the presence of moisture, chemically react with calcium hydroxide at ordinary temperatures to form compounds possessing cementitious properties. Raw or calcined natural pozzolans (sometimes used to replace a portion of cement in mass concrete) vary widely in their composition but will reduce the total heat evolved by a value up to one half of the replacement percentage figure [20].

Fly ash, also classified as a pozzolan, may range in fineness before processing from below to well above that of portland cement. When used in concrete, a minimum fineness limit considerably above the portland cement value is imposed. The reaction between the glass in the fly ash and the lime in the cement is particularly sensitive to heat, and adiabatic curing of concrete containing a fly ash replacement will serve to increase substantially the chemical activity of the fly ash component.

In a study of portland blast-furnace slag cements, Klieger and Isberner [21] found no consistent difference in heat of hydration up to three days of age between Type I and Type IS cements, and, at later ages up to one year, the Type IS cements exhibited slightly to moderately lower heat generation characteristics.

The term, chemical admixtures, as is used here, refers to materials added in relatively small amounts to mortar or concrete during mixing to modify some characteristic of the product. Generally accelerating, retarding, and to some extent, water reducing admixtures will affect the rate of hydration of the cement, but the processes are not well delineated and will differ significantly with different cements and proprietary products. There is little if any evidence that air-entraining admixtures influence either the rate or the total amount of heat generated for a specific cement or cement blend.

Test Methods

The most widely used method for determining the heat of hydration of a hydraulic cement over long periods of time is by heat of solution, ASTM Test Method for Heat of Hydration of Hydraulic Cement (C 186). In this procedure, the heat of hydration of a hydraulic cement is determined by measuring the heat of solution of the dry cement and the heat of solution of a separate portion of the cement that has been partially hydrated for 7 and 28 days, the difference between these values being the heat of hydration for the respective hydrating period. The procedure requires several corrections and Lea [6] discusses a number of possible errors. Most significant to the user, other than laboratory techniques, is that the storage temperature is standardized at 23°C (73°F), which is usually not representative of the environment experienced in a concrete construction. A standard paste w/c of 0.40 is required, but this ratio will result in an understatement of the heat generated where higher ratios are encountered in construction. The procedure is not recommended for pozzolan blends nor for slag cements in which a portion of the test sample usually remains insoluble at the end of the test procedure.

The Carlson-Forbrich vane conduction calorimeter, developed prior to 1940, determined the amount of heat developed by a 0.40 w/c cement paste by measuring the rate of heat flow from the paste receptacle through cooling vanes into a water bath. Because of difficulties in controlling extraneous heat transfer, the test period usually was limited to three days. The method is sensitive to the initial starting temperature and is neither adiabatic nor isothermal.

Monfore and Ost [22] developed a refined calorimeter for measuring early rates of heat liberation that ideally is suited for laboratory research. In this calorimeter, the temperature rise of a sample of hydrating cement weighing up to 8 g can be held to less than 0.5°C, thus producing nearly isothermal conditions at any temperature level. The test usually is limited to three-day duration, but the results include the immediate heat of hydration (0 to 1 h). The method is well suited to determining the effects of additions and admixtures to the cements, as well as properties of specific cement compounds or other similar materials.

The Corps of Engineers [10] Method of Test for Temperature Rise in Concrete (CRD-C 38) covers a procedure for determining the temperature rise in concrete under adiabatic conditions primarily due to heat liberated on hydration of cement. The test specimen is a 762- by 762-mm (30- by 30-in.) cylinder that is sealed, insulated, and maintained in a temperature environment corresponding to its own temperature history for a period of 28 days. The advantages of this method are: (a) cement-pozzolan blends may be used, (b) the aggregates proposed for use in the prototype structure may be incorporated in the test, and (c) the temperature rise approximates the nearly adiabatic conditions existing in the interior of a massive concrete structure. A minor difficulty is maintaining an adequate insulation condition at the later stages when cement hydration rate is very low.

The Bureau of Reclamation uses Procedure No. 4911, for determining temperature rise of concrete; this procedure can be found in Ref 13. In this procedure, temperature rise tests are conducted on 318-kg (700-lb) concrete specimens sealed in 546- by 546-mm (21½- by 21½-in.) cylindrical metal containers and placed in a calorimeter chamber. The temperature of the air within the chamber is maintained at the same temperature as the specimen by automatic control equipment, which keeps an electrical resistance thermometer exposed to the chamber air in balance with a resistance thermometer in a well extending to the center of the concrete specimen. As a result, any heat generated by the cement or pozzolan or both results in a temperature rise in the concrete and a corresponding rise in chamber temperature.

SIGNIFICANCE OF THERMAL PROPERTIES

Heat Generation and Temperature Rise

The heat resulting from the exothermic water-cement reaction over a given time interval is expressed in terms of temperature rise when the specific heat value for the paste, mortar, or concrete mixture are known. While the range of values for specific heat of concrete is relatively narrow, the heat generation characteristics of the cementitious materials available for use in concrete are quite broad. When positive control over temperature rise is desired, upper limits on the heat of hydration of the cement may be imposed through ASTM Specification for Portland Cement (C 150) and ASTM Specification for Blended Hydraulic Cements (C 595). Specification limits for these cementitious materials and typical ranges for Types I and III not subject to specification restriction are indicated in Table 13, with values in calories per gram.

In lieu of established specification products, partial replacement of the cement in a mixture by natural or processed pozzolans or one of the several types of fly ashes will reduce significantly both the hydration rate and the total amount of heat generated.

The importance of heat generation and disposal of that heat has long been recognized by designers of large dams. Upon final completion, these structures should be monolithic and in intimate contact with the foundation and abutments in order to achieve the design stress distribution and stability. Cracks that disrupt the stress pattern and decrease stability are caused principally by thermal tensile stress created by concrete volume changes associated with a temperature decline as heat is dissipated. While several design and construction practices are available to deal with temperature changes that tend to occur, restricting the amount of heat that causes the temperature rise is a fundamental and certain scheme for mitigating the thermal stress problem.

Waugh and Rhodes [23] have listed experiences at several large mass concrete dams constructed by the Bureau of Reclamation, the Corps of Engineers, and the Tennessee Valley Authority. The favorable trend toward lower peak concrete temperatures has allowed the use of larger monolith lengths without serious consequences. The temperature rise at Glen Canyon and Dworshak Dams was limited to 14°C (25°F).

Biological shielding structures of concrete are widely used with nuclear reactors. Such shields are up to 2 m (6 ft) in thickness and generally require heavy aggregates that are responsible for cement contents up to 350 kg/m^3 (590 lb/yd^3). Peak temperatures in the concrete of as much as 36°C (65°F) above the initial placing temperature may be expected.

Pier footings and heavily reinforced raft foundation slabs up to 2 m thick and totaling 2000 m^3 of concrete with cement contents of 400 kg/m^3 (675 lb/yd^3) have been successfully placed with no cracking detected [5]. Peak interior concrete temperatures about 50°C (90°F) above initial concrete placing temperature required exterior insulation to control gradients and cooling rates for prevention of thermal cracking.

Heat Flow

The rate at which heat flows into, through, or out of a concrete structure is governed by the thermal conductivity of the concrete. The ease or difficulty with which the concrete undergoes temperature change, as a result of heat loss or gain, depends also on the heat capacity and is measured by the thermal diffusivity of the concrete.

Concrete in massive structures is usually placed in horizontal lifts from 0.75 to 2.3 m (2.5 to 7.5 ft) in thickness; the shallower lifts facilitate the loss of heat at locations where restraint is high. Control of thermal stresses by restricting lift thickness and placement frequency is especially effective for concrete that must be cast at a high initial temperature.

In exceptionally large concrete structures, principally arch and gravity dams, embedded pipe cooling is used frequently to remove much of the heat generated during hydration of the cement. Water is circulated through pipe coils placed at the bottom of each new lift of concrete. Spacing of the cooling pipes, initial temperature of the water, rate of water circulation, duration of the cooling operation, and capacity of the refrigeration-pumping plant are determined primarily from the thermal properties of the concrete, including heat generation characteristics of the cement.

The practice of precooling concrete materials so as to depress the initial temperature of the concrete at the time of placing is resorted to frequently in massive concrete construction. This practice reduces the peak temperature that subsequently will be attained. A maximum placing

TABLE 13—Specification Limits for Types I and II.

	7 Days	28 Days
Type I (C 150)	63 to 88	83 to 109
Type II (C 150)	70	80
Type III (C 150)	84 to 95	101 to 107
Type IV (C 150)	60	70
Type IS (C 595)	70	80
Type IP (C 595)	70	80
Type P (C 595)	60	70

temperature of 10°C (50°F), which is specified frequently for large concrete dams, is usually well below the ambient air temperature during the normal construction season. This low placing temperature reduces the rate of cement hydration initially, but during the period the concrete is at a temperature below air temperature, heat is gained from the surrounding air. However, this period of time is short as compared to the time the concrete is warmer than the surroundings.

Besides being important in the dissipation of internal heat, thermal properties of concrete contribute to damping ambient cyclic variations. Few buildings, residential or industrial, are designed without consideration for thermal insulation. The lightweight aggregate concrete and foamed or cellular concretes provide a measurable degree of insulation for the interior, in addition to their lower dead-weight structural advantage. In consideration of heat and energy costs, designers now utilize the damping effect, expressed by thermal diffusivity, of leveling out the cyclic air temperature variation. Normal-weight concrete ideally combines strength and durability with a favorable thermal diffusivity property that is most effective in reducing the heating and cooling requirements of the building interior.

Volume and Length Changes

All concrete elements and structures are subject to volume change [24]. These volume changes may be due to heat generation and subsequent cooling, shrinkage, creep or stress relaxation, or other mechanisms. When these changes are restrained, either by internal or external forces, tensile, compressive, or flexural strains will develop. The magnitude of these strains is determined primarily by the coefficient of expansion and the amount of temperature change. Cracking occurs when the magnitude of these strains exceeds the strain capacity of the concrete.

Restrained volume changes that cause tensile strains are a primary concern in mass concrete structures, particularly in the first few days after the placement of the concrete when the tensile capacity of the concrete can be quite low. External restraint is caused by bond or frictional forces between the concrete and the foundation or underlying lifts. The degree of external restraint depends upon the stiffness and strength of the concrete and restraining material and upon the geometry of the section. Internal restraint is caused by temperature gradients within the concrete. The warmer concrete in the interior of the lift provides restraint as the concrete in the periphery of the lift cools due to heat transfer to its surroundings. The degree of internal restraint depends upon the quantity of heat generated, the thermal properties of the concrete, and thermal boundary conditions.

The prediction of stresses, strains, and cracking in mass concrete at early times presents special problems, because many of the properties of concrete depend on the degree of hydration of the cementitious materials. The rate of hydration of the cementitious materials is affected by the type of materials used and by the temperature and moisture history during the period of hydration. At the same time, the internal environment of mass concrete is affected by the hydration of the cementitious materials. Elevated temperatures generated by hydration are maintained for long periods of time in the center of mass concrete structures and affect mechanical properties that are essential in determining the stress/strain condition of the concrete such as elastic modulus, compressive strength, creep, and volumetric changes associated with hydration.

The Corps of Engineers has adapted a general-purpose, heat transfer and structural analysis finite element program with an easily implemented user-defined material model to the problem of thermal analysis of mass concrete structures [25,26]. The heat-transfer capability of this program uses the finite element method to numerically solve the governing differential equation for heat transfer. Material properties necessary for input to the program are thermal conductivity, specific heat, a mathematical description of the applied heat flux (adiabatic temperature rise), and density. Calculations are carried out in time steps to model the incremental construction of mass concrete structures. The temperature-time history obtained in the heat-transfer analysis is used as the loading in a stress analysis. The effects of time-dependency and temperature on modulus, strength, and creep compliance along with an interactive cracking criteria for concrete are incorporated into the calculations with the user-supplied material model [27].

The thermal and incremental construction analysis procedures have been used to develop specific recommendations to limit thermal-related cracking during construction of mass concrete projects [24,28]. A number of parameters may be controlled to limit cracking caused by restrained volume changes. These parameters can be generally classified as either material or construction parameters. Concrete material parameters include (a) heat generation; (b) thermal properties (coefficient of expansion, specific heat, and conductivity); (c) mechanical properties (strength, modulus of elasticity, and tensile strain capacity); and (d) time-dependent deformations (shrinkage and creep or stress relaxation). Construction parameters include (a) lift height; (b) time between placement of lifts; (c) concrete placement temperature; (d) ambient temperature; (e) use of insulation; (f) use of cooling coils; and (g) monolith geometry (section thickness, monolith length, location and size of inclusions such as galleries, culverts, etc.).

In addition to new construction, thermal stress analysis procedures have also been used to address thermal cracking problems in rehabilitation of existing navigation lock walls [29,30]. Improvements in material properties and new design and construction procedures resulting from these studies have significantly reduced the extent of cracking in lock walls resurfaced with cast-in-place concrete.

For arch dams and similar massive structures containing joints that must be subsequently closed by grouting, the product of temperature drop and coefficient of expansion (or contraction) should be such that joint openings of at least 1 to 2 mm (0.04 to 0.08 in.) will occur. A minimum opening of this magnitude is required to permit successful and complete grouting.

The design of prestressed concrete reactor vessels should include consideration of stresses or strains originating from temperature gradients during operation as well as during shutdowns. Thermal conductivity at higher temperature levels and the resulting thermal expansion or contraction can be determined analytically for introduction into the stress and strain patterns.

The principal thermal property to be considered in rigid pavement design is coefficient of thermal expansion under the cyclic variations in ambient temperature. Spacing with the width of expansion joints is determined mainly by the amount of thermal expansion caused by the amplitude of the ambient temperature cycle and solar heating of the pavement slab.

Conventional building construction, principally rigid frame or continuous structures, warrant consideration of thermal expansion coefficients as they may affect either joints or stresses.

Thermal incompatibility is a primary concern in repair of concrete with polymer mortars. Depending on the type of polymer, the coefficient of expansion for unfilled polymers is 6 to 14 times higher than that for conventional portland cement concrete. Adding fillers or aggregate to polymers will improve this thermal mismatch, but the coefficient of expansion for the polymer-aggregate combinations will still be about 1.5 to 5 times more than that of concrete. The modulus of elasticity of the polymer and the thickness of the repair will also affect the magnitude of the stresses caused by thermal incompatibility. Thick repairs with stiff materials present the greatest potential for failure.

Winter Concreting and Insulation

Newly placed concrete must be maintained at temperature levels that will facilitate hydration and development of minimum strength requirements. Protection is required to prevent freezing of the uncombined free water in saturated new concrete, and accelerating the hydration process by the addition of calcium chloride or similar admixture will reduce the duration of protection required. The American Concrete Institute Recommended Practice for Cold Weather Concreting (ACI 306) [31] sets forth guidance for concrete placement at low temperatures. Surface insulation is also utilized to avoid development of steep thermal gradients that would result in large differential length changes, tensile strain, and possible development of cracks.

REFERENCES

[1] "Thermal Properties of Concrete," Boulder Canyon Project Final Reports, *Bulletin No. 1*, Part VII, U.S. Bureau of Reclamation, Denver, CO, 1940.

[2] Jakob, M. and Hawkins. G. A., *Elements of Heat Transfer and Insulation*, Wiley, New York, 1950.

[3] Lentz, A. E. and Monfore, G. E. "Thermal Conductivities of Portland Cement Paste, Aggregate and Concrete Down to Very Low Temperatures," *Bulletin 207*, Research Department, Portland Cement Association, Skokie, IL, 1966.

[4] Brewer, H. W., "General Relation of Heat Flow Factors to the Unit Weight of Concrete," *Journal*, Research and Development Laboratories, Portland Cement Association, Vol. 9, No. 1, Jan. 1967.

[5] Campbell-Allen, D. and Thorne, C. P., "The Thermal Conductivity of Concrete," *Magazine of Concrete Research*, Cement and Concrete Association, Wexham Springs, Slough, UK, Vol. 15, No. 43, March 1963 (and subsequent discussions).

[6] Lea, F. M., *The Chemistry of Cement and Concrete*, 3rd ed., Chemical Publishing Co., New York, 1970.

[7] Neville, A. M., *Properties of Concrete*, 2nd ed., Wiley, New York, 1973.

[8] Lentz, A. E. and Monfore, G. W., "Thermal Conductivity of Concrete at Very Low Temperatures," *Journal*, Research and Development Laboratories, Portland Cement Association, Vol. 7, No. 2, May 1965.

[9] *Temperature and Concrete*, SP-25, American Concrete Institute, Detroit, MI, 1971.

[10] *Handbook for Concrete and Cement*, U.S. Army Engineer Waterways Experiment Station, Vicksburg, MS, Aug. 1949 (with quarterly supplements).

[11] Mass Concrete, ACI 207.1R.87, *ACI Manual of Concrete Practice, Part I*, American Concrete Institute, Detroit, MI, 1993.

[12] Harmathy, T. Z. and Allen, L. W., "Thermal Properties of Selected Masonry Unit Concrete," *Journal*, American Concrete Institute; Proceedings, Vol. 70, No. 2, Feb. 1973, pp. 132–142.

[13] *Concrete Manual, Part 2*, 9th ed., U.S. Department of the Interior, Bureau of Reclamation, Denver, CO, 1992.

[14] Powers, T. C., "The Physical Structure and Engineering Properties of Concrete," *Bulletin 90*, Research Department, Portland Cement Association, Skokie, IL, July 1958.

[15] *Behavior of Concrete Under Temperature Extremes*, SP-39, American Concrete Institute, Detroit, MI, 1973.

[16] Monfore, G. E. and Lentz, A. E., "Physical Properties of Concrete at Very Low Temperatures," *Bulletin 145*, Research Department, Portland Cement Association, Skokie, IL, 1962.

[17] Philleo, R. E., "Some Physical Properties of Concrete at High Temperatures," *Bulletin 97*, Research Department, Portland Cement Association, Skokie, IL, Oct. 1958.

[18] Verbeck, G., "Energetics of the Hydration of Portland Cement," *Proceedings*, Fourth International Symposium on the Chemistry of Cement, National Bureau of Standards, Monograph 43, Vol. 1, Washington, DC, 1960.

[19] Verbeck, G., "Cement Hydration Reactions at Early Ages," *Journal*, Research and Development Laboratories, Portland Cement Association, Vol. 7, No. 3, Sept. 1965.

[20] "Investigation of Cement-Replacement Materials, Report No. 1," Corps of Engineers, Waterways Experiment Station, Vicksburg, MS, MP No. 6-123, April 1955.

[21] Klieger, P. and Isberner, A. W., "Laboratory Studies of Blended Cements-Portland Blast-Furnace Slag Cements," *Journal*, Research and Development Laboratories, Portland Cement Association, Vol. 9, No. 3, Sept. 1967.

[22] Monfore, G. E. and Ost, B., "An 'Isothermal' Conduction Calorimeter for Study of the Early Hydration Reactions of Portland Cements," *Journal*, Research and Development Laboratories, Portland Cement Association, Vol. 8, No. 2, May 1966.

[23] Waugh, W. R. and Rhodes, J. A., "Control of Cracking in Concrete Gravity Dams," *Journal*, Power Division, American Society of Civil Engineers, New York, PO 5, Oct. 1959.

[24] "Prediction of Creep, Shrinkage, and Temperature Effects in Concrete Structures," ACI 209R-82, *ACI Manual of Concrete Practice*, ACI Committee 209, American Concrete Institute, Detroit, MI, 1993.

[25] Bombich, A. A., Norman, C. D., and Jones, H. W., "Thermal Stress Analyses of Mississippi River Lock and Dams 26 (R)," Technical Report SL-87-21, U.S. Army Engineer Waterways Experiment Station, Vicksburg, MS, July 1987.

[26] Bombich, A. A., Garner, S., and Norman, C. D., "Evaluation of Parameters Affecting Thermal Stresses in Mass Concrete," Technical Report SL-91-2, U.S. Army Engineer Waterways Experiment Station, Vicksburg, MS, Jan. 1991.

[27] Garner, S. B. and Hammons, M. I., "Development and Implementation of Time-Dependent Cracking Material Model for Concrete," Technical Report SL-91-7, U.S. Army Engineer Waterways Experiment Station, Vicksburg, MS, April 1991.

[28] Garner, S., Hammons, M., and Bombich, A., "Red River Thermal Studies, Report 2, Thermal Stress Analysis," Technical Report SL-90-8, U.S. Army Engineer Waterways Experiment Station, Vicksburg, MS, Dec. 1991.

[29] Norman, C. D., Campbell, R. L., Sr., and Garner, S., "Analysis of Concrete Cracking in Lock Wall Resurfacing," Technical Report REMR-CS-15, U.S. Army Engineer Waterways Experiment Station, Vicksburg, MS, Aug. 1988.

[30] Hammons, M. I., Garner, S. B., and Smith, D. M., "Thermal Stress Analyses of Lock Wall, Dashields Locks, Ohio River," Technical Report SL-89-6, U.S. Army Engineer Waterways Experiment Station, Vicksburg, MS, June 1989.

[31] "Cold Weather Concreting (ACI 306R-88)," *ACI Manual of Concrete Practice, Part 2*, ACI Committee 306, American Concrete Institute, Detroit, MI, 1993.

Pore Structure and Permeability

Nataliya Hearn,[1] R. Douglas Hooton,[1] and Ronald H. Mills[1]

PREFACE

In previous editions of this manual, there was a chapter on pore structure written by George Verbeck. The increased awareness of the role of transport mechanisms on concrete durability has prompted the inclusion of permeability in the chapter title. The introduction and section on the summary of porosity have been taken from previous editions that were originally written by George Verbeck and revised by W. L. Dolch in *ASTM STP 169B*. The bulk of this chapter was adopted from Nataliya Hearn's Ph.D. thesis [1].

INTRODUCTION

The pores, or voids, in concrete consist of pores in the hardened cement paste, entrained or entrapped air voids, and voids in the pieces of aggregate. Other void spaces such as honeycombing, and bleeding channels and pockets are also important but are not treated in detail here, because they are the result of poor practice and are not inherent to properly prepared concrete. In addition, concretes deliberately designed to contain large, interconnected voids (such as no-fines or free-draining concretes) are not addressed in this chapter. The porosity of the aggregate is treated extensively in Chapter 38 of this publication.

Most of the important properties of hardened concrete are related to the quantity and the characteristics of the various types of pores in the concrete. The engineering properties, such as strength, durability, shrinkage, creep, permeability, and ionic diffusion are directly influenced or controlled by the relative amounts of the different types and sizes of pores.

The pores can exert their influence on the properties of the concrete in various ways. As regards the strength and elasticity of the concrete, it is primarily the total volume of the pores that is important, not their size or continuity. The permeability is influenced by the volume, size, and continuity of the pores. Shrinkage, at least that part of drying shrinkage that is reversible, is largely a function of changes in surface energy at the solid-to-pore interface and, therefore, depends upon the nature of the solid surface and the total surface area of the pore system. Irreversible drying shrinkage may involve capillary phenomena. The resistance of concrete to freezing and thawing and deicer scaling is controlled by the volume and the characteristics of air voids. Therefore, it is not surprising that there has been considerable interest in developing ways of measuring and characterizing the pore structure and permeability of concrete and in elucidating the various mechanisms by which they influence the properties of the concrete.

POROSITY OF CONCRETE

The pores formed in the original plastic concrete are either water- or gas-filled. After the concrete has hardened, the water-filled pores may tend to dry, or the air-filled pores may tend to become water saturated, depending upon the history of the concrete, the external moisture conditions, and the dimensions of the concrete member.

Hardened concrete that is properly proportioned, mixed, and placed consists of hardened cement paste, aggregate, and relatively small air voids. If these air voids are the result of the use of an added air-entraining agent, they are termed entrained air. If they are the result of the inevitable inability to effect complete consolidation of the plastic concrete, they are termed entrapped air.

The rate of mass transfer through a porous solid and its mechanical properties are influenced by its pore system. Figure 1 illustrates various possible pore systems. Cementitious materials are best represented by the high-porosity/low-permeability category.

The constituents of concrete, that is, hardened cement paste (HCP) and aggregates, are characterized by widely different porosities but similar permeabilities (Table 1). A very high-quality cement paste has a porosity of about 25% by volume, and when it is incorporated in concrete the overall porosity reduces to about 7% by volume. Corresponding figures for material of average quality are 50% for paste and 15% for concrete. Few normal aggregates have porosities greater then 5% by volume.

The initial porosity of concrete is determined by the sum of the volume of mixing water, intentionally entrained air, and accidental voids due to incomplete compaction (aggregate voids are discussed in Chapter 38). As the cement reacts with water, the new solids (cement hydrate) occupy space partly within the original grain boundary

---

[1] Assistant Professor, associate professor, and professor emeritus, respectively, Department of Civil Engineering, University of Toronto, Ontario, Canada M551A4.

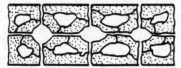 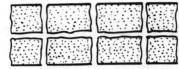

FIG. 1—Illustration of permeability and porosity (after Bakker [2]).

TABLE 1—Comparison of Permeability and Capillary Porosity Between Well-Hydrated Cement Paste and Natural Rocks [3].

Rock		Permeability Coefficient, m/s (rock and paste)	Cement Paste	
Type	Porosity, %		Porosity, %	w/c
Sandstone	4.3	1.7×10^{-11}	30	0.71
Limestone	3.1	8.0×10^{-13}	28	0.66
Fine-grained marble	1.8	3.3×10^{-14}	15	0.48
Dense trap	0.6	3.5×10^{-15}	6	0.38

and partly in the original water-filled space between cement grains and around aggregates. Thus, the porosity of concrete may be classified as follows:

1. Porosity of the aggregates.
2. Water- and air-filled voids after consolidation and final set.
3. Water- and air-filled voids after partial hydration of the cement.

Porosity of Components

Microstructure of Hardened Cement Paste

The discussion of pore structure of HCP would be incomplete without a brief review of the microstructure. Powers et al. [4] aptly noted that, "Knowledge of the structure of cement paste is essential to an understanding of almost all concrete properties of technical importance." The properties of concrete, such as strength, creep, shrinkage, and permeability, are dependent upon the microstructural behavior of the HCP. The understanding of this behavior enables the prediction of concrete's response to various environments. However, the actual microstructural makeup of the cement paste is not well understood, because its major constituent, the calcium silicate hydrate (C-S-H) gel, is almost an amorphous solid with varied chemical composition and generally unresolved morphology (Table 2). Three major competing schools of thought have evolved to explain concrete's behavior, through microstructural models of C-S-H: namely, Powers', Feldman and Sereda's, and the Munick model.

Powers' model

Structure—This model describes gel as a rigid substance made up of colloidal size particles, with constant porosity of 28% by volume (Fig. 2). The particles consist of several layers having a bond structure similar to that of clay. No long-range order between these particles exists. The particles have a large specific surface area, in the order of 200 m²/g, as determined by application of the Brunauer-Emmett-Teller (BET) method [6] using water vapor isotherms.

Water Classification—Water in the hardened cement paste (Fig. 3) is classified into nonevaporable (which is chemically bound water) and evaporable (which can be removed by drying at some arbitrary drying condition such as, P-drying, D-drying, or at 105°C [5]). Evaporable water includes water held in the capillaries and water adsorbed on the poorly crystalized layers of the gel.

Interparticle Bonds—The particles are held together by surface forces with sparse ionic-covalent bonds providing permanent links.

Shrinkage and Swelling Mechanism—The shrinkage and swelling in the hardened cement paste is attributed to the variation in the disjoining pressure (0 to 100% relative humidity (RH)) (Fig. 4), which is further augmented by the meniscus effects above 45% RH. In Fig. 4, the adsorption of water molecules (interlayer water) causes disjoining pressure between particles thus attenuating solid-

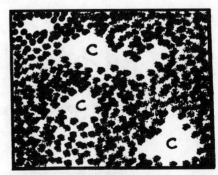

FIG. 2—Simplified model of paste structure. Solid dots represent gel particles; interstitial spaces are gel pores; spaces such as those marked C are capillary cavities [6].

TABLE 2—Summary of Properties of the Hydration Products of Portland Cement Compounds [5].

Compound	Specific Gravity	Crystallinity	Morphology in Pastes	Typical Crystal Dimensions in Pastes	Resolved by[a]
C-S-H	2.3 to 2.5[b]	very poor	spines; unresolved morphology low porosity striated material	1×0.1 μm (less than 0.01 μm)	SEM
Calcium hydroxide (CH)	2.24	very good	long slender prismatic needles	0.01 to 0.1 mm	OM, SEM
Ettringite	~1.75	good	thin hexagonal plates	10×0.5 μm	OM, SEM
Monosulfoaluminate	1.95	fair-good	irregular "rosettes"	$1 \times 1 \times 0.1$ μm	SEM

[a] OM = optical microscopy; SEM = scanning electron microscopy.
[b] Depends on water content.

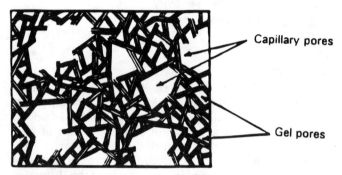

FIG. 3—Model of paste structure based on fibrous or platy gel [6].

to-solid attraction; reduction of RH relaxes disjoining pressure and the opposite effect-dilation-occurs when RH increases. Powers attributes irreversible shrinkage to the formation of new primary bonds following migration of adsorbed water. These are not subsequently broken by rewetting.

Creep Mechanism—Creep in the cement paste is modeled in a similar way to shrinkage and swelling. The applied load creates an energy gradient causing migration of the adsorbed water from the hindered adsorption areas (refer to Fig. 4) by a time-dependent diffusion process. As the water is squeezed out, solid separation is reduced, resulting in time-dependent deformation known as creep. Irreversible creep is due to the formation of primary bonds between surfaces when they are brought close together by the combined effect of mechanically induced strain and migration of adsorbed water.

Basis for the Model—The model is based on a classification of evaporable pore water according to whether it sustains a meniscus ($p/p_s > 0.45$) or exists as discrete molecules ($p/p_s < 0.45$).

Criticism of the Model—The major criticism of the model is its interpretation of water-sorption isotherms. It is believed by some (Feldman, Sereda, and others) that the interaction of water vapor with C-S-H in adsorption experiments is so strong that the cement paste is unstable; and that the surface areas thus obtained are too high, because both the internal and the external surface areas of the gel layers are measured. The arbitrary classification of water is also disputed, because D-drying or P-drying (and certainly drying at 105°C) is said to remove some of the structural (chemically bound) water. Even with these possible shortcomings, Powers' model has survived over 30 years of criticism and is still the only model that can quantitatively calculate volumetric composition of HCP [9].

Feldman and Sereda's model

Structure—This model [10] consists of irregularly arranged, layered, crumpled sheets two to four water molecules thick (Fig. 5).

Water Classification—In addition to capillary water, two types of water in the gel are identified: interlayer, hydrate water (zeolitic) that may be regarded as a structural and chemical component, and physically adsorbed water. The pore volume of the gel is equal to the difference between the volume of evaporable water and the interlayer water. The classification of the various types of water is done by examining scanning adsorption isotherms (Fig. 6). The

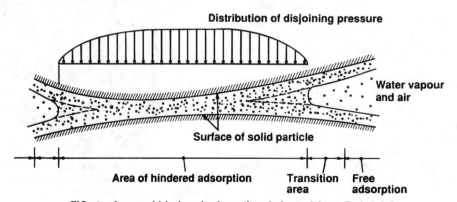

FIG. 4—Areas of hindered adsorption (adapted from Ref 7) [8].

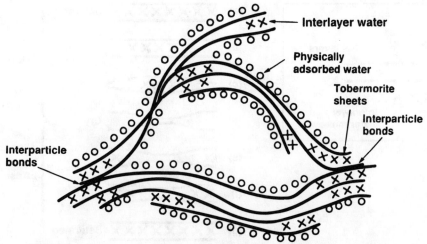

FIG. 5—Feldman and Sereda model of hydrated portland cement [11].

movement of water into and out of the C-S-H layers is considered reversible.

Interparticle Bonds—The interparticle bonds result from solid contact caused by the bridging together of surfaces. There is no atomic regularity in these bonds, and they can be broken and remade. Their strength lies between the strength of the weak van der Waals' bond and the strong ionic-covalent bonds.

Shrinkage and Swelling Mechanism—Shrinkage and swelling is attributed to the movement of the interlayer water (Fig. 7). On drying between 100 to 35% RH, variation in surface energy and capillary tension causes shrinkage. However, most of the shrinkage occurs below 35% RH and is due to the removal of the interlayer water. On rewetting, the same mechanisms are at work, with 80% of expansion attributed to the re-entry of the interlayer water [10].

Creep Mechanism—Initially, creep was attributed, as in the case of shrinkage, to the displacement of the interlayer water due to the externally applied load. In later studies, however, a revised theory was put forward, namely, that: "creep is a manifestation of the gradual crystallisation or aging process of the layered material, resulting in further layering. Water movement, although occurring, is not the major mechanism. Other processes, such as slippage and micro-cracking are also present [12]."

Basis for the Model—The model recognized the layered structure of C-S-H and is based on adsorption isotherms obtained using nitrogen (N_2) as the adsorbate, and the reversible movement of the interlayer water.

Criticism of the Model—The model is not quantitative so that calculations of the gel volumetric properties are indeterminate. As the structure of C-S-H is uncertain, there is still doubt which sorption isotherm, water-vapor or N_2, is more representative. There was some evidence that C-S-H particles are not layered in structure [13] but the concept of a layered structure is generally recognized. Some experiments have shown (Soroka [8] after Powers and Brownyad [7]) that: "only 10 to 20 percent of the total deformation of individual particles caused by a change of the amount of interlayer water can be measured as a volume change in the structure as a whole."

However, this model has received support from many including Taylor [14].

The Munich model

Structure—The Munich model, developed by Wittman [15,16], is a physical model describing C-S-H as, "a three dimensional network of amorphous gel particles forming a xerogel" [17].

Water Classification—No specific classification of water is made, except that water is considered to be strongly attracted to solid surfaces (so that all water is adsorbed water), and that its presence influences all the essential properties of the HCP (Fig. 8).

Interparticle Bonds—The predominant bonds are ionic-covalent, with van der Waals' forces playing an important role, especially at the lower RH levels, when the particles of the xerogel are close together.

Shrinkage and Swelling Mechanism—Shrinkage and swelling [18] is considered to be due to three mechanisms of water-solid interaction (Fig. 9):

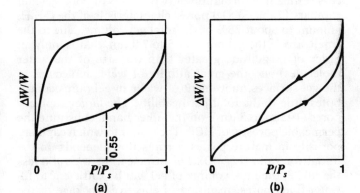

FIG. 6—Sorption isotherms for (A) interlayer water and (B) adsorbed water [8].

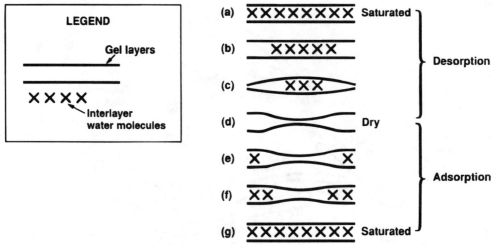

FIG. 7—Simplified model for exit and entry of water into the layers of the C-S-H gel [11].

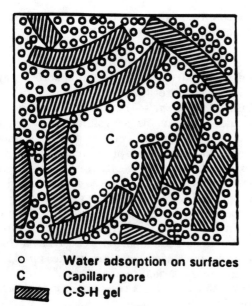

FIG. 8—Schematic representation of C-S-H by Munich model [5].

1. change of surface free energy,
2. disjoining pressure, and
3. capillary pressure.

Creep Mechanism—The creep mechanism is not directly related to the gel water. However, changes in the RH affect the magnitude of creep by weakening particle interaction at lower RH. At higher RH, an increase in the number of xerogel particles taking part in the creep process (creep centers) results in a high increase in creep deformation [17].

Basis for the Model—The basis for this model is the study of the surface free energy of the C-S-H gel and the disjoining pressures between the particles.

Criticism of the Model—The model is still fairly new, and more work needs to be done to bridge the gap between the observed microstructure of C-S-H and the mechanical behavior of the HCP.

Prediction of the behavior of concrete in terms of the microstructure of CSH remains a matter of considerable controversy. However, what is beyond dispute is that water held at low vapor pressures ($p/p_s < 0.45$) is extremely interactive with the solid phase and with externally applied mechanical load. Because this water is so strongly attracted to the solid, it may be regarded as a fluid of such high viscosity that the space it occupies does not contribute to permeable porosity.

Pores in HCP

For practical purposes, the porosity of HCP depends on the initial w/c ratio, the degree of compaction, the degree of hydration, and the presence of supplementary cementing materials (mineral admixtures). The w/c ratio determines the initial porosity, while the degree of hydration determines the extent to which the original pores become filled with new solid. When water is mixed with cement, the clinker grains begin to hydrate. According to Powers [19], the hydrate is approximately 1.6 times the absolute volume of its constituents or 2.1 times the bulk volume that includes pore volume of the C-S-H. This process results in a formation of two classes of pore: gel and capillary (Table 3). Gel pores, characteristic of the C-S-H, amount to about 28% of the total gel volume. Due to the small size of the gel pores (1.5 to 2.0 nm, that is, only an order of magnitude greater than the size of the water molecules) and the great affinity of water molecules to the gel surfaces, movement of water in gel pores contributes little to the total permeability. The larger capillary pores (10 nm to 5 μm), on the other hand, constitute the permeable porosity of HCP. The extent to which capillary porosity in mature pastes permits flow depends on the initial w/c ratio (Fig. 10). For instance, for w/c ratio less than 0.38, the bulk volume of gel will be sufficient to fill water-filled pores, resulting in effective blockage of the capillaries. While for initial w/c ratios above 0.7, even full hydration would not result in total filling and segmenta-

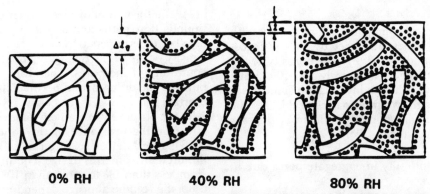

FIG. 9—Simplified model of the three-dimensional network of amorphous C-S-H particles. In the dry state (0% RH), all particles of the xerogel are compressed by surface tension. At 40% RH, adsorbed water films reduce surface tension and as a result all particles expand. This leads to macroscopic expansion, Δl_0. At higher RH, disjoining pressure of the adsorbed water separates some contact points in the xerogel and hence further swelling occurs, that is, Δl_π [17].

TABLE 3—Size Distribution of the Various Pores (Adapted From Ref 20).

Pore Type	Size Range, nm
Interparticle spacing between C-S-H sheets	1 to 3
Capillary voids	10 to 1000
Entrained air bubbles	1×10^3 to 5×10^4
Entrapped air voids	1 to 3×10^6

tion of the capillaries (Tables 4 and 5). The segmentation of pores implies impermeability even in concretes of high w/c ratios. The results in Table 6 are relevant to well-cured laboratory specimens; structural or mass concrete, even at lower w/c, may not attain the levels of hydration at which segmentation of capillaries occurs. However, recent advances in mix design and increased attention to curing are likely to produce segmentation in field concrete as well as laboratory specimens.

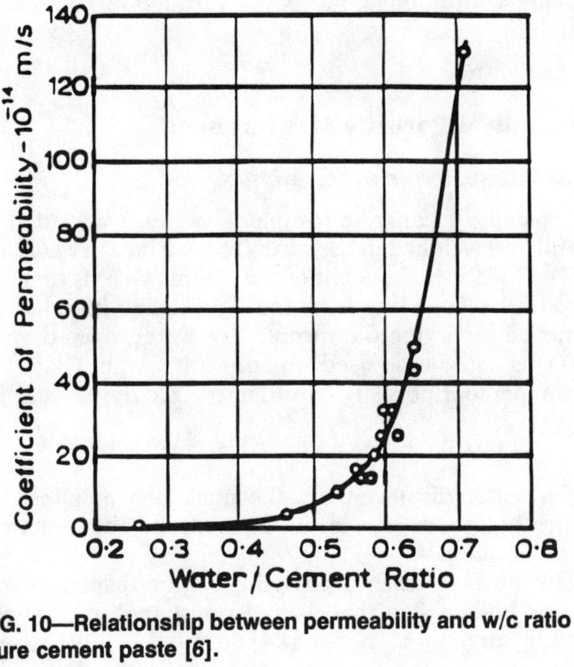

TABLE 4—Reduction of Permeability of Cement Paste (w/c Ratio = 0.7) With the Progress of Hydration [21].

Age, days	Coefficient of Permeability, K (m/s)
Fresh	2×10^{-6}
5	2×10^{-10}
6	2×10^{-10}
8	2×10^{-11}
13	2×10^{-12}
24	2×10^{-12}
Ultimate	2×10^{-13} (calculated)

TABLE 5—Effect of Age of Cement Paste on Permeability Coefficient (w/c = 0.51) [5].

Age, days	K_p, m/s
Fresh paste	10^{-5} independent of w/c
1	10^{-8}
3	10^{-9}
4	10^{-10} capillary pores
7	10^{-11} interconnected
14	10^{-12}
28	10^{-13}
100	10^{-16} capillary pores
200 (maximum hydration)	10^{-18} discontinuous

FIG. 10—Relationship between permeability and w/c ratio for mature cement paste [6].

TABLE 6—Approximate Age Required to Produce Maturity at Which Capillaries Become Segmented [6].

Water/Cement Ratio by Mass	Time Required
0.40	3 days
0.45	7 days
0.50	14 days
0.60	6 months
0.70	1 year
Over 0.70	impossible

Permeability of the Aggregate

The pore system of most aggregates, which is dealt with in detail in Chapter 38 is best described by the low porosity, low permeability model of Fig. 1. The porosity of the aggregates is usually under 3% and rarely over 10%, however their permeability, although low, may in some cases approximate that of high w/c ratio pastes (Table 7). The relatively high permeability/porosity ratio in aggregates relative to cement pastes results from:

1. a high degree of continuity in aggregate pores due to jointing and fissures,
2. a significant fraction of impermeable porosity and disconnected pores in HCP, and
3. a coarser pore distribution in aggregates.

Permeability of the Paste-Aggregate Interface

Because of the water films that surround aggregate particles during mixing, the paste within about 20 μm from the aggregate surface has an effectively higher w/c. This porous zone can be augmented by trapped bleed water. Also, calcium hydroxide and ettringite form at early ages in this zone and result in a weak, porous layer in the hardened concrete. In the presence of supplementary cementing materials (mineral admixtures) such as fly ash, blast furnace slag, or silica fume, the thickness of this porous interfacial zone tends to decrease through reaction with the calcium hydroxide with the supplementary materials to form C-S-H. This is thought to be responsible for higher strength and lower permeability of such concretes.

Mixing-and-Placing Porosity

Mixing-and-placing porosity either is purposely introduced to improve concrete's performance, such as air-entrainment, or stems from faulty practices allowing excessive bleeding, or voids such as entrapped air or honeycombing due to incomplete compaction. Adequate workability is the key to reduction of unintentional voids. Workability can be controlled by quality and characteristics of aggregate, w/c ratio, and judicious use of admixtures. Table 3 compares the sizes of the various pore types, indicating the possible contribution of the accidental flaws to increased permeability. Potentially, entrained air, entrapped air, or honeycombing (large irregular voids due to poor compaction) are less damaging than bleeding, as they are isolated and do not form continuous flow paths in the cement matrix. Bleeding has more serious effects [22]. The formation of bleed water channels creates continuous flow paths and as the upward movement of the bleed water is impeded by aggregate particles, zones of low density occur below the aggregates. Both effects enhance continuity of the pores. The latter, in particular, contributes to the difference in permeability between concrete, mortar, and pure cement paste.

Those purposely entrained voids that have a significant effect on the resistance of the concrete to freezing and thawing and deicer scaling range from a few to several hundred micrometres in size. The air voids may constitute from less than 1% to more than 10% of the concrete volume, the volume and size depending upon several factors including the amount of air-entraining agent used, size distribution of fine aggregate, concrete consistency, duration of mixing, and so on [23].

The small air voids having a significant effect on concrete durability can be seen and studied using a microscope at a magnification of approximately 50 to 150 diameters. The volume of the air voids in hardened concrete and the characteristics of the air-void system (void size, voids per inch of traverse, and spacing factor) can be determined in accordance with ASTM Practice for Microscopical Determination of Air-Void Content and Parameters of the Air-Void System in Hardened Concrete (C 457).

Porosity in Hardened Concrete

Porosity occurring due to the exposure of concrete to curing conditions and to the environment is in the form of continuous channels due to a lack of hydration or due to microcracking in the HCP and in the mortar surrounding the coarse aggregate. Microcracking can result from drying shrinkage, carbonation shrinkage, thermal shrinkage, and externally applied loads. The cracks are larger than most capillary cavities (Table 3) and generally provide continuous flow paths throughout the cement matrix.

Methods of Porosity Measurement

Liquid Displacement Techniques

Typically, a concrete sample is vacuum-saturated in a liquid and weighed in the saturated, surface dry condition (Ws). The buoyant weight of the sample (Wb) is then measured by suspending it in the liquid. The liquid is then removed by drying the sample (for water, oven drying at 110°C is commonly used) to constant weight (Wd).

The percent porosity can then be calculated as follows

$$\text{Porosity} = (Ws - Wd)/(Ws - Wb) \times 100\%$$

For water displacement, the mass of water lost from saturation to drying is an estimate of the evaporable water content.

The porosity measured using water displacement is always higher than that using isopropanol, methanol, or even by mercury intrusion [24] or initial helium displace-

TABLE 7—Comparison Between Permeabilities of Rocks and Cement Pastes [6].

Type of Rock	Coefficient of Permeability, m/s	Water/Cement Ratio of Mature Paste of the Same Permeability
Dense trap	2.47×10^{-14}	0.38
Quartz diorite	8.24×10^{-14}	0.42
Marble	2.39×10^{-13}	0.48
Marble	5.77×10^{-12}	0.66
Granite	5.35×10^{-11}	0.70
Sandstone	1.23×10^{-10}	0.71

ment. This indicates that water can enter into space that other fluids cannot and this is likely the interlayer space described by Feldman and Sereda [10].

Mercury Intrusion Porosimetry

The early use of mercury intrusion porosimetry (MIP) for cementitious systems was reported by Winslow and Diamond [25]. Because of the ease of test and the wide range of effective pore sizes that can be measured (200 μm to 2 nm), this test has become very widely used to study the pore structure of concrete. Unfortunately, it suffers from a number of problems when applied to the complex pore structure in cementitious systems.

The method of sample drying can influence pore-size distributions [24] as can sample-size effects [26] and selection of contact angle [27]. The biggest problem is with the assumption that the voids are conical in shape and decrease in diameter from the surface to the core. The presence of so-called ink bottle pores, which HCP almost certainly has, will not be correctly measured in the recorded pore-size distribution.

Supplementary cementing materials such as slag or fly ash, which often hydrate after the initial pore structure is established, were postulated by Bakker [2] to result in blockages of capillary pores without a huge reduction of total porosity. These would certainly be examples of ink bottle pores. Feldman [28] using double intrusions of mercury showed that this indeed was the case with blended cement pastes where, unlike portland cement pastes, the mercury forced its way through these blockages resulting in damage to the pore system.

A number of problems with MIP and other techniques are reviewed by Diamond [29] and Feldman [30]. Interesting insights into new interpretations of MIP data are given by Winslow [31].

Other Methods

Other methods, which are not discussed in detail, include capillary condensation, BET nitrogen or water vapor adsorption, helium inflow, image analysis, low-angle X-ray scattering, and nuclear magnetic resonance. These methods have been reviewed by Diamond [29] and by Feldman [30] and with the exception of image analysis are largely research techniques.

The problem with using image analysis to automate ASTM C 457 has been in distinguishing between entrained air voids in the paste, and cracks as well as voids in the aggregates. Better sample preparation techniques and shape factor analysis, which would automatically separate out the spherical entrained air voids, are needed before ASTM C 457 can be fully automated.

Summary of Porosity

In summary, concrete can be visualized as consisting of a heterogeneous mixture of components, each component having its own characteristic pores. In terms of the other pores in the concrete, the air voids, normally the coarsest of all, may constitute from less than 1 to more than 10% of the total volume of the concrete. Approximately 75% of the concrete is aggregate, frequently heterogeneous, with an internal pore volume varying from almost 0 to 20% or more (most commonly about 1 to 5%), the pores ranging from relatively fine to coarse. The cement paste component usually contains both extremely fine gel pores and the coarser but submicroscopic capillary spaces [32].

Although these various pores and voids in concrete influence the physical properties of water contained therein, the properties of the separate types of pores are not sufficiently different to permit their complete identification in concrete. The combined porosity of concrete (including the air voids) can be represented by the total capacity for evaporable water between the stages of complete saturation and dryness—dry except for the combined or nonevaporable water content of the cement hydrate [32].

MASS TRANSFER IN CONCRETE

Material characteristics that govern the mass transfer of fluids include pore structure, moisture content, and boundary conditions. In concrete, the movement of water is controlled by cracking and the properties of the hardened cement paste (HCP) because in most cases, the aggregates are, by comparison, impervious. Even though aggregates are less permeable than HCP, the cement aggregate interface is generally more permeable than either constituent. Therefore, the inclusion of aggregate effectively increases the permeability of HCP. In concrete, the natural porosity of the paste (as determined by the initial w/c ratio) is augmented by the voids, cracks, and low-density or so-called "transition zones" at the aggregate-paste interface. Moreover, the movement of water in HCP is complicated by changes in its pore structure due to continued hydration and changing solubility of its constituents.

The key to this variation in the pore structure is the presence of water, which is often divided into evaporable (W_e) and nonevaporable (W_n). In the examination of mass transfer of water, only the W_e component is significant. The W_e is divided into adsorbed and interlayer water (W_{gel}) and mobile (W_m) (capillary condensed) water (Fig. 11). The way in which water is held depends on the pore structure. Figure 12 shows the state of water at various levels of relative vapor pressure for normal cured (high surface area) and steam cured (low surface area) HCP. Changes in the boundary conditions, such as pressure, temperature, or chemical potential will affect the state of water held in the HCP. The adjustment to such changes, in order to restore thermodynamic equilibrium, is controlled through humidity flux, which includes adsorption, surface diffusion, vapor diffusion, and bulk flow (Fig. 13).

Adsorption

In the Powers [19] model, adsorption occurs at all vapor pressures but principally at $p/p_s < 0.45$. As menisci cannot exist at $p/p_s < 0.45$ [19], the water held by the porous structure below this pressure is all adsorbed water that is attracted to the solid at different energy levels. Depending on surface energy of the solid and relative vapor pressure, adsorption is not limited to a single mono-layer, and fol-

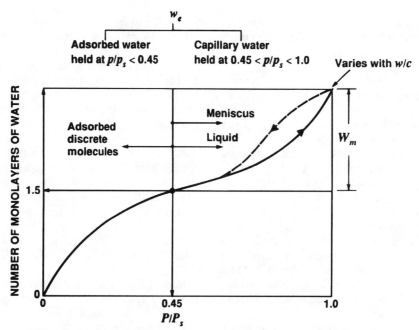

FIG. 11—Distribution of evaporable water in HCP (adapted from Ref 7).

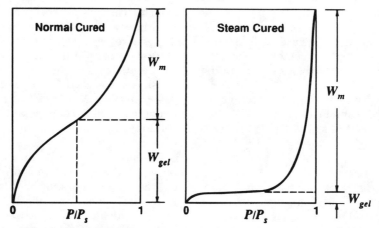

FIG 12—The effect of pore structure on the amounts of W_m and W_{gel} (adapted from Ref 7).

lows the BET model of multi-layer adsorption, with sequential layers having weaker ties with the adsorbate (Fig. 14).

The surface area of hydrate is variously given as 210 m^2/g by water vapor adsorption, 50 m^2/g by nitrogen adsorption, and 750 m^2/g by small-angle X-ray scattering [34]. At low relative vapor pressures, mass transfer of water vapor occurs by molecular migration rather than by coherent flow [35]. The hydrates with high surface energy, high surface area, and low porosity capture the traveling molecules. Peer [36] has shown that pores with separation distances of 10^{-7} m and shape factor (length/diameter) of 2 cannot contribute significantly to mass transfer of water through cement paste. The transport of discrete molecules in such restricted space takes place by surface diffusion.

Surface Diffusion

The mass transfer through porous media of highly adsorbing gases has been shown [37] to be significantly greater than that of nonadsorbing ones. This excess flux has been attributed to surface flow of the adsorbed molecules also known as film transfer. Water vapor molecules are highly adsorbing because of polarity, and the hydrophillic nature of HCP. The adsorbed molecules, no longer in a gaseous state, are so strongly attracted to the bounding solid that they have a density of about 0.9 and assume the characteristics of ice IV, a fluid of very high viscosity [19]. The movement of the surface layers is achieved by satisfying thermodynamic equilibrium of the surrounding surfaces, so that the loosely bound top molecular layers of the condensate may slide over the underlying layer, thus

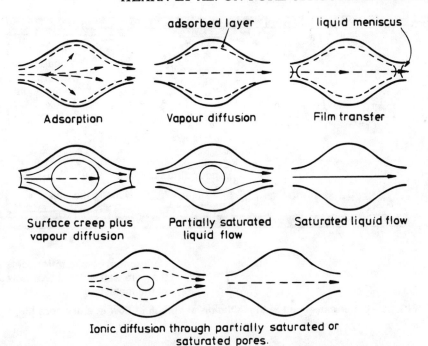

FIG. 13—Idealized model of movement of water and ions within concrete [33] adapted from [38].

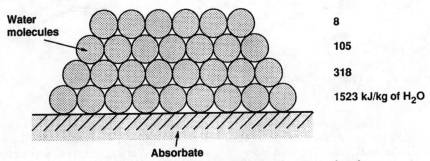

FIG. 14—Energy of surface adsorption of water molecules.

wetting bounding surfaces and hence causing mass transfer (Fig. 15). Change in the boundary conditions, such as an increase of vapor pressure, will result in the buildup of adsorbate layers until capillaries begin to fill and vapor diffusion and subsequently bulk flow become predominant transport mechanisms.

Vapor Diffusion

Vapor diffusion of water occurs in accordance with Fick's law

$$Q = D_p \frac{dp}{dx} \quad (1)$$

where

Q = mass transport rate (g/m²s),
D_p = diffusion coefficient (m²/s), and
dp/dx = vapor pressure potential gradient (m/m).

Vapor diffusion plays a considerable part in the overall moisture flux, as the actual transfer of a particular molecule from Point A to Point B is not necessary for moisture movement from A to B.

Rose [38] and Mills [35,39] contended that the continuous film of moisture in a partially saturated sample of HCP constituted a short circuit allowing rapid moisture transfer without coherent flow. The concept envisages a starting condition where the atmosphere on the upstream side, the menisci in pores of the solid, and the atmosphere on the downstream side are in quasi-equilibrium. This means that the number of water molecules condensing onto the meniscus surface is exactly equal to the number that will evaporate from that surface (Fig. 16). Suppose that the relative humidity of the atmosphere upstream is increased while that downstream remains static. The hydrostatic tension of condensed water then reduces and rapidly creates disequilibrium downstream with the result that water molecules are ejected in order to equalize hydraulic tension. If Powers [19] is correct in his statement

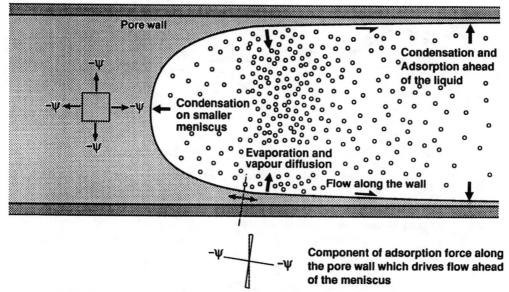

FIG. 15—Water movement at the initiation of bulk fluid flow in a dry pore [36].

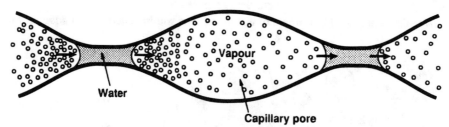

FIG. 16—"Short circuit" mass transfer by vapor diffusion.

that hydraulic tension is equal throughout as soon as a meniscus forms, then the actual linear dimension of the flow path is not as significant as it is for Darcian flow.

Bulk Flow

Bulk flow describes movement of liquid through an unsaturated or saturated matrix, or both. In unsaturated material with one end exposed to water and the other to the atmosphere, the bulk of moisture flux will be transmitted by capillary tension (absorption). Even in the case of penetration flow, where water is supplied under pressure into an unsaturated matrix, capillary tension is still the main transport mechanism. In concrete, where pore sizes are small, ranging by three orders of magnitude from 1 to 10^3 nm, the effect of external pressure on flow is negligible

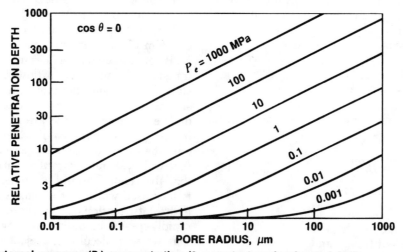

FIG. 17—Influence of external pressure (P_e) on penetration (θ = contact angle of water) [40].

compared to the flow due to capillary tension (Fig. 17 —showing the ratio between penetration of water with external pressure and without external pressure as a function of capillary radius [40]). Capillary movement relies on the difference in pressure over the upstream and downstream surfaces of the menisci, in concrete pores, to transfer water mass. The curvature of the meniscus depends on the vapor pressure, and becomes zero once saturation level is reached. At this stage, external pressure, such as a head of water, is required to drive water through a saturated matrix and Darcian flow governs.

In unsaturated concrete, initial water absorption is linearly related to the square root of time ($\sqrt{t}$), with the slope of the line defining sorptivity of the material. The mathematical expressions describing this process vary widely in the literature. Table 8 gives examples of the equations used to define absorption. This variation is largely due to the premises of the derivations, which either stem from the extension of Darcy's permeability to unsaturated flow, or from the physics of the capillary rise.

In saturated concrete, Darcy's permeability describes the steady flow of water through pores

$$\frac{dq}{dt}\frac{1}{A} = k\frac{\Delta h}{L} \qquad (2)$$

where

$\frac{dq}{dt}$ = rate of flow(m³/s),
A = cross sectional area of the sample (m²),
k = coefficient of permeability (m/s),
Δh = drop in hydraulic head across the sample (m), and
L = thickness of the specimen (m).

Darcy's equation [47] of flow was developed for water flow through sand. It defines permeability (k) as the velocity of fluid per unit hydraulic gradient. This equation implies that a linear relationship exists between flow velocity and hydraulic gradient. This linearity has been found to hold not only for soils, but for porous media in general, including concrete with cement at stated hydration levels. Ruettgers [48] confirmed this linear relationship for concrete by measuring flow rates over a wide range of applied pressures. Powers [5] clearly demonstrated the validity of Darcy's law for hardened cement paste for various samples at several temperature levels (0.2 to 27°C).

Summary

Three stages of moisture front propagation in concrete can be defined:

TABLE 8—Examples of Mathematical Equations (Adapted from Emerson [41]).

Authors	Equation		Notation
Bamforth et al. [42]	$x = \frac{r}{2}\left(\frac{P_0 t}{\eta}\right)^{1/2}$	x	= distanced traveled
		r	= radius (mean) of capillaries
		P_0	= driving pressure
		t	= time
		η	= viscosity of water
	$K = \frac{r^2 v \rho g}{8\eta}$	K	= permeability
		r	= pore radius
		v	= porosity
		ρ	= density
		g	= acceleration due to gravity
		η	= viscosity
	$x = \sqrt{\frac{2kT(P_1 - P_2)}{v\rho g}}$	x	= penetration distance
		k	= permeability
		T	= penetration time
		$(P_1 - P_2)$	= pressure difference
		v	= porosity
		ρ	= acceleration due to gravity
Fagerlund [43]	$t = mz^2$	t	= time
		m	= resistance to water penetration
		z	= depth of penetration
Hall [44]	$i = St^{1/2}$	i	= cumulative absorption
		S	= sorptivity
		t	= time
Ho and Lewis [45]	$i = St^{1/2}$	i	= volume of water absorbed per unit area of inflow surface
		S	= sorptivity
		t	= time
Vuorinen [46]	$x = (2Kht)^{1/2}$	x	= depth of penetration
		K	= coefficient of permeability
		h	= hydraulic head
		t	= time

1. for partial vapor pressures below 0.45, before a meniscus is formed, the moisture movement is controlled by adsorption and surface diffusion;
2. for partial vapor pressures between 0.45 and close to 1, moisture transfer is achieved through vapor diffusion and capillary tension; and
3. in saturated or nearly-saturated material, moisture transfer is mainly due to laminar flow, controlled by viscosity and defined by Darcy's law.

Flow into unsaturated concrete may be supported by all of the preceding mechanisms. The complexity of the pore structure of concrete, its variation with mix proportions, curing and conditioning, complicates numerical analysis even when only one flow mechanism is examined. Moreover, the presence or introduction of water into the concrete's pore system complicates analysis even further, because processes such as swelling, continued hydration, and dissolution and precipitation of alkalies result.

Methods of measuring multi-phase moisture flux in concrete suffer from variability or simplistic modeling divorced from the real environment. The following sections discuss procedures of measuring mass transfer in concrete and their advantages or disadvantages over direct measurement of saturated flow.

PERMEABILITY TESTING AND STANDARDS

Gas

Gas Flow

Pressure-induced gas flow through a sample appears to follow Darcy's law, and the porous solid is often characterized by the intrinsic permeability coefficient, K.

$$\frac{dq}{dt} = KA \frac{\rho g}{\mu} \frac{dh}{dL} \tag{3}$$

where

ρ = density of the permeant (kg/m^3),
g = acceleration due to gravity (9.81 m/s^2),
μ = viscosity of the permeant (Pa·s), and
K = intrinsic permeability coefficient (m^2).

Equation 3 incorporates Carman's findings (which show an inverse relationship between the rate of flow and viscosity of the permeant) with Darcy's equation. The intrinsic permeability coefficient thus depends only on the microstructural properties of the porous medium and is supposed to be independent of the properties of the permeant.

The test can be rapid and reproducible, but only if the moisture condition of the concrete sample is known and carefully controlled. As the moisture content may vary from surface dry to nearly saturated, so will intrinsic permeability, because water is an effective barrier to gas flow. Estimates of the permeable porosity at various moisture levels can be made by estimating the wetted surface area (that is, the surface area as reduced by vapor adsorption) and calculating the hydraulic mean radius. The moisture level, however, must be constant throughout the specimen. Conditioning of the test samples to a constant moisture content is a lengthy process, especially for thick samples. Though oven drying is a quick and uniform method of conditioning, it alters the microstructure and causes extensive cracking. Moreover, there is no direct relationship between gas and water permeabilities. For materials of low porosity, gas permeability is always greater than that of water. This discrepancy was attributed by Klinkenberg [49] to the gas-slippage effect. He proposed the following relationship between gas and liquid permeabilities for a porous medium

$$K_a = K\left(1 + \frac{b}{P_m}\right) \tag{4}$$

where

K_a = gas permeability (m^2),
K = actual intrinsic permeability (m^2),
b = constant (m), and
P_m = mean pressure (m).

Bamforth [50] recently derived a relationship, using Eq 4, for concrete. There is, however, some doubt regarding the application of Eq 4 to cementitious materials, because of changes to the microstructure, including cracking, that accompany drying. The alteration of pore structure and flow can be minimized by drying techniques such as freeze-drying or solvent replacement using isopropanol [1, 24, 51].

Gas Diffusion

Gas diffusion measures the concentration of a chosen gas (usually oxygen) across the sample, due to exposure at the upstream face. Typically, a nitrogen stream is initially maintained at equal pressure on either side of the specimen. A slight over-pressure of oxygen is applied at the upstream face and the resultant gas transfer is detected by means of a gas chromatograph on the downstream side [33]. This technique is useful for dense concretes, and has also been used to predict carbonation rates and reinforcement corrosion rates. The limitations discussed with regard to gas flow also apply to gas diffusion testing and interpretation because water in a partially saturated system has the same effect as a solid barrier.

Depth of Carbonation

Depth of carbonation is taken as a measure of the rate of reaction between carbon dioxide from the atmosphere and mainly calcium hydroxide

$$Ca(OH)_2 + CO_2 = CaCO_3 + H_2O$$

although C-S-H is also affected. The rate of reaction depends on the diffusion of carbon dioxide into the concrete, and the carbonation front is said to be revealed by the color change resulting from the application of phenolphthalein indicator onto the freshly exposed concrete cross section. This test, of course, detects the advance of a pH change and indicates acidity from all possible sources, not exclusively carbonation. This test is most useful for rapid on-site evaluation where the depth of acidity together with the age of the structure is used to estimate the rate of carbonation and the possible onset of corrosion in reinforcing steel.

Water

Water-Vapor Diffusion

Water-vapor diffusion refers to the movement of water vapor across the sample driven by a difference in partial pressure between the upstream and downstream faces. The common methods of testing water vapor diffusion are either by measuring the rate of drying of a sample, or by creating a vapor pressure gradient and then monitoring moisture transfer [35]. The latter is achieved by using a dry cup filled with desiccant on the downstream side of the specimen and storing the whole in a controlled humidity environment. The movement of moisture through the sample is monitored by periodic weighing of the sample and the desiccant. The diffusion coefficient, D_p, based on Fick's law, is calculated using Eq 1. The water vapor transmission has its most practical application when the sample is in contact with water on one side and a desiccant on the other. These conditions, under which both saturated and unsaturated flow take place, are common in concrete water retaining structures. Not much research has been done in this area, but the intractability of analyzing the combined effect of saturated flow, capillary action, and vapor diffusion is well known [36].

Absorption and Rate of Absorption

Most concrete is only partly saturated and the initial ingress of water and dissolved salts is dominated, at least initially, by capillary absorption rather than either water permeability or ion diffusion. A wide variety of water absorption tests on concrete have been developed. These tests measure weight gain of a sample, volume of water entering the sample, depth of penetration, or a combination thereof, by either complete immersion of dry samples in water, or exposing only one face to water, or spraying the specimen surface with water. Absorption is either measured at a single, arbitrary time or by measuring the rate of absorption. Although in all these tests the absorption process is proportional to the square root of time, the sorptivity varies a great deal. The absorption tests have been standardized in many countries (for example, in Great Britain: BS1881 Test for Determining the Initial Surface Absorption of Concrete; in North America: ASTM Test Method for Specific Gravity, Absorption, and Voids in Hardened Concrete (C 642) and ASTM Test Method for Evaluating the Effectiveness of Materials for Curing Concrete (C 1151); and in Australia: AS1342), however, the differences in the test limits and procedures create considerable variation in sorptivity measurements [52].

In ASTM C642, the water absorbed by an oven-dried specimen is measured after 48-h immersion or after such immersion followed by 5 h in boiling water. The ratio of the water absorbed to the dry weight is the absorption. In ASTM C1151, the absorptivity of 10 by 25-mm diameter, methanol-dried cores, are measured after 60-s contact of one surface with water. The difference in absorptivity between the top, cured surface and the bottom, molded surface is taken as a measure of the effectiveness of curing.

In British Standard 1881, a closed reservoir is sealed onto the concrete surface. The reservoir is flooded with water and sealed except for an open-ended capillary tube that is used to monitor the volume of surface absorption after 10, 30, 60, and 120 min.

It has been suggested [41] that the best approach is that used by Gummerson, Hall, and Hoff [53], that defines absorption by sorptivity obtained from the relationship of the volume of water absorbed per unit area of suction surface and the square root of the absorption time (see equation in Table 8). Their test measures the reliable and easily obtained absorbed-water volume, rather than depth or distance of penetration, which is often subjective and requires splitting of the test specimens. Moreover, the amount per unit area of absorbed water is independent of the area of suction surface for uniaxial flow. A rate of absorption test, based on that of Hall [54] is being considered by the ASTM C9 Permeability Task Group. This method has been used by Hooton [55] to evaluate the effects of curing. A typical, sorptivity plot is shown in Fig. 18.

Capillary absorption tests, besides measuring degree of imperviousness, also give an indication of mean pore radius and porosity of the sample. The Concrete Society [33] in their summary has provided the following equations for determination of mean pore radius and porosity

$$r = \sqrt{\frac{4\mu}{P_0}} \cdot \frac{d}{\sqrt{t}} \qquad (5)$$

where

r = mean pore radius (m);
μ = viscosity of water (Pa · s);
P_0 = atmospheric pressure (Pa);
d = depth of penetration, for example, from a splitting test (m); and
t = time (s).

$$v = \frac{1000\,M}{Ad} \qquad (6)$$

where

v = porosity
M = gain in mass (g),
A = area of penetration (m²), and
d = depth of penetration (m).

The major difficulty in the assessment of absorption tests is the determination of the initial moisture condition, which affects sorptivity measurements [36,53]. Moreover, in poorly compacted concrete, where entrapped air voids do not get filled with water, the absorption may be lower than for a thoroughly compacted sample.

Water Penetration

The water penetration test, standardized in Germany as the DIN 1048 test, involves subjecting one end of the unsaturated concrete to a pressure head. The measure of water penetration is achieved either by measuring the

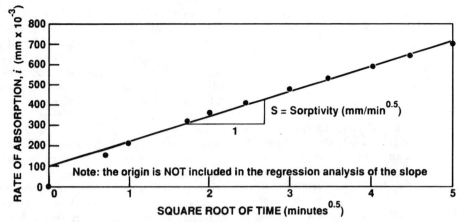

FIG. 18—Typical plot of rate of absorption versus the square root of time [55].

volume of water entering the sample or by splitting open the cylinder and measuring the average depth of discoloration (due to wetting) taken as equal to the depth of penetration. As discussed earlier in the section on Bulk Flow, the usual external hydraulic forces in structures are small compared to that of capillary tension. In penetration tests where artificially high pressures are applied, the equations presented in Table 8 must be modified. Reinhardt [40] presented the combined effects of capillary tension (P_c) and hydraulic pressure (P_e) on the fluid penetration coefficient (B), in the following form

$$B = \frac{r}{2}\left(\frac{P_e + P_c}{\eta}\right)^{1/2} \quad (7)$$

where B is equivalent to sorptivity (S), as defined by Hall [54] (Table 8).

Besides the DIN 1048 test, several in-situ tests utilize water penetration or absorption or both for estimating the quality of concrete, such as Figg's tests [56] and several modified versions (for example, see Refs 54, 57, and 58).

The water penetration results are often complicated by the initial moisture state of the specimen and the nonuniformity of moisture distribution. Moreover, the microstructural characteristics in concrete change with the introduction of water, modifying the pore size distribution of the matrix.

Saturated Flow

Pressure-induced water flow through a saturated sample follows Darcy's law (Eq 2). The test is performed on a water-saturated specimen and involves subjecting one end of the sample to a pressure head. The measurement of the outflow enables the determination of the permeability coefficient, k (described in Eq 2), or intrinsic permeability coefficient, K, which are related by

$$K = \frac{\mu}{\rho g} k \quad (8)$$

where

ρ = density of the permeant (kg/m^3),
g = gravitational acceleration (9.81 m/s^2), and
μ = viscosity (Pa · s)

so that conversion factors can be calculated for the various fluids at various temperatures. For water at 23°C, for example

$$k = 9.75 \times 10^6 K \quad (9)$$

One of the major objections of the saturated water flow test is that the boundary conditions are not representative of the usual concrete environment. Even so, it has drawn considerable attention over the past century (from Hyde and Smith [59] to Hearn and Mills [60]).

The major attractions of this test are:

1. Saturated flow gives intrinsic permeability as defined by Darcy's law.
2. Saturated flow has only one fluid transport mechanism—flow of water through voids—while unsaturated flow encompasses diffusion, absorption, capillary, and saturated flows.
3. The conditioning process for saturated flow (water saturation of the sample) does not damage the existing microstructure.

The major disadvantages of saturated water permeability testing are:

1. Potential problems with saturation of specimens [1,61,62,63].
2. Decrease in flow with the progress of the test (the self-sealing phenomenon [1]).

Ionic Diffusion

While concrete can be penetrated by many aggressive ions because of the extent of reinforcement corrosion damage due to deicing and marine salts, the main concern with ionic diffusion is chloride ion migration into concrete. Chloride diffusion of in-situ concrete is usually determined by analyzing a concrete core at successive depths, thus establishing a chloride ion concentration profile. The diffusion coefficient, D_c, is calculated using Fick's second law (Eq 10) after fitting an equation to the chloride profile in order to first establish an estimate of the surface chloride concentration, Cc

$$C(x,t) = Cc[1 - erf(x/2\sqrt{Dc \cdot t})] \quad (10)$$

where

$C(x,t)$ = chloride concentration at depth x (m) at age t (years),
Cc = surface chloride concentration (%), and
erf = error function (from standard tables).

Other methods involve the use of a concrete slice as a membrane between two salt solutions. One of these methods uses the concentration gradient of a salt as a driving potential across a thin slice of paste or mortar. The rate of diffusion is monitored by the analyses of the ion concentration change on the lower concentration side [64,65]. Once steady state has been established, often taking several weeks, Fick's first law (Eq 1) can be used to calculate the ionic diffusion coefficient

$$Q = D_c \frac{dc}{dx} \qquad (11)$$

where

Q = mass transport rate (mol/m²s),
D_c = diffusion coefficient (m²/s), and
dc/dx = concentration gradient of the ion (mol/m³/m).

A rapid method, known as the Rapid Chloride permeability test, AASHTO T 277-83 and also ASTM Test Method for Electrical Indication of Concrete's Ability to Resist Chloride Ion Penetration (C 1202), has become widely used in North America and involves measurement of the total charge passed in 6 h across a sample sandwiched between NaOH and NaCl solutions. The driving force is provided by 60-V of direct current applied between the two sides of the sample. In this test, the conductivity of the saturated sample, including the effects of all dissolved ions, is measured without emphasis on a particular ion. It is also not clear whether there is a connection between the charge passed and permeability, although this test has some similarity to the electro-osmosis used to test the permeability of clays [66]. Recently, another method has been developed that combines the best features of the two aforementioned chloride tests [67]. A 12-V d-c potential across a sample immersed between chloride and NaOH (neutral) solutions provides the driving force, and the neutral solution is periodically titrated to determined the concentration change of chloride ions. The test is relatively fast and provides information regarding diffusion of a particular salt.

POROSITY, PERMEABILITY, AND MECHANICAL PROPERTIES

In concrete structures, all of the components of the total porosity contribute at least in part to mass transport through the material. In the laboratory, where many factors determining porosity can be controlled, the key elements affecting permeability have been established. As with pastes, the w/c ratio has been found [68] to have a considerable effect on permeability. It has been argued, however, that the amount of mixing water per cubic metre of the mix is a more significant parameter than w/c ratio [69–71], as mixes of the same w/c ratio but of different water and therefore cement contents produce different porosities in concrete. Norton and Pletta [72] have shown that for concrete, the cement to void ratio gives a better correlation (Fig. 19 a and b) with permeability than does w/c ratio. Such findings are reasonable, as the grading of the aggregate and consistency will affect the cement-voids ratio, but not w/c ratio. Most studies, however, have sought correlation between w/c ratio and permeability mostly because it is convenient and easy. Figure 20 [1] combines the results of several such studies [73]. The data of the figure include only those concrete specimens that were moist cured for at least 28 days and tested without any previous drying. This procedure was followed in order to have samples of relatively uniform hydration, and to eliminate variation in porosities produced by microcracking, and so potentially reduce the scatter, although the scatter in the data even with these precautions is high. A similar collection of data for HCP (Fig. 21) shows a reduced band of scatter, indicating that introduction of aggregate into cement matrix does have an effect on permeability. The data for HCP show good correlation between permeability and w/c ratios above 0.60. However, this correlation is of limited use, as commonly used w/c ratios in research and practice are in the range of 0.40 to 0.60; here, the scatter is high. The results of individual studies have produced directly proportional correlations between permeability and w/c ratios. Inter-study data comparisons have produced high scatter, indicating limited value to such findings.

Because of the difficulty in obtaining reliable water permeability data, attempts have been made to predict permeability from pore structural parameters. An extensive review was made by Brown et al. [23]. Pore structural parameters have included total porosity, hydraulic mean radius, critical pore radius, threshold pore size, and the porosity larger than some arbitrary pore size [74–78]. While general relationships have been developed, there is typically too much scatter of the data for these to be useful for predictive purposes.

The studies on the effect of aggregate type and grading on permeability have not been consistent. Work by Collins [79] and Norton and Pletta [72] have shown correlation between permeability and the size and grading of the aggregate. The results from Norton and Pletta [72] in Fig. 22 are misleading because the largest variations in flow occur for aggregate to cement (a/c) ratios above 8, which is outside the practical range of mix proportions. For the a/c ratio below 8, the variation of flow was minimal. Dhir et al. [80] found no significant difference between permeabilities of concrete with a common w/c when using aggregate sizes between 5 and 20 mm, but all of the specimens were certainly cracked by oven drying. Such apparently contradictory results can often be attributed to variations in the compaction of mixes ranging from lean to rich. For instance, in Collins' [79] research, all of the specimens were vibrated at the same speed for a constant 30 s, so that the observed increase in permeability, as the mix became leaner, may actually be related to insufficient or excessive compaction.

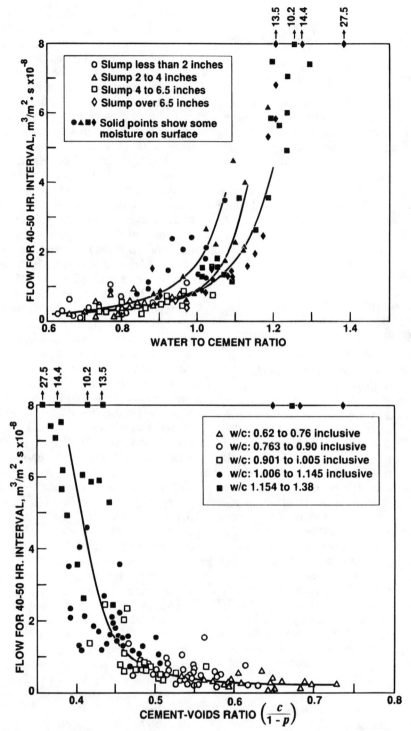

FIG. 19—(a) Effect of w/c ratio on flow [72]; and (b) Effect of cement-voids ratio and w/c ratio on flow [72].

The effects of admixtures on permeability remain obscure [80], mainly because of the number and variety of commercial products used in the industry. It is, however, clear that admixtures that improve workability during mixing and casting are likely to improve compaction and therefore the density and water tightness of concrete.

Hydration and subsequent treatment have been shown in all relevant studies to have a significant effect on permeability. The extent of hydration, especially early hydration, establishes the extent to which the initial pore volume gets filled up with the hydration products. The subsequent treatment establishes the severity of crack formation in the matrix due to shrinkage, thermal, or load stresses. Figure 23 taken from Dhir et al. [80] demonstrates the effects of hydration and curing regimes on the permeability of concrete. Many researchers [46,68,81,82] have

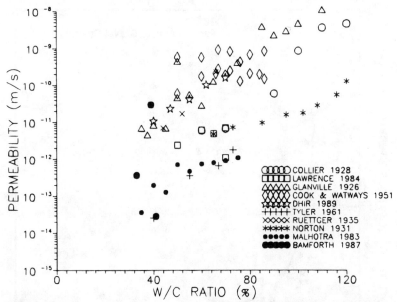

FIG. 20—Correlation between w/c and permeability of concrete using data from several studies [1].

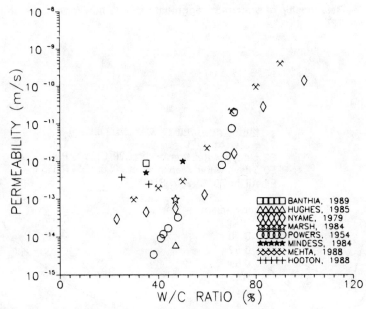

FIG. 21—Correlation between w/c and permeability of HCP using data from several studies [1].

shown that early drying leads to an irreversible increase in permeability. In Collier's [83] study, some of the specimens were placed in water three days after demolding, and their permeability after 28 days of curing was two orders of magnitude greater than the permeability of companion specimens that were placed in water immediately after demolding.

For decades, the acceptance of concrete has been based on strength alone. The importance of permeability to the performance of concrete, however, was not overlooked, as many researchers attempted to correlate permeability with strength. The rationalization behind such attempts was based on the fact that strength, as permeability, is dependant on the w/c ratio; degree of hydration, size, type, and grading of the aggregate; and the extent to which the cement matrix is flawed, that is, by poor compaction or cracking. It is common experience that for the same level of hydration and porosity, dry concrete is stronger than saturated concrete in spite of extensive cracking of the HCP. Thus when concrete is dried, strength increases and permeability also increases. However, permeability is affected to a much larger extent by poor curing than is strength. In the laboratory environment, many studies have shown at least a vague inverse proportionality

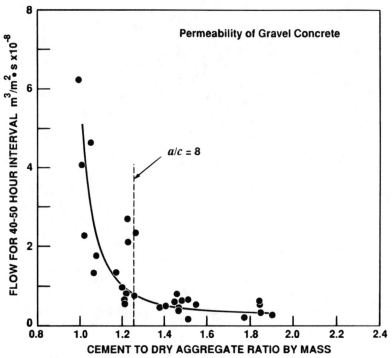

FIG. 22—Relationship of cement-dry aggregate ratio to flow for mixes with (2 to 4 in.) slump [72].

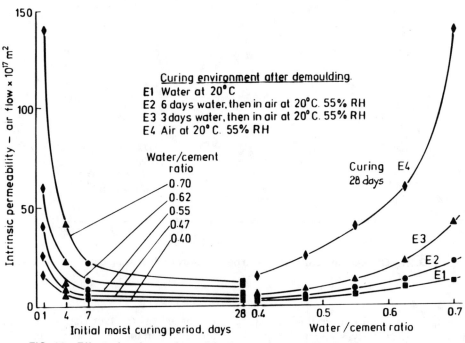

FIG. 23—Effect of curing and conditioning on permeability [80].

between strength and permeability. The amalgamation of the laboratory results (Fig. 24) obscures the inverse proportionality somewhat. The large scatter in Fig. 24 can mainly be attributed to three factors. First, the strength depends on the quality of paste (that is, on w/c ratio), while permeability (for equal strength) depends on the volume of paste in concrete, thus it is possible to have a wide range in permeability for constant strength. Second, the inverse proportionality between strength and permeability does not hold after drying of the specimen, especially severe drying. Strength increases, due to the new bond formation between collapsed gel layers, even though drying causes severe microcracking. Permeability also increases, due to the widening and interconnection of the

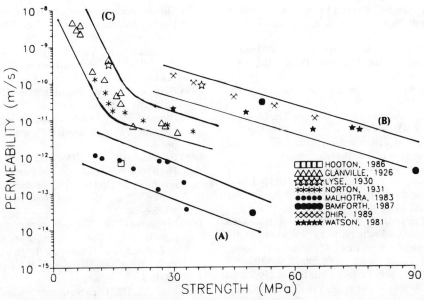

FIG. 24—Correlation between strength and permeability of concrete using data from several studies [1].

capillaries as the gel layers collapse, and new flow paths form due to microcracking. And third, the comparison between strength and permeability is made with data that is not absolute, in the sense that w/c ratio or cement content are, because the results for strength and permeability are influenced by the testing procedures especially in the case of permeability. This produces an additional source of scatter.

Data in Fig. 24 can be divided as follows according to the conditioning regime of the sample and the date of the study:

(a) wet-cured samples tested in recent studies,
(b) dry samples tested in recent studies, and
(c) wet-cured samples tested in old studies.

Three distinct groups of data evolved (Fig. 24). The variation between the older (Group C) and more recent studies (Group A) for wet-cured samples indicates that changes, over the last several decades, in cement composition and fineness, together with casting and testing procedures may have affected the strength-permeability relationship. In general, attempts at correlating strength with permeability have failed and the consensus is that strength alone cannot be used to predict the permeation characteristics of the concrete.

Extensive work has been done in correlating mix design variables with the permeability of concrete. General trends have emerged indicating low permeability for the richer, well-compacted mixes that require not only a certain level of cement content, but also an appropriate fine to coarse ratio of the aggregate. The importance of moist curing for developing a tight structure, especially for young concrete, and increased permeability due to drying, has been also demonstrated. The existing scatter, however, is high, even for painstakingly prepared HCP samples [4,74,84]. Such scatter indicates that either (1) the heterogeneity of the cementitious materials is uncontrollable, or (2) the relationship of water permeability to the general water-cement interaction is poorly understood. Possibility (1) is unlikely. The scatter appears in both concrete and paste; and although concrete is heterogeneous, many properties, such as strength, can be controlled in the mix design, with individual studies producing reasonable correlations between permeability and design variables. Explanation (2), however, has not been addressed. The water-cement interaction in the physical and chemical sense is not well understood; even the contribution of various types of porosities to the overall flow is unclear. It has also been shown [74] that permeability does not maintain a constant value during the length of the test and can substantially decrease with time [1]. The scatter in published data indicates that before a clear correlation between permeability and mix design variables can be achieved, the understanding of the interaction between water and cement needs to be addressed.

REFERENCES

[1] Hearn, N., "Saturated Permeability of Concrete as Influenced by Cracking and Self-Sealing," Ph.D. thesis, University of Cambridge, Cambridge, UK, 1993.

[2] Bakker, R. F. M., "Permeability of Blended Cement Concretes," SP-79, Vol. 2, V. M. Malhotra, Ed., American Concrete Institute, Detroit, MI, 1983, pp. 589–605.

[3] Young, J. F., "A Review of the Pore Structure of Cement Paste and Concrete and its Influence on Permeability," SP-108, Proceedings, Permeability of Concrete, D. Whiting, Ed., American Concrete Institute, Detriot, MI, 1988, pp. 1–18.

[4] Powers, T. C., Copeland, L. E., and Mann, H. M., "Flow of Water in Hardened Portland Cement Pastes," Special Report 40, Highway Research Board, Washington, DC, 1959, pp. 308–323.

[5] Mindess, S. and Young, J. F., Concrete, Prentice-Hall Inc., New York, 1981.

[6] Powers, T. C., "Structure and Physical Properties of Hardened Portland Cement Paste," *Journal*, American Ceramic Society, Vol. 41, 1958, pp. 1–6.

[7] Powers, T. C. and Brownyard, T. L., "Studies of the Physical Properties of Hardened Portland Cement Paste," Research and Development Bulletin 22, Portland Cement Association, 1948; reprinted from *Journal*, American Concrete Institute, *Proceedings*, Vol. 43, 1947.

[8] Soroka, I., *Portland Cement Paste and Concrete*, The MacMillan Press Ltd., New York, 1979.

[9] Hansen, T. C., "Physical Structure of Hardened Cement Paste, Classical Approach," *Materials and Structures*, RILEM, Vol. 19, No. 114, 1986, pp. 423–436.

[10] Feldman, R. F. and Sereda, P. J., "A New Model of Hydrated Cement and its Practical Implications," *Engineering Journal*, Canada, Vol. 53, 1970, pp. 53–59.

[11] Feldman, R. F. and Sereda, P. J., "A Model for Hydrated Portland Cement Paste as Deduced from Sorption-Length Change and Mechanical Properties," *Materials and Structures*, RILEM, Paris, Vol. 1, No. 6, 1968, pp. 509–520.

[12] Feldman, R. F., "Density and Porosity Studies of Hydrated Portland Cement," *Cement Technology*, Vol. 3, No. 1, 1972, pp. 5–14.

[13] Grudemo. A., "The Microstructure of Hardened Cement Paste," *Proceedings*, 4th International Symposium on the Chemistry of Cement, Washington, DC, Vol. 2, 1960, pp. 615–658.

[14] Taylor, H. F. W., *Cement Chemistry*, Academic Press Ltd., London, 1990.

[15] Wittman, F. H., "Interaction of Hardened Cement Paste and Water," *Journal*, American Ceramic Society, Vol. 56, No. 8, 1973, pp. 409–415.

[16] Wittman, F. H. and Englert, G., "Water in Hardened Cement Paste," *Matériaux et Constructions*, Vol. 1, No. 6, 1968, pp. 535–546.

[17] Wittman, F. H., "Trends in Research on Creep and Shrinkage of Concrete," *Cement Production and Use*, J. P. Skalny, Ed., Engineering Foundation, New York, No. 79-08, 1979, pp. 143–162.

[18] Wittman, F. H. and Ferraris, C. F., "Shrinkage Mechanism of Hardened Cement Paste," *Cement and Concrete Research*, Vol. 17, 1987, pp. 453–464.

[19] Powers, T. C., "Physical Properties of Cement Paste," *Proceedings*, 4th International Symposium on the Chemistry of Cement, Washington, DC, 1960, Vol. 2, pp. 577–613.

[20] Mehta, P. K., *Concrete: Structure, Properties and Materials*, Prentice-Hall Inc., New York, 1986.

[21] Powers, T. C., Copeland, L. E., Haynes, J. C., and Mann, H. M., "Permeability of Portland Cement Pastes," *Journal*, American Concrete Institute, Vol. 51, 1954, pp. 285–298.

[22] Mills, R. H., "The Influence of Stress on the Permeability of Concrete," *Proceedings*, FIP-CPCI Symposium on Concrete and Pressure Storage Vessels, Calgary, Alberta, Canada, Aug. 1984, pp. 43–47.

[23] Brown, P. W., Shi, D., and Skalny, J. P., "Porosity/Permeability Relationships," *Concrete Science II*, J. P. Skalny and S. Mindess, Eds., American Ceramic Society, Westerville, OH, 1991, pp. 83–109.

[24] Day, R. L. and Marsh, B. K., "Measurement of Porosity in Blended Cement Pastes," *Cement and Concrete Research*, Vol. 18, 1988, pp. 63–73.

[25] Winslow, D. N. and Diamond, S., "Mercury Porosimetry Study of the Evolution of Porosity in Portland Cement," *Journal of Materials*, Vol. 5, 1970, pp. 564–585.

[26] Hearn, N. and Hooton, R. D., "Sample Mass and Dimension Effects on Mercury Intrusion Porosimetry Results," *Cements and Concrete Research*, Vol. 22, 1992, pp. 970–980.

[27] Shi, D. and Winslow, D. N., "Contact Angle and Damage During Mercury Intrusion Into Cement Paste," *Cement and Concrete Research*, Vol. 15, 1985, pp. 645–654.

[28] Feldman, R. F., "Pore Structure Damage in Blended Cements Caused by Mercury Intrusion," *Journal*, American Ceramic Society, Vol. 67, 1984, pp. 30–33.

[29] Diamond, S., "Methodologies of PSD Measurement in HCP: Postulates, Peculiarities and Problems," *Pore Structure and Permeability of Cementitious Materials*, L. R. Roberts and J. P. Skalny, Eds., Materials Research Society, Vol. 137, 1989, pp. 83–92.

[30] Feldman, R. F., "The Porosity and Pore Structure of Hydrated Portland Cement Pastes," *Pore Structure and Permeability of Cementitious Materials*, L. R. Roberts and J. P. Skalny, Eds., Materials Research Society, Vol. 137, 1989, pp. 59–74.

[31] Winslow, D. N., "Some Experimental Possibilities with Mercury Intrusion Porosimetry," *Pore Structure and Permeability of Cementitious Materials*, L. R. Roberts and J. P. Skalny, Eds., Materials Research Society, Vol. 137, 1989, pp. 23–103.

[32] Verbeck, G., "Pore Structure," *Significance of Tests and Properties of Concrete and Concrete Making Materials*, ASTM STP 169B, American Society for Testing and Materials, Philadelphia, 1978, pp. 262–274.

[33] "Permeability Testing of Site Concrete," Technical Report No. 31, Concrete Society, UK, 1988.

[34] Winslow, D. N. and Diamond, S., "Specific Surface of HPCP as Determined by Small-Angle X-Ray Scattering," *Journal*, American Ceramic Society, Vol. 57, No. 5, 1974, pp. 193–197.

[35] Mills, R. H., "Mass Transfer of Water Vapour Through Concrete," *Cement and Concrete Research*, Vol. 15, 1985, pp. 74–82.

[36] Peer, L. B. B., "Water Flow into Unsaturated Concrete," Ph.D. Thesis, University of Cambridge, Cambridge, UK, 1990.

[37] Barrer, R. N., *The Solid-Gas Interface*, E.A., Flood, Ed., Vol. II, Dekker, New York, 1967, pp. 557–609.

[38] Rose, D. A., "Water Movement in Unsaturated Porous Materials," *Materials and Structures*, RILEM, Paris, Bulletin No. 29, Dec. 1965, pp. 119–123.

[39] Mills, R. H., "Gas and Water Permeability of Concrete for Reactor Buildings—Small Specimens," Research Report INFO 0188-1, Atomic Energy Control Board, Ottawa, Ontario, Canada, 1986.

[40] Reinhardt, H. W., "Transport of Chemicals Through Concrete," *Materials Science of Concrete III*, J. P. Skalny, Ed., American Ceramic Society, Westerville, OH, 1992, pp. 209–241.

[41] Emerson, H., "Mechanisms of Water Absorption by Concrete," *Proceedings*, Protection of Concrete, R. K. Dhir and J. W. Green, Eds., E. & F. N. Spon, Dundee, 1990, pp. 689–700.

[42] Bamforth, P. B., Pocock, D. C., and Robery, P. C., "The Sorptivity of Concrete," *Proceedings*, Our World in Concrete and Structures, C. I. Premier Pte. Ltd., Singapore, 27–28 Aug. 1985.

[43] Fagerlund, G., "On the Capillarity of Concrete," *Nordic Concrete Research*, The Nordic Concrete Federation, Oslo, Publication No. 1, 1982, pp. 6.1–6.20.

[44] Hall, C., "Water Movement in Porous Building Materials—I," *Building and Environment*, Vol. 12, 1977, pp. 117–125.

[45] Ho, D. W. S. and Lewis, R. K., "Water Penetration into Concrete—A Measure of Quality as Affected by Material Composition and Environment," *Proceedings*, Symposium on Concrete, Perth, Australia, Oct. 1983, pp. 20–21.

[46] Vuorinen, J., "Applications of Diffusion Theory to Permeability Tests on Concrete, Part 1: Depth of Water Penetration into Concrete and Coefficient of Permeability," *Magazine of Concrete Research*, Vol. 37, No. 132, 1985, pp. 145–152.

[47] Darcy, H., "Determination of the Law of the Flow of Water Through Sand," Les fontaines publiques de la ville de Dijon, Victor Dalmont, Paris, 1856, pp. 590–594 (the English translation appears in *Physical Hydrology*, R. A. Freeze and W. Buck, Eds., Hutchinson Ross Publishing Co. Stroudsburg, PA).

[48] Ruettgers, A., Vidal, E. N., and Wing, S. P., "An Investigation of the Permeability of Mass Concrete with Particular Reference to Boulder Dam," *Journal*, American Concrete Institute, *Proceedings*, Vol. 31, 1935, pp. 389–416.

[49] Klinkenberg, L. J., "The Permeability of Porous Media to Liquids and Gases," *Drilling and Production Practice*, American Petroleum Institute, New York, 1941, pp. 200–214.

[50] Bamforth, P. B., "The Relationship Between Permeability Coefficients for Concrete Obtained Using Liquid and Gas," *Magazine of Concrete Research*, Vol. 39, No. 138, 1987, pp. 3–11.

[51] Kumar, A. and Roy, D. M., "Pore Structure and Ionic Diffusion in Admixture Blended Portland Cement Systems," *Proceedings*, 8th International Congress on the Chemistry of Cement, Rio de Janeiro, Vol. 4, 1986, pp. 72–79.

[52] Guger, H., "Water Absorption of Concrete," *Monier Rocla Technical Journal*, No. 75, Nov. 1983.

[53] Gummerson, R. J., Hall, C., and Hoff, W. D., "Water Movement in Porous Building Materials—II," *Building and Environment*, Vol. 15, 1980, pp. 101–108.

[54] Hall, C., "Water Sorptivity of Mortars and Concretes: A Review," *Magazine of Concrete Research*, Vol. 41, No. 147, 1989, pp. 51–61.

[55] Hooton, R. D., Mesic, T., and Beal, D. L., "Sorptivity Testing of Concrete as an Indicator of Concrete Durability and Curing Efficiency," *Proceedings*, 3rd Canadian Symposium on Cement and Concrete, R. F. Feldman, Ed., Ottawa, Ontario, Canada, 1993, pp. 264–278.

[56] Figg, J. W., "Methods of Measuring the Air and Water Permeability of Concrete," *Magazine of Concrete Research*, Vol. 25, No. 85, 1973, pp. 213–219.

[57] Richards, P. W., "A Laboratory Investigation of the Water Permeability and Crushing Strength of Concrete Made With and Without Pulverised-Fuel Ash, as Affected by Early Curing Temperature," Cement and Concrete Association, Slough, UK, 1982, Advanced Concrete Technology Project 82/9, No. 59, 1982.

[58] Pihlajavaara, S. E. and Paroll, H., "On the Correlation Between Permeability Properties and Strength of Concrete," *Cement and Concrete Research*, Vol. 5, 1985, pp. 321–327.

[59] Hyde, G. W. and Smith, W. J., "Results of Experiments Made to Determine the Permeability of Cements and Cement Mortars," *Journal*, Franklin Institute, Philadelphia, Vol. 128, No. 3, Sept 1889, pp. 199–207.

[60] Hearn, N. and Mills, R. H., "A Simple Permeameter for Water or Gas Flow," *Cement and Concrete Research*, Vol. 21, 1991, pp. 257–261.

[61] Day, R. L., Joshi, R. C., Langan, B. W., and Ward, M. A., "Measurement of the Permeability of Concretes Containing Fly Ash," *Proceedings*, 7th International Ash Utilization Symposium, Orlando, USDOE Report DOE/METC-85/6018, Vol. 2, 1985, pp. 811–821.

[62] Hooton, R. D., "What is Needed in a Permeability Test for Evaluation of Concrete Quality," *Pore Structure and Permeability of Cementitious Materials*, L. R. Roberts and J. P. Skalny, Eds., Materials Research Society, Vol. 137, 1989, pp. 141–150.

[63] Denno, G., "The Permeability of High Strength Lightweight Aggregate Concrete," Ph.D. thesis, Imperial College, London, 1992.

[64] Page, C. L. and Vennesland, O., "Pore Solution Composition and Chloride Binding Capacity of Silica-Fume Cement Pastes," *Materials and Structures*, RILEM, Paris, Vol. 16, No. 91, 1983, pp. 19–25.

[65] Buenfeld, N. R. and Newman, J. B., "The Permeability of Concrete in a Marine Environment," *Magazine of Concrete Research*, Vol. 36, No. 127, 1984, pp. 67–76.

[66] Mitchell, J. K., *Fundamentals of Soil Behaviour*, Wiley, New York, 1976, pp. 353–359.

[67] Detwiler, R. J., Kjellsen, K. O., Gjørv, O. E., "Resistance to Chloride Intrusion of Concrete Cured at Different Temperatures," *ACI Materials Journal*, Vol. 88, No. 1, Jan.–Feb. 1991, pp. 19–24.

[68] Glanville, W. H., "The Permeability of Portland Cement Concrete," *Building Research*, UK, Technical Paper No. 3, 1931 (reprinted from 1926).

[69] Markestad, A., "An Investigation of Concrete in Regard to Permeability Problems and Factors Influencing the Results of Permeability Test," Cement and Concrete Research Institute, Norwegian Institute of Technology Report, STF 65A 77027, 1977.

[70] Mills, 1991 (personal communication).

[71] Hooton, R. D., "High Strength Concrete as a By-Product of Design for Low Permeability," Concrete 2000, R. K. Dhir and M. R. Jones, Eds., 1993, Chapman and Hall, Dundee, Vol. 2, pp. 1627–1638.

[72] Norton, P. and Pletta, D., "Permeability of Gravel Concrete," *Journal*, American Concrete Institute, *Proceedings*, Vol. 27, 1931, pp. 1093–1132.

[73] Cook, H. K., "Permeability Tests of Lean Mass Concrete," *Proceedings*, American Society for Testing and Materials, Philadelphia, Vol. 51, 1951, pp. 1156-1165.

[74] Nyame, B. K. and Illston, J. M., "Relationship Between Permeability and Pore Structure of Hardened Cement Paste," *Magazine of Concrete Research*, Vol. 33, No. 116, 1981, pp. 139–146.

[75] Mehta, P. K. and Manmohan, D., "Pore Size Distribution and Permeability of Hardened Cement Pastes," *Proceedings*, Seventh International Congress on the Chemistry of Cement, Paris, No. VIII, 1980, pp. VII 1–5.

[76] Roy, D. M., "Relationships Between Permeability, Porosity, Diffusion and Microstructure of Cement Pastes, Mortars and Concrete at Different Temperatures," *Pore Structure and Permeability of Cementitious Materials*, L. R. Roberts and J. P. Skalny, Eds., Materials Research Society, Vol. 137, 1989, pp. 179–189.

[77] Garboczi, E. J., "Permeability, Diffusivity and Microstructural Parameters; A Critical Review," *Cement and Concrete Research*, Vol. 20, 1990, pp. 591–601.

[78] Hooton, R. D., "Permeability and Pore Structure of Cement Pastes Containing Fly Ash, Slag, and Silica Fume," *Blended Cements, ASTM STP 897*, G. Frohnsdorff, Eds., American Society for Testing and Materials, Philadelphia, 1986, pp. 128–143.

[79] Collins, J. F., Jr., "Permeability of Concrete Mixtures: Part I Literature Review," *Journal of Civil Engineering for Practising and Design Engineers*, Vol. 5, 1986, pp. 579–638.

[80] Dhir, R. K., Hewlett, P. C., and Chan, Y. N., "Near Surface Characteristics of Concrete: Intrinsic Permeability," *Magazine of Concrete Research*, Vol. 41, No. 147, June 1989, pp. 87–97.

[81] Tyler, I. L. and Erkin, B., "A Proposed Simple Test Method for Determining the Permeability of Concrete," *Journal*, Portland Cement Association, Research and Development Laboratories, Vol. 3, No. 3, 1961, pp. 2–7.

[82] Lawrence, C. D., "Measurement of Permeability," *Proceedings*, 8th International Congress on the Chemistry of Cements, Rio de Janeiro, Vol. 5, 1986, pp. 29–34.

[83] Collier, I. L., "Permeability of Concrete," *Proceedings*, American Society for Testing and Materials, Philadelphia, Vol. 28, No. 2, 1928, pp. 490–496.

[84] Marsh, B. K., "Relationships Between Engineering Properties and Microstructural Characteristics of Hardened Cement Paste Containing Pulverized-Fuel Ash as a Partial Cement Replacement," Ph.D. Thesis, Hatfield Polytechnic, UK, 1984.

Chemical Resistance of Concrete

G. W. DePuy[1]

PREFACE

This is a revision of Chapter 24, "Chemical Resistance of Concrete," prepared by Lewis H. Tuthill, that appeared in the original edition of *ASTM STP 169* and in subsequent editions of *ASTM STP 169A* and *ASTM STP 169B*. We are greatly indebted to Lewis H. Tuthill for his contributions to concrete technology as exemplified by his chapter. He has been a leader in the unending quest for excellence in the production of quality concrete, and his wisdom has greatly increased our understanding of how to make concrete better and more durable in exposure to severe environmental conditions. His contributions are well known throughout the concrete industry. The quality and engineering soundness of his work is indicated by the fact that many of his recommendations for making durable and chemically resistant concrete, and in particular those for making sulfate-resistant concrete, were developed many years ago but remain still intact and fully in force today. The contributions of the many other pioneers and leaders in the study of the chemical resistance of concrete, too numerous to adequately recognize, are also respectfully recognized. Special thanks are given to Doug Craft and Andy Murphy, chemists with the Bureau of Reclamation, for their helpful explanations on basic and applied chemistry, and to the ASTM reviewers of the manuscript for their many constructive comments that were incorporated into this chapter.

INTRODUCTION

Under ordinary conditions, concrete is a very durable material and may be expected to have a long service life. In addition to being a strong and versatile material with good constructability and low cost, the durability of concrete is often one of the main reasons for concrete being the material of choice. In most situations concrete is a stable material, and well-made concrete resists deterioration for many years. However, there are instances where concrete is subjected to chemical attack by aggressive substances, and the durability of concrete may become a concern.

The common forms of chemical attack discussed are: leaching by soft water, acid attack, sulfate attack, carbonation, exposure to industrial chemicals, and exposure to saline water. Two other types of destructive chemical reactions, alkali-aggregate reaction (Chapters 32 and 33) and corrosion of reinforcing steel and other embedded metals (Chapter 18), are special cases and are not included in this discussion. Chemical reactions per se in concrete are not necessarily harmful; some may be beneficial, such as pozzolanic reactions. Other reactions may be beneficial in some circumstances but harmful at other times, as is the case with carbonation. Most harmful reactions involve chemicals in a liquid or gaseous form. Most chemicals in a dry and solid form are not harmful to concrete until they are brought into contact with water or other solvents.

The two questions most commonly faced by engineers with regard to concrete exposed to aggressive agents are how long will the structure stand up to the exposure and what can be done to prolong the service life? The answers to these questions require information on the specific exposure conditions and chemical reactions so that appropriate protective strategies can be developed. The more commonly occuring situations involving exposure to aggressive chemicals have been extensively investigated over the years and some general guidelines have been developed for dealing with these situations; however, the attack in a specific situation is a complex matter and may vary greatly depending on the variables governing the reaction. This is particularly true in unusual situations where concrete is exposed to high temperatures or where the performance of concrete is especially critical, as in hazardous waste containment or nuclear waste storage [1]. In these instances, the general guidelines that have served so well in the past may not adequately cover the situation and it becomes increasingly more important that the engineer have a good understanding of the chemical reactions, factors governing the rate of reaction, and mechanisms causing deterioration in concrete.

There are three main types of mechanisms involved in the deterioration of concrete from chemical reactions: (1) purely mechanical expansive forces arising from the formation of expansive reaction products or products that have a larger volume than the original materials, (2) chemical alteration that weakens or softens concrete constituents, (3) and dissolution and leaching of soluble components.

---

[1] Supervisory materials research engineer, Bureau of Reclamation, U.S. Department of the Interior, Denver, CO 80225.

Concrete may be thought of as a man-made rock, and in this context, concrete structures exposed to the elements are subject to the same natural chemical weathering processes (dissolution, oxidation, hydrolysis, and acid hydrolysis) that attack rock [2]. In industrial exposures, concrete faces a wider variety of chemical reactions, ranging from dissolution (dissolving of a solid substance into a solvent such as water) and decomposition (dissociation of one or more compounds and chemical combination into other compounds, such as redox and acid-base reactions), to ion exchange (substitution of one ion for another in a molecular structure that may change the physical properties of the compound).

The rate and severity of attack is governed by reactivity of the reactants, supply of the reactants, and severity of exposure conditions. The main factors affecting reaction rates are concentration of reactants, the rate at which the reactants are brought together, and temperature. The reaction is also affected by other factors, such as effects of other substances that alter, inhibit, or promote the reaction; the physical configuration of the reaction site; the nature of the reaction products; and whether the reaction goes to equilibrium or completion. Some reactions, such as dissolution and leaching may tend to accelerate future deterioration; other reactions form products that produce a physical blocking or passivation effect and tend to impede further deterioration.

The supply of aggressive agents for reaction is governed by the velocity the reactants are brought into contact with concrete and the penetration of the reactants into the concrete, which is generally considered in terms of permeability [3] and diffusion. Permeability, the penetration of a substance through fractures, voids and capillaries in concrete, generally is linearly related to pressure and viscosity of the aggressive agent. Diffusion, the passage of molecules of an aggressive substance through the molecules of the cement hydration products, is a function of the concentration gradient and is generally related to the square root of time. The velocity of the aggressive agent coming into contact with concrete is a key factor in surface reactions.

Although both the hydrated cement paste and aggregate particles may be involved in harmful chemical reactions, most attention will be given to reactions with cement hydration products. The main cement hydration products are calcium silicate hydrates and calcium hydroxide, and may include varying amounts of other reaction products such as calcium aluminate hydrates and calcium sulfoaluminate.

The portland cement hydration product of greatest significance to chemical resistance is calcium hydroxide $(Ca(OH)_2)$, also referred to as lime, hydrated lime, or portlandite. Calcium hydroxide constitutes about 25 to 30% of the hydrated cement and is more soluble in water and chemically reactive than the other hydration products. It reacts readily with acids and with carbonate and sulfate ions. One strategy to improve the chemical resistance of portland cement is to minimize the calcium hydroxide content of the hydrated cement by using a "low-lime" cement (contains more dicalcium silicate and less tricalcium silicate), or by adding pozzolans to the concrete mixture that will react with calcium hydroxide to form more stable calcium silicate hydrates.

Three general strategies are used to improve the service life of structures under the threat of chemical attack. These are the design of the structure to avoid or minimize contact with aggressive agents, the production of high-quality impermeable concrete, and the provision of a physical barrier to exclude aggressive agents.

The best strategy starts with the design of the structure. This would include avoiding or minimizing contact with aggressive agents, providing good drainage to intercept or eliminate the water migrating through concrete or ponding behind structures, using contraction and construction joints that are water tight and frequent enough to prevent intermediate cracking, and attention to details to control cracking. Design measures may also include the provision of sacrificial concrete (using sections with added thickness to compensate for the chemical attack).

The intrinsic durability of concrete may also be improved by:

1. Production of high-quality, well-consolidated, impermeable concrete that requires selection of the appropriate materials, proper mixture design, and good quality control and workmanship. This involves good workmanship throughout the entire production process from batching and mixing, placing and consolidation, finishing and curing to produce a crack-free concrete with a hard and dense surface.
2. In addition to good workmanship, attention must be given to the selection and proportioning of concrete materials, which includes:
   a. sufficient cement;
   b. good mix design to obtain a workable and compactable mix;
   c. a low water-cement (w/c) ratio (usually 0.40 or lower that can be easily provided by using a water-reducing admixture) to reduce shrinkage, capillarity, and permeability;
   d. blending the cement with pozzolan or slag, such as 20 to 30% of an active fly ash or a well-dispersed silica fume, to minimize permeability; and
   e. selection of a well-graded, hard, clean, physically sound, and chemically resistant aggregate.
3. Special surface treatments to produce a hard, impervious surface.
4. Use of chemically resistant materials, such as chemical-resistant cement and aggregates, or polymer concretes.
5. Use of an impermeable barrier such as chemically resistant membranes and protective coatings [4–6].

SCALING, EFFLORESCENCE, AND LEACHING

Concrete exposed to water is subject to leaching and the deposition of mineral deposits. Leaching and deposition are related effects produced by the dissolution and precipitation of minerals in water from any source, including groundwater, surface water, or water contained in or transported by concrete structures [7–12]. Leaching refers to the chemical dissolution and removal of water-soluble constituents in concrete. The deposition of mineral deposits includes both scaling or the formation of incrustation, and efflorescence.[2] Scale deposits are generally hard, layered precipitates of calcium sulfate, calcium carbonate, or magnesium salts, such as produced by the heating or evaporation of water containing dissolved salts. Efflorescence refers to a white, powdery to hard deposit (commonly calcium carbonate, but may consist of other salts), such as is formed when water containing dissolved minerals migrates through concrete and subsequently evaporates on exposed surfaces.

Mineral deposits may form on exposed surfaces where water seeps through concrete along joints and cracks of concrete structures, such as dam faces, pipelines and tunnels, retaining walls and abutments, and building foundations. These deposits may be objectionable for several reasons—they are unsightly, and they may adversely affect the performance of a structure such as by reducing the effectiveness of heat exchanger surfaces or by obstructing the passage of water in drain holes and water tunnels [13,14]. The deposits are sometimes taken as evidence of leaching in the concrete, but the leaching in such cases is often relatively minor.

Leaching may erode concrete either externally on exposed surfaces, as in the case of water flowing in canals and pipelines, or internally in the case of water migrating through cracks and joints in a concrete structure. Relatively cold and pure water from mountain streams and reservoirs, and product water from desalting plants are capable of leaching concrete and create the sandy appearance sometimes observed in some concrete conduits, canals, and flumes. Internal leaching by water seeping through cracks and joints in concrete is sometimes suspected of creating voids and weakening the concrete.

The principal constituent of concrete usually involved with leaching or mineral deposition is $Ca(OH)_2$ (calcium hydroxide). $Ca(OH)_2$ is soluble to a certain extent in water and is the main cement hydration product susceptible to leaching. The other principal hydration products (calcium silicate hydrates and calcium aluminate hydrates) are much less soluble. The observable differences in resistance to leaching due to the use of different types of portland cements are very minor and usually attributed to differing amounts of $Ca(OH)_2$ formed from the hydration of the cement. Other factors affecting the permeability of the concrete, such as water/cement ratio and the presence of micro fractures and cracks, are more important.

$Ca(OH)_2$ dissolves in water by dissolution and by acid-base reaction in acidic water. The amount of $Ca(OH)_2$ that can be dissolved is dependent upon the pH of the water, temperature, and the concentration of other soluble materials in the water. The pH of natural water low in dissolved solids is related to the absorption of carbon dioxide (CO_2) gas from the air. The solubility of CO_2 in water increases with pressure and decreases with temperature. On this basis, cold water is more aggressive towards concrete than warm water as calcium hydroxide is more soluble in cold water than in hot water.

The effects of other dissolved salts in water are major factors governing the dissolution of lime in concrete by water. Hard water (water containing dissolved calcium, magnesium, and iron salts) will not dissolve calcium hydroxide to any appreciable extent, but soft water (water relatively free of the dissolved salts) readily dissolves calcium hydroxide and may become quickly saturated with calcium.

Saturated lime water is in a delicate equilibrium condition, and slight changes in conditions can quickly result in the formation of a precipitate. A common example is that of lime-saturated water migrating through concrete. When the water becomes exposed to the atmosphere, calcium carbonate frequently precipitates on the surface of the concrete. The precipitation occurs as a result of a shift in equilibrium of CO_2 in solution due to a decrease in pressure, a probable increase in temperature, and evaporation.

Leaching and deposition processes usually tend to either increase or decrease with time, rather than continue on at a more or less constant rate. Most commonly, the processes tend to diminish. The rate of increase or decrease may be taken as evidence of whether the problem is becoming worse or is solving itself. At times, leaks are observed to plug themselves through the formation of calcium carbonate precipitates—the so-called autogenous crack healing sometimes observed in concrete pipes and canals.

The tendency of water to dissolve or to form a precipitate of a calcium salt is an issue related to the degree of saturation of the salt in a particular situation, and the other factors governing solubility. A common indicator of the tendency of a particular water to dissolve or precipitate calcium carbonate is the Langlier Index that takes into account the calcium content of the water, pH, total alkalinity, dissolved solids, and temperature [15]. This is applicable to concrete, as calcium usually precipitates in the form of calcium carbonate. More recently, chemical equilibrium models, such the U.S. Geological Survey's WATEQ model, have been developed to evaluate mineral saturation [16,17]. A more definitive answer would require a chemical mass balance approach [18–20].

Questions on the cause of mineral deposits, or on whether leaching is damaging to a particular concrete structure, are difficult to answer as usually there is insufficient information on the chemical equilibrium condition of the water before and after exposure to the concrete. An indication of the potential for the water to dissolve lime could be obtained by testing the water immediately upstream of the structure to obtain a Langlier, or preferably a WATEQ saturation index. This may be difficult to do as it would require sampling of the water upstream of the structure and also at the location where the water seeps through the concrete. Drilling and coring may also be required to determine if leaching has weakened the concrete; however, some caution in interpretation should be exercised since internal voids can also be produced by

---

[2] In regards to concrete technology, the term "scaling" is used more often in an entirely different sense to refer to the physical process of the removal of thin layers of surface concrete.

266 TESTS AND PROPERTIES OF CONCRETE

defective workmanship in the construction of the structure. In most cases, the appearance is worse than the damage, and very little good-quality concrete has actually been destroyed or made unserviceable by leaching of lime [21].

The Bureau of Reclamation investigated the effects of warm to hot distilled water on portland-cement concrete in a study on the use of concrete in desalting plants, and concluded that concrete was not suitable in exposure to flowing distilled water [22]. The studies showed that concrete does not resist leaching in warm to hot distilled water, and that severity of leaching increases as temperature increases. The use of up to 23% fly ash (ASTM Specification for Fly Ash and Raw or Calcined Natural Pozzolan for Use as a Mineral Admixture in Portland Cement Concrete (C 618), Class F) was not effective in reducing the leaching by warm to hot distilled water.

In addition to the general steps to improve chemical resistance, other means to consider include:

1. Mineral deposits
   a. Stop the water seeps and leaks.
   b. Removal by physical or chemical means (such as sulfamic acid, or using a protective coating as an aid).
2. Leaching
   a. Use sacrificial concrete.
   b. Stop or divert the flow of water.
   c. Pretreat the water to saturate it with calcium before it comes into contact with the concrete.
   d. Use of special cements, such as aluminous cement, "low-lime" portland cement (contains more dicalcium silicate and less tricalcium silicate), a polymer modified cement, or a polymer concrete.
   e. Use of mineral admixtures blended with portland cement (may not be effective against hot distilled water), such as ground blast furnace slag, pozzolans, such as 20 to 30% of a good natural pozzolan or fly ash, or silica fume.

SULFATE RESISTANCE

Sulfate Attack

Sulfate attack, sometimes called "sulfate corrosion," is a particularly severe type of deterioration resulting from chemical reactions occurring when concrete is exposed to water containing a sufficiently high concentration of dissolved sulfates ($SO_4^=$). Sulfate attack has been reported in many parts of the world, including western United States, the northern Great Plains area of the United States and Canada, Spain, Great Britain, the Middle East, and in industrial situations where concrete has been exposed to solutions containing sulfates. The attack is particularly prevalent in arid regions where naturally occurring sulfate minerals are present in the water and ground in contact with the structures. The necessary conditions for sulfate attack are known and preventive measures can be taken to eliminate or minimize the risk of sulfate attack.

Without taking adequate precautions, concrete structures such as floors, foundations, drainage pipe, and lower parts of canal structures may completely disintegrate in only a very few years when exposed to water containing dissolved sulfates. The severity of the attack can be affected by the presence of other dissolved substances in the water, but generally increases as the concentration of sulfates in the water increases, and becomes even more severe if the concrete is subjected to frequently alternating periods of wetting and drying. On the other hand, concrete exposed to dry sulfate-bearing soils will not be attacked.

The source for the sulfates involved in the chemical reaction may be either external or internal to the concrete [23–25]. In the case of external attack, the sulfates may originate from groundwater or from sulfates leached from adjacent soil as the reaction progresses from the surface into the interior of the concrete. In the case of internal attack, the sulfates may come from minerals in the aggregates, sulfates dissolved in the mix water, admixtures and additives, and possibly from sulfates in the portland cement. It is therefore necessary to have a chemical analysis of the ingredients of the concrete and of the groundwater and soil surrounding the structure in order to assess the probability of sulfate attack.

The gypsum normally added to portland cement to control setting characteristics is not sufficient to cause sulfate attack. Portland cement normally contains gypsum, which is added in amounts of about 2.5 to 4% SO_3 to control setting characteristics. This gypsum is consumed during the normal course of the cement hydration process.

The main constituents of hardened portland-cement paste susceptible to sulfate attack are the hydration products of the tricalcium aluminate ($3CaO \cdot Al_2O_3$) phase. Other cement hydration products may also be attacked in more severe sulfate environments.

The major factors governing the sulfate reactions are: (1) the permeability of the concrete to water, which is a control on the rate at which the reactants are brought together; (2) the composition of the hydrated phases of the cement paste, as not all cement hydration products are equally susceptible to reaction with sulfates; (3) and the chemical composition and concentration of the sulfate-laden water, which affects the course and the severity of the reaction. There are several different sulfate reactions depending on the composition of the sulfate water. The course of the reaction and the mechanism for deterioration may vary according to the situation. As the reaction proceeds inward into the concrete, the composition of the pore water may change and alter the course of the reaction.

Although sulfate attack has been extensively investigated, it is still not completely understood. The situation is not as straightforward as might be expected since several different chemical reactions rather than a single reaction may be involved. Some controversy still remains over the mechanism responsible for producing expansion and deterioration of the concrete. Most of the knowledge on sulfate attack has been developed from laboratory studies involving relatively simple chemical systems of pure materials, yet in actual practice the situation is more complicated as the exposures may involve varying combinations of impure materials in different proportions. In such cases, the water may contain other aggressive anions and cations that attack hydrated cement constituents by chemical reactions other than those associated strictly with the sulfate anion. At times these reactions may impede the attack, and at other times may aggravate the attack.

Deterioration due to sulfate attack is generally attributed to chemical decomposition of certain portland-cement hydration products after hardening, and formation of an expansive reaction product [26–32]. This produces expansive forces and a subsequent physical disruption of the concrete. Sulfate solutions may also react with the cement paste to form products that have little cementing value, and thereby turn the concrete into mush [29,33].

The main reaction product involved in sulfate attack is calcium sulfoaluminate. Calcium sulfoaluminate forms during the normal hydration of portland cement, and also forms as a reaction product of sulfate ion with calcium hydroxide and calcium aluminate hydrate. Calcium sulfoaluminate occurs in two forms, depending largely upon the supply of sulfate: a low sulfate form called calcium monosulfoaluminate ($3CaO \cdot Al_2O_3 \cdot CaSO_4 \cdot 12H_2O$), and a high sulfate form called calcium trisulfoaluminate or ettringite ($3CaO \cdot Al_2O_3 \cdot 3CaSO_4 \cdot 32H_2O$). The monosulfate form may be converted to ettringite when exposed to sulfate solutions, and is potentially destructive.

Ettringite is an expansive reaction product and is usually considered as the main destructive force in sulfate attack. However, there is some controversy over the formation of ettringite and the deterioration of the concrete [30,34,35]. The presence of ettringite alone is insufficient evidence of a destructive sulfate attack, as ettringite is also a secondary reaction product from the hydration of portland cement. If ettringite crystals are found in fractures, there may be a question of whether the fracture was caused by crystal growth or if the crystals happened to grow in the space provided by the fractures. Ettringite as a normal cement hydration product is usually not considered to be destructive. There is also a lack of correlation between the amount of ettringite formed and the amount of expansion observed. Furthermore, it appears that there is more than one type of ettringite, and not all types are expansive [33,36,37]. The mode of formation may also have an effect, that is, ettringite formed by solid-state reactions is more expansive than that formed through solution [30,38]. Mechanisms for the expansion of ettringite have been reviewed by Cohen [39] and Ping and Beaudin [40].

Thaumasite ($CaCO_3 \cdot CaSO_4 \cdot CaSiO_3 \cdot 15H_2O$) is structurally similar to ettringite and was observed by Erlin and Stark in 1965 in concrete attacked by sulfate solutions [41]. Since then, other instances of thaumasite in deteriorated portland cement concrete, soil cement, and plaster attacked by sulfates have been reported [42–47]. Thaumasite may form as a conversion product from ettringite from carbonation and silicon substitution, or may form directly under favorable conditions when there is a supply of alumina, calcium silicates or free silica gel, sulfate, and carbonate [43,46]. It would appear that the presence of C_3A is not necessarily a prerequisite for thaumasite formation, and therefore such precautions as a Type V cement would not necessarily provide protection against the formation of thaumasite. Thaumasite forms preferentially at low temperatures due to the higher solubility of calcium salts at low temperatures, and appears to form rather quickly at temperatures around 5°C [43,46].

In some cases, other reaction products are formed such as gypsum ($CaSO_4 \cdot 2H_2O$) and silica gel. Gypsum is also an expansive reaction product, but there is some question as to whether the formation of gypsum produces the same disruptive effect as is attributed to ettringite [37]. Even so, gypsum formation under sulfate attack conditions is associated with producing a deterioration effect on concrete as it involves the conversion of cementitious constituents into products having little cementing properties [29].

The American Concrete Institute (ACI) divides sulfate attack into two general categories: (1) a sulfate reaction with calcium hydroxide to form gypsum (sometimes called gypsum attack, gypsum corrosion, and acid type of sulfate attack), and (2) a sulfate reaction with calcium aluminate hydrate to form ettringite (sometimes called sulfoaluminate attack or corrosion) [5,6]. In more severe forms of the reaction, other constituents of concrete are susceptible to sulfate attack, such as calcium silicate hydrate, and even some types of aggregates particles, such as weathered feldspars.

The type of chemical reaction is related to the cations associated with the sulfates in solution, which most commonly are CA^{++}, Na^+, K^+, and Mg^{++}. The cations affect the solubility of sulfate minerals and the concentration of the sulfates in solution, and also the course of the reaction. With extremely high concentrations of sulfate, the reaction may intensify into acid attack that forms a different reaction product [10,37]. These reactions will be discussed in the following paragraphs. The course of the reaction and destructive effects are also affected by the presence of other dissolved constituents in the system that can tend to either ameliorate or intensify the destructiveness of the reaction. Sulfate attack is also affected by environmental factors such as temperature and wetting and drying.

In addition to the chemical form of attack involving chemical reactions, a purely physical form of sulfate attack is sometimes included under the general category of sulfate attack [48]. Crystallization pressures resulting from the precipitation of sulfate salts, such as Na_2SO_4 or $MgSO_4$, and subsequent growth of crystals, as may occur when concrete is exposed to sulfate solutions under wetting and drying conditions, can cause destruction of concrete. This mechanism is employed in ASTM Test Method for Soundness of Aggregates by Use of Sodium Sulfate or Magnesium Sulfate (C 88).

Calcium Sulfate Reaction

Calcium sulfate (most commonly occurring as gypsum, $CaSO_4 \cdot 2H_2O$) is not highly soluble in water, and sulfate attack on concrete in contact with soils containing gypsum may range from negligible to severe, depending upon the amount of gypsum dissolved in water in contact with the concrete. In concentrations approaching saturation (2000 mg/L water), calcium sulfate becomes reactive with calcium aluminate hydrate ($4CaO \cdot Al_2O_3 \cdot 19H_2O$), the main hydration product of the calcium aluminate phase in portland cement. The reactions of calcium sulfate with calcium aluminate hydrate and calcium monosulfoaluminate have been described [10,28,31,33,49–51]:

$$3(CaSO_4 \cdot 2H_2O) + 4CaO \cdot Al_2O_3 \cdot 19H_2O + 17H_2O \rightarrow 3CaO \cdot Al_2O_3 \cdot 3CaSO_4 \cdot 32H_2O \quad (1)$$
gypsum + calcium aluminate hydrate + water → ettringite

$$CaSO_4 \cdot 2H_2O + 3CaO \cdot Al_2O_3 \cdot CaSO_4 \cdot 12H_2O + 18H_2O \rightarrow 3CaO \cdot Al_2O_3 \cdot 3CaSO_4 \cdot 32H_2O \quad (2)$$
gypsum + calcium monosulfoaluminate + water → ettringite

Sodium Sulfate Reaction

Sodium sulfate is potentially more destructive than gypsum as it is more highly soluble (40 000 mg/L water) than calcium sulfate, and as it enters into two types of reactions with the hydrated cement phases in the hardened cement paste. Sodium sulfate reacts with calcium hydroxide as well as the calcium aluminate hydrate phases. At lower concentrations (SO_4 content less than 1000 mg/L), sulfate reacts with the hydrated calcium aluminate phases to produce ettringite and subsequent deterioration; at higher concentrations, sodium sulfate reacts with calcium hydroxide to produce gypsum [10]. Gypsum formation results in a volume increase, but there is some controversy as to whether the formation of gypsum produces a harmful expansion in concrete. The evidence for gypsum producing expansion and deterioration of the concrete does not appear as straightforward as does the case for ettringite formation [37].

The sodium sulfate reactions have been variously described [10,31,49,52,53]

$$Na_2SO_4 + Ca(OH)_2 + 2H_2O \rightarrow Ca_2SO_4 \cdot 2H_2O + NaOH$$
sodium sulfate + calcium hydroxide + water → gypsum + sodium hydroxide

and (3)

$$Ca_2SO_4 \cdot 2H_2O + 4CaO \cdot Al_2O_3 \cdot 19H_2O \rightarrow 3CaO \cdot Al_2O_3 \cdot 3CaSO_4 \cdot 32H_2O + Ca(OH)_2$$
gypsum + calcium aluminate hydrate → ettringite + calcium hydroxide

$$3Na_2SO_4 + 2(4CaO \cdot Al_2O_3 \cdot 19H_2O) + 14H_2O \rightarrow 3CaO \cdot Al_2O_3 \cdot 3CaSO_4 \cdot 32H_2O + 2Al(OH)_3 + 6NaOH \quad (4)$$
sodium sulfate + calcium aluminate hydrate + water → ettringite + aluminum hydroxide + sodium hydroxide

$$2Na_2SO_4 + 3CaO \cdot Al_2O_3 \cdot CaSO_4 \cdot 12H_2O + 2Ca(OH)_2 + 2H_2O \rightarrow 3CaO \cdot Al_2O_3 \cdot 3CaSO_4 \cdot 32H_2O + 2NaOH \quad (5)$$
sodium sulfate + calcium monosulfoaluminate + calcium hydroxide + water → ettringite + sodium hydroxide

Magnesium Sulfate Reaction

Magnesium compounds are potentially more destructive to concrete as the magnesium ion is capable of completely replacing the calcium in hydrated portland cement. Magnesium sulfate is highly soluble (70 000 ppm) and can form more highly concentrated sulfate solutions than sodium sulfate. Strong solutions of magnesium sulfate are capable of reacting with the calcium silicate hydrate phases as well as calcium hydroxide and calcium aluminate hydrate phases. The concentration of magnesium sulfate may affect the course of the reaction [10]:

1. At low concentrations (less than 3200 mg SO_4/L—or less than 4000 mg $MgSO_4$/L), the attack is characterized by ettringite formation. However, in the continued presence of $MgSO_4$, ettringite is eventually decomposed to gypsum, magnesium hydroxide, and hydrated alumina [31].

2. At intermediate concentrations (between 3200 and 6000 mg SO_4/L or 4000 to 7500 mg $MgSO_4$/L), the reaction is characterized by ettringite and gypsum formation. Deterioration and cracking may slow down and may be hardly perceptible.

3. At high concentrations (above 6000 mg SO_4/L or above 7500 mg $MgSO_4$/L), the reaction is characterized by the formation of magnesium hydroxide, gypsum, and silica gel; ettringite does not form. Deterioration becomes very severe as the magnesium concentration increases.

The various magnesium sulfate reactions have been described [10,29,31,50,54]

$$MgSO_4 + Ca(OH)_2 + 2H_2O \rightarrow CaSO_4 \cdot 2H_2O + Mg(OH)_2 \quad (6)$$
magnesium sulfate + calcium hydroxide + water → gypsum + magnesium hydroxide

$$3MgSO_4 + 3CaO \cdot SiO_2 \cdot 6H_2O \rightarrow 2CaSO_4 \cdot 2H_2O + 3Mg(OH)_2 + 2Al(OH)_3 \quad (7)$$
magnesium sulfate + calcium aluminate hydrate → gypsum + magnesium hydroxide + aluminum hydroxide

$$3MgSO_4 + 3CaO \cdot Al_2O_3 \cdot CaSO_4 \cdot 12H_2O \rightarrow 4CaSO_4 \cdot 2H_2O + 3Mg(OH)_2 + Al_2O_3 \cdot 3H_2O \quad (8)$$
magnesium sulfate + calcium monosulfoaluminate hydrate → gypsum + magnesium hydroxide + aluminum hydroxide

$$3MgSO_4 + 3CaO \cdot SiO_2 \cdot nH_2O + zH_2O \rightarrow 3CaSO_4 \cdot 2H_2O + 3Mg(OH)_2 + 2SiO_2 \cdot xH_2O$$
magnesium sulfate + calcium silicate hydrate + water → gypsum + magnesium hydroxide + silica gel

and (9)

$$Mg(OH)_2 + SiO_2 \cdot xH_2O \rightarrow 4MgO \cdot SiO_2 \cdot yH_2O$$
magnesium hydroxide + silica gel → magnesium silicate hydrate

Magnesium hydroxide has a beneficial effect as it is highly insoluble and forms surface deposits, which tend to block pores and hinder the further penetration of sulfate solutions into the concrete.

Other Reactions

The most common types of sulfate attack in natural environments are from calcium, sodium, magnesium, and occasionally potassium sulfate. Potassium sulfate reacts with concrete in a manner similar to that of sodium sulfate. In some industrial situations, concrete may be exposed to other types of sulfate solutions and to bisulfate solutions (HSO_4^-). Bisulfate solutions tend to be more acidic and can contribute to acid attack as well as sulfate attack. The aggressiveness of various sulfate salts towards concrete is partly related to solubility to water, in that sulfates of very low solubility (barium sulfate and lead sulfate) are relatively harmless. Ammonium sulfate is highly soluble in water (~800 g/L) and is highly aggressive towards concrete. Sulfates of copper, manganese, aluminum, iron, cobalt, and zinc, which are moderately soluble in water, are also aggressive towards concrete [55,56].

In low concentrations, chlorides have been shown to decrease the sulfate resistance of concrete [57,58], yet in high concentrations chloride is thought to temper sulfate attack as chloride increases the solubility and decreases the stability of ettringite and gypsum [10,31,36,59]. Chloride ions react with hydrated aluminates and monosulfoaluminate to form chloroaluminates.

There is conflicting evidence concerning the effects of solutions containing mixtures of sulfate salts and other dissolved substances. In some cases, sulfate solutions containing chloride, magnesium, sodium bicarbonate, and carbon dioxide, are highly aggressive towards concrete, yet seawater, which contains these substances, is not as aggressive as might be expected based upon magnesium and sulfate contents [31,50,56,60].

Control of Sulfate Attack

The widespread occurrence and destructiveness of sulfate attack led to many investigations over the years into the mechanism for the deterioration and means to combat the reaction. There have been many studies in many parts of the world and an extensive body of knowledge on sulfate attack has been developed. It is not possible to adequately recognize the many contributions and only a few of the contributions can be mentioned in passing.

Some of the earlier studies in the United States include those conducted by the Metropolitan Water District [26], the U.S. Department of Agriculture and the University of Minnesota [61], the Portland Cement Association [27,29,62–64], the California Division of Highways [39], the Bureau of Standards [65], the Corps of Engineers [66,67], the Bureau of Reclamation [57,68–71], and at the University of California for the Bureau of Reclamation [72].

Extensive studies have been conducted elsewhere in the world such as in Canada, England, France, Russia, South Africa, Australia, Belgium, Sweden, Hungary, and Israel, (for example, see Refs 10,28–31,73–76 for brief summaries, and more recently Refs 32,33,35,37,53,77–84).

Most of these investigations seem to reach (or confirm) the same three basic conclusions on the resistance of concrete to sulfate attack:

1. The permeability and quality of concrete is of great importance to sulfate resistance; therefore, it is essential to have a high quality, impermeable concrete. This requires good workmanship (workable mixes, good consolidation, a hard finish, and good curing), and to use a relatively rich mix with a low water/cement ratio.
2. The deterioration in most cases shows a relationship with the composition of the cement. The deterioration generally increases as the amount of tricalcium aluminate (C_3A) in the cement increases. However, some exceptions have been noted: in some cases, low C_3A cements were observed to have poor sulfate resistance, some moderately high C_3A cements were observed to have good sulfate resistance, and in several instances zero C_3A cements have shown poor sulfate resistance [32,74].
3. The deterioration also shows some relationship to $Ca(OH)_2$ in the hydrated cement paste, as $Ca(OH)_2$ enters into the chemical reactions involved in sulfate attack. Therefore, it should be possible to improve the sulfate resistance by controlling the amount of $Ca(OH)_2$ in the cement paste, either by limiting the amount formed from hydration or by converting the $Ca(OH)_2$ to a more chemically resistant CSH. $Ca(OH)_2$ may be converted to CSH through the addition of SiO_2, either in an active form as a pozzolan or slag, or as pulverized quartz and steam curing.

The three main strategies for improving resistance to sulfate solutions are: (1) making a high quality, impermeable concrete; (2) using a sulfate-resistant cement; and (3) using pozzolans or slag [5,6]. The service life of concrete also may be increased by protecting the concrete from exposure to sulfates. These measures include limiting the sulfate content of the materials used to make the concrete, designing the structure so as to minimize exposure to sulfate solutions, and using impermeable barriers to prevent sulfate solutions from coming into contact with the concrete. Impermeable barriers are not recommended as a long-term solution to an aggressive sulfate condition as there is no guarantee the barriers will perform effectively over a long term.

It essential to have a high-quality impervious concrete to improve sulfate resistance [82,83,85,86]. This involves using a low water-cement ratio and a high cement content in the mix, and also good workmanship and construction techniques. This includes good mixture proportioning and selection of materials, consolidation of the concrete during placement, and good finishing and curing to produce a hard and dense surface.

The sulfate resistance of portland cements may be improved by limiting the C_3A content in the cement clinker, by limiting the amount of $Ca(OH)_2$ formed in the hydrated cement paste, or by converting the $Ca(OH)_2$ to a more stable form either as CSH or $CaCO_3$. Carbonation of the surface layer, and air entrainment also may aid by reducing the permeability of concrete and improving sulfate resistance [65,87–89]. The use of calcareous fillers has been reported as having a beneficial effect due to a reaction with C_3S to form carboaluminate ($C_3A \cdot CaCO_3 \cdot 12H_2O$) [90]. It may be noted that the development of high alumina cements (which contain calcium aluminates in forms other than C_3A) was in part stimulated by a search for sulfate-resistant cements. The high alumina cements generally have very good sulfate resistance [31,91].

One means to reduce the C_3A content of the cement is to convert the C_3A phase to C_4AF and CF by the addition of iron oxide during the cement manufacturing process. But as the hydrated ferrite phases are ultimately susceptible to sulfate attack in severe environments, specifications for highly sulfate-resistant cements place a limit on the total amount of C_3A and C_4AF + CF.

The $Ca(OH)_2$ content of the hardened cement paste can be controlled to a certain extent by increasing the amount of C_2S and decreasing the C_3S phase in the portland cement as the C_2S phase liberates less $Ca(OH)_2$ than C_3S during hydration; however, the evidence does not clearly indicate that increasing the relative proportion of C_2S with respect to C_3S is effective in controlling sulfate resistance [31].

The $Ca(OH)_2$ formed during the hydration of portland cement can be converted to a more stable CSH form through a chemical reaction with reactive forms of SiO_2. This can be accomplished by adding finely ground SiO_2 to the mix followed by high pressure steam curing at 100°C or higher, or by combining portland cement with active siliceous mineral admixtures such as natural or calcined pozzolans, fly ash, silica fume, or slags. In regards to heat-treated or low-temperature steam-cured concrete (around 75 or 80°C), it should be noted that some observations have been reported of an apparent delayed formation of ettringite, which in some cases has been suspected of causing deterioration in cast concrete products [25,92–95]. In some cases, alkali aggregate reaction and freeze-thaw deterioration are also suspected of being involved in the deterioration.

Pozzolans and slags improve sulfate resistance by chemically reacting with $Ca(OH)_2$ to form CSH and by reducing the permeability of the concrete. Slags have been shown to be effective when substituted for at least 50% of the cement as long as the Al_2O_3 content is limited [96]. However, the addition of a pozzolan or slag does not automatically guarantee sulfate resistance, and tests should be made to determine effectiveness in improving sulfate resistance.

There are some apparent inconsistencies regarding the effectiveness of pozzolans in improving sulfate resistance. Not all pozzolans are effective in improving sulfate resistance. The apparent inconsistencies are due in part to differences in the composition and fineness of the pozzolans, and also to the amount of pozzolan and the amount and type of cement in the mixture. Some differences in performance may depend upon whether a particular pozzolan is used as a replacement or as an addition to the cement. Other differences may arise depending upon the criteria selected to base performance (comparison with other mixes or against an absolute scale).

Pozzolans are used for several different purposes, and often used as a means to reduce the amount of cement in a mix. Consequently, pozzolans are customarily thought of in terms of a cement replacement, but in regards to improving sulfate resistance, it may be better to think of pozzolans in terms of an addition to the cement rather than as a replacement (that is, maintain a high cement content in the mix) [97].

A pozzolan specification according to ASTM C 618 alone does not guarantee effectiveness for sulfate resistance [98], but at the risk of over simplification, the following generalizations may be made regarding the effectiveness of pozzolans classified according to ASTM C 618 for improving sulfate resistance:

1. Class N pozzolans are variable but generally good. The least effective are usually those causing appreciably higher mixing water requirements. Best results are more likely to result if the replacement is limited to 15% [71].
2. Most Class F fly ashes improve the sulfate resistance of Type II and Type V cements [99–101], and generally are more efficient in improving sulfate resistance than Class C ashes [102]. However, some ashes with a high alumina content are not effective in improving sulfate resistance.
3. Low calcium Class C fly ashes are often good; but high calcium Class C ashes are variable, often poor, and may reduce sulfate resistance [71,97,99,101,103–105]. Some Class C ashes must be used at replacement levels greater than 75% to achieve sulfate resistance [101].
4. Silica fume is generally very effective in improving sulfate resistance [102,106–109], perhaps even more so than fly ash or ground slag.

Two general approaches have been taken for the specification of a sulfate-resistant cements: (1) base specification on the chemical composition of the cement such as ASTM Specification for Portland Cement (C 150); or (2) on performance tests such as ASTM Test Method for Length Change of Hydraulic-Cement Mortars Exposed to a Sulfate Solution (C 1012), ASTM Test Method for Potential Expansion of Portland-Cement Mortars Exposed to Sulfate (C 452), or Bureau of Reclamation (USBR 4908) Procedure for Length Change of Hardened Concrete Exposed to Alkali Sulfates. There are some cases where a specification based on chemical composition alone would exclude a sulfate-resistant cement. In these cases, performance tests may be specified in lieu of a chemical composition specification. Performance tests should be specified for combinations of portland cements and pozzolans or slags.

Portland cement with a C_3A content below 5% is generally considered to be sulfate resistant [29,84]. ASTM C 150 designates Type II and Type III cements containing no more than 8% C_3A as having moderate sulfate resistance, and Type V cements (contain less than 5% C_3A and 25% $C_4AF + 2(C_3A)$ or $C_4AF + C_2F$), and Type III cements containing no more than 5% C_3A, as having high sulfate resistance. Type I cements with pozzolans or slag, generally are considered to have moderate sulfate resistance.

Studies have been made for the development of specifications for sulfate-resistant fly ashes. These have included specifications based on chemical composition and mineralogical composition [110]. For example, an "R value" based on chemical composition of fly ash (where R = %CaO − 5 ÷ %Fe_2O_3) has been proposed as a method to estimate sulfate resistance [98,111,112]. For replacement mixes in the 15 to 25% range, R values below 3 indicate improved sulfate resistance. The R value has been questioned as it does not directly take into account the alumina content of the fly ash; however, the R value indirectly allows for alumina since alumina combined with lime is available for attack and not when combined with iron [112].

Specific recommendations for sulfate resistance cited by Tuthill are fairly well established and have shown little change [26,113]:

1. When long-range durability of concrete is at stake in an aggressive environment, it is more important to specify limits that will provide the most sulfate resistance, rather than to require composition limits that most manufacturers can meet.
2. Use a portland cement with a low C_3A content [61,114].
3. Use ample amounts of cement. The lower the C_3A content of the cement and the richer the mix, the better will be the resistance of concrete to sulfate attack. An additional sack of Type V sulfate-resistant cement will increase resistance at least 50%. However, an additional sack of Type II cement will not provide additional sulfate resistance equal to that provided by Type V cement with 4.5% C_3A, unless the Type II cement contains less than 5.5% of C_3A. Extending the curing period provided prior to sulfate exposure also improves sulfate resistance.
4. Use a low w/c ratio. A w/c ratio below 0.50 is recommended for mild exposures, and 0.45 or below for moderate to severe exposures [4,6,58]. A w/c of 0.40 or lower is even better and is preferable in extreme exposures. At the very low w/c ratios, the effects from the composition of the cement are less apparent [83].
5. If the pozzolan is to be used on the basis of a cement replacement, the pozzolan should be substituted for 15 to 30% by weight of the cement, provided the pozzolan is proven effective in increasing sulfate resistance of Type II and Type V cement. Silica fume is particularly effective in improving sulfate resistance. Slag, when substituted at least 50% by weight of cement, has been shown to be effective in improving sulfate resistance as long as the Al_2O_3 content of the slag is low [96].[3]
6. A low-lime and low-aluminate cement, having less than 50% C_3S and less than 12% $C_3A + C_4AF$, in which less than 4% is C_3A, makes the most sulfate-resistant portland cement.
7. Air entrainment is generally recognized as being beneficial to sulfate resistance, especially when air entrainment is used to advantage in reducing the water-cement ratio [87,88,102].
8. Precast units such as concrete pipe and block can be made appreciably more resistant by introducing a period of several weeks of drying following good curing. Presumably, carbonation occurring during this period contributes to improved sulfate resistance, as it reduces permeability.
9. Steam curing at temperatures above 212°F (100°C) increases the sulfate resistance of concrete, preferably at temperatures of 177°C (350°F), with silica additions to the concrete mix [61,74,115].
10. Calcium chloride reduces sulfate resistance [57].

Two aspects of controlling sulfate attack are identification of the severity of the exposure conditions and selection of appropriate precautionary measures. Table 1 [5,116] shows the severity of sulfate attack to be expected from the sulfate concentrations encountered in both soil and in water, and recommended precautionary measures. These recommendations are generally consistent with those given by the Bureau of Reclamation [58], the Corps of Engineers, and the Portland Cement Association [102].

The table is intended for use in concrete construction under ordinary service life conditions, but may not necessarily be reliable for the ultimate protection of concrete in exposures to exceptionally severe conditions such as magnesium-sulfate exposures, high temperature exposures, or for mega year service life, such as may be encountered in severe industrial exposures or hazardous waste containment [37,117]. For example, water containing magnesium sulfate in concentrations of more than 6000 ppm SO_4 (equivalent to 0.75% $MgSO_4$) is capable of attacking all portland cement hydration products regardless of cement type or combination of pozzolan or slag [10,37]. In this case, a special cement should be used. Therefore, some judgement must be exercised as the table lists only the concentration of the sulfate anion (SO_4) as a criterion for potential severity of sulfate attack and does not include cations that also have an effect on the conditions and severity of the reaction. It is also possible to have cases where a potentially destructive sulfate attack condition may exist even though a sulfate analysis indicates the ground water or soil to have a low sulfate content. These are situations where the exposure conditions may cause the water in contact with the structure to become more concentrated in sulfates due to ponding or evaporation [37], or where under severe wetting and drying conditions, water may evaporate and crystals of sulfate salts may grow and physically attack the concrete [29,118].

Tests for Sulfate Resistance

Tests for sulfate resistance are made to study the sulfate resistance of different cements and various combinations of materials. These tests include field exposures under

---

[3] The w/c rules for concrete made with ordinary portland cement are also equally applicable to blended cements.

TABLE 1—Recommendations for Concrete Exposed to Sulfate Solutions.[a]

Relative Degree of Sulfate Attack	Water Soluble Sulfate[b] in Soil, %	Sulfate[b] in Water, ppm	Recommended Type of Cement	Normal-Weight Aggregate Concrete, water-cement ratio (max)	Lightweight Aggregate Concrete, minimum compressive strength, (f'_c)
Mild[c]	0.00 to 0.10	0 to 150	Type I or II	...	...
Moderate[d]	0.10 to 0.20	150 to 1500	Type II, IP(MS), IS(MS)[e]	0.50	3750
Severe[f]	0.20 to 2.00	1500 to 10 000	Type V[g]	0.45	4250
Very severe[f]	over 2.00	10 000 or more	Type V + pozzolan or slag[h]	0.45	4250

[a] Table after ACI 201 "Guides to Durable Concrete" [5], ACI 318 "Building Code Requirements for Reinforced Concrete and Commentary" [116], PCA "Design and Control of Concrete Mixtures" [102], and USBR, *Concretes Manual* [58].
[b] Sulfate expressed as SO_4 is equivalent to 1.2 × sulfate expressed as SO_3.
[c] If the corrosion of reinforcing steel is of concern and chlorides or other depassivaing agents are present, a lower water-cement ratio may be necessary to reduce corrosion potential (see Chapter 4 in ACI 201 "Guide to Durable Concrete" [5])
[d] Includes seawater exposure.
[e] Or a blend of Type I cement and a ground granulated blast furnace slag or a pozzolan that has been determined by tests to give equivalent sulfate resistance. In addition, ACI 318 also permits P(MS), I(PM)(MS), and I(SM)(MS) cement types.
[f] Use of calcium chloride admixture is not permitted.
[g] Or a blend of Type II cement and a ground granulated blast furnace slag or a pozzolan that has been determined by tests to give equivalent sulfate resistance. (ASTM [113] recommends use of Type V or Type V + extra cement or an effective pozzolan.)
[h] Use a pozzolan or slag that has been determined by tests to improve sulfate resistance when used in concrete containing Type V cement.

natural conditions, and laboratory tests of concrete or mortar specimens exposed to artificial and generally accelerated exposure conditions. As the exposure conditions vary, some differences in the rate and severity of attack may be expected. Laboratory tests involve exposing mortar or concrete specimens to sulfate, either internally by adding the sulfate to the mixture, or externally by soaking the specimen in a solution containing sulfate. Sulfate exposure conditions are accelerated by using high sulfate concentrations, by using magnesium sulfate solutions, or by using a wet-dry exposure cycle. A variety of test methods have been devised [10,25,68,84,89,119–121].

ASTM C 452 involves adding gypsum to a mortar bar (1 by 1 by 10-in. or 25 by 25 by 250-mm size prism) in amounts so that the total sulfur trioxide content of the specimen is 7% by mass. The specimens are stored in water and length changes are measured over time. The test is considered suitable for portland cements, but not for blended cements or blends of portland cement with pozzolans or slags. Results of this test generally conform with the long-term behavior of cements in concrete exposed to aggressive sulfates and with predictions based on the C_3A content of the cement. This test is used to establish that a sulfate-resisting portland cement meets the performance requirements of ASTM C 150.

ASTM C 1012 involves expansion of mortar specimens in a 5% Na_2SO_4 solution (50 g/L) at 73.4°F or 23°C and measuring length change. The test is considered suitable for portland cements, blends of portland cements with pozzolans or slags, and blended hydraulic cements. The test does not simulate sulfate attack by solutions of sulfate compositions other than that used, and if evaluation of behavior in exposure to a sulfate solution of a given composition is desired, then that solution should be used in the test.

The Bureau of Reclamation Procedure 4908 [122] describes three methods for testing 3 by 6-in. (76 by 152-mm) cylindrical specimens. Method A consists of continuous soaking of test specimens in a 2.1% Na_2SO_4 solution at room temperature; Method B consists of continuous soaking of test specimens in a 10% Na_2SO_4 solution at room temperature; and Method C consists of alternating wetting and drying tests of specimens soaked in 2.1% Na_2SO_4 solution. Methods B and C produce failure in about one-sixth the time required in Method A. An expansion of 0.5% is considered failure in the test. A failure criterion of a 40% reduction in dynamic modulus of elasticity has also been used.

ACID ATTACK

As concrete is an alkaline material, it is subject to deterioration in exposure to acid solutions. Fortunately, most natural waters are neutral or only slightly acidic and do not present a problem. However, natural waters can become acidic, either through natural processes or from contamination, and attack concrete. Natural processes for acidifying water include the decay of organic matter in marshes and swamps, and absorption of gases and pollutants from the atmosphere such as CO_2 and SO_2. Natural waters can also be contaminated by acid mine drainage, and agricultural and industrial wastes. In industrial exposures, concrete may be exposed to a wide variety of acidic solutions.

Acid solutions attack concrete through dissolution and acid-base chemical reactions. The reaction attacks concrete on the exposed surface and works inward, and generally involves removal of soluble reaction products. The acid attacks cement hydration products and, in some cases, may also attack certain types of aggregate particles. The hydration product most susceptible to reaction with acid is $Ca(OH)_2$, which is readily attacked and dissolved by even relatively mild acid solutions. More aggressive acid solutions react with other cement hydration products, and ultimately all cement hydration products may be attacked. As an example, ASTM Test Method for Portland

Cement Content of Hardened Hydraulic-Cement Concrete (C 1084) is based on dissolution of all hydrated cement paste in hydrochloric acid. Certain types of aggregates, such as limestone and dolomite that contain carbonate minerals, react with acids.

The two main factors involved in the chemical reactivity of an acid are the concentration and strength of the acid. The concentration of a particular acid solution refers to the weight of acid per unit volume of water and is commonly expressed either in terms of normality or molarity.

The strength of an acid is determined by its tendency to dissociate in water and form protons (H^+ ions). The intrinsic property of a particular acid to dissociate in water is indicated by the dissociation constant, K_a, for the acid, which also may be expressed in terms of pK_a (defined as the negative logarithm of the equilibrium constant, K_a, for the acid, $pK_a = -\log K_a$). Strong acids have a large dissociation constant and a small pK_a. Strong acids ionize completely in water; weak acids do not completely ionize. For example, battery acid is concentrated H_2SO_4, is highly ionized ($pK_a = 1.92$), and is a strong acid. Vinegar is dilute acetic acid, dissociates to a much lesser extent ($pK_a = 4.5$), and is a weak acid.

The chemical reactivity of an acid is due to the hydrogen ion (H^+), and reactivity increases as the H^+ concentration increases. The H^+ concentration of a particular solution is related to both the strength and concentration of the acid in the solution, and is commonly expressed in terms of pH (originally defined as equal to the negative logarithm of the H^+ concentration). A 1-normal solution of a strong acid (one gram equivalent weight of a completely dissociated acid per litre of water) theoretically has a pH of 0, and a 0.1-normal solution has a pH of 1. A neutral solution has a pH of 7. A decrease of a full pH unit represents a 10 times increase in the concentration of the acid, that is, an acid solution of pH 4 is 100 times more concentrated than an acid solution of pH 6.

pH measurements are commonly used as an indicator of the aggressiveness of a solution, but may be misleading. For example, it is possible for a dilute solution of a strong acid to have the same pH as a concentrated solution of a weak acid. Both solutions may initially attack concrete at about the same rate, but on an equal volume basis, a given volume of the weak acid ultimately will be more destructive to concrete than an equal volume of the stronger acid as the weaker solution contains more acid [55].

The rate and extent of attack is governed by several factors in addition to the strength and concentration of acid. These factors include exposure conditions (exposure to static or running solutions), the transport of the acid into the concrete (diffusion and permeation), the nature of the reaction product formed, and in some cases, the rate at which the reaction product is removed. Flowing solutions are more aggressive than static solutions as more acid is brought into contact with the concrete and as the flowing solutions tend to remove the reaction products. The extent of the attack is governed by the rate the acid penetrates into the concrete. Acids that form soluble reaction products are generally more aggressive than acids that form insoluble reaction products. Most acids react to form soluble products; oxalic and phosphoric acids are exceptions and form insoluble reaction products that are not readily removed [6]. Insoluble reaction products, such as SiO_2 gel, $CaCO_3$, or $CaSO_4$, may tend to seal the concrete surface from the acid and impede further reaction of the acid with the concrete.

Hydrochloric acid (muriatic acid), nitric acid, and sulfuric acid are among the more common acids and are highly reactive with concrete. Sulfuric acid is particularly aggressive since the sulfate ions it contains are also available for sulfate attack reactions. Dilute hydrochloric acid has been used to clean concrete surfaces and for architectural purposes in making exposed aggregate surfaces; however, hydrochloric acid is not recommended for these purposes as the reaction is difficult to control and as it is difficult to make sure the surfaces have been washed clean and all acid removed. Organic acids range from highly to slightly reactive. Lactic acid, acetic acid (vinegar), citric acid, and tartaric acid are highly to moderately reactive and oxalic acid is only slightly reactive. Oxalic acid is sometimes beneficial as it has been used as a floor treatment to improve the resistance of the floors to other weak organic acids. The literature contains tabulations on the exposure of concrete to various acids [4,31,55,123].

The leaching of concrete ascribed to soft water is largely due to carbonic acid (H_2CO_3) which forms when the water absorbs carbon dioxide (CO_2) from the atmosphere. Such waters are quite aggressive towards concrete [9,10,12,124,125]. The effect of CO_2 on the pH of soft water is quite complex [126,127]. The amount of CO_2 that can be dissolved in water is directly proportional to the partial pressure of CO_2 in the atmosphere and inversely proportional to temperature. As the dissolved CO_2 content of the water increases, the CO_2 combines with water to form carbonic acid. Carbonic acid dissociates to produce H^+ in two steps. In the first step, H_2CO_3 dissociates to form $H^+ + HCO_3^-$, lowering the pH of the water. In the second step, as more CO_2 is dissolved and the pH decreases, the HCO_3^- ionizes to $H^+ + CO_3^=$, which further lowers the pH.

Concrete sewer lines are susceptible to acid attack although the sewage itself is usually neutral or only slightly acidic. The deterioration of concrete in sewer lines is observed above the water line, rather than below where the sewage is in direct contact with the concrete. Problems of acid attack above the water level in sewers arise indirectly from the bacterial decomposition of sewage, which produces hydrogen sulfide gas. The gas rises and combines with oxygen and with moisture condensed on upper surfaces of the sewer conduit to form sulfuric acid where it reacts with the cement paste [128]. Acid corrosion problems in the crown portions may be reduced by design or operating modifications that result in the sewers running full or with ventilation at higher velocities. Lower temperatures also tend to reduce the production of acid-forming hydrogen sulfide gas. When sewage is stagnant or moves slowly, bacterial creation of the sulfide may be too rapid for the sulfide to be oxidized by air dissolved in the sewage. But if industrial wastes are discharged into sewage lines, the situation may significantly change and present a special problem.

Engineers are often faced with questions concerning whether concrete will be resistant to acidic solutions of varying concentrations. As acid solutions are commonly described in terms of pH, most engineers are interested in rules of thumb regarding the resistance of concrete to acids at various pH levels. Although pH by itself is not necessarily a good indicator of aggressiveness, most investigators seem to feel above pH 5.5 the prospective severity of acid attack is zero to very slight, from about pH 4 to 5.5 acid attack may range from slight to somewhat aggressive if the acid is replenished and some precautionary measures may be advisable (such as chemically resistant slag cements or supersulfated cement), and below pH 3.5 acid attack may be severe and the concrete should be protected [4,6,10,55,129–131]. For neutral to moderately acidic natural waters, the American Water Works Association recommends using an Aggressiveness Index, which is a function of pH, calcium hardness, and total alkalinity [132].

Testing for acid resistance usually consists of soaking test specimens in acid solutions followed by periodic testing. Most tests are run under static conditions, and may show some inconsistencies as the acid is consumed during the test. ASTM Test Method for Chemical Resistance of Mortars (C 267) consists of immersing test specimens in the particular solution under investigation over a period of time. Specimens are removed periodically and tested for visual appearance, weight loss, and loss of compressive strength. Some feel such tests should be conducted on a depth of scaling basis rather than weight loss.

Surface treatments or protective coatings are sometimes prescribed for concrete exposed to various acids, but usually there is no guarantee as to their long-term effectiveness [4,6,123]. If the exposure is severe, it is better to use an acid-resistant cement than to rely on a protective coating system.

SEAWATER AND BRINES

The fact that plain concrete (nonreinforced) of good quality can have good resistance to seawater is amply demonstrated by the many examples of structures showing good service life after many years exposure to seawater. Examples of good performance of concrete made using seawater as mix water can also be found. But there are also many examples of concrete deterioration in exposure to seawater. Most of these examples involve the deterioration of poor quality, permeable concrete or the corrosion of reinforcing steel. The permeability of concrete is one of the key factors in seawater attack [59].

There are several mechanisms involved in the deterioration: leaching and chemical attack of the cement paste by seawater, corrosion of steel reinforcement, wetting and drying effects, freezing and thawing, and wave action [8,9,31,38,59,106,133–136]. Biological attack has also been identified [137].

The type of attack on concrete depends on the exposure conditions of the concrete to seawater. Generally speaking, concrete exposed above the high tide line is subject to attack from corrosion of reinforcement and freezing and thawing; concrete in the tidal zone or splash zone is susceptible to cracking and spalling from wetting and drying, frost action, corrosion of reinforcement, and erosion of material from chemical reactions by wave action; and below the water line by chemical reactions between seawater and cement paste that soften the concrete and make it susceptible to erosion [59,138]. The mechanisms for deterioration are complicated as the reactions involved in seawater attack may vary according to the exposure and temperature conditions and according to the depth of penetration of the various seawater constituents into the concrete.

Seawater contains about 3.5% dissolved salts, mainly magnesium sulfate and sodium chloride. The most commonly described reaction of seawater with cement paste is sulfate attack although other reactions involving chloride and magnesium ions and carbonation also occur (see discussion of $MgSO_4$ reactions under the section on sulfate attack) [10,31]. The reaction is somewhat different than ordinary sulfate attack in that the reactions seem to be more complex, the concrete generally deteriorates from erosion and loss of concrete constituents rather than by expansion and cracking, and the chemical composition of the cement seems to play only a minor role if at all [31,50,65]. The observed deterioration does not seem to be related to the presence of ettringite per se. Two types of ettringite have been observed: one type is associated with Type I and II cements and is considered expansive[4] and the other type is associated with Type V cement and does not appear to be expansive [36].

Most investigators observe that seawater is not as destructive to concrete as the total sulfate content of the water would indicate, and suggest this is due to the effects of dissolved chloride on the reactions. The role of chloride appears to be quite complex. Chloride ions penetrate cement paste more readily than sulfate and other ions, with the result that the course of chemical attack may vary with depth within the concrete according to rate the reactive species penetrate into the concrete. Chloride and dissolved CO_2 in the seawater also act to dissolve $Ca(OH)_2$ in the cement paste. Leaching and subsequent erosion of concrete is more severe in seawater as calcium hydroxide and gypsum are more soluble in seawater than in ordinary water.

Chloride reactions appear to have three main effects on the cement paste: lowering of pH and dissolution of $Ca(OH)_2$ and gypsum, reaction with calcium aluminate to form chloroaluminate and suppression of the formation of calcium sulfoaluminate, and enlargement of pores and increase in the permeability of the cement paste [31,38,139,140]. Chlorides tend to reduce the pH of the system, and as the pH is lowered, the solubility of $Ca(OH)_2$ is increased. The leaching of $Ca(OH)_2$ is only very slight at pH 13, but increases as pH is lowered [139]. Ettringite is also more soluble in solutions containing chloride [140]. The presence of chloride favors the formation of chloro-

---

[4] This does not mean all ettringite in concrete made with Types I and II cements has caused expansion.

aluminate over ettringite; however, in the presence of sufficient sulfate the chloroaluminate will convert to ettringite.

The main line of defense against seawater attack is making good quality impervious concrete. This generally means using a high cement content, low water/cement ratio, good consolidation, and workmanship. The use of pozzolans such as fly ash and silica fume, or slag cements generally help to make dense and impermeable concrete and are of help [10,31,73,141,142]. In very severe cases, special cements such as high alumina cements and polymer concretes work well [143].

The question of the durability of concrete in exposure to brines and seawater, especially at high temperatures, arises in industrial situations, hazardous waste containment, and in saline water conversion projects involving the use of concrete structures.

The durability of concrete exposed to distilled water and synthetic seawater brine at normal and elevated temperatures has been investigated fairly extensively by the U.S. Bureau of Reclamation in a study on the use of portland-cement concrete in saline water distillation plants [22,144]. These investigations showed that conventional portland-cement concrete is not affected by exposure to seawater for four years at temperatures up to 200°F (93°C), and showed only slight surface deterioration at exposures up to 250°F (121°C). Conventional portland cement concrete is considered suitable for use in exposures ranging up to 200°F (93°C), and up to 250°F (121°C) if sacrificial concrete is provided. Concrete exposed to seawater at 290°F (143°C) for 40 months showed surface deterioration to a depth of about 15 mm and was not considered suitable for long-term use at this temperature. Hot seawater exposure studies performed on concrete made with a crushed limestone coarse aggregate showed that the limestone is susceptible to dissolution when directly exposed to hot brines, such as on sawed surfaces, and that the concrete undergoes abnormally large expansions. The use of limestone aggregates in such exposures is not recommended. However, concrete made with natural siliceous aggregates did not show signs of deleterious sulfate attack.

CARBONATION

The reaction of carbon dioxide gas with hydrated portland cement, called "carbonation," is known to produce both beneficial and detrimental effects in concrete (the reactions of carbon dioxide dissolved in water is discussed in the sections on Scaling, Efflorescence, and Leaching and Acid Attack). Carbon dioxide gas reacts with calcium hydroxide in the hydrated cement paste to form calcium carbonate [31,51,145–149]. All calcium silicate hydration products are susceptible to carbonation, and over long periods of time will be ultimately converted to calcium carbonate and hydrated silica [51,147,148,150]. The basic reaction is

$$CO_2 + Ca(OH)_2 \rightarrow CaCO_3 + H_2O$$
carbon dioxide + calcium hydroxide → calcium carbonate + water

The long-term exposure of concrete to the atmosphere often results in shrinkage of the concrete and sometimes cracking if the shrinkage is great enough. The shrinkage is generally attributed to two main causes: (1) a gradual loss of moisture (drying shrinkage), (2) and carbonation. In situations where the long-term volume stability of concrete is important, carbonation becomes of interest not only in studies to separate the two causes for shrinkage, but also as a means to control shrinkage. Intentional carbonation of freshly placed concrete is an effective means to control shrinkage and also to improve strength, hardness, and imperviousness of concrete [151].

Moisture is an important factor in carbonation. Studies have indicated that carbonation does not occur in dry cement or cement at 100% relative humidity (RH) [148]. Apparently, at 100% RH, the moisture blocks carbon dioxide from passing through the pores. The optimum conditions for carbonation appear to be around 50% RH.

Carbonation is of concern in Europe in regard to the deterioration of concrete balconies, facades, parapets, etc., although these problems do not appear to be quite so wide spread in North America [152]. The deterioration is caused by the corrosion of reinforcing steel. Carbonation reaction consumes calcium hydroxide, destroys the passification effect of calcium hydroxide in preventing the corrosion of reinforcing steel, and lowers the pH to a point where corrosion readily occurs. Carbonation also causes shrinkage of the cement paste that may produce cracking or surface crazing of concrete [149,153–155]. Carbonation often acts in concert with other destructive mechanisms, such as chemical attack or freeze-thaw action [156].

Both fresh and hardened concrete are susceptible to carbonation, and processes have been developed to use carbonation as a treatment to produce beneficial effects. The treatments are reported to harden surfaces, increase compressive strength, increase imperviousness, and reduce shrinkage [146,151,157]. The process has been used as a means to prevent crazing, decrease permeability and improve frost resistance, control dusting and increase wear resistance of the surface, and to control aggressive chemical reactions including further carbonation and alkali-aggregate reaction.

ATTACK BY OTHER CHEMICALS

Concrete has good resistance to attack by dry chemicals, dry soils containing chemicals, most alkalies, mineral oils, neutral salts and many salts of strong bases, although it is susceptible to varying degrees of attack by other chemicals not already covered in this chapter. These include acid-producing substances, ammonia salts, magnesium salts, concentrated solutions of strong bases such as NaOH, chlorides, sulfates, nitrates, salts of weak bases, animal wastes, vegetable and animal oils, fats, sugar, glycol, glycerol, phenol, and creosote. The literature contains fairly comprehensive lists of various substances and their effects on concrete [4,6,10,31,55,123,158–160].

Such generalities on the resistance or susceptibility of concrete to chemical attack should be taken with some caution as the length of exposure, temperature, exposure conditions, concentration, and reactivity of the particular chemical must also be considered. Some chemicals ordinarily considered as not particularly aggressive towards concrete may react very slowly under prolonged exposure conditions or at elevated temperatures, and progressively attack concrete over a period of many years, and ultimately may prove to be aggressive towards concrete. Some substances, which by themselves may be harmless, may react with other substances in the environment and become aggressive. Examples of these materials are vegetation that may decompose to form organic acids and sulfides that may oxidize to form sulfuric acid. The severity and rate of attack may vary greatly and an actual test under anticipated exposure conditions is recommended to assess the potential for deterioration.

Vegetable and animal oils usually contain fatty acids that may be oxidized to form more acid during the exposure and become even more aggressive towards concrete. Many oils, fats, and glycerides are saponified by the lime in hydrated cement to form alcohols that can react with more lime to further deteriorate the concrete. On the other hand, drying oils (linseed oil and tung oil), undergo oxidization and polymerize when exposed to air. These oils form hardened films and have been used as protective coatings on concrete.

Many salts are not particularly harmful to plain portland cement concrete under ordinary circumstances, but in some particular situations may cause damage in other ways, such as by accelerating the corrosion of reinforcing steel or by "salt attack" (corrosion is discussed in the chapter on Embedded Metals and Materials Other Than Reinforcing Steel). Salt attack includes both "salt scaling" and "salt weathering." These forms of salt attack are caused by destructive physical processes rather than by a strictly chemical type of attack, and therefore will only be mentioned briefly.

Salt scaling refers to spalling, cracking, and deterioration of concrete surfaces associated with the use of deicing salts, usually sodium or calcium chloride, on concrete in freezing weather [161–168]. The attack is quite rapid and destructive to concrete surfaces. The deterioration appears to be primarily caused by frost action (see the chapter on Freezing and Thawing), but at times other phenomena may also be involved, such as corrosion of reinforcing steel, leaching, carbonation, osmotic pressures, and salt heaving. Nonair-entrained concrete is quite susceptible to salt scaling.

Remedial measures are similar to those recommended for freeze-thaw resistance: use a high quality, dense concrete made with air entrainment and a water/cement ratio of 0.45 or less. Some sealers and surface treatments have also shown promise in reducing salt scaling [163,169]. Pozzolans, including silica fume and fly ash, may also be effective. Particular combinations of materials should be tested to determine their effectiveness [167,170,171], such as by ASTM Test Method for Scaling Resistance of Concrete Surfaces Exposed to Deicing Chemicals (C 672).

Salt weathering is a destructive physical process commonly occurring in arid climates, coastal regions, and in wet-dry exposures. Deterioration of concrete by salt weathering has been relatively limited and has been reported only infrequently [48,87,172,173] although the process is fairly widespread. The subject has been explored more thoroughly as a geologic weathering process [174,175], and also in regards to historic preservation and the weathering of historic and cultural artifacts [176,177]. Most investigators attribute the mechanism of the attack to expansive pressures generated by the formation of salt crystals in fractures and pores [34,118,178]; others feel other mechanisms may be involved, such as osmotic, surface tension, and adsorption effects that cause swelling and shrinkage [179]. The crystal growth mechanism for salt attack has been fairly well established [35,178,180–183].

REFERENCES

[1] Barret, P. and Glasser, F. P., "Proceedings of Symposium D of the E-MRS Fall Meeting on Chemistry of Cements for Nuclear Applications," *Cement and Concrete Research*, Vol. 22, Nos. 2/3, March/May 1992.

[2] Raiswell, R. W., Brimblecombe, P., Dent, D. L., and Liss, P. S., *Environmental Chemistry*, Wiley, New York, 1980.

[3] Roy, D. M., Shi, D., Scheetz, B., and Brown, P. W., "Concrete Microstructure and Its Relationships to Pore Structure, Permeability, and General Durability," *Durability of Concrete, SP-131*, American Concrete Institute, Detroit, MI, 1992.

[4] "Effect of Various Substances on Concrete and Guide to Protective Treatments," PCA Concrete Information Sheet, Portland Cement Association, Skokie, IL, 1989.

[5] "Guide to Durable Concrete," *ACI Manual of Concrete Practice, Part 1*, ACI 201.2R-77, American Concrete Institute, Detroit, MI, 1992.

[6] "Proposed Revision of: Guide to Durable Concrete," *ACI Materials Journal*, ACI 201.2R, Sept.-Oct. 1991.

[7] Terzaghi, R., "Concrete Deterioration Due to Carbonic Acid," *Journal*, Boston Society of Civil Engineers, Vol. 36, No. 2, April 1949.

[8] Terzaghi, R., "Leaching of Lime in Concrete" (Discussion), *Journal*, American Concrete Institute, Vol. 21, No. 5, Feb. 1950.

[9] Mather, K., "Leaching of Lime from Concrete" (Discussion), *Journal*, American Concrete Institute, Vol. 21, No. 6, Feb. 1950.

[10] Biczok, I., *Concrete Corrosion and Concrete Protection* (English ed.), Chemical Publishing Co., New York, 1967.

[11] Stumm, W. and Morgan, J. J., *Aquatic Chemistry*, 2nd ed., Wiley, New York, 1981.

[12] Koelliker, E., "Zur Hydrolytischen Zersetzung von Zementstein und zum Verhalten von Kalkzuschlag bei der Korrosion von Beton durch Wasser (On the Hydrolytic Decomposition of Cement Paste and the Behavior of Calcareous Aggregate during the Corrosion of Concrete by Water)," *Betonwerk Fertigteil-Technik*, Vol. 4, 1986.

[13] Prokopovich, N. P., "Calcareous Deposits in Clear Creek Tunnel, California," *Bulletin*, Association of Engineering Geologists, Vol. XIV, No. 2, Spring 1977.

[14] Schaffer, A., "Deposition of Calcium Corbonate in Foundation Drain Holes," *REMR Bulletin*, Corps of Engineers Waterways Experiment Station, Vol. 6, No. 1, Feb. 1989.

[15] Langlier, W., "The Analytical Control of Anti-Corrosion Water Inhibition," *Journal*, American Water Works Association, Vol. 69, 1936.

[16] Ball, J. W., Nordstrom, D. K., and Zachman, D. W., "WATEQ4F—A Personal Computer Fortran Translation of

the Geochemical Model WATEQ2 with Revised Data Base," U.S. Geological Survey Open File Report 87-50, Reston, VA, 1987.

[17] *Standard Methods for the Examination of Water and Waste-Water*, 17th ed., A. E. Greenburg, J. J. Conners, and D. Jenkins, Eds., APHA-AWWA-WPCF, American Public Association, Washington, DC, 1990.

[18] Craft, D., "Evaluation of Mineral Dissolution at Deer Flats Embankments—Arrowrock Division—Boise Project, Idaho, Using Mass Balance and a Chemical Equilibrium Model," Applied Sciences Technical Memorandum 89-3-1, Bureau of Reclamation, Denver, Co, 1989.

[19] Garrels, R. M. and MacKenzie, F. T., "Origin of the Chemical Compositions of Some Springs and Lakes," *Equilibrium Concepts in Natural Water Chemistry*, ACS Advances in Chemistry Series No. 67, American Chemical Society, Washington, DC, 1967.

[20] Plummer, L. N. and Back, W., "The Mass Balance Approach: Applications to Interpreting the Chemical Evolution of Hydrologic Systems," *American Journal of Science*, Vol. 280, 1980.

[21] Tuthill, L. H., "Resistance to Chemical Attack," *Concrete and Concrete-Making Materials, ASTM STP 275*, American Society for Testing and Materials, Philadelphia, 1966.

[22] Graham, J. R., Backstrom, J. E., Redmond, M. C., Backstrom, T. E., and Rubenstein, S. R., "Evaluation of Concrete for Desalination Plants," Report REC-ERC-71-15, Bureau of Reclamation, Denver, CO, March 1971.

[23] Al-Rawi, R. S., "Internal Sulfate Attack in Concrete Related to Gypsum Content of Cement with Pozzolan Addition," *Technology of Concrete When Pozzolans, Slags, and Chemical Admixtures Are Used*, Facultad de Ingenieria Civil, Universidad Autonoma de Neuvo Leon, Mexico, 1985.

[24] Ouyang, C., Nanni, A., and Chang, W. F., "Internal and External Sources of Sulfate Ions in Portland Cement Mortar: Two Types of Chemical Attack," *Cement and Concrete Research*, Vol. 18, 1988.

[25] Grabowski, E., Czamecki, B., Gillott, J. E., Duggan, C. R., and Scott, J. F., "Rapid Test of Concrete Expansivity Due to Internal Sulfate Attack," *ACI Materials Journal*, Sept.-Oct. 1992.

[26] Tuthill, L. H., "Resistance of Cement to the Corrosive Action of Sodium Sulfate Solution," *Journal*, American Concrete Institute; *Proceedings*, Vol. 33, 1936.

[27] McMillan, F. R., Stanton, T. E., Taylor, I. L., and Hansen, W. C., "Long-Time Study of Cement Performance in Concrete—Chapter 5, Concrete Exposed to Sulfate Soils," *ACI SP Dec. 1949*; reprinted as *Bulletin 30*, Portland Cement Association, Dec. 1949.

[28] Bogue, R. H., *The Chemistry of Portland Cement*, 2nd ed., Reinhold Publishing Corp., New York, 1955.

[29] Hanson, W. C., "Attack on Portland Cement Concrete by Alkali Soils and Waters—A Critical Review," *Highway Research Record 113*, Highway Research Board, Washington, DC, 1966.

[30] Hanson, W. C., "Chemistry of Sulphate-Resisting Portland Cement," *Performance of Concrete*, Thorvaldson Symposium, E. G. Swenson, Ed., University of Toronto Press, Toronto, Canada, 1968.

[31] Lea, F. M., *The Chemistry of Cement and Concrete*, Chemical Publishing Co., New York, 1971.

[32] Kalousek, G., Porter, L. C., and Harboe, E. M., "Past, Present, and Potential Developments of Sulfate-Resisting Concrete," *Journal of Testing and Evaluation*, Vol. 4, No. 5, Sept. 1976.

[33] Mehta, P. K., "Mechanism of Sulfate Attack on Portland Cement Concrete—Another Look," *Cement and Concrete Research*, Vol. 13, 1983.

[34] Chatterji, S. and Jensen, A. D., "Efflorescence and Breakdown of Building Materials," Nordic Concrete Research Publication, Oslo, Norway, No. 8, 1989.

[35] Cabrera, J. G. and Plowman, C., "The Mechanism and Rate of Attack of Sodium Sulphate Solution on Cement and Cement/PFA Pastes," *Advances in Cement Research*, Vol. 1, No. 3, July 1988.

[36] Kalousek, G. and Benton, E. J., "Mechanism of Seawater Attack on Cement Paste," *Journal*, American Concrete Institute; *Proceedings*, Vol. 67, Feb. 1970.

[37] Cohen, M. D. and Mather, B., "Sulfate Attack on Concrete—Research Notes," *ACI Materials Journal*, Vol. 88, No. 1, Jan.-Feb. 1991.

[38] Mather, B., "Effects of Sea Water on Concrete," *Highway Research Record No. 113*, Highway Research Board, Washington, DC, 1966.

[39] Cohen, M. D., "Theories of Expansion in Sulfoaluminate-Type Expansive Cements: Schools of Thought," *Cement and Concrete Research*, Vol. 13, 1983.

[40] Ping, X. and Beaudin, J. J., "Mechanism of Sulfate Expansion," *Cement and Concrete Research*, Vol. 22, 1992.

[41] Erlin, B. and Stark, D. C., "Identification and Occurrence of Thaumasite in Concrete," *Highway Research Record 113*, Highway Research Board, Washington, DC, 1965.

[42] Lukas, W., "Betonzerstörung Durch SO_3—Angriff Unter Bildung von Thaumasit und Woodfordit," (Concrete Deterioration From SO_3—Attack With Formation of Thaumasite and Woodfordite), *Cement and Concrete Research*, Vol. 5, 1975.

[43] van Aardt, J. H. P. and Visser, S., "Thaumasite Formation: A Cause of Deterioration of Portland Cement and Related Substances in the Presence of Sulphates," *Cement and Concrete Research*, Vol. 5, 1975.

[44] Gouda, G. R., Roy, D. M., and Sarkar, A., "Thaumasite in Deteriorated Soil-Cements," *Cement and Concrete Research*, Vol. 5, 1975.

[45] Regourd, M., "Physico-Chemical Studies of Cement Pastes, Mortars, and Concretes Exposed to Sea Water," *Performance of Concrete in Marine Environment*, SP-65, American Concrete Institute, Detroit, MI, 1980.

[46] Crammond, N. J., "Thaumasite in Failed Cement Mortars and Renders From Exposed Brickwork," *Cement and Concrete Research*, Vol. 15, 1985.

[47] Berra, M. and Baronio, G., "Thaumasite in Deteriorated Concretes in the Presence of Sulfates," *Concrete Durability*, SP-100, Katharine and Bryant Mather International Conference, American Concrete Institute, Detroit, MI, 1987.

[48] Reading, T. J., "Physical Aspects of Sodium Sulfate Attack on Concrete," SP-77, George Verbeck Symposium on Sulfate Resistance of Concrete, American Concrete Institute, Detroit, MI, 1982.

[49] Stanton, T. E. and Meder, L. C., "Resistance of Cements to Attack by Sea Water and by Alkali Soils," *Journal*, American Concrete Institute; *Proceedings* Vol. 34, 1938.

[50] Mindess, S. and Young, J. F., *Concrete*, Prentice-Hall, Englewood Cliffs, NJ, 1981.

[51] Roy, D. M., "Mechanism of Cement Paste Degration Due to Chemical and Physical Factors," *Proceedings*, 8th International Symposium on Chemistry of Cement, Rio de Janeiro, Vol. 1, 1986.

[52] Heller, L. and Ben-Yair, M., "Effect of Sulphate Solutions on Normal and Sulphate-Resisting Portland Cement," *Journal of Applied Chemistry*, Vol. 14, Jan. 1964.

[53] Lawrence, C. D., "Sulphate Attack on Concrete," *Magazine of Concrete Research*, Vol. 42, Dec. 1990.

[54] Bonen, D. and Cohen, M. D., "Magnesium Sulfate Attack on Portland Cement Paste—II Chemical and Mineralogical Analyses," *Cement and Concrete Research*, Vol. 22, No. 4, July 1992.

[55] Kuenning, W. H., "Resistance of Portland Cement Mortar to Chemical Attack-A Progress Report," *Highway Research Record 113*, Highway Research Board, Washington, DC, 1966.

[56] Bensted, J., "Chemical Considerations of Sulphate Attack," *World Cement Technology*, May 1981.

[57] Smith, F. L., "Effect of Calcium Chloride Additions on Sulfate Resistance of Concretes Placed and Initially Cured at 40° and 70°F Temperatures," Report No. C-900, Bureau of Reclamation Concrete Laboratory, Denver, CO, 7 July 1959.

[58] *Concrete Manual*, 8th ed., Bureau of Reclamation, Denver, CO, 1975.

[59] Mehta, K., "Durability of Concrete in Marine Environments—A Review," *Performance of Concrete in Marine Environment*, SP-65, American Concrete Institute, Detroit, MI, 1980.

[60] Eglinton, M. S., *Concrete and Its Chemical Behavior*, Thomas Telford, London, 1987.

[61] Miller, D. W. and Manson, P. W., "Long Time Tests of Concretes and Mortars Exposed to Sulfate Waters," Technical Bulletin No. 194, University of Minnesota, Minneapolis, MN, May 1951.

[62] Lerch, W. and Ford, C. L., "Long-Time Study of Cement Performance in Concrete. 3. Chemical and Physical Tests of the Cements," *Journal*, American Concrete Institute; *Proceedings*, Vol. 44, 1948.

[63] Dahl, L. A., "Cement Performance in Concrete Exposed to Sulfate Soils," *Proceedings*, American Concrete Institute, Vol. 46, 1950.

[64] Verbeck, G. J., "Field and Laboratory Studies of the Sulfate Resistance of Concrete," *Bulletin 227*, Portland Cement Association Research Department, Skokie, IL, 1967.

[65] Bates, P. H., Phillips, A. J., and Wig, R. J., "Action of the Salts in Alkali Water and Sea Water on Cements," Bureau of Standards Technical Paper 12, National Bureau of Standards, Washington, DC, 1913.

[66] Mather, B., "Field and Laboratory Studies of the Sulphate Resistance of Concrete," *Performance of Concrete*, E. G. Swenson, Ed., University of Toronto Press, Toronto, Canada, 1968.

[67] Mather, K., "Tests and Evaluation of Portland and Blended Cements for Resistance to Sulfate Attack," *Cement Standards—Evolution and Trends*, ASTM STP 663, P. Mehta, Ed., American Society for Testing and Materials, Philadelphia, 1978.

[68] Higgenson, E. C. and Glanz, O. J., "The Significance of Tests on Sulfate Resistance of Concrete," *Proceedings*, American Society for Testing and Materials, Vol. 53, 1953.

[69] Smith, F. L., "Resistance to Sulfate Attack on Concrete Using Special Cements as Compared with Concrete Using Type V Sulfate-Resisting Cement," Report No. 968, Bureau of Reclamation Concrete Laboratory, Denver, CO, 18 May 1961.

[70] Bellport, B., "Combating Sulfate Attack on Concrete on Bureau of Reclamation Projects," *Performance of Concrete*, University of Toronto Press, Toronto, Canada, 1968.

[71] Kalousek, G., Porter, L. C., and Benton, E. J., "Concrete for Long-Term Service in Sulfate Environment," *Cement and Concrete Research*, Vol. 2, No. 1, 1972.

[72] Davis, R. E., Hanna, W. C., and Brown, E. H., "Cement Investigations for Boulder Dam—Results of Tests on Mortars up to Age of 10 Years," *Journal*, American Concrete Institute; *Proceedings*, Vol. 43, 1946.

[73] Miller, D. G., Manson, P. W., and Chen, T. H. R., "Annotated Bibliography on Sulfate Resistance of Portland Cement, Concretes and Mortars," Paper 758, University of Minnesota Miscellaneous Journal Service, 1952.

[74] Bouge, R. H., "The Studies on Volume Stability of Portland Cement Paste," PCA Fellowship Paper No. 55, Portland Cement Association, Skokie, IL, 1949.

[75] Hurst, W. D., "Experience in the Winnipeg Area with Sulphate-Resisting Cement Concrete," *Performance of Concrete*, Thorvaldson Symposium, E. G. Swenson Ed., University of Toronto Press, Toronto, Canada, 1968.

[76] Thorvaldson, T., "Chemical Aspects of The Durability of Cement Products," *Proceedings*, 3rd International Symposium on the Chemistry of Cement, Cement and Concrete Association, London 1954.

[77] de Sousa Coutinho, A., "Aspects of Sulfate Attack on Concrete," *Cement, Concrete, and Aggregates*, Vol. 1, No. 1, 1979.

[78] Harboe, E. J., "Longtime Studies and Field Experiences with Sulfate Attack," SP-77, George Verbeck Symposium on Sulfate Resistance of Concrete, American Concrete Institute, Detroit, MI, 1982.

[79] Mather, K., "Factors Affecting Sulfate Resistance of Mortars," *Proceedings*, 7th International Congress on the Chemistry of Cements, Paris, 1980.

[80] Mather, K., "Current Research in Sulfate Resistance at the Waterways Experiment Station," SP-77, George Verbeck Symposium on Sulfate Resistance of Concrete, American Concrete Institute, Detroit, MI, 1982.

[81] Tuthill, L. H., "Lasting Concrete in a Sulfate Environment," *Concrete International*, Dec. 1988.

[82] Stark, D., "Long Time Study of Concrete Durability in Sulfate Soils," SP-77, George Verbeck Symposium on Sulfate Resistance of Concrete, American Concrete Institute, Detroit, MI, 1982.

[83] Stark, D., "Durability of Concrete in Sulfate-Rich Soils," *PCA R&D Bulletin RD 097*, Portland Cement Association, Skokie, IL, 1989.

[84] Morales, R. T. "Los Cementos Portland de Moderada Resistencia Sulfatica: Metodos Acelerados de Ensayo para Determinar las Bases para su Caracterizacion y Control," ICCET/CSIC Monograph Series #399, Eduardo Torroja Institute for Construction Engineering Sciences, Madrid, Spain, Dec. 1989, USBR Translation No. 2183, "Moderate Sulfate-Resisting Portland Cements: Accelerated Test Methods for Determining the Bases for their Characterization and Control," Bureau of Reclamation, Denver, CO, Sept. 1991.

[85] Price, G. C. and Peterson, R., "Experience with Concrete in Sulphate Environments in Western Canada," *Performance of Concrete*, E. G. Swenson, Ed., University of Toronto Press, Toronto, Canada, 1968.

[86] Hamilton, J. J. and Handegord, G. O., "The Performance of Ordinary Portland Cement Concrete in Prairie Soils of High Sulfphate Content," *Performance of Concrete*, Thorvaldson Symposium, E. G. Swenson, Ed., University of Toronto Press, Toronto, Canada, 1968.

[87] Stanton, T. E., "Durability of Concrete Exposed to Sea Water and Alkali Soils—California Experience," *Journal*, American Concrete Institute; *Proceedings*, Vol. 44, 1948.

[88] Verbeck, G. J., "Field and Laboratory Studies of the Sulphate Resistance of Concrete," *Performance of Concrete*, Thorvaldson Symposium, E. G. Swenson, Ed., University of Toronto Press, Toronto, Canada, 1968.

[89] Osborne, G. J., "The Sulphate Resistance of Portland and Blastfurnace Slag Cement Concretes," *Durability of Concrete, SP 126*, American Concrete Institute, Detroit, MI, 1991.

[90] Soroka, I., *Portland Cement Paste and Cement*, Chemical Publishing Co., New York, 1979.

[91] Robson, T. D., "Aluminous Cement and Refractory Castables," *The Chemistry of Cements*, Vol. 2, H. F. W. Taylor, Ed., Academic Press, New York and London, 1964.

[92] Tepponen, P. and Eriksson, B. E., "Damages in Concrete Railway Sleepers in Finland," *Nordic Concrete Research*, No. 6, 1987.

[93] Heinz, D. and Ludwig, U., "Mechanism of Secondary Ettringite Formation in Mortars and Concretes Subjected to Heat Treatment," *Concrete Durability, SP-100*, Vol. 2, American Concrete Institute, 1987.

[94] Shayan, A. and Quick, G. W., "Relative Importance of Deleterious Reactions in Concrete: Formation of AAR Products and Secondary Ettringite," *Advances in Cement Research*, Vol. 4, No. 16, Oct. 1991/1992.

[95] Siedel, H., "Secondary Ettringite Formation in Heat Treated Portland Cement Concrete: Influence of Different W/C Ratio and Heat Treated Temperatures," *Cement and Concrete*, Vol. 23, 1993, pp. 453–461.

[96] Hooton, R. D. and Emery, J. J., "Sulfate Resistance of a Canadian Slag Cement," *ACI Materials Journal*, Nov.–Dec. 1990.

[97] Ellis, W. E., Jr., "For Durable Concrete, Fly Ash Does Not 'Replace' Cement," *Concrete International*, July 1992.

[98] Dunstan, E. R., "Fly Ash and Fly Ash Concrete," Report No. REC-ERC-82-1, Bureau of Reclamation, Denver, CO, May 1984.

[99] Dikeou, J. T., "Fly Ash Increases Resistance of Concrete to Sulfate Attack," *Research Report No. 23*, Bureau of Reclamation, Denver, CO, 1975.

[100] Dunstan, E. R., "Performance of Lignite and Subbituminous Fly Ash in Concrete-A Progress Report," Report No. REC-ERC-76-1, Bureau of Reclamation, Denver, CO, Jan. 1976.

[101] von Fay, K. F. and Pierce, J. S., "Sulfate Resistance of Concretes with Various Fly Ashes," *Standardization News*, Dec. 1989.

[102] Kosmatka, S. H. and Panarese, W. C., "Design and Control of Concrete Mixtures," *PCA Engineering Bulletin*, 13th ed., Portland Cement Association, Skokie, IL, 1988.

[103] Mehta, P. K., "Studies on Chemical Resistance of Low Water/Cement Ratio Concretes," *Cement and Concrete Research*, Vol. 8, 1986.

[104] Wong, G. and Poole, T., "The Effect of Pozzolans and Slags on the Sulfate Resistance of Hydraulic Cement Mortars," *Concrete Durability, SP-100*, American Concrete Institute, Detroit, MI, 1987.

[105] Tikalsky, P. J., Carrasquillo, R. L., and Snow, P. G., "Sulfate Resistance of Concrete Containing Fly Ash," *Durability of Concrete, SP-131*, American Concrete Institute, Detroit, 1992.

[106] Mather, B., "Concrete in Sea Water," *Concrete International*, March 1982.

[107] Sellevold, E. J. and Nilsen, T., "Condensed Silica Fume in Concrete—A World Review," *Supplementary Cementing Materials for Concrete*, V. M. Malhotra, Ed., Ministry of Supply and Services, Canada, 1987.

[108] Hooton, R. D., "Some Aspects of Durability with Condensed Silica Fume in Pastes, Mortars and Concretes," *CANMET International Workshop on Condensed Silica Fume in Concrete*, Montreal, Canada, 1987.

[109] Cohen, M. D. and Bentur, A., "Durability of Portland Cement—Silica Fume Pastes in Magnesium and Sodium Sulfate Solutions," *ACI Materials Journal*, Vol. 85, No. 3, May–June 1988.

[110] Mehta, P. K., "Effect of Fly Ash Composition on Sulfate Resistance of Cement," *ACI Journal*, Nov.-Dec. 1986.

[111] Dunstan, E. R., "Possible Method of Identifying Fly Ashes That Will Improve Sulfate Resistance," *Cement, Concrete and Aggregates*, Vol. 2, No. 1, American Society for Testing and Materials, Philadelphia, 1980.

[112] Dunstan, E. R., "Sulfate Resistance of Fly Ash Concrete—The R-Value," *Concrete Durability, SP 100-103*, Katherine and Bryant Mather International Conference, Vol. 2, American Concrete Institute, Detroit, MI, 1987.

[113] Tuthill, L. H., "Resistance to Chemical Attack," *Significance of Tests and Properties of Concrete and Concrete-Making Materials, ASTM STP 169B*, American Society for Testing and Materials, Philadelphia, 1978.

[114] Lerch, W., "Significance of Tests for Sulfate Resistance," *Proceedings*, American Society for Testing and Materials, Vol. 61, 1961.

[115] Thorvaldson, T., Vigfusson, V. A., and Wolochow, D., "Studies of the Action of Sulfate on Portland Cement," II "Steam Curing of Portland Cement Mortar and Concrete as a Remedy," *Canadian Journal of Research*, Vol. 1, 1929, pp. 359–384.

[116] "Building Code Requirements for Reinforced Concrete and Commentary," *ACI Manual of Concrete Practice, Part 3*, Chapter 4, "Durability Requirements," ACI 318R-89, American Concrete Institute, Detroit, MI, 1990.

[117] Clifton, J. R. and Knab, L. I., "Service Life of Concrete," NISTIR 89-4086, U. S. Department of Commerce, National Institute of Standards and Technology, Gaithersburg, MD, June 1989.

[118] Winkler, G. M. and Singer, P. C., "Crystallization Pressure of Salts in Stone and Concrete," *Bulletin*, Geological Society of America, Vol. 83, Nov. 1972.

[119] Patzias, T., "Evaluation of Sulfate Resistance of Hydraulic-Cement Mortars by the ASTM 4012 Test Method," *SP 100-108, Concrete Durability*, Katherine and Bryant Mather International, Conference, American Concrete Institute, Detroit, MI, Vol. 2, 1987.

[120] "Concrete in Sulphate-bearing Soils and Groundwaters," BRE Digest 250, Building Research Establishment, Garston, Herts, UK, 1986.

[121] Almedia, I. R., "Resistance of High Strength Concrete to Sulfate Attack: Soaking and Drying Tests," *Durability in Concrete, SP-126*, American Concrete Institute, Detroit, MI, 1991.

[122] "Length Change of Hardened Concrete Exposed to Alkali Sulfates," *Concrete Manual*, Part 2, 9th ed., USBR 4908-92 U.S. Bureau of Reclamation, Denver, CO, 1992.

[123] "A Guide to the Use of Waterproofing, Dampproofing, Protective, and Decorative Barrier Systems for Concrete," *ACI Manual of Concrete Practice, Part 5*, ACI 515.1R-79, American Concrete Institute, Detroit, MI, 1990.

[124] Grube, H. and Rechenberg, W., "Durability of Concrete Structures in Acidic Water," *Cement and Concrete Research*, Vol. 19, 1989.

[125] James, A. N., "Preliminary Field Studies of Rates of Dissolution of Hydrated Cement," *Magazine of Concrete Research*, Vol. 41, No. 148, Sept. 1989.

[126] "Watershed and Lake Processes Affecting Surface Water Acid-Base Chemistry, Acidic Deposition: State of Science and Technology, Report 10," National Acid Precipitation Assessment Program, Oak Ridge, TN, Sept. 1990.

[127] Pankow, J. F., "Chemistry of Dissolved CO_2," Chapter 9, *Aquatic Chemistry Concepts*, Lewis Publishers, Chelsea, UK, 1991.

[128] Hammerton, C., "The Corrosion of Cement and Concrete," *The Surveyor*, 1944, p. 587.

[129] Woods, H., "Durability of Concrete Construction," American Concrete Institute and Iowa State University Press, Ames, IA, 1968.

[130] Gutt, W. H. and Harrison, W. H., "Chemical Resistance of Concrete," Current Practice Sheet 3PC/10/1 No. 37, *Concrete*, Vol. 11, No. 5, May 1977.

[131] Harrison, W. H., "Durability of Concrete in Acidic Soils and Waters," *Concrete*, 1986.

[132] "Aggressive Water—Assessing the Problem," *Journal*, American Water Works Association, May 1980, p. 262.

[133] Conjeaud, M. L., "Mechanism of Sea Water Attack on Cement Mortar," SP-65, *Performance of Concrete in Marine Environment*, American Concrete Institute, Detroit, MI, 1980.

[134] Terzaghi, R., "Concrete Deterioration in a Shipway," *Journal*, American Concrete Institute, Detroit, MI, Vol. 44, June 1948.

[135] Wakeman, C. M., Dockweiler, E. V., Stover, H. E., and Whiteneck, L. L., "Use of Concrete in Marine Environments," *Journal*, American Concrete Institute, Detroit, MI, Vol. 54, April 1958.

[136] Mehta, P. K. and Haynes, H. H., "Durability of Concrete in Seawater," *Journal*, Structural Division, American Society of Civil Engineers, New York, Aug. 1975.

[137] Beslac, J., "Durability of Concrete to Marine Environment," SP-126, *Durability of Concrete*, American Concrete Institute, Detroit, MI, 1991.

[138] Tibbetts, D. C., "Performance of Concrete in Sea-Water: Some Examples from Halifax, N. S.," *Performance of Concrete*, Thorvaldson Symposium, E. G. Swenson, Ed., University of Toronto Press, Toronto, Canada, 1968.

[139] Geogout, P., Revertegat, E., and Moine, G., "Action of Chloride Ions Hydrated Cement Pastes: Influence of the Cement Type and Long Time Effect of the Concentration of Chlorides," *Cement and Concrete Research*, Vol. 22, 1992.

[140] Ftikos, C. and Parissakes, G., "A Study on the Effect of Some Ions Contained in Seawater on Hydrated Cement Compounds," *Concrete Durability*, SP-100, American Concrete Institute, Detroit, MI, 1987.

[141] Malhotra, V. M., Carette, G. G., and Brewer, T. W., "Durability of Concrete in Marine Environment Containing Granulated Blast Furnace Slag, Fly Ash, or Both," *Performance of Concrete in Marine Environment*, SP-65, American Concrete Institute, Detroit, MI, 1980.

[142] Suprenant, B. A., "Designing Concrete for Exposure to Seawater," *Concrete Construction*, Dec. 1991.

[143] Gjorv, O. E., "Long-Time Durability of Concrete in Seawater," *Journal; Proceedings*, American Concrete Institute, Vol. 68, Jan. 1971.

[144] Backstrom, J. E. and Graham, J. R., "Evaluation of Concrete and Related Materials for Desalination Plants," Executive Summary No. 35, Bureau of Reclamation, Denver, CO, June 1971.

[145] Ying-yu, L. and Qui-dong, W., "The Mechanism of Carbonation of Mortars and the Dependence of Carbonation on Pore Structure," *Concrete Durability*, SP–100, American Concrete Institute, Vol. 2, 1987.

[146] Groves, G. W., Rodway, D. I., and Richardson, I. G., "The Carbonation of Hardened Cement Pastes," *Cement Research*, Vol. 3, No. 11, July 1990.

[147] Papadikas, V. G., Vayenas, C. G., and Fardis, M. N., "Physical and Chemical Characteristics Affecting the Durability of Concrete," *ACI Materials Journal*, March–April 1991.

[148] Verbeck, G., "Carbonation of Hydrated Portland Cement," *Cement and Concrete*, ASTM STP 205, American Society for Testing and Materials, Philadelphia, 1958.

[149] Hunt, C. M. and Tomes, L. A., "Reaction of Hardened Portland Cement Paste with Carbon Dioxide," *NBS Journal of Research—A. Physics and Chemistry*, Vol. 66A, No. 6, Nov.-Dec. 1962.

[150] Steinour, H. H., "The Ultimate Products of the Carbonation of Portland Cement," Research and Development Division, Portland Cement Association, Chicago, IL, April 1957.

[151] Malinowski, R., "Vacuum Carbonation of Lightweight Aggregate Concrete," *Nordic Concrete Research*, No. 2, Dec. 1983.

[152] Bickley, J. A., "Potential for Carbonation of Concrete in Canada," SP-122, Paul Klieger Symposium on Performance of Concrete, American Concrete Institute, Detroit, MI, 1990.

[153] Powers, T. C., "A Hypothesis on Carbonation Shrinkage," *Journal*, Research and Development Labs, Portland Cement Association, Vol. 4, No. 2, May 1962, (Research Department Bulletin 146).

[154] Kamimura, K., Sereda, P. J., and Swenson, E. G., "Changes in Weight and Dimensions in the Drying and Carbonation of Portland Cement Mortars," *Magazine of Concrete Research*, Vol. 17, No. 50, March 1965.

[155] Ho, D. W. S. and Lewis, R. K., "Carbonation of Concrete and its Prediction," *Cement and Concrete Research*, Vol. 17, 1987.

[156] Sarkar, S. L., Chandra, S., and Rodhe, M., "Microstructural Investigation of Natural Deterioration of Building Materials in Gothenburg, Sweden," *Materials and Structures*, Vol. 125, Aug./Sept. 1992.

[157] Leber, I. and Blakley, F. A., "Some Effects of Carbon Dioxide on Mortars and Concrete," *Proceedings*, American Concrete Institute, Vol. 53, 1957.

[158] Kleinlogel, A., *Influences on Concrete*, Ungar, New York, 1950.

[159] Orchard, D. F., *Concrete Technology*, Vol. 1, 3rd ed., Wiley, New York, 1971.

[160] Gutt, W. H. and Harrison, W. H., "Chemical Resistance of Concrete," *Concrete*, Vol. 11, No. 5, May 1977.

[161] Verbeck, G. J. and Klieger, P. F., "Studies of 'Salt' Scaling of Concrete," *Highway Research Bulletin 150*, Highway Research Board, Washington, DC, 1957.

[162] Klieger, P., "Curing Requirements for Scale Resistance of Concrete," *Bulletin RX082*, Research Department, Portland Cement Association, 1957.

[163] Timms, A. G., "Resistance of Concrete Surfaces to Scaling Action of Ice-Removal Agents," *Highway Research Bulletin 128*, Highway Research Board, Washington, DC, 1958.

[164] "Effect of Cement Content on Salt Scaling Resistance of Concrete," Ontario Ministry of Transportation Report EM-28, 1979.

[165] Sayward, J. M., "Salt Action on Concrete," U.S. Army Corps of Engineers Cold Regions Research and Engineering Laboratory Special Report 84-25, Hanover, NH, Aug. 1984.

[166] Gagne, R., Pigeon, M., and Aitcin, P.-C., "Deicer Salt Scaling Resistance of High Performance Concrete," SP-122, Paul Klieger Symposium on Performance of Concrete, American Concrete Institute, Detroit, MI, 1990.

[167] Sellevold, E. J. and Farstad, T., "Frost/Salt-Testing of Concrete: Effect of Test Parameters and Concrete Moisture History," *Nordic Concrete Research*, No. 10, Dec. 1991.

[168] Marchand, J., Boisvert, J., Pigeon, M., and Isabelle, H. L., "Deicer Salt Scaling Resistance of Roller-Compacted Concrete Pavements," *Durability of Concrete, SP-126*, American Concrete Institute, Vol. 1, 1991.

[169] Yamasaki, R. S., "Coatings to Protect Concrete Against Damage by De-Icer Chemicals," *Journal of Paint Technology*, Vol. 39, No. 509, June 1967.

[170] "RILEM Recommendation: Methods of Carrying Out and Reporting Freeze-Thaw Tests on Concrete with De-Icing Chemicals," *Materials and Structures, Research and Testing*, Vol. 10, No. 58, July-Aug. 1977.

[171] Zaman, M. S., Ridgway, P., and Ritchie, A. G. B., "Prediction of Deterioration of Concrete Due to Freezing and Thawing and to Deicing Chemical Use," *Journal*, American Concrete Institute, Jan.-Feb. 1982.

[172] Novak, G. A. and Colville, A. A., "Efflorescent Mineral Assemblages Associated with Cracked and Degraded Residential Concrete Foundations in Southern California," *Cement and Concrete Research*, Vol. 19, 1989.

[173] Rzonca, G. F., Pride, R. M., and Colin, D., "Concrete Deterioration, East Los Angeles County Area: Case Study," *ASCE Journal of Performance of Constructed Facilities*, Vol. 4, No. 1, Feb. 1990.

[174] Wellman, H. W. and Wilson, A. T., "Salt Weathering, a Neglected Geological Erosive Agent in Coastal and Arid Environments," *Nature*, No. 4976, 13 March 1965.

[175] Beaumont, P., "Salt Weathering on the Margin of the Great Kavir, Iran," *Bulletin*, Geological Society of America, Vol. 79, Nov. 1968.

[176] Heath, M., "Polluted Rain Falls in Spain," *New Scientist*, 18 Sept. 1986.

[177] Smith, B., Whalley, B., and Fassina, V., "Elusive Solution to Monumental Decay," *New Scientist*, June 1988.

[178] Hanson, W. C., "Crystal Growth as a Source of Expansion in Portland-Cement Concrete," *Proceedings*, American Society for Testing and Materials, Vol. 63, 1963.

[179] Larson, E. S. and Nielsen, C. B., "Decay of Bricks Due to Salt," *Materials and Structures/Materiaux et Constructions*, Vol. 213, 1990.

[180] Becker, G. F. and Day, A. L., "The Linear Force of Growing Crystals," *Proceedings*, Washington Academy of Sciences, Vol. 7, 24 July 1905.

[181] Taber, S., "The Growth of Crystals Under External Pressure," *American Journal of Science*, Vol. XLI, 1916.

[182] Correns, C. W., "Growth and Dissolution of Crystals Under Linear Pressure," *Discussions*, Faraday Society, No. 5, London, UK, 1949.

[183] Brown, L. S., "Mechanism of Seawater Attack on Cement Pastes," Discussion by Kalousek and Benton, *Journal*, American Concrete Institute, Detroit, MI, Aug. 1970.

Resistance to Fire and High Temperatures

Peter Smith[1]

PREFACE

The fundamentals of concrete performance at high temperatures, including fire resistance, were well established by the time the previous version of this chapter was written by the same author. Accordingly, major revisions have not been necessary, though new issues are addressed, such as those arising from the use of high-strength concrete. Relevant references have been updated, but no attempt has been made to extend the 267-item bibliography of Ref *1*, current to 1975, since electronic databases covering the interim are now readily accessible.

INTRODUCTION

One of the main reasons why portland cement concrete is so widely used in building construction is that it can help satisfy the cardinal need for public safety in the face of the hazards of fire better than most of its competitors. Concrete is incombustible and a reasonable insulator against the transmission of heat. These qualities alone help confine the fire and limit the extent of the damage. Though the surface may crumble or spall, more often than not the essential engineering properties of the body of the concrete remain intact. However, the main role of concrete in a fire is to protect any embedded steel for as long as possible against a rise in temperature to the point where its physical properties are reduced significantly, causing excessive structural deflections that might lead ultimately to collapse.

Every building code contains minimum fire protection requirements based on a combination of knowledge of the physical properties and past experience of the behavior of various building materials when exposed to fire, and upon fire endurance ratings specified by the survival times of specific structural assemblies or components in standard laboratory fire tests. Usually only the ambient temperature regime to a maximum is controlled in these tests. Therefore, such tests do not provide much information about the effect of specific high temperatures on the properties of concrete, its constituent materials, or reinforcing steel except in a very general way. To improve fire resistance in building design, or to assess the condition and possibilities of repair of a structure damaged by fire, more needs to be known about the thermal and mechanical properties of steel and concrete at elevated temperatures, or those residual after slow or quench cooling. It is relatively easy to determine the residual properties by standard test methods and the results provide much of the information needed to determine what can be saved after a fire. However, experiments to determine the same properties at sustained or cyclical high temperature require special equipment and measuring techniques. Yet this kind of information has become increasingly in demand for two main reasons. First, it is needed for computer-based modeling techniques of heat flow and structural behavior that offer great promise for extending the results of fire tests on discrete assemblies to the more general case, and of demonstrating the beneficial effects of restraint and support from unaffected cooler parts of the structure. Second, advanced industrial applications, in particular for nuclear reactors, require a greater knowledge of the physical properties of various types of concrete when subject to complex, sustained or repetitive, mechanical and thermal stress regimes at moderately high temperatures.

Some industrial and military applications require special concretes that are resistant to specific service temperature regimes; fortunately for most constructions, three of the most important factors are within the control of the designer and specifier prior to placing concrete. These are: the structural system and design details chosen, the depth of cover provided to any embedded steel, and the type of coarse aggregate specified. By selection of the best combination of these for a particular application, significant improvements in resistance to high temperatures, including fire, can be achieved.

FIRE TESTING, ENDURANCE STANDARDS, AND MODELING

In North America, the fire resistance of building components is usually measured, and fire endurance is specified, on the basis of ASTM Method for Fire Tests of Building Construction and Materials (E 119). A standard fire exposure is defined by a time-temperature curve in which 538°C (1000°F) is attained in 5 min, 927°C (1700°F) in 1 h, and 1260°C (2300°F) in 8 h (assuming an end point has not

---

[1] Director (retired), Research and Development Branch, Ontario Ministry of Transportation, Downsview, Ontario, Canada M3M 1J8.

yet been reached). Exposure to the furnace temperature is on one side only for slabs and walls, beams are exposed from beneath and from the sides, and columns are exposed on all faces. Minimum specimen sizes (areas or lengths exposed) and a moisture condition requirement prior to testing are specified and superimposed loads and restraint may be added to simulate in-service conditions.

End point criteria in the test are intended to simulate modes of failure in an actual fire. They include development of a passage for flames or gases through to the unheated side, increase in temperature of steel or the unexposed surface due to heat transmission, and failure to support load after hosing down. Fire endurance is then stated in terms of time to reach whichever end point occurs first (such as 1, 2, 3, or 4-h rating). The appropriate fire endurance requirements, necessary to ensure safety in a particular set of circumstances, are called forward in the applicable building code along with related design criteria. It must be emphasized that the prime purpose of fire tests and related criteria is to ensure safety of the building and occupants in case of fire and not to ensure structural survival or ease of repair after the fire (though obviously the more fire resistant the construction is, the more likely that it will be salvageable).

The requirements and significance of standard fire test procedures and the influence of structural design and material parameters on fire endurance have been debated for over 100 years, with improvements being introduced as the development of knowledge and experience dictated. Uddin and Culver [1] cite 149 historic references covering the period from 1884 to 1961, when an American Concrete Institute (ACI) Symposium was held. The proceedings of that symposium provide a good opening to present-day thinking on fire resistance through four comprehensive papers. Carlson [2] looked at the effect of moisture, specimen size, restraint, surface temperature criteria, and radiant heat. Benjamin [3] reported on thermal factors arising from moisture content and type of aggregate, on structural factors such as concrete strength, cover to reinforcing steel, spalling, and lateral restraint as affecting fire endurance ratings. Sheridan [4] discussed concrete as a protection for structural steel. Troxell [5] addressed the special problems of prestressed concrete sections. Since then, researchers and fire testing laboratories around the world have worked actively to evaluate new types of construction and improve fundamental knowledge.

The establishment of fire endurance ratings by means of standard fire tests is both expensive and time consuming. The results are specific and particular to the component or assembly tested and cannot be extended with confidence to include even comparatively minor variations from the configuration tested. Lastly, it is difficult during the test to simulate the beneficial effects of restraint by cooler parts of a concrete structure that are clear of the actual fire. Accordingly, with the advent of computers, mathematical models have been developed for the heat flow process in a variety of structural concrete units to establish temperature distributions and, taking material and structural parameters into account, to then predict fire endurance and structural behavior at high temperatures.

The report of ACI Committee 216 [6] is definitive on fire endurance of concrete elements, and it is beyond the scope of this chapter to further discuss fire-resistant structural design.

FACTORS INFLUENCING BEHAVIOR AND AFFECTING TEST METHODS

Interest in the high-temperature behavior of structural concretes for most applications starts at a lower bound temperature of 100°C (212°F) and immediately above as free water starts to be driven off. Generally speaking, the engineering properties and behavior of concrete at these temperatures vary by only a few percentage points from those measured at room temperatures. However, above about 149°C (300°F), the progressive continuum of cement dehydration reactions, thermal incompatibilities between paste and aggregate, and eventual physical-chemical deterioration of the aggregate leads to high thermal stresses, microcracking, and a rapid worsening in most mechanical properties of structural value. However, the end of structural usefulness of a particular concrete does not occur suddenly at a specific temperature.

In their review, Uddin and Culver [1] list six interrelated material properties as having particular influence on the high-temperature performance of structural concrete. They are: (1) type of aggregate, (2) free moisture in the concrete, (3) stress levels in both steel and concrete, (4) cover over reinforcing steel, (5) modulus of elasticity, and (6) thermal conductivity, diffusivity, specific heat, etc. As will be discussed, other factors, such as different types of normal cements, are of much less importance. Most of the factors interrelate and the format has been adopted of first considering the influences of component materials, then the mechanical and thermal properties of normal concretes and, finally, concretes in special applications. Within the space available, it is not possible to reproduce the many graphical displays of test results in the references discussed. All the essential ones covering steel properties, mechanical and thermal properties of concrete, and temperature distribution across various sections can be found in Ref 6.

Throughout the discussion, experimental techniques to measure various properties will be identified or, where this is not essential, left to the details provided by the referenced sources. Two references [7,8], not later discussed, are specific to experimental techniques and support a general observation that a property that can easily be determined at normal temperatures by standard tests often can only be determined at elevated temperatures with difficulty, that is, by modifying the normal procedure or developing a new or indirect method through considerable ingenuity. As a result, many of the differences reported in measured values of various properties may be ascribed to differences in test conditions. However, while the actual numbers may differ, there is a usually good agreement on the general trends reported.

Two tests have proved useful in obtaining an understanding of the observed behavior of concrete at high temperatures. Thermogravimetry tests provide insight into chemical reactions involving loss of weight while dilatometry tests that measure change of length may detect other reactions not necessarily accompanied by loss of weight. Harmathy and Allen [9], among other investigators, used these to examine the overall stability of a variety of concretes and to pin-point specific chemical and physical changes that occur with increasing temperature. Cement dehydration or the decomposition of carbonate aggregate and the chemical stability of lightweight or siliceous aggregate concretes are very evident in their loss-of-weight test data. One important result from the change-of-length test is to show that the α-β quartz transformation at 573°C (1063°F), while certainly causing an increase in volume in siliceous aggregates, did not cause a corresponding change in the concrete. This Harmathy and Allen attributed to the accommodating plasticity of the cement paste at high temperatures. The overall pattern of their length change data confirms the complex superposition of dehydration shrinkage in the cement paste and expansion of the aggregate as being a major factor in high-temperature performance of concrete. The corresponding cooling curves demonstrate that most of the reactions and events that occur during heating are irreversible.

INFLUENCE OF CEMENT PASTE COMPONENT

Choosing one portland cement over another will do little to improve the properties of concrete at high temperature. High alumina cements are used in refractory concretes but may not be appropriate for structural purposes because of the conversion phenomena that may occur under normal service conditions.

While a case may be made for using fly ash or blast furnace slag cements mainly because these produce lesser amounts of calcium oxide as a result of dehydration in the heat of a fire, in practice these are little used for such benefit alone. At lower temperatures such as those that might be experienced in mass concrete in nuclear reactors, Nasser and Marzouk [10] attributed an increase in compressive strength and decrease in the modulus of elasticity in the temperature range of 121 to 149°C (250 to 300°F) to the formation of tobermorite that is much stronger than tobermorite gel. After six months in a temperature range of 177 to 232°C (350 to 450°F), there was a great reduction in structural properties of fly ash concrete, thought to be due to the formation of crystalline alpha dicalcium silicates that are poor binders. Fly ash as it affects lightweight concrete is discussed under that section.

Silica fume additions have been shown to increase the risk of explosive spalling in very high strength concrete [11]. In the same investigation, it was noted that when a melamine-based high range water reducer was incorporated, it decomposed releasing ammonia gas. Otherwise, admixtures appear to have no unusual effects.

The development of distress and the change for the worse in the thermal and mechanical properties of portland cement pastes that occur with increasing temperature are the result of an uninterrupted series of physical-chemical reactions that are accompanied by shrinkage and microcracking. Desorption of what is commonly called evaporable water first takes place as the temperature increases. Then, and overlapping, chemically bound water (nonevaporable) is released progressively from the complex system of low crystalline order, calcium silicate hydrates, and other hydrates in the cement paste and from the calcium hydroxide crystals formed when the cement originally hydrated.

The dehydration of calcium hydroxide itself to form the oxide is a relatively simple process occurring above 400°C (752°F). Subsequently, the calcium oxide, after cooling and in the presence of moisture, will rehydrate to calcium hydroxide with a disruptive 14% increase in volume. On the other hand, the dehydration of the silicate hydrates and other compounds present, through any intermediate stages, is more complicated and gradual and not fully understood. Dehydration is complete at 800°C (1472°F) and above. The stage reached at any given temperature is dependent also upon the rate of heating since all the reactions are not instantaneous. Most of these decompositions are irreversible so damage to concrete from this cause at sustained high temperatures is essentially permanent, and the practical end point is probably about 538°C (1000°F). After exposure to sustained temperatures of 649 to 816°C, (1200 to 1500°F), concrete is friable, porous, and, after cooling, usually can be taken apart with the fingers. Many of these events are evident in thermogravimetric, dilatometric, and differential thermal analysis [9,12].

Zoldners' review of experimental thermal data [13] shows that the variation of thermal properties of the cement paste with temperature, in particular thermal volume change, is a very important component to the overall behavior of concrete because of incompatibility with the same properties in the aggregates. There appear to be two components to thermal volume change of paste prior to decomposition: (1) a true thermal expansion that is essentially constant and reversible, and (2) an apparent thermal expansion that is a hygrothermal contraction dependent on the internal transportation of moisture between various states in the capillaries and gel pores, resulting in shrinkage that is irreversible. Thermal expansion is additive to shrinkage due to dehydration, with the result that many investigators have confirmed that cement paste expands up to 100°C (212°F), and then with further heating and loss of moisture, it contracts rapidly to more than cancel the initial expansion. Above 500°C (932°F) shrinkage again changes to expansion. Actual values of the apparent coefficient of contraction or expansion are time dependent as is dehydration, one of the causative mechanisms. Thermal conductivity of cement paste is also dehydration dependent; the lower the moisture content the lower the thermal conductivity. At 750°C (1382°F), where the paste is much drier and more porous, thermal conductivity is only half its value at room temperature.

In addition to the effects of thermal incompatibilities, the decrease in evaporable and chemically bound water with progressive dehydration of the paste modifies the physical bonds existing initially in the concrete and promotes microcracking. The resulting influence for the worse in the mechanical properties of concrete with increasing temperature is clear, though separate data related to the paste component alone are sparse.

INFLUENCE OF AGGREGATE COMPONENT

Unlike the situation with cements, the selection of one aggregate over another can have a great influence on the resistance of concrete to high temperature. Though not entirely justified by fact, many building codes and insurance rating handbooks group concretes into two or four classes of fire endurance rating depending on the aggregate used.

Crushed firebrick and fused aluminum oxide (corundum) and other special aggregates used in refractory concretes perform best. Then in descending order of fire endurance come expanded slags, shales, slates, and clays; air-cooled slag; basic, finely grained igneous rocks such as basalt; calcareous; and siliceous aggregates. Attention naturally centers on the reasons for the significantly different behavior of calcareous and siliceous aggregates that are the most widely used.

Carbonate aggregates are decomposed chemically by heat whereas siliceous aggregates in general are not. Between 660 to 979°C (1220 to 1795°F), calcium carbonate breaks down into calcium oxide with the release of carbon dioxide. Magnesium carbonate is likewise decomposed between 741 and 838°C (1365 and 1540°F). Both reactions are endothermic, thus absorbing heat and delaying temperature rise in the concrete. The calcined material is also less dense and hence a better insulator. Furthermore, the carbon dioxide, escaping from the surface of the concrete in considerable volume, forms an inert insulative layer thus further retarding temperature rise in the concrete.

On the other hand, though siliceous aggregates may be chemically stable, they have serious deficiencies in their physical properties at high temperatures. Most striking is the transformation of the quartz crystal from the α to β polymorph at 573°C (1063°F) with an increase in volume. At higher temperatures, other internal volume changes occur and the crystal form may be metastable. Furthermore, quartz has a much greater coefficient of thermal expansion than most other rocks up to 600°C (1112°F). The consequence in concrete is much greater thermal incompatibility between the cement paste and aggregate and hence greater internal thermal stresses. Many people have ascribed a greater tendency to spalling for these reasons, in particular, the crystal form transformation, though, as later discussed, other mechanisms dominate.

The desirable thermal properties of rock as aggregate are low thermal conductivity that delays temperature rise, a thermal expansion as close to that of the paste as possible in order to minimize development of parasitic thermal stresses, and a high specific heat to absorb heat. Zoldners [13] reviewed studies at CANMET that give exhaustive data on rock thermal expansions up to 1000°C (1832°F) and confirm the dominance of mineral composition. For most rocks, the rate of thermal elongation is many times greater above 500°C (932°F) than at room temperatures and abrupt increases or decreases may occur at specific temperatures, as notably happens with siliceous rocks. Mirkovich [14] found the greatest decrease in thermal conductivity occurs with those rock types that show the greatest thermal expansions, that is, quartz, sandstone, and granite, while anorthosite, basalt, and limestone show relatively little change. This connection between thermal conductivity and expansion is ascribed by Zoldners [13] to the fact that higher thermal expansion stresses cause more microcracking of crystals and loosening of grains resulting in increased rock porosity and hence decreased thermal conductivity. The "ideal" aggregate from a thermal conductivity point of view seems to require an amorphous microstructure and porous macrostructure, a specification best met by lightweight aggregates.

INFLUENCE OF EMBEDDED STEEL AND STRUCTURAL SYSTEMS

A prime objective in the design of fire-resistant structural concrete is to prevent collapse for as long as possible. This requires keeping any embedded steel as cool as possible for as long as possible to avoid loss of tensile yield strength.

Normal reinforcing steel shows a linear decrease in yield strength with increase in temperature up to about 370°C (700°F) totaling about 15%. After that, the decrease is more rapid and reaches 50% by about 593°C (1100°F). Less than 20% of the original yield strength remains at 760°C (1400°F). Prestressing wire and strand reach the 50% loss of yield strength at a lower temperature, about 423°C (800°F), and some stress relaxation losses may remain after cooling. These temperatures, 593°C (1199°F) and 423°C (800°F), are taken as one of the end points in standard fire tests. The length of time before the steel reaches a critical temperature, defined as that at which the moment capacity is reduced to the applied moment, in a fire depends on the amount of cover to the steel and on the insulative properties of the concrete. The influence of cover is striking. For the same normal-weight concrete exposed in a standard fire test for 1 h, steel with only 25-mm (1-in.) cover would reach over 427°C (800°F) whereas, with 50-mm (2-in.) cover, the steel temperature would be about 204°C (400°F). With 75-mm (3-in.) cover, the steel would probably be hardly warm to the touch. Similar orders of limitation on steel temperature with increased cover are applicable to other periods of fire exposure [6]. Steel that has remained encased in concrete will recover much of its lost strength on cooling and only exposed reinforcement that has become distorted and twisted will usually need replacing or supplementing. Fire protection requirements relating to structural steel encased in concrete are covered in Refs 4 and 6.

Both the coefficient of thermal expansion and the thermal conductivity of steel influence the performance of the surrounding concrete in a fire. The coefficient of thermal expansion of steel varies with temperature in a dissimilar manner to that of concrete at very high temperatures, but is generally close to that of concrete where compatibility is critical at lower temperatures. Changes in the coefficient of thermal expansion of steel at high temperatures are related to changes in the phase and crystal composition of the steel. The coefficient increases with temperature up to 649°C (1200°F), decreases to zero in the range up to 816°C (1500°F), and then increases again. The thermal conductivity of steel is much greater than that of concrete and sometimes after a fire a thin, whitish, porous layer may be evident surrounding reinforcing steel embedded in otherwise sound concrete. Usually, this happens if part of the bar has become exposed to the heat source (as by spalling), and the steel temperature then rapidly becomes high enough to dehydrate the cement paste. This effect and the difference in coefficients of thermal expansion may adversely affect the local bond of concrete to steel at higher temperatures. Up to about 316°C (600°F), the bond strength between steel and concrete is the same as, or slightly higher than, at room temperature. Thereafter, it declines and work by Diederichs and Schneider [15] and others shows that the bond strength at elevated temperatures may be an important structural consideration. The use of epoxy-coated reinforcing steel as a defence against corrosion raised the question of the effect of the organic coating on fire resistance, in particular on bond strength. Fire tests [16] and the author's experience of a bridge fire show that the epoxy coating adds little overall risk.

It should be emphasized that there is no one critical temperature or condition at which embedded steel (normal reinforcement, prestressed, or structural steel) suddenly fails. The dead and live load intensity, structural indeterminacy, continuity, and member restraint all influence the outcome in terms of the distortion, deflection, and ultimate collapse of a member. A combination of established engineering principles, computer-based modeling of temperature-structural regimes, and applicable material properties promise that the rational design of concrete structures for a particular high-temperature resistance may become as routine as any other design task.

EFFECT OF EXPERIMENTAL CONDITIONS ON MECHANICAL PROPERTIES

The observed strength of the same concrete when heated to a particular temperature, or after heating and cooling, depends on the circumstances and superimposed conditions. Variables include:

1. whether the concrete is restrained or not during heating,
2. whether free moisture is contained or not during heating,
3. whether the concrete is subject to thermal cycling,
4. whether it is tested while hot or after cooling, and
5. whether it is quenched or allowed to cool slowly.

In addition, each concrete, depending on the behavior of the cement paste and aggregates, will have different strength and moduli values at different temperatures.

Compressive and flexural concrete strengths at elevated temperatures have been studied widely, tensile and shear strength less so. Bond strength has been discussed in the preceding section. For purposes of a general understanding, only the more recent and significant studies have been referenced. Access to earlier work is by Ref 1.

EFFECT ON COMPRESSIVE STRENGTH

Because compressive strength is one of the prime qualities by which concrete is judged, it is probably the most investigated property at high temperature. In simply supported members, differences in the strength of concrete in the compressive zone have relatively minor consequences and do not affect fire endurance. However in the design of complex frames and continuous or restrained members, the compressive strength is important and specific data are needed if fire endurance is to be improved by these means.

Various techniques have been used to modify the compressive strength testing procedure used at normal temperatures, that is, ASTM Test Method for Compressive Strength of Cylindrical Concrete Specimens (C 39) to permit testing at high temperatures in universal testing machines. Usually, the specimen is heated by means of an electrical or gas furnace that jackets a concrete cylinder varying in size from 50 mm (2 in.) diameter and 100 mm (4 in.) length to 100 mm (4 in.) by 200 mm (8 in.). To maintain uniform temperature conditions, load is applied from the testing machine platens through cement-asbestos disks above and beneath steel cylinders abutting flat machined faces of the specimens. Where length changes as well as strength are to be measured, then extension rods or optical devices may be used to bring the gage points outside the furnace. Temperatures are measured and controlled by surface or embedded thermocouples.

Using such apparatus and 75 by 150-mm (3 by 6-in.) cylinders, Abrams [17] determined the compressive strength of concrete between 93 and 871°C (200 to 1600°F) for concretes in the strength range of 23 to 45 MPa (3300 to 6500 psi) that contain carbonate, siliceous, and lightweight (expanded shale) aggregates. The testing regimes included heating without load and testing while hot, heating at three stress levels and testing while hot, and testing after slowly cooling.

In summary, his results showed: (1) carbonate and sanded lightweight aggregate concrete retained more than 75% of original strength up to 649°C (1200°F) when heated unstressed and tested hot; for siliceous aggregate concrete, the corresponding temperature was 427°C (800°F); (2) when loaded during heating, compressive strengths were 5 to 25% higher but were not affected by the applied stress level; (3) residual strengths after cooling were a little lower than the corresponding hot compressive strength; and (4) the original strength of the concrete had little effect on the percent reductions in strength observed.

Turning to Malhotra [18], who tested 50 by 100-mm (2 by 4-in.) concrete (10 mm (^{3}/$_8$-in.) siliceous aggregate) cylinders with maturities of from 7 to 42 days conditioning at 24°C (75°F) and 55 to 60% relative humidity, provides the following observations on some additional factors: (1) aggregate/cement ratio had little bearing on changes in compressive strength up to 204°C (400°F); however, above this temperature lean mixes showed a proportionally smaller reduction in compressive strength than occurred with richer mixtures, and (2) water/cement (w/c) ratios in the range from 0.37 to 0.65 had little influence on the order of reduction of compressive strength of concrete up to 593°C (1100°F). His tests also confirmed those of Abrams [17] with respect to the benefit of the concrete being under compressive stress (in the order of design stress) when heated. He also recorded an additional reduction of compressive strength of about 20% in specimens tested after cooling.

If concrete specimens are quenched, as may happen to concrete that is hosed down with water during a fire, then as Zoldners [19] showed, up to about 500°C (932°F), there is a greater reduction in residual compressive strength than when slowly cooled. This he attributed to thermal shock. He also noted a harder outer shell in quenched cylinders due to rehydration of the cement, and that at higher temperatures this largely eliminated the difference in strength between quenched and slowly cooled specimens.

Zoldners' [19] results include the effect of four different aggregate types on compressive strength at elevated temperatures. They were a gravel comprised mainly of a mixture of metamorphic and granitic igneous rock, a relatively pure sandstone, a high-calcium fine-grained limestone, and an expanded slag. His conclusions provide three additional pieces of information in addition to generally agreeing with Abrams [17].

1. The crystalline igneous and metamorphic rock gravel aggregate deteriorated more rapidly than did the limestone aggregate concrete; at 400°C (752°F), only 85% of the initial strength remained as compared with over 95% for the limestone.
2. The sandstone aggregate concrete showed a significant strength gain up to about 300°C (572°F) after which deterioration was rapid.
3. Above 525°C (977°F), the expanded slag aggregate concrete out-performed the others whereas below that temperature it was equal or inferior in compressive strength.

EFFECT ON FLEXURAL STRENGTH

The most common procedure for determining flexural strength appears to be a modification of three-point loading of small beams along the lines of ASTM Test Method for Flexural Strength of Concrete (Using Simple Beam with Third-Point Loading (C 78). The decreases observed in flexural strength are not necessarily of the same order and magnitude as those for compressive strength of concrete of the same mix proportions and containing the same constituents [19]. Speaking generally, the decline of flexural strength is much greater than that of compressive strength. For example, Sullivan and Poucher [20] found the flexural strength of a siliceous aggregate concrete at 400°C (752°F) varied from 25% to 0% of the original strength. Zoldners [19] obtained similar results, but more hopefully found only a 50% flexural strength reduction with limestone aggregate concrete at 400°C (752°F) (a reduction level that he found to occur at over 600°C (1112°F) in the case of compression). Again, quenching resulted in greater strength losses than occurred when specimens were tested hot.

EFFECT ON MODULUS OF ELASTICITY, POISSON'S RATIO, AND BULK MODULUS

The ratio of stress to strain is an important structural concrete design parameter. At high temperatures, the value of E, the modulus of elasticity, may be obtained dynamically as by Philleo [21], or by displacement measurements as by Sullivan and Poucher [20] on concrete specimens heated within a furnace. In general terms, there is a decrease in E in the order of 10 to 15% between initial values at room temperature and those at 100°C (212°F). E then remains relatively constant up to 300°C (572°F) after which a progressive decline sets in so that by 538°C (1000°F) it is only half or less than half of that before heating. Sullivan and Poucher [20] noted in their tests, made up to 400°C (752°F) that for a given mortar or concrete (with siliceous aggregates), the value of E after cooling was essentially the same as that prevailing at the high test temperature, thus determining that the reduction in E with increasing temperature is a permanent one. Philleo [21] noted that the value of E depended on initial w/c ratio, curing regime, and age. Between unheated values for calcareous gravel aggregate concrete ranging from 3.20 to 7.34×10^{-6} and final values at 760°C (1400°F) of from 0.91 to 2.30×10^{-6}, some benefit was found in low w/c ratios and air drying. Marechal [22] looked at the variation of E with temperature for expanded clay, crushed rock aggregate (mainly porphyry and quartzite), and gravel aggregate (siliceous and siliceous limestone) at one cement factor. He found the reduction in this modulus to be aggregate dependent, in the preceding order, up to 400°C (752°F).

Philleo [21] was able to compute Poisson's ratio because he measured both flexural and torsional frequencies of vibration. Because of sensitivity complications in the test procedure, the results were erratic, but a general tendency to decline with increasing temperature was apparent. Marechal [22] also calculated Poisson's ratio, but did so directly from the longitudinal axial load and measured transverse strain data and a similar decline was found after the evaporable water had been driven off. He made further calculations for the bulk modulus, K, which declined with temperature in a curve paralleling that for E.

EFFECT ON CREEP

Data on inelastic behavior of concrete at high temperature are essential for nuclear applications, since long-term behavior under complex mechanical thermal loadings is involved, and is important for fire resistance design if the effects of restraint are to be taken into account. If inelastic behavior is not taken into account and if only data secured on small specimens are used, over-estimations of the affected parameters will occur. A first step in overcoming the continuing elusiveness of creep at high temperature was to devise apparatus for measuring it as reported by Cruz [23].

A study of Wang [24] of moderately high-strength concretes 38 and 46 MPa (5500 and 6700 psi) containing quartz gravel aggregate over the temperature range from 93 to 427°C (200 to 800°F) showed that the shape of creep-time curves was the same as at room temperature, but the creep rate was higher when the concrete was subject to both high temperature and a high stress-to-strength ratio. He also found that (1) the latter affects the creep rate more than the temperature; (2) there was a nonlinear relationship between creep and stress level; and (3) with a lower w/c ratio, creep was less. Sullivan and Poucher [20] attributed the fact that creep rates for concrete above 200°C (392°F) were higher than those for mortar, while the reverse was true below 200°C (392°F), to the development of microcracking between aggregate and paste. At 300°C (572°F), they measured creep rates in concrete at three times those at 125°C (257°F), and above 400°C (752°F), they observed that creep increased very rapidly at low stress levels as a plastic stage was reached. Dependence of creep on moisture content is stressed by Marechal [25]. Higher creep rates occur in saturated or sealed specimens for which a fully accepted explanation does not exist.

In addition to creep tests, Cruz [23] also carried out exploratory stress relaxation tests under constant strain over a 5-h period. Stress was reduced by 2, 32, and 74% at 24, 316, and 649°C (75, 600, and 1200°F), respectively.

Investigations of concrete under transient high-temperature conditions, including time-dependent effects such as creep and relaxation, confirm that a very complex pattern of behavior probably exists [26].

DETERMINING THERMAL PROPERTIES

The basic thermal properties of concern to the behavior of concrete at elevated temperatures are: thermal conductivity (the ability of the material to conduct heat, a ratio of the flux of the heat to the temperature gradient), thermal diffusivity (the rate at which temperature changes can take place), and the two more widely understood parameters of specific heat and the coefficient of thermal expansion. Reference 13 provides experimental values but, as will be discussed, there are difficulties with many of the test procedures at elevated temperatures.

Thermal Conductivity, Diffusivity, and Specific Heat

Thermal conductivity and diffusivity of a mixture such as concrete are not additives of the properties of the constituents and there are difficulties in experimental determination at a given elevated temperature. These are due principally to accompanying moisture migration-desorption, though other physical-chemical changes also contribute. Thompson [27] discusses the shortcomings of steady-state methods such as ASTM Test Method for Steady-State Heat Flux Measurements and Thermal Transmission Properties by Means of the Guarded-Hot-Plate Apparatus (C 177) or ASTM Test Method for Steady-State Thermal Performance of Building Assemblies by Means of the Guarded Hot Box (C 236) and indicates that meaningful determination would only be possible after moisture equilibrium is reached, a self-defeating state of affairs. Transient methods, such as the hot wire method, also present problems. Experimental errors are greater, since the hot wire does not present a surface of constant flux and the initial temperature rise in the wire in a region of high conductivity will be less than in a region of low conductivity, both of which exist side by side in a heterogeneous material such as concrete. However, Harmathy [28] offers two variable state methods that appear to overcome problems associated with other methods. One is based on fitting a temperature distribution curve used (assuming linear heat flow) to determine all thermal properties, and the other on the time-temperature rise response to heat pulses to determine diffusivity. Both are accompanied by the necessary mathematical interpretations.

Difficulties also plague direct measurement of true specific heat at elevated temperatures. Equilibrium at the test temperature is disturbed in making the determination by standard physical methods, and chemical dehydration reactions take place in the cement phase that may add or subtract heat. Such latent heat effects, while overridden to some extent by the greater contribution of the aggregate fraction to the overall apparent specific heat of concrete, must be taken into account if reasonably accurate results are to be attained in heat flow model studies. It is therefore not surprising that data from Harmathy [12] Harmathy and Allen [9], and other investigators show considerable dispersion in values for the specific heat of concrete but, in general, specific heat appears to be insensitive to the aggregate used or to the mix proportions of the concrete. Values reported in Ref 9 for a range of concretes over a temperature range from normal ambient to 650°C (1202°F) show a slight trend to higher values with higher temperature. A high specific heat, including the contribution from moisture present, has the advantage of taking up more heat and delaying temperature rise in a fire.

Fortunately, thermal conductivity (k), thermal diffusivity (h), and specific heat (c) are simply related by the expression: $k = hcp$; where p is the density.

Conductivity and diffusivity hence vary in step and, with the relatively small variation in specific heat with temperature, density of the concrete is an important overall parameter when low density is due to air voids. This is most apparent between a dry lightweight concrete and a normal-weight concrete. Moisture content has a significant role in heat conduction in concrete until dehydration is complete after which, and with decomposition of the constituents, the now dry, porous, normal-weight concrete becomes a much better heat insulator than at lower temperatures. The greater the moisture content the greater the thermal conductivity according to data summarized by Zoldners [13]. For normal-weight concretes, both thermal conductivity and diffusivity show considerable decrease at high temperature. Harada et al. [29] cite a 50% decrease at 800°C (1472°F) as compared with thermal conductivity at room temperature. For a lean concrete as compared with a rich one, a 20% increase in thermal conductivity was noted assuming both were consolidated fully. From an examination of 16 concretes made with Type 1 cement and a range of aggregates and processes used in the manufacture of concrete masonry units, Harmathy and Allen [9] confirm this order of reduction and find that all thermal diffusivities became approximately equal at about 600°C (1112°F) whereas at lower temperatures there was considerable difference between lightweight aggregate and normal-weight aggregate concretes. The lightweight aggregates showed lower diffusivities and less reduction with increase in temperature than did concretes of normal weight. Their data on conductivity, while showing a similar trend, did not merge the aggregate groups at the upper temperature.

The three thermal properties so far discussed (conductivity, diffusivity, and specific heat) are important to overall behavior of concrete at elevated temperature because they essentially determine the inward progression of temperature rise into concrete and hence the extent of damage that may occur to steel or concrete in a fire. They are also important in heat flow calculations necessary in the design of structures for performance at elevated service temperatures as, for example, in nuclear reactors where significant temperature gradients are a feature of normal operations. On the microscale, they influence the thermal incompatibilities that occur between the cement paste and aggregates due to differential thermal volume changes.

THERMAL VOLUME CHANGE AND THERMAL INCOMPATIBILITIES

The thermal expansion of the paste is highly dependent on moisture content while that of the aggregate is highly dependent on mineral composition. The observed thermal expansion in concrete is an additive of those of the components. There are three aspects of volume change in concrete at elevated temperatures to consider. As well as true thermal volume changes, volume changes due to chemical reactions, for example, decomposition of calcium hydroxide or rehydration of the lime, and changes in physical form of the component materials, for example, the transition in quartz crystals, also occur. The term "thermal incompatibility" is used widely to describe those differential volume changes that result in differential strains and hence stresses between aggregate particles and paste and between the constituent minerals of each. Unless relieved by creep or plasticity in the paste, these parasitic thermal stresses may be very high.

Looking at Philleo's data [21] for a calcareous aggregate concrete shows thermal expansion is linear up to about 260°C (500°F), then after a transition period there is a further linearity to about 427°C (800°F), but at a higher coefficient. Above 538°C (1000°F), the coefficient is again lower, probably due to a permanent dilation as a result of decomposition. Philleo's experimental technique seems fairly typical of those used to secure such data. The horizontal displacement between two wires hung vertically from gage points on a horizontal concrete beam is measured at the test temperatures, clear of the furnace containing the specimen.

Zoldners [13] summarizes data from other investigations that show that the thermal coefficient of expansion of concrete is in approximate proportion to the weighted value of those of the constituent aggregates once initial shrinkage of the cement paste due to dehydration has occurred. The coefficient of thermal expansion of concrete with increasing temperature thus tends to approach asymptotically that of the rock type in the aggregate [29]. A point of particular interest in this data is that the expansion curve for limestone aggregate concrete is very close to that of mild steel over the temperature range up to 700°C (1292°F). Andesite, the "best" natural rock aggregate concrete Harada et al. [29] investigated, had about half the coefficient of thermal expansion of limestone or sandstone concrete.

Zoldners' own data [19] show a very rapid expansion above about 400°C (752°F) for a predominantly metamorphic and igneous gravel aggregate concrete and for a sandstone aggregate concrete, as compared with limestone aggregate concrete and, in the case of an expanded slag or shale lightweight concrete, a slight overall contraction with an almost constant coefficient up to 900°C (1652°F). In all cases an initial shrinkage up to about 300°C (572°F) is noted. These data are for only one cement factor, one w/c ratio, and one aggregate/cement ratio. However, Philleo's work [21] did include three cement factors and corresponding w/c ratios. Above 427°C (800°F), these variables and initial curing (moist or air dried) had little effect; however, below 260°C (500°F), all these variables affected the coefficient values to a greater extent. Specimens having a higher w/c ratio and lower cement factor tended to have the higher coefficients.

Crispino [30] pursued an investigation to find the best combination of thermally compatible materials that would give good performance as a structural concrete for nuclear reactor pressure vessels at high temperature under thermal cycling. As compared with concretes containing limestone aggregate and a variety of portland and high-alumina cements, he found barytes aggregate with portland cement far superior.

The importance of thermal incompatibilities and the resulting thermal strains and stresses cannot be overemphasized. While the cement paste is shrinking as dehydration proceeds, the aggregate is expanding. After dehydration is complete, both are expanding, though at different rates. As a result, there is a clear, if not as yet fully explained, effect on the mechanical properties of concrete in which, as will be discussed, moisture effects may play a substantial role. Considerable hypothesizing continues toward a full explanation of the behavior of concrete at all temperatures with respect to thermal properties. Further reading is provided by Dougill [31].

THERMAL CYCLING

Since most investigations in the past have addressed fire resistance, it is not surprising that the effect of more than one thermal cycle has only really been addressed after interest arose in using concrete in nuclear reactor vessels. Both mechanical and thermal properties are affected [30].

The effect of thermal cycling on strength has been investigated by Campbell-Allen and Desai [32] and Campbell-Allen et al. [33] who compared the compressive strengths of limestone, expanded shale, and firebrick aggregate concretes up to 302°C (575°F) and noted a progressive deterioration with increasing number of thermal cycles though most of the loss occurred in the first few cycles. Similar effects are observed with flexural strengths and modulus of elasticity attributable to incompatible dimensional changes between cement paste and aggregate and to observed microcracking [20,32,33]. The interrelationship of thermal cycling, moisture content, and thermal volume change is also important in nuclear applications of concrete. Polivka et al. [34] examined concrete that allowed moisture a controlled escape or sealed it in, over durations up to 14 days of thermal exposure cycling up to 143°C (300°F). They showed that:

1. For a constant moisture condition (sealed) of the concrete, its thermal coefficient of expansion decreased during thermal cycling to 143°C (300°F). A similar decrease was observed for concretes that successively dried out during thermal exposure.
2. After the first thermal cycle (21 to 143 to 21°C) (70 to 300 to 70°F) of the concrete in a moist condition, a significant residual expansion was observed, and this expansion successively increased with the number of cycles. This was also true for concrete specimens that were permitted to lose moisture during thermal exposure. A larger permanent expansion was observed for concretes of high moisture contents than for those of lower moisture contents.

In examining the significance of data from thermal cycling and sustained high-temperature tests, the nature of the thermal regime must be considered since many of the reactions influencing the measured thermal and mechanical properties including the internal transport or the expulsion of water, are dependent on time or the rate of temperature increase.

REVIEW OF BEHAVIOR MECHANISMS AND INFLUENCE OF MOISTURE CONTENT

The mechanical and thermal performance of concrete at elevated temperatures is determined by complex, interrelated physical-chemical behavior and strongly influenced by moisture content. The stresses from thermal loadings due to overall expansion or contraction, when added to those arising from mechanical loadings, determine the imposed total stress levels in the concrete. However, thermal stresses on a smaller scale between particles determine the resisting strength, compressive or tensile, of the concrete and the effect on other mechanical properties such as the various moduli. Thermal strain incompatibilities and the resulting parasitic thermal stresses are thus critical to performance of concrete at high temperature, especially where thermal cycling occurs.

Thus, investigators have explained their observations on compressive strength changes with temperature in general terms of the changes in physical properties of the cement and aggregate and the thermal incompatibilities between these components with temperature. From petrographic examination, Abrams [17] draws attention to microcracking and deterioration in the aggregate-to-paste bond. The advantage in compressive strength where the concrete is preloaded in compression is ascribed to the suppression of thermal microcracking in the cement paste, and the increase in compressive strength at moderately high temperatures has been attributed by Zoldners [19] to "a heat-stimulated cement hydration and densification of the cement gel due to the loss of absorbed water."

From his pulse velocity measurements, Zoldners [19] attributed the greater reductions in flexural strength as compared with compressive strength at the same temperature to microcracking. Whereas small defects would have little effect on compressive strength, they would immediately show up as a reduction in flexural strength.

Since the reductions in modulus of elasticity, Poisson's ratio, and bulk modulus, K, with increase in temperature follow the general pattern for reduction in compressive strength, these have usually likewise been put down to thermal incompatibilities, microcracking, and relaxation of bonds.

Lankard et al. [35] reviewed existing data secured on specimens at elevated temperatures from which moisture could escape, made their own tests on both sealed and unsealed specimens up to 260°C (500°F), and provided this summary:

1. First, and probably most important, the influence of heat exposure per se on the structural properties of concrete is critically dependent on the moisture content of the concrete at the time of heating and is quite different for concrete in which the moisture is free to evaporate as compared to concrete sealed against moisture loss.
2. For concrete that is heated slowly at atmospheric pressure, the primary factor influencing changes in the structural properties is the loss of free water. Partial loss of chemically combined water (dehydration of hydrated cement phases) occurs above 121°C (250°F) and primarily affects the flexural strength.
3. For concrete slowly heated at saturated steam pressures, the primary factor influencing changes in the structural properties is the hydrothermal reactions that can occur in the cement phase and that convert the original highly cementitious calcium silicate hydrates into crystalline, lime-rich calcium silicate hydrates producing lower strength.
4. Deterioration of structural properties under both moisture conditions (sealed and unsealed) can be expected from any test variable that can produce large temperature differentials in the concrete on heating (for exam-

ple, quench heating and cooling, use of very large specimens, etc.).

While their data strongly support the role of moisture content in influencing mechanical properties of concrete at moderately high temperatures, they do provide somewhat of a challenge to the views expressed by others on the role of microcracking.

For unsealed specimens, they noted the microcrack system was no worse after careful heating, possibly not as bad as that in the original concrete. This observation helps to explain that compressive strength decline is not significant below 260°C (500°F) (and in some cases may even increase) provided all the free water is removed on heating. It also adds to the usual explanation of the development of flaws and stress concentrations around flaws to account for the reduction in tensile strength being greater than that for compressive strength. In this case, they considered that the benefit of free water removal is offset by the loss of chemically bound water that reduces the critical stress requirements for tensile failure.

The removal of free water also appears to account for the observed independence of the reduction in modulus of elasticity with temperature at moderately high temperatures. However, once dehydration proceeds to a significant loss of other forms of water and to decomposition of the cement hydration products, and once aggregate deterioration and thermal incompatibilities increase with increasing temperature, then these latter mechanisms appear to account fully for the observed behavior. But it does appear that the presence or lack of free water is the controlling parameter at lower temperatures. Polivka et al. [34] established that at 143°C (300°F) a maximum reduction in strength and stiffness, in the order of 75 and 55%, respectively, occurred at an intermediate moisture content corresponding to about a 50% loss of the free water in the concrete.

Most of the data so far discussed on the thermal and mechanical properties of concrete have been secured by testing small unsealed specimens. The justifications for this are that most tests at normal temperature are made on small specimens and, in a fire, moisture is usually free to be progressively driven off from the concrete, sometimes accompanied by spalling in which moisture expulsion plays a significant role. However, nuclear reactor shields and primary containment vessels present a different set of conditions in which the concrete essentially is sealed against moisture loss. The prestressed concrete sections tend to be massive and concrete may be hot and sealed by a steel liner on one side, even where it is cool and open on the other. In hot sealed areas, the exposure conditions may be quite akin to autoclaving mature concrete and, with chemical and mineralogical changes in the component phases of the concrete and restriction on the mobility of moisture, there is a greater reduction with temperature in the physical properties of concrete than occurs when the moisture is freely driven off.

In "saturated steam" conditions, Lankard et al. [35] found a continual and much greater decline in compressive strength and modulus of elasticity up to 260°C (500°F). Compressive strength at the upper temperature was only 50% of original while the modulus of elasticity fell to only 31% of the original value. In observations of the composition of the hydrated cement phases under these sealed heated conditions, they determined from X-ray analysis the presence of new, more highly crystalline, probably lime-rich calcium silicate hydrates, some of which, if they occur in autoclaved products, are known to lead to lower strengths.

The amount of free water in concrete also has a major influence on thermal properties. The contraction of cement paste is due to the expulsion of free water and, while water is a relatively poor direct conductor of heat as compared with aggregate, it is a many times better conductor of heat than is the air that would be occupying voids, capillaries, or pores if the water was not present.

Because it influences almost every parameter of concrete behavior, including spalling, control and uniformity of moisture content in the concrete prior to fire tests has been of long standing concern. This led to investigations and proposals to modify ASTM E 119 by preconditioning the specimens [36]. Natural drying has been replaced with a specified conditioning regime of relative humidity and temperature and moisture content at the start of the test. While it is still not possible to cover every circumstance of possible moisture condition in a standard test, the present requirements are an improvement and save time and money.

Many of the disparities in the results of physical tests for strength and other properties in the investigations referenced are probably explained by lack of control of the moisture parameter and the temperature gradient in the concrete. However, this apparent lack of control is the real situation when mature or young concrete is exposed to fire and sometimes it spalls and sometimes it does not.

SPALLING AND CRACKING

Spalling of concrete surfaces, particularly at an arris, is an evident feature of most fire-damaged concrete. The worst consequence of this process is that it opens up fresh surfaces to the heat source, thus making for more rapid progression of heat into the concrete and increasing the probability of embedded steel being adversely affected, especially if it becomes exposed. There are two possible causes of such spalling: thermal incompatibilities and moisture entrapment. For many years, direct thermal causes, in particular such things as the 2.4% increase in volume of quartz at 573°C (1063°F), were considered a main contributing factor. However, the quartz volume change may not be as significant as earlier thought. Experienced investigators such as Harmathy and Allen state they have never identified spalling due to this cause [9]. It seems that by the temperature necessary for quartz expansion to occur the plasticity of the paste is great enough to accommodate the expansion, though sudden pop-outs of siliceous aggregate particles at or near the surface will likely occur. Moisture entrapment now is identified [37,38] as predominant, at least in the earlier stages of spalling deterioration. A commonly held view is that the disruptive pressure causing spalling is the result of rapid heating with sufficient moisture present so that, even after the evaporable water escapes, closely bound water in capillaries and gel pores will produce saturated or superheated steam with explosive force. However, this does not appear to be an adequate explanation, and it is worthwhile to examine a more plausible explanation in some detail.

Harmathy [38] (including his work with Shorter) developed an explanation called "moisture clog spalling" along the following lines. As moisture is desorbed in a layer adjacent to the heated surface, the vapor migrates toward the colder region and is readsorbed. As the thickness of the dry layer increases, a completely saturated layer builds up at some distance in from the exposed face and later a sharply defined front forms between the two layers. As the temperature of the exposed surface increases, a very steep temperature gradient develops across the dry layer resulting in a high rate of heat flow, intensified desorption at the frontal plane, and vapor pressure buildup. Because the concrete to the rear is saturated, the vapor can only leave through the dry layer, expanding and meeting increasing flow resistance as it goes. If the permeability of the material is low, the pressure build up at the front continues until a tensile spall occurs separating the dry layers from the rest.

In establishing an overall picture of the effects of moisture on the fire resistance of concrete, Harmathy [38] stressed two very important practical points. First, high moisture contents are conducive to spalling but if spalling did not take place (because, if resistance of the pores to moisture flow was not too high, the existing large vapor pressure differential would move the "moisture clog region" towards the colder region and the pressure buildup would level off), then the moisture in the concrete is most beneficial. Second, the absorption of heat associated with the desorption of moisture checks the rise in temperature in the concrete.

Data on the thermal properties of cement paste, aggregates, and concrete emphasize the thermal contraction and expansion strain incompatibilities that develop with increasing temperature, within and between the constituents, and between the hotter and cooler parts of the concrete itself. The resulting thermal stresses may be sufficient to induce cracking. Thermal cracking may range from crazing on the surface and microcracking to deeper and substantial spalling or cracks of structural significance at locations where overall structural deformations or thermal stresses exceed the tensile capacity of the concrete. Planes of weakness may also develop at the level of the reinforcing steel, because of thermal shock, if the surface of the concrete is quenched with water.

Selvaggio and Carlson [39] in fire tests of prestressed concrete beams observed some cases of longitudinal tensile splitting through the full depth of the flanges and webs when the temperature at the level of the steel tendons was as low as 149°C (300°F). By analysis, they established that the differential thermal expansion between concrete and embedded steel can be a source of tangential tensile (bursting) stresses of considerable magnitude. Two concrete factors were considered to contribute to lack of resistance to these stresses, (1) a high modulus of elasticity, and (2) reduced tensile and compressive strength of concrete when simultaneously exposed to tensile and compressive stresses (from the longitudinal prestress), together with factors originating from the steel arrangement and moisture expulsion.

Sometimes it appears as though additional spalling occurs as the concrete cools down. While the normal contractions are opposed by a volume increase if moisture is present to rehydrate the calcium hydroxide, these processes are not rapid and most likely the apparent spalling is simply the falling away of already loosened pieces of concrete.

INVESTIGATION AND REPAIR OF FIRE DAMAGE

Once the structure is safe to enter after the fire, a first step is to try to obtain an estimate of the temperature reached and its duration and hence a first estimate of the likely remaining properties of the concrete. Many circumstantial clues may be found by sifting through the debris and identifying the remains of common objects that have known melting points. The appearance of the concrete can also provide important evidence [40,41]. Permanent color changes may accompany dehydration of the cement paste. As temperature increases, the color of siliceous or limestone aggregate concrete changes through pink or red to gray at about 600°C (1112°F) to buff at about 900°C (1652°F). Harmathy [42] proposes thermogravimetric and dilatometric testing of samples secured within a day or two of the fire as a means of determining a temperature history inwardly into the concrete. The approach is based on the fact that the continuum of reactions in the cement paste remains frozen at the temperature level reached (for at least a few days) and, by comparison with a reference sample of unaffected concrete or standard data, this temperature marker can be identified. In 1980, Placid [43] proposed a test based on thermoluminescence that requires only small samples and is sensitive to the thermal exposure experienced rather than just the maximum temperature reached. More recently, Riley [44] has explored the potential of the petrographic analysis of thin sections to assess damage.

Abrams and Erlin [45] provide a method based on examining the microstructure and microhardness of prestressing steel for estimating the temperature it reached in a fire and hence its remaining tensile strength. Though the effect of the temperature on normal reinforcing steel is not so critical, similar metallographic methods may be useful.

Once the likely temperature regime has been established, the available data on the engineering properties of the concrete and steel, at either the elevated temperature reached or after cooling, can be incorporated in the appropriate structural analysis to determine the mode of failure, if the structure should be demolished or can be repaired.

Confirmation of the residual engineering properties of the concrete and steel after cooling can be determined by securing samples for standard laboratory testing. Standard methods for in situ nondestructive testing such as ultrasonic pulse velocity measurements can be used to help delineate the damaged areas and compare the properties of the affected concrete with that untouched by the fire. Where doubt still exists after the structural analysis or repairs have been completed, confirmatory full-scale load tests can be made.

Smith [46] described the investigative, material, and structural evaluation techniques and restoration methods available to examine concrete damaged in construction fires. Twenty-five years later he evaluated the performance of the remedial work undertaken and reviewed developments in investigative techniques over the intervening years [47]. Numerous accounts of investigations and repair of fire damage to mature concrete have been published. The account by Fruchtbaum [48] in 1941 remains as a classic, and Ref 41 provides an excellent general review.

LIGHTWEIGHT AGGREGATE CONCRETE

The thermogravimetric, dilatometric, and thermal studies made by Harmathy and Allen [9] on concrete used in masonry units confirm the reasons why concretes made from lightweight aggregates have superior heat resistance to normal-weight concretes. Lightweight aggregates (man-made expanded shales, clays, slates, slags, perlites, or micas and natural pumice or scoria) have significantly lower thermal diffusiveness (up to 600°C (1112°F)) and thermal conductivities (at all temperatures). The capacity of the aggregate to retain moisture and its chemically stable vesicular and glassy (amorphous) composition are also advantageous. Abrams [17], among others, points out that the compatibility in relative stiffnesses between the cement paste and lightweight aggregates is also beneficial. He further noted a good bond strength between paste and aggregate.

Zoldners [19] (with Wilson in subsequent work) looked at the residual engineering properties of lightweight concretes after exposure to temperatures of 300, 500, 700, and 1000°C (572, 932, 1292, and 1832°F) and after cycling 5 or 25 times from 100 to 300°C (212 to 572°F). They concluded from ultrasonic pulse velocity, compressive, and flexural strength determinations made after cooling that all the lightweight (expanded shale and slag) concrete stood up to the effects of high temperature better than did semi-lightweight concrete. They also found that, although there was no advantage in using Type 1S (blended portland blast-furnace slag) cement as compared with a normal Type 1 cement, the addition of microfillers (fly ash, silica flour, and calcined shale dust) as 20 or 40% cement replacements did improve the heat resistance of the concretes up to 500°C (932°F). In all cases where microfillers were used, the improvement in residual flexural strength was in the range of 10 to 20% though similar improvements were not always noted in compressive strength. The pulse velocity measurements, taken to be an indicator of porosity and microcracking in the matrix, generally supported the conclusions drawn from the strength data. Residual strengths for all lightweight concretes were reduced to about 80% by heating to 300°C (572°F) and to about 45% by heating to 700°C (1292°F) as compared with the unheated reference concretes. After heating to 1000°C (1832°F), residual strengths were negligible. Since failure is largely because of the deterioration of the cement paste matrix, by using high alumina cement rather than portland or blended cements, insulating lightweight refractory concretes good up to about 950°C (1742°F) can be made. Very lightweight (cellular, vermiculite, or perlite insulating) concretes, 480 kg/m^3 (30 pcf), as used in roof or other building applications, enhance fire endurance [6].

HEAVYWEIGHT AGGREGATE CONCRETE

Heavyweight concrete is used for radiation shielding in nuclear reactors. This shielding is usually in the shape of a hollow cylinder or sphere. Under moderately high operating temperatures, thermal gradients induce moisture migrations, and complex triaxial stresses can occur. In summarizing, the results of compressive strength of portland cement concrete containing limonite, serpentine, magnetite and steel punchings, and combinations thereof, Desov et al. [49] reported reductions of up to 40% in hot tested specimens after heating to 350°C (662°F) for between 3 and 7 h. As with normal-weight concrete, these effects were ascribed to differences in thermal expansion between the cement paste and aggregates, and to the dehydration of the paste.

VERY HIGH STRENGTH CONCRETE

Most of the data and their significance so far discussed have been from tests on concretes in the normal strength range used in general constructions. Very high strength concretes ranging in strength from 70 MPa (10 000 psi) to as high as 170 MPa (25 000 psi) have been used, or proposed for use, in tall buildings or off-shore structures where enhanced mechanical properties lead to cost effectiveness. It has been observed in laboratory investigations that very high strength concretes, especially if saturated, have a tendency to spall explosively when heated rapidly [11,50]. These concretes are more "brittle" with a denser cement paste matrix obtained by use of high cement factors, silica fume, fly ash, and superplasticizers. Since the cement paste matrix will carry higher loads than in normal concretes, maintaining a more homogeneous stress distribution with the aggregate, the decomposition of the cement paste with temperature rise is a critical factor in fire resistance. Coupled with the slower drying out of moisture due to the less porous microstructure, this increases the risk of damage from fire at an early age. Otherwise the mechanical properties appear to decline with increasing temperature in a similar trend to normal strength concretes. Continuing research into the behavior of very high strength concrete, in particular for load-bearing columns, its prime use, is certainly needed in the interest of public safety.

REFRACTORY CONCRETE

Concretes used to withstand prolonged high temperatures, for example, in monolithic linings for kilns, can be made with high alumina cements and common refractory materials, such as crushed firebrick, as aggregate [51]. This type of concrete will be stable and retain considerable strength at dry heat temperatures of up to 1300°C (2372°F) because ceramic bonds have replaced the hydraulic bonds that were lost during desiccation. By using aggregates such as fused aluminum oxide with a cement that is essentially calcium aluminate, service temperatures for a refractory concrete can be as high as 1800°C (3272°F). The report of ACI Committee 547 [52] provides a definitive discussion of refractory concrete properties for those having deeper interest.

ELEVATED TEMPERATURE COUPLED WITH BLAST

At the time of the introduction of jet aircraft into service, there was considerable concern that the hot blast might be damaging to concrete in engine test cells, warm-up aprons, and runway ends. Immediately at the jet exhaust, temperatures of 677°C (1250°F) occur and this may be as high as 1927°C (3500°F) where afterburners are used. The velocity of the attendant hot blast has been estimated at 1067 m/s (3500 ft/s) and impingement on the concrete was particularly exaggerated by the downward inclination and low slung engines on many early jet aircraft. Tail and higher wing mountings in more recent planes have largely removed the problem from the practical sphere and a 1953 paper by Bishop [53] remains definitive.

The launching of rockets from pads or underground silos for space exploration or military purposes presents another aspect of the hot blast problem and is one for which all the answers have not yet reached the public domain. However, it is known that the solutions tried include sacrificial concrete and protective shields including ceramic ablative shields. However, the conditions though extreme, involving temperatures of about 2760°C (5000°F) and velocities over 2438 m/s (8000 ft/s), last for only a few seconds and the choice between sacrificing the concrete or trying to shield it is one of cost and convenience.

CONCLUDING REMARKS

Despite inherent deficiencies in performance at elevated temperatures arising from dehydration and thermal incompatibilities, concrete has a long-standing and justified reputation as a fire-resistant material, notwithstanding a long-recognized need for better means of testing and specifying endurance [1,54]. Development of temperature regime-structural behavior models and greater understanding of thermal and mechanical properties are leading to significant improvements in specification and design against the extremes of fire and for moderately high temperature applications such as nuclear reactors. While striking advances have been made, full physical-chemical, thermodynamic explanations that tie together all the aspects of the very complex visco-elastic, moisture-dependent behavior pattern of concrete at elevated temperatures are still awaited.

REFERENCES

[1] Uddin, T. and Culver, C. G., "Effects of Elevated Temperatures on Structural Members," *Journal*, Structural Division; *Proceedings*, American Society of Civil Engineers, Vol. 101, No. ST7, July 1975, pp. 1531–1549; Errata ST11, Nov. 1975, pp. 2463–2464; Discussion ST3, Vol. 102, March 1976, p. 685; ST6, Vol. 102, June 1976, p. 1268; and ST12, Vol. 102, Dec. 1976, p. 2376.

[2] Carlson, C. C., "Fire Endurance Testing Procedures," *SP-5*, American Concrete Institute, Detroit, MI, 1962, pp. 3–22.

[3] Benjamin, I. A., "Fire Resistance of Reinforced Concrete," *SP-5*, American Concrete Institute, Detroit, MI, 1962, pp. 25–39.

[4] Sheridan, R. R., "Fire Resistance with Concrete as Protection," *SP-5*, American Concrete Institute, Detroit, MI, 1962, pp. 43–55.

[5] Troxell, G. E., "Fire Resistance of Prestressed Concrete," *SP-5*, American Concrete Institute, Detroit, MI, 1962, pp. 59–86.

[6] "Guide for Determining the Fire Endurance of Concrete Elements," 216 R-89, American Concrete Institute, Detroit, MI, 1989.

[7] Bertero, V. V., Bresler, B., and Polivka, M., "Instrumentation and Techniques for Study of Concrete Properties at Elevated Temperatures," *SP-34*, American Concrete Institute, Detroit, MI, 1972, pp. 1377–1419.

[8] Purkiss, J. A. and Dougill, J. W., "Apparatus for Compression Tests on Concrete at High Temperatures," *Magazine of Concrete Research*, Vol. 25, No. 83, June 1973, pp. 102–108.

[9] Harmathy, T. Z. and Allen, L. W., "Thermal Properties of Selected Masonry Concrete Units," *Journal V70*, American Concrete Institute, Feb. 1973, pp. 132–142.

[10] Nasser, K. W. and Marzouk, H. M., "Properties of Mass Concrete Containing Fly Ash at High Temperatures," *ACI Journal*, April 1979, pp. 537–550.

[11] Hertz, K. D., "Danish Investigations on Silica Fume Concretes at Elevated Temperatures," *ACI Materials Journal*, Vol. 89, No. 4, July–August 1992.

[12] Harmathy, T. Z., "Thermal Properties of Concrete at Elevated Temperatures," *Journal of Materials*, Vol. 5, No. 1, March 1970, pp. 47–74.

[13] Zoldners, N. G., "Thermal Properties of Concrete under Sustained Elevated Temperatures," *SP-25*, American Concrete Institute, Detroit, MI, 1971, pp. 1–32.

[14] Mirkovich, V. V., "Experimental Study Relating Thermal Conductivity to Thermal Piercing of Rocks," *International Journal of Rock Mechanics in Mining Sciences*, England, Vol. 5, 1968, pp. 205–218.

[15] Diederichs, V. and Schneider, V., "Bond Strength at High Temperatures," *Magazine of Concrete Research*, Vol. 33, No. 115, June 1981.

[16] Lin, T. D., Zwiers, R. I., Shirley, S. T., and Burg, R. G., "Fire Tests of Concrete Slab Reinforced with Epoxy-Coated Bars," *ACI Structural Journal*, Vol. 86, No. 2, March–April 1989, pp. 156–162.

[17] Abrams, M. S., "Compressive Strength of Concrete at Temperatures of 1600°F," *SP-25*, American Concrete Institute, Detroit, MI, 1971, pp. 33–58.

[18] Malhotra, H. L., "The Effect of Temperature on the Compressive Strength of Concrete," *Magazine of Concrete Research*, Vol. 8, No. 23, 1956, pp. 85–94.

[19] Zoldners, N. G., "The Effect of High Temperatures on Concrete Incorporating Different Aggregates," *Proceedings*, American Society for Testing and Materials, Vol. 60, 1960, pp. 1087–1108.

[20] Sullivan, P. J. and Poucher, M. P., "The Influence of Temperature on the Physical Properties of Concrete and Mortar in

the Range 20°C to 400°C," *SP-25*, American Concrete Institute, Detroit, MI, 1971, pp. 103–135.

[21] Philleo, R., "Some Physical Properties of Concrete at High Temperatures," *Journal*, American Concrete Institute; *Proceedings*, Vol. 54, No. 10, April 1958, pp. 857–864.

[22] Marechal, J. C., "Variations in the Modulus of Elasticity and Poisson's Ratio with Temperature," *SP-34*, American Concrete Institute, Detroit, MI, 1972, pp. 495–503.

[23] Cruz, C. R., "Apparatus for Measuring Creep of Concrete at High Temperature," *Journal*, Portland Cement Association, Research and Development Laboratory, Vol. 10, No. 3, Sept. 1968, pp. 36–42.

[24] Wang, C. H., "Creep of Concrete at Elevated Temperatures," *SP-27*, American Concrete Institute, Detroit, MI, 1971, pp. 387–400.

[25] Marechal, J. C., "Creep of Concrete as a Function of Temperature," *SP-34*, American Concrete Institute, Detroit, MI, 1972, pp. 547–564.

[26] Schneider, U., "Physical Properties of Concrete under Nonsteady State Condition," *Proceedings*, VII, Federation Internationale de la Precontrainte, Congress, New York, 1974.

[27] Thompson, N. E., "A Note on Difficulties of Measuring the Thermal Conductivity of Concrete," *Magazine of Concrete Research*, Vol. 20, No. 62, March 1968, pp. 45–49.

[28] Harmathy, T. Z., "Variable-State Methods of Measuring the Thermal Properties of Solids," *Journal of Applied Physics*, Vol. 35, No. 4, April 1964.

[29] Harada, T., Takeda, J., Yamane, S., and Furumura, F., "Strength, Elasticity and Thermal Properties of Concrete Subjected to Elevated Temperatures," *SP-34*, American Concrete Institute, Detroit, MI, 1972, pp. 377–406.

[30] Crispino, E., "Studies of the Technology of Concretes Under Thermal Conditions," *SP-34*, American Concrete Institute, Detroit, MI, 1972, pp. 443–479.

[31] Dougill, J. W., "Some Effects of Thermal Volume Changes on the Properties and Behaviour of Concrete," *Proceedings*, International Conference on the Structure of Concrete, London, 1968.

[32] Campbell-Allen, D. and Desai, P. M., "The Influence of Aggregate on the Behaviour of Concrete at Elevated Temperatures," *Nuclear Engineering and Design* (Amsterdam), Vol. 6, No. 1, 1967, pp. 65–77.

[33] Campbell-Allen, D., Low, E. W. E., and Roger, H., "An Investigation of the Effect of Elevated Temperatures on Concrete for Reactor Vessels," *Nuclear Structural Engineering* (Amsterdam), Vol. 1, No. 2, 1965, pp. 382–388.

[34] Polivka, M., Bertero, V. V., and Gjorv, O. E., "The Effect of Moisture Content on the Mechanical Behaviour of Concrete Exposed to Elevated Temperatures," *Proceedings*, International Conference on Mechanical Behaviour of Material, Vol. IV, 1972, pp. 203–213.

[35] Lankard, D. R., Birkimer, D. L., Fondriest, F. F., and Snyder, M. J., "Effects of Moisture Content on Structural Properties of Portland Cement Concrete Exposed to Temperatures up to 500°F," *SP-25*, American Concrete Institute, Detroit, MI, 1971, pp. 59–102.

[36] Abrams, M. S. and Orels, D. L., "Concrete Drying Methods and Their Effect on Fire Resistance," *Moisture in Materials in Relation to Fire Tests*, ASTM STP 385, American Society for Testing and Materials, Philadelphia, 1964, pp. 52–73.

[37] Meyer, O. C., "Explosive Spalling Occurring in Reinforced and Prestressed Concrete Structural Members of Normal Dense Concrete Under Fire Attack," (in German), Deutscher Ausschuss für Stahlbeton, H248, Einst and Sohn, Berlin, 1975.

[38] Harmathy T. Z., "Effect of Moisture on the Fire Endurance of Building Elements," *Moisture in Materials in Relation to Fire Tests, ASTM STP 385*, American Society for Testing and Materials, Philadelphia, 1964, pp. 74–94.

[39] Selvaggio, S. L. and Carlson, C. C., "Fire Resistance of Prestressed Concrete Beams, Study B, Influence of Aggregate and Load Intensity," *Journal*, Portland Cement Association, Research and Development Laboratory, Vol. 6, No. 1, Jan. 1964, pp. 41–64.

[40] Bessey, G. E., "Investigation on Building Fires, Part 2—The Visible Changes in Concrete or Mortar Exposed to High Temperature," National Building Studies, Technical Paper 4, Her Magesty's Stationary Office, London, 1950.

[41] "Concrete Structures after Fires, Evaluating Fire Damage to Concrete Structures and Reinstating Fire Damaged Structures," three-part series on "Repair of Fire Damage," *Concrete Construction*, Vol. 17, Nos. 4, 5, and 6, April, May, June 1972.

[42] Harmathy, T. Z., "Determining the Temperature History of Concrete Constructions Following Fire Exposure," *Journal*, American Concrete Institute, Vol. 65, No. 11, Nov. 1968, pp. 959–1064.

[43] Placid, F., "Thermoluminescence Test for Fire Damaged Concrete," *Magazine of Concrete Research*, Vol. 32, No. 111, June 1980, pp. 112–116.

[44] Riley, M., "Assessing Fire Damaged Concrete," *Concrete International*, American Concrete Institute, Vol. 13, No. 6, June 1991, pp. 60–64.

[45] Abrams, M. S. and Erlin, B., "Estimating Post Fire Strength and Exposure Temperature of Prestressing Steel by a Metallographic Method," *Journal*, Portland Cement Association, Research and Development Laboratory, Vol. 9, No. 3, Sept. 1967, pp. 23–33.

[46] Smith, P., "Investigation and Repair of Damage to Concrete Caused by Formwork and Falsework Fire," *Journal*, American Concrete Institute; *Proceedings*, Vol. 60, No. 11, Nov. 1963, pp. 1535–1566.

[47] Smith, P., "Old Flames Speak Well of Young Concrete," *SP-92*, American Concrete Institute, Detroit, MI, 1986, pp. 33–44.

[48] Fruchtbaum, J., "Fire Damage to General Mills Building and Its Repair," *Journal*, American Concrete Institute; *Proceedings*, Vol. 37, No. 3, Jan. 1941, pp. 201–252.

[49] Desov, A. E., Nekrasov, K. D., and Milovaniov, A. G., "Cube and Prism Strength of Concrete at Elevated Temperatures," *SP-34*, American Concrete Institute, Detroit, MI, 1972, pp. 423–434.

[50] Castillo, C. and Durrani, A. J., "Effect of Transient High Temperature on High Strength Concrete," *ACI Materials Journal*, Vol. 87, No. 1, Jan.–Feb. 1990, pp. 47–53.

[51] Zoldners, N. G., Malhotra, V. M., and Wilson, H. S., "High-Temperature Behaviour of Aluminous Cement Concrete Containing Different Aggregate," *Proceedings*, American Society for Testing and Materials, Philadelphia, Vol. 63, 1963, pp. 966–995.

[52] "Refractory Concrete State-of-the-Art," 547 R-79 (83)(87) Report, American Concrete Institute, Detroit, MI, 1987.

[53] Bishop, J. A., "The Effect of Jet Aircraft on Air Force Pavements; Investigation Conducted by the Bureau of Yards and Docks," *Proceedings*, American Society of Civil Engineers, New York, Separate 317, Oct. 1953.

[54] Menzel, C. A., "Tests of the Fire Resistance and Thermal Properties of Solid Concrete Slabs and Their Significance," *Proceedings*, American Society for Testing and Materials, Vol. 43, 1943, pp. 1099–1153.

Air Content and Unit Weight of Hardened Concrete

Kenneth C. Hover[1]

PREFACE

Samuel Helms discussed the unit weight and air content of hardened concrete in each of the three preceding editions of this ASTM Special Technical Publication. This present edition makes much use of Helms' clear and concise work, particularly in the area of unit weight. The discussion of air content is more extensive here, however, largely in response to the great deal of new research since the last major update. It has also become evident that the further in time that the industry has come from the landmark contributions of the early workers in air-void systems and frost resistance in concrete, the less understood are the fundamental principles laid down in the late 1940s and early 1950s. Following the pattern set by Helms, the present edition therefore reviews much of this key literature to explain the origins of today's state of the art. Finally, this edition appears when there is a proliferation of the use of the ASTM Practice for Microscopical Determination of Air-Void Content and Parameters of the Air-Void System in Hardened Concrete (C 457) microscopical analysis procedure, which was once exclusively performed by a small cadre of experts. While expansion of the use of the test has produced more data, it has in some cases produced more confusion where less experienced operators have used nonstandardized equipment and procedures. The chapter has therefore been written in part to point out pitfalls and sources of error in performing and interpreting the test.

PART I: AIR CONTENT

Introduction

The air content within a given volume of concrete is the cumulative volume of a large number of air voids of multiple sizes ranging from microscopic bubbles to larger, irregularly shaped air pockets. These voids have their origins in the air initially trapped among the dry constituents, dissolved air in the water, and the air introduced as a result the kneading and folding action of mixing and trapped in the mix while depositing the concrete in the forms [1].

The presence of these air voids initially suspended in the mix water among the solid particulate constituents of the concrete influences the workability, consistency, bleeding tendency and yield of the fresh concrete, and the density, strength, and frost resistance of the hardened concrete. The most significant of these effects is the influence of air voids in mitigating the damaging effects of freezing and thawing of absorbed water. The magnitude and utility of these effects depends on not only the total volume of these air voids, but on the entire size-distribution of voids and on their dispersion throughout the concrete. The beneficial consequences of air voids are obtained by using air-entraining admixtures and appropriate concrete production and construction procedures to encourage the retention of the smallest voids, and, by using effective placing and consolidation procedures to reduce the number and volume of the largest voids.

While the concrete is still plastic, however, the air voids have an opportunity to move, become larger or smaller, coalesce, or change shape. Some voids can be removed from the concrete entirely. The total air content and other air-void characteristics therefore depend on the stage in the mixing, transport, placement, and consolidation processes at which the measurement was taken. Once the concrete has hardened, however, permanent void spaces remain, preserving the size and shape of those air voids present at the time of setting. All air voids remaining within a given concrete mass, regardless of their size, shape, or origin, are often referred to as the "air-void system." In order to evaluate the characteristics of the air-void system, it is presently necessary to obtain a sample of the hardened concrete, perform a statistical analysis on a fraction of the exposed voids observed under a microscope, and to estimate the relevant characteristics to various degrees of accuracy and precision. The significance of such sampling, testing, and analysis, and the interpretation of the results obtained are the subjects of this chapter.

Influence of Air Content on the Behavior and Performance of Concrete

Introduction

Incorporating air voids into concrete influences the behavior of the material in both the fresh and hardened states. The magnitude and degree of utility of these influences depends on the total volume, sizes, and dispersion of the voids, and on the material properties of the concrete.

---

[1] Professor of Structural Engineering, Cornell University, Ithaca, NY 14853.

Workability

A large number of microscopically small voids, well-dispersed throughout the cement paste, will generally increase workability at any given water content. This is because sufficiently small air bubbles can separate solid particles in the mix [2], acting as air-cushions to reduce inter-particle friction. In proportioning concrete mixes, one can take advantage of this increased workability by reducing the water content by 6 to 12%, depending on aggregate gradation and other constituents, while maintaining slump [3].

Cohesion

When air-entraining admixtures are used to stabilize the smaller air voids, a mutual attraction develops between microscopic bubbles and the grains of portland cement. This attraction not only "anchors" the air bubbles to inhibit their loss from the fresh concrete due to buoyancy, but imparts a beneficial cohesion to the mix that resists segregation, settlement, and bleeding [1]. In some mixes, this can also cause air-entrained concrete mixes to become "sticky" with increased adhesion to construction equipment and increased drag on finishing tools.

Unit Weight and Yield

By displacing heavier components in the mix, the air voids reduce the unit weight or density of the mix. As described in Chapter 9 and Part II of this chapter, the unit weight test, ASTM Test Method for Unit Weight, Yield, and Air Content (Gravimetric) of Concrete (C 138), can therefore be a useful means of measuring air content in fresh concrete.

Strength Reduction

By occupying space between the cement grains, creating a more porous cement paste, and reducing the density of the hardened concrete, air voids can reduce the strength of concrete. Thus, at equal water/cement ratios, the impact of air content decreases compressive strength, with various rules of thumb equating each 1% increase in air content to a 3 to 5% reduction in 28-day compressive strength [1]. (Strength reductions of up to 10% per 1% air have been reported [4].) However, the water-reducing effect of the air voids permits a reduced water content in mixes intentionally incorporating air, allowing such mixes to have a reduced water/cement ratio without increasing the cement content. The lower water/cement ratio characteristic of well-designed mixes using an air-entraining admixture can compensate in part or in full for the strength-reduction effect of the air voids [5].

Frost Resistance

Primary among the multiple benefits of intentionally incorporating air into concrete mixtures is the increased frost-resistance afforded to the hardened cement paste. This is accomplished through several mechanisms as described in the next section.

Nature of Concrete and the Mechanism of Freeze-Thaw Damage

Mechanism of Freeze-Thaw Damage

Concrete is fundamentally a porous material composed of coarse and fine aggregate particles of varying porosity, held together by a hardened cement paste whose porosity depends on the original water/cement ratio, the effectiveness of consolidation, and on subsequent curing conditions. Water can be absorbed into the pores in the aggregate particles and into the "capillary pores" of the hardened cement paste [6,7].

These capillary pores can be saturated to various degrees, as influenced by residual, non-evaporated mix water or the subsequent absorption of water from the environment. At sufficiently low temperatures, this water freezes and expands in volume by 9%. When absorption of water has filled the capillaries to the point where the remaining empty pore space cannot accommodate expansion of the ice, the volume of the hardened concrete itself will be forced to expand with accompanying tensile or bursting stresses. Depending on the rate at which the freezing takes place and whether salt water or deicing agents are present, the expansive pressure can originate in the actual expansion of ice, the movement of unfrozen water escaping the advancing ice front [8], or in osmotic pressure caused by differential salt concentrations [6,7,9]. Since these pressures are generated with the onset of freezing and relaxed with the advent of thawing, multiple freeze-thaw cycles "fatigue" the concrete. Whether physical damage occurs depends on the level of stress and the frequency of these fatigue cycles, and on the strength of the concrete. Air voids are intentionally incorporated into the cement paste to reduce (but not eliminate) the stresses generated with each freeze-thaw cycle, and thus increase the number of cycles to failure.

Influence of Air Voids

Air voids in concrete intersect the network of capillary pores in the hardened cement paste. Since the largest of the capillary pores are typically smaller than the minimum diameter of an air void (0.010 to 0.020 mm or 0.0004 to 0.0008 in.) absorbed water remains in the smaller capillaries, moving towards the air voids only at subfreezing temperatures or when under pressure. Under normal conditions, the air voids therefore remain empty and able to accept either ice or unfrozen water when a freezing cycle begins.

Upon freezing, water and ice move towards the air voids accompanied by a pressure that increases with the required distance of travel [8,10,11]. This pressure is minimized by a well-dispersed system of air voids intersecting the capillary network at many points, providing a minimum travel distance from any point in the paste to the nearest air void.

Just how far the water or ice can travel without generating damaging pressures depends on degree of saturation, rate of freezing, porosity, permeability, degree of hydration of the hardened cement paste, viscosity of the water, and the tensile strength of the concrete [8–11]. For any given cement paste and set of environmental conditions, one can theoretically compute a critical distance (Powers' term was "critical thickness") beyond which movement of ice or water will generate excessive pressures. Damage results when absorbed water in a portion of the concrete must move further than this critical distance to arrive at the nearest air void. Powers calculated this theoretical distance as 0.25 mm (0.010 in.) "or thereabouts" for commonly encountered pastes and freezing conditions [11].

As a consequence of the critical distance concept, any given air void can provide frost protection to only the hardened cement paste falling within a sphere of influence radiating outwards from the periphery of the void. The zone of protected paste occupies a "shell" surrounding the air void with a thickness approximately equal to the critical distance. (Among the mathematical details of the development of the "protected paste shell" concept is a reduction in the critical distance due to the curvature of the bubble surface, and the limiting condition that the volume of freezable water in the shell must be no greater than the volume of the bubble [10,12]. Shell thickness is therefore not independent of bubble size, but tends to be so as the bubbles become larger.)

Requirements for an effective air-void system

To effectively provide frost resistance, the air-void system must have a total volume of air voids that equals or exceeds the volume of water or ice not accommodated by empty space in the capillary pore system. (Unsaturated concrete would need less air than saturated concrete, and concrete that is continuously dry would need no air at all.) Of equal importance, the air voids must be dispersed throughout the cement paste so that nearly all of the paste is within the protective shell of one or more air voids [8,10–12]. As will be discussed, the precise requirements for volume, dispersion, and spacing of the voids depend on both the concrete and on the environment. Meeting such requirements will not eliminate the stress that accompanies freezing but will maintain it at tolerable levels.

Principle of achieving dispersion and an acceptably small bubble spacing

Achieving dispersion and an acceptably small spacing for a given air content requires that the air-void system be comprised of a large number of small bubbles. As the same volume of air is subdivided into smaller and smaller bubbles, the total bubble surface area expands geometrically, as does the cumulative volume of the protected paste shells around the bubbles.

Influence of factors other than air volume and air-void dispersion

The volume of freezable water initially present in the pores, or the "degree of saturation" will depend on the initial porosity and age of the concrete, curing conditions, and environmental exposure. Similarly, the velocity at which the water or ice must move through the capillaries enroute to the air voids depends on the rate of freezing, another environmental variable. As pointed out earlier, rapid freezing conditions probably induce water movement, while slow-freeze conditions are more likely to induce the movement of ice [6,8–11]. Mattimore and Arni et al. [13,14] demonstrated reduced freeze-thaw damage at slower rates of freezing, while Flack [15] produced data showing rate of freezing effects to be more complicated.

It can also be shown that the pressure developed as water moves towards the air voids depends on the permeability of the paste and on the viscosity of the water [6,8–11]. The additional component of osmotic pressure is influenced chiefly by the presence of deicing salts.

Details of the interaction of air-void geometry, material properties of the paste, and environmental conditions are described in detail by multiple researchers [6–12,16–20].

Influence of Frost Resistance of Aggregates

Freezing and thawing damage to concrete can result from the mechanisms of paste destruction described, or from the expansion of absorbed water in the aggregates, or both. The air voids protect only the hardened cement paste and do not improve the inherent frost resistance of the aggregates. As Powers described it, concentrating on air voids "ignores 75% of the problem" of the frost resistance of concrete (assuming 75% of the volume of the material is composed of coarse and fine aggregates) [11]. Obtaining frost-resistant concrete therefore requires the selection of frost-resistant aggregates combined with the incorporation of appropriately sized and dispersed air voids in the paste. The selection or testing of frost-resistant aggregates is discussed in other chapters of this publication.

Origin and Geometric Characteristics of Air-Void Systems in Concrete

Introduction

With or without the use of an air-entraining admixture, air is unavoidably present in fresh concrete as a remanent of the air initially present in and among the dry mix ingredients [8], and subsequently incorporated by the kneading and folding action of mixing and as a result of transporting and depositing the concrete in the forms [1,16]. While the regular action of controlled mixing can stimulate formation and entrapment of a fairly well-defined gradation of predominantly small air voids, the more random activity of depositing the concrete in the forms tends to trap randomly sized voids, generally much larger than those created in the mixing process.

Air-void systems without benefit of an air-entraining admixture

In ordinary concrete mixed without the stabilizing effects of an air-entraining admixture, a large number of the smaller air voids initially formed in the mixing process are lost from the system, leaving primarily larger voids in the hardened concrete. The problem with this is not that the larger air voids do not provide frost resistance; a protected paste shell exists around the periphery of all air voids in concrete regardless of their size or origin [9]. The problem is that unless the air content is extraordinarily high, the large voids offer too little volume of protected paste. (The volume of protected paste relative to the volume of the air voids themselves increases for smaller air voids.) Large voids are therefore not ineffective, they are simply less effective in providing frost resistance than a much greater number of smaller voids occupying the same total air volume.

Function of Air-Entraining Admixtures

Air-entraining admixtures stabilize the smaller air voids in the mix by promoting their formation, retention, and dispersion. These admixtures, related to soaps, detergents, and the general class of chemicals known as "surface active agents" initially reduce the surface tension of water, promoting the formation of smaller bubbles. They also provide an attractive force between the bubbles and the cement grains to resist air loss due to buoyancy. Finally, the electrical charge effects responsible for anchoring the bubbles to the cement grains cause the bubbles to repel one another, discouraging coalescence [1,21–24]. As pointed out by Whiting and Stark, however, [1] "entrained air is produced by the mechanical stirring, kneading, and infolding actions of the concrete mixer"; the air-entraining admixture merely stabilizes these bubbles once they are formed.

Air-void system resulting from the use of an air-entraining admixture

As reported by Powers, "Comparison of sections of concrete made of the same materials, with and without an air-entraining agent, indicates that the voids present when an entraining agent is not used are also present when the agent is used" [11]. The consequence of using an air-entraining admixture is therefore to augment the coarser or "natural" air voids with a large number of the smaller voids stabilized during the mixing process. The resulting air-void system is therefore a "composite" [11] of the smaller voids stabilized and trapped in the concrete during mixing combined with the much larger voids trapped in the concrete as a result of handling and placing.

The term "entrained air" as conventionally used is intended to apply to those voids trapped during mixing and stabilized by the air-entraining admixture; the term "entrapped air" is generally intended to apply to those voids trapped in the concrete during handling and placement. Given that these larger, so-called "entrapped" air voids reduce the density and strength of the concrete to a greater degree than is justified by their limited contribution to frost resistance [4,25], it is normally advantageous to remove them from the fresh concrete by appropriate consolidation techniques.

The Composite Air-void Gradation

While not implied in the literature, a common industry misinterpretation of the terms entrained and entrapped is that air voids in concrete are of two sizes only: large "entrapped" voids and small "entrained" voids. It is observed that the voids actually occupy a broad gradation of sizes ranging from 10 μm to several or hundreds of millimetres—a size range from smallest to largest of more than a factor of 1000. As shown in Fig. 1, the air voids occupy the broadest range of sizes of all the constituents of concrete.

Despite this continuous gradation of voids, the simplest way to describe the air-void gradation is to report the

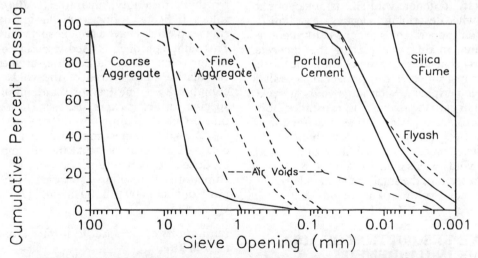

FIG. 1—Example gradation curves of concrete constituents. Vertical axis represents the mass or volume "finer than" the size indicated on the horizontal axis. Mass or volume percent is shown as a fraction of the total mass or volume of that particular constituent. Coarse and fine aggregate gradation bands are from ASTM C 33. Cement gradations are examples of a coarse and fine grinds. Fly ash and silica fume examples are taken from Ref 76. In the coarser of the two air-void gradations, 90% of the total air volume is contained in voids larger than 1 mm in diameter. In the fine air-void gradation, 90% of the total air content is contained in voids smaller than 1 mm in diameter. An air-entraining admixture is normally required to stabilize the air bubbles in such a mix.

number or volume of voids larger than, and smaller than, some predetermined size. ASTM Terminology Relating to Concrete and Concrete Aggregates (C 125) sets a size criterion of 1 mm (0.04 in.) and defines "entrapped" air voids as those above this nominal limit, and "entrained" air as those below. Underlying this definition by size may be the observation that retention of a significant number of air bubbles much less than 1 mm in diameter is unlikely without the stabilizing influence of an air-entraining admixture. Alternatively, air voids of greater than 1 mm in diameter or so can be present with or without the influence of an air-entraining admixture [11].

The Shape of Air Voids

While the smaller air voids in concrete are generally observed to be spherical in shape or nearly so, the larger voids are often seen to be nonspherical and irregularly formed. The observation that void shape becomes increasingly irregular with void size is observed in many other physical systems [26], and is however not necessarily related to the use of an air-entraining admixture or to the effectiveness of the air voids in providing frost protection.

Caveat concerning terminology

As described earlier, the terms "entrained air void" and "entrapped air void" as defined in ASTM C 125 have specific meaning only in regard to the size and shape of the voids. These same terms can be misleading when used to imply the origin, evolution, or effectiveness of the voids. As has been discussed, "air-entraining admixtures" do not strictly "entrain" air into fresh concrete. Further, the smaller voids classified as the entrained air voids are in fact the result of air having been "entrapped" in the mix and stabilized via the admixture. The term "entrapped" when used correctly to denote void size is analogous to the term "coarse" when describing an aggregate gradation. Finally, large or "entrapped" voids are the natural consequence of batching, mixing, and placing the concrete. Large or irregularly shaped voids are not altogether without benefit to frost resistance, since a protected paste shell exists around their periphery. Such large voids are relatively inefficient in regard to proving frost resistance, however, since their protected paste volume as a fraction of the void volume itself is many times less than for the smaller, more efficient voids. The advantage of the more efficient, smaller voids was stated by Powers [8], "For a given degree of protection, the smaller the air-filled cavities the smaller the total volume of air required."

QUANTITATIVE DESCRIPTION OF AIR-VOID SYSTEMS IN HARDENED CONCRETE—THE AIR-VOID SYSTEM PARAMETERS

Measures of Air Content

The air content of hardened concrete is most commonly expressed as a percentage of the combined volume of all constituents of the concrete including the total volume of air. Klieger [5,27] found that it can be more informative to express the air volume as a percentage of the mortar volume (that is, mortar volume = concrete volume minus the volume of coarse aggregate), observing that frost resistance consistently resulted from 9 ± 1% air content in the mortar. Given that air voids contribute to the frost-resistance of the hardened cement paste only, it can be useful to express the air content as a percentage of the paste (volume of cementitious materials, water, and air). When calculating the spacing factor of the air-void system (as will be discussed), the air content must be expressed as a fraction of the "air-free" paste, by dividing the total air volume by the volume of cement and water only.

Measures of Air-Void Size and Size-Distribution

As pointed out earlier, the air-void gradation can range from 20 μm to more than 20 mm—a spread of more than three orders of magnitude. While Willis and Lord [28,29] developed mathematical techniques for estimating the complete gradation curve for air voids in concrete, characterization of void size is currently done much more simply and approximately. The size of the air voids is typically defined by a statistical parameter known as the "specific surface," based on the ratio of total air void surface area to total air volume.

The Specific Surface

The specific surface (α) of the air-void system is analogous in some ways to the "fineness" of cement, which is expressed as the estimated total cement surface area per unit mass of the cement (300 m^2 of cement surface per kilogram of cement, for example), ASTM Test Method for Fineness of Portland Cement by the Turbidimeter (C 115) and ASTM Test Method for Fineness of Portland Cement by Air Permeability Apparatus (C 204). Higher fineness values indicate a finer, smaller-grained cement since smaller particles have a greater surface area per unit mass. This same approach is used to describe the gradation of the air voids, where the specific surface is defined as the cumulative surface area of the voids divided by their cumulative volume (rather than cumulative mass). The specific surface is expressed as surface area per unit volume resulting in units of mm^2/mm^3 or 1/mm (in.^{2}/in.^{3}) or (1/in.); higher values imply a finer air-void system. It is observed, for example, that the air-void system typically produced in concrete incorporating an air-entraining admixture will result in a specific surface of 25 to 45 mm^2/mm^3 (or about 600 to 1100 in.^{2}/in.^{3}) [11,17], see also ASTM C 457.

Limitations of the specific surface alone as an index to void size

The specific surface of air voids and the fineness of cement are two examples of the industry's need to characterize the behavior of multi-sized systems with a single, numerical size-index. The "fineness modulus" of sand (see ASTM C 125) is another example. In each of these cases where a single number is used to represent an entire gradation, critical information is lost concerning the range of sizes included. While single-number indices may imply an "average" fineness of sorts, they provide no information about the actual number of particles or air voids with a fineness near the average value. Such indices do not uniquely define the size distribution, since multiple, broadly different size distributions could be described by the same index.

Powers recognized this limitation of specific surface, citing its utility only for air-void distributions without "extremely coarse voids" [11]. When void systems vary considerably in the breadth or overall shape of their size distributions, the specific surface loses much of its value as a comparative index, as do any other indices calculated from the specific surface. Among air-void systems with similar distributions of void sizes, however, the specific surface can be a useful indicator of the relative void fineness.

Air-Void Dispersion and Spacing

As discussed earlier, only those saturated portions of cement paste within the "critical distance" of an air void will be able to tolerate the pressure generated during freezing. While both the air content (expressed as the volume of air per unit volume of air-free paste) and the air-void size (as characterized by the specific surface) will jointly be used to estimate the number of air voids within a volume of paste, it remains to assess the dispersion or spacing of those voids and so determine whether substantial portions of the paste are sufficiently close to one or more air voids.

Characteristics of Actual Geometric Arrangement

As observed under the microscope, actual air voids in concrete are randomly dispersed in regard to both size and location [16,30,31], see also ASTM C 457. Mather [32] observed an inhomogeneity in bubble size and dispersion such that, "Many samples show variations from area to area, and one cannot escape the conclusion that different parts of such concrete would behave differently with respect to resistance to freezing and thawing."

Spacing Factor

Mathematically rigorous approaches are available for determining the relative proportion of the hardened cement paste within the beneficial zone of influence of one or more air voids [12,20,33,34]. A less accurate but far simpler approach has been adopted by the industry, however, in which one computes a theoretical maximum distance from any point in the paste to the nearest air void. Under the assumption that a majority of the maximum paste-to-void spacings in the paste are less than this computed value, the so-called "spacing factor," generally designated by $\bar{L}$, can serve as an index to the effectiveness of the air-void system in contributing to frost resistance.

The simplified model of air-void spacing: Power's spacing factor

Powers developed the concept of the spacing factor as a simplified approach to the complex mathematics of the actual distribution of air-void spacing in concrete [8,10,11]. The relevant equations for determining the spacing factor are given in ASTM C 457. To those not familiar with the derivation of these equations, their apparent complexity can imply a level of mathematical rigor beyond the intent of the developer, as the relationships bypass the complexity of the real air-void system and substitute a very simplified model of reality.

The basis for the utility of the spacing factor is that it takes into account the combined influence of the "total" air content, total air-free paste content, and representative void size on the spatial distribution of the air voids. To compute the spacing factor in accordance with the ASTM equations, one needs to first determine the air content, paste content, and specific surface of the air voids. No measurements are required or performed relative to observed distances between voids or from points in the paste to the nearest air void.

After having obtained the relevant input data, the first simplifying step is to replace the multi-sized voids in the actual system with a system of single-sized voids. The void size in the simplified system is chosen to have the same total volume and same total void surface area as in the actual concrete. As pointed out by Willis [28] and by Powers [10] the "number" of air voids in the actual versus simplified systems may be significantly different.

Next, the random dispersion of air voids in the actual cement paste is replaced in the model with a geometrically regular pattern of single-sized voids arranged in a uniform, three-dimensional grid. As described by Walker [25], "Consider a hypothetical set of air voids, all one size, arranged in paste in a cubic, three-dimensional array. Every void is equidistant from six other voids, and imaginary lines connecting the voids are mutually perpendicular" (see Fig. 2). The dimensions of the grid system are determined so that the volume of the air voids relative to volume of "grid space" between them is equivalent to the measured volume of air relative to the measured volume of hardened cement paste. The spacing factor is then determined as half of the greatest distance between any two adjacent air voids. These assumptions and the simplified geometry just described are embodied in the standard equation for spacing factor (ASTM C 457-90, Eq 13).

When it has been determined that air content in the actual system is more than about 23% of the air-free paste volume (normally occurring in only high air content/low paste concretes), an even simpler model is used to determine the spacing factor (ASTM C 457-90, Eq 12). In this case, one simply assumes that the total paste volume is evenly distributed over the combined surface area of the air voids like a uniformly thick coating. The thickness of this hypothetical paste layer is assumed to be equivalent to the spacing factor [10,11]. This simplified model likewise ignores the complexity of the actual void sizes and random dispersion.

Approximate nature of the spacing factor

Due to the nature of the simplifying assumptions in constructing this model, Powers himself pointed out that the method "does not give actual spacing" [11]. ASTM C 457 reports that the spacing factor is "related to the maximum distance in the cement paste from the periphery of an air void." Philleo [34] summarized two primary limitations of the Powers spacing factor, "first, being derived from total void volume and surface area, it [Powers spac-

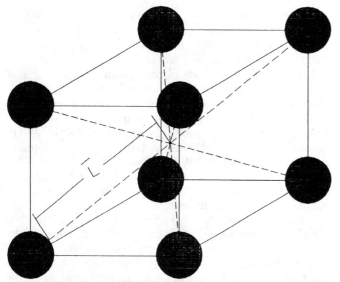

FIG. 2—Air voids arranged in a simple cubic lattice. Calculation of the spacing factor (L) is based on the assumptions that all air voids are (1) the same size, and (2) arranged in a simple cubic lattice where each void is equidistant from its nearest neighbor. The ratio of the volume of the unit cube to the volume of the air voids is set to be equal to the ratio of the paste volume to the total air volume in the paste. The spacing factor (L) is the distance along the interior diagonal from the center of the unit cube to the periphery of the nearest void, representing the farthest that water or ice would have to travel through paste to get to the nearest air void in the hypothetical air-void system.

ing factor] is strictly applicable only to concretes with similar void size distributions and, second, even if the average spacing is determined accurately, the performance of a particular concrete might be determined by a much larger spacing factor that occurs less than half the time. Thus two concretes with the same calculated spacing factor might not behave the same."

METHODS OF EVALUATING THE GEOMETRY OF THE AIR-VOID SYSTEM IN HARDENED CONCRETE

Introduction

As previously discussed the frost-resistance of concrete depends on environmental conditions, the frost-resistance of the aggregates, material properties of the hardened cement paste, and the air-void system. Once the concrete has hardened, however, frost resistance is normally estimated on the basis of actual measurements of the air-void system, coupled with assumptions concerning material properties and environmental conditions. The current procedure for evaluating the air-void system of hardened concrete is the ASTM C 457.

ASTM C 457 Microscopical Analysis

ASTM C 457 was originally proposed by Brown and Pierson [35] on the basis of earlier developments in geological sciences, with modifications for use on concrete. The mathematical foundation for the technique was laid by Powers [8,10], Willis [28], and Lord and Willis [29], and key modifications were made by Mielenz and Wolkodoff [19] based on work by Chayes [36].

A sample of hardened concrete is sawn to expose a plane surface that is then ground to provide an "extremely smooth and plane section" suitable for microscopic inspection (ASTM C 457). The operator causes the prepared sample to move in a systematic pattern under the crosshairs of a microscope, and makes a series of measurements or countings on only those air voids that come into a narrow field of view. Statistical estimates of air content, paste content, air-void size distribution, and spatial dispersion of voids are computed on the basis of these measurements.

Sampling

Samples submitted for microscopical analysis are often cores extracted from concrete in service, or may be hardened samples cast specifically for the purpose. ASTM C 457 recommends at least three samples per determination when the tests are being performed for "referee purposes or to determine the compliance of hardened concrete with requirements of specifications for the air void system." Actual performance of the test, however, requires only one sample, provided the minimum area requirements of ASTM C 457 can be met by the single specimen.

Specimen Preparation

When observing the specimen through the microscope, the operator must be able to clearly distinguish among the various constituents, with the air voids appearing as sharply defined depressions in the surface. (These depressions are made more visible under the microscope by sidelighting with a harsh, low-angle light.) The less distinct the edges of these depressions the less accurate are the measurements, and sharpness of the void edges depends on the quality of the surface preparation prior to microscopic evaluation.

The current procedure recommends grinding the surface with successively smaller abrasive grits, although alternative methods have been used with success [30,37–39]. The end result should be an "extremely smooth" surface, although Brown and Pierson [35] and Mather [40] recommended stopping short of "polishing."

Conduct of the Test

ASTM C 457 describes two alternative methods, (1) the Rosiwal linear traverse technique (ASTM C 457 Procedure A), and (2) the modified point count method (ASTM C 457, Procedure B). The nature of the measurements differs somewhat between the two methods as will be described, but the ultimate results are essentially the same. Both methods are statistical survey procedures, however, in which inferences about the air-void system of the sample as a whole are made on the basis of measurements of only a fraction of the air voids visible on the specimen surface, which are in turn only a fraction of the total number of air voids present in the sample. For example, a typical 100 mm (4 in.) diameter core taken from concrete in which an air-entraining agent was used may contain something on the order of 20 million air voids. Of these only about 1000 will be evaluated under the microscope.

The inspected surface is a two-dimensional plane cut and ground from the three-dimensional concrete. Three-dimensional air voids intersected by this plane are represented by depressions of various shape and depth in the plane of the sample. Irregularly shaped air voids in the concrete appear as depressions with an irregular outline, while those voids that were spherical or nearly so are represented by depressions that are clearly circular. The diameters of these circles are only indirectly related to the diameters of the original voids, however, as the plane of the surface could have randomly cut the original void at any point [28,29]. Only a coincidental and unlikely cut at mid-depth of a void would leave a circle with the diameter of the original void. A small circular depression on the specimen surface could therefore be the two-dimensional artifact of a small void cut near its center, or of a much larger void sliced near its periphery.

It can be demonstrated mathematically, however, that small circles are more likely to have originated in small rather than large spheres, which permits the use of statistical theory combined with a sufficiently large number of observations to project an estimate of the original bubble sizes [16,28,29,35,37,41]. Using these principles, one could painstakingly measure the diameters of the circles on the plane of the specimen and from these measurements project an estimate of the distribution of the original bubble sizes. As will be discussed, a simpler operation is performed in lieu of measuring the diameters of the depressions.

Linear Traverse Method

The linear traverse method requires that a series of "preferably parallel and equally spaced" lines be traced across the surface of the specimen as shown in Fig. 3 [35]. (In practice, this is done by moving the specimen under the microscope in a regular pattern.) These traverse lines randomly intercept a fraction of the depressions on the plane surface, and the number and length of these intercepts are recorded as the test continues. No attempt is made to direct the lines through the diameters of the circles; the intercepts are therefore random "chord lengths" or "chord intercepts." A typical test may consist of total traverse length of 1400 to 4000 mm per specimen distributed over about 50 to 1600 cm^2 (7 to 250 in.^{2}) of sample surface depending on aggregate size (see ASTM C 457), and can take 2 to 6 h of microscopic observation to complete.

The total air content is estimated from the cumulative length of chord intercepts across air voids, divided by the total length of the traverse, multiplied by 100 to convert to percentage measure (Eq 4 of ASTM C 457-90). Paste or aggregate content is similarly estimated by dividing the cumulative chord intercept across the constituents of interest by the total traverse length.

As previously discussed, it is possible to estimate the distribution of air-void sizes from the distribution of diameters of the depressions on the plane of the specimen. This can also be done, although with less certainty, from the distribution of the chord intercepts. Neither of these steps are generally taken, however, as void-size distributions are usually neglected in favor of determining the specific surface, α. It has been shown that α is readily obtained

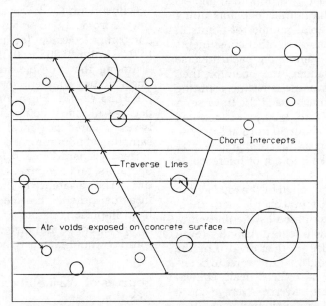

FIG. 3—Schematic diagram of linear traverse procedure. The microscope moves in parallel, closely spaced paths across the prepared concrete surface. Chord intercepts are measured that do not coincide with diameters of the air voids. Note also that the size of the two-dimensional void depressions does not necessarily indicate the spherical diameter of the voids.

directly from the average of the lengths of the individual chord intercepts [28,29,37].

The cumulative length of chord intercepts divided by the number of chords intercepted produces the average chord intercept, $\bar{l}$, (ASTM C 457-90, Eq 6). As demonstrated by Willis [28] the average chord intercept or average chord length is statistically related to the specific surface of the air-void distribution via the expression

$$\alpha = 4/\bar{l} \text{ (ASTM 457-90, Eq 8)}$$

Having obtained values for the air content, paste content, and specific surface, one computes the spacing factor in accordance with Eqs 12 or 13 of ASTM C 457-90.

One additional air-void system parameter in common use is the number of voids encountered per unit length of traverse, or the "void frequency," generally designated by n. The greater this number, the more voids per unit volume, making the void frequency a useful index. Since the number of voids encountered is equivalent to the number of chords intercepted, the void frequency is equal to the number of chords divided by the total length of traverse. ASTM C 457, Eq 7, relates void frequency and air content to average chord length.

Modified Point Count Method

The alternative to the linear traverse method is the modified point count technique in which a series of traverse lines are superimposed on the sample surface in a manner similar to that just described in Fig. 3. Rather than measuring the individual and cumulative chord lengths, however, regularly spaced stops are made and the constituent directly under the cross hairs is identified at each stop [19]. (Regularly spaced stops along parallel, equally spaced traverse lines form a rectangular grid of points, Fig. 4.) The air content is estimated as the number of points falling on an air void divided by the total number of points in the grid. The volume proportions of all other constituents are estimated similarly. After completing each traverse line in this point-to-point manner, the operator then reverses the direction of travel over the same line, counting the total number of air voids intersecting the traverse. A typical test may consist of 1000 to 2400 points counted per specimen, distributed over about 50 to 1600 cm^2 (7 to 250 in.^{2}) of sample surface depending on aggregate size (see ASTM C 457), and can take 2 to 6 h of microscopic observation to complete.

The computed air content multiplied by the total length of traverse provides an estimated total chord length, theoretically equivalent to that determined by the linear traverse method. This "cumulative chord length" (even though no chords were recorded) is then divided by the total number of voids encountered on the traverse to determine an "average chord length," also theoretically equivalent to that determined by linear traverse. Average chord length may also be determined from void frequency. This average chord length is then used as explained earlier to estimate specific surface and subsequently the spacing factor.

Precision and Bias

The discussion of precision and bias that accompanies ASTM C 457 includes the results of two studies on precision [30]. In tests sponsored by the ASTM subcommittee responsible for the test method and using one set of prepared specimens, it was estimated that in 95% of all cases the expected difference between two independent measurements of air content on a single specimen would be less than or equal to 0.82% air if the two tests were performed in the same laboratory. The expected difference would be less than or equal to 1.16% air if tested in two different laboratories. In Sommer's independent tests [30], the expected difference between two measurements of air content on the same specimen would be less than 1.61% air within the same laboratory, and less than 2.01% air if performed in different laboratories. Variations in estimated air content would be greater than these reported values if based on analyses of different samples from the same batch.

ASTM C 457-90 reports Sommer's precision data for spacing factor, showing that at the 95% confidence level two subsequent measurements of spacing factor within the same laboratory could vary by as much as 22.6% of the value of the spacing factor. If the studies were done in different laboratories, the variation could be as much as 56.9%.

Langan and Ward [42] conducted a study in which two prepared specimens were sent to various labs for evaluation by various operators within those labs. Results reported for air content ranged between 6.14 and 9.45% air for one of the specimens, and between 2.3 and 2.89% on the other. For the specimen with the higher air content, the specific surface ranged from 21.9 to 27.6 mm^2/mm^3 (556 to 701 in.^{2}/in.^{3}), and values for spacing factor ranged from 0.079 to 0.170 mm (0.003 to 0.007 in.). For their specimen with the lower air content, the specific surface ranged from 21.9 to 46.5 mm^2/mm^3 (556 to 1180 in.^{2}/in.^{3}), and computed values for spacing factor ranged from 0.135 to 0.284 mm (0.005 to 0.011 in.).

Among the multiple studies of the variabilities and uncertainties inherent in the ASTM C 457 procedures [19,31,35,42-48], Pleau and Pigeon [48] have presented perhaps the most comprehensive. None of these studies have replicated the typical industrial conditions of random sampling of nonuniform concrete combined with variable surface preparation and local variations in procedures, operators, and equipment. ASTM C 457 therefore advises that, "The variability of the test method would be higher in actual practice for specimens sampled and prepared from in-place concrete since additional variation due to sample selection and surface preparation in different laboratories would increase the coefficient of variation."

Sources of Variability and Uncertainty in Test Results

The discussion on precision and bias has demonstrated that the results of the microscopical analysis procedure are not only variable from one sample to another, one

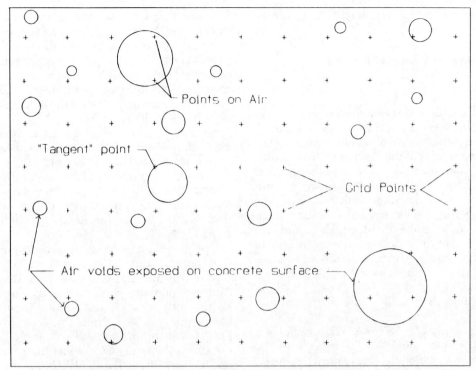

FIG. 4—Schematic diagram of modified point count procedure. The microscope moves in a grid pattern over the prepared concrete surface. Air content is determined by the number of points identified as falling on air voids divided by the total number of points counted. Note that points falling on the edge between air and paste, for example, can complicate the procedure.

laboratory to another, and one operator to another, but being statistical estimates the results are uncertain as well. Specific sources of this variability and uncertainty are detailed subsequently.

Inherent Statistical Uncertainty

The linear traverse and modified point count procedures are based on random statistical sampling of a small fraction of the air voids within a small fraction of the concrete so that test results are statistical estimates based on a limited number of observations. The uncertainty in these statistical estimates depends on the inherent variability of the concrete, the area of concrete sampled, the area of cement paste available, the number of voids measured, and on the breadth of the distribution of void sizes [31,47,48].

Multiple studies have been conducted on these sources of uncertainty, independent from the variability caused by operational factors [19,29–31,35,42,47,48]. While these references should be consulted for details, it is observed that the uncertainty in air content is primarily a function of the length of the traverse or the number of points counted [19,35,42]. The longer the traverse or the greater number of points the more accurate will be the estimate of air content.

To increase the probability of having traversed sufficient concrete surface to have a representative estimate of air and paste content, ASTM C 457 requires a minimum length of traverse for Procedure A, and a minimum number of points for Procedure B. Both these minimum values increase with increasing nominal or observed maximum aggregate size, recognizing that mixes with larger aggregates normally have a reduced paste content [19,35]. Note that whether a sufficiently large number of voids has been intercepted to permit a reasonable estimate of void size depends not only on the traverse length but on the void frequency. For concrete with a low air content and a correspondingly small number of air voids, fewer total voids will have been intercepted even when the ASTM C 457 minimum traverse length or number of points has been completed. This condition tends to increase the uncertainty in the estimates of specific surface and spacing factor. It is of interest to note that a microscopical analysis of hardened concrete is most often called-for when a low air content is suspected; yet the lower the air content the less certain are the results of the analysis. Snyder et al. [47] have shown that when using the linear traverse method this problem may be minimized by running the traverse beyond the ASTM minimum traverse length as necessary until about 1000 voids have been intercepted.

Fundamental statistical uncertainty in estimating the specific surface depends on both the uncertainty in the air content and the uncertainty in establishing the average chord length or void frequency, while the uncertainty in estimating the spacing factor depends on the combined uncertainties in air content, paste content, and average chord length [42,47].

Procedural Sources of Uncertainty and Variability

Sampling

If the concrete placement being evaluated were homogeneous throughout, the accuracy of the microscopical analysis would still depend on the number of samples examined. Since the results of the test are statistical estimates of the true values, the error inherent in the reported values "decreases exponentially" with the number of samples [31]. For example, at the 95% confidence level, the error in estimating the specific surface can drop from 25% to about 15% by examining three rather than one specimen, assuming that the concrete is homogeneous. In actual structures, the characteristics of the air-void system vary with location due to variations in batching, mixing, placement, consolidation, and finishing [49–52], making the issue of sampling substantially more influential and complex.

No guidance is provided by ASTM C 457 for applying the results from a single sample to the balance of the volume of concrete in question. In this context, however, Simon et al. [49] have shown broad variations in air content, specific surface, and spacing factor among multiple samples taken from within 125 to 250 mm (5 to 10 in.) of one another within the same concrete slabs. Such variations can be intensified by the localized effects of consolidation [49,51].

Surface Preparation

Faulty or incomplete specimen preparation has been cited as a source of error by many researchers [16–19,30,31,35,37,53]. Roberts and Gaynor [53] have reported that the effects of specimen preparation alone can skew the results of the ASTM C 457 test by as much as 3% on air content. Sommer reports similar errors stemming from surface preparation [30].

Level of Magnification of the Microscope

At higher levels of magnification the operator can see and measure smaller voids. While such small voids have little influence on total air content, they have significant impact on specific surface, and therefore on spacing factor [54]. While ASTM C 457 requires a minimum magnification of ×50, Bruere [55] reported that in comparison to tests performed at ×120, "Low magnifications (40× and 60×) gave erroneous results since very small bubbles could not be seen clearly." Sommer reported that spacing factor was always decreased by magnification of more than ×50, and a magnification of ×100 instead of ×50 can be assumed to reduce an $\bar{L}$ of 0.25 mm (0.010 in.) by at least 10% [30]. Langan and Ward reported the identical effect [42]. In light of such observations, Pleau et al. [31] suggested a minimum magnification of ×100 and a maximum of ×125.

Operator Subjectivity

The most sensitive and sophisticated piece of equipment required in the microscopical analysis of air-void systems in concrete is a human brain, with human eyes as input devices and the human frame as a support system. Each of these is severely taxed by the tedious and time-consuming process described. Despite the mechanically rote aspects of the test (which can be streamlined with modern equipment), it is the human operator who makes the decisions about where to stop and start the individual measurements of chord length, and whether the constituent under the crosshairs is paste, coarse aggregate, fine aggregate, or an air void. In many cases, these decisions are not easy, such as when the crosshairs or line of traverse is nearly tangent to a void, when poor surface preparation makes boundaries unclear, or when the presence of aggregates or pozzolans mimic the appearance of an air void. Detailed discussions are found in Ref 31 and in the text and notes in ASTM C 457.

Operator bias can be particularly important in the modified point count procedure when the crosshairs fall directly on the edge of an air void. Counting this as an "air" or "non-air" point is a subjective choice that can significantly affect the test results. There have been multiple suggestions for keeping this error from accumulating [31]. It can be meaningful to simply keep track of the questionable points and report their number along with the number of clearly identifiable points.

Given the subjective nature of these decisions and the need for informed judgement, it is clear that operators must be well-trained and well-experienced. In comparing test results obtained with experienced operators, Rodway [56] reported "erratic" results from the inexperienced operators using modified point count procedure, interpreting this as a confirmation of Langan and Ward's previous findings relative to operator variability [42]. Mielenz [19] reported that "the results of the linear traverse are adequately reproducible provided the operator is trained properly." In one test series, Pleau et al. [31] documented operator subjectivity as the cause of variability in computed spacing factor, with an inexperienced operator reporting about 0.145 mm (0.006 in.) and three experienced operators averaging about 0.210 mm (0.008 in.) on the same specimen. While emphasizing the "paramount importance" of training new operators, Pleau et al. conclude that "an intrinsic error due to the operator's subjectivity will always remain."

Arbitrary Deletion of Large Voids

ASTM C 457 states that "no provision is made for distinguishing among entrapped air voids, entrained air voids, and water voids. Any such distinction is arbitrary, because the various types of voids intergrade in size, shape, and other characteristics" (ASTM C 457-90, paragraph 5.3). Some testing agencies have nevertheless found it useful to distinguish at least between the coarse, so-called "entrapped" air voids and the finer, so-called "entrained" air voids, [30,57,58]. Other agencies have determined such practices to produce misleading results [43]. One justification for discounting the larger voids is to avoid their possible impact on skewing the specific surface as discussed earlier. To delete large voids, however, is to artificially modify the recorded air-void size distribution and calculated specific surface, and to delete the contribution made by the large voids to air content and protected paste volume. The effect of discounting the larger chords is consistently to decrease the reported value of air content and to increase the reported value of specific surface.

Calculations

Several potential errors in the void system calculations are worth noting. First, the equation for spacing factor requires a value for the paste content, and the accuracy of the computed value will depend, in part, on the accuracy of the paste content. In some cases the computed value for the spacing factor is either so great or so small that appropriate conclusions may be drawn without much concern for an accurate paste content. In less obvious cases, an accurate estimate of paste content is essential, and may be determined from either the linear traverse or the point count technique. (However, Pleau et al. [31] have observed that microscopical examination always underestimates paste content.) Some laboratories use the paste content determined from the mix design or reported batch weights or, in some cases, will merely assume a value for paste content.

A closely related source of error is the common mistake of using a value for paste content that includes the volume of the air itself. The value required in computing spacing factor is the fractional volume of "air-free" paste.

Image Analysis Techniques

Brief mention is made of the promise that computer-based image analysis techniques can improve the accuracy and reduce the time and effort required to perform the microscopical analysis. Image-based methods preceded the current techniques [37,59] and have been subsequently attempted to various effectiveness [57,60]. While multiple semi-automated systems are presently marketed, only those in which a human operator discerns the air voids in accordance with ASTM C 457 meet the requirements of that method. It is almost certain that sophisticated equipment will replace the current method in time. Depending on the direction taken by the appropriate technology, it is likely that the critical issues of sampling and surface preparation will remain. It may be that more rapid turn-around on test results or a reduced cost per test will permit a larger number of samples to be examined or each sample examined more thoroughly, with a resulting decrease in overall uncertainty and variability.

Comparing Air-Void System Parameters in Fresh and Hardened Concrete

Total air content is the only air-void system parameter that can be directly compared between the fresh and hardened concrete. This is because of the fundamental inability of any of the standard test methods including ASTM Test Method for Air Content of Freshly Mixed Concrete by the Pressure Method (C 231), ASTM Test Method for Air Content of Freshly Mixed Concrete by the Volumetric Method (C 173), and ASTM C 138 (unit weight) to provide information about void size or dispersion in fresh concrete. These standard test methods indicate the total air content only, regardless of the size or distribution of the air voids. (Developing a test method that can detect void size in the fresh concrete is an acknowledged, high-priority research need in the industry.)

When differences exist between the mixing, handling, placing, or consolidation of the concrete sampled for the determination of air content in the fresh state, and the concrete sampled for later microscopic determination of air content, differences in the two test results are expected. This is because the air voids can change in size, number, shape, and volume within the fresh concrete, and can be removed from the system entirely under the influence of mixing, vibration, pressure, and temperature [1,50,61–65]. In fact, the reason for vibrating the concrete is to remove air from the concrete, and not all of the air removable by the vibrator is that which became trapped after mixing and discharge in the process of placing the concrete in the forms [19,49,51,66,67].

The procedure for the ASTM C 231 pressure method therefore alerts the user that "The air content of hardened concrete may be either higher or lower than that determined by this method." The ASTM C 231 procedure goes on to explain that the magnitude and direction of the difference between fresh and hardened air contents will depend upon "the methods and amount of consolidation effort applied to the concrete from which the hardened concrete specimen is taken; uniformity and stability of the air bubbles in the fresh and hardened concrete; accuracy of the microscopic examination, if used; time of comparison; environmental exposure; stage in the delivery, placement and consolidation processes at which the air content of the unhardened concrete is determined, i.e., before or after the concrete goes through a pump and other factors."

INTERPRETATION OF TEST RESULTS

Interpreting the Results of Tests on Hardened Concrete

General Comments

A more detailed discussion of criteria for obtaining frost resistance is included in Chapter 16 of this publication. It is useful to keep the following principles in mind, however, when interpreting the air-void system parameters obtained using the procedures already described.

1. The test results themselves are variable and subject to uncertainty, complicating a reliable inference concerning frost resistance.
2. The criteria against which the results are to be compared can be equally uncertain. Natesaiyer [12,20] has shown that while present criteria (to be discussed) can identify concrete that is almost certainly frost-resistant and concrete that is almost certainly non-frost-resistant, there exists a broad, marginal zone or "gray area" in which frost resistance is difficult to judge.
3. Some of the difficulty in interpreting frost resistance on the basis of air-void system parameters that are neither "clearly acceptable" nor "clearly unacceptable" is due

308 TESTS AND PROPERTIES OF CONCRETE

to the fact that frost resistance depends also on the mix proportions and material properties of the concrete, and on environmental exposure.

4. The incorporation of a beneficial air-void system in concrete does not eliminate the pressure caused by freezing, but merely reduces it to tolerable levels. In a practical sense, frost damage is minimized, not eliminated entirely, by the incorporation of air [56].
5. The frost resistance of concrete in-service depends on its properties in-place. Measures of air content in the fresh concrete, or air-void system parameters determined on concrete samples not handled, placed, consolidated, or finished in a representative manner may lead to incorrect conclusions about frost resistance.

Air Content

Klieger recognized that the necessary air content was dependent on mix proportions [5,27]. In two independent studies, he demonstrated that when the coarse aggregate volume (which is not protected by air voids in concrete) is ignored, the optimum air content for frost resistance was consistently 9% ± 1% of the volume of the mortar. Klieger's observations are reflected in the mix design recommendations of ACI 211.1 [3] in which the suggested air content increases as the nominal maximum coarse aggregate size decreases. This is not because smaller coarse aggregate particles are less frost-resistant (the reverse is generally true) [68]. It is because the ACI 211 mix design method will correctly result in a higher mortar content in mixes using smaller coarse aggregates. This conclusion was independently reached by Siggelokow [69]. As Saucier pointed out, however, while a particular value of air content is necessary to provide frost resistance for a given concrete, obtaining such an air content is not by itself sufficient because of the need to obtain an appropriately sized and distributed air-void system [65].

Specific Surface

While frost-resistant concrete is generally characterized by values of specific surface greater than 25 mm²/mm³ (or about 600 in.²/in.³), Neville [70] reports frost resistance for certain concrete at values as low as about 16 mm²/mm³ (400 in.²/in.³.) Specific surface is merely an indicator of bubble size, however, providing no information about volume or dispersion. One could have an air-void system composed of acceptably small bubbles as indicted by specific surface, but there may be too few of these bubbles or they may be nonuniformly spaced so as to leave large gaps of unprotected paste.

Void Frequency

For frost-resistant concrete, the number of voids encountered per unit length of traverse is generally expected to be "one to one and a half times the numerical value of air content" [71,72]. Thus, if the air content is 5%, one would expect 5 to 7.5 voids per inch. This rule of thumb arises from the algebraic relationships among the air content, void frequency, and specific surface; when void frequency is about equal to the air content, specific surface is about 16 mm²/mm³ (400 in.²/in.³). When void frequency is 1.5 times air content, the specific surface is about 25 mm²/mm³ (600 in.²/in.³). Void frequency and specific surface are therefore not independent variables.

Spacing Factor

Various agencies and organizations, such as ACI Committee 212 on concrete admixtures, have recommended a spacing factor of 0.008 in. (0.200 mm) as being indicative of frost-resistant concrete [71]. While it is likely that concrete with a spacing factor in this range or smaller will be frost resistant, it is not equally likely that concrete with a larger spacing factor will necessarily be non-frost resistant. Interpretation of frost resistance on the basis of spacing factor is not without difficulty, therefore, in spite of the fact that of the indices available, spacing factor is the most commonly used.

Multiple studies correlating freeze-thaw durability with computed spacing factor have shown a scattered but general trend of increased durability with decreasing spacing factor [12,20,43,72-75]. Ivey and Torrans [73] concluded that for conventional concretes, "the transition between durable and nondurable concrete seems to be somewhere between an L of 0.008 and 0.010 in. [0.20 to 0.25 mm]." This general assessment appears to remain valid, and coincides with Powers original proposal that "for typical concretes and environmental conditions spacing factors not exceeding 0.250 mm (0.010 in.) or thereabouts" were generally indicative of frost-resistant concrete [10]. In this proposal, Powers also recognized that the spacing factor required for frost resistance depended on material and environmental factors.

PART II: UNIT WEIGHT

Introduction

The term "unit weight" as used here refers to the weight per unit volume of hardened concrete. The term unit weight when used to describe the weight of fresh concrete per unit volume, as determined by ASTM (C 138), is discussed in Chapter 10 of this publication. While the unit weights of fresh and subsequently hardened samples of a particular mix are expected to be related, the values are not expected to be identical. The relationship between the fresh and hardened unit weights for a particular concrete mixture will depend on the mix proportions and characteristics of the aggregate, the degree of consolidation, sampling, volume changes, age, and curing.

Significance of Unit Weight as a Characteristic of Concrete

Unit Weight

In some cases, the unit weight of hardened concrete, per se, is critical to the performance of the structure or facility. Examples include setting a maximum unit weight requirement when lightweight aggregate concrete is used to limit structure self-weight, or setting a minimum unit weight requirement when self weight is to be maximized for structural stability or for nuclear shielding. As described by ASTM Test Method for Unit Weight of Structural Lightweight Concrete (C 567), "The test for air-dried weight of concrete determines whether design weight requirements have been met."

Uniformity or Consistency of Materials, Construction Operations, and Testing

Consistent unit weights normally indicate consistency in all phases of concreting operations. This is because the unit weight of hardened concrete is a function of the unit weights of the initial ingredients, mix proportions, initial and final water content, air content, degree of consolidation, degree of hydration, volume changes, and subsequent gain or loss of water, among other factors. Dependence on these factors makes unit weight an effective indicator of the uniformity of raw materials, mixing, batching, placing, sampling, and testing. A significant change in unit weight signals a change somewhere in the process.

For example, if the unit weight is seen to vary among samples of hardened concrete that had been cast at the point of concrete delivery, variability in the constituent materials or proportions could be indicated. On the other hand, in this situation variable unit weights could mean nonuniform batching or mixing, or nonuniform casting of test specimens. Routine weighing of standard compressive strength specimens before testing is recommended to quickly approximate unit weight and get an indication of sample uniformity.

Alternatively, unit weight tests performed on samples of hardened concrete extracted from the structure can be useful for indicating segregation, nonuniform consolidation, or other problems. Because this useful information is so readily obtained, routine unit weight testing for all cores extracted in the field is often recommended, regardless of the primary purpose for obtaining those cores.

Voids Content

In dried, hardened concrete all of the internal voids are air-filled, including capillary pores in the hardened cement paste, voids in the aggregate particles, bleed water channels, water gain voids, microcracks, and air voids intentionally or unintentionally incorporated into the mixture. The volume of some, but not all, of these voids can be determined by the weight of absorbed water when a dried specimen is immersed for a period of time. Those voids that communicate directly with the exterior surface of the sample or are connected to the surface via capillary channels, absorb water upon immersion. These are called the "permeable pores," and represent the pore space measured by drying followed by immersion. Other voids, which include a portion of the capillary void system, some of the aggregate pores, and a fraction of the system of air voids, are termed "impermeable pores." These spaces do not fill with water upon immersion and cannot be measured by these techniques. Because one cannot discriminate among the various types of voids present when determining voids content by absorption methods, the final result is termed "total permeable voids." For a given set of raw materials of fixed unit weights, the lower the unit weight of the mixture the greater will be the voids content. Helms [77] has documented an empirical relationship between oven-dry unit weight and voids content.

The total volume of permeable voids of a sample is related to porosity, a basic characteristic of concrete that influences many of its properties. Numerous studies, beginning with those of Feret and continuing more recently to Popovics, have explored these relationships [78–83]. In fact, Feret [78] established a clear relationship between the strength of mortars and the voids/cement ratio (volume of voids in the sample divided by the volume of the cement) well before the more currently recognized relationship was established between strength and water/cement ratio. Data published by the U.S. Bureau of Reclamation confirmed that the relationship between voids and strength applied equally well to a wide variety of concrete mixes [84]. Helms [77] has reported a similar relationship even when lightweight aggregate concretes were tested.

Permeability

While the porosity of concrete can be related to unit weight and voids content, permeability depends not only on the total pore or flow channel volume, but also on the connectivity, tortuosity, and hydraulic characteristics of the channels. Nevertheless, correlations have been demonstrated between certain types of permeability tests and unit weight of hardened concrete [85–87]. Such relationships, even if only empirical in nature, suggest a linkage between the unit weight and durability, since the durability of concrete can be related to the degree to which water, water vapor, oxygen, and carbon dioxide can permeate the concrete.

Degree of Consolidation

Analogous to routine measurements in soil mechanics and bituminous materials, one can estimate the degree of consolidation from unit weight measurements of hardened concrete. Degree of consolidation is generally expressed as the ratio between the observed unit weight and some value taken for the "maximum" or "optimum" unit weight. Olsen [88] defined degree of consolidation as the unit weight of the hardened concrete divided by the unit weight of the fresh concrete. In Whiting's work [85], the maximum unit weight was that obtained from hardened samples that had been consolidated on a vibrating table. (It is interesting to note that within Whiting's laboratory testing program the standard deviation on measured unit weight was approximately 40 kg/m^3 (2.5 lb/ft^3), which may approximate a minimum uncertainty for unit weight tests.) Whiting went on to correlate the degree of consolidation (or "relative unit weight") to compressive strength, bond strength, and rapid chloride permeability. Bisaillon and Malhotra [87] also demonstrated the benefits of improved consolidation in reducing the hydraulic permeability of concrete.

Thermal, Acoustic, and Nuclear Shielding Properties

Unit weight is a key parameter in defining the ease of transmission of energy through the concrete, and when such properties are of interest it may be more appropriate to measure unit weight than compressive strength. Valore and Brewer separately [89–90] demonstrated the relationship between density of hardened concrete and its ability to transmit heat. Both the modulus of elasticity of concrete and the ultrasonic pulse velocity can be shown to be dependent on density. For that reason, it is often necessary to determine the unit weight of concrete in order to interpret the results of dynamic modulus or the pulse velocity tests.

Although the topic of heavyweight, nuclear-shielding concrete is beyond the scope of this chapter, the key issue in attenuating the transmission of atomic particles is to put as many atomic nuclei in the path of the radiation as possible. This means that, in general, the denser the mass, the better is its shielding ability. The unit weight test is a simple means of determining whether the required density has been achieved.

Inferring Batch Weights and Composition

Just as the unit weight test can lead to estimates of the voids content, some information about the composition of the balance of the sample is theoretically possible as well. The effects of composition on unit weight will be discussed.

Typical Values

Figure 5 [77,91,92] displays the approximate range of unit weights and air contents represented by aggregates, concrete, and cementitious materials. The range is bounded at the high-density end with steel shot, steel punchings, and magnetite aggregate used for radiation shielding concretes and counterweights, with unit weights above 5000 kg/m^3 (above 310 lb/ft^3). The low-density end of the range is occupied by cellular concretes with unit weights less than 1000 kg/m^3 (60 lb/ft^3) and with air contents above 25%.

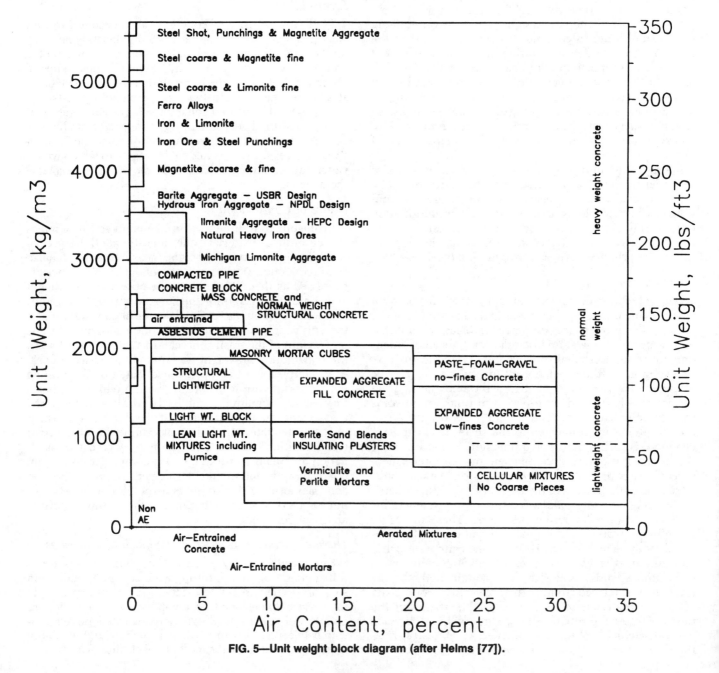

FIG. 5—Unit weight block diagram (after Helms [77]).

TABLE 1—Observed Average Unit Weight (adapted from Ref 96).

Maximum Size of Aggregate, mm (in.)	Average Values			Unit Weight, kg/m³ (lb/ft³) Specific Gravity of Aggregate, SSD				
	Air Content, %	Water, kg/m³ (lb/yd³)	Cement, kg/m³ (lb/yd³)	2.55	2.60	2.65	2.70	2.75
19 (³/₄)	6.0	168 (283)	336 (566)	2200 (137)	2230 (139)	2260 (141)	2290 (143)	2330 (145)
38 (1¹/₂)	4.5	145 (245)	291 (490)	2260 (141)	2290 (143)	2340 (146)	2370 (148)	2405 (150)
76 (3)	3.5	121 (204)	242 (408)	2310 (144)	2357 (147)	2390 (149)	2440 (152)	2470 (154)
152 (6)	3.0	97 (164)	167 (282)	2360 (147)	2389 (149)	2440 (152)	2470 (154)	2520 (157)

Litvin and Fiorato [93] have independently prepared the "Lightweight Aggregate Spectrum" that also graphically depicts concretes with compressive strengths ranging from 0.7 to 40 MPa (100 to 6000 psi) with concrete unit weights from 240 to 2000 kg/m³ (15 to 120 lb/ft³).

Methods of Determining the Unit Weight of Hardened Concrete

General

The measurement of density of hardened concrete is based on procedures that are simple and direct, using samples such as molded specimens, drilled cores, or portions taken from hardened structures and trimmed to cylinders or prisms of sufficient size to be representative. Hardened samples are dried or otherwise conditioned, and weight and volume measurements are taken. The specimens are then immersed in water so that the permeable pores are filled. Weight measurements are repeated after immersion and from these data one determines the volume of the permeable pores and weight/volume relationships for the dry and saturated specimens. It is reasonable to require unit weight by displacement measurement to be reported to the nearest 1.6 kg/m³ (0.1 lb/ft³) and within an accuracy of 0.1%. Reference is made to the detailed provisions of ASTM C 567 and ASTM Test Method for Specific Gravity, Absorption, and Voids in Hardened Concrete (C 642). (Exceptions to the general simplicity of density measurements are the nuclear methods described in ASTM Test Method for Density of Unhardened and Hardened Concrete in Place by Nuclear Methods (C 1040).)

More advanced, yet nonstandard techniques for determining the density of hardened concrete include mercury pycnometry [94], although such methods are more appropriate for very small samples of hardened concrete, mortar, or paste (that is, generally less than 20 cm³).

Procedural Factors Influencing the Results

In principle, the unit weight of hardened concrete is obtained just as the unit weight, specific gravity, absorption, etc. are obtained for aggregates, and the same concerns apply to proper moisture conditioning. This is made more difficult, however, when dealing with concrete samples that are many times larger than aggregate particles, and therefore come to moisture equilibrium more slowly. Other concretes are more absorptive than normal-weight aggregates and therefore can take on water at a faster rate.

The moisture condition to be used for weight determination depends on the purpose of the measurement and the in-service condition of the concrete in question. For example, when specimens are obtained by extracting cores from a structure, care is required to retain the "as-is" moisture condition. Since the density of lightweight aggregate concrete in-service is a critical issue, ASTM C 567 makes special note of establishing in-service and "equilibrium" moisture conditions in the sample.

In both ASTM C 567 and C 642, the volume of the sample is determined by the "displacement method," in which the difference between weighing in air and weighing in water is attributed to the buoyant effect of the water, which in turn is related to the density of the water and the displaced volume. In informal testing, the volume of the specimen is frequently calculated from the physical dimensions of the sample. This latter approach can be useful only when the shape of the specimen is highly regular and the dimensions accurately obtained. Even with standard concrete cylinders, the unevenness of uncapped ends and the tendency to out-of-roundness can introduce significant errors.

The issue of sample size is briefly addressed in ASTM C 642, pointing out the need to have a representative sample. This is more difficult than it may at first appear, particularly given the variations in degree of compaction that exist vertically and horizontally within a given concrete member. Such variations are well documented in work by Simon et al. [49] and by Kagaya et al. [95].

Influence of Composition on Unit Weight

Effect of Aggregate Density

Since aggregate can occupy 60 to 80% of the volume of most concrete mixes and the specific gravity of normal-weight coarse and fine aggregate is approximately one-third greater than that of the hardened cement paste, the aggregate has considerable influence on unit weight of hardened concrete. The data shown in Table 1 were collected by the U.S. Bureau of Reclamation [96] and show a 6 to 7% variation in unit weight for mixes using aggre-

TABLE 2—Effect of Paste Content and Water/Cement Ratio on Unit Weight.

Water/Cement Ratio	Specific Gravity of Paste	Unit Weight of Paste, kg/m³	Unit Weight of Paste, lb/ft³
0.30	1.90	1900	119
0.40	1.68	1680	105
0.50	1.48	1480	92
0.60	1.34	1340	84
0.70	1.22	1220	76

gates of various specific gravities but identical proportions.

Effect of Paste Content

Cook [94] reported the dry specific gravity of hardened cement pastes at various water/cement ratios after 56 days of continuous wet cure. The results are shown in Table 2, in which it is clear that the unit weight of hardened cement paste decreases as water/cement ratio increases. Since the densities of normal aggregates usually range from about 2560 to 2800 kg/m^3 (160 to 175 lb/ft^3), it is clear that a reduction of paste volume by substitution of aggregate will increase the unit weight. Further data from the Bureau of Reclamation [97] demonstrates this, in which mixes with a 43% paste content had a unit weight of 2206 kg/m^3, while a similar mix with a 22% paste content had a unit weight of 2533 kg/m^3 (137 and 155 lb/ft^3, respectively).

Effect of Air Content

Unit weight is reduced with an increase in the relative volume of pore space in the concrete sample, regardless of the origin of the pores. Therefore, all other factors remaining constant, unit weight will decrease with an increase in air volume, regardless of the size, shape, or distribution of the air voids. The unit weight test cannot discern air-void size, nor can it differentiate between those voids constituting the desirable air-void system in the hardened concrete and spaces such as capillary pores, bleed water channels, aggregate voids, etc. For these reasons, one has to exercise judgement in using measurements of the the unit weight of hardened concrete to evaluate air content.

REFERENCES

[1] Whiting, D. and Stark, D., "Control of Air Content in Concrete," National Cooperative Highway Research Program Report No. 258, Transportation Research Board, Washington, DC, 1983.

[2] Kennedy, H. L., "The Function of Entrained Air in Concrete," and Discussion by M. Spindel, *Journal*, American Concrete Institute, June 1943; *Proceedings*, Vol. 39, pp. 529–542.

[3] "Recommended Practice for Proportioning Normal and Heavyweight Concrete Mixtures," ACI 211.1, American Concrete Institute, Detroit, MI, 1991.

[4] Gutmann, P. F., "Bubble Characteristics as they Pertain to Compressive Strength and to Freeze-Thaw Durability," *ACI Materials Journal*, Sept.–Oct. 1988; *Proceedings*, Vol. 85, pp. 361–366.

[5] Klieger, P., "Effect of Entrained Air on Strength and Durability of Concrete Made with Various Maximum Sizes of Aggregates," *Proceedings*, Highway Research Board, Vol. 31, 1952, pp. 177–201.

[6] Powers, T. C., "Frost Resistance of Concrete at Early Ages," *Proceedings*, RILEM Symposium: Wintertime Concreting, Theory and Practice, Copenhagen, Feb. 1956, The Danish Institute of Building Research Special Report, Copenhagen, 1956, Section C.

[7] Powers, T. C. and Helmuth, R. A., "Theory of Volume Changes in Hardened Portland Cement Paste During Freezing," *Proceedings*, Highway Research Board, Vol. 32, 1953, pp. 285–297.

[8] Powers, T. C., "A Working Hypothesis for Further Studies of Frost Resistance of Concrete," *Journal*, American Concrete Institute, Feb. 1945; *Proceedings*, Vol. 41, pp. 245–272.

[9] Powers, T. C., "Basic Considerations Pertaining to Freezing-And-Thawing Tests," *Proceedings*, American Society for Testing and Materials, Vol. 55, 1955, pp. 1132–1155.

[10] Powers, T. C., "The Air Requirement of Frost-Resistant Concrete," *Proceedings*, Highway Research Board, Vol. 29, 1949, pp. 184–202.

[11] Powers, T. C., "Void Spacing as a Basis for Producing Air Entrained Concrete," *Journal*, American Concrete Institute, May 1954; *Proceedings*, Vol. 50, pp. 741–759.

[12] Natesaiyer, K. C. and Hover, K. C., "The Protected Paste Volume of Air Entrained Cement Paste; Part I," *Journal of Materials in Civil Engineering*, Vol. 4, No. 2, May 1992, pp. 166–184.

[13] Mattimore, H. S., "Durability Tests of Certain Portland Cements," *Proceedings*, Highway Research Board, National Research Council, Washington, DC, Vol. 16, 1936, pp. 135–166.

[14] Arni, H. T., Foster, B. E., and Clevenger, R. A., "Automatic Equipment and Comparative Test Results for the Four ASTM Freezing-And-Thawing Methods for Concrete," *Proceedings*, American Society for Testing and Materials, Philadelphia, Vol. 56, 1956, pp. 1229–1256.

[15] Flack, H. L., "Freezing and Thawing Resistance of Concrete as Affected by the Method of Test," *Proceedings*, American Society for Testing and Materials, Philadelphia, Vol. 57, 1957, pp. 1077–1095.

[16] Mielenz, R. C., Wolkodoff, V. E., Backstrom, J. E., and Flack, H. L., "Origin, Evolution, and Effects of the Air Void System in Concrete. Part 1—Entrained Air in Unhardened Concrete," *Proceedings*, American Concrete Institute, Vol. 55, 1958, pp. 95–121.

[17] Backstrom, J. E., Burrows, R. H., Mielenz, R. C., and Wolkodoff, V. E., "Origin, Evolution, and Effects of the Air Void System in Concrete. Part 2—Influence of Type and Amount of Air-Entraining Agent," *Proceedings*, American Concrete Institute, Vol. 55, 1958, pp. 261–517.

[18] Backstrom, J. E., Burrows, R. H., Mielenz, R. C., and Wolkodoff, V. E., "Origin, Evolution, and Effects of the Air Void System in Concrete. Part 3—Influence of Water-Cement Ratio and Compaction," *Proceedings*, American Concrete Institute, Vol. 55, 1958, pp. 359–375.

[19] Mielenz, R. C., Wolkodoff, V. E., Backstrom, J. E., and Burrows, R. H., "Origin, Evolution, and Effects of the Air Void System in Concrete. Part 4—The Air Void System and Job Concrete," *Proceedings*, American Concrete Institute, Vol. 55, 1958, pp. 507–517.

[20] Natesaiyer, K. C. and Hover, K. C., "Protected Paste Volume of Air Entrained Cement Paste; Part II," *Journal of Materials in Civil Engineering*, Vol. 5, No. 2, May 1992, pp. 170–186.

[21] Dodson, V., *Concrete Admixtures*, Van Nostrand Reinhold, New York, 1990.

[22] Dolch, W. L., "Air Entraining Admixtures," *Concrete Admixtures Handbook*, V. S. Ramachandran, Ed., Noyes Publishers, Park Ridge, NJ, 1984, pp. 269–302.

[23] Bennet, K., "Air Entraining Admixtures for Concrete," *Concrete Admixtures*, M. R. Rixom, Ed., Longman, NY, 1977, pp. 37–48.

[24] Rixom, M. R. and Mailvaganam, N. P., *Chemical Admixtures for Concrete*, E&F.N. Spon, London and New York, 1986.

[25] Walker, H. N., "Formula for Calculating Spacing Factor for Entrained Air Voids," *Cement, Concrete, and Aggregates*, Vol. 2, No. 2, Winter 1980, pp. 63–66.

[26] Clift, R., Grace, J. R., and Weber, M. E., *Bubbles, Drops, and Particles*, Academic Press, New York, 1985.

[27] Klieger, P., "Effect of Entrained Air on Strength and Durability of Concrete Made with Various Size Aggregates," *Bulletin No. 128*, Highway Research Board, Washington, DC, 1956.

[28] Willis, T. F., Discussion of Powers. T. C., "The Air Requirement of Frost-Resistant Concrete," *Proceedings*, Highway Research Board, Vol. 29, 1949, pp. 203–211.

[29] Lord, G. W. and Willis, T. F., "Calculation of Air Bubble Size Distribution from Results of a Rosiwal Traverse of Aerated Concrete," *ASTM Bulletin No. 177*, American Society for Testing and Materials, Philadelphia, Oct. 1951, pp. 56–61.

[30] Sommer, H., "The Precision of the Microscopical Determination of the Air Void System in Hardened Concrete," *Cement, Concrete, and Aggregates*, Vol. 1, No. 2, 1979, pp. 49–55.

[31] Pleau, R., Plante, P., Gagné, R., and Pigeon, M., "Practical Considerations Pertaining to the Microscopical Determination of Air Void Characteristics of Hardened Concrete (ASTM C 457 Standard)," *Cement, Concrete, and Aggregates*, Vol. 12, No. 2, Summer 1990, pp. 3–11.

[32] Mather, K., "Measuring Air in Hardened Concrete (LR-49-5), a continued discussion of Brown, L. S. and Pierson, C. U., 'Linear Traverse for Measurement of Air in Hardened Concrete,'" *Proceedings*, American Concrete Institute, Vol. 49, 1952, pp. 61–64.

[33] Larson, T. D., Cady, P. D., and Malloy, J. J., "The Protected Paste Volume Concept Using New Air Void Measurement and Distribution Techniques," *Journal of Materials*, Vol. 2, No. 1, March 1967, pp. 202–224.

[34] Philleo, R. E., "A Method for Analyzing Void Distribution in Air-Entrained Concrete," *Cement, Concrete, and Aggregates*, Vol. 5, No. 2, Winter 1983, pp. 128–130.

[35] Brown, L. S. and Pierson, C. U., "Linear Traverse for Measurement of Air in Hardened Concrete," *Proceedings*, American Concrete Institute, Vol. 47, 1950, pp. 117–123.

[36] Chayes, F., *Petrographic Modal Analysis, An Elementary Statistical Appraisal*, Wiley, New York, 1956.

[37] Warren, C., "Determination of Properties of Air Voids in Concrete," *Bulletin No. 70*, Highway Research Board, Washington, DC, 1953, pp. 1–10.

[38] Ray, J. A., "Preparation of Concrete Samples for Petrographic Studies," *Proceedings*, 6th International Conference on Cement Microscopy, International Cement Microscopy Association, Duncanville, TX, pp. 292–303.

[39] Balaguru, P. and Ramakrishnan, V., "Chloride Permeability and Air-Void Characteristics of Concrete Containing High Range Water Reducing Admixture," *Cement and Concrete Research*, Vol. 18, 1988, pp. 401–414.

[40] Mather, K., discussion of Brown, L. S. and Pierson, C. U., "Linear Traverse for Measurement of Air in Hardened Concrete," *Proceedings*, American Concrete Institute, Vol. 47, 1950, pp. 124–2–124–5.

[41] Whiting, D. and Schmitt, J., "A Model for Deicer Scaling Resistance of Field Concretes Containing High-Range Water Reducers," *Superplasticizers and Other Chemical Admixtures in Concrete, SP-119*, V. M. Malhotra, Ed., American Concrete Institute, Detroit, MI, 1989, pp. 343–360.

[42] Langan, B. W. and Ward, M. A., "Determination of the Air Void System Parameters in Hardened Concrete—An Error Analysis," *Journal*, American Concrete Institute, Vol. 83, No. 6, Nov.–Dec. 1986; *Proceedings*, Vol. 83, pp. 943–952.

[43] Walker, H. N., "Correlation of Hardened Concrete Air Void Parameters Obtained by Linear Traverse with Freeze-Thaw Durability as Found by ASTM C 666," *Cement, Concrete, and Aggregates*, Vol. 6, No. 1, Summer 1984, pp. 52–55.

[44] Amsler, D. E., Eucker, A. J., and Chamberlin, W. P., "Techniques for Measuring Air Void Characteristics of Concrete," Research Report No. 11, Engineering Research and Development Bureau, New York State Department of Transportation, Albany, NY, Jan. 1973.

[45] Reidenouer, D. R. and Howe, R. H., "Air Content of Plastic and Hardened Concrete," Research Project No. 73-1, Bureau of Materials, Testing and Research, Pennsylvania Department of Transportation, Harrisburg, PA, Dec. 1974.

[46] Burg, G. R. U., "Slump Loss, Air Loss, and Field Performance of Concrete," *Journal*, American Concrete Institute, Vol. 80, No. 4, July–Aug. 1983; *Proceedings*, Vol. 80, pp. 332–339.

[47] Snyder, K., Hover, K., and Natesaiyer, K., "An Investigation of the Minimum Expected Uncertainty in the Linear Traverse Technique," *Cement, Concrete, and Aggregates*, Vol. 13, No. 1, Summer 1991, pp. 3–10.

[48] Pleau, R. and Pigeon, M., "Precision Statement for ASTM C 457 Practice for Microscopical Determination of the Air-Void Content and Other Parameters of the Air Void System in Hardened Concrete," *Cement, Concrete, and Aggregates*, Vol. 14, No. 2, Summer 1992, pp. 118–126.

[49] Simon, M. J., Jenkins, R. B., and Hover, K. C., "The Influence of Immersion Vibration on the Air Void System in Concrete," *Durability of Concrete*, G. M. Idorn International Symposium, *SP 131*, American Concrete Institute, Detroit, MI, 1992, pp. 99–126.

[50] Powers, T. C., *The Properties of Fresh Concrete*, Wiley, New York, 1968.

[51] Stark, D. C., *Effects of Vibration on the Air Void System and Freeze-Thaw Durability of Concrete*, (RD092.01T), Portland Cement Association, Skokie, IL, 1986.

[52] Rodway, L. E., "Void Spacing in Exposed Concrete Flatwork," *Concrete International, Design & Construction*, Vol. 1, No. 2, Feb. 1979, pp. 83–89.

[53] Roberts, L. W. and Gaynor, R. D., discussion by L. W. Roberts and R. D. Gaynor on Ozyildirim, C., "Comparison of Air Contents in Fresh and Hardened Concretes Using Different Airmeters," *Cement, Concrete, and Aggregates*, Vol. 13, No. 1, Summer 1991, pp. 11–17.

[54] Willis, T. F., discussion of Brown, L. S. and Pierson, C. U., "Linear Traverse for Measurement of Air in Hardened Concrete," *Proceedings*, American Concrete Institute, Vol. 47, 1950, pp. 124–1 to 124–2.

[55] Bruere, G. M., "The Effect of Type of Surface Active Agent on Spacing Factors and Surface Areas of Entrained Bubbles in Cement Pastes," *Australian Journal of Applied Science*, Vol. 11, No. 2, June 1960, pp. 288–294.

[56] Rodway, L. E., "Effect of Air Entraining Agent on Air Void Parameters of Low-And High-Calcium Fly Ash Concrete," *Cement, Concrete, and Aggregates*, Vol. 10, No. 1, Summer 1988, pp. 35–38.

[57] Whiting, D., "Air Contents and Air Void Characteristics in Low-Slump Dense Concretes," *Journal*, American Concrete Institute, Sept.–Oct. 1985, pp. 716–723.

[58] Ozyildirim, C., "Comparison of Air Contents in Fresh and Hardened Concretes Using Different Airmeters," *Cement, Concrete, and Aggregates*, Vol. 13, No. 1, Summer 1991, pp. 11–17.

[59] Verbeck, G. J., "The Camera-Lucida Method for Measuring Air Voids in Hardened Concrete," *Journal*, American Concrete Institute, May 1947; *Proceedings*, Vol. 43, pp. 1025–1040.

[60] Mullen, W. G. and Waggoner, C. K., "Air Content of Hardened Concrete from Comparison Electron Microscope Photographs," *Cement, Concrete, and Aggregates*, Vol. 2, No. 1, Summer 1980, pp. 43–49.

[61] Pigeon, M., Plante, P., and Plante, M., "Air Void Stability, Part I: Influence of Silica Fume and Other Parameters," *ACI Materials Journal*, Vol. 86, No. 5, Sept.–Oct. 1989, pp. 482–490.

[62] Plante, P., Pigeon, M., and Saucier, F., "Air Void Stability, Part II: Influence of Superplasticizers and Cement," *ACI Materials Journal*, Vol. 86, No. 6, Nov.–Dec. 1989, pp. 581–589.

[63] Saucier, F., Pigeon, M., and Plante, P., "Air Void Stability, Part III: Field Tests of Superplasticized Concretes," *ACI Materials Journal*, Vol. 87, No. 1, Jan.–Feb. 1990, pp. 3–11.

[64] Pigeon, M., Saucier, F., and Plante, P., "Air Void Stability, Part IV: Retempering," *ACI Materials Journal*, Vol. 87, No. 3, May–June 1990, pp. 252–259.

[65] Saucier, F., Pigeon, M., and Cameron, G., "Air Void Stability, Part V: Temperature, General Analysis, and Performance Index," *ACI Materials Journal*, Vol. 88, No. 1, Jan.–Feb. 1991, pp. 25–36.

[66] Crawley, W. O., "Effect of Vibration on Air Content of Mass Concrete," and discussion by C. E. Wuerpel, *Journal*, American Concrete Institute, June 1953; *Proceedings*, Vol. 49, pp. 909–919.

[67] Higginson, E. C., "Some Effects of Vibration and Handling on Concrete Containing Entrained Air," *Journal*, American Concrete Institute, Sept. 1952; *Proceedings*, Vol. 49, pp. 1–12.

[68] Cordon, W. A., *Freezing and Thawing of Concrete—Mechanisms and Control*, ACI Monograph No. 3, American Concrete Institute, Detroit, MI, 1966.

[69] Siggelokow, J., "The Air Quantity in the Design of Air-Entraining Concrete (LR 46–39)," *Journal*, American Concrete Institute, Dec. 1949; *Proceedings*, Vol. 50, pp. 300–301.

[70] Neville, A., *Properties of Concrete*, 3rd ed., Pitman Press, London, 1982.

[71] "Chemical Admixtures for Concrete (ACI 212.3R-91)," ACI Committee 212, American Concrete Institute, Detroit, 1991, pp. 7–10; *ACI Manual of Concrete Practice*, Part I, 1992, pp. 212.3R-7 to 212.3R-10.

[72] Fears, F. K., "Correlation between Concrete Durability and Air-Void Characteristics," *Air Voids in Concrete and Characteristics of Aggregates, Bulletin 196*, Highway Research Board, Washington, DC, 1958, pp. 17–28.

[73] Ivey, D. L. and Torrans, P. H., "Air Void Systems in Ready Mixed Concrete," *Journal of Materials*, Vol. 5, No. 2, June 1970, pp. 492–522.

[74] Pigeon, M. and Gagné, F. C., "Critical Air Void Spacing Factors for Low Water-Cement Ratio Concretes with and without Condensed Silica Fume," *Cement and Concrete Research*, Vol. 17, 1987, pp. 896–906.

[75] Aïtcin, P. C. and Vezina, D., "Resistance to Freezing and Thawing of Silica Fume Concrete," *Cement, Concrete, and Aggregates*, Vol. 6, No. 1, Summer 1984, pp. 38–42.

[76] Mehta, P. K. and Monteiro, P. J. M., *Concrete Structure, Properties, and Materials*, Prentice Hall, Englewood Cliffs, NJ, 1993.

[77] Helms, S. B., "Air Content and Unit Weight," *Significance of Tests and Properties of Concrete and Concrete-Making Materials, ASTM STP 169B*, American Society for Testing and Materials, Philadelphia, 1978, pp. 435–461.

[78] Feret, R., "Sur la Compacite (Density) des Mortars Hydrauliques," Vol. 4, 1892; *Annales des Ponts et Chausses* and *Bulletin de la Societe d'Encouragement pour l'Industrie Nationale*, Vol. 2, 1897, p. 1604.

[79] Feret, R., discussion of "The Laws of Proportioning Concrete," *Transactions*, American Society of Civil Engineers, New York, Vol. 59, 1907, p. 154.

[80] Talbot, A. N. and Richart, F. E., "The Strength of Concrete—Its Relation to the Cement, Aggregates, and Water," *Bulletin No. 137*, Engineering Experiment Station, University of Illinois, Urbana, 1923.

[81] Morris, M., "The Mortar Voids Method of Designing Concrete Mixtures," *Journal*, American Concrete Institute, Sept. 1932; *Proceedings*, Vol. 29, p. 9.

[82] Thornburn, T. H., "The Design of Concrete Mixes Containing Entrained Air," *Proceedings*, American Society for Testing and Materials, Vol. 49, 1949, p. 921.

[83] Popovics, S., "Prediction of the Effect of Porosity on Concrete Strength," *Journal*, American Concrete Institute; *Proceedings*, Vol. 82, No. 2, 1985, pp. 136–146.

[84] *Concrete Manual*, 8th ed., U.S. Department of the Interior, Bureau of Reclamation, Denver, CO, 1975, p. 39.

[85] Whiting, D., Seegebrecht, G. W., and Tayabji, S., "Effect of Degree of Consolidation on some Important Properties of Concrete," *Consolidation of Concrete, SP-96*, S. Gebler, Ed., American Concrete Institute, Detroit, MI, 1987, pp. 125–160.

[86] Whiting, D., "Permeability of Selected Concretes," *Permeability of Concrete, SP-108*, D. Whiting and A. Walitt, Ed., American Concrete Institute, Detroit, MI, 1988, pp. 195–222.

[87] Bisaillon, A. and Malhotra, V. M., "Permeability of Concrete Using a Uniaxial Water-Flow Method," *Permeability of Concrete, SP-108*, D. Whiting and A. Walitt, Eds., American Concrete Institute, Detroit, MI, 1988, pp. 175–194.

[88] Olsen, M. P. J., "Energy Requirements for Consolidation of Concrete During Internal Vibration," *Consolidation of Concrete, SP-96*, S. Gebler, Ed., American Concrete Institute, Detroit, 1987, pp. 179–196.

[89] Valore, R. C., "Insulating Concretes," *Journal*, American Concrete Institute; *Proceedings*, Vol. 14, No. 2, 1956, pp. 509–532.

[90] Brewer, H. W., "General Relation of Heat Flow Factors to the Unit Weight of Concrete," *Journal*, Portland Cement Association Research and Development Laboratories, Skokie, IL, Vol. 9, No. 1, Jan. 1967, pp. 48–60.

[91] Helms, S. B., "Air Content and Unit Weight," *Significance of Tests and Properties of Concrete and Concrete-Making Materials, ASTM STP 169*, American Society of Testing and Materials, Philadelphia, 1955, p. 208.

[92] Helms, B., "Air Content and Unit Weight," *Significance of Tests and Properties of Concrete and Concrete-Making Materials, ASTM STP 169A*, American Society for Testing and Materials, Philadelphia, 1966, pp. 309–325.

[93] Litvin, A. and Fiorato, A. E., "Lightweight Aggregate Spectrum," *Concrete International, Design & Construction*, Vol. 3, No. 3, 1981, p. 49.

[94] Cook, R. A., "Advanced Mercury Porosimetry for the Investigation of Cement Paste, Mortar, and Concrete," Ph.D. thesis, Cornell University, Ithaca, NY, Aug. 1992.

[95] Kagaya, M., Tokuda, H., and Kawakami, M., "Experimental Considerations on Judging Adequacy of Consolidation in Fresh Concrete," *Consolidation of Concrete, SP-96*, S. Gebler, Ed., American Concrete Institute, Detroit, MI, 1987, pp. 161–178.

[96] *Concrete Manual*, 7th Ed., U.S. Department of the Interior, Bureau of Reclamation, Denver, CO, p. 34.

[97] "Cement and Concrete Investigations," *Bulletin 4*, U.S. Department of the Interior, Bureau of Reclamation, Boulder Canyon Project; as noted by Helms, S. B., "Air Content and Unit Weight," *Significance and Properties of Concrete and Concrete-Making Materials, ASTM STP 169B*, American Society for Testing and Materials, Philadelphia, 1978, p. 441.

Analyses for Cement and Other Materials in Hardened Concrete

William G. Hime[1]

PREFACE

The chemical analysis of concrete was covered in *ASTM STP 169* by H. F. Kriege. Dr. L. John Minnick provided an excellent discussion of methods for the analysis of concrete in *ASTM STP 169A*. The present writer extended the discussions in *ASTM STP 169B* to include analytical methods for other concrete components. The present edition addresses enacted and proposed changes to the ASTM procedures for cement content analyses, and extends the discussions of analyses of concrete for other components. Accordingly, the title has been broadened from the "Cement Content" of the prior two editions.

INTRODUCTION

Analyses of concrete for cement content are properly requested by investigators who understand the limitations of the test and of the information obtained. Because water-cement ratio (or, as applicable, water-cementitious ratio and voids-cementitious ratio) more directly influences such concrete properties as strength and shrinkage, the cement content information alone often fails to adequately disclose the cause for the problem at hand. It is the purpose of this chapter to outline the most frequently used cement content analyses procedures, and to point out their proper use. Additionally, supplementary analytical methods for other components of concrete are briefly discussed and sources of information about them are provided.

Minnick [1], in an earlier edition of this book, *ASTM STP 169A*, primarily discussed the ASTM Test for Cement Content of Hardened Portland Cement Concrete (C 85) that was in effect from 1954 to 1987.

Both Minnick, in that book, and Hime [2] in a later edition, *ASTM STP 169B*, pointed out the many sources of error in that procedure—errors that often led analysts, especially those who were less experienced or capable, to report considerably lower cement contents than were actually present. Hime proposed a modification of the test that was subsequently adopted as the ASTM Test Method for Portland-Cement Content of Hardened Hydraulic-Cement Concrete (C 1084). This modification, as discussed later, usually leads to either slight underestimations or large overestimations of cement content. As such, it should not cast undue doubt on concrete of proper cement content, or on its producer.

Although the procedures of analysis for portland cement have improved, concrete has become more complex, with the present-day product often incorporating admixtures such as fly ash, silica fume, slag, and calcium salts (such as chlorides, nitrites, and nitrates), most of which can lead to erroneously high results.

Further, since the methods presented in this chapter are for portland-cement content, care must be used in interpreting the results since the actual cement employed in the project may be a blended hydraulic formulation conforming to ASTM Specification for Blended Hydraulic Contents (C 595) or a "bag-equivalent" mixture where a ready-mix company may sell, for example, a "five-bag mix" that contains four bags of portland cement and 50 lb of fly ash.

For these reasons, cement content analyses should usually be made in conjunction with a petrographic study that can not only provide an independent estimate of cement content, but can also direct the analyst's attention to compositional features of the concrete components that may dictate use of a particular analysis procedure.

OUTLINE OF CEMENT CONTENT ANALYSIS PROCEDURES

Analytical procedures for portland cement utilize chemical or physical differences between the cement and the aggregates, and they are performed by chemists or petrographers. The procedures, in general, are outlined in the following paragraphs.

Chemical Methods

Because portland cement has an elemental composition similar to that of many types of aggregate, and because in concrete it has undergone chemical reaction, there is no satisfactory method of analyzing hardened concrete for portland cement, per se. As discussed by Lawrence in Chapter 13, the situation is less complicated for fresh concrete, where neither hardening nor reaction has significantly progressed.

The generally recognized methods for cement content of hardened concrete take advantage of certain potential

---

[1] Principal, Erlin, Hime Associates Division of Wiss, Janney, Elstner Associates, Inc., Northbrook, IL 60062.

differences between portland cement and aggregate: concentrations of three chemical constituents—calcium, silicates, and sulfates—and solubility of the silicates.

Calcium is the major element in portland cement and is relatively constant in concentration, about 60 to 66% calculated as the oxide (CaO). It is not a significant component of most siliceous aggregates, but it is the major component of limestone and dolomite. ASTM C 1084 provides a procedure for calcium analyses.

Acid-soluble silica is also a major, well-regulated component of portland cement, being present at a concentration of about 20 to 22%, calculated as silicon dioxide (SiO_2). It is not a significant component of most calcareous aggregates such as limestone or dolomite, and even the components of many siliceous aggregates are not acid soluble, at least initially. Unfortunately, some siliceous aggregates such as feldspars can react with portland cement paste during residence in the concrete to gradually become acid soluble, thus leading to error even when analytical data for the original aggregate is used to correct for interference. ASTM C 1084 provides a procedure for soluble silica analyses.

Because sulfate is the unique substance that is present in portland cement but not in virtually any concrete aggregate, analyses for it have been the method of choice of some laboratories, However, sulfate concentrations of a given type of cement from a single manufacturer can vary significantly, and, from manufacturer to manufacturer, or type to type, they vary substantially. Except for concrete employing a known cement, cement content errors may be 50% or more using sulfate procedures. Further, because the sulfate content of cement is only about 3%, errors of only 0.1% in the sulfate analysis translate to over 3% in cement content.

Selective dissolution by alcoholic maleic acid is, at this writing, the subject of a proposed ASTM procedure for cement content and is the method utilized by many laboratories. Maleic acid is weaker than the hydrochloric acid employed in the ASTM C 1084 procedures and dissolves hydrated portland cement but usually little or none of any limestone (calcium carbonate) or dolomite aggregates, thus permitting its use not only with most calcareous aggregates, but also with many siliceous aggregates. The fact that the maleic acid solution does not dissolve calcium carbonate can be a disadvantage: carbonation of concrete insolubilizes much of the calcium component of the cement. Thus, analyses of porous or highly fractured concrete, or even of an aged powdered concrete sample, will underestimate cement content to the extent of carbonation. Further, for reasons not yet understood, some laboratories consistently achieve poor results using the procedures presently proposed.

The maleic acid test also presents difficulty in discarding the used solutions. Governmental regulations in this respect may impose costly procedures.

Petrographic Procedures

Studies of concrete using the petrographic microscope, ASTM Practice for Petrographic Examination of Hardened Concrete (C 856) and the procedures of ASTM Practice for Microscopical Determination of Air-Void Content and Parameters of the Air-Void System in Hardened Concrete (C 457) can allow a determination of the "paste" (water, unhydrated cement, and hydrated cement) content of concrete. From such features as color, hardness, and size of crystals, experienced concrete petrographers can often estimate cement content or water-cement ratio, or both.

Regardless of ability to precisely determine cement content, petrographic studies are invaluable in disclosing potential interferences to each of the cement content procedures. Further, and most importantly, they may reveal the cause for the problem that led to the request for the cement content analysis. For example, low strength may be due to high air content, to cracking due to alkali-silica reaction, or to scaling due to improper finishing.

Other Cement Content Procedures

Several instrumental methods for cement content determinations have been proposed, but have found little general use or acceptance because they are applicable only to particular concrete mixtures, such as those that do not contain a calcareous aggregate. For example, nuclear methods that determine total calcium, or total silicon, have been proposed by Covaut and Poovey [3] and Iddings et al. [4].

PROCEDURE DETAILS

Sample Selection

Selection of a proper sample is of extreme importance in attaining meaningful cement content data, and is often not achieved. Although ASTM C 85 (replaced by ASTM C 1084) required three ten-pound concrete pieces, experience indicated that a smaller sample, carefully chosen, can be sufficient and will present fewer laboratory problems. As indicated in ASTM C 1084, a single core may be satisfactory for the cement content determination. It should have a diameter of at least four times the maximum aggregate size and be taken through the entire depth of a concrete slab. For thicker concrete members, the selection must consider that concrete composition may vary with depth or structural feature. In such cases, several cores analyzed separately will disclose the degree of uniformity of cement distribution.

A second core or sample, or a larger core, is generally also required to perform a determination of unit weight, to allow petrographic examination, or to allow horizontal sampling for components such as chloride, so that concentrations can be determined as a function of depth.

Sample Preparation

If a second core or sample has not been obtained, it is desirable to cut off (longitudinally) a portion of the sample for unit weight determination, petrographic examination, or further analyses as might be suggested by the problem or the cement content results. The major portion should be crushed into quarter-inch size (or as directed), quartered, and a portion pulverized as required. Since very fine particles interfere with many of the analysis procedures, the sample should be pulverized for short intervals, with sieving after each so that most particles just pass the designated sieve.

Significant amounts of iron may be introduced by some pulverization and grinding equipment. A magnet may be passed through the sample to remove the iron before analysis if a petrographer confirms that such a process does not remove aggregate components such as magnetite.

ASTM C 1084 Analysis

ASTM C 1084 requires analyses for acid-soluble silica by a prescribed procedure. In contrast to ASTM C 85, ice-cold hydrochloric acid is employed to dissolve the cement, thus minimizing dissolution of silicates in aggregates, fly ash, or slag. It is useful to examine the acid-insoluble residue so that the presence of fly ash may be revealed.

A procedure is also given for the calcium determination, but it is lengthy and difficult. Alternatively, any determination method may be employed if the method can be shown to provide accurate analyses of NIST (NBS) Standard Cements. Atomic absorption spectroscopy or X-ray emission spectroscopy is employed by many laboratories to determine calcium rapidly.

Maleic Acid Analyses

Maleic acid determinations of cement content were first suggested by Tabikh [5], with modifications by Clemena [6], Pistilli [7], and Marusin [8]. These vary primarily in the concentration of the maleic acid solution and the length of digestion time. All are designed to minimize solution of both calcareous and siliceous aggregates, but to allow nearly complete dissolution of the portland cement component. For a particular aggregate, one of the versions of the test may be better than the others, but most laboratories do not have the luxury of allowing a test variance for each sample. The procedure by Marusin is probably the present method of choice.

It is imperative that the analyst minimize carbonation of the sample because carbonation converts soluble cement hydration products to insoluble calcium carbonate. If highly fractured concrete is the subject of the analysis, either a petrographer should confirm that carbonation is minimal or another procedure should be employed.

Cement-Type Analysis

At times an investigation requires a determination of the type of cement used in the concrete: for example, concern has been expressed that a sewer has not been made with the specified Type V cement. Dissolution of the cement paste by the maleic acid procedure provides an avenue to a successful approach if the aggregate is insoluble and fly ash is not a component of the concrete. The maleic acid is then destroyed by evaporating with nitric acid, and aluminum and iron determined by chemical or instrumental methods. From this data the tricalcium aluminate (C_3A) content may be determined and compared to ASTM Specification for Portland Cement (C 150) limits.

Similar methods may be used to determine if the cement was "low-alkali," but interference by the aggregate may occur if it contains alkalies (for example, a feldspar), and will occur if fly ash is present.

Of course, no study of the concrete will reveal if the cement satisfied most other chemical or any physical requirements.

Calculations and Report

Perhaps the greatest error in the cement content analysis is the assumption that the cement content determined represents the cement content delivered. Two factors can cause serious divergence between determined and delivered cement factor. First, because the calculation is in terms of pound per cubic yard, poor consolidation of the concrete leads to a lowered cement content. And second, even under good placement practice, cement concentration may vary with height of placement (for example, in columns and walls), or with location in heavily reinforced structures. To partially account for such occurrences, ASTM C 1084 provides two methods for density determination, oven-dry and saturated surface-dry. The latter relates better to as-delivered concrete.

ANALYSIS OF MORTAR AND GROUT

The cement content methods detailed earlier have been employed by various laboratories to analyze mortars or grouts, usually on the mistaken belief that "mortar" is concrete made without coarse aggregate. Masonry mortars and grouts almost always contain, in addition to portland cement and sand, hydrated lime or finely ground limestone. The latter is the major component of most masonry cements. Therefore, analyses based on calcium overestimate portland cement content, analyses based only on soluble silica underestimate total cementitious material, and analyses based on the maleic acid procedure usually cannot differentiate between the cementitious components.

Mortar analyses should generally be made only in conjunction with microscopic studies by an experienced petrographer. With the guidance of the petrographer, successful analyses are often possible by determining portland cement through the soluble silica data, and hydrated lime or calcium carbonate from the calcium data after subtracting off the portland cement contribution. Such analyses are in error to the extent that the aggregate contains soluble calcium or silica. A procedure is outlined by Erlin and Hime [9].

DETERMINATION OF OTHER CONCRETE CONSTITUENTS

The analysis of concrete for cement content alone usually provides insufficient information to adequately determine cause for distress. In many cases, the determination of the presence of cement additives, concrete admixtures, and of water content, may provide more useful information. These determinations are outlined in the following paragraphs.

Additives and Admixture

Organic additives to cement and admixtures to concrete are often present in concentrations of a few thousandths of a percent. In general, appropriate extraction and spectroscopic analysis procedures allow determination of most of the commercially available products.

Spectroscopic procedures applicable to the active organic components of most commercial products were described by Connolly et al. [10], and earlier by Hime et al. [11]. Unfortunately, there have been few publications in this area since ASTM STP 169B. In addition to the methods of Connolly and Hime, methods have been presented by other investigators for specific compounds or admixture types:

Lignosulfonates—Kroome [12], Reul [13], Rixom [14]
Sugars—Shima and Mishi [15], Rixom [14]
Polycarboxylic Acids—Frederick and Ellis [16], Rixom [14]
Triethanolamine—Connolly and Hime [17], Reul [13], Rixom [14]
Vinsol Resin—Kroome [12], Reul [13]
Retarders—Halstead and Chaiken [18], Reul [13], Rixom [14]
Waterproofers—Reul [13], Rixom [14]

Procedures for "mineral" admixtures such as fly ash, silica fume, and slag are almost completely proprietary to a producer or an analytical laboratory. If the particular fly ash or slag used in a concrete is available for analysis, some element not present in significant amounts in the cement or aggregate may be found, thus allowing elemental analyses for it in the concrete. Among the elements found useful in this connection for some fly ashes and slags are barium, manganese, and titanium. Petrographic examination usually will allow approximation of the fly ash or slag concentration, generally as related to the cement content. No generally accepted procedure for silica fume content has been developed. Indeed, except by use of electron microscopy, successful methods for detection of silica fume in concrete have not been reported.

For the determination of chloride in concrete, procedures employ either water extraction or acid extraction. The former relates best to present-time corrosion potential, but the latter is more conservative. In either case, currently used procedures generally pulverize the concrete so that chloride embedded in the aggregate is determined, thus often over-estimating the potential for corrosion. Recently adopted ASTM procedures for acid-soluble chloride in cement, ASTM Test Methods for Chemical Analysis of Hydraulic Cement (C 114), and in concrete, ASTM Test Methods for Acid-Soluble Chloride in Mortar and Concrete (C 1152), are widely used, but most governmental agencies use AASHTO Sampling and Testing for Total Chloride Ion in Concrete and Concrete Raw Materials (T-260) for "total" (acid-soluble) and water-soluble chloride. The AASHTO use of the word "total" is improper because chloride-containing organic chemicals are not determined by the procedure. "Total" chloride may be determined by fusion of the sample in calcium oxide and then use of an acid-dissolution procedure.

For nitrite, a component of a commercial corrosion inhibitor, colorimetric procedures are useful, but specific methods have not been published.

Air Voids

The air void content of hardened concrete is determined by the procedures of ASTM C 457. Details are discussed by Hover in Chapter 28 of this book.

Aggregates

The aggregate content of concrete is usually determined by the linear traverse or point count procedures of ASTM C 457, which follow those of Polivka et al. [19] and Axon [20]. Aggregate content can also be determined as the residue from the maleic acid cement content procedure. With some siliceous aggregates, the aggregate content is determined by an insoluble residue procedure such as in ASTM C 114.

Water

The determination of the water content of hardened concrete is readily made by ignition to 1000°C and correction for carbon dioxide by any usual procedure. In contrast, there is no generally applicable method for the much more useful determination of the amount of free water that was present in the fresh concrete. For non-air-entrained concrete, the procedures of Blackman [21] and Brown [22] can be employed, but are difficult to perform. An estimation procedure by Axon [20] employs microscopic techniques. Many petrographers estimate water-cement ratio by petrographic techniques and experience. Substitution of cement content data in that water-cement ratio value allows at least an estimate of water content.

General Reference Works

Method of analysis for cement content as presented in the two previous editions of this book provide considerable information about the effect of aggregate on the calcium and silica procedures, and other useful historical background. General methods for analysis of cement and concrete were presented by Hime [23]. Methods of analysis of concrete and other cement products are presented in books by the Society of Chemical Industry [24], and Figg and Bowden [25]. The latter contains a bibliography of 193 references.

REFERENCES

[1] Minnick, L. J., "Cement Content," Significance of Tests and Properties of Concrete and Concrete-Making Materials, ASTM STP 169A, American Society for Testing and Materials, Philadelphia, 1966, pp. 326–339.

[2] Hime, W. G., "Cement Content," *Significance of Tests and Properties of Concrete and Concrete Making Materials, ASTM STP 169B*, American Society for Testing and Materials, Philadelphia, 1978, pp. 462–470.

[3] Covault, D. O. and Poovey, C. E., "Use of Neutron Activation To Determine Cement Content of Portland Cement Concrete," *Publication 1022*, Highway Research Board, Washington, DC, 1962, pp. 1–29.

[4] Iddings, F. A., Arman, A., Perez, A. W., Kiessel, D. W., and Woods, W., "Nuclear Techniques for Cement Determination," *HRB Bulletin 268*, Highway Research Board, Washington, DC, 1969, pp. 118–130.

[5] Tabikh, A. A., Balchunas, M. J., and Schaefer, D. M., "A Method for the Determination of Cement Content in Concrete," *Highway Research Record 370*, Highway Research Board, Washington, DC, 1971, pp. 1–7.

[6] Clemena, G. G., "Determination of the Cement Content of Hardened Concrete by Selective Solution," Final Report, VHRC 72-R7, Virginia Highway Research Council, Charlotteville, VA, Sept. 1972.

[7] Pistili, M. F., "Cement Content of Hardened Concrete," presented at Cement Chemist's Seminar, Portland Cement Association, Skokie, IL, 1976.

[8] Marusin, S. L., "Use of the Maleic Acid Method for the Determination of the Cement Content of Concrete," *Cement, Concrete, and Aggregates*, Winter 1981, pp. 89–92.

[9] Erlin, B. and Hime, W. G., "Methods for Analyzing Mortar," *Proceedings*, Third North American Masonry Conference, The Masonry Society, University of Texas at Arlington, Arlington, TX, Paper 63, 1985, pp. 1–6.

[10] Connolly, J. D., Hime, W. G., and Erlin, B., "Analysis for Admixtures in Hardened Concrete," *Admixtures, Concrete International 1980*, The Construction Press, Lancaster, UK, 1980, pp. 114–129.

[11] Hime, W. G., Mivelaz, W. F., and Connolly, J. D., *Analytical Techniques for Hydraulic Cement and Concrete, ASTM STP 395*, American Society for Testing and Materials, Philadelphia, 1966, pp. 132–134.

[12] Kroome, B., "A Method of Detecting and Determining Lignin Compounds in Mortars and Concrete," *Magazine of Concrete Research*, Vol. 23, No. 76–76, June–Sept. 1971, pp. 132–134.

[13] Reul, H., "Detection of Admixtures in Hardened Concrete," *Zement-Kalk-Gips*, Vol. 33, No. 3, 1980 (translation of article).

[14] Rixom, R. R., *Chemical Admixtures for Concrete*, E. & F. N. Span, London, 1978.

[15] Shima, I. and Nishi, T., "Determination of Saccharose in Hydrated Cement," *Semento Gijutsu Nempo*, Vol. 17, 1963, pp. 106–109; *Chemist Abstract*, Vol. 61, 1964, 5344b.

[16] Frederick, W. L. and Ellis, J. T., "Determination of a Polyhydroxy Carboxylic Acid Retarder in Hardened Concrete," *Materials Research and Standards*, Vol. 8, March 1968, pp. 14–18.

[17] Connolly, J. D. and Hime, W. G., "Analysis of Concrete for Triethanolamine," *Cement and Concrete Research*, Vol. 6, 1976, pp. 741–746.

[18] Halstead, W. J. and Chaiken, B., "Water-Reducing Retarders for Concrete Chemical and Spectral Analysis," *Public Roads*, Vol. 31, No. 6, 1961, pp. 126–135.

[19] Polivka, M., Kelly, J. W., and Best, C. H., "A Physical Method for Determining the Composition of Hardened Concrete," *Papers on Cement and Concrete, ASTM STP 205*, American Society for Testing and Materials, Philadelphia, 1956, pp. 135–152.

[20] Axon, E. O., "A Method of Estimating the Original Mix Composition of Hardened Concrete Using Physical Tests," *Proceedings*, American Society for Testing and Materials, Philadelphia, Vol. 62, 1962, pp. 1068–1080.

[21] Blackman, J. S., "Method for Estimating Water Content of Concrete at the Time of Hardening," *Proceedings*, American Concrete Institute, Vol. 50, 1954, pp. 533–541.

[22] Brown, A. W., "A Tentative Method for the Determination of the Original Water-Cement Ratio of Hardened Concrete," *Journal of Applied Chemistry*, Vol. 7, Oct. 1957, pp. 565–572.

[23] Hime, W. G., "Cements, Mortars and Concrete," *Encyclopedia of Industrial Chemical Analysis*, Vol. 9, Wiley, New York, 1970, pp. 94–155.

[24] "Analysis of Calcareous Materials," *SCI Monograph No. 18*, Society of Chemical Industry, London, 1964.

[25] Figg, J. W. and Bowden, S. R., *The Analysis of Concretes*, Her Majesty's Stationery Office, London, 1971.

30

Nondestructive Tests

V. Mohan Malhotra[1]

PREFACE

The subject of nondestructive testing of concrete was covered in *ASTM STP 169A* and *ASTM STP 169B*. The chapter in *ASTM STP 169* was entitled "Dynamic Tests" and was authored by E. A. Whitehurst, whereas the chapter in *ASTM STP 169B* was entitled "Nondestructive Tests," and was authored by E. A. Whitehurst and V. M. Malhotra. This chapter updates the information presented in the previous publications and, in addition, includes the discussion of new ASTM standards on the subject.

INTRODUCTION

In the inspection and testing of concrete, the use of nondestructive testing (NDT) is relatively new. The slow development of these testing methods for concrete is due to the fact that, unlike steel, concrete is a highly nonhomogenous composite material, and most concrete is produced in ready-mixed plants and delivered to the construction site. The in-place concrete is, by its very nature and construction methods, highly variable, and does not lend itself to testing by traditional NDT methods as easily as steel products.

Notwithstanding the preceding, there has been considerable progress in the development of NDT methods for testing concrete in recent years. A number of these methods have been standardized by the American Society for Testing Materials (ASTM), the International Standards Organization (ISO), and the British Standards Institute (BSI).

For the purposes of this chapter, tests that are identified generally as nondestructive may be subdivided into two main types. The first type includes those identified as sonic and pulse velocity tests that involve the determination of the resonant frequency and the measurement of the velocity of a compressional pulse traveling through the concrete. Also included in this category are stress wave tests for locating the flaws or discontinuities that may be present or measuring thickness of concrete. The second type includes those tests that are used to estimate strength properties and include the surface hardness, penetration, pullout, break off, maturity, pulloff, and combined methods. Some of these methods are not truly nondestructive because they cause some surface damage that is generally insignificant.

RESONANT AND PULSE VELOCITY METHODS

Resonant Frequency Methods

Natural frequency of vibration is a dynamic property of an elastic system and is primarily related to the dynamic modulus of elasticity and density in the case of a vibrating beam. Therefore, the natural frequency of vibration of a beam can be used to determine its dynamic modulus of elasticity. Although the relationship between the two is valid for homogenous solid media that are isotropic and perfectly elastic, they may be applied to concrete when the size of a specimen is large in relation to the size of its constituent materials.

For flexural vibrations of a long, thin unrestrained rod, the following equation or its equivalent may be found in any complete textbook on sound [1]

$$N = \frac{m^2 k}{2\Pi L^2} \sqrt{\frac{E}{\rho}} \quad (1)$$

and solving for E

$$E = \frac{4\Pi^2 L^4 N^2 \rho}{m^4 k^2} \quad (2)$$

where

E = dynamic modulus of elasticity,
ρ = density of the material,
L = length of the specimen,
N = fundamental flexural frequency,
k = radius of gyration of the section about an axis perpendicular to the plane of bending ($k = t/\sqrt{12}$ for rectangular cross section where t = thickness), and
m = a constant (4.73 for the fundamental mode of vibration).

The dynamic modulus of elasticity can also be computed from the fundamental longitudinal frequency of vibration of an unrestrained specimen, according to the following equation [2]

---

[1] Program principal, Advanced Concrete Technology Program, Energy, Mines and Resources Canada, Canada Centre for Mineral and Energy Technology, Ottawa, Canada K1A 0G1.

$$E = 4L^2\rho N^2 \qquad (3)$$

Equations 1 and 3 were obtained by solving the respective differential equations for the motion of a bar vibrating: (1) in flexure in the free-free mode, and (2) in the longitudinal mode.

Thus, the resonant frequency of vibration of a concrete specimen directly relates to its dynamic modulus of elasticity. If the concrete undergoes degradation, the modulus of elasticity will be altered and so will the resonant frequency of the specimen. Therefore, monitoring the change in resonant frequency allows one to infer changes in the integrity of concrete.

The method of determining the dynamic elastic moduli of solid bodies from their resonant frequencies has been in use for the past 45 years. However, until recently, resonant frequency methods had been used almost exclusively in laboratory studies. In these studies, natural frequencies of vibration are determined on concrete prisms and cylinders to calculate the dynamic moduli of elasticity and rigidity, the Poisson's ratio, and to monitor the degradation of concrete during durability tests.

The resonant frequency method was first developed by Powers [3] in the United States in 1938. He determined the resonant frequency by matching the musical tone created by concrete specimens, usually 51 by 51 by 241-mm prisms, when tapped by a hammer, with the tone created by one of a set of orchestra bells calibrated according to frequency. The error likely to occur in matching the frequency of the concrete specimens to the calibrated bells was of the order of 3%. The shortcomings of this approach, such as the subjective nature of the test, are obvious. But this method laid the groundwork for the subsequent development of more sophisticated methods.

In 1939, Hornibrook [4] refined the method by using electronic equipment to measure resonant frequency. Other early investigations on the development of this method included those by Thomson [5] in 1940, by Obert and Duvall [2] in 1941, and by Stanton [6] in 1944. In all the tests that followed the work of Hornibrook, the specimens were excited by a vibrating force. Resonance was indicated by the attainment of vibrations having maximum amplitude as the driving frequency was changed. The resonant frequency was read accurately from the graduated scale of the variable driving audio oscillator. The equipment is usually known as a sonometer and the technique is known as forced resonance.

The forced resonance testing apparatus as described in ASTM Test Method for Fundamental Transverse, Longitudinal, and Torsional Frequencies of Concrete Specimens (C 215) consists primarily of two sections, one generates mechanical vibrations and the other senses these vibrations [7]. The principal part of vibration generation section is an electronic audio-frequency oscillator that generates audio-frequency voltages. The oscillator output is amplified and fed to the driver unit for conversion into mechanical vibrations.

The mechanical vibrations of the specimen are sensed by a piezoelectric transducer. The transducer is contained in a separate unit and converts mechanical vibrations to electrical voltage of the same frequencies. These voltages are amplified for the operation of a panel-mounted meter that indicates the amplitude of the transducer output. As the frequency of the driver oscillator is varied, maximum deflection of the meter needle indicates when resonance is attained. Visible indications that driving frequency equals the fundamental frequency can be obtained easily through the use of an auxiliary cathode-ray oscilloscope. This is because a resonance condition can be established at driving frequencies that are fractions of the fundamental frequency.

Some skill and experience are needed to determine the fundamental resonant frequency using a meter-type indicator because several resonant frequencies may be obtained corresponding to different modes of vibration. Specimens having either very small or very large ratios of length to maximum transverse direction are frequently difficult to excite in the fundamental mode of transverse vibration. It has been suggested that the best results are obtained when this ratio is between three and five.

The supports for the specimen under test should be of a material having a fundamental frequency outside the frequency range being investigated, and should permit the specimen to vibrate without significant restriction. Ideally, the specimens should be held at the nodal points, but a sheet of soft sponge rubber is quite satisfactory and is preferred if the specimens are being used for freezing and thawing studies.

The fundamental transverse vibration of a specimen has two nodal points, at distances from each end of 0.224 times the length. The vibration amplitude is maximum at the ends, about three fifths of the maximum at the center, and zero at the nodal points. Therefore, movement of the pickup along the length of the specimen and observation of the meter reading will show whether the specimen is vibrating at its fundamental frequency. For fundamental longitudinal and torsional vibrations, there is a point of zero vibration (node) at the midpoint of the specimen and the maximum amplitude is at the ends.

Sometimes in resonance testing of prismatic concrete specimens, two resonant frequencies may appear that are close together. Kesler and Higuchi [8] believed this to be caused by a nonsymmetrical shape of the specimen that causes interference due to vibration of the specimen in some direction other than that intended. Proper choice of specimen size and shape should practically eliminate this problem; for example, in a specimen of rectangular cross section the problem can be eliminated by vibrating the specimen in the direction parallel to the short side.

In performing resonant frequency tests, it is helpful to have an estimate of the expected fundamental frequency. Approximate ranges of fundamental longitudinal and flexural resonant frequencies of standard concrete specimens have been given by Jones [9].

Calculation of Dynamic Moduli of Elasticity and Rigidity and Poisson's Ratio

The dynamic moduli of elasticity and rigidity (or shear modulus of elasticity) and the Poisson's ratio of the concrete can be calculated by equations given in ASTM C 215.

These are modifications of theoretical equations applicable to specimens that are very long in relation to their cross section and were developed and verified by Pickett [10], and Spinner and Teftt [11]. The corrections to the theoretical equations involve Poisson's ratio and are considerably greater for transverse resonant frequency than for longitudinal resonant frequency. For example, a standard 102 by 102 by 510-mm prism requires a correction factor of about 27% at fundamental transverse resonance, as compared with less than 0.5% at fundamental longitudinal resonance [12]. The longitudinal and flexural modes of vibration give nearly the same value for the dynamic modulus of elasticity. The dynamic modulus of elasticity may range from 14.0 GPa for low-quality concretes at early ages to 48.0 GPa for good-quality concrete at later ages [13]. The dynamic modulus of rigidity is about 40% of the modulus of elasticity [14]. It should be mentioned that more input energy is needed for longitudinal resonance and, therefore, transverse resonance mode is used more often in laboratory investigations.

Other Methods of Resonant Frequency Testing

A new method for determining fundamental frequencies has been proposed by Gaidis and Rosenberg [15] as an alternative to the forced resonance method. In this method, the concrete specimen is struck with a small hammer. The impact causes the specimen to vibrate at its natural frequencies. Hence the technique is known as impact resonance. The amplitude and frequency of the resonant vibrations are obtained using a spectrum analyzer that determines the component frequencies via the fast Fourier transform. The amplitude of the specimen response versus frequency is displayed on the screen of a frequency analyzer, and the frequencies of major peaks can be read directly.

In operation, the pick-up accelerometer is fastened to the end of the specimen with microcrystalline wax, and the specimen is struck lightly with a hammer. The output of the accelerometer is recorded digitally by the waveform analyzer and the recorded signal is processed to obtain the frequency response. On the resulting amplitude versus frequency curve, a dot marker (cursor) may be moved to coincide with the peak, and the frequency value of the peak is displayed on the screen. The advantages of this method over the forced-resonance procedure are the greater speed of testing, the capability of testing specimens having a wide range of dimensions, and the ability to measure readily the longitudinal frequency. However, the initial high cost of equipment appears to be a disadvantage. This impact resonance procedure was adopted by ASTM in 1991 as an alternative to the existing procedure. The various modes of vibration are obtained by the proper location of the impact point and an accelerometer.

Damping Properties of Concrete

Damping is the property of a material causing free vibrations in a specimen to decrease in amplitude as a function of time. Several investigators, particularly Thomson [5], Obert and Duvall [2], Kesler and Higuchi [16], Shrivastava and Sen [17], and Swamy and Rigby [14], have shown that certain properties of concrete can be related to its damping ability.

There are several methods of determining the damping characteristics of a material, but two common methods used for concrete are [18]:

1. The determination of logarithmic decrement, δ, that is the natural logarithm of the ratio of any two successive amplitudes in the free vibration of the specimen.
2. Calculation of the damping constant, Q, from the amplitude versus resonance curve of the test specimen.

The measurement of the damping properties of concrete specimens has not been standardized by ASTM, but research continues toward gaining a better understanding of the significance of these measurements.

Factors Affecting Resonant Frequency and Dynamic Modulus of Elasticity

Several factors influence the resonant frequency measurements, the dynamic modulus of elasticity, or both. These include the influence of mixture proportions and properties of aggregates, specimen size effect, and the influence of curing conditions. These have been discussed in detail elsewhere [18].

Standardization of Resonant Frequency Methods

ASTM C 215 was published in 1947 and since then has been revised seven times. The last revision to this standard was in 1991, when the impact-resonance procedure was added as an alternative to the forced resonance procedure.

The significance and use statement of the resonant frequency method as given in ASTM C 215 is as follows:

3.1 This test method is intended primarily for detecting significant changes in the dynamic modulus of elasticity of laboratory or field test specimens that are undergoing exposure or weathering or other types of potentially deteriorating influences.

3.2 This test method may be used to assess the uniformity of field concrete, but it should not be considered as an index of compressive or flexural strength nor as an adequate test for establishing the compliance of the modulus of elasticity of field concrete with that assumed in design.

3.3 The conditions of manufacture, the moisture content, and other characteristics of the test specimens materially influence the results obtained.

3.4 Different computed values for the dynamic modulus of elasticity may result from widely different resonant frequencies of specimens of different size and shapes of the same concrete. Therefore, comparison of results from specimens of different sizes or shapes should be made with caution.

Limitations and Usefulness of Resonant Frequency Methods

Although the basic equipment and testing procedures associated with the resonant frequency techniques have been standardized in various countries, and commercial testing equipment is easily available, the usefulness of the tests is seriously limited because of the following two points.

1. Generally, these tests are carried out on small-sized specimens in a laboratory rather than on structural members in the field because resonant frequency is affected considerably by boundary conditions and the properties of concrete. The size of specimens in these tests is usually 152 by 305-mm cylinders or 76 by 76 by 305-mm prisms.
2. The equations for the calculation of dynamic elastic modulus involve "shape factor" corrections. This necessarily limits the shape of the specimens to cylinders or prisms. Any deviation from the standard shapes can render the application of shape factor corrections rather complex.

Notwithstanding the preceding limitations, the resonance tests provide an excellent means for studying the deterioration of concrete specimens subjected to repeated cycles of freezing and thawing and to deterioration due to aggressive media. The use of resonance tests in the determination of damage by fire and the deterioration due to alkali-aggregate reaction have also been reported by Chefdeville [19] and Swamy and Al-Asali [20].

The resonant frequency test results are often used to calculate the dynamic modulus of elasticity of concrete but the values obtained are somewhat higher than those obtained with standard static tests carried out at lower rates of loading and to higher strain levels. The use of dynamic modulus of elasticity of concrete in design calculations is not recommended.

Various investigators have published correlations between the strength of concrete and its dynamic modulus of elasticity. Such correlations should not be used to predict strength properties of concrete unless similar relationships have been developed in the laboratory for the concrete under investigation [18].

Pulse Velocity Method

The application of pulse transmission techniques to the testing of concrete is believed to have had its origin with Obert [21]. Tests were made on concrete replacement pillars in mines and involved the use of two geophones, two high-gain amplifiers, and a camera with a moving film strip. Two holes, approximately 6.09 m (20 ft) apart vertically, were drilled into the pillars. The geophones were placed in the backs of the holes and the holes filled with cotton waste. A hammer blow was struck at the base of the pillar, and at the same time the camera lens was opened and the moving film strip exposed. After the film was developed, the transit time of the impulse in traveling from one geophone to the other was determined by measuring the distance between the two signals on the film, the speed of motion of the film having been controlled carefully. The velocity of the stress pulse could then be calculated.

Long and Kurtz [22] reported performing somewhat similar experiments with a seismograph in which the longitudinal velocity of the stress pulse created by a single impact was measured between arbitrarily placed geophones. They stated that only very limited experiments of this nature had been conducted but the method appeared to hold great promise providing the apparatus could be adapted to measure much shorter time intervals than could be measured with the seismograph.

Long et al. [23] undertook further investigations along these lines and in 1945 reported on the instrument and technique that resulted from their work. The apparatus consisted of two vibration pickups (in the form of phonograph cartridges), two amplifiers, two thyratron tube circuits, and a ballistic galvanometer circuit. The concrete was struck with a hammer approximately in line with the two pickups. The propagating pulse actuated the first pickup, the voltage from which energized the first thyratron and started a flow of current through the galvanometer. When the energy impulse reached the second pickup, the voltage from its amplifier ionized the second thyratron and cut off the flow of current. The deflection of the galvanometer was directly proportional to the time required for the wave to travel the distance between the two pickups.

In a discussion of this paper, the substitution of an electronic interval timer for the ballistic galvanometer was suggested. This device consists of a capacitor that begins to charge when the first thyratron is ionized and stops charging when the second is ionized and a vacuum-tube voltmeter measures the charge. The meter may be calibrated directly in units of time, thus eliminating the necessity for computations involving the magnitude of the current flowing through the galvanometer. This device was found to be more reliable than the ballistic galvanometer for field use.

Subsequent investigations in North America and Europe have resulted in the development of a number of other devices quite similar in most respects. These include the Micro-timer developed by the U.S. Bureau of Reclamation, and the Condenser Chronograph developed by the Danish National Institute of Building Research.

In 1946, the Hydro-Electric Power Commission of Ontario, Canada, in an effort to develop a technique for examining cracks in monolithic concrete structures, began a series of studies that resulted in the construction of an instrument known as the Soniscope. The device developed by Leslie and Cheesman [24] consists basically of a stress transmitter using piezoelectric crystals, a similar pulse receiver, and electronic circuits that actuate the pulse transmitter, provide visual presentation of transmitted and received signals on a cathode-ray tube, and accurately measure the time interval between the two.

The physical and electrical features of the Soniscope passed through several stages of improvement, and a number of these instruments have been built by various laboratories in the United States and Canada. The instrument was used extensively in Canada [24,25] and the United States [26].

During approximately the same time that the Soniscope was being developed in Canada and the United States, work of a similar nature was being conducted in England. These investigations resulted in the development of an instrument known as the Ultrasonic Concrete Tester. This instrument and its applications have been described at length by Jones [27,28]. The Ultrasonic Concrete Tester differed from the Soniscope primarily in the higher frequency of the transmitted pulse and the pulse repetition rate that was about three times greater than that of the

Soniscope. These changes improved the accuracy of measurement on very small specimens but limited the usefulness of the instrument for field testing, since the high frequencies suffer much greater attenuation in passing through concrete than do the lower ones. The maximum range of the Ultrasonic Concrete Tester was about 2 m, whereas that of the Soniscope in testing reasonably good concrete was 15 m or more.

In recent years, portable, battery-powered ultrasonic testing units have become available worldwide. One of the units available in the United States is called the V-Meter [11].[2]

The pulse travel time is displayed in three numerical digits that can be varied for three different ranges: (1) 0.1 to 99.9 μs with an accuracy of 0.1 μs, (2) 1 to 999 μs with an accuracy of 1 μs and (3) 10 to 9990 μs with the accuracy of 10 μs.

Whereas the use of the resonant frequency tests has been restricted primarily to the evaluation of specimens undergoing natural or artificial weathering, pulse velocity techniques have been applied to concrete for many purposes and, in most areas of investigation, only limited agreement has been reached concerning the significance of test results. The quantity measured by all of these instruments is the travel time of stress pulses passing through the concrete under test. If the path length between transmitter and receiver is known, or can be determined, the velocity of the pulse can be computed. It is in the interpretation of the meaning of this velocity and in its use for determining various properties of concrete that agreement is incomplete. The technique is as applicable to in-place concrete as to laboratory-type specimens, and results appear to be unaffected by the size and shape of the concrete tested, within the limits of transmission of the instrument employed, provided care is taken when testing very small specimens. This, of course, is a highly desirable attribute and, in many respects, makes the pulse velocity techniques more useful than those involved in resonant frequency testing.

Because of the fundamental theoretical relationship between pulse velocity techniques and resonant frequency techniques, there is a strong inclination for users of the pulse technique to endeavour to compute the dynamic modulus of elasticity from the results of the tests. Theoretically, such values of modulus should be the same as those determined by resonant frequency tests upon the same specimens. It has been shown that on some occasions this is true and on others it is not [29]. Because of these unexplainable differences, most of those experienced in the use of pulse velocity techniques are inclined to leave their results in the form of velocity without attempting to calculate elastic moduli therefrom.

If the modulus of elasticity is to be computed from the pulse velocity, the relationship generally recommended is

$$E = V^2 \rho \frac{(1 + \mu)(1 - 2\mu)}{(1 - \mu)} \quad (4)$$

where

E = dynamic modulus of elasticity,
V = longitudinal pulse velocity,
ρ = mass density, and
μ = Poisson's ratio.

This equation relates modulus to pulse velocity and density in an infinite medium and presumably should apply only to mass concrete. However, the experience of most investigators has been that, even for very small laboratory specimens, this relationship gives better results than do those applying to either slabs or long slender members. Leslie and Cheesman [24] have suggested that best results are obtained if, for concretes having unit weights in excess of approximately 2240 kg/m^3, the value of Poisson's ratio is assumed to be 0.24.

The use of pulse velocity techniques has been suggested for evaluating the strength of concrete, its uniformity, its setting characteristics, its modulus of elasticity, and the presence or absence of cracks within the concrete. There appears to be little question of the suitability of such techniques to determine the presence, and to some extent the magnitude, of cracks in concrete, although it has been suggested that if the cracks are fully water-filled their locations may be more difficult to ascertain. In all of the other fields of investigation, independent investigators have reported widely different degrees of success through the use of these techniques [26].

A comprehensive report of the early experiences of users of pulse velocity techniques in the United States and Canada may be found in Ref 30. Recently, Sturrup et al. [31] have published data dealing with the experiences of Ontario Hydro, Toronto, in the use of pulse velocity for strength evaluation.

Experiences in the use of pulse velocity measurements for evaluating concrete quality have been reported in many other countries, notably Great Britain and Russia. In some instances, investigators in these countries have appeared to have greater confidence in the use of such techniques for acceptance testing than has been the case in the United States.

It is generally agreed that very high velocities of >4570 m/s are indicative of very good concrete and that very low velocities of <3050 m/s are indicative of poor concrete. It is further agreed that periodic, systematic changes in velocity are indicative of similar changes in the quality of the concrete. Beyond these areas of agreement, however, it appears that the investigator must have a rather intimate knowledge of the concrete involved before attempting to interpret velocities as measures of strength or other properties of the concrete. This is particularly true if the aggregate involved is a lightweight aggregate.

Standardization of Pulse Velocity Method

ASTM Committee C9 initiated the development of a test method for pulse velocity in the late 1960s. A tentative standard was issued in 1968. A standard test method was issued in 1971 and no significant changes have been made

---

[2] This is the same unit that is manufactured in England and is known as PUNDIT.

in the standard since then [10]. The significance and use statement of the test method, as given in the ASTM Test Method for Pulse Velocity Through Concrete (C 597), is as follows:

1. This test method may be used to advantage to assess the uniformity and relative quality of concrete, to indicate the presence of voids and cracks, to estimate the depth of cracks, to indicate changes in the properties of concrete, and in the survey of structures, to estimate the severity of deterioration or cracking.

Note 1: Moisture content of concrete can affect pulse velocity.

2. The results obtained by the use of this test method should not be considered as a means of measuring strength nor as an adequate test for establishing compliance of the modulus of elasticity of field concrete with that assumed in design.

Note 2: When circumstances permit, a velocity-strength (or velocity-modulus) relationship may be established by the determination of pulse velocity and compressive strength (or modulus of elasticity) on a number of samples of a concrete. This relationship may serve as a basis for the estimation of strength (or modulus of elasticity) by further pulse-velocity tests on that concrete.

3. The procedure is applicable in both field and laboratory testing regardless of size or shape of the specimen within the limitations of available pulse-generating sources.

Note 3: Presently available test equipment limits path lengths to approximately 50 mm minimum and 15 m maximum, depending, in part, upon the frequency and intensity of the generated signal. The upper limit of the path length depends partly on surface conditions and partly on the characteristics of the interior concrete under investigation. The maximum path length is obtained by using transducers of relatively low vibrational frequencies (10 to 20 kHz) to minimize the attenuation of the signal in the concrete. (The resonant frequency of the transducer assembly, that is, crystals plus backing plate, determines the frequency of vibration in the concrete). For the shorter path lengths where loss of signal is not the governing factor, it is preferable to use vibrational frequencies of 50 kHz or higher to achieve more accurate transit-time measurements and hence greater sensitivity.

Stress Wave Propagation Methods

In recent years, considerable research has been undertaken in Canada [32] by CANMET and in the United States [33,34] by NIST to develop methods to determine flaws and discontinuities in concrete. These methods are classified as stress wave propagation methods and the common feature of the various methods under development is that stress waves are introduced into concrete, and their surface response is monitored using a receiving transducer connected to a digital data acquisition system capable of performing frequency analysis. One of the most attractive features of the stress wave propagation methods is that access to only one surface of concrete is required.

Two main types of stress wave methods being used are the echo method and the impulse response method.

Echo Method

In the echo method, a stress pulse is introduced into concrete by a transducer and its reflection by flaws or discontinuities is monitored by the same transducer that transmits the wave or by a separate transducer located near the transmitting transducer. In the former case, the technique is called pulse-echo and in the latter case the technique is known as pitch-catch. The receiver output is displayed on an oscilloscope. If the compressional wave speed is known, time-domain analysis, that is, measuring the travel time of the pulse to and from the flaw in the concrete, can be used to determine the depth of the flaw or discontinuity.

In the impact-echo method, instead of a transducer, mechanical impact is used to generate a lower frequency stress wave that can penetrate into concrete. The stress waves are reflected by the discontinuities, and the surface motion caused by the arrival of reflected waves is measured by a receiving transducer and recorded on a digital oscilloscope. Frequency analysis of the recorded signal permits measurement of the depth of the reflecting interface.

These methods are being used for thickness measurements, flaw detection, and integrity testing of piles.

Impulse Response Method

This method is somewhat similar to the impact-echo method. An instrumented hammer is used to generate a stress wave. The arrival of the reflected echos causes the concrete surface to vibrate, and the velocity of this vibration is measured by a transducer located near the point of impact. A dynamic signal analyzer is used to analyze the force-time history of the impact and the recorded velocity. From the analysis, information is obtained about the location of the reflecting interface and the dynamic stiffness of the test object. This method is primarily being used to test the integrity of the piles.

These wave propagation methods are still under development and therefore have not been standardized by ASTM.

SURFACE HARDNESS METHODS

The increase in the hardness of concrete with age and strength has led to the development of test methods to measure this property. These methods consist of the indentation type and those based on the rebound principle. The indentation methods consist principally of impacting the surface of concrete by means of a given mass having a given kinetic energy and measuring the width or depth or both of the resulting indentation. The methods based on the rebound principle consist of measuring the rebound of a spring-driven hammer mass after its impact with concrete.

Rebound Method

In 1948, a Swiss engineer, Ernst Schmidt [35–37] developed a test hammer for measuring the hardness of concrete by the rebound principle. The Schmidt rebound hammer is principally a surface hardness tester with little apparent theoretical relationship between the strength of concrete and the rebound number of the hammer. However, within limits, empirical correlations have been established between strength properties and the rebound number. Further, Kolek [38] has attempted to establish a correlation between the hammer rebound number and the hardness as measured by the Brinell method.

Description

The Schmidt rebound hammer weighs about 1.8 kg and is suitable for use both in a laboratory and in the field. The main components include the outer body, the plunger, the hammer mass, and the main spring. Other features include a latching mechanism that locks the hammer mass to the plunger rod and a sliding rider to measure the rebound of the hammer mass. The rebound distance is measured on an arbitrary scale marked from 10 to 100. The rebound distance is recorded as a "rebound number" corresponding to the position of the rider on the scale.

Method of Testing

To prepare the instrument for a test, the plunger is released from its locked position by pushing the plunger against the concrete and slowly moving the body away from the concrete. This causes the plunger to extend from the body and the latch engages the hammer mass to the plunger rod. The plunger is then held perpendicular to the concrete surface and slowly the body is pushed towards the test object. As the body is pushed, the main spring connecting the hammer mass to the body is stretched. When the body is pushed to the limit, the latch is automatically released, and the energy stored in the spring propels the hammer mass toward the plunger tip. The mass impacts the shoulder of the plunger rod and rebounds. During rebound, the slide indicator travels with the hammer mass and records the rebound distance. A button on the side of the body is pushed to lock the plunger in the retracted position, and the rebound number is read from the scale.

The test can be conducted horizontally, vertically upward or downward, or at any intermediate angle. Due to different effects of gravity on the rebound as the test angle is changed, the rebound number will be different for the same concrete and require separate calibration or correlation charts [39,40].

Correlation with Strength

Each hammer is furnished with correlation curves developed by the manufacturer using standard cube specimens. However, the use of these curves is not recommended because material and testing conditions may not be similar to those in effect when the calibration of the instrument was performed.

A typical curve established by Zoldners [41] for limestone aggregate concrete is shown in Fig. 1. This curve was based on tests performed at 28 days using different concrete mixtures.

Although the rebound hammer provides a quick, inexpensive means of checking the uniformity of concrete, it has serious limitations and these must be recognized. The results of the Schmidt rebound hammer test are affected by:

1. smoothness of test surface;
2. size, shape, and rigidity of specimens;
3. age of test specimens;
4. surface and internal moisture conditions of the concrete;
5. type of coarse aggregate;
6. type of cement (portland, high alumina, super sulfated);
7. type of mold; and
8. carbonation of the concrete surface.

These limitations are discussed in detail elsewhere (Ref 18, Chapter 1).

To gain a basic understanding of the complex phenomena involved in the rebound test, Akashi and Amasaki [42] have studied the stress waves in the plunger of a rebound hammer at the time of impact. Using a specially designed plunger instrumented with strain gages, the authors showed that the impact of the hammer mass produces a large compressive wave, σi, and a large reflected stress wave, σr, at the center of the plunger. The ratio, $\sigma r/\sigma i$, of the amplitudes of these waves and the time, T, between their appearance was found to depend upon the surface hardness of concrete. The rebound number was found to be approximately proportional to the ratio of the two stresses and was not significantly affected by the moisture condition of the concrete [42].

Carette and Malhotra [43] have investigated the within-test variability of the rebound hammer test at test ages of one to three days and have studied the ability of the test to determine early-age strength development of concrete for formwork removal purposes. The rebound tests were performed at one, two, and three days on plain concrete slabs, 300 by 1270 by 1220-mm in size. Also, companion cylinders and cores taken from the slabs were tested in compression.

From the analyses of the test data, the authors concluded that because of the large within-test variation, the rebound hammer test was not a satisfactory method for reliable estimates of strength development of concrete at early ages.

According to Kolek [38] and Malhotra [39,40] there is a general correlation between compressive strength of concrete and the hammer rebound number. However, there is a wide degree of disagreement among various researchers concerning the accuracy of the estimation of strength from the rebound readings and the empirical relationship. Coefficients of variation for estimated compressive strength for a wide variety of specimens averaged 18.8% and exceeded 30% for some groups of specimens. The large deviations in strength can be narrowed down considerably by developing a proper correlation curve for the hammer, which allows for various variables discussed earlier. By consensus, the accuracy of estimation of compressive strength of test specimens cast, cured, and tested under laboratory conditions by a properly calibrated hammer lies between ± 15 and ± 20%. However, the probable accuracy of estimation of concrete strength in a structure is ± 25%.

Greene [44] and Klieger et al. [45] have established correlation relationships between the flexural strength of concrete and the hammer rebound number. They have found

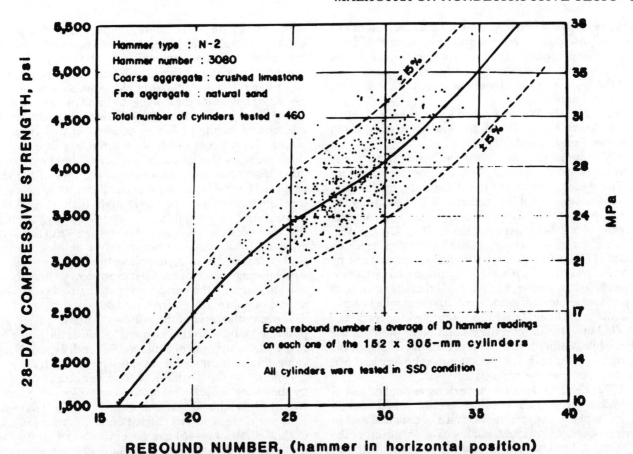

FIG. 1—Relationship between 28-day cylinder compressive strength and rebound number for limestone aggregate concrete obtained with Type N2 hammer (from Ref 41).

that the relationships are similar to those obtained for compressive strength, except that the scatter of the results is greater. Further, they found that the rebound numbers for tests conducted on the top of finished surface of a beam were 5 to 15% lower than those conducted on the sides of the same beam.

Limitations

The limitations of the Schmidt hammer should be recognized and taken into account when using the hammer. It cannot be over stressed that the hammer must not be regarded as a substitute for standard compression tests but as a method for determining the uniformity of concrete in the structures, and comparing one concrete against another.

The rebound method has won considerable acceptance, and standards have been issued both by the ASTM and ISO and by several other countries for determining the rebound number of concrete.

Standardization of Rebound Test Method

ASTM Test Method for Rebound Number of Hardened Concrete (C 805) was revised in 1985; the significance and use statement of the test method as given in ASTM C 805 is as follows:

4.1 The rebound number determined by this method may be used to assess the uniformity of concrete in-situ, to delineate zones or regions (areas) of poor quality or deteriorated concrete in structures, and to indicate changes with time in characteristics of concrete such as those caused by the hydration of cement so that it provides useful information in determining when forms and shoring may be removed.

4.2 This test method is not intended as an alternative for strength determination of concrete.

4.3 Optimally, rebound numbers should be correlated with core testing information. Due to the difficulty of acquiring the appropriate correlation data in a given instance, the rebound hammer is most useful for rapidly surveying large areas of similar concretes in the construction under consideration.

Probe Penetration Test

Penetration resistance methods are based on the determination of the depth of penetration of probes (steel rods or pins) into concrete. This provides a measure of the hardness or penetration resistance of the material that can be related to its strength [18].

The measurement of concrete hardness by probing techniques was reported by Voellmy [46] in 1954. Apart from the data reported by Voellmy, there is little other published work available on these tests, and they appear to have received little acceptance in Europe or elsewhere. Perhaps the introduction of the rebound method around 1950 was one of the reasons for the failure of these tests to achieve general acceptance [18].

In the 1960s, the Windsor probe test system was introduced in North America, and this was followed by a pin penetration test in the 1980s.

The Windsor probe test system was advanced for penetration testing of concrete in the laboratory as well as in-situ. The device was meant to estimate the quality and compressive strength of in-situ concrete by measuring the depth of penetration of probes driven into the concrete by means of a powder-actuated driver. The development of this technique was closely related to studies reported by Kopf [47]. Results of the investigations carried out by the Port of New York Authority were presented by Cantor [48] in 1970. Meanwhile, a number of other organizations had initiated exploratory studies of this technique [49–51] and a few years later, Arni [52,53] reported the results of a detailed investigation on the evaluation of the Windsor probe, while Malhotra [54–56] reported the results of his investigations on both 150 by 300-mm cylinders and 610 by 610 by 200-mm concrete slabs.

In 1972, Klotz [57] stated that extensive application of the Windsor probe test system had been made in investigations of in-place compressive strength of concrete and in determinations of concrete quality. The Windsor probe had been used to test reinforced concrete pipes, highway bridge piers, abutments, pavements, and concrete damaged by fire. In the 1970s, several U.S. federal agencies and state highway departments reported investigations on the assessment of the Windsor probe for in-situ testing of hardened concrete [58–62]. In 1984, Swamy and Al-Hamed [63] in the United Kingdom, published results of a study on the use of the Windsor probe system to estimate the in situ strength of both lightweight and normal-weight concretes.

Description of Test

The Windsor probe consists of a powder-actuated gun or driver, hardened alloy-steel probes, loaded cartridges, a depth gauge for measuring the penetration of probes, and other related equipment. The probes have a tip diameter of 6.3 mm, a length of 79.5 mm, and a conical point. Probes of 7.9 mm diameter are also available for the testing of lightweight aggregate concrete. The rear of the probe is threaded and screws into a probe-driving head that is 12.7 mm in diameter and fits snugly into the bore of the driver. The probe is driven into the concrete by the firing of a precision powder charge [7,18]. For testing of relatively low-strength concrete, the power level can be reduced by pushing the driver head further into the barrel.

The method of testing is relatively simple and is given in ASTM Test Method for Penetration Resistance of Hardened Concrete (C 803). The area to be tested must have a brush finish or a smooth surface. To test structures with coarse finishes, the surface first must be ground smooth in the area of the test. Briefly, the powder-actuated driver is used to drive a probe into concrete. If flat surfaces are to be tested, a suitable locating template is used to provide a 178-mm equilateral triangular pattern, and three probes are driven into the concrete, one at each corner. The exposed lengths of the individual probes are measured by a depth gage. The manufacturer also supplies a mechanical averaging device for measuring the average exposed length of the three probes fired in the triangular pattern. The mechanical averaging device consists of two triangular plates. The reference plate with three legs slips over the three probes and rests on the surface of the concrete. The other triangular plate rests against the tops of the three probes. The distance between the two plates, giving the mechanical average of exposed lengths of the three probes, is measured by a depth gage inserted through a hole in the center of the top plate. For testing structures with curved surfaces, three probes are driven individually using the single probe locating template. In either case, the measured average value of exposed probe length may then be used to estimate the compressive strength of concrete by means of appropriate correlation data.

The manufacturer of the Windsor probe test system has published tables relating exposed length of the probe with compressive strength of concrete. For each exposed length value, different values for compressive strength are given, depending on the hardness of the aggregate as measured by the Mohs' scale of hardness. The tables are based on empirical relationships established in his laboratory. However, investigations carried out by Gaynor [50], Arni [52], Malhotra [54–56] and several others [51,59,64–66] indicate that the manufacturer's tables do not always give satisfactory results. Sometimes they considerably overestimate the actual strength and in other instances they underestimate the strength. It is therefore, imperative for each user of the probe to correlate probe test results with the type of concrete being used. A practical procedure for developing such a relationship is described elsewhere [67].

The correlation published by several investigators for concretes made with limestone gravel, chert, and traprock aggregates are shown in Fig. 2. Note that different relationships have been obtained for concretes with aggregates having similar Moh's hardness numbers.

There is no rigorous theoretical analysis of the probe penetration test available. Such analysis may, in fact, not be easy to achieve in view of the complex combinations of dynamic stresses developed during penetration of the probe and the heterogeneous nature of concrete. The test involves a given initial amount of kinetic energy of the probe that is absorbed during penetration, in large part through crushing and fracturing of the concrete and in lesser part through friction between the probe and the concrete. Penetration of the probe causes the concrete to fracture within a cone-shaped zone below the surface with cracks propagating up to the surface (Fig. 3).

The probe penetrations relate to some strength parameter of the concrete below the surface, which makes it possible to establish useful empirical relationships between the depth of penetration and compressive strength.

There appears to have been no systematic attempts to determine the relative influences of these factors that could affect the probe penetration test results. However, it is generally agreed that the largest influence comes from

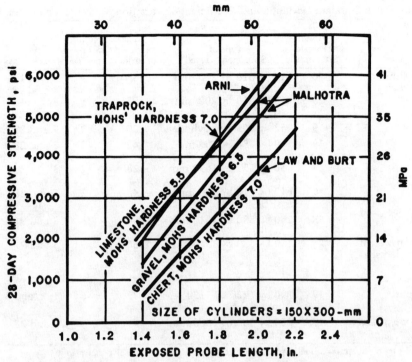

FIG 2—Relationship between exposed probe length and 28-day compressive strength of concrete as obtained by different investigators (from Ref 67).

the coarse aggregate. Apart from its hardness, the type and size of coarse aggregate used have been reported to have a significant effect on probe penetration [52,63,65]. The considerable differences shown in Fig. 2 tend to support the important influence of the aggregate type. However, other parameters such as mixture proportions, moisture content, curing regime, and surface conditions are likely to have affected these correlations to some extent and could explain some of the observed differences.

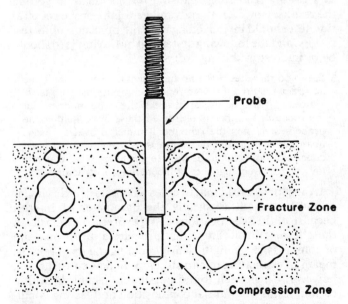

FIG. 3—Typical failure of mature concrete during probe penetration (from Ref 67).

The within-batch variability in the probe test results, as obtained by various investigators, is shown in Table 1. Variability is reported in terms of standard deviations and coefficients of variation with values in the latter case being calculated from the exposed length of the probe readings, although more correctly they should be based on the embedded lengths of the probe. These data show that for concrete with a maximum aggregate size of 19 mm, a typical value for the within-test coefficient of variation (based on depth of penetration) is about 5% [68]. Statistically, for such concrete, the minimum number of individual penetration tests required to ensure that the average penetration is known with the same degree of confidence as the average standard cylinder strength (assuming a coefficient of variation of 4% based on two cylinders) would be three. This number, however, would not ensure that the in-situ strength is known with the same degree of confidence, since, obviously, the preceding within-test coefficient of variation for the probe penetration test does not take into account the uncertainty of the correlation relationship, which also affects the reliability of the estimated strength.

An increasingly important area of the application of nondestructive techniques is in the estimation of early-age strength of concrete for the determination of safe form-removal times. Relatively little information has been published in regard to the performance of the probe penetration test at early ages. However, by the late 1970s, it had been reported that the probe penetration test was probably the most widely used nondestructive method for the determination of safe stripping times [61]. One main advantage cited was the great simplicity of the test: "One

TABLE 1—Within-Batch Standard Deviation and Coefficient of Variation of Probe Penetration Measurements.

Investigation Reported by	Type of Aggregate Used	Maximum Aggregate Size, mm	Type of Specimens Tested	Total Number of Probes	Number of Probes per Test	Age of Test, days	Average Standard Deviation, mm	Average Coefficient of Variation %[a]
Arni [52]	gravel, limestone, trap rock	50	410 by 510 by 200-mm slabs	136	9	3, 7, 14, and 28	3.62	7.1
		25	410 by 510 by 200-mm slabs	198	9	3, 7, 14, and 28	2.66	5.4
Malhotra [67]	limestone	19	152 by 305-mm cylinders	20	2	7 and 28	3.14	7.7
	gravel	19	150 by 150 by 200-mm slabs	48	3	7 and 28	1.37	3.4
		19	150 by 150 by 1690-mm prisms	28	2	35	1.57	3.4
		19	610 by 610 by 200-mm slabs	48	3	7 and 28	2.21	5.5
Gaynor [50]	quartz	25	150 by 580 by 1210-mm walls	384	16	3 and 91	4.05	...
	semi-lightweight expanded shale as coarse aggregate	25	150 by 580 by 1210-mm walls	256	9	3 and 91	4.30	...
Carette, Malhotra [43]	limestone	19	300 by 1220 by 1220-mm slabs	72	6	1, 2, and 3	2.52	8.2 (5.4)
Keiller [64]	limestone, gravel	19	250 by 300 by 1500-mm prisms	45	3	7 and 28	1.91	3.5

[a] Based on exposed length of probe, except for value in parentheses that is based on depth of penetration.

simply fires the probes into the concrete and compares penetration to previously established criteria. If the probes penetrate too far, the contractor knows the concrete is not yet strong enough" [61].

Carette and Malhotra [42] have investigated in the laboratory the within-test variability at the ages of one to three days of the probe penetration test, and the ability of the test to indicate the early-age strength development of concrete for formwork removal purposes. The penetration tests were performed at one, two, and three days on plain concrete slabs, 300 by 1220 by 1220-mm in size, along with compression tests on standard cylinders and cores taken from the slabs. Excellent correlations between compressive strength and probe penetration were observed at these ages for each concrete. From the analysis of the test data, the authors concluded that unlike the rebound method, the probe penetration test can estimate the early-age strength development of concrete with a reasonable degree of accuracy, and thus can be applied to determine safe stripping times for the removal of formwork in concrete construction.

Probe Penetration Test versus Core Testing

The determination of the strength of concrete in a structure may become necessary when standard cylinder strength test results fail to comply with specified values, or the quality of the concrete is being questioned because of inadequate placing or curing procedures. It may also be required in the case of older structures where changes in the quality of the concrete must be investigated. In these instances, the most direct and common method of determining the strength of concrete is through drilled core testing; however, some nondestructive techniques such as the probe penetration test have been gaining acceptance as a means to estimate the in-situ strength of concrete [61,69,70].

It has been claimed that the probe penetration test is superior to core testing and should be considered as an alternative to the latter for estimating the compressive strength of concrete. It is true that the probe test can be carried out in a matter of minutes, whereas cores, if from exposed areas and if they have to be tested in accordance with ASTM Test Method for Obtaining and Testing Drilled Cores and Sawed Beams of Concrete (C 42), must be soaked for 40 h; also, the cores may have to be transported to a testing laboratory, causing further delay in getting the results. However, the advantages of the probe penetration test should be judged against the precision of its test results, and the following statement by Gaynor [50] should be of interest in this regard:

> Based on these tests, the probe system does not supply the accuracy required if it is to replace conventional core tests. However, it will be useful in much the same manner that the rebound hammer is useful. In these tests, neither the probe system nor the rebound hammer provides precise quantitative estimates of compressive strength of marginal concretes. Both should be used to locate areas of relatively low- or relatively high-strength concretes in structures.

It must be stressed that in cases where standard cylinder or cube strength is strictly the parameter of interest because of the specifications being expressed primarily in these terms, the core test that provides a direct measure of compressive strength clearly remains the most reliable means of estimating in-situ strength. In new construction, however, it has been found possible to establish, within certain limits of material composition and testing conditions, relationships between probe penetration and strength that are accurate enough so that the probe test can be used as a satisfactory substitute for the core test [70].

In the early 1980s, a survey of concrete testing laboratories in Canada and the United States indicated that the Windsor probe penetration technique was the second most often used method for in-situ strength testing of concrete (see ASTM C 42). The survey included methods such as rebound hammer, probe penetration, pullout, pulse velocity, maturity, and cast-in-place cylinder. In terms of reliability, simplicity, accuracy, and economy, the probe test was given one of the best combined ratings.

Advantages and Disadvantages of the Probe Penetration Test

The probe penetration test system is simple to operate, rugged, and needs little maintenance except for occasional cleaning of the gun barrel. The system has a number of built-in safety features that prevent accidental discharge of the probe from the gun. However, wearing of safety glasses is required. In the field, the probe penetration test offers the main advantages of speed and simplicity, and that of requiring only one surface for the test. Its correlation with concrete strength is affected by a relatively small number of variables that is an advantage over some other methods for in-situ strength testing.

However, the probe test has limitations that must be recognized. These include minimum size requirements for the concrete member to be tested. The minimum acceptable distance from a test location to any edges of the concrete member or between two given test locations is of the order of 150 to 200 mm, while the minimum thickness of the members is about three times the expected depth of penetration. Distance from reinforcement can also have an effect on depth of probe penetration especially when the distance is less than about 100 mm [71]. The importance of aggregate type and aggregate content on the correlation is emphasized.

As previously indicated, the uncertainty of the estimated strength value, in general, is relatively large and the test results may lack the degree of accuracy required for certain applications. The test is limited to a certain range of strength (<40 MPa), and the use of two different power levels to accommodate a larger range of concrete strength within a given investigation complicates the correlation procedures. Finally, as noted earlier, the test causes minor damage to the surface that generally needs to be repaired.

Pin Penetration Test

In the late 1980s, Nasser and Al-Manaseer [72,73] reported the development of a simple pin penetration test for the determination of early-age strength of concrete for removal of concrete formwork. Briefly, this apparatus consists of a device that grips a pin having a length of 30.5 mm, a diameter of 3.56 mm, and a tip machined at an angle of 22.5°. The pin is held within a shaft that is encased within the main body of the tester. The pin is driven into the concrete by a spring that is mechanically compressed when the device is prepared for a test. The spring is reported to have a stiffness of 49.7 N/mm and stores about 10.3 J of energy when compressed.

When ready for testing, the apparatus is held against the surface of concrete to be tested, and a triggering device is used to release the spring forcing the pin into the concrete. Following this, the apparatus is removed, and the small hole created in the concrete is cleared by means of an air blower. A depth gage is used to measure the penetration depth.

The test, though simple in concept, has limitations. The pin penetrates only a small depth into the concrete, and therefore the results can be seriously affected by the conditions of the material at the surface. The test results are invalid when an aggregate particle is struck. The simplicity of the test makes it possible to obtain as many readings as necessary at little extra cost. The equipment is rather heavy for field use and, because of the nature of the spring mechanism, cannot be used for concrete with strength greater than about 30 MPa. Calibration of the equipment is important to ensure a consistent level of energy and its frequent verification may be necessary.

Standardization of Penetration Resistance Techniques

ASTM Committee C9 initiated the development of a standard for the probe penetration technique in 1972 and a tentative test method covering its use was issued in 1975. A standard test method designated ASTM C 803 was issued in 1982, the latest revision was issued in 1990, and the pin penetration method was added as an alternative penetration test. The significance and use statement of the test method as given in the 1990 standard is as follows:

1. This test method may be used to assess the uniformity of concrete, to delineate zones of poor quality or deteriorated concrete in structures, and to indicate in-place strength development of concrete.
2. For a given concrete and a given test apparatus, a relationship between penetration resistance and strength can be experimentally established and used to assess in-situ concrete strength. Such a relationship may change with curing and exposure conditions, type and size of aggregate, and level of strength developed in the concrete.

Note 1: Results of penetration resistance testing with probes have been correlated with results of strength tests of drilled cores, or cast specimens, or both. Results of penetration testing with pins have been correlated with strength test results of cast specimens. Since penetration results may be affected by the nature of the formed surfaces (for example, wooden forms versus steel forms), correlation testing should be performed on specimens with formed surfaces similar to those to be used during construction. Statistical correlation studies of penetration tests have been used as the basis for estimating the strength of similar concrete.

3. Steel probes are driven with a high-energy, powder-actuated driver and probes may penetrate some aggregate particles. Probe penetration resistance is affected by concrete strength as well as the nature of the coarse aggregate. Steel pins are smaller in size than probes and are driven by a low energy, spring-actuated driver. Pins are intended to penetrate the mortar fraction only, therefore a test in which a pin strikes a hard, dense, coarse aggregate particle is disregarded.

PULLOUT TEST

A pullout test, by using a dynamometer and a reaction bearing ring, measures the force required to pullout from concrete a specially shaped insert whose enlarged end has been cast into the concrete. Because of its shape, the insert is pulled out with a cone of the concrete. The concrete is simultaneously in tension and in shear, the generating lines of the cone are defined by the key dimensions of the insert and bearing ring (Fig. 4). The pullout force is then related to compressive strength by means of a previously established relationship.

The pullout techniques, though in use in the former Union of Soviet Socialist Republics [74] since 1935, are relatively new elsewhere [75,76]. In 1944 in the United States, Tremper [77] reported results of laboratory studies dealing with pullout tests covering strengths up to 35.2 MPa. In 1968, Tassios [78], in Greece, reported the development of a nail pullout test.

In the 1970s Kierkegaard-Hansen [79], in Denmark, and Richards [80], in the United States, advocated the use of pullout tests on structural concrete members. A number of researchers including Malhotra [75], Carino [81], and others [76] have published data dealing with laboratory studies and field testing. A pullout system known as LOK-Test based on Kierkegaard-Hansen's work is now available commercially. Carino [81] has summarized analytical and experimental studies to gain an understanding of the fundamental failure mechanisms of the pullout test and he concluded as follows:

1. It is understood that the pullout test subjects the concrete to a nonuniform, three-dimensional state of stress. It also has been demonstrated that there are at least two circumferential crack systems involved: a stable primary system which initiates at the insert head at about ⅓ of the ultimate load and propagates into the concrete at a large apex angle; and a secondary system which defines the shape of the extracted cone. However, there is not a consensus on the failure mechanism at the ultimate load. Some believe that ultimate load occurs as a result of compressive failure of concrete along a line from the bottom of the bearing ring to the top face of the insert head. This could explain why good correlation exists between pullout strength and compressive strength. Others believe that the ultimate failure is governed by aggregate interlock across the secondary crack system, and the ultimate load is reached when sufficient aggregate particles have been pulled out of the matrix. In the latter case, it is argued that good correlation exists between pullout strength and compressive strength because both properties are controlled by the strength of the mortar.

2. Despite the lack of agreement on the exact failure mechanism, it has been shown that pullout strength has good correlation with compressive strength of concrete, and that the test has good repeatability.

The main advantage of the pullout test is that it provides a direct measure of the in situ strength of concrete. The method is relatively simple and testing can be done in the field in a matter of minutes.

An example of the correlation between compressive strength and pullout strength or load is shown in Fig. 5. Such correlations depend on the geometry of the pullout test configuration and the coarse aggregate characteristics.

A major disadvantage of the pullout test is that minor damage to the concrete surface must be repaired. However, if a pullout force corresponding to a given minimum strength is applied without failure, it may be assumed that a minimum strength has been reached in the concrete and the pullout insert need not be stressed to failure. Another disadvantage is that the standard pullout tests have to be planned in advance, and unlike other in-situ tests, cannot be performed at random after the concrete has hardened.

In order to overcome the second disadvantage mentioned, new techniques have been proposed. These include the core and pullout (CAPO) test and the drilled-in pullout tests [18]. In the CAPO test, a hole is drilled into concrete and a special milling tool is used to under cut a 25-mm-

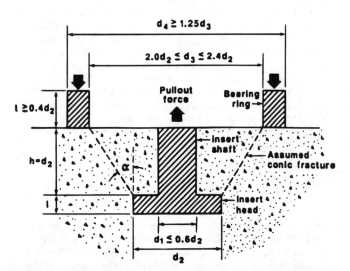

FIG. 4—Schematic cross section of pullout test (from ASTM C 900).

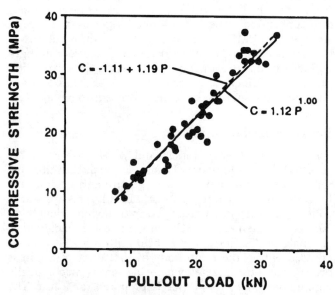

FIG. 5—Correlation data by Khoo and best-fit linear and power function relationship (from Ref 81).

diameter slot at a depth of 25 mm. An expandable steel ring is placed into the hole, and the ring is expanded into the slot using a special device. The entire assembly used to expand the ring is then pulled out of the concrete using the LOK test loading system. However, the test has found very limited acceptance because of the high variability associated with the test data in the field tests [81]. Variability may be reduced assuring a flat surface beneath the bearing ring.

The development of drilled-in pullout tests have been reported by the Building Research Establishment (BRE) (United Kingdom) [82] and by Mailhot et al. [83]. In the BRE tests, an anchor bolt is placed in a hole drilled in hardened concrete, and a pullout force is applied to cause failure in concrete. As with the CAPO test there is a high degree of variability in the test results.

Another pullout method studied by Mailhot et al. [83] involved epoxy grouting a 16-mm-long threaded rod to a depth of 38 mm in a 19-mm-diameter hole. After the epoxy had cured, the rod was pulled out using a tension jack reacting against a bearing ring. Once again, the within-test variability was reported to be high.

Standardization of Pullout Tests

ASTM Committee C9 initiated the development of a standard for the pullout tests in the early 1970s, and a standard was issued in 1982 and was revised in 1987. The revised standard is designated as ASTM Test Method for Pullout Strength of Hardened Concrete (C 900). The current standard deals with the use of cast-in-place inserts, but work is in progress to develop a standard for drilled-in tests.

The significance and use statement as given in ASTM C 900 is reproduced here:

1. For a given concrete and a given test apparatus, the pullout strength is related to other strength test results. Such strength relationships depend on the configuration of the embedded insert, bearing ring dimensions, depth of embedment, and level of strength development in that concrete. Prior to use, this relationship must be established for each system and each new combination of concreting materials. Such relationships tend to be less variable where both pullout and other test specimens are of consistent size and cured under similar conditions.
2. Pullout tests are used to determine when the in-place strength of concrete has reached a specified level so that, for example:
   a) post-tensioning may proceed;
   b) forms and shores may be removed; or
   c) winter protection and curing may be terminated.
3. When planning pullout tests and analyzing test results, consideration should be given to the normally expected decrease of concrete strength with increasing height within a given concrete placement in a structural element.

BREAK-OFF METHOD

In 1979 Johansen [84] reported the development of a method to determine the in-situ strength of concrete. The method known as "break-off" has found some acceptance in the Scandinavian countries and has been standardized by ASTM as ASTM Test Method for the Break-Off Number of Concrete (C 1150). Briefly, the test consists of breaking-off an in-situ cylindrical concrete specimen at a failure plane parallel to the finished surface of the concrete element. The break-off stress at failure is then related to the strength of concrete.

The principle of the break-off test is illustrated in Fig. 6. A plastic sleeve with an annular seating ring is inserted in fresh concrete to form a cylindrical test specimen and a counter bore. After the concrete has hardened, the sleeve is removed and a special loading mechanism is placed in the counter bore. A hand-operated pump is used to generate a force at the uppermost section of the cylinder so as to break from the concrete mass. The test result is reported as a break-off number, that is, the maximum pressure recorded by the gage measuring the hydraulic pressure in the loading mechanism. In hardened concrete, in cases where the plastic sleeve has not been installed, a concrete coring machine with a specially shaped coring drill bit may be used to drill similarly shaped test specimens.

Carlsson et al. [85], Dahl-Jorgenson and Johansen [86] and Naik [87] investigated the use of the break-off test and have published correlations between the break-off number and the compressive and flexural strength. Earlier investigations performed at CANMET had indicated that the test results obtained by this method had a high degree of variability, but according to Naik [87] and others, the changes in the design of the equipment has overcome this problem.

Standardization of Break-Off Test

The significance and use statement of ASTM C 1150 is as follows:

4.1 The break-off number determined by this test method may be used to assess the in-place strength of concrete, and delineate zones, regions, or areas of varying quality or deteriorated concrete in structures.
4.2 Prior to using this test method for determining in-place strength, a correlation relationship between the break-off number and the concrete strength should be established. Since such a correlation may vary with type and size of aggregates and method of specimen preparation, a relationship may be developed to take these and other variables into account. This relationship must be established for each new combination of concrete-making materials. In developing such relationships, care must be taken to ensure that the break-off specimens and the strength test specimens undergo similar curing histories up to the time of the test.
4.3 The break-off test may be used to evaluate the in-place concrete in order to:
   4.3.1 Determine if formwork or reshoring can be removed,
   4.3.2 Test if concrete meets break-off number specifications,
   4.3.3 Determine when prestressing strands may be cut to release the prestressing force,
   4.3.4 Determine if concrete has sufficient strength to all post-tensioning to proceed,
   4.3.5 Estimate efficiency of curing techniques, and

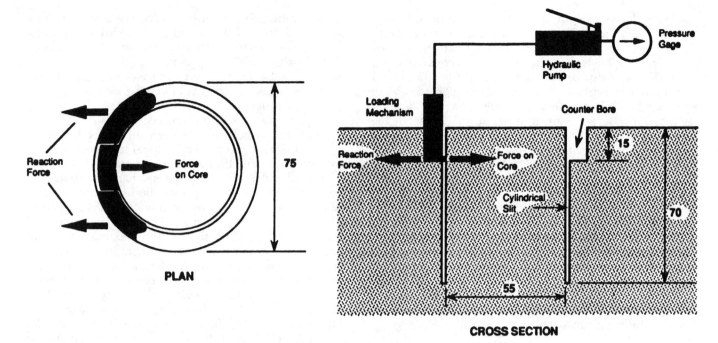

FIG. 6—A schematic of the break-off method (from ASTM C 1150).

4.3.6 Evaluate the effects of exposure to environmental or chemical attack.

4.4 When planning the break-off test and analyzing test results, consideration should be given to, (1) the normally expected decrease of concrete strength with increasing height within a given concrete placement in a structural element, and (2) locations with less favorable curing conditions prior to form removal.

4.5 Break-off tests are not recommended for concrete with a nominal maximum aggregate size greater than 25 mm. The within test variability of the break-off test has been found to increase in concrete with larger aggregate size.

4.6 The cylindrical break-off specimens may be kept and used for additional testing. The break-off test shall not be performed on concrete that is at a temperature of less than −5°C. Prior to starting a testing program the break-off tester must be calibrated according to the manufacturer's procedure (using the calibrator force gage and calibration diagram provided with the test unit) to ensure a consistent relationship.

MATURITY METHOD

It is well known that the compressive strength of well-cured concrete increases with time. However, the increase in strength is governed by many factors other than curing time, the most important being the concrete temperature and the availability of moisture. The combined effect of time and temperature has been studied by several investigators since 1904, but no hypothesis was formulated in early years. Then, in the 1950s, the concept of maturity was advanced by McIntosh [88], Nurse [89], Saul [90], and others [91–93], and strength-maturity relationships were published. Maturity was defined as the product of time and temperature above a datum temperature of −10°C (14°F).

In 1956, Plowman [92] examined relationships between concrete strength and maturity and attempted to establish a rational basis for the datum temperature used for maturity to calculate the maturity index. He defined the datum temperature for maturity as the temperature at which the strength gain of hardened concrete ceases. From his investigations, Plowman concluded that the datum temperature was −12.2°C (10°F). These earlier studies have been extensively reviewed by Malhotra [93].

Maturity Functions

Maturity functions are mathematical expressions to convert the temperature history of concrete to an index indicative of its strength development.

The Nurse-Saul maturity function is as follows

$$M(t) = = \Sigma(T_a - T_0)\Delta t \quad (5)$$

where

$M(t)$ = the temperature-time factor at age t, degree-days or degree-hours;
Δt = a time interval, days or hours;
T_a = average concrete temperature during time interval, Δt, °C; and
T_0 = datum temperature, °C.

It has been shown that the Nurse-Saul function does not accurately represent time-temperature effects because it is based on the assumption that the rate of strength development is a linear function of temperature.

In order to overcome some of the limitations of the Nurse-Saul maturity function, the maturity function proposed by Freiesleben Hansen and Pederson [94] can be used to compute equivalent-age at a specified temperature as follows

$$t_e = \Sigma e^{-[Q((1/T_a) - (1/T_s))]}\Delta t \qquad (6)$$

where

t = equivalent age at a specified temperature T_s, days or hours;
Q = activation energy divided by the gas constant, K;
T_a = average temperature of concrete during time interval Δt, K;
T_s = specified temperature, K; and
Δt = time interval, days or hours.

Equation 6 is based on the Arrhenius equation that describes the influence of temperature on the rate of chemical reaction.

Byfors [95], Naik [96], and Carino [97] have shown that the equivalent age maturity function based on Arrhenius equation accounts better for the effects of temperature on strength gain and is applicable over a wider temperature range than the Nurse-Saul maturity function.

One of the first engineering applications of the maturity method was by Swenson [98] in Canada who used it in forensic investigations to estimate the strength gain of concrete in structures.

The maturity method has been used to estimate the in-situ strength of concrete during construction. In the United States, Hudson and Steele [99] used the maturity approach to predict potential strength of concrete based upon early-age tests. Their results have been incorporated into ASTM Method for Developing Early Age Compression Test Values and Projecting Later Age Strengths (C 918). Malhotra [100] attempted to relate compressive strengths using accelerated-strength tests with the maturities for these tests.

The maturity of in-situ concrete can be monitored by thermocouples or by instruments called maturity meters. Basically, these maturity meters monitor and record the concrete temperature as a function of time using thermocouples or thermistors embedded in the fresh concrete and connected to strip-chart recorders or digital data recorders. The temperature-time factor or equivalent age is automatically computed and displayed. Also, disposable mini-maturity metres based on the Arrhenius equation have also become available [97]. These primarily consist of a glass tube containing a liquid that has an activation energy for evaporation that is similar to the activation energy for strength gain in concrete; the amount of evaporation from the capillary tube at a given time is indicative of strength development in the concrete.

Standardization of the Maturity Method

ASTM Committee C9 initiated the development of a standard on the maturity method in 1984, and a standard was issued in 1987. This standard is designated as ASTM Practice for Estimating Concrete Strength by the Maturity Method (C 1074).

The significance and use statement as given in ASTM C 1074 is reproduced here:

1. This procedure can be used to estimate the in-place strength of concrete to allow the start of critical construction activities such as: (1) removal of formwork and reshoring; (2) post-tensioning of tendons; and (3) termination of cold weather protection.
2. This procedure can be used to estimate strength of laboratory specimens cured under non-standard temperature conditions.
3. The major limitations of the maturity method are: (1) the concrete must be maintained in a condition that permits cement hydration; (2) the method does not take into account the effects of early-age concrete temperature on the long-term ultimate strength; and (3) this method needs to be supplemented by other indications of the potential strength of the concrete mixture.
4. The accuracy of the estimated strength depends on properly determining the maturity function for the particular materials used.

PULLOFF TESTS

In the mid-1970s, researchers in the United Kingdom developed a surface pulloff test to estimate the in situ strength of concrete [101]. Although some test data have been published, the test has found little acceptance outside the United Kingdom.

Briefly, the pulloff test involves bonding a circular steel disk to the surface of the concrete under test by means of any epoxy resin. Prior to this bonding, sandpaper is used to abrade the surface of the concrete to remove laitance, followed by degreasing the surface using a suitable solvent. After the epoxy has cured, a tensile force is then applied to the steel disk. Because the tensile strength of the bond is greater than that of concrete, the latter fails in tension. From the area of the disk and the force applied at failure, it is possible to obtain a measure of the tensile strength of concrete. The pulloff test is still in its infancy compared with other in-situ tests and, as of 1992, has been standardized only in the United Kingdom.

The pulloff test is relatively simple to perform, and stress at failure is a direct measure of the tensile strength of concrete. The test gives reproducible results and does not require planning in advance of placing the concrete. However, the serious disadvantage of the test is that it is confined to the testing of the surface layers of concrete.

COMBINED METHODS

In Europe in general and in Romania in particular, the use of more than one in situ/nondestructive testing technique to improve the accuracy of prediction of strength parameters of concrete has gained some credibility [102,103]. Some researchers have suggested the use of rebound hammer and pulse velocity techniques, while others have suggested the use of other combined methods. The proponents of this approach claim that the use of two methods, each measuring a different property, can overcome the limitation associated with the use of one method. Some case histories have been published supporting this claim; data disputing this claim are also available.

The use of more than one method may provide useful information in some instances but its general use is not advocated because of economy and time requirements, and the possible marginal increase in the accuracy of predicting compressive strength.

CONCLUDING REMARKS

Considerable progress has been made over the past two decades in the development and use of nondestructive methods for estimating strength of concrete. A number of these methods have been standardized by ASTM, ISO, and other organizations, and several new methods are in the process of being standardized. However, research is needed to standardize those tests that determine properties other than strength. Radar scanning and impact/pulse echo techniques appear to be the most promising of these type of tests.

In-situ/nondestructive tests may not be considered as a replacement for the standard cylinder test, but should be considered as additional techniques. When performed in conjunction with standard core tests, they can provide additional information and reduce the number of cores required for testing.

Unless comprehensive correlations have been established between the strength parameters to be predicted and the results of in-situ/nondestructive tests, the use of the latter to predict strength properties of concrete is not recommended.

REFERENCES

[1] Rayleigh, J. W., *Theory of Sound*, 2nd ed., Dover Press, New York, 1945.
[2] Obert, L. and Duvall, W. I., "Discussion of Dynamic Methods of Testing Concrete with Suggestions for Standardization," *Proceedings*, American Society for Testing and Materials, Vol. 41, 1941, p. 1053.
[3] Powers, T. C., "Measuring Young's Modulus of Elasticity by Means of Sonic Vibrations," *Proceedings*, American Society for Testing and Materials, Vol. 38, Part II, 1938, p. 460.
[4] Hornibrook, F. B., "Application of Sonic Method to Freezing and Thawing Studies of Concrete," *ASTM Bulletin*, No. 101, American Society for Testing and Materials, Dec. 1939, p. 5.
[5] Thomson, W. T., "Measuring Changes in Physical Properties of Concrete by the Dynamic Method," *Proceedings*, American Society for Testing and Materials, Vol. 40, 1940, p. 113; also, discussion by T. F. Willis and M. E. de Reus, pp. 1123–1129.
[6] Stanton, T. E., "Tests Comparing the Modulus of Elasticity of Portland Cement Concrete as Determined by the Dynamic (Sonic) and Compression (Secant at 1000 psi) Methods," *ASTM Bulletin*, No. 131, American Society for Testing and Materials, Dec. 1944, p. 17; also, discussion by L. P. Witte and W. H. Price, pp. 20–22.
[7] Malhotra, V. M., "Testing of Hardened Concrete: Nondestructive Methods," Monograph No. 9, American Concrete Institute, Detroit, MI, 1976, p. 52.
[8] Kesler, C. E. and Higuchi, Y., "Problems in the Sonic Testing of Plain Concrete," *Proceedings*, International Symposium on Nondestructive Testing of Material and Structures, Vol. 1, RILEM, Paris, 1954.
[9] Jones, R., *Non-Destructive Testing of Concrete*, Cambridge University Press, London, 1962.
[10] Pickett, G., "Equations for Computing Elastic Constants from Flexural and Torsional Resonant Frequencies of Vibration of Prisms and Cylinders," *Proceedings*, American Society for Testing and Materials, Vol. 45, 1945, p. 846.
[11] Spinner, S. and Tefft, W. E., "A Method of Determining Mechanical Resonance Frequencies and for Calculating Elastic Moduli from these Frequencies," *Proceedings*, American Society for Testing and Materials, Vol. 61, 1961, p. 1221.
[12] Jones, R., "The Effect of Frequency on the Dynamic Modulus and Damping Coefficient of Concrete," *Magazine of Concrete Research* (London), Vol. 9, No. 26, 1957, p. 69.
[13] Orchard, D. F., *Concrete Technology, Practice*, Vol. 2, Wiley, New York, 1962, p. 181.
[14] Swamy, N. and Rigby, G., "Dynamic Properties of Hardened Paste, Mortar, and Concrete," *Materials and Structures/Research and Testing*, Paris, Vol. 4, No. 19, 1971, p. 13.
[15] Gaidis, J. M. and Rosenberg, M., "New Test for Determining Fundamental Frequencies of Concrete," *Cement Concrete Aggregates*, Vol. 8, No. 2, 1986, p. 117.
[16] Kesler, C. E. and Higuchi, Y., "Determination of Compressive Strength of Concrete by using its Sonic Properties," *Proceedings*, American Society for Testing and Materials, Vol. 53, 1953, p. 1044.
[17] Shrivastava, J. P. and Sen, B., "Factors Affecting Resonant Frequency and Compressive Strength of Concrete," *Indian Concrete Journal*, Bombay, Vol. 37, No. 1, 1963, p. 27, and Vol. 37, No. 3, 1963, p. 105.
[18] *Handbook of Nondestructive Testing of Concrete*, V. M. Malhotra and N. J. Carino, Eds., CRC Press, Boca Raton, FL, 1991.
[19] Chefdeville, J., "Application of the Method toward Estimating the Quality of Concrete," *RILEM Bulletin*, Special Issue-Vibrating Testing of Concrete, Paris, No. 15, Aug. 1953, Part 2, p. 61.
[20] Swamy, R. N. and Al-Asali, M. M., "Engineering Properties of Concrete Affected by Alkali-silica Reaction," *ACI Materials Journal*, Vol. 85, No. 5, 1988, p. 367.
[21] Obert, L., "Measurement of Pressures on Rock Pillars in Underground Mines," RI, 3521, U.S. Bureau of Mines, 1940.
[22] Long, B. G. and Kurtz, H. J., "Effect of Curing Methods Upon the Durability of Concrete as Measured by Changes in the Dynamic Modulus of Elasticity," *Proceedings*, American Society for Testing and Materials, Vol. 43, 1943, p. 1051.
[23] Long, B. G., Kurtz, H. J., and Sandenaw, T. A., "An Instrument and a Technique for Field Determination of the Modulus of Elasticity of Concrete (Pavements)," *ACI Proceedings*, Vol. 41, 1945, p. 11.
[24] Leslie, J. R. and Cheesman, W. J., "An Ultrasonic Method of Studying Deterioration and Cracking in Concrete Structures," *Journal*, American Concrete Institute, 1949; *ACI Proceedings*, Vol. 53, 1953, p. 1043.
[25] Whitehurst, E. A., "A Review of Pulse Velocity Techniques and Equipment for Testing Concrete," *Proceedings*, Highway Research Board, Vol. 33, 1954, p. 226.
[26] Whitehurst, E. A., "Soniscope Tests Concrete Structures," *ACI Proceedings*, Vol. 47, 1951, p. 433.
[27] Jones, R., "The Non-Destructive Testing of Concrete," *Magazine of Concrete Research*, No. 2, June 1949.

[28] Jones, R., "The Testing of Concrete by an Ultrasonic Pulse Technique," *Proceedings*, Highway Research Board, Vol. 32, 1953, p. 258.

[29] Cheesman, W. J., "Dynamic Testing of Concrete with the Soniscope Apparatus," *Proceedings*, Highway Research Board, Vol. 29, 1949, p. 176.

[30] "Effects of Concrete Characteristics on the Pulse Velocity—A Symposium," *Bulletin No. 206*, Highway Research Board, Washington, DC, 1958

[31] Sturrup, V. R., Vecchio, F. J., and Caratin, H., "Pulse Velocity as a Measure of Concrete Compressive Strengths," *ACI SP 82*, V. M. Malhotra, Ed., American Concrete Institute, Detroit, MI, 1984, pp. 201–228.

[32] Cummings, N. A., Seabrook, P. T., and Malhotra, V. M., "Evaluation of Several Nondestructive Testing Techniques for Locating Voids in Grouted Tendon Ducts," *ACI SP-128*, V. M. Malhotra, Ed., American Concrete Institute, Detroit, MI, 1991, pp. 47–68.

[33] Sansalone, M. and Carino, N. J., "Stress Wave Propagation Methods," *Handbook on Nondestructive Testing of Concrete*, CRC Press, Boca Raton, FL, 1991, pp. 275–304.

[34] Carino, N. J., "Recent Developments in Nondestructive Testing of Concrete," *Advances in Concrete Technology*, V. M. Malhotra, Ed., CANMET, Ottawa, Canada, 1992, pp. 281–328.

[35] Schmidt, E., "The Concrete Test Hammer (Der Beton-Prufhammer)," *Schweizer Baublatt*, Zurich, Vol. 68, No. 28, 1950, p. 378.

[36] Schmidt, E., "Investigations with the New Concrete Test Hammer for Estimating the Quality of Concrete" (Versuche mit den neuen Beton-Prufhammer zur Qualitatsbestimmung des Beton), *Schweizer Archiv für angewandte Wissenschaft und Technik*, Solothurn, Vol. 17, No. 5, 1951, p. 139.

[37] Schmidt, E., "The Concrete Sclerometer," *Proceedings*, International Symposium Non-destructive Testing on Materials and Structures, Vol. 2, RILEM, Paris, 1954, p. 310.

[38] Kolek, J., "An Appreciation of the Schmidt Rebound Hammer," *Magazine of Concrete Research*, London, Vol. 10, Nos. 28 and 27, 1958.

[39] Malhotra, V. M., "Non-destructive Methods for Testing Concrete, Mines Branch," *Monograph No. 875*, Department of Energy, Mines and Resources, Ottawa, Canada, 1968.

[40] Malhotra, V. M., "Testing of Hardened Concrete: Nondestructive Methods," *Monograph No. 9*, American Concrete Institute, Detroit, MI, 1976, p. 188.

[41] Zoldners, N. G., "Calibration and Use of Impact Test Hammer," *ACI Proceedings*, Vol. 54, No. 2, 1957, p. 161.

[42] Akashi, T. and Amasak, S., "Study of the Stress Waves in the Plunger of a Rebound Hammer at the Time of Impact," *ACI SP-82*, V. M. Malhotra, Ed., American Concrete Institute, Detroit, MI, 1984, p. 17.

[43] Carette, G. G. and Malhotra, V. M., "In-situ Tests: Variability and Strength Prediction of Concrete at Early Ages," *ACI SP-82*, V. M. Malhotra, Ed., American Concrete Institute, Detroit, MI, 1984, p. 111.

[44] Greene, G. W., "Test Hammer Provides New Method of Evaluating Hardened Concrete," *ACI Proceedings*, Vol. 51, No. 3, 1954, p. 249.

[45] Klieger, P., Anderson, A. R., Bloem, D. L., Howard, E. L., and Schlintz, H., Discussion, "Test Hammer Provides New Method of Evaluating Hardeneded Concrete," Gordon W. Greene, *ACI Proceedings*, Vol. 51, No. 3, 1954, p. 256–1.

[46] Voellmy, A., "Examination of Concrete by Measurements of Superficial Hardness," *Proceedings*, International Symposium on Non-destructive Testing of Materials and Structures, RILEM, Paris, Vol. 2, 1954, p. 323.

[47] Kopf, R. J., "Powder Actuated Fastening Tools for Use in the Concrete Industry, Mechanical Fasteners for Concrete," *ACI SP-22*, American Concrete Institute, Detroit, MI, 1969, p. 55.

[48] Cantor, T. R., "Status Report on Windsor Probe Test System," *Proceedings*, 1970 Annual Meeting, Highway Research Board, Committee A2-03, Mechanical Properties of Concrete, Washington, DC, 1970.

[49] Freedman, S., "Field Testing of Concrete Strength," *Modern Concrete*, Vol. 14, No. 2, 1969, p. 31.

[50] Gaynor, R. D., "In-Place Strength of Concrete-A Comparison of Two Test Systems," *Proceedings*, 39th Annual Convention, National Ready Mixed Concrete Association, New York, 28 Jan. 1969; NRMCA Technical Information Letter No. 272, 4 Nov. 1969.

[51] Law, S. M. and Burt, W. T., III, "Concrete Probe Strength Study," Research Report No. 44, Research Project No. 68-2C(b), Louisiana HPR (7), Louisiana Department of Highways, Baton Rouge, LA, Dec. 1969.

[52] Arni, H. T., "Impact and Penetration Tests of Portland Cement Concrete," *Highway Research Record*, No. 378, 1972, p. 55.

[53] Arni, H. T., "Impact and Penetration Tests of Portland Cement Concrete," Rep. No. FHWA-RD-73-5, Federal Highway Administration, Washington, DC, 1973.

[54] Malhotra, V. M., "Preliminary Evaluation of Windsor Probe Equipment for Estimating the Compressive Strength of Concrete," Mines Branch Investigation Report IR 71-1, Department of Energy, Mines and Resources, Ottawa, Canada, Dec. 1970.

[55] Malhotra, V. M. and Painter, K. P., "Evaluation of the Windsor Probe Test for Estimating Compressive Strength of Concrete," Mines Branch Investigation Report IR 71-50, Department of Energy, Mines and Resources, Ottawa, Canada, 1971.

[56] Malhotra, V. M., "Evaluation of the Windsor Probe Test for Estimating Compressive Strength of Concrete," *Materials and Structures*, RILEM, Paris, Vol. 7, No. 37, 1974, pp. 3–15.

[57] Klotz, R. C., "Field Investigation of Concrete Quality Using the Windsor Probe Test System," *Highway Research Record*, Vol. 378, 1972, p. 50.

[58] Keeton, J. R. and Hernadez, V., "Calibration of Windsor Probe Test System for Evaluation of Concrete in Naval Structures," Technical Note N-1233, Naval Civil Engineering Laboratory, Port Hueneme, CA, 1972.

[59] Clifton, J. R., "Non-Destructive Tests to Determine Concrete Strength—A Status Report," NBSIR 75-729, National Bureau of Standards, Washington, DC.

[60] Bowers, D. G. G., "Assessment of Various Methods of Test for Concrete Strength," Connecticut Department of Transportation/Federal Highway Administration, Dec. 1978; available through National Technical Information Service, NTIS No. PB 296317, Springfield, VA.

[61] Bartos, M. J., "Testing Concrete in Place," *Civil Engineering*, American Society of Civil Engineers, 1979, p. 66.

[62] Strong, H., "In-Place Testing of Hardened Concrete with the Use of the Windsor Probe," New Idaho Test Method T-128-79, Division of Highways, State of Idaho, 1979, p. 1.

[63] Swamy, R. N. and Al-Hamed, A. H. M. S., "Evaluation of the Windsor Probe Test to Assess In-Situ Concrete Strength," *Proceedings*, Institute of Civil Engineering, Part 2, June 1984, 1967.

[64] Keiller, A. P., "A Preliminary Investigation of Test Methods for the Assessment of Strength of In-Situ Concrete," Tech-

nical Report No. 551, Cement and Concrete Association, Wexham Springs, Sept. 1982.

[65] Bungey, J. H., *The Testing of Concrete in Structures*, Chapman and Hall, New York, 1982.

[66] Keiller, A. P., "Assessing the Strength of In-Situ Concrete," *Concrete International*, Feb. 1985, p. 15.

[67] Malhotra, V. M. and Carette, G. G., "Penetration Resistance Methods," *Handbook on Nondestructive Testing of Concrete*, CRC Press, Boca Raton, FL, 1991, pp. 19–39.

[68] "In-place Methods for Determination of Strength of Concrete," *ACI Journal of Materials*, Vol. 85, No. 5, 1988, p. 446.

[69] Malhotra, V. M. and Carette, G. G., "In-Situ Testing for Concrete Strength," CANMET Report 79-30, Energy, Mines and Resources Canada, Ottawa, Canada, May 1979.

[70] Kopf, R. J., Cooper, C. G., and Williams, F. W., "In-Situ Strength Evaluation of Concrete Case Histories and Laboratory Investigations," *Concerete International*, March 1981, p. 66.

[71] Lee, S. L., Tam, C. T., Paramasivam, P., Ong, K. C. G., Swaddiwudhipong, S., and Tan, K. H., "Structural Assessment in In-Situ Testing and Interpretation of Concrete Strength," Department of Civil Engineering, National University of Singapore, July 1988.

[72] Nasser, K. W. and Al-Manaseer, A., "New Non-Destructive Test for Removal of Concrete Forms," *Concrete International*, Vol. 9, No. 1, 1987, p. 41.

[73] Nasser, K. W. and Al-Manaseer, A., "Comparison of Non-Destructive Testers of Hardened Concrete," *ACI Journal of Materials*, Vol. 84, No. 5, 1987, p. 374.

[74] Skramtajew (sic), B. G., *ACI Proceedings*, Vol. 34, No. 3, Jan.–Feb. 1938, pp. 285–303; Discussion, pp. 304–305.

[75] Malhotra, V. M., "Evaluation of the Pull-out Test to Determine Strength of In-Situ Concrete," *Materials and Structures*, RILEM, Vol. 8, No. 43, Jan.–Feb. 1975, pp. 19–31.

[76] Bickley, J. A., "Evaluation and Acceptance of Concrete Quality by In-place Testing," *ACI SP-82*, V. M. Malhotra, Ed., American Concrete Institute, Detroit, MI, 1984, pp. 95–110.

[77] Tremper, B., "The Measurement of Concrete Strength by Embedded Pull-Out Bars," *Proceedings*, American Society for Testing and Materials, Vol. 44, 1944, pp. 880–887.

[78] Tassios, T. P., "A New Nondestructive Method of Concrete Strength Determination," *Publication No. 21*, National Technical University, Athens, 1968.

[79] Kierkegaard-Hansen, P., "Lok-Strength," *Nordisk Betong*, No. 3, 1975, p. 19.

[80] Richards, O., "Pull-out Strength of Concrete," *Proceedings*, Research Session, Annual Meeting, American Concrete Institute, Dallas, TX, 1972.

[81] Carino, N. J., "Pullout Tests," *Handbook on Nondestructive Testing of Concrete*, V. M. Malhotra and N. J. Carino, Eds., CRC Press, Boca Raton, FL, 1991, pp. 39–82.

[82] Chabowski, A. J. and Bryden-Smith, D. W., "Assessing the Strength of Concrete of In-Situ Portland Cement Concrete by Internal Fracture Tests," *Magazine of Concrete Research*, Vol. 32, No. 112, 1980, p. 164.

[83] Mailhot, G., Bisaillon, A., Malhotra, V. M., and Carette, G., "Investigations into the Development of New Pullout Techniques for In-situ Strength of Determination of Concrete," *ACI Journal*, Vol. 76, No. 2, 1979, p. 1267.

[84] Johansen, R. "In-situ Strength of Concrete, the Breakoff Method,"*Concrete International*, 1979, p. 45.

[85] Carlsson, M., Eeg, I. R., and Janer, P., "Field Experience in the Use of the Break-off Tester," *In Situ/Nondestructive Testing of Concrete, ACI SP-82*, V. M. Malhotra, Ed., American Concrete Institute, Detroit, MI, 1984, pp. 277–292.

[86] Dahl-Jorgenson, E. and Johansen, R., "General and Specialized Use of the Break-off Concrete Strength Test Method," *In Situ/Nondestructive Testing of Concrete*, ACI SP-15, V. M. Malhotra, Ed., American Concrete Institute, Detroit, MI, 1984, pp. 293–308.

[87] Naik, T., "The Break-off Test Method," *Handbook on Nondestructive Testing of Concrete*, V. M. Malhotra and N. J. Carino, Eds., CRC Press, Boca Raton, FL, 1991, pp. 83–100.

[88] McIntosh, J. D., "Electrical Curing of Concrete," *Magazine of Concrete Research*, Vol. 1, No. 1, 1949, pp. 21–28.

[89] Nurse, R. W., "Steam Curing of Concrete," *Magazine of Concrete Research*, Vol. 1, No. 2, 1949, pp. 79–88.

[90] Saul, A. G. A., "Principles Underlying the Steam Curing of Concrete of Atmospheric Pressure," *Magazine of Concrete Research*, Vol. 2, No. 6, 1951, pp. 127–140.

[91] Bergstrom, S. G., "Curing Temperature, Age and Strength of Concrete," *Magazine of Concrete Research*, Vol. 5, No. 14, 1953, pp. 61–66.

[92] Plowman, J. M., "Maturity and the Strength of Concrete," *Magazine of Concrete Research*, Vol. 8, No. 22, 1956, pp. 13–22.

[93] Malhotra, V. M., "Maturity Concept and the Estimation of Concrete Strength," Information Circular IC 277, Department of Energy, Mines and Resources, Ottawa, Canada, Nov. 1971.

[94] Freiesleben Hansen, P. and Pedersen, E. J., "Maturity Computer for Controlled Curing and Hardening of Concrete," *Nordisk Beton*, Vol. 1, No. 19, 1977.

[95] Byfors, J., "Plain Concrete at Early Ages," Report 3:80, Swedish Cement and Concrete Research Institute, Stockholm, 1980.

[96] Naik, T. R., "Maturity Functions, Concrete Curred During Winter Conditions," *Temperature Effects on Concrete, ASTM STP 858*, T. R. Naik, Ed., American Society for Testing and Materials, Philadelphia, 1985, p. 107.

[97] Carino, N. J., "The Maturity Method," *Handbook on Nondestructive Testing of Concrete*, V. M. Malhotra and N. J. Carino, Eds., CRC Press, Boca Raton, FL, 1991, pp. 101–146.

[98] Swenson, E. G., "Estimation of Strength of Concrete," *Engineering Journal*, Canada, Vol. 50, No. 9, 1967, pp. 27–32.

[99] Hudson, S. B. and Steele, G. W., "Developments in the Prediction of Potential Strength of Concrete from Results of Early Tests," *Transportation Research Record*, No. 558, Transportation Research Board, Washington, DC, 1975.

[100] Malhotra, V. M., "Maturity Strength Relations and Accelerated Strength Testing," Canada Mines Branch Internal Report, MPI (P) 70-29, Department of Energy, Mines and Resources, Ottawa, Canada, 1970.

[101] Long, A. E. and McMurray, A., "The Pull-off Partially Destructive Test," *ACI SP-82*, V. M. Malhotra, Ed., American Concrete Institute, Detroit, MI, 1984, pp. 327–350.

[102] Samarin, A., "Combined Methods," *Handbook on Nondestructive Testing of Concrete*, V. M. Malhotra and N. J. Carino, Eds., CRC Press, Boca Raton, FL, 1991, pp. 189–202.

[103] Malhotra, V. M., "In-Situ Nondestructive Testing of Concrete-A Global Review," *ACI SP-82*, V. M. Malhotra, Ed., American Concrete Institute, Detroit, MI, 1984, pp. 1–16.

PART IV
Concrete Aggregates

Petrographic Evaluation of Concrete Aggregates

Richard C. Mielenz[1]

PREFACE

Petrographic examination of concrete aggregate was included in the three previous editions of ASTM Technical Publication, *ASTM STP 169*, *ASTM STP 169A*, and *ASTM STP 169B*, under the present authorship. The construction and content of this chapter is like that of the former presentations except that a substantial addition treats petrographic examination of recycled concrete for use as aggregate in new construction. The properties of recycled concrete are covered in some detail so as to aid the petrographer in decision on investigation of constructions to be demolished with intent to produce concrete for recycling and on observations that should be made as a part of the examination of the finished aggregate. Experience and literature pertinent to examination of all types of concrete aggregate are cited as appropriate.

INTRODUCTION

Petrographic examination of concrete aggregate is visual examination and analysis in terms of both lithology and properties of the individual particles. The procedure requires use of a hand lens and petrographic and stereoscopic microscopes. Less commonly, X-ray diffraction, differential thermal analysis, or electron microscopy is used to supplement the examination using optical microscopes. By petrographic examination, the relative abundance of specific types of rocks and minerals is established; the physical and chemical attributes of each, such as particle shape, surface texture, pore characteristics, hardness, and potential chemical reactivity, are described; coatings are identified and described; and the presence of contaminating substances is determined, each in relation to proposed or prospective conditions of service in concrete constructions.

As will be discussed subsequently, petrographic examination contributes in several ways to the investigation, selection, testing, and control of quality of aggregates. Consequently, the method is progressively being applied more widely. Since 1936, all aggregates used in concrete construction by the Bureau of Reclamation, United States Department of the Interior, have been examined petrographically as part of the basis for their selection [1,2]. The method has been applied similarly by the Corps of Engineers since before 1940 [3]. Coarse aggregates proposed for use in either portland-cement concrete or bituminous concrete are examined by petrographic methods in laboratories of the Ministry of Transportation–Ontario, and specifications governing acceptance are based upon the results [4]. Petrographic examination is performed also by several other agencies of the U.S. government and state departments of transportation, and may be obtained through some testing and engineering laboratories.

In 1952, ASTM Tentative Recommended Practice for Petrographic Examination of Aggregates for Concrete (C 295) was accepted, and was adopted as a standard in 1954. Minor modifications were made subsequently, most recently in 1985. The document has been published as the Guide for Petrographic Examination of Aggregates for Concrete in the *1991 Annual Book of ASTM Standards*. Petrographic examination is also cited in ASTM Specification for Concrete Aggregates (C 33).

The abundant data obtained and the rapidity with which petrographic examination can be completed justify more general use of the method in investigation, selection, manufacture, and use of concrete aggregates. However, in all cases, the petrographer should be supplied with available information on the conditions of service to which the concrete is to be exposed.

This paper summarizes the objectives and applications of petrographic examination of aggregates with reference to gravel, sand, crushed stone, blast-furnace slag, the most common types of lightweight aggregates, and aggregates produced by recycling of hardened concrete from constructions. Special attention has been given to developments in this field since publication of *ASTM STP 169B* in 1979. Techniques of the examination are treated briefly because instructions have been published [3–11] and are included in ASTM C 295.

PURPOSE OF THE PETROGRAPHIC EXAMINATION OF CONCRETE AGGREGATES

Preliminary Determination of Quality

Preliminary petrographic examination of concrete aggregate is performed either in the field or in the laboratory as an adjunct to geologic examination, exploration,

[1] President, Richard C. Mielenz, P. E., Inc., Gates Mills, OH 44040-9706.

mapping, and sampling. The examination assists the geologist or materials engineer in determining the extent to which consideration of an undeveloped deposit is justified. Also, the preliminary petrographic examination indicates the relative quality of aggregates from alternate sites. By revealing variations in the material, examination of exposures or cores from pilot drill holes establishes the minimum program of exploration and sampling necessary for acceptance or rejection of the deposit.

Establishing Properties and Probable Performance

Petrographic examination is primarily a supplement to the acceptance tests. Probable performance of concrete aggregate is estimated in two general ways by petrographic examination. First, the examination reveals the composition and physical and chemical characteristics of the constituents. From this information, the probable response of the aggregate to such phenomena as attack by cement alkalies, freezing-thawing, wetting-drying, heating-cooling, or high temperatures usually can be estimated. The rapidity with which the petrographic examination predicts potential alkali reactivity of aggregate is especially valuable because of the widespread occurrence of this condition and the long time commonly required by realistic tests of concrete or mortar.

Second, petrographic examination establishes the fundamental nature of aggregates so that aggregates from unfamiliar sources can be compared with aggregates upon which information is available. This application is discussed subsequently.

Correlation of Samples with Aggregates Previously Tested or Used

Detailed petrographic examination is the only procedure that permits comparison and correlation of samples with aggregates previously used or tested. Thus, data and experience previously obtained by use and long-time tests of similar aggregates can be applied in the selection of materials proposed for current work, even though the materials come from new sources.

By relating the sample to aggregates previously used in construction, aggregate indicated to be unsound by standard tests may be found adequate, or conversely, aggregate indicated to be sound in standard tests might be found unsatisfactory for the intended use.

For example, gravel in certain deposits near Jackson, Michigan, meets usual specification requirements for soundness, abrasion resistance, and content of soft particles, yet produces objectionable popouts in pavements after exposure for two winters (Fig. 1). Petrographic examination of the gravel and of individual particles producing popouts during service has identified the unsound rock types and the critical minimum size required for failure. Examination of proposed materials will demonstrate the presence or absence of such particles and will thus indicate, in the light of other data, whether the aggregate should be accepted, rejected, or subjected to special tests.

FIG. 1—Popouts produced by claystone, shale, and chert in concrete pavement near Jackson, Michigan (courtesy of Michigan Department of Transportation).

Petrographic examination of aggregate may be a part of an investigation of concrete constructions as described in the ASTM Practice for Examination and Sampling of Hardened Concrete in Constructions (C 823) and ASTM Practice for Petrographic Examination of Hardened Concrete (C 856). For example, the examination may be applied to determine the source of aggregate used in the constructions, such as whether it was obtained from an approved or not approved source.

Selecting and Interpreting Other Tests

All of the properties of aggregates influencing performance of concrete in service ordinarily are not evaluated by test before selection of the aggregate to be used in the work, primarily because of cost and time required. Such factors as thermal properties and volume change with wetting-drying rarely are determined. Other properties, such as chemical reactivity or effect of the aggregate on the freeze-thaw resistance of concrete, usually are not determined. Consequently, it is worthwhile to apply a procedure by which the relative significance of such properties can be determined and the need for supplementary quantitative tests indicated.

Petrographic examination aids interpretation of other tests. For example, are the particles identified as clay lumps in accordance with ASTM Test Method for Clay Lumps and Friable Particles in Aggregates (C 142) indeed clay lumps or are they merely friable or pulverulent particles of some other composition? What is the cause of unexpected failure of concrete specimens in freezing and thawing? Is it the presence of unsound particles that do not disintegrate in the sulfate soundness test, yet expand in freezing and thawing? Is it the result of an alkali-aggregate reaction? Is failure in the soundness or abrasion test the result of complete breakdown of a small proportion of unsound or soft particles or partial disintegration of the greater proportion of the aggregate?

Detection of Contamination

Petrographic examination is the best method by which potentially deleterious and extraneous substances can be detected and determined quantitatively. Inadvertent contamination with natural substances, industrial products, or wastes, such as overburden or from trucks or railroad cars not properly cleaned of previous cargo, may decrease the quality of aggregate markedly. Contamination by containers may invalidate samples. Such substances as clay, soil, coal, vegetable matter, chemical fertilizers, petroleum products, or refractories containing free calcium or magnesium oxides are especially important. Incomplete processing of synthetic aggregates may contaminate the finished product with raw or partly fired materials or coal.

Undesirable substances inherent in the material, such as coatings, clay, plant remains, coal, and soluble salts, are detected easily and can be determined quantitatively by petrographic methods or by other procedures that can be selected most definitively pursuant to petrographic identification.

Determining Effects of Processing

Petrographic examination aids in production and processing of aggregate. The relative merit of alternative processing methods and equipment can be determined quickly by comparison of the original material with the processed aggregate. Comparison can be based on particle shape, content of unsound or chemically reactive constituents, removal of coatings, or production of rock dust.

The feasibility of beneficiation by removal of unsound or deleterious constituents depends on the properties of the particles and their abundance. By petrographic examination, the undesirable particles can be identified and separated. Their properties can be evaluated and compared with properties of the remainder of the aggregate. If the undesirable particles are unusually soft, friable, dense, lightweight, or high in magnetic susceptibility, separation may be feasible on a commercial scale.

Petrographic examination can be used to control the manufacture of synthetic aggregate, such as expanded shale or clay, perlite, slag, and other types. Microscopical examination will reveal the presence of raw or underburned materials, alkali-reactive phases, and contaminants, such as coal (Fig. 2). X-ray diffraction analysis is the most dependable means to identify and determine quantitatively the proportion of crystalline phases, such as periclase (MgO), free lime (CaO), and clays whose crystal structure was not destroyed by the calcination. Differential thermal analysis can be used to estimate the effective temperature to which clays, shales, slate, and similar materials have been fired; the trace of the differential thermal curves for the raw and fired product will coincide above the effective temperature achieved in the firing operation.

FIG. 2—Coal in sintered clay before crushing (courtesy of Bureau of Reclamation, U.S. Department of the Interior).

PERFORMANCE OF THE PETROGRAPHIC EXAMINATION

Samples for Petrographic Examination

Samples of aggregate for petrographic examination should be representative of the source. Recommended procedures for sampling and for preparation of the sample for analysis are covered by ASTM Practices for Sampling Aggregates (D 75), ASTM C 295, and technical literature [12].

Examination in the Field

Petrographic analysis of samples in the field is usually qualitative or only semiquantitative because lack of facilities makes detailed work difficult. However, detailed examination in the field may be warranted if aggregate retained on the 75-mm (3-in.) sieve is to be used in the work, inasmuch as a representative sample of this size fraction may weigh a hundred or more kilograms and transportation of the sample to the laboratory is costly. Chips from cobbles not adequately identified in the field should be taken to the laboratory for further study.

If the deposit or rock exposure is variable, samples should be selected from each zone, and detailed notes made at the site should relate each sample to a particular zone and portion of the deposit or rock ledge. The relative proportion of unsound, fractured, or chemically deleterious materials should be estimated from measurements made at exposures. These notes and the results of the tests on the samples will be the basis for operation of the deposit inasmuch as it may be desirable to waste or avoid zones or portions containing inferior, unsuitable, or alkali-reactive materials.

Examination in the Laboratory

Petrographic examination of aggregate in the laboratory may be brief or detailed. Brief examination indicates the relative merit of materials from alternate sources and supplies justification for abandonment or continued investigation of undeveloped deposits or other exposures. The preliminary examination should not replace the quantitative analysis included in the program of acceptance tests.

Samples supplied to the laboratory comprise: (a) granular materials, such as gravel, sand, crushed stone, slag, or synthetic aggregate; (b) stone in quarried blocks and irregularly shaped pieces or drilled cores; and (c) fragments, sawed pieces, or drilled cores of concrete that contain the aggregate (coarse, fine, or both) that is of interest. For fully graded granular materials, the examination should be performed on at least three size fractions included in the aggregate. The sample of each size fraction should comprise at least 150 particles. For natural sand and gravel and crushed stone, the minimum representative samples are shown in Table 1.

Samples of the aggregate for petrographic examination are obtained by sieving in accordance with the ASTM Method for Sieve Analysis of Fine and Coarse Aggregates (C 136) and ASTM Test Method for Material Finer Than 75-μm (No. 200) Sieve in Mineral Aggregates by Washing (C 117). The fractions are quartered or, for fine aggregate, split repeatedly on an appropriate riffle.

Details of the procedure are outlined by K. Mather and B. Mather [3] and in ASTM C 295.

During the analysis, helpful information on identity and physical condition can be obtained by recording such features as: (1) friability or pulverulence in the fingers; (2) resonance when struck; (3) ease of fracturing; (4) nature of the fracture surface and fracture fillings; (5) odor on fresh fracture; (6) color and its variation; (7) internal structure, such as porosity, granularity, seams, and veinlets; (8) reaction to water, such as absorption of droplets on fresh fracture, evolution of air on immersion, capillary suction against the tongue, slaking, softening, or swelling; and (9) differential attack by acids or other media.

Fractions retained on the 600-μm (No. 30) sieve are best identified, examined, and counted under the stereoscopic microscope. The analysis can be made conveniently by traversing a representative portion of each fraction by means of a mechanical stage and microscope assembly like that employed in the point-count procedure according to ASTM Practice for Microscopical Determination of Air-Void Content and Parameters of the Air-Void System in Hardened Concrete (C 457) (Fig. 3). The mineralogic composition of finer fractions usually is determined most easily and accurately in immersion oils or thin sections under the petrographic microscope. However, better continuity

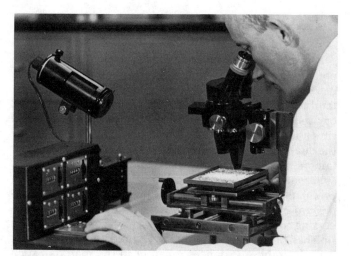

FIG. 3—Performing petrographic examination of fine aggregate with stereoscopic microscope, mechanical stage, and tally counter.

in description of physical characteristics of the particles is obtained if analysis of fractions passing the 600-μm (No. 30) sieve and retained on the 150-μm (No. 100) sieve is performed under the stereoscopic microscope.

Thin sections occasionally are necessary in examination of natural aggregate. They usually are employed in the study of quarried stone. However, preparation of plane sections by sawing and lapping, with or without etching or staining, commonly is preferable because of the large area made available for examination or analysis. Such examination should be supplemented by microscopical examination of grain mounts in immersion oils. Thin sections or polished surfaces, supplemented by use of oil immersion mounts of granular material, usually are used in the analysis of blast-furnace slag aggregates. X-ray diffraction and differential thermal analysis may be required to identify or determine quantitatively constituents that are very finely divided or dispersed through particles of the aggregate (Fig. 4).

Any of the previously noted types of coarse or fine aggregate can be examined while enclosed in hardened concrete, using broken fragments, sawed and lapped sections, and microscopical thin sections, as required and appropriate.

Ordinarily, petrographic analysis of concrete aggregates is performed on at least three coarse fractions and five fine fractions. The results may be used to compute the lithologic composition of the aggregate in any gradation comprising the analyzed size fraction. The bulk composition of a concrete sand may vary widely with fineness modulus. Occasionally, one analysis may be performed on a graded aggregate. For such analysis, the composition by count may differ greatly from the composition by mass.

Numerical results of the analysis may be expressed by mass or count of particles retained on the 600-μm (No. 30) sieve, but for the finer fractions, the results are based only upon count of grains unless a correction factor is applied. Consequently, consistent analyses of all fractions of fine aggregate and coarse aggregate can be reported

TABLE 1—Minimum Representative Sample Size.

Size Fraction, mm (in.)	Weight of 150 Particles
90 to 38 (3 to 1½)	13 kg (20 lb)
38 to 19 (1½ to ¾)	4.3 kg (9.5 lb)
19 to 9.5 (¾ to ⅜)	0.6 kg (1.3 lb)
9.5 to 4.7 (⅜ to ³⁄₁₆)	0.17 kg (0.38 lb)
4.75 to 2.36 (No. 4 to 8)	7.5 g
2.36 to 1.18 (No. 8 to 16)	1.1 g
1.18 to 0.60 (No. 16 to 30)	0.14 g
0.60 to 0.30 (No. 30 to 50)	0.017 g
0.30 to 0.15 (No. 50 to 100)	0.0033 g

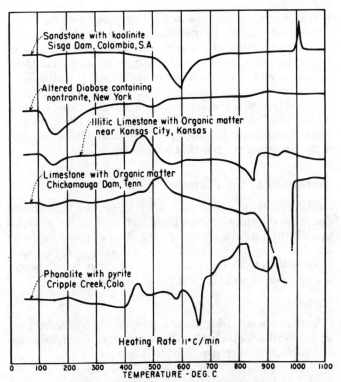

FIG. 4—Typical differential thermal analysis (DTA) records obtained on concrete aggregates. Kaolinite is indicated by the endotherm (downward shift) at 990 to 1025°C. Nontronite is revealed by endotherms at 100 to 350°C and 450 to 550°C. Illite produces the small endotherms at 100 to 200°C and 500 to 615°C. Organic matter produces large exotherms (upward shifts) at 430 to 500°C or 440 to 600°C. Pyrite (ferrous sulfide, FeS_2) develops a marked exotherm at 400 to 485°C (courtesy of the Bureau of Reclamation, U.S. Department of the Interior).

Observations Included in the Petrographic Examination

In reporting the results of the petrographic examination, the petrographer should supply information on the following subjects as necessary for evaluation of the aggregate for service under the anticipated conditions of exposure in the concrete:

Mineralogic and lithologic composition
Particle shape
Surface texture
Internal fracturing
Coatings
Porosity, permeability, and absorption
Volume change, softening, and disintegration with wetting-drying
Thermal properties
Strength and elasticity
Density
Hardness
Chemical activity
    Solubility
    Oxidation
    Hydration
    Carbonation
    Alkali-silica reactivity
    Alkali-carbonate reactivity
    Sulfate attack on cementitious matrix
    Staining
    Cation-exchange reactions
    Reactions of organic substances
    Effects of contaminants

The significance of these properties is discussed by Rhoades and Mielenz [7,8], Dolar-Mantuani [11], Swenson and Chaly [9], Hansen [13], Mielenz [14], and others [15–19]. Sarkar and Aitcin [20] emphasize the need for petrographic examination of aggregates for very high-strength concrete, especially regarding internal texture and structure of the particles.

A simple test to detect forms of pyrite and similar ferrous sulfides that are likely to oxidize while enclosed in concrete was devised by Midgley [21] and is discussed more generally by Mielenz [14]. Shayan [22] reported popout formation and deterioration of the surface of a concrete floor by oxidation and hydration of pyrite included in exposed particles of coarse aggregate.

Soles [23] found that pyritiferous dolomite used as coarse aggregate produces distress when the concrete was heated to about 150°C (300°F) but less than 300°C (575°F). The deterioration is a result of expansion of the particles as oxidation and hydration of the pyrite and crystallization of iron and calcium sulfates took place. The present author has observed similar effects when compact pyritiferous shales constitute appreciable proportions of gravel and sand coarse and fine aggregate. These observations are pertinent to the evaluation of aggregates to be incorporated in concrete that will be subjected to sustained high temperatures in the indicated range, such as structural members and slabs that are included in or support furnaces, kilns, and the like.

only by count. Analysis by count is the more appropriate determination technically because the influence of particles of given type upon performance of concrete in test or service depends primarily upon their frequency and distribution in the mass. When applied to coarse aggregate only, the petrographic analysis is more rapid if the relative proportion of the several rock types or facies is determined by mass. However, the accuracy and precision of analysis by mass are affected adversely by unavoidable loss of portions of the particles when they must be broken to allow identification of the composition and description of the physical condition. Specifications on lithologic composition preferably are expressed as percentage by count of particles.

Details of calculating and reporting of results of petrographic examination and analysis of concrete aggregates are summarized in ASTM C 295. The results of analysis of composition of aggregates enclosed in hardened concrete, such as by point-count analysis on sawed and lapped surfaces, are expressed as percent by volume of the concrete or percent by volume of total aggregate. Analysis of aggregate enclosed in concrete usually is confined to determination of the proportional volume of individual selected types of particles relative to the proportional volume of the total aggregate (see subsequent discussion).

Stark and Bhatty [24] concluded that significant amounts of alkalies (sodium and potassium) can be released from certain alkali-bearing aggregates when in contact with a saturated solution of calcium hydroxide and that such released alkalies engage in expansive alkali-silica reaction in concrete. The release appears to involve ion-exchange phenomena and possibly dissolution of silicates within the aggregate. Sources of alkalies include glassy volcanic rocks, plagioclase feldspars, potassium feldspar (microcline), and feldspathic and quartzose sand. The release of potassium from phlogopite mica [25] and from particles of bentonite clay in natural sand [26] has been shown to promote an alkali-carbonate reaction in concrete. Similar effects can be produced by montmorillonite-type clays and zeolites from which sodium or potassium released by cation exchange may engage in alkali-silica reaction involving other siliceous constituents of the aggregate [27].

Schmitt [28] recently summarized information on micas as constituents of fine aggregate in concrete, citing work done in England whereby it was concluded that for each 1% of mica in fine aggregate by mass of total aggregate, the compressive strength of the concrete is reduced as much as 5%. Frost resistance is reduced and drying shrinkage is increased by increasing content of biotite. He commented also on applicable specifications on allowable quantities of mica in coarse and fine aggregate. Occurrence of mica in concrete sands from numerous sources in the continental United States is reported by Gaynor and Meininger [29].

Higgs [30] reported a correlation of methylene blue adsorption and the amount of smectite (montmorillonite-type) clay in natural sand that was the cause of undue slump loss and increased drying shrinkage of concrete, citing work by Davis et al. [31]. Elsewhere, he described the effect of chlorite in aggregate relative to its effect on the freeze-thaw durability of concrete [32]. The critical criteria required for poor durability were found to be chlorite content of volcanic rocks in excess of 20% and distribution of the mineral in seams, clots, and fissures in contrast to a disseminated condition in the rock.

Petrographic criteria useful in detecting dolomitic rocks that are potentially susceptible to a deleterious degree of the alkali-carbonate reaction in concrete are described by Dolar-Mantuani [11], Hadley [33], Gillott [34], and others [35–38]. Dolar-Mantuani [39] described the alkali-silica reactivity of argillites and graywackes. Bachiorrini and Montanaro [40] reported the occurrence of alkali-silica reaction in non-dolomitic carbonate rocks containing disseminated, finely divided silicate minerals. The alkali reactivity of strained quartz and quartzose rocks has been described and discussed by Buck [41] and others [6,11,42]. Investigators of concrete affected by alkali-silica reaction in many structures in Nova Scotia have concluded that alkali-silica reaction in which strained quartz is the primary alkali-reactive constituent has been augmented substantially by an alkali-aggregate reaction involving vermiculite that is present in phyllites, argillites, and graywackes that are widespread in natural aggregates of that region [43].

It is intended that the preceding properties be determined qualitatively by observation of mineralogic composition and texture and internal structure of the particles by petrographic techniques and simple tests, such as those noted, as appropriate. When other tests are available, or if the particles can be compared petrographically with previously tested materials, the properties can be evaluated semiquantitatively or quantitatively. The petrographer should participate in the formulation of any supplementary studies that are contemplated.

Condition of the Particles

The following classification of properties is useful in cataloging the physical and chemical condition of particles constituting an aggregate. Physical condition is defined by three terms: (1) satisfactory, (2) fair, and (3) poor. Chemical stability in concrete is designated by two terms: (1) innocuous and (2) potentially deleterious.

Satisfactory

Particles are hard to firm and relatively free from fractures, capillary absorption is very small or negligible, and the surface texture is relatively rough.

Fair

Particles exhibit one or two of the following qualities: firm to friable, moderately fractured, capillary absorption small to moderate, surface relatively smooth and impermeable, very low compressibility, coefficient of thermal expansion approaching zero or being negative in one or more directions.

Poor

Particles exhibit one or more of the following qualities: friable to soft or pulverulent, slake when wetted and dried, highly fractured, capillary absorption high, marked volume change with wetting and drying, combine three or more qualities listed under "fair."

Innocuous

Particles contain no constituents that will dissolve or react chemically to a significant extent with constituents of the atmosphere, water, or hydrating portland cement while enclosed in concrete or mortar under ordinary conditions of service in constructions. Particles are stable at high temperature or decompose without expansion.

Potentially Deleterious

Particles contain one or more constituents in significant proportion that are known either to react chemically under conditions ordinarily prevailing (or applicable in the present instance) in portland-cement concrete or mortar in such a manner as to produce significant expansion, interfere with the normal course of hydration of portland cement, or supply substances that might produce harmful effects upon mortar or concrete. By extrapolation, this category is extended to include individual constituents that produce notable expansion under conditions that are expected in the proposed work, such as in concrete to be exposed to high temperatures as a part of the planned service regimen, of which a prime example is quartz and highly quartzose rock types that are subject to disruption as the quartz crystal expands about 2.4% by volume while inversion takes place from the α- to the β-polymorph at 573°C (1063°F) [44].

Coatings should be reported and evaluated separately because coatings usually are confined to portions of a deposit and, for crushed stone, the nature and abundance of coatings vary with processing methods and equipment. Schmitt [28] recently summarized information concerning petrographic features of aggregate coatings and their significance.

Similarly, particle shape should be considered apart from other aspects of physical quality because particle shape commonly is subject to control or modification by processing of the aggregate. Unless otherwise defined by applicable specifications, particles whose length is five or more times their width should be designated as "elongated pieces," and those having a ratio of width to thickness greater than five should be designated as "flat pieces." The measurement should be made in accordance with definitions of these terms as they appear in ASTM Terminology Relating to Concrete and Concrete Aggregates (C 125). Particle shape relative to coarse aggregate is discussed by Sarker and Aitcin [20], noting in particular the unfavorable effect of elongated or flaky pieces on workability of concrete. See Gaynor and Meininger [29] and Mass [45] for helpful discussions of the significance of the particle shape of concrete sands.

Petrographic Examination of Aggregates in Hardened Concrete

General

Aggregates enclosed in hardened concrete from structures or pavements in service or from test specimens can be subjected to petrographic examination for any of many reasons, such as initially to identify the nature of the aggregates in older structures under investigation or to determine the involvement of the aggregates in either satisfactory performance or deleterious activity. The following features can be investigated during such examination:

1. Analysis of composition
   Identification and qualitative or quantitative determination of proportions of constituents
   Proportions of physically unsound or chemically reactive constituents
   Contamination by foreign substances
   Identification of sources of coarse and fine aggregate
   Proportioning of blended aggregates
2. Proportions of coarse and fine aggregates
   Sand-aggregate ratio
3. Segregation
   Homogeneity of distribution of coarse and fine aggregates
   Differential distribution of lightweight and heavyweight particles; coarse and fine aggregates
   Concentration of unsound particles at exterior surfaces
4. Effects of attrition during handling of aggregates and mixing of concrete
   Rounding of edges and corners
   Coatings of dust of fracture
   Aggregate fines within the cement-paste matrix
5. Coatings on aggregate particles
   Composition and physical characteristics
   Frequency and extent
   Deleterious effects
6. Unsoundness of constituents during service or test exposure
   Relationship of lithology and particle size to popouts
   Involvement in scaling, D-cracking
   Thermal expansion at high temperature in service exposure or fire
   Decarbonation of carbonate aggregates in high-temperature service or fire
7. Chemical reactivity of constituents during service or test exposure
   Alkali-silica reaction
   Alkali-carbonate reaction
   Dissolution of soluble constituents
   Oxidation and hydration of ferrous sulfides in normal service
   Oxidation and hydration of ferrous sulfides in high-temperature service
   Hydration of free lime (CaO) or magnesia (MgO) in slag aggregates or contaminants
   Staining of cement paste matrix by organic matter, sulfides, or other substances

Proportions of Coarse and Fine Aggregates

The proportions of coarse and fine aggregates within hardened concrete can be determined fairly readily on sawed and lapped sections by means of the microscopical point-count method or the linear-traverse method in general accord with the requirements of ASTM C 457. The task is straightforward if the lithology of the coarse and fine aggregates are unambiguously distinctive such that no mutually common constituents are included in the respective materials. The particles of coarse or fine aggregates are accordingly identified by composition and relegated to the appropriate category, regardless of particle size. The area to be traversed should be at least three times that shown in Table 1 of ASTM C 457 for the respective nominal maximum size of aggregate; the method is of doubtful practicality for concrete containing aggregate whose nominal maximum size is larger than 37.5 mm (1½ in.) [46].

For concrete containing natural or crushed gravel coarse aggregate and manufactured sand or natural sand or a crushed stone coarse aggregate and a natural sand, wherein one or more constituents are found in both the coarse and fine aggregates, other considerations are required. For example, in southern Michigan, gravels and sands contain high proportions of dolomites and calcitic dolomites originating in geologic formations that are operated as sources of crushed stone coarse aggregate for use in structural concrete and pavements. Although the natural aggregates characteristically are rounded because of the only moderate hardness of the rock, the particles of

348 TESTS AND PROPERTIES OF CONCRETE

crushed stone typically are rounded at edges and corners by attrition incidental to processing and handling of the aggregate and mixing of concrete. In this instance, the distinction of coarse and fine particles can be approximated by classifying as coarse aggregate all particles whose cross section on sawed and lapped surfaces include one or more dimensions greater than 3/16 in., the opening of the U.S. Sieve No. 4 (4.75-μm), the remainder being classified as fine aggregate.

The measurement is made conveniently if the reticle of one eyepiece of the stereoscopic microscope includes a scale which, when calibrated, will demark the 4.75-μm (3/16-in.) dimension in the field of view. Of course, the dimensions of the cross section so revealed are not a measure of the size of the particle. Sections of coarse aggregate adjacent to edges and corners of the particle may be less than 4.75 μm (3/16 in.) in diameter and the maximum dimension of flat or elongated particles of fine aggregate may be greater than 4.75 μm (3/16 in.). However, these effects tend to be self-compensating. Completely angular and irregularly shaped particles of requisite lithology whose size is less than 4.75 μm (3/16 in.) would appropriately be relegated to the coarse-aggregate fraction if the aggregate is a combination of a crushed stone and natural sand.

These determinations permit estimation of the sand-aggregate ratio, an important factor in proportioning of concrete for optimum workability.

Alkali Reactivity of Aggregate Constituents

Examination of hardened concrete from service or test exposure allows determination of alkali reactivity of siliceous or dolomitic rock types through observations made on sawed and lapped sections or fracture surfaces. The manifestations of reactivity include such features as rim formation, microcracking within the particles and adjacent cement-paste matrix, and the presence of secondary deposits. Darkened or clarified rims within the peripheral border of aggregate particles are the initial and most frequent indication of cement-aggregate reaction. The presence of such rims requires determination that the rims are, in fact, a consequence of processes occurring within the concrete rather than a result of prior weathering.

A decision on this matter is made readily if a bonafide sample of the unused aggregate is available for separate study. Peripheral rims occurring adjacent to fractured faces of crushed stone or crushed natural aggregate can be taken as being a product of cement-aggregate reaction. Also, rims produced by cement-aggregate reaction will be seen to thin or disappear where the aggregate particle is bordered by a void space [47]. A condition that is more frequent and more readily discovered than the relationship between peripheral rims and adjacent void space is the thinning or disappearance of the peripheral reaction rim along the bottom side of the aggregate particle as cast in the concrete, a circumstance related to bleeding and settlement of the fresh concrete, presumably producing a real or incipient separation at this location or a more dilute cement paste of higher water-cement ratio and lower alkalinity.

Summarizing the Petrographic Examination

Tables 2 through 6 exemplify a variety of forms in which the petrographic analysis may be reported. All are based upon samples received as a part of engineering investigations. The tabulations always are accompanied by appropriate discussions and supplementary descriptions. Table 2 is in the form that has been employed by the Bureau of Reclamation [48]. It departs somewhat from the format recommended in ASTM C 295, but the inclusion of pertinent descriptions facilitates interpretation of the analysis. The tabulation could be simplified by cataloging the rock types into major and secondary classifications, such as "granite: fresh, moderately weathered and internally fractured, and deeply weathered." The summary of physical and chemical quality is included in a separate tabulation.

Table 3 is the analysis of the sand produced with the gravel whose composition is summarized in Table 2. The format conforms with ASTM C 295. Table 4 is similar except that the designations of quality are not used because they are inappropriate for the description of synthetic lightweight aggregates.

Table 5 is an analysis of a sample representing a commercial crushed stone coarse aggregate passing the 37.5-mm (1½-in.) sieve. The analysis was obtained because the aggregate apparently delayed or prevented development of specified strength by the concrete under certain conditions. In the tabulation, the denotation of "innocuous" and "deleterious" is restricted to potential deleterious alkali-silica reactivity because the significance of the sulfides and organic matter in the stone was not evaluated.

Table 6 is an example of petrographic analysis of aggregates in hardened concrete, namely, a crushed stone coarse aggregate and a natural sand containing particles of shale.

PETROGRAPHIC EXAMINATION OF NATURAL AGGREGATES

Examination of Natural Aggregates in the Field

Sand and gravel result from weathering, natural abrasion, and impacting of rock and the deposition of the resulting particles along streams, in lakes or marine basins, or by wind or glaciers on the earth's surface. Consequently, sand and gravel are more or less complex mixtures of different kinds of rocks and minerals. Moreover, deposits of sand and gravel usually vary vertically by stratification and laterally because of the lenticular nature of zones or strata. The concrete-making qualities of the aggregate are influenced by these changes. Examination in the field also should reveal the variability of the sand and gravel with reference to unsound or deleterious particles, interstitial clay, and organic matter.

Deposits of sand and gravel commonly are changed by deposition of mineral matter from ground water or by weathering of particles. Examination in the field should indicate the lateral and vertical extent and the physical nature and composition of the coatings. Areas of the deposit that are free from coatings and zones that are

TABLE 2—*Example of Tabulation of a Petrographic Analysis of Gravel.*

_____ Plant, _____ Company, Near Denver, Colorado

Sample No. _____

Rock Types	Size Fraction, % by weight[a]			Description of Rock Types	Physical Quality	Chemical Quality
	1½ to ¾ in.	¾ to ⅜ in.	⅜ to 3/16 in.			
Granite	29.5	40.0	48.6	medium- to fine-grained, rounded to fragmental	satisfactory	innocuous
Weathered granite	12.0	17.7	17.2	fractured, weathered, rounded to fragmental	fair	innocuous
Deeply weathered granite	...	0.5	...	fractured, slightly friable, rounded to fragmental	poor	innocuous
Coarse-grained granite	6.4	6.1	8.4	pink, rounded, includes some free quartz	satisfactory	innocuous
Fractured coarse-grained granite	...	0.9	...	pink, rounded, includes some free quartz	fair	innocuous
Rhyolite porphyry	0.8	0.2	1.1	microcrystalline, porphyritic, white to brown	satisfactory	innocuous
Andesite porphyry	2.2	1.2	0.1	microcrystalline, porphyritic, tan, to green	satisfactory	innocuous
Weathered andesite porphyry	...	0.3	...	as above, fractured and weathered	fair	innocuous
Basalt	0.2	...	0.6	weathered, fractured, black, microcrystalline	fair	innocuous
Diorite	0.4	0.4	0.1	medium- to fine-grained, hard, massive	satisfactory	innocuous
Granite gneiss	32.2	14.3	15.8	hard, banded, fine- to medium-grained	satisfactory	innocuous
Weathered gneiss	10.3	7.8	2.7	as above, fractured to slightly friable	fair	innocuous
Deeply weathered gneiss	0.2	...	...	as above, intensely fractured to friable	poor	innocuous
Schist	2.2	2.3	2.2	hornblende schists, hard, rounded	satisfactory	innocuous
Fractured schist	...	0.4	0.2	as above, fractured	fair	innocuous
Quartzite	2.8	6.2	2.0	fine-grained, hard, massive to schistose	satisfactory	innocuous
Milky quartz	0.6	1.0	0.5	massive, hard, brittle, dense, smooth	fair	innocuous
Quartzose sandstone	...	0.1	...	fine-grained, massive, firm to hard	satisfactory	innocuous
Ferruginous sandstone	...	0.4	...	porous, brown, platy, quartzose	fair	innocuous
Shale	...	0.2	0.5	soft, absorptive, rounded, gray	poor	innocuous
Rhyolite	0.2	...	...	cryptocrystalline, porphyritic, pink to gray	satisfactory	deleterious

NOTE—Conversion factors—1 in. = 25.4 mm and 1 lb = 0.45 kg.

[a] Based upon analysis of 19.0 lb of 1½ to ¾-in., 2.7 lb of ¾ to ⅜-in., and 0.80 lb of ⅜ to 3/16-in. aggregate.

so heavily coated as to preclude processing as aggregate should be delineated.

Weathering of gravel and sand after formation of the deposit is common on terraces and at lower levels along existing stream channels, glacial deposits, and outwash. The examination should define the extent and distribution of such weathering.

Close observation of gravel and sand exposed on the surface of the deposit commonly will reveal unsound particles that slake or fracture with freezing-thawing or wetting-drying (Fig. 5). Identification of such particles will aid evaluation of the petrographic examination performed in the laboratory. In interpreting the effects of natural freezing and thawing, the consequences of particle size should be evaluated [49]. Representative specimens of particles affected by freezing and thawing should be packaged separately and transmitted to the laboratory with samples of the gravel and sand. Water-soluble salts in coatings or ground water also may be revealed by efflorescence at or near the surface of the deposit. Their presence forewarns of the need for their identification and quantitative determination of their concentration in the aggregate.

Examination of Natural Aggregates in the Laboratory

Samples and data from the field should be examined to determine: (1) the abundance of individual lithologic or mineralogic types; (2) the abundance of particles in various conditions of alteration and internal texture and structure and degrees of chemical reactivity; (3) the composition, frequency, abundance, and physical nature of

TABLE 3—Example of Tabulation of a Petrographic Analysis of Natural Sand.
_____ Plant, _____ Company Near Denver, Colorado
Sample No. _____

Constituents	Amount, as Number of Particles, %												
	In Size Fractions Indicated[a]							In Whole Sample[b]					
								Physical Quality			Chemical Quality		
	No. 4–8	No. 8–16	No. 16–30	No. 30–50	No. 50–100	No. 100–200	Passing No. 200	S[c]	F[c]	P[c]	I[c]	D[c]	T[c]
Granite and granite gneiss	34.9	33.6	21.0	4.8	...	...	...	13.5	1.0	0.6	15.1	...	15.1
Pegmatite	34.2	28.7	2.0	...	...	...	...	8.2	0.8	0.2	9.2	...	9.2
Rhyolite tuff	0.8	0.9	0.3	...	...	...	...	0.3	...	...	...	0.3	0.3
Basalt	...	1.0	...	...	...	...	...	0.2	...	...	0.2	...	0.2
Sericite schist	2.4	0.7	...	9.5	11.2	1.1	...	4.7	...	...	4.7	...	4.7
Quartz and quartzite	24.6	41.2	59.2	65.0	43.0	48.8	28.8	51.6	...	...	51.6	...	51.6
Feldspar	2.6	3.3	16.7	11.1	18.7	23.8	10.5	...	11.4	...	11.4	...	11.4
Claystone	0.5	...	...	...	...	1.2	...	...	...	0.2	0.2	...	0.2
Chalcedonic chert	...	0.6	...	3.3	...	...	...	1.2	...	...	...	1.2	1.2
Mica	...	...	0.8	6.3	10.2	8.3	51.1	...	3.8	...	3.8	...	3.8
Hornblende, garnet, zircon, etc.	...	...	...	...	16.9	16.8	9.6	2.3	...	...	2.3	...	2.3
Total	100.0	100.0	100.0	100.0	100.0	100.0	100.0	82.0	17.0	1.0	98.5	1.5	100.0

[a] Based on count of 500 particles in each sieve fraction.
[b] Based on gradation of the sample received and on the distribution of constituents by size fractions shown at the left above.
[c] S = satisfactory; F = fair; P = poor; I = chemically innocuous; D = potentially chemically deleterious; T = total of constituent in whole sample.

TABLE 4—Example of Tabulation of a Petrographic Analysis of Expanded Clay Aggregate.
_____ Plant, _____ Company
Sample No. _____

Constituent	Amount, as Number of Particles, percent									Remarks
	In Size Fractions Indicated[a]								In Whole Sample[b]	
	3/4 to 3/8 in.	3/8 to 3/16 in.	No. 4–8	No. 8–16	No. 16–30	No. 30–50	No. 50–100	Passing No. 100		
Black to gray, vesicular particles	44.6	30.2	36.2	44.1	46.5	55.0	62.2	64.4	42.4	vesicular, hard
Black to gray, vesicular particles	13.1	1.7	...	...	...	...	...	...	1.2	vesicular, friable
Red to tan, vesicular particles	26.1	29.5	20.2	25.0	15.4	24.0	21.8	20.6	23.5	vesicular, hard
Red to tan, vesicular particles	7.7	1.8	...	...	...	...	...	...	0.8	vesicular, friable
Red to brown, brick-like particles	4.9	23.6	32.3	22.0	25.6	12.4	7.8	4.9	21.3	not vesicular, firm to fragile
Gray to pink, brick-like particles	2.8	9.8	7.7	8.8	9.2	3.0	1.6	1.1	7.1	not vesicular, fragile, many slake in water
Gray to pink, brick-like particles	0.6	1.0	0.4	...	...	...	...	...	0.4	not vesicular, friable, many slake in water
Gray to black, friable particles	...	1.7	2.0	...	0.8	0.3	...	...	1.0	not vesicular, contain coal
Granite	0.2	0.4	0.4	...	1.4	2.8	4.7	7.0	1.4	hard, dense
Sandstone	...	0.2	...	...	...	...	...	...	0.1	hard, fine-grained
Coal	...	0.1	0.8	0.1	1.1	2.5	1.9	2.0	0.8	hard to friable
Total	100.0	100.0	100.0	100.0	100.0	100.0	100.0	100.0	100.0	

NOTE—1 in. = 25.4 mm.
[a] Based on count of 500 particles in each size fraction.
[b] Based on gradation of the sample received, and on the distribution of constituents by size fractions shown at left above.

coatings; (4) particle shape and surface texture; and (5) the possible qualitative contribution of particles of the several types to properties of concrete (Tables 1 and 2).

Natural aggregate may contain more than 20 rock and mineral types. Consequently, petrographic examination commonly is time consuming. Based upon the similarity in composition and the probable performance in concrete, two or more rock types commonly can be combined into a single category with considerable savings in time and without loss in the validity of the analysis. For example, granites, quartz diorites, granodiorites, and quartz monzonites—or rhyolites, dacites, and latites—of similar composition and physical condition might be combined, thus eliminating the need for tedious examination sufficient to effect a separation.

Soft, friable, and altered particles may be original constituents of sand and gravel, or may be developed by weathering in the deposit. Weathering in the deposit is

TABLE 5—Example of Tabulation of a Petrographic Analysis of Crushed Stone Coarse Aggregate.[a] _____ Plant, _____ Company
Sample No. _____

Rock Type or Facies	Description of the Rock Type or Facies	Amount, % by weight[b]					
		Physical Quality			Chemical Quality		
		S[c]	F[c]	P[c]	I[c]	D[c]	T[c]
Dense dolomitic limestone	gray to buff, contains sparse organic matter with pyrite and marcasite	56.2	...	...	56.2	...	56.2
Soft dolomitic limestone	gray to buff, soft to friable, slightly porous, sparse organic matter with pyrite and marcasite	...	33.9	...	33.9	...	33.9
Soft, organic dolomitic limestone	same as above except containing one or more seams of iron sulfides and organic matter	...	4.2	...	4.2	...	4.2
Limestone	white to gray, coarse- to medium-grained	2.4	...	...	2.4	...	2.4
Laminated limestone	laminated, fine-grained, iron sulfides and organic matter abundant	...	2.1	...	2.1	...	2.1
Chalcedonic limestone	white to gray, particles of zones of chalcedony evident	1.0	...	...	...	1.0	1.0
Chalcedonic chert	conchoidal fracture, dense	...	0.1	...	...	0.1	0.1
Sandstone	includes also grains of quartz and feldspar	0.1	...	...	0.1	...	0.1
Total		59.7	40.3	...	98.9	1.1	100.0

[a] Sample graded in accordance with specifications of the _____ State Highway Dept. for 1½ in. to No. 4 aggregate for concrete highway pavement (1 in. = 25.4 mm).
[b] Based upon analysis of 25.9 lb of aggregate split from the sample (1 lb = 0.45 kg).
[c] S = satisfactory; F = fair; P = poor; I = not deleteriously alkali reactive; D = potentially deleteriously alkali reactive; T = total of constituent in the sample.

TABLE 6—Microscopical Analysis of Aggregate in Hardened Concrete[a]
Core No. 1—Street pavement in front of _____.
Washington Street, _____, Michigan
Core No. 2—Sidewalk, SW corner of _____ Street and _____ Street, _____, Michigan

Item	Core No. 1	Core No. 2
Coarse aggregate, % by volume of total aggregate	54.99	54.70
Sand fine aggregate, % by volume of total aggregate	45.01	45.30
Shale particles, % by volume of sand fine aggregate	11.35	8.96

[a] Microscopical point-count method, ASTM C 457, based on the following criteria:

Core No.	Number of Counts	Area Traversed, in.[2b]
1	1944	35.9
2	2838	24.9

[b] Maximum area available in submitted samples (1 in. = 25.4 mm).

FIG. 5—Cobbles of argillaceous limestone disrupted by freezing-thawing in the deposit, near Charlesvoix, Michigan. Observation of natural disintegration forewarns of possible difficulty in service as concrete aggregate (courtesy of Michigan Department of Transportation).

especially significant because the alteration affects most or all particles, possibly causing softening and absorptivity in the superficial portion of the particles. This action decreases both the bond with the cementitious matrix and the strength, durability, and volume stability of the concrete.

Coatings on gravel and sand vary from minute spots and films to a cement that produces zones of sandstone or conglomerate in the deposit. Coatings usually are composed of fine sand, silt, clay, and calcium carbonate; but organic matter, iron oxides, opal, manganiferous substances, alkali- and alkali-earth sulfates, and soluble phosphates have been identified [8,28]. The petrographic examination should reveal the composition, abundance, physical properties, probable potential chemical reactivity, and the ease with which the coatings are removed by impact and abrasion.

Gravel and sand potentially susceptible to the alkali-silica reaction occur along many important rivers in the United States and have been responsible for serious distress in many concrete structures of all types, including highway pavements (Fig. 6). The location of many known deposits of alkali-reactive natural aggregate in the western United States was reported by Holland and Cook [50]. The known alkali-reactive substances are the silica minerals, opal, chalcedony, tridymite, cristobalite, and strained or very finely divided quartz; glassy to cryptocrystalline rhyolites, dacites, latites, and andesites, and their tuffs; at least certain artificial siliceous glasses; and certain phyllites, slates, argillites, and graywackes (Fig. 7). Occasional basalts contain glass whose index of refraction is less than 1.535 and palagonite may be partly altered to opal; such types are potentially alkali reactive. Fraser et al. [51] have provided a detailed study of composition of chalcedonic cherts in relation to their reactivity with alkaline solutions. Alkali-reactive dolomitic rocks may occur as constituents of natural aggregates. Any aggregate containing a significant proportion of these substances is potentially deleteriously alkali reactive in concrete in service.

PETROGRAPHIC EXAMINATION OF CRUSHED STONE

Examination of Stone in the Field

Rock formations are massive or stratified; the strata can occur in any attitude relative to the horizontal; and rock may vary widely in porosity, hardness, toughness, degree of fracturing, or chemical reactivity. For example, the alkali reactivity of dolomitic limestones typically varies

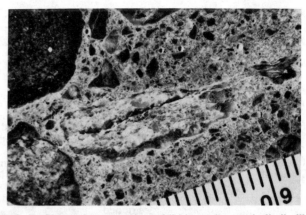

FIG. 7—Metasubgraywacke exhibiting advanced alkali-silica reaction in mass concrete, including darkened peripheral rim, internal microcracking, and radial cracking of enclosing mortar matrix. The internal microcracks are characteristically open in the interior of the particle and closed at the periphery. Note white deposits in external microcracks. A millimetre scale is shown.

widely even within the same quarry when deleterious facies are present [33,34].

Rock formations commonly contain zones of faulting, jointing, or local shearing, within which the materials are fractured or chemically decomposed. Certain zones may contain deleterious or unsound substances, such as chal-

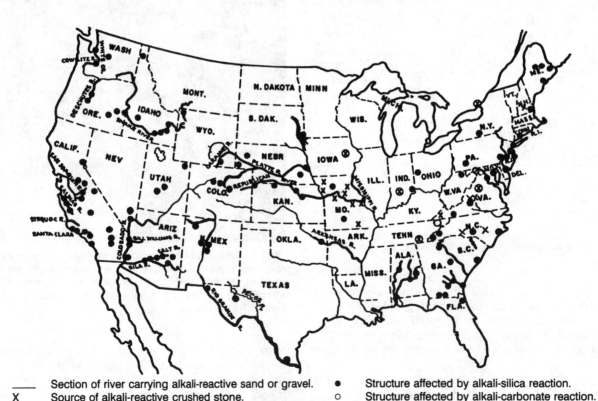

FIG. 6—Location of structures in the United States (other than Alaska and Hawaii) known to be affected by alkali-aggregate reaction in concrete, sources of sand or gravel subject to alkali-silica reaction, and sources of crushed stone subject to alkali-carbonate reaction.

cedonic or opaline chert and clay or shale in limestone or dolomite, or zones of hydrothermal alteration in igneous or metamorphic formations in which sulfides, zeolites, clays, or clay-like minerals are prominent. Especially in warm and humid areas, rock formations commonly are fractured, leached, and partially decomposed by weathering near the surface to varying depths, depending on localized differences in fracture, porosity, or lithology.

These features should be discovered by geologic and petrographic examination of natural exposures, quarried faces or other excavations, and drilled core. Petrographic examination in the field should include exposures of geologic formations and stone used in the area for fill, ballast, riprap, and masonry. Excavation of trenches or test pits may be required. This survey may reveal lithologic varieties that have failed during the natural exposure to freezing-thawing or wetting-drying or to oxidation and hydration (Fig. 8). The significance of the size of fragments should be recognized in any evaluation of such observations [49]. Disintegration of rock on natural exposures commonly does not coincide with results of the sulfate soundness test, ASTM Test Method for Soundness of Aggregates by Use of Sodium Sulfate or Magnesium Sulfate (C 88) [52].

Shale, claystone, and argillaceous rocks, including deeply altered basalts and diabases (traprock) containing nontronite, may slake and fracture during brief exposure even though they appear sound when excavated. Similar effects occur if the zeolites leonardite-laumontite constitute a significant proportion of the rock or occur as seams or veinlets in the formation; they are especially common in granodiorites, quartz diorites, quartz gabbros, and anorthosites. Soluble salts usually will be revealed by efflorescence at exposed surfaces.

Note that the term "crushed stone" covers aggregates produced by processing of boulders, cobbles, and the like as they may be derived from deposits of gravel and sand, where substantially all of the faces of the particles have resulted from the crushing operation (ASTM C 125). Field examination of such sources is analogous to that applied to natural aggregates as discussed previously.

Examination of Crushed Stone in the Laboratory

As is indicated earlier, crushed stone aggregates commonly are complex petrographically and not only when they are derived by crushing of boulders and cobbles recovered from gravel. Consequently, they should be examined in the same detail as required for natural aggregates (Table 5).

Petrographic examination of stone in the form of quarried blocks or irregularly shaped pieces should include inspection of the entire sample. The examination should establish the relative abundance of individual rock types or facies (Fig. 9). Specimens representative of each type should be obtained by sawing or coring of typical pieces; these specimens provide for special tests, detailed petrographic examination, and reference. Procedures for obtaining representative samples are described by Abdun-Nur [12].

Quantitative petrographic analysis can be made on calcareous rocks by preparing plane surfaces by sawing and lapping, followed by etching. Thirty seconds of immersion in 10% hydrochloric acid (HCl) will provide adequate differential attack on calcite and dolomite and will expose noncalcareous constituents, such as clay, chert, pyrite or marcasite, and siliceous sand or silt. The etched surface can be analyzed by point-count or linear traverse in general accordance with procedures given in ASTM C 457. This technique can be applied to calcareous aggregates in hardened concrete (Fig. 10).

Quantitative analysis can be made similarly on prepared surfaces of quartzose and feldspathic rocks following etching by 1½ to 2 min immersion in hydrofluoric acid (HF) (Fig. 11).

FIG. 8—Disintegration of argillaceous facies of limestone used as riprap at Chickamauga Dam, Tennessee. Observation of such disintegration indicates unsoundness of stone in at least portions of a quarry (courtesy of the Bureau of Reclamation, U.S. Department of the Interior).

FIG. 9—Crushed limestone aggregate containing seams and pieces of illite shale, Webster Dam, Kansas. The physical quality was indicated as follows: satisfactory, 61%; fair, 34%; poor, 5% by mass (courtesy of the Bureau of Reclamation, U.S. Department of the Interior).

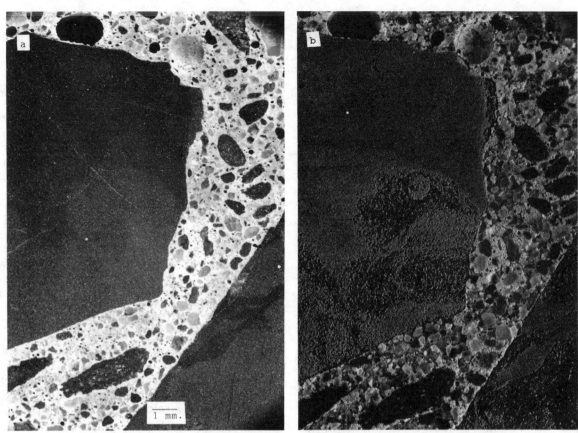

FIG. 10—Concrete containing dolomitic limestone coarse aggregate (a) before and (b) after etching for 30 s in 10% HCl. Scattered crystals in the etched areas on the lapped surface of the aggregate particles are dolomite.

If the sample of stone was submitted for crushing tests in the laboratory, the manufactured aggregate produced should be examined to determine the effects of natural fracturing, jointing, and internal texture on particle size and shape; frequency of fractures and seams within the particles; distribution of unsound or deleterious substances in the size fractions; and the abundance and composition of crusher dust. These features should be correlated with the processing equipment and methods employed.

Inspection of the stone prior to processing is important because only thus can the examination of the finished aggregate be interpreted fully. For example, if unsound or deleterious particles constitute 10% of the finished aggregate, was this proportion derived from approximately one piece in ten of the original sample or does a typical piece of the stone contain approximately 10% of unsound material? The former possibility suggests that the quarry should be examined to determine whether the unsound zones can be avoided or wasted; the latter suggests that the material should not be used as aggregate in concrete for permanent construction.

If the samples are in the form of drilled core, the entire length of the core should be examined and compared with logs available from the driller and geologist. Special attention must be given to sections in which core loss was high or complete, inasmuch as such zones commonly represent fractured, altered, or otherwise unsound rock. The core should be examined by means of the hand lens, stereoscopic microscope, and petrographic microscope, as necessary, to establish variations in lithology; frequency and intensity of fracturing; content of clay or other soft materials, regardless of rock type; and presence of chemically reactive substances. The examination is facilitated by sawing and lapping of the core along the length, with or without etching or staining. These observations should be correlated from hole to hole so that the variation in lithology or quality of the rock, both in depth and laterally, is established.

The quality of the rock to be expected from the formation represented by the cores also will be revealed by petrographic examination of aggregate produced from cores in the laboratory.

PETROGRAPHIC EXAMINATION OF BLAST-FURNACE SLAG

Blast-Furnace Slag

Blast-furnace slag is the nonmetallic product, consisting essentially of silicates and aluminosilicates of calcium and other cations, that is developed simultaneously with iron in a blast furnace (see ASTM C 125). Properties of blast-

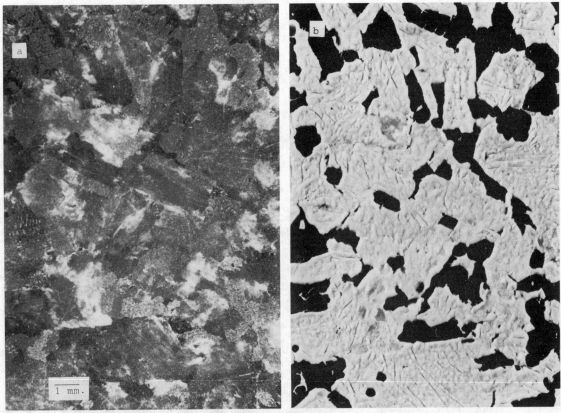

FIG. 11—Augite gabbro (a) before and (b) after 1½ min etching in concentrated HF. White areas are plagioclase feldspar, dark areas are augite in the etched surface.

furnace slag as concrete aggregate have been reported by Gutt et al. [53]. Three general types of blast-furnace slag are used for concrete aggregate; namely, air-cooled slag, granulated slag, and lightweight slag [54]. Petrographic examination of lightweight or expanded slag will be discussed in a later section.

Performance of the Petrographic Examination of Blast-Furnace Slag

The procedure for petrographic examination of blast-furnace slag is not included specifically in ASTM C 295. However, the instructions provided for examination of ledge rock, crushed stone, and manufactured sand are applicable. In addition to the indicated microscopical methods, examination of polished and etched surfaces in reflected light is a valuable technique, being preferred by some petrographers because the surfaces are easy to prepare in sizes larger than standard thin sections, the two-dimensional aspect simplifies quantitative estimation of composition, and microchemical tests can be used to identify various phases in the surface under study.

Air-cooled slag is more or less well-crystallized, depending primarily upon the method of disposal employed at the steel plant. Such slag crushes to angular and approximately equidimensional pieces, the surface texture of which is pitted, rough, or conchoidal. Crystals range from submicroscopic to several millimetres in size. Abrasion resistance relates to glass content and the condition of internal stress [54].

Well-granulated slag is substantially all glass. Crystals occur individually or in clusters scattered through the glass matrix. Incipient crystallization produces brown or opaque areas in thin sections (Fig. 12).

The petrographic examination of blast-furnace slag aggregate should include description of the various types of slag as well as of contaminating substances. The slag constituent usually can be segregated into two or more varieties, depending upon particle shape, surface texture, color, vesicularity, crystallinity, or the presence of products of weathering. Evaluation of effects of weathering is especially important when old pits or deposits or fill are being exploited. Each type should be studied in sufficient detail to assure identification of potentially deleterious compounds.

More than 20 compounds have been identified in blast-furnace slag (Table 7), but even well-crystallized slag rarely contains more than five compounds (Table 8). The typical constituent of blast-furnace slag is melilite, a compound of variable composition between akermanite ($2CaO \cdot MgO \cdot 2SiO_2$) and gehlenite ($2CaO \cdot Al_2O_3 \cdot SiO_2$). Pseudowollastonite and anorthite are of common occurrence. High-lime blast-furnace slags usually contain one or more forms of dicalcium silicate (α, β, or γ forms of $2CaO \cdot SiO_2$). Magnesium-containing blast-furnace slags commonly include monticellite, forsterite, or merwinite. Calcium sulfide

TABLE 7—Compounds Occurring in Blast-Furnace Slag.[a]

Compound	Chemical Formula	Compound	Chemical Formula
Gehlenite	$2CaO \cdot Al_2O_3 \cdot SiO_2$	oldhamite	CaS
Akermanite	$2CaO \cdot MgO \cdot 2SiO_2$	ferrous sulfide	FeS
Pseudowollastonite	$\alpha CaO \cdot SiO_2$	manganous sulfide	MnS
Wollastonite	$\beta CaO \cdot SiO_2$	spinel	$(Mg, Fe)O \cdot Al_2O_3$
Bredigite	$\alpha'2CaO \cdot SiO_2$	anorthite	$CaO \cdot Al_2O_3 \cdot 2SiO_2$
Larnite	$\beta 2CaO \cdot SiO_2$	periclase	MgO
γ-dicalcium silicate	$\gamma 2CaO \cdot SiO_2$	lime	CaO
Olivine	$2(Mg, Fe)O \cdot SiO_2$		
Merwinite	$3CaO \cdot MgO \cdot 2SiO_2$	cristobalite	SiO_2
Rankinite	$3CaO \cdot 2SiO_2$	calcium aluminate	$CaO \cdot Al_2O_3$
Monticellite	$CaO \cdot MgO \cdot SiO_2$	cordierite	$2MgO \cdot 2Al_2O_3 \cdot 5SiO_2$
Pyroxene			
Diopside	$CaO \cdot (Mg, Fe)O \cdot 2SiO_2$	sillimanite	$Al_2O_3 \cdot SiO_2$
Enstatite	$MgO \cdot SiO_2$	mullite	$3Al_2O_3 \cdot 2SiO_2$
Clinoenstatite	$MgO \cdot SiO_2$	madisonite	$2CaO \cdot 2MgO \cdot Al_2O_3 \cdot 3SiO_2$

[a] Compiled from several sources, primarily Nurse and Midgley [56], McCaffery et al. [62], and American Concrete Institute Committee 201 [60].

(CaS) almost always is present in small proportion. Sulfides of manganous manganese and ferrous iron are common. Properties and techniques for identification of these compounds are summarized by Rigby [55], Nurse and Midgley [56], Insley and Frechette [57], and Snow [58], and in standard works on mineralogy. X-ray diffraction methods are necessary if crystalline phases are submicroscopic and are a great aid to a petrographer who is developing experience independently in this field.

Several constituents of blast-furnace slag may be deleterious to the performance of concrete. Sulfides released into the cement-paste matrix produce innocuous green staining of the interior of the concrete. Occasionally, a mottled aspect is produced on the surface of damp concrete [14], but the coloration fades to a homogeneous shade of gray on drying and exposure to the atmosphere. Presence of colloidal sulfides is suggested by yellow or brown coloration of the glass phase [53]. Gypsum (calcium sulfate dihydrate, $CaSO_4 \cdot 2H_2O$) commonly forms in blast-furnace slag by weathering and may result in sulfate attack on the cement-paste matrix of concrete in service. Granulated slag is most susceptible to the formation of gypsum during weathering exposure.

Inversion of β-dicalcium silicate to the γ-dicalcium silicate, with accompanying 10% increase in volume of the crystal, causes "dusting" or "blowing" of slag [54,59,60]. The inversion ordinarily takes place before the slag has cooled, and the disintegrated material is removed by screening during the production of coarse aggregate. In less severe occurrences, the disintegration takes place slowly, producing pieces that are partly or wholly weak and friable, and thus unsuitable for concrete aggregate. This action of dicalcium silicate can be avoided by maintaining a ratio of CaO to SiO_2 in the slag sufficiently low to prevent formation of the compound, or by chilling the molten slag so that the compound does not crystallize [53]. If air-cooled slag is poured in thin layers, rapid cooling ordinarily arrests the compound in the β-modification. Dicalcium silicate can be identified microscopically in slag by special techniques [59]. Scattered crystals of β-dicalcium silicate commonly are stable in blast-furnace slag and therefore are innocuous.

Free lime (CaO) and magnesia (MgO) are extremely rare as constituents of air-cooled blast-furnace slag and are not likely to form either as a primary phase or devitrification product in granulated blast-furnace slag [56,61].[2] Nevertheless, their absence or presence and abundance should be established by petrographic examination. These compounds are deleterious because of the increase in solid volume resulting from hydration and carbonation in place.

Cristobalite (SiO_2) has been reported as a constituent of blast-furnace slag [62]. This compound is potentially subject to the alkali-silica reaction in portland-cement

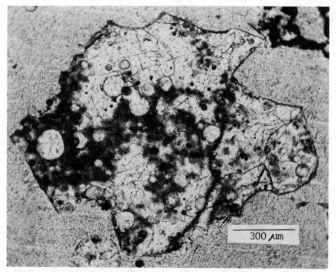

FIG. 12—Photomicrograph of granulated blast-furnace slag. Note the vesicles (bubbles) in glass phase (white). The dark areas are concentrations of microcrystalline melilite and merwinite (courtesy of the Bureau of Reclamation, U.S. Department of the Interior).

[2] Unlike blast-furnace slags, basic open-hearth slags commonly contain free oxides that are subject to hydration and carbonation in portland-cement concrete, namely, free lime (CaO) and magnesiowustite, a solid solution of magnesium oxide (MgO), manganese oxide (MnO), and ferrous oxide (FeO).

TABLE 8—Most Frequently Occurring Combinations of Compounds of CaO·MgO·Al₂O₃·SiO₂ Produced by Crystallization of Blast-Furnace Slag.[a]

Flux Stone	C₂AS	C₂MS₂	C₂S	CS	C₃S₂	C₃MS₂	MA	CMS₂	CMS	CAS₂	M₂S	MS	MgO
Limestone	X	X	X		X								
	X	X	X			X							
	X	X				X	X						
	X	X					X			X			
	X	X		X						X			
Dolomite		X		X				X		X			
		X						X		X	X		
		X					X			X	X		
		X					X		X	X			
		X				X	X		X				
						X	X		X				X
										X	X	X	

[a] After Nurse and Midgley [56].
[b] Key: C₂AS = gehlenite ⎫ melilite
         C₂MS₂ = akermanite ⎭
         C₂S = dicalcium silicate
         CS = wollastonite or pseudowollastonite
         C₃S₂ = rankinite
         C₃MS₂ = merwinite
         MA = spinel
         CMS₂ = diopside
         CMS = monticellite
         CAS₂ = anorthite
         M₂S = forsterite
         MS = enstatite
         MgO = periclase

concrete; if identified, its abundance should be determined.

The glass phase of blast-furnace slag is not deleteriously reactive with cement alkalies.

Contaminating substances whose presence or absence should be determined by petrographic methods are metallic iron, iron carbide, coke, and incompletely fused fluxstone. The last is important because hydration and carbonation of free lime and magnesia may produce expansion of the concrete and popouts. Metallic iron and iron carbides rust by oxidation and hydration if exposed at the surface of the concrete.

PETROGRAPHIC EXAMINATION OF LIGHTWEIGHT CONCRETE AGGREGATES

Expanded Clay, Shale, and Slate

Petrographic examination of lightweight aggregate should include segregation of the particles into as many categories as required to describe the sample adequately (Table 4). Particles of expanded material may be segregated on the basis of particle shape, surface texture, development of a coating or "skin," and vesicularity. Incompletely expanded particles and particles not expanded should be distinguished from vesicular ones. Such particles usually represent sandstone or siltstone that was intercalated within the clay, shale, or slate. They should be separated on the basis of porosity, absorptivity, density, friability, softness, occurrence of efflorescence, and reaction to water (softening, slaking, or swelling). The petrographic examination supplements standard tests in distinguishing clayey particles from "clay lumps" determined in accordance with ASTM C 142. Other materials to be identified and determined quantitatively are underburned or raw material, coal, and miscellaneous rock particles. The composition and content of raw materials commonly can be established most easily and accurately by X-ray diffraction or differential thermal analysis.

The individual types of particles also should be analyzed petrographically to establish the presence of free lime or magnesia. These compounds result mainly from decomposition of calcium and magnesium carbonates in the feed during firing. They may produce distress or popouts in concrete or concrete products unless the aggregate is water- or steam-cured prior to use [63]. Hard-burned magnesia may not be accommodated by such treatment.

Petrographic examination also assists in the selection of raw materials, development of manufacturing equipment and methods, and process control.

Industrial Cinders

Industrial cinders used as concrete aggregate are the residue from high-temperature combustion of coal and coke in industrial furnaces. Petrographic examination should determine the physical nature of the cinder particles on the basis of composition, friability, softness, particle shape, and surface texture. Particular attention should be given to identification of sulfides, sulfates, coal, and coke.

Expanded Blast-Furnace Slag

Expanded blast-furnace slag is produced by carefully controlled intermingling of molten slag and water or steam in one of several ways [54]. The petrographic examination should describe the aggregate in terms of the nature of the expanded particles, including their particle shape, surface texture, friability or softness, effects of weathering in stockpiles, and content of contaminating substances, such as dense slag. The petrographic examination is especially important when fill or waste deposits are later contemplated for use as a source of concrete aggregate.

FIG. 13—Typical basaltic scoria aggregate. The pieces are black, gray, and reddish brown. Note the rounded vesicles (courtesy of the Bureau of Reclamation, U.S. Department of the Interior).

Pumice, Scoria, Tuff, and Volcanic Cinders

Pumice, scoria, tuff, and volcanic cinders used for lightweight aggregate are naturally occurring porous or vesicular lava and ash. Pumice is a very highly porous and vesicular rock composed largely of natural glass drawn into approximately parallel or loosely intertwined fibers and tubes. Scoria is a highly porous and vesicular rock in which the vesicles typically are rounded or elliptical in cross section, the interstitial glass occurring as thin films (Fig. 13). Tuff is a general term designating consolidated volcanic ash of any lithologic type or physical character. Volcanic cinder is a loose accumulation of highly vesicular (scoriaceous) fragments of lava, predominantly ranging from 4 to 32 mm (0.16 to 1.26 in.) in diameter.

Petrographic examination of these types of aggregate includes segregation of the particles on the basis of particle shape, surface texture, porosity or vesicularity, fracturing, friability or softness, weathering, specific gravity, secondary deposits in voids, coatings, and potential alkali reactivity. Extraneous or contaminating substances are primarily dense particles of volcanic rock and organic matter. In production of lightweight aggregate, two types of volcanic materials occasionally are intermixed for economy or to control gradation or unit weight. The type and relative proportion of the materials can be established by petrographic examination.

Volcanic glass with an index of refraction less than 1.535 is potentially deleteriously reactive with cement alkalies; glass whose index of refraction is in the range of 1.535 to 1.570 probably is alkali reactive. Opal, chalcedony, tridymite, and cristobalite are also common alkali-reactive constituents of volcanic aggregates. These minerals are especially common as alteration products of volcanic rocks of acidic to intermediate composition. However, opal may occur as an alteration product of palagonite in basaltic lavas.

In spite of alkali-silica reaction and formation of alkalic silica gel in highly porous vesicular lavas and tuffs, expansion and cracking of the enclosing concrete usually is prevented by the abundant voids into which the hydrating gel can escape without development of excessive stress in the mortar. For example, a highly porous rhyolite tuff from Hideaway Park, Colorado, containing abundant tridymite caused only 0.041% expansion of a high-alkali cement mortar during one year of moist storage in accordance with ASTM Test Method for Potential Alkali Reactivity of Cement-Aggregate Combinations (Mortar-Bar Method) (C 227), yet the specimen contained abundant alkalic silica gel [64]. A similar but dense rhyolite tuff from near Castle Rock, Colorado, produced an expansion of 0.400% under the same conditions.

Perlite

When heated rapidly to fusion, certain obsidians and pitchstones release gases that, being trapped within the molten glass, vesiculate the rock and cause disruption into small pieces. The product is known commercially as perlite.

Petrographic examination should indicate the composition of the aggregate in terms of particle shape, surface texture, rock classification, density, friability or fragility, and potential alkali reactivity. Perlite may contain particles of dense volcanic rock or individual crystals.

Being composed of rhyolitic volcanic glass, typical perlite is potentially reactive with cement alkalies, although significant expansion may not occur because of the porosity of the particles. However, laboratory tests demonstrate that certain perlites produce significant expansion of mortar stored in accordance with ASTM C 227 in combination with either high-alkali or low-alkali cement [65,66]. Such volume change will not necessarily cause structural distress if appropriately accommodated in the design of the constructions or concrete product, such as masonry units or precast panels.

Exfoliated Vermiculite

Exfoliated vermiculite is produced by rapid heating of the micaceous mineral, vermiculite. Release of combined water expands the crystals—like an accordion—increasing the volume to as much as 30 times its original size. The degree of expansion varies widely, depending upon mineralogic properties and purity and the conditions of firing.

During petrographic examination, the particles of vermiculite are segregated by degree of expansion, elasticity or brittleness of the flakes, and fragility of the expanded crystals. These differ significantly within individual samples from some sources, especially from marginal deposits where the vermiculite grades into hydrobiotite or biotite. Also to be reported is the intermixture of the vermiculite with particles of rocks and minerals occurring with the vermiculite in the ore deposit.

Diatomite

Crushed and sized natural diatomite typically is soft, porous, absorptive, and ranges from firm to pulverulent.

Finely divided opal and opaline skeletons of diatoms are the predominant constituents. Fine sand, silt, clay, and volcanic ash are present in widely differing proportions. At least certain diatomites produce significant expansion of mortars stored in accordance with ASTM C 227, with both high- and low-alkali cement [65,66].

PETROGRAPHIC EXAMINATION OF RECYCLED CONCRETE FOR USE AS CONCRETE AGGREGATE

Recycled Concrete

Recycled concrete is defined in the report of the American Concrete Institute (ACI) Committee 117 on Cement and Concrete Terminology [67] as "hardened concrete that has been processed for reuse, usually as aggregate," and in ASTM C 33 as "crushed hydraulic-cement concrete." Such reclaimed concrete from structures and pavements was used widely in Great Britain and Germany following World War II and, more recently, it has been investigated in the United States, Great Britain, Canada, and elsewhere as a partial or complete replacement for conventional aggregates for purposes of economy, substitution for natural or crushed stone aggregates in short supply, and environmental benefits.

Buck [68] concluded that use of recycled concrete as aggregate in new concrete is feasible and may become routine. Forster [69] reported on extensive studies of such applications by state departments of transportation. Committee 37-DRC, Reunion Internationale des Laboratoires D'Essais et de Recherches sur les Materiaux et les Constructions (RILEM), reported on the state-of-the-art in this technology for the period of 1945 to 1985 [70].

The following summary of the properties of concrete containing recycled concrete as aggregate may assist the petrographer in decisions on: (1) features to be considered in examination of constructions to be demolished with the intent that the concrete will be reclaimed for production of concrete aggregate, and (2) observations that should be included in petrographic examination of samples of such materials.

Compressive Strength

Hansen and Narud [71] concluded that the compressive strength of concrete containing recycled concrete is controlled largely by the water-cement ratio of the original concrete when other conditions are essentially identical. When the water-cement ratio of the original concrete is equal to or lower than that of the new concrete containing the recycled aggregate as coarse aggregate only, they observed that the strength of the new concrete is equal to or higher than that of the original concrete, and vice versa, provided the fine aggregate is a natural sand or manufactured sand of suitable quality.

In evaluating the probable influence of recycled concrete from a given source on the strength of new concrete, the compressive strength of the concrete to be recycled is a more practical measure of quality than is the water-cement ratio because the water-cement ratio usually is not known and is not readily determined in practice.

A substantial reduction in compressive strength may result when the conventional fine aggregate is replaced in whole or in part by fine aggregate derived from the recycled concrete. The RILEM report [70] concludes that all material below 2 mm in recycled concrete should be screened out and wasted.

Water Requirement

Use of recycled concrete decreases workability of fresh concrete, increases the water requirement for a given consistency, increases drying shrinkage at a given water content, and reduces the modulus of elasticity at a given water-cement ratio [72]. These effects are greatest when the recycled concrete is used as both coarse and fine aggregate.

Freeze-Thaw Resistance

Widely varying results have been obtained in tests of freeze-thaw resistance of concrete containing recycled concrete as aggregate. The results relate to many factors, including use of recycled concrete as coarse aggregate alone or as both coarse and fine aggregate, the quality of the original concrete in terms of water-cement ratio and parameters of the air-void system, frost resistance of the aggregates included in the recycled concrete, and presence or absence of purposeful air entrainment in the new concrete containing the recycled concrete. In any event, there is no reason to believe that the recycled concrete should not be required to meet generally accepted standards for quality of the cementitious matrix, air entrainment, and soundness of aggregates when the new concrete is to be exposed to severe conditions of weathering.

Presence of Chemical or Air-Entraining Admixtures

Available data [73] indicate that, when used at generally recommended rates in the original concrete, plasticizing, set-retarding, or air-entraining admixtures in recycled concrete have no significant effect on slump, air content, or setting time of the new concrete or compressive strength of the new concrete after hardening. High concentrations of water-soluble chloride ion in recycled concrete have been shown to contribute to accelerated corrosion of steel embedments included in the new concrete. It is not expected that the chloride-containing admixtures used at ordinary rates as normal-setting or accelerating admixtures in the recycled concrete will influence the setting time of the new concrete.

Presence of Mineral Admixtures

Mineral admixtures, such as fly ash, natural pozzolans (raw or calcined), or silica fume, included in recycled concrete are unlikely to modify the properties of the new concrete.

Unsound Recycled Concrete

Prospective sources of recycled concrete may be unsound or have been rendered unsound in service, namely, presence of finely porous aggregate that is susceptible to disruption by freezing and thawing, deterioration by sulfate attack, damage by fire or service at high temperatures, presence of alkali-silica or alkali-carbonate reaction, and so on. Tests and experience show that concrete disrupted by D-cracking can be successfully recycled as concrete aggregate, presumably in part because of the reduction in maximum size of the offending coarse aggregate, use of recycled concrete as coarse aggregate only, air entrainment of both the original and the new concrete, reduction of the water-cement ratio, and introduction of fly ash as a partial replacement of the portland cement [69].

Concrete affected by a harmful degree of alkali-aggregate reactivity should be rejected as a source of recycled concrete unless it can be shown that the expansive processes have terminated, the new concrete will serve under conditions in which the internal relative humidity will be maintained at very low levels, say 70% or less, or that no deleterious effects are found in appropriately extended tests of the proposed concrete mixtures. The possibility of introduction of additional alkalies into the particles of recycled concrete during the service exposure of the new concrete should be recognized, such as from the cementitious binder of the new concrete or by penetration of alkaline solutions from an external source, such as sea water, alkaline soils, deicing chemicals, or industrial media.

The utility of concrete damaged by fire or high-temperature service must be examined individually. Portions of the constructions may be suitable for such use. Of particular concern is recycled concrete that contains aggregates including disseminated pyrite (ferrous sulfide, FeS_2), such as certain dolomites and firm shales, where the concrete in service was maintained for long periods at temperatures in the range 150 to 300°C (370 to 575°F). As noted previously [23], pyritiferous dolomite or shales in aggregate can produce distress when the concrete is heated in that range due to expansion of the particles by oxidation and hydration of the pyrite and subsequent crystallization of iron and calcium sulfates. These observations are pertinent to evaluation of recycled concrete derived from concrete exposed to temperatures in this intermediate range, such as structural members adjacent to kilns and furnaces or constituting the frame of such facilities. They pertain also to use of such recycled aggregates in new concrete intended for these applications.

Chemical Contamination or Radioactivity

Concrete contaminated by noxious, toxic, or radioactive substances should be rejected for use as recycled aggregate. An otherwise desirable source may be salvaged by elimination of excessively contaminated portions of the concrete if they can be isolated and disposed of separately, such as portions of pavements or marine structures in which high concentrations of chloride or water-soluble sodium are found.

Contamination by Bituminous Materials

Recycled concrete containing asphaltic materials, such as bituminous concrete, may or may not affect air entrainment of the new concrete [69]. Enhanced entrainment of air may require use of an air-detraining admixture, but elimination of the contaminant is preferable. The proportions of such materials can be brought to acceptable levels by specifications on removal of overlayments, joint fillers, and the like during the course of demolition and processing of the recycled aggregate.

Metallic Contaminants

Metallic particles or fragments that survive the crushing and screening operations and magnetic separation usually will be in acceptable proportions in the finished recycled concrete. Ferrous particles may cause staining where they lie adjacent to exposed surfaces, especially in the presence of appreciable concentrations of water-soluble chloride salts. Small particles of aluminum metal or galvanized iron may produce blisters or shallow cracking on surfaces of fresh or "green" concrete as a result of the release of hydrogen formed by the interaction with the alkalies within the cementitious matrix.

Glass as a Contaminant

Fragments of plate glass, bottle glass, or glassware may be present in demolished concrete structures and difficult to avoid or to extract during production of recycled aggregate. As scattered particles lying adjacent to exposed surfaces, they can produce popouts and local cracking as a result of the alkali-silica reaction, and unsightly and annoying bulges may develop on floors where the eruptions are confined by flexible coverings. The effects of the alkali-silica reaction may develop in the presence of either low- or high-alkali cements.

Contamination by Brickwork

The report of RILEM Committee 37-RDC [70] concludes that up to 5% by mass of fragmented brick masonry usually can be tolerated in building rubble for production of recycled-concrete aggregate. On the other hand, highly porous, fired-clay brick may be susceptible to disruption when particles of such brick lie adjacent to surfaces that are exposed to freezing and thawing while wet. The report recommends that particles of brick rubble having a density less than 1.95 g/cm³ be rejected as lightweight particles in accordance with ASTM C 33.

Refractory bricks having a high content of crystalline magnesia (MgO, periclase) present special problems, such as where recycled concrete is obtained from facilities that include furnaces or kilns lined by such masonry. Large-size fragments of such bricks included in conventional aggregates are known to create cracking, spalling, and popouts as deep as 50 mm (2 in.) in portland-cement structural concrete, pavements, and tunnel linings.

Miscellaneous Contaminants

A Japanese standard for use of recycled concrete limits various impurities as shown in Table 9 [70].

Stringent limits are required on the allowable concentration of water-soluble sulfates in recycled concrete, such as that derived from gypsum plaster or plasterboard. The RILEM Committee report [70] recommends that the sulfate content of the new concrete be limited to 4%, expressed as the SO_3 content by mass of the entire concrete, and that the same limit be applied to the recycled-concrete aggregate. The Committee recommended further that stringent limits be placed on allowable amounts of

TABLE 9—Recommended Limits on Contaminants for New Concrete to be Subjected to Wetting and Drying or Freezing and Thawing as Reported by Committee 37-RDC, International Union of Testing Laboratories for Materials and Structures.

Material	Maximum Allowable % by Volume of Recycled Concrete[a]
Lime plaster	7
Gypsum plaster	3
Soil	5
Wood	4
Asphalt	2
Vinyl acetate paint	0.2

[a] Amount that reduced compressive strength 15% compared to the control concrete containing recycled concrete.

the contaminants listed in Table 9 for new concrete to be subjected to wetting and drying or freezing and thawing. Noting that organic substances, such as paint, may entrain excessive amounts of air, they recommend a limit of 0.15% by mass of the recycled-concrete aggregate for organic particles.

Examination of Prospective Sources of Recycled Concrete in the Field

The petrographer can assist in evaluation of prospective sources of recycled concrete by examining the intact structures or pavements at the site of demolition prior to initiation of procedures for aggregate production. This evaluation might constructively designate those locations or portions of concrete placements that are unsuitable or of questionable quality for use as concrete aggregate, such as portions highly contaminated by sea water, industrial chemicals or wastes, or radioactivity, and where extensive fenestration, installations of gypsum plaster or wallboard, or applications of paints or other coatings or overlayments will make difficult the securing of aggregate containing acceptable levels of deleterious substances. Similar avoidance of difficulties may relate to installations of lightweight and porous brickwork or potentially deleterious refractory bricks.

Such inspection should reveal the existence of portions of the installations that are seriously affected by the alkali-silica or alkali-carbonate reactions or severe corrosion of steel embedments. In each instance, use of the concrete in production of recycled concrete aggregate would be brought into question.

Performance of Petrographic Examination of Recycled Concrete

ASTM C 295 does not contemplate petrographic examination of recycled-concrete aggregate. Nevertheless, the procedures for examination of ledge rock, crushed stone, lightweight aggregates, and manufactured sand may be employed as appropriate.

The petrographic examination should be carried out with recognition of the possibility that the aggregate was derived from rubble that comprises two or more classes of concrete of differing composition and condition, possibly from more than one source, and may include contaminants of divergent types.

Objectives of the examination may include evaluation of processing procedures and facilities as well as determination of composition, quality, and anticipated performance of the aggregate in one or more concretes to be subjected to differing levels of attrition, impact, and load and to various types of environmental exposure.

The following features should be observed, recorded, and reported relative to the effectiveness of operations that constitute the production of the recycled concrete:

Grading
Particle shape
Surface texture
Angularity
Coatings of dust of fracture
Variability among nominally similar production lots
Contamination by foreign materials

The following features should be observed, recorded, and reported relative to the composition and condition of the particles:

Hardness (hard, moderately hard, weak, soft)
Relative proportion of mortar and coarse aggregate
Bond of mortar to coarse aggregate
Characteristics of the cement-paste matrix (firmness, absorptivity, bond to fine aggregate, luster on fresh fracture, presence of a mineral admixture, estimated water-cement ratio or water-to-cementitious materials ratio, microcracking)
Air-void content and parameters of the air-void system
Alteration of the cementitious matrix (carbonation, leaching, sulfate attack, staining)
Secondary chemical deposits in included aggregate, air voids, cracks, and cement-aggregate interface
Nature of coarse and fine aggregates (natural, crushed natural, crushed stone, slag, calcined or sintered lightweight aggregate, other)
Lithologic composition of aggregate and foreign materials
Quality of aggregates (sound, unsound, alkali reactive, thermal stability, cement-aggregate reactivity other than alkali-aggregate reaction)
Damage to original concrete during service (freezing before setting, freeze-thaw, leaching, sulfate attack, in situ hydration of iron sulfides, fire damage, high-temperature exposure, other)

Techniques of the petrographic examination directed to these objectives are covered by ASTM C 295, C 457, and C 856. Where the examination, other information, or instructions by the purchaser or supervisor show the need for partial chemical analysis of the recycled concrete or portions thereof, the chemical analytical work and petrographic examination should be closely coordinated and the data mutually exchanged.

Because of its great effect on the properties and performance of concrete containing recycled concrete, Hansen and Narud [71] determined the volumetric ratio of coarse aggregate and mortar in the recycled concrete used in their tests by casting specimens of a mixture of the aggregate and a binder identified by them as "red cement." Presumably, satisfactory specimens could be prepared with the portland cement blended with a suitable red iron-oxide pigment so as to allow easy distinction between the particles of recycled concrete and the binder. After hardening of the specimen, the relative proportions of coarse aggregate and mortar in the recycled concrete were determined on sawed and lapped surfaces by the linear traverse method. No details were provided as to the explicit criteria applied to distinguish sections of coarse aggregate from those of larger particles of fine aggregate.

This writer has employed the point-count method of ASTM C 457 in analysis of ordinary portland-cement concrete, distinguishing coarse aggregate from fine aggregate by classifying as coarse aggregate all particles that include a dimension greater than 4.75 mm (3/16 in.) in the plane of the section (see the previous section on Petrographic Examination of Aggregates in Hardened Concrete).

In any event, the method of Hansen and Narud can be employed to determine the relative proportions of coarse aggregate and mortar and of coarse aggregate, fine aggregate, and cementitious matrix. The same type of specimen can be used to determine air-void content and other parameters of the air-void system of the recycled concrete in accordance with ASTM C 457, these being features important to the performance of new concrete subjected to moderate or severe conditions of weathering exposure.

As noted previously, detection of rock types or other particles susceptible to the alkali-silica reaction or the alkali-carbonate reaction in recycled concrete presents two general problems, namely: (1) if pre-existing manifestations of one or both of these reactions are found, will they proceed sufficiently to damage the new concrete; and (2) if no evidence of these reactions is seen, will the composition and environmental exposure of the new concrete be such as to initiate and propagate these reactions to a deleterious extent? These questions require careful analysis of all factors concerned, including the kind and amount of alkali-reactive aggregate, the concentration of available alkalies in the recycled concrete, the previous extent of development of distress, the availability of alkalies in the new concrete (including those derived from the cement, mineral admixtures, and other sources), and the availability of moisture to the new concrete during the service exposure.

CONCLUSION

Petrographic examination should be included in the investigation and testing of concrete aggregate for use in permanent construction and in the manufacture of concrete products and precast structural elements. Applied in the field, the method aids exploration and sampling and permits preliminary evaluation of materials from alternative sources. Detailed examination of aggregates in the laboratory supplements the standard acceptance tests, especially by (1) detecting adverse properties, (2) comparing the aggregate with aggregates for which service records or previous tests are available, (3) explaining results of tests and justifying special tests as required, (4) detecting contamination, and (5) determining the efficiency and relative merit of processing and manufacturing methods for aggregate production.

Validity of the results depends upon the training and experience of the petrographer. However, with proper training and the adoption of uniform techniques and nomenclature, subjective elements in the examination are not significant.

The method can be applied effectively to sand, gravel, crushed stone, slag, natural and synthetic lightweight aggregates, and aggregates produced from recycled concrete.

REFERENCES

[1] Mielenz, R. C. and Witte, L. P., "Tests Used by the Bureau of Reclamation for Identifying Reactive Concrete Aggregates," *Proceedings*, American Society for Testing and Materials, Philadelphia, Vol. 48, 1948, pp. 1071–1103.

[2] DePuy, G. W., "Petrographic Investigations of Concrete and Concrete Aggregates at the Bureau of Reclamation," *Petrography Applied to Concrete and Concrete Aggregates, ASTM STP 1061*, B. Erlin and D. Stark, Eds., American Society for Testing and Materials, Philadelphia, 1990, pp. 32–46.

[3] Mather, K. and Mather B., "Method of Petrographic Examination of Aggregates for Concrete," *Proceedings*, American Society for Testing and Materials, Philadelphia, Vol. 50, 1950, pp. 1288–1312.

[4] Roberts, C. A., "Petrographic Examination of Aggregate and Concrete in Ontario," *Petrography Applied to Concrete and Concrete Aggregates, ASTM STP 1061*, B. Erlin and D. Stark, Eds., American Society for Testing and Materials, Philadelphia, 1990, pp. 5–31.

[5] Mielenz, R. C., "Petrographic Examination of Concrete Aggregates," *Bulletin*, Geological Society of America, Vol. 57, 1946, pp. 309–318.

[6] Mielenz, R. C., "Petrographic Examination of Concrete Aggregate," *Proceedings*, American Society for Testing and Materials, Philadelphia, Vol. 54, 1954, pp. 1188–1218.

[7] Rhoades, R. F. and Mielenz, R. C., "Petrography of Concrete Aggregate," *Proceedings*, American Concrete Institute, Vol. 42, 1946, pp. 581–600.

[8] Rhoades, R. F. and Mielenz, R. C., "Petrographic and Mineralogic Characteristics of Aggregates," *Mineral Aggregates, ASTM STP 83*, American Society for Testing and Materials, Philadelphia, 1948, pp. 20–48.

[9] Swenson, E. G. and Chaly, V., "Basis for Classifying Deleterious Characteristics of Concrete Aggregate Materials," *Proceedings*, American Concrete Institute, Vol. 52, 1956, pp. 987–1002.

[10] Mielenz, R. C., "Petrographic Examination of Concrete Aggregate to Determine Potential Alkali Reactivity," *Highway Research Report 18-C*, Highway Research Board, National Research Council, Washington, DC, 1958, pp. 29–35.

[11] Dolar-Mantuani, L., *Handbook of Concrete Aggregates*, Noyes Publications, Park Ridge, NJ, 1983.

[12] Abdun-Nur, E. A., "Random Sampling Made Easy," *Concrete International*, American Concrete Institute, Vol. 3, No. 9, 1981, pp. 66–70.

[13] Hansen, W. C., "Anhydrous Minerals and Organic Materials, as Sources of Distress in Concrete," *Highway Research Record No. 43*, Highway Research Board, National Research Council, Washington, DC, 1964, pp. 1–7.

[14] Mielenz, R. C., "Reactions of Aggregates Involving Solubility, Oxidation, Sulfates or Sulfides," *Highway Research Record No. 43*, Highway Research Board, National Research Council, Washington, DC, 1964, pp. 8–18.

[15] Smith, P., "Investigation and Repair of Damage to Concrete Caused by Formwork and Falsework Fire," *Proceedings*, American Concrete Institute, Vol. 60, 1963, pp. 1535–1566.

[16] Stark, D., "Alkali-Silica Reactivity: Some Reconsiderations," *Cement, Concrete, and Aggregates*, Vol. 2, 1980, pp. 92–94.

[17] Robinson, R. F., "Lithologic Characteristics of Concrete Aggregates as Related to Durability," *Cement, Concrete, and Aggregates*, Vol. 5, 1983, pp. 70–72.

[18] Aitcin, P. C. and Mehta, P. K., "Effect of Coarse Aggregate Characteristics on Mechanical Properties of High-Strength Concrete," *Materials Journal*, American Concrete Institute, Vol. 87, 1990, pp. 103–107.

[19] Dobie, T. R., "Correlating Water-Soluble Alkalies to Total Alkalies in Cement—Considerations for Preventing Alkali-Silica Popouts on Slabs," *Alkalies in Concrete, ASTM STP 930*, V. H. Dodson, Ed., American Society for Testing and Materials, Philadelphia, 1985, pp. 46–57.

[20] Sarkar, S. L. and Aitcin, P. C., "The Importance of Petrological, Petrographical, and Mineralogical Characteristics of Aggregates in Very High-Strength Concrete," *Petrography Applied to Concrete and Concrete Aggregates, ASTM STP 1061*, B. Erlin and D. Stark, Eds., American Society for Testing and Materials, Philadelphia, 1990, pp. 129–144.

[21] Midgley, H. G., "The Staining of Concrete by Pyrite," *Concrete Research*, Vol. 10, 1958, pp. 75–78.

[22] Shayan, A., "Deterioration of a Concrete Surface Due to the Oxidation of Pyrite Contained in Pyritic Aggregate," *Cement and Concrete Research*, Vol. 18, 1988, pp. 723–730.

[23] Soles, J. A., "Thermally Destructive Particles in Sound Dolostone Aggregate from an Ontario Quarry," *Cement, Concrete, and Aggregates*, Vol. 4, 1982, pp. 99–102.

[24] Stark, D. and Bhatty, S. Y., "Alkali-Silica Reactivity: Effect of Alkalies in Aggregate on Expansion," *Alkalies in Concrete, ASTM STP 930*, V. H. Dodson, Ed., American Society for Testing and Materials, Philadelphia, 1986, pp. 16–30.

[25] Grattan-Bellew, P. E. and Beaudoin, J. J., "Effect of Phlogopite Mica on Alkali-Aggregate Expansion in Concrete," *Cement and Concrete Research*, Vol. 10, 1980, pp. 789–797.

[26] Buck, A. D., Kennedy, T. B., Mather, B., and Mather, K., "Alkali-Silica and Alkali-Carbonate Reactivity of a South Dakota Sand," Miscellaneous Paper 6-530, U.S. Army Corps of Engineers, Waterways Experiment Station, Vicksburg, MS, 1962.

[27] Mielenz, R. C. and King, M. E., "Physical-Chemical Properties and Engineering Performance of Clays," *Bulletin 1969*, Division of Mines, Department of Natural Resources, State of California, San Francisco, CA, 1955, pp. 196–250.

[28] Schmitt, J. W., "Effects of Mica, Aggregate Coatings, and Water-Soluble Impurities on Concrete," *Concrete International*, American Concrete Institute, Vol. 12, No. 12, 1990, pp. 54–58.

[29] Gaynor, R. D. and Meininger, R. C., "Evaluating Concrete Sands," *Concrete International*, American Concrete Institute, Vol. 5, No. 12, 1983, pp. 53–60.

[30] Higgs, N. B., "Methylene Blue Testing of Smectite as Related to Concrete Failure," *Petrography Applied to Concrete Aggregates, ASTM STP 1061*, B. Erlin and D. Stark, Eds., American Society for Testing and Materials, Philadelphia, 1990, pp. 47–54.

[31] Davis, R. E., Mielenz, R. C., and Polivka, M., "Importance of Petrographic Analysis and Special Tests Not Usually Required in Judging the Quality of a Concrete Sand," *Journal of Materials*, American Society for Testing and Materials, Philadelphia, Vol. 2, No. 3, 1967, pp. 461–479.

[32] Higgs, N. B., "Chlorite: A Deleterious Constituent With Respect to Freeze-Thaw Durability of Concrete Aggregates," *Cement and Concrete Research*, Vol. 17, 1987, pp. 793–804.

[33] Hadley, D. W., "Alkali Reactivity of Carbonate Rocks—Expansion and Dedolomitization," *Proceedings*, Highway Research Board, National Research Council, Washington, DC, Vol. 40, 1961, pp. 462–474.

[34] Swenson, E. G. and Gillott, J. E., "Alkali Reactivity of Dolomitic Limestone Aggregate," *Concrete Research*, Vol. 19, 1967, pp. 95–104.

[35] Smith, P., "15 Years of Living at Kingston With a Reactive Carbonate Rock," *Highway Research Record No. 525*, Transportation Research Board, Washington, DC, 1974, pp. 23–27.

[36] Ryell, J., Chojnacki, B., Woda, G., and Koniuszy, D., "The Uhthoff Quarry Alkali Carbonate Rock Reaction: A Laboratory and Field Performance Study," *Highway Research Record No. 525*, Transportation Research Board, Washington, DC, 1974, pp. 43–54.

[37] Ozol, M. A. and Newlon, H. H., Jr., "Bridge Deck Deterioration Promoted by Alkali-Carbonate Reaction: A Documented Example," *Highway Research Record No. 525*, Transportation Research Board, Washington, DC, 1974, pp. 55–63.

[38] Buck, A. D., "Control of Reactive Carbonate Rocks in Concrete," Technical Report C-75-3, U.S. Army Corps of Engineers, Waterways Experiment Station, Vicksburg, MS, 1975.

[39] Dolar-Mantuani, L., "Alkali-Silica-Reactive Rocks in the Canadian Shield," *Highway Research Record No. 268*, Transportation Research Board, Washington, DC, 1969, pp. 99–117.

[40] Bachiorrini, A. and Montanaro, L., "Alkali-Silica Reaction (ASR) in Carbonate Rocks," *Cement and Concrete Research*, Vol. 8, 1988, pp. 731–738.

[41] Buck, A. D., "Alkali Reactivity of Strained Quartz as a Constituent of Concrete Aggregate," *Cement, Concrete, and Aggregates*, Vol. 5, 1983, pp. 131–133.

[42] Cortelezzi, C. R., Maiza, P., and Pavlicevic, R. E., "Strained Quartz in Relation to Alkali-Silica Reaction," *Petrography Applied to Concrete and Concrete Aggregates, ASTM STP 1061*, B. Erlin and D. Stark, Ed., American Society for Testing and Materials, Philadelphia, 1990, pp. 145–158.

[43] Gillott, J. E., Duncan, M. A., and Swenson, E. G., "Alkali-Aggregate Reaction in Nova Scotia. IV. Character of the Reaction," *Cement and Concrete Research*, Vol. 3, 1973, pp. 521–535.

[44] Zoldners, N. G., "Effect of High Temperatures on Concrete Incorporating Different Aggregates," *Proceedings*, American Society for Testing and Materials, Philadelphia, Vol. 60, 1960, pp. 1087–1108.

[45] Mass, G. R., "Aggregate Particle Shape Considerations," *Concrete International*, American Concrete Institute, Vol. 5, No. 8, 1983, pp. 23–29.

[46] Mielenz, R. C., "Diagnosing Concrete Failures," Stanton Walker Lecture Series on the Material Sciences, Lecture No. 2, University of Maryland, College Park, MD, 1964.

[47] Mielenz, R. C., "Petrography Applied to Portland-Cement Concrete," *Reviews in Engineering Geology*, Geological Society of America, Vol. 1, 1962, pp. 1–38.

[48] *Concrete Manual*, Bureau of Reclamation, U.S. Department of the Interior, 8th ed., 1975.

[49] Verbeck, G. J. and Landgren, R., "Influence of Physical Characteristics of Aggregates on Frost Resistance of Concrete," *Proceedings*, American Society for Testing and Materials, Philadelphia, Vol. 60, 1960, pp. 1063–1079.

[50] Holland, W. Y. and Cook, R. H., "Alkali Reactivity of Natural Aggregates in Western United States," *Mining Engineer*, 1953, p. 991.

[51] Fraser, G. S., Harvey, R. D., and Heigold, P. C., "Properties of Chert Related to Its Reactivity in an Alkaline Environment," Circular 468, Illinois State Geological Survey, Urbana, IL, 1972.

[52] Loughlin, G. F., "Usefulness of Petrology in the Selection of Limestone," *Rock Products*, Vol. 31, 1928, p. 50.

[53] Gutt, W., Kinniburgh, W., and Newman, A. J., "Blast-Furnace Slag as Aggregate for Concrete," *Concrete Research*, Vol. 19, 1967, pp. 71–82.

[54] Josephson, G. W., Sillers, F., Jr., and Runner, D. G., "Iron Blast Furnace Slag Production, Processing, Properties, and Uses," *Bulletin No. 479*, Bureau of Mines, U.S. Department of the Interior, Washington, DC, 1949.

[55] Rigby, G. R., *The Thin Section Mineralogy of Ceramic Materials*, British Ceramic Research Association, 2nd ed., 1953.

[56] Nurse, R. W. and Midgley, H. G., "The Mineralogy of Blast Furnace Slag," *Silicates Industriels*, Vol. 16, No. 7, 1951, pp. 211–217.

[57] Insley, H. and Frechette, V. D., *Microscopy of Ceramics and Cements*, Academic Press, Inc., New York, 1955.

[58] Snow, R. B., "Identification of CaO·MgO Orthosilicate Crystals, Including Merwinite ($3CaO·MgO·2SiO_2$), Through the Use of Etched Polished Sections," Technical Publication No. 2167, American Institute of Mining, Metallurgical, and Petroleum Engineers, New York, 1947.

[59] Parker, T. W. and Ryder, J. F., "Investigations on "Falling" Blast Furnace Slag," *Journal*, Iron and Steel Institute (London), Vol. 146, No. II, 1942, pp. 21P–51P.

[60] "Report of Committee 201 on Blast Furnace Slag as Concrete Aggregate," *Proceedings*, American Concrete Institute, Vol. 27, 1930–31, pp. 183 and 661.

[61] Stutterheim, N., "The Risk of Unsoundness Due to Periclase in High Magnesia Blast Furnace Slag," Monograph 43, *Proceedings*, 4th International Symposium on the Chemistry of Cement, National Bureau of Standards, Washington, DC, Vol. 2, 1960, pp. 1035–1040.

[62] McCaffery, R. S., Jr., Oesterle, R., and Schapiro, L., "Composition of Iron Blast Furnace Slag," Technical Publication No. 19, American Institute of Mining, Metallurgical, and Petroleum Engineers, New York, 1927.

[63] Nordberg, B., "The Basalt Rock Company Story," *Rock Products*, Vol. 57, 1954, p. 104.

[64] Mielenz, R. C., Greene, K. T., and Benton, E. J., "Chemical Test for Reactivity of Aggregates With Cement Alkalies; Chemical Processes in Cement-Aggregate Reaction," *Proceedings*, American Concrete Institute, Vol. 44, 1948, p. 193–221.

[65] Hickey, M. E., Cordon, W. A., and Price, W. H., "Properties of Concrete Made with Typical Lightweight Aggregates—Housing and Home Finance Agency Research Program," Laboratory Report No. C-385, Bureau of Reclamation, U.S. Department of the Interior, Denver, CO, 1948.

[66] Price, W. H. and Cordon, W. A., "Tests of Lightweight Aggregate Concrete Designed for Monolithic Construction," *Proceedings*, American Concrete Institute, Vol. 45, 1949, pp. 581–600.

[67] "Cement and Concrete Terminology," *Publication SP-19*, (ACI 116R-90), Committee 116, American Concrete Institute, Detroit, MI, 1990.

[68] Buck, A. D., "Recycled Concrete as a Source of Aggregate," *Proceedings*, American Concrete Institute, Vol. 74, 1977, pp. 212–219.

[69] Forster, S. W., "Recycled Concrete as Aggregate," *Concrete International*, American Concrete Institute, Vol. 8, No. 10, 1986, pp. 34–40.

[70] Hansen, T. C., "Recycled Concrete and Recycled Aggregate," *Materials and Structures* (RILEM), Vol. 19, 1986, pp. 201–246.

[71] Hansen, T. C. and Narud, H., "Strength of Recycled Concrete Made from Crushed Concrete Coarse Aggregate," *Concrete International*, American Concrete Institute, Vol. 5, No. 1, 1983, pp. 79–83.

[72] Nixon, P. J., "Recycled Concrete as an Aggregate for Concrete—A Review," *Materials and Structures*, International Union of Testing and Research Laboratories for Materials and Structures (RILEM), Vol. 11, 1978, pp. 371–378.

[73] Hansen, R. C. and Hedegard, S. E., "Properties of Recycled Aggregate Concrete as Affected by Admixtures in Original Concretes," *Proceedings*, American Concrete Institute, Vol. 81, 1984, pp. 21–26.

Alkali-Silica Reactions in Concrete

David Stark[1]

PREFACE

This chapter is the third of a series dealing with chemical reactions of aggregates in concrete. The first version, entitled "Chemical Reactions," was written by W. C. Hansen and included in *ASTM STP 169A*, published in 1966. It dealt with several types of chemical reactions known at the time to occur in concrete, including alkali-silica reactivity. The second version was written in 1978 by Sidney Diamond for *ASTM STP 169B*, and was entitled "Chemical Reactions Other Than Carbonate Reactions." It specifically excluded alkali-carbonate reactivity and emphasized alkali-silica reactivity. The present version is focused entirely on alkali-silica reactivity.

The shifting emphasis in this chapter since 1966 particularly reflects the predominating concern with alkali-silica reactivity in present-day construction and also reflects our updated knowledge of how to deal more effectively with the problem. Accordingly, more detail and emphasis than previously included are provided for the practicing technologist responsible for providing long-term durable concrete.

INTRODUCTION

For many years, the selection of aggregates for use in concrete was based primarily on physical characteristics such as grading, particle shape, hardness, density, and "cleanness." Likewise, such characteristics were the major, if not the only, consideration of the engineer in achieving long-term specified concrete strengths and volume stability. Virtually no attention was given to the chemical or mineralogical composition of the aggregate, despite the known fact that concrete is a highly alkaline system in which pore solutions usually exceed pH values of 13.

In the late 1930s, Stanton, of the California Division of Highways, detected a previously unknown deleterious chemical reaction between pore solutions in the concrete and certain types of siliceous aggregates [1]. He determined that resulting expansion could lead to abnormal cracking, reduction in strength, and early in-service failure of the concrete structure. This phenomenon became known as alkali-silica reactivity (ASR).

By the 1940s, ASR was known to have developed in widespread areas of the United States and through the following decades has been identified in most areas of the world. The fact that aggregates must be evaluated for chemical stability in concrete as well as for physical competence thus became evident with costly realization. In response to this situation, ASTM has developed standards for evaluation of materials to avoid deleterious ASR. Salient features of ASR, together with a summary of existing methods of ameliorating ASR, are given in this chapter.

SYMPTOMS OF ALKALI-SILICA REACTIVITY

Symptoms of ASR are typified by the expansion of concrete, usually accompanied by cracking [2]. Expansions without discernible or abnormal cracking may be indicated by closing of joints, relative movement or displacement of concrete members, or abnormal movement or misalignment of machinery embedded in, or anchored on, the concrete. Creep of the concrete may initially obscure evidence of expansive ASR but, commonly, the reaction is carried to the extent of inducing cracking on exposed surfaces as well as at internal reaction sites. Resulting patterns of distress depend on the magnitude and orientation of restraint to movement within the concrete. Where restraint is more or less uniform and at low levels, interconnected "pattern" or "map" cracks develop that define polygonal areas more or less similar to those that develop in mud flats upon drying. These cracks may remain open for indefinite periods of time or become filled with secondary reaction products, depending on ambient exposure conditions. Prolonged drying tends to result in persistently open cracks, while frequent wetting combined with drying results in filling of the cracks. Crack width also determines the degree of crack filling.

Most concrete members are subjected to nonuniform restraint in one form or another. Embedded reinforcing steel is a common form of internal restraint tending to cause cracks to form parallel to the steel. That is, greatest expansion develops normal to the orientation of the steel. In reinforced support columns or parapet walls, the predominant orientation of cracking will be "longitudinal," interconnected by less prominent random cracking. Restraint may exist external to the concrete member as well, such as in pavement concrete where abutting panels

---

[1] Senior principal scientist, Materials Research and Consulting, Construction Technology Laboratories, Inc., Skokie, IL 60077.

provide restraint that results in cracking predominantly in the longitudinal direction.

Triaxial restraint sufficient to overcome the expansive forces associated with ASR, even though the restraint may be nonuniform, apparently is capable of preventing progressive cracking due to ASR. Mass concrete, such as in large dams, may exhibit random or map cracking on external surfaces where restraint is minimal, but fail to develop associated cracking tens of feet inside the concrete unit. The mass of the concrete, confinement by abutments, plus relatively uniform conditions (for example, temperature and relative humidity) appear to minimize distress associated with ASR.

Ambient conditions also may serve to accentuate cracking associated with ASR. Once cracks develop due to expansion, significant shrinkage due to drying may occur at greater depths, thereby further opening cracks. However, these effects are relatively shallow and may extend only a few centimetres below the exposed surface.

Cracking resulting from ASR may be only cosmetic in effect, or actually lead to impairment of performance. In the latter case, other forces are generally found to be present, such as traffic loading, freezing and thawing, etc.

On a finer scale, ASR is uniquely characterized by a glassy clear to white alkali-silica gel reaction product that is found within reacted aggregate particles, in air voids, and in cracks. The presence of such gel indicates that ASR has occurred but not necessarily that it has caused abnormal expansion or distress.

Reaction rims usually occur on reacted aggregate particles and are associated with ASR although, on natural surfaces of sand and gravel particles, they can be confused with rims due to weathering prior to use as aggregate.

MECHANISM OF REACTIONS AND DISTRESS IN CONCRETE

Voluminous literature [3] has been produced dealing with all aspects of ASR, including mechanisms of reaction and distress. Review of these data [4] indicates that the Powers and Steinour [5] model best fits published results. Powers and Steinour proposed that safe reactions are explained as those in which sufficient calcium ion is available in alkaline pore solutions to produce limited-swelling high-calcium-content gel. Conversely, when less calcium ion is available, a higher alkali content gel forms, which has a greater capacity to swell with absorption of moisture.

The initial surface reaction in this model is assumed to occur in contact with the pore solution in the cement paste, which is saturated with respect to calcium hydroxide and therefore produce essentially a non-swelling reaction product. For a safe reaction to continue, it is assumed that consumption of alkali in the reaction results in greater solubility and availability of calcium at subsequent reaction sites, thereby continuing to produce high calcium non-swelling gel reaction product. These safe reactions were believed to begin and continue without expansion if the initial alkali concentration is not greater than that produced by cements of about 0.6% equivalent sodium oxide (Na_2O), or about 0.4 normal sodium hydroxide ($0.4N$ NaOH) concentration in pore solutions at 0.50 water-cement ratio (w/c) by mass.

According to Powers and Steinour, an expansive reaction occurs when alkali concentrations in pore solutions are higher, thus reducing the concentration of calcium ion. Accordingly, a low-calcium, high-alkali unlimited-swelling gel reaction product is formed. In this case, the reacted layer in the aggregate becomes too thick, and calcium ion can not diffuse to the reaction site at a sufficient rate to prevent formation of a swelling gel. The diameter of the hydrated calcium ion is substantially larger than that of the hydrated sodium or potassium ion, with the result that they diffuse much more readily than the calcium ion to the internal reaction sites within a particle of reactive aggregate. The lower solubility of calcium in solution and the lower diffusion rate of calcium to reaction sites within the particle both serve to reduce its participation in the reaction. However, if the amount of reactive silica, or its fineness, is increased sufficiently, the alkali content of the solution is rapidly reduced to a safe level so that a higher calcium, limited-swelling gel can form. Obviously, innumerable intermediate-alkali and corresponding calcium ion concentrations, diffusion rates, permeabilities of reactive aggregates, moisture availabilities, etc. exist and result in intermediate conditions in which ASR develops.

One other aspect of the Powers and Steinour model is the diffusion of silica away from reaction sites and out of reacted particles. They proposed that silica was able to diffuse out through the reaction layer in the aggregate particle and that expansion-controlling processes were diffusion rates of alkali and calcium to reaction sites. Chatterji [6] suggested that calcium hydroxide plays an additional role in ASR in that it can cause or increase expansions by impeding the escape of dissolved silica from reactive particles. In this circumstance, calcium reacts to form calcium-alkali silica gel that blocks escape of other reaction products out of the particle.

As might be expected, ASR gels vary widely in composition, depending not only on circumstances just indicated, but on other factors such as distance from the reaction site and movement along cracks [7]. Electron microprobe and X-ray studies suggest that ASR gels are two-phase composites consisting of the swelling alkali-silicate hydrate phase of relatively narrow composition and a limited swelling calcium-alkali silicate hydrate phase. The latter varies in both proportion in the gel and composition and, being calcium-bearing, provides some rigidity to the paste structure. Viscosity of the gel is a major factor affecting swelling pressures generated in concrete.

Types of Reactive Aggregate

Potentially deleteriously reactive rock types probably exist in every state in the United States and in most countries throughout the world. Many natural siliceous rock types are known to have reacted deleteriously with the highly alkaline solutions in concrete. Probably the most commonly used of such aggregates are those that contain non-crystalline or poorly or imperfectly crystalline silica. These include opaline and chalcedonic silica that are found in such rocks as chert, flint, siliceous shale, as secondary fillings in voids in, for example, basalt, and as interstitial and interlayer material in rock types such as sandstone. Reactive silica of this type also is found as microscopic veinlets in otherwise innocuous rock types such as limestone and dolomite. The most highly reactive rock types known to the writer are shales that contain opaline diatoms, which have been observed to react deleteriously and produce distress within one day of mixing the concrete.

Deleteriously reactive silica also is present in metamorphic rocks where it is classified as strained quartz and microcrystalline quartz. In these cases, crystal structures are variously distorted, thus rendering them susceptible to deleterious ASR. This imperfectly crystallized quartz has been recognized as the reactive component of such metamorphic rocks as gneiss, schist, metagraywacke, and quartzite. These rock types are comparatively slowly reactive and are known to have exhibited deleterious reactivity with high-alkali cements only after five or more years of service.

The other general reactive component of siliceous aggregates is the glassy to cryptocrystalline matrix of volcanic rocks of approximately rhyolitic to andesitic composition. This material constitutes a major proportion of these rock types, and is commonly altered by weathering that facilitates penetration of alkaline solutions into the aggregate particle. Weathered volcanic rock is particularly troublesome in that it has produced ASR-associated distress when used with cements with less than 0.60% equivalent Na_2O. Obsidian and pumice, which are very dense and very porous volcanic glasses, respectively, also are known to have reacted deleteriously in concrete. Slightly metamorphosed volcanic rocks, such as metarhyolite, also have been found to produce expansive ASR.

Moisture Availability and Environmental Effects

As indicated previously, moisture must be sufficiently available in the concrete to permit expansive ASR to occur. Laboratory data indicate that relative humidity (RH) values in the concrete must exceed about 80%, referenced to 23 ± 1°C, before expansion due to ASR can develop [8]. This condition is easily met in outdoor slab-on-grade concrete, regardless of climate, including desert conditions. Interior portions of elevated concrete members, such as bridge decks, support columns, and girders, also exceed the 80% RH level, at least on a cyclic basis even where prolonged severe drying conditions prevail. Except for near-surface concrete in elevated structures, concrete in damp climates can be expected to be sufficiently damp to support expansive ASR on an almost continuous basis. Near-surface concrete experiences major RH fluctuations that support expansive ASR on an intermittent basis, regardless of climate. Where high water-cement ratios are used, residual mixing water may be sufficient to support expansive ASR in mass concrete, even in interior air-conditioned exposures.

METHODS OF IDENTIFYING POTENTIALLY REACTIVE AGGREGATE

ASTM has available three methods intended to identify potentially reactive siliceous aggregates. They are:

1. ASTM Test Method for Potential Alkali Reactivity of Cement-Aggregate Combinations (Mortar-Bar Method) (C 227),
2. ASTM Test Method for Potential Reactivity of Aggregates (Chemical Method) (C 289), and
3. ASTM Guide for Petrographic Examination of Aggregates for Concrete (C 295).

ASTM also has under consideration a test that is rapid and shows promise of greater reliability than either ASTM C 227 and C 289.

Another test, ASTM Test Method for Potential Volume Change of Cement-Aggregate Combinations (C 342), was intended to identify expansive cement-aggregate reactions involving so-called sand-gravel aggregates found primarily in eastern Colorado, Kansas, Nebraska, and southeastern Wyoming. The basis for this test was the belief that the cause of distress observed in concrete made with these aggregates was more or less unique to those areas and so deserved the special test procedure. The writer believes that alkali-silica reactivity is the major cause of distress and that the test is of little additional value. The use and limitations of the other tests are described subsequently.

ASTM C 227—Mortar Bar Method

ASTM C 227 has been an ASTM standard since 1950. Mortar-bar tests were initially conceived by Stanton as a means of identifying safe or unsafe cement-aggregate combinations and, in this sense, simulated what might actually occur in concrete [1]. They are made using the aggregate in question, sized to a fine aggregate grading. If a coarse aggregate is to be evaluated, it must be crushed, sized, and washed to the prescribed grading. A high-alkali cement, approximately 1.0% equivalent Na_2O, or the contemplated job cement, is used in the test. Four companion mortar bars are made and stored over water in a sealed container held at 38 ± 2°C for the duration of the test period. A zero reference comparator reading is made the day following casting, and biweekly to bimonthly comparator readings are made to a suggested test period of six months.

The Appendix of ASTM Specification for Concrete Aggregates (C 33), presently states that in ASTM C 227, cement-aggregate combinations showing expansions greater than 0.10% at six months "usually should be considered capable of harmful reactivity." The Appendix also states that combinations showing expansions of at least 0.05% at three months should be considered "potentially capable of harmful reactivity." During the 1950s, the 0.05% and 0.10% expansion levels were applied to test ages of six and twelve months, respectively. This was a stricter test criterion.

Forty years of experience with ASTM C 227 has revealed shortcomings that, in some cases, have resulted in misleading conclusions with eventual deterioration of concrete structures. The suggested expansion criteria appear to be valid for rock or mineral types arbitrarily defined as rapidly reactive aggregates, such as opaline-bearing materials and weathered glassy to cryptocrystalline volcanics of rhyolitic to andesitic composition, when they are tested with cement with 1.0% alkali. ASTM C 33 leaves open to the engineer's discretion the option of extending the test period to, for example, 12 or more months. However, extending the test period to more than 12 months still results in failure to define so-called slowly reactive aggregates as potentially deleteriously reactive in field structures. These aggregate types include certain gneisses, schists, graywackes, and metavolcanics. Even when used with cements with 1.0% alkali, expansions for these rock types may reach no more than 0.05% after two to three, or more, years [9]. Obviously, such a time frame is not practical in these cases.

A further limitation resulting from ASTM C 227 results in a recommendation to use low-alkali cement (not more than 0.60% equivalent Na_2O) if an aggregate is judged to be potentially deleteriously reactive. Field and laboratory studies [9] have revealed that certain aggregates, such as glassy volcanics, are deleteriously reactive with low as well as high-alkali cements.

More recently, leaching of alkalies has been found to be a serious limitation on expansion [10] in ASTM C 227. Thus, an aggregate may be judged to be innocuous when used with a cement of certain alkali level when, in fact, it may otherwise have been deleteriously reactive.

Overall, ASTM C 227 is useful for identifying certain aggregates as deleteriously reactive. However, cases exist where the procedure falls short of requirements.

ASTM C 289—Chemical Method

A second procedure used to identify whether an aggregate is potentially deleteriously reactive is ASTM C 289, commonly referred to as the Quick Chemical Test. This method was originally developed by Mielenz and co-workers at the U.S. Bureau of Reclamation [11]. In this procedure, the aggregate in question is sized to the 300- to 150-μm sieves, then three 25-g samples of the material are immersed in $1N$ NaOH solution in sealed containers held at 80 ± 1°C for 24 h. The resulting suspensions are then filtered, and the filtrates analyzed for reduction in the original hydroxyl ion concentration caused by reaction with the test sample, and for the amount of silica dissolved in the solution. The result is plotted on a graph where dissolved silica in millimoles per litre is plotted logarithmically on the abscissa and reduction in alkalinity is plotted on a linear scale on the ordinate. A curve is drawn on the graph in which "Aggregate Considered Innocuous" is located on one side of the curve, and "Aggregate Considered Deleterious" or "Potentially Deleterious" are located on the other. Innocuous aggregates are considered to be those that produce little or no reduction in alkalinity, or those for which a significant reduction in alkalinity is accompanied by relatively little dissolution of silica. Location of the innocuous-deleterious curve on the graph presumably is based on field performance and on ASTM C 227 mortar-bar test results, with the mortar bar criterion being 0.10% or greater linear expansion.

ASTM C 289 appears to fairly reliably identify rapidly reactive aggregates susceptible to expansive ASR when used with high-alkali cements. However, like ASTM C 227, this test fails to identify many slowly reactive aggregates, such as gneiss, schist, or quartzite, whose alkali reactivity results from strained or microcrystalline quartz. In these cases, both the amount of dissolved silica and the reduction in alkalinity are very low.

ASTM C 295—Petrographic Examination

A more reliable ASTM standard for identifying potentially reactive aggregate is ASTM C 295, a guide for petrographic examination of aggregates for concrete. This can be done rapidly but is somewhat subjective in nature, and depends on the experience and capabilities of the petrographer examining the aggregate. Fundamentally, the results depend on the judgment of the petrographer in contrast to fixed measuring procedures for classifying the aggregate. Its major use is for alerting the materials engineer to the presence of potentially reactive rock and mineral types and is highly recommended for use in conjunction with other tests and field performance evaluations.

Rapid Mortar-Bar Expansion Test

It is evident from the preceding discussion that limitations exist in current versions of ASTM C 227 and C 289 for identifying known slowly reactive aggregates, such as gneiss, schist, and quartzite. A test developed in South Africa at the National Building Research Institute [12] was designed to rapidly identify slowly as well as highly reactive aggregates, based on field performance record. However, it is not a test of cement-aggregate combinations. It was identified as ASTM Proposed Test Method for Accelerated Detection of Potentially Deleterious Expansion of Mortar Bars Due to Alkali-Silica Reaction (P 214).

Aggregate for the P 214 test is sized and washed to meet the specified grading, then cast into mortar bars of 0.50 w/c and stored in a moist cabinet at 23 ± 2°C. After one day, the specimens are removed from the molds, a comparator reading is obtained, then stored for 24 h in water brought to 80 ± 2°C. An additional comparator reading is obtained, then the bars are transferred to 1 N NaOH solution and again stored at 80 ± 2°C in sealed containers for 14 days. Periodic comparator readings, for example, at 1, 3, 7, 10, and 14 days, are made, while being careful to prevent cooling and drying during the taking of the readings. Thus, total time required between making the mortar bars and obtaining the final measurement of length change is 16 days. Test criteria are suggested as follows.

1. Expansions in NaOH solution greater than 0.20% indicate potentially deleteriously reactive aggregate.
2. Expansions between 0.10 and 0.20% are uncertain but are known to include aggregates that have reacted deleteriously in field concrete.

3. Expansions not greater than 0.10% represent innocuous aggregates.

There are several important points in the procedure that must be emphasized. First, the alkali content of the cement used in making the mortar bars is comparatively unimportant because NaOH in the immersion solution diffuses into the mortar bars and exerts overwhelming control on reactions. Second, the test result indicates only that the aggregate is potentially reactive, not that it will react deleteriously in field concrete. Cement alkali level exerts control on field performance. Third, the test has successfully identified, in conformance with field observations, the deleterious nature of reactivity with the slowly reactive aggregates, such as certain gneisses, schists, and quartzites. Overall, the procedure is rapid and compensates for but may overstate shortcomings found in ASTM C 227 and C 289.

At the present time, two ASTM procedures, C 295 and P 214, together with careful examination of field performance, collectively appear to be the most reliable indicators of potential for deleterious ASR. At the time of this writing (1992), evaluating aggregates for reactivity is in a state of uncertainty, but the writer believes the three procedures listed here will emerge, perhaps with some modification, as the most reliable for identifying potential for expansive ASR.

AVOIDING EXPANSIVE ALKALI-SILICA REACTION

Three general methods are used today to prevent abnormal expansions due to ASR. They are:

1. avoid use of the reactive aggregate,
2. limit the cement alkali level, and
3. use pozzolans or other admixtures.

The most feasible method for a particular job will depend on economics and the local availability of suitable materials. In some cases, combinations of the various methods have been used. Each of the three methods is discussed next.

Avoid Use of Reactive Aggregate

At first glance, it would seem a simple matter to use innocuous aggregates to avoid expansive ASR. However, this presupposes that the "innocuous" aggregate has a proven service record and has been evaluated properly using pertinent ASTM standards. If field service record is to be used, assurance must be made that the prospective aggregate from a commercial or lithologic source is of similar petrographic character as that on which the service record is based. If quarried stone is being considered, for example, one must be careful to evaluate the same rock strata, particularly where bedding is convoluted or inclined. If water-laid deposits from a given source of sand or gravel are being considered, one must be sure to evaluate materials from, for example, the same river terrace level. Petrologic character of the source material can vary significantly not only vertically from bed to bed but also laterally. Variations often develop in mineralogic composition, proportion, and particle size of individual rock types in the prospective aggregate source.

Use of the field service record also requires that cement alkali level be known for the existing concrete, regardless of whether low or high alkali cement was used previously and will be used in the new construction. Aggregate from a particular source that performed innocuously when used with a reactive aggregate may not perform safely when combined with an innocuous aggregate from some other source.

A second major factor in evaluating field service record is the environment to which the concrete is exposed. This includes temperature, moisture accessibility, and restraint. Elevated temperatures and ready availability of moisture are known to exacerbate alkali-silica reactivity, whereas, restraint or confining pressure reduces expansion and reactivity-induced distress.

Deleterious ASR has been observed in concrete structures in all habitable climates, including hot arid as well as cool damp regions where the concrete is exposed directly to the ambient outdoor conditions. Thus, moisture usually is sufficiently available, either continuously or cyclically, to support expansive ASR, and that temperature change, in a practical sense, serves only to change the rate of reaction. Commonly, increases in ambient outdoor temperature are accompanied by reductions in relative humidity with possible offsetting effects on ASR. For example, increase in temperature may increase the rate of chemical reaction, but drying would remove water that might otherwise be available for absorption by, and resulting expansion of, ASR gel. Water-cement ratio also will affect reactivity, with residual mixing water possibly being the only source of moisture available to support expansive reactions.

A third factor that has emerged in the assessment of field performance is extraneous sources of alkali, such as sodium chloride (NaCl) deicer salts, alkali salts occurring naturally in soils, for example, sodium sulfate (Na_2SO_4), industrial solutions (for example, NaOH), and seawater. Simple calculations can show that many times the quantity of alkali present in the cement can diffuse into concrete from external sources, increase hydroxyl ion concentration, and possibly convert an otherwise innocuous cement-aggregate combination into a deleterious one. Thus, the observer must consider all of these factors in his assessment of field performance relating to potential for deleterious ASR.

Results of considerable laboratory research have alluded to so-called pessimum (intermediate) proportions of reactive aggregate at which maximum expansion rates and levels develop. This appears to be largely of academic interest except to explain some case histories in which all of the aggregate was reactive but no damage resulted while, in related cases, only a relatively small proportion of reactive material was used and damage resulted. Attempting to specify aggregates for field concrete on the basis of avoiding pessimum ratios should be avoided.

Limit Cement Alkali Level

As noted earlier, cement alkali level is a major factor determining expansion due to ASR. In general, the higher the cement alkali level, the greater will be expansion due to reactivity. Simply specifying low-alkali cement (less than 0.60% equivalent Na_2O) to avoid expansive ASR has not always been successful. In some cases, cement alkali levels as low as 0.45 to 0.50% have failed to prevent expansive reactivity where partially weathered glassy to cryptocrystalline volcanic rock of rhyolitic to andesitic composition constitute no more than 5 to 10% of the aggregate. The best indicator presently available to determine a safe cement alkali level is field performance record where all pertinent factors are properly considered. Another option in this case is to use a pozzolan or mineral admixture meeting ASTM requirements as discussed next.

Use Admixed Materials

Finely ground admixed materials may be included in concrete to permit the safe use of otherwise potentially reactive cement-aggregate combinations. These materials are intended to interfere with the normal course of the reaction by complexing alkalies to reduce hydroxyl ion concentrations (pH) in concrete, or by altering diffusion rates of alkali and lime to reaction sites, thereby resulting in the formation of harmless nonexpansive reaction products. The most widely used material for this purpose is fly ash obtained from power plants burning coal. Other, less commonly used, materials include ground granulated blast-furnace slag, finely ground volcanic glass, and silica fume. In addition, lithium hydroxide solutions, which have been found to be extremely effective in reducing expansions in experimental work, are contemplated for the future. The use and testing of these materials are discussed next.

Fly Ash

Two categories of fly ash are recognized in ASTM Specification for Fly Ash and Raw or Calcined Natural Pozzolan for Use as a Mineral Admixture in Portland Cement Concrete (C 618): Class F and Class C. For purposes of preventing expansive ASR, the most important distinction is the minimum permissible proportion of $SiO_2 + Al_2O_3 + Fe_2O_3$ which, for Class F ash is 70%, and for Class C ash is 50%. Class C ashes usually have greater lime (CaO) contents—often exceeding 10% and even 20%. High lime contents adversely affect the capability of a fly ash to prevent expansive reactivity. An optional requirement of fly ash is a 1.5% maximum limit on "available" alkalies as determined by methods described in ASTM Method for Sampling and Testing Fly Ash or Natural Pozzolans for Use as a Mineral Admixture in Portland Cement Concrete (C 311). The writer believes the latter restriction is important because total alkali content of cement + fly ash may result in the equivalent of very high cement alkali levels that may convert an otherwise innocuous cement-aggregate combination into a deleterious one.

ASTM provides several testing procedures to determine the suitability of a fly ash for controlling expansive ASR in concrete. The ASTM Test Method for Effectiveness of Mineral Admixtures or Ground Blast-Furnace Slag in Preventing Excessive Expansion of Concrete Due to the Alkali-Silica Reaction (C 441) involves replacing 25% of the cement, by mass, with an amount of fly ash equal to the volume of cement replaced. Alkali content of the cement should be about 1.0%. Highly reactive Pyrex glass is used as fine aggregate in a specified grading, and the water content is gaged to meet a flow requirement. Companion mortars are made with and without the fly ash. Test storage is over water in sealed containers held at 38 ± 2°C. Expansion using the fly ash is determined at 14 days, and the absolute expansion level of the test mortar is evaluated.

ASTM C 618 requires expansion of the test mortar to not exceed 0.020% at 14 days for mortars containing either Class F or Class C ash, but does not provide a minimum limit for acceptance of the fly ash on the basis of percent reduction in expansion of the mortar containing the fly ash compared with the control mortar. However, ASTM C 441 requires the control mortar to expand at least 0.250% at 14 days and, considering the 0.020% maximum expansion limit of ASTM C 618 for the test mortar, the corresponding reduction in expansion for acceptance would be 92%, which is overly conservative. ASTM C 618 provides a minimum reduction in expansion of 75% for Class N (raw or calcined natural) pozzolan. This figure, although still quite rigid, has been used as a practical guideline for fly ash. Values between 60 and 75% also are known to have been used for acceptance levels for fly ash even though validating data for this criterion range are not known.

ASTM C 441 has been criticized for a number of reasons. First, Pyrex glass is used as the standard reactive aggregate. It contains appreciable alkali that, it has been argued, interferes with proper evaluation of the pozzolan. Of much greater importance is the fact that Pyrex glass reacts significantly faster than many fly ashes while, in field concrete containing natural aggregate, the reverse is probably true. For this same reason, meeting the ASTM C 618 expansion criterion of 0.020% at 14 days for the test mortar will be difficult, if not impossible, to meet even though the fly ash might effectively suppress expansive alkali-silica reaction [13].

Raw or Calcined Natural Pozzolan

Remarks presented for fly ash apply generally to raw or calcined natural pozzolans, such as some volcanic ashes, opaline shales, and diatomaceous earths. However, ASTM C 618 recommends 75% reduction in expansion after 14 days under ASTM C 441 test conditions, as noted earlier. The 0.020% expansion at 14 days applies to this category of pozzolan as well.

Ground Granulated Blast-Furnace Slag (GGBFS)

Ground granulated blast-furnace slag (GGBFS) is occasionally used to prevent expansive ASR. ASTM C 441 is used to evaluate slag as well as fly ash, except that up to 50%, instead of 25%, of the portland cement is replaced by slag equal to the volume of cement replaced. ASTM C 989 requires that reduction in expansion be at least 75%, which is the same as that for natural pozzolan.

The mechanism by which slags inhibit expansive reactivity is purported to be different than that by which fly ashes prevent expansion. From laboratory investigations [14], it was concluded that slag hydration in portland cement-water systems serves to reduce diffusion rates of alkali to aggregate reaction sites, thereby minimizing the rate of ASR. This would have the effect of producing a potentially less expansive or non-expansive gel reaction product. This explanation needs further confirmation.

Silica Fume

Silica fume is occasionally used to mitigate expansive ASR. It appears to function in much the same manner as fly ash in reducing expansions due to reactivity. That is, the calcium silicate hydrate reaction product complexes alkali, thereby reducing the hydroxide-ion concentration in the pore solution in the concrete.

Lithium Salts

McCoy and Caldwell [15] first reported the exceptional capability of lithium salts to inhibit expansive ASR. They found that small threshold quantities of any of several such salts could eliminate expansion when used with high-alkali cement and Pyrex glass. Later work [16] verified these findings using highly reactive commercially available aggregate. Here it was recommended that lithium hydroxide (LiOH) salt be used because it is soluble in tapwater, unlike lithium fluoride or lithium carbonate, thereby permitting better dispersal in large concrete batches. It also has been included, to a limited extent, in experimental field concrete pavement. Currently, there are no specification guidelines for its use, but the experimental laboratory work suggests a dosage rate of about 1:1 LiOH · H_2O : equivalent Na_2O in the portland cement.

An important advantage of lithium salts over other admixed materials is the fixed composition of the salt, thus requiring no testing in this respect. ASTM presently does not address this means of preventing expansive ASR.

CONCLUDING REMARKS

The concrete technologist has at his disposal the knowledge and guidelines to produce concrete that is not susceptible to deleterious ASR. ASTM provides various test procedures and recommendations that, if used prudently with other information noted in this chapter, should ensure deleterious reactions are avoided in future construction.

REFERENCES

[1] Stanton, T., "Expansion of Concrete Through Reaction between Cement and Aggregate," American Society of Civil Engineers, New York, Dec. 1940, pp. 1781–1811.

[2] Stark, D., *Handbook for The Identification of Alkali-Silica Reactivity In Highway Structures*, SHRP-C/FR-91-101, Strategic Highway Research Program, National Research Council, Washington, DC, 1991.

[3] Diamond, S., "Alkali Aggregate Reactions in Concrete: An Annotated Bibliography 1939–1991," Strategic Highway Research Program, SHRP C/VWP-92-601, National Research Council, Washington, DC, 1992.

[4] Helmuth, R., "Alkali-Silica Reactivity in Concrete—An Overview of Research," Strategic Highway Research Program, SHRP C/FR-92, National Research Council, Washington, DC, in publication.

[5] Powers, T. C. and Steinour, H., "An Interpretation of Published Researches on the Alkali-Aggregate Reaction," *Journal*, American Concrete Institute; *Proceedings*, Vol. 51, 1955, pp. 497–516 and 785–812.

[6] Chatterji, S., "Mechanism of Alkali-Silica Reaction and Expansion," *Proceedings*, 8th International Conference on Alkali-Aggregate Reaction in Concrete, K. Okada, S. Nishibayashi, and M. Kawamura, Eds., 8th ICAAR Local Organizing Committee, The Society of Materials Science, Kyoto, Japan, 1989, pp. 101–105.

[7] Moranville-Regourd, M., "Products of Reaction and Petrographic Examination," *Proceedings*, 8th International Conference on Alkali-Aggregate Reaction, K. Okada, S. Nishibayashi, and M. Kawamura, Eds., 8th ICAAR Local Organizing Committee, The Society of Materials Science, Kyoto, Japan, 1989, pp. 445–456.

[8] Stark, D., "The Moisture Condition of Field Concrete Exhibiting Alkali-Silica Reactivity," *Durability of Concrete, Proceedings*, Second International Conference, Montreal, Canada, SP-126, Vol. II, American Concrete Institute, Detroit, MI, 1991, pp. 974–987.

[9] Stark, D., "Alkali Silica Reactivity: Some Reconsiderations," *ASTM Cement, Concrete, and Aggregates*. Vol. 2, No. 2, 1980, pp. 92–94.

[10] Rogers, C. A. and Hooton, R. C., "Leaching of Alkalies in Alkali-Aggregate Reaction Testing," *Proceedings*, 8th International Conference on Alkali-Aggregate Reaction in Concrete, K. Okada, S. Nishibayashi, and M. Kawamura, Eds., 8th ICAAR Local Organizing Committee, The Society of Materials Science, Kyoto, Japan, 1989, pp. 327–332.

[11] Mielenz, R. C., Greene, K. T., and Benton, E. J., "Chemical Test for Reactivity of Aggregates with Cement Alkalies: Chemical Processes in Cement-Aggregate Reaction," *Journal*, American Concrete Institute; *Proceedings*, Vol. 44, 1948, pp. 193–219.

[12] Oberholster, R. E. and Davies, G., "An Accelerated Method for Testing the Potential Alkali Reactivity of Siliceous Aggregates," *Cement and Concrete Research*, Vol. 16, 1986, pp. 186–189.

[13] Stark, D., Portland Cement Association, Skokie, IL, unpublished data, 1993.

[14] Bakker, R. F. M., "About the Cause of the Resistance of Blastfurnace Cement Concrete to the Alkali-Silica Reaction," *Proceedings*, Fifth International Conference on Alkali-Aggregate Reaction in Concrete, S252/29, NBRI of the Council for Scientific and Industrial Research, Pretoria, South Africa, 1981.

[15] McCoy, W. J. and Caldwell, A. G., "New Approach to Inhibiting Alkali-Aggregate Reaction," *Journal*, American Concrete Institute; *Proceedings*, Vol. 47, 1951, pp. 693–706.

[16] Stark, D., "Lithium Salt Admixtures—An Alternative Method to Prevent Expansive Alkali-Silica Reactivity," *Proceedings*, 9th International Conference on Alkali-Aggregate Reactions in Concrete, London, 1992, pp. 1017–1025.

Alkali-Carbonate Rock Reaction

Michael A. Ozol[1]

PREFACE

The subject of alkali-carbonate rock reactivity (ACR) was discussed for the first time as a distinct and separate chapter-length topic in *ASTM STP 169B* by H. N. Walker [1]. Prior to that, W. C. Hansen, in *ASTM STP 169A* in 1966 [2], discussed the reactivity of dolomitic limestones, along with the reactivity of other rocks and minerals, under the general heading of chemical reactions. Previous editions of *Significance of Tests and Properties of Concrete and Concrete Making Materials (ASTM STP 169*, 1956) and its predecessors *ASTM STP 22A*, in 1943, and *ASTM STP 22*, in 1935, did not treat the subject of deleterious carbonate rock reactions because such reactions were not recognized as a distinct problem until identified as such by Swenson in 1957 [3], based on his investigations of expanded, cracked, and buckled concrete in the vicinity of Kingston, Ontario, Canada.

In addition to the now numerous occurrences of reactive carbonate rock in concrete in Ontario, in the United States there have been affected concretes in Virginia, West Virginia, Kentucky, Missouri, Tennessee, Iowa, Illinois, Indiana, and New York; and elsewhere in the world in Iraq, Bahrain, England, and China.

INTRODUCTION

When a portland cement concrete mixture is designed and placed, it is expected that the coarse and fine aggregate will not participate in any chemical reactions with components of the cement paste that will cause the aggregate particles to crack or to be otherwise dimensionally unstable.

Historically, there was a time when that expectation was satisfactorily fulfilled, or at any rate believed to have been satisfactorily fulfilled, because for certain instances of concrete deterioration, the reasons for that deterioration were ascribed to causes other than deleteriously reactive aggregates.

The satisfactorily innocuous, or "inert," behavior of aggregates was, at that time, favored by such factors as:

1. Continued availability of aggregate materials from "old" sources, those materials that had performed without problems over years of consumption.
2. The general availability of low-alkali cement that was furnished as a matter of course during the years when kiln dust was not included in the product.

As demand for concrete increased, in large part to meet the requirements of transportation-related construction, new sources for aggregates were necessarily developed. It should be noted that new sources include previously unused or undeveloped strata or portions of old operating limestone quarries as well as new "greensite" quarry sources. And, at around the same time, in the late 1960s and early 1970s, in response to economic and environmental constraints, cement kiln dust, which is richer in alkalies than the cement product, began to be returned to the product. Thus, the preceding Factors 1 and 2 no longer favored so much the satisfactorily innocuous behavior of aggregates with respect to alkalies in the cement, and instances of deleterious behavior became more numerous.

The following ASTM documents are directly applicable to the subject of alkali-carbonate reactivity and are listed with a notation as to their particular relevance.

ASTM Standard Specification for Concrete Aggregates (C 33)

Directs, in Paragraph 10.2, that deleteriously reactive aggregates shall not be used in moisture-exposed concrete made with cement containing more than 0.60% total alkalies and, in Appendix X1.1.5, describes the petrographic characteristics of potentially reactive carbonate rocks and, in part, how to evaluate them.

ASTM Guide for Petrographic Examination of Aggregates for Concrete (C 295)

States in Paragraph 4.5 that petrographic examinations should identify and call attention to potentially alkali-carbonate reactive constituents, determine their amounts, and recommend additional tests to determine whether those amounts are capable of deleteriously affecting concrete. In Paragraphs 10.1 and 11.1, reference is made to examination of drilled core and ledge rock. These topics are of particular importance since it is more often than not the case that alkali-carbonate reactive rock comprises

---

[1] Consultant, geologist/petrographer, Concrete and Materials, Baltimore, MD 21209.

a relatively small stratigraphic portion of a producing quarry.

ASTM Descriptive Nomenclature for Constituents of Natural Mineral Aggregates (C 294)

In Paragraph 21.1, describes the characteristics, texture, and composition of carbonate rocks that are potentially alkali-reactive, in essentially the same language as is in ASTM C 33, Appendix X1.1.5.

ASTM Test Method for Potential Alkali Reactivity of Carbonate Rocks for Concrete Aggregates (Rock Cylinder Method) (C 586)

Directly measures the expansion of small cylinders of rock during exposure in sodium-hydroxide solution. The rock sample is secured in accordance with the applicable requirements of ASTM D 75 (cited later) that are in Appendix X2.2. Positive test results cannot, by themselves, be used as an accurate predictor of the damage that the rock will cause in concrete. ASTM C 1105 (see next paragraph) should be used to evaluate the behavior of the rock in concrete.

ASTM Test Method for Length of Change of Concrete Due to Alkali-Carbonate Rock Reaction (C 1105)

Directly measures the expansion produced in concrete prisms by the alkali-carbonate reaction, using up to 19.0 mm (¾ in.) maximum size aggregate, and the job cement.

ASTM Practice for Examination and Sampling of Hardened Concrete in Constructions (C 823)

Provides general and specific guidance for evaluating constructions in the field and for obtaining samples for laboratory examination as by ASTM C 856 (see next paragraph), with which it may be used in an interactive way, with ASTM C 856 providing preliminary information for detailed and rigorous application of ASTM C 823 to provide a second set of samples for petrographic examination by ASTM C 856. ASTM C 823 is of particular relevance when evaluating an aggregate source for alkali-carbonate reactivity based on its service record in previous constructions. In that connection, it is important to have information on the alkali content of the cement used, and the cement content of the concrete in the constructions.

ASTM Practice for Petrographic Examination of Hardened Concrete (C 856)

Provides the general method for how to go about examining field and laboratory concrete chiefly microscopically to establish whether alkali-carbonate reactions have taken place, and gives some specific criteria for recognizing the effects of alkali-carbonate reactions.

ASTM Practices for Sampling Aggregates (D 75)

Gives the procedure for obtaining samples for examination and testing by ASTM C 295, C 1105, C 586, and any other tests and examinations from which general conclusions regarding the aggregate may be drawn. It should be noted that ASTM C 586 is best, and most conveniently, done on ledge rock samples that will permit the preparation of three mutually perpendicular small rock cylinders. Information regarding obtaining ledge rock samples is in Appendix X2.2 of ASTM D 75. Alternatively, in the event that ledge rock samples cannot be obtained, it is possible that the small cores (1.38 in., 35 mm) needed for ASTM C 586 can be drilled out of larger particles, for example, 2 in. (50 mm) or larger, of aggregate obtained from stock pile or conveyor belt.

ASTM Method for Reducing Field Samples of Aggregate to Testing Size (C 702)

Samples taken in accordance with ASTM D 75 may be reduced in size by quartering or by use of a mechanical splitter as described in ASTM C 702.

TYPES OF CARBONATE ROCK REACTIONS

Although various reactions involving carbonate rock, both dolomitic and nondolomitic, embedded in concrete or mortar have been recognized, only one, the expansive dedolomitization reaction that is the principal subject of this chapter, is of importance with respect to having produced significant damage in concrete constructions. This reaction may or may not produce rim zones on particles of reacting rock in concrete.

Other reactions, causing the production of rim zones and other internal changes in carbonate rocks of various compositions are of scientific interest. Some other reactions are discussed in Refs 1, 4, and 5.

EXPANSIVE DEDOLOMITIZATION REACTION

The expansive dedolomitization reaction occurs between alkaline solutions (almost always sodium and potassium hydroxides derived from the cement[2]) in the cement paste matrix, and limestone aggregate particles of particular but nevertheless somewhat variable compositions and textures. However, a feature common to all such limestones is that there will be small dolomite crystals present in the matrix of the rock that will react with the alkaline solutions in a manner such that the dolomite crystals, of original composition $CaMg(CO_3)_2$, become other minerals and compounds; hence, the term dedolomi-

---

[2] Alkalies from deicing materials such as sodium chloride (NaCl) may augment the alkalies supplied by the cement.

tization, as represented, simply, here and as discussed in greater detail in subsequent sections.

calcium magnesium carbonate (Dolomite)
+ alkali hydroxide solution → magnesium hydroxide
+ calcium carbonate + alkali carbonate

The chief physical result of dedolomitization, which result distresses and damages concrete, is that the reaction causes the affected aggregate particles to expand and crack and thereby to extend the cracks into the enclosing mortar matrix.

MANIFESTATIONS OF DISTRESS

Concrete that is affected by the alkali-carbonate rock reaction expands and cracks. Depending on the perfection of the petrological characteristics of the rock that promotes expansion, and the convergence of the environmental and materials-related factors that promote expansion, important among which are availability of moisture and alkali content in the concrete, the concrete may in a matter of months exhibit severe surface cracking accompanied by closing of joints, as was reported in Canada by Swenson [3]. Pattern cracking after three years of exposure is shown in Fig. 1.

When all of the factors favoring alkali-carbonate reaction combine to produce near-ideal conditions for causing the reaction, "spectacular expansion and deterioration" as reported by Rogers [7] can occur within three years. In that instance, the concrete in sidewalks, curbs, and gutters expanded about 1.2% (Fig. 2) with consequent compressive failures, blow-ups, and shoving of adjacent asphalt pavement. It is reported that the alkali content of the cement was about 1.0% Na_2O equivalent, and the cement content of the concrete was about 300 kg/m³ (500 lb/yd³).

FIG. 2—Concrete slab affected by alkali-carbonate reaction (upper portion of photo) that has moved 9 cm to the right relative to the concrete in the lower portion of the photo that has nonreactive aggregate. Expansion was measured at 1.2% in three years, Ontario, Canada [35].

FIG. 1—Typical cracking of concrete slab by alkali-carbonate reaction after three years, Ontario, Canada [6].

In ACR-affected field concrete, there is no definitive feature of the concrete or of the geometry of the cracking that precisely identifies the cause of the cracking as being alkali carbonate-rock reaction. (In contrast, gel deposits, if present, may provide a strong indication that cracking is due to alkali-silica reaction but, not exclude the possibility that both reactions can be present.) The crack geometry, or pattern, produced is the response of the concrete to internal expansive force based on its particular size and shape and the direction of maximum moisture availability. Generally, expansion of ordinary, plain, cast-in-place concrete as in sidewalks, floors, decks, slabs, and footings where there is a moisture gradient from top to bottom or from side to side results in pattern cracking similar to that shown in Fig. 1. Where moisture conditions are uniform or where the concrete elements are thick, or of low surface area-to-volume ratio, bulk expansion of the concrete is evidenced by closing of joints, extrusion of filler material, shoving or buckling of adjacent materials, or crushing of adjacent weaker concretes. Cracks may be prominent or faint—visible only following wetting of the surface to produce some contrast—and concrete between the cracks can be hard and intact.

Conclusive evidence that the observed distress has been caused by alkali-carbonate reaction is best obtained by detailed examination of the concrete as by ASTM C 856, during which it will be observed that the aggregate exhibits the characteristic microscopic texture and composition of alkali-reactive carbonate rock (see the section on Characteristics of Alkali-Reactive Carbonate Rocks), and that internal cracking in the aggregate extends into the mortar.

Reinforced and post-tensioned concrete, precast and prestressed concrete, and other kinds of prefabricated concrete can develop their own individualized cracking responses to internal expansion that may not be similar to those produced in slabs on grade. For example, elongate prestressed or post-tensioned pieces typically develop long cracks parallel to the direction of tensioning or prestressing.

In at least one particular type of exposure condition, the manifestation of distress due to ACR does not involve gross expansion of the concrete and widespread cracking, but instead, the damage is confined to the surface region and is the result of the application of deicing salts. For example, a parking deck investigated by the author containing reactive carbonate aggregate throughout was deeply scaled exposing numerous cross-fractured coarse aggregate particles in the scaled surface. Only the coarse aggregate particles in the near-surface region had cracked, extending cracks into the mortar to connect with similar cracks from other reactive coarse aggregate particles, with the result that the surface concrete was lost. Although the deck has reactive aggregate throughout, the alkali content of the concrete, in the absence of deicing salts, is insufficient to sustain the reaction and to develop the overall expansion and tensile cracks producing the map, or pattern cracking, that is usually observed.

Smith [8] has presented experimental data showing the exacerbating effect of sodium chloride (NaCl) on alkali-carbonate reactivity. Using concrete prisms, made with a 1.1% alkali cement and a known reactive aggregate, stored in saturated NaCl solution at 70°F (21°C), he recorded about 0.19% expansion at 300 days. The same concrete stored in pure water at the same temperature showed 0.12% at 300 days.

Using reactive aggregate from the same quarry (Pittsburgh quarry at Kingston, Ontario) about 27 years later, Alasali et al. [9] recorded comparable results, measuring 0.23% expansion at 275 days on concrete prisms made with 1.13% alkali cement stored in 5% NaCl at 23°C. A similar prism stored in water at 38°C measured 0.155% expansion at 275 days.

COMPARISON OF ALKALI CARBONATE ROCK REACTION WITH ALKALI-SILICA REACTION AS REGARDS GENERATION OF CRACK DAMAGE

In alkali silica reaction (ASR), the first damage to the concrete can be produced by both (a) internal cracking of aggregate particles extending cracks into the paste and mortar, and (b) extension of existing cracks and production of new cracks as a consequence of the migration and subsequent expansion of the ASR gel, initially produced by the reacting particles, away from the particles and into cracks and voids in the concrete.

In contrast, direct damage due to alkali carbonate reaction is produced by the cracking, and extension of cracks into the mortar, generated within the reactive coarse aggregate particle itself.

The initial, or direct, damage produced in both ASR and ACR may then be exacerbated by the action of cyclical freezing and thawing. And, it should be recognized that, although bulk expansions of concrete produced solely by the alkali-carbonate reaction may be lower than those produced in severe alkali-silica reaction, the ultimate damage to the concrete can be of the same order as produced by ASR if the concrete is exposed in an environment where cyclical freezing and thawing occurs [10]. That is, the initial cracks produced by the alkali-carbonate reaction are then exploited by saturation of the cracks and subsequent freezing.

CHARACTERISTICS OF ALKALI-REACTIVE CARBONATE ROCKS

Petrographic

Carbonate rocks participating in the alkali-carbonate reaction have a particular mineralogical composition, and a microscopic texture and structure that is characterized by relatively larger, rhombic crystals of dolomite [$CaMg(CO_3)_2$] set in a finer-grained matrix of calcite [$CaCO_3$], clay, and (commonly) silt-sized quartz. The characteristic composition is that in which the carbonate-mineral components of the rock consist of substantial amounts of both dolomite and calcite, and the dilute hydrochloric acid (HCl) insoluble residue contains a significant amount of clay.

The basic theme of the reactive alkali carbonate rock microtexture wherever it is found is the discrete dolomite rhombs in a background mass. The basic theme is observed in several variations in dolomitic limestones that are known to be alkali-carbonate reactive rocks, Fig. 3.

The rhombic crystals of dolomite occurring in the matrix of the reactive rock may be relatively sparsely distributed and appear to be "floating" in the background (Fig. 3, Field A) or may be more crowded together with dolomite rhombs adjacent or touching (Fig. 3, Field E).

Many of the discrete "floating" dolomite rhombs set in the matrix are sharp-edged euhedral sections while others may occur with rounded corners and a more irregular, less pronounced, rhombic shape. The typical average size of the rhombs is approximately 25 to 30 μm maximum dimension with the largest being approximately 50 to 75 μm. The smallest of the rhombic sections may be very small, at the limit of optical resolution, because of the "sectioning effect" caused in the preparation of thin-sections or opaque polished sections.

The grain sizes of the calcite and clay matrix in which the dolomite rhombs are set is typically, 2 to 6 μm for the calcite with smaller clay particles disseminated throughout. It is often the case that silt-sized quartz grains are also disseminated throughout the matrix.

The typical texture that is described here is found in those reactive carbonate rocks that are identified as early expanders, showing rock prism expansions (ASTM C 586) of a few tenths percent in a matter of weeks, and field manifestations in concrete within perhaps one year after construction.

A modification of the typical texture is found in reactive carbonate rocks identified as late expanders that may not show noticeable rock prism test expansions until approximately 25 weeks (rarely, not before one year) and that may not show manifestations in concrete after five years in moist storage. In those rocks, the typical feature of dolomite rhombs in a clayey fine-grained matrix is consistent, but the matrix is coarser-grained and is composed of interlocking dolomite grains together with calcite, clay, and, more commonly than not, silica minerals. Those differences between early and late expanders are reflected in the bulk compositions, as shown in Table 1, and in differences in internal textural restraint, or rigidity, as investigated by Hilton [11,12] and discussed subsequently. Whether the possibility of late (rock prism) expanders causing distress in field concrete after many years of exposure should be disregarded is an open question.

The best technique for identifying the characteristic reactive texture is by use of thin sections with a petrographic microscope. It is also possible, and often convenient for rapid survey of a large number of thin sections, to observe the thin sections with a stereo-microscope. If a dark-field type of illumination is available for the stereo scope, it is usually useful. If it is not available, some sort of illumination to enhance the contrast of the dolomite rhombs in the thin section can be "jerry-rigged" by adjustment of the angle of illumination and the background against which the section is observed.

Alternatively, the reactive texture can be identified in opaque section in concrete or rock samples. To do it this way usually requires a highly polished specimen and manipulation of the illumination to get the contrast and reflectance just right. A light etching of the specimen surface with dilute HCl can be very helpful.

Carbonate rocks displaying the characteristic reactive texture can comprise a substantial or a small portion of a bed in a limestone quarry. Or, the reactive texture may occur as veins within a limestone bed that does not otherwise display the reactive texture (Fig. 4).

The occurrence of the reactive carbonate rock texture is a reliable diagnostic guide since there appears to be no known exception to the relationship that all limestones that have the characteristic texture and composition will react or dedolomitize, in an alkaline environment, and all limestones that dedolomitize in an alkaline environment have the characteristic texture and composition.

Alkali-carbonate reactive coarse aggregate particles in concrete may or may not display reaction rims. The formation of reaction rims on ACR rock particles is incidental to their propensity for reaction and generation of crack damage. However, the presence of reaction rims calls for closer inspection of those coarse aggregate particles, since often, but not always, their presence is indicative of alkali-carbonate reaction.

Broadly speaking, it appears to be the case geologically that the conditions for formation of alkali-expansive carbonate rocks are more restricted than for other limestones, and they are not abundant in the volume of limestone that exists on earth. They may occur in carbonate rocks of any age but, thus far, have been found most often in carbonate rocks of Ordovician age.

It is clear from the microtexture of reactive carbonate rocks, with the commonly sharp-edged rhombs, and occurrence as veins of reactive texture, that that texture is of replacement, or possibly diagenetic, origin rather than one resulting from primary deposition. As such, it is reasonable to infer that, since the production of the reactive texture requires special conditions subsequent to deposition and not obtaining as components of the ordinary depositional regime for limestones, then there will be less of it in the total limestone section—as field experience thus far seems to indicate.

To put reactive carbonate rocks in perspective with reactive siliceous rocks, reactive siliceous rocks are far more abundant in the overall category of siliceous aggregates (crushed stone together with sand and gravel) that might be used in concrete than are reactive carbonate rocks in the category of all carbonate rocks. And, generally, but not entirely, the carbonate aggregates that participate in alkali-carbonate reaction are crushed stones from limestone quarries. This is in contrast to alkali-silica reactivity where reactive siliceous components can derive from both gravel and crushed stone sources. That is, except near their bedrock source, and under conditions of rapid erosion and deposition, limestone-bearing gravels, considering earth's gravels as a whole, are not as abundant as siliceous gravels because limestones are less able than siliceous rocks to survive the weathering and transportation necessary to put them in gravel deposits as hard and sound particles suitable for use as concrete aggregate. Consequently, alkali-carbonate reactive particles in gravel deposits, because such particles are limestones, are less likely to be found in gravels than alkali-silica reactive particles. However,

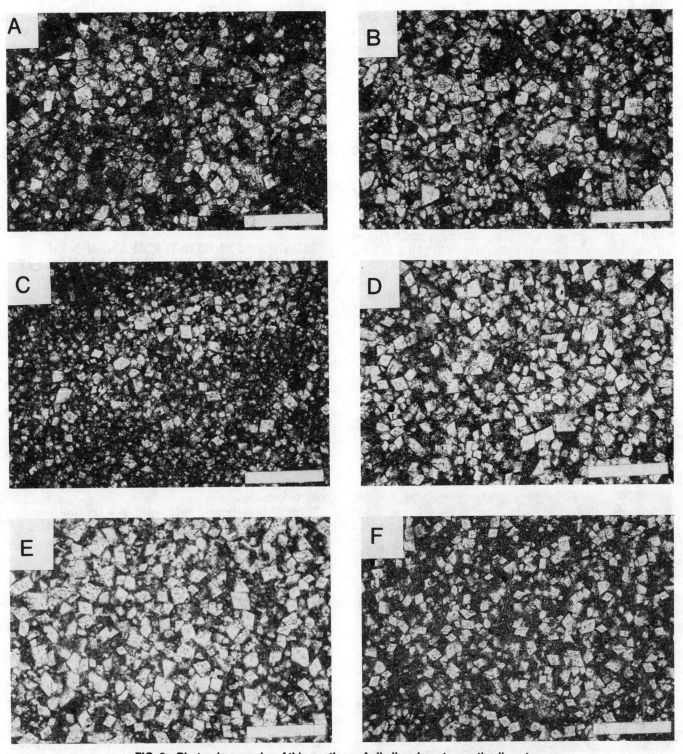

FIG. 3—Photomicrographs of thin sections of alkali-carbonate reactive limestones from quarries in eastern Canada. Scale bar is 0.25 mm. Fields A, B, and C are from different quarries. Fields D, E, and F are from the same quarry, D and F from the same bed [35].

gravel deposits containing reactive carbonate rock exist and are reported to have been used in concrete in England in the M 50 Motorway[3] and have been found and used in concrete, causing damage, in Ontario, Canada, near their bedrock source.[4]

Chemical and Mineralogical Composition

Typical compositions of early and late expanding alkali-reactive expansive carbonate rocks are shown in Table 1, with their geographic location and publication reference indicated.

In an extended study of the composition of reactive and nonreactive limestones, versus their expansion in the Rock Cylinder and Concrete Prism Expansion Tests, Rogers [7] found that the compositions of potentially expansive rocks fell within a fairly distinct field as shown in Fig. 5. In fact, the empirical correlations depicted in Fig. 5 are the basis for a quick chemical screening test used in Ontario (MTC LS-615) (see *Determination of Potential Alkali Carbonate Reactivity of Carbonate Rocks by Chemical Composition*).

A similar distinction of the potentially reactive field was obtained when plotting CaO:MgO versus insoluble residue, Fig. 6.

The information presented in Figs. 5 and 6 and as reported by previous workers [16,19,20,21] establishes that alkali reactive carbonate rocks do conform to systematic, geochemical and petrological realms and are not widely variable with respect to their amounts of carbonate, clay, and siliceous minerals, but, rather, have a characteristic composition with respect to those components.

FACTORS AFFECTING EXPANSION OF CONCRETE WITH REACTIVE CARBONATE ROCK AND FACTORS AFFECTING EXPANSION OF REACTIVE CARBONATE ROCK BY ITSELF

A suite of carbonate rocks that have reactive texture and that exhibit varying degrees of (unrestrained) expansion in the rock cylinder test (ASTM C 586) may exhibit a very different expansion ranking when tested "in restraint" in the rock cylinder test [11,12] or when restrained by incorporation in concrete [19]. Based on Hilton's work [11] the degree to which reactive carbonate rocks will cause expansion in concrete is related, on the one hand, to the restraint imposed by the concrete, and, on the other, to the volume of dolomite in the rock, up to a point, and the internal textural restraint, of rigidity, of the carbonate rock in question.

A reactive carbonate rock with low textural restraint can be highly expansive unrestrained, but when restrained, either experimentally or within concrete, it is more compressible and less expansive.

In Hilton's experimental work [11,12], the most expansive rock, unrestrained in ASTM C 586, expanded 7.8% at 14 weeks and restrained in a steel frame in 1N sodium hydroxide (NaOH), expanded 0.275% at 14 weeks. The rock has 58.6% by volume dolomite, 23.5% calcite, and 5.63% porosity. Petrographically, the rock has a high volume of dolomite rhombs tightly packed but with abundant voids between the masses of tightly packed rhombs (Fig. 7).

The least expansive sample did not expand at all after eight weeks in 1N NaOH. It is a typical "crystalline" dolomite composed of a mosaic of equant dolomite grains in a tight virtually nonporous structure (Fig. 8). It has 94% dolomite by volume, < 6% calcite, and 0.18% porosity. It did begin to expand after 40 weeks' continuous soaking in the NaOH solution.

A rock, representative of several, with the typical alkali-carbonate reactive texture (Fig. 9) and similar to Fig. 3 (Fields A, B, and F), expanded unrestrained 1.4% at 14 weeks, and restrained 0.31%. It has 32.4% by volume dolomite, 36.7% calcite, and 1.25% porosity.

The rock with the lesser value of unrestrained expansion expanded more under restraint than the rock with five

TABLE 1—Composition of Early and Late Expansive Carbonate Rocks (Adapted From Ref *1*).

	Acid Insoluble Residue, %	Dolomite % of total carbonate
Kingston, Ontario, early expanders [*3,13,14*]	5 to 15	about 50
Iowa, Illinois, and Indiana early expanders [*15,16*]	10 to 20	40 to 60
Virginia early expanders [*17,18*]	13 to 29	46 to 73
Gull River, Ontario, late expanders [*14*]	21 to 49	75 to 87
Virginia late expander [*19*]	33	>90

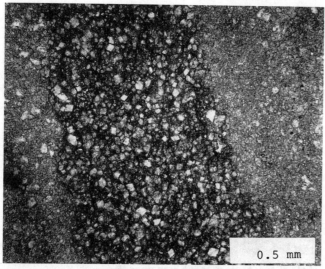

FIG. 4—Vein of reactive carbonate texture in limestone displaying alternating bands of reactive and nonreactive texture. Northern Virginia, field is approximately 2 mm across. The same type of structure can also exist on a larger scale, from bed to bed, or within a bed, with the alternating bands being several inches thick.

[3] W. J. French, personal communication.
[4] C. A. Rogers, personal communication.

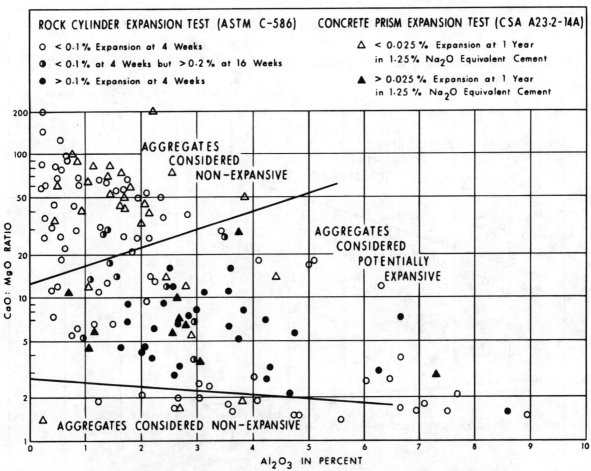

FIG. 5—CaO:MgO ratio versus Al_2O_3 for a suite of alkali-carbonate reactive and nonreactive carbonate rocks [7].

times greater unrestrained expansion, and almost two times greater dolomite volume.

Thus, carbonate rocks with different or equivalent, chemical/mineralogical composition could have vastly different expansion characteristics in concrete if their structural fabrics differ to such an extent that their elastic properties are quite different.

The petrological factors leading to the perfection of the lithology for producing expansion in concrete are: (1) clay content, or insoluble residue content, in the range of 5 to 25%; (2) calcite to dolomite ratio of approximately 1:1; (3) increase in dolomite volume up to the point at which interlocking texture becomes a restraining factor; and (4) small size of the discrete "floating" dolomite rhombs.

The expansion of concrete containing alkali-reactive carbonate rocks is promoted by: (1) increasing coarse aggregate size; (2) moisture availability; (3) higher temperature; (4) high alkali content of the concrete and high pH of the liquid phase in cement pores; (5) high proportion of reactive stone in the coarse aggregate; and (6) lower concrete strength [19,22].

MECHANISM OF REACTION AND EXPANSION

Following the first recognition of the alkali-carbonate reaction by Swenson [3], work by Hadley [4], and later investigators, showed that the main chemical reaction that occurred in the rock was that the dolomite $[CaMg(CO_3)_2]$ decomposed, or dedolomitized, to calcite $(CaCO_3)$ and brucite $Mg(OH)_2$, as represented by the following reaction in which M represents an alkali element, such as potassium, sodium, or lithium.

$$CaMg(CO_3)_2 + 2MOH \rightarrow Mg(OH)_2 + CaCO_3 + M_2CO_3$$

A further reaction that occurs in concrete is that the alkali carbonate produced in the initial reaction may then react with the $Ca(OH)_2$, produced as a normal product of cement hydration, to regenerate the alkali hydroxide. For example

$$Na_2CO_3 + Ca(OH)_2 \rightarrow 2 NaOH + CaCO_3$$

Although the starting and end products of the reaction have been well documented and establish the overall bulk chemical and mineralogical changes that take place, the dedolomitization reaction, as such, does not obviously translate into a mechanical explanation for the expansion, because the reactions proceed as a volume-for-volume replacement without the direct production of some "extra" volume to generate expansion. The explanation was sought in behavior of the reactive rock and alkali system with respect to indirect mechanisms and secondary reaction products that had been identified in various investiga-

380 TESTS AND PROPERTIES OF CONCRETE

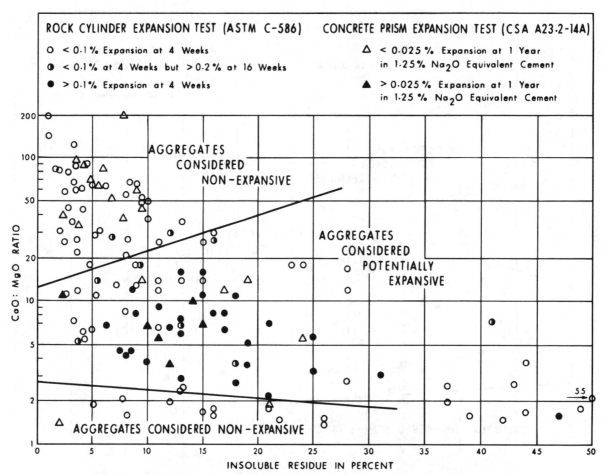

FIG. 6—CaO:MgO ratio versus insoluble residue for a suite of alkali-carbonate reactive and nonreactive carbonate rocks [7].

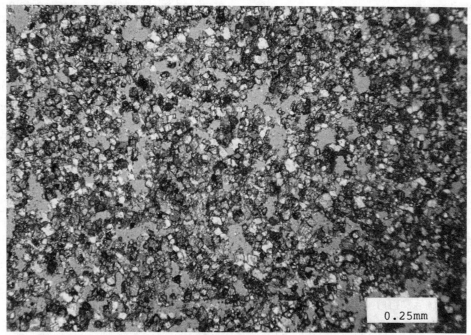

FIG. 7—Expansive rock from Missouri, 7.8% in 14 weeks in ASTM C 586. High content of dolomite rhombs in porous matrix.

tions. That is, the primary dedolomitization reaction was considered to be the prerequisite that enabled other processes to work to produce the expansion.

An idea common to some proposed mechanisms is that, in one way or another, water uptake, absorption, and incorporation is important in producing the expansion.

Sherwood and Newlon [23] proposed that expansion could be related to the formation of the hydrated, relatively high-volume, low-density, secondary mineral products gaylussite and buetschliite. Feldman and Sereda [24] found trace amounts, in reacted rock, of a material that expands when it absorbs water. Walker [25] and Buck [26] found minerals of the hydrotalcite-sjogrenite group, which minerals may possibly help promote water absorption and consequent expansion.

Swenson and Gillott [21] propose that dedolomitization exposes "active" clay minerals that are present as inclusions, in the rocks that they studied, within the dolomite euhedra. Exchange sites on the clay surface adsorb sodium ions, and water uptake by the "new" clay results in swelling. Thus, it is the "active" clay released during dedolomitization that contributes to the expansion.

Clay inclusions within the dolomite rhombs of expansive rocks have not been universally reported. They have not been observed, for example, in expansive rocks from Virginia.

Ming-Shu Tang et al. [27] conclude that the expansion due to ACR is caused by ions and water molecules migrating into restricted spaces and by the growth and rearrangement of the products of dedolomitization, particularly brucite. Therefore, that the expansion results directly from dedolomitization and not from water uptake of clay mineral inclusions within dolomite rhombs. Using STEM and EDAX to observe the locations of the reaction products in expanded samples of Kingston, Ontario, reacted rock, they found that the brucite produced by the reaction occurred in a 2-µm-thick ring of parallely oriented 25Å, crystals with space between them surrounding the euhedral dolomite rhombs of the typical reactive texture. The brucite crystals were, in turn, surrounded by the calcite and clay particles of the matrix. The calcite of the dedolomitized rhomb remained within the interior of the dedolomitized crystal.

The authors conclude that the expansion is caused by the combined process of: (1) the migration of alkali ions and water molecules into the restricted space of the calcite/clay matrix surrounding the dolomite rhomb, (2) migration of those materials into the dolomite rhomb, and (3) the growth and rearrangement of the dedolomitization products, especially brucite exerting crystallization pressure, around the outsides of the rhombs.

The rhombs maintain their original dimensions, supported by the remaining calcite framework, but the $MgCO_3$ component of the original dolomite rhomb is now transformed into the brucite that accumulates in parallel layers around the periphery of the rhomb. The original rhomb, together with the brucite layer, now needs more space and generates an expansive force that translates into cracking of the coarse aggregate particle.

The role of the clay minerals in the fabric of the rock, which may contribute to the expansion mechanism, is to provide mechanical pathways to the dolomite rhombs by disrupting the structural framework of the rock, thus weakening the carbonate matrix and, by that means, helping to promote expansion. Thus, the presence of clays is not essential to the chemistry of the dedolomitization reaction but is an enabling factor, mechanically promoting expansion when dedolomitization occurs.

Based on subsequent theoretical and experimental work, Tang et al. [28] and Deng and Tang [29] concluded that both the limit and the rate of the dedolomitization reaction and expansion depend directly on the pH value of the pore water in the concrete. And, at pH lower than 12, the reaction, and consequent expansion, proceeds very slowly or not all.

FIG. 8—Nonexpansive rock, in ASTM C 586, from Virginia. Structural framework of equant dolomite grains with no discrete "floating" rhombs and no interstitial matrix.

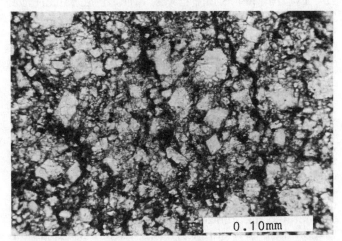

FIG. 9—Expansive rock from northwest Virginia with typical reactive texture. Expansion in ASTM C 586 was 1.4% at 14 weeks.

EVALUATING THE POTENTIAL FOR ALKALI CARBONATE REACTIVITY

Field Service Record

Field performance information on concrete incorporating the aggregate under consideration can furnish direct answers to the questions that the laboratory evaluation tests attempt to answer indirectly. That is, tests on the rock, in or out of concrete, are intended to provide information on which, in essence, to base a prediction on what will happen if the rock is used in concrete construction. However, if "good" information is available from the field then, logically, that is best, and indirect information to predict what can be observed directly is second best.

Provided certain essential information, noted later, is available and certain exposure conditions are fulfilled, an evaluation of the potential for aggregate reactivity based on the investigation of concrete structures in the field can be the most unequivocal and economic way of accomplishing the evaluation.

When investigating and sampling a concrete structure (ASTM C 823) known to contain aggregate from the source in question, for example, by prior petrographic examination, the alkali content of the cement and the cement content per cubic yard must be known. In general, if the cement used in the concrete was low alkali cement, less than 0.6% total alkalies as per ASTM Specification for Portland Cement (C 150), then the concrete cannot provide information on the propensity for reaction of the aggregate with high alkali cement. However, in view of the facts, (1) the most reactive carbonate rocks can produce deleterious expansion with cement of 0.4% total alkalies [19] and (2) the more important measure is the alkali content per unit volume of concrete, and not the alkali content of the cement per se. Then even if the cement used was less than 0.6% total alkalies, if the cement content of the concrete was high, the alkali level per unit volume may equal or exceed that of concrete in which the cement content was low but the alkali content of the cement was high. A threshold guideline, based on pounds of sodium oxide (Na_2O) equivalent per cubic yard, must be established for judging whether the aggregate should be considered to have been used in a "high" or "low" alkali concrete.

If the structure shows no distress and it has been established that the alkali content of the concrete would be sufficient to cause reaction if the aggregate is potentially reactive, then the effects of aggregate size, proportion of reactive aggregate, and moisture availability to the concrete must also be considered singly and in combination before concluding that the aggregate is non-deleteriously expansive.

Larger-sized reactive aggregate causes greater expansion than smaller-sized reactive aggregate. A reactive aggregate that has not caused distress with "high" alkali cement in small top size may cause distress when used in the larger size with the same cement.

The greater the proportion of reactive aggregate of the total aggregate, the greater the expansion. A reactive aggregate that has not caused distress as a small portion of the coarse aggregate in a structure with "high" alkali level, may cause distress when used in larger proportion with the same cement.

Potentially reactive aggregates will not cause deleterious amounts of expansion if the concrete is protected and in a relatively dry condition. Where exposure to moisture also includes exposure to sodium-chloride-containing deicing salts, which are known to exacerbate ACR, then the conclusion that the aggregate is non-deleteriously expansive is enhanced.

Concrete Prism Expansion Test

The surest measure of the potential for deleterious expansion of a carbonate aggregate in concrete is the measurement of the length change (for percent expansion) of concrete prisms, as by ASTM C 1105 or Canadian Standard CSA A23.2-14A.

ASTM C 1105 is intended primarily for evaluating a particular cement-aggregate combination prior to use in concrete, although, as noted, it can be used for investigational or research purposes (Paragraphs 4.5 and 6.3).

CSA A23.2-14A, Potential Expansivity of Cement Aggregate Combinations (Concrete Prism Expansion Method), because of its stipulations regarding use of the same gradation and same (high) alkali content, can be considered to be a test aimed at maximizing the conditions for detection of potentially expansive rocks and for facilitating comparisons among test results.

Both methods use similarly sized test specimens 75 by 75 by 300 to 450 mm (CAN) or 285 mm (ASTM), and storage conditions of 100% humidity and 23°C. However, ASTM C 1105 specifies that if the job-proposed aggregate has material larger than 19 mm (¾ in.), it shall not be used; based on the assumption that the plus 19-mm material does not differ in composition, lithology, and presumed expansive characteristics from the minus 19-mm material. If, based on information from petrographic examination or rock cylinder tests (ASTM C 586) or both, the plus 19-mm material differs from the minus 19-mm material in those respects, then instructions are given for testing the coarser fraction. In ASTM C 1105, the job cement is used when the test is used for its primary purpose.

In CSA A23.2-14A, the coarse aggregate is separated on 20-, 14-, 10-, and 5-mm sieves, oversize and undersize material discarded, and equal masses of the three sizes recombined. The alkali content of the cement, as per the current edition of the method (1990) shall be 1.0 ± 0% Na_2O equivalent but raised by the addition of NaOH to the mix water to 1.25% Na_2O equivalent by mass of cement. In the earlier (1977) version of the Canadian standard, the alkali content of the cement was 0.9 ± 0.1% Na_2O equivalent raised by NaOH to 1%. The change to 1.25% alkalies was made to accelerate expansion, in furtherance of the objective to identify potentially expansive rocks, rather than to reflect job conditions. The change was concluded to be necessary in the light of field experience and research, for example, Rogers [7], that showed that the prior specification requirements regarding alkali content and maximum expansion did not exclude rocks that, in the course of time, caused deleterious expansion in the field.

Expansion limits for ASTM C 1105 have not been adopted in ASTM C 33 but are suggested in the Appendix

TABLE 2—Concrete Prism Test, Expansion Limits, and Alkali Content.

Test	Na₂O Equivalent Alkali Content	Deleterious Expansion Limit at Indicated Time		
		3 months	6 months	1 year
ASTM C 1105	job cement	0.015	0.025	0.030
CSA CAN3 A23.2-14A	1.25%	0.01	...	0.025[a]

[a] For concrete not exposed to freezing and thawing or to the application of deicing chemicals, the limit is 0.04% at any age.

to ASTM C 1105 and are, therefore, nonmandatory. As such, they place greater responsibility on the specifier of the concrete to understand the factors promoting expansion due to alkali-carbonate reactivity versus the service environment of the concrete.

Expansion limits for the CSA prism test are specified and are tabulated in Table 2 with the suggested ASTM C 1105 limits.

Work by Rogers and Hooton [30] calls attention to the fact that the specified storage conditions for both the Canadian Standard and ASTM concrete prism tests allow leaching of the alkalies from the specimens resulting in reduced expansion. A known reactive aggregate produced approximately 0.15% expansion at 52 weeks in the moist room at 23°C, still well above the specification limit of 0.025%, but replicate prisms stored over water at 38°C in a sealed box produced greater than 0.3% expansion at 52 weeks. Determination of the water-soluble alkalies remaining in the prisms at 130 weeks indicated a loss of 63% versus 42% for the moist room versus the sealed box storage, respectively. The authors recommend that the solution to detecting alkali leaching problems during storage is the use of a reference reactive aggregate of known performance. If the reference aggregate expansion is not within prescribed limits, the test results on the unknown would be invalid, or suspect.

Rock Cylinder Expansion Test

The expansion of small (approximately 1.4-in. length by 0.4-in. diameter, 35.5 by 10 mm) cylinders of carbonate rock that are continuously immersed in 1N NaOH solution has been used as a direct indication of expansive alkali-carbonate reactivity of the rock since the method, originally using square cross-section specimens, was first described and used by Hadley in 1961 [16]. And since its earliest use, the correlation between rock cylinder (or rock prism) expansion and concrete prism expansion has been substantiated (Figs. 10 and 11). However, note in Fig. 11 the spread of the six rock prism expansion values corresponding to the concrete prism expansion value (average of 3) and, in Fig. 10 that the rock prism expansion values plotted are the average of 20 specimens.

Thus, despite good correlations between the "average" values of rock prism and concrete prism expansions, due to the variability inherent in the rock, the differences that can exist between restrained (in concrete) and unrestrained (in ASTM C 586) behavior, and the other factors governing expansion in concrete (for example, w/c, total water soluble alkalies, paste to aggregate ratio, etc.), the situation is such that ASTM C 586 is a relatively rapid indicator of potential expansive reactivity, and an effective tool for screening aggregate sources, but it does not predict in-concrete expansion of the rock and is not used as an acceptance test.

ASTM C 586 periodically measures the length change of small, elongate, cylinders of carbonate rock (or, as permitted by the method, square prismatic specimens of approximately the same size) that are continuously immersed in 1N NaOH solution. One test specimen per sample of rock is the minimum required by the method. It should be taken normal to the bedding if bedding is discernible. If bedding is not discernible, then three mutually perpendicular specimens are made. In general, even if bedding is discernible, it is desirable to prepare three

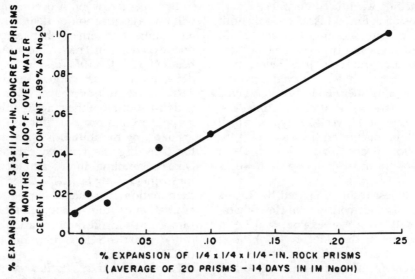

FIG 10—Expansion of small rock prisms in alkaline solution correlates well with expansion of concretes using these rocks as aggregate [4].

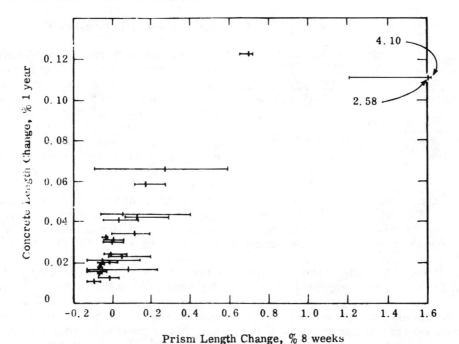

FIG 11—Relationship between one-year length change of (ASTM C 157) concrete prisms (average of three) with high alkali cement (0.95%) and length change of (ASTM C 586) rock prisms at eight weeks. Average and range of six rock prism length changes are shown for each sample [19].

mutually perpendicular specimens; one perpendicular to the bedding and two at right angles parallel to the bedding. Of three such specimens, in either case, the one exhibiting the greatest expansion is the one on which to base the test result.

An expansion of 0.10% at 28 days is considered to indicate the presence of material with a potential for deleterious expansion when used in concrete. For example, as reported by Rogers [7], most quarry faces sampled in Ontario, Canada, that gave average rock cylinder expansions of greater than 0.1% at four weeks also caused excessive expansion and cracking when used in concrete.

Although 0.1% expansion at four weeks is the only limit cited in ASTM C 586, an alternate limit of 0.20% at 16 weeks is also considered applicable by several researchers in view of the fact that some rocks contract before they begin to expand [19,20]. For example, a delayed expander tested by Newlon, Ozol, and Sherwood [19] contracted 0.1% at four weeks but expanded to 0.25% at 16 weeks, eventually peaking at 0.5% at 44 weeks. A typical alkali carbonate reactive rock producing on the order of 0.1% concrete prism expansion at one year with high alkali cement, might expand on the order of 0.5% at four weeks in ASTM C 586.

In the suite of sample test results reported by Rogers [7] (see Figs. 5 and 6), there were considerably fewer samples in the category "< 0.1% at 4 weeks but > 0.2% at 16 weeks" than there were in the category "> 0.1% at 4 weeks."

In view of the requirements of specimen selection and the occasional irregular, or late expanding, behavior of samples in the rock cylinder test, it is strongly recommended that all aspects of the testing program and interpretation of the results be guided by petrographic information.

Petrographic Examination

Petrographic examination of dolomitic carbonate rocks (ASTM C 295) proposed for use as coarse aggregate in concrete will detect the presence of the characteristic reactive lithologic texture. And, the characteristic reactive texture, as described in ASTM C 294 and discussed elsewhere in this chapter, when it occurs in a dolomitic limestone will undergo reaction or dedolomitization in an alkaline environment. Whether, when tested, the reaction will produce expansion of a rock clinder (ASTM C 586) or concrete beam (ASTM C 1105), insofar as the rock is concerned, depends on the perfection of the type lithology, as regards ASTM C 586, and, as regards ASTM C 1105, on the amount and distribution of the reactive lithology within the stratigraphic section proposed for use as concrete aggregate.

Following the initial use of petrographic examination for detecting the reactive carbonate rock texture (see Quarry Sampling) in order to identify candidate samples for further testing by ASTM C 586 or C 1105, petrographic examination is valuable during successive steps of the evaluation to guide the understanding of the behavior of samples in, and the interpretation of results from, the rock cylinder test (ASTM C 586) and the concrete prism test (ASTM C 1105).

Petrographic examination speaks directly to the concerns of Paragraphs 10.1, 10.2, and 10.3, "Interpretation of Results" of ASTM C 586. And, although not specifically called for in ASTM C 1105, petrographic examination of the concrete prisms at the conclusion of the test can furnish information helpful for interpretation of results at, or near, the test limits. The Canadian Standard notes the usefulness of petrographic examination at this stage.

Determination of Potential Alkali Carbonate Reactivity of Carbonate Rocks by Chemical Composition

The Ontario, Canada, Ministry of Transportation and Communications has developed a quick chemical screening test for the determination of potential alkali-carbonate reactivity of carbonate rocks by chemical composition (MTC LS 615), which is also under consideration as a Canadian standard.

The test is based on work by Rogers [7] and uses the relationships shown in Figs. 5 and 6 that are based on a correlation between expansion of concrete prisms, and rock cylinders, and performance of aggregates in field structures, and the CaO:MgO ratio of the limestone versus the alumina (Al_2O_3) content or the insoluble residue content.

In Figs. 5 and 6, using either Al_2O_3 or insoluble residue, regions are delineated in which the (compositions of) potentially expansive carbonate rocks cluster. Rocks whose analyses are in these regions should be considered to be potentially deleterious until their nondeleterious level of expansion is demonstrated by service record or by a concrete prism expansion test.

Other Methods and Approaches

Miniature Rock Prisms

Grattan-Bellew [31] and Tang and Lu [32] using miniature rock prisms, 8 by 15 mm of known reactive aggregate, in alkaline solution, and monitored by a linear differential transformer, measured expansions within days. In the latter paper, the authors used 1N potassium hydroxide (KOH) solution and temperatures of 20, 60, and 80°C to accelerate expansion to 14 days. The 80°C specimen expanded to over 1% at 11 days, at which point it cracked, the two-day expansion for the 60 and 80°C samples were 0.20 and 0.38%, respectively.

Autoclaved Concrete Microbars

Tang et al. [33] present a rapid method for determining the reactivity of carbonate rock based on the expansion of autoclaved concrete microbars that is based on a similar previously reported method for alkali-silica reaction [34]. Using 2 by 2 by 6-cm bars fabricated from 5 to 10-mm size aggregate, and 1.5% Na_2O equivalent portland cement at a 1:1.5 cement/aggregate ratio and 0.3 water/cement ratio, the bars are demolded following one-day moist-chamber curing, then exposed to 100°C steam for 4 h, and then autoclaved in 10% KOH solution at 150°C for 6 h. The expansion of the typical Kingston, Ontario, rock was about 0.175%, and other reactive rocks from various localities in China expanded between about 0.11 and 0.16%. The authors state that the method may be especially suitable for detecting delayed expanders, and that results can be obtained in three days.

Freezing and Thawing in NaCl Solution

Rogers [7], in comparing unconfined freeze-thaw (F-T) loss after five cycles in 3% NaCl solution versus loss in the $MgSO_4$ soundness test, noted that rocks that gave concrete prism expansions greater than 0.025% at one year also had high, generally greater than 18%, unconfined F-T losses while at the same time giving acceptable losses in the sulfate soundness test. He considers that unconfined F-T loss in NaCl solution may show promise as a screening test for alkali-carbonate expansivity.

SELECTION OF SAMPLES FOR TESTING AND FOR CONCRETE AGGREGATE

Quarry Sampling

Sampling of the stratigraphic section in limestone quarries for detecting the presence of alkali-carbonate reactive rock by petrographic examination should recognize that the characteristic reactive texture and composition are only subtly and indirectly related to the primary factors of sedimentation that produce bedding. The reactive texture is of secondary, or replacement, origin and may or may not conform faithfully to lithologic or stratigraphic units, or individual beds as observable in the field. The amount of reactive texture and the expansivity of that texture may vary laterally and vertically within beds or within a sequence of beds comprising a definable, or mappable, stratigraphic unit, to the same extent that the variations in rock prism (ASTM C 586) expansions within a given bed, or lithology, may be as large or larger than those between lithologies [19].

Sampling for the reactive lithology should therefore be guided by petrographic examination using as many thin or polished sections as necessary to delineate the reactive rock and to select samples for further tests. And, because of the variability that can exist, any thin or polished section that shows even a minor area of reactive texture should, at the first sampling stage (Stage I) require the field sampling unit that it represents to be considered suspect [19].

The objective of the first (Stage I) sampling is to establish the presence or absence of reactive rock within a quarry. It is understood that the establishment of 100% absolute absence requires examination of every bed. Lithology and structure should be considered in a way that samples are located with due regard to their relationship to the quarry production horizons, and in sufficient detail to permit a calculated "weighting" for comparison with the maximum "allowable" reactive content of the total aggregate that can be established by subsequent testing using ASTM C 1105. In Stage I sampling, a minimum of five and an average of approximately ten samples should be taken in a quarry. Each lithology representing more than 10% of the production should be sampled. Lithologies representing less than 10% can be combined and sampled at random locations in the combined thickness. Where the face exhibits relatively thin beds of varying lithology, it should be sampled at a minimum of ten systematically located stations [22].

If Stage I sampling establishes the presence of reactive rock within a quarry, then the objective of the Stage II sampling will be to identify as precisely as the limitations of time and budget will permit, the precise beds and their amounts of reactive material. Perhaps from 50 to 100 samples may be required. Previous experience suggests that the results of the more extensive and finer-scale sampling can end up delineating a lesser volume of reactive material in the quarry face than may have been indicated from the results of the Stage I sampling [22].

USING COARSE AGGREGATE FROM A QUARRY WITH POTENTIALLY EXPANSIVE ROCK

If potentially expansive rock is present in a quarry, the following courses of action should be considered.

If the stratigraphy of the producing section is favorable, avoid the reactive rock by selective quarrying. If the reactive rock cannot be totally avoided, utilize selective quarrying such that the aggregate from the produced section, when tested in concrete by ASTM C 1105, does not produce deletrious expansion. If the "natural" dilution approach is infeasible, dilute the produced aggregate with non-reactive stone such that the resulting mixture does not produce a deleterious level of concrete expansion by ASTM C 1105.

Research and experience with a group of Virginia quarries indicated that, for those sources, an acceptable level of dilution was reached when the reactive rock did not exceed 20% of the coarse aggregate, 20% of the fine aggregate, or 15% of the total aggregate, if both fractions contained reactive material [22].

In the concrete, use the smallest aggregate size that is economically feasible and the lowest alkali cement, bearing in mind that cement with total Na_2O equivalent alkali at the 0.4% level can produce alkali-carbonate reaction, and that alkali content per unit volume of the concrete is the important measure.

In the use of the concrete in constructions that will contain diluted, or otherwise minimized, amounts of reactive aggregate and alkali, restrict availability of moisture to the concrete by barriers or by design of the construction if possible.

If no mitigating approach is feasible, prohibit the use of the reactive rock in portland-cement concrete.

Acknowledgment

The author is grateful to Christopher A. Rogers, Ministry of Transportation, Downsview, Ontario, Canada, who kindly provided important graphs, photographs, and technical information.

REFERENCES

[1] Walker, H. N., "Chemical Reactions of Carbonate Aggregates in Cement Paste," *Significance of Tests and Properties of Concrete and Concrete-Making Materials, ASTM STP 169B*, American Society for Testing and Materials, Philadelphia, 1978, pp. 722–743.

[2] Hansen, W. C., "Chemical Reactions," *Significance of Tests and Properties of Concrete and Concrete-Making Materials, ASTM STP 169A*, American Society for Testing and Materials, Philadelphia, 1966, pp. 487–496.

[3] Swenson, E. G., "A Reactive Aggregate Undetected by ASTM Tests," *Bulletin No. 226*, American Society for Testing and Materials, 1957, pp. 48–51.

[4] Hadley, D. W., "Alkali Reactivity of Dolomitic Carbonate Rocks," *Proceedings*, Symposium on Alkali-Carbonate Rock Reactions, Highway Research Board, Record No. 45, 1964, pp. 1–20.

[5] Mather B., "Developments in Specification and Control," *Cement-Aggregate Reactions, Transportation Research Record 525*, Transportation Research Board, Washington, DC, 1974, pp. 38–42.

[6] *Field Trip Guidebook*, 7th International Conference on Alkali-Aggregate Reaction, Ontario Ministry of Transportation, Engineering Materials Office, Materials Information Report 99, Ottawa, Canada, 1986.

[7] Rogers, C. A., "Evaluation of the Potential for Expansion and Cracking of Concrete Caused by the Alkali-Carbonate Reaction," *Cement, Concrete, and Aggregates*, Vol. 8, No. 1, Summer 1986, pp. 13–23.

[8] Smith, P., "Learning to Live with a Reactive Carbonate Rock," *Proceedings*, Symposium on Alkali Carbonate Rock Reactions, Record No. 45, Highway Research Board, 1964, pp. 126–133.

[9] Alasali, M. M., Malhotra, V. M, and Soles, J. A., "Performance of Various Test Methods for Assessing the Potential Alkali Reactivity of Some Canadian Aggregates," *ACI Materials Journal*, Vol. 88, No. 6, Nov.–Dec. 1991, pp. 613–619.

[10] Ozol, M. A. and Newlon, H. H., Jr., "Bridge Deck Deterioration Promoted by Alkali-Carbonate Reaction: A Documented Example," *Cement-Aggregate Reactions, Transportation Research Record 525*, Transportation Research Board, National Research Council, Washington, DC, 1974, pp. 55–63.

[11] Hilton, M. H., "Expansion of Reactive Carbonate Rocks under Restraint," *Cement-Aggregate Reactions, Transportation Research Record 525*, Transportation Research Board, National Research Council, Washington, DC, 1974, pp. 9–21.

[12] Hilton, M. H., "The Effects of Textural and External Restraints on the Expansion of Reactive Carbonate Aggregates," Progress Report No. 7b, Virginia Highway Research Council, Charlottesville, VA, Sept. 1968.

[13] Swenson, E. G. and Gillott, J. E., "Characteristics of Kingston Carbonate Rock Reaction," *Concrete Quality Control, Aggregate Characteristics and the Cement Aggregate Reaction*, Bulletin No. 275, Highway Research Board, Washington, DC, 1960, pp. 18–31.

[14] Dolar-Mantuani, L. M. M., "Expansion of Gull River Carbonate Rocks in Sodium Hydroxide," *Proceedings*, Symposium on Alkali-Carbonate Rock Reactions, Record No. 45, Highway Research Board, Washington, DC, 1964, pp. 178–195.

[15] Hadley, D. W., "Alkali-Reactive Carbonate Rocks in Indiana—A Pilot Regional Investigation," *Proceedings*, Symposium on Alkali-Carbonate Rock Reactions, Record No. 45, Highway Research Board, Washington, DC, 1964, pp. 196–221.

[16] Hadley, D. W., "Alkali Reactivity of Carbonate Rocks—Expansion and Dedolomitization," *Proceedings*, Highway Research Board, Vol. 40, 1961, pp. 462–474.

[17] Newlon, H. H. and Sherwood, W. C., "An Occurrence of Alkali-Reactive Carbonate Rock in Virginia," *Carbonate*

Aggregate and Steam Curing of Concrete, Bulletin No. 355, Highway Research Board, Washington, DC, 1962, pp. 27–44.

[18] Sherwood, W. C. and Newlon, H. H., Jr., "A Survey for Reactive Carbonate Aggregates in Virginia," Record No. 45, Highway Research Board, Washington, DC, 1964, pp. 222–233.

[19] Newlon, H. H., Ozol, M. A., and Sherwood, W. C., "Potentially Reactive Carbonate Rocks," *An Evaluation of Several Methods for Detecting Alkali-Carbonate Reaction*, Progress Report No. 5, Virginia Highway Research Council, 71-R33, Charlottesville, VA, May 1972.

[20] Ryell, J., Chojnacki, B., Woda, G., and Koniuszy, Z. D., "The Uhthoff Quarry Alkali Carbonate Rock Reaction: A Laboratory and Field Performance Study," *Transportation Research Record 525*, Transportation Research Board, National Research Council, Washington, DC, 1974, pp. 43–54.

[21] Swenson, E. G. and Gillott, J. E., "Alkali-Carbonate Rock Reaction," *Proceedings*, Symposium on Alkali-Carbonate Rock Reactions, Highway Research Record No. 45, Highway Research Board, Washington, DC, 1964.

[22] Newlon, H. H. Jr., Sherwood, W. C., and Ozol, M. A., Potentially Reactive Carbonate Rocks, Progress Report No. 8., A Strategy for Use and Control of Carbonate Rocks including an Annotated Bibliography of Virginia Research, VHRC 71-R41, Virginia Highway Research Council, Charlottesville, VA, June 1972.

[23] Sherwood, W. C. and Newlon, H. H. Jr., "Studies on the Mechanisms of Alkali-Carbonate Reaction, Part I. Chemical Reactions," *Highway Research Record No. 45*, Highway Research Board, Washington, DC, 1964, pp. 41–56.

[24] Feldman, R. F. and Sereda, P. J., "Characteristics of Sorption and Expansion Isotherms of Reactive Limestone Aggregate," *Journal*, American Concrete Institute; *Proceedings*, Vol. 58, No. 2, 1961, pp. 203–214.

[25] Walker, H. N., "Reaction Products in Expansion Test Specimens of Carbonate Aggregate," *Transportation Research Record 525*, National Research Council, Washington, DC, 1974.

[26] Buck, A. D., "Control of Reactive Carbonate Rocks in Concrete," Technical Report C-75-3, U.S. Army Engineer Waterways Experiment Station, Corps of Engineers, Vicksburg, MS, 1975.

[27] Tang, M., Liu, Z., and Han, S., "Mechanism of Alkali-Carbonate Reaction," *Concrete Alkali-Aggregate Reactions, Proceedings*, 7th International Conference, 1986, Ottawa Canada, P. E. Grattan-Bellew, Ed., Noyes Publications, Park Ridge, NJ, 1987.

[28] Tang, M., Liu, Z., Lu, Y., and Han, S., "Alkali Carbonate Reaction and pH Value," *IL Cemento*, Vol. 88, July-Sept. 1991, pp. 141–150.

[29] Deng, M. and Tang, M., "Mechanism of Dedolomitization and Expansion of Dolomitic Rocks," Department of Silicate Engineering, Nanjing Institute of Chemical Technology, Nanjing, China, 1992; to be published in *Cement and Concrete Research*.

[30] Rogers, C. A. and Hooton, R. D., "Reduction in Mortar and Concrete Expansion with Reactive Aggregates due to Alkali Leaching," *Cement, Concrete, and Aggregates*, Vol. 13, No. 1, Summer 1991, pp. 42–49.

[31] Grattan-Bellew, P. E., "Evaluation of Miniature Rock Prism Test for Determining the Potential Alkali-Expansivity of Aggregates," *Cement and Concrete Research*, Vol. 11, 1981, pp. 699–711.

[32] Tang, M. and Lu, Y., "Rapid Method for Determining the Alkali Reactivity of Carbonate Rock," *Concrete Alkali-Aggregate Reactions, Proceedings*, 7th International Conference, 1986, P. E. Grattan-Bellew, Ed., Noyes Publications, Park Ridge, NJ, 1987, pp. 286–287.

[33] Tang, M., Deng, M., Lan, X., and Han, S., "Studies on Alkali-Carbonate Reaction," Department of Silicate Engineering, Nanjing Institute of Chemical Technology, Nanjing, Jiengsu, China, 1992.

[34] Tang, M., Han, S., and Zhen, S., "A Rapid Method for Identification of Reactivity of Aggregate," *Cement and Concrete Research*, Vol. 13, No. 3, 1983, pp. 417–422.

[35] Rogers, C. A., *Alkali Aggregate Reactions, Concrete Aggregate Testing and Problem Aggregates in Ontario—A Review*, Ontario Ministry of Transportation and Communications, Engineering Materials Office, Report EM-31, 5th rev. ed., 1985.

Degradation Resistance, Strength, and Related Properties

Richard C. Meininger[1]

PREFACE

The intent of this chapter is to be useful to those in the field and office who specify aggregates and who are involved in inspection and quality assurance activities. This chapter concentrates on tests relating to degradation of aggregates—both wet and dry tests—and includes a discussion of the effect of aggregate strength on concrete. Also, the effect of aggregates on wear and frictional properties of concrete in service is briefly covered.

The previous edition of this chapter [1] was longer. It included material concerning research on strength and abrasion properties of aggregates and a review of North American and European methods and research. It also included additional references and discussion of the mineralogic nature of the breakdown of aggregates in tests and information on hardness, impact, and toughness. It is recommended that engineers or researchers interested in these additional topics refer to that earlier edition. In earlier editions of this book, *ASTM STP 169* and *ASTM STP 169A*, this topic was covered by D. O. Woolf in his chapter on "Toughness, Hardness, Abrasion, Strength, and Elastic Properties."

INTRODUCTION

Aggregates for use in concrete, and to a large extent the individual constituent particles, must possess a reasonably high degree of inherent strength, tenacity, and stability to resist, without detrimental degradation, the static and dynamic stresses, impacts, and wearing actions to which it may be exposed both in concrete production operations and, ultimately, in concrete in service. In many uses of concrete, the roughest treatment to which an aggregate may be subjected in terms of mechanical forces and attrition is in the concrete production process. Actions involved in production are fairly predictable; those in service may be less predictable. In end uses such as beams, columns, covered slabs, walls, footings, and other mostly structural or architectural elements, the only real strength property needed of the aggregate after the concrete is in place is that necessary to give the concrete enough strength to resist the distributed service loads. Other uses may expose aggregate near or at the surface to a variety of localized impact and abrasive stresses that will be of overriding importance in aggregate evaluation and selection. Examples include pavements and slabs exposed to heavy traffic and hydraulic structures subject to eroding forces of moving water and sediment material.

Some recognition must be given to the consideration of an aggregate as a whole versus the properties of the individual particles present. In many crushed stone aggregates and some sands and gravels, the lithology of the particles and, consequently, their mechanical properties show little variation. In other cases, great variations may exist among the mineral and rock particles involved. A small percentage of weak particles in most instances is not objectionable, and may be thought of as an impurity or deleterious substance, and therefore not properly the subject of this discussion. These weaker or less abrasion resistant particles allowed in small percentages may cause a slight increase of fines during mixing or may contribute to occasional surface imperfections not particularly harmful to overall performance. In other instances, eliminating even very small percentages of friable or weak particles is desired. An example is high-velocity hydraulic structures where high surface strength and very accurate surface alignment is needed to resist cavitation damage that can spread rapidly once started.

Concrete carrying a distributed static or dynamic load, and not subjected to local stress concentrations, requires aggregate that will bond with the surrounding cement paste permitting the transfer of stress through the aggregate particles. Here, the particles are expected to possess enough strength and rigidity to carry the stresses without mechanical breakdown or excessive deformation—for usual aggregates and typical stress levels in concrete, this is not a problem. The rock, mineral, or synthetic materials making up aggregates are usually much stronger in compression or tension than the concrete; and the weakest link in the system is the bond between the paste and aggregate. In some cases, however, aggregate materials are used that possess compressive strength of the same order of magnitude as the concrete in which they are used. Examples are lightweight aggregates, marine limestones used in Florida, and volcanic cinders sometimes used in concrete. These weaker materials are used because of economic reasons or to produce concrete of lighter unit weight. However, high-strength aggregate materials will probably be

---

[1] Vice president of Research, National Aggregates Association, National Ready Mixed Concrete Association, 900 Spring St., Silver Spring, MD 20910.

required in the production of very high-strength concrete, which is presently used in the range 52.5 MPa (7500 psi) to 140 MPa (20 000 psi).

Impact, abrasion, scuffing, attrition, and other wearing actions, either in a dry environment or in the presence of water, can be an important factor. These loadings or actions usually are applied to the surface of the concrete so that only the properties of the aggregate at or near the surface are of any substantial importance in determining whether the long-term performance of the concrete will be acceptable. In some cases, special aggregates, concrete mixtures, or coatings may be used at the surface to improve the surface properties locally.

Surface actions are of two basic modes—impact and rubbing. Impact is where hard particles or objects impinge against the concrete surface with enough momentum to cause shattering, yielding, or debonding of aggregate particles. Materials with good impact resistance are said to possess good toughness. In the rubbing or scratching mode, damage can be inflicted by the movement of particles or objects on the surface under enough load to cause indentation, for the relative hardness involved, and therefore cause scratching or gouging as movement occurs. Scratches and gouges from relatively large particles or objects are on a macroscopic scale and are termed wear. In other words, the amount of material removed by such action, if repeated over and over again, will wear away a significant quantity of surface mortar exposing more and more coarse aggregate with time. Rubbing or scratching action from much smaller particles—on a microscopic scale—does not have the capacity to remove much material, but it can cause a polishing action at the surface. In highway uses, and occasionally other types of service, aggregates that allow rapid polishing can contribute to poor frictional properties of wet surfaces.

APPLICABLE ASTM STANDARDS

The ASTM standards given here fall under the general purview of this chapter. Some relate directly to aggregate properties, and others to properties of concrete in which the performance of different aggregates can be compared.

1. ASTM Specification for Concrete Aggregates (C 33)
2. ASTM Test Method for Resistance to Degradation of Small-Size Coarse Aggregate by Abrasion and Impact in the Los Angeles Machine (C 131)
3. ASTM Test Method for Resistance to Degradation of Large-Size Coarse Aggregate by Abrasion and Impact in the Los Angeles Machine (C 535)
4. ASTM Test Method for Degradation of Fine Aggregate Due to Attrition (C 1137)
5. ASTM Test Method for Aggregate Durability Index (D 3744)
6. ASTM Test Method for Abrasion Resistance of Concrete by Sandblasting (C 418)
7. ASTM Test Method for Abrasion Resistance of Concrete (Underwater Method) (C 1138)
8. ASTM Test Method for Abrasion Resistance of Concrete or Mortar Surfaces by the Rotating-Cutter Method (C 944)
9. ASTM Test Method for Abrasion Resistance of Horizontal Concrete Surfaces (Procedure A—Revolving Disks; Procedure B—Dressing Wheels) (C 779)
10. ASTM Test Method for Accelerated Polishing of Aggregates Using the British Wheel (D 3319)
11. ASTM Test Method for Insoluble Residue in Carbonate Aggregates (D 3042)
12. ASTM Test Method for Skid Resistance of Paved Surfaces Using a Full-Scale Tire (Locked Wheel Skid Trailer) (E 274)
13. ASTM Test Method for Skid Resistance of Paved Surfaces Using North Carolina State University Variable—Speed Friction Tester (E 707)
14. ASTM Practice for Accelerated Polishing of Aggregates or Pavement Surfaces Using a Small-Wheel, Circular Track, Polishing Machine (E 660)

ASTM C 33 for concrete aggregates includes a modular format where many of the required properties of coarse aggregates are varied depending on severity of exposure to weather and the end use of the concrete. However, due to the lack of enough significantly explicit data relating concrete performance with degradation in the Los Angeles Machine (ASTM C 131 or C 535), the "abrasion" limit in ASTM C 33 has been set at 50% for all categories. (Note: ASTM C 131 and C 535 now properly refer to "degradation" in the tests by "abrasion and impact in the Los Angeles machine;" however, the terms: "abrasion," "abrasion loss," and "LA wear" are often used to refer to the percentage loss in these tests and "L.A. abrasion" as an abbreviated name for the tests.) In addition, judgment provisions such as the following have been included in many specifications to allow use of known satisfactory materials: "Coarse aggregates having test results exceeding the limits specified . . . may be accepted provided that concrete made with similar aggregate from the same source has given satisfactory service . . . or in the absence of a demonstrable service record provided that the aggregate produces concrete having satisfactory relevant properties . . ." when tested in the laboratory. ASTM C 33 specifies requirements for a wide range of aggregate sizes. The determination of which grading to use in ASTM C 131 or C 535, to test an aggregate product, is based on the requirement that degradation in the L.A. machine is to be determined using the grading most nearly corresponding to the grading to be used in the concrete. Crushed air-cooled blast-furnace slag is excluded from the L.A. abrasion requirement in ASTM C 33. In its place, a minimum compact unit weight is required.

The test methods listed here are discussed in subsequent sections relating to the properties measured. Other than the percent loss limits for degradation by abrasion in the L.A. machine, as measured by ASTM C 131 or C 535, there are generally no other specified limits with respect to the effect of aggregate as measured in the other tests unless there are special requirements limiting degradation of aggregates in handling and mixing, limiting abrasion resistance of concrete surfaces, or requirements pertaining to the frictional properties of pavement surfaces after being subjected to traffic [2].

DEGRADATION OF COARSE AGGREGATE (LOS ANGELES ABRASION)

ASTM C 131 and C 535 are accepted and used almost universally in the United States as specification qualification tests for coarse aggregate for concrete. They are also used widely around the world to evaluate aggregates.

ASTM C 131 was first adopted as a tentative standard test in 1937. At first, it provided for testing only 37.5 or 19.0-mm (1½ or ¾-in.) maximum size aggregates (Gradings A and B). Later 9.5-mm (⅜-in.) and 4.75-mm (No. 4) sizes were added (Gradings C and D). In 1947, a revision was prepared to add three more gradings for large aggregate sizes—75-mm (3-in.), 50-mm (2-in.), and a second narrower 37.5-mm (1½-in.) maximum size grading (Gradings E, F, and G)—with an increased sample size from 5000 to 10 000 g and a doubling of the number of revolutions from 500 to 1000 for these larger sizes. This was an attempt to obtain generally corresponding percentages of wear for each grading when a material of uniform quality was tested. Such uniformity of results was usually obtained for Gradings A, B, C, and D, within a reasonable margin of error. For the larger gradings, however, agreement was found in some instances and contrary data in other cases. To clarify the situation in 1964 and 1965, the larger sizes were deleted from ASTM C 131 and a new test method was adopted, ASTM C 535, for them.

Briefly, the method is a dry abrasion and impact test that involves placing a sample of aggregate and a charge of steel balls into a hollow steel cylinder that is 508 mm (20 in.) long and 711 mm (28 in.) in diameter, is closed on the ends, and has a single rigid steel shelf extending 89 mm (3½ in.) into the chamber. The cylinder is rotated at 30 to 33 rpm for 500 revolutions. Percent loss is obtained as the difference between the original sample weight and the amount of the final sample coarser than the 1.70-mm (No. 12) sieve, expressed as a percentage of the original weight. Often the loss is determined at 100 and 500 revolutions to give an indication of the rate of loss. For a material degrading at a uniform rate, the ratio of the loss after 100 revolutions to the loss after 500 revolutions should not greatly exceed 0.2; a small percentage of very soft, friable, or brittle material or a large number of flat or elongated pieces might cause this ratio to increase. A nominal 5000-g sample is used for the gradings given in Table 1. The required charge of nominal 420-g steel spheres placed in the drum with the aggregate sample is as follows:

Grading A = 12 steel spheres,
Grading B = 11 steel spheres,
Grading C = 8 steel spheres, and
Grading D = 6 steel spheres.

The precision statement in ASTM C 131 is based on 19.0-mm (¾-in.) maximum size aggregate samples distributed to a large number of laboratories throughout the United States by the AASHTO Materials Reference Laboratory. The samples were tested by both ASTM and AASHTO procedures, which differ only in minor respects for this particular test. Multilaboratory coefficient of variation was found to be 4.5%, and single operator coefficient of variation was 2.0%. These data are based on samples ranging in loss from 10 to 45%. In another precision investigation within one state, the data show a higher level of variability on samples having a loss of 13 to 18% [3].

The same equipment is used for ASTM C 535. Gradings 1, 2, and 3 are given in Table 2. The required charge of steel spheres is 12 for all three gradings. Note that 10 000-g samples are used and that 1000 revolutions of the drum are used. Loss is again determined using the 1.70-mm (No. 12) sieve, and an indication of uniformity can be obtained by comparing the loss after 200 revolutions with that after 1000 revolutions. The ratio should not be greatly different from 0.2 for uniform material. Unfortunately, no data are available to use in the development of a precision statement for ASTM C 535.

Caution should be exercised in comparing values obtained from ASTM C 131 and C 535. They may give different results for similar materials. For this reason, specification limits for larger sizes may not necessarily be the same as those judged to be needed for the smaller sizes of coarse aggregate.

Impact probably causes more loss, at least during early revolutions of the drum before any fines are created that might tend to cushion the impact forces. To some extent, harder rocks and minerals, even though they are strong,

TABLE 1—ASTM C 131 Aggregate Test Sample Gradings.

Sieve Size (square openings)		Weight of Indicated Sizes, g			
		Grading			
Passing	Retained on	A	B	C	D
37.5 mm (1½ in.)	25.0 mm (1 in.)	1250 ± 25	...	...	...
25.0 mm (1 in.)	19.0 mm (¾ in.)	1250 ± 25	...	...	...
19.0 mm (¾ in.)	12.5 mm (½ in.)	1250 ± 10	2500 ± 10	...	...
12.5 mm (½ in.)	9.5 mm (⅜ in.)	1250 ± 10	2500 ± 10	...	...
9.5 mm (⅜ in.)	6.3 mm (¼ in.)	...	...	2500 ± 10	...
6.3 mm (¼ in.)	4.75 mm (No. 4)	...	...	2500 ± 10	...
4.75 mm (No. 4)	2.36 mm (No. 8)	...	...	...	5000 ± 10
Total		5000 ± 10	5000 ± 10	5000 ± 10	5000 ± 10

TABLE 2—ASTM C 535 Aggregate Test Sample Gradings.

Sieve Size, mm (in.) (square openings)		Weights of Indicated Sizes, g		
		Grading[a]		
Passing	Retained on	1	2	3
75 (3)	63 (2½)	2500 ± 50	...	...
63 (2½)	50 (2)	2500 ± 50	...	...
50 (2)	37.5 (1½)	5000 ± 50	5000 ± 50	...
37.5 (1½)	25.0 (1)	...	5000 ± 25	5000 ± 25
25.0 (1)	19.0 (¾)	...	...	5000 ± 25
Total		10 000 ± 100	10 000 ± 75	10 000 ± 50

[a] Gradings 1, 2, and 3 correspond, respectively, in their size distribution to Gradings, E, F, and G in the superseded ASTM C 131–55, Test for Abrasion of Coarse Aggregate by Use of the Los Angeles Machine, which appears in the 1961 *Book of ASTM Standards*, Part 4.

tend to shatter more than softer materials that can better absorb the force. On the other hand, softer minerals will be more susceptible to surface wearing. The product of the degradation action in that case may be more of a dust rather than the larger angular pieces resulting from the shattering. Also, dust produced early in the test may provide a cushioning effect, giving possible misleading results.

Smith [4] used coarse aggregates in concrete with L.A. abrasion losses ranging from 13 to 39%. He found no significant correlation between abrasion loss and the abrasion of concrete containing the aggregates using three different test procedures: (1) Davis steel ball, (2) dressing wheel, and (3) the Ruemelin shot blast apparatus. The most significant correlations with concrete abrasion resistance were with concrete strength and water-cement ratio. A soft limestone did decrease the concrete abrasion resistance at strengths below 55 MPa (8000 psi). Above that strength, no coarse aggregate effect could be detected.

The loss in the L.A. abrasion test has been correlated by Woolf [5] with the strength of concrete prepared with a wide variety of aggregates. However, laboratory studies reported by Bloem and Gaynor [6] failed to show such a relationship with strength. Jumper [7] and Walker and Bloem [8] conducted research where aggregates of high and low abrasion loss were blended to produce a range of values. They did show some correlation of reductions in abrasion values with increased strength. Jumper blended aggregate to obtain loss values from 42 to 58%. Over that range, the decrease in compressive strength for constant cement factor concrete was from about 27 to 23 MPa (3900 to 3300 psi). Walker and Bloem did not detect any dependence of concrete compressive strength on abrasion loss, but they did find that high abrasion losses did lower flexural strength with the aggregates studied. Bartel and Walker [9] and Walker [10] showed a slight drop-off of compressive and flexural strength with increased abrasion losses. Correlation of L.A. abrasion with strength is inconclusive.

With respect to the freezing and thawing durability of aggregates tested in concrete specimens, Gaynor and Meininger [11] did not find a relationship between concrete durability and abrasion loss at 500 revolutions or 100 revolutions. This study included 56 coarse aggregates, ranging in abrasion loss from about 15 to 65%. This and other freezing and thawing data is also reported in Ref 12.

Kohler [13], in a comprehensive study reported to the Highway Research Board compared the results of aggregates tested by the German Impact Test, the L.A. abrasion test, the older Deval Abrasion Test, the British Impact Test (British Standard 812), and a modified Marshall Impact Test. He found that even though the methods use different loads and procedures, the tests characterize "the same or at least similar properties of the aggregates." Good correlation was found between all the tests run in one laboratory. High variation of results was found for the L.A. abrasion and German Impact Tests when tests were run in different laboratories, suggesting that further strict standardization of procedures may be necessary.

In spite of the drawbacks of the L.A. abrasion test, it is a fairly fast test with known variability; and as a general test, it does provide a means of identifying coarse aggregate that may degrade readily during handling.

WET DEGRADATION AND ATTRITION TESTS

There has been increased interest in abrasion and attrition procedures run in a wet environment. The California durability index test for fine and coarse aggregate, which measures the tendency of an aggregate to break down into plastic clay-like fines, has been standardized as ASTM D 3744. Some aggregates degrade in a different manner in the presence of water than that exhibited in dry test methods [14]. For concrete aggregates, the principal value of such tests are identification of aggregates that may create undesirable fines during handling, batching, and mixing that could affect concrete properties such as strength and water requirement. These tests also have significance for base courses and aggregate surfaced roads where the possibility of interparticle movement in the presence of water may occur over the period of service of the material.

Liu [15] reported on research that led to the development of an underwater abrasion test method (ASTM C 1138) designed to assess the abrasive effect of waterborne gravel, rocks, and other debris traveling over concrete surfaces in hydraulic structures. He concluded that the results of this test were correlated with concrete strength and the hardness of the aggregate. However, the underwater abrasion of concrete was not related to the L.A. abrasion of the coarse aggregate.

Highway Research Board Circular 144 [16] indicates use by state transportation agencies of various test methods relating to aggregate durability (other than freezing and thawing). The L.A. abrasion (a dry test) was used by the great majority; however, a number also indicated use of some type of wet abrasion or attrition procedure including: Durability Index, Washington Degradation Procedure, Idaho Degradation, Nevada Air Test, the Oregon Air Test, and modified L.A. abrasion tests using water. A few relied on petrographic analysis to identify mineral constituents that are associated with rapid weathering. Almost all agencies reported some degradation problems; in most cases, the problems were termed moderate.

Breese [17] gives a history of the development of wet abrasion tests in the Western United States. He experimented with different procedures designed to detect comminution degradation and alteration degradation. He concluded that Nevada had a number of degradation susceptible aggregates and that the California durability index test is not as harsh as some of the other procedures. Ekse and Morris [18] compared properties of fines produced from a standard L.A. abrasion test and an extended 4-h test without the steel balls. The aggregates were crushed basalt, a pit-run river gravel, and a crushed river gravel all from sources in the State of Washington. Plasticity index of the fines produced from the extended test were respectively 6.8, 2.0, and 1.0. Fines from the standard L.A. abrasion test were found to be nonplastic.

Minor [19] describes the development of the Washington degradation test that was prompted by the poor performance of several basalt crushed stones and volcanic gravels that gave losses in the standard L.A. abrasion test ranging from 12 to 50. These aggregates contained altered minerals that contributed to rapid weathering. In studies at the Pennsylvania DOT and at Purdue University in Indiana [20,21] to help identify the properties of shales that might help determine suitability for its use in embankment or as a granular material, the Washington degradation test was one of the tests that interested these researchers. In addition to the Washington test, good indications of performance were the ultrasonic disaggregation test and the ethylene glycol soak test used by the Corps of Engineers (CRD-C 148, Method of Testing Stone for Expansive Breakdown on Soaking in Ethylene Glycol.) At Purdue, the Swiss Hammer was found to be a potentially useful tool in identifying hard shales. High rebound readings were obtained for two shales that showed less than 40% loss in the L.A. abrasion test and that gave good results in the Washington Test.

The durability index for fine and coarse aggregate (ASTM D 3744) was developed to measure breakdown during construction and normal use under traffic conditions. It is intended to complement the sand equivalent, cleanness value, L.A. abrasion, and sulfate soundness procedures. There is little correlation between the durability index and the L.A. abrasion test.

The durability index tests starts with a washed sample and then measures the quality of the fines generated from interparticle abrasion during a wet agitation period. Figure 1 shows durability index values versus L.A. abrasion. The precision of the fine and coarse durability tests is

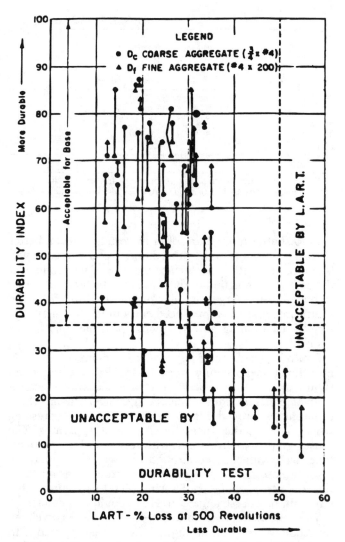

FIG. 1—L.A. abrasion loss versus California durability index [23].

given in ASTM D 3744 with the standard deviation increasing with decreasing durability index values. The single-operator standard deviation is about 1.5 to 3.5 and the multi-laboratory standard deviation is about 2 to 5.

Goodewardane [22] in work with the Washington degradation test looked into sources of variation. He concluded that both surface attrition and leaching of clays from the aggregate pores were involved in the test. The value obtained is very dependent on the surface area of the charge; finer samples will indicate more degradation. Aggregate quality improved with repeated agitation cycles as found by Hveem and Smith [23]. Weathered basalt was found to produce lower durability results than fresh basalt.

Degradation or attrition of aggregates during mixing and agitating of concrete has been the subject of several investigations. In studies by Slater [24], and more recently by Gaynor [25] and by McKisson [26], it was identified that during prolonged mixing the fine aggregate with its much greater surface area tended to break down by attrition much more readily than the coarse aggregate. In 2 h of mixing in a laboratory mixer, in one case, and 2 h and

40 min, in another case, little change was noted in the coarse aggregate grading. The amount of minus 150-μm (No. 100) material generated from the fine aggregate depended on aggregate hardness. A fine aggregate known to break down, caused an increase in minus 150-μm (No. 100) material of almost 12% in 2 h in one study. In the other study, a soft limestone sand caused an increase in minus 150-μm (No. 100) size of about 6% in 160 min. Hard quartz sand only caused an increase of 1% in the same 160-min mixing period.

A number of tests have been developed to measure the propensity of fine aggregates to degrade due to attrition during concrete mixing [27–30]. In 1990, ASTM C 1137 was approved. The method is based on work reported in the preceding references using a vane-type attrition chamber, mounted in a drill press and turned at 800 rpm. Results of the test are given in terms of the amount of minus 74-μm (No. 200) material generated and the decrease in fineness modulus as a result of the 6-min attritioning period. Meininger [29], in the work conducted at the NAA-NRMCA Joint Research Laboratory, found that the grading used in the test affects the results. Therefore, fine aggregates should be compared in one of the alternative test gradings given in ASTM C 1137 with fineness moduli ranging from 2.15 to 3.33; aggregates tested in the coarser gradations tended to have more minus 75-μm (No. 200) material generated and a greater decrease in fineness modulus. Some sources of fine aggregate, which were known to cause slump and air loss in concrete, and in some cases, strength problems, generated 23% or more 75-μm (No. 200) material during the test run with the coarse grading.

The Micro-Deval test is being used in Canada (Quebec and Ontario) for the evaluation of fine and coarse aggregate [30,31]. It is a wet attrition test developed in France during the 1960s. It is based on equipment used in the grinding industry.

For fine aggregate, the test was modified by the Ministry of Transportation in Ontario (MTO) to use a 700-g sample of sand washed to remove material passing the 75-μm (No. 200) sieve. The sample is placed in a steel jar mill with 1250 g of steel balls and 750 mL of water and rotated at 100 rpm for 15 min. Loss is expressed as the percentage of minus 75-μm (No. 200) material generated in the test. In the Ontario study, the Micro-Deval test was found to correlate with sulfate soundness and water absorption of fine aggregate. They also included the ASTM C 1137 attrition test and a similar Ontario (MTO) attrition test with a more robust impeller and a reduction in the rpm from 850 to 390 rpm to reduce splashing problems. Table 3 shows a comparison of the three tests, both amount of fines produced and the coefficient of variation. It was found that the Micro-Deval test had the lowest within-laboratory variability, and it was relatively insensitive to fine aggregate grading. It is being used in Ontario to evaluate fine aggregate.

In a similar Ontario study of coarse aggregate tests, the Micro-Deval coarse aggregate test (Quebec BNQ 2560-070182) was evaluated and compared to three British Standard Tests (BS 812) for aggregate impact value, polished stone value, and aggregate abrasion value. In the Micro-Deval test for coarse aggregate, 500 g of a graded sample is placed with 5000 g of 9.5 mm (⅜ in.) steel balls and 2.5 L of water in a 5-L steel jar mill that is rotated at 100 rpm for 2 h and the loss is measured by the amount passing the 1.18-mm (No. 16) sieve. Ontario Test Methods (MTO LS-614) for unconfined freeze-thaw of coarse aggregate and the magnesium sulfate soundness and Los Angeles abrasion tests were also included. The Micro-Deval test correlated with sulfate soundness, but it had lower variation, see Fig. 2. It was concluded that the Micro-Deval and unconfined freeze-thaw tests approximate the deterioration due to weathering and that, when used in conjunction with water absorption and petrographic examination, they are better predictors of field performance for marginal aggregates. Also, the British aggregate impact value test was seen as a practical substitute for the L.A. abrasion test, see Fig. 3.

Higgs [28] reported another fine aggregate wet attrition test using a paint shaker to attrition a 1000-g regraded sample of fine aggregate. He investigated a number of basalt fine aggregates and found that those causing rapid slump loss in concrete yielded more than 8 or 9% minus

TABLE 3—Within-Laboratory Variability of Attrition and Micro-Deval Tests.[a]

	Natural Sand		Shaley Limestone Screenings	
	Mean Loss	Coefficient of Variation	Mean Loss	Coefficient of Variation
Micro-Deval	13.8%	1.9%	23.4%	1.1%
ASTM attrition	13.3%	11.0%	38.8%	6.8%
MTO attrition	4.1%	14.1%	20.1%	5.4%

[a] Number of replicates = 10.

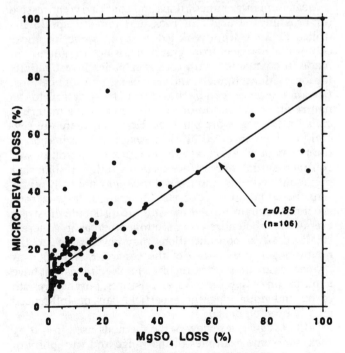

FIG. 2—Correlation of Micro-Deval abrasion against magnesium sulfate soundness of coarse aggregates [31].

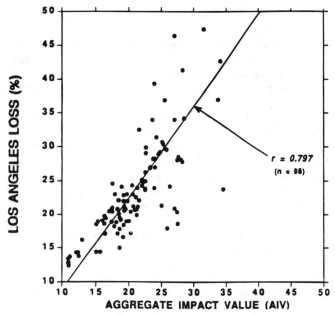

FIG. 3—Relationship between loss in the Los Angeles test and aggregate impact value of coarse aggregates [31].

75-μm (No. 200) material in the shaker test. These sands also had high magnesium sulfate soundness losses. One sand that produced 6% fines in the test caused some problems on a construction project with increased fines in concrete. This test method has been standardized by the U.S. Army Corps of Engineers as CRD-C141.

STRENGTH OF AGGREGATE MATERIAL

Measurements of unconfined and confined compressive (or crushing) strength of rock and mineral specimens drilled or sawn from rock ledges, or in some instances cobbles or boulders from gravel deposits, are quite often made in connection with rock mechanics investigations for foundations, tunnels, and various mining applications. Certain inferences can be drawn from these data as to the nature and distribution of the strengths of the materials constituting aggregate particles. However, there are several problems involved in translating this information to the behavior of aggregate in concrete: (1) conditions of confinement in concrete are certainly different than that of unconfined tests, and triaxial tests may not truly represent the situation in concrete; (2) strength tests of rock specimens usually involve measures to distribute the stress evenly over the entire cross section, whereas in concrete localized stress concentrations may develop due to the nonhomogeneous nature of the concrete; and (3) rock properties are most often reported for specimens that have a minimum of flaws, cracks, and fissures, but the strength of an individual aggregate particle can be influenced greatly by such defects.

Udd [32] gives a review of the methods used in testing rock and some examples of the range and variability of results. Compressive strength increases with confinement and both elastic and plastic deformations occur. There are many sources for data on compressive strength of rock and in some cases tensile strength or modulus of elasticity. Bond [33] shows compressive strengths of small cubes and cylinders cut from rock material. For limestone, the values ranged from 90 to 270 MPa (13 000 to 39 000 psi); for traprock, 105 to 235 MPa (15 000 to 34 000 psi); and for granite, 85 to 275 MPa (12 000 to 40 000 psi). These values represent ranges from a number of sources. He also reports compressive strength results from one gravel source ranging from 165 to 235 MPa (24 000 to 34 000 psi). The great bulk of the strength data available is for stone materials where specimens of the required size can be obtained and usually the material is less variable. Gravel deposits, on the other hand, present difficult problems in attempting to measure strength. Strength specimens can only be cut from sizes larger than used in many products and the lithology of many gravel deposits is quite variable, necessitating the testing of a large number of specimens to get meaningful results. The assumption can be made that the strength of the rock materials in gravels is similar to that of the rock materials from which the gravels were derived.

Tests of two types of rocks used as aggregate by Iyer et al. [34] yielded, for a limestone source, compressive strengths of 105 to 200 MPa (15 000 to 29 000 psi) and tensile strengths of 2.1 to 6.2 MPa (300 to 900 psi), and for a quartzite source, compressive strengths of 340 to 470 MPa (49 000 to 68 000 psi) and tensile strengths of 7.6 to 15.8 MPa (1100 to 2300 psi). Kaplan [35] tested gravel, limestone, granite, and basalt aggregates and obtained compressive strengths ranging from about 160 MPa (23 000 psi) for the gravel and the granite on the low end to 305 MPa (44 000 psi) for the basalt on the high end. The tensile strength measured for the gravel, limestone, and granite ranged from about 6.2 to 11.7 MPa (900 to 1700 psi). The basalt had a tensile strength of 15.2 MPa (2200 psi). Of the almost 1000 compressive strengths reported by Woolf, for rock from which crushed stone coarse aggregates are made, in his state-by-state summary [36], only about 60 results indicated compressive strength levels less than 70 MPa (10 000 psi).

From the foregoing, it is apparent that the great majority of the aggregate-making rock materials are strong in compression, over 70 MPa (10 000 psi), and like all brittle materials relatively weak in tension. There is also a proportion of aggregates used for concrete that are weak enough, less than 70 MPa (10 000 psi), that their strength characteristic may be an important limitation on the properties of concrete made with these materials. Kaplan [35], in using a number of strong aggregate materials in concrete, concluded that "no relationship between the strength of the coarse aggregate and the strength of concrete was established. This finding should not be taken to mean that aggregates of low strength will not affect concrete strength." Johnston [37] using a 48 MPa (6900 psi) sandstone coarse aggregate in concrete and Collins and Hsu [38] using a 28 MPa (4000 psi) oolite coarse aggregate in concrete were both able to produce concrete over 40 MPa (6000 psi) at 28 days age. These data suggest, as does broad experience with lightweight structural aggregates and experience using Florida limestone [39], volcanic cinders [40], and coral in concrete [41], that aggregate strength need only be of the same order of magnitude as the concrete strength needed. Tanigawa [42] in his model analysis with various aggregate and mortar strengths concluded that the best results in terms of delay in formation of bond cracks at the coarse aggregate-mortar interface were obtained when the aggregate and mortar were about the same strength.

It is recognized that coarse aggregate-to-mortar bond is extremely important to concrete strength and that a reasonably high-strength aggregate with excellent bonding characteristics is necessary in the production of very high-strength concrete over 55 MPa (8000 psi). The strength of individual fine aggregate particles is apparently not too important as long as some minimum level is achieved, principally so that it does not break down and create excess fines and surface area during concrete production. Bond is not as important with fine aggregate because of its large surface area-to-volume ratio. The influence of fine aggregate on concrete strength is almost exclusively through its shape, texture, and grading effects on mixing water demand needed for workability.

The significance of strength tests of aggregates is not well established. There are no ASTM test methods or specification requirements in this area. The usual practice is to evaluate the aggregates' performance in concrete to see if it will produce concrete of the needed strength. Aggregate strength may be an indicator of other aggregate properties, but relationships have not been established.

ELASTIC PROPERTIES

Modulus of elasticity and related values such as the Poisson's ratio of concrete aggregates are not included in aggregate specifications and there are no ASTM aggregate test methods designed to yield this information. The influence of aggregate source on concrete modulus of elasticity is determined normally by testing concrete mixtures containing each of the aggregates. As is the case for compressive strength, a good deal of information from rock mechanics investigations is available concerning rock deposits [43–45], but little is known about the elastic properties of gravel or sand particles other than what can be inferred from tests of parent rock material. A number of investigators have also determined rock modulus of elasticity in connection with researches of aggregate and concrete properties [34,35,37,38]. There has been a good deal of activity in the more theoretical area of developing equations to predict the resulting modulus of elasticity of a composite material from a knowledge of the elastic properties of the aggregate and the matrix [46–50].

Figure 4 gives an indication of the range of Young's modulus of elasticity for various rock types. Both static and dynamic E data are included. Generally, the dynamic values are higher, and there is a good linear relationship between dynamic and static values determined in the laboratory. A direct linear relationship has also been found between rock compressive strength and modulus of elasticity [44,45].

Both in compression and tension, stress-strain plots for stone [43] and for cement paste [37,51] are normally fairly straight lines indicating that the aggregate and matrix are reasonably elastic. Concrete and mortar, on the other hand, have a curved stress-strain plot when the stress exceeds about 30% of ultimate strength. This is due to the formation of bond cracks and slipping at the aggregate paste interface. With a curved stress-strain plot the modulus of elasticity computed for the concrete will vary depending on which of the several recognized definitions is used. Because of this and because unknown factors can affect the influence of the paste-aggregate bond on the stress-strain curve, there is no simple relationship between aggregate modulus of elasticity and concrete modulus of elasticity.

It does not always hold that an increase in aggregate modulus of elasticity will increase the E of the concrete even if the properties of the paste are held relatively constant. An example of this is contained in Houghton's paper [52] where natural rounded aggregate with an E of about 41 400 MPa (6×10^6 psi) produced concrete with E ranging from 28 000 to 34 000 MPa (4 to 5×10^6 psi) at 90 days compared to a quarried aggregate with an E of about 62 000 MPa (9×10^6 psi) that produced concrete with an E of 21 000 to 28 000 MPa (3 to 4×10^6 psi). Hirsch [50], however, shows data where aggregates with E values of about 14, 34, 62, 76, and 207×10^3 MPa (2, 5, 9, 11, and 30×10^6 psi) were used in different proportions in concrete indicating "that the modulus of elasticity of concrete is a function of the elastic moduli of the constituents. An increase or decrease in the modulus of either the aggregate or matrix constituent will produce a corresponding effect on the concrete." The equations advanced by Hirsch and others are complex and include, of necessity, various assumptions.

Less work has been done on the tensile modulus of elasticity. Johnston [37,53] gives values of tensile modulus of elasticity of rock. The values are somewhat lower than the corresponding compression values. Houghton [52] discusses the concept of predicting the tensile strain capacity of mass concrete so that tensile strains caused by thermal volume changes can be accommodated by the concrete without cracking. Relationships with age for tensile strength and modulus of elasticity enter into the determination of strain capacity for rapidly loaded concrete. Creep is a third factor that enters into the estimate when stress is applied over long periods.

HARDNESS, WEAR, FRICTIONAL PROPERTIES

"Hardness is the single most important characteristic that controls aggregate wear," according to Stiffler [54]. Mineral and rock materials when subjected to rubbing, scratching, and gouging actions, particularly with the aid of an abrasive, are worn down due to the scratching and pitting that occurs. Stiffler has observed that, unlike metals, which are just scratched when subjected to the movement of an abrasive on the surface under a load, minerals are pitted as well. This is apparently due to mineral grains

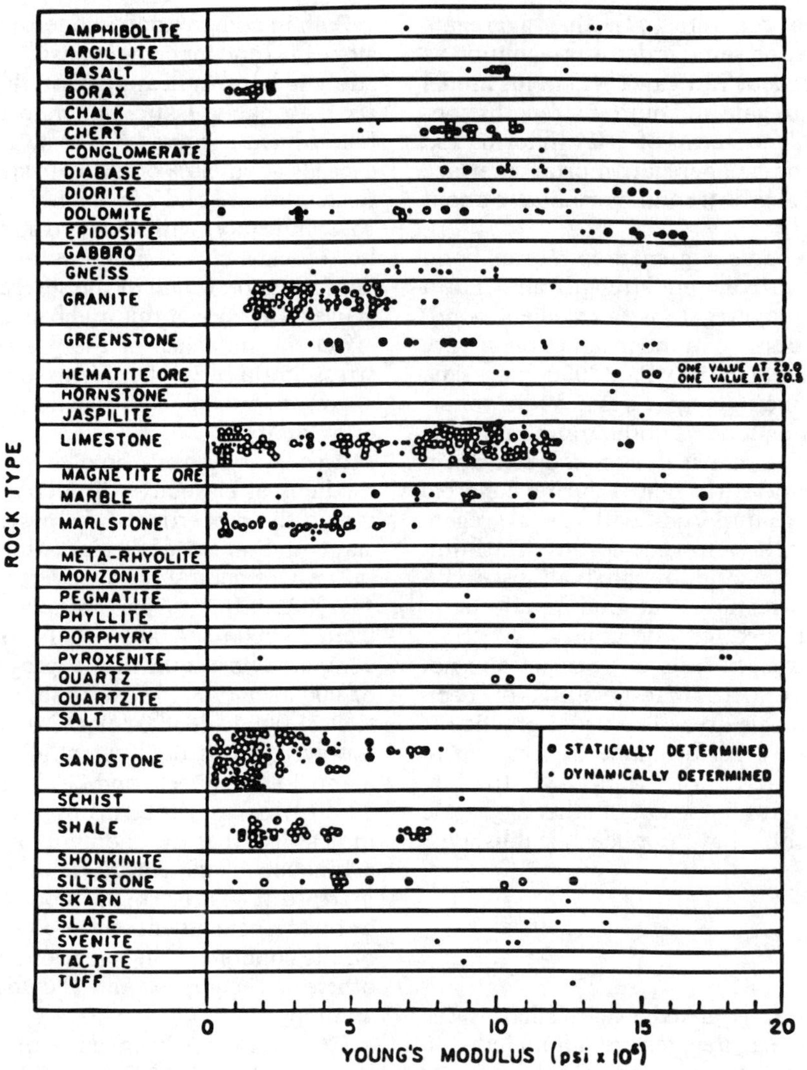

FIG. 4—Relationship between rock type and Young's modulus; 1 psi × 10⁶ = 6895 MPa [44].

or particles being pulled from the matrix. There are many potential ways of measuring the hardness [55].

1. Indention Hardness—Vickers hardness and Rockwell hardness [56].
2. Rebound Hardness—Shore sclerascope and Swiss rebound hammer.
3. Scratch Hardness—Moh's hardness scale.
4. Wear or Abrasion Hardness—Dorry hardness where a core of rock is subjected to wear with an abrasive on a revolving horizontal wheel [57] and British aggregate abrasion value (British Standard 812).
5. Rate of cutting or drilling.

Hardness is an often-used term in describing desirable aggregate properties, but there has been little agreement as to how measurements should be made and whether hardness should be a property that is specified for concrete aggregates. At one time, ASTM had a scratch hardness test, but it was withdrawn because of the high variability of the results, no way to correct the test, and the inability to correlate results with performance. ASTM C 33 has no limits on soft particles and no other provisions referring to hardness.

As a general rule, concrete strength has been found to be the most significant factor in rate of wear of concrete [58]; but fine and coarse aggregate may be an important factor in some instances for concrete surfaces subjected to heavy traffic or abrasive forces, particularly lower-strength concrete. Initially, the hardness of the fine aggregate is important. The coarse aggregate will become involved only if there is enough loss of surface to expose a significant amount of the coarse particles.

Polishing is a special form of wear where abrasive size is quite small, such as typical road grit at 10 to 40 μm; and the action is such that any texture such as existing pits, gouges, or scratches are smoothed and polished gradually. This can happen particularly to exposed cement paste and the top surfaces of fine or coarse aggregate particles. Wear is "waste or diminish by continual attrition, scraping, percussion or the like" [59]. It is any removal of particles that produces shape changes on a microscopic or macroscopic scale [54].

The mechanism of concrete wear can vary [4]. For example, in concrete abrasion tests using the steel balls or dressing wheel, harder aggregates retard abrasion loss and stand out from the surface. A softer limestone aggregate, on the other hand, wears along with the paste in these tests. The reverse is true for a shot blast test where the limestone cushioned the abrasive shot and decreased the wear compared to more brittle aggregate.

Wear of highway pavements is due in large measure to the fine mineral abrasive present on the pavements. Studded tires have also been an important factor in road wear. With the phasing out of the studs in many areas, this factor has diminished.

In Canada and the United States, studded tire wear was studied intensively. Smith and Schonfeld [60] found that in an area where almost a third of the passenger vehicles were equipped with studs, over the winter of 1969-70, wear on both portland cement concrete and bituminous concrete approached 5 to 8 mm in one season. Pavements with aggregates of similar hardness to the matrix showed uniform wear. With harder aggregates, the matrix was preferentially worn down around them until the particles were dislodged by the studs because of lack of embedment. It was determined that studded tires did not change the frictional properties much. In concrete containing a sand with a substantial amount of soft minerals, pavements with traprock and limestone coarse aggregates wore at about the same overall rate even though the mechanism was somewhat different. When a 100% silica sand was used with the harder traprock, wear of both the matrix and the overall pavement was reduced.

Keyser [61] found that the age of the mortar and whether or not limestone coarse aggregate was used were significant factors in studded tire wear studies. Other aggregate types, size of coarse aggregate, and concrete compressive strength was not significant for the materials used. Preus [62] in reporting test track studies showed that studs produced more than 100 times as much wear as regular tires even when sand and salt are applied to the surface. Rosenthal et al. [63] show photos of stud wear marks. They confirmed that studs tend to skid over hard aggregate and leave grooves in softer material.

In a review of pavement wear testing, Stiffler [54] reports that sliding movement between rubber and pavement of up to 6.4-mm (¼-in.) occur in normal rolling tires. Therefore, many pavement-wear researchers, as well as ASTM D 3319 (British wheel) and ASTM E 660 (circular track polishing machine), utilize a wearing mode of rubber on pavement specimens, or rubber on mineral aggregate samples, with an abrasive supplied to simulate road grit. Polishing occurs when fine abrasive is used and the surface is made up of materials of similar hardness. Regeneration of the surface can occur when two components of differing hardness are present and particles of the weaker, softer material wear faster causing the harder material to protrude and eventually be undercut and torn out leaving an unpolished surface.

Frictional properties of pavement surfaces in wet weather depends on microtexture (amplitude less than 0.5 mm) and, also, on macrotexture (amplitude more than 0.5 mm) if significant speeds are involved. The definitions for texture are taken from ASTM Definitions of Terms Relating to Traveled Surface Characteristics (E 867). Macrotexture is controlled by concrete finishing operations, and it is important in removing excess water from between the tire and pavement. Microtexture is controlled by the polishing tendency of the exposed cement paste or aggregate surfaces. Aggregate particles individually do not have skid resistance, but in a pavement may contribute frictional properties affecting the skidding propensity for vehicles traveling on that pavement. Meyer [64], in using a number of concrete finishing textures, silica gravel and limestone coarse aggregates, and silica or lightweight fine aggregate, found good frictional properties in all cases. The lightweight fines did wear faster.

In other studies where calcareous fine aggregates were used in concrete, low frictional properties have been found [65]. For concrete pavements exposed to normal traffic wear that does not expose much coarse aggregate during the life of the pavement, it is the fine aggregate that tends to control the polishing rate and microtexture [66]. Colley et al. [65] in tests at the Portland Cement Association showed, after polishing, that the energy needed to turn a rubber tire against a fixed concrete specimen decreased markedly as the siliceous particle count in the fine aggregate was decreased below 25%. Mullen and Dahir [67] studied the wearing and polishing characteristics of a number of aggregate sources. They found no general correlation of properties for all aggregates. For a granite aggregate, there was an inverse relationship between wear resistance and frictional properties; and, for sandstone, synthetic aggregate, and one mountain gravel, friction increased as absorption and surface capacity of the particles increased. For carbonate aggregates, a direct relationship between sand-size, acid-insolubility, residue, and friction properties was implied.

Some highway agencies have advocated the use of ASTM D 3042 as a tool in selecting fine or coarse carbonate aggregates for use in surface coarses. Insoluble material may be clay or siliceous material. A higher insoluble residue retained on the 75-μm sieve indicates a higher percentage of harder and perhaps more polish resistant minerals. ASTM D 3319 (British wheel) is similar to the polishing value test determination in British Standard 812 for coarse aggregate. The procedure involves polishing oriented coarse aggregate particles held by an epoxy backing using a rotating rubber tire running against the specimens that are mounted around the perimeter of a second wheel. Abrasive and water are fed onto the tire-specimen interface at a constant rate. Degree of polish is measured using ASTM Method of Measuring Surface Frictional Properties Using the British Pendulum Tester (E 303). A number of other laboratory procedures for evaluating pavement materials and mixtures for polishing are currently in use.

Franklin and Calder [66] report results of frictional properties research using several types of fine aggregate that are in decreasing order of performance: (1) calcined bauxite fines with both good polishing and abrasion resistance; (2) a gritty sandstone material with good polish resistance, but poor abrasion resistance; (3) a flint sand with poor polish resistance, but good abrasion resistance; and (4) a carbonate fine aggregate with both poor polishing and abrasion performance.

Weller and Maynard [68] at the British Road Research Laboratory developed an accelerated wear machine for pavement samples. They use a dry wearing cycle of 50 h with a small flint gravel as an abrasive followed by 5 h of wet polishing with a fine emery abrasive. The most important characteristic of the sands tested is hardness. Harder sands stand out from the surface and show higher frictional properties after dry wearing than after wet polishing. Conversely, for soft dolomite and limestone fine aggregates, the particles were worn flush with the cement paste during the dry wearing cycle; and, since portland cement paste polishes more during the dry cycle and gains frictional properties during wet polishing, the friction numbers of the pavement surfaces containing the soft fine aggregate were lowest after dry wearing and improved during wet polishing.

IMPACT ON SPECIFICATIONS

Tests that measure directly degradation resistance, strength, or toughness characteristics of an aggregate are not generally referenced in specifications for concrete aggregates in the United States. The one exception is the L.A. abrasion test for coarse aggregate that is used in ASTM C 33 and ASTM Specification for Aggregates for Radiation-Shielding Concrete (C 637) as well as in a large number of agency specifications.

There appears to be a difference in the approach taken in the United States and Canada with respect to using aggregate strength and impact tests in specifications as compared to European and British practices that do utilize tests such as the crushing value, 10% fines value, and impact value as potential specification tests [69]. In North America, the practice is to first, rely more on past service record and judgment for those aggregates that have performed satisfactorily; and second, to test the properties of concrete made with the aggregate in question if special abrasion, impact, or strength properties are required. These methods can be used to evaluate the performance of alternative aggregate sources as well as various mix proportions. With respect to aggregate strength, ASTM Specification for Lightweight Aggregates for Structural Concrete (C 330) and ASTM Specification for Lightweight Aggregates for Concrete Masonry Units (C 331), require the aggregate to be of sufficient strength to yield certain minimum compressive and tensile splitting values when used in concrete.

With respect to an aggregate's effect on frictional properties of pavements, many state highway agencies rely on friction trailer performance history of highway surfaces and assign aggregate classifications on that basis. Others use petrographic or acid solubility data or both to classify aggregates, and a few states use the British wheel (ASTM D 3319) as a specification tool for aggregates for highway surfaces [70].

CONCLUDING REMARKS

There is a good deal of history and data available concerning the measurement of physical properties of aggregates and attempts to relate these properties with concrete properties. However, the evaluation and testing of concrete containing alternative materials remains the best approach to assuring performance when special abrasion resistance, strength, toughness, or frictional properties are needed.

REFERENCES

[1] Meininger, R. C., "Aggregate Abrasion Resistance, Strength, Toughness, and Related Properties," *Significance of Tests and Properties of Concrete-Making Materials*, ASTM STP 169B, American Society for Testing and Materials, Philadelphia, 1978.

[2] "Guide for Use of Normal Weight Aggregates in Concrete," ACI Committee 221, Report ACI 221R-89, American Concrete Institute, Detroit, 1989.

[3] Benson, P. E. and Ames, W. H., "The Precision of Selected Aggregate Test Methods," *TRB Record No. 539*, Transportation Research Board, Washington, DC, 1975, p. 85.

[4] Smith, F. L., "Effect of Aggregate Quality on Resistance of Concrete to Abrasion," *Cement and Concrete*, ASTM STP 205, American Society for Testing and Materials, Philadelphia, 1958, p. 91.

[5] Woolf, D. O., "The Relation Between Los Angeles Abrasion Test Results and the Service Records of Coarse Aggregates," *Proceedings*, Highway Research Board, Washington, DC, 1937, p. 350.

[6] Bloem, D. L. and Gaynor, R. D., "Effects of Aggregate Properties on Strength of Concrete," *Journal*, American Concrete Institute, Detroit, MI, Oct. 1963.

[7] Jumper, E. A., Herbert, J. D., and Beardsley, C. W., "Rattler Losses Correlated with Compressive Strength of Concrete," *Journal*, American Concrete Institute, Detroit, MI, Jan. 1956, p. 563.

[8] Walker, S. and Bloem, D. L., "Effect of Soft Sandstone in Coarse Aggregate on Properties of Concrete," *Series 88 and 107*, National Sand and Gravel Association, Washington, DC, NSGA TIL No. 72, 1949.

[9] Bartel, F. F. and Walker, S., "Concrete-Making Properties of Gravels from Southwestern United States," *Series 62*, National Sand and Gravel Association, Washington, DC, NSGA TIL No. 48, 1946.

[10] Walker, S., "The Flexural Strength of Concrete," *Convention Talk*, National Sand and Gravel Association, Washington, DC, 1944.

[11] Gaynor, R. D. and Meininger, R. C., "Investigation of Aggregate Durability in Concrete," *HRB Record No. 196*, Highway Research Board, Washington, DC, 1967, p. 25.

[12] Gaynor, R. D., "Laboratory Freezing and Thawing Tests—Method of Evaluating Aggregates?" Circular No. 101, National Aggregates Association, Silver Spring, MD, Nov. 1967.

[13] Kohler, G., "Comparative Investigations of International Test Methods for Small-Sized Coarse Aggregates," *HRB Record No. 412*, Highway Research Board, Washington, DC, 1972.

[14] Melville, P. L., "Weathering Study of Some Aggregates," *Proceedings*, Highway Research Board, Vol. 28, 1948, p. 238.

[15] Liu, A. C., "Abrasion Resistance of Concrete," *Journal*, American Concrete Institute, Detroit, MI, Sept.-Oct. 1981, p. 341.

[16] Hendrickson, L. G. and Shumway, R. D., "Analysis of Questionnaire on Aggregate Degradation," Circular No. 144, Highway Research Board, Washington, DC, July 1973.

[17] Breese, C. R., "Degradation Characteristics of Selected Nevada Mineral Aggregates," Engineering Report No. 4, Civil Engineering Department, University of Nevada, Reno, NV, 1966.

[18] Ekse, M. and Morris, H. C., "A Test for Production of Plastic Fines in the Process of Degradation of Mineral Aggregates," *Symposium on Road and Paving Materials, ASTM STP 277*, American Society for Testing and Materials, Philadelphia, 1959, p. 122.

[19] Minor, C. E., "Degradation of Mineral Aggregate," *Symposium on Road and Paving Materials, ASTM STP 277*, American Society for Testing and Materials, Philadelphia, 1959, p. 109.

[20] Reidenouer, D. R., Geiger, E. G., Jr., and Howe, R. H., "Shale Suitability-Phase-Final Report," Pennsylvania Department of Transportation, Bureau of Materials, Testing, and Research, Harrisburg, April 1974.

[21] Chapman, D. R., "Shale Classification Tests and Systems—A Comparative Study," Report No. 75-11, Joint Highway Research Project, Purdue University, West Lafayette, IN, June 1975.

[22] Goodewardane, K., "Behavior of Aggregate in the Washington Degradation Test," *Journal of Testing and Evaluation*, Vol. 5, No. 1, Jan. 1977, p. 16.

[23] Hveem, F. N. and Smith, T. W., "A Durability Test for Aggregates," *HRB Record No. 62*, Highway Research Board, Washington, DC, 1964, p. 119.

[24] Slater, W. A., "Tests of Concrete Conveyed from a Central Mixing Plant," *Proceedings*, Part 2, American Society for Testing and Materials, Philadelphia, 1931, p. 510.

[25] Gaynor, R. D., "Effect of Prolonged Mixing on the Properties of Concrete," *Publication No. 111*, National Ready Mixed Concrete Association, Silver Spring, MD, 1963.

[26] McKisson, R. L., "Degrading of Aggregate by Attrition in a Concrete Mixer," Report No. C-1308, Bureau of Reclamation, Denver, CO, 1969.

[27] Davis, R. E., Mielenz, R. C., and Polivka, M., "Importance of Petrographic Analysis and Special Tests Not Usually Required in Judging Quality of Concrete Sand," *Journal of Materials*, Vol. 2, No. 3, Sept. 1967, p. 461.

[28] Higgs, N. B., "Montmorillonite in Concrete Aggregate Sands," *Bulletin*, Association of Engineering Geologists, Vol. 12, Winter 1975, p. 57.

[29] Meininger, R. C., "Aggregate Degradation During Mixing," *Convention Talk*, National Ready Mixed Concrete Association, Silver Spring, MD, 1977.

[30] Rogers, C. A., Bailey, M. L., and Price, B., "Micro-Deval Test for Evaluating the Quality of Fine Aggregate for Concrete and Asphalt," *Record No. 1301*, Transportation Research Board, 1991, p. 68.

[31] Senior, S. A. and Rogers, C. A., "Laboratory Tests for Predicting Coarse Aggregate Performance in Ontario," *Record No. 1301*, Transportation Research Board, 1991, p. 97.

[32] Udd, J. E., "Physical Properties of Rocks and Probable Applications," *Application of Advanced Nuclear Physics to Testing Materials, ASTM STP 373*, American Society for Testing and Materials, Philadelphia, 1965, pp. 114–126.

[33] Bond, F. C., "Crushing Tests by Pressure and Impact," *Technical Publication No. 1895*, American Institute of Mining and Metallurgical Engineers, 1946.

[34] Iyer, Rahn, and Remarkrishnam, "Durability Tests on Some Aggregates for Concrete," *Journal*, Construction Division, American Society of Civil Engineers, New York, Sept. 1975, p. 593.

[35] Kaplan, M. F., "Flexural and Compressive Strength of Concrete as Affected by the Properties of Coarse Aggregate," *Journal*, American Concrete Institute, Detroit, MI, May 1959, p. 1193.

[36] Woolf, D. O., "Results of Physical Tests of Road-Building Aggregate," Bureau of Public Roads, Washington, DC, 1953.

[37] Johnston, C. D., "Strength and Deformation of Concrete in Uniaxial Tension and Compression," *Magazine of Concrete Research*, March 1970, p. 5.

[38] Collins, J. J. and Hsu, T. T. C., "Properties of Florida-Oolite Concrete," *Journal of Testing and Evaluation*, March 1977, p. 141.

[39] McCaulley, D. B., Mittelacher, M., Mross, J. L., Roebuck, J. P., and Winemberg, R., "Florida Aggregates in Construction," Technical Report 90-01, Florida Concrete and Products Association and the Concrete Materials Engineering Council, Orlando, 1990.

[40] Ross, A. W. and Kienow, K. K., "Arizona Volcanic Cinder Concrete—A Comparative Study," *Bulletin No. 9*, University of Arizona, Engineering Experiment Station, Tucson, AZ, 1959.

[41] Lorman, W. R., "Coral and Coral Concrete," Technical Report No. 068, U.S. Naval Civil Engineering Laboratory, 1960.

[42] Tanigawa, Y., "Model Analysis of Fracture and Failure of Concrete as a Composite Material," *Cement and Concrete Research*, Pergamon Press, Elmsford, NY, 1976, p. 679.

[43] Brandon, J. R., "Rock Mechanics Properties of Typical Foundation Rocks," Report REC-ERC-74-10, Bureau of Reclamation, Denver, CO, 1974.

[44] Clark, G. B., "Deformation Moduli of Rock," *Testing Techniques for Rock Mechanics, ASTM STP 402*, American Society for Testing and Materials Philadelphia, 1966, p. 133.

[45] Hartley, A., "A Review of the Geological Factors Influencing the Mechanical Properties of Road Surface Aggregates," *Quarterly Journal of Engineering Geology*, UK, Vol. 7, No. 1, 1974, p. 69.

[46] Lott, J. E. and Kesler, C. E., "Crack Propagation in Plain Concrete," *HRB Special Report 90*, Highway Research Board, Washington, DC, 1966, p. 204.

[47] Ko, K. C. and Haas, C. J., "The Effective Modulus of Rock as a Composite Material," *International Journal of Rock Mechanics and Mineral Sciences*, Pergamon Press, Elmsford, NY, 1972, p. 531.

[48] Hobbs, D. W., "The Dependence of the Bulk Modulus, Young's Modulus, Shrinkage and Thermal Expansion of Concrete Upon Aggregate Volume Concentration," Technical Report TRA 437, Cement and Concrete Association, London, Dec. 1969.

[49] Hansen, T. C., "Influence of Aggregate and Voids on Modulus of Elasticity of Concrete, Cement Mortar, and Cement Paste," *Journal*, American Concrete Institute, Detroit, MI, Feb. 1965, p. 193.

[50] Hirsch, T. J., "Modulus of Elasticity of Concrete Affected by Elastic Moduli of Cement Paste Matrix and Aggregate," *Journal*, American Concrete Institute, Detroit, MI, March 1962, p. 427.

[51] Shah, S. P. and Winter, G., "Inelastic Behavior and Fracture of Concrete," *SP-20*, American Concrete Institute, Detroit, MI, 1968, p. 5.

[52] Houghton, D. L., "Determining the Tensile Strain Capacity of Mass Concrete," *Journal*, American Concrete Institute, Detroit, MI, Dec. 1976, p. 691.

[53] Johnston, C. D., "Deformation of Concrete and Its Constituent Materials in Uniaxial Tension," *HRB Record No. 324*, Highway Research Board, Washington, DC, 1970, p. 18.

[54] Stiffler, K., "Mineral Wear in Relation to Pavement Slipperiness," Report J-15, Pennsylvania State University Joint Road Friction Program and Automotive Safety Research Program, 1967; also, Stiffler, K., "Relation Between Wear and Physical Properties of Roadstones," *HRB Special Report 101*, Highway Research Board, Washington, DC, 1969.

[55] Davis, H. E., Troxell, G. E., and Wiskocil, C. T., *The Testing and Inspection of Engineering Materials*, McGraw-Hill, New York, 1964.

[56] Harvey, R. D., Fraser, G. S., and Baxter, J. W., "Properties of Carbonate Rocks Affecting Soundness of Aggregate," Illinois Minerals Note 54, Illinois State Geological Survey, Champaign, IL, Feb. 1954.

[57] Woolf, D. O., "Methods for the Determination of Soft Pieces in Aggregate," *Crushed Stone Journal*, Sept. 1951, p. 13.

[58] Scripture, E. W., Jr., Benedict, S. W., and Bryant, D. E., "Floor Aggregates," *Journal*, American Concrete Institute, Detroit, MI, Dec. 1953, p. 305.

[59] Prior, M. E., "Abrasion Resistance," *Significance of Tests and Properties of Concrete and Concrete-Making Materials, ASTM STP 169A*, American Society for Testing and Materials, Philadelphia, 1966.

[60] Smith, P. and Schonfeld, R., "Studies of Studded-Tire Damage and Performance in Ontario During the Winter of 1969–70," *HRB Record No. 352*, Highway Research Board, Washington, DC, 1971, p. 1.

[61] Keyser, J. H., "Resistance of Various Types of Bituminous Concrete and Cement Concrete to Wear by Studded Tires," *HRB Record No. 352*, Highway Research Board, Washington, DC, 1971, p. 16.

[62] Preus, C. K., "Discussion," *HRB Record No. 352*, Highway Research Board, Washington, DC, 1971, p. 31.

[63] Rosenthal, P., et al., "Evaluation of Studded Tires," *HRB-NCHRP Report No. 61*, Highway Research Board, Washington, DC, 1969.

[64] Meyer, A. H., "Wearability of P C Concrete Pavement Finishes," *Transportation Engineering Journal*, American Society of Civil Engineers, New York, Aug. 1974, p. 719.

[65] Colley, B. E., Christensen, A. P., and Nowlen, W. J., "Factors Affecting Skid Resistance and Safety of Concrete Pavements," *HRB Special Report No. 101*, Highway Research Board, Washington, DC, 1969, p. 80.

[66] Franklin, R. E. and Calder, A. J. J., "The Skidding Resistance of Concrete—The Effect of Materials Under Site Conditions," Laboratory Report 640, Transportation and Road Research Laboratory, UK, 1974.

[67] Mullen, W. G. and Dahir, S. H. M., "Skid Resistance and Wear Properties of Aggregates for Paving Mixtures," North Carolina State University Highway Research Program Interim Report, June 1970 (PB 197 625).

[68] Weller, D. E. and Maynard, D. P., "Influence of Materials and Mix Design on the Skid Resistance Value and Texture Depth of Concrete," Report No. LR 334, British Road Research Laboratory, London, 1970.

[69] Pike, D. C., *Standards for Aggregates*, Ellis Horwood Ltd., Chichester, UK, 1990.

[70] Meininger, R. C., Report of ASTM Committee D-4 Survey of Aggregate Requirements for Highway Surfaces of State Highway Agencies, National Aggregates Association, Technical Information Letter No. 386, Aug. 1986.

Grading, Shape, and Surface Properties

Joseph E. Galloway, Jr.[1]

PREFACE

The subject of grading has been included in the three previous editions of *ASTM STP 169* and all authored by W. H. Price.

The discussion of Shape and Surface Properties has also appeared in previous editions of this STP. The most recent in *ASTM STP 169B* was authored by M. A. Ozol and titled "Shape, Surface, Texture, Surface Area, and Coatings." Those chapters on this subject in *ASTM STP 169* and *ASTM STP 169A* were authored by B. Mather and were titled "Shape, Surface Texture, and Coatings."

This chapter is a condensation of the previously referenced works and discusses grading with some new trends being considered. The information on shape and surface properties is more general and does not delve into the technical discussions contained in the previous editions. If the reader is concerned about those technical geological discussions, they are encouraged to refer to Chapter 35 of the *ASTM STP 169B* edition.

INTRODUCTION

The three attributes of concrete mixtures discussed in this chapter may be considered benign features because even a well-trained eye may not be able to discern changes unless they are in the extreme. However, even slight changes can have an effect on the characteristics of concrete mixtures.

Grading changes are perhaps more prevalent than shape and surface texture, in the case of a coarse aggregate, because of its natural tendency to segregate during stockpiling, transporting, and other forms of handling.

Therefore, attention to every detail of these phases from the time of processing to the time of introduction into the concrete mixture is very important and can lessen problems later in the concrete-making process.

Shape and surface texture properties are not as likely to change, but the changes that can occur will be discussed in this chapter. Unlike grading, these two attributes are not within the direct control of the user.

The items discussed here pertain to normal, lightweight, and heavyweight aggregates, because the aggregate as well as other ingredients fit together as absolute volumes in the concrete mixture and in no other way [1]. The effect these changes have will be discussed in this chapter.

GRADING

Definition

Grading is simply the frequency distribution of the particle sizes of a given aggregate.

This distribution is given in certain ranges for each sieve size and each size is usually assigned an arbitrary number. ASTM Specification for Concrete Aggregates (C 33) gives a wide range for each sieve to accumulate different localities and to allow for economical production considerations. These size numbers are widely used but other specifiers may have a different number assigned for virtually the same material. Until the early 1920s, aggregate "sizes" were concocted by almost everyone specifying or producing aggregate; but, in 1937 ASTM Committee D-4 adopted ASTM Classification for Sizes of Aggregate for Road and Bridge Construction (D 448) that listed standard sizes of aggregate in an attempt to bring uniformity within the industry.

These grading sizes are reflected to a great degree in ASTM C 33, Table 2, that is largely used today by specifiers of concrete mixtures.

Test Method

The basic document for determining grading is ASTM Method for Sieve Analysis for Fine and Coarse Aggregate (C 136). This is done basically by separating the material on a nest of sieves meeting the requirements of ASTM Specifications for Wire-Cloth Sieves for Testing Purposes (E 11) and determining the percent of each size present. Note that each successive sieve is approximately one-half the opening of its predecessor and all openings are expressed in SI units. Using such a system and employing a log scale, lines on a graph can be spaced at constant intervals to represent the successive sizes, and the fineness modulus can be computed [1].

The fineness modulus (FM) is an empirical figure derived by adding the cumulative percentages retained on a specified series of sieves and dividing by 100. These sieves are, in millimetres, 150 (6 in.), 76 (3 in.), 37.5 (1½ in.), 19 (¾ in.), 9.5 (⅜ in.), 4.75 (No. 4), 2.36 (No. 8), 1.18 (No. 16), 0.6 (No. 30), 0.3 (No. 50), and 0.15 (No. 100).

---

[1] Consultant, Mechanicsville, VA 23111-1314.

A typical computation for the FM of a fine aggregate is shown in Table 1.

The FM of a coarse aggregate is computed in a similar way. It should be noted that all sieves in the series are always considered. Obviously, in the case of a coarse aggregate, 100% would be retained on the smaller sieve sizes so a typical coarse aggregate, Size 57 from ASTM C 33, would have an FM of about 7.00. However, the FM on coarse aggregate is not usually of any value, but for fine aggregate where the eye cannot discern changes in grading, it is of value. ASTM C 33 states that the FM shall not vary from the base by more than 0.20 for a given source.

Significance

Since the fine-aggregate FM is used in computing the mixture design in the solid volume process, changes beyond the 0.20 value would call for a redesign of the mixture. This fact is often overlooked and problems with workability are searched for in other areas or, in many cases, the actual FM was not determined originally, but assumed.

While ASTM C 33 contains fine-aggregate grading limits, succeeding paragraphs modify these limits considerably even to the point of discarding the grading limits, provided similar aggregate from the same source has a demonstrated performance record or, in the absence of such record, if it can be shown that concrete of the class specified made with the aggregate in question has shown equal relevant properties. There are many qualifying adjectives in these statements and only someone qualified to judge these equals should make the decision.

As previously noted, the grading limits in ASTM C 33 are very broad to accommodate a wide area of conditions. In any particular area, adjustments should be made to accommodate the local conditions. It is not meant to prohibit the concrete designer from modifying the grading range, that is, specifying more than one coarse aggregate size to be used or any other modification, provided it can be done in an economical manner.

Once the ranges on each sieve are agreed upon by the supplier and the user, whether they be ASTM C 33 ranges or some other, the supplier should strive to produce to the midpoint of that range. Production at either extreme will statistically cause the product to fall out of the specification limits and result in problems with the concrete mixture.

TABLE 1—Computation for the Fineness Modulus.

Sieve Size, mm (No.)	Percent Retained	Cumulative Percent Retained
4.75 (4)	1	1
2.36 (8)	15	16
1.18 (16)	23	39
0.6 (30)	19	58
0.3 (50)	16	74
0.15 (100)	15	89
−0.15 (100)	11	
		277
277 divided by 100 = 2.77 FM		

What then is the ideal grading?

The influence of aggregate on the properties of concrete has been extensively discussed in the technical literature during the past 125 years and many methods for arriving at "optimum" or "ideal" gradings have been presented. However, none of these has been accepted as being universally applicable because of economic considerations, differences in particle shape and texture of the aggregates, and the effects of entrained air and amount of cementitious material contained in the concrete [2–25].

Experience has demonstrated that either very fine or very coarse sand, or coarse aggregate having a large deficiency or excess of any size fraction, is usually undesirable, although aggregates with discontinuous or gap grading have sometimes been used to advantage [26–28]. These, however, are special circumstances and should only be considered by someone familiar with gap grading.

A well-graded material is the closest to ideal with a representative amount on each standard sieve size listed in that size specification. The scarcity of a particular sieve size could result in poor workability and even poor durability of the concrete.

The following paragraphs speak to coarse and fine aggregates, but the trend is to consider the aggregate as a composite in the concrete mix.

There is work underway in ASTM to deemphasize the designer's need to stay with one coarse aggregate size. That size (usually No. 57) can easily lead to a gap-graded aggregate. The proposed change would allow any grading the specifier chose and would not necessarily conform to one of the sizes in ASTM C 33.

Figure 1 is a computer generated distribution chart of an aggregate meeting the median requirements of ASTM C 33, Size 57, and the concrete fine aggregate (notice the peak-valley arrangement) [29].

It is obvious that if the aggregate had a grading more closely related to that in Fig. 2 [29], space between coarse aggregate particles would be filled with aggregate rather than mortar, thereby causing a more workable and more durable concrete.

This can be accomplished by adding another size coarse aggregate containing those particle sizes that were deficient.

A proposed standard practice by the SHRP program titled "Packing-Based Concrete Aggregate Proportioning" will result in the most densely packed system and minimum void content for the aggregates under consideration [30].

Coarse Aggregate

The importance of grading has long been recognized. The original issue of ASTM C 33 in 1921 had very few requirements other then grading. The coarse aggregate was not designated by size but rather by maximum size [31]. This is still the "starting point" of selecting the coarse aggregate "size" to be used and is dependent on thickness of section, spacing of reinforcement, availability, economics and placement procedures (aggregates greater than 63 mm (2½ in.) become more difficult to pump).

The terms, maximum size and nominal maximum size, should be understood. ASTM Terminology Relating to

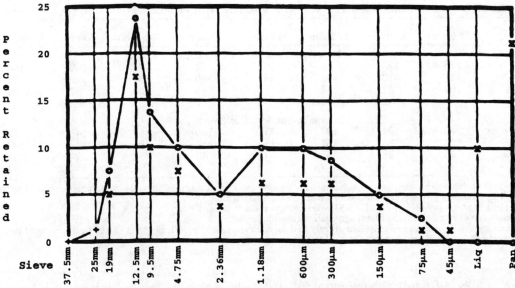

FIG. 1—Particle distribution for a typical gap-graded aggregate mixture.

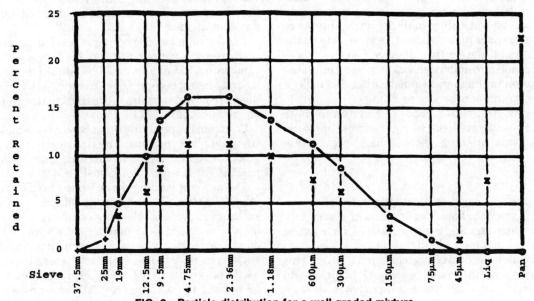

FIG. 2—Particle distribution for a well-graded mixture.

Concrete and Concrete Aggregates (C 125) defines the two terms as the smallest sieve opening through which the entire amount of aggregate is required to pass and the smallest sieve opening through which the entire amount of aggregate is permitted to pass, respectively.

The importance of maximum size is illustrated in Table 2 that shows the relationship to aggregate size and the amount of water and cement needed. Simply put, the smaller the aggregate size the more surface area is present for a given value and, therefore, the more mortar needed in the mix to surround the aggregate particle. There are limits on both ends where this does not hold true.

For compressive strengths below 21 MPa (3046 psi), the cementitious content of the concrete can be progressively reduced as the maximum-size aggregate is increased while maintaining a constant water/cement ratio and strength. At the 28 MPa (4061 psi) level, no cementitious material can be saved by using aggregate above 76 mm (3 in.) in size. As the strength level is increased, smaller maximum-size aggregates must be used for the most efficient use of cement with 37.5 mm (1½ in.) for the 35 MPa (5076 psi) level, 19 mm (¾ in.) for the 42 MPa (6092 psi) level, and 9.5 to 12.5 mm (⅜ to ½ in.) for strength above 42 MPa (6092 psi) [28,32].

In the production of high-strength concrete above 42 MPa (6092 psi), there seems to be no advantage in using maximum-size aggregate below 12.5 mm (½ in.) [33,34]. A large particle of aggregate has less area for bonding with the mortar than an equal weight of smaller aggregate particles, and it is for this reason that it is not possible to obtain high strengths with large maximum-size aggregate [35]. Also, it is not possible to obtain strengths above a

TABLE 2—Mortar Requirements for Workable Concrete With Various Maximum Aggregate Sizes.

Maximum Size Aggregate	4.75 (sand)	9.5 mm	12.5 mm	19 mm	25 mm	37.5 mm	50 mm	76 mm	150 mm
			Non-Air-Entrained Concrete						
Water, kg/m³[a]	276	228	216	202	192	177	168	160	139
Cement, kg/m³[b]	520	430	406	382	360	334	316	302	262
Fine aggregate, %[c]	100	62	54	49	44	40	37	34	28
Entrapped air, %[d]	6	3	2.5	2	1.5	1	0.5	0.3	0.2
			Air-Entrained Concrete						
Water, kg/m³[a]	245	201	192	177	169	157	148	139	118
Cement, kg/m³[b]	462	380	362	334	318	296	279	262	222
Fine aggregate, %[c]	100	58	50	45	40	37	34	31	25
Total air, %[d]	13	8	7	6	5	4.5	4	3.5	3

[a] Approximate amount of mixing in kilograms per cubic metric required for 75-mm slump with well-shaped angular coarse aggregate. Quantities listed can be reduced significantly through the addition of water-reducing admixture.
[b] Cement required in kilograms per cubic metre for 0.53 water/cement ratio by weight.
[c] Approximate percentage of fine aggregate of total aggregate by absolute volume.
[d] Recommended average total percentage of entrained air required for frost resistance from Table 5.3.3 of ACI Recommended Practice for Selecting Proportions for Normal and Heavy Weight Concrete (ACI 211.1).

practical level of 35 MPa (5076 psi) with the entrained air contents listed in Table 2 [36].

Because it is not practical to produce aggregate that is perfectly separated, ranges are given in grade specifications. There will be some finer material due to breakage and sometimes coarser material due to screen wear or the use of screens slightly larger than the specified size. These deviations are not normally significant when compared to the segregation problem mentioned earlier due to poor stockpiling or handling practices or both.

Therefore, proper attention needs to be given to those procedures to avoid problems in the concrete mixture. Obviously, the less handling, the less likely segregation will occur.

Fine Aggregate

Fine-aggregate grading has a much greater effect on workability of concrete than does coarse-aggregate grading [37]. With the water and cement (and, in some cases, other mineral admixtures), the fine aggregate comprises the matrix in which the coarse aggregate must "live." This matrix needs to coat the coarse-aggregate particles and retain sufficient fluidity for placement purposes.

Thus, the fine aggregate cannot be too coarse or harshness, bleeding, and segregation will occur. At the same time, if it is too fine, the additional surface areas will require additional water and also result in segregation.

The grading in ASTM C 33 is usually satisfactory but, as previously mentioned, deviations are permitted because certain areas do not have native material containing these sizes.

In addition to the grading requirement, ASTM C 33 also prohibits more than 45% passing any one sieve and retained on the next consecutive sieve shown in the standard grading. It also limits its FM to between 2.3 and 3.1.

Working in conjunction with the fine-aggregate grading is its particle shape that will be discussed later.

Regardless of the grading used, consistency is most important. If the grading varies considerably, problems with workability will probably result. Thus, good control calls for a constant check on the grading and computation of the FM. This is particularly true with fine aggregate where changes can occur without it being obvious. The material may well stay within the grading limits but the FM vary more than the 0.2 allowed in ASTM C 33. Changes have been noted over a range as much as 1.0 in FM in a day's production [38].

For the high cement contents used in the production of high-strength concrete, it has been found that a coarse sand with a FM of around 3.0 produced the best workability and highest compressive strength [33]. In general, manufactured sands require more fines than natural sands for equal workability [21].

Concrete for pumping must be very workable, with high fine-aggregate contents. The fine aggregate must be well-graded and somewhat on the fine side, with 15 to 30% passing the 0.3-mm (No. 50) sieve.

The amount passing the 0.3-mm (No. 50) and 0.15-mm (No. 100) sieves have a greater influence on workability, surface texture, and bleeding of concrete. The lower limit of 10% in ASTM C 33 may be satisfactory where placing conditions are easy or where mechanical finishing is used such as in pavements. However, where hand finishing is used and a smooth texture is desired, at least 15% should be passing the 0.3-mm (No. 50) sieve and a minimum of 3% passing the 0.15-mm (No. 100) sieve [39].

Grading Effect on Air Entrainment

Table 2 shows the reduction in fine-aggregate percentages that may be realized through the use of air entrainment in concrete. In addition to a reduction in the amount of fine aggregate that may be accomplished for a given set of materials and grading, it is sometimes possible to use a coarser or finer grading, or a jump grading, and obtain concrete of satisfactory workability with the proper amounts of entrained air. ASTM C 33 recognizes the effect of entrained air on workability and permits a smaller percentage of fines for air-entrained concrete [1].

Fine-aggregate particles passing the 0.6-mm (No. 30) to 0.15-mm (No. 100) sieves entrain more air than either the finer or coarser particles. Therefore, fluctuations in these sieve sizes can affect the air entrainment of the concrete. Significant amounts passing the 0.15-mm (No. 100) sieve will cause a significant reduction in air content [39].

Specification

As noted throughout this discussion, ASTM C 33 is the document used to specify grading by most architects and engineers.

However, other agencies such as the American Railway Association and the various State transportation departments may have their own grading specifications. Many of these specifications are very similar to ASTM C 33 but may call a particular grading by a different size number.

SHAPE

Definition

The concept of particle shape incorporates three geometrical ideas; namely, sphericity, roundness, and form, that are distinct and separately definable properties in the abstract or mathematical sense. They may be linked properties in the geological sense, in that a process that affects the expression of one property may, concurrently, promote or inhibit development of the others.

Sphericity is a measure of how nearly equal are the three axes or dimensions of a particle, based on the degree to which the volume of a particle fills the volume of a circumscribed sphere whose diameter is the maximum dimension of the particle.

Roundness is a measure of the sharpness of the edges and corners of a fragment, or the degree to which the contour of a particle fits the curvature of the largest sphere that can be contained within the particle.

Sphericity and roundness can be visualized conveniently, but not defined rigorously, by the analogy of an irregular solid within which is a sphere of the largest possible size and around which is a sphere of the smallest possible size. The spheres may or may not be concentric. The congruence of the particle boundary to the inner sphere is an indication of roundness and its coincidence to the outer sphere, the sphericity. When the spheres coincide, the particle itself is a sphere with both a roundness and sphericity of 1—the maximum value.

The form of Folk [40] or the alternative shape factor of Asehenbrenner [41] is a measure of the relationship between the three dimensions of a particle based on ratios between the proportions of the long, medium, and short axes of the particle, or the smallest circumscribing ellipsoid. Form or shape factor distinguishes between particles of the same numerical sphericity but of different axial proportions.

A more thorough discussion of how these dimensions interrelate is contained in Chapter 35 of *ASTM STP 169B* by Ozol [42].

The shape is influenced by the natural breakage of the particle during mining (blasting) operations, by the type of crushing equipment used, or by the processing techniques such as the speed of feed to the crusher.

A desirable shape is one that is round or a near perfect cube. Poorly shaped aggregate is more difficult to define and depends largely on the specification being followed.

One generally accepted definition of a flat particle is one in which the width or length exceed the thickness by some ratio, usually 3:1. In other words, it is approaching the shape of a coin.

An elongated particle is one where the length exceeds the width by some ratio, usually 3:1. Its shape is approaching that of a rod.

Some specifications may use a ratio of 5:1. Most specifications will use one of these ratios to describe undesirable shapes.

Test Method

There are no ASTM test methods at the present time for determining particle shape directly. An indirect method that establishes a particle index is ASTM Test Method for Index of Aggregate Particle Shape and Texture (D 3398). This method is under the jurisdiction of ASTM Committee D-4 and is used mostly for research work on asphalts.

There is work underway to establish a void content test method in ASTM for fine aggregate that will be an indirect method of measuring particle shape. It basically follows a method originally described by The National Crushed Stone Association in the late 1950s [43]. It consists of dropping the test material from a prescribed height into a vessel of known volume and comparing the resultant weight with the theoretical weight of the vessel's solid volume.

With regard to coarse aggregate, ASTM Test Method for Flat or Elongated Particles in Coarse Aggregate (D 4791) may be used. Some state transportation agencies follow a similar method used by the U.S. Army Corps of Engineers [44]. These methods require that the technician measure each individual particle in a set of proportional calipers set up according to the specification on a 2:1, 3:1, or 5:1 basis. While the method is effective in determining the weight or number of particles above the prescribed ratio, the method becomes tedious and time consuming and can lead to carelessness or lethargy on the part of the operator.

There are other methods that have been developed (see *ASTM STP 169B*, p. 594 [42]), however, the one's described herein are those in more common use.

Significance

The shape of the aggregate particle influences the fresh and hardened concrete. If all particles were cubes, we would have an ideal shape factor. As the particles become more elongated or flatter, the ease with which they move in the mixing and handling processes become increasingly more difficult.

This is especially critical in pumped concrete where workability is of extreme importance.

If the number of poorly shaped particles is not too great, the workability problems may be overcome with the use of water-reducing admixtures. The question then becomes what quantity is considered too great? ASTM C 33 does not have any requirements with regard to this attribute. Some other specifying agencies such as State transportation agencies do have limits generally ranging between 8 and 20% maximum allowable on a 3:1 ratio.

One of the problems related to particle shape is the particle shape changing after trial mixes are made or during the course of the project. This increase in workability problem is solved many times on the job by the addition of unwanted and unauthorized water. Therefore, when workability problems occur, aggregate particle shape should be one of the areas to be investigated.

In the hardened concrete, the particle orientation may influence compressive and flexural strengths, elasticity, and the distribution of stresses. For instance, if such a thin, flat particle was oriented in the hardened concrete where outside stresses were introduced on that particle, the strength of the concrete might be significantly lowered because of the lack of this thin particle's ability to carry the stress. Thus, the importance of trying to find a source of relatively round to cubical-shaped aggregates is vital.

As previously mentioned, should the source be marginal with respect to this property, the use of designing techniques, including the use of admixtures, can still produce a quality end product. This may become necessary because in some areas locating good naturally occurring aggregates is becoming more difficult.

Rounded gravels that were rounded by eons of rolling to lower elevations along stream or river banks will not be replaced in the foreseeable future. The rapid development and therefore large use of good gravel aggregate in the last 50 years, coupled with stricter zoning laws, noise and dust pollution laws, stricter permit requirements, and use of land, particularly in urban areas, for more economic gain than quarrying, have joined together to limit the use of what is still left of the natural deposits.

Therefore, the use of crushed stone and, more critically, the use of synthetic or reclaimed material as aggregates will become more prevalent, and the use of less than ideal-shaped material will increase [45].

Sphericity increases with size, although the relationship in naturally occurring materials is less pronounced than with crushed materials [42]. In a study of the shape of various, as-supplied coarse aggregates (two gravels and three crushed stones), Conway [46] found that the sphericity in all cases increased with increasing sieve size [42]. Similarly, coarser grains round easier than fine ones and roundness increases with size [47].

The form and sphericity of aggregates are primarily the result of the degree of anisotropy of the material and the original shape of the particle, that is, the shape prior to the outside influence of blasting, handling, sizing, and transporting [42].

In the case of gravels, the original particle shape is controlled by larger-scale structural features of the bed rock, such as joints, fractures, faults, and bedding planes, that was then exploited by glacial action, stream action, and mechanical (frost) and chemical weathering. In the case of natural sands, the original particle shape, that is, its shape prior to transportation, was probably most strongly controlled by its mineral grain shape in the parent material.

With crushed stone, the analogous "original" particle shape is that produced by blasting of the bedrock. Those shapes are a complex function of the influence of the large-scale structural features of the bedrock but, additionally, are a function of the size and frequency of the cracks initiated by the action of shock waves on the smaller-scale flaws and discontinuities of the rock on the order of magnitude of the grain sizes. Most of the shot rock requires further size reduction by crushing, and the final shapes of the particles are influenced strongly by preferred orientations of any sort contributing to anisotropy. Rocks with bedding, cleavage schistosity, shale partings, mineral cleavage, etc. may yield particles more elongated or flattened than equidimensional. Equigranular rocks with no preferred orientations tend to yield more cubical particles on crushing. The characteristics of crushing equipment also influence the particle shape of the rock being crushed. Generally, the greater the reduction ratio, the more flat or elongated the product—slightly more so if the machine is of the compression (jaw, gyrator, or cone crusher) type. That effect is less pronounced with impact- or impeller-type machines, which would produce more nearly cubical particles at equivalent throughputs.

Rounding of sand and gravel particles is chiefly the result of those geologic factors that were of secondary importance in contributing to their form and sphericity; that is, the abrasion, attrition, chipping, rubbing, and (possibly) solution incident to their transportation and to the site of deposition. The roundability of particular mineral rock fragments depends directly on their hardness and toughness and inversely on the presence of cleavage or cracks, which tend to induce fracturing, negating what rounding had been accomplished [42].

The shape then becomes an important factor with regard to all aspects of the concrete and, while universally recognized methods of measurement are not at hand, the characteristic should be taken into account in the concrete design.

Specification

ASTM C 33 does not have any limitation on particle shape for either fine or coarse aggregates. As mentioned in the text, work is underway on a test method for fine aggregate and, if adopted, one should consider that limitations would follow in the specification. No such consideration is underway for coarse aggregate.

SURFACE TEXTURES

Definition

The surface texture of an aggregate particle is the degree to which the surface may be defined relative to arbitrary numbers as being rough or smooth (loosely referring to the height of the asperities) or coarse grained or fine grained (loosely referring to the spacing of the grains). Additional elements of surface texture, not easily incorporated into a concise comprehensive definition, include the lateral and vertical irregularity of the roughness; that is, the statistics of the height distribution of the population of asperities and their frequency of occurrence over an area, as well as their morphology.

Two independent geometric properties are the bias component of surface texture: (1) the degree of the surface relief, also called roughness or rugosity; and (2) the amount of surface area per unit of dimensional or projected area. The latter property, although it is the ratio of areas, has been defined by Wenzel [48,49] as the roughness factor, $R = A/a$, where A is the true (real) surface area, and a is the apparent or project area. A rough surface (in the sense of degree of relief) does not, ipso facto, have more surface area than a smooth surface or equivalent dimensions—it may or may not. The two properties are not related functionally but depend on the relative amplitude and frequency of the asperities on the surfaces being compared [42].

The criterion by which one surface is designated rougher than another, in terms of their reliefs, is variously defined depending on the intended use of the information, but almost invariably involves some measure of the deviation of a profile of the surface from a hypothetical reference surface [42].

Test Method

There are several methods for measuring surface texture [50]. These are research methods. There are no laboratory operational procedures except that the tests measuring shape by one of the flow tests are indirectly influenced by the particle texture. For example, a rough-textured particle will rub against other similar particles and not flow as readily as smooth particles. Therefore, any flow test using time as a part of the measurement will be affected by the relative textures.

Significance

In work done by Kaplan [51], it was concluded that surface texture has no appreciable effect on the workability of concrete. It is obvious that it certainly does not affect the workability of the concrete to the degree that grading and particle shape do.

However, other work by Kaplan [52] found that among the factors of angularity, texture, flakiness, and elongation indexes, the texture had the largest effect on compressive strength. This probably can be attributed to the extra mechanical interlock and increased surface area available for bond in the rough texture.

Patten [53] investigated the relative contributions of adhesion and "keying" (or mechanical interlock) by compressive and tensile strength measurements of concrete at five different ages in which the same coarse aggregate had been used in both a "bondless" (or surface-treated) and an untreated condition. The object was to eliminate adhesive bonds without affecting the mechanical interlock (or physical keying) between the aggregate and the mortar. The coarse aggregate was coated with a mold release agent. The elimination of adhesion between the mortar and the coarse aggregate reduced compressive strength by an average of about 23%. Compared to the control, the strength of the bondless concrete decreased with age from 19% difference (or reduction) at seven days to 27% difference at six months. Tensile strength was less affected, ranging from 6 to 28% difference, with an average reduction of about 17%; but in this case, the strength of the bondless concrete increased with age as a percent of the control. At six months, in the high-strength series, the splitting tensile strength of the concrete with the bondless aggregate was 92% of the strength of the concrete with the untreated aggregate; whereas at seven days, it was 79%. Investigations along similar lines have been conducted by Darwin and Slate [54] and others [55,56].

Studies focusing further on the components of the adhesive force making up the percent contribution of adhesion to bond strength endeavor to test the interface in as pure a state of tension as possible to avoid mechanical contributions. Various specimen configurations and techniques for measuring aggregate-cement bond strength have been reviewed by Alexander et al. [57]. Under conditions of tension, the total adhesive force is the product of: (1) the specific adhesion, that is, the strength of the adhesive force per unit area of surface from whatever physicochemical mechanism it derives; and (2) the amount of surface area available for bonding over which the adhesive force acts. Different rock types have been observed to have different tensile bond strengths to cement paste [58–60]. A possible explanation is that the bond is chemical and its specific adhesion differs significantly according to the particular chemistry of the mineral surface [61]. Alternative possibilities are that (1) the bond is chemical but its specific adhesion on the smallest unit area basis does not differ greatly between minerals, or (2) the bond is physical [62], deriving from the same sorts of forces that hold materials together in general. From either of these possibilities, an alternative hypothesis to explain differences in bond strength between different rock types is that surfaces of different lithologies have different roughness factors [42] and therefore present different true surface areas available for bonding [63].

Coatings

These are defined as any material adhering to the particle whether it be foreign or fine material of the same mineralogy as the parent particle. Some coatings are a result of mining or processing procedures or may be due to a natural weathering of the rock.

In any case, coatings are for the most part undesirable because they interfere with the bond between the aggregate and cement paste.

One of the more common forms of coatings occurs in stockpiles where the material may be stored in a wet condition (in the case of gravel) and dust from passing traffic or wind-borne dust coats the top layer. Another common problem can be the carelessness of an operator loading the material (usually with a front-end loader) into trucks for transportation to the concrete holding bins. If he is not careful, the loader scoop gets too low and he picks up part of the soil on which the stockpile rests. This, in turn, contaminates the aggregate and, in the case of clay or plastic soils, adheres to the material.

Fortunately, washing the aggregate usually removes coatings and, in the case of gravels that have been washed during the grading process, only careless contamination after processing presents a problem. In the case of crushed stone that is not generally washed, any coatings are usually the dust resulting from fracture and may be eliminated during the processing. If the coating is something that is tightly adhering, washing or scrubbing may be necessary.

Most crushed-stone plants now use a dust suppressant for ecological reasons, therefore, cautions should be taken so that this material does not increase the chances of contamination as noted for wet gravel.

Soluble coatings also present a problem because they became dissolved in the matrix with undesirable results. One common coating is iron sulfide that stains the concrete. Chlorides and sulfates can also cause staining, efflorescence, and premature corrosion of imbedded reinforcing steel.

ASTM C 33 contains limits for the amount of material finer than the 75-μm (No. 200) sieve for fine and coarse aggregate. This amount is usually determined during the sieving procedure unless the coating is tightly adhering in which case the sample is washed in accordane with the Test Method for Materials Finer than the 75-μm (No. 200) Sieve in Mineral Aggregates by Washing (ASTM C 117).

Specification

ASTM C 33 is silent with regard to requirements for surface texture, probably due to the fact that there is no easy way to measure texture and the fact that when texture reaches the point of interfering with the production of quality concrete other factors will have come into play.

CONCLUSIONS

1. Grading has a significant effect on the concrete mixture proportioning and workability. The gradings listed in ASTM C 33 are generally satisfactory and may be used in several different combinations to achieve desired results. The main concern is inconsistency of the grading and the ability to detect changes particularly with fine aggregate. The fineness modulus is one way to maintain a check on the fine aggregate grading that is not discernable by a casual observation of the grading.

Segregation of coarse aggregate during the handling process is a major problem. Choosing sizes that do not range at the extremes of the grading scale will help. Many times, two coarse aggregate sizes are preferable, and can be recombined at the time they are introduced into the mixer or can be added separately, depending on the economics involved.

Variations in the grading can change the surface area that is to be coated with mortar and, because this is done during the design and trial mix stages, variations in the grading (surface area) change the amount of optimum mortar and can lead to the addition of unwanted water.

Continuous attention to grading is of paramount importance.

2. Particle shape also has a significant effect on both fresh and hardened concrete. Flat and elongated particles do not "roll" as well as rounded or cubical particles during the mixing process and therefore require more water for the same consistency. If the water-cement ratio is to be maintained (and this is usually mandatory), then more cement needs to be used for the same strength, thus becoming uneconomical.

The fine aggregate particle shape is just as important as the coarse aggregate particle shape. A 1% increase in void content as measured by the NCSA method [43] requires about one additional gallon of water per cubic yard for the same slump, all other things being maintained.

In the hardened concrete, the orientation of the coarse particle could affect the stresses in the concrete mass leading to reduced strength.

There are no standard ASTM test methods at this time that measure the particle shapes of either fine or coarse aggregates. Therefore, attention must be directed to void content tests for fine aggregate and a discernable eye for coarse aggregate.

3. The texture of aggregate particles has no appreciable effect on workability but does affect compressive strength. This is probably because the rough texture presents a better mechanical interlock with the matrix and, with an increased surface area, there is more interface with which the mortar may react.

Unfortunately, a rough texture increases the propensity to retain undesirable coatings. At the same time, unless the coating has a strong adherence, it may loosen during processing or handling. If such adherence is strong, washing or scrubbing may be necessary.

Coatings can result as an inherent feature of the parent rock, but more often coatings are a result of carelessness during the stockpiling or handling phases.

Coatings can prevent the matrix from adhering to the aggregate particle and may be chemically reactive to the point of staining the concrete. Also, coating fines may simply mix into the matrix, not affecting the bond, but adding fines to the matrix.

Therefore, clean, uncoated aggregate is important to ensure a good quality concrete.

REFERENCES

[1] Price, W. H. "Chapter 34—Grading," *Significance of Tests and Properties of Concrete and Concrete-Making Materials, ASTM STP 169B*, American Society for Testing and Materials, 1978, p. 573.

[2] Wright, W. H., *Treatise on Mortars*, 1845.

[3] Feret, R., Bulletin de la Societe' d' Encouragement pour l'Industrie Nationale, 1897.

[4] Mercer, L. B., *The Law of Grading for Concrete Aggregates*, Melbourne Technical College Press, Australia, 1951.

[5] Fuller, W. B. and Thompson, S. E. "The Laws of Proportioning Concrete," *Transactions*, American Society of Civil Engineers, Vol. 59, 1907, p. 67.

[6] Wig, R. J., Williams, G. M., and Gates, E. R., "Strength and Other Properties of Concrete as Affected by Materials and Methods of Preparation," Technical Paper No. 58, National Bureau of Standards, Washington, DC.

[7] Talbot, A. N. and Richart, F. E., "The Strength of Concrete and Its Relation to the Cement, Aggregate, and Water," *Bulletin No. 137*, University of Illinois, Ubana, Oct. 1923.

[8] Edwards, L. N., "Proportioning of Mortars and Concrete by Surface Area of Aggregates," *Proceedings*, American Society for Testing and Materials, Vol. 18, Part 2, 1918, p. 235.

[9] Young, R. B., "Some Theoretical Studies on Proportioning Concrete by the Method of Surface Areas of Aggregates," *Proceedings*, American Society for Testing and Materials, Vol. 19, Part 2, 1919, p. 444.

[10] Abrams, D. A., "Design of Concrete Mixtures," *Bulletin No. 1*, Structural Materials Research Laboratory, Lewis Institute, Chicago.

[11] Swayze, M. A. and Gruenwald, E., "Concrete Mix Design—A Modification of the Fineness Modulus Method," *Journal*, American Concrete Institute, March 1947; *Proceedings*, Vol. 43, p. 829.

[12] Walker, S. and Bartel F., discussion of paper by Swayze and Gruenwald, *Journal*, American Concrete Institute, Dec. 1947; *Proceedings*, Vol. 43, 1947, pp. 844–851.

[13] Weymouth C. A. G., "Effects of Particle Interference in Mortars and Concretes," *Rock Products*, 25 Feb. 1933, p. 26.

[14] Weymouth, C. A. G., "A Study of Fine Aggregate in Freshly Mixed Mortars and Concretes," *Proceedings*, American Society for Testing and Materials, Vol. 38, Part 2, 1938, p. 354.

[15] Weymouth, C. A. G., "Designing Workable Concrete," *Engineering News–Record*, 29 Dec. 1938, p. 818.

[16] Walker, S., "Relations of Aggregates to Concrete," *Bulletin No. 2*, National Sand and Gravel Association, Washington, DC, 1928.

[17] *Standard Guide Specifications for Concrete*, U.S. Army Corps of Engineers, Washington, DC, 1963.

[18] Walker, S. and Bloem, D. J., "Effect of Aggregate Size on Properties of Concrete," *Journal*, American Concrete Institute, Sept. 1960.

[19] Kellerman, W. F., "Effect of Type and Gradation of Coarse Aggregate Upon Strength of Concrete," *Public Roads*, June 1929.

[20] Blanks, R. F., Vidal, E. N., Price, W. H., and Russell, F. M., "The Properties of Concrete Mixes," *Journal*, American Concrete Institute, April 1940; *Proceedings*, Vol. 36, p. 433.

[21] Tyler, I. L., *Concrete of Norris Dam*, Bulletin, Engineering and Construction Department, Tennessee Valley Authority, Sept. 1937.

[22] Gilkey, H. J., "Size, Shape, Surface Texture, and Grading of Aggregates," *Report on Significance of Tests of Concrete and Concrete Aggregates (Second Edition), ASTM STP 22A*, American Society for Testing and Materials, Philadelphia, 1943, p. 92.

[23] Higginson, E. C., Wallace, G. B., and Ore, E. L., "Effect of Aggregate Size on Properties of Concrete," *Symposium on Mass Concrete*, SP-6, American Concrete Institute, Detroit, MI.

[24] Price, W. H., "Grading and Surface Area (of Aggregates)," *Significance of Tests and Properties of Concrete and Concrete Aggregates, ASTM STP 169*, American Society for Testing and Materials, Philadelphia, 1955, p. 274.

[25] Price, W. H., "Grading and Surface Area (of Aggregates)," *Significance of Tests and Properties of Concrete and Concrete-Making Materials, ASTM STP 169A*, American Society for Testing and Materials, Philadelphia, 1966, p. 404.

[26] Mercer, L. B. "Gap Grading for Concrete Aggregates," *Journal*, American Concrete Institute; *Proceedings*, Vol. 39, Feb. 1943, p. 309.

[27] Li, S., "Proposed Synthesis of Gap-Graded Shrinkage-Compensating Concrete," *Journal*, American Concrete Institute, Oct. 1967; *Proceedings*, Vol. 64, 1967, p. 654.

[28] *Concrete Manual*, 8th ed., Bureau of Reclamation, U.S. Department of the Interior, Washington, DC.

[29] Shilstone, J. M., Sr. in *Concrete Products*, June 1991.

[30] SHRP Program No. 5, working document of the SHRP committee.

[31] ASTM Specification for Concrete Aggregates (C 33-21T), *Annual Book of ASTM Standards*, American Society for Testing and Materials, Philadelphia, 1921.

[32] Cordon, W. A. and Gillespie, H. A., "Variables in Concrete Aggregates and Portland Cement Paste Which Influence the Strength of Concrete," *Journal*, American Concrete Institute, Aug. 1963; *Proceedings*, Vol. 60, 1963, p. 1029.

[33] Schmidt, W. and Hoffman, E. S., "9000 psi Concrete—Why? Why Not?" *Civil Engineering*, Vol. 45, No. 5, May 1975, p. 48.

[34] Mather, K., "High-Strength, High-Density Concrete," *Journal*, American Concrete Institute; *Proceeding*, Vol. 62, Aug. 1965, p. 951.

[35] "Tentative Interim Report on High Strength Concretes," *Journal*, American Concrete Institute; *Proceedings*, Vol. 64, Sept. 1967, p. 556.

[36] Gaynor, R. D., *High Strength Air Strength Air Entrained Concrete*, Publication No. 17, National Sand and Gravel Association and National Ready Mixed Concrete Association, Silver Spring, MD, March 1968.

[37] "Guide and Recommendations on Aggregates for Concrete for Large Dams," *Bulletin 18*, Commission Internationale Des Grand Barrages, Paris, France.

[38] Study by Virginia Department of Transportation, Charlottsville, VA, 1960.

[39] *Design and Control of Concrete Mixtures*, Portland Cement Association, 13th ed., Charlottesville, VA, p. 34.

[40] Folk, R. L., *Petrology of Sedimentary Rocks*, University Station, Austin, TX, 1968.

[41] Ashenbrenner, B. C., "A New Method of Expressing Particle Sphericity," *Journal of Sedimentary Petrology*, Vol. 26, 1956, pp. 15–31.

[42] Ozol, M. "Chapter 35—Shape, Surface Texture, Surface Area, and Coatings," *Significance of Tests and Properties of Concrete and Concrete-Making Materials, ASTM STP 169B*, American Society for Testing and Materials, Philadelphia, 1978, p. 586.

[43] Gray, J. E. and Bell, J. E., "Stone Sand," *Engineering Bulletin No. 13*, National Crushed Stone Association, 1964.

[44] "Method of Test for Flat and Elongated Particles in Coarse Aggregate," *Handbook for Concrete and Cement*, Designation CRD-C 119-48-42-53, U.S. Army Engineers Waterways Experiment Station, Vicksburg, MS, 1949, 1952, 1953.

[45] *Living with Marginal Aggregates, ASTM STP 597*, American Society for Testing and Materials, Philadelphia, 1976.

[46] Conway, C. D., "Measurement of Shape Characteristics of Coarse Aggregates," Virginia Highway Research Council, Charlottesville, VA, 1968.

[47] Pettijohn, F. J., *Sedimentary Rocks*, 2nd ed., Harker and Brothers, New York, 1957.

[48] Wenzel, R. N., "Resistance of Solid Surfaces to Wetting by Water," *Industrial and Engineering Chemistry*, Vol. 28, 1936, p. 988.

[49] Wenzel, R. N., "Surface Roughness and Contact Angle," *Journal of Physical and Colloid Chemistry*, Vol. 53, 1949, p. 1466.

[50] Ozol, M., "Chapter 35—Shape, Surface Texture, Surface Area, and Coatings," *Significance of Tests and Properties of Concrete and Concrete-Making Materials, ASTM STP 169B*, American Society for Testing and Materials, Philadelphia, 1978, pp. 605–607.

[51] Kaplan, M. F., "The Effects of Properties of Coarse Aggregates on the Workability of Concrete," *Magazine of Concrete Research*, Vol. 10, No. 29, Aug. 1958, pp. 63–74.

[52] Kaplan, M. F., "Flexural and Compressive Strength of Concrete as Affected by the Properties of Coarse Aggregate," *Journal*, American Concrete Institute, Vol. 30, No. 11, May 1959, pp. 1193–1208.

[53] Patten, B. J. F., "The Effects of Adhesive Bond Between Coarse Aggregate and Mortar on the Physical Properties of Concrete," *Civil Engineering Transactions*, The Institution of Engineers, Australia, 1973.

[54] Darwin, D. and Slate, F. O., "Effect of Paste-Aggregate Bond Strength of Behavior of Concrete," *Journal of Materials*, Vol. 5, No. 1, March 1970, pp. 86–98.

[55] Shah, S. P. and Chandra, S., "Critical Stress Volume Change and Microcracking of Concrete," *Journal*, American Concrete Institute, Vol. 65, No. 9, Sept. 1968, pp. 770–781.

[56] Nepper-Christensen, P. and Nielsen, T. P. H., "Modal Determination of the Effect of Bond Between Coarse Aggregate and Mortar on the Compressive Strength of Concrete," *Journal*, American Concrete Institute, Vol. 66, No. 1, Jan. 1969, pp. 69–72.

[57] Alexander, K. M., Wardlaw, J., and Gilbert, D. J., "Aggregate-Cement Bond, Cement Paste Strength and the Strength of Concrete," *Proceedings*, International Conference on the Structure of Concrete, London, Sept. 1965, pp. 59–81.

[58] Hsu, T. T. C. and Slate, F. O., "Tensile Bond Strength between Aggregate and Cement Paste or Mortar," *Journal*, American Concrete Institute; *Proceedings*, Vol. 60, No. 4, April 1963.

[59] Valenta, O., "The Significance of the Aggregate-Cement Bond for the Durability of Concrete," *Colloque International*, Durabilite des Betons, Prague, Nakladatelstoi Ceskoslovenske Akademie ved, 1961, pp. 53–87.

[60] Alexander, K. M., "Strength of the Cement Aggregate Bond," *Journal*, American Concrete Institute, Vol. 56, No. 5, Nov. 1959, pp. 377–390.

[61] Farran, J., "Mineralogical Contribution to the Study of Adhesion Between the Hydrated Constituents of Cement and the Embedded Material," *Revue des Materiaux de Construction*, July–Aug. and Sept. 1956, pp. 155–172 and pp. 191–209.

[62] Chatterji, S. and Jeffery, J. W., "The Nature of the Bond Between Different Types of Aggregates and Portland Cement," *Indian Concrete Journal*, Vol. 45, Aug. 1971, pp. 346–349.

[63] Ozol, M. A., "The Portland Cement Aggregate Bond—Influence of Surface Area of the Coarse Aggregate as a Function of Lithology," Report No. 71-R40, Virginia Highway Research Council, Charlottesville, VA, June 1972.

36

Soundness, Deleterious Substances, and Coatings

Stephen W. Forster[1]

PREFACE

A great deal of the material in this chapter is based on the work of the authors in the two previous editions of *ASTM STP 169*. D. L. Bloem authored the section on soundness and deleterious substances in *ASTM STP 169A* and L. Dolar-Mantuani did the chapter of the same name in *ASTM STP 169B*. I am indebted to them for their earlier work, most of which remains very valid today. For *ASTM STP 169C*, the subject of aggregate coatings is grouped with soundness and deleterious substances in this chapter. The discussion of coatings by M. A. Ozol in *ASTM STP 169B* was used as a basis for the inclusion on that subject here. In contrast to previous editions, the chapter has been organized to concentrate on relevant ASTM standards and their use in dealing with these three aspects of aggregate quality.

INTRODUCTION

Standards Development

During the past 100 plus years, many tests have been developed to assess the quality and suitability of aggregates for use in concrete. Mather and Mather [1] provide some historical background on the early use and evaluation of aggregate materials. This paper presents a discussion of a series of aggregate quality tests; including a brief overview of each method, the characteristics of the aggregate that are being assessed, and the meaning and application of the test results. After a section on nomenclature, the remainder of the chapter is arranged into three major headings according to the characteristic being discussed; that is, soundness, deleterious substances, and coatings. A final section presents a summary and conclusions based on the preceding discussions.

NOMENCLATURE

Terminology has developed over the years that is used in the description and assessment of aggregate quality. As is commonly the case, the same term can have different meanings to different people. In order to make the rest of the chapter clearer, the relevant terms will be discussed and defined here.

Soundness

Bloem [2] used "soundness" as a general term to describe the durability characteristics of an aggregate as a whole. On the other hand, some literature limits the terms sound and unsound to chemical properties. In the scope section of the ASTM Test Method for Soundness of Aggregates by Use of Sodium Sulfate or Magnesium Sulfate (C 88), it states, "This test method covers the testing of aggregates to estimate their soundness when subjected to weathering action in concrete or other applications." This statement indicates that soundness is an aggregate's resistance to weathering; whether bound in concrete or not. Thus, soundness includes resistance to wetting and drying, heating and cooling, freezing and thawing, or any combination thereof. This will be the context to which soundness applies in this chapter.

Deleterious Substances

Bloem [2] used "deleterious substances" to describe individual particles or contaminants that are detrimental to the aggregate's use in concrete. Rhoades and Mielenz [3] and Mielenz [4,5] used deleterious and its antonym, innocuous, to refer to chemical properties of aggregate particles. Swenson and Chaly [6] developed a detailed classification of harmful substances based on the published literature, and they used deleterious to mean any quality of an aggregate that is harmful to concrete. Their definition is correct, however, this chapter is devoted to deleterious "substances," herein defined as any material in the aggregate that is detrimental to concrete. For instance, Table 3 of ASTM Specification for Concrete Aggregates (C 33), specifically lists the following as deleterious substances:

clay lumps and friable particles,
chert (less than 2.40 saturated surface dry (SSD) specific gravity),
material finer than the 75-μm (No. 200) sieve, and
coal and lignite.

Coatings

Ozol [7] described "coatings" as adhering materials that may be strongly or weakly cemented to the aggregate parti-

[1] Research geologist and team leader, Pavements Division, Federal Highway Administration, HNR-20, McLean, VA 22101.

cle's surface. This description is certainly appropriate for this chapter, although the use of "bound" rather than "cemented" may better express the range of adherence included. Coatings may be found on sand, gravel, crushed stone, or artificial aggregate, and may be produced by either chemical or physical processes.

SOUNDNESS

Although individual particles may behave differently in concrete, it is desirable to have general criteria for the overall rating and acceptance of an aggregate source for use in concrete. It is convenient to test representative aggregate samples to obtain data that correlate well with the quality of concrete made from the aggregate. No individual test method accomplishes this. However, with respect to aggregate soundness, as defined earlier, ASTM C 88 is most frequently applied. This test, and others that have been used to evaluate soundness, are discussed subsequently.

Sulfate Soundness Test

Description

The sulfate soundness test, ASTM C 88, was first published as a tentative test method in 1931, and finally approved as a standard method in 1963. As noted earlier, the scope of the method indicates that the test is a means to estimate the soundness of the aggregate in concrete or other applications. The test is conducted by repeated immersions of the aggregate sample in saturated solutions of sodium or magnesium sulfate alternating with oven drying to precipitate the salt in permeable pore spaces. Expansive forces are exerted by rehydration of the salt when the sample is re-immersed. This expansion is said to simulate the expansion of water upon freezing. Magnesium sulfate is generally more destructive than sodium sulfate. After a fixed number of cycles of soaking and drying, the amount of material that has been lost from the original particles is estimated. This is done by weighing the coarse aggregate size fractions retained as a result of hand sieving on sieves with openings five-sixths the size of those in the original sieves used to separate the sample. For fine aggregate, this material loss is determined by sieving using the same sieves and method as originally used. The loss is equal to the weighted sum of the difference between the original and after test size fraction weights.

Examination of the test samples after testing reveals that the distress caused by the procedure may be manifested in a number of ways. Aggregate particles may show flaking, splitting, crumbling, granular disintegration, or some combination thereof. To better categorize the type of distress that occurred after completing the required number of cycles, the sample fraction retained on the 19.0 mm (¾ in.) sieve is examined visually with or without the aid of magnification. Affected particles are categorized by type of distress, and the number of particles in each category recorded for comparison with a count taken on the number of plus 19.0 mm (¾ in.) particles in the original test sample. Although the method calls for examining only the plus 19.0 mm (¾ in.) size fraction, examination of the other sizes may help define the type of distress that is occurring. For further insight as to the type and progression of distress that is occurring, the test samples may be visually examined after each cycle. Shaley particles, with closely spaced planes of weakness, may split a number of times during the test without greatly reducing the sieve sizes of the particles. This phenomenon should be watched for, and noted in the report if detected.

In reporting the results of the test, a number of items should be included. The actual test results will provide: the kind of solution used; the mass of each sample fraction before and after test; the percent loss for each size fraction; the weighted percent loss for each size fraction and for the fine and coarse aggregate portions; and for the plus 19.0 mm (¾ in.) size fractions, the number of particles before the test and the number of particles in each class of distress after the test. Information on the source of the sample is needed, and additional information on the aggregate, such as a general description and past performance, would prove useful in interpretation of the test results. Since the method currently allows flexibility in the number of cycles run, this must also be reported.

Precision and Bias

Before attempting to interpret the results of the soundness test (or any test), available information on the "precision" of the test (that is maximum allowable differences between the results of two tests, properly conducted either by a single operator or by two different laboratories); and the "bias" of the test (a systematic error in the results that is inherent in the test method) must be carefully considered. In the case of the sulfate soundness test, since the soundness loss can be defined only in terms of the test method, bias cannot be determined. On the other hand, precision has been investigated and determined in a number of studies.

Precision is normally defined in terms of two variables: (1) the coefficient of variation and (2) the difference between two tests. Precision is further broken down into within-laboratory, single-operator precision (repeatability) or between-laboratory precision (reproducibility). Early studies [8–11] indicated that numerous portions of the sulfate soundness test procedures and equipment were not well enough defined to yield results with good precision, particularly between laboratories. Improvements suggested by Wuerpel [12], Paul [13], and Woolf [14,15], among others, continued to be added to the method in the interest of better precision. Even with the continual search for and addition of improvements, the precision of the method remains poor. ASTM Subcommittee C9.20 recently conducted a carefully controlled round-robin test series on the soundness test, in yet another effort to better define the causes of poor precision. Although many laboratories feel that they produce repeatable results within their individual laboratories, the round-robin series indicated continued poor precision within as well as between laboratories [16]. The failure to get results of even comparable magnitude appears to stem from two groups of factors:

(1) differences in materials or procedures used to conduct the test that are allowed under the current standard method; (2) deviations from the prescribed method, either due to misunderstanding of the requirements or modifications made to the method over the years by the laboratory conducting the test. Table 1 lists some of the main factors affecting precision. Precision was reported to improve [16] with tighter control on a number of these factors noted in Table 1. As with any test method, obtaining a representative test sample is critical to getting good results. The difficulty in obtaining a representative sample is especially high for sand and gravel deposits of variable composition.

In order to improve the precision of the results, several avenues are open; the most obvious being to more closely define the conditions under which the test may be conducted and the equipment and materials used in the test. Bloem [17] investigated the impact of using various options available in the method and discovered a number of trends. For instance, fine aggregates have a loss from the test about twice as large when magnesium sulfate is used as when sodium sulfate is used. For coarse aggregate, magnesium sulfate is also usually more severe, but the trend is not so clear, and can be reversed.

As would be expected, as the number of cycles specified to be conducted increases, the loss also increases. Ten cycle losses average about 50% higher than five cycle losses for both coarse and fine aggregate tested with either type of salt. Precision indices given in the latest version of ASTM C 88 indicate that much better precision can be obtained with magnesium sulfate than with sodium sulfate (see Table 2). These results may be due to magnesium sulfate having a solubility with a lower temperature sensitivity than that of sodium sulfate. In spite of the apparent advantage of magnesium sulfate, laboratories that have traditionally used sodium sulfate are reluctant to switch because direct comparison to past test results would then be lost.

Interpretation of Sulfate Test Results

In order to correctly interpret the results of this test, its significance must be clearly understood. As stated in the definition, soundness is an aggregate's resistance to weathering. How accurately the sulfate test measures soundness is the subject of debate that has continued unabated since the inception of the test. The test is most often related to that portion of weathering due to frost action.

There have been a number of explanations of the process that occurs during the sulfate test that causes the disintegration of the aggregate particles [8–10,12,18]. There is general agreement that the cycles of soaking in the sulfate solution and oven drying cause the salt to accumulate in the cracks and pores in the aggregate particles. After some variable number of cycles, depending on the susceptibility of the aggregate particle, the quantity of salt in these spaces becomes more than can be accommodated during hydration in the next soaking phase. At this point, disruptive forces are exerted within the particle. Magnesium sulfate results in a greater loss than sodium sulfate because a saturated solution of it contains more material, both in mass and volume [2].

Evidence suggests that there are other forces acting on the particles in addition to those exerted by crystal growth. Experiments run with distilled water in place of the solution resulted in significant disintegration and loss, indicating that wetting and drying or heating and cooling are factors in the test results [10]. For carbonate aggregates, Hudec and Rodgers [19] found that disintegration was influenced by the quantity of water adsorbed on the particles. This result points to wetting and drying as being a factor in the test results. It has been found that continuing the oven drying beyond the time required to dehydrate the sulfate crystals results in greater disintegration of the aggregate, again indicating the influence of heating and cooling.

During the early development and experiments with the test, it was suggested that the resulting disintegration was more the result of the number of pores in the aggregate than their size and distribution [11,20–25]. It should be pointed out, however, that this early work was mostly done prior to the era of air-entrained concrete and the development of some understanding of the role that the characteristics of a void system can have on freeze-thaw resistance. Various authors have pointed out that there is little theoretical basis for equating/relating the results of the sulfate soundness test with freeze-thaw disintegration/resistance in the field. However, in a statistical analysis of test results on 70 samples, Vollick and Skillman [26] did find limited correlation between the sulfate soundness test results and the freeze-thaw resistance of concrete containing the same coarse aggregates. The amount of scatter in the data indicates that there certainly were other major influencing factors. Porosity of concrete, porosity of aggregates, and freeze-thaw resistance of concrete are discussed in other chapters of this volume.

TABLE 1—Factors Affecting the Precision of the Sulfate Soundness Test (ASTM C 88).

Design of containers used for immersing samples
Temperature of the solution
Temperature variation of the solution during test
Chemical grade of the sulfate used
Purity of the water used to make the solution
Age of the solution
Temperature of samples at time of immersion after oven drying (should match solution temperature)
Temperature variation and rate of drying in oven

TABLE 2—Precision Indices for the Sulfate Soundness Test (ASTM C 88).

	Coefficient of Variation (1S%), %	Difference Between Two Tests (D2S%), % of Average
Multilaboratory		
Sodium sulfate	41	116
Magnesium sulface	25	71
Single-operator		
Sodium sulface	24	68
Magnesium sulfate	11	31

NOTE — For further explanation of 1S% and D2S%, see ASTM Practice for Preparing Precision and Bias Statements for Test Methods for Construction Materials C 670.

In view of the precision of the test method as just discussed, and the uncertainty as to what performance characteristics of the aggregate the test actually relates, it can easily be seen that interpretation of the results must proceed with caution. One suggestion was to accept aggregates that passed the test, but not to reject those that fail it [15]. The reasoning is that the test is severe and, therefore, aggregates that survive it are certainly "good," but those that deteriorate under test may or may not be good. This theory may hold for some aggregate uses, but not for others. For example, some aggregates that fail the sulfate soundness test have shown good freeze/thaw resistance in portland-cement concrete, while conversely, other aggregates that have performed very well in the sulfate soundness test cause concrete to fail under freeze-thaw conditions [26].

In summary, interpreters of the soundness test results must proceed with caution. A low soundness loss will "usually" indicate a durable aggregate, however, collaborative evidence from other tests, or ideally prior service records in the same application, lend more credence to any conclusions reached.

Specification Limits

In view of the preceding discussions on precision and interpretation of results, the difficulty in setting specification limits for the soundness test can easily be seen. Normally, specifications provide some flexibility to the user to consider other evidence when making final determination of the suitability of an aggregate. ASTM C 33 allows use of fine aggregate that has exceeded the specified limits of the soundness test results if "concrete of comparable properties, made from similar aggregate from the same source, has given satisfactory service when exposed to weathering similar to that to be encountered" or, lacking a service record, "provided it gives satisfactory results in concrete subject to freezing and thawing tests." For coarse aggregate, ASTM C 33 similarly allows the use of aggregate that has exceeded specified limits of the soundness test results "provided that concrete made with similar aggregate from the same source has given satisfactory service when exposed in a similar manner to that to be encountered; or in the absence of a demonstrable service record, provided that the aggregate produces concrete having satisfactory relevant properties."

These statements thoroughly demonstrate that the authors of this specification well realized the shortcomings of the sulfate soundness test to unfailingly discriminate between sound and unsound aggregate for use in concrete. The way the specification is written allows the specifiers to apply the soundness test as an acceptance test to the degree of strictness they are comfortable with.

Other Tests as Indicators of Soundness

Because of the imperfect prediction of soundness by the sulfate soundness test, a number of other tests have been used in lieu of, or to supplement, the results of the sulfate test. The more commonly used tests are briefly discussed subsequently. These tests often address only one aspect of soundness as defined; that is, resistance to either freezing and thawing, wetting and drying, or heating and cooling.

Absorption Tests

It has been suggested that results of ASTM Test Method for Specific Gravity and Absorption of Coarse Aggregate (C 127) and ASTM Test Method for Specific Gravity and Absorption of Fine Aggregate (C 128) might be useful as indicators of the soundness of aggregate. Originally, it was felt that since absorption was a direct measure of accessible pore space in the aggregate, it would be closely related to the freeze-thaw behavior of the aggregate in concrete [24,25]. This relationship has not proven to be reliable; however, absorption can serve as an initial general indicator of the soundness tendency of an aggregate, prior to the completion of more reliable, definitive tests.

Freeze-Thaw Testing of Unconfined Aggregate

Since the sulfate soundness test is often considered a simulation of the action on aggregate particles due to freezing and thawing, a logical extension of the soundness test is the testing of loose (unconfined) aggregate by actual freeze-thaw action. Various cycling rates, sample conditioning, number of cycles, and types of additives to the freezing water have been tried in an attempt to more closely duplicate actual field performance in concrete (see, for example, Ref 27). All the variations met with limited success in this regard, for the most part no better than the sulfate soundness test, itself. New York State, however, does feel comfortable using a laboratory freeze-thaw test for the acceptance of aggregates.

Freezing and Freeze-Thaw Tests in Concrete

Application of freeze-thaw tests to aggregate contained in concrete of the proposed mix design is the most satisfactory laboratory method of predicting the field performance of an aggregate in concrete exposed to freeze-thaw conditions. There are a number of ASTM freeze-thaw tests currently on the books, including ASTM Test Method for Resistance of Concrete to Rapid Freezing and Thawing (C 666), ASTM Test Method for Critical Dilation of Concrete Specimens Subjected to Freezing (C 671), and ASTM Practice for Evaluation of Frost Resistance of Coarse Aggregates in Air-Entrained Concrete by Critical Dilation Procedures (C 682). ASTM C 666 has optional Procedures A and B, that, along with the conditioning of the test specimens prior to the start of the test, can have a major effect on the results. Sample preparation and conditioning can also have a significant effect on the results of the other tests. These effects all are related to the moisture content of the concrete (and aggregates) during testing. Even nonfrost resistant concrete will not be adversely affected by freezing tests if it is not critically saturated. This may occur because of air storage of the specimens prior to test, or other steps in sample preparation and curing. It is generally agreed that ASTM C 666 Procedure B (freezing in air) is less severe than Procedure A (freezing in water), perhaps because of the potential water loss from the specimens during the freezing step.

ASTM C 682 points out several areas of concern when interpreting test results that bear repeating. Many of these cautions apply equally well to the other methods. First, as noted earlier, minor changes in the conditioning of the specimens can have major effects on the test results. Therefore, repeatability should be one of the factors in choosing a conditioning procedure. Sorting the aggregate by a petrographer into homogeneous fractions prior to testing will improve the precision of the test results. Even if it is decided to test the heterogeneous aggregate sample as a whole, identifying the lithologies present, condition, and percentage of each fraction will provide insight for the interpretation of the test results. Any one test cannot serve as the definitive answer for all aggregates. A combination of tests and evaluation by a trained petrographer is the best approach. This thought will be revisited at the end of this chapter.

The freezing and freeze-thaw methods are discussed in detail in another chapter of this volume.

Other Tests

A number of other tests are or have been used as indicators of aggregate soundness. These tests include the copper nitrate test as described by Dolar-Mantuani [28]. This test, conducted by immersing the test sample in copper nitrate solution overnight, has a similar effect on shaley or argillaceous particles as the sulfate soundness test. Other tests that may be indicative of soundness and that are described elsewhere in this volume include the two impact and abrasion tests using the Los Angeles machine (ASTM Test Method for Resistance to Degradation of Small-Size Coarse Aggregate by Abrasion and Impact in the Los Angeles Machine (C 131) and ASTM Test Method for Resistance to Degradation of Large-Size Coarse Aggregate by Abrasion and Impact in the Los Angeles Machine (C 535)); ASTM Test Method for Lightweight Pieces in Aggregate (C 123); and ASTM Test Method for Clay Lumps and Friable Particles in Aggregates (C 142).

Petrographic Examination for Soundness

The evaluation of an aggregate sample by a trained, experienced petrographer both independent of, and in concert with, other tests is an invaluable tool in correctly predicting the soundness of an aggregate source. ASTM Guide for Petrographic Examination of Aggregates for Concrete (C 295) provides a good outline for conducting a systematic evaluation. The lithology (rock type) and state of weathering are two important factors that must be considered when evaluating the soundness of an aggregate. The procedure is discussed in detail in Chapter 22 of this publication. Beyond the specifics of the practice, the basic lesson to be learned from the general approach of the petrographic examination, and something equally applicable to any type of evaluation, is to "take advantage of all available pertinent information prior to reaching a conclusion."

As was noted earlier, ASTM C 88 requires the qualitative examination of the sample particles larger than 19.0 mm (¾ in.) for categorizing the types of distress present. (Examination of the smaller sizes will often add additional insight on the aggregate's performance). If conducted by a qualified petrographer, this examination, in concert with a petrographic examination of the sample prior to testing, will provide great insight as to the causes for the aggregate being susceptible to the conditions of the soundness test. Knowing the causes of distress in the affected particles, and their overall percentage in the aggregate source, a better prediction of the actual performance of the aggregate in concrete in service can be made, as well as recommendations on the need and practicality of beneficiation of the aggregate.

DELETERIOUS SUBSTANCES

As noted in the preceding section on nomenclature, a deleterious substance (as used in this chapter) refers to any material in the aggregate that is detrimental to concrete. According to ASTM C 33, deleterious substances include the following categories: clay lumps and friable particles, chert (less than 2.40 SSD specific gravity), material finer than the 75-μm (No. 200) sieve, and coal and lignite. Each of these categories will be briefly described.

Description

Clay Lumps

The term, clay lumps (or clay balls), refers to lumps of clay to fine sand-sized particles that are present during and after the aggregate processing. The lumps would have to be mechanically broken up to be effectively dispersed. If they survive the aggregate processing procedures, they also will usually survive the mechanical action associated with the mixing and placing of concrete. They will, however, be subject to some surface abrasion during the mixing and, as a consequence, often contribute additional, unaccounted-for fines to the mix.

Friable Particles

Friable particles are those aggregate pieces that have little bond between the mineral grains that compose the aggregate particles. The individual minerals may or may not be soft; however, the fact that the bond between grains is weak results in the aggregate particles breaking down into smaller pieces or rapidly losing grains, or both, from their surface during the mechanical action of processing or concrete mixing and placing. Examples of friable aggregates are a sandstone composed of quartz grains weakly cemented together by a clay or calcite matrix, and a poorly indurated shale.

Chert (SSD Specific Gravity Less Than 2.40)

Based on empirical evidence, chert particles with an SSD specific gravity of less than 2.40 are considered to be usually objectionable as aggregate in concrete due to frost susceptibility and the resulting cracking of the concrete or the formation of popouts at the concrete surface. Where individual aggregate particles are composed of some lightweight chert and the parent rock (limestone or dolomite), the particles may be above the 2.40 specific gravity but still cause these problems due to the chert. These aggregates will therefore have to be examined carefully.

Material Finer Than the 75 μm (No. 200) Sieve

Particles that pass the 75-μm (No. 200) sieve are generally referred to as clay and silt. This includes the fine material that can occur in sand and gravel deposits as well as the dust of fracture that may result from the crushing and mechanical processing of aggregate. If this material adheres to the coarser aggregate particles during processing, it can create problems if it provides too much unanticipated fines to the concrete during mixing or results in a poor coarse aggregate-to-paste bond in the hardened concrete. If this material is essentially free of clay "minerals" or shale (as is usually the case with dust of fracture), it may be present in greater amounts without adverse effect on the concrete. A small amount of dust of fracture may even be of benefit to strength for lean concrete mixes.

Coal and Lignite

Coal and lignite may occur as discrete coarse particles in the aggregate or as fine material disseminated throughout. If not derived through contamination from another source, coal and lignite are often associated with shale, something not normally used in concrete; however, coal and lignite may also be found associated with other rock types.

Other Deleterious Substances

Other deleterious substances not specifically mentioned earlier include lightweight pieces other than chert, soft particles, and organic impurities other than coal and lignite. Lightweight pieces include highly porous aggregates that float on a liquid of density 2.0. These particles often lack durability. Soft particles, as distinguished from friable particles, are composed of minerals that are soft and therefore are very susceptible to abrasion and wear. They are often the result of long-term (in a geologic sense) weathering and chemical alteration of originally durable rock. Organic particles other than coal and lignite include plant roots, twigs, and other vegetable and animal material.

Effect of Deleterious Substances on Concrete

When clay lumps survive the processing of the aggregate, and also the mixing and placing of the concrete, they are subject to breakdown during wetting and drying, and freezing and thawing of the hardened concrete. This can result in popouts near the concrete surface and also the appearance of a pock-marked surface where the clay lumps have weathered away.

Friable particles and soft pieces are easily broken down into smaller particles or create additional fines. If the deterioration of the friable particles continues once the concrete is in place, it may result in freeze-thaw damage to the concrete. If soft particles are present in sufficient percentages, a lowering of the concrete strength and durability will result. Soft or friable particles exposed at the concrete surface will certainly lower its abrasion resistance and will often quickly weather away leaving a pitted or pock-marked surface.

Low specific gravity chert particles are susceptible to frost action due to their internal pore structure. Their deterioration in concrete can result in cracking of the concrete or popouts if the aggregates are located just below the concrete surface.

An excessive amount of minus 75-μm material in the aggregate results in high water demand by the concrete for mixing, placing, and finishing. Once in place, the excess water increases drying shrinkage and the likelihood of cracking. The excess water also leads to lower strengths. If the fines include one or more of the swelling clay minerals (montmorillonite, smectite, or nontronite), volume change during wetting and drying in the hardened state can be increased, with the formation of microcracks (see ASTM C 295, Ref 29, and Ref 30). If the fines remain adhered to the coarse aggregate during mixing, a poor coarse aggregate-to-paste bond may develop, resulting in lower strengths. As noted earlier, if the minus 75-μm material is essentially free of clay minerals and shale, higher percentages may be allowable (see ASTM C 33); however, the total amount of this material in the coarse and fine aggregate must be considered. A small percentage of this material may have strength, workability, and density benefits for lean concrete mixes.

Finely divided organic material may retard the hardening of the concrete and reduce the concrete strength. This material may also interact with or affect the effectiveness of any air-entraining or other admixtures used in the concrete. Coal and lignite are usually present in larger-sized pieces and often result in pits or popouts at the surface of the concrete due to subsequent weathering. Staining of the surface may also result from coal/lignite. Brown lignite coal, wood, tall oil residues, and pine oil contain materials that readily disperse in the alkaline solution present in concrete and may cause dark brown stains and zones of weakness [31].

Tests for Deleterious Substances

Clay Lumps and Friable Particles

Testing for these two classes of deleterious substances is combined in one ASTM test, ASTM C 142. This method is applied to an aggregate sample after it has undergone a washed gradation (ASTM Test Method for Material Finer Than 75-μm (No. 200) Sieve in Mineral Aggregates by Washing (C 117)) to determine the amount of minus 75-μm material in the aggregate. As a result, any clay lumps or friable particles (or portions thereof) that break down in the wet sieving process will not be measured in the ASTM C 142 test. The test procedure itself consists of a 24-h soak of the sample in distilled water (after the sample has undergone the ASTM C 117 test) followed by manual manipulation by the fingers to break any susceptible particles into smaller sizes. After manipulation, the sample is wet-sieved over a designated sieve according to the size range of the original sample. The percent of clay lumps and friable particles is the change in mass of the sample due to the soaking, manipulation, and sieving processes.

Chert (SSD Specific Gravity Less Than 2.40) and Other Lightweight Pieces

Due to its pore structure, chert that has the potential for physical disruption of concrete under freeze-thaw conditions usually has an SSD specific gravity of less than 2.40, and this characteristic is used as a distinguishing criteria in ASTM C 33. It should be noted that this criteria applies only when the chert is present in the aggregate in small amounts; if the aggregate is predominantly chert, its potential effect on the durability of the aggregate must be determined by other means, preferably prior service record in similar concrete, structure, and exposure. ASTM C 123 is used to separate aggregate pieces of different specific gravity. The procedure employs a heavy liquid that can be mixed to possess the specific gravity required as the separation value (2.40 for chert, 2.0 for coal and lignite). Then it is simply a matter of placing the test aggregate in the heavy liquid, skimming off the lightweight (reject) pieces that float, and calculating their percentage of the total sample.

It has been reported [32] that sand-sized particles of chert (which is detrimental to the durability of concrete in coarse aggregate sizes) do not damage concrete containing them when it is subjected to ASTM C 666. The vulnerability of the chert to freeze-thaw damage depends on a number of interrelated factors, including the characteristics of the pore system and the particle size, as well as the characteristics of the surrounding concrete paste. As is often the case, directly comparable service records are the surest means to predict performance.

If there is any doubt about the composition of the lightweight pieces identified by ASTM C 123, the particles should be examined by a petrographer. In addition to the detrimental materials noted earlier and deeply weathered soft (non-durable) aggregate particles, certain man-made aggregates and natural volcanic rocks that perform well in concrete may float in the test.

Material Finer Than 75-μm (No. 200) Sieve

ASTM C 117 is the method used to determine the amount of this material in an aggregate sample. The test involves washing the material over a 75-μm (No. 200) sieve, and any material carried through the sieve by the wash water (or less frequently dissolved and removed in the water) is the amount of minus 75-μm material in the sample. The precision statement for ASTM C 117 indicates that for coarse aggregate, the results of two properly conducted tests should differ by no more than 0.28% for a single operator or 0.62% for two different laboratories. Similarly, for fine aggregate, the difference should be no greater than 0.43% for a single operator or 0.82% for two different laboratories.

Other tests have been developed that are reported to indicate the amount or proportion of clay-sized fines in concrete aggregate [29]. Of these, the ASTM Test Method for Sand Equivalent Value of Soils and Fine Aggregate (D 2419) is the most notable. By means of water suspension and settlement procedures, this test provides a rapid indication of the relative proportion of "clay-like or plastic fines and dusts" in a fine aggregate or soil. Tests conducted by Gaynor [33] on 130 samples showed, however, that there was little correlation between the amount of minus 75-μm material in the sample and the results of the sand-equivalent test. Others have reported some correlation between the sand-equivalent test and the water demand of a concrete mix [31]. In order to determine the mineralogy of the minus 75-μm material, and whether it is potentially detrimental clay minerals or dust of fracture or other material, it may be necessary to run an X-ray diffraction or other analysis technique. A petrographer should be consulted in this regard.

Soft Particles

Although often confused, soft particles and friable particles are distinct. As noted earlier, the component mineral grains that make up a soft particle are soft (often weathered), whereas the component grains of a friable particle may or may not be soft; rather it is the weak bond holding the grain together that defines a friable particle. Discontinued ASTM Test for Scratch Hardness of Coarse Aggregate Particles (C 235) (later included as ASTM C 851) was used for some time as a means of identifying soft particles. Deletion of ASTM C 851 from the *ASTM Book of Standards* followed a number of years of debate in committee meetings as to the merits of the test. Percentages of soft particles in an aggregate seemed to have little correlation with concrete strength. This low correlation may have been due to the difficulty in differentiating between soft particles and friable particles using the test method.

Two test methods, ASTM C 131 and ASTM C 535, have also been used as a measure of the presence and amount of soft particles. As with the scratch test, the results should be interpreted carefully since many factors other than the softness of the particles can influence the test results. Petrographic examination of the material will help in identifying its characteristics that led to the test results obtained.

Organic Impurities

ASTM Test Method for Organic Impurities in Fine Aggregates for Concrete (C 40) is a simple screening test for determining the presence of organic material. The procedure consists of immersing a given amount of fine aggregate in a standard solution of sodium hydroxide and noting the color change in the solution after 24 h due to reaction of the solution with any organic material present in the aggregate. The color of the solution is compared to a standard solution or standard color glass plates to qualitatively determine the amount of organic matter. Since the test solution reacts with all types of organic matter, including bits of wood and twigs that do not affect concrete strength when present in small amounts, it is only a screening test. Aggregates resulting in colors darker than the standard should be further evaluated by other tests such as mortar strength and setting time. ASTM Test Method for Effect of Organic Impurities in Fine Aggregate on Strength of Mortar (C 87) compares strength results of mortar samples containing the subject aggregate as received with strength results of mortar samples made using the subject aggregate after it has been washed to remove the organic matter.

In addition to strength and setting time, some organic matter may reinforce or detract from the efficiency of admixtures. Air entrainment is a particular area of concern [34] in that the organic material may act as an air-entraining agent of variable efficiency.

COATINGS

Description

Naturally Occurring Coatings

Naturally occurring coatings are those that are deposited on the surfaces of the aggregate particles by means of natural processes. Within this context, this class of coatings is limited to occurrence in sand and gravel deposits, since only this type of aggregate source has the aggregate particle surfaces exposed for the requisite period of time for these natural processes to take place. The chemical and mineralogical composition of the coating can vary over a wide range. The two most important chemically precipitated coatings are calcium carbonate and silica. Deposition of these coatings often results from solution of the minerals from one portion of the deposit (by percolating groundwater) and redeposition in another (usually lower) portion of the deposit. Since the source of the coatings is the particles in the deposit itself, the coatings and the aggregate particles on which they precipitate are often of the same composition. Other possible chemically precipitated coatings include gypsum, other sulfates, iron oxide, opal, and phosphates. There also are physically deposited coatings in sand and gravel deposits, namely, a layer or layers of silt and clay-sized particles. These coatings are also transported by the moving groundwater in the sand and gravel deposit, but migrate as solid particles rather than being dissolved, transported in solution, and then precipitated as in the chemical coatings process.

Artificially Generated Coatings

Coatings on aggregate particles may also result from one or more of the steps necessary to produce an aggregate from a source material. These coatings usually consist of fine materials of the same composition as the particles they coat because they are derived from these particles through impact and abrasion during handling. Loading, crushing, unloading sizing, and stockpiling all produce a certain amount of fines (often called "dust of fracture"), the amount dependent on the characteristics of the source material being processed. A hard, polish, and wear-resistant source material will produce only small amounts of fines during processing; usually not nearly enough to produce any undesirable coating of the larger aggregate pieces. A soft, wear, and polish-susceptible aggregate source material, on the other hand, may produce enough fine material to form a coating on the coarse aggregate. This coating may have to be dealt with either by additional aggregate washing to remove it or, if the coating of fines is shown to be removed from the aggregate during mixing, by compensation for the additional fines in the mix during the design of the concrete.

Identification

The usual approach used for identifying the composition and extent of coatings is a petrographic examination, ASTM C 295. As noted earlier, sand and gravel deposits may have either chemically or physically deposited coatings, whereas a crushed stone can have only a physically deposited coating. This practice allows a wide latitude in the equipment and procedures used depending on the nature of the material being identified and other characteristics to be determined. If the coatings are considered to contain materials (such as opal, gypsum, organic material, or easily soluble salt) likely to be deleterious when the aggregate is used in concrete, the coatings must be positively identified by whatever necessary means. The strength of the bond of the coating to the particles should also be determined, first qualitatively during the petrographic examination, and then, if necessary, by conducting strength tests on concrete containing the aggregate, including post-test examination of the break surfaces.

Effect of Coatings on Concrete

From a physical standpoint, aggregate coatings may be detrimental to the strength properties of concrete if the bond of the cement paste to the coating is greater than the bond of the coating to the aggregate particles. When this occurs, the concrete will fail at the coating/aggregate interface at a smaller load than it would otherwise fail. In this regard, Goldbeck [35] reported that flexural strength reductions of up to 1.5%, and compressive strength reductions of up to 2%, can occur for every percent of dust of fracture included in the aggregate. As noted elsewhere, if the dust of fracture occurs not as a coarse aggregate coating, but dispersed in the mix in small quantities, it may not be detrimental to the concrete and may in fact be of benefit to some properties of the plastic and hardened concrete.

Soluble or chemically reactive coatings will cause the same deleterious results in the concrete as they would if included in the concrete in any other form, such as coarse or fine aggregate particles. Certain siliceous coatings (opal, for example) may react with the alkalis in the cement, depending on the amount of alkalis present, other components in the mix, and additional factors [36]. The products of this reaction are susceptible to swelling in the presence of moisture, producing tensile stresses in the concrete that can lead to cracking. Coatings composed of iron compounds may result in staining of the concrete surface, as well as a weakening of the aggregate/paste bond mentioned earlier. Chlorides and sulfates can cause staining and efflorescence at the concrete surface and may also lead to corrosion of reinforcing steel [7].

In summary, aggregate coatings may be innocuous or deleterious to concrete depending on: their solubility in the concrete and their effect on any chemical reaction that might occur; their bond to the aggregate relative to the cement paste's bond to them; and the effect that any fine particles derived from them have on the properties and characteristics of the plastic and hardened concrete. Examination of specimens (both before and after strength testing) made during the mix design process will often help to verify conclusions reached about coatings during the petrographic examination of the aggregate.

CONCLUSIONS

An aggregate's "soundness" may be best equated to its resistance to all aspects of weathering when used in concrete or other applications. Weathering may include any combination of heating and cooling, wetting and drying, or freezing and thawing. Because it is so complex, no one test adequately measures the soundness of an aggregate, although several tests certainly provide good insight as to how an aggregate will behave. Results of ASTM C 88 are often considered to be an aggregate's soundness without considering all that the term implies. ASTM C 88 certainly can be a useful indicator of soundness and, when used along with other information (preferably including a prior service record), a reliable assessment of soundness can be made.

A deleterious substance, as defined herein, refers to any material in the aggregate (other than coatings) that is detrimental to the concrete in which the aggregate is used. By convention, the term, deleterious substances, usually refers to accessory or minor constituents in the aggregate rather than major components. Deleterious substances include clay lumps and friable particles, low SSD specific gravity chert, fine material, coal and lignite, and others. The effect of deleterious substances on concrete, and the tests used to determined their presence and amount, will depend on the type of substance under consideration and the structure and environment in which the concrete will be placed.

Coatings may be subdivided into naturally occurring coatings and those generated by artificial means through processing of the aggregate. Because of the wide variety of materials included under coatings, evaluation by a qualified petrographer using ASTM C 295 is the most practical approach. Following identification of the coatings involved, specific tests may be prescribed depending on the identity, nature, and quantity of the coating.

In all three of these areas of aggregate quality assessment, the surest means of prediction of performance of an aggregate in concrete is the availability of a prior service record. To be most reliable, the service record should involve the same aggregate (often just the same source is not specific enough), the same concrete mix design and components, incorporation into the same type of structure, and should be subject to the same environment. Usually, one or more of these components of the service record is different from the intended use, therefore, some interpretation will be necessary. It is at this point that examination of test results, a petrographic evaluation, and good engineering judgment must be employed.

REFERENCES

[1] Mather, K. and Mather, B., "Aggregates," *The Heritage of Engineering Geology; The First Hundred Years*, Centennial Special Vol. 3, Geological Society of America, Boulder, CO, 1991, pp. 323–332.

[2] Bloem, D. L., "Soundness and Deleterious Substances," *Significance of Tests and Properties of Concrete and Concrete-Making Materials, ASTM STP 169A*, American Society for Testing and Materials, Philadelphia, 1966, pp. 497–512.

[3] Rhoades, R. and Mielenz, R. C., "Petrography of Concrete Aggregates," *Proceedings*, American Concrete Institute, Vol. 42, 1946, p. 581.

[4] Mielenz, R. C., "Petrographic Examination Concrete Aggregate," *Proceedings*, American Society for Testing and Materials, Vol. 54, 1954, pp. 1188–1218.

[5] Mielenz, R. C., "Petrographic Examination," *Significance of Tests and Properties of Concrete and Concrete-Making Materials, ASTM STP 169A*, American Society for Testing and Materials, Philadelphia, 1966, pp. 381–403.

[6] Swenson, E. G. and Chaly, V., "Basis for Classifying Deleterious Characteristics of Concrete Aggregate Materials," *Proceedings*, American Concrete Institute, Vol. 52, 1956, pp. 987–1002; *Journal*, May 1956.

[7] Ozol, M. A., "Shape, Surface Texture, Surface Area, and Coatings," *Significance of Tests and Properties of Concrete and Concrete-Making Materials, ASTM STP 169A*, American Society for Testing and Materials, Philadelphia, 1966, pp. 584–628.

[8] Garrity, L. V. and Kriege, H. F., "Studies of Accelerated Soundness Tests," *Proceedings*, Highway Research Board, Vol. 15, 1935, p. 237.

[9] Walker, S. and Proudley, C. E., "Studies of Sodium and Magnesium Sulfate Soundness Tests," *Proceedings*, American Society for Testing and Materials, Vol. 36, Part 1, 1936, p. 327.

[10] Wuerpel, C. E., "Factors Affecting the Testing of Concrete Aggregate Durability," *Proceedings*, American Society for Testing and Materials, Vol. 38, Part 1, 1938, p. 327.

[11] Walker, S. and Bloem, D. L., "The Problem of Deleterious Particles in Aggregates," *Circular No. 35*, National Sand and Gravel Association, Silver Spring, MD, 1950.

[12] Wuerpel, C. F., "Modified Procedure for Testing Aggregate Soundness by Use of Magnesium Sulfate," *Proceedings*, American Society for Testing and Materials, Vol. 39, 1939, p. 882.

[13] Paul, I., "Magnesium Sulfate Accelerated Soundness Test on Concrete Aggregates," *Proceedings*, Highway Research Board, Vol. 12, 1932, p. 319.

[14] Woolf, D. O., "Improvement in the Uniformity of the Accelerated Soundness Test of Coarse Aggregate," *Bulletin No. 213*, American Society for Testing and Materials, Philadelphia, Jan. 1953, p. 42.

[15] Woolf, D. O., "An Improved Sulfate Soundness Test for Aggregates," *Bulletin No. 213*, American Society for Testing and Materials, Philadelphia, April 1956, p. 77.

[16] ASTM Subcommittee C09.03.05, Methods for Testing and Specifications for Physical Characteristics of Concrete Aggregates, minutes for the years 1990 and 1991, American Society for Testing and Materials, Philadelphia.

[17] Bloem, D. L., "Sulfate Soundness Test Relationship." Report to Subcommittee III-e, ASTM Committee C9, American Society for Testing and Materials, Philadelphia, 23 Sept. 1958.

[18] McCown, V., "The Significance of Sodium Sulfate and Freezing and Thawing Tests on Mineral Aggregates," *Proceedings*, Highway Research Board, Vol. 11, 1931, p. 312.

[19] Hudec, P. P. and Rogers, C. A., "The Influence of Rock Type, Clay Content and Dominant Cation on the Water Sorption Capacity of Carbonate Rocks," Abstract of paper presented

at Geological Society of Canada, Annual Meeting, Edmonton, May 1976.
[20] Woolf, D. O., "Relation Between Sodium Sulfate Soundness Tests and Absorption of Sedimentary Rock," *Public Roads*, Dec. 1927, p. 225.
[21] Cantrill, C. and Campbell, L., "Selected Aggregates for Concrete Pavement Based on Service Records," *Proceedings*, American Society for Testing and Materials, Vol. 39, 1939, p. 937.
[22] Adams, A. and Pratt, H. A., "A Comparison of Absorption and Soundness Tests on Main Sands," *Proceedings*, American Society for Testing and Materials, Vol. 45, 1945, p. 771.
[23] Mather, K., "Relation of Absorption and Sulfate Test Results of Concrete Sands," *Bulletin No. 144*, American Society for Testing and Materials, Philadelphia, Jan. 1947, p. 26.
[24] Bloem, D. L., "Review of Current and Projected Researches," Report to Annual Convention, National Sand and Gravel Association and National Ready Mixed Concrete Association, Chicago, Feb. 1960.
[25] Sweet, H., "Chert as a Deleterious Constituent in Indiana Aggregates," *Proceedings*, Highway Research Board, Vol. 20, 1940, p. 599.
[26] Vollick, C. A. and Skillman, E. I., "Correlation of Sodium Sulfate Soundness of Coarse Aggregate with Durability and Compressive Strength of Air-Entrained Concrete," *Proceedings*, American Society for Testing and Materials, Vol. 52, 1952, p. 1159.
[27] Brink, H. R., "Rapid Freezing and Thawing Test for Aggregate," *Bulletin No. 201*, Highway Research Board, Washington, DC, 1958.
[28] Dolar-Mantuani, L., *Handbook of Concrete Aggregates*, Noyes Publications, Park Ridge, NJ, 1983.
[29] Temper. B. and Haskel, W. E., "The Effect of Clay on the Quality of Concrete Aggregates," California Highways and Public Works, Vol. 34, Nov.–Dec. 1955; *Highway Research Abstracts*, Vol. 26, No. 2, Feb. 1956, p. 30.
[30] Lyse, I., "Tests Indicate Effect of Fine Clay in Concrete," *Engineering News Record*, Vol. 113, 23 Aug. 1934, p. 233.
[31] Dolar-Mantuani, L., "Soundness and Deleterious Substances," *Significance of Tests and Properties of Concrete and Concrete-Making Materials, ASTM STP 169A*, American Society for Testing and Materials, Philadelphia, 1966, pp. 744–761.
[32] Bloem, D. L., "Factors Affecting Freezing-and-Thawing Resistance of Chert Gravel Concrete," *Research Report No. 18*, Highway Research Board, Washington, DC, 1963, pp. 48–60.
[33] Gaynor, R. D., "Investigation of Concrete Sands," *Technical Information Letter No. 266*, National Sand and Gravel Association, Silver Spring, MD, 13 May 1968, p. 24.
[34] MacNaughton, M. F. and Herbich, J. B., "Accidental Air in Concrete," *Proceedings*, American Concrete Institute, Vol. 51, 1955–1956, pp. 273–284.
[35] Goldbeck, A. T., "A Digest of a Report on Effect of Stone Dust on the Properties of Concrete," *Proceedings*, American Society for Testing and Materials, Vol. 29, Part 1, 1929, p. 301.
[36] Hansen, W. C., "The Chemical Reactions," *Significance of Tests and Properties of Concrete and Concrete-Making Materials, ASTM STP 169A*, American Society for Testing and Materials, Philadelphia, 1966, pp. 487–496.

Unit Weight, Specific Gravity, Absorption, and Surface Moisture

Robert Landgren[1]

PREFACE

This chapter describes the background and application of elemental tests for the basic aggregate properties of unit weight, specific gravity, and absorption. Except for the discussion of the validity tests for aggregate absorption, this chapter summarizes the information on the same subjects from the original article by Timms [1] in *ASTM STP 169* and the revised articles by Brink and Timms [2] in *ASTM STP 169A* and by Mullen [3] in *ASTM STP 169B*.

INTRODUCTION

Aggregate in a stockpile is a collection of various sizes of irregularly shaped particles often with rough or convoluted surfaces. Aggregate particles almost always have internal pores that will hold water. Because aggregate particles do not fit together well, spaces called voids are left between them. The aggregate pile probably contains water in the pores of individual particles and on particle surfaces, even in desert conditions.

In order to use aggregates as the major volumetric ingredient of a concrete mixture, five factors should be resolved:

1. The volume of voids, so they may be filled with paste and the aggregate particles coated with a paste film.
2. The volume of solids, to bulk-out the concrete mixture.
3. Aggregate specific gravity, a useful factor in concrete mixture proportioning and control.
4. Absorption, because of its effect upon specific gravity.
5. Aggregate surface moisture, which constitutes part of the mixing water.

Aggregates used in concrete are identified by their physical and chemical properties. Aggregate properties are defined by standard specifications and methods of test regulated by such agencies as The American Society for Testing and Materials, American Association of State Highway and Transportation Officials, and others.

For most aggregates, the physical properties that lend themselves readily to determination are specific gravity, unit weight, absorption, and grading. Grading and unit weights are determined easily for almost all aggregates. For some porous or lightweight aggregates, the specific gravity and absorption may not be readily determined because of surface roughness and high porosity. Fortunately, it is possible to arrive at workable concrete mixture proportions for these aggregates through the use of alternative means, such as "specific gravity factors" [4].

Aggregate physical properties vary with aggregate source, mineralogy, size, grading, shape, and surface texture. For a given aggregate, these values are relatively stable. When these properties change significantly, it will be necessary to reproportion the concrete mixture.

The primary variable in concrete production is the aggregate moisture content. This value usually is not required until after mixture proportions have been established. Corrections for changes in aggregate moisture content are made routinely without changing the basic concrete mixture.

AGGREGATE CONFIGURATION

It is generally recognized that aggregate particles come in various sizes that are described for engineering identification purposes by separation on square-opening sieves. It is also recognized that aggregates having the same grading may vary widely in shape, angularity, and surface texture or roughness. Differently shaped aggregates may be rounded, subangular, or angular.

Aggregate particle surfaces vary from smooth to rough and irregular. Mineralogy and porosity of the source rock as well as natural polishing contribute to the final surface texture. Generally, natural sands or gravels have smooth surfaces. They are abraded and polished by wind, water, or glacial transportation. Crushed or manufactured aggregates generally have rougher surfaces and greater angularity than natural sands and gravels. Crushed, highly porous particles, such as slags or other lightweight aggregates have extremely rough surfaces. Shape and surface texture of aggregate particles have been discussed by Mather [5] using the terms sphericity, roundness, smoothness, and roughness as descriptors. Ozol [6] presents a more detailed discussion of this subject.

Shape and texture affect aggregate mobility, that is, the ease with which aggregate particles move past one another when manipulated. Aggregate mobility affects the energy required for compaction either of concrete or of concrete aggregate when making a standard test for aggregate unit weight. Visual evidence of differences in aggregate mobility between particles of different shapes and texture is

---

[1] Senior engineer, Wiss, Janney, Elstner Associates, Inc., Northbrook, IL; presently, Cody, WY 82414.

shown in Fig. 1, where it is possible to form a pile of unconfined crushed stone several particles deep while unconfined glass marbles can be piled only one particle deep. The mobility of an aggregate and the ratio of aggregate voids volume to solids volume affects the workability of fresh concrete as well as the proportion of coarse aggregate that may be used in a mixture. Goldbeck and Gray [7], in their landmark "b/b_0" procedure for concrete mixture proportioning, applied packing concepts to concrete technology. Their measure of packing density was the dry rodded unit weight of coarse aggregate.

If the maximum size of aggregate and the range of individual sizes of aggregate are considered as elements of aggregate configuration, then both affect unit weight significantly and slightly affect absorption, free or surface moisture, and specific gravity. The last three properties change because small particles have larger surface areas per unit volume, greater exposure of pore openings to the entrance of water, and short pore lengths. Small highly porous particles of lightweight aggregate or slag have less pore volume and higher specific gravities than larger particles of the same aggregate.

Aggregate shape and grading affect packing of particles. It is well known that a compaction of particles containing a range of particle sizes has a voids content smaller than a compaction of one-size particles. For example, a compaction of one-size spheres has a void volume of less than 30% of the total aggregate volume while that percentage may exceed 50% for compacted one-size angular particles. When graded particles with several sizes are compacted, small particles fit into the voids between larger particles. Classic studies in this area are those by Fuller and Thompson [8] and by Weymouth [9].

In concrete, aggregate particle surfaces must be covered with a discrete paste layer. Concrete using small aggregates with large surface areas per unit volume will generally require more paste than comparable concrete utilizing large aggregates with a smaller surface area. Figure 2 shows that the surface area of cubes per unit volume increases as cube size decreases. This change in surface area with size is even more pronounced if the aggregate particles are flat and elongated.

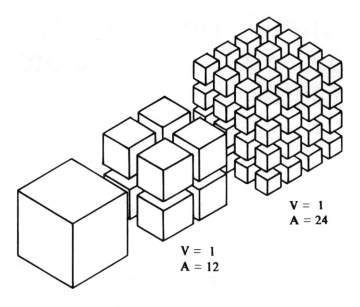

FIG. 2—Surface area versus particle size at equal volumes.

UNIT WEIGHT AND VOIDS

Unit weights and measures of bulk products, such as grain or aggregates, have been with us almost since recorded history [10]. Bulk product particles do not fit together perfectly with the result that voids are left between the particles. Voids are affected by the shape of the particles, the size distribution of the particles, compaction effort, and the orientation or packing arrangement of the particles as discussed previously. For all aggregate packings, unit weight and voids are complimentary; when one increases, the other decreases.

Aggregate unit weight is defined by Brink and Timms [2] as "the weight of a unit volume of representative particles." Methods to determine unit weight and void content are given in ASTM Test Method for Unit Weight and Voids

FIG. 1—Aggregate mobility affected by shape and surface texture.

in Aggregate (C 29). That standard defines void content as a percentage relating the ratio of void volume to the volume of a unit weight measuring container just filled with the granular material.

ASTM C 29 provides for determination of unit weights of aggregates with a maximum size of 6 in. (150 mm) and smaller. The volume of the required measuring container decreases with aggregate size, with the minimum measure size being 3½ ft³ for a 6 in. maximum size coarse aggregate, ½ ft³ for a 1½-in. (40-mm) maximum size coarse aggregate, and 1/10 ft³ for fine aggregate.

ASTM C 29 also provides for compacted and loose unit weight determinations. Compaction usually is accomplished by rodding aggregates that are 1½ in. (40 mm) maximum size and smaller. Aggregates larger than that cannot be compacted effectively by rodding, hence "jigging" compaction of these aggregates is required. Weak, friable lightweight aggregate may be broken by rodding, hence unit weight measures are simultaneously filled and the aggregates compacted by gently shoveling the material into a standard measure to obtain a "dry-loose unit weight." Whether the unit weight is compacted or loose, the measure is filled in three layers of aggregate dried to essentially constant weight at 110 ± 5°C (230 ± 9°F). The first two layers are leveled with the fingers and the third layer placed to overflowing. The top surface is leveled using a straightedge or the fingers in such a way as to balance projections and depressions in the upper surface.

The unit weight, either "dry rodded" or "loose" is directly useful in proportioning concrete mixtures where the b/b_0 method [7] or its variations [11,12] are used. For a given source of normal-weight aggregate, changes in unit weight indicate changes in angularity, grading, or both. Unit weight changes in lightweight aggregate may signal changes in aggregate specific gravity, hence are particularly useful in showing the need for possible corrections to lightweight aggregate concrete mixtures. Unit weight measurements are also essential for calculations of grout quantities required for preplaced aggregate concrete where work must be completely underground, underwater, or obscured from visual control of grout level [3].

Aggregates are classified in terms of unit weight into categories such as nonstructural or insulating lightweight, structural lightweight, air cooled slag, normal weight, and heavyweight. Table 1 contains such a classification.

Aggregate void contents are calculated with equations given in ASTM C 29. For aggregate of a given specific gravity, voids will vary inversely as the unit weight. For a particular coarse aggregate, concrete unit weights tend to be greatest when grout or mortar has a higher specific gravity than the aggregate, especially if the b/b_0 value is less than 1.

Some rules of thumb relate aggregate shape, grading, and compaction as follows:

1. Rounded particles pack more closely and have fewer voids than angular particles.
2. Graded aggregates pack more densely than one-sized aggregates because small particles fill the voids between larger particles.

TABLE 1—Unit Weight Classification For Aggregates.

Aggregate Classification	Unit Weight Range, lb/ft³ (kg/m³)	Compaction Mode	Unit Weight Source
Insulating	6 to 12 (96 to 196)	dry loose	ASTM C 332[a]
Lightweight for masonry	55 to 70 (880 to 1120)	dry loose	ASTM C 331[a]
Lightweight for concrete	55 to 70 (880 to 1120)	dry loose	ASTM C 330[a]
Air cooled slag	>70 (>1120)	compacted	ASTM C 33[a]
Normal weight	75 to 110 (1200 to 1760)	compacted	PCA[b], ASTM C 29[c]
Heavyweight	110 to 290 (1760 to 4640)	compacted	PCA[b]

[a] ASTM Specification for Lightweight Aggregates for Insulating Concrete (C 332); ASTM Specification for Lightweight Aggregates for Concrete Masonry Units (C 331); ASTM Specification for Lightweight Aggregates for Structural Concrete (C 330); ASTM Specification for Concrete Aggregates (C 33), and ASTM C 29.
[b] *Design and Control of Concrete Mixtures*, Portland Cement Association, Skokie, IL, July 1968.
[c] Inferred from balance capacity required by ASTM C 29.

3. Voids decrease with compaction effort that brings about the most favorable particle orientation.

Unit weight procedures described here are laboratory methods determining aggregate properties useful in concrete proportioning and control. Be careful when applying laboratory dry unit weights to aggregate volumes elsewhere, for example, when calculating volumes of aggregate stockpiles. Aggregate stockpiles are usually wet and not uniformly compacted. Dampness causes masses of aggregate to "bulk" or expand to a volume larger than that which the same aggregate would have if it were in a dry condition. The surface tension of water in the damp aggregate binds individual particles together, making it more difficult to compact or consolidate the aggregate than it would if the aggregate particles were dry.

AGGREGATE WATER ABSORPTION

Absorption is the process by which the pores inside aggregate particles imbibe water. The term "absorption" is also used to describe the percent by weight of water absorbed inside the pores of an initially dry aggregate after a 24-h period of soaking in water. Absorption test procedures are given in ASTM Test Method for Specific Gravity and Absorption of Coarse Aggregate (C 127) or ASTM Test Method for Specific Gravity and Absorption of Fine Aggregate (C 128). Because ASTM absorption values are determined after a fixed 24-h period of water immersion, absorption may be regarded as an aggregate property that is a function of aggregate porosity and pore size. Absorption and other factors that are a function of aggregate porosity are critical in determining freeze-thaw durability of an aggregate and of concrete made with the aggregate [13].

The weight of water absorbed by the aggregate is the difference between the weight of the oven-dry aggregate and the "saturated but surface dry" (SSD) weight of the aggregate after it has been soaked in water for 24 h. ASTM C 127 requires that saturated coarse aggregate surface dryness is accomplished by wiping the aggregate particles with an absorbent cloth. In the ASTM C 128 method, water-saturated fine aggregate is reduced to surface dryness by air drying the aggregate until interparticle cohesion produced by surface moisture is lost.

Figure 3 shows factors differentiating three aggregate moisture conditions: oven-dry, SSD, and wet. ASTM absorption testing involves oven-dry and SSD conditions only. The primary difference between these two moisture states is a pore-filling combination of water and entrapped air. The moisture condition of aggregates in field stockpiles usually brackets that of the SSD state. Most stockpiled aggregates are wet; a few are "air-dry" and contain less pore moisture than the SSD condition.

The weight of absorbed water in the pores of an aggregate will increase with time of immersion of the initially dry aggregate in water. Figure 4 shows the absorption rate of a representative granitic material. During the initial phases of absorption, the inrush of water into aggregate pores compresses air originally occupying the pore space, forcing air bubbles out from the larger pore channels. These processes occur quickly, and initial absorption rates are extremely high. Eventually, air trapped deep inside aggregate particles can only be displaced by dissolving that air in pore water and transporting it out of the aggregate by slow diffusion through the pore water. This accounts for the rapid initial absorption of pore water and the slow, but measurable absorption long after the dry aggregate is immersed in water.

In normal concrete operations, the amount of water absorbed in the pores of stockpiled aggregates might approximate the 24-h absorption determined by standard ASTM methods for measuring absorption. However, it is probable that most active aggregate stockpiles contain a significantly higher percentage of absorbed water than that determined by the ASTM procedure because the stockpiled aggregate usually has been kept wet for long periods of time before it is incorporated in the concrete mix.

Alternative methods of measuring concrete aggregate absorption that have been proposed will be discussed later.

AGGREGATE SPECIFIC GRAVITY

The terms specific gravity and density describe the relationship between the weight and volume of a substance. Because ASTM C 127 and C 128 use the term "specific gravity" to express the ratio of the weight of an aggregate to the weight of an equal volume of water, specific gravity will be used here.

The top illustrations of Fig. 5 are a schematic of the "weight in air and weight in water" procedure used by ASTM C 127 to determine specific gravities of coarse aggregate. The difference between these two weights is the buoyancy afforded the aggregate by water equal in volume to that of the aggregate. Figure 5 also shows the pycnometer (fixed volume) method required by ASTM C 128 to measure specific gravity of fine aggregate. The difference between (a) the weight of the aggregate plus the pycnometer filled to the mark only with water and (b) the weight of the pycnometer containing aggregate and filled to the mark with water is determined. That difference is the weight of water displaced by the aggregate. Both

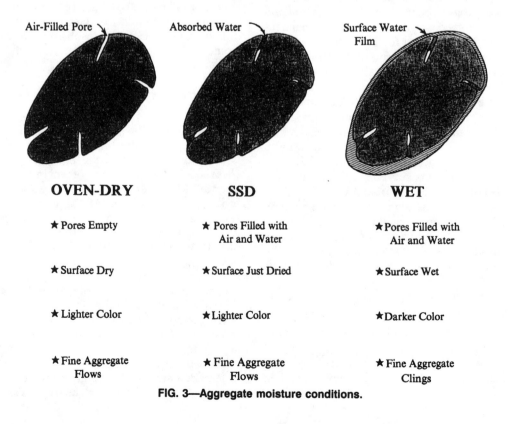

FIG. 3—Aggregate moisture conditions.

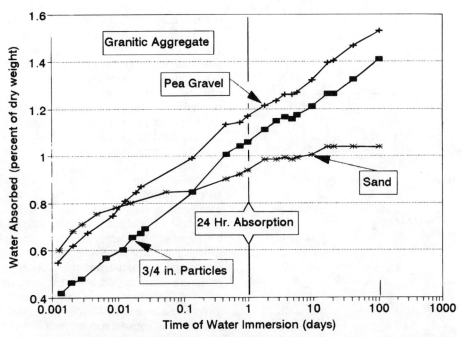

FIG. 4—Semi-log, long-term water absorption as water replaces entrapped air in pores.

methods yield exactly the same information, that is, the weight of water displaced by the concrete aggregate. ASTM specific gravities must be determined on initially oven-dry aggregates that have been immersed in water for 24 h.

The presence of water-filled pores inside aggregate particles (Fig. 3) complicates specific gravity determination. ASTM C 127 and C 128 define three types of specific gravity values. These specific gravity values have the following significance:

1. "Apparent specific gravity" is the ratio of the oven-dry weight of an aggregate to the weight of an equal volume of water. The aggregate volume represented by the apparent specific gravity is the volume of aggregate solids plus the volume of any air entrapped in aggregate pores after 24 h of water saturation. The apparent specific gravity can be determined absolutely and is not affected by inaccuracies in measurement of aggregate absorption.
2. "Aggregate bulk specific gravity" is the ratio of the oven-dry weight of the aggregate to the weight of a volume of water equal to the "bulk volume" of the aggregate particles. Bulk volume may be described as the volume of aggregate solids plus the volume of water absorbed by the aggregate pores plus the volume of air entrapped in the aggregate pores. Inaccuracies in the measurement of aggregate absorption will affect the value of the calculated aggregate bulk specific gravity.
3. "Aggregate bulk specific gravity (saturated-surface dry basis)" is the ratio of the SSD weight of the aggregate to the weight of a volume of water equal to the "bulk volume" of the aggregate, as described previously. Inaccuracies in the measurement of aggregate absorption will affect the value of the calculated aggregate bulk specific gravity (SSD basis).

Aggregate specific gravity values measured after 24 h of water absorption are significant properties of the aggregate. Aggregates with moisture contents different than that found by 24 h of absorption will have different specific gravities. Equations relating specific gravities with absorption are given in Appendix to ASTM C 127 and C 128.

AGGREGATE SURFACE MOISTURE

Aggregate surface moisture may be defined as all moisture in the aggregate except that absorbed inside the pores of individual aggregate particles. Aggregate surface moisture is a critical factor in concrete operations. Surface moisture on batched concrete aggregates is part of the water in the concrete mixture. Aggregate surface moisture compensation is necessary for proper mixture control. Concrete fine aggregate generally contains the greatest quantity of surface moisture, since the voids between particles are small and capable of retaining large percentages of surface moisture.

In concrete plants, sand surface moisture is usually monitored by moisture meters installed on batching equipment. Proper placement and maintenance of moisture meter equipment can permit accurate, continuous measurement of aggregate surface moisture, provided other plant operations, such as bin-filling and stockpiling are done properly. One critical feature in plant management is to utilize sand stocks that have been inactive long enough that most excess water has drained from the pile. Concrete mix water control may suffer if sand stockpiles are too wet or are nonuniformly wet.

Moisture meter performance should be checked routinely. ASTM standard methods of doing this are the ASTM

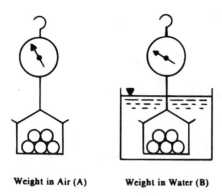

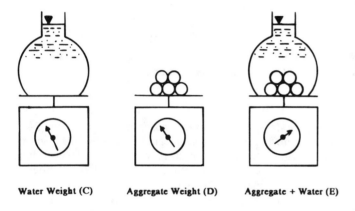

FIG. 5—Specific gravity test alternatives.

Test Method for Total Moisture Content of Aggregate by Drying (C 566) and the less widely used ASTM Test Method for Surface Moisture in Fine Aggregate (C 70).

ASTM C 566 covers the determination of total evaporable water in an aggregate by forced drying using either a hot plate, a hot plate together with alcohol additions to the aggregate, or a microwave oven. The last two methods accelerate drying but can be dangerous; observe the precautions given in the standard if contemplating their use. The weight of surface moisture in the aggregate sample is determined as the total weight of evaporated water minus the calculated weight of water absorbed in the aggregate pores. Utilize the percentage of water actually absorbed in the pores of aggregate from typical plant stockpiles in that calculation. As discussed previously, aggregate from a stockpile that has been wet a long time may contain significantly more absorbed water than the same aggregate that has been soaked in water for 24 h and then tested for absorption by the methods of ASTM C 128 or C 127.

The water-immersion test method of ASTM C 70 determines aggregate surface moisture quickly and with reasonable accuracy provided the actual bulk specific gravity (SSD basis) of typical stockpiled material is used in the calculation. Stockpiled aggregate may have a bulk specific gravity that is different than that determined for dry aggregate immersed in water only 24 h.

Expedient methods not accepted by ASTM have been utilized to determine sand surface moisture. One such method is the proprietary "Speedy Moisture Test" in which gas pressure generated by the reaction between acetylene crystals and the surface moisture of a weighed quantity of wet aggregate is determined and related to surface moisture percentage.

Another non-standard meter calibration procedure is an elaboration of ASTM C 566. In the "frying pan moisture" test, a weighed sample of wet sand in a pan on a hot plate is lightly heated and stirred continuously until incipient color change and decreasing interparticle cohesion signal approaching surface dryness and prompt removing the pan from the burner. The slower surface moisture evaporation from the cooling aggregate sample permits determination of the precise time at which surface moisture appears to be lost and the sample of aggregate approximating surface dryness can be weighed. At this point, the percentage of surface moisture in terms of dry aggregate can be calculated on the basis of an assumed aggregate absorption. Alternatively, the sample can be further dried to constant weight, at which time the dry aggregate can be weighed and the approximate water absorption percentage of the aggregate calculated. A patient, proficient technician can utilize the methods just described to determine the weight of an aggregate sample in the wet, surface dry, and dry condition with an accuracy sufficient for the control of concrete batching.

Tests of moist aggregate displacement in water have been used successfully to adjust directly the batching of aggregate and mix water when changes occur in the total moisture content of aggregates [14–16]. These procedures work well because displacement methods accurately measure total volume of wet aggregate. Consequently, other factors remaining the same, they can detect changes in aggregate volume caused by changes in aggregate surface moisture and permit adjustments to these changes to be calculated.

THE VALIDITY OF ABSORPTION MEASUREMENTS

Several alternative procedures have been proposed to determine surface-dryness of an aggregate and from that the aggregate absorption [16–18]. Two of these alternatives [17,18] may be significant. Two other procedures, which are volume change methods, have been proposed both as procedures for concrete mixture control and for independently determining aggregate absorption [15,16]. These two methods evidently provide adequate concrete mixture control, but their validity in determining aggregate absorption is highly questionable.

Black's [15] absorption test procedure is similar to that proposed by Saxer [16]; both measured volume changes occurring in a water-aggregate suspension from the time of immersion of a dry aggregate through 24 h of immersion. Saxer's method was more refined because he utilized absorption rate linear extrapolation to calculate the considerable amount of water absorbed by the aggregate in the interval between the time when the aggregate was immersed and the first volumetric reading could be taken. Even with such sophisticated procedures, when compared with ASTM absorption methods, Saxer's method underestimated absorption percentages. Black ignored such early absorption, presumably causing even greater differences between absorption by his method and those of the ASTM procedures.

Black and Saxer based their absorption methods on the assumption that the displacement volume of a dry aggregate could be measured or calculated from measurements made quickly after water is introduced into a pycnometer containing the dry aggregate. If it is possible to measure the original bulk volume of the dry aggregate, decreases in bulk volume of the aggregate due to absorption of water inside the aggregate pores can be determined easily and accurately. Then the absorbed water content of the aggregate can be calculated for a particular time after aggregate immersion. While such an approach is theoretically valid, it probably will not work in actual practice.

The problem with the Black-Saxer approach is that the initial rate of absorption of dry aggregate is so great that it seems virtually impossible to measure by pycnometer the volume of the dry aggregate at the precise instant when water has contacted the aggregate surfaces but no water has penetrated into the aggregate pores.

The difficulty in attaining this objective is indicated by the data of Fig. 4, which was obtained by measuring aggregate displacement volume changes in a pycnometer. In those tests, baseline 24-h absorptions were determined by ASTM C 127 and C 128. Absorbed water contents of the aggregate at other times were calculated with the assumption that changes in displacement volume of the aggregate from that determined after 24 h of aggregates immersion were caused by changes in the absorbed water content of the aggregates. The operator performing those tests assembled the pycnometer, filled it with water, removed entrapped air bubbles from the system and weighed the pycnometer in slightly less than 2 min. The absorption measured after this short immersion time was almost half that measured after 24 h of immersion. This data shows that the initial rate of aggregate water absorption is extremely high. It also suggests that attempts to measure aggregate absorption by immersion procedures similar to those of Black and Saxer will not succeed unless it is possible to measure dry aggregate displacement volumes within milliseconds after aggregate surfaces come in contact with water. Such methods are not now available.

Saxer inferred and Black concluded that determining aggregate absorption by displacement methods gave valid absorptions because the absorptions and specific gravities obtained by these methods were capable of maintaining good control of concrete mixtures monitored by these methods. This assumption is not correct because of the singular relationship that exists between bulk specific gravity and absorption values.

As stated earlier, calculated bulk specific gravities are a function of absorption. If an incorrect absorption were used in the specific gravity calculation from data obtained by an actual specific gravity test, that absorption and specific gravity could still be used to calculate the absolute volumes of aggregate and net mix water for a concrete mixture. Both calculated absolute volumes would be incorrect. However, the sum of those incorrect absolute volumes would equal the true absolute volume of the aggregate plus mix water. Stated another way, if absorption measurements produced an incorrect absorption value that was too large and that absorption value were utilized to calculate the aggregate bulk specific gravity, the absolute volume of aggregate calculated with that specific gravity would be slightly higher and the calculated volume of mix water would be slightly lower than the real values. For both the aggregate and mix water, the volumetric error would be the same and would equal the calculated volume of absorbed water minus the actual volume of absorbed water in the aggregate pores.

As indicated previously, Black used water-displacement methods to test for aggregate absorption and specific gravity and applied those values for purposes of concrete control. His success with this method was probably due to consistent, frequent testing of mixture components. While there is a question about the accuracy of individual test values, Black's method probably did maintain consistent concrete quality. That is a basic aim of concrete quality control. However, it should be emphasized that water-cement ratios based on defective aggregate absorption data will be incorrect. That may be the greatest practical defect of Black's method.

Two other unique procedures to determine aggregate absorption have been proposed. One of them [4,17] purged surface moisture from initially wet aggregate and "surface-dried" it by centrifuging the aggregate for a fixed period of time. This method, originally developed for lightweight aggregate, gave normal-weight fine aggregate absorption percentages approximately 1¼% higher than ASTM C 128 values. Little data is available to establish the validity of this test method.

Hughes and Familli [18] forced water-saturated air through a compaction of aggregate particles. Surface dryness was defined as a specific break in the curve relating time of drying and aggregate weight. Their automatic method gave coarse aggregate absorption values approximately 0.2 to 0.3% lower than ASTM C 127 values. Some rather convincing arguments are advanced concerning the validity of this test method.

The possibility that the ASTM absorption measurement procedures are incorrect must be addressed. Intuitively, the ASTM C 127 and C 128 approaches to measure aggregate absorption appear correct; they monitor real manifestations of wet aggregate surface moisture and arrive at a point of aggregate surface-dryness when those manifestations disappear. In the case of the coarse aggregate (ASTM C 127), surface moisture is blotted with a towel. In the case of a sand (ASTM C 128), surface moisture is removed by air drying until aggregate particles lose the cohesion afforded by the surface moisture. Furthermore, at or near the ASTM points of surface dryness the aggregates begin to change appearance and color, presumably because one is no longer looking at the particles through a film of surface water.

Intuition aside, there is little experimental evidence to confirm or discount the precision of the ASTM absorption measurement. One study of the subject [19], with a restricted scope and limited accuracy, suggests that ASTM absorption methods are moderately accurate. Hughes and Familli [18] question ASTM and similar British standard methods because the absorption of a graded aggregate did not match the absorption calculated for that grading from the absorptions of individual sizes of that graded aggregate. That observation, and the surface-drying method proposed by those authors to provide such a correlation, makes their approach worthy of more attention than it apparently has received to date.

It is obvious that certain aggregates must respond to ASTM aggregate surface-drying methods differently than most of our concrete aggregates. For example, the pores of a normal weight coarse aggregate are so fine they will not lose absorbed water to a towel during surface drying. Conversely, a few lightweight aggregates [19] have such large interconnected pores that toweling the aggregate surfaces can remove some of the internal pore water. As to the ASTM C 128 test for fine aggregate absorption, concrete sands have different surface textures ranging from smooth, rounded river-sands to irregular, crushed "manufactured sands." Particle mobility, illustrated for coarse materials in Fig. 1, must have similar variable effects upon the collapse of near-surface dried sands once the sand cone is raised from the compacted sand during the test for surface dryness. Manufactured sands, which hold their shape better than rounded sands when compacted, may well have lower measured absorptions simply because of their lack of particle mobility.

SUMMARY

The accuracy of ASTM absorption tests is uncertain. However, present ASTM absorption/specific gravity procedures have been utilized for a very long time; this author can attest to the fact that they have been virtually unchanged for the last 45 years. An enormous database of information derived from these ASTM tests has been amassed upon aggregate absorption and upon concrete water-cement ratio, the values of which depend upon absorption measurements. Consequently, unless past or future research provides realistic alternatives to the methods of ASTM C 127 and C 128, it would be best to stay with these old tried and, probably nearly, true ASTM methods.

REFERENCES

[1] Timms, A. G., "Weight, Density, Absorption, and Surface Moisture," *Significance of Tests and Properties of Concrete and Concrete-Making Materials*, ASTM STP 169, American Society for Testing and Materials, Philadelphia, 1956, p. 297.

[2] Brink, R. H. and Timms, A. G., "Weight, Density, Absorption, and Surface Moisture," *Significance of Tests and Properties of Concrete and Concrete-Making Materials*, ASTM STP 169A, American Society for Testing and Materials, Philadelphia, 1966, p. 432.

[3] Mullen, W. G., "Weight, Density, Absorption, and Surface Moisture," *Significance of Tests and Properties of Concrete and Concrete-Making Materials*," ASTM STP 169B, American Society for Testing and Materials, Philadelphia, 1978, p. 629.

[4] "Standard Practice for Selecting Proportions for Structural Lightweight Concrete," ACI 211.2-91, *ACI Manual of Concrete Practice*, Part 1, American Concrete Institute, Detroit, MI, 1992.

[5] Mather, B., "Shape, Surface Texture, and Coatings," *Significance of Tests and Properties of Concrete and Concrete-Making Materials*, ASTM STP 169A, American Society for Testing and Materials, Philadelphia, 1966, p. 415.

[6] Ozol, M. A., "Shape, Surface Texture, Surface Area, and Coatings," *Significance of Tests and Properties of Concrete and Concrete-Making Materials*, ASTM STP 169B, American Society for Testing and Materials, Philadelphia, 1978, p. 584.

[7] Goldbeck, A. T. and Gray, J. E., "A Method of Proportioning Concrete for Strength, Workability, and Durability," *Engineering Bulletin No. 11*, National Crushed Stone Association, 1942.

[8] Fuller, W. B. and Thompson, S. E., "The Laws of Proportioning Concrete," *Transactions*, American Society of Civil Engineers, New York, Dec. 1907, p. 67.

[9] Weymouth, C. A. G., "Effect of Particle Interference in Mortars and Concrete," *Rock Products*, 25 Feb. 1933, p. 26.

[10] *Bible*, King James Version, Luke 6:38.

[11] Design and Control of Concrete Mixtures, 13th ed., Portland Cement Association, Skokie, IL, 1990.

[12] Standard Practice for Selecting Proportions for Normal, Heavyweight and Mass Concrete, ACI 211.1-91, *ACI Manual of Concrete Practice*, Part 1, American Concrete Institute, Detroit, MI, 1992.

[13] Dolch, W. L., "Porosity," *Significance of Tests and Properties of Concrete and Concrete-Making Materials*, ASTM STP 169B, American Society for Testing and Materials, Philadelphia, 1978, p. 646.

[14] Landgren, R., Hanson, J. A., and Pfeifer, D. W., "An Improved Procedure for Proportioning Mixes of Structural Lightweight Concrete," *Journal*, PCA Research and Development Laboratories, Portland Cement Association, Skokie, IL, May 1965, p. 47.

[15] Black, R. W., "The Determination of Specific Gravity Using the Siphon-Can Method," *Cement, Concrete, and Aggregates*, American Society for Testing and Materials, Summer 1986, p. 46.

[16] Saxer, E. L., "A Direct Method of Determining Absorption and Specific Gravity of Aggregates," *Rock Products*, May 1956, p. 78.

[17] Wills, M. H., "Lightweight Aggregate Particle Shape Effect on Structural Concrete," *Journal*, American Concrete Institute, Detroit, MI, March 1974, p. 134.

[18] Hughes, B. P. and Famili, H., "Absorption of Concrete Aggregates, Part 1—Absorption of Concrete Aggregates, Part 2—Saturated Air Techniques for Determining the Absorption of Aggregates," RR/C8, Birmingham University, Great Britain, 1971.

[19] Landgren, R., "Determining the Water Absorption of Coarse Lightweight Aggregates for Concrete," *Proceedings*, American Society for Testing and Materials, Philadelphia, 1964, p. 846.

The Pore System of Coarse Aggregates

Douglas Winslow[1]

PREFACE

A chapter on the pore systems of aggregates has appeared in each of the predecessors to this volume. The first, *ASTM STP 169*, contained a chapter entitled "Porosity and Absorption" by D. W. Lewis and W. L. Dolch. The following revisions, *ASTM STP 169A* and *ASTM STP 169B*, had chapters titled "Porosity" by W. L. Dolch. The continued presence of a chapter devoted solely to the void space in aggregates is indicative of the continued importance of the topic.

INTRODUCTION

The pore, or void, space within the individual particles of an aggregate is probably their most important single feature. Virtually all aggregates have pore space that can affect their incorporation in concrete and its subsequent performance. This makes the aggregate's pore space of universal interest and importance. Before discussing specific aspects of the pore space, it is appropriate to consider some general points concerning this crucial aggregate feature.

In general, the pore space in an object can be divided into two parts: that which is accessible from the surface, and that which is completely isolated by surrounding solid. It is likely that the vast majority of the pore space in an aggregate is of the former type. Further, while the isolated pore volume in an aggregate is pertinent to the aggregate's thermal properties, elastic modulii, etc., it will not play a part in the main topic of interest in this chapter, that is, durability. Therefore, the isolated pore space will not be considered further. Or, to put it another way, this pore space will be considered as a part of the solid volume, and the term "pore volume" will be restricted here to that which is accessible from the exterior.

The pore space is a potential source of difficulty when the aggregate is used. Therefore, it is desirable to use aggregates in which this space is minimized. However, economic issues frequently act to force one to use aggregates with appreciable volumes of pore space. Thus, it is important to know how to assess quantitatively the pore space. And, to know how much pore space can be tolerated.

The two principal, quantitative assessments of pore space are its volume and its "size." The measurement of the volume of a pore space is usually reasonably unambiguous. The size is more complicated. This term refers to some linear dimensionality of the cross section of the pore space at some point. The cross section is almost always so irregular that one cannot assign a simple, rigorous geometric definition to the idea of size. Therefore, size is generally defined by the experimental technique used to determine it, and by how the analysis of that technique is modeled. Nevertheless, the concept of pore size has been found to be useful.

One frequently refers to the "pores" in an aggregate. This may make for simple grammar, but it conveys the unfortunate implication that an aggregate contains a number of discrete pore spaces that can be identified, counted, etc. It is likely that the vast majority of the pore space in most aggregates forms a continuous, albeit tortuous, system having many points of egress at the aggregate's surface. That is, the pore space in a typical piece of aggregate is probably composed of only a single pore that possesses a varying size from place to place. Thus, for example, when one speaks of "pores having a size of X," it is more appropriate to think of the volume of that portion of the continuous pore system where the cross section is characterized by a size parameter of X. In this chapter, the discussion will be of the pore space, or system, so as not to further the erroneous implication that a number of discrete "pores" exist in most aggregates.

The pore space will likely have a cross section that varies with position. The size may range from the order of single atoms to as much as eight orders of magnitude larger. The majority of this size range is of importance to concrete technology. This presents one with the problem of quantitatively characterizing a complicated pore space having a wide range of cross-sectional sizes. A complete characterization would include both the volume and sizes of the pore space. The complexity, range of sizes, and need for rapid and economical test methods have combined to force compromises in characterization. Thus, many of the test methods that will be discussed, and their interpretations, will be more or less limited to a partial characterization of the pore system of aggregates.

BASIC CONCEPTS AND DEFINITIONS

One must distinguish between pore space lying within aggregate particles (intra-particle) and pore space that is

---

[1] School of Civil Engineering, Purdue University, West Lafayette, IN 47907.

between particles (inter-particle). This chapter addresses only the intra-particle pore space lying inside an imaginary, flexible membrane surrounding the particle. It should be noted that ASTM Test Method for Unit Weight and Voids in Aggregate (C 29) does not, in fact, test for the void space "in" the aggregate. It (Eq 4 in ASTM C 29) determines the void space "between" the aggregate particles. The test method has a misleading title.

There may be some ambiguity between pore space immediately adjacent to external surfaces and major depressions or irregularities in the external surface. A functional differentiation might be to consider the pore space of an aggregate particle to be that non-solid volume that will not be filled with cement hydration products when the aggregate is used in portland-cement concrete.

Parameters Related to Pore Volume

If the volume of the pore space is V_p and the volume of the solid part of the particle is V_s, then one can define several useful pore parameters.

$$\text{Porosity} = \frac{V_p}{V_p + V_s} \quad (1)$$

and

$$\text{Void Ratio} = \frac{V_p}{V_s} \quad (2)$$

The porosity and the void ratio are expressions of the volume of pore space associated with the aggregate particles; there is a relationship between the two. Generally, porosity is the more common term in aggregate technology where the pore volume remains essentially constant in service. Void ratios are used more in soil technology where the pore volume may change appreciably.

When water is present in the pore space, its amount may be associated with the pore volume. One can express the amount of water in several ways. If the volume of this water, V_w, has a mass M_w, and M_s is the mass of solid, then

$$\text{Absorption} = \frac{M_w}{M_s} \quad (3)$$

and

$$\text{Degree of Saturation} = \frac{V_w}{V_p} \quad (4)$$

In some other fields, the absorption is called the water or moisture content.

With a porous material, it is possible to define a variety of densities. The details of these are important to an understanding of some of the techniques used to measure the volume of the pore space in an aggregate. Two important densities are

$$\text{True Density} = \frac{M_s}{V_s} \quad (5)$$

and

$$\text{Bulk Density} = \frac{M_s}{V_s + V_p} \quad (6)$$

The true density is also sometimes called the solid density or the skeletal density. It is the only density possessed by nonporous solids. The bulk density, as defined in Eq 6, includes the intra-particle pore volume, and excludes the inter-particle pore space. In some other technical fields, bulk density is defined as including both kinds of pore space, but in concrete aggregates Eq 6 is the usual definition.

In some pycnometric techniques for measuring the solid volume, one uses a liquid to infiltrate the pore space. Usually, one is unsuccessful in completely filling the pore volume, and a portion is left filled with air, V_a. This leads to another density as

$$\text{Apparent Density} = \frac{M_s}{V_s + V_a} \quad (7)$$

Note that, to the extent that one is successful in filling more of the pore volume, and reducing V_a, the apparent density approaches the true density. In fact, when one wishes to determine the true density, one often actually measures an apparent density after going to great lengths to ensure that V_a is minimized.

There is a set of specific gravities that corresponds to these densities. These are defined as the appropriate density divided by the density of water.

Parameters Related to Pore Size

The preceding parameters all have some application in defining the volume of the pore space. They give no information concerning the size of the cross section of that space. Unless the pore space has a simple geometry, there will be no unique, simply defined measure of its size. This is, of course, true of solid volume as well as pore volume. There is, for example, no simple way to describe the size of a piece of aggregate any more than the size of its pore space.

What is done in the face of this difficulty is to let the technique being used to measure the size also define what is meant by the size. This is, again, the same thing that is done with solid volume. For example, the size of an aggregate piece is usually defined by the distance between the wires in the sieve through which it either does, or does not, pass.

The most common pore sizing techniques model the pores as having a circular cross section. Thus, in the vast majority of cases, the pore size is reported as either the radius or diameter of the model cross section with the diameter being the more straightforward expression. When one uses these techniques, one is really finding the diameter of a hypothetical model pore space that would respond to the measurement technique in the same way as does the actual pore space.

There are a number of techniques that allow the measurement of pore size. They will be discussed in a following section on Experimental Methods. Of these, the most used provide one with a series of pore diameters and, associated with each, a pore volume representing that portion of the pore space that has either larger or smaller sizes. Such a series of pairs of pore size and volume probably represent the most complete description of a pore system that is experimentally available today.

The preferred way to display information of this sort is as a cumulative distribution. This distribution has an abscissa (the conventional X-axis) displaying the pore sizes, generally on a logarithmic scale because of the wide range in their magnitudes. The ordinate (the conventional Y-axis) of the distribution shows the pore volume having either larger or smaller sizes; both types of ordinates are in common usage.

The experimental data leading to a distribution are obtained in a cumulative, or integral, fashion. Unfortunately, these data are sometimes differentiated for presentation. In the process, information is lost and distorted, and nothing is gained. Differentiation of these experimental data, in common with the differentiation of all experimental data, introduces, and either greatly magnifies or suppresses local fluctuations. These may appear in the resulting curve out of all proportion to their actual importance. Thus, peak heights in a differential plot may be artificially high or low. Further, a differential presentation makes it impossible to determine the total pore volume involved, or the fraction of the total existing in any size range of interest. For these reasons, differential pore size distributions are not desirable, and should be avoided.

If one wants to avoid the difficulty of ascribing a size to a pore volume, one can use a parameter that is associated with some sort of "average" size of the total space. This, of course, results in a decrease in the information that one has about the pore system in the same way that averages obscure information about most systems.

The most common of these parameters is the surface area of the boundary between the pore space and the solid. This area is properly expressed on some specific basis such as the solid mass, M_s. Alternately, it can be given per unit of solid volume, V_s, pore volume, V_p, or bulk volume, V_b. When this is done, it is called the specific surface area, or simply the specific surface, and has units of area/mass or area/volume as appropriate. Frequently, for simplicity, the specific surface area is simply called the surface area with the specific basis being understood by the units. The general trend between surface area and pore size is that the former increases as the later decreases, as

$$\text{Specific Surface Area} \propto \frac{1}{\text{Pore Size}} \quad (8)$$

The pore size in Eq 8 is an extremely biased average size associated with the pore system. It is overwhelmingly influenced by the smallest-sized portion of the pore space. Therefore, when that space has sizes spanning an order of magnitude or more, the surface area gives information on only the smallest-sized portion of the total pore space. For example, the smallest 10% of the pore volume of most aggregates possesses about 75% of the surface area. Or, viewed the other way round, typically 90% of the pore volume contributes only something like 25% of the surface area.

Another parameter that is sometimes used to describe a pore system is the hydraulic radius. It originated in the analyses of fluid flow, but it can be viewed also as a purely geometric parameter of a pore system. It is defined, at a point in the pore system, as the cross-sectional area divided by the perimeter of the pore space. As such, this is a point parameter of the pore space, and does not require a particular geometric model of the pore space for its definition. In general, the hydraulic radius varies in proportion to the pore size as

$$\text{Hydraulic Radius} \propto \text{Pore Size} \quad (9)$$

For example, the hydraulic radius of a circular cross section is one-quarter of the diameter of the section.

Usually, one does not have area and perimeter data available from point to point. However, one can define an average hydraulic radius for an entire pore system as

$$\text{Average Hydraulic Radius} = \frac{\text{Pore Volume}}{\text{Surface Area}} \quad (10)$$

In Eq 10, it is important that both the pore volume and the surface area be expressed on the same basis. Since one generally determines the pore volume and surface area of an entire pore system, and not at discrete points, it is appropriate to refer to this system-wide parameter as the average hydraulic radius. This parameter is also referred to as the hydraulic mean radius.

Note that this average hydraulic radius, Eq 10, is just the reciprocal of the specific surface area expressed on a unit pore volume basis. This means that the average hydraulic radius can convey no more information than does the surface area when the two parameters are determined with compatible experimental techniques.

Other Pore System Parameters

One can infer something about a combination of pore volume and pore size from the permeability of an aggregate. A general reference to permeability is Ref 1. Consider a piece of aggregate as a plug in a pipe through which a fluid is forced in lamellar flow under a pressure differential, ΔP. The plug has a cross-sectional area, normal to the pipe, of A, and a length of L. The volume of fluid flowing per unit time, Q, through the plug will be directly proportional to the area, A, and the pressure gradient causing the flow, $\Delta P/L$, and inversely proportional to the viscosity of the fluid, η. This is Darcy's law, and the constant that makes these proportionalities an equality is defined as the permeability, K, or

$$Q = K \times [A] \times \left[\frac{\Delta P}{L}\right] \times \left[\frac{1}{\eta}\right] \quad (11)$$

As one reduces either the pore volume or size, one finds a smaller flow rate and associated permeability. There are several expressions of permeability in terms of pore system parameters. One of these is the Kozeny equation

$$K = \frac{\text{Porosity}^3}{K_k \times \text{Surface Area}^2} \quad (12)$$

where K_k is the Kozeny constant, and the surface area is expressed on a unit bulk volume basis. The Kozeny equation demonstrates that a reduced pore volume (reduced porosity) or a smaller pore size (greater surface area), or both, will reduce the permeability.

EXPERIMENTAL METHODS

There are existing ASTM standard test methods that directly measure some of the important aspects of an aggregate's pore system. In some cases, these standard tests have been developed for porous materials other than concrete aggregates. Nevertheless, if they are useful in assessing some other pore system, they should be equally useful for aggregates. In addition, some standard methods can be adapted to yield information on the pore space. Finally, there are pore parameters that cannot be determined with ASTM methods or adaptations. It may be necessary to employ tests from all of these categories to characterize properly an aggregate's pore system.

Techniques Involved with Pore Volume Measurement

The most straightforward way to determine pore volume is to measure the volume of a liquid that fills the pore space. Unfortunately, this technique usually fails to detect all of the pore volume, because the liquid fails to occupy the total pore space. On some occasions, this failure may be because the molecules of the liquid are too large to enter some of the pore space. Much more common causes of incomplete filling are pockets of trapped gas, and a failure to wait long enough for the liquid to flow into all of the pore volume.

The most common ASTM test that makes this sort of measurement is ASTM Test Method for Specific Gravity and Absorption of Coarse Aggregate (C 127). In the absorption portion of this method, one allows water to permeate the pore system for 24 h. If the aggregate is initially oven dried, this test typically results in only 50 to 80% of the pore volume being filled. Thus, this absorption is a poor measure of pore volume. (If the aggregate is tested in the naturally wet state, more of the pore space will be filled. Nevertheless, one remains unsure of the total pore volume.)

A more rigorous method might be based on an adaptation of ASTM Test Methods for Apparent Porosity, Liquid Absorption, Apparent Specific Gravity, and Bulk Density of Refractory Brick and Shapes by Vacuum Pressure (C 830). This technique involves the prior evacuation of the pore space, and water introduction, while the aggregate is under vacuum, followed by a release of the vacuum and the imposition of a modest pressure. This eliminates the problem of trapped gas pockets. The completeness of the pore filling then depends upon how long one is willing to wait. The standard method calls for a period of 1 h. This is probably not long enough for many aggregates, and one should repeat the test with varying soak times until the maximum amount of water has entered the pore system. The imposition of a super-ambient pressure will not significantly speed the infiltration, since the natural capillary pressure causing the inflow is already sufficient.

The preceding technique provides one with a measure of pore volume that must be put on some specific basis to be useful. A simple basis is the aggregate mass as is done when calculating the absorption (ASTM C 127). Another basis is the aggregate volume as is done when finding the porosity (ASTM C 830). This, however, requires an additional measurement of either the bulk volume, V_b, or the solid volume, V_s. Measuring either of these volumes presents additional difficulties.

One can measure the bulk volume directly if one can obtain a regularly shaped piece of aggregate, such as a cylindrical core or a prism. This method has much to recommend it. It is simple, accurate, and inexpensive, and should not be overlooked when it is possible.

Another approach is pycnometric. One displaces a nonwetting liquid with the aggregate and determines the bulk volume. This technique requires that the liquid not enter the pore system, and the common choice is mercury. This technique is complicated in that it requires a special pycnometer that avoids the entrapment of exterior gas bubbles and that allows for the fact that aggregates float in mercury. A suitable system is presented in ASTM Test Method for Bulk Density and Porosity of Granular Refractory Materials by Mercury Displacement (C 493).

Another approach to measuring bulk volume is based on the buoyancy of a porous object immersed in a liquid. ASTM C 127 provides for such a measurement by determining the bulk specific gravity of an aggregate. The result is easily converted to a bulk density, and this to a bulk volume. This method has a great deal to recommend it, with the only real difficulty being the necessity to obtain a sample mass with water inside the pore system but not on the surface. This requires surface drying and introduces some uncertainty. The use of larger aggregate pieces, when possible, will help, since sample volume rises with the cube of size while surface rises only with the square.

An alternate approach to providing a basis for the pore volume is to determine the solid volume. The most direct approach here is, again, pycnometric. When one wants the solid volume, one uses a fluid that does enter the pore space, and finds the volume of that portion of the aggregate not penetrated by the fluid. If the fluid completely penetrates the pore system, the volume not penetrated will be the solid volume. The best choice for the fluid is a gas (not a vapor) that does not react with the solid. Nitrogen and helium are common choices, and there are a number of commercially available gas pycnometers for this type of measurement.

A less desirable choice of fluid is a liquid because it will less easily fill the pore space. One might use a procedure similar to ASTM Test Method for Specific Gravity of Soils (D 854) or ASTM Test Method for Density of Hydraulic Cement (C 188) to obtain a value for the specific gravity of the solid portion of the aggregate. This can, again, be converted readily to a solid density and, hence, a solid volume.

These methods depend for their accuracy on the liquid penetrating all of the pore space. It may be desirable to grind an aggregate to make the pore space more readily accessible to the liquid. Grinding will not destroy a significant portion of the pore space in most aggregates, since the size of this space is typically much smaller than the post-grinding particle size. It is also desirable to use a modest vacuum to remove entrapped air.

If one has gone the route of measuring both bulk and true densities, then a simple relationship exists for the porosity, as

$$\text{Porosity} = 1 - \frac{\text{Bulk Density}}{\text{True Density}} \quad (13)$$

However, one must be aware of the potential errors in determining both of these densities that have been discussed.

All of these procedures leave the experimenter with a variety of ways of determining the pore, solid, and bulk volumes. Many combinations can be used, and there is no "best" way to go about it. One must select techniques that will yield a sufficiently accurate result for the application in question while minimizing one's effort.

Image analysis offers a completely different approach to determining porosity that avoids these difficulties while, unfortunately, introducing its own problems. It is necessary to assume that the pore space is randomly distributed throughout the aggregate. This is generally true of many naturally occurring aggregates, but some samples of sedimentary aggregates may require special attention. Good general discussions on the subject are found in Refs 2 and 3. For image analysis, one prepares an adequately polished, flat surface of the aggregate on which the pore space can be distinguished clearly. The technique is an adaptation of ASTM Practice for Microscopical Determination of Air-Void Content and Parameters of the Air-Void System in Hardened Concrete (C 457).

One can use either a point, lineal, or areal analysis of the surface to obtain the porosity. The choice depends solely on the preference of the experimenter. In the case of point counting, one examines a sufficiently large number of total points, P_t, on the surface, and notes the number of points falling in the pore space, P_p. A lineal analysis involves traversing a sufficiently long total line length, L_t, and also measuring the length of that portion of the line passing through the pore space, L_p. Finally, in an areal analysis, one measures a sufficiently large total area, A_t, and also the area occupied by the pore space, A_p.

The calculation of porosity, $V_p/(V_p + V_s)$, stems from the following relationships with the point, lineal and areal fractions, as

$$\frac{V_p}{V_p + V_s} = \frac{V_p}{V_t} = \frac{P_p}{P_t} = \frac{L_p}{L_t} = \frac{A_p}{A_t} \quad (14)$$

where

V_t = total sample volume,
P_p = number of points falling in the pore space,
P_t = total number of points examined,
L_p = length of line falling in the pore space,
L_t = total length of line examined,
A_p = area of the pore space, and
A_t = total area examined.

The old problem of tedium with image analysis has been largely overcome by the development of computer systems that automatically determine any of the three fractions given in Eq 14. The difficulty with this method is that all of the pore space must be resolved in the imaging system. This requires careful preparation of the analyzed surface. Also, many aggregates have a pore size smaller than the resolution of a light microscope, and one may need to use a scanning electron microscope, or an electron microscope, or a combination of microscopes, to obtain adequate images.

Techniques Involved with Pore Size Measurement

One of the earliest techniques used to determine the size of the cross section of a pore space involves the condensation of a vapor in that space. A good general reference to this, and other pore sizing techniques, is Ref 4. The method is based on a development by Kelvin [5] in which he demonstrated that a vapor will condense in a pore space at a partial pressure (less than the saturation pressure, P_O) associated with the size of that space. The controlling equation, for a circular pore cross section, is

$$P_p = e^{-\left[\frac{4\gamma \cos\Theta M}{RTd}\right]} \quad (15)$$

where

P_p = partial pressure of vapor,
γ = surface tension of condensed vapor,
Θ = contact angle of condensed vapor on solid,
M = molar volume of condensed vapor,
R = gas constant,
T = absolute temperature at which condensation occurs, and
d = diameter of pore space.

All of the factors in the exponent except d are functions of the vapor and of experimental conditions. If they are grouped into a constant, K, and Eq 15 is rearranged, the relationship between the partial pressure and pore size is more readily seen as

$$d = \frac{K}{\ln P_p} \quad (16)$$

The experiment consists of exposing a sample to a succession of partial pressures, and measuring the amount of vapor that condenses at each pressure. It is described in ASTM Test Method for Determination of Nitrogen Adsorption and Desorption Isotherms of Catalysts by Static Volumetric Measurements (D 4222). The amount of condensed vapor is converted to the pore volume, and the partial pressure is converted to the corresponding pore size. The procedure, when nitrogen is the condensing vapor, is described in ASTM Practice for Calculation of Pore Size Distributions of Catalysts from Nitrogen Desorption Isotherms (D 4641). The standard practice can be easily adapted to the use of other vapors. There are a number of commercially available instruments that will perform this experiment and reduce the data to a pore size distribution automatically.

The major disadvantage of this technique is that it is limited to pore sizes smaller than about 0.1 μm (1000 Å) when using most conventional vapors. For larger sizes, the condensing pressure, P_p, is nearly equal to P_O. It is experimentally difficult to control pressures in the near vicinity of P_O and the method loses its utility. Because of its limited size range, this technique is not used much with aggregates. However, it may be useful for materials having a pore volume with predominantly small sizes.

The most common technique for determining the pore size of aggregates is based on the forced intrusion of a liquid that does not voluntarily enter the pore space. Any nonwetting liquid can be used, but the common one is mercury. The pressure required to force the intrusion is inversely proportional to the size of the pore space being intruded. The Washburn equation [6] defines the pressure-size relationship, for circular cross sections, as

$$P = \frac{-4\gamma\cos\Theta}{d} \quad (17)$$

where

P = pressure causing intrusion,
γ = surface tension of mercury,
Θ = contact angle of mercury on solid, and
d = diameter of entrance to pore space.

Since γ and Θ are constants for a given experiment, one can recast Eq 17, in a form similar to Eq 16, as

$$d = \frac{K}{P} \quad (18)$$

The crucial difference between Eqs 16 and 18 is that, in Eq 18, the pore size is not such a sensitive function of the pressure. Therefore, the method is applicable to a much broader range of pore sizes. Indeed, one can measure sizes over whatever range one can generate, and accurately control, pressures. Typically, this translates to sizes between about 2 nm (20 Å) and 500 µm (0.5 mm). This enormous range-in-size advantage is the principal reason that intrusion is the most used method for determining pore size. Another advantage of intrusion over vapor condensation is that the experiment is much faster.

Briefly, the experiment consists of surrounding an aggregate sample with mercury and then forcing it into the pore space by pressuring. The pressure is converted to a pore size. And, the volume of mercury intruded is converted to the pore volume. The method is described in ASTM Test Method for Determination of Pore Volume and Pore Volume Distribution of Soil and Rock by Mercury Intrusion (D 4404). Instruments for performing the experiment are available from several manufacturers. These provide computer automation of the experiment and the reduction of the data to a pore-size distribution.

Image analysis also has techniques for determining a pore-size distribution. These, again, require that one prepare a suitable plane surface passing through the aggregate, and that one be able to obtain an image of the entire pore system. They involve measuring the length of a sufficient number of chords intersecting the pore space so that one can find the chord-length distribution. One then uses statistical procedures to determine the pore-size distribution of a pore system that would have such an intersecting chord-length distribution. The procedure requires considerable computation. However, modern computer-based systems have made this approach much more palatable. Further, this is perhaps the only technique that can measure truly large pore sizes (greater than 1 mm). Also, it is faster than the vapor condensation approach, and does not entail the use of high pressures common in mercury intrusion.

The concept of specific surface area provides an extremely biased estimate of the "average" pore size. Still, it gives some information on pore size. By far the most used technique for measuring the surface area is by measuring the amount of vapor needed to cover the surface. This is done by exposing the aggregate to a vapor at a series of partial pressures and measuring the amount that is adsorbed by the surface. (These partial pressures are much lower than those required by the Kelvin analysis, Eqs 15 and 16.)

A difficulty comes from the fact that most vapors do not neatly adsorb into a single layer. Rather, in some places they pile up into multi-layers even before other regions are covered by the first layer. The most common analysis that rectifies this problem is the B.E.T. theory [7] that permits one to find the amount of vapor that would just fill a single layer. This amount, and a knowledge of the size of the vapor's molecules, permits the calculation of the surface area.

This technique is described in several standards. These have been developed for the study of catalysts, but they are equally applicable to aggregates. Among these are: ASTM Test Method for Surface Area of Catalysts (D 3663), ASTM Test Method for Single-Point Determination of Specific Surface Area of Catalysts Using Nitrogen Adsorption by Continuous Flow Method (D 4567), and ASTM Test Method for Determination of Low Surface Area of Catalysts by Multipoint Krypton (D 4780). Instruments are available commercially to perform the sorption measurements and to analyze the results to yield a specific surface area.

The specific surface area can also be determined by image analysis. One prepares a plane surface passing through the aggregate. Then, one measures the number of times that a sufficiently long line on the surface intersects the pore-solid interface. The surface area, per unit bulk volume of material, is calculated as

$$S_b = 2 \times \frac{N}{L_t} \quad (19)$$

where

S_b = surface area per unit bulk volume,
N = number of line intersections with pore-solid boundary, and
L_t = total length of line examined.

Again, this approach needs an appropriate image of the pore system. However, it is unburdened by the assumptions of the vapor sorption technique. And, it is uniquely valuable in the case of an aggregate with an exceedingly small surface area, that is, one with only large pore sizes. (It is difficult to measure the small amount of vapor adsorbed by such aggregates.)

Techniques Involving the Measurement of Other Pore System Parameters

After the various measurements described previously, permeability is probably the most commonly measured parameter of a pore system. The experiment is, in principle, straightforward. One causes a fluid to flow through a shaped piece of aggregate. One determines the appropriate factors on the right side of Eq 11, and measures the flow rate. Then Eq 11 is used to calculate the permeability.

There are two main experimental difficulties with which one must deal. One is that the flow must be restricted to the interior of the aggregate. The sample must be confined so that no flow is possible at the sample-holder interface as this will give wildly inaccurate results. The second difficulty is that the flow rates may be small in the cases of some aggregates and fluids. Therefore, one may need a means of making accurate measurements of small amounts of flow. ASTM Test Method for Permeability of Rocks by Flowing Air (D 4525) describes the procedure when air is the fluid. Other fluids may be used with suitable modification of the equipment.

SIGNIFICANCE OF THE PORE SYSTEM ON AGGREGATE PROPERTIES

The void space within concrete aggregates affects many of the aggregate's properties. Among these are the thermal and elastic properties of both the aggregate and concrete made with the aggregate. However, these are addressed specifically in other chapters of this book. In addition, the void space acts as a reservoir into which water may flow during the mixing of plastic concrete. The extent to which this happens affects the proportions of constituents used in the mix. Again, this question of absorption is specifically addressed elsewhere. What remains of great significance to be discussed here is the relationship between an aggregate's pore system and its durability in a freezing environment.

The durability problems that may come from an aggregate are, at heart, caused by an almost unique property of water: its expansion upon solidification. Under the right circumstances, this expansion can generate sufficiently large internal pressures to crack concrete. This problem is not unique to pieces of aggregate in concrete. The danger of cracking exists for all materials that possess a void system that can imbibe water. The potential for cracking is governed solely by the amount of water present in the void space. This is, in turn, governed only by the properties of the void space. The chemical nature of most porous solids is of no significance in this regard.

When a sufficiently saturated void space is completely sealed, the surrounding solid will always be cracked by the pressure that is developed upon freezing. This pressure generates a tensile stress in the aggregate of several hundred MPa (several tens of thousands of psi). For the simplest of systems, this critical level of saturation is about 91.7%. For the more complex pore structures of most aggregates, this critical level may be somewhat different, but it is still in the vicinity of 90%. In a heterogeneous material, such as concrete, it is only necessary that some regions, or pieces of aggregate, reach this critical level to precipitate failure. Thus, a gross assessment of degree of saturation may well be misleading.

The freezing of water in an aggregate does not take place simultaneously throughout its pore system. Freezing will generally occur in different places at different times because of one or more of the following reasons: (1) the approach of a freezing front from one direction, (2) the difference in pore size from place to place, or (3) the presence of differing concentrations of salts in the pore fluid. For whatever reason, the nonsimultaneous freezing causes water to move ahead of critically saturated regions that become frozen. This flow can be a source of disruptive pressure also. The hydraulic pressure generated by the flow will be greater in regions of smaller-sized pore space and along longer flow paths. In some cases, the pressure will be sufficient to cause rupture of the surrounding solid.

Thus, the freezing of water in an aggregate's pore system can create disruptive pressures in the aggregate either because of the simple expansion of the water upon freezing or because of the hydraulic pressure of the liquid flow induced by the ice formation, or both. Further, water expelled from an aggregate can generate disruptive pressures in the surrounding paste. The propensity of an aggregate for causing difficulties by any of these mechanisms is associated with its pore volume and with the permeability of its pore system, and with nothing else.

These types of failures have lead to a general classification of aggregates into three groups [8]. The first, and most benign, are those aggregates that have such a small pore volume that the expansion of water in their pore systems results in strains in the surrounding solid that are too small to cause cracking. Such aggregates are said to accommodate the pressure elastically. Aggregates with typical strengths and modulii must have a porosity less than about 0.5% to fall within this category. Such aggregates cannot cause distress in a freezing environment.

A second class of aggregates is that in which the hydraulic pressure generated by water flowing ahead of a freezing front in a critically saturated pore system is sufficient to crack the surrounding rock itself. In order for the pressure to be high enough, such aggregates must have a sufficiently low permeability and a long enough flow path.

The third class differs from the second in that it has a permeability that is sufficiently high so that the hydraulic pressure never reaches the crack inducing level within the aggregate particle. This class may cause distress by forcing too much water into the surrounding, and less permeable, paste and so generating disruptive pressures there.

In the latter two classes, the potential for distress is governed by the aggregate's ability to become critically saturated. If it does become so saturated, then the potential for and type of distress is governed by the permeability of the aggregate and the length of the flow path within the aggregate.

Leaving permeability aside for the moment, a shorter flow path will be associated with a smaller aggregate size. This will reduce both the intra-aggregate hydraulic pressure, and the amount of water that is expelled into the paste. In either case, the potential for distress is reduced with smaller maximum aggregate sizes. However, reducing aggregate size always brings with it the need for a higher cement content in the mix and results in more expensive concrete.

For a given aggregate size, the ability of an aggregate to become critically saturated, and the aggregate's permeability, are both functions of the aggregate's pore-size distribution. If one knew the quantitative relationships between the distribution and these two critical properties of an aggregate, one would be able to assess the potential for distress of any aggregate. Unfortunately, there has been little research into these relationships.

The importance of permeability is that it dictates whether a problem will develop in the aggregate itself, or in the surrounding paste. In addition, permeability may have some effect on the speed with which a critical level of saturation is reached in an aggregate's pore system. However, aggregates are surrounded by a usually much less permeable paste matrix. It is likely that the rate at which water passes through this matrix will be the controlling factor in the availability of water to an aggregate's pore system. And, exterior concrete has many years in which its aggregate can reach critical saturation. Thus, permeability may dictate the location of the distress, and the rate at which it develops, but it does not control the incidence of the problem.

The important point is that no aggregate-induced problem will develop anywhere if the aggregate does not become sufficiently saturated. Thus, the important question is that of the pore system's water retention. It is the ability of an aggregate to reach the critical saturation, regardless of how long permeability may delay it, that is of utmost importance to its durability potential.

One can lay out a few generalities about an aggregate's pore system and its level of saturation. The pore space in an aggregate is in competition with the pore space in the surrounding paste for the retention of water. Smaller pore sizes have a higher capillary potential, and are more likely to retain water. Thus, one expects that a sufficiently large pore size in the aggregate will be relatively benign regardless of the pore volume since regions with this size will usually not be saturated. Thus, one finds that highly porous aggregates with a large pore size, such as many slags, do not cause distress.

Further, the freezing point of water is depressed when it is contained in a pore space of sufficiently small cross section. The smaller the section, the greater will be the depression. Thus, one may expect that there will be a cross section that is small enough to prevent freezing at the lowest temperatures that are typically encountered by exterior concrete. This means that the volume of pore space having a sufficiently small cross section will be unimportant in determining the distress potential of an aggregate.

The preceding arguments leave one with the conclusion that only the volume of pore space having cross sections larger than some minimum, and smaller than some maximum, would be associated with durability. The determination of the volume of pore space having sizes lying between two limits generally requires that a pore size distribution be determined.

A surprisingly small number of researchers have attempted to correlate any aspect of a pore size distribution with the durability of aggregate. In one study, the volume of pore space having diameters between 0.04 and 0.2 µm was found to be associated with durability [9]. In this work, a less durable aggregate had a greater pore volume in this size range. Another investigation found that a lack of durability was associated with a large pore volume having diameters less than 0.1 µm [10]. This study did not specifically look for a lower limit for the critical pore size range. A third study found that the volume and median diameter of the pore space having a diameter greater than 0.0045 µm was important [11]. This finding gives rise to an expected durability factor (EDF) that is related to the pore system parameters as

$$\text{EDF} = \frac{A}{\text{PV}} + B \times \text{MD} + C \qquad (20)$$

where

EDF = expected durability factor,
PV = pore volume with diameter > 0.0045 µm,
MD = median diameter of pore volume with diameter > 0.0045 µm, and
A, B, and C = constants.

An EDF of greater than 50 was found to be associated with durable aggregates.

In addition, other porous materials subject to wetting and freezing have been found to exhibit the same general pore size-distress relationship. For example, one investigation of bricks found that a lack of durability was associated with the an increased pore volume between 0.01 and 0.1 µm [12]. Another study of paving brick durability found that increased pore volume smaller than 3 µm was detrimental, and that pore space with a size larger than 3 µm was actually beneficial [13].

Clearly, these few studies do not agree on the exact size range that is critical. This is, perhaps, not surprising, since only a modest number of samples have been studied to date. However, an important point to be made concerning the limited research that has been carried out is that all of the studies have concluded that durability is related to the volume of a segment of the pore space lying within certain size limits. No researcher has found the contrary.

The preceding list of studies is not long. This paucity of work is, in itself, an informative finding. When one considers the magnitude of the durability problem, it is surprising that so little work has been directed toward the heart of the aggregate durability issue. Too often, tests that do not measure the important aspects of the pore system have predominated even in the face of arguments against their efficacy.

The author has recently surveyed ten midwestern U.S. state highway departments. All of the states have severe aggregate freeze-thaw durability problems apparent in their concrete roads. Only one of these ten currently uses any pore parameter measurement in the selection of its aggregates. And, that one uses only the aggregate absorption. Thus, one must conclude that there is virtually no attempt being made to use aggregate pore parameters to select aggregates and screen out potentially nondurable ones.

What these agencies are generally using is laboratory freeze-thaw testing of concrete beams made with aggregates. These tests are either the A or B variant of ASTM Test Method for Resistance of Concrete to Rapid Freezing and Thawing (C 666), or some variation typically concerning aggregate pre-treatment. However, concrete highways continue to be constructed in these states that display distress stemming from nondurable aggregates. Thus, it is evident that this test is not providing information that allows the agencies to discriminate between durable and nondurable aggregates.

Therefore, one is left with a paradox concerning the significance of the pore systems of coarse aggregates. There seems to be no disagreement that it is the pore system that governs durability. Yet, virtually no pore-system parameters are being measured on a regular basis by aggregate users, and pore-system oriented research is relatively rare. Meanwhile, a large amount of experimentation continues on non-pore-system parameters. This state of affairs continues even though, if one is to judge by pavement condition, the current approaches are not providing significant guidance to aggregate users.

CLOSURE

A broad variety of ASTM test methods exist for quantifying the pore system of aggregates. Some of these tests are specifically aimed at aggregates, and some can be readily adapted from their intended use on other porous materials. Only a few of these tests bear directly on the critical points controlling an aggregate's ability to resist freeze/thaw distress in concrete. Of these few, none are being used to screen aggregates for potential difficulties. The problem of concrete distress from nondurable aggregates continues to be prevalent. Only the application of appropriate test procedures and interpretations holds any current hope of alleviating this problem.

REFERENCES

[1] Dullien, F. A. L., *Porous Media: Fluid Transport and Pore Structure*, Academic Press, New York, 1979.
[2] *Quantitative Microscopy*, R. Dehoff and F. Rhine, Eds., McGraw-Hill, New York, 1968.
[3] Underwood, E., *Quantitative Stereology*, Addison-Wesley, Reading, MA, 1970.
[4] Gregg, S. J. and Sing, K. S. W., *Adsorption, Surface Area and Porosity*, 2nd ed., Academic Press, New York, 1982.
[5] Thomson, W. (Lord Kelvin), *Philosophical Magazine*, Vol. 42, 1871, pp. 448–452.
[6] Washburn, E. W., *Proceedings*, National Academy of Science, Vol. 7, 1921, pp. 115–116.
[7] Brunauer, S., Emmett, P. H., and Teller, E., *Journal*, American Chemical Society, Vol. 60, 1938, pp. 309–319.
[8] Verbeck, G. and Landgren, R., *Proceedings*, American Society for Testing Materials, Vol. 60, 1960, pp. 1063–1079.
[9] Marks, V. and Dubberke, W., *Research Record 853*, Transportation Research Board, 1982, pp. 25–30.
[10] Shakoor, A., West, T., and Scholer, C., *Bulletin*, Association of Engineering Geologists, Vol. 19, 1982, pp. 371–384.
[11] Kaneuji, M., Winslow, D., and Dolch, W. L., *Cement and Concrete Research*, Vol. 10, 1980, pp. 433–442.
[12] Nakamura, M., *Bulletin*, American Ceramic Society, Vol. 67, 1988, pp. 1964–1965.
[13] Winslow, D., *Journal of Testing and Evaluation*, Vol. 19, 1991, pp. 29–33.

39
Thermal Properties of Aggregates
D. Stephen Lane[1]

PREFACE

The chapter on Thermal Properties of concrete aggregates authored by H. K. Cook appeared in the three previous editions of *Significance of Tests and Properties of Concrete and Concrete Aggregates*, *ASTM STP 169* in 1956, *ASTM STP 169A* in 1966, and *ASTM STP 169B* in 1978. Since much of the discussion in the previous editions remains appropriate, the current edition is a revision of the previous work rather than a rewrite. This revision presents typical values of thermal properties for various common rocks and minerals and describes a means for calculating such values from existing data sources. A revised method for determining the coefficient of thermal expansion and the extent of anisotropy of rocks is referenced as well as a new method for measuring conductivity using a conventional needle probe.

INTRODUCTION

Aggregates typically compose 65 to 70% or more of the volume of concrete. Because the thermal properties of composite materials such as concrete are dependent on the thermal properties of their constituents, the aggregates have a great influence on the thermal properties of concrete. The degree to which thermal properties of concrete aggregates are of concern to the user depends upon the nature of the concrete structure and the exposure to which it is subjected. Knowledge of thermal properties are important in the design of massive structures, such as dams, where thermal and volume stability of the composite are important; in concretes exposed to extreme temperatures or temperature cycles, where the relative properties of the individual constituents may be important; and in lightweight concrete structures where insulating value of the composite is a primary factor.

The thermal properties of aggregates that are of general concern are the coefficient of thermal expansion, conductivity, diffusivity, and specific heat. Of these, the thermal expansion of aggregates is by far the most important because of the importance of the thermal expansion of concrete in the design of all concrete structures. The conductivity, diffusivity, and specific heat of aggregates are usually of concern only in special instances. The most obvious of these is massive structures where the dissipation of heat generated by cement hydration is important to control thermal cracking. Other instances include insulating concrete and the durability of concrete exposed to cyclical temperature changes. The design engineer should carefully evaluate the circumstances of a particular concrete application to estimate the importance that thermal properties of aggregates will have on concrete performance.

The attention that has been given to the thermal properties of concrete aggregates is not as great as that given to other properties of aggregates. This is probably because in many instances the effects do not appear to be as important as the effects of other properties of aggregates and, consequently, the significance of thermal properties is not as apparent. The need for a better understanding of the effects of the thermal characteristics of aggregates, particularly the coefficient of thermal expansion on the durability of concrete, has been expressed by Allen [1], Woolf [2], Scholer [3], and others.

The discussion in this paper is primarily confined to the significance of tests and thermal properties of the aggregates. While this necessarily requires discussion of the thermal properties of the concrete as affected by the aggregates, the significance of tests and thermal properties of concrete is discussed in Chapter 24 of this publication.

GENERAL CONCEPTS OF THERMAL PROPERTIES

Thermal properties of rocks are related directly to the thermal properties of the constituent minerals in proportion to their relative abundance. A reasonably good approximation of the value of thermal expansion, specific heat, conductivity, or diffusivity of a rock can be calculated if the mode (mineral content, percent by volume) of the rock is known using the following formula [4]

$$Pt(\text{rock}) = S(n_1^* Pt_1 + n_2^* Pt_2 + n_3^* Pt_3 \ldots) \quad (1)$$

where

n_1 = percent by volume, Mineral 1;
Pt_1 = thermal property value of Mineral 1; and
S = solidity (1 − decimal porosity).

The mode of an aggregate can be determined petrographi-

[1] Research scientist, Virginia Transportation Research Council, Charlottesville, VA 22923.

cally using ASTM Guide for Petrographic Examination of Aggregates for Concrete (C 295). The value thus calculated for a given aggregate will not be as accurate as could be measured. However, given the considerable care and skill necessary to accurately measure these properties, a calculated value may prove acceptable for many general applications. Robertson [4] has compiled a large volume of data on the thermal properties of rocks and minerals.

COEFFICIENT OF THERMAL EXPANSION

Numerical Values for Thermal Coefficients of Expansion

The coefficient of thermal expansion of aggregates is directly dependent on the coefficients of the constituent minerals and thus on the mineral composition of the rock. Zoldners [5] has reported coefficients of thermal expansion for common rock-forming minerals and some common rocks that are presented in Table 1. Although values for the thermal coefficients of expansion of aggregates from specific locations are contained in many of the references listed and are available from many other sources, Refs 4 to 16 list more values than most of the others.

As will be noted from the table, quartz has the highest coefficient of thermal expansion of the common minerals, while calcite and high calcium feldspars exhibit the lowest. Thus, the coefficients of rocks vary to a large degree in direct proportion to their quartz content.

While most minerals will exhibit some degree of anisotropy by expanding more in a direction parallel to one crystallographic axis than another, this is generally not of significance except when the differences are great. The most notable example is calcite, which can have a linear thermal coefficient of expansion as great as $25.8 \times 10^{-6}/°C$ parallel to its C-axis and as low as $-4.7 \times 10^{-6}/°C$ perpendicular to this direction. Potassium feldspars are another group of minerals exhibiting rather extreme anisotropy.

TABLE 1—Average Linear Coefficient of Thermal Expansion of Some Common Rocks and Minerals [5].

Mineral	$\times 10^{-6}/°C$
Quartz	11.5 to 12
Orthoclase, microcline	6.5 to 7.5
Pyroxenes, amphiboles	6.5 to 7.5
Olivine	6 to 9
Albite	5 to 6
Calcite	4.5 to 5
Oligoclase, andesine	3 to 4
Labradorite, Bytownite	3 to 4
Anorthite	2.5 to 3

Rock	$\times 10^{-6}/°C$
Quartzite, silica shale, chert	11.0 to 12.5
Sandstones	10.5 to 12.0
Quartz sands and pebbles	10.0 to 12.5
Argillaceous shales	9.5 to 11.0
Dolomite, magnesite	7.0 to 10.0
Granites and gneisses	6.5 to 8.5
Syenite, andesite, diorite, phonolite, gabbro, diabase, basalt	5.5 to 8.0
Marbles	4.0 to 7.0
Dense, crystalline, porous limestones	3.5 to 6.0

For this reason, the cubical expansion of rocks and minerals is not always related directly to the linear expansion, and this possibility should be kept in mind when investigating the thermal properties of aggregates.

The paper by Meyers [17] is concerned completely with thermal coefficient of expansion of portland cement rather than aggregates. It is included because a discussion of the thermal properties of concrete aggregates would not be complete without some information on the thermal properties of the cements with which they are used. The coefficient of thermal expansion of hydrated cement pastes may range from $10.8 \times 10^{-6}/°C$ to $16.2 \times 10^{-6}/°C$ and mortars from about $7.9 \times 10^{-6}/°C$ to $12.6 \times 10^{-6}/°C$. Mitchell [7] has found that at early ages or at certain critical saturations the linear thermal coefficient of expansion of cement pastes may be somewhat higher than those reported earlier. He reports values as high as $22.3 \times 10^{-6}/°C$ for some samples of neat cement specimens.

An average value for the linear thermal coefficient of expansion of concrete may be taken as $9.9 \times 10^{-6}/°C$, but the range may be from about $5.8 \times 10^{-6}/°C$ to $14.0 \times 10^{-6}/°C$, depending upon the type and quantities of the aggregates, the mixture proportions, and other factors.

Effect of Thermal Expansion

The significance of the coefficient of thermal expansion of aggregates with respect to its effect on concrete is twofold. The primary, or at least better understood, effect is that on the volume change of concrete. Aggregates with high coefficients of thermal expansion will produce concretes of low thermal volume stability and vice versa, all other things equal. The thermal expansion of concrete should be accounted for in the design of the structure and is further discussed in Chapter 23 on Volume Change in this publication. The other effect is the development of internal stresses as a result of large differences between the coefficients of thermal expansion of the various components of concrete subjected to temperature extremes or thermal cycling and the impact this may have on concrete durability.

There seems to be fairly general agreement that the thermal expansion of the aggregate has an effect on the durability of concrete, particularly under severe exposure conditions or under rapid temperature changes. There is less agreement on whether more durable concrete will be produced with an aggregate of high thermal coefficient of expansion similar to that of the cement paste or with an aggregate of low thermal coefficient.

In one of the earliest papers on the subject, Pearson [18] attributes a case of rapid concrete deterioration to the use of an aggregate of low thermal coefficient. The aggregate, a dolomitic marble with a coefficient of thermal expansion of $3.6 \times 10^{-6}/°C$ was reported to be very resistant to freezing and thawing deterioration. The concrete was severely cracked after the first winter, having been subjected to very low temperatures and to severe frost action. Given the aggregate's freezing and thawing durability and because the deterioration occurred much more rapidly than would have been expected from freezing and thawing action alone, other causes were suspected. Pearson speculated that tensile stresses developed when the matrix contracted three to four times that of the coarse aggregate at extremely low winter temperatures. These stresses were sufficient to cause fine cracking of the concrete providing pathways for water into the concrete and subsequent freezing and thawing damage. Pearson conducted a laboratory investigation [19] with concretes containing aggregates of both low and high thermal coefficients. The concretes were subjected to 100 thermal cycles between −28.9 and 21.3°C before being subjected to freezing-and-thawing tests. Concretes containing the aggregates of low thermal coefficient failed much more rapidly in the freezing-and-thawing test than the concretes containing aggregates of high thermal coefficient.

Callan [20,21] statistically analyzed 78 combinations of aggregate in concrete with respect to durability in freezing and thawing and differences in thermal expansion between the coarse aggregate and the mortar. He concludes that large differences between the coefficients of expansion of coarse aggregate and mortar may considerably reduce the durability of concrete from that predicted by the results of the usual acceptance tests. He suggests that differences between these coefficients exceeding 5.4 × 10^{-6}/°C should be considered large enough to warrant caution in the selection of materials for highly durable concrete exposed to temperature extremes. Swenson and Chaly [22] include some discussion of the effect of thermal coefficient of expansion of the aggregates on concrete durability and caution against the possible deleterious effects if large differences are observed. Smith [23] presents calculations to indicate the potential magnitude of physical incompatibility of the matrix and aggregate in concrete based on differences in thermal expansion. Kennedy and Mather [24], in attempting to correlate laboratory-accelerated freezing and thawing with natural weathering at Treat Island, Maine, state among other conclusions that, while there appears to be a correlation between the resistance of concrete to freezing and thawing and differences in thermal expansion between the coarse aggregate and the mortar, the correlation is probably usually of lesser importance than other characteristics of the concrete.

Walker, Bloem, and Mullen [25], on the other hand, report the results of heating and cooling concrete specimens over the temperature range of 4.4 to 60°C at various rates. They found that "changes in temperature were destructive to the concrete with sudden changes in temperature being much more severe than slower ones; and concretes having higher coefficients of expansion were less resistant to temperature changes than concretes with lower coefficients." It also was determined that the thermal coefficients of expansion of concrete and mortar containing different aggregates varied approximately in proportion to the thermal coefficient and quantity of aggregate in the mixture.

The findings of Walker et al. [25] stand in contrast to those of Pearson [18,19], Callan [20,21], and the others [22–24] that aggregates with low coefficients of thermal expansion compared to the cement paste may reduce concrete durability. One explanation for the difference in findings is that Walker's study did not subject specimens to temperatures below the freezing point. Pearson, however, used temperatures as low as −28.9°C. Koenitzeer [26] subjected several concretes and the aggregates used in them over temperature ranges of −12.8 to 26.7°C, 29.4 to 88°C, and −12.8 to 88°C, in both moist and dry conditions. One of Koenitzer's conclusions was that the elastic and thermal expansion properties for any one material vary with the conditions of test, the greatest variation being caused by freezing.

However, the U.S. Bureau of Reclamation produced concrete at Grand Coulee Dam using a predominately basalt aggregate with a thermal expansion of about 7.2 × 10^{-6}/°C that showed an extremely high resistance to freezing and thawing. The Bureau of Reclamation also found [8] in the Kansas-Nebraska area that replacement of part of the aggregate with limestone, which had a thermal expansion averaging 4.5 × 10^{-6}/°C, greatly increased the durability of the concrete in which it was used. Although the addition of limestone was for the purpose of inhibiting alkali-aggregate reaction and the improvement in durability probably was primarily because of the inhibition of this reaction, the low thermal expansion apparently introduced no adverse effects.

More recently, Venecanin [27,28] has renewed interest in the durability of concrete produced with aggregates of low thermal coefficient of expansion. He has derived equations [27] to evaluate the potential stresses developed between binders and aggregates of different thermal coefficients and attributed abnormal cracking in several bridges to a limestone aggregate with low thermal coefficient and a high degree of anisotropy [28]. The bridges are in areas subject to extreme subfreezing winter temperatures. Other researchers [29,30] following Venecanin's lead have concluded that limestone aggregates that exhibit extreme anisotropic thermal characteristics can adversely affect concrete durability even in environments not subject to sub-freezing temperatures.

The preceding discussion has dealt with the effects of thermal expansion of aggregates over temperature ranges that can occur under natural exposure conditions. Zoldners [5] indicates that thermally stable aggregates such as fine-grained rocks of anorthositic composition are best suited for concretes that will be subjected to extreme elevated temperatures. For such applications, aggregates containing quartz should be avoided since at a temperature of 572.7°C quartz changes state and suddenly expands 0.85%, usually producing a disruptive effect at the surface of concrete in which it is used. Endell [31] has reported the results of experiments to determine the structural and expansion changes of concrete aggregates with temperatures up to 1200°C. These are considered to be highly specialized conditions and are not discussed farther here, but are discussed further in another paper in this publication.

METHODS OF DETERMINING THERMAL EXPANSION OF AGGREGATES

Several ingenious methods have been developed for determining the thermal coefficient of expansion of coarse aggregate. The majority of the test methods are based on the measurement of linear expansion over a temperature range. This range is usually 55°C or more because the change in unit length per degree is extremely small, and the multiplication of the change over a substantial temperature range greatly increases the facility and precision of the determination. However, the size of a representative specimen that can be obtained from a coarse aggregate rapidly approaches a practical maximum. Except in the case of a crushed aggregate of sufficient uniformity to permit obtaining a larger specimen that will be representative of the sand sizes, this means of multiplying the change in unit length is not available for determining the linear expansion of fine aggregate.

The method described by Willis and DeReus [32] allows measurements to be made over a considerable temperature range with the use of an optical lever. The optical lever has also been used to obtain additional multiplication of the length change. The specimens used by Willis and DeReus were 25.4-mm-diameter cores, 50 mm long, drilled from the aggregate specimens to be tested and placed in a controlled-temperature oil bath with a range of 2.78 ± 1.7°C to 60 ± 2.8°C. The vertical movement of the specimen as the temperature was varied was measured by reading the image reflected by the mirror of the optical lever, having a 25.4-mm lever arm, on a vertical scale placed 6.1 m from the mirror, by means of a precise level. It is reported that consideration of the possible errors involved in the measurements indicates that the calculated coefficients are probably accurate to $\pm 3.6 \times 10^{-7}/°C$.

Another method uses the interferometer described by Merritt [33] and modified by Saunders [34]. Detailed descriptions of the apparatus, the preparation of specimens, and the test procedure are also given by Johnson and Parsons [10]. A third method for determining the thermal coefficient of expansion of coarse aggregate is that developed by the Corps of Engineers [21,35] in which an SR-4 strain gage is bonded to a prepared piece of aggregate and readings are taken over a temperature range of 1.7 to 57.2°C. This method requires that the piece of aggregate be sliced in three mutually perpendicular directions, two of these directions to lie in the major structural plane of the rock, if such a plane can be located. The strain gages are then mounted so as to measure strain in each of the three directions. The purpose of this requirement is to determine if anisotropy or preferred crystal orientation exist. Recently, Venecanin [28] has reported on a similar, but more elaborate setup, where strain gages are mounted to obtain measurements parallel to the edges and in both diagonal directions on each of the six faces of a cube of rock.

Mitchell [7] describes a method that employs specimens from 25.4 to 76.2 mm in size coated with wax and held in fulcrum-type extensometer frames. Measurements are made with electromagnetic strain gages with electronic indicators while the specimen is immersed in a circulating ethylene glycol solution held at the desired temperature.

Because of the size and usually heterogenous nature of fine aggregate, none of the preceding methods are readily adaptable to the determination of the coefficient of expansion of this type of material. The usual approach has been to determine the linear expansion of mortar bars containing the fine aggregate. Of course, the results obtained include the effects of the length change contributed by the cement. Verbeck and Hass [36] have developed a dilatometer method for determining the thermal coefficient of expansion that is particularly adaptable for use with fine aggregate. The method determines the cubical thermal coefficient of expansion from which the linear expansion may be calculated. The apparatus consists of a 1-L dilatometer flask to which is attached a capillary-bulb arrangement containing electrical contacts spaced over a calibrated volume. The flask is filled with aggregate and water, and the apparatus is allowed to come to equilibrium at one of the controlling electrical contacts. The equilibrium temperature is noted, and the procedure is repeated at the other electrical contact. After proper calibration, the only measurements required are the weight of the water placed in the flask and the temperature needed to produce an expansion equivalent to the volume between the electrical contacts.

THERMAL CONDUCTIVITY, THERMAL DIFFUSIVITY, AND SPECIFIC HEAT

Thermal conductivity, thermal diffusivity, and specific heat are largely interrelated, and all three normally are determined only for concrete as used in massive structures. The Bureau of Reclamation [37] indicates the application of these data in connection with computing concrete placement temperatures and designing cooling systems, and in other thermal calculations aimed at reducing thermal volume change and thus cracking in large dams. The same type of measurements and calculations would apply equally to other massive structures. Thermal conductivity is also of importance in lightweight concrete for insulating purposes. It has been indicated by some [38–40] that thermal diffusivity may have an important effect on concrete durability.

Thermal conductivity, measured as the rate of heat flow through a body of unit thickness and unit area with a unit temperature difference between two surfaces, normally is expressed in calculations as watts per metre kelvin (W/mK).

Thermal diffusivity is defined as the thermal conductivity divided by the specific heat and density and is a physical property of the material that determines the time rate of change of temperature of any point within a body. Its units are square metres per second (m^2/s).

Specific heat is the amount of heat required to raise the temperature of a unit mass of material one unit of temperature. Its units are joules per kilogram kelvin (J/kgK).

Thermal Conductivity

The thermal conductivity of aggregates varies in large measure with their mineral content. Quartz is one of the more important minerals to consider in evaluating the conductivity of an aggregate as it exhibits a relatively high conductivity. The feldspars on the other hand are also important because of their abundance and relatively low conductivity. The porosity of an aggregate also plays a significant role in determining its conductivity because air is a very good insulator. However, the moisture content of the aggregate has a tremendous effect on thermal conductivity. Tyner [41] reports that in a 1:5 mixture of Florida limerock concrete an increase of moisture from 0 to 5% increases the conductivity by 23%. Although one of the more useful properties of lightweight concrete is its low thermal conductivity, Davis and Kelly [42] state that "the presence of a small amount of moisture in the interior of a lightweight concrete greatly increases its thermal conductivity; hence, under conditions of continuous or intermittent exposure to moisture, if a high degree of insulation is desired, an aggregate (and concrete) of relatively low absorption should be used." Kluge, Sparks, and Tuma [43] and Price and Condon [44] also have found pronounced reductions in the thermal conductivity of concrete containing lightweight aggregate, but they indicate that the reduction seems to be influenced more by the reduction in the density of the concrete than by the characteristics of the aggregate. Thermal conductivity values for some common rock types reported by Clark [45] are presented in Table 2.

For most heavy-weight aggregates used in radiation shielding, thermal conductivities will be high. The exception is barite, a sulfate mineral with a thermal conductivity somewhat lower than normal concrete aggregate [46].

Thermal Diffusivity

As mentioned earlier, thermal diffusivity is directly proportional to conductivity. The thermal diffusivity of a normal-weight aggregate is thus primarily dependent on its quartz content. Investigations have indicated that the normal diffusivity of the aggregate may have an influence on the durability of the concrete in which it is used. Thomson [38] states that for a given body with specified boundary conditions the thermal stresses depend on certain physical properties of the materials. In a homogenous body, such physical properties as thermal conductivity, specific heat, and the density of the material influence the temperature distribution and the thermal stresses during the transient period only in a certain combination known as the thermal diffusivity. If in a mixture such as concrete the thermal diffusivities and conductivities are the same for each material, the body can be thought of as being thermally homogenous. Since a difference in diffusivities would result in different rates of diffusion of heat through the aggregate and cement, it is believed that such a combination would result in higher thermal stresses than those existing in homogenous bodies. Nothstine [39] and Weiner [40] have reported the results of their approach to the problem. Weiner's work was instigated by the failure of a gravel concrete, exposed to natural freezing and thawing accompanied by thermal shock and characterized by bond failure and internal expansion. He attributes the failure to the relatively high thermal coefficient of expansion of the concrete, which is responsible for surface stress, and to the diffusivity of the gravel that, being higher than the mortar, responds more quickly to temperature changes, resulting in differential volume change. Fox and Dolch [47] in an investigation of four limestones found a large change in the thermal diffusivity with a relatively small degree of saturation. The increases in diffusivity ranged from 20 to 59% for saturations of less than 5%. The authors of the references cited essentially agreed that the thermal diffusivity of the aggregates apparently has an effect on the durability of concrete, but that further work is needed to determine the significance of the effect and to find a practical means for using this knowledge to improve concrete durability. Calculated diffusivities for some rocks are presented in Table 3.

Specific Heat of Aggregates

Specific heat is of considerable importance in connection with the calculations involved in the control of placement temperatures and the limiting of thermal volume change of mass concrete. The specific heat of the aggregate contributes materially to the specific heat of the concrete [37]. The specific heat of rocks can be calculated from the mineral composition of the rock using Eq 1. The effect of temperature on specific heat is considerable and should be considered. Table 4 presents values of specific heat for some common minerals and rocks.

Methods of Test for Conductivity, Diffusivity, and Specific Heat

Methods are available for the direct determination of thermal conductivity, thermal diffusivity, and specific

TABLE 2—Thermal Conductivity Values for Some Common Rocks (Data From Clark [45]).

Rock	Thermal Conductivity (W/mK)	
	Mean	Range of Values
Quartzite	6.7	5.9 to 7.4
Dolomite	4.6	4.0 to 5.0
Gneiss, parallel to foliation	3.5	2.6 to 4.4
Granite, quartz monzonite	3.3	2.8 to 3.6
Granite	3.2	2.6 to 3.8
Granodiorite (California)	3.2	2.9 to 3.5
Diabase	3.0	2.6 to 3.4
Amphibolite	2.9	2.6 to 3.8
Granodiorite (Nevada)	2.8	2.6 to 2.9
Gneiss, perpendicular to foliation	2.6	2.0 to 3.6
Limestone	2.6	2.0 to 3.0

TABLE 3—Calculated Diffusivities of Some Common Rocks (Data From Robertson [4]).

Rock	Diffusivity, $10^{-6} m^2/s$
Basalt	0.9
Marble	1.0
Limestone	1.1
Gabbro	1.2
Sandstone	1.3
Rhyolite	1.6
Peridotite	1.7
Quartzite	2.6
Dolomite	2.6

TABLE 4—Specific Heat of Some Common Minerals and Rocks (Data From Robertson [4]).

Mineral	10^3 J/kgK	
	27°C	127°C
Fayalite	0.64	0.72
Quartz	0.75	0.88
Pyroxene	0.75	0.92
Microcline, Sanidine	0.77	0.85
Anorthite	0.77	0.88
Philogopite	0.77	0.92
Forsterite	0.77	0.95
Albite	0.80	0.92
Muscovite	0.81	0.95
Calcite	0.86	0.95
Dolomite	0.86	0.98

Rock	0°C	50 to 65°C	200°C
Diabase	0.70	...	0.87
Granite	0.65	...	0.95
Granite	0.80	0.77	0.95
Granodiorite	0.70	...	0.95
Quartzite	0.70	0.77	0.97
Diorite	0.71	0.81	0.99
Gabbro	0.72	...	0.99
Slate	0.71	...	1.00
Marble	0.79	0.85	1.00
Limestone	...	0.83	...
Limestone	...	1.00	...
Granitic gneiss	0.74	0.79	1.01
Basalt	0.85	...	1.04

heat. However, as a matter of practical convenience, it is customary to determine diffusivity and specific heat and calculate conductivity or to determine conductivity and specific heat and calculate diffusivity. This is possible because the formula includes all three values, and knowing the values for any two, and the density, permits solving the equation for the unknown property. The formula is

$$k = hcp \qquad (2)$$

where

k = thermal conductivity in W/mK,
h = thermal diffusivity in m^2/s,
c = specific heat in J/kgK, and
p = density in kg/m^3.

Whether conductivity is determined directly and diffusivity calculated, or vice versa, is largely a matter of the most convenient equipment setup available and the preference of the laboratory doing the work. Some of the test methods included in the references to this paper are based on determinations made on concrete specimens, but, in most cases, they may be used with aggregates if properly modified with respect to specimen size and shape.

The Corps of Engineers [35], Thomson [38], and Fox and Dolch [47] describe methods for the direct determination of diffusivity of stone and concrete. All of the methods depend basically on obtaining time-temperature differential curves between the temperatures at the center and the surface of a specimen by starting at essentially equilibrium temperature, then changing the surface temperature, and plotting the time-temperature curve until equilibrium conditions are obtained at the new surface temperature.

Depending upon the degree of accuracy desired, refinements can be made by grinding the specimen to a sphere, for example, and by refining the instrumentation. The determination is by no means routine in nature and requires careful experimentation, precise equipment, and a capable operator.

Recently, Sass, Kennelly, Smith, and Wendt [48] have described the use of a conventional needle probe to measure the thermal conductivity of machined, drilled, or unconsolidated rock material. This method offers a simplified and versatile means for measuring the conductivity of aggregates. Thermal conductivity can also be measured directly by ASTM Test Method for Steady-State Heat Flux Measurements and Thermal Transmission Properties by Means of the Guarded-Hot-Plate Apparatus (C 177) or similar methods [41,49]. However, the equipment is designed for use with a flat specimen, and consequently is not easily adapted to measure the conductivity of such materials as concrete aggregates. The Bureau of Reclamation [50] has developed a method for the determination of the thermal conductivity of concrete by the use of a 203 by 406-mm (8 by 16-in.) hollow cylindrical concrete specimen. This procedure probably could be used for coarse aggregate provided a specimen of the required size and shape could be fabricated from a large rock specimen.

The specific heat of aggregate usually is determined by a procedure known as the method of mixtures [35]. It is a calorimetric procedure wherein the net heat required to raise the temperature of a specimen of known weight a given amount is measured.

References 44 through 59 provide additional background information on the theory and mathematics of thermal tests. Other references may be found appended to many of the references cited here.

CONCLUSIONS

Test methods are available that when used with proper attention to procedure, specimen size and shape, instrumentation, and technique are entirely adequate for the determination of the thermal properties of aggregates.

There appears to be no doubt that the normal properties of aggregates, particularly thermal expansion, have an effect on the durability and other qualities of concrete. Investigations reported to date do not present a clear-cut picture of the effects that might be expected, and some aspects of the problem are controversial. The ultimate solution must be based on the performance of aggregates of known thermal properties in concrete, and, as is normal in this field of investigation, the major difficulty is to separate the effects of the thermal properties of the aggregates from the numerous other variables existing in the concrete. There is a real need for additional research work on the subject, both to resolve existing controversy and to improve concrete further as a construction material.

REFERENCES

[1] Allen, C. H., "Influence of Mineral Aggregates on the Strength and Durability of Concrete," *Mineral Aggregates*,

ASTM STP 83, American Society for Testing and Materials, Philadelphia, 1948, p. 153.
[2] Woolf, D. O., "Needed Research," *Mineral Aggregates, ASTM STP 83*, American Society for Testing and Materials, Philadelphia, 1948, p. 221.
[3] Scholer, C. H., "Durability of Concrete," *Report on Significance of Tests of Concrete and Concrete Aggregates, ASTM STP 22A*, American Society for Testing and Materials, Philadelphia, 1943, p. 29.
[4] Robertson, E. C., "Thermal Properties of Rocks," Open-File Report 88-441, U.S. Geological Survey, Reston, VA, 1988.
[5] Zoldners, N. G., "Thermal Properties of Concrete Under Sustained Elevated Temperatures," *Temperature and Concrete*, SP 25, American Concrete Institute, Detroit, MI, 1971.
[6] Hallock, W., "Preliminary Notes on the Coefficients of Thermal Expansion of Certain Rocks," *Bulletin No. 78*, U.S. Geological Survey, Reston, VA, 1891.
[7] Mitchell, L. J., "Thermal Expansion Tests on Aggregates, Neat Cements and Concrete," *Proceedings*, American Society for Testing and Materials, Vol. 53, 1953, p. 963.
[8] "A Study of Cement-Aggregate-Incompatibility in the Kansas-Nebraska Area," Report C-964, U.S. Bureau of Reclamation, Denver, CO.
[9] Griffith, J. H., "Thermal Expansion of Typical American Rocks," *Bulletin No. 128*, Iowa Engineering Experiment Station, Ames, IA, 1936.
[10] Johnson, W. H. and Parsons, W. H. "Thermal Expansion of Concrete Aggregate Material," *Journal of Research*, RP 1578, National Bureau of Standards, Vol. 32, 1944, p. 101.
[11] Parsons, W. H. and Johnson, W. H., "Factors Affecting the Thermal Expansion of Concrete Aggregate Materials," *Journal*, American Concrete Institute, April 1944; *Proceedings*, Vol. 40, p. 457.
[12] "Laboratory Investigations of Certain Limestones Aggregates for Concrete," Technical Memorandum No. 6-371, Waterways Experiment Station, U.S. Army Corps of Engineers, Vicksburg, MS, Oct. 1953.
[13] "Test Data on Concrete Aggregates in Continental United States," Technical Memorandum No. 6-370, Vols. 1–5, Waterways Experiment Station, U.S. Army Corps of Engineers, Vicksburg, MS, Sept. 1953.
[14] Rhodes, R. and Mielenz, R. C., "Petrography of Concrete Aggregate," *Journal*, American Concrete Institute, June 1946; *Proceedings*, Vol. 42, p. 581.
[15] Rhodes, R. and Mielenz, R. C. "Petrographic and Mineralogic Characteristics of Aggregates," *Mineral Aggregates, ASTM STP 83*, American Society for Testing and Materials, Philadelphia, 1948, p. 20.
[16] *Concrete Manual*, 7th ed., U.S. Bureau of Reclamation, Denver, CO, 1963, p. 60.
[17] Meyers, S. L., "Thermal Coefficient of Expansion of Portland Cement," *Industrial and Engineering Chemistry*, Vol. 32, Aug. 1940, p. 1107.
[18] Pearson, J. C., "A Concrete Failure Attributed to Aggregate of Low Thermal Coefficient," *Journal*, American Concrete Institute, June 1942; *Proceedings*, Vol. 38, pp. 36–41.
[19] Pearson, J. C., "Supplementary Data on the Effect of Concrete Aggregate Having a Low Thermal Coefficient of Expansion," *Journal*, American Concrete Institute, Sept. 1943; *Proceedings*, Vol. 40, p. 33.
[20] Callan, E. J., "Thermal Expansion of Aggregates and Concrete Durability," *Journal*, American Concrete Institute, Feb. 1952; *Proceedings*, Vol. 48, pp. 504–511.
[21] Callan, E. J., "The Relation of Thermal Expansion of Aggregates to the Durability of Concrete," *Bulletin No. 34*, Waterways Experiment Station, U.S. Army Corps of Engineers, Vicksburg, MS, Feb. 1950.
[22] Swenson, E. G. and Chaly, V., "Basis for Classifying Deleterious Characteristics of Concrete Aggregate Materials," *Journal*, American Concrete Institute, May 1956; *Proceedings*, Vol. 52, p. 987.
[23] Smith, G. M., "Physical Incompatibility of Matrix and Aggregate in Concrete," *Journal*, American Concrete Institute, March 1956; *Proceedings*, Vol. 52, p. 791.
[24] Kennedy, T. B. and Mather, K., "Correlation Between Laboratory Accelerated Freezing and Thawing and Weathering at Treat Island, Maine," *Journal*, American Concrete Institute, Oct. 1953; *Proceedings*, Vol. 50, p. 141.
[25] Walker, S., Bloem, D. L., and Mullen, W. G., "Effects of Temperature Changes on Concrete as Influenced by Aggregates," *Journal*, American Concrete Institute, April 1952; *Proceedings*, Vol. 48, p. 661.
[26] Koenitzer, L. H., "Elastic and Thermal Expansion Properties of Concrete as Affected by Similar Properties of the Aggregate," *Proceedings*, American Society for Testing and Materials, Vol. 36, Part 2, 1936, p. 393.
[27] Venecanin, S. D., "Durability of Composite Materials as Influenced by Different Coefficients of Thermal Expansion of Components," *Durability of Building Materials and Components, ASTM STP 691*, P. J. Sereda and G. G. Litvan, Eds., American Society for Testing and Materials, Philadelphia, 1980, pp. 179–192.
[28] Venecanin, S. D., "Thermal Incompatibility of Concrete Components and Thermal Properties of Carbonate Rocks," *ACI Materials Journal*, American Concrete Institute, Vol. 87, No. 6, 1990.
[29] Al-Tayyib, A. J., Baluch, M. H., Sharif, A. M., and Mahamud, M. M., "The Effect of Thermal Cycling on the Durability of Concrete Made From Local Materials in the Arabian Gulf Countries," *Cement and Concrete Research*, Vol. 19, 1989, pp. 131–142.
[30] Baluch, M. H., Al-Nour, L. A. R., Azad, A. K., Al-Mandil, M. Y., and Pearson-Kirk, D., "Concrete Degradation Due to Thermal Incompatibility of Its Components," *Journal of Materials in Civil Engineering*, Vol. 1, No. 3, 1989, pp. 105–118.
[31] Endell, K., "Experiments on Concrete Aggregate Materials to Determine Structural and Linear Expansion Changes with Temperatures up to 1200°C," Ernst and Sohn, Berlin; (revised) *Tonindustrie-Zeitung*, Vol. 53, 1929, p. 1472; *Ceramic Abstracts*, Vol. 9, 1930, p. 159; *Chemical Abstracts*, Vol. 24, p. 2265.
[32] Willis, T. F. and Dereus, M. E., "Thermal Volume Change and Elasticity of Aggregates and Their Effect on Concrete," *Proceedings*, American Society for Testing and Materials, Vol. 39, 1939, p. 919.
[33] Merritt, G. E., "The Interference Method of Measuring Thermal Expansion," *Journal of Research*, RP 515, National Bureau of Standards, Vol. 10, 1933, p. 591.
[34] Saunders, J. B., "Improved Interferometric Procedure with Application to Expansion Measurements," *Journal of Research*, RP 1227, National Bureau of Standards, Vol. 23, 1939, p. 179.
[35] *Handbook for Concrete and Cement*, Waterways Experiment Station, U.S. Army Corps of Engineers, Vicksburg, MS, 1949.
[36] Verbeck, G. J. and Haas, W. E., "Dilatometer Method for Determination of Thermal Coefficient of Expansion of Fine and Coarse Aggregate," *Proceedings*, Highway Research Board, Vol. 30, 1950, p. 187.
[37] "Cement and Concrete Investigations," Boulder Canyon Project Final Reports, Part 7; "Thermal Properties of Concrete," *Bulletin No. 1*, Bureau of Reclamation, Denver, CO, 1940.

[38] Thomson, W. T., "A Method of Measuring Thermal Diffusivity and Conductivity of Stone and Concrete," *Proceedings*, American Society for Testing and Materials, Vol. 40, 1940, p. 1073; Discussion, p. 1081.

[39] Nothstine, L. V., "Thermal Diffusivity and Modulus of Elasticity in Relation to the Durability of Concrete," Kansas State College, Manhattan, KS, 1940.

[40] Weiner, A., "A Study of the Influence of Thermal Properties on the Durability of Concrete," *Journal*, American Concrete Institute, May 1947; *Proceedings*, Vol. 43, p. 997.

[41] Tyner, M., "Effect of Moisture on Thermal Conductivity of Limerock Concrete," *Journal*, American Concrete Institute, Sept. 1946; *Proceedings*, Vol. 43, p. 9.

[42] Davis, R. E. and Kelly, J. W., "Lightweight Aggregates," *Mineral Aggregates, ASTM STP 83*, American Society for Testing and Materials, Philadelphia, 1948, p. 160.

[43] Kluge, R. W., Sparks, M. M., and Tuma, E. C., "Lightweight-Aggregate Concrete," *Journal*, American Concrete Institute, May 1949; *Proceedings*, Vol. 45, p. 625.

[44] Price, W. H. and Cordon, W. A., "Tests of Lightweight-Aggregate Concrete Designed for Monolithic Construction," *Journal*, American Concrete Institute, April 1949; *Proceedings*, Vol. 45, p. 581.

[45] Clark, S. P., Jr., "Thermal Conductivity," *Handbook of Physical Constants, GSA Memoir 97*, S. P. Clark, Jr., Ed., Geological Society of America, Boulder, CO, 1966, pp. 461–482.

[46] Witte, L. P. and Blackstrom, J. E., "Properties of Heavy Concrete Made with Barite Aggregates," *Journal*, American Concrete Institute, Sept. 1954; *Proceedings*, Vol. 51, p. 65.

[47] Fox, R. G., Jr. and Dolch, W. L., "A Technique for the Determination of a Thermal Characteristic of Stone," *Proceedings*, Highway Research Board, Vol. 30, 1950, p. 180.

[48] Sass, J. H., Kennelly, J. P., Jr., Smith, E. P. and Wendt, W. E., "Laboratory Line-Source Methods for the Measurement of Thermal Conductivity of Rocks Near Room Temperature," Open-File Report 84-91, U.S. Geological Survey, Reston, VA, 1984.

[49] "Projected Testing Method for Conductivity of Materials," *Bulletin No, 19*, International Association of Testing and Research Laboratories for Materials and Structures, RILEM, Paris, Nov. 1954, p. 3.

[50] *Materials Laboratory Procedures Manual*, U.S. Bureau of Reclamation, Denver, CO, 1951.

[51] Shack, A., *Industrial Heat Transfer*, Wiley, New York, 1933 (translated from the German by Hans Goldschmidt and E. P. Partridge).

[52] McAdams, W. H., *Heat Transmission*, McGraw-Hill, New York, 1942.

[53] Carslaw, H. S. and Jaeger, J. C., *Conduction of Heat in Solids*, Oxford University Press, London, 1947.

[54] Clark, H., "The Effects of Simple Compressing and Wetting on the Thermal Conductivity of Rocks," *Transactions*, American Geophysical Union, Part 3, 1941, p. 543; *Chemical Abstracts*, Vol. 34, 1942, p. 2820.

[55] Clark, H. and Birch, F., "Thermal Conductivities of Rocks and Its Dependence Upon Temperature and Composition," *American Journal of Science*, Vol. 238, 1940, p. 529; *Chemical Abstracts*, Vol. 34, 1940, p. 7796.

[56] Ingersoll, L. R., Zobel, O. J., and Ingersoll, A. C., *Heat Conduction with Engineering and Geological Applications*, McGraw-Hill, New York, 1948.

[57] Jakob, M., *Heat Transfer*, Wiley, New York, 1949.

[58] Niven, C. D., "Thermal Conductivities of Some Sedimentary Rocks," *Canadian Journal of Research*, Vol. 18, 1940, p. 132; *Chemical Abstracts*, Vol. 28, 1934, p. 7455.

[59] Kingery, W. D. and McQuarrie, M. C., "Concepts of Measurements and Factors Affecting Thermal Conductivity of Ceramic Materials," *Journal*, American Ceramic Society, Vol. 37, No. 2, Part 2, Feb. 1954, p. 67.

PART V
Other Concrete Making Materials

Hydraulic Cements—Physical Properties

Leslie Struble[1] and Peter Hawkins[2]

PREFACE

Physical properties of hydraulic cements are covered for the first time in this edition of *ASTM STP 169*. The chapter covers the full range of physical properties—fineness, consistency, setting properties, strength, and durability. The emphasis is on significance of standards and how they might be improved.

INTRODUCTION

Hydraulic cements are manufactured products that find their principal uses in concrete and related construction materials. When cement and water are mixed, they undergo various chemical reactions that gradually change the mixture from a plastic (or fluid), which can be molded or cast, into a rigid solid, capable of bearing substantial compressive loads. Thus, cement and its reactions with water are largely responsible for most of the key aspects of concrete—its workability, set, strength, creep, shrinkage, and durability.

Hydraulic cements have a wide range of physical properties. The objective of this chapter is to review those properties that are included in the standard specifications for hydraulic cements; that is, Tables 3 and 4 of the ASTM Specification for Portland Cement (C 150), Tables 2 and 3 of the ASTM Specification for Blended Hydraulic Cements (C 595), and Table 3 of the ASTM Specification for Expansive Hydraulic Cement (C 845). The properties are summarized here in Table 1 and their test methods in Table 2. The emphasis throughout this chapter is on the significance of the physical properties, limits in their use, and ways in which their tests could be improved.

Some physical properties (for example, activity index) are used to control quality during cement production. Others are needed for determination of some other cement property (for example, density to calculate fineness). Most of the properties, however, attempt to predict how cement influences the performance of concrete. For those properties, their significance lies in how well they predict the concrete behavior. Unfortunately, predicting this behavior is often very difficult. The relationships between cement properties and concrete behavior are often not understood, at least not quantitatively nor in detail. The development of new or improved test methods continues to be very difficult in the absence of this basic understanding, which will only be gained through research.

There are several good references on cement and concrete, many of which were used in preparing this chapter. Particularly useful were two textbooks on concrete [1,2], one on cement chemistry [3], and a publication of the American Concrete Institute on cement [4]. Some of the ideas discussed here were introduced recently in a separate paper [5].

FINENESS

Of the various physical properties that relate to cement quality, probably the most important and widely used is fineness. Because cement is a ground material, it has an inherently broad particle size distribution (Fig. 1). The

TABLE 1—Summary of Physical Properties Specified for Hydraulic Cements.

Physical Property	Portland Cement (ASTM C 150)	Blended Cement (ASTM C 595)	Expansive Cement (C 845)
Fineness	s	r,m	...
Density	...	...	r
Activity index	...	m	...
Consistency	...	s,r	...
Set	s,o	s	s
Heat of hydration	o	s	...
Volume change			
Drying shrinkage	...	s	...
Expansion	c	...	s
Autoclave expansion and contraction	s	s	...
Strength	s	s	s
Durability			
Air Content	s	s	s
Alkali reactivity	...	o,m	...
Sulfate expansion	o	s	...

KEY:
s = specified.
m = specified for constituent materials.
c = specified under certain conditions.
o = optional specification.
r = value must be reported but no limit has been specified.

---

[1] Assistant professor, Department of Civil Engineering, University of Illinois, Urbana, IL 61801-2352.

[2] Director of Cement Laboratory, California Portland Cement Co., Colton, CA 92324.

TABLE 2—Summary of Test Methods for Physical Properties of Hydraulic Cements.

Physical Property	ASTM Test Method
Fineness	
Air permeability	C 204
Turbidimeter	C 115
Sieving	C 184 (Nos. 100 and 200, dry)
	C 786 (Nos. 50, 100, 200, wet)
	C 430 (No. 325, wet)
Density	C 188
Activity index	C 311
Consistency	
Water requirement	included in C 109
Normal consistency	C 187
Set	
Time of set	C 266 (Gillmore)
	C 191 (Vicat)
	C 807 (modified Vicat)
False set	C 451
Heat of hydration	C 186
Volume Change	C 157
Expansion	C 1038
Restrained expansion	C 806
Autoclave expansion	C 151
Strength	C 109
Optimum SO_3	C 563
Durability	
Air content	C 185
Alkali reactivity	C 227 (using Pyrex glass)
Sulfate expansion	C 452 (for portland cement)
	C 1012 (for blended cement)

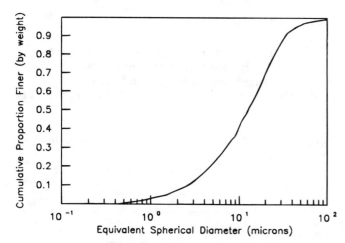

FIG. 1—Particle size distribution of a typical portland cement.

median particle diameter of Type I or II portland cement is typically about 10 to 20 mm, but particles range in size from a few tenths of a micrometre to 50 mm or more. Type III portland cement is typically finer than Type I, and Types IV and V portland cements are often coarser. Blended cements are also broad in their particle size distribution, may cover a somewhat different size range, and are often bimodal.

Significance

Fineness is a very important physical property for cement. Hydration rate is a function of fineness, so both strength and permeability are influenced by fineness. Increasing the fineness substantially increases the rate of hydration, a common strategy for meeting the faster strength gain specified for a Type III portland cement. Not only does strength increase at a faster rate, but also permeability is reduced at a faster rate when fineness is increased. Finer cements adsorb chemical admixtures more rapidly, often requiring higher admixture dosages. Finer cements also generate higher temperatures during hydration.

Fineness also affects workability. The particle size distribution controls the density with which cement particles pack, therefore influencing the fluidity (viscosity) and, indirectly, the strength.[3] With increased fineness generally comes a higher proportion of submicron-sized particles which are prone to flocculate. Unless prevented through the use of dispersing admixtures (water-reducers or superplasticizers), flocculation substantially increases the water demand of the cement, either reducing slump or, if a higher water content is used, reducing strength.

Until recently, there have been few parameters in the grinding process that could be used to control the particle size distribution. Recent grinding technology utilizing high-efficiency classifiers allows better control of the distribution [7]. Helmuth and colleagues [8–10] have recently shown that controlling the particle size distribution of cement, particularly by limiting the coarse particles (>20 mm) and the fine particles (<2 mm), can substantially improve both performance and energy efficiency during grinding. This particle size control requires specialized mechanical particle separation equipment, which is gradually becoming available. It is anticipated that this technology will find increasing application in the future.

Standards

There are three standard tests to measure cement fineness—air permeability, turbidimeter, and sieving. The first, ASTM Test Method for Fineness of Portland Cement by Air Permeability Apparatus (C 204), allows measurement of the specific surface area, called the Blaine surface area after the inventor of the apparatus. The second test, ASTM Test Method for Fineness of Portland Cement by the Turbidimeter (C 115), uses the Wagner apparatus to measure the particle size distribution by sedimentation. In the third, ASTM Test Method for Fineness of Hydraulic Cement by the 45-μm (No. 325) Sieve (C 430), powder is allowed to pass through a standard-sized sieve, assisted by a water spray.[4]

---

[3] Optimizing the particle size distribution to achieve very dense packing is the basis for DSP, a new high-performance cement-based material [6]. DSP is an acronym for Densified with Small Particles, a material developed in Denmark in the early 1980s. It is a castable product made from cement, silica fume, superplasticizer, and water at a low water-to-solid ratio, often as low as 0.18.

[4] Other tests are available using coarser sieves: ASTM Test Method for Fineness of Hydraulic Cement by the 300-μm (No. 50), 150-μm (No. 100), and 75-μm (No. 200) Sieves by Wet Methods (C 786); and ASTM Test Method for Fineness of Hydraulic Cement by the 150-μm (No. 100) and 75-μm (No. 200) Sieves (C 184).

Fineness is specified for most hydraulic cements. For all types of portland cement except III and IIIA, a minimum specific surface value is specified: either 280 m^2/kg by the air permeability test or 160 m^2/kg by the turbidimeter test. For blended cements, fineness (both the amount retained on the 45-μm sieve and the specific surface by the air permeability method) is listed as a physical requirement and must be included if the purchaser requests certification, but no fineness limits are specified. In addition, in blended cements containing pozzolan, no more than 20.0% of the pozzolan may be retained on a 45-μm sieve. For expansive cements, there is no fineness specification.

Improvements

The influence of fineness on hydration rate is well established, and there is a large body of literature relating Blaine surface area and such properties as rate of strength gain and heat of hydration. Likewise, the Blaine surface area has been a very reliable parameter for use in controlling particle size distribution during cement production. However, there are some problems with the Blaine specific surface. One is that the Blaine values are somewhat lower than the values calculated from particle size distributions or measured using conventional sorption techniques.[5] Also, surface area does not fully describe fineness. The full particle size distribution curve should be considered when assessing the effects of fineness on cement and concrete behavior. Two cements may have the same Blaine but quite different particle size distributions. The Blaine surface area is easy to measure, but new instruments based on sedimentation or light scattering, as described [11] in a recent ASTM Committee C1 symposium, offer convenience in measuring the full particle size distribution; efforts are now underway within ASTM Committee C1 to develop standard test methods using such methods. The Blaine surface area is convenient to use because it is a single parameter, whereas the full particle size distribution is difficult to apply in correlations with other parameters. One solution is to use the Rosin Ramler distribution [3,7], which describes the particle size distribution in terms of two parameters, the average particle size and the breadth of the distribution.

Given the major influence of fineness on cement and concrete behavior, it is surprising that fineness is not specified either for blended cement or for expansive cement. One reason is undoubtedly that the test methods for air permeability (ASTM C 204) and sedimentation test (ASTM C 115) were developed for use with portland cement, and need to be modified to use on blended cement and expansive cement.

DENSITY

The density of hydraulic cement, though not in itself a particularly significant property and not specified, is used to calculate fineness in ASTM C 204 and C 115, and is used to calculate concrete mix proportions by the absolute volume method [12]. Density is measured using the ASTM Test Method for Density of Hydraulic Cement (C 188), which utilizes a Le Chatelier flask filled with kerosine. The density of portland cement is usually assumed to be 3.15 Mg/m^3. Blended cement and expansive cement may vary substantially in chemical composition, so density must be measured. There is no specification regarding density of blended cement, and, for expansive cement, there is only the requirement that density be measured and reported.

The standard test method using the Le Chatelier flask is accurate and inexpensive. However, it tends to be time-consuming and uses large quantities of kerosine, which requires special disposal. The standard test allows use of alternate methods, and pycnometers[6] are now available that provide rapid and reproducible density measurements and entail no special disposal.

Though the density of portland cement is usually assumed to equal 3.15 Mg/m^3, it in fact ranges rather widely, from as low as 3.05 Mg/m^3 to as high as 3.25 Mg/m^3. This variation is enough to affect the cement volume used in calculating concrete mix proportions. It would be better to measure density of portland cement more frequently and to report the measured density. Unfortunately, the difficulties with the present standard method (discussed earlier) make it unlikely that routine measurement of density would be specified.

ACTIVITY INDEX

The activity index is discussed in a separate chapter of this manual (Mineral Admixtures) because it is also specified for mineral admixtures. The test is described in ASTM Method for Sampling and Testing Fly Ash or Natural Pozzolans for Use as a Mineral Admixture in Portland-Cement Concrete (C 311). The activity index is simply the compressive strength of mortar using a blend of 80% (by mass) portland cement and 20% (by mass) pozzolan relative to the compressive strength of mortar using the plain portland cement. The test for slag is modified slightly from that described in ASTM C 311 for pozzolan; these modifications are described in ASTM C 595. The specification for slag or pozzolan in ASTM C 595 stipulates that the activity index at 28 days be a minimum of 75%.

CONSISTENCY

Consistency refers to the flow behavior of a fresh mixture. It falls under rheology, the study of flow. Consistency is specified for blended cement, though not for portland or expansive cements. It must be measured and adjusted to a standard value when testing for compressive strength, autoclave expansion, setting time, and premature stiffening (discussed later). The tests do not actually measure

---

[5] Sorption of a gas such as nitrogen or water vapor is commonly used to measure the specific surface area of solids.

[6] For example, AcuPyc 1330 from Micromeritics, Norcross, GA.

consistency, but rather measure the amount of water required to produce some standard consistency.

Consistency measures (for example, slump of concrete, flow of mortar, and penetration of paste) are empirical. Fundamental rheological measures (for example, yield stress and plastic viscosity) exist but are not used for cement and concrete except perhaps in research studies. Tattersall and Banfill [13] and later Struble [5] have argued that fundamental measures of consistency have clear advantages over empirical measures. In particular, it is important to be able to measure both yield stress and viscosity, whereas slump and flow relate largely to yield stress [13]. Cement paste, mortar, and concrete are fluid at high stress, but highly viscous or even solid at low stress. Therefore, behavior during processes that involve high stress, such as consolidation using vibration, is expected to correlate with viscosity; whereas behavior during processes that involve low stress, such as settling of aggregate due to gravity, is expected to correlate with yield stress.

The consistency of cement paste depends on the water-to-cement ratio (w/c) and on various aspects of the cement: fineness, flocculation, and rate of hydration reactions. The consistency of mortar or concrete depends on the consistency of the cement paste plus various aspects of the aggregate: quantity, grading, and shape.

Significance

The significance of cement consistency is twofold. Strength (for some cements), autoclave expansion, setting time, and premature stiffening tests are measured at stipulated consistency rather than w/c, and the test is used to determine what w/c provides the specified consistency. More importantly, cement consistency is generally assumed to affect concrete workability. Concrete workability (slump) is assumed to correlate with paste consistency at the same w/c and including the same mineral and chemical admixtures. Unfortunately, this correlation has eluded researchers, probably due (at least in part) to the highly empirical nature of both the cement and concrete consistency test methods.

Standards

As noted previously, consistency tests actually measure the amount of water required to produce a standard consistency. There are two tests: flow and penetration resistance.

The flow test is included in the ASTM Test Method for Compressive Strength of Hydraulic Cement Mortars (Using 2-in. or 50-mm Cube Specimens) (C 109). This test establishes the amount of water required in mortar to obtain a flow of 105 to 115 on a standard flow table.

The other test, the ASTM Test Method for Normal Consistency of Hydraulic Cement (C 187), involves penetration resistance. Normal consistency is measured on paste using the Vicat apparatus equipped with plunger rather than needle. A stiff consistency is reached when the plunger penetrates to a point 10 ± 1 mm below the original surface. This test is used in preparing pastes to measure setting time (discussed later) and, at a slightly different consistency, to measure premature stiffening (also discussed later).

Consistency (or water requirement) using standard flow (ASTM C 109) is only specified for certain blended cements. However, the water requirement using constant flow must be determined as part of the compressive strength test for all blended cements and expansive cements (any cement other than portland and air-entraining portland which are tested at fixed water contents).[7]

Improvements

The consistency tests allow selection of w/c for use in various other tests, and are satisfactory for this purpose. However, the tests and specifications concerning consistency are particularly confusing and need clarification. For example, the test for water requirement to provide standard flow would be better as a separate test, not part of the strength test. The relationship, assuming one exists, between cement consistency and concrete workability would be much more clear if consistency tests were structured to measure consistency at some standard w/c, rather than the amount of water to provide some standard (and arbitrary) consistency. Such an approach would facilitate the use of cement consistency tests to study the influence of other factors on concrete workability (for example, chemical or mineral admixtures, temperature, time). Finally, cement consistency is an important contributor to concrete workability, so consistency specifications should be developed for all hydraulic cements.

Most importantly, test methods are needed that provide fundamental measures of consistency based on rheology. Consistency should be described using more than one rheological parameter (at the very least, yield stress and plastic viscosity). Unfortunately, there are major technical difficulties in measuring rheological properties of paste, mortar, and concrete, and these difficulties will continue to limit development of improved standards in this area.

SET

Set refers to the transformation of cement paste, mortar, or concrete, from a fluid material to a rigid solid. Set is a gradual and progressive change controlled by hydration of the cement. Therefore, it is logical to expect that concrete would set at the same time as cement paste or mortar

---

[7] The same approach is used for mortar-bar expansion tests: ASTM Test Method for Effectiveness of Mineral Admixtures or Ground Blast-Furnace Slag in Preventing Excessive Expansion of Concrete Due to the Alkali-Silica Reaction (C 441) (flow 100 to 115); ASTM C 227 (flow 105 to 120, procedure modified slightly); and ASTM Test Method for Potential Volume Change of Cement-Aggregate Combinations (C 342) (flow 100 to 115). A similar approach is used for various other tests, sometimes with a slight modification in the flow level: ASTM Test Method for Early Volume Change of Cementitious Mixtures (C 827), ASTM C 185, ASTM Test Method for Bleeding of Cement Pastes and Mortars ASTM Test Method for Drying Shrinkage of Mortar Containing Portland Cement (C 596), ASTM C 1012, and ASTM C 157.

prepared using the same w/c. Two setting times are recognized, initial set and final set. These times are arbitrary, in that they do not correspond exactly to any specific change in properties nor to any specific levels of hydration reaction.

It is particularly important to show that the cement is not prone to premature stiffening. Cements occasionally show premature stiffening, either false set or flash set. In false set, the cement stiffens rapidly soon after mixing but regains its fluidity if remixed. False set may result from hydration of calcium sulfate hemihydrate ($C\bar{S}H_{1/2}$) to form gypsum ($C\bar{S}H_2$). It is also thought that false set can result from the formation of excess ettringite ($C_3A \cdot 3C\bar{S} \cdot H_{32}$) soon after mixing. This phenomenon has not been satisfactorily demonstrated. Ettringite precipitation may not actually cause false set, but instead contribute to its severity. Finally, false set may result from flocculation of cement particles, rendering the mixture highly thixotropic in its rheological behavior. Flash set, on the other hand, occurs when the level of calcium sulfate is not sufficient to retard hydration of tricalcium aluminate (C_3A), allowing the formation of substantial calcium aluminate hydrate or calcium monosulfoaluminate hydrate ($C_3A \cdot C\bar{S} \cdot H_{12}$). In the case of flash set, fluidity cannot be regained on remixing.

Significance

It is very important to predict and control setting time during concrete processing so concrete remains workable for a sufficient time that it can be placed and consolidated, but not for such a long time that finishing or form removal is excessively delayed. It is particularly important to prevent premature stiffening of the concrete due to false set or flash set. Because concrete set is controlled by reactions of the cement and water, perhaps modified by admixtures, it is expected that setting times of concrete may be measured using cement paste or mortar. The tests are used to ensure that the cement does not produce abnormal setting times or to test the response of a particular combination of cement and chemical admixture. Limits in setting time of paste and mortar also help assure overall concrete performance.

Standards

The several laboratory tests for setting time rely on measuring penetration resistance of cement paste or mortar. In the ASTM Test Method for Time of Setting of Hydraulic Cement Paste by Gillmore Needles (C 266), set is the time elapsed from mixing until the paste supports either the initial or the final needle without appreciable indentation. The Gillmore apparatus is shown in Fig. 2. The test is run at a normal consistency (ASTM C 187, as described previously). In the ASTM Test Method for Time of Setting of Hydraulic Cement by Vicat Needle (C 191), normal consistency paste is allowed to rest for 30 min, then its resistance to penetration by a 1-mm needle is measured at specified time intervals. This apparatus is shown in Fig. 3. Initial set occurs when the penetration is 25 mm, and final set occurs when there is no visible penetration. Finally, the ASTM Test Method for Time of Setting of Hydraulic Cement Mortar by Vicat Needle (C 807) utilizes specified water and cement contents, while the sand content is adjusted to obtain a specified consistency (measured using the modified Vicat plunger). The procedure is similar to that of the previous test; the specimen is allowed to rest for 30 min, then its resistance to penetration of 2-mm needle is measured at specified time intervals. Set has occurred when the penetration is 10 mm. This test is specified for expansive cements.

For measuring setting time of paste, there is no obvious reason to prefer the Vicat test over the Gillmore test (or

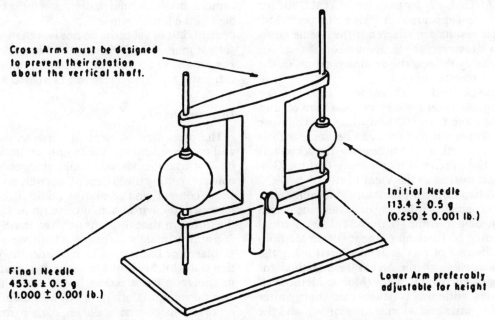

FIG. 2—Gillmore apparatus and test specimen (adopted with permission from ASTM C 266).

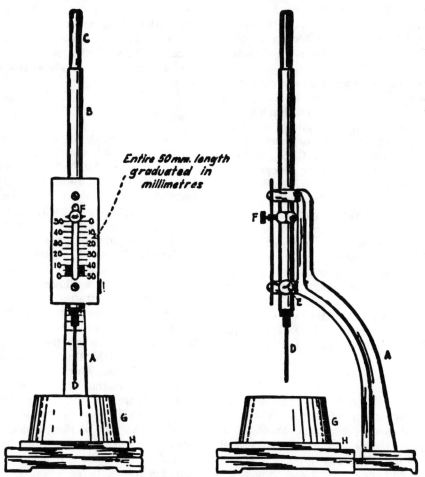

FIG. 3—Vicat apparatus (adopted with permission from ASTM C 191).

vice versa), except perhaps personal preference. Of the two Vicat tests (ASTM C 191 for paste and ASTM C 807 for mortar), the paste test appears to provide better precision, probably because results are affected if the needle lands on a sand grain. However, the large number of tests for setting time probably reflects the arbitrary nature of the test methods and specifications.

Setting time is specified for all hydraulic cements. For portland cement, initial set must be not less than 60 min and final set not more than 600 min using the Gillmore test (ASTM C 266); or time of set must be not less than 45 min and not more than 375 min using the Vicat test (ASTM C 191). The purchaser may specify that the Gillmore test be used instead of the Vicat test.

There are also two tests for premature stiffening (false set or flash set) of cement. Stiffening of paste is measured using the Vicat needle according to the ASTM Test Method for Early Stiffening of Portland Cement (Paste Method) (C 451), and stiffening of mortar is measured using the Vicat plunger according to the ASTM Test Method for Early Stiffening of Portland Cement (Mortar Method) (C 359). In both tests, stiffening is determined shortly after mixing (5 min for paste and 11 min for mortar), and the penetration may be measured after remixing to help differentiate between false set and flash set.

There is an optional false set specification for portland cement, in which final penetration must be a minimum of 50% of the initial penetration (ASTM C 451). For blended cement, initial set must be not less than 45 min and final set not more than 7 h using the Vicat test (ASTM C 191). For expansive cement, the minimum time of set is 90 min using the modified Vicat test (ASTM C 807).

Improvements

The presumed correlation between paste (or mortar) and concrete setting time is only tentative, and can only be verified under standard laboratory conditions, because concrete setting times depend not only on cement content, water content, and admixture addition, but also on shear history and environmental conditions. It is assumed that the different time limitations given in ASTM C 191 and C 266 approximately correspond with one another and with similar time limitations in the concrete. While it appears that the arbitrary limits in general give satisfactory results in concrete, more research could lead to more specific and meaningful definitions of initial and final set based on reactions occurring during early hydration, especially in the presence of admixtures (which can substantially affect the reaction kinetics). For methods dealing with premature stiffening, the limitations are even more arbitrary, since there are no guidelines to judge how and when an occurrence of premature stiffening in a cement may be manifested in concrete. Premature stiffening can also be drastically affected by cement and water contents, admixture presence, type of mixing, and environmental conditions in the concrete.

There are several improvements that could be made in setting tests. For example, it should be possible to reduce the number of tests and specifications concerning setting time. In addition, set is not a very reproducible measurement. The correlation between set of cement paste and set of concrete is not as strong as one would expect, perhaps due (at least in part) to the arbitrary way in which set is defined. Rheological parameters might provide a way to define set that is not so arbitrary, and that would allow separating the effects of hydration reactions from the effects of flocculation.

HEAT OF HYDRATION

Concrete structures generate a substantial amount of heat; if not dissipated, this heat produces a substantial increase in the concrete temperature. Concrete that is subject to large temperature changes will crack, especially if the paste and aggregate have substantially different coefficients of thermal expansion. The heat is a direct result of the cement hydration reactions which are exothermic and occur most rapidly during the first hours and days after mixing. Thus, it is important to use cement with a moderate or low heat of hydration when constructing large concrete structures from which heat is not rapidly dissipated or during hot weather construction. Preventing excessive temperature rise is the basis for specification of a Type II or IV portland cement.

The ASTM Test Method for Heat of Hydration of Hydraulic Cement (C 186) measures the difference in heat of solution of dry cement and cement that has been hydrated for either 7 or 28 days. A maximum heat of hydration using this test is an optional requirement for Type II or Type IV portland cement (70 cal/g for Type II, 60 cal/g for Type IV at seven days, and 70 cal/g for Type IV at 28 days).

The heat of hydration provides a convenient way to monitor hydration, using either isothermal or adiabatic calorimetry. A calorimetry curve of the type shown in Fig. 4 could be used to estimate heat of hydration at any specific age as an alternate method to the heat of solution measurement (ASTM C 186), if a suitable test method were developed. The advantage of using a calorimeter is that information would also be obtained concerning other aspects of hydration (for example, the time at which the induction period ends, which correlates approximately with initial set).

VOLUME CHANGE

Concrete can undergo changes in volume (shrinkage or expansion) for a number of reasons: load (creep), drying and wetting cycles, expansive reactions that may occur in the cement, and expansive reactions that may occur between cement and aggregate. Creep and drying shrinkage are expected; these result from the porous nature of the cement hydration products. They are to some extent restrained in concrete by the aggregate, and must be taken into account during the design of concrete structures. Excessive or unexpected volume changes may occur due to specific reactions in the cement (unsoundness) or to reactions between cement and aggregate and must be avoided. Reactions between cement and aggregate fall under the category of durability problems (discussed later), while reactions that occur during cement hydration are measured using the tests described here.

Drying shrinkage is specified only for certain blended cements (P and PA), for which it is limited to a maximum of 0.15%. It is measured using ASTM Test Method for Length Change of Hardened Hydraulic Cement Mortar and Concrete (C 157). Although the test purports to mea-

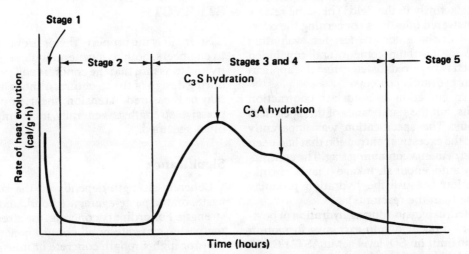

FIG. 4—Isothermal calorimetry curve of a typical portland cement (reprinted with permission from Ref 1).

sure expansion or contraction due to any cause other than applied force or temperature change, it is clearly intended to measure drying shrinkage. As noted earlier, all hydraulic cements are prone to drying shrinkage. It is known that composition and fineness of the cement can influence drying shrinkage, though the effects may not be large. Therefore, it would seem worthwhile to have a shrinkage specification for all hydraulic cements.

Expansion may occur in portland cement or in blended cement due to certain hydration reactions after the cement has set. Several reactions may cause expansion, in particular the hydration of free lime (CaO) to form calcium hydroxide [Ca(OH)$_2$], the hydration of periclase (MgO) to form magnesium hydroxide [Mg(OH)$_2$], and the formation of excess $C_3A \cdot 3C\bar{S} \cdot H_{32}$ through reaction of C_3A or of $C_3A \cdot C\bar{S} \cdot H_{12}$ with $C\bar{S}H_2$. These reactions are slow, so they are accelerated in the laboratory by testing at an elevated pressure and temperature.

The tendency to expand due to hydration of CaO or MgO is measured using the ASTM Test Method for Autoclave Expansion of Portland Cement (C 151). The specification is a maximum expansion of 0.80% for portland cement, and 0.50% for blended cement. It is not obvious why the allowed expansion is lower for blended cement than for portland cement, or why the expansion is lower for blended cement than it is for mineral admixtures in concrete (0.8% expansion according to the ASTM Specification for Fly Ash and Raw or Calcined Natural Pozzolan for Use as a Mineral Admixture in Portland Cement Concrete (C 618)). Although there is a specification for blended cement using this test method, the method at present is only applicable to portland cement.

It is assumed that excessive autoclave expansion corresponds to unsoundness in concrete. However, a recent literature review [14] on the efficacy of the autoclave test uncovered examples where such correspondence was not observed in cases involving blended cements containing fly ash. In one case, a blended cement containing fly ash and high free-CaO clinker gave an acceptable autoclave expansion, yet produced disruptive expansion in the field; in another case, a blended cement containing a high free-CaO Type C fly ash gave excessive autoclave expansion, yet performed satisfactorily in the field. The same review posed several unanswered questions concerning the overall appropriateness of the autoclave test for evaluating unsoundness even for portland cement, and concluded that further research is required to determine the applicability of this test to hydraulic cements.

There is also a specification on autoclave contraction for blended cements, but the significance of this specification is not obvious. The specification was apparently imposed to prevent the excessive contraction that had been reported in blended cements containing slag. The contraction may be merely autogenous shrinkage (that is, shrinkage due to hydration because the hydration products occupy less volume than the reactants).

As previously noted, expansion during hydration of portland cement may also occur due to excessive formation of $C_3A \cdot 3C\bar{S} \cdot H_{32}$. The limit on SO$_3$ level set in ASTM C 150 (and discussed in the chapter on chemical composition) prevents this. However, for some portland cements the SO$_3$ required to optimize strength or drying shrinkage is greater than allowed by this limit. In that case, the higher level of SO$_3$ may be used as long as the cement does not develop expansion greater than 0.020% at 14 days when tested using the ASTM Test Method for Expansion of Portland Cement Mortar Stored in Water (C 1038).

The alkali-silica reaction is a cause of expansion in concrete, and is discussed in the section on durability, but there is the possibility of a similar expansion in blended cements due to reaction between alkalies from the cement and silica in the pozzolan. To ensure that such a reaction does not cause deleterious expansion in concrete, a mortar expansion test for pozzolan for use in blended cement is included as part of ASTM C 595. This method tests combinations of pozzolan and cement or clinker using the ASTM Test Method for Potential Alkali Reactivity of Cement-Aggregate Combinations (Mortar-Bar Method) (C 227). The test is usually used to detect the alkali-silica reaction between cement and aggregate (discussed in the section on durability). When run with a non-reactive aggregate, as described here, it tests for expansive reactions between pozzolan and cement. The test is run at various levels of pozzolan between 2.5% and 15% in order to detect excessive expansion due to an adverse proportion of pozzolan in combination with portland cement. This test method would be easier to understand if it were described more fully in a separate method, rather than included as part of the blended cement specification. It should be noted that there is no similar limit for fly ash used as a mineral admixture in concrete (ASTM C 618).

The situation is somewhat different for expansive cements, whose object is to use expansive reactions to offset the drying shrinkage and the resulting tensile cracking of ordinary concrete. Expansion is measured using the ASTM Test Method for Restrained Expansion of Expansive Cement Mortar (C 806). To comply with the expansive hydraulic cement specification (ASTM C 845), expansion must be within the range 0.04% to 0.10% at 7 days and not be more than 115% of its 7-day value at 28 days.

STRENGTH

Strength is the property that is probably most important to engineers, both as a general indicator of concrete quality and to assure that the concrete will perform as intended during design of the structure. Although concrete strength may be measured in tension, shear, or compression, compressive strength is generally most important and most often specified.

Significance

Concrete strength depends on the strength of cement paste, on the paste-aggregate bond, and on the aggregate strength. For ordinary concrete, the strengths of paste and the paste-aggregate bond control concrete strength. It is only for high-strength concrete or unusually low-strength aggregate that aggregate characteristics become important.

The strength of both paste and the paste-aggregate bond depends largely on paste porosity. Porosity is reduced (and strength is increased) by initially packing particles very densely and by filling remaining voids with hydration products. So strength is increased by reducing the w/c ratio and by increasing the degree of hydration. It is the latter parameter that may be influenced considerably by the composition and microstructure of the cement, and both parameters are influenced by fineness.

Neville [15] reported a reasonably good linear correlation between concrete strength and strength of mortar at the same w/c and degree of hydration. It is reasonable, therefore, to suppose that concrete strength can be predicted from the strength of cement paste or mortar. But the evidence for such a direct relationship has generally been lacking, as noted recently by Struble [5], and even in ASTM C 109 there is a caution against using mortar results to predict concrete strength. However, Gaynor [16] more recently showed a good correlation between mortar and concrete strength, concluding that concrete strength may be predicted from mortar strength, w/c, and air content of the concrete. The appreciable variability in strength data must be considered when attempting to predict cement performance in concrete, and more than a single batch of mortar must be tested.

The nature of the fracture process has important implications to mortar and concrete strength. When concrete is loaded, small cracks develop at the interface between cement paste and aggregate. The cracks may develop because this interface is weaker than the bulk paste or the aggregate, or they may exist in concrete before it is loaded due to drying shrinkage or thermal stresses. The development and growth of these cracks reduce concrete strength. Therefore, it is expected that fracture mechanics, the area of research concerned with development and growth of cracks, will provide a better understanding of strength and fracture (as discussed later).

The significance of mortar strength is broader, however, than its relationship to concrete strength. Many of the potential problems in cement hydration affect its strength. Furthermore, permeability of hydrated cement is a function of w/c and degree of hydration just as is strength, so permeability decreases as strength increases, and many aspects of durability are improved (as will be discussed further). Thus, strength provides an excellent indication of the overall quality of hydrated cement, mortar, or concrete.

Standards

There are several tests for mortar strength (compressive, tensile, and flexural strength),[8] though the specifications utilize only ASTM C 109. This laboratory test allows measurement of compressive strength of mortar prepared using a graded standard sand[9] and water adjusted to provide either a specified w/c or a specified flow. The specified w/c is used for portland cement, either 0.485 for portland cement or 0.460 for air-entraining portland cement. A specified flow of 105 to 115 is used for all other cements. Cubes are moist-cured 24 h in their molds, then stripped and immersed in lime-saturated water until they are tested.

Minimal compressive strength requirements are specified for all hydraulic cements—portland cement, blended cement, and expansive cement. However, these strength levels are easily met by cements with a wide variety of chemical composition and fineness levels.

In all cases, the minimum average compressive strength measured using ASTM C 109 is specified. Most types of portland cement have compressive strength requirements at three and seven days, though some cement types have 1- and 28-day requirements. For Type I, the seven-day specified strength is 19.3 MPa. The values are slightly lower for Type II cement, even lower for Type IV and Type V cements, and higher for Type III cement. Values are lower for each corresponding air-entraining cement. There is also an optional 28-day specification for Type I and Type II. Likewise, strengths are specified at 3, 7, and 28 days for most blended cements, though some types do not have three-day specifications. For most general-purpose blended cements (Types IS and IP), the seven-day strength level is 19.3 MPa, similar to the level specified for Type I portland cement, while levels specified for moderate-sulfate resistant cements (MS) are slightly lower, and levels specified for portland-pozzolan cement (Type P) and slag cement (Type S) are much lower. For expansive cements, minimum strength levels are specified at 7 and 28 days; the seven-day level is 14.7 MPa, somewhat lower than that for Type I portland cement. In all cases, these strength levels are relatively low. Measured strength values are typically much higher than the levels specified.

Improvements

As noted previously, the existing strength specifications are well below strength levels obtained in current practice. In fact, these strength specifications do not represent acceptable minimum strength levels, and they are so low as to be of limited use. It would be an improvement if the strength specifications were raised. By raising the specifications to levels more consistent with standard practice, cement users would get a product that provides acceptable minimum strength levels.

Even more important is the issue of strength uniformity. Strength specifications for all hydraulic cements would be enhanced if they were modified to include some restriction on variability. In this way, the cement user would be guaranteed a uniform performance as well as a specified minimum strength level.

---

[8] Other strength tests include ASTM Test Method for Compressive Strength of Hydraulic Cement Mortars (Using Portions of Prisms Broken in Flexure) (C 349); ASTM Test Method for Tensile Strength of Hydraulic Cement Mortars (C 190), now discontinued; ASTM Test Method for Flexural Strength of Hydraulic Cement Mortars (C 348); and ASTM Test Method for Evaluation of Cement Strength Uniformity from a Single Source (C 917).

[9] A natural silica sand from Ottawa, IL, that conforms to ASTM Specification for Standard Sand (C 778).

As previously discussed, it has been shown that mortar and concrete strength are well correlated. Therefore, the caution in ASTM C 109 against using mortar results to predict concrete strength should probably be deleted from the test method.

Although we use strength in compression to describe the performance of cement and concrete, failure (that is, development and growth of cracks) typically occurs in tension. Concrete is a brittle material, with a tensile strength only about one tenth of its compressive strength. Other types of strength, tensile and shear, are generally influenced by the same parameters that control compressive strength—in particular, the w/c and the degree of cement hydration. Therefore it is assumed that specifications and tests for compressive strength are applicable to other types of strength. This is a simplifying assumption, in that compressive strength is much easier to measure. However, this assumption is not without risk. It is known that the ratio of tensile strength to compressive strength is not constant, but rather is affected by a number of parameters, from air drying during aging (which reduces tensile strength more than it does compressive strength) to air entrainment (which reduces compressive strength more than it does tensile strength). While it is probably not necessary to have additional test methods and specifications for tensile strength of cement, it is certainly necessary to appreciate the differences between tensile and compressive strength.

As discussed earlier, fracture mechanics is the area of research concerned with failure due to cracking. This is an active area of research and should provide a better understanding of the processes of cracking in concrete, mortar, and paste. Although there is currently no ASTM test for fracture mechanics parameters of concrete, fracture mechanics tests have been developed that are suitable for use as standard test methods for these materials [17]. The ACI Committee on Fracture Mechanics of Concrete recently concluded that fracture mechanics now (after intense research during the last decade) appears ripe for applications in design practice [18]. However, considerable work remains to be done before concrete strength (or strength-related performance of concrete structures) can be predicted from measurements of fracture mechanics parameters, and it is not yet clear whether fracture mechanics parameters of concrete may be estimated from parameters measured on its constituents.

OPTIMUM SO$_3$ CONTENT

The effect of SO$_3$ content of cement on both strength and drying shrinkage was established by Lerch as early as 1946 [19]. The proportion of SO$_3$ in the cement required for optimum performance may be different for mortar compared with that for concrete, and may change with age of the material. The mechanism of this phenomenon is not well understood, but it seems likely that the proportion of sulfate affects the degree of porosity during early aluminate hydration reactions, though it is modified by subsequent silicate hydration. Optimum performance presumably corresponds to minimum porosity at any age, and the increase in SO$_3$ requirement for optimum performance at later ages supports this contention. Some cements, however, show a regression of their SO$_3$ requirements at ages of 28 days, and this is more difficult to explain. Further research is required for this important parameter.

The current test for optimum SO$_3$ content is the ASTM Test Method for Optimum SO$_3$ in Portland Cement (C 563). This method only evaluates optimum SO$_3$ for 24-h strength, and does not address drying shrinkage. Efforts are underway within ASTM Committee C1, however, to extend the method to later ages. Further research is needed to broaden the scope even further—to consider concrete, including the effects of admixtures, and to include drying shrinkage.

DURABILITY

There are several causes of concrete deterioration that relate in part to properties of the cement: freeze-thaw deterioration, alkali-silica expansion, and sulfate reaction. Tests are available to predict the influence of the specific cement on each of these deterioration reactions.

Air Content

Significance

The air content of concrete controls its ability to withstand cycles of freezing and thawing. Without entrained air, most concrete will deteriorate after only a few freeze-thaw cycles. Air is obtained in concrete using an air-entraining admixture, a surfactant that lowers the surface tension of water and thereby stabilizes air bubbles. Whether the added air is sufficient to protect against freeze-thaw damage depends not only on the volume of air, but also on the size and distribution of the bubbles throughout the paste portion of the concrete.

An air-entraining admixture may be added to the concrete or as part of the cement (called an air-entraining cement). ASTM recognizes several air-entraining cements (designated A), both portland cement and blended cement.

The volume of entrained air in concrete depends not only on the dosage of air-entraining agent (whether added directly to the concrete or as part of an air-entraining cement), but also on various other parameters of the concrete and how it is mixed. It should be noted that even a nonair-entraining cement may influence the air content in concrete. Finely ground cements entrain less air than do coarsely ground cements. Use of finely ground slag or pozzolan in blended cements also reduces the air content and necessitates a larger dosage of air-entraining admixture. Most importantly, fly ash in a blended cement may reduce the air content due to adsorption of air-entraining agent, in particular by carbon impurities in fly ash.

Gaynor [16] addressed the correlation of air content in mortar and concrete. Correlations were good (standard errors of estimated air content in concrete were 0.3% to 0.8%), especially considering the inherent variability of air content tests. But he noted that air contents in mortar are higher than air contents in concrete and that many nonair-entrained cements show mortar air contents greater than 10%.

Concrete air content is also important because air (whether entrained or entrapped) reduces strength. The rule of thumb is that every 1% increase in air content reduces strength of concrete by about 5% [2]. Thus, it is important that no more air than necessary be entrained.

Standards

The potential of cement to entrain air is determined using the ASTM Test Method for Air Content of Hydraulic Cement Mortar (C 185). In this test, the air content of mortar is measured by the difference between measured and calculated volume, the calculated volume depending on the proportion and density of each mortar constituent.

Both minimum and maximum air contents are specified for each air-entraining hydraulic cement. For nonair-entraining cements (portland and blended), only a maximum air content is specified. The minimum air content assures that the air-entraining cement provides an acceptable air content in concrete for resistance to freeze-thaw deterioration. The maximum air content assures that no unacceptable loss of strength will occur due to entrained air.

The specified air content is the same for portland cement and blended cement. The maximum air content for nonair-entraining cement is 12%. For air-entraining cements, the minimum air content is 16% and the maximum air content is 22%. There is no specification on air content for expansive cement.

Improvements

There should be a specified air content for expansive cement, for the same reasons there are specifications for the other hydraulic cements.

Alkali-Silica Reaction

Another common cause for concrete deterioration is the alkali-silica reaction, a reaction in which certain forms of silica (amorphous or reactive crystalline forms), generally part of the aggregate, react with the highly alkaline pore fluid in concrete to produce an alkali-silica gel. This gel may sorb water and swell, causing expansion or cracking of the concrete.

Alkalies (sodium and potassium) in concrete are typically derived from the cement. Deleterious expansion is usually prevented through use of a low-alkali cement, an optional specification for portland cement. Expansion is generally reduced by using a blended cement; pozzolan reduces the alkalinity of the pore fluid, and slag may serve the same function. Not every blended cement is effective in reducing expansion, so specific cements must be tested explicitly.

Significance

Tests are used to determine whether a specific blended cement or the pozzolan or ground slag used in a blended cement is successful in preventing deleterious alkali-silica expansion when combined with reactive aggregate in mortar. These tests are only significant, however, in-so-far as alkali-silica expansion of mortar bars corresponds to the occurrence of deterioration in concrete. There is much concern within ASTM Committee C9 about the validity of mortar-bar tests, and efforts are underway to improve the tests for predicting alkali-silica reaction in concrete.

Standards

Of the several tests relating to alkali-silica reaction, the one used to assess performance of blended cement is ASTM C 227. In this test, crushed Pyrex[10] glass is used as the reactive aggregate.[11]

Blended cement (ASTM C 595) includes an optional specification for mortar expansion due to alkali-silica reaction: a maximum of 0.020% expansion at 14 days and 0.060% expansion at eight weeks. These expansion levels, while reasonable, are different from those listed as excessive in the ASTM Specification for Concrete Aggregates (C 33)—0.05% at three months or 0.10% at six months.

Similarly, the alkali reactivity of pozzolan for use in pozzolan-modified (PM) blended cement must be measured to assure that the pozzolan reduces expansion to an acceptable level. In this test (described as part of ASTM C 595), blends of various proportions of clinker and pozzolan are tested according to ASTM C 227 using a sand judged to be nonreactive. The maximum expansion for all blends after 91 days is 0.05%.

There are currently no physical tests or specifications regarding the alkali-silica reaction for portland cement or for expansive cement, though research is currently underway to develop such standard test methods and specifications for portland cement.

Improvements

The level of expansion specified for blended cement (ASTM C 595) is different than the level specified for aggregate (ASTM C 33). It seems more reasonable that these levels should be the same, in that they apply the same test method to the same deterioration process.

A more substantial concern regarding these alkali-silica tests is their use of Pyrex glass as a standard reactive aggregate. Studies have shown that Pyrex is not suitable as a standard aggregate (for example, Ref *20*), and the current research to develop a performance test and specification for portland cement is exploring alternatives for the standard reactive aggregate.

Sulfate Reaction

Attack of concrete by sulfate ions is a particular problem where the sulfate concentration is high, such as in the western part of the United States. The attack involves reactions between cement hydration products and sulfate ions, reactions that are expansive and cause cracking and deterioration. The primary reaction is between $C_3A \cdot \bar{C}S \cdot H_{12}$ and sulfate ion to produce $C_3A \cdot 3\bar{C}S \cdot H_{32}$. To prevent this

---

[10] Pyrex Glass No. 7740 from Corning Glass Works, Corning, NY.

[11] Otherwise, the approach is basically the same as ASTM C 441.

deterioration, it is necessary to limit the amount of $C_3A \cdot \bar{C}S \cdot H_{12}$ in the hydrated cement, usually by reducing the amount of C_3A in the cement, as in a Type II or V portland cement or a blended cement. It is also important to reduce the permeability of the cement paste, generally by reducing its w/c.

Significance

The resistance of a particular cement to sulfate attack is important because it relates directly to the performance of concrete when exposed to sulfate ions.

Standards

The resistance of portland cement to sulfate attack is measured using the ASTM Test Method for Potential Expansion of Portland Cement Mortars Exposed to Sulfate (C 452). Excess sulfate is added to the portland cement in the form of gypsum, and length changes of mortar bars prepared from this cement-gypsum mixture are measured. This was found to be a satisfactory test for predicting the performance of portland cement in concrete exposed to sulfate ions, but not for blended cement [21]. Therefore an alternative test was developed for blended cements, the ASTM Test Method for Length Change of Hydraulic-Cement Mortars Exposed to a Sulfate Solution (C 1012). In this test, bars are immersed in a sulfate solution and their length changes measured.

Sulfate expansion is an optional requirement for Type V portland cement; the maximum expansion at 14 days using ASTM C 452 is 0.040%. A sulfate expansion requirement is also stipulated for moderate sulfate-resisting blended cements (designated MS); the maximum expansion using ASTM C 1012 is 0.10% at 180 days.

DISCUSSION OF TESTS FOR PHYSICAL PROPERTIES

More knowledge is needed about the fundamental mechanisms responsible for the various properties of cement paste, mortar, and concrete. Many of the tests described here are empirical and need a better fundamental basis. The need for better fundamental understanding was recently noted for sulfate deterioration [22], but applies equally well to most aspects of cement and concrete behavior. In some cases, the fundamental knowledge is available but just needs to be applied in standards (as in the case of consistency), but, in most cases, additional research is needed.

There are many benefits of testing paste or mortar rather than concrete. Tests using paste or mortar are generally preferred because they are simpler than tests using concrete. There is no need to develop a standard coarse aggregate. The effects of mineral and chemical admixtures on concrete performance may be studied using paste or mortar. But the results from paste or mortar do not necessarily correlate with concrete, and further research is needed to understand how to make such correlations.

Many tests developed for portland cement have been applied to blended cement and expansive cement without modification. Some tests are satisfactory for all hydraulic cements, but many tests (for example, fineness, sulfate deterioration) are not. It is important to modify existing tests or develop new tests for blended cement and expansive cements.

One aspect of concrete performance that we do not currently measure, and which depends largely on properties of the cement, is permeability. Permeability, or the related parameter diffusivity, refers to the rate at which fluid, or ionic species, moves through the material. In hardened paste, mortar, or concrete, such movement takes place through the small pores, either capillary pores or the much smaller gel pores. Thus, permeability is in many respects the inverse of strength, in that it is reduced either by lowering the w/c or by increasing the degree of hydration.

As with strength, the permeability of concrete depends, at least broadly, on the permeability of the paste portion of the concrete. Permeability of paste depends on the volume and interconnectedness of pores. Concrete permeability also depends on permeability of the aggregate and of the paste-aggregate interface, and may be increased by microcracking in the paste or at the interface.

The rate of most concrete deterioration processes, such as those discussed here, depends directly on permeability or diffusivity. Despite its importance, there is currently no standard test method by which to measure concrete permeability. Several laboratories are carrying out concrete permeability measurements, so perhaps a standard test method will be developed in the near future. Because permeability is so important to concrete durability, it is an important area for further research and for the development of new standard test methods for both cement paste and concrete.

Creep is another important property of hydrated cement that is not currently measured. Concrete deforms under load, and these creep deformations must be measured or predicted and incorporated into design of concrete structures. Creep, like drying shrinkage, occurs in the hydrated paste and is restrained by aggregate. Creep depends on the level and duration of load. Like shrinkage, permeability, and strength, creep also depends on the pore structure of the hydrated paste, the w/c, the degree of hydration, and the moisture content. There is a general inverse relationship between ultimate specific creep and compressive strength in concrete. Thus, it would seem useful to measure creep of cement paste or mortar in order to determine the influence of the specific cement on creep in concrete.

CONCLUSIONS

For most of the physical properties of cement, their significance lies in how well they relate to or predict the behavior of cement in concrete. The relationships between cement properties and behavior of cement in concrete in some cases are understood only in a general way, and in other cases are not understood at all. Some of the properties allow us to predict the behavior in concrete with confidence—fineness, heat of hydration, expansion, and durability. Others allow us to predict behavior only in a general way, and the precise nature of the relationship is not yet understood and additional research is needed—consistency, set, drying shrinkage, strength. There are several aspects of concrete behavior that are clearly influenced by cement, but for which no cement tests exist—fracture mechanics parameters, permeability, creep.

There are several tests that should be improved to take advantage of better knowledge or of recent technical developments—fineness, consistency, heat of hydration. Other tests need to be modified or new tests developed for use with blended cements or hydraulic cements. Finally, tests and specifications that are not logical or consistent should be improved.

REFERENCES

[1] Mindess, S. and Young, J. F., *Concrete*, Prentice-Hall, Inc., Englewood Cliffs, NJ, 1981.

[2] Mehta, P. K. and Monterro, P. J. M., *Concrete Structure, Properties, and Materials*, Second Edition, Prentice-Hall, Inc., Englewood Cliffs, NJ, 1993.

[3] Taylor, H. F. W., *Cement Chemistry*, Academic Press, London, 1990.

[4] "Guide to the Selection and Use of Hydraulic Cements," *Manual of Concrete Practice*, Committee 225, American Concrete Institute, Detroit, MI, 1991.

[5] Struble, L., "The Performance of Portland Cement," *ASTM Standardization News*, Vol. 20, 1992, pp. 38–45.

[6] Bache, H. H., "Densified Cement/Ultra-Fine Particle Based Materials," *Proceedings*, Second International Conference on Superplasticizers in Concrete, National Research Council of Canada, 1981, p. 33.

[7] Johansen, V., "Cement Production and Cement Quality," *Materials Science of Concrete I*, J. Skalny, Ed., American Ceramic Society, Westerville, OH, 1989, pp. 27–72.

[8] Helmuth, R. A., "Energy Conservation Potential of Portland Cement Particle Size Distribution Control," Phase I, NTIS No. C00-4269-1, 1978.

[9] Helmuth, R. A., Whiting, D. A., and Gartner, E. M., "Energy Conservation Potential of Portland Cement Particle Size Distribution Control," Phase II, NTIS No. DE 86-001926, 1984.

[10] Tresouthick, S. W., "Energy Conservation Potential of Portland Cement Particle Size Distribution Control," Phase III, NTIS No. DE 91-007537-38, 1985–1986, pp. 42–44.

[11] Malghan, S. G. and Lum, L.-S. H., "An Analysis of Factors Affecting Particle-Size Distribution of Hydraulic Cements," *Cement, Concrete, and Aggregates*, Vol. 13, 1991, pp. 115–120.

[12] "Standard Practice for Selecting Proportions for Normal, Heavyweight, and Mass Concrete," *Manual of Concrete Practice*, Committee 211, American Concrete Institute, Detroit, MI, 1991.

[13] Tattersall, G. H. and Banfill, P. F. G., *The Rheology of Fresh Concrete*, Pitman Advanced Publishing Program, Boston, MA, 1983.

[14] Helmuth, R. A., private communication to ASTM Subcommittee C01.31 on Volume Change, Nov. 1987.

[15] Neville, A. M., *Properties of Concrete*, Pitman Publishing, London, 1973.

[16] Gaynor, R. D., "Cement Strength and Concrete Strength—An Apparition or a Dichotomy?" *Cement, Concrete and Aggregates*, Vol. 15, 1993, pp. 135–144.

[17] "Determination of Fracture Parameters (K^s_{Ic} and $CTOD_c$) of Plain Concrete Using Three-Point Bend Tests," *Materials and Structures*, Vol. 23, 1990, pp. 457–460.

[18] "Fracture Mechanics of Concrete: Concepts, Models and Determination of Material Properties, ACI 446. 1R-XX," *Concrete International*, Vol. 12, 1990, pp. 67–70.

[19] Lerch, W., "The Influence of Gypsum on the Hydration and Properties of Portland Cement Pastes," *Proceedings*, American Society for Testing and Materials, Vol. 46, 1946, pp. 1252–1292.

[20] Struble, L. and Brockman, M., "Standard Aggregate Materials for Alkali-Silica Reaction Studies," *Proceedings*, 8th International Conference on Alkali-Aggregate Reaction, K. Okada, S. Nishibayashi, and M. Kawamura, Eds., 8th ICAAR Local Organizing Committee, Kyoto, Japan, 1989, pp. 433–437.

[21] Mather, K., "Tests and Evaluation of Portland and Blended Cements for Resistance to Sulfate Attack," *Cement Standards—Evolution and Trends*, P. K. Mehta, Ed., American Society for Testing and Materials, Philadelphia, 1978, pp. 74–86.

[22] Cohen M. D. and Mather, B., "Sulfate Attack on Concrete—Research Needs," *ACI Materials Journal*, Vol. 88, 1991, pp. 62–69.

Hydraulic Cement—Chemical Properties

Sharon M. DeHayes[1]

PREFACE

This chapter on the chemical properties of hydraulic cements is new to this edition of the "Significance of Tests and Properties of Concrete and Concrete-Making Materials." The members of ASTM Committee C1 would like to thank the editors for the opportunity to include information about the composition of hydraulic cements in this publication.

INTRODUCTION

This chapter will describe the ASTM specifications and standard test methods covering the composition and analysis of hydraulic cements. More importantly, the effects of chemical properties on concrete performance will be described because this area is of more concern to the user than a detailed discussion of testing procedures.

The audience is assumed to be those who use or specify cement to make ready-mixed concrete and other types of concrete products. The information may also be of interest to laboratory personnel, students, and researchers new to the field of cement chemistry and concrete technology. The objective of this chapter is to provide sufficient information about the chemical properties of hydraulic cements to enable the user to interpret ASTM specifications and understand the technical documentation provided by cement manufacturers. A better understanding of the chemical properties of cement will also aid in evaluating and predicting the performance of cement in concrete. For the reader desiring more detailed information on cement chemistry, there are a number of textbooks and articles that provide more complete coverage of the subject [1–3].

Hydraulic cement is defined in ASTM Terminology Relating to Hydraulic Cement (C 219) as "a cement that sets and hardens by chemical interaction with water and that is capable of doing so under water." This chapter will cover two classes of hydraulic cements: both portland and blended hydraulic cements will be discussed, although emphasis will be given to portland cement because there are more constraints on chemical composition in ASTM specifications for these cements. Masonry, expansive, and other portland-cement-based materials will not be discussed, although many of the principles related to chemical composition apply to special cements.

PORTLAND CEMENT

According to the definition in ASTM C 219, portland cement is "a hydraulic cement produced by pulverizing portland-cement clinker, and usually containing calcium sulfate." Portland-cement clinker is made by heating a finely ground raw mix in a large kiln, to temperatures of about 1500°C. A raw mix is made by combining calcium, silica, alumina, and iron-bearing materials to achieve the correct chemical composition. As the raw mix is heated, carbon dioxide and water are driven off, and calcium silicates and aluminates are formed during the clinkering reactions. The calcium silicates, along with two other phases, aluminate and ferrite, are responsible for the strength and setting characteristics of portland cement. The final step in the manufacturing process requires crushing and grinding the clinker nodules to make the powder that is sold commercially as portland cement. Calcium sulfate is added during the grinding step to control the early hydration reactions.

The properties of portland cement can be controlled to some extent by altering the chemical composition of the raw materials and the resulting clinker. The performance of portland cement will also be influenced by the manufacturing process; burning and cooling conditions can affect clinker reactivity, and finish grinding controls the particle size distribution (effects of which are discussed in Chapter 45). This discussion will focus on the chemical composition, with attention to the requirements in the ASTM Specification for Portland Cement (C 150).

Chemical Composition of Portland Cement

The chemical elements found in portland cement are generally expressed as oxides; SiO_2, Al_2O_3, Fe_2O_3, and CaO are the four main oxides found in clinker and cement. The actual compounds responsible for hydraulic activity are expressed as sums of oxides. One of the major components in cement is tricalcium silicate, a combination of calcium oxide (CaO), and silicon dioxide (SiO_2). The formula for tricalcium silicate is written Ca_3SiO_5 using standard chemical nomenclature, but the cement chemist

[1] Director of Technical Services, California Portland Cement Company, Glendora, CA 91740.

expresses it as $3CaO \cdot SiO_2$, indicating that the compound is composed of three parts calcium oxide and one part silicon dioxide. To further simplify, individual oxides are abbreviated to a single letter. The abbreviations used to describe the major and minor phases in portland-cement clinker are listed Table 1.

Using these abbreviations, $3CaO \cdot SiO_2$ becomes C_3S. The primary phases are C_2S ($2CaO \cdot SiO_2$), C_3A ($3CaO \cdot Al_2O_3$), and C_4AF ($4CaO \cdot Al_2O_3 \cdot Fe_2O_3$). These chemical formulas can be threatening to the nonchemist, but it is quite possible to comprehend and discuss the chemical properties of cement referenced in ASTM standards by using only the simplified abbreviations.

Phase Composition

Although the composition of cement is expressed as oxides, it is important to note that the mineralogical composition of cement is not adequately described by a list of oxide weight percentages. Table 2 shows the chemical analysis results for a Type I/II cement. In addition to the oxide results, the mineralogical composition is described. In cement chemistry, the term "phase composition" is often used to indicate the true mineralogical composition of clinker and cement. In this paper, chemical composition refers to the oxide analysis, and phase composition will be used when discussing the compounds found in clinker and cement. The single-letter abbreviations will only be used in the name of the phases, or compounds. The oxides are listed in reports following the approximate order in which they are determined by classical methods of chemical analysis.

TABLE 1—Cement Chemist's Shorthand.

Oxides	Abbreviations
SiO_2	S
Al_2O_3	A
Fe_2O_3	F
CaO	C
MgO	M
SO_3	$\bar{S}$
Na_2O	N
K_2O	K
CO_2	$\bar{C}$
H_2O	H

TABLE 2—An Example of a Typical Portland-Cement Composition.

Oxide	Percentage
SiO_2	22.0
Al_2O_3	4.3
Fe_2O_3	3.1
CaO	64.5
SO_3	2.6
POTENTIAL PHASE COMPOSITION	
C_3S	54
C_2S	22
C_3A	6
C_4AF	9

The oxide analysis in Table 2 shows 22.0% total SiO_2, but the SiO_2 is distributed in the two major calcium silicate phases, C_3S and C_2S. Similarly, the cement contains 64.5% CaO, but it is distributed in each of the major phases; C_3S, C_2S, C_3A, and C_4AF. There may also be about 1.0% free CaO, lime that did not combine completely during the clinkering reactions. In addition, about 2% of the CaO will be from calcium sulfate added during the grinding process.

Bogue Calculations

Determination of the oxide composition of cement is fairly straightforward, but the measurement of the actual phase composition is more difficult. In order to correlate chemical composition and cement performance, it is necessary to know how much of each phase is present. Historically, the Bogue calculation has been used to calculate the "potential" phase composition. Developed by Bogue in 1929, this procedure is based on a number of assumptions [4]. There are many variations of the Bogue calculations, but the following formulae are used in ASTM C 150 specifications.

$$C_3S = 4.071CaO - 7.600SiO_2 - 6.718Al_2O_3 \quad (1)$$
$$- 1.430Fe_2O_3 - 2.852SO_3$$

$$C_2S = 2.867SiO_2 - 0.7544C_3S \quad (2)$$

$$C_3A = 2.650Al_2O_3 - 1.692Fe_2O_3 \quad (3)$$

$$C_4AF = 3.043Fe_2O_3 \quad (4)$$

The use of these equations assumes that the clinker reactions go essentially to completion, and also that the four major phases are pure C_3S, C_2S, C_3A, and C_4AF. Equations 3 and 4 are modified if the ratio of Al_2O_3 to Fe_2O_3 is more than 0.64, because a solid solution—expressed as $ss(C_4AF + C_2F)$—will be formed instead of C_3A and C_4AF.

$$ss(C_4AF + C_2F) = 2.100Al_2O_3 + 1.702Fe_2O_3 \quad (5)$$

$$C_3S + 4.071CaO - 7.600SiO_2 - 4.479Al_2O_3 \quad (6)$$
$$- 2.859Fe_2O_3 - 2.852SO_3$$

The sum of the four major compounds will be less than 100%, because the Bogue calculation does not account for minor phases and the incorporation of impurities into the major phases. The version of the Bogue calculations used in ASTM C 150 also does not correct for the amount of free CaO present in the cement that can lead to high values for C_3S. Another source of error arises because ASTM C 150 includes TiO_2 and P_2O_5 in the weight percentage of Al_2O_3.

Although based on many assumptions, the Bogue calculations have the advantage of being easy to calculate. When used appropriately, the results can be effectively used to compare cement properties. This is especially true when evaluating cement produced at a single plant where the raw materials and burning conditions do not change radically over time. However, the Bogue calculations give only the potential phase composition; one should not assume that the results are always close to the true phase composition.

Incorporation of Impurities in Clinker Phases

It has been shown that there are a number of ways to express the chemistry of cement, and that it is necessary to differentiate oxide and phase composition. As mentioned previously, the Bogue equation does not take into account the incorporation of impurities into the major phases. In commercial clinker, the phases will contain impurities, and these impurities will affect performance. In general, impure phases are more reactive than the pure compounds. The C_3S may contain up to 2% MgO, and will also incorporate aluminum and iron. Sulfur, sodium, and potassium, and many other elements have been identified within the C_3S crystal structure [5]. Alite is the mineral name used to identify impure C_3S, which includes all C_3S in commercial clinker. The arrangement of impurities in the C_3S crystals, as well as the heating and cooling conditions of manufacturing, lead to a variety of polymorphic forms (polymorphs have the same chemical composition, but a different crystal structure). Monoclinic and triclinic are two of the polymorphs discussed in the literature. The mineral names of the clinker phases are shown in Table 3.

Belite, the mineral form of C_2S found in clinker, can incorporate aluminum, iron, and other elements into its crystal structure. Like alite, belite exists in a number of polymorphic forms, and the β-polymorph of belite is the predominant form in commercial clinker. The C_3A as aluminate in clinker can contain iron and silicon. The alkalies also influence the aluminate crystal structure, stabilizing the orthorhombic form (also called alkali-aluminate). The principal type of aluminate in clinker is referred to as cubic. Ferrite contains significant amounts of silicon and magnesium.

Clinker contains other impurities—TiO_2, P_2O_5, SrO, Mn_2O_3, and various sulfur compounds are examples. These do not usually have a significant impact on performance, but Jeknavorian and Hayden have reported one example where excessive amounts of zinc caused severe retardation problems [6]. Chloride, fluoride, and trace metals can also affect cement performance. Studies on impurities in clinker phases are numerous, and the possible effects on cement performance are much more complex than this brief introduction.

Hydration Reactions

Hydration refers to the chemical reactions that take place when cement is mixed with water. The modeling of hydration and characterization of hydration products are active fields of research, and the reader is referred to a number of references for more detailed information [7–9]. There are a few reactions that should be reviewed in order to better understand the effects of composition on cement performance.

The calcium silicates react with water to form C-S-H, a calcium silicate hydrate gel that is responsible for the strength development of hydraulic cement paste. For example, C_3S undergoes the following reaction

$$C_3S + (y + z)H_2O \rightarrow C_xSH_y + zCH \qquad (7)$$

where $x + z = 3$, but x, y, and z are not necessarily integers [3]. The actual composition of C-S-H may vary, and hyphens are used between the oxide abbreviations to show that the composition is indefinite. A byproduct of the hydration reaction is the formation of CH, calcium hydroxide. The C_2S also hydrates to form C-S-H and CH. The quantity of CH produced by C_2S hydration is one third the amount from C_3S hydration.

The C_3A participates in the reactions affecting setting (early stiffening with no substantial development of compressive strength) and early strength gains. In the presence of gypsum, the following reaction occurs

$$C_3A + 3C\overline{S} + 26H \rightarrow C_6AS_3H_{32} \qquad (8)$$
(aluminate + gypsum + water → ettringite)

The C_3A can also participate in many other reactions, with and without sulfate. All of the hydration reactions involving the clinker phases are exothermic, and the amount of heat generated is an important property of the cement. These hydration reactions and compounds are not specifically part of ASTM cement standards, but their chemistry forms the basis of many test methods designed to predict the performance of cement: performance with respect to rheology, strength, and durability.

Contributions of the Major Phases to Cement Properties

It is the properties of the major phases when they hydrate that allow us to control the performance of cement by changing the chemical composition. Following is a summary of these properties.

C_3S

Alite hydrates rapidly and is responsible for early strength and early setting characteristics. Taylor reports that approximately 70% of the C_3S in portland cement will have reacted by 28 days, and virtually 100% by one year [1]. Portland cements with higher C_3S will generally show higher strengths through about seven days. The C_3S hydration also generates more heat than C_2S hydration, and the heat of hydration will be related to C_3S content. ACI Committee Report 225 includes a discussion on heat of hydration [10].

C_2S

Belite hydrates more slowly than alite, and contributes mostly to later age strength. Taylor reports 30% complete hydration at 28 days for C_2S, and 90% completion at one year [1]. The C_2S hydration generates less heat than C_3S, and also produces less calcium hydroxide.

TABLE 3—Nomenclature for Major Clinker Phases.

Name of Pure Compound	Mineral Name	Shorthand
Tricalcium silicate	alite	C_3S
Dicalcium silicate	belite	C_2S
Tricalcium aluminate	aluminate (celite)	C_3A
Tetracalcium aluminoferrite	ferrite (felite)	C_4AF

C_3A

The aluminate phase influences setting and early strengths. Gypsum or other sulfate forms are needed to control the C_3A reaction rates, which also generate significant heat of hydration. The C_3A hydration products may also participate in the reactions leading to sulfate attack [11].

C_4AF

The contributions of C_4AF are not well-understood, but it may function similarly to C_3A. Cement color is influenced by the composition and amount of the iron-containing phase. A recent report by Chiesi, Myers, and Gartner suggests that the strength-producing capabilities of C_4AF can be enhanced with the use of chemical additives [12].

CHEMICAL REQUIREMENTS IN ASTM C 150

The chemical requirements for portland cement are summarized in Table 4. This table shows the compositional limits for Types I, II, III, IV, and V. ASTM C 150 also includes Types I-A, II-A, and III-A; the A indicates that an air-entraining admixture has been interground during the manufacturing process. The chemical requirements are the same for regular and air-entrained cements. In addition to air-entraining additives, ASTM C 150 permits the addition of water, calcium sulfate, and processing additions conforming to ASTM Specification for Processing Additions for Use in the Manufacture of Hydraulic Cements (C 465) (usually grinding aids and pack-set inhibitors). If requested, the manufacturer must supply in writing all relevant information about the air-entraining and processing additions used in processing.

The limitations in Table 4 are based on the oxides, as well as the potential compound composition. The different types of cement have different oxide limitations because the composition of the cement is known to affect the resulting properties of concrete. Type I, a general purpose cement, has the fewest restrictions, whereas Types II, III, IV, and V have more limits. The chemical properties limited for all types of cement will be discussed first.

TABLE 4—Standard Chemical Requirements for Portland Cements (From ASTM C 150, Table 1).

Cement Type	I	II	III	IV	V
SiO_2, min, %	...	20.0	...	...	...
Al_2O_3, max, %	...	6.0	...	...	...
Fe_2O_3, max, %	...	6.0	...	6.5	...
MgO, max, %	6.0	6.0	6.0	6.0	6.0
SO_3, max, %					
when $C_3A \leq 8\%$	3.0	3.0	3.5	2.3	2.3
when $C_3A > 8\%$	3.5	...	4.5	...	...
LOI, max, %	3.0	3.0	3.0	2.5	3.0
Insol Res, max, %	0.75	0.75	0.75	0.75	0.75
C_3S, max, %	...	...	...	35	...
C_2S, min, %	...	...	...	40	...
C_3A, max, %	...	8	15	7	5
$(C_4AF + 2(C_3A))$ or $(C_4AF + C_2F)$, max, %	...	...	...	...	25

Limitations Common to All Types of Portland Cement

MgO

All types show a 6% limit on MgO. The MgO is limited because of concern about expansion that can occur if free MgO hydrates, leading to disruption. The MgO is one of two chemical properties limited in the original ASTM cement specification proposed in 1902. The maximum percentage was first set at 4%, but was raised to 5% in 1926 and then 6% in 1976 [13].

Insoluble Residue

All types are subject to a maximum limit on insoluble residue, the material that cannot be dissolved in strong acid or alkaline solutions. Insoluble residue was first limited to 0.85% in 1916 and was lowered to 0.75% in 1941 [13]. The purpose of this limit is to prevent adulteration or contamination of cement with siliceous and argillaceous components. Insoluble residue may result from raw materials that did not combine completely in the burning process, or possibly from contamination during clinker reclaiming. It is usually a silicate or alumino-silicate material. All portland cements contain a small amount of insoluble residue, and the most common source is from silicate impurities in the calcium sulfate added during the finish grinding step. The 0.75% limit on insoluble residue must be considered when determining the level of acceptable purity of calcium sulfate sources for portland-cement manufacture.

Loss on Ignition

Commonly abbreviated LOI, loss on ignition is the weight percentage lost when portland cement is heated at 950°C. Set at a maximum of 3.0% today (2.5% for Type IV cement), LOI was added to the specification in 1917 to prevent the addition of carbonate minerals such as limestone and dolomite [13]. Surface and absorbed moisture, as well as carbon dioxide, will contribute to LOI. The primary source of LOI in modern cements is from water combined in calcium sulfate ($CaSO_4 \cdot 2H_2O$). For example, a portland cement containing 5% calcium sulfate has a theoretical LOI of about 1.5%, if only the combined water from calcium sulfate is contributing to LOI. In practice, some of the combined water from calcium sulfate is lost through dehydration during the mill-grinding step. Additional moisture may be picked up from the atmosphere and from water sprays used to cool grinding mills.

Another source of LOI is moisture picked up from the clinker components during storage as well as during grinding. For example, free lime (CaO) not combined during burning is hygroscopic and readily absorbs water. During storage, particularly if stored outside and exposed to rain, free CaO hydrates to form $Ca(OH)_2$ and it may absorb CO_2 from the atmosphere to form $CaCO_3$. These reactions contribute to a higher LOI. Other clinker components, particularly C_3A and C_3S are subject to prehydration during storage. These reactions can be quite complex and may affect the cement performance [14].

SO_3

All types of portland cement have a limit on the percentage of allowable SO_3, although the limits differ for Types II, III, and V. The SO_3 limit was one of the original two chemical specifications, fixed at 1.75% in 1904 [13]. The SO_3 is simply a way to express the amount of sulfate in the cement. The SO_3 may be present in the form of calcium sulfate, hemihydrate, anhydrite, and many other forms. Clinker SO_3 in the form of alkali sulfates will be included in the measured SO_3 level and will reduce the amount of SO_3 that can be added. The SO_3 is added to regulate the initial setting and hardening reactions that take place during hydration. The amount of SO_3 allowed is directly related to the fineness and composition of the cement, particularly the C_3A content as shown in Table 5. Type V cement has the lowest allowable SO_3 content, a function of the low C_3A level in Type V cement. This same rationale holds for the SO_3 limits on Types I, II, and III. Type III has the highest SO_3 limit because it usually has the highest C_3A content and the highest fineness. The SO_3 limits are all set at a maximum level; there are no restrictions on the minimum SO_3 content. However, it will almost always be necessary to add some form of sulfate in order to meet the physical requirements in ASTM C 150.

Optimum SO_3

Careful examination of Table 1 in ASTM C 150 shows a note concerning the SO_3 limits. Note B allows addition rates of SO_3 to exceed the maximum in the table if it can be shown that the compressive strength will be improved at the higher SO_3 levels. The present method for determining optimum SO_3 is ASTM Test Method for Optimum SO_3 in Portland Cement (C 563), a compressive strength evaluation based on 24-h test results. It must also be shown that the added SO_3 does not cause expansion greater than 0.020% as demonstrated by ASTM Test Method for Expansion of Portland Cement Mortar Stored in Water (C 1038). There may be different optimum SO_3 contents for early strength than for strength at later ages. Optimum SO_3 may vary for other properties such as setting time and shrinkage, and the optimum SO_3 in concrete may be influenced by the addition of admixtures or heat treatment. Optimum SO_3 testing for a particular cement should be repeated if there is a change in manufacturing conditions or chemical composition of the clinker.

ASTM C 150 CHEMICAL REQUIREMENTS FOR DIFFERENT TYPES OF CEMENTS

By varying the chemical composition, it is possible to change the properties of the cement. Examples of the phase composition, calculated by the Bogue method, for Types I through V are given in Table 6. Reference to Table 4 shows that there are only a few limitations on phase content. Even though the chemical requirements are not listed for each oxide, a manufacturer's certification will normally show the complete chemical analysis, along with the calculated phase composition. The test certificate (also called mill report or mill certificate) is a good starting point for correlating cement properties with concrete performance characteristics.

Type I Portland Cement

Type I is to be used when the special properties specified for any other type are not required. According to ASTM C 150, Type I has no restrictions on the major oxides or the phase composition. A Type I cement can have a C_3S content of 20 or 70%, but typically will have 55 to 60% C_3S. Even though ASTM C 150 does not limit C_3S, there are theoretical and practical limitations. In addition, the chemical composition cannot preclude meeting physical requirements such as strength and setting time, discussed in Chapter 45. Cement composition is influenced by the composition of available raw materials. For example, a plant with readily available high purity limestone and a very expensive silica source would find it more economical to make a high C_3S cement (C_3S uses three parts calcium oxide, the major component in limestone, compared to only two parts for C_2S). Economies of raw material sourcing must be balanced with manufacturing costs—it takes more energy to produce C_3S than C_2S. Therefore, cement compositions tend to be more alike throughout the country than might be expected. Chemical composition will also be affected by market demands. As the construction industry strives for higher early strengths, there may be a tendency toward higher C_3S and C_3A contents. When composition is changed to maximize properties such as strength, changes to other properties may also occur.

Type II Portland Cement

Type II has the most restrictions, that is, SiO_2, Al_2O_3, and Fe_2O_3 are limited, in addition to the limits on insoluble residue, LOI, and SO_3. Type II cement is designed for general use when moderate sulfate resistance or moderate heat of hydration is desired. The 20.0% minimum SiO_2 limits the amount of C_3S a Type II cement can contain, even though there are no requirements for C_3S. The limits on Al_2O_3 and Fe_2O_3 restrict the amounts of C_3A and C_4AF that can be made, leading to lower heat of hydration. There is also a limit of 8% on C_3A, for the purpose of reducing susceptibility to sulfate attack. As long as the

TABLE 5—Relationship Between C_3A and SO_3 (From ASTM C 150, Table 1).

Cement	C_3A, %	SO_3, %
Type III	> 8	4.5
Type I	> 8	3.5
Type II	≤ 8	3.0
Type IV	≤ 7	2.3
Type V	≤ 5	2.3

TABLE 6—Examples of Phase Composition for ASTM C 150 Portland Cements.

Cement Type	I	II	III	IV	V
C_3S, %	55	50	60	30	50
C_2S, %	20	25	15	45	30
C_3A, %	10	7	12	5	4
C_4AF, %	10	12	8	12	12

physical requirements are met, Type II cements also meet the chemical requirements for Type I. In many areas of the country, Type II cements have replaced Type I as the most common cement and are often referred to as Type I/II.

Type III Portland Cement

Type III, for use when high early strength is desired, has no additional restrictions other than a 15% C_3A maximum. While many Type III cements have a C_3A content close to the maximum, it is also acceptable to have a very low C_3A content because there is no minimum specified in ASTM C 150. Type III is allowed higher SO_3 than Types I and II, even if they have the same C_3A level. This is because Type III generally has a greater surface area (to meet the higher strength requirements), and more SO_3 may be required to control a finer ground cement. Recall from the discussion of optimum SO_3, that SO_3 limits in ASTM C 150, Table 1, can be exceeded if certain other conditions are met.

Type IV Portland Cement

Type IV is restricted to less than 35% C_3S, a minimum 40% C_2S, and C_3A can be no more than 7%. Designed for low heat of hydration, Type IV is not readily available in the United States. The use of pozzolans and granulated blast furnace slags, either as mineral admixtures or as components of blended cements, have reduced the demand for Type IV cement. There may be renewed interest in low-heat cements for applications such as roller-compacted concrete.

Type V Portland Cement

Type V, for use when high sulfate resistance is desired, has a maximum limit of 5% C_3A and 25% for the quantity ($C_4AF + 2(C_3A)$). The C_3A has been identified as the component in cement that can react with sulfate to lead to deleterious expansion, but research has shown that limiting C_3A may not provide sufficient protection against sulfate attack [15]. The C_3A limit for Type V in ASTM C 150 does not apply when the sulfate expansion limit, an optional physical requirement, is specified.

Optional Chemical Requirements

C_3A and C_3S

The optional chemical requirements in Table 2 of ASTM C 150 apply only when specifically requested. Optional limits on C_3A for Type III are 8% for moderate sulfate resistance and 5% for high sulfate resistance. For Type II, the sum of C_3S and C_3A may be limited to 58% to help ensure moderate heat of hydration.

Low Alkali Limit

The optional requirement for low-alkali cement limits the alkali content to 0.60% maximum, calculated as ($Na_2O + 0.658K_2O$). This formula expresses the total alkali content as equivalent percentage of Na_2O. A review of the history of the alkali limit is given by Frohnsdorff et al. [16]. Briefly, the alkali limit is intended to eliminate deleterious expansive action arising from alkali silica reaction (ASR). Discussed in more detail in other chapters, ASR occurs when cement alkalies and reactive silica components in aggregate form a gel that leads to cracking and expansion. Continuing developments in accelerated physical testing for ASR and the discovery of reactive aggregates previously thought to be innocuous may lead to reexamination of the alkali limits.

BLENDED HYDRAULIC CEMENTS

Blended hydraulic cements consist of two or more inorganic constituents (one of which is normally portland cement or portland-cement clinker) that separately or in combination contribute to the strength-gaining properties of the cement. The most common materials used in blended cements are blast-furnace slag and fly ash. Natural pozzolans and microsilica (also called condensed silica fume) are examples of other constituents in blended cements. Blended cements may be made by intergrinding the constituents with portland cement clinker and calcium sulfate, or they can be made by blending the constituents with portland cement.

A pozzolan is a material that has little or no hydraulic activity of its own, but it acts as a hydraulic cement when mixed with water and CaO. In blended cements, portland cement is the source of CaO for the pozzolanic reactions ($Ca(OH)_2$ is produced when portland cement hydrates). Pozzolans are high in SiO_2 and may also contain significant amounts of Al_2O_3. Fly ash, microsilica, and natural pozzolans are examples of pozzolanic materials. Blast-furnace slag may have latent hydraulic activity as well as pozzolanic properties.

ASTM C 595 SPECIFICATION FOR BLENDED HYDRAULIC CEMENTS

There are five major classifications of blended cements, and some of these classes are further subdivided, based on the percentage of portland cement in the mixture. The five main types of blended hydraulic cements are

1. portland blast-furnace slag cement (Type IS),
2. portland-pozzolan cement (Types IP and P),
3. slag cement (Type S),
4. pozzolan-modified portland cement (Type I (PM)), and
5. slag-modified portland cement (Type I(SM)).

There are only a few chemical requirements for blended cements, and the rationale for most of these requirements is similar to that for ASTM C 150. Table 7 summarizes the chemical requirements. The MgO is limited to 5.0% for all portland-pozzolan and pozzolan-modified portland cements, but there is no limit on MgO for slag cements. The SO_3 limits for all types of blended cements may be exceeded if it can be demonstrated that the optimum SO_3 is higher than the limit. Insoluble residue is limited to 1.0% for slag cements, but there is no limit for pozzolan-

TABLE 7—Chemical Requirements for Blended Hydraulic Cements (From ASTM C 595, Table 1).

Cement Type	I(SM) I(SM)-A IS, IS-A	S, SA	I(PM) I(PM)-A P, PA, IP, IP-A
MgO, max, %	...	...	5.0
SO_3, max, %	3.0	4.0	4.0
Sulfide sulfur, max, %	2.0	2.0	...
Insoluble residue, max, %	1.0	1.0	...
Loss on ignition, max, %	3.0	4.0	5.0
Water-soluble alkali, max, %	...	0.03	...

containing cements. Loss on ignition varies from 3.0 to 5.0%. The source of LOI for blended cements is the same as for portland cements, although pozzolans such as fly ash may contain unburned carbon that is measured in the LOI determination (carbon can affect the efficiency of air-entraining admixtures). There is no total alkali limit for blended cements, but water-soluble alkali for slag cements is limited to 0.03% to prevent staining of limestone aggregate.

The chemical composition of pozzolanic materials cannot always be directly related to the performance of concrete, but examination and tracking of chemical properties can provide a means to check product uniformity. In addition to the chemical requirements in ASTM Specification for Blended Hydraulic Cements (C 595), the standards dealing with the individual types of finely divided mineral admixtures provide useful guidelines about the composition of these materials. These are summarized in the next section.

CHEMICAL COMPOSITION OF POZZOLANIC MATERIALS AS MINERAL ADMIXTURES IN CONCRETE

Fly Ash and Natural Pozzolans

ASTM Specification for Fly Ash and Raw or Calcined Natural Pozzolan for Use as a Mineral Admixture in Portland Cement Concrete (C 618) gives the requirements for fly ash and natural pozzolans for use in concrete. Class N, raw or calcined natural pozzolans, include materials such as diatomaceous earths, opaline cherts and shales, tuffs and volcanic ashes or pumicites, and some calcined shales and clays. The sum of SiO_2, Al_2O_3, and Fe_2O_3 must equal at least 70%, and there are also limits on SO_3, moisture, and LOI. Class F, fly ash produced from burning anthracite or bituminous coal, must also meet the 70% minimum sum of SiO_2, Al_2O_3, and Fe_2O_3. The sum of these three oxides for Class C fly ash produced from burning lignite or sub-bituminous coals, must equal 50% or more. These fly ashes normally contain CaO and have some cementitious properties in addition to being pozzolanic. Helmuth has summarized the chemical composition and properties of a variety of fly ashes [17].

Ground Granulated Blast-Furnace Slag

Blast-furnace slag is the nonmetallic product formed during iron production, consisting of silicates and aluminosilicates of calcium. The major oxides, SiO_2, Al_2O_3, CaO, and MgO, constitute 95% or more of the total oxides [18]. If water-quenched, blast-furnace slag has a glass content of greater than 95%; pelletized blast-furnace slag has a lower glass content. Air-cooled slag has little hydraulic activity, but may be used as aggregate. Attempts to relate the reactivity of granulated blast-furnace slag to chemical composition, such as the ratio $(CaO + MgO + Al_2O_3)/SiO_2$, have not been adequate, but these constraints are included in some specifications. ASTM Specification for Ground Iron Blast-Furnace Slag for Use in Concrete and Mortars (C 989) uses the slag-activity index, based on actual compressive strength test results to characterize slags.

Microsilica

Microsilica, also called condensed silica fume, is not yet covered by an ASTM specification, but a standard is in the draft form. Microsilica is a by-product from the reduction of quartz (or quartz and iron) with coal to produce silicon and ferro-silicon alloys. Microsilica condenses from the gaseous phase, and the particles are spherical, amorphous, and have a very high surface area [19]. Microsilica from silicon alloy production usually contains more than 90% SiO_2, and ferrosilicon alloy production produces microsilica with varying amounts of Fe_2O_3. Small amounts of other oxides may also be present. Microsilica functions as a very active pozzolan, because the very fine SiO_2 particles quickly react with $Ca(OH)_2$ and water to form C-S-H.

TEST METHODS FOR THE COMPOSITION OF HYDRAULIC CEMENTS

ASTM Test Methods for Chemical Analysis of Hydraulic Cement (C 114) provides directions for chemical analysis procedures for both portland cements and blended hydraulic cements. The scope of ASTM C 114 states that any method may be used for analysis of hydraulic cement, as long as it can be demonstrated that the method achieves the required levels of precision and bias. ASTM C 114 does provide a set of specific test procedures for analyzing cement. These procedures are separated into Reference Test Methods and Alternative Test Methods. Reference test methods are "long accepted wet chemical test methods which provide a reasonably well-integrated basic scheme of analysis for hydraulic cements." Reference test methods can be performed in any reasonably equipped chemical laboratory and do not require expensive analytical instrumentation. Generally, satisfactory results can only be obtained by experienced analysts, and ASTM C 114 requires that individual analysts demonstrate their ability to achieve acceptable precision and bias when using the reference test methods.

Demonstration of Precision and Bias

The competency of an individual analyst is determined by comparing the analyst's duplicate results for an SRM cement to the maximum permissible variation in results shown in Table 8. Table 8 has been adapted from Table 1 in ASTM C 114 that provides information on oxides in addition to the ones included in Table 8 of this report. An SRM cement is a standard reference material cement, prepared and certified by the National Institute of Standards Technology (NIST, formerly NBS). Cement SRMs cover a range of compositions, and certified values are provided. In addition to being used to demonstrate competency of analysts and test methods, SRMs are used as calibration standards for instrumental methods.

Table 8 also provides a convenient means to assess the acceptability of optional test methods (also called rapid methods). A scheme for qualifying these test methods was developed and became part of ASTM C 114 in 1977. Briefly, the qualification procedure involves analysis of at least seven SRM samples; two rounds of tests on nonconsecutive days are required. Six of the seven differences between duplicates obtained for any single component shall not exceed the limits in Column 2 of Table 8. Also, the average of the duplicates must not exceed the limits in Column 3, again for six of the seven SRMS. Test methods must be requalified at least every two years, or when there is substantial evidence that the test method is not performing in accordance with Table 1 in ASTM C 114. Such evidence may come from a comparison of results with the average in the CCRL Proficiency Sample program. Many companies also conduct their own interlaboratory sample exchange programs to monitor chemical analysis test procedures. Additional details and restrictions on the qualification procedure are given in ASTM C 114 in the section titled Performance Requirements for Rapid Test Methods.

Instrumental Methods for Cement Analysis

Table 1 in ASTM C 114 has facilitated the application of modern instrumental methods to cement analysis, and a manufacturer can use any qualified method, as long as the procedure used is noted on the manufacturer's certification. Examples of test methods used successfully to analyze hydraulic cements are discussed in *ASTM STP 985* [20]. Methods included are X-ray fluorescence, atomic absorption spectrophotometry, and a spectrophotometric scheme. The most common of these methods will be discussed further.

X-Ray fluorescence (XRF)

Both wavelength and energy dispersive X-ray methods have been used successfully to analyze portland cements, but wavelength is the more popular technique in cement plant quality control laboratories. XRF is well-suited to analysis in a cement plant because it can be used to obtain chemical analyses of quarry and imported raw materials, clinker, and portland cement. Analysis time is greatly reduced compared to classical wet methods, and varying degrees of automation are now in use. Three methods are used for sample preparation: pressed powder, fused pellets, and fusion followed by grinding. ASTM C 114 allows the use of any method, as long as acceptable precision and bias can be demonstrated.

Other Rapid Methods

Atomic absorption involves dissolution of the cement prior to elemental analysis. Used successfully in many laboratories, it is subject to error for the major elements because a number of dilution steps are required. Inductively coupled plasma (ICP) spectrometry can also be used for complete analysis. Specialized techniques, such as ion chromatography, can also be used for selected elements. For more information on analysis of anhydrous cement, refer to the American Ceramic Society's yearly publication of *Cements Research Progress*.

Selective Dissolution

Selective dissolution covers a variety of procedures that attempt to separate a component or phase from the composite. Methods for free CaO fall into this classification, as do procedures such as the maleic acid/methanol (M/M) extraction. The M/M extraction, or a variation using salicylic acid, can be used to remove the calcium silicates [21–23]. This then concentrates the aluminate, ferrite, periclase, and calcium sulfate phases. These techniques are sometimes used as precursors to other methods of analysis such as quantitative X-ray diffraction.

Significance of Test Results

Table 8 also provides a means for the user to evaluate the significance of test results. When comparing two cements, Table 8 can be used as a guideline. For example, the SO_3 content for Cement A is 2.70% and 2.75% for Cement B. Examination of Table 8 shows that a difference of 0.10% is allowed for duplicate analyses of the same sample, a good indication that the SO_3 contents of Cements A and B are not significantly different. This type of examination may be helpful when determining if a cement that is close to a specification limit should be rejected. The values in Table 8 apply only to within-lab variation, but between-lab variation can be inferred because all laboratories are comparing their results to reference standards.

Looking at another example, it is possible to get an approximation of the variation that can occur between two

TABLE 8—Maximum Permissible Variations in Results (From ASTM C 114, Table 1).

Component	Maximum Difference Between Duplicates	Max Difference of the Average of Duplicates from SRM values (+/−)
SiO_2	0.16	0.2
Al_2O_3	0.20	0.2
Fe_2O_3	0.10	0.10
CaO	0.20	0.3
MgO	0.16	0.2
SO_3	0.10	0.1
LOI	0.10	0.10
Na_2O	0.03	0.05
K_2O	0.03	0.05

laboratories, both of which have demonstrated acceptable analysis of SRM cements. If Lab A shows a consistent bias of +0.05% in the analysis of Na_2O, and Lab B shows a consistent bias of −0.05%, they may differ by as much as 0.10%. Unlike the previous SO_3 example, a difference of 0.10% in Na_2O content is not trivial, especially if the cement is near the optional 0.60% alkali limit. Fortunately, it is unlikely that laboratories will show a consistent bias at the maximum allowable range. However, even a bias of 0.02 or 0.03% can be important for plants operating near the limit for the minor oxides.

The preceding examples show that bias in test results can have a significant impact on the minor oxides, but testing variation for the major oxides can also influence the potential phase composition calculated by the Bogue equations. For example, errors or bias of a few tenths of a percent in measuring SiO_2, Al_2O_3, and CaO can cause differences of more than 5% in the calculated C_3S content. This example illustrates the need for cement manufacturers and others who test cement to establish a quality assurance program for analytical procedures. One should use caution when comparing the composition of two or more cements, especially when the chemical analyses were not performed in the same laboratory.

QUANTITATIVE PHASE ANALYSIS

Currently, there are several task groups working within ASTM Subcommittee C01.23 on Compositional Analysis of Hydraulic Cements to develop standard test methods for measuring the phase composition of cements. The impetus for this effort comes from the growing shift away from prescription to performance-based specifications. In addition, there is a desire to correlate cement properties with performance in concrete, requiring a more definitive characterization of cement structure.

Microscopical Techniques

Optical microscopy has long been used to quantitatively determine mineralogical composition in geological samples, and the technique can be applied to the measurement of clinker phase content. Work in the current task group is focusing on developing a method that is economical, that can be used by trained laboratory technicians, and that has acceptable levels of precision and bias. Methods for quantitative clinker microscopy with the light microscope have been reviewed by Campbell and Galehouse [24]. Determination of phase content by optical microscopy is considered a direct method since phases are identified and counted to determine the volume percentage. The disadvantages of optical microscopy include (1) it is difficult to obtain a representative sample, (2) it is time-consuming to count at least 2000 to 3000 points to get a statistical sampling, (3) the method is subject to errors of misidentification of phases, (4) the precision of the method can vary with the clinker microstructure (that is, size of crystals), and (5) the method may not be easily applied to ground cement.

The use of optical microscopy to evaluate clinker reactivity and predict strength performance of portland cement has been the subject of numerous papers reported in the proceedings of the International Cement Microscopy Association (ICMA). One procedure is the Ono Method, named after Dr. Yoshio Ono from Japan, where a small amount of ground clinker or cement powder is evaluated to determine kiln burning and cooling conditions [25]. The crystal observations can then be correlated to potential strength gain. This type of compositional analysis is not presently covered by ASTM standards, but it is an example of methods that might be investigated in the future.

Scanning electron microscopy (SEM) and other electronic imaging techniques have been successfully applied to characterizing portland-cement clinker, portland cements, and blended cements [26]. The SEM, if equipped with an X-ray analyzer, can be used to perform chemical analysis of specific crystals. X-ray mapping produces an image of the elemental composition, and also shows the distribution of the elements throughout the crystal structure.

Quantitative X-Ray Diffraction (QXRD)

X-ray diffraction has been used for many years to characterize cements and is useful for identifying phases in cement. It can also be used for quantitative phase analysis, but the difficulties are numerous. The status of QXRD analysis applied to cement and clinker is discussed by Struble in a report prepared for a 1991 ASTM symposium on the characterization of hydraulic cements [27]. Struble describes the problems associated with QXRD, in general, and the specific difficulties associated with analysis of clinker and cement. To date, progress has been made by the ASTM task group in developing a standard test method for MgO, aluminate, and ferrite. Methods for the calcium silicate phases are expected to be more difficult because of the difficulty in preparing appropriate standards.

Even though standard ASTM test methods for QXRD have not yet been finalized for cement and clinker, the future is promising. Advances in X-ray diffraction automation and computer software have made it possible to collect and analyze large amounts of data, and these improvements will help to overcome some of the challenges inherent to the technique. Experienced diffractionists have successfully applied QXRD analysis to cement and clinker, reporting accuracy levels of better than 5% for C_3S and 2% for C_3A [1]. Some laboratories may find that good results can be obtained by combining microscopic and XRD techniques. It is very difficult to differentiate aluminate from ferrite in some clinkers, and current procedures for QXRD may already be capable of achieving greater accuracy than optical microscopy. Conversely, the calcium silicates are usually easily identified by microscopy and an experienced microscopist may be able to achieve improved accuracy levels compared to present QXRD methods. Combined techniques of XRD and microscopy have also been used to successfully characterize periclase (free MgO).

Other Compositional Considerations

Carbonate Additions

A third task group within Subcommittee C01.23 is concerned with the measurement of carbonate additions to both portland and blended cements. Suggested methods for measuring carbonate additions include thermogravimetric analysis (TGA), split-LOI procedures, and the use of a carbon/carbonate analyzer. The purpose is to measure the amount of CO_2, and then back calculate to the addition rate for the carbonate mineral (usually calcite or dolomite).

SO_3 Form

The SO_3 content was discussed in the section on ASTM C 150, but it is also important to determine the form of sulfate. For example, when natural gypsum is added to clinker and ground in the finish mill, enough heat may be generated to dehydrate the gypsum to varying degrees. Natural gypsum may also contain anhydrite, and there is an increasing use of calcium sulfate byproducts as a source of SO_3. The form and amount of the sulfate phases can affect the hydration reactions. As previously mentioned, gypsum and other sulfates can also react during storage, perhaps leading to cement flowability problems. There is interest in sulfate form and amount because reports during the last several years have shown that these parameters can affect a cement's interaction with chemical admixtures, particularly water reducers [28].

Calcium sulfates can be identified by X-ray diffraction, optical and electron microscopy, and by differential thermal analysis/thermogravimetric analysis (DTA/TGA) methods. There is a need for reexamination of these techniques and the development of more accurate procedures for the quantitative measurement of the calcium sulfate phases.

Alkali Sulfates

In addition to forms of calcium sulfate, there is the concern about the presence of alkali sulfates. The most commonly occurring alkali sulfates are arcanite, apthitalite, and calcium langbeinite. Hydration of alkali sulfates during cement storage can lead to lumping and flowability problems. Alkali sulfates can affect the strength characteristics of cement, reportedly improving early strengths but giving lowered 28-day and later strengths. Changes in cement alkali levels have also been reported to affect the effectiveness of certain air-entraining admixtures [28].

Trace Elements

A review of the literature on cement composition will show hundreds of references dealing with trace elements (primarily trace metals, but chloride, fluoride, and rare earth metals are also included). Covered are effects on burnability of cement raw materials, effects on kiln refractory, effects on clinker reactivity and cement performance, and methods for measuring trace elements. Interest in recent years has increased because of the environmental concerns. A recently completed survey of trace elements in portland cements and cement kiln dust has been published by the Portland Cement Association [29]. This reference also provides information about the sources and effects of trace elements. Of particular interest is the increased attention being given to trace elements as a result of the growing practice of using waste materials as both raw materials and fuels in the clinker manufacturing process. Barger [30] presented a thorough discussion on the utilization of waste solvent fuels, including a description of trace metal, hydrocarbon, and chloride balances. Analysis of cements produced with waste solvent fuels, using XRD and optical microscopy techniques, showed no detrimental effects.

Testing for trace elements has brought a number of techniques to the cement quality control laboratory. Discussion of these methods is beyond the scope of this paper, but they include atomic absorption (AA), inductively coupled plasma (ICP), gas chromatography (GC), mass spectroscopy, and many other instrumental techniques for analysis of metals, halides, hydrocarbons, and other organics. Improved procedures for rapid analysis of heating values for fuels are also being developed.

Concerns of the User

Cement manufacturers may be interested in pursuing alternate raw materials and fuels for economic reasons, but maintaining product uniformity and quality should be the prime concern of the knowledgeable user. Significant changes in raw materials or fuels may affect properties like alkali levels, sulfate form, and even reactivity of the major phases [31]. In turn, physical properties such as setting time, bleeding rate, strength gain, and cement-admixture interaction may be affected. Data is just now beginning to be published about the effects of alternate materials on cement quality, so users should maintain close communication with their suppliers. Consideration should also be given when developing ASTM standards to ensure that specifications address all relevant properties of cement, both chemically and physically.

PERFORMANCE VERSUS PRESCRIPTION STANDARDS

Improved methods are needed to predict the performance of hydraulic cement in concrete, and many subcommittee and task group activities in ASTM are working on this issue. Ideally, there would be no need to specify limits on chemical composition because there would be rapid tests to correctly predict performance. However, all of the properties that contribute to the performance of hydraulic cement have yet to be identified, and adequate performance tests have not been developed for the properties we know to be significant. Furthermore, even accelerated tests take up to two weeks or more to produce results, especially with respect to durability. It is likely that the standards process will continue to evolve, and prescription limits on chemical composition will be modified or eliminated as better performance test methods are developed. Is it conceivable that the matter will go full circle, and tomorrow's performance tests will be replaced with better prescriptions? If a hydraulic cement can be completely characterized, chemically and physically, will it be necessary to run compressive strength or setting time tests? If we know the mechanisms of sulfate attack, will it be necessary to subject mortar bars to accelerated curing regimes?

The composition of hydraulic cements may become more complex as environmental issues such as the concern for global warming dictate increased uses of materials other than portland cement and the pozzolanic materials we are familiar with today. A necessary part of the evolution toward improved ASTM standards is the development of better methods to characterize the properties of hydraulic cements, leading to more useful predictive models.

REFERENCES

[1] Taylor, H. F. W., *Cement Chemistry*, Academic Press Inc., San Diego, CA, 1990.

[2] Lea, F. M., *The Chemistry of Cement and Concrete*, Chemical Publishing Company, Inc., New York, 1971.

[3] Bye, G. C., *Portland Cement. Composition, Production, and Properties*, Pergammon Press Inc., Elmsford, NY, 1983.

[4] Bogue, R. H., *The Chemistry of Portland Cement*, 2nd ed., Reinhold, New York, 1955.

[5] Hofmanner, F., *Microstructure of Portland Cement Clinker*, Holderbank Management and Consulting, Ltd., Holderbank, Switzerland, 1973.

[6] Jeknavorian, A. A. and Hayden, T. D., "Troubleshooting Retarded Concrete: Understanding the Role of Cement and Admixtures Through an Interdisciplinary Approach," *Cement, Concrete, and Aggregates*, Vol. 13, No. 2, Winter 1991, pp. 103–108.

[7] Mindess, S. and Young, J. F., *Concrete*, Prentice-Hall, Inc., Englewood Cliffs, NJ, 1981, pp. 76–111.

[8] Gartner, E. F. and Gaidis, J. M., "Hydration Mechanisms, I," *Materials Science of Concrete I*, J. Skalny, Ed., American Ceramic Society, Inc., Westerville, OH, 1989, pp. 95–126.

[9] Jawed, I., Skalny, J., and Young, J. F., "Hydration of Portland Cement," *Structure and Performance of Cements*, P. Barnes, Ed., Elsevier Science Publishing Co., Inc., New York, 1983, pp. 237–318.

[10] "Guide to the Selection and Use of Hydraulic Cements," *Manual of Concrete Practice*, ACI Committee 225, American Concrete Institute, Detroit, MI, 1991.

[11] Mather, K., "Factors Affecting Sulfate Resistance of Mortars," *Proceedings*, 7th International Congress on the Chemistry of Cement, V., Paris, 1980, pp. 580–585.

[12] Chiesi, C. W., Myers, D. F., and Gartner, E. M., "Relationship Between Clinker Properties and Strengths Development in the Presence of Additives," *Proceedings*, Fourteenth International Conference on Cement Microscopy, Costa Mesa, CA, International Cement Microscopy Association, Duncanville, TX, 1992, pp. 388–401.

[13] Weaver, W. S., "Committee C-1 on Cement—Seventy-Five Years of Achievement," *Cement Standards—Evolution and Trends, ASTM STP 663*, P. K. Mehta, Ed., American Society for Testing and Materials, Philadelphia, 1978, pp. 3–15.

[14] Johansen, V., "Cement Production and Cement Quality," *Materials Science of Concrete I*, J. Skalny, Ed., American Ceramic Society, Inc., Westerville, OH, 1989, pp. 32–34.

[15] Stark, D., *Durability of Concrete in Sulfate-Rich Soils*, Portland Cement Association, Skokie, IL, 1989.

[16] Frohnsdorff, G., Clifton, J. R., and Brown, P. W., "History and Status of Standards Relating to Alkalies in Hydraulic Cements," *Cement Standards—Evolution and Trends, ASTM STP 663*, P. K. Mehta, Ed., American Society for Testing and Materials, Philadelphia, 1978, pp. 16–34.

[17] Helmuth, R. A., *Fly Ash in Cement and Concrete, SP040T*, Portland Cement Association, Skokie, IL, 1987.

[18] "Ground Granulated Blast-Furnace Slag as a Cementitious Constituent in Concrete," *Manual of Concrete Practice*, Committee C 226, American Concrete Institute, Detroit, MI, 1991.

[19] Roberts, L. R., "Microsilica in Concrete, I," *Materials Science of Concrete I*, J. Skalny, Ed., American Ceramic Society, Inc., Westerville, OH, 1989, pp. 197–222.

[20] *Rapid Methods for Chemical Analysis of Hydraulic Cement, ASTM STP 985*, R. F. Gebhardt, Ed., American Society for Testing and Materials, Philadelphia, 1988.

[21] Javellana, M. P. and Jawed, I., "Extraction of Free Lime in Portland Cement and Clinker by Ethylene Glycol," *Cement and Concrete Research*, Vol. 12, 1982, pp. 399–403.

[22] Tabikh, A. A. and Weht, R. J. "An X-ray Diffraction Analysis of Portland Cement," *Cement and Concrete Research*, Vol. 1, 1971, pp. 317–328.

[23] Struble, L. J., "The Effect of Water on Maleic Acid and Salicylic Acid Extraction," *Cement and Concrete Research*, Vol. 15, 1985, pp. 631–636.

[24] Campbell, D. H. and Galehouse, J. S., "Quantitative Clinker Microscopy with the Light Microscope," *Cement, Concrete, and Aggregates*, Vol. 13, No. 2, Winter 1991, pp. 94–96.

[25] Ono, Y., "Microscopical Observation of Clinker for the Estimation of Burning Condition, Grindability, and Hydraulic Activity," *Proceedings*, Third International Conference on Cement Microscopy, International Cement Microscopy Association, Houston, TX, 1981, pp. 198–210.

[26] Stutzman, P. E., "Cement Clinker Characterization by Scanning Electron Microscopy," *Cement, Concrete, and Aggregates*, Vol. 13, No. 2, Winter 1991, pp. 109–114.

[27] Struble, L. J., "Quantitative Phase Analysis of Clinker Using X-Ray Diffraction," *Cement, Concrete, and Aggregates*, Vol. 13, No. 2, Winter 1991, pp. 97–102.

[28] Dodson, V., *Concrete Admixtures*, Van Nostrand Reinhold, New York, 1990.

[29] *An Analysis of Selected Trace Metals in Cement and Kiln Dust, SP109T*, Portland Cement Association, Skokie, IL, 1992.

[30] Barger, G. S., "Utilization of Waste Solvent Fuels in Cement Manufacturing," presented at 1991 Engineering Foundation Meeting, Potosi, MI.

[31] Hills, L. M. and Tang, F. J., "The Effect of Alternate Raw Materials on Clinker Microstructure and Cement Performance: Two Case Studies," *Proceedings*, Fourteenth International Conference on Cement Microscopy, Costa Mesa, CA, International Cement Microscopy Association, Duncanville, TX, 1992, pp. 412–426.

Mixing and Curing Water for Concrete

James S. Pierce[1]

PREFACE

Walter J. McCoy, then Director of Research for Lehigh Portland Cement Company, wrote the first version of this chapter for *ASTM STP 169*. For *ASTM STP 169A*, Mr. McCoy revised his chapter and added information on typical municipal water analyses, tolerable concentrations of impurities, the effects of sugar in mixing water, and the effects of water hardness on concrete air content. He also added several new references. For *ASTM STP 169B*, Mr. McCoy, then Director of Cement Technology for Master Builders, made only minor changes to the *ASTM STP 169A* version of the chapter. This current version is essentially Mr. McCoy's (now retired) chapter with minimal updating. There has been very little new technology published regarding mixing and curing water for concrete.

INTRODUCTION

This chapter is concerned primarily with the significance of quality tests of various types of waters for mixing and curing concrete and makes no attempt to include the effect of the quantity of mixing water. The quality of water is important because poor-quality water may adversely affect the time of setting, the strength development, or cause staining. Almost all natural waters, fresh waters, and waters treated for municipal use are satisfactory as mixing water for concrete if they have no pronounced odor or taste. Because of this, very little attention is usually given to the water used in concrete, a practice that is in contrast to the frequent checking of the admixture, cement, and aggregate components of the concrete mixture. In fact, most of the references appear to be outdated, but they still represent the bases for modern concrete technology with regard to water for mixing and curing.

MIXING WATER

A popular criterion as to the suitability of water for mixing concrete is the classical expression, "If water is fit to drink it is all right for making concrete." This does not appear to be the best basis for evaluation, since some waters containing small amounts of sugars or citrate flavoring would be suitable for drinking but not mixing concrete [1], and, conversely, water suitable for making concrete may not necessarily be fit for drinking [2]. In an attempt to be more realistic, some concrete specifications writers attempt to ensure that water used in making concrete is suitable by requiring that it be clean and free from deleterious materials. Some specifications require that if the water is not obtained from a proven satisfactory source, the strength of concrete or mortar made with the questionable water should be compared with similar concrete or mortar made with water known to be suitable. The U.S. Army Corps of Engineers, in addition to a general description of acceptable water requirements [3], also states that if the pH of water is between 6.0 and 8.0 and the water is free from organic matter, it may be regarded as safe for use in mixing concrete. (An exception to this is the case where potassium or sodium salts or other natural salts are present in excessive amounts.) This standard also states that if water is of questionable quality before the water is judged to be acceptable, it should be tested in mortar cubes for which the average 7- and 28-day compressive strengths must equal at least 90% of those of companion test specimens made with distilled water [4]. Other than comparative tests of this type, no special test has been developed to determine the quality of mixing water, and, hence it is difficult to judge the fitness of water for use in concrete.

Guidance on mixing water quality is also available using the ASTM Specification for Ready-Mixed Concrete (C 94) and the American Association of State Highway and Transportation Officials (Designation T26). In ASTM C 94, precautions are given to exercise care if the water contains substances that discolor it, or make it smell or taste unusual. When water quality is questionable, service records of concrete made with the questionable water should be examined. If service records are not available or not conclusive, then the water quality should be clarified by comparing mortar cube compressive strengths and times of setting with mortars made with the water in question and with distilled water. ASTM C 94 also requires mortar strengths to be a minimum of 90% of the strength of cubes made with distilled water and the time of setting in the test mortar should not be more than 1 h quicker nor more than 1½ h later than the time of setting when distilled water is used.

---

[1] Chief, Materials Engineering Branch, U.S. Bureau of Reclamation, Denver, CO 80225–0007.

This specification also permits the use of wash water from mixer washout operations for mixing water, as long as it meets maximum concentration criteria for chloride as Cl (500 ppm), sulfate as SO_4 (3000 ppm), alkalies as Na_2O equivalent (600 ppm), and total solids (50 000 ppm).

Caution must be exercised when groundwater supplies are tapped for mixing water supplies. The rule of thumb about being potable is still only a guide, and before critical concrete operations are undertaken, comparative tests for strength and time of setting should be completed. Periodic water analyses are recommended.

The two principal questions regarding mixing water appear to be: "How do impurities in the water affect the concrete?" and "What degree of impurity is permissible?" The following discussion is a summary of available information on these two items.

Effects of Impurities in Mixing Water

The most extensive series of tests on this subject was conducted by Abrams [5]. Approximately 6000 mortar and concrete specimens representing 68 different water samples were tested in this investigation. Among the waters tested were sea and alkali waters, bog waters, mine and mineral waters, and waters containing sewage and solutions of salt. Tests with fresh waters and distilled water were included for comparative purposes. Time of setting tests and cement and concrete strength tests from three days to 2.33 years were conducted for each of the various water samples. Some of the more significant conclusions based on these data are as follows.

1. The time of setting of portland cement mixtures containing impure mixing waters was about the same as those observed with the use of clean fresh waters with only a few exceptions. In most instances, the waters giving low relative compressive strength of concrete caused slow setting, but, generally speaking, the tests showed that time of setting is not a satisfactory test for suitability of a water for mixing concrete.
2. None of the waters caused unsoundness of the neat portland cement pat when tested over boiling water.
3. In spite of the wide variation in the origin and type of waters used, most of the samples gave good results in concrete due to the fact that the quantities of injurious impurities present were quite small.
4. The quality of mixing water is best measured by the ratio of the 28-day concrete or mortar strength to that of similar mixtures made with pure water. Waters giving strength ratios that are below 85%, in general, should be considered unsatisfactory.
5. Neither odor nor color is an indication of quality of water for mixing concrete. Waters that were most unpromising in appearance gave good results. Distilled waters gave concrete strengths essentially the same as other fresh waters.
6. Based on a minimum strength-ratio of 85% as compared to that observed with pure water, the following samples were found to be unsuitable for mixing concrete: acid water, lime soak water from tannery waste, carbonated mineral water discharged from galvanizing plants, water containing over 3% of sodium chloride or 3.5% of sulfates, and water containing sugar or similar compounds. The concentration of total dissolved solids in these waters was over 6000 ppm except for the highly carbonated water that contained 2140 ppm total solids. These data support one of the principal reasons for the general suitability of drinking water, for few municipal waters contain as much as 2000 ppm of dissolved solids and most contain far less than 1000 ppm. Very few natural waters other than seawater contain more than 5 000 ppm of dissolved solids [6].
7. Based on the minimum strength-ratio of 85%, the following waters were found to be suitable for mixing concrete: bog and marsh water, waters with a maximum concentration of 1% SO_4, seawater (but not for reinforced concrete), alkali water with a maximum of 0.15% Na_2SO_4 or NaCl, water from coal and gypsum mines, and wastewater from slaughterhouses, breweries, gas plants, and paint and soap factories.

Many of the specifications for water for mixing concrete, especially those requiring that it be potable, would have excluded nearly all of the aforementioned waters, but contrary to this rather general opinion, the test data show that the use of many of the polluted types of water did not result in any appreciable detrimental effect to the concrete. The important question is not whether impurities are present, but do impurities occur in injurious quantities? It should be noted that the conclusions on suitable waters cited earlier were based entirely on tests of specific samples from the indicated sources, and it should not be assumed that all waters of the type described would be innocuous when used as mixing water.

Typical analyses of municipal water supplies as reported by the U.S. Geological Survey are given in Table 1. Although dated, there is no reason to believe that typical analyses have changed much from those in Table 1. Collins [7] reported that these analyses represent public water supplies used by about 45% of the cities of the United States that have a population of more than 20 000. These analyses indicate they would be acceptable sources for mixing water. However, when investigating a municipal source, analyses for several periodic tests should be examined to determine if comparisons with distilled water are

TABLE 1—Typical Analyses of City Water Supplies (PPM)[a].

Analysis Number	1	3	5	6	7
Silica (SiO_2)	2.4	12.0	10.0	9.4	22.0
Iron (Fe)	0.14	0.02	0.09	0.2	0.08
Calcium (Ca)	5.8	36.0	92.0	96.0	3.0
Magnesium (Mg)	1.4	8.1	34.0	27.0	2.4
Sodium (Na)	1.7	6.5	8.2	183.0	215.0
Potassium (K)	0.7	1.2	1.4	18.0	9.8
Bicarbonate (HCO_3)	14.0	119.0	339.0	334.0	549.0
Sulfate (SO_4)	9.7	22.0	84.0	121.0	11.0
Chloride (Cl)	2.0	13.0	9.6	280.0	22.0
Nitrate (NO_3)	0.54	0.1	13.0	0.2	0.52
Total dissolved solids	31.0	165.0	434.0	983.0	564.0

[a] Taken from Collins [7].

needed. Such an examination will also provide a perspective on the variability of inorganic and organic compounds present. The presence of certain compounds may not create a serious deleterious effect, but the variability may cause varying effects on the efficiency of air-entraining admixtures, the time of setting, and the strength development.

A concrete manual [8] published in Denmark in 1944 points out that humic acid and other organic acids should be avoided because their presence means a danger to the stability of concrete. Most organic compounds will have an effect on time of setting and comparative tests should be conducted to evaluate the effect.

An article appearing in a 1947 British publication [9] discusses the harmful effects of using acid waters in concrete and claims that the harmful effects of organic acid are not evident as soon as those of mineral acids, while deleterious salts have a greater effect on early-age strengths than at later ages.

Kleinlogel [10] stated that mixing water should not contain humus, peat fiber, coal particles, sulfur, or industrial wastes containing fat or acid.

An article in Ref 11 contains a tabulation of maximum limits for impurities in mixing water that is summarized in Table 2.

The limit for suspended particles in Table 2 agrees with the requirements of the U.S. Bureau of Reclamation [12] that has a turbidity limit of 2000 ppm for mixing water.

Sugar is probably the organic contaminant that causes the most concern. Steinour [13] explained that, although sugar as a contaminant in the field has gained a bad reputation as a retarder and strength reducer, these judgments need qualification. Laboratory tests have shown that, although smaller amounts retard the setting, they increase the strength development. With larger amounts, the setting is further retarded and early strengths such as those for two and three days (or even seven days) are severely reduced. The later strengths, however, are increased or at least not affected adversely, provided proper curing is maintained as required. With still larger amounts, the cement becomes quick-setting and strengths are reduced markedly for 28 days and probably permanently. The amount of sugar that can cause these different effects varies with the other factors involved, such as the composition of the cement, cement content, and ambient conditions.

Use of Seawater in Mixing Concrete

In addition to the supporting reference previously mentioned in Abrams' paper [5], the English article [9] also states that seawater with a maximum concentration of salts on the order of 3.5% does not appreciably reduce the strength of concrete, although it may lead to corrosion of reinforcement.

A paper by Liebs [14] contains the results of comparative 7-, 28-, and 90-day compressive strength tests of concrete mixed with fresh water and with seawater. The data show that the seawater concrete had about 6 to 8% lower strengths than the fresh water concrete. No efflorescence was observed. The article in Ref 11 pointed out that concrete made with seawater may have higher early strength than normal concrete, but strengths at later ages (after 28 days) may be lower. Steinour [13] stated that the use of seawater may cause a moderate reduction in ultimate strength, an effect that can be avoided by the use of a higher cement content. He also noted that concrete in which seawater is used as mixing water is sound but its use may cause efflorescence or dampness, and in reinforced concrete, the risk of corrosion of the steel is increased. Seawater definitely should not be used for making prestressed concrete.

Hadley [15] points out that the seawater was used in the concrete for the foundation of the lighthouse at the extremity of the Los Angeles breakwater that was built by the U.S. Army Corps of Engineers in 1910 and that 25 years later it was examined and found to be in good condition with sharp-edged corners and no disintegration. There are several references in the literature that indicate that salt water has been used in mixing plain concrete without incurring trouble at later periods. Much of the concrete for the Florida East Coast Railway was mixed with seawater with no detrimental effect due to its use [16]. Most engineers are of the opinion that seawater should not be used for mixing reinforced concrete; however, Dempsey [17] describes construction of military bases in Bermuda using coral aggregate and concludes that seawater seems to be satisfactory for making reinforced concrete and causes no problem beyond an acceleration in stiffening of the mixture. No harmful effect on the durability of reinforced concrete had occurred at the end of four years. Nonetheless, extreme caution is urged when mixing water for reinforced concrete is selected. If the water contains salts, the residual salts in the concrete when combined with air and moisture will result in some corrosion.

Effects of Algae in Mixing Water on Air Content and Strength of Concrete

A rather extensive series of laboratory tests reported by Doell [18] showed that the use of water containing algae had the unusual effect of entraining considerable quanti-

TABLE 2—Tolerable Concentrations of Impurities in Concrete Mixing Water.

Impurity	Maximum Tolerable Concentration
1. Sodium and potassium carbonates and bicarbonates	1000 ppm
2. Sodium chloride	20 000 ppm
3. Sodium sulfate	10 000 ppm
4. Calcium and magnesium bicarbonates	400 ppm of bicarbonate ion
5. Calcium chloride	2% by mass, of cement in plain concrete
6. Iron salts	40 000 ppm
7. Sodium iodate, phosphate, arsenate, and borate	500 ppm
8. Sodium sulfide	100 ppm warrants testing
9. Hydrochloric and sulfuric acids	10 000 ppm
10. Sodium hydroxide	0.5%, by mass, of cement if set not affected
11. Salt and suspended particles	2000 ppm

ties of air in concrete mixtures with an accompanying decrease in strength. The data in Table 3 were extracted from Doell's paper and are based on tests with 19.0-mm (¾-in.) maximum-size aggregate concrete having a water/cement ratio of 0.5 and a slump of 40 to 75 mm (1.5 to 3 in.), with a constant ratio of coarse to fine aggregate.

In addition to the detrimental effect on strength, one of the important aspects of these data is that considerable quantities of air can be entrained in concrete by the use of mixing water containing algae.

Effect of Hardness of Mixing Water on Air Content of Concrete

Wuerpel [19] reported a series of air determination tests with waters of various degrees of hardness that shows that the air content was not affected by the hardness of the water.

CURING WATER

There are two primary considerations with regard to the suitability of water for curing concrete. One is the possibility that it might contain impurities that would cause staining, and the other is that it might contain aggressive impurities that would be capable of attacking or causing deterioration of the concrete. The latter possibility is unlikely, especially if water satisfactory for use in mixing concrete is employed. In some instances the staining or discoloration of the surface of concrete from curing water would not be objectionable. The most common cause of staining is usually a relatively high concentration of iron or organic matter in the water; however, relatively low concentrations of these impurities may cause staining, especially if the concrete is subjected to prolonged wetting by runoff of curing water from other portions of the structure [3].

Test data from the Corps of Engineers [20], show that there is not a consistent relationship between dissolved iron content and degree of staining. In some cases, 0.08 ppm of iron resulted in only a slight discoloration and in other cases, waters with 0.06 ppm of iron gave a moderate rust-colored stain, while 0.04 ppm produced considerable brownish-black stain. Generally speaking, the conditions of these tests were such as to accentuate the staining properties of the water, since considerably more water was evaporated over a unit area than would be the case in most instances in the field.

TABLE 3—Effect of Algae in Mixing Water on Air Content and Strength of Concrete.

Mixture Number	Algae in Mixing Water, %	Air in Concrete, %	Compressive Strength, 28 days, MPa (psi)
10	none control	2.2	33.3 (4830)
8	0.03	2.6	33.4 (4840)
7	0.09	6.0	27.9 (4040)
5	0.15	7.9	22.8 (3320)
9	0.23	10.6	17.8 (2470)

With respect to organic impurities in water, it is virtually impossible to determine from a chemical analysis if the water would cause objectionable staining when used for curing concrete. It is advisable to use a performance-type test procedure, such as Designation CRD-C 401 of the Corps of Engineers [21]. This method outlines three procedures for evaluating the staining properties of water proposed for use in curing concrete. The Preliminary Method is intended for use in selecting sources that are worthy of more complete investigation and consists of evaporating 3000 mL of the test water in the concave area formed by the impression of a 100-mm (4-in.) watch glass in the surface of a neat white cement or plaster of Paris specimen. The Complete Method can be used to evaluate those sources that the Preliminary Method indicate to be promising. In the Complete Method, 11 dm^3 (3 gal) of test water drip on a mortar specimen exposed to heat lamps and forced air circulation. The Field Method is intended as a means of evaluating the water finally selected for use and involves the curing of a 1.9-m^2 (20-ft^2) slab of concrete with the test water for at least 28 days with maximum exposure to the sun with the test slab placed at a slight angle sufficient to keep it in a wet condition with a minimum runoff. The test results by each of these three methods are evaluated by visual observation.

The Corps of Engineers' Standard Practice for Concrete [22] clearly states that there must be no permanent staining of surfaces where appearance is important. For these surfaces, the contractor has the option of using nonstaining water or of cleaning the surface after completion of moist curing. No cleaning is required for surfaces that will subsequently be stained when the structure is in service.

SUMMARY

Mixing Water

The significance of the foregoing information presented indicates that any naturally occurring or municipal water supply suitable for drinking purposes can be used as mixing water for concrete and that most naturally occurring waters ordinarily used for industrial purposes are satisfactory. Many waters that upon casual examination would be judged to be unsuitable because of color, odor, or contamination with impurities as in the case of marsh water, alkaline sulfate waters, and water containing industrial wastes could be found to be satisfactory when tested in mortar or concrete since, in many instances, the strength would be greater than 90% of the strength of comparative specimens made with pure waters. In the case of seawater, a strength reduction ranging from 8 to 15% can be expected depending on job conditions; however, seawater ordinarily is not recommended for use as mixing water in reinforced concrete. The hardness of water usually does not affect air content of concrete; however, with certain anionic and non-ionic admixtures, additional dosage may be required to obtain the desired air content. Algae in mixing water; however, can entrain air and significantly reduce strength.

Curing Water

It is improbable that a water used for curing would attack concrete if it were of the type suitable for use as mixing water. Organic matter or iron in the curing water can cause staining or discoloration of concrete, but this is rather uncommon especially where a relatively small volume of water is used; however, the suggested performance tests [21] will determine if a water possesses any potential staining qualities.

REFERENCES

[1] Clair, M. N., "Effect of Sugar on Concrete in Large Scale Trial," *Engineering News-Record*, March 1929, p. 473.

[2] Neville, A. M. and Brooks, J. J., *Concrete Technology*, Longman Group UK Limited, Essex, UK, 1987, pp. 74–75.

[3] "Requirements for Water for Use in Mixing or Curing Concrete," CRD-C 400, *Handbook for Concrete and Cement*, U.S. Army Corps of Engineers, Vicksburg, MS, 1963.

[4] "Test Method for Compressive Strength of Mortar for Use in Evaluating Water for Mixing Concrete," CRD-C 406, *Handbook for Concrete and Cement*, U.S. Army Corps of Engineers, Vicksburg, MS, 1979.

[5] Abrams, D. A., "Tests of Impure Waters for Mixing Concrete," *Proceedings*, American Concrete Institute, Vol. 20, 1924, pp. 442–486.

[6] Proudley, C. E., "Effect of Alkalies on Strength of Mortar," *Public Roads*, Vol. 5, 1924, pp. 25–27.

[7] Collins, W. D., "Typical Water Analyses for Classification with Reference to Industrial Use," *Proceedings*, American Society for Testing and Materials, Vol. 44, 1944, pp. 1057–1061.

[8] Plum, N. H., Christiani, and Nielsen, "Concrete Manual," *Bulletin No. 39*, Copenhagen, Denmark, 1944.

[9] "Water for Making Concrete," Title No. 44-197, "Job Problems and Practice," *Proceedings*, American Concrete Institute, Vol. 44, 1948, pp. 414–416; reprinted from "Water for Making Concrete," *Concrete and Constructional Engineering*, London, Vol. 42, No. 10, 1947, p. 295.

[10] Kleinlogel, A., *Influences on Concrete*, Frederick Ungar Publishing Co., New York, 1950, p. 158.

[11] "Requirements of Mixing Water for Concrete," *Indian Concrete Journal*, March 1963, pp. 95, 98, and 113.

[12] *Concrete Manual*, 8th ed., U. S. Department of the Interior, Bureau of Reclamation, Washington, DC, 1981, p. 70.

[13] Steinour, H. H., "Concrete Mix Water—How Impure Can It Be?" *Journal*, Portland Cement Association, Research and Development Laboratories, Skokie, IL, Vol. 2, No. 3, 1960, pp. 32–50.

[14] Liebs, W., "The Change of Strength of Concrete by Using Sea Water for Mixing and Making Additions to Concrete," *Bautechnik*, 1949, pp. 315–316.

[15] Hadley, H., "Letter to Editor," *Engineering News-Record*, May 1935, pp. 716–717.

[16] "Job Problems and Practice," *Proceedings*, American Concrete Institute, Vol. 36, 1940, pp. 313–314.

[17] Dempsey, J. G., "Coral and Salt Water as Concrete Materials," *Proceedings*, American Concrete Institute, Vol. 48, 1951, p. 165.

[18] Doell, B. C., "The Effect of Algae Infested Water on the Strength of Concrete," *Proceedings*, American Concrete Institute, Vol. 51, 1954, pp. 333–342.

[19] Wuerpel, C. E., "Influence of Mixing Water Hardness on Air Entrainment," *Proceedings*, American Concrete Institute, Vol. 42, 1946, p. 401.

[20] Mather, B. and Tye. R. V., "Requirements for Water for Use in Mixing or Curing Concrete," Waterways Experiment Station Technical Report No. 6-440, U.S. Army Corps of Engineers, Vicksburg, MS, Nov. 1956, p. 36.

[21] "Method of Test for the Staining Properties of Water," CRD-C 401, *Handbook for Concrete and Cement*, U.S. Army Corps of Engineers, Vicksburg, MS, 1975.

[22] *Engineer Manual, Standard Practice for Concrete*, EM 1110-2-2000, U.S. Army Corps of Engineers, Washington, DC, 1985.

Curing and Curing Materials

Ephraim Senbetta[1]

PREFACE

The subject of curing materials was first discussed by John Swanberg in *ASTM STP 169* in 1955. In 1966, C. E. Proudley wrote about curing materials in *ASTM STP 169A*. The discussion on curing materials in *ASTM STP 169B* was by Rodger Carrier. This paper attempts to summarize the extent to which the state of the art has changed over the years since 1955.

INTRODUCTION

Fresh concrete is converted into a solid mass when the cementitious materials hydrate. The degree to which the concrete achieves its full potential strength and durability depends on the degree of hydration of the cementitious materials. The actions taken to ensure proper hydration after the concrete is placed constitute the curing of the concrete.

The need for curing concrete has been recognized since the inception of concrete, and much has been written on the subject. The American Concrete Institute (ACI) has had a committee that deals with the curing of concrete, Committee 308, since 1936. Similar committees have also been operating for sometime now within ASTM and other organizations. In spite of all these activities, as far as curing materials for cast-in-place concrete under normal atmospheric conditions are concerned, there have not been significant breakthroughs or significant technological developments for a long time. The methods and types of materials for curing concrete that are in use today have been in use for more than 40 years.

The first edition of this publication, *ASTM STP 169* [1], that was published in 1955 had a chapter on curing materials that described the same types of curing methods and materials that are used today. The new piece of information in that particular publication was the development of a new tentative ASTM Test Method for Water Retention Efficiency of Liquid Membrane-Forming Curing Compounds and Impermeable Sheet Materials for Curing Concrete (C 156) that was published in 1940, and a new ASTM Specification for Liquid Membrane-Forming Compounds for Curing Concrete (C 309), published in 1953. There was also a mention of the use of calcium chloride on the surface of fresh concrete and integrally for curing. This is no longer recommended.

In 1966, *ASTM STP 169A* [2] was published, and it too had a chapter on curing materials. The chapter contained significantly more information and discussion about the importance of curing compared to what was written in the 1955 edition. In addition, there was a mention of the lack of methods for identifying the effects of poor curing, after the fact, but nothing new was reported about new curing materials or methods.

Twelve years later, in 1978, *ASTM STP 169B* [3] was published. The chapter on curing materials in this publication provided significantly more information about the effect of curing on concrete properties. It also discussed the need for monitoring the effectiveness of the curing of concrete in the field, and it made a brief mention of a new technology that was developed for this purpose. Unfortunately, this monitoring device that was known as the Relative Humidity Button did not receive wide acceptance. Other than that, there was essentially no mention of new developments regarding curing materials or methods.

This chapter, which is being written approximately 14 years after the publication of *ASTM STP 169B*, will unfortunately not contain much new information about advances made in the development of curing materials or methods. This is because as stated earlier, there have not been any noteworthy developments or changes in the way cast-in-place concrete is cured in nearly half a century. However, information on new developments concerning testing and monitoring methods and activities related to product improvements to comply with environmental regulations will be discussed. In addition, a review of the subject of curing in a general way with emphasis on the reasons why concrete should be cured and on the limitations of the current test methods and specifications governing curing materials will be presented. There will also be a brief mention of how to select curing materials. The chapter will conclude by outlining what needs to be done in the area of curing of concrete to make sure it is practiced more faithfully than it has been and by providing suggestions for future new technological developments that may help to ensure proper curing of concrete.

THE EFFECT OF CURING ON CONCRETE PROPERTIES

The capillary porosity of cement paste has a profound effect on both the strength and durability of concrete. The

---

[1] Technical director, Master Builders, Inc., Cleveland, OH 44122.

two major factors that affect capillary porosity are water/cement (w/c) ratio and the degree of hydration of the cement. For a given w/c ratio, the capillary porosity of paste is essentially the space that was originally occupied by the mixing water. As hydration of cement progresses, the space that was originally occupied by water is gradually filled with hydration products. Therefore, the capillary porosity of cement paste is the greatest shortly after the initial mixing of the concrete before any significant cement hydration occurs and it is the least when nearly all the cement is hydrated. Studies by Powers [4] have shown that the solid materials in fully hydrated hardened cement paste normally occupy 45 to 60% of the total volume of the paste and the average porosity is 40 to 55% including both capillary and gel pores.

Figure 1 illustrates the effect of w/c ratio and degree of hydration on capillary porosity of cement paste. The degree of hydration values, that is, amounts of cement that have hydrated, are assumed to range from 50 to 90%. Note that the capillary porosity values in the figure are calculated results based on what is known about changes in the volume of paste constituents during cement hydration [5]. The profound effect of porosity on strength is shown on Fig. 2 [6]. As stated earlier, applied curing affects the near-surface region of concrete more so than the interior because the surface is affected more by the ambient conditions than the interior. Therefore, depending on the quality and effectiveness of the applied cure, a porosity gradient that decreases with depth can be expected. The concrete's strength can also be expected to increase as the porosity decreases. Because of this consideration, the use of flexural strength test as a test for the efficiency of curing materials has been suggested [7]. In general, as long as a strength test determines the bulk property of concrete, as would be the case with compressive strength, the test would not be a sufficiently sensitive indicator of the effect of curing because both the surface and the interior of the concrete are contributing to the strength.

The effect of capillary porosity on durability is more pronounced than it is on strength because as capillary porosity increases, permeability increases. Powers [8] has illustrated that the coefficient of permeability of paste with w/c ratio of 0.645 to be 190 times larger after one day of curing than the same paste when cured for three days. Permeability has a profound effect on durability of concrete because as the permeability decreases, the ingress of water or any harmful solutions into the concrete is reduced. The effect of curing on durability is discussed in greater detail by this author in another publication [9].

Curing also has an effect on other physical properties of concrete. A study [10] has shown that the development of modulus of elasticity of concrete is much faster than either compressive or tensile strength at early ages. It was also pointed out that the greatest increase of static modulus of elasticity in the first three days of curing occurred between 6 and 12 h after the test specimens were cast.

In addition to the effect on strength and durability that has just been discussed, curing practice focusing on temperature control has been shown to have a very significant effect on reducing cracking [11]. This involves monitoring maximum temperature, maximum temperature differences in the concrete, and the use of the maturity method to predict strength gain before form removal.

THE EFFECT OF ATMOSPHERIC CONDITIONS ON CURING

The need for taking deliberate steps to ensure proper curing of concrete is dependent on the ambient atmospheric conditions. In certain environments where the relative humidity and temperature are favorable, no deliberate action needs to be taken to cure the concrete. Usually, conditions are such that some action is required.

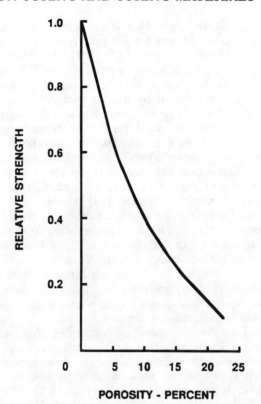

FIG. 2—Effect of porosity on strength.

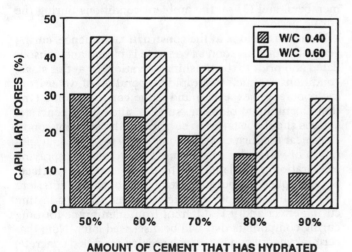

FIG. 1—Effect of water/cement ratio W/C and degree of hydration on capillary porosity.

Although there are no standard definitions, general ambient conditions can be put into three temperature categories. These are hot, (32°C) (90°F) and above; cold, (10°C) (50°F) and below; and normal, (10 to 32°C) (50 to 90°F). Within these broad categories, the severity of the environment is affected by relative humidity and wind speed. Therefore, by determining the ambient temperature, relative humidity, wind speed, and concrete temperature, the rate of evaporation of water from the exposed surface of fresh concrete can be estimated. A nomograph [12] has been commonly used for this purpose. However, the nomograph should be used with caution because, among other things as Mather [13] pointed out, it was developed for conditions where a surface is covered with free water. It does not apply to evaporation from porous materials that contain water within them but have no continuous surface layer of water. Thus, it is important to keep in mind that unless the conditions on which the nomograph is based are met, the evaporation rate estimated from the nomograph can, in some cases, be underestimated by a significant amount.

Two of the three temperature conditions, cold and hot, call for particular attention during curing. According to ACI Committee 306, a cold weather condition is described as a period of three or more consecutive days when the average daily air temperature is less than (5°C) (40°F) and when the air temperature is less or equal to (10°C) (50°F) for more than half of any 24-h period [14]. The major concerns in cold weather conditions are: damage due to freezing at early ages, that is, before the concrete attains compressive strength of at least (3.5 MPa) (500 psi); reduced rate of strength development for form removal or for handling of structural loads at a given age; and damage due to thermal stresses as a result of rapid surface cooling creating a large temperature gradient between the surface and the interior of concrete. In addition to preventing moisture loss, keeping concrete warm by using insulation or a heat source may be the primary action taken during the curing period under cold weather conditions. Although in cold weather conditions the focus is on temperature control, it has been shown that temperature difference between the concrete and air can cause excessive drying that may result in cracking [15]. This is particularly the case when hot water is used for mixing the concrete to reduce retardation thereby magnifying the temperature difference between the concrete and the ambient air. This, in turn, accelerates evaporation of water from the concrete surface drying the surface and possibly causing plastic shrinkage cracking. Therefore, in addition to the common concerns during cold weather mentioned earlier, it is important to guard against rapid surface drying and the subsequent plastic shrinkage cracking due to large temperature differences. Water curing in cold weather conditions should be avoided or used with caution to guard against the concrete freezing in a saturated condition when protection is removed at the end of the curing period. The protection of the curing period during cold weather is generally longer than what it would be under normal temperature conditions since cement hydration is temperature dependent. Cement hydration also varies with the type of cement, the cement factor, and in response to any accelerating admixtures that may have been used. General guidelines for the recommended duration of curing under the different conditions are provided by ACI Committee 306 report and specification, ACI 306R-88 [14].

Hot weather conditions are normally considered more damaging to concrete because of the potential drying of concrete at early ages as a result of high temperatures that may be accompanied by low relative humidity and wind. The primary concern under these conditions is preventing evaporation of water. Control of concrete temperature is generally accomplished by using ice as part of the mixing water. High concrete temperature has no particular adverse effect on the curing of the concrete as long as there is no large temperature gradient in the concrete. Limiting concrete temperatures is therefore usually done to control mixing water content and its effect on strength, durability, drying shrinkage, and slump loss. However, for those situations where thermal cracking can be a potential problem under specific environmental and construction conditions, Concrete Curing Tables have been prepared by researchers at the University of Pennsylvania [16]. According to the researchers, the tables will be helpful in predicting "... four conditions in the curing process: satisfactory curing conditions, risk of early freezing or inadequate strength development, risk of too large a thermal gradient between the slab interior and the surrounding concrete, and risk of too high a temperature within the slab." The tables will take into account factors such as concrete mix type, wind velocity, type of surface protection, and elapsed time between placement and removal of surface protection. More information about curing in hot weather conditions can be found in ACI Committee 305 report, ACI 305R-89 [17].

CURING METHODS AND MATERIALS FOR MOISTURE RETENTION

The methods and materials employed in the curing of concrete vary depending on (1) the type of concrete, (2) the type of structure including the orientation of the structural members, and (3) on the ambient conditions during the curing period.

If one was to look at the construction sequence, curing generally starts as soon as concrete is placed and consolidated and prior to the finishing operation. If, at this stage, conditions are such that rapid evaporation of water from the concrete is expected, and if the concrete is the type that has minimal bleeding, such as silica fume concrete, actions should be taken to either alter the ambient conditions near the surface of the concrete by fogging, or evaporation of water can be prevented or reduced by spraying evaporation retardant on the fresh concrete surface. Spray-applied evaporation retardant does not interfere with the finishing of concrete nor does it have a lasting curing effect and it is different from membrane-forming curing compounds that will be discussed later. Note that certain types of concrete, such as latex-modified concrete, may not require any action to prevent drying beyond the first 24 h.

After concrete is placed and all other subsequent operations such as finishing and texturing are completed, the curing of the concrete can proceed by either keeping the concrete moist or by sealing the surface to prevent evaporation. For formed concrete placements, the forms will provide adequate protection against moisture loss as long as they remain in place. After removal, other curing procedures may be needed depending on the age of the concrete at the time of form removal.

Moist curing may involve ponding, continuous sprinkling, or covering with materials such as burlap, sand, or straw that are kept continuously wet. This approach will not only prevent the loss of moisture from the concrete surface but it can also furnish additional moisture to the surface of the concrete as the mixing water is used in the hydration of cement. The disadvantages of this approach include the effort it requires to keep the concrete wet continuously, ensuring the availability of an adequate supply of water, and providing for proper drainage of excess water during continuous sprinkling. In addition, for certain types of structures and structural members that are vertical or inclined, the method may not be practical. More information about this method of curing can be found in the ACI Committee 308 document, ACI 308-81 [12].

The second approach for curing concrete involves sealing the surface, and it is applied by covering the concrete surface with reinforced waterproof paper, plastic sheet material, or membrane-forming curing compounds. Leaving formwork in place may also provide curing, provided that the form is not absorbent and it does not have gaps that may allow moisture to leave the concrete. Of the preceding methods, the use of membrane-forming curing compounds is probably the most common and perhaps the most convenient. The use of waterproof paper and plastic sheets requires more labor and keeping the materials in place is not always easy, particularly in windy environments. Certain structural members, for example vertically oriented, may not lend themselves to curing with sheet materials. Combinations of different materials for curing concrete, such as the use of wet burlap covered with plastic sheet, or covering with wet burlap initially for a day or two and then applying curing compound, may also be used.

Membrane-forming curing compounds are liquid materials that can either be sprayed, rolled, or brushed on the surface of concrete. After the curing compound is applied, it is intended to form a continuous film that completely seals the surface restricting evaporation. These materials have different chemical compositions and can be water or solvent based. Curing compounds should be applied to the surface of fresh concrete as soon as the bleed water evaporates from the surface of the concrete. The amount of material that is applied on the surface is usually recommended by the manufacturer. However, if none is specified, for testing purposes the coverage rate is usually 5.0 m^2/L (200 ft^2/gal) when applied to relatively smooth surfaces. As the surface texture of concrete increases, a greater amount of curing compound should be specified. The increased application rate is to compensate for the greater surface area of textured surfaces and for the inevitable nonuniform coverage due to the hills and valleys.

Curing compounds may have a variety of colors. Some may contain fugitive dye that provides a visual indication of uniform coverage of the concrete surface and that disappears in a few days. Others are white-pigmented to reduce heat buildup in concrete by reflecting the rays of the sun. And there are those that are pigmented to match the color of the concrete they are applied on, usually to meet architectural requirements for colored concrete.

It is important to keep in mind that some curing compounds, although effective, may yellow after they are applied and exposed to the elements. In some cases, this may not be acceptable for aesthetics reasons. One also needs to keep in mind that there may be a bond or adhesion problem if coatings or toppings are to be placed on top of concrete cured with curing compounds. Removing curing compounds at the end of the curing period is often difficult. However, curing compounds usually wear off from the concrete by natural weathering over extended periods of time, or by traffic action, and may leave a blotchy appearance due to nonuniform wear.

The popularity of the use of curing compounds is a result of their cost effectiveness, convenience, and simplicity. Although there is a standard specification for membrane-forming curing compounds, ASTM C 309, there is some controversy or concern with the test procedure for evaluating their effectiveness. A detailed discussion of this matter is provided in the next section.

TEST METHODS AND SPECIFICATIONS

Although there are no specifications for the moist curing methods, there are standard specifications for sealing materials that prevent evaporation of water. ASTM Specification for Sheet Materials for Curing Concrete (C 171) was first published in 1942. It includes requirements for minimum tensile strength, minimum elongation, minimum reflectance (for white materials), and moisture retention. The tensile strength requirement is to guard against tearing when the material is handled during use and a minimum reflectance is specified to ensure a minimum level of heat absorption from the rays of the sun. The moisture retention requirement that calls for moisture loss through the material of no more than 0.55 kg/m^2 of surface when tested according to ASTM Test Method for Water Retention by Concrete Curing (C 156) is interesting. If a material meets all the other requirements there is no reason why any measurable moisture should go through the membrane unless the membrane has pinholes, is torn, or the surface of the test sample is not covered properly. The limit was set to be comparable to the liquid membrane-forming curing compounds rather than on the capabilities of sheet materials.

The standard specification for liquid membrane-forming curing compounds, ASTM C 309, was adopted in 1953. The specification includes requirements for drying time, reflectance, and moisture retention. The moisture retention requirement, as stated earlier, is limited to evaporation loss of no more than 0.55 kg/m^2 of surface in 72 h when the testing is done according to ASTM C 156. This is, by far, the most important requirement, but the test method has been controversial. The requirement for a given minimum reflectance, which applies to white-pigmented curing compounds, is related to the prevention of heat buildup in concrete. The maximum drying time requirement is to guard against the concrete having a slippery surface and to make sure the curing compound does not track off when walked on. Studies [18,19] have shown that the results of the moisture retention test method have poor reproducibility, and the precision statement in the standard itself indicates this. A survey of the state highway departments in the United States revealed that 17% of the states that use the test method regard it unreliable and 22% of the states have modified it in order to improve it [20]. Factors that contribute to the lack of reproducibility of test results include temperature; humidity; wind in test cabinets; the times of application of curing compounds on test specimens, that is, how long after the specimens are cast; and specimen preparation. These concerns have led to the development of a new test method that is discussed subsequently.

If meeting the ASTM C 309 specification requirements is not a guarantee for having a curing compound that will do the job, one may then wonder how one goes about selecting an effective curing compound. This task is not as simple as it may seem in the absence of reliable quantitative evaluation methods. Key questions that must be asked to ensure the curing compound chosen is the right one for the job can be found in a paper published in *Concrete International* in 1989 [21].

SUMMARY OF NEW DEVELOPMENTS

Although, as stated earlier, there have not been significant changes in the methods and materials for curing concrete, some progress has been made that has the potential of improving the curing of concrete and protecting the environment.

Most of the activities regarding curing materials have probably been in improving the volatile organic compound (VOC) content of solvent-based curing compounds. As concern for the environment grows, the federal government has been enacting laws to reduce air pollutants. The first of these laws is the Clean Air Act of 1963. It has since been amended a number of times and the Environmental Protection Agency now addresses issues of hazardous air pollutants, atmospheric ozone, and volatile organic compounds nationally. To comply with these regulations, manufacturers of curing compounds have been reducing the VOC content of curing compounds, in many cases switching to water-based materials including hydrocarbon resin and wax emulsions in water.

There has also been research work to explore alternate approaches for curing concrete. One such method is the use of microwave energy to cure concrete rapidly and uniformly without sacrificing some of the ultimate strength that is normally associated with accelerated cement hydration [22].

One new development is the adoption of a new test procedure for determining the effectiveness of materials for curing concrete. It is intended to be an alternate to ASTM C 156 that has had questionable reliability. The new test procedure, ASTM Test Method for Evaluating the Effectiveness of Materials for Curing Concrete (C 1151), was adopted in 1990, and is based on evaluation of the absorptivity of mortar test specimens. Absorptivity is a measure of the degree of the pore structure development of the paste fraction of the mortar as a result of cement hydration. As cement hydrates, capillary porosity decreases, which in turn reduces the absorptivity. The validity of the new test method has been verified by its excellent correlation with abrasion tests, and its reproducibility has been shown to be high. A detailed explanation of this approach can be found in a thesis on which the test procedure is based [23]. With this test method, curing materials can be evaluated under any desired atmospheric condition and for any length of curing time. The method is also conducive to evaluating the effect of curing on any part of concrete including the near-surface region that is the most vulnerable. It has been shown that the top 2.5 cm (1 in.) of concrete is the zone that is affected the most by curing [22].

An important aspect of enforcement of any specification is having tools for monitoring the achievement of the specified desirable results. Current curing specifications simply rely on specifying the materials to be used and how and when they should be used. The assumption being, if these requirements are followed, good quality concrete with the desirable properties is produced. Unfortunately, without a method for verifying what is being assumed, the motivation for adhering to the specifications and for doing what needs to be done suffers. One new method for assessing concrete curing efficiency is the use of gas permeability apparatus for testing concrete surface permeability [24,25]. This method is claimed to be a nondestructive and a rapid method of determining the air permeability of the near-surface layer of concrete within approximately 10 min. Other monitoring methods are being looked at as they become available. In his report to ACI Committee 308, on curing of concrete, Hover [26] has indicated that devices that have been used for agricultural applications to estimate evaporation rate from crops in the field may be used for monitoring evaporation from concrete surfaces during the curing period.

NEEDS FOR FUTURE WORK

To improve the curing of concrete, new enforceable specifications, reliable testing, and field curing monitoring methods, new innovative materials, and effective educational programs are needed to convey the value of curing and the methods by which good curing can be achieved.

At the time of this writing, there is no standard specification for curing concrete except for its brief mention in other specifications such as the ACI Specification for Structural Concrete for Buildings [27]. A complete stand-alone curing specification that can be incorporated into job specifications can have a significant impact, particularly if it includes requirements for field curing monitoring and the desired properties of concrete at the end of the specified curing period. ACI Committee 308 is currently working on this.

Reliable curing monitoring methods that can be used by inspectors at job sites, and new test procedures for evaluating the properties of concrete at the end of the curing period to determine the effectiveness of the curing, are also needed. These approaches can provide the necessary motivation for paying particular attention to the curing of the concrete and for making sure the end product is acceptable.

From the curing materials point of view, the primary means of curing, as stated earlier, is preventing the drying of concrete by covering the surface of the concrete with cloth, paper, plastic, or a membrane-forming curing compound. The effectiveness of these materials depends on the quality of the materials; the time of application of the materials, that is how soon after the concrete is placed; and how well they are applied and kept intact and in place for the specified length of time. Each of these steps is a potential problem area. Curing materials of the future may include materials that are added to concrete as an admixture during mixing whose purpose is to prevent evaporation of the mix water and perhaps to also regulate the temperature of the concrete. Approaches such as this may make the curing of the concrete virtually effortless.

Another essential component of future developments to improve the curing of concrete is the development of effective educational programs to enable people to develop a real appreciation for the value of curing. Unless one understands the need for curing and its profound impact on the properties of concrete, the steps prescribed to cure the concrete merely become requirements that one must satisfy without concern for the end result.

It is this author's opinion that, in the construction of concrete structures, curing of the concrete continues to be the most neglected step. A great deal of concrete the durability problems are a result of poor concrete surface properties because of the lack of adequate curing. The next edition of this publication will hopefully include significant advances in both curing materials and curing methods that will enhance the strength and durability of concrete.

REFERENCES

[1] Swanberg, J. H., "Curing Materials," *Significance of Tests and Properties of Concrete and Concrete Aggregates, ASTM STP 169*, American Society for Testing and Materials, Philadelphia, 1955, pp. 361–365.

[2] Proudley, C. E., "Curing Materials," *Significance of Tests and Properties of Concrete and Concrete-Making Materials, ASTM STP 169A*, American Society for Testing and Materials, Philadelphia, 1966, pp. 522–529.

[3] Carrier, R. E., "Curing Materials," *Significance of Tests and Properties of Concrete and Concrete-Making Materials, ASTM STP 169B*, American Society for Testing and Materials, Philadelphia, 1978, pp. 774–786.

[4] Powers, T. C., "The Nature of Concrete," *Significance of Tests and Properties of Concrete and Concrete-Making Materials, ASTM STP 169A*, American Society for Testing and Materials, Philadelphia, 1966, p. 67.

[5] Neville, A. M., *Properties of Concrete*, 3rd ed., Pitman Publishing, London, 1981, pp. 26–35.

[6] Neville, A. M., "Properties of Concrete—An Overview," *Concrete International*, Vol. 8, No. 2, 1986, p. 22.

[7] Taylor, W. H., *Concrete Technology and Practice*, Fourth Ed., McGraw Hill, New York, 1977, p. 267.

[8] Powers, T. C., "The Physical Structure and Engineering Properties of Concrete," *PCA Bulletin 90*, Portland Cement Association, Skokie, IL, 1958, p. 17.

[9] Senbetta, E. and Malchow, G., "Studies on Control of Durability of Concrete Through Proper Curing," *ACI SP-100*, American Concrete Institute, Detroit, MI, Vol. 1, 1987, pp.73–87.

[10] Ohiokun, F. A., et al., "Rates of Development of Physical Properties of Concrete at Early Ages," *Transportation Research Record No. 1284*, Transportation Research Board, Washington, DC, 1990, pp. 16–21.

[11] Gotfredsen, H.-H. and Idorn, G. M., "Curing Technology at the Faroe Bridges, Denmark," *Properties of Concrete at Early Ages, ACI SP-95*, American Concrete Institute, Detroit, MI, 1986, pp. 17–31.

[12] "Standard Practice for Curing Concrete," *ACI Manual of Concrete Practice*, ACI 308-81 (Revised 1986), American Concrete Institute, Detroit, MI, 1991, pp. 308–1–308–11.

[13] Mather, B., "Curing of Concrete," *ACI SP-104*, American Concrete Institute, Detroit, MI, 1987, p. 147.

[14] "Cold Weather Concreting," *ACI Manual of Concrete Practice*, ACI 306R-88, American Concrete Institute, Detroit, MI, 1991, pp. 306R–1–306R–23.

[15] Senbetta, E. and Bury, M. A., "Control of Plastic Shrinkage Cracking in Cold Weather," *Concrete International*, Vol. 13, No. 3, 1991, pp. 49–53.

[16] "Concrete Curing Tables Help Determine Curing Risks at Various Temperatures," *FOCUS*, Strategic Highway Research Program, Washington, DC, Jan. 1992, p. 6.

[17] "Hot Weather Concreting," *ACI Manual of Concrete Practice*, ACI 305R-89, American Concrete Institute, Detroit, MI, 1991, pp. 305R–1–305R–17.

[18] Leitch, H. C. and Laycraft, N. E., "Water Retention Efficiency of Membrane Curing Compounds," *Journal of Materials*, Vol 6, No. 3, Sept. 1971, pp. 606–616.

[19] Senbetta, E. "Tightening Concrete Curing Specifications," *The Construction Specifier*, Dec. 1989, pp. 53–57.

[20] Senbetta, E. "Concrete Curing Practices in the United States," *Concrete International*, Vol. 10, No. 11, Nov. 1988, pp. 64–67.

[21] Senbetta, E., "Curing Compound Selection," *Concrete International*, Vol. 11, No. 2, Feb. 1989, pp. 65–67.

[22] Brodwin, E., "Rapid and Uniform Curing of Concrete Products Using Microwave Energy," *Cementing the Future*, (Northwestern University), Evanston, IL, Vol. 3, No. 2, Fall 1991.

[23] Senbetta, E., "Development of a Laboratory Technique to Quantify Concrete Curing Quality," Ph.D. thesis, Purdue University, West Lafayette, IN, 1981.

[24] Schonlin, K. and Hilsdorf, H., "Evaluation of the Effectiveness of Curing of Concrete Structure," *ACI SP-100*, Vol. 1, 1987, pp. 207–226.

[25] Cabrera, J. G., et al., "An Assessment of Concrete Curing Efficiency Using Gas Permeability," *Magazine of Concrete Research*, Vol. 41, No. 149, 1989, pp. 193–198.

[26] Hover, K., "Report on Monitoring the Effectiveness of Curing Procedure," prepared for ACI Committee 308, American Concrete Institute, Detroit, MI, Feb. 1991.

[27] "Specification for Structural Concrete for Buildings," *ACI Manual of Concrete Practice, Part 3*, ACI 301-89, American Concrete Institute, Detroit, MI, 1991, pp. 301–23–301–24.

44

Air-Entraining Admixtures

Paul Klieger[1]

PREFACE

Air-entraining admixtures were first discussed in *ASTM STP 169*, in 1956, by Carl E. Wuerpel of the U.S. Army Corps of Engineers. Subsequent chapters by the present writer appeared in *ASTM STP 169A* in 1966, and in *ASTM STP 169B* in 1978.

INTRODUCTION

The use of intentional air entrainment in concrete is a well-established means for greatly enhancing the ability of concrete to resist the potentially destructive effect of freezing and thawing. Its use should be mandatory when concrete is to be exposed to such an environment, particularly when chemical deicers are being used, as on pavements and bridge decks.

A thorough survey of the early development of air entrainment is presented by Gonnerman [1]. The following paragraph from Gonnerman's report is of particular significance:

> These projects (test roads constructed in 1935-1937) revealed no relationship between surface scaling and composition of the cement, but they did show clearly that portland cement that inadvertently contained "crusher oil" reduced surface scaling as did many of the blends of portland and natural cement that contained tallow added during grinding of the natural cement. Laboratory tests disclosed that the beneficial effect of the crusher oil and tallow was due entirely to the additional air entrapped in the concrete by these air-entraining agents.

Other investigators [2,3] came to similar conclusions.

In these early instances, the air entrainment was not intentional but resulted from the presence of the crusher oil or the use of the tallow as a grinding aid during the production of the cement. These were the forerunners of materials called air-entraining additions, now used to produce air-entraining cements. Materials similar to presently used additions are called air-entraining admixtures when added with the other concrete ingredients at the time of mixing, the more widely used method for obtaining intentionally entrained air. This chapter is concerned with this class of materials.

What are these materials; how do they function, both as to the process of entraining air and enhancing durability; how can they be specified and tested to ensure adequate performance? These are some of the questions that we will attempt to answer in this chapter.

Definitions

It will be helpful to define certain of the terms used in discussing air entrainment in mortars and concretes. This listing is by no means complete, but the following definitions appear appropriate.

1. Air Entrainment—The introduction of air in the form of discrete air voids or bubbles dispersed throughout the mixture as a result of the use of air-entraining materials.
2. Entrained Air—The air, made up of discrete air voids, that becomes part of a mixture during the process of air entrainment.
3. Entrapped Air Voids—Air voids not resulting from intentional air entrainment. Such voids are larger than those resulting from intentional air entrainment and are at times referred to as natural air voids.
4. Entrained Air Voids—Air voids resulting from the use of intentional air entrainment. Such voids are generally spherical in shape and considerably smaller than the natural air voids.
5. Air-Entraining Admixture—A material added to cementitious mixtures at the time materials are batched for mixing, the use of which results in intentional air entrainment. (See ASTM Terminology Relating to Concrete and Concrete Aggregates (C 125) for the definition of an admixture.)
6. Air-Entraining Addition—Air-entraining material interground with, or to be interground with, hydraulic cement. (See ASTM Terminology Relating to Hydraulic Cement (C 219) for the definition of an addition.)
7. Air-Entraining Agent—A material the use of which results in intentional air entrainment when included in a mixture; a term that should be used only when it is intended to refer to materials that can be used both as air-entraining additions or admixtures.

MATERIALS USED AS AIR-ENTRAINING ADMIXTURES

There are many materials capable of functioning as air-entraining admixtures. In an extensive evaluation pro-

[1] Consultant, Concrete and Concrete Materials, Northbrook, IL 60062.

gram, the Bureau of Public Roads [4] separated 27 commercial air-entraining admixtures submitted for test into the following classifications: (1) salts of wood resins (pine wood stumps), (2) synthetic detergents (petroleum fractions), (3) salts of sulfonated lignin (paper pulp industry), (4) salts of petroleum acids (petroleum refining), (5) salts of proteinaceous materials (processing of animal hides), (6) fatty and resinous acids and their salts (paper pulp and animal hide processing), and (7) organic salts of sulfonated hydrocarbons (petroleum refining). Of these 27 materials, 17 were liquids, five were powders, four were in flake form, and one was semisolid.

Rixom [5] in 1978 provides an excellent description of the chemistry of air-entraining admixtures and discusses their function in freshly mixed and hardened concretes. Dodson [6] contains a detailed discussion and up-dating of information on air-entraining admixtures as of 1990. Dodson notes that most of the modern air-entraining admixtures are anionic in character because these produce stable air voids; cationic materials are costly and questionable as to stability of air voids; and nonionic materials do not impart stability to air voids.

FUNCTION OF ENTRAINED AIR IN FRESHLY-MIXED AND HARDENED CONCRETE

The major reason for the use of intentionally entrained air is to provide concrete with a high degree of resistance to freezing and thawing, particularly when chemical deicers are used. (The discussion to follow will also be applicable to the use of air-entraining cements, in which the air-entraining agent or material is used as an addition during the grinding of the cement clinker.) There are numerous other advantages, also, to the use of intentionally entrained air. For example, plasticity and workability are increased, enabling a reduction in water content. Uniformity of placement and consolidation can be achieved more readily, thus reducing segregation, and bleeding is reduced. These and other advantages are discussed in detail by Lerch [7] and Bruere [8].

To achieve the improvement in frost resistance, the intentionally entrained air must have certain characteristics. In addition to the proper total volume of air, the size and distribution of the air voids must be such as to provide efficient protection to the cement paste. Powers' [9–11] contributions to the understanding of how entrained air functions in providing increased frost resistance have been outstanding.

This work by Powers developed the concept of internal hydraulic pressure created by the resistance to flow or movement of excess water volume produced during the freezing process as being the mechanism responsible for distress. To keep this internal pressure below the tensile or rupture strength of the paste, Powers showed that the air voids must be well-distributed throughout the matrix (cement-water paste component) and sufficient in number so that each void provides protection to the cement paste surrounding it, and the protected volumes overlap to leave no unprotected paste.

Later work by Powers and Helmuth [12] indicated that, in addition to the generation of hydraulic pressure during freezing, another important factor may be the diffusion of gel water to capillary cavities contributing to the growth of ice bodies in these cavities resulting in the development of additional expansive forces.

Powers [10] developed the concept of void spacing factor to characterize an air-void system and analyzed laboratory freezing and thawing data available at the time to show that the void spacing factor for frost resistance should be about 0.01 in. (0.25 mm) or less. The void spacing factor is defined by Powers as the average maximum distance from a point in the cement paste to the nearest air void. This is an indication of the distance water would have to travel, during the freezing process, to reach a protective air void. Work by Mielenz et al. [13] indicates that an upper limit of about 0.006 to 0.008 in. (0.152 to 0.203 mm) is required for extreme exposures. Extensive freezing and thawing tests by Klieger [14,15] provided further substantiation of the void spacing factor concept. These tests called attention to the need for different volumetric air-content requirements for concretes made with different maximum sizes of coarse aggregate. Table 1, adapted from Ref 14, shows this effect of maximum size of aggregate on the optimum air content along with void spacing factors. Although the total air contents of the mixtures shown in Table 1 vary through a wide range, the air content of the mortar fraction is essentially constant at about 9%. As the maximum size of coarse aggregate increases and workability and cement content held constant, less mortar is required in the mixture; therefore, change in total concrete air content with change in maximum size of coarse aggregate is to be expected.

TABLE 1—Air-Void Parameters at Optimum Air Contents of Concretes (Adapted From Tables 17 and 18 of Ref 14).

Maximum Size of Aggregate		Cement Content, 5½ bags/yd³		$\bar{L}$		Cement Content, 7 bags/yd³		$\bar{L}$	
in.	(mm)	Optimum Air Content, %[a]	Mortar Air Content, %	in.	(mm)	Optimum Air Content, %	Mortar Air Content, %	in.	(mm)
2½	(63)	4.5	9.1	0.007	(0.18)	4.5	9.2	0.007	(0.18)
1½	(37.5)	4.5	8.5	0.008	(0.20)	4.5	8.4	0.008	(0.20)
¾	(19)	5.0	8.3	0.009	(0.23)	5.5	9.2	0.007	(0.18)
⅜	(9.5)	6.5	8.7	0.011	(0.28)	7.0	9.6	0.008	(0.20)
No. 4	(4.75)	9.0	9.0	0.012	(0.30)	10.0	10.0	0.008	(0.20)

[a] Optimum air contents determined from the relationship between expansion during 300 cycles of freezing and thawing and air contents of concretes.

Most of the early field and laboratory work on air-entrained concretes dealt with paving-type concrete in which the coarse aggregate was generally about 1½ in. (38 mm) maximum size. In concretes without intentional air entrainment, the air content may range up to as high as 1 or 2% by volume. However, this air is composed of entrapped air voids that are too large to be effective with respect to improving frost resistance or workability. The provision of an additional 3% of intentionally entrained air by the use of an air-entraining admixture will provide an air-void system well-distributed throughout the matrix and containing a sufficient number of air voids to meet the void spacing factor requirements for adequate resistance to freezing and thawing, even in the presence of deicing chemicals.

The importance of size and distribution of air voids, as contrasted with total volume of voids alone, can be seen in the results of a study made some years ago in the laboratories of the Portland Cement Association. Air-entrained concretes were prepared using an acceptable proprietary air-entraining admixture and four nonproprietary materials that exhibited a potential for entraining air. A nonair-entrained concrete was also included in these tests. In addition to the determination of air content of the freshly mixed concrete as described in ASTM Test Method for Air Content of Freshly Mixed Concrete by the Pressure Method (C 231), the air content, void spacing factor, specific surface, and number of voids per lineal inch of traverse were determined on the hardened concretes as described in ASTM Practice for Microscopical Determination of Air-Void Content and Parameters of the Air-Void System in Hardened Concrete (C 457). Additional refinements of the technique enabled the determination of the total number of air voids per unit volume of concrete. The results of these measurements and the performance of the concretes when frozen and thawed while immersed in water are shown in Table 2. It is apparent from these data that an air-entraining admixture must not only be capable of entraining some volume of air but also that the air void system must be characterized by a large number of small, well-distributed air voids in order to provide a high degree of frost resistance.

Specifications and control tests will continue to be based on the volume of air entrained in the concrete, rather than on the size and distribution of the air voids in the cement-paste matrix, until a means is developed for readily determining other air-void parameters directly on the freshly mixed concrete in the field. It would be highly desirable to have a test method available that could provide a measure of size and distribution of air voids within a few minutes after completion of mixing. Not having such a test method available, it is indeed fortunate that concretes having total air contents in the range of the optimum air contents shown in Table 1 will have the size and distribution of air voids that will provide adequate resistance to freezing and thawing when the air-entraining admixture used meets the requirements of ASTM Specification for Air-Entraining Admixtures for Concrete (C 260).

FACTORS INFLUENCING AMOUNT AND CHARACTER OF ENTRAINED AIR IN FRESHLY-MIXED CONCRETE

General

Mielenz et al. [13] and Bruere [8,16–18] have made significant contributions to the understanding of the mechanism by which air-entraining admixtures function and the influence of a number of different variables. These contributions have withstood the test of time and are still pertinent. Mielenz and his co-workers dealt extensively with the origin, evolution, and effects of the air-void system in concrete. They showed that in the concentrations normally used in concrete, air-entraining admixtures are adsorbed at air-water interfaces and that the surface tension of the water is decreased about 25%. This adsorption at air-water interfaces produces a "film" of air-entraining admixture that influences the air-retention properties of discrete air voids formed during mixing. For some air-entraining admixtures, the calcium salt of the active constituent in the admixture may be only slightly soluble in water. In such instances, the film at the air-water interface may include a precipitated solid or gelatinous film enclosing each air void. Admixtures that produce a relatively insoluble precipitate in portland-cement concrete include sodium soaps of wood resin, such as neutralized Vinsol resin and sodium abietate; sodium soaps of lignin derivatives, rosin, or fatty acid; or triethanolamine salts of sulfonic acid. Many calcium salts of sulfonic acids are soluble in water and many air-entraining admixtures in which the surface-active constituent is a sulfonate do not form such precipitated films around the air voids.

The amount and character of the air entrained in concrete is influenced by numerous factors, some of which

TABLE 2—Influence of Air-Void Characteristics on Resistance of Concrete to Freezing and Thawing.

Air Entraining Admixture	Air Content (C 231), %	Air Content, %	Number of Voids/in.	Void Spacing Factor (L), in.	Void Spacing Factor (L), (mm)	Specific Surface, in.^{2}/in.^{3} of air	Number of Voids/in.^{3} Concrete, millions	Freezing and Thawing Cycles for 0.10% Expansion
None	1.8	1.1	0.8	0.031	(0.79)	302	0.08	19
A	6.0	4.1	4.0	0.013	(0.33)	387	0.11	29
B	6.0	4.1	4.9	0.010	(0.25)	480	0.22	39
D	5.0	3.2	3.3	0.013	(0.33)	416	0.26	82
E	5.8	3.5	5.1	0.009	(0.23)	577	0.78	100
F[a]	5.2	3.9	9.6	0.006	(0.15)	990	3.79	550

[a] A commercial air-entraining admixture meeting the requirements of ASTM C 260.

are: (1) concentration of the air-entraining admixture and its influence on surface tension; (2) time of mixing; (3) speed of mixing, that is, rate of shearing action; (4) consistency of mortar or concrete mixture; (5) temperature of the mixture; (6) water-cement ratio and water content of the mixture; and (7) the gradation of the solids in the mixture, including the cement. Mielenz et al. [13] have theoretically concluded that both the total volume of air and the size distribution of the air voids can change in the unhardened concrete due to interchange of air between air voids and dissolution of air. Their tests in air-water foams and dilute cement pastes tended to corroborate these conclusions. Bruere [18], however, has shown that such changes do not take place to any significant degree in air-entrained pastes after cessation of mixing, although such interchange is of significance during the mixing process and may be for a few minutes thereafter.

Further information on the influence of numerous variables on the characteristics of air entrained in freshly mixed concrete is presented in the following paragraphs.

Type and Amount of Air-Entraining Admixture

Mielenz and his co-workers theorize that the type of organic ingredient in the air-entraining admixture influences the amount and character of entrained air voids by its effect on: (1) surface tension, (2) the elasticity of the film at the air-water interface, (3) transmission of air across the air-water interface, and (4) adhesion of the air voids to particles of cement or aggregate. All of these factors will be operative during the mixing operation.

At the same volumetric air content, different air-entraining admixtures will produce air-void systems having different specific surfaces, number of air voids per unit volume, and void spacing factors. Part of the difference occurs during the mixing operation and part during handling, placing, and compaction. The potential importance of these differences with respect to freezing and thawing durability was illustrated by the data in Table 2.

Increasing the amount of acceptable air-entraining admixtures will increase the volume of air entrained, increase the specific surface of the air-void system, and decrease the void spacing factor.

Fine Aggregate

Changes in grading of sand may alter the volume and nature of air in the mortar [19]. An appreciable increase in the quantity of very fine particles of sand will decrease the amount of entrained air and may reduce the maximum and median size of the individual air bubbles [20]. An appreciable increase in the quantity of the middle sizes of sand will tend to increase the air in the mortar. Sand gradation is of more importance in leaner mixes. In the richer mixes, the influence of gradation on entrained air is not as marked.

Cement

As the cement content increases, the air entraining potential of an admixture will tend to diminish, and an increase in the fineness of cement will result in a decrease of the air entrained in the mortar [21,23]. Some regular (nonair-entraining) cements naturally entrain more air than others, and these require less air-entraining admixture to develop a given mortar air content. Soluble alkalies in cements will influence the required amount of air-entraining admixture dosage [24]. At fixed dosage levels, higher air contents will be obtained with higher alkali cements. Therefore, if more than one cement is being used and soluble alkali contents differ significantly, care must be exercised to adjust dosages when necessary.

Water

Increase in water-cement ratio is likely to result in an increase in air content. Although the volume of air entrained may increase, the specific surface of the air voids generally decreases and the void spacing factor increases [13]. Nondegradable detergents present in water can result in excessively high and variable air contents.

Consistency

Within the normally used range, increase in initial slump is accompanied by an increase in air content in concrete mixtures [7,21,23]. Work by Klieger [15] indicates that the optimum mortar air content remains at about 9.0%.

Chemical Admixtures

Whiting and Stark [25] note that chemical admixtures such as water-reducers, retarders, and accelerators may increase air content somewhat when used in normal, recommended dosages, requiring adjustments in dosages of air-entraining admixtures. They state further that high-range water reducers will generally result in an altered void size distribution of the entrained air, manifested by higher void spacing factors than those normally considered acceptable for adequate resistance to freezing and thawing. Nevertheless, excellent freeze-thaw durabilities have been obtained both in laboratory and field concretes with these admixtures.

Supplementary Cementing Materials

Supplementary cementing materials, such as pozzolans (fly ash, silica fume, etc.) and ground granulated blast-furnace slags, generally require increased amounts of air-entraining admixtures to attain the proper volume of entrained air [26–28]. Klieger and Gebler [29] called attention to the effect of certain fly ashes in reducing the stability of the entrained air void system as a function of time after mixing. Further details concerning this phenomenon and techniques for control are described in an ACI Committee Report [30].

Mixing

The effect of mixing action on the amount of air entrained varies with the type and the condition of the mixer [7,31]. The amount of air entrained by any given mixer will decrease appreciably as the blades become worn, or as the mixing action is impaired if hardened mortar is allowed to accumulate in the drum and on the blades. An increase in entrained air will occur if the mixer is loaded to less than rated capacity, and a decrease will result from overloading the mixer. A stationary mixer, a paving mixer, and a transit mixer may develop significant differences in the volume of air entrained in a given concrete mixture. The air content will increase with increased time of mixing up to about 2 min in stationary [7] or paving mixers (and to about 15 min in most transit mixers), after which the air content may remain approximately constant for a considerable period before it begins to drop off. The reduction in air may result from an increase in very fine particles in the mixture with prolonged mixing action, from an increase in the ratio of air-escape to foam-generation in the latter portion of the mixing period, from adsorption of the chemical by unburned carbon in fly ashes, or from adsorption of the chemical on rapidly hydrating aluminate phases of the cement. The air void system, as characterized by specific surface and void spacing factor, does not usually appear to be harmed by prolonged agitation. Different air-entraining admixtures may require significantly different mixing periods to reach maximum air content.

Temperature

For a constant amount of air-entraining admixture, less air will be entrained at 100°F (38°C) than at 70°F (21°C) and more will be entrained at 40°F (4.4°C). In other words, everything else being equal, air entrainment varies inversely with temperature [7,22].

Vibration

Intensive internal vibration applied to freshly mixed concrete will cause air voids to rise to the surface and be expelled. The larger natural voids are most readily expelled [13,32]. Moderately small air voids may tend to work upward if the vibration is intense and prolonged. There is increasing evidence, however, that the critically important spacing of small entrained-air voids in the matrix is disturbed very little, even by intense vibration. If vibration is applied as it should be, with just enough intensity and duration to effect consolidation, and if the mixture is designed properly, removal of the effective portion of the entrained air will not occur. In some instances, externally applied vigorous vibration may cause an increase in air content. However, in this case, the added air is in the form of relatively large natural voids.

General Comments

One of the most frequent and pronounced causes for variations in air content results from variation in the amount, type, or condition of the air-entraining admixture itself. This last cause, as with many of the other variations in concrete, is a function in turn of the alertness and adequacy of the control and inspection given the work.

Since air-entraining admixtures are generally incompatible with other admixture types, care must be exercised to prevent them coming in direct contact while being dispensed into the concrete during mixing.

Any influence that would maintain or even actually improve the distribution of the air voids (number, or spacing factor) within the matrix or increase the ratio of air-boundary surface to air volume (specific surface) would be desirable. It is certain that an improved surface-to-volume ratio would reduce considerably the total volume of air needed to be entrained compared with what is now considered normal for optimum results.

METHODS FOR DETERMINING AIR-VOID CHARACTERISTICS

Freshly Mixed Concrete

Since parameters such as specific surface and void spacing factor of entrained air are more reliable criteria of effectiveness than the volume of entrained air, the ideal test would be one that could measure these characteristics in the freshly mixed concrete. Unfortunately, no such method has yet been developed, although a considerable amount of effort is currently underway to do so. For field control purposes, presently available test techniques can determine, with reasonable accuracy, only the volume of entrained air in the freshly mixed concrete.

The gravimetric method described in ASTM Test Method for Unit Weight, Yield, and Air Content (Gravimetric) of Concrete (C 138) is a simple test to make. It is based on the relationship of actual-to-calculated air-free unit weight of the concrete and produces reasonably accurate results when aggregates of relatively uniform specific gravity are used. However, errors attributable to inadequate sampling may be introduced where the specific gravities of the fine and coarse aggregate differ materially or where either aggregate is itself composed of particles of materially differing densities.

The most widely used method is that described in ASTM C 231, that is a procedure for determining the air content of freshly mixed concrete by application of pressure. Based on Boyle's law, that at a constant temperature the volume of given quantity of any gas varies inversely with the pressure to which the gas is subjected, the method is generally adequate for use with all ordinary types of mortar or concrete containing reasonably dense aggregate. Large errors will be introduced where highly vesicular or porous aggregates are used due to the impracticability of differentiating between the air in the aggregate particles and the entrained air in the paste.

The volumetric method described in ASTM Test Method for Air Content of Freshly Mixed Concrete by the Volumetric Method (C 173) is most useful for determining the air contents of concretes made with lightweight aggregates. This technique eliminates the possibility of significant errors in differentiating between air in the aggregate particles and air in the paste.

Hardened Concrete

The important characteristics of the entrained-air voids can most readily be determined in hardened concrete by microscopic examination of sawed and ground surfaces of a sample of the hardened concrete. ASTM C 457 is a procedure for determining the total air volume, specific surface, and spacing factor of air voids by either a linear traverse method or a modified point-count method. Additional refinements of the linear traverse equipment enable the measurement of chord-size distribution of air voids from which the total number of air voids per unit volume of concrete can be calculated, as shown in Table 2.

Mielenz and his co-workers [13] show the results of measurements of air-void characteristics of cores taken from a wide variety of structures. Although such measurements are time consuming, they can provide reassuring evidence of the effectiveness of the air-entraining admixture in providing the desired air-void system.

STATUS OF CURRENT SPECIFICATIONS AND TEST RESULTS FOR AIR-ENTRAINING ADMIXTURES

It was realized rather quickly (in the early 1940s) in the development of air entrainment that the size and distribution of the air voids could be expected to be of major importance with respect to the effectiveness of the entrained air in enhancing durability. Since there was no ready and reliable means for determining these air-void characteristics and since there was a need to evaluate the influence of these air-entraining materials on other concrete properties, a performance-type specification (ASTM C 260) was developed in 1950 by ASTM. It remains essentially the same as it appeared initially.

ASTM C 260 currently evaluates the effects that any given air-entraining admixture under test may exert on the bleeding, compressive and flexural strength, resistance to freezing and thawing, and the length change on drying of a concrete mixture, all in comparison with a similar concrete mixture containing a reference air-entraining admixture. The methods by which these effects may be tested are given in ASTM Methods of Testing Air-Entraining Admixtures for Concrete (C 233). The criteria of ASTM C 260 afford assurance that if, under the conditions of the specified mixtures and conditions in ASTM C 233, the particular sample of the admixture under test exerts satisfactory influence on certain properties of the laboratory concrete, it will be reasonable to expect that the quantity of the air-entraining admixture represented by the sample will develop satisfactory air-entrainment in field concrete.

The testing required by ASTM C 260 reduces simply to a very indirect method of determining whether: (1) the particular admixture under test will produce relatively stable air voids that will become widely dispersed throughout the matrix of field mortar or concrete so as to produce an air-void system having the proper characteristics for enhancing durability, and (2) that the admixture contains nothing that will have a deleterious effect on other properties of such mixtures. This specification, with its attendant test methods, has provided a means for evaluating air-entraining admixtures on a performance basis. The wide variety chemically of materials that can function as satisfactory air-entraining admixtures precludes the inclusion of chemical requirements.

A further consideration that is receiving attention is the possibility that the freezing and thawing tests, which are probably the most costly and time-consuming part of the testing procedure, can be supplemented by an examination of the characteristics of the air-void system produced by the admixture under test and a comparison with the system produced by the reference admixture. The CSA Standard A23.1, Concrete Materials and Methods of Concrete Construction (Canada), has adopted this alternative and considers the concrete to have a satisfactory air-void system if the average of all tests of the hardened concrete shows a void spacing factor not exceeding 0.009 in. (0.23 mm) and no single test greater than 0.010 in. (0.26 mm).

ASTM C 457 provides the means for such a comparison; however, further information is needed on the accuracy and reproducibility of the method. Such information is being developed by ASTM Committee C9, which is responsible for the various ASTM specifications and test methods to which this chapter has referred. Until that time, however, ASTM C 260 requires that the job performance of an air-entraining admixture must still be based on the direct measurement of "total air content" and on its effect on other readily measured properties, such as slump, bleeding, and strength. It is fortunate that air-entraining admixtures generally produce satisfactory air-void systems in the concrete at the total air contents shown in Table 1.

REFERENCES

[1] Gonnerman, H. F., "Air-Entrained Concrete, A Look at the Record," *Consulting Engineer*, Oct. 1954, pp. 52–61.

[2] Swayze, M. A., "More Durable Concrete With Treated Cement," *Engineering News-Record*, Vol. 126, 19 June 1941, pp. 946–949.

[3] Hansen, W. C., "History of Air-Entraining Cements," *Proceedings*, American Concrete Institute, Vol. 58, pp. 243–245; also, *Concrete Briefs*, Aug. 1961.

[4] Halstead W. J. and Chaiken, B., "Chemical Analysis and Sources of Air-Entraining Admixtures for Concrete," *Public Roads*, Vol. 27, No. 12, Feb. 1954, pp. 268–278.

[5] Rixom, M. R., "Chemical Admixtures for Concrete," Wiley, New York, 1978.

[6] Dodson, V. H., *Concrete Admixtures*, Van Nostrand Reinhold, New York, 1990.

[7] Lerch, W., "Basic Principles of Air-Entrained Concrete," Research Laboratories, Portland Cement Association, Skokie, IL, 1950.

[8] Bruere, G. M., "Relative Importance of Various Physical and Chemical Factors on Bubble Characteristics in Cement Paste," *Australian Journal of Applied Science*, Vol. 12, No. 1, March 1961, pp. 78–86.

[9] Powers, T. C., "A Working Hypothesis for Further Studies of Frost Resistance of Concrete," *Journal*, American Concrete Institute, Feb. 1945; *Proceedings*, Vol. 41, pp. 245–272; Discussion by R. D. Terzaghi, D. McHenry, H. W. Brewer, A. Collins, and T. C. Powers, *Journal*, American Concrete Insti-

tute, Nov. 1945, Supplement; *Proceedings*, Vol. 41, pp. 272–1 to 272–20; *Bulletin No. 5A*, Research Department, Portland Cement Association, Skokie, IL.
[10] Powers, T. C., "The Air Requirement of Frost Resistant Concrete," *Proceedings*, Highway Research Board, Vol. 29, 1949, pp. 184–211; *Bulletin No. 33*, Research Department, Portland Cement Association, Skokie, IL.
[11] Powers, T. C., "Void Spacing as a Basis for Producing Air-Entrained Concrete," *Journal*, American Concrete Institute, May 1954; *Proceedings*, Vol. 50, pp. 741–760; *Bulletin No. 49*, Research Department, Portland Cement Association, Skokie, IL.
[12] Powers, T. C. and Helmuth, R. A., "Theory of Volume Changes in Hardened Portland Cement Paste During Freezing," *Proceedings*, Highway Research Board, Vol. 32, 1953; *Bulletin No. 46*, Research and Development Laboratories, Portland Cement Association, Skokie, IL.
[13] Mielenz, R. C., Wolkodoff, V. E., Backstrom, J. E., and Flack, H. L., "Origin, Evolution, and Effects of the Air Void System in Concrete. Part 1—Entrained Air in Unhardened Concrete," *Journal*, American Concrete Institute, July 1958; *Proceedings*, Vol. 55; "Part 2—Influence of Type and Amount of Air-Entraining Agent," *Journal*, American Concrete Institute, Aug. 1958; *Proceedings*, Vol. 55; "Part 3—Influence of Water-Cement Ratio and Compaction," *Journal*, American Concrete Institute, Detroit, MI, Oct. 1958, and *Proceedings*, Vol. 55.
[14] Klieger, P., "Effect of Entrained Air on Strength and Durability of Concrete Made with Various Maximum Sizes of Aggregate," *Proceedings*, Highway Research Board, Vol. 31, 1952, pp. 177–201; *Bulletin No. 40*, Research and Development Laboratory, Portland Cement Association, Skokie, IL.
[15] Klieger, P., "Further Studies on the Effect of Entrained Air on the Strength and Durability of Concrete Made with Various Maximum Sizes of Aggregate," *Proceedings*, Highway Research Board, Vol. 34, 1955; *Bulletin No. 77*, Research and Development Laboratory, Portland Cement Association, Skokie, IL.
[16] Bruere, G. M., "Effect of Type of Surface-Active Agent on Spacing Factors and Surface Areas of Entrained Bubbles in Cement Pastes," *Australian Journal of Applied Science*, Vol. 11, No. 2, 1960, pp. 289–294.
[17] Bruere, G. M., "Air-Entrainment in Concrete," *Australian Journal of Applied Science*, Vol. 11, No. 3, 1960, pp. 399–401.
[18] Bruere, G. M., "Rearrangement of Bubble Sizes in Air-Entrained Cement Pastes During Setting," *Australian Journal of Applied Science*, Vol. 13, No. 3, 1962, pp. 222–227.
[19] Scripture, E. W., Jr., Hornibrook, F. B., and Bryant, D. E., "Influence of Size Grading of Sand on Air Entrainment," *Journal*, American Concrete Institute, Detroit, MI, Nov. 1948; *Proceedings*, Vol. 45, pp. 217–228.
[20] Mather, K., "Crushed Limestone Aggregates for Concrete," *Mining Engineering*, Oct. 1953, pp. 1022–1028; *Technical Publication No. 36164*, American Institute of Mining, Metallurgical, and Petroleum Engineers.
[21] Scripture, E. W., Jr., and Litwinowicz, F. J., "Some Factors Affecting Air Entrainment," *Journal*, American Concrete Institute, Detroit, MI, Feb. 1949; *Proceedings*, Vol. 45, pp. 433–442.
[22] Scripture, E. W., Jr., Benedict, S. W., and Litwinowicz, F. J., "Effect of Temperature and Surface Area of the Cement on Air Entrainment," *Journal*, American Concrete Institute, Detroit, MI, Nov. 1951; *Proceedings*, Vol. 48, pp. 205–210.
[23] Wright, P. J. F., "Entrained Air in Concrete," *Proceedings*, Institute of Civil Engineers, London, Vol. 2, Part 1, May 1953, pp. 337–358.
[24] Greening, N. R., "Some Causes for Variation in Required Amount of Air-Entraining Agent in Portland Cement Mortars," *Journal*, PCA Research and Development Laboratories, May 1967; *Bulletin No. 213*, Research and Development Laboratories, Portland Cement Association, Skokie, IL.
[25] Whiting, D. and Stark, D., "Control of Air Content in Concrete," National Cooperative Highway Research Program Report 258, Transportation Research Board, Washington, DC, May 1983.
[26] Mather, B., Discussion of "Investigations Relating to the Use of Fly Ash as a Pozzolanic Material and as an Admixture in Portland-Cement Concrete," *Proceedings*, American Society for Testing and Materials, Vol. 54, pp. 1159–1164.
[27] Malhotra, V. M. and Carette, G. G., "Silica Fume Concrete—Properties, Applications, and Limitations," *Concrete International*, May 1983.
[28] "Ground Granulated Blast-Furnace Slag as a Cementitious Constituent in Concrete," ACI 226.1R-87, American Concrete Institute, Detroit, MI, 1987 (now ACI 233).
[29] Gebler, S. and Klieger, P., "Effect of Fly Ash on the Air-Void Stability of Concrete," *Fly Ash, Silica Fume, Slag and Other Mineral By-Products in Concrete*, SP-79, American Concrete Institute, Detroit, MI, 1983.
[30] "Use of Fly Ash in Concrete," ACI 226.3R-87, American Concrete Institute, Detroit, MI, 1987 (now ACI 232).
[31] Scripture, E. W., Jr. and Litwinowicz, F. J., "Effects of Mixing Time, Size of Batch, and Brand of Cement on Air Entrainment," *Journal*, American Concrete Institute, May 1949; *Proceedings*, Vol. 45, pp. 653–662.
[32] Higginson, E. C., "Some Effects of Vibration and Handling on Concrete Containing Entrained Air," *Journal*, American Concrete Institute, Sept. 1952; *Proceedings*, Vol. 49, pp. 1–12.

45

Chemical Admixtures

Bryant Mather[1]

PREFACE

The first edition of this compilation published in 1956 did not have a chapter on "Chemical Admixtures" since at that time there was no ASTM standard covering chemical admixtures. The ASTM Specification for Chemical Admixtures for Concrete (C 494) was issued in 1962. Hence, in the second edition, issued in 1966, there was a chapter prepared by Dr. Bruce Foster. The chapter in *ASTM STP 169B* [1], published in 1978, was an updating of Dr. Foster's paper prepared by me.

This discussion is limited to certain features of chemical admixtures regarded as most appropriate to the scope of this volume. For a more comprehensive review of knowledge in this field, reference should be made to the following works as may be relevant.

a. In 1968, RILEM published the *Proceedings* of the International Symposium on Admixtures for Mortar and Concrete that was held in Brussels [2].
b. In 1989, the American Concrete Institute (ACI) published the sixth report of its Committee 212 on Admixtures for Concrete, entitled "Chemical Admixtures for Concrete" (ACI 212.3R) [3]. This committee, which was organized in 1943, has functioned in close, but informal, liaison with the Transportation Research Board Committee A2E05 on Chemical Additions and Admixtures for Concrete and with Subcommittee C09.03.08 on Test Methods and Specifications for Admixtures for Concrete of ASTM Committee C-9 on Concrete and Concrete Aggregates. Symposia, bibliographies, reports of research, and the state-of-the-art reviews sponsored by all three groups have been studied and used, with benefit, by all three.
c. In 1984, the *Concrete Admixtures Handbook: Properties, Science, and Technology*, edited by V. S. Ramachandran, was published [4].

INTRODUCTION

Although, in a broad sense, all concrete admixtures are chemicals, by convention, in concrete technology, the term "chemical admixture" is restricted to water-soluble substances. Most chemical admixtures react chemically with the cement in concrete. The reports of unfavorable behavior of some admixtures with certain cements and under certain conditions of use are counterbalanced by a record of successful use under controlled conditions in many concreting operations. However, when experience with specific admixture-cement combinations under similar job conditions is not available, tests with specific materials should precede a decision for use in construction.

When two or more admixtures are added to a given concrete mixture, they should be added separately during the mixing operation unless they have been shown to be compatible when added to the concrete in a single operation [5].

DEFINITIONS

ACI 116R-90 [6] includes the following definitions:

admixture—a material other than water, aggregates, hydraulic cement, and fiber reinforcement, used as an ingredient of concrete or mortar, and added to the batch immediately before or during its mixing.

admixture, accelerating—an admixture that causes an increase in the rate of hydration of the hydraulic cement, and thus shortens the time of setting, or increases the rate of strength development, or both.

admixture, air-entraining—an admixture that causes the development of a system of microscopic air bubbles in concrete, mortar, or cement paste during mixing.

admixture, retarding—an admixture that causes a decrease in the rate of hydration of the hydraulic cement, and lengthens the time of setting.

admixture, water-reducing—an admixture that either increases slump of freshly mixed mortar or concrete without increasing water content or maintains slump with a reduced amount of water, the effect being due to factors other than air entrainment.

admixture, water-reducing (high-range)—a water-reducing admixture capable of producing large water reduction or great flowability without causing undue set retardation or entrainment of air in mortar or concrete.

ASTM STANDARDS

The 1993 edition of the *Annual Book of ASTM Standards* includes the following standards:

(a) ASTM Specification for Admixtures for Shotcrete (C 1141) and ASTM Test Method for Time of Setting of

[1] Director, Structures Laboratory, U.S. Army Engineer Waterways Experiment Station, Vicksburg, MS 39180-6199.

Portland-Cement Pastes Containing Accelerating Admixtures for Shotcrete by the Use of Gillmore Needles (C 1102) are discussed in Chapter 55 on Shotcrete.

(b) ASTM Specification for Foaming Agents Used in Making Preformed Foam for Cellular Concrete (C 869) and ASTM Test Method for Foaming Agents for Use in Producing Cellular Concrete Using Preformed Foam (C 796) are discussed in Chapter 48 on Lightweight Concrete.

(c) ASTM Specification for Pigments for Integrally Colored Concrete (C 979) is discussed in Chapter 46 on Mineral Admixtures.

(d) ASTM Specification for Air-Entraining Admixtures for Concrete (C 260) and ASTM Methods of Testing Air-Entraining Admixtures for Concrete (C 233) are discussed in Chapter 44 on Air-Entraining Admixtures.

(e) Admixtures intended to retard corrosion of embedded metal are covered in Chapter 18 on Corrosion of Metal.

(f) ASTM Specification for Chemical Admixtures for Concrete (C 494), ASTM Specification for Calcium Chloride (D 98), and ASTM Specification for Chemical Admixtures for Use in Producing Flowing Concrete (C 1017), together with admixtures to increase pumpability of concrete and to minimize washout in concrete placed underwater, are discussed here.

There are specifications of the American Association of State Highway and Transportation officials corresponding to ASTM D 98 and C 494 having designations M 144 and M 194, respectively.

Other types of admixtures discussed in ACI 212.3R [3] include gas-forming admixtures; grouting admixtures; expansion-producing admixtures; bonding admixtures; flocculating admixtures; fungicidal, germicidal, and insecticidal admixtures; dampproofing admixtures; permeability-reducing admixtures; and chemical admixtures to reduce alkali-aggregate reaction.

TYPES OF MATERIALS AND THEIR ACTION IN CONCRETE

Set-Retarding and Water-Reducing Admixtures

The materials referred to as conventional set-retarding and water-reducing admixtures in the United States were classified by Prior and Adams [7] into four classes: (1) lignosulfonic acids and their salts, (2) modifications and derivatives of lignosulfonic acids and their salts, (3) hydroxylated carboxylic acids and their salts, and (4) modifications and derivatives of hydroxylated carboxylic acids and their salts. In each of these, the primary component has both water-reducing and set-retarding properties. In the formulation of products of Classes 2 and 4, these admixtures may be modified by the addition of other components to give various degrees of retardation, no significant change in setting time, or acceleration, while at the same time preserving the water-reducing properties. They also may be modified by the addition of an air-entraining admixture.

Odler [8] reported that a series of additional compounds was effective as water-reducing or plasticizing agents, or both, among them: acrolein; polyglycerol; a combination of anthranylic acid, formaldehyde, and phenol; polyethylene-oxide with a molecular weight from 8×10^5 to 6.7×10^6; polyacrylic and polymethacrylic acid or their alkali metal or ammonium salts; admixture of triethanolamine and gallic acid; glutamic acid amide; and a compound with the formula $(NaO_2) \cdot CH(SO_3Na) \cdot CHR(CO_2Na)$ where R is H or CH_3.

The water reduction resulting from the use of conventional water-reducing admixtures was reported by Vollick [9] to range from 5 to 15%. A part of the water reduction found with lignosulfonate water-reducers may be the result of the air entrained by these materials. In addition to varying with the particular cement employed, the amount of water reduction with a given admixture is also influenced by dosage, cement content, type of aggregate, and the presence of other admixtures, such as air-entraining agents or pozzolans. Water-reducing admixtures are effective with all types of portland cement, portland blast-furnace slag cement, portland-pozzolan cement, and high-alumina cement.

The retardation of setting time brought about by retarders is dependent not only upon the particular cement with which they are used, but also upon the temperature, dosage, and other factors. Overdosage may produce excessive setting times of 24 h or more, but in such cases, if the concrete finally sets and has been protected from drying, ultimate strengths developed may be satisfactory if forms are left in place for a sufficient length of time. Severe overdosage of lignosulfonate admixtures may produce excessive air contents and consequently reduced strength. With either lignosulfonate or hydroxylated carboxylic acid admixtures, the degree of retardation can be controlled by varying the dosage, provided that the allowable air content with lignosulfonate materials is not exceeded. Classes 2 and 4 water reducers may be formulated to give no retardation or to produce acceleration, both effects being produced by the incorporation of a catalyst or an accelerator.

The extent of any modification of setting time will depend upon the properties of the cement to which an admixture is added. The behavior of concrete containing water-reducing retarders appears to be dependent even more upon the properties of the cement. Polivka and Klein [10] reported that the effectiveness of water-reducing retarders, both from a water-reducing standpoint and from a retarding standpoint, was greater with cements of low alkali and low C_3A content. Tuthill et al. [11] and those who discussed their paper have reported excessive retardation when retarders were used with certain cements, which could be overcome by an increase in the sulfate content of the cement. Often, the excess retardation was accompanied by stiffening, even though the cement without the water-reducing retarder showed no early stiffening and had a normal setting time. Seligman and Greening [12] have reported on work that provides information on the chemistry of this phenomenon.

Bruere [13] and Dodson and Farkas [14] have noted that the effect of water-reducing retarders depends in some cases on the time at which the retarder is added to the mixture. Dodson and Farkas [14], working with cements little affected by the addition of either lignosulfonate or hydroxylated carboxylic admixtures, found that the efficiency of the admixtures as retarders, and their capacity to enhance air entrainment and reduce water requirement, was increased greatly by a 2-min delay in their addition after mixing commenced.

Lignosulfonate water-reducing retarders usually entrain 2 to 3% of air when used in normal dosages. The hydroxylated carboxylic admixtures do not entrain air. However, both classes of material enhance the effectiveness of air-entraining admixtures from the standpoint of volume of air produced, so that less air-entraining admixture is required when added to concrete containing one of these other admixtures.

Water-reducing admixtures that entrain air reduce bleeding, the reduction being due to the entrained air and the lower water content. Hydroxylated carboxylic acid-based water-reducing retarders have been reported to increase the rate and amount of bleeding. Such bleeding, which reduces effective water-cement ratio (w/c), may be responsible for a portion of the strength increase observed with the use of these materials.

Contrary to expectations, water-reducing retarders usually have not been found effective in reducing slump loss resulting from substantial delays in placing mixed concrete. As pointed out earlier, the use of retarders with some cements may actually produce an early stiffening. However, the addition of a water-reducing admixture will give a high initial slump with the same w/c and permit more slump loss before concrete becomes unworkable.

Usually compressive strength is increased 10 to 20% by use of a water-reducing admixture based on lignosulfonate or a salt of a hydroxylated carboxylic acid [2]. The percentage of strength gain at three and seven days is usually higher than that at 28 days, while tests up to five years in duration have shown a continued strength benefit.

As pointed out earlier, lignosulfonate water-reducers normally entrain some air. The portion of the water reduction attributable to the air content can, with suitable adjustment of the amount of sand, compensate largely or wholly for the loss of strength due to the entrained air. The balance of the water reduction when using a lignosulfonate water reducer, and the reduction produced by a hydroxylated carboxylic water reducer, are effective in increasing the 28-day strength over that which would be produced with similar concrete without admixtures. Further, this increase in strength is generally greater than would be predicted from the reduction in w/c. The permeability of concrete is not changed significantly except that the concrete matures more rapidly, and hence may be affected less by failure to provide proper curing.

The available data on the effect on drying shrinkage are conflicting, probably because of the influence of variations in test methods employed. Tremper and Spellman [15], using concrete specimens of 3 by 3-in. (approximately 75 by 75-mm) cross section made from a blend of Type II cements, moist cured for seven days, and dried at 50% relative humidity for 28 days, found drying shrinkages of 8 to 17% greater than similar concrete without admixtures for one hydroxylated carboxylic acid and two lignosulfonates. The percent increase in shrinkage over the control concrete was usually higher for shorter than for longer drying periods. In assessing the importance of these figures, it should be kept in mind that they are comparable in magnitude with effects that may be introduced through choice of cement, choice of maximum aggregate size, contamination of aggregate, and choice of aggregate source.

While the air-void spacing obtained with water-reducing retarders is slightly greater than that for an equivalent amount of entrained air produced by typical air-entraining admixtures, the performance of concrete containing water-reducing retarders, as measured by freezing and thawing tests, often has been found to be better than concrete of the same air content, but without the water-reducing retarders. This increase might be the result of the reduction in w/c.

The usual water-reducing retarders, unless modified by the addition of chlorides, have not been found to bring about corrosion. There is some evidence that retarding admixtures, when used in conjunction with chlorides, reduce the increased electrical conductivity that normally would result from use of the chlorides, and hence the resulting corrosion [16]. Laboratory tests have shown some small improvement in sulfate resistance, through the use of water-reducing retarders [17].

These chemical admixtures have been shown to have little, if any effect on the total heat liberated during the hydration of cement [17,18], but they may have a pronounced effect upon the rate at which the heat is liberated. When a water-reducing admixture is used to increase workability of concrete as an alternative to using more cement and water at a given w/c, the cement content of the resulting concrete is lower than it would otherwise have been, the heat evolved is proportionately less, and so is the temperature rise.

High-Range Water-Reducing Admixtures (HRWR)

High-range water-reducing (HRWR) admixtures were added to ASTM C 494 in 1980 and their use to produce flowing concrete is covered by ASTM C 1017, adopted in 1985. They have been used to produce high-strength concrete by taking advantage of the opportunity they provide to lower the w/c more than generally has been practical with more traditional water-reducing admixtures or to produce "flowing" concrete, that is, concrete of much greater slump. It has been reported that some of these products provide the alteration of properties that characterizes their use for a relatively short period of time. Such admixtures may be added to ready-mixed concrete after arrival at the site for placement. The early data may be located beginning with the work in the United Kingdom [19], Germany [20,21], or the United States [22].

Three American Concrete Institute Symposia have been held on HRWR. The first was in 1978 and yielded *ACI SP-62* in 1979 which contained 20 papers; the second in 1981 yielded *ACI SP-68* containing 30 papers; the third in 1989 yielded *ACI SP-119* with 33 papers [23–25].

Most investigators [22,26–29] have found, in laboratory tests, that the addition of HRWRs to air-entrained concrete may increase the spacing factor and decrease the specific surface area of the air-void system. However, early reports of a reduction in frost resistance of such concretes [22,26] have not been substantiated by later research. Nevertheless, it would be prudent to evaluate the effect a specific high-range water reducer on the frost resistance of a concrete mixture if this is a significant factor and if the manufacturer cannot supply results of such an evaluation.

Zakka and Carrasquillo [30] report results of tests of concrete made with and without four HRWR admixtures, two of which were described as "first generation" and two as "second generation." The latter were said to "allow extended workability and can be added at the batching plant." The first generation products included one of the naphthalene type and one of the melamine type. The second generation products were obtained from two manufacturers but the chemical basis was not stated. None of the 26 stated conclusions differ as a function of whether the admixture was naphthalene or melamine based. The conclusions, in general, reflect anticipated relationships among measurable properties of concrete that would exist whether or not the concrete contained a chemical admixture.

Accelerating Admixtures

The accelerating admixture that has found greatest use is calcium chloride [31]. The organic material, triethanolamine, also has been used extensively but usually in combination with other materials, and little published information is available on its properties [15,18,32]. Therefore, the discussion of accelerators will be primarily concerned with calcium chloride.

The addition of recommended amounts of calcium chloride has been found to reduce the water requirement by a small, but definite, amount over that required to produce the same slump, with no calcium chloride added. The magnitude of the effect on setting time of the addition of calcium chloride depends not only upon the dosage, but also upon the particular cement, the temperature, and other factors. The recommended maximum dosage has a substantial effect on setting time at normal temperatures and can produce a very rapid set at high temperatures, as is also the case with a very large dosage at normal temperatures.

Calcium chloride does not entrain air but its use enhances the effectiveness of air-entraining admixtures so less air-entraining admixture is required for a given air content of concrete. Calcium chloride may result in early stiffening and in many cases, therefore, is not added until after mixing has commenced.

Concrete containing calcium chloride shows maximum strength gain at early ages [33]. Concrete with 2% calcium chloride has often been found to be stronger at one year than similar concrete without the admixture. Accelerators are particularly effective at very early ages and at relatively low temperatures. Flexural strength generally is improved, but the effect is less pronounced than that on compressive strength, and some reduction in flexural strength at later ages may be noted. The permeability of concrete is not changed significantly, except that the concrete matures more rapidly, and hence may be affected less by failure to provide proper curing.

The available data on the effect on drying shrinkage are conflicting, probably because of the influence of variations in test procedures employed. Tremper and Spellman [15], using concrete specimens of 3 by 3-in. (approximately 75 by 75-mm) cross section made from a blend of Type II cements, moist cured for seven days, and dried at 50% relative humidity for 28 days, found drying shrinkages 30% greater for concrete with calcium chloride than without. The percent increase in shrinkage over the control concrete was usually higher for shorter than for longer drying periods. The relative shrinkage where calcium chloride was added was found to be dependent on the SO_3 content of the cement used. Other investigators [10,17,33,34], using various procedures and drying times, have reported drying shrinkage figures that are usually lower than those of Tremper and Spellman. These results were discussed by Mather [35].

In assessing the importance of these figures, it should be kept in mind that they are comparable in magnitude with effects that may be introduced through choice of cement, choice of maximum aggregate size, contamination of aggregate, and choice of aggregate source.

The use of calcium chloride has been reported to decrease somewhat the resistance of concrete to sulfates [21]. Concrete containing calcium chloride liberates heat earlier, as would be expected from the earlier strength development. Calcium chloride has been found to improve the early resistance of concrete to freezing and thawing, either in the presence or absence of ice-removal salts, but to reduce somewhat the eventual durability of the fully cured concrete [17,33,36].

The addition of calcium chloride to reinforced concrete has not been found to contribute significantly to corrosion of the reinforcing. If the concrete is proportioned properly, consolidated properly to form a continuous contact with the steel, and of adequate cover thickness, no corrosion problems normally are encountered. However, chlorides should not be used in prestressed concrete [37–39]; in concrete where stray currents are present [16]; in concrete in which dissimilar metals are imbedded, such as aluminum conduit and steel reinforcement [40]; or where galvanized forms are to be left in place [41]. The use of calcium chloride in concrete made with Type V cement was reviewed [42] and it was concluded that its use was not harmful. Also, calcium chloride may bring about corrosion where elevated temperature is employed during curing [43]. By contrast, another accelerator, stannous chloride [43], when properly used, was found not to contribute to corrosion.

APPLICATIONS

Water-Reducing Admixtures

Water-reducing admixtures may be used with no change in cement content and slump to produce concrete with a lower w/c; with no change in cement content and w/c to produce a higher slump; or with reduced cement content and unchanged w/c and slump. In the first case, the usual benefits accruing from the use of a lower w/c normally will be obtained, and, in many cases an increase in strength greater than normally produced by the reduction in w/c alone may result. In the second application, easier placing of concrete, or a higher slump with delayed placing of the concrete, may be obtained. In the third application, a reduced cost should result.

Retarders

Retarders may be used to delay the setting of concrete during hot weather, or to extend the vibration limit so that large members can be cast and consolidated without cold joints and without damage to the freshly placed concrete due to settlement of forms as concreting proceeds [44].

Accelerators

Accelerators have their primary application in cold weather concreting where they may be used to permit earlier starting of finishing operations, and in certain cases, the application of insulation; reduce the time required for curing; and permit earlier removal of forms or loading of the concrete. Accelerators cannot be used as antifreeze agents since, at the allowable dosages, the freezing point will be lowered less than 2°C.

Dosage

In the case of water-reducing and retarding admixtures, dosages required to produce specific results usually are recommended by the manufacturers. Variation in the dosage often can be made to obtain the desired concrete properties under particular job conditions. In other cases, the manufacturer may change the formulation of the admixture to suit the conditions under which it will be used. Increase in the dosage in a multipurpose material to obtain one particular effect might not be feasible because, as an example, too much or too little air entrainment might result.

Calcium chloride usually is added in amounts of 1 to 2% by mass of the cement, and the latter figure should not be exceeded.

SPECIFICATIONS

A specification for a concrete admixture should serve several purposes.

1. It should provide test methods and specification limits by which the material to be tested may be judged as to its general ability to perform the functions for which it is purchased. Tests for strength, water requirement, and setting time measure properties important in the use of chemical admixtures.
2. Specifications should provide test procedures and limits against which the material may be judged from the standpoint of not producing deleterious properties in the concrete. Excessive drying shrinkage, poor resistance to freezing and thawing, and excessive setting time are examples of such properties.
3. The specification should provide procedures whereby the performance of a particular admixture with cement and aggregate from particular sources and to be used under specific job conditions can be assessed.
4. A specification should provide a ready means of identifying materials in successive shipments, both from the standpoint of composition and concentration, to give the user some assurance that the material being used is uniform and is the same as that which was tested.

ASTM C 494 recognizes each of these needs. A test procedure is outlined using a blend of cement, aggregates of specified grading, and mixture proportions based on ACI Standard 211.1, and recommends that, where practical, tests be made using the cement, aggregates, and air-entraining admixture proposed for the specific work. It contains recommendations that can serve to assure that a particular admixture previously subjected to all the specification tests and found satisfactory when tested with specific aggregates, air-entraining agent, and blend of cement would perform satisfactorily with the other concrete ingredients on a particular job. The materials covered by ASTM C 494 are often found particularly useful at temperatures well below or above the laboratory temperature of 23°C (73°F) required by the specification, and a few tests at temperatures anticipated in the field might be profitable.

The specification requires the manufacturer to recommend, upon request, appropriate test procedures for establishing the equivalence of materials from different lots. Qualitative, and to some extent quantitative, compositional analysis of water-reducing retarders may be obtained by infrared or ultraviolet absorption spectroscopy [45,46]. One alternative to this type of test consists of watching the rate of hardening, water requirement, and slump in the job concrete as use of a new lot of admixture is started, while another involves laboratory concrete tests using the new batch and a reserve sample from the initial batch [47].

The complete set of tests prescribed in ASTM C 494 is extensive, and requires an extended period of time, hence it is unlikely that it will be performed more than a few times for any given admixture, and perhaps never for specific lots of materials used in construction work. These tests are desirable and necessary to establish the potential value of an admixture, and the test data should be available to the purchaser, but less extensive tests might well be used for assessing the performance of an admixture with the specific materials and conditions of given concrete work. Since its introduction, the specification has been undergoing constant improvement and clarification, and further revisions no doubt will be made.

In establishing specifications for concrete admixtures, allowance must be made not only for the statistical variation of test results on the concrete containing the admixture, but also that of the control concrete without admixture. For example, if it is intended that a water-reducing admixture increase the 28-day compressive strength by 20%, and the limit therefore is set at 120% of the control, an admixture which actually, on the average, will produce an increase of 20% in strength would fail to meet the specification requirement 50% of the time. The limits given in Table 1 of ASTM C 494 take this factor into account by lowering the strength requirements and the water requirement by approximately 10%. The specification requirement for Type B (retarding) admixture that the strength of treated concrete at any age be at least 90% of that of the untreated control has the objective of requiring that there be no sacrifice in strength due to use of the admixture. However, under the requirements of the specification, an admixture that actually caused a 10% reduction in strength would be rejected, on the average, only 50% of the time. The rate of rejection for true strengths above or below this point depends on the standard deviation of the test data and the number of test specimens. Protection to the consumer against acceptance of inferior products can be increased by improving test procedures so that the variance is lowered, increasing the number of specimens tested, or both.

The balance between probability of accepting an inferior product and rejecting a satisfactory one can be adjusted within limits, but substantial improvement can be achieved only by further reduction in the variance through better testing and test methods, or the use of more specimens.

OTHER CHEMICAL ADMIXTURES

Pumping Aids

In the beginning of this chapter, a listing was given of current ASTM specifications for chemical admixtures and references to the other chapters where information on several of them is given. Chemical admixtures not currently covered by ASTM specifications include pumping aids. Valore [48] reviewed this subject for *ASTM STP 169B*; much of the following is a revision of what he wrote.

Pumpable concrete must be sufficiently fluid to be moved through changes in shape and direction of lines without a buildup of excessive pressure. Also it must be sufficiently stable or cohesive to resist separation or segregation of its constituents under the pressures of pumping [49]. Forced bleeding of water or cement paste from concrete under pressure is an example of segregation that, in excess, can make pumping impossible. Workable and cohesive pumpable concrete possesses a sufficient volume of cement paste to act as a lubricant between the concrete and walls of the line to reduce frictional resistance. Pumping can be impaired, however, when the amount of sub-sieve fines is excessive. If it is assumed that the density of dry portland cement is 1440 kg/m³ (90 lb/ft³), it has been shown that, in laboratory trials, when the volume of cement was equal to or greater than the aggregate interstitial void volume, the concrete mixtures generally were pumpable. Exceptions were mixtures in which the volumes of total fines, consisting of cement plus aggregate passing the 150-μm (No. 100) sieve were excessively high.

Pumping aids for concrete are admixtures whose sole or primary function is the improvement of pumpability; they will not normally be used in concrete that is not pumped or in concrete that can be readily pumped. Their primary purpose is to overcome difficulties of obtaining satisfactory aggregate interstitial void volume and individual void size when using available materials. Pumpability admixtures can make some, but not all, unpumpable concretes pumpable, or can improve pumpability of marginally pumpable concretes. The principle involved in virtually all of the admixtures marketed for improving the pumpability of concrete is that of thickening or increasing the viscosity of water. Since the objective is to inhibit forced bleeding of water from cement paste or permeation of paste through interstitial aggregate voids, permeability of the void system is effectively reduced by increasing the viscosity of the permeating fluid. The objective is to prevent dewatering of the paste under pressure of pumping. Dewatering has two cumulatively adverse effects: it decreases mobility of the concrete and depletes the lubricating fluid.

The function of the pumpability admixture is to impart water retentivity to the cement paste under forces tending to separate water. The desired inhibition of segregation must not be obtained at a cost of decreased mobility. Under optimum conditions, these agents impart a degree of thixotropy to suspensions of mineral solids in water: stability of the suspension at rest and compliability when worked upon.

The performance of a given admixture can change drastically when dosage rates, cement composition, mixing temperature and time, and other factors change. An example is provided by polyethylene oxide. When used in small amounts of 0.01 to 0.05% of cement mass, it improves pumpability. Smaller amounts make water slipperier; larger amounts produce thickening that may or may not disappear upon prolonged mixing. Other examples are provided by synthetic polyelectrolytes that act as flocculents or thickeners, depending upon dosage levels. It would appear to be highly undesirable to induce flocculation and increase bleeding in pumped concrete. Factors to consider in the use of emulsions (paraffins, polymers) is whether they function in the desired way in cement paste by remaining stable or by breaking of the emulsion. Many of the thickening agents cause entrainment of air; to control air content, a defoamer (for example, tributyl phosphate) may need to be used, especially for higher concentrations of pumpability admixture in mortars and concretes. Many of the synthetic and natural organic thickening agents retard the setting of portland cement pastes. For dosages of methyl or hydroxyethyl cellulose of 0.1% or more, by mass of portland cement, retardation may be substantial. In any case, the particular concrete system in which a pumpability admixture is incorporated must be evaluated in terms of side effects upon the fresh and hardened concrete in addition to assessing the effectiveness of the admixture in performing its intended function.

ASTM Test Method for Bleeding of Cement Pastes and Mortars (C 243) based on a carbon tetrachloride displacement method developed by Valore et al. [50] provides data on initial bleeding rate and bleeding capacity and was used successfully, when modified, in determining the bleeding characteristics of concrete. Ritchie [51] adapted the method to testing of concrete mixtures. In this and the Valore et al. [50] studies, it was concluded that only by obtaining bleeding data for concrete and for the corresponding cement paste can a complete picture of the potential dewatering of concrete be seen, including effects of aggregate interference in causing internal bleeding.

Browne and Bamforth [49] described a pressure bleeding test and apparatus for the purpose of establishing pumpability of concrete. The concrete sample is subjected to a pressure of 3.5 MPa (500 psi) by a piston, actuated by a hydraulic jack. The volume of bleeding water collected at 140 s (V_{140}), minus that collected at 10 s (V_{10}), is said to correlate directly with pumpability for concrete of a given slump. This was interpreted as showing that concretes that dewater quickly (in the first 10 s) under pressure tend to be unpumpable. The authors developed a calibration curve of ($V_{140} - V_{10}$) versus slump that forms a line of demarcation between zones of "pumpability" and "unpumpability." Results of this work tend to negate the principles involved in using flocculating admixtures to improve pumpability by increasing initial bleeding rates and decreasing bleeding capacities.

Gray [52] developed a laboratory procedure for comparing pumpability of concrete mixtures that involved pumping concrete samples upward through a 1.2-m (4-ft) vertical section of pipe with a curved offset of 305 mm (12 in.) and measuring the pressure required to move the concrete.

Tanaka [53] and Ozaki and Emura [54] presented results of pumping tests on lightweight aggregate mortars and concretes, respectively. Test mortars were made with natural sands and fine lightweight aggregates, and included mixtures containing methyl cellulose, paraffin emulsion, and fly ash. Evaluation of pumpability was based on the rate of pumping in litres per minute. The admixtures used were found to reduce segregation; rates of pumping were reduced when methyl cellulose or paraffin emulsion were used and increased when fly ash was used in mortar mixtures at dosage levels required to prevent segregation. Lightweight mortars were more difficult to pump than natural sand mortars. In pumping tests of lightweight concretes, pressures were measured at several stations in a U-shaped pumping circuit about 45 m (150 ft) in length. Slump and flow were measured before and after pumping. The testing setup provided a good indication of pumpability in the field but, because of the space required, it is not well-suited for use as a laboratory method for evaluating admixtures.

Admixtures for pumping of concrete might be evaluated indirectly by means of determination of their effects on initial bleeding rates and bleeding capacities of cement pastes, mortars, or concretes under pressure, using methods now available.

Ragan [55] evaluated the pressure bleed method as proposed by Browne and Bamforth [49] and the checklist analysis as proposed by Anderson [56]. He evaluated 14 mixtures in the laboratory and studied field concrete at two projects. He concluded that the pressure bleed test will indicate relative pumpability if the pump to be used is poorly maintained and leaky; but it will declare pumpable mixtures as nonpumpable using a well-maintained pump. The checklist approach was regarded as conservative and was recommended for consideration for use.

OTHERS

Other sorts of chemical admixtures not covered by ASTM specifications include (a) freeze-protection admixtures, (b) extended setting control admixtures, and (c) anti-washout admixtures.

Senbetta and Scanlon [57] discuss results of tests of concrete containing a freeze-protection admixture that will "prevent the concrete from freezing at a temperature of 20°F (-7°C)" in which the requirements of ASTM C 494 for Types C and E were met. Since the admixture contains sodium thiocyanate, it was evaluated for effects on corrosion of steel and found not to promote corrosion at a dosage of 8L/100 kg of cement (120 fl oz/cwt). They also reported that concrete in which an extended setting control admixture was used and reactivated after 18-h storage, when tested for resistance to freezing and thawing, gave equally good results as concrete not so treated, and complied with ASTM C 494. They reported that tests of concrete containing an anti-washout admixture gave equally good results as the concrete not so treated in the underwater abrasion test, in freezing and thawing, and in sulfate resistance.

The same extended setting control materials reported on by Senbetta and Scanlon [57] were studied by Senbetta and Dolch [58] for effects on paste as revealed by X-ray diffraction, thermogravimetric analysis, differential thermal analysis, scanning electron microscopy, and determination of non-evaporable water content, surface area, and pore size distribution. No significant differences between treated and untreated pastes were noted.

Neeley [59] reported on evaluation of concrete mixtures for use in underwater repairs. He included concrete made with five different anti-washout admixtures. He found that all had a beneficial effect but none stood out as better than the others.

Acknowledgment

I appreciate the cooperation of the authorities at the U.S. Army Engineer Waterways Experiment Station and the Headquarters, U.S. Army Corps of Engineers, that permitted me to prepare and present this paper for publication.

REFERENCES

[1] Mather, B., "Chemical Admixtures," *Significance of Tests and Properties of Concrete and Concrete-Making Materials, ASTM*

STP 169B, American Society for Testing and Materials, Philadelphia, 1978, pp. 823–835.

[2] *Proceedings*, RILEM-ABEM International Symposium on Admixtures for Mortar and Concrete, Brussels, 1967, RILEM, Paris, 1968; also "General Reports," *Materials and Structures, Research and Testing*, RILEM, Vol. 1, No. 2, 1968, pp. 75–149.

[3] "Chemical Admixtures for Concrete," *Materials Journal*, ACI 212.3R, American Concrete Institute, Vol. 86, 1989, pp. 297–327; also, ACI *Manual of Concrete Practice*.

[4] Ramachandran, V. S., *Concrete Admixtures Handbook: Properties, Science, and Technology*, Noyes Publications, Park Ridge, NJ, 1984.

[5] Foster, B., "Chemical Admixtures," *Significance of Tests and Properties of Concrete and Concrete-Making Materials*, ASTM STP 169A, American Society for Testing and Materials, Philadelphia, 1966, pp. 556–564.

[6] *Cement and Concrete Terminology*, ACI 116R-90, American Concrete Institute, Detroit, MI, 1990.

[7] Prior, M. E. and Adams, A. B., "Introduction to Producers' Papers on Water-Reducing Admixtures and Set-Retarding Admixtures for Concrete," *Effect of Water Reducing and Set Retarding Admixtures on Properties of Concrete*, ASTM STP 266, American Society for Testing and Materials, Philadelphia, 1960, p. 170.

[8] Odler, I., "Admixtures," *Cements Research Progress 1975*, Cements Division, American Ceramic Society, Columbus, OH, 1976, pp. 83–95.

[9] Vollick, C. A., "Effect of Water-Reducing Admixtures and Set-Retarding Admixtures on the Properties of Plastic Concrete," *Effect of Water-Reducing and Set-Retarding Admixtures on Properties of Concrete*, ASTM STP 266, American Society for Testing and Materials, Philadelphia, 1960, p. 180.

[10] Polivka, M. and Klein, A., "Effect of Water-Reducing Admixtures and Set-Retarding Admixtures as Influenced by Portland Cement Composition," *Effect of Water-Reducing Admixtures and Set-Reducing Admixtures on Properties of Concrete*, ASTM STP 266, American Society for Testing and Materials, Philadelphia, 1960, p. 124.

[11] Tuthill, L. H., Adams, R. F., Bailey, S. N., and Smith, R. W., "A Case of Abnormally Slow Hardening Concrete for Tunnel Lining," *Journal*, American Concrete Institute; *Proceedings*, Vol. 57, No. 9, 1961, p. 1091; discussion by K. E. Palmer, p. 1828, and B. Tremper, p. 1831.

[12] Seligman, P. and Greening, N. R., "Studies of Early Hydration Reactions of Portland Cement by X-ray Diffraction," *Record No. 62*, Highway Research Board, Washington, DC, 1964, p. 80.

[13] Bruere, G. M., "Importance of Mixing Sequence When using Set-Retarding Agents with Portland Cement," *Nature*, London, Vol. 199, 1963, pp. 32–33.

[14] Dodson, V. H. and Farkas, E., "Delayed Addition of Set-Retarding Admixtures to Portland Cement Concrete," *Proceedings*, American Society for Testing and Materials, Vol. 64, 1964, pp. 816–829.

[15] Tremper, B. and Spellman, D. L., "Shrinkage of Concrete—Comparison of Laboratory and Field Performance," *Record No. 3*, Highway Research Board, Washington, DC, 1963, p. 30.

[16] Kondo, Y., Takeda, A., and Hideshima, S., "Effects of Admixtures on Electrolytic Corrosion of Steel Bars in Reinforced Concrete," *Journal*, American Concrete Institute; *Proceedings*, Vol. 56, No. 4, 1959, pp. 299–312.

[17] Wallace, G. B. and Ore, E. L., "Structural and Lean Mass Concrete as Affected by Water-Reducing, Set-Retarding Agents," *Effect of Water-Reducing and Set-Retarding Admixtures on Properties of Concrete*, ASTM STP 266, American Society for Testing and Materials, Philadelphia, 1959, p. 38.

[18] Newman, E. S., Blaine, R. L., Jumper, C. H., and Kalousek, G. L., "Effects of Added Materials on Some Properties of Hydrating Portland Cement Clinkers," *Journal of Research*, National Bureau of Standards, Vol. 30, 1943, p. 281.

[19] Hewlett, P. and Rixom, R., "Superplasticised Concrete," *Concrete*, The Concrete Society, London, 1976, pp. 39–42.

[20] Sasse, H. R., "Water-Soluble Plastics as Concrete Admixtures," *Proceedings*, 1st International Congress on Polymer Concretes, Paper No. 2 London, 1975.

[21] "Flieszbeton," *Zement Taschenbuch*, Bauverlag GmbH, Verein Deutsche Zementwerke e. V., Wiesbaden-Berlin, 1976/77, pp. 327–350.

[22] Tynes, W. O., "Investigation of Proprietary Admixtures," Technical Report No. C-77-1, U.S. Army Engineer Waterways Experiment Station, Vicksburg, MS, 1977.

[23] *Superplasticizers in Concrete*, ACI SP–62, V. M. Malhotra, Ed., American Concrete Institute, Detroit, MI, 1979.

[24] *Developments in the Use of Superplasticizers*, ACI SP–68, V. M. Malhotra, Ed., American Concrete Institute, Detroit, MI, 1981.

[25] *Superplasticizers and Other Chemical Admixtures on Concrete*, ACI SP–119, V. M. Malhotra, Ed., American Concrete Institute, Detroit, MI, 1989.

[26] Mather, B., "Tests of High-Range Water-Reducing Admixtures," *Superplasticizers on Concrete*, ACI SP–62, American Concrete Institute, Detroit, MI, 1979, pp. 157–166.

[27] Schutz, R. J., "Durability of Superplasticized Concrete," presented at the International Symposium on Superplasticizers in Concrete, Ottawa, Canada, May 1978.

[28] Whiting, D., "Effects of High-Range Water Reducers on Some Properties of Fresh and Hardened Concrete," *Research and Development*, Bulletin No. RD-61-01T, Portland Cement Association, Skokie, IL, 1979.

[29] Litvan, G. G., "Air Entrainment in the Presence of Superplasticizers," *Journal*, American Concrete Institute; *Proceedings*, Vol. 80, No. 4, 1983, pp. 326–331.

[30] Zakka, Z. A. and Carrasquillo, R. L., "Effects of High-Range Water Reducers on the Properties of Fresh and Hardened Concrete," Report 1117-3F, Center for Transportation Research, University of Texas at Austin, Austin, TX, 1989.

[31] Vollmer, H. C., "Calcium Chloride in Concrete," *Bibliography No. 13*, Highway Research Board, Washington, DC, 1952.

[32] Mather, B., "Effects of Three Chemical Admixtures on the Properties of Concrete," Miscellaneous Paper No. 6–123, Report No. 3, U.S. Army Corps of Engineers, Vicksburg, MS, 1956.

[33] Shideler, J. J., "Calcium Chloride in Concrete," *Journal*, American Concrete Institute; *Proceedings*, Vol. 48, No. 7, 1952, p. 537.

[34] Grieb, W. E., Werner, G., and Woolf, D. O., "Water-Reducing Retarders for Concrete," *Public Roads*, Vol. 31, 1961, p. 136.

[35] Mather, B., "Drying Shrinkage—Second Report," *Highway Research News*, No. 15, Washington, DC, 1964, pp. 34–38.

[36] Klieger, P., "Curing Requirement for Scale Resistance of Concrete," *Bulletin No. 150*, Highway Research Board, Washington, DC, 1958.

[37] Evans, R. H., "Effects of Calcium Chloride on Prestressing Steel and X-Rays for Anchorage Lengths," *Proceedings*, World Prestressed Conference, San Francisco, CA, 1957.

[38] Monfore, G. E. and Verbeck, G. J., "Corrosion of Prestressed Wire in Concrete," *Journal*, American Concrete Institute; *Proceedings*, Vol. 57, No. 5, 1960, pp. 491–516.

[39] Godfrey, H. T., "Corrosion Tests on Prestressed Concrete Wire and Strand," *Journal*, Prestressed Concrete Institute, Vol. 5, No. 1, 1960, pp. 45, 48–51.

[40] Wright, T. E., "An Unusual Case of Corrosion of Aluminum Conduit in Concrete," *The Engineering Journal*, Vol. 38, 1955, p. 1357.

[41] Mange, C. E., "Corrosion of Galvanized Steel in Contact with Concrete Containing Calcium Chloride," 13th Annual Conference, National Association of Corrosion Engineers, St. Louis, *Proceedings*, American Concrete Institute, Vol. 54, Nov. 1957, p. 431.

[42] Mather, B., "Calcium Chloride in Type V-Cement Concrete," *Durability of Concrete, SP-131*, G. M. Idorn International Symposium, J. Holm and M. Geiker, Eds., American Concrete Institute, Detroit, MI, 1992, pp. 169–178.

[43] Vivian, H. E., "Some Chemical Additions and Admixtures in Cement Paste and Concrete," *Proceedings*, 4th International Symposium on Chemistry of Cement, Monograph No. 43, National Bureau of Standards, Washington, DC, 1960, pp. 909–923.

[44] Schutz, R. J., "Setting Time of Concrete Controlled by the Use of Admixtures," *Journal*, American Concrete Institute; *Proceedings*, Vol. 55, No. 7, 1959, pp. 769–781.

[45] Swenson, E. G. and Thorvaldson, T., "Detection of Lignosulfonate Retarder in Cement Suspensions and Pastes," *Effect of Water-Reducing and Set-Retarding Admixtures on Properties of Concrete, ASTM STP 266*, American Society for Testing and Materials, Philadelphia, 1960, p. 159.

[46] Halstead, W. J. and Chaiken, B., "Water-Reducing Retarders for Concrete—Chemical and Spectral Analyses," *Bulletin 310*, Highway Research Board, Washington, DC, 1962, p. 33.

[47] Tuthill, L. H., Adams, R. F., and Hemme, J. M., Jr., "Observations in Testing and Use of Water-Reducing Retarders," *Effect of Water-Reducing and Set-Retarding Admixtures on Properties of Concrete, ASTM STP 266*, American Society for Testing and Materials, Philadelphia, 1960, p. 97.

[48] Valore, R. C., Jr., "Pumping Aids for Concrete," *Significance of Tests and Properties of Concrete and Concrete-Making Materials, ASTM STP 169B*, American Society for Testing and Materials, Philadelphia, 1978, pp. 860–872.

[49] Browne, R. D. and Bamforth, P. B., "Tests to Establish Concrete Pumpability," *Journal*, American Concrete Institute; *Proceedings*, Vol. 74, No. 5, 1977, pp. 193–203.

[50] Valore, R. C., Jr., Bowling, J. E., and Blaine, R. L., "The Direct and Continuous Measurement of Bleeding in Portland Cement Water Mixtures," *Proceedings*, American Society for Testing and Materials, Vol. 49, 1949, pp. 891–908.

[51] Ritchie, A. G. B., "Stability of Fresh Concrete Mixes," *Proceedings*, American Society of Mechanical Engineers; *Journal*, Construction Division, Vol. 92, No. CO 1, 1966, pp. 17–35.

[52] Gray, J. E., "Laboratory Procedure for Comparing Pumpability of Concrete Mixtures," *Proceedings*, American Society of Civil Engineers, Vol. 62, 1962, p. 964.

[53] Tanaka, I., "On the Artificial Lightweight Aggregate Mortar-Mesalite Mortar," *Journal of Research*, Onoda Cement Company, Vol. 21, 1969, p. 93 (in Japanese).

[54] Ozaki, Y. and Emura, K., "Lightweight Aggregate Concrete Placing with Pump Method," *Journal of Research*, Onoda Cement Company, Vol. 20, No. 73, 1968, p. 62 (in Japanese).

[55] Ragan, S. A., "Evaluation of Tests for Determining the Pumpability of Concrete Mixtures," MP SL-81-29, U.S. Army Engineer Waterways Experiment Station, Vicksburg, MS, 1981.

[56] Anderson, W. H., "Analyzing Concrete Mixtures for Pumpability," *Journal*, American Concrete Institute; *Proceedings*, Vol. 74, No. 9, 1977, pp. 447–451.

[57] Senbetta, E. and Scanlon, J. M., "Effects of Three New Innovative Chemical Admixtures on Durability of Concrete," Supplementary Papers, Second CANMET/ACI International Conference on Durability of Concrete, Montreal, Canada, 1991, pp. 29–48.

[58] Senbetta, E. and Dolch, W. L., "The Effects on Cement Paste of Treatment with an Extended Set Control Admixture," *Cement and Concrete Research*, Vol. 21, Pergamon Press, New York, Oxford, Seoul, Tokyo, 1991, pp. 750–756.

[59] Neeley, B. D., "Evaluation of Concrete Mixtures for Use in Underwater Repairs," Technical Report REMR-CS-18, (NTIS AD A193 897), U.S. Army Engineer Waterways Experiment Station, Vicksburg, MS, 1988.

46

Mineral Admixtures

Craig J. Cain[1]

PREFACE

This writer is much indebted to E. C. Higginson [1] who prepared the effort on this chapter on mineral admixtures in 1966 and to L. H. Tuthill [2] who wrote the chapter in *ASTM STP 169B* in 1978. Their work provides the background for this version. I have borrowed from their ideas and specific words. However, this chapter will present less discussion of the use of mineral admixtures in concrete and more on the characteristics of the mineral admixtures and the significance and use of their acceptance tests. For discussion of the use of mineral admixtures, the reader is referred to two reports from former Committee 226 on Admixtures for Concrete of the American Concrete Institute (ACI). These reports, ACI 226. 1R-87 "Ground Granulated Blast-Furnace Slag as a Cementitious Constituent of Concrete" [3] and ACI 226.3R-87 "Use of Fly Ash in Concrete" [4] were published in 1987. Updated reports on these two materials are currently being prepared by ACI Committees 233 and 232, respectively. Donald W. Lewis also made significant contributions with information and comments on the section in this chapter on slag. This chapter will deal with the following mineral admixtures: fly ash, natural pozzolans, ground granulated blast-furnace slag, and silica fume.

INTRODUCTION

This chapter will touch on the history of the use and interest in mineral admixtures. It will chronicle the paths that the various specifications for mineral admixtures have taken through ASTM task groups, subcommittees, and committees. A discussion of tests will follow at that point in the chapter.

MINERAL ADMIXTURES

Definitions

An admixture is defined "as a material other than water, aggregates, hydraulic cement and fiber reinforcement used as an ingredient of concrete or mortar and added to the batch immediately before or during its mixing," according to ASTM Terminology Relating to Concrete and Concrete Aggregates (C 125). Mineral admixtures include any essentially water insoluble material other than cement or aggregate that is used as an ingredient for concrete. According to Mielenz [5], "mineral admixtures include finely divided materials that fall into four types: those that are (*a*) cementitious, (*b*) pozzolanic, (*c*) both cementitious and pozzolanic, and (*d*) those that are nominally inert chemically." They include natural materials, processed natural materials, and artificial materials. They are finely divided and therefore form pastes to supplement portland cement paste, in contrast to soluble admixtures that act as chemical accelerants or retardants during the hydration of portland cement or otherwise modify the properties of the mixture.

Cementitious Materials Not Discussed

Cementitious materials such as natural cements, hydraulic limes, portland-pozzolan cements, and slag cements were discussed as mineral admixtures in the previous versions. They will not be dealt with in this chapter because specifications for those materials are not under the jurisdiction of the ASTM subcommittee dealing with mineral admixtures. Even though the aforementioned materials might have been considered mineral admixtures at one time, the greater availability of fly ash and slag has made their use in this category negligible.

Low-Reactivity Admixtures Not Discussed

Such materials as ordinary clay, ground quartz, ground limestone, bentonite, hydrated lime, and talc were once used to improve workability and reduce bleeding in concrete. However, they are admixtures with low reactivity. Now, more suitable and available mineral admixtures such as fly ash and slag have largely taken their place for these purposes. Therefore, such materials will not be dealt with further in this chapter.

POZZOLANIC MATERIALS

History and Use

Although slag cements and natural pozzolans had been used in concrete in local areas for many years, it was the

---

[1] Board chairman, American Fly Ash Company, Naperville, IL 60563-9338; also, chairman, ASTM Subcommittee C09.24 on Fly Ash, Slag, Mineral Admixtures, and Supplementary Cementitious Materials.

use and acceptance of fly ash in concrete in the national market in the late 1940s and early 1950s that created the need for national specifications for mineral admixtures. It was recognized that the use of fly ash in concrete would have the following benefits:

(a) improved workability,
(b) lower heat of hydration,
(c) lower cost concrete,
(d) improved resistance to sulfate attack,
(e) improved resistance to alkali-silica reactions, [Type F]
(f) higher long-term strength,
(g) opportunity for higher strength concrete,
(h) equal freeze-thaw durability,
(i) lower shrinkage characteristics, and
(j) lower porosity and improved impermeability.

Good results were being obtained with fly ash [6,7], and therefore there was a need for an ASTM specification for its use. The preparation of a specification for fly ash for use in portland-cement concrete was begun in 1948 by ASTM Committee C-9 on Concrete and Concrete Aggregates. This work was done by Subcommittee III-h dealing with all admixtures, chemical and mineral, later designated as Subcommittee C09.03.08. In early 1953, the Subcommittee proposed two standards on fly ash. One of these dealt with the methods of testing and was published as ASTM Tentative Test Methods for Sampling and Testing Fly Ash or Natural Pozzolans for Use as a Mineral Admixture in Portland-Cement Concrete (C 311-53T). The other proposed standard dealt with specifications for fly ash with various chemical and physical limits. This latter proposal met with serious opposition from some members of the cement industry and was not approved by Committee C-9 until several changes were made. In point of fact, these changes lumped all fly ash into one class where a class distinction would have allowed the better fly ashes better recognition. It was published as ASTM Tentative Specification for Fly Ash as Admixture in Portland Cement Concrete (C 350-54T) and covered only the use of what is now designated as Class F fly ash in concrete "where the use of increased quantities of fine material may be indicated to promote workability and plasticity." The fly ash was to be treated only as a portion of the fine aggregate and not as a partial replacement or substitute for portland cement. Notes of caution on the use of fly ash in concrete were inserted in the early versions of the specification and dealt with the amount of sulfur, magnesium oxide, and effect on autoclave expansion. Experience has shown them to be of less concern than expected, but many of the reflected limits are still in the specification.

ASTM C 350 was revised repeatedly based on new experience and research. In 1960, the scope of ASTM C 350 was extended to cover fly ash as a pozzolan in concrete acknowledging the prospect that the use of fly ash in the mixture might result in a reduced amount of portland cement.

In 1957, ASTM Tentative Specification for Raw or Calcined Natural Pozzolans for Use as Admixture in Portland Cement Concrete (C 402-57T) was published. At that time differences between natural pozzolans and fly ash were recognized by different requirements on drying shrinkage, loss on ignition, fineness, and reactivity with cement alkalies. Then in 1968, ASTM C 350 was combined with ASTM C 402 resulting in ASTM Specification for Fly Ash and Raw or Calcined Natural Pozzolan for Use as a Mineral Admixture in Portland Cement Concrete (C 618). Three classes of mineral admixture were dealt with in three columns. Class F was the fly ash from ASTM C 350. Class N was the natural pozzolan from ASTM C 402. A new Class S having the same requirements as Class F allowed for other materials that could not meet the requirements for natural pozzolans, but could contribute desirable qualities to the concrete mixture.

In 1977, Class C fly ash was added to ASTM C 618 to recognize it as a material with cementitious characteristics compared to those of the original Class F. There was no apparent use of Class S pozzolan in the construction industry, so the designation was dropped from ASTM C 618 in 1980.

In 1980, ASTM Committee C-9 formed a new subcommittee, ASTM C09.03.10 on Fly Ash, Slag, Mineral Admixtures, and Supplementary Cementitious Materials, with Robert E. Philleo as chairman. The number designation of the subcommittee was changed to C09.24 in 1992. The scope of ASTM C09.24 is "to develop and maintain test methods and specifications for finely divided mineral materials other than cement and pigments, used in substantial proportions in concrete." This new subcommittee now had no concern for chemical admixtures and instead had new interest in slag and eventually silica fume.

Classification of Fly Ash

In 1977, when Class C fly ash was added to ASTM C 618 there was recognition of the fact that this was a new material made available because utilities had begun to burn larger quantities of low-sulfur coal. This fly ash varies from the Class F fly ash around which the earlier version of the specification had been written. The new Class C has cementing characteristics of its own. It has higher measured quantities of lime or calcium oxide (CaO) and is less of a pozzolan. It has greater hydraulicity and makes a greater contribution to early strength over that obtained from the Class F fly ash. However, there are not yet any acceptable tests for measuring these attributes. Therefore, the only differentiation made between the two fly ashes in ASTM C 618 is coal source and chemistry. ASTM C 618 states that Class F fly ash is "normally produced from burning anthracite or bituminous coal" and Class C fly ash is "normally produced from lignite or subbituminous coal." This distinction has been controversial. The subcommittee has not been able to make a better one.

Chemical Requirements for Fly Ash

ASTM C 618, Table 1, gives the chemical requirements for fly ash and natural pozzolan mineral admixtures. There are limits on the sum of the three elements expressed as compounds of silicon dioxide (SiO_2) plus aluminum oxide (Al_2O_3) plus iron oxide (Fe_2O_3). There are also limits on the sulfur trioxide (SO_3), moisture content, and loss on ignition. ASTM C 311, specifying the methods to be used in the determinations, refers to ASTM Test Methods for Chemical Analysis of Hydraulic Cement (C 114) with some modifications. Rapid and instrumental methods may be employed similar to those in ASTM C 114. The earliest versions of ASTM C 618 assumed that the pozzolanic contribution from fly ash had to come from its makeup of SiO_2, Al_2O_3, and Fe_2O_3. The minimum requirement placed on these three constituents is based on that assumption. However, variations in the total of those three constituents for any fly ash have not been seen to correlate directly with results in concrete. At one point in the development of the specification, there was a suggestion to remove this requirement with the rationale that it served only to define the material as fly ash. At that time, the expense and delay in running these tests was significant and not altogether worthwhile. Now, with quicker instrumental chemical testing procedures, the determination of the total SiO_2, Al_2O_3, and Fe_2O_3 is not a concern.

With respect to the test method for Fe_2O_3, ASTM C 311 refers to ASTM C 114 for determining the iron content of hydraulic cements with high insoluble residues. The U.S. Army Corp of Engineers (USAE) Waterways Experiment Station (WES) found the procedure in effect up to 1989 to be lengthy, labor intensive, and in one case, at least, apparently underestimated the iron content [8]. An alternative method that is more rapid and efficient in decomposing the sample, but at least as precise, was developed by the WES group and was adopted in the 1989 edition of ASTM C 114.

The only specified difference between Classes F and C is the total of SiO_2 plus Al_2O_3 plus Fe_2O_3. Class F has always required a minimum of 70% and this limit served well for many years. Class C requires a minimum of 50%. Actually, this is the subcommittee's way of recognizing that Class C may have 20% more CaO content than Class F with the corollary result of lowering the sum of the other three oxides. This is the only chemical difference recognized between Classes F and C in the specification. Any material passing the Class F requirement will also pass as Class C. Therefore, this is not a helpful distinction.

The greater reactivity of Class C fly ash over Class F has been attributed to greater CaO content. To date, the subcommittee has not found a more satisfactory method to measure the amount of CaO than by determining the three oxides discussed earlier. The procedure in ASTM Test Methods for Chemical Analysis of Limestone, Quicklime, and Hydrated Lime (C 25) for determining quicklime was tried. The measurement of temperature rise of a fly ash mixed with water and the measurement of the pH of a fly ash mixture are two other methods to measure reactivity, and therefore CaO, that were tried and rejected.

Although the CaO content of a fly ash effects the activity in significant ways, the results are not direct since the CaO may be in different compounds in various fly ashes. Excessive autoclave expansion has been experienced in the laboratory when the CaO was in the form of free lime. The hydration of that free lime takes place quickly enough in plastic concrete, so it has not been a problem in the field. But, when such a fly ash is used in a relatively dry mix, like shotcrete or for concrete products, there might not be enough moisture to hydrate the free lime before the product takes its initial set, therefore, delayed volume change could be harmful.

Sulfur in fly ash is measured and reported as SO_3. The first editions of this specification limited the SO_3 to 3%. Subsequent research showed that a fly ash with up to 17% SO_3 did not give significant delayed volume changes or setting problems. The SO_3 limits for fly ash were raised to 5% maximum. Higher amounts of sulfur can contribute to efflorescence, so when they are present, testing for discoloration can be considered.

The moisture content is limited to 3% maximum because the material would become sticky and hard to handle if too much moisture were encountered. The test is easy and inexpensive to run and the values determined are needed in other parts of the test methods.

The limit on loss on ignition (LOI), now set at a maximum of 6%, was originally set at 12% maximum. At that time, many of the fly ashes available in the eastern United States could not pass limits much lower, and they were giving satisfactory results in concrete. With the advent of better combustion in electric power plants and with the coming of Class C fly ash to the market, it was found that many agencies were dropping the limit to 6% maximum. Subcommittee C09.03.08 was divided in the action to lower the limit. The maximum LOI was already at 6% for Class C fly ash. The note under ASTM C 618, Table 1, was added providing for a 12% maximum if acceptable performance records were provided. Everyone concedes, however, that the lower the loss on ignition, the easier will be the control of air entrainment content in the concrete. Class C fly ash usually is below 1% LOI. Agencies frequently lower this maximum when they are sure there are competitive materials available in their marketplace.

Loss on ignition is generally assumed to be slightly higher than, but directly related to, carbon content. When the specification was first written, the determination of LOI was chosen because it was an available test method from ASTM C 114 and was preferable to the determination of carbon. LOI is run at 750°C rather than 950°C, the temperature given in ASTM C 114. Currently, instrumental tests for carbon are frequently run instead of LOI and are quicker and more desirable.

Loss on ignition or carbon content uniformity is important. Higher carbon will dilute the beneficial components of the material and, therefore, detract from strength at least in a small way. It will generally make the product coarser and increase the water demand of the concrete mixture. Most importantly, it will make control of air entrainment more difficult. In some mixtures, small changes in carbon content have significant effects on air entrainment. Effects on air entrainment may also be due to the form or compound in which the carbon is found.

Supplementary Optional Chemical Requirement for Fly Ash

Table 1A from ASTM C 618 shows that the only supplementary optional chemical requirement for fly ash deals with the maximum limit on available alkalies. This limit should be imposed only when there is a potential alkali-aggregate reaction in the concrete in which the fly ash is to be used. Some agencies use this limit for all concrete and in so doing might limit the availability of good fly ash for their projects.

Up until 1983, magnesium oxide (MgO) was limited in the optional requirements. Originally, there was concern that the presence of MgO would indicate the presence of periclase in the fly ash. There have been only a few scattered situations where research has found small quantities of periclase in fly ash. The autoclave soundness test will detect periclase should it ever be a concern.

There has been much criticism of ASTM C 618 because some people think it has irrelevant requirements. Generally, those are thought to be in the chemical requirements just discussed. All of us in ASTM would like to get away from the prescriptive type of chemical requirements shown in these two tables. Philleo [9] suggested that "all the compositional limits in the current draft could be dropped with no adverse effect on the user since all the properties intended to be controlled by these limits—pozzolanicity, hydraulicity, soundness, and prevention of alkali-aggregate reaction—may be assured by physical tests also provided in the draft specification." However, tests for these chemicals help ensure that the product under test is like the material used in any previous evaluation and help define the product. As ASTM C 311 states, "the chemical component determinations and the limits placed on each, do not predict performance in portland cement concrete, but collectively help describe composition and uniformity of the mineral admixture."

Physical Requirements for Fly Ash

These requirements are set forth in Table 2A of ASTM C 618 and are listed as fineness, strength activity index, water requirement, soundness, and uniformity requirements. This chapter will deal with them in that order.

Fineness of fly ash is important in its contribution to a concrete mixture. Although there is not a direct correlation, the water needed for workability and the relative chemical contribution of the fly ash are both dependent on fineness. The first versions of this specification used specific surface as measured by the Blaine method to measure fineness. Then, mean particle diameter was substituted in 1965. The consensus went back to specific surface in 1968 along with percent retained on 45 μm (No. 325) sieve. A maximum of 20% retained was the original limit. This was changed to 34% maximum when the procedure in ASTM Test Method for Fineness of Hydraulic Cement by the 45-μm (No. 325) Sieve (C 430) began to call for the use of the new electroformed sieves. These changes are mentioned to indicate the difficulty experienced measuring fineness in fly ash in a way that correlates with results in concrete. In the Blaine test, higher carbon can give readings indicating higher surface area that would assume a finer product when actually the material is coarser or less fine. The percent retained test does not measure the fine material that is actually contributing to the concrete reaction, but it generally acts as an accurate indicator of the amount of fine material present. With that in mind, it was found that the multiple factor shown in ASTM C 618, Table 2A, combined fineness and LOI in a manner that correlated with results in concrete. This relationship does not apply to Class C fly ash because the LOI is generally below 1% and is, therefore, too low to act as a multiplier.

The strength activity index test with lime has been controversial for some time. The requirement was dropped for Class C fly ash in 1985. Class C fly ash makes its own contribution of lime to the mixture, so this test was irrelevant. However, it has been the only 7-day strength test for all of the classes of mineral admixture and was used for that reason alone. Some problems with the lime test were answered when more control over the fineness of the test lime was required. The test is still felt to be as much a test of lime as of mineral admixture. There is a current effort to remove the requirement for all classes.

During 1989 and 1990, the strength activity index test with portland cement received extensive changes in procedure and evaluation. The replacement of portland cement is done by mass rather than volume to reflect field usage. The curing temperature has been reduced also to reflect laboratory conditions. The most important addition has been a 7-day alternate with the previous 28-day requirement. Using the new 7-day activity index with portland cement allows for deletion of the 7-day lime test.

Some criticism of the physical requirements is due to the fact that the strength activity index test with portland cement does not directly correlate with performance in concrete. Since it is performed with a mortar mixture with a test cement, it should not necessarily be expected to forecast results in a concrete mixture. Fly ash will react differently with different cements and with different proportions. The strength activity index test with portland cement is meant as a product acceptance test on a nationwide basis and not as a guide to mixture designs for specific projects.

The water requirement limit gives some measure of the contribution to workability that the fly ash will impart to the concrete mixture. This limit will depend on fineness, carbon, or LOI, and, to a lesser extent, on glassy spherical particles in the fly ash. The limit is set to make the user aware of that contribution to workability.

Soundness, as measured by the autoclave expansion test, will indicate any possible periclase as well as free lime if those materials are present in deleterious quantities. The drier consistency of the mortar used in the soundness test will not allow for hydrating lime when it is in the fly ash in a so-called "hard burned" state. Some Class F fly ash has failed this test due to the lime, even though the lime would have been adequately hydrated when used in normal plastic concrete. With respect to periclase or MgO, we are reminded that up to 1980 when ASTM C 618, Table 1A, limited MgO to 5% maximum, it allowed for a higher percent if the fly ash passed the autoclave test. Note c to ASTM C 618, Table 2, states that "If the mineral admixture will constitute more then 20% by weight of the cementitious material in the project mix design, the test specimens for autoclave expansion shall contain that anticipated percentage. Excessive autoclave expansion is highly significant in cases where water to mineral admixture and cement ratios are low, for example, in block or shotcrete mixes."

The uniformity of the fineness is also specified since it is an easily determined indicator of a change in the product. Such a change is important because it might effect water demand in the concrete. The other uniformity requirement in Table 2 of ASTM C 618 is on specific gravity. Currently, there is consideration of changing the words "specific gravity" to "density." Whichever is used, a change of 5% in the uniformity of specific gravity or density is not thought to be significant either as an indicator of product change or in effect on concrete mixture design.

Supplementary Optional Physical Requirements for Fly Ash

ASTM C 618, Table 2A, lists the optional physical requirements as multiple factor, drying shrinkage, uniformity requirements, and reactivity with cement alkalies. The multiple factor has already been discussed in its relationship to fineness.

Increase of drying shrinkage is a requirement because some natural pozzolans will cause an unacceptable increase in drying shrinkage. Fly ash will generally reduce drying shrinkage.

The test procedure for uniformity of air-entraining agent demand determined at 18% air in ASTM C 618, Table 2A, has recently been challenged as not being sensitive enough. The subcommittee will consider a lower air content that would make the test more sensitive. In the meantime, this optional test will deal with the concern that some fly ash causes problems with control of entrained air in the concrete mixture.

The reactivity with cement alkalies requirement test has not been satisfactory. The required "reduction of mortar expansion" does not apply at all to fly ash because results are not attainable or dependable. The required "mortar expansion" at 14 days of a maximum percent of 0.020 is too low for many otherwise acceptable fly ashes and has, therefore, served to cause the rejection of materials that might have been helpful in correcting the alkali-silica reaction in concrete. The 0.020 maximum expansion was set assuming low-alkali portland cements could meet that limit. The use of Pyrex glass as aggregate with modern cements has given undependable results.

A new test procedure measuring the contribution of fly ash in correcting the alkali-silica reaction is being considered by ASTM. Low alkali cement would be used in a control test to determine an expansion level that would or should be acceptable in the concrete. If the use of fly ash in a mortar test mixture can then ensure that a maximum level of expansion is not exceeded even with a cement of higher alkali content than that used in the control, then that fly ash will be not only acceptable but beneficial in the concrete. There is some feeling that this optional requirement should be made mandatory, but that discussion will follow ASTM acceptance of this new procedure and limit.

Test Requirements Not in the Specification

Various other tests have been proposed and used in research. These have been considered for use in these specifications, but they were not generally applicable or definitive enough to include. These dealt with, among other things, a discernment between Classes F and C, heat of hydration, sulfate resistance, and effect on air entrainment.

This chapter has already suggested that the distinction between Classes F and C fly ash is not properly drawn in ASTM C 618. The subcommittee has worked with a test for hydraulic index in hopes of being able to forecast the possible contribution of a pozzolan to the strength of a concrete mixture. This seemed at least to be one factor that could be isolated. Such a test would help discern between Classes F and C. No such test procedure has proven totally acceptable because the complex contribution to a concrete mixture made by a pozzolan with various cements, in various proportions, at various temperatures, and in various types of concrete mixtures has not allowed any correlations that are acceptable. Manz [10] suggested a performance test measuring strength with no cement in the mixture. If such a mixture gave a measurable strength, the fly ash would be assumed to be Class C, if not, Class F. Various other strength tests of mortars or even of mixtures without aggregate were tried. Temperature rise tests were also tried without conclusive results.

These same temperature rise tests on mortars have been used for information on heat of hydration with varying degrees of success. However, in measuring both contribution to strength and heat of hydration, it is more advisable to test a complete concrete mixture rather than a laboratory mortar.

Another test that is not in the specification deals with the contribution of fly ash to the sulfate resistance of the concrete. Dunstan [11] has proposed an R factor equation for sulfate resistance as follows: $R = (CaO - 5)/Fe_2O_3$. This uses the oxide analysis percentages. The fly ash should improve sulfate resistance when R is less than 1.5. This proposal has been controversial because some studies have shown that fly ashes with an R factor greater than 1.5 have contributed to sulfate resistance when used in proper proportions in the concrete mixture.

The foam-index test is a quick test to show possible changes in the amount of air-entraining agent required when using fly ash in the concrete. Since it deals with one cement and only the source of fly ash under test, the results apply only to that single combination.

Quality Assurance

Some of the tests discussed earlier could be used as part of a quality assurance program, inasmuch as ASTM C 618 and ASTM C 311 cannot be expected to be a complete quality assurance program. A quality assurance program can better be done by a producer of fly ash than by a user, since the producer will know more about the factors that affect change and will therefore know the tests that really apply to his situation. Fly ash is a by-product of coal combustion in an electric steam power plant and acceptability is therefore more a matter of quality assurance than quality control. Some agencies, like the U.S. Army Corps of Engineers once used sealed storage and complete testing before shipment. In 1984, that organization began an effort to avoid the consequent cost and delay by qualifying sources of pozzolan based on prior testing history and producer product certification [12]. Such a program might use some tests not discussed in this chapter that a producer has found to be indicative of possible changes in product. The ACI Committee report [4] discusses some of those tests that cannot be used in a specification like ASTM C 618 but can be used effectively for quality assurance.

Natural Pozzolans

Since 1968, when ASTM Specification for Natural Pozzolans in Concrete (C 402) was combined with the fly ash specification, ASTM C 350, to form the current ASTM C 618, there have been no changes to the requirements for natural pozzolans that were not brought on because of experiences with fly ash. The maximum LOI was lowered from 12% to 10%. It should be noted that the maximum water requirement is 115% of control in contrast to the two fly ash classes at 105%. The new strength activity index test with portland cement was evaluated for natural pozzolans in the same way that it was for fly ash. At both 7 and 28 days, the correlations were satisfactory for the materials in the round-robin series. It is of note that there was some difficulty getting samples of natural pozzolan to put in the series. None were being produced in the United States at that time.

In ASTM C 618, Table 2A, there is still an optional requirement under reactivity with cement alkalies that natural pozzolans reduce mortar expansion a minimum of 75% of control. This applies only to natural pozzolans because fly ash cannot be expected to meet this requirement with any degree of certainty. If the changes in the alkali-silica test procedure and requirements discussed under the fly ash section are made, they will be applied to natural pozzolans. All of the other requirements for natural pozzolans are the same as for the two classes of fly ash, and statements made with respect to tests under the fly ash section of this chapter apply to natural pozzolans as well.

SLAG

History and Use

The term "slag" as used in this chapter and as defined in ASTM Specification for Ground Granulated Blast-Furnace Slag for Use in Concrete and Mortars (C 989) is meant to include each of the modifiers used in the specification title. According to ASTM C 125, it is "the glassy, granular material formed when molten blast-furnace slag produced as a by-product in the making of iron is rapidly chilled as by immersion in water." The slag is "the nonmetallic product, consisting essentially of silicates and aluminosilicates of calcium and other bases, that is developed in the molten condition simultaneously with iron in a blast furnace." The granulated material is then ground to cement fineness. The slag specified in ASTM C 989 may be used for blending with portland cement to produce a cement that meets the requirements of ASTM Specification for Blended Hydraulic Cements (C 595) or it may be added as a separate ingredient in concrete or mortars. This chapter will deal with slag in the latter use as separately ground slag that is used as a mineral admixture, one that is both cementitious and pozzolanic. Lewis [13] has pointed out the advantages of using slag added as a separate ingredient in concrete as follows:

(a) higher ultimate strengths with a tendency toward lower early strengths,
(b) higher ratio of flexural to compressive strengths,
(c) improved refractory properties,
(d) lower coefficients of variation in strengths,
(e) improved resistance to sulfates and seawater,
(f) lowered expansion from alkali-silica reaction,
(g) lower temperature rise due to lower heat of hydration,
(h) better finish and lighter color,
(i) equivalent durability in freezing and thawing, and
(j) decreased porosity and chloride penetration.

For many years the hydraulic properties of slag have been utilized in blends with cements and lime. Such blends have been used in France, Germany, and the United States since the 1800s. Intensive investigations of the blended material were conducted in South Africa beginning in the 1960s. This new knowledge and experience led to more recognition of the benefits of adding separately ground slag at the mixer over its use as a blended material at the mill.

In the United States, the ASTM specification for separately ground slag was originally developed under the jurisdiction of Committee E-38 on Resource Recovery in 1982. By starting in ASTM Committee E-38, the task group working on the specification was able to avoid unnecessary prescription specifications and, instead, pioneer with meaningful performance criteria. That task group succeeded in getting realistic test requirements that had a minimum of long drawn out testing and at the same time included all reasonably effective slags. The task group based its findings on: (a) extensive tests of a wide range of ground slags from North America, Japan, England, and South Africa (the last two with extensive use records as admixtures); and (b) all available literature. Once it was developed, jurisdiction for this standard (ASTM C 989) was changed to ASTM Committee C-9 on Concrete and Concrete Aggregates in November of 1982.

Classification

Factors that determine the cementitious properties of slag are: (a) chemical composition, (b) alkali concentration of the reacting system, (c) glass content of the slag, (d) fineness of the slag, and (e) temperature during the early phase of the hydration process [74]. These factors all affect one another in a complex manner. Many investigators have tried to relate hydraulicity of a slag, or the slag's contribution to compressive strength, to its various chemical components as measured by the quantities of various oxides. Lea [15] has commented on the usefulness of such empirical formulae: "Composition moduli are convenient for the rapid control of slag quality, since regular chemical analysis of the slag is necessary in any event in the control of the pig iron production process, but the only fixed guide is the strength developed in cements."

Glass Content

It is generally recognized that glass content of granulated slag is a primary factor in determining its contribution to a concrete mixture or its hydraulicity. The granulation process in the slag processing procedure is fundamental. Rapid chilling or quenching of the molten slag inhibits the formation of a crystalline structure. Consequently, this quick cooling is necessary so that the slag will be composed largely of glass or amorphous material. One method for determining glass content is by microscopic count. As important as glass content is in slag, investigators have not developed clear-cut relationships between glass content and the strength contribution of slag.

Slag Activity Test with Portland Cement

ASTM C 989 has gone directly to an evaluation of a slag's contribution to the strength of a mortar after studies of other methods for hydraulicity rating. Three grades of slag are covered in the specification. These grades are based on the ratio of the strength of mortar cubes (using only portland cement) to the strength of mortar cubes (made with a 50% slag–50% cement blend) made in accordance with ASTM Test Method for Compressive Strength of Hydraulic Cement Mortars (Using 2-in. or 50-mm Cube Specimens) (C 109). The three grades are designated as Grade 80, Grade 100, and Grade 120. The reference cement used must have a minimum 28-day strength of 35 MPa (5000 psi) and an alkali content between 0.6 and 0.9%. It was found that performance of the slag might vary significantly with other portland cements and that the standard cement provided the best differentiation between grades of slag. Proportioning of mixtures for field use should be based on tests with project cement.

Laboratory strength test procedures can be misleading in evaluating a slag's performance in the field. Because the test involves curing at room temperatures and the average field temperature of concrete placed in the United States is higher than room temperature, slag is penalized by the lower laboratory temperature. Slag is more sensitive to temperature differences than is portland cement, because slag hydration is accelerated to a greater extent by increases in curing temperature and it is retarded more at lower temperatures. Slag's strength contribution to concrete in the field will often be greater than shown in laboratory tests.

It should also be noted that slag is not usually given proper credit for its contribution to better workability in a concrete mixture. Although the ASTM C 109 procedure does adjust for water requirement by equalizing flow, that test and the slump test do not measure the contribution to workability, placeability, or ease of consolidation that slag will make. In concrete, this contribution results in less water required and in better concrete overall. The smooth surface texture and particle shape of the ground slag account for this benefit [16].

Other Physical Requirements

ASTM C 989, Table 1, sets forth the slag activity index limits discussed earlier. In addition, that table sets forth a fineness requirement of a maximum of 20% retained on a 45-μm (No. 325) sieve. Specific surface by air permeability is to be determined and reported, but no limits are set for that parameter. One of the benefits of grinding slag separately, as compared to intergrinding with cement at the mill, is the opportunity to grind the slag finer. Slag is harder to grind than portland cement. An interground blend will result in the cement being finer than the slag, which is the opposite to what is desired. By grinding the two materials separately, each can be ground to its optimum fineness that will result in the slag being finer than the portland cement and that is more desirable. Idorn concludes that the fineness of the slag should be about 5000 to 5500 cm_2/g and a portland cement should be about 3800 cm^2/g [17].

Quality Control Test

In 1985, ASTM Committee C-9 gave approval to ASTM Test Method for Hydraulic Activity of Ground Slag by Reaction with Alkali (C 1073). This method originally was designated ASTM E 1085 because its preparation had been started when the originating task group was under Committee E-38, but that number was quickly changed by ASTM to a C number so it could be found in the construction volumes of the test methods. ASTM C 1073 states, "This test method can be used as a quality-control test for slag production from a single source after adequate correlation with tests stipulated in Specification C 989." The test is helpful in guiding fineness level required and "in evaluating the hydraulic activity of slags from different sources."

Use of Appendixes

ASTM C 989 has three appendixes. They are X1. Contribution of Slag to Concrete Strength; X2. Sulfate Resistance; and X3. Effectiveness of Slag in Preventing Excessive Expansion of Concrete Due to Alkali-Aggregate Reaction. This nonmandatory information is helpful in interpreting the specification and in evaluating the slag when any of the preceding factors are important to a user of concrete. When ASTM C 989 was originally approved and published in 1982, the ACI Report 226.1R-87 [3] "Ground Granulated Blast-Furnace Slag as a Cementitious Constituent of Concrete," which also deals with this information, was not available. Some of the material in the appendixes is now in the ACI report, but the appendixes are helpful additions.

Chemical Requirements

Sulfide sulfur is limited to 2.5% maximum and sulfate ion, reported as SO_3, is limited to 4.0% maximum in ASTM C 989, Table 2. This is to avoid any excess amounts of these products in the concrete although there is no apparent evidence that this would affect the slag reactions. Some feared that sulfides would produce sulphuric acid and corrode rebars (although it has never happened), and others felt that SO_3 might affect the cement and provide the effect of an over-sulfated cement. The task group did not feel that the specification would be approved without some limits on sulfur. Also, ASTM C 989 provides for the determination of chloride content of the slag but has no limits placed.

SILICA FUME

History and Use

Some of the first field experimentation with silica fume in concrete in the United States was done in Kentucky in 1982. In January 1985, ASTM Subcommittee C09.03.10 organized a task group to develop a specification for silica fume as a mineral admixture in concrete. According to Richter [18], silica fume, also known as condensed silica fume, microsilica, and amorphous silicon dioxide, can be used to make concrete with the following desirable properties:

1. reduce the heat of hydration,
2. retard alkali-aggregate reaction,
3. reduce freeze-thaw effects and water erosion,
4. produce high strength,
5. increase sulfate resistance, and
6. reduce permeability.

Silica fume is almost as fine as tobacco smoke. Along with this fineness, it is mostly amorphous and is spherical in shape. The ASTM task group at first thought to add silica fume as a new class or column in the existing ASTM C 618. Test procedures from ASTM C 311 were to be used where possible and modified when needed. Prescription and performance specifications were both to be used when they gave desirable information. By late 1986, it was decided not to try to incorporate silica fume into ASTM C 618. Still, much of the thinking from that specification as well as ASTM C 311 has guided the preparation of the ASTM Standard Specification for Silica Fume for Use in Portland Cement Concrete and Mortar (C 1240).

Chemical Requirements

Silicon reported as silicon dioxide (SiO_2) is usually the active material in silica fume. The limit was set at a minimum of 85% when Richter at the Waterways Experiment Station [18] found that, with 16 different fumes with a minimum of 85% SiO_2, no other oxide except potassium oxide (K_2O) that was at 2.27% was over 2%. According to Richter, "most research has been done on silica fume produced from the production of silicon or ferro-silicon alloys. Since very little is known about the other types at the present time, the Canadian specification and ASTM are limiting silicon as SiO_2 to 85% and higher" [19]. The originating task group from ASTM decided to report elements from the chemical analysis as conventional oxides even though in some fumes none or very little of that oxide is present. Moisture content is limited to 3% maximum to avoid problems in handling. Most of the 6% LOI is due to "foreign materials" such as carbon and wood chips. These foreign materials do not come from contamination, but are present from the process that produces silica fume. The optional requirement on available alkalies comes from the other pozzolan experience in ASTM C 618. There is no limit on SO_3 because it is very low or non-existent in silica fume.

Physical Requirements

Under physical requirements, there is an oversize provision that 10% of the material may be retained on the 45-μm (No. 325) sieve. All of the silica fume will pass the 45-μm (No. 325) sieve. This retention on the 45-μm (No. 325) sieve would be an indication of foreign materials such as carbon and wood chips. With respect to the accelerated pozzolanic activity index test, it should be noted that specimens are cured at 65°C. This is not in accordance with the new and current ASTM C 311. The round-robin work done for this specification was performed before ASTM C 311 was changed in 1989. The 85% of control was set with the old test procedure at higher temperatures in mind. The note at the bottom of Table 3 in the specification points out that "Accelerated pozzolanic activity index is not to be considered a measure of the compressive strength of concrete containing the silica fume. This is a measure of the reactivity of a given silica fume with a given cement and may vary as to the source of both the silica fume and the cement."

The originating task group felt there should be uniformity requirements on density. In order to determine density, a volume displacement method was found and used in the qualifying round-robin testing. The task group also felt a need for a uniformity requirement on percent retained or "oversize." The percent retained test has a note saying, "Care should be taken to avoid the agglomeration of extremely fine material."

The task group developing the silica fume specification also felt a need to measure particle size or specific surface area even after rejecting the air-permeability test because of a wide variance of results between operators and laboratories. They decided to put the measurement of specific surface area by The Brunauer, Emmett, and Teller (BET), nitrogen adsorption method in the optional physical requirements table, even though it was voted down originally because of cost and the fact that few laboratories had the necessary equipment. Each of the other optional physical requirements have their obvious purposes and do not need further comments. All of these tests will be evaluated as more experience with them is gained.

CONCLUSION

In conclusion, the mineral admixtures dealt with in this chapter have properties that are beneficial in their use in concrete. The many tests used in the ASTM specifications discussed in this chapter are significant in controlling and defining the aforementioned properties and their use in the concrete industry. As L. H. Tuthill [1] said in his conclusion to this chapter in 1976, "Much has been done in recent years toward the development and refinement of the use of mineral admixtures. Through improved technology and a thorough knowledge of proper design of concrete mixtures to meet specific needs, the use of many admixtures previously used has been eliminated. In addition, this improved technology has resulted in more economical and advantageous uses of mineral admixtures, which have provided better concrete at lower cost. Based upon present technological advances, it is anticipated that the future will continue to show marked improvements in concrete technology, which will undoubtedly include further refinements in the use of mineral admixtures."

REFERENCES

[1] Higginson, E. C. "Mineral Admixtures," *Significance of Tests and Properties of Concrete and Concrete-Making Materials, ASTM STP 169A*, American Society for Testing and Materials, Philadelphia, 1966, pp. 543–555.

[2] Tuthill, L. H. "Chapter 46—Mineral Admixtures," *Significance of Tests and Properties of Concrete and Concrete-Making Materials, ASTM STP 169B*, American Society for Testing and Materials, Philadelphia, 1978, pp. 804–822.

[3] "Ground Granulated Blast-Furnace Slag as a Cementitious Constituent of Concrete," *Manual of Concrete Practice, Part 1—Materials and General Properties of Concrete*, ACI 226.1R-87, Committee 226, Admixtures for Concrete, American Concrete Institute, Detroit, MI, 1987.

[4] "Use of Fly Ash in Concrete," *Manual of Concrete Practice, Part 1—Materials and General Properties of Concrete*, ACI 226.3R-87, Committee 226, Admixtures for Concrete, American Concrete Institute, Detroit, MI, 1987.

[5] Mielenz, R. C., "Mineral Admixtures-History and Background," *Concrete International*, Aug. 1983.

[6] Davis, R. E., Carlson, R. W., Kelly, J. W., and Davis, H. E., "Properties of Cements and Concrete Containing Fly Ash," *Journal*, American Concrete Institute; *Proceedings*, Vol. 33, May–June 1937, pp. 577–612.

[7] Lamond, J. F. "Twenty-Five Years' Experience Using Fly Ash in Concrete," *Proceedings*, First CANMET/ACI Conference on Fly Ash, Silica Fume, Slag & Other Mineral By-Products in Concrete, SP-79 Volume I, Montebello, Canada, American Concrete Institute, Detroit, MI, 1983.

[8] Richter, R. E. and Poole, T. S., "Chemical Determination of Iron Content of Cements and Pozzolans," unpublished files of Task Group 4, ASTM Subcommittee C09.24, American Society for Testing and Materials, Philadelphia.

[9] Philleo, R. E. "Recent Developments in Pozzolan Specifications," *Proceedings*, Second CANMET/ACI Conference on Fly Ash, Silica Fume, Slag & Natural Pozzolans in Concrete, Madrid, Spain, American Concrete Institute, Detroit, MI, 1986.

[10] Manz, O. E. "Review of International Specifications for Use of Fly Ash in Portland Cement Concrete," *Proceedings*, First CANMET/ACI Conference on Fly Ash, Silica Fume, Slag & Other Mineral By-Products in Concrete, SP-79, Vol. I, Montebello, Canada, American Concrete Institute, Detroit, MI, 1983.

[11] Dunstan, E. R., Jr., "A Spec Odyssey—Sulfate Resistant Concrete For The 80's," *Proceedings*, Verbeck Memorial Symposium of Sulfate Resistance, American Concrete Institute Annual Meeting, 2–7 March 1980.

[12] Lamond, J. F. and Faber, J. H. "Fly Ash Quality Assurance and the Corps of Engineers 'Pozzolan Quality Management System'," *Proceedings*, Seventh International Ash Symposium and Exposition, Vol. I, Technical Information Center, Office of Scientific and Technical Information, U.S. Department of Energy, Washington, DC, May 1985.

[13] Lewis, D. W., Discussion of "Admixtures for Concrete," (ACI 212.1R-81), *Concrete International: Design & Construction*, Vol. 27, No. 5, May 1982, pp. 64–65.

[14] "Ground Granulated Blast-Furnace Slag," *Transportation Research Circular-Admixtures and Ground Slag for Concrete*, No. 365, Transportation Research Board/National Research Council, Washington, DC, Dec. 1990.

[15] Lea, F. M., *The Chemistry of Cements and Concrete*, 3rd ed., Edward Arnold Ltd., London, 1970.

[16] Fulton, F. S., "The Properties of Portland Cements Containing Milled Granulated Blastfurnace Slag," Portland Cement Institute Monograph, The Portland Cement Institute, Johannesburg, South Africa, 1974.

[17] Idorn, G. M., *The Effect of Slag Cement in Concrete*, NRMCA Publication No. 167, National Ready Mixed Concrete Association, Silver Spring, MD, April, 1983.

[18] Richter, R. E. "Development of ASTM Specification For Silica Fume," unpublished files of Task Group 4, ASTM Subcommittee C09.24, American Society for Testing and Materials, Philadelphia, 1988.

[19] Canadian Specification for Silica Fume, CAN/CSA-A23.5-M86, Supplementary Cementing Materials, Canadian Standards Association, Toronto Canada, July 1986.

PART VI
Specialized Concretes

47

Ready Mixed Concrete

Richard D. Gaynor[1]

PREFACE

This chapter on Ready Mixed Concrete has been a part of the original *ASTM STP 169* publication since Stanton Walker authored it in 1955 and 1966. Richard Gaynor authored the chapter in 1978. He has been a member of the ASTM Subcommittee C 09.40 responsible for ready mixed concrete specifications C 94 and C 685 since the early 1960s and was its chairman from 1974 through 1990.

This chapter follows the organization of ASTM Specification C 94 on Ready Mixed Concrete and provides the authors' perspective of the technology underlying the specification.

INTRODUCTION

The American Society for Testing and Materials (ASTM) has two specifications covering ready-mixed concrete: ASTM Specification for Ready-Mixed Concrete (C 94) and ASTM Specification for Concrete Made by Volumetric Batching and Continuous Mixing (C 685). This paper will cover both specifications, but the emphasis will be on ASTM C 94 because of its much greater use.

ASTM C 94 and C 685 are specifications for the material and concrete, as manufactured and delivered to a purchaser in a freshly mixed and unhardened state. They form the basis for a contract between a manufacturer and a purchaser [1,2].

HISTORY OF THE INDUSTRY

The first concrete mixed off-site and delivered to the job may have been furnished in 1913, but it was not a recognizable industry in the United States until the late 1920s when the first revolving drum truck mixers were developed [3]. ASTM C 94 was first published in 1933. The volumetric batching with continuous mixing specification, ASTM C 685, was published in 1971. In 1933, the ready mixed concrete industry used about 5% of the portland cement used in the United States. In the early days, the competition was with site-mixed concrete. However, as measured by the amount of total cement used in the United States, the industry's use grew from about one third to two thirds in the period from 1950 to 1975. In 1990 the industry used 72.4% of the cement [4]. At that time, the estimated United States ready mixed concrete production was 253 million cubic yards (193 million cubic meters).

The 1978 edition of this chapter on ready mixed concrete [5] estimated that there were 5000 companies in the industry and that few companies operated in more than one or two geographical areas. In the past few years, the industry has changed dramatically and companies have expanded their markets through acquisition of existing companies [6]. Although there are no reliable statistics, consolidation has taken place on both a national and a state or regional basis. The total number of companies has decreased to perhaps 3700. In recent years, the historical pattern that 10% of the companies produce 50% of the concrete has changed and now 6 or 7% of companies produce 50% of the concrete. Generally, these companies operate more than 100 truck mixers. At the present time, small companies operating less than 15 trucks account for more than half the number of companies and produce less than 10% of the concrete.

This trend in the development of larger companies also fosters greater technical sophistication and greater use of performance specifications. Increasingly, the ready mixed concrete producer is the concrete technology expert. He relies less and less on the highway department and other specifiers. His involvement in local American Concrete Institute Chapters and his state associations have helped greatly.

SPECIFICATIONS

ASTM C 94 and C 685 are specifications for the material ready mixed concrete. As such, the specifications address the separate and joint responsibilities of the various parties in a typical job—the owner, specifying agency, contractor, concrete producer, and testing agency. The ASTM C 94 and C 685 specifications are used in a number of rather different situations.

In major public and private construction, they are incorporated by reference in the job specifications. Here, the basic contract is between an owner or a specifier and a contractor. The concrete producer is a material supplier who agrees in his contract with the job contractor, to comply with ASTM C 94 or C 685.

---

[1] Executive vice president, National Ready Mixed Concrete Association, Silver Spring, MD 20910.

512 TESTS AND PROPERTIES OF CONCRETE

At the other end of the scale, in small jobs, these ASTM specifications form the basis for an agreement to furnish material for incidental construction, perhaps to a homeowner.

A third and important use is the protection of the public interest by incorporation in public building codes. In this instance, the requirements, which are binding on the owner, are then passed down to the contractor and finally to the concrete producer through a purchase order from the contractor. Inevitably, circumstances often dictate that job specification requirements differ from those in ASTM C 94 and C 685. Therefore, both specifications include the statement that if there are differences, the purchaser's specifications shall govern.

Since the ready mixed concrete producer functions as a supplier of materials and often has no binding contract with the owner, there is a need for close cooperation between the contractor and the concrete producer. In 1965, the Associated General Contractors and the National Ready Mixed Concrete Association adopted a "Joint Statement of Responsibilities" that has been found useful over the years and has been reviewed by both associations periodically and republished without change [7]. The statement addresses the separate and joint responsibilities of each party and is helpful in defining the traditional roles of each.

Basis of Purchase

An early section in both ASTM C 94 and C 685 defines a cubic yard of concrete as the basis of purchases and then describes the method of testing and calculation of the volume of fresh concrete. The volume of concrete is defined as the total weight of the batch divided by the concrete unit weight as determined by ASTM Test Method for Unit Weight, Yield, and Air Content (Gravimetric) of Concrete (C 138). The unit weight used is the average of the results of three tests made on separate samples from different loads. Although the unit weight method, ASTM C 138, permits tests in ¼ ft³ (0.19 m³) air meter bases, ASTM C 94 requires tests in standard ½ ft³ (0.38 m³) unit weight buckets.

Yield or Volume In-Place

A note explains that the volume of concrete may be less than expected due to waste, spillage, over-excavation, spreading of forms, settlement in forms, or some loss of air. Because the quantity of concrete actually used on a job will be greater than that calculated from plan dimensions, yield tests must be made early in the job and appropriate adjustment made in batch proportions if deficiencies are found.

When yield is confirmed by the unit weight test, another possible source of error is in the weights of materials batched. Scale calibration should be checked at three- to six-month intervals. If these checks are made, the concrete discharged from the truck should be within 1 or 2% of that determined by the standard yield test. Seldom do the contractors' initial estimates agree that closely with the amount delivered by the concrete producer. The practice of ordering full loads by many contractors practically ensures that 2 to 3% more will be ordered than is needed. Also, a 6-in. (150-mm) slab that is ¼ in. (6 mm) too thick will require 4% more concrete, and deflection of bar joist construction, which is not shored, can produce slabs that average a full inch too thick. In general, the ACI 117 standard on construction tolerances [8] permits slabs, beams, walls, and columns up to +½ in. (12 mm) thicker than required.

In a "Concrete Mobile" or a mobile mixer under ASTM C 685, a check on the yield is made by weighing the concrete discharged in a given number of revolutions of the cement feeder and then determining the unit weight of that concrete by ASTM C 138. The volume of concrete discharged is then the weight concrete discharged per revolution times the number of revolutions divided by the weight per cubic foot or cubic yard.

Truck Mixer Hold Back

Generally, it is realized that when concrete is batched into a clean, washed-out truck mixer, it will take 1 to 2% of a capacity batch to coat the drum and blades. This material is principally mortar with negligible coarse aggregate. Since mixers are washed out and wash water and solids discharged only at the end of the day, this does not significantly affect the volume of concrete delivered if the truck delivers four or five batches each day.

However, when a 1 yd³ (0.76 m³) batch is mixed in a clean 10 yd³ (7.6 m³) capacity truck mixer, the effect on concrete proportions can be dramatic. Compared to the batched proportions, if the amount of mortar retained is equal to 2% of the weight of a capacity batch, the cement content of the concrete discharged will drop from 600 to 470 lb/yd³ (356 to 280 kg/m³), and the sand as a percentage of the total aggregate will decrease from 36 to 26%. The batch is harsh and has low-cement content and low strength.

The solution is to increase cement, sand, and water weights up to about 40% in such small truck-mixed batches. If a clean tilting central mixer is used, the effect on proportions is much more dramatic since the gross volume of the drum is about 25% larger than a truck mixer, the surface to be coated is much greater, and the hold back perhaps 50% greater.

ORDERING INFORMATION

Both ASTM specifications include the fundamental elements of prescription and performance specifications and contain the basic information needed by both parties. However, the purchaser often adds additional requirements. Sometimes this is an attempt to ensure quality, perhaps because of past bad experiences or in an effort to provide characteristics that were not readily available in standard mixes. The growing number of types of chemical and mineral admixtures permit the concrete producer to produce concrete with properties quite different from the "standard mixes" in common use a few years ago. Some of the newer applications are flowing concrete, anti-wash-out concrete, corrosion-inhibited concrete, low-permeability corrosion-resistant concrete, and high-performance concrete. Concrete with three or even five admixtures is becoming more common. Control of such mixtures has become a significant challenge for concrete producers.

The Ordering Information section requires the purchaser to specify the size of coarse aggregate, slump, air content, unit weight of structural lightweight concrete, and one of three options; A, B, or C. The options are:

Option A is the performance format where the purchaser specifies strength and the producer selects proportions of ingredients.

Option B is the prescription format where the purchaser specifies cement content, maximum water content, and admixtures.

Option C is a mixed format where the purchaser specifies minimum cement content, required strength, and admixtures, if required.

Water-Cement Ratio in Specifications

The A, B, and C options do not include a requirement for a maximum water-cement ratio even though Option B does include both cement and water content. The principal reason that it is not included is the difficulty of actually measuring the water-cement ratio in practice. From the users point of view, a water-cement ratio is needed to ensure durability, that is, resistance to freezing and thawing, deicer salts, sulfate solutions, or intrusion of chlorides in reinforced concrete. The maximum water ratios usually cited are those required in the "Building Code Requirements for Reinforced Concrete" (ACI 318) [9] or "Specifications for Structural Concrete for Buildings" (ACI 301) [10] that range from 0.50 to 0.40. In many high-performance concretes, it is possible to produce concretes with water cementitious material ratios as low as 0.30 or even 0.28 and values as low as 0.35 are being specified in parking structures for resistance to intrusion of chlorides.

The concrete producer has difficulty conforming to a maximum water-cement ratio since the amount of water necessary to produce a given slump will vary with delivery time which depends on traffic delays and variations in the job-placing operations. Additionally, the moisture content of the aggregates will vary from batch to batch making accurate measurement difficult. Another difficulty is that the procedures used by specifiers to determine compliance with maximum water-cement ratio specifications is rarely defined; and if every batch is to be checked, the production schedule will be disrupted.

In practice, specifiers use a number of systems of enforcing maximum water-cement ratio specifications. One of the simplest is to require submission of laboratory trial batch data demonstrating compliance with the specified maximum water-cement ratio and then relying on routine strength tests to determine compliance. This system breaks down if the specified strength is not reasonably consistent with the strength obtained at the desired water-cement ratio. Increasingly large agencies are requiring 2 to 4 yd^3 (.76 to 1.5 m^3) trial batches instead of laboratory trial batches to improve the accuracy of the mix approval process.

If the water-cement ratio is to be strictly enforced in the field, then a specified maximum water-cement ratio is inappropriate and as unrealistic as an absolute minimum strength specification!

On well-controlled jobs, the standard deviation of the water-cement ratio ranges from 0.02 to 0.03. A concrete producer would have to furnish concrete with an average water-cement ratio of as low as 0.35 or even lower to avoid batches that have calculated water ratios greater than 0.40!

The writer's opinion on water-cement ratio specifications is that they should not be used and do not, in themselves, provide better assurance of "durability" than other much more reliable, more accurately measured characteristics such as strength. This issue is one that has been under considerable discussion, particularly in recent years [11,12]. The purchaser is better served by specifying concrete with adequately high-compressive strength and, of course, air content if the durability concern is freezing and thawing or deicer scaling.

Under all three of the optional methods of specifying proportions in ASTM C 94 and C 685, if the purchaser requests it, the manufacturer is required to furnish proportions of ingredients that will be used. This information is related to that required to be furnished on the delivery ticket.

This system of requiring submittal and "approval" of concrete proportions is firmly embedded in specifications and codes in the United States. It started prior to about 1940 when virtually all concrete was furnished as prescription or cement content mixes—not strength.

Today, most mixes have specified design strengths. Once mix proportions are identified, it is difficult to reduce cement content without generating opposition or suspicion from specifiers or contractors. The net result is that there is little incentive for the concrete producer to implement a quality control program that is designed to control strength at the specified level, and the concrete is furnished with more cement and much higher strength than is actually needed.

If the producer was free to vary proportions to produce the required performance, the incentive would exist to invest in sophisticated quality control based on accelerated strength testing and increased emphasis and monitoring of the quality and uniformity of the process at all stages for the consistent materials, batching, delivery, placing, and testing. Adoption of performance-based requirements and elimination of prescriptive mix proportions are necessary before concrete suppliers will have the incentive to adopt more formal quality control systems.

There is, in the American Concrete Institute (ACI) Building Code ACI 318, a provision that permits a producer to recalculate the required over-design when he has accumulated 15 tests and to make adjustments in proportions to conform to the required average strength level. This is at least a beginning for performance specifications.

MATERIALS

This section includes references to the commonly used ASTM specifications for cement, aggregates, admixtures, and water. A number of materials such as fibers, expansive cement, corrosion inhibitors, and silica fume are not included, either because there is no ASTM specification or because there is a feeling that the use of the material requires different batching, mixing, delivering, or testing procedures than those for "normal" concretes.

Water Quality

Water that is clear and clean without objectionable color, smell, or taste is acceptable. Other waters are subject to tests for strength and time of set.

Because of increasingly restrictive governmental regulations on disposal of wash water, yard runoff, and returned concrete, producers are increasingly reusing wash water as mixing water and are considering incorporation of returned concrete or the partially hydrated cement from returned concrete in freshly mixed concrete.

The specifications permit the use of water from mixer wash-out operations as mixing water in concrete if the water does not substantially reduce strength or affect time of set. Additional limits are set on the amount of chlorides, sulfates, alkalies, and total solids in the water. Research and industry experience with the use of wash water has been that effects on product quality are not significant or important [13–16]. As currently written, ASTM C 94 would permit up to 50 000 ppm of solids in mixing water. Typically, this permits up to about 15 lb/yd^3 (9 kg/m^3) of dry solids. Available data indicate that up to about twice this amount of hydrated cement solids can be permitted without detrimentally affecting concrete properties [17].

A note in ASTM C 94 directs attention to the fact that the amount of air-entraining admixture required to produce the required air content may increase and even double, if these admixtures are added to high pH wash water before the water is batched into the mixer. The air-entraining admixtures should be added either with the sand or with an increment of clean mixing water. Although it does not appear that similar problems are experienced when other commonly used accelerators, retarders, or water reducers are batched into or at the same time as wash water, this possibility should not be ignored with the increasingly sophisticated admixture systems that are being developed and used.

SLUMP AND AIR CONTENT

The section on slump in ASTM C 94 contains tolerances in two different specification formats. It also establishes a 30-min period after arrival on the job during which the producer is responsible for the slump. The job addition of high-range water reducers to produce flowing concrete has created a number of field control problems when the slump of the flowing concrete is subject to strict slump control and testing. A preferred procedure is to specify the slump of the concrete before addition of the high-range water reducer and to accept the fact that the slump of the flowing concrete will vary.

The section on air content contains a table of recommended air contents taken from ACI Committee 211 [18]. The delivery tolerance is ±1.5% air.

There have been a number of problems that have surfaced recently in the control and measurement of air content. Over ten years ago, Meininger [19,20] described tests that indicated that some fly ashes contain carbon that adsorbs air-entraining agent and reduces concrete air content during delivery. More recently, Ozyildirim [21] and Hover [22] have made studies to determine if the ASTM pressure meter accurately measures the air content of plastic concrete and whether the pressure method provides a reasonable estimate of the amount to be expected in hardened concrete. These issues are the subject of other chapters in this book. However, the answer generally appears to be that ASTM Test Method for Air Content of Freshly Mixed Concrete by the Pressure Method (C 231) does provide the desired measurement and that the air content of hardened concrete is often either lower or higher for a multitude of reasons.

Another issue is the observation that much of the initial air can be lost during pumping, particularly with the newer long boom pumps with 5-in. (125-mm) lines. The author believes that this occurs when a section of the boom is essentially vertical and concrete slides down from its weight and develops a vacuum in the pump line. The air bubbles then expand and fail to reform when the concrete drops out the end of the pump line or impacts an elbow in the boom [23]. The solution has been to insert some resistance in the line. This can be as simple as inserting a loop in the flexible hose or laying a length of hose on the deck. Under normal conditions, the loss of air in pumping should not exceed 1 or 1.5%.

Another problem is the development of surface delaminations and blisters when air-entrained concrete is steel troweled. This occurs in industrial floors with mechanical and vibrating screeds that are mechanically troweled. Although a number of factors are involved, the solution is to avoid the use of entrained air in industrial floors or, at least, to keep the air content below 3 or 4%.

BATCHING AND MEASURING MATERIALS BY ASTM C 94

ASTM C 94 requirements for measuring materials recognize both individual scales and hoppers for weighing a single material and cumulative scales and an associated hopper for weighing more than one material. This system, and the terminology that is used, developed from the use of dial scales where the material weights were accumulated as materials were weighed. With the increasing use of computerized batching equipment, it has now become possible to have the computer do the subtraction and print the weights of the individual materials in a cumulative batcher, including a recognition of the zero reading or tare. A note is needed in ASTM C 94 to explain the fact that cumulative tolerances apply in cumulative batchers irrespective of the format used to report the batch weights.

Cement and Pozzolans

Cement and pozzolans may be weighed cumulatively on a single scale provided the fly ash is weighed after the cement. The weighing tolerance is ±1% of the required cumulative weight. For small batches where the cumulative weight is less than 30% of scale capacity, the tolerance is from −0 to +4%. Since fly ash tends to flow through small cracks and openings, the ash is required to be weighed after cement. This tends to ensure the correct cement weight and, if anything, an excess of fly ash. Cumulative weighing of cement and fly ash also has the advantage that the batcher tends to rathole when it discharges into the mixer and the fly ash blends with the cement as it is loaded. There have been a few instances when individual separate batchers were used for cement and fly ash and the fly ash wound up in one part of the batch, separate from most of the cement. When separate batchers are used, which is rare, the batcher discharge must blend the two materials.

Although it is not addressed in ASTM C 94, many concrete plants contain several cement and fly ash silos; and there is always the possibility that material can be placed in the wrong silo. Producers generally use different colored fill pipes, signs, and distinctively colored bills of lading to distinguish between materials. In some cases, it is possible to locate fill pipes at different plant locations. Some producers use keys and locks on fly ash pipes and control access to the keys. Because fly ash tends to flow freely through cracks, some specifications [24,25] do not permit common walls between multi-compartment cement and fly ash silos. The space between these double-walled bins needs to be free-draining with access provided for inspection.

Aggregates

Aggregates can be weighed either in cumulative or individual weigh batches. The basic batching tolerance in cumulative batches is 1% of the cumulative weight and 2% of the required amount in individual batchers. For cumulative weights less than 30% of scale capacity, the tolerance is ±0.3% of scale capacity or ±3% of the required cumulative weight, whichever is less. This means that at less than 10% of scale capacity, the 3% tolerance will govern.

Mixing Water

ASTM C 94 defines mixing water as any water added to the batch plus surface moisture on aggregates, ice, or liquid admixtures. The tolerance on batching added water is ±1% of the total water content, not a percentage of that being batched. The tolerance on the total water is ±3% of the total water content. This does mean that if a water meter is accurate to about 10 lb (4.5 kg) or 1 gal (3.78 L), and the nominal total water content is 280 lb/yd^3 (166 kg/m^3), then the minimum batch size that can be batched is about 3 yd^3 (2.3 m^3). The NRMCA Checklist [26] establishes a tolerance of ±1.5% on added water and is considered more realistic. When water is weighed, the present ASTM C 94 tolerances are more easily achieved; but the ultimate importance to concrete quality is questionable.

Admixtures

Admixtures are rarely batched by weight, except when they are added in prepackaged fixed amounts. Volumetric dispenser systems are becoming more sophisticated and capable of being integrated into plant automation. Admixtures are required to be batched to within ±3% of the desired amount or plus or minus the amount required per sack of cement, whichever is greater.

It should be noted that on small batches of lean concrete the admixture batching accuracy may be as large as ±25% of the amount batched. Even in a 10-yd^3 (7.6-m^3) capacity batch, if the dosage rate is ½ oz/100 lb (0.3 cm^3/kg) and the dispenser is accurate to ±1 oz (30 cm^3), the overall batching accuracy will be ±5%. However, variations of this size would not change the air content, setting time, or concrete strength. In a 2-yd^3 (1.5-m^3) load, the effects will be significant.

BATCHING PLANT

The requirements of ASTM C 94 for the batching plant are relatively basic and straightforward. Scales are required to be accurate to ±0.4% of the capacity of the scale when tested with standard weights. A recent revision of ASTM C 94 will require testing the scale at its quarter points. Over the past 10 years, *NIST Handbook 44* [27] has been extensively revised and, in 1991, the reference in ASTM C 94-90 was being updated. The scale accuracy requirements of the NRMCA Check List [26] are ±0.2% of capacity and those of the Concrete Plant Manufacturer's Bureau (CPMB) [25] for new equipment are ±0.1% of capacity.

Scales in concrete plants generally consist of a container supported at the four corners and transmitting reduced loads through a system of levers to a calibrated beam or springless dial-indicating device. Load cells are beginning to be incorporated in the supporting lever system to measure the load.

Use of load cells to support the batchers has the potential to simplify a scale by eliminating much of the lever system and the dial scale that has typically been used. Load cells have not been widely accepted since the dial scale provides a mechanical backup in case of a failure of the electronic load cell system in the middle of a pour.

ASTM C 94 does not define or require different types of batching controls. The CPMB Standards [25] provide a consistent terminology for plant control systems.

A manual control is one that is operated manually and is dependent on the operators' visual observation of the scale or meter. A semi-automatic control is one that, when started, stops automatically when the required weight has been reached. A semiautomatic interlocked control is similar but contains provisions to prevent discharge of the device until the material is within tolerances.

An automated control starts and stops automatically and includes interlocks to:

(a) prevent charging until the scale returns to zero,
(b) prevent charging if the discharge gate is open,
(c) prevent discharging if the charging gate is open, and
(d) prevent discharge until the material is within tolerances.

A batching system consists of the required combination of individual batchers. The CPMB Standards define the following types:

1. Manual—a combination of manual batchers except that water or admixture batchers may be semiautomatic or automatic.
2. Partially automatic—includes at least one automatic or semiautomatic batcher. Interlocks are optional.
3. Semiautomatic batching—a system of semiautomatic interlocked or automatic batchers and volumetric devices where the interlocks, other than those required for individual batchers, are optional.
4. Automatic batching—requires a combination of automatic devices:

   (a) that must start with a single starting mechanism, except water or admixture not batched at the same time, and may have separate starting devices;
   (b) where each batcher must return to zero within tolerance and reset to start; and
   (c) where discharge of any ingredient may not start until individual batchers have returned to zero and all weighed ingredients have been batched within tolerance.

The development of computerized batching systems and control panels have made great strides, and the cost of such systems continues to decline. These have greatly simplified the process of recording batch weights and delivery tickets.

Recorders

The CPMB Standards define both digital and graphical recorders. Graphical recorders were once popular for large jobs where one mix design was produced for extended periods. This made it easy to identify mistakes and malfunctions. More recently, digital recorders have become increasingly popular, particularly in computerized batching systems. They print delivery tickets and can provide information for billing and inventory control. The CPMB Standards require that if target weights, simulated weights, or other than actual batch weights are recorded, they must be clearly identified. Digital recorders must reproduce the scale reading within 0.1% of scale capacity or within one increment on volumetric devices.

MIXING CONCRETE

There are three types of mixing operations defined in ASTM C 94:

a. Central mixing where concrete is mixed in a central plant mixer and delivered to the job in a revolving drum truck, an agitator or nonagitating unit.
b. Shrink mixing where concrete materials are blended in a central plant mixer with the mixing completed in a revolving drum truck mixer.
c. Truck mixing where ingredients are loaded into a revolving drum truck mixer for mixing and delivery.

The Plant Mixer Manufacturers Division Standards (PMMD) [28] define four principal types of concrete plant mixers:

1. Tilting mixers—revolving drum mixers that discharge by tilting. This is the type used in most ready mixed concrete plants. Standard sizes range from 2 to 15 yd³ (1.5 to 11.5 m³). The rated capacity ranges from 30 to 45% of gross drum volume.
2. Vertical shaft mixers—These mixers have an annular mixing compartment with rotating blades or paddles. The mixing compartment or pan may rotate or not. Generally, they discharge through a door or hatch in the bottom of the "pan." They were popular 25 years ago because of their low overall height and rapid mixing, but because of rapid wear are little used today.
3. Nontilting drum mixers—These are revolving drum horizontal axis nontilting mixers. They are little used today.
4. Horizontal shaft mixer—These have a horizontal cylindrical mixing compartment with blades or paddles rotating about the horizontal axis. A number of new designs have been introduced recently although the general concept is not new.

In addition to the recognized PMMD types, there has been considerable interest in several other high-energy mixing designs that are reported to provide more efficient mixing at higher energy levels, presumably shear rates, that result in more efficient dispersion of cement and produce significant cement savings [29].

Central Mixing

The choice between central or truck mixing depends on a large number of factors. The technical advantage of central mixing is that it provides centralized control of the mixing process and requires a less-skilled truck mixer operator. However, the CPMB study of economic factors [30,31] shows that the decision will depend on a number of other factors, including the market area, market volume, blade life in trucks, and truck utilization. Although there are no recent data on the percentage of the ready mixed concrete produced by central mixing, it is likely that it is in the middle or lower end of the historical range of 20 to 25%.

The principal advantages of truck mixing are lower capital investment, lower plant heights, lower electrical costs, and somewhat greater flexibility when long deliveries are required in rural areas. With special loading sequences designed to keep cement essentially dry until the concrete is mixed at the job site, loss of slump, and use of retempering water can be avoided [32].

In the mid 1960s, the Federal Highway Administration (FHWA) [33,34] conducted mixing efficiency tests of tilting drum central mixers. Generally, they demonstrated that 45 to 90 s mixing times were feasible if care were taken to blend or ribbon-load all ingredients as they entered the mixer. The importance of the loading sequences is likely much less if, as in most ready mixed concrete operations, the concrete will be transported in a revolving drum truck mixer. As will be noted later, loading sequences to ensure uniform mixing in truck mixers must not attempt to achieve uniform ribbon loading or blending of ingredients.

Although most State Departments of Transportation (DOTs) permit 60 or 90 s mixing time in central mixers, ASTM C 94 still requires a minimum of 1 min mixing for the first cubic yard and an additional 15 s for each additional cubic yard. This minimum can be reduced if mixing uniformity criteria are met after shorter periods.

Shrink Mixing

Early in the development of the ready mixed concrete industry, shrink mixing was designed to permit hauling a larger batch in a truck mixer. The idea was to partially mix and shrink the volume of concrete before it was placed in a truck mixer for final mixing. Only a small percentage of producers use the system today.

Truck Mixing

Two general types of inclined axis truck mixers are in use today: the traditional rear discharge unit, and the newer front discharge unit. Because the front discharge unit requires a special truck chassis, it tends to be a significantly more expensive unit than a rear discharge unit. With a rear discharge unit, the mixer can be more easily positioned on the truck chassis to comply with the truck weight laws of the various states.

Many contractors prefer front discharge units because the truck driver can drive into the job with little direction from contractor personnel, control chute movement, and discharge from within without leaving the truck cab.

The requirements of ASTM C 94 for truck mixers are that the volume of mixed concrete not exceed 63% of the gross drum volume for truck or central mixed concrete or 80% for central mixed concrete.

The extra carrying capacity of central mixed concrete is much less of a consideration with present weight laws and today's 9 to 11 yd^3 (7 to 8.5 m^3) mixers. When units were only 5 or 6 yd^3 (4 to 4.5 m^3) capacity, the angle of inclination of the drum was greater and mixers would hold 80% of the gross drum capacity, even in hilly areas.

The larger drums with larger drum openings and lower angles of inclination used today are not able to transport the rated agitator capacity without spillage.

Front discharge units tend to have much larger gross drum volumes than rear discharge units for two principal reasons. In a quick stop, a front discharge unit is much more likely to spill concrete and the extended cylindrical section over the truck cab tends to be relatively ineffective in mixing concrete. The result is that the manufacturer's rated mixing capacity may be less than half the gross drum volume.

ASTM C 94 requires revolution counters on truck mixers and that the mixer be capable of mixing concrete in 70 to 100 revolutions. The limits set for mixing uniformity include tests for:

1. air-free unit weight of concrete,
2. air-free unit weight of mortar,
3. air content,
4. slump,
5. coarse aggregate content,
6. and 7-day compressive strength.

Acceptable performance requires compliance with five of the six tests. Two samples are taken after discharging approximately 15 to 85% of the load. For central mixers, the samples can be taken during discharge or directly from the mixer at points approximately equidistant from the front and rear of the load.

Slump tests of samples taken after the discharge of 15 and 85% of the load can be made as a quick test of the probable degree of mixing uniformity.

The specifications also require mixers to be examined or weighed routinely to detect accumulations of hardened concrete.

Generally, it has been found that when blade wear or accumulations of hardened concrete have become significant enough to affect mixing uniformity, discharge performance will also have deteriorated enough to be noticeable, particularly with moderately low slump concrete.

CONTROL OF THE ADDITION OF WATER

ASTM C 94 recognizes that concrete loses slump with the passage of time and that either water will have to be added to restore slump or the slump on initial mixing will have to be higher than that required at the job. The rate at which slump is lost depends on a great number of factors, including concrete temperature, properties of the cement, and the admixtures used. The literature on this issue is extensive [35–38].

Under ASTM C 94, within the limits set by the maximum water-cement ratio, water can be added only once on arrival at the job. Because of the difficulty of measuring aggregate moisture and inevitable traffic delays, concrete is generally shipped from the plant at a slump less than the specified maximum. Repeated tempering or retempering, especially during discharge, should not be permitted. However, initial tempering to obtain the desired slump is routinely necessary. The "one" addition of water permitted in ASTM C 94 should not be taken too strictly and the driver should be permitted to adjust the amount added over a period of perhaps 3 to 5 min on arrival at the job site.

The specification requires 30 revolutions at mixing speed to ensure incorporation of the water. Tests by NRMCA confirm that the 30 revolutions are necessary if mixing is at less than about 15 rpm [39]. However, if mixing is at 22 to 25 rpm as few as 5 or 10 revolutions will be sufficient.

The specifications further require that discharge be completed within 90 min after the cement is wetted or before the drum has completed 300 revolutions, whichever comes first. Both can be waived by the purchaser if the concrete can be placed without the addition of water. The 90-min time limit has been one of the most controversial requirements in the specification. Recently, an attempt was made to allow 90 min to the start of discharge with the provision that the concrete could be used as long as the slump was acceptable for placement and no water was added. The proposal was not accepted by the ASTM Subcommittee. However, in the author's opinion, it should have been. Both field and laboratory data demonstrate that concrete strengths tend to improve with time, but *only* when water is not added.

Clearly, the 90-min time limit is too conservative when the concrete temperature is less than about 70°F (21°C) [40]. At higher temperatures, the time limit can only be justified by a concern that the prohibition against water addition, after the adjustment on arrival at the job site, can not be or will not be enforced.

In recent years, the 300 revolution limit has created difficulty when high-range water-reducing admixtures are added at the job site. This is likely to happen if the slump of the concrete is specified and tested both before and after addition of the high-range water reducer (HRWR). When this is done, the HRWR must be added carefully in several increments to avoid exceeding the maximum slump and a large number of drum revolutions will accumulate.

A better solution for job site additions is to bring the concrete to the necessary 2 or 3 in. (~50 to 75 mm) slump on arrival at the job and then add a designated amount of HRWR and accept the fact that the final slump will vary, perhaps from 6 to 10 in. (~150 to 250 mm). Recognize that the accuracy of the slump test deteriorates at these high slump levels.

Except for a very few soft aggregates, the 300 revolution limit is of no practical consequence. With soft aggregates that are subject to grinding, the effect will be to decrease slump or increase the mixing water requirement. Sand is more subject to grinding than coarse aggregate because of its large surface area. The 300 revolution limit was developed many years ago when mixers were powered by separate mixer engines and had only one basic drum speed, about 6 rpm. The limit on revolutions tended to control delivery time.

The ASTM specifications recognize that many of the requirements such as the 90-min time limit can be waived by the specifier. This is the type of thing that should be discussed at a pre-job conference with the specifier, contractor inspection agency, and concrete producer. A sample Pre-pour Conference Agenda developed originally by the D.C. Ready Mixed Concrete Producers is available [41].

Volumetric Batching and Continuous Mixing

The Concrete Mobile is the only equipment available in the United States that produces concrete under ASTM C 685. The equipment is truck or trailer mounted and consists of bins for sand, coarse aggregate, cement, and water. The aggregate bins have a longitudinal belt at the bottom that moves material to calibrated gates that control the rate of material flow. The aggregate belt is geared to a rotary vane feeder in the cement bin. Water and admixtures are controlled with valves and flow meters. Production rate is inversely proportional to cement content. The materials are fed into a rubber-lined inclined chute that contains a mixing auger. The auger mixes and elevates the concrete. The units are available in sizes from 4 to 12 yd^3 (3 to 9 m^3). The mixer holds about 2 ft^3 (0.06 m^3) of concrete and mixing time is 15 or 20 s. Slump is readily adjustable and the unit can be started and stopped as concrete is needed. ASTM C 685 requires calibration at six-month intervals.

The unit is versatile and has been used in a variety of work, generally where the amount of concrete is relatively limited. In somewhat larger work where requirements are less than about 40 yd^3 (30 m^3) per hour, it has been used as a job-site mixing plant.

The outstanding advantage of the unit is that it produces freshly mixed concrete. Therefore, strengths and other properties are improved. The units do encounter problems with false setting cements because of the very short mixing time.

SAMPLING AND TESTING

Testing Laboratories

Both ASTM C 94 and C 685 contain requirements that the individual who samples and tests the concrete be qualified and knowledgeable in the proper conduct of the test procedures required. This requires an ACI Concrete Field Testing Technician, Grade I certification, or perhaps an equivalent process. The concept of requiring demonstrated knowledge of the test procedures has grown rapidly. By 1991, over 25 000 ACI Grade I Field Testing Technicians had been certified. The ACI and the National Institute for Certification of Engineering Technicians (NICET) programs are becoming increasingly popular.

For many years, testing laboratories have been required to conform to ASTM Practice for Use in the Evaluation of Testing and Inspection Agencies as Used in Construction (E 329). In 1991, a change was accepted to require conformance with ASTM Practice for Laboratories Testing Concrete and Concrete Aggregates for Use in Construction and Criteria for Laboratory Evaluation (C 1077). In the past, conformance with ASTM E 329 has rarely been enforced by users of either ASTM C 94 or C 685. To assure reliable testing, building codes and specifiers must require laboratories that conform to ASTM C 1077.

Within ASTM Committee C9, a principal objection to both ASTM E 329 and C 1077 is that they require a registered professional engineer in charge of the laboratory. At the present time, there are three national agencies that inspect or certify concrete testing laboratories:

1. AASHTO—American Association of State and Highway Transportation Officials in Washington, District of Columbia.
2. A2LA—American Association of Laboratory Accreditation, in Gaithersburg, Maryland.

3. NVLAP—National Voluntary Laboratory Accreditation Program, in Gaithersburg, Maryland.

Although the AASHTO program is new and growing, none of the programs has been widely accepted or required by specifying authorities. The one exception is that of the Concrete Materials Engineering Council, in Orlando, Florida. The program is required by the Florida Building Codes and is supported by the concrete industry, Engineering Testing Laboratories, and the Florida Department of Transportation. In 1990, 65 laboratories throughout the state were certified and accredited to comply with ASTM C 1077.

Sampling

Under ASTM C 94, concrete samples, except those for uniformity testing, are required to be taken in accordance with ASTM Practice for Sampling Freshly Mixed Concrete (C 172). Generally, this means the sample should be taken at two or more regularly spaced intervals "during discharge of the middle portion of the batch." In practice, this is seldom done and the sample is taken as a single increment near the start of discharge. The procedure in ASTM C 685 permits sampling any time after at least 2 ft^2 (0.06 m^3) has been discharged.

Sampling at a single point during discharge should be permitted in ASTM C 94 and C 172. The risk is that when cement is the last ingredient loaded in the mixer and the concrete is not well mixed, the first concrete discharged can have high strength and will not be representative of the majority of the batch.

Compressive Strength Testing

ASTM C 94 requires air and slump tests when strength specimens are made. If either falls outside the specifications, a retest is required before the concrete is considered to have failed.

Both ASTM C 94 and C 685 require that strength specimens be made in accordance with ASTM Test Methods of Making and Curing Concrete Test Specimens in the Field (C 31), cured under standard moist curing procedures, and tested by ASTM Test Method for Compressive Strength of Cylindrical Concrete Specimens (C 39). A test is defined as the average of results from two cylinders made from the same sample and tested at the same age. ASTM C 31 permits use of cylinders smaller than 6 by 12 in. (150 by 300 mm) only when required by the project specifications. Increasingly, 4 by 8 in. (100 by 200 mm) cylinders are being used for concretes with specified strengths greater than about 8000 psi (55 MPa). This is because very few compression testing machines are available with load capacities greater than 300 000 lb (136 000 kgf) that is, 10 600 psi (73 MPa) on a 6 by 12 in. (100 by 300 mm) cylinder.

Several years ago, an unsuccessful attempt was made to modify the section of ASTM C 94 that required discarding the result of a test of a single cylinder if it "shows definite evidence, *other than low strength*, of improper sampling, handling, curing, or testing." A proposal was made to set a limit on the range of pairs of cylinders that are averaged for a test and permit discarding the low value.

The average within-test coefficient of variation of cylinder tests, for strength levels less than about 8000 psi (55 MPa), is about 2% [42,43], and 4% is a large value.

When the range between two cylinders exceeds 5.5% of their average more than 1 time in 20, the test should be considered suspect.

1. If the higher of the two values is more like the other tests of this concrete, and the 7-day result is normal, discard the lower result. If neither cylinder is unusually high or low, average the two and accept the result.
2. And whether or not the lower or higher test is disregarded, consider the possibility that the sampling and testing may be poor (see Table 1 [44]). Testing with a within-test coefficient of variation of 4% will have one range in three exceeding 5.5%! Improve testing.

Table 1 can be used to evaluate cylinder testing data, but recognize that whether a single low test is sufficiently unusual to discard the result depends on "normal" quality of the testing on that job. What this means is that when 15, 30, or 60 pairs of cylinders have been tested, calculate the ranges and convert them to a coefficient of variation.

$$C_v = (R)(0.8865)/x$$

where

C_v = coefficient of variation, %;
R = average range, psi;
0.8865 factor for a range of two cylinders;
0.5907 factor for a range of three cylinders; and
x = average strength.

If the coefficient of variation is much over 2%, realize that many, if not most, jobs (laboratories) can do better. Note also that if you have less than 60 pairs of results, the frequency of large ranges can be larger than indicated in Table 1. With only ten results, the ranges in Table 1 would be about 10% larger.

There is some concern that the within-test coefficient of variation may be larger for very high strength concretes;

TABLE 1—Distribution of Ranges of 2 or 3 Cylinders that are Averaged to Constitute "A Test."

	Within Test C_v, %[a]			
	2% = Average Lab		4% = Poor Lab	
Frequency	n = 2	n = 3	n = 2	n = 3
	RANGE SHOULD NOT EXCEED % OF AVERAGE			
1:10	4.7	5.8	9.3	11.6
1:20	5.5	6.3	11.1	12.5
1:100	7.3	8.2	14.6	16.5
1:1000	9.3	10.1	18.6	20.3
	RANGE SHOULD NOT BE LESS THAN, % AVERAGE			
1:10	0.4	1.2	0.7	2.5
1:5	0.7	1.8	1.4	3.6

[a] Population value.

but in the writer's opinion, this is the result of problems of capping, curing, and testing these concretes. With careful attention, it should be possible to obtain a within-test coefficient of variation of 2%. However, sulfur mortar caps must be less than 3/32 in. (2.4 mm) thick and probably should not be permitted on concrete with strength greater than about 10 000 psi (70 MPa). Neat cement caps are preferred and often ground ends are not sufficiently flat to give optimum results.

In 1989, ASTM C 94 and C 685 were revised to conform to the two acceptance criteria used in the ACI 318 building code for almost 20 years:

1. the average of any three consecutive strength tests should be equal to or greater than the specified strength, $f'c$; and
2. no individual strength test should be more than 500 psi (3.4 MPa) below the specified strength.

A table provides advice on the "over-design" needed to meet these requirements, depending on the standard deviation expected. The values given have been calculated from the equations given in ACI 318–89 [9], with correction for instances where the standard deviation is calculated from less than 30 tests.

Failure to Meet Strength Requirements

A section in both specifications requires that, if the concrete was properly tested, the manufacturer and purchaser confer to see if they can agree on what adjustments, if any, should be made. If they do not agree, then a decision is to be made by a panel of three engineers.

NRMCA *Publication 133* [45] outlines an orderly and deliberate process for determining the strength of the concrete in the structure and for developing information on the assignment of responsibility for deficiencies, if any. The practice suggests that if the cause of the low strength was improper testing, the party responsible should bear the cost of the investigation.

CLOSURE

In the earlier version of this chapter, the author expressed the belief that great strides would be made in the use of formal quality systems. The British have made significant progress in implementing their "Quality Scheme" for Ready Mixed Concrete [46,47].

The development and acceptance of quality control in the ready mixed industry that was expected when this chapter was originally written almost 15 years ago has not occurred in the United States. However, once again, new concepts that hold great promise are in the offing. New computerized quality control software is available. The U.S. Department of Commerce Malcolm Baldrige Award and the development of "total quality management" concepts with greater worker commitment and productivity with customer focus could change the ready mixed concrete industry. Traditional quality control will be only a small, but important process to measure the success of the quality management system. The FHWA and AASHTO along with industry are about to approve a "National Policy on Highway Quality" that, among other things, would reduce adversarial relationships between specifiers, contractors, and the industry. The Associated General Contractor's Association has a "partnering program" that attempts to avoid court claims by fostering agreements that will settle job difficulties on the job promptly at the lowest possible level of management. All of these promise to increase productivity and increase efficiency. If we can get everyone to accept responsibility for quality, not just the quality-control department, things will change greatly! [48].

REFERENCES

[1] "Cement and Concrete Terminology, SP-19 (90)/ACI 116R01-90," *ACI Manual of Concrete Practice, Part 1*, American Concrete Institute, Detroit, MI 1991, p. 116R-1.

[2] "3273 Ready Mixed Concrete," *Standard Industrial Classification Manual*, U.S. Department of Commerce, Bureau of the Census, Washington, DC, 1987, p. 169.

[3] *Pictorial History of the Ready Mixed Concrete Industry*, National Ready Mixed Concrete Association, Silver Spring, MD, 1964, p. 44.

[4] "Cement in 1990," *Mineral Industry Surveys*, U.S. Department of the Interior, Bureau of Mines, Washington, DC, July 1991.

[5] Gaynor, R. D., "Ready-Mixed Concrete," *Significance of Tests and Properties of Concrete and Concrete-Making Materials*, ASTM STP 169B, American Society for Testing and Materials, Philadelphia, 1978, pp. 471–502.

[6] Fredericks, N. J., "Foreign Ownership of U.S. Ready Mix Firms," *Concrete Products*, June 1991, pp. 27–29.

[7] "Joint Statement of Responsibilities," National Ready Mixed Concrete Association and Associated General Contractors of America, Silver Spring, MD, Jan. 1980, p. 1.

[8] "Standard Specifications for Tolerances for Concrete Construction and Materials (ACI 117-90)," *ACI Manual of Concrete Practice, Part 2*, American Concrete Institute, Detroit, MI, 1991, p. 117–1.

[9] "Building Code Requirements for Reinforced Concrete (ACI 318-89) and Commentary—ACI 318R-89," *ACI Manual of Concrete Practice 1991, Part 3*, American Concrete Institute, Detroit, MI, 1991.

[10] "Specifications for Structural Concrete for Buildings (ACI 301-89)," *ACI Manual of Concrete Practice 1991, Part 3*, American Concrete Institute, Detroit, MI, 1991.

[11] Mather, B., "How to Make Concrete That Will Be Immune to the Effects of Freezing and Thawing," Paul Klieger Symposium on Performance of Concrete in Aggressive Environments, ACI Convention, 29 Oct. 1989, pp. 1–24.

[12] Barton, R. B., "Water-Cement Ratio is Passe," *Concrete International*, Nov. 1989, p. 75–7.

[13] Gaynor, R. D.,"C 94 and C 685 Requirements for Quality of Mixing Water," unpublished memorandum to ASTM Subcommittee C09.03.09 Methods of Testing and Specifications for Ready Mixed Concrete, American Society for Testing and Materials, Philadelphia, 16 June 1975, p. 9.

[14] Parker, L. C. and Slimak, M. W., "Waste Treatment and Disposal Costs for the Ready Mixed Concrete Industry," *Journal*, American Concrete Institute, Vol. 74, No. 7, July 1977, pp. 281–287.

[15] Pistilli, M. F., Peterson, C. F., and Shah, S. P., "Properties and Possible Recycling of Solid Waste from Ready Mixed Concrete," *Cement and Concrete Research*, Vol. 5, No. 3, May 1975, p. 249.

[16] Ullman, G. R., "Re-use of Wash Water as Mixing Water," Technical Information Letter No. 298, National Ready Mixed Concrete Association, Silver Spring, MD, 30 March 1973, p. 4.

[17] Meininger, R. C., "Recycling Mixer Wash Water—Its Effect on Ready Mixed Concrete," Technical Information Letter No. 298, National Ready Mixed Concrete Association, Silver Spring, MD, 30 March 1973, p. 7.

[18] "Standard Practice for Selecting Proportions for Normal, Heavyweight, and Mass Concrete (ACI 211.1–89)," *ACI Manual of Concrete Practice, Part 1*, American Concrete Institute, Detroit, MI, 1991.

[19] Meininger, R. C., "Use of Fly ASH in Air-Entrained Concrete—Report of Recent NSGA-NRMCA Research Laboratory Studies—Series J 153," Technical Information Letter No. 381, National Ready Mixed Concrete Association, Silver Spring, MD, 3 April 1981.

[20] Helmuth, R., "Fly Ash in Cement and Concrete," *SP040.01T*, Portland Cement Association, Skokie, IL, 1987, pp. 79–82.

[21] Ozyildirim, C., "Comparison of the Air Contents of Freshly Mixed and Hardened Concretes," *Cement, Concrete and Aggregates*, Vol. 13, No. 1, Summer 1991, pp. 11–24.

[22] Hover, K. C., "Some Recent Problems With Air-Entrained Concrete," *Cement, Concrete, and Aggregates*, Vol. 11, No. 1, Summer 1989, pp. 67–72.

[23] Gaynor, R. D., "Summer Problem Solving," *Concrete Products*, June 1991, p. 11.

[24] "Civil Works Construction Guide Specification CW-03305," U.S. Army Corps of Engineers, Office of the Chief of Engineers, Washington, DC, Jan. 1985, pp. 3–28.

[25] "Concrete Plant Standards," ninth revision, Concrete Plant Manufacturers Bureau, Silver Spring, MD, 1 Jan. 1990.

[26] "Plant Certification Check List," *Publication QC 3*, fourth revision, National Ready Mixed Concrete Association, Silver Spring, MD, 1 Jan. 1984.

[27] Specifications, Tolerances, and Other Technical Requirements for Weighing and Measuring Devices, *NIST Handbook 44*, U.S. Department of Commerce, National Institute of Standards and Technology, Washington, DC, 1990, pp. 1–274.

[28] "Concrete Plant Mixer Standards," sixth revision, Plant Mixer Manufacturers Division, Silver Spring, MD, 24 April 1990.

[29] Mass, G. R., "Premixed Cement Paste," *Concrete International*, Nov. 1989, pp. 82–85.

[30] "CPMB Does Cost Survey on Central Versus Transit Mixing," *Concrete Products*, Nov. 1974, p. 41.

[31] "A Study of Economic Factors of Central Mixing in the Production of Ready Mixed Concrete," *CPMB Publication 103*, Concrete Plant Manufacturers Bureau, Silver Spring, MD, 1976.

[32] Meininger, R. C., "Study of ASTM Limits on Delivery Time," *Publication 131*, National Ready Mixed Concrete Association, Silver Spring, MD, Feb. 1969, pp. 1–17.

[33] Bozarth, F. M., Granley, E. C., and Grieb, W. E., "A Study of Mixing Performance of Large Central Plant Concrete Mixers," *Research and Development Report*, U.S. Department of Commerce, Bureau of Public Roads, Office of Research and Development, Washington, DC, July 1966.

[34] Bozarth, F. M., "Case Study of Influence of Imbalances in Charging of Cement and Water on Mixing Performance of an 8-Cubic Yard Central Plant Mixer," U.S. Department of Transportation, Federal Highway Administration, Bureau of Public Roads, Washington, DC, July 1967.

[35] Gaynor, R. D., "Effects of Prolonged Mixing on Properties of Concrete," *Publication 111*, National Ready Mixed Concrete Association, Silver Spring, MD, 1963, pp. 1–18.

[36] Meininger, R. C., "Study of ASTM Limits on Delivery Time," *Publication 131*, National Ready Mixed Concrete Association, Silver Spring, MD, 1969, pp. 1–17.

[37] Ravina, D., "Retempering a Prolonged-Mixed Concrete with Admixtures in Hot Weather," *Journal*, American Concrete Institute, Vol. 72, No. 6, June 1975, pp. 291–295.

[38] Beaufait, F. W. and Hoadley, P. G., "Mix Time and Retempering Studies on Ready Mixed Concrete," *Journal*, American Concrete Institute, Vol. 70, No. 12, Dec. 1973, p. 810; and Discussion, *Journal*, American Concrete Institute, Vol. 73, No. 4, April 1976, p. 233.

[39] Gaynor, R. D. and Mullarky, J. I., "Mixing Concrete in a Truck Mixer," *Publication 148*, National Ready Mixed Concrete Association, Silver Spring, MD, Jan. 1975, pp. 1–14.

[40] Gaynor, R. D., Meininger, R. C., and Khan, T. S., "Effect of Temperature and Delivery Time on Concrete Proportions," *Temperature Effects on Concrete, ASTM STP 858*, American Society for Testing and Materials, Philadelphia, June 1985; also *Publication No. 171*, National Ready Mixed Concrete Association, Silver Spring, MD.

[41] "Pre-Pour Conferences," *Concrete Construction*, March 1985, Vol. 30/No. 3, pp. 265–268; also in *Publication QC 2*, National Ready Mixed Concrete Association,

[42] "Simplified Version of the Recommended Practice for Evaluation of Strength Test Results of Concrete, (ACI 214.3R-88)" *ACI Manual of Concrete Practice, Part 2*, American Concrete Institute, Detroit, MI, 1991, p. 214.3R–1.

[43] Ahari, H. E., Discussion of the report by ACI Committee 214, "Proposed Revision of ACI 214-65: Recommended Practice for Evaluation of Strength Test Results of Concrete," *Journal*, American Concrete Institute, Detroit, MI, Jan. 1977, pp. 39–40.

[44] Harter, L., *Order Statistics and Their Use in Testing and Estimation*, Vol. 1, Aerospace Research Laboratories, U.S. Air Force, NTIS Document Acquisition No. ADA058262,

[45] "In-Place Strength Evaluation—A Recommended Practice," *Publication 133-79*, Committee on Research, Engineering and Standards, National Ready Mixed Concrete Association, Silver Spring, MD, 1970, revised 1979.

[46] Dewar, J. D. and Anderson, R., "Appendix 1: Quality Scheme for Ready Mixed Concrete," *Manual of Ready-Mixed Concrete*, Blackie and Son LTD, London, p. 219.

[47] "BRMCA Authorization Scheme, Part IV of BRMCA Code for Ready Mixed Concrete," British Ready Mixed Concrete Association, London, May 1975.

[48] "Introduction to Total Quality Management and Its Application to Total Quality Management in the Construction Industry," *DCQ Forum*, Design and Construction Quality Institute, Washington, DC, Fall 1991, pp. 1–94.

Lightweight Concrete and Aggregates

Thomas A. Holm[1]

PREFACE

Lightweight concrete and aggregates were first discussed by R. E. Davis and J. Kelly in the 1956 edition of *ASTM STP 169*. The *ASTM STP 169A* and *ASTM STP 169B* editions were authored by D. W. Lewis. The general presentation of this paper is similar to earlier chapters; however, additional information on elastic properties of lightweight aggregates, as well as strength making, durability, and placement characteristics of lightweight concrete is included to reflect the current state of the art. This edition also includes new discussions relative to the contact zone and internal curing as well as revisions to ASTM methods for calculating the equilibrium density of structural lightweight concrete adopted by ASTM since the publication of *ASTM STP 169B* in 1978.

CLASSIFICATION OF LIGHTWEIGHT AGGREGATES AND LIGHTWEIGHT AGGREGATE CONCRETES

ASTM Standards provide requirements for lightweight aggregates that are used in structural masonry units and insulating types of concrete. Structural and insulating lightweight aggregate concretes are broadly divided into three groups based upon their use and physical properties. Unit weight, thermal conductivity and compressive strength ranges normally associated with each class of concrete are summarized in (Table 1).

This chapter addresses concretes where weight reduction is achieved through the use of lightweight aggregates and does not include cellular or foam concrete, where lighter weight is developed primarily by inclusion of large amounts of air or gas through foaming-type agents. No-fines concretes with very large, unfilled interstitial voids produced with aggregate content deficient in fine aggregates are also excluded from this review, which restricts discussion to the predominant forms of lightweight aggregate concretes based upon inorganic lightweight aggregates.

[1] Vice president of Engineering, Solite Corporation, Richmond, VA 23261.

STRUCTURAL-GRADE LIGHTWEIGHT AGGREGATE AND STRUCTURAL LIGHTWEIGHT AGGREGATE CONCRETE

Structural-grade lightweight concretes generally contain aggregates made from pyroprocessed shales, clays, slates, expanded slags, expanded fly ash, and those mined from natural porous volcanic sources. Minimum compressive strength of structural-grade lightweight aggregate concrete has, in effect, been jointly established by the ASTM Specification for Lightweight Aggregates for Structural Concrete (C 330) and the Standard Building Code for Reinforced Concrete (ACI 318) [1] which requires that: "Structural concrete made with lightweight aggregate; the air-dried unit weight at 28 days is usually in the range of 1440 to 1850 kg/m³ (90 to 115 lb/ft³) and the compressive strength is more than 17.2 MPa (2500 psi)." This is a definition, not a specification and project requirements may permit equilibrium unit weights up to 1900 kg/m³ (120 lb/ft³). Although structural concrete with equilibrium unit weights from 1450 to 1920 kg/m³ (90 to 120 lb/ft³) are often used, most lightweight aggregate concrete used in structures have equilibrium unit weights between 1600 to 1760 kg/m³ (100 and 110 lb/ft³).

Structural-grade lightweight aggregates are produced in manufacturing plants from raw materials including suitable shales, clays, slates, fly ashes, or blast furnace slags. Naturally occurring lightweight aggregates are mined from volcanic deposits that include pumice and scoria types. Pyroprocessing methods include the rotary kiln process (a long, slowly rotating, nearly horizontal cylinder lined with refractory materials similar to cement kilns); the sintering process wherein a bed of raw materials including fuel is carried by a traveling grate under ignition hoods; and the rapid agitation of molten slag with controlled amounts of air or water. No single description of raw material processing is all-inclusive and the reader is urged to consult local lightweight aggregate manufacturers for physical and mechanical properties of lightweight aggregates and the concrete made with them.

Increased usage of processed lightweight aggregates is evidence of environmentally sound planning, as these products utilize materials with limited structural applications in their natural state, thus minimizing construction industry demands on finite resources of natural sands, stones, and gravels.

ASTM C 330 requires fine lightweight aggregates used in the production of structural lightweight concrete to be properly graded, with 85 to 100% passing the 4.75 mm

TABLE 1—Lightweight Aggregate (LWA) Concrete Classified According to Use and Physical Properties.[a]

Class of Lightweight Aggregate Concrete	Type of Lightweight Aggregate used in Concrete	Typical Range of Lightweight Concrete Unit Weight	Typical Range of Compressive Strength	Typical Range of Thermal Conductivities
Structural	Structural-grade LWA C 330	(1440 to 1840) 90 to 115 air dry	(>17) (>2500)	not specified in C 330
Structural/ Insulating	Either structural C 330 or insulating C 332 or a combination of C 330 and C 332	(800 to 1440) (50 to 90) air dry	(3.4 to 17) (500 to 2500)	C 332 from (0.22) (1.50) to (0.43) (3.00) oven dry
Insulating	Insulating-grade LWA C 332	(240 to 800) (15 to 50) oven dry	(0.7 to 3.4) 100 to 500	C 332 from (0.065) (0.45) to (0.22) (1.50) oven dry

[a] Unit weights are in (kg/m^3) (lb/ft^3), compressive strengths in (MPa) (psi), and thermal conductivity in (W/m · °K) (Btu · in./h · ft^2 · °F).

(^{3}/$_{16}$ in.) screen with a dry loose bulk density less than 1120 kg/m^3 (70 lb/ft^3). Four coarse aggregate gradations are provided for use in structural lightweight concrete with maximum dry loose bulk density limited to 880 kg/m^3 (55 lb/ft^3). Combined fine and coarse aggregate formulations must not exceed a maximum dry loose unit weight of 1040 kg/m^3 (65 lb/ft^3). Tests are conducted in accordance with ASTM Test Method for Unit Weight and Voids in Aggregate (C 29) using the shoveling procedure.

INSULATING-GRADE LIGHTWEIGHT AGGREGATES AND INSULATING LIGHTWEIGHT CONCRETES

Very light nonstructural concretes, employed primarily for high thermal resistance, incorporate low-density low-strength aggregates such as vermiculite and perlite. With low unit weights, seldom exceeding 800 kg/m^3 (50 lb/ft^3), thermal resistance is high. These concretes are not intended to be exposed to the weather and generally have a compressive strength range from about 0.69 to 6.89 MPa (100 to 500 psi).

ASTM Specification for Lightweight Aggregates for Insulating Concrete (C 332) limits thermal conductivity values for insulating concretes to a maximum of 0.22 W/m·K (1.50 Btu · in. · /h · ft^2 °F) for concrete having an oven-dry density of 800 kg/m^3 (50 lb/ft^3) or less, and to 0.43 W/m · K (3.0 Btu · in. · /h · ft^2 °F) for those weighing up to 1440 kg/m^3 (90 lb/ft^3). Lighter concretes are those made with Group I aggregates (perlites and vermiculite), while higher unit weights result from the use of Group II aggregates (expanded shales, expanded slags and natural lightweight aggregates).

Thermal conductivity values may be determined in accordance with ASTM Test Method for Steady-State Thermal Performance of Building Assemblies by Means of a Guarded Hot Box (C 236) and ASTM Test Method for Steady-State Heat Flux Measurements and Thermal Transmission Properties by Means of the Guarded-Hot-Plate Apparatus (C 177). Oven-dried specimens are used for both thermal conductivity and unit weight tests on the insulating concretes. Moisture content of insulating materials directly affects both the thermal conductivity and unit weight, but to varying degrees. A 1% increase in moisture content will increase unit weight by an equivalent 1% but may increase thermal conductivity by as much as 5 to 9% [2]. Use of oven-dried specimens provides an arbitrary basis for comparison but clearly does not duplicate in-service applications. The controlled test conditions serve to permit classification of materials and to provide a standardized reference environment.

STRUCTURAL/INSULATING LIGHTWEIGHT AGGREGATE CONCRETES

Widespread industrial applications that call for "fill" concretes require modest compressive strengths with densities intermediate between the structural- and insulating-grade concretes. These concretes may be produced with high air mixes with structural-grade lightweight aggregate, with sanded insulating lightweight aggregate mixes, or with formulations incorporating both structural- and insulating-grade lightweight aggregates. Compressive strengths from 3.4 to 17 MPa (500 to 2500 psi) are not uncommon with thermal resistance less than concretes containing only insulating-grade lightweight aggregate.

LIGHTWEIGHT AGGREGATE PROPERTIES

Internal Structure of Lightweight Aggregates

Lightweight aggregates develop low particle specific gravity because of the cellular pore system. Cellular structure within the particles is normally developed at high temperatures by formation of gases due to the reaction of heat on certain raw material constituents coincident with incipient fusion causing gas expansion to be trapped in the viscous, pyroplastic mass. Strong, durable lightweight aggregates are produced when small-size, well-distributed, noninterconnected pores are enveloped in a continuous, crack free, vitreous phase [3] (Fig. 1).

PARTICLE SHAPE AND SURFACE TEXTURE

Depending on the source and the method of production, lightweight aggregates exhibit considerable differences in

FIG. 1—Contact zone—structural lightweight concrete from 30-year-old bridge deck, W.P. Lane Memorial Bridge over the Chesapeake Bay, Annapolis, Maryland: compression strength 24 MPa (3500 psi); density 1680 kg/m³ (105 lb/ft³).

particle shape and texture. Shapes may be cubical, rounded, angular, or irregular. Textures may range from fine pore, relatively smooth skins to highly irregular surfaces with large exposed pores. Particle shape and surface texture directly influence workability, coarse-to-fine aggregate ratio, cement content requirements, and water demand in concrete mixes, as well as other physical properties.

SPECIFIC GRAVITY

The specific gravity of an aggregate is the ratio between the mass of a quantity of the material and the volume occupied by the individual particles contained in that sample. This volume includes the pores within the particles but does not include the voids between the particles. In general, the volume of the particles is determined from the volume displaced submerged in water when penetration of water into the particles during the test is limited by previous saturation. Specific gravity of individual particles depends both on the specific gravity of the poreless vitreous material and the pore volume within the particles, and generally increases when particle size decreases. The specific gravity of the pore-free vitreous material may be determined by pulverizing the lightweight aggregate in a jar mill and then following procedures used for determination of the specific gravity of cement in the ASTM Test Method for Density of Hydraulic Cement (C 188).

BULK UNIT WEIGHT OF LIGHTWEIGHT AGGREGATES

Aggregate bulk unit weight is defined as the ratio of the mass of a given quantity of material and the total volume occupied by it. This volume includes the voids between as well as within the particles. Unit weight is a function of particle shape, density, size, gradation, and moisture content, as well as the method of packing the material (loose, vibrated, rodded) and varies not only for different materials, but for different sizes and gradations of a particular material. Table 2 summarizes the maximum unit weights for lightweight aggregates listed in ASTM C 330, ASTM Specification for Lightweight Aggregates for Concrete Masonry Units (C 331), and ASTM C 332. Minimum unit weights for perlite and vermiculite are also provided to limit over-expanded, weak particles that would break down in mixing.

Density of insulating concrete is usually determined in an over-dry condition, using oven-dry weight and dimensions of specimens associated with to those subjected to either thermal conductivity or compressive strength tests. Density of insulating concretes made with perlite or vermiculite aggregates or cellular concretes may range from 240 to 800 kg/m³ (15 to 50 lb/ft³), while those made with other types of lightweight aggregates are usually in the range of 800 to 1440 kg/m³ (50 to 90 lb/ft³).

TOTAL POROSITY

Void content (within particle pores and between particles' voids) can be determined from measured values of particle specific gravity and bulk unit weight. If, for example, measurements on a sample of lightweight coarse aggregate are:

1. bulk dry loose unit weight, 770 kg/m³ (48 lb/ft³);
2. particle specific gravity, 1400 kg/m³; and
3. specific gravity of poreless vitreous material, 2500 kg/m³;

TABLE 2—Requirements of ASTM C 330, C 331, and C 332 for Dry Loose Unit Weight of Lightweight Aggregates.

Aggregate Size and Group	Maximum Unit Weight, kg/m³ (lb/ft³)	Minimum Unit Weight, kg/m³ (lb/ft³)
ASTM C 330 AND C 331		
fine aggregate	(1120) (70)	...
coarse aggregate	(880) (55)	...
combined fine and coarse aggregate	(1040) (65)	...
ASTM C 332		
Group 1		
Perlite	(196) (12)	(120) (7.5)
Vermiculite	(160) (10)	(88) (5.5)
Group 2		
fine aggregate	(1120) (70)	...
coarse aggregate	(880) (55)	...
combined fine and coarse aggregate	(1040) (65)	...

then the fractional pore volume of an individual particle is

$$\frac{2500 - 1400}{2500} = 0.44$$

and the fractional interstitial void volume (between particles) is

$$\frac{1400 - 770}{1400} = 0.45$$

For this example, total porosity (pores and voids) would then equal

$$[0.45 + (0.44 \times 0.55)] = 0.69$$

GRADATION

Gradation requirements are generally similar to those provided for normal-weight aggregate with the exception that lightweight aggregate particle size distribution permits a higher weight through smaller sieves. This modification recognizes the increase in specific gravity typical for the smaller particles of most lightweight aggregates, and that while standards are established by weights passing each sieve size, ideal formulations are developed through volumetric considerations.

An exception to the procedures of ASTM Method for Sieve Analysis of Fine and Coarse Aggregates (C 136) requires reduction of the weight of fine aggregate sample tested according to the lightweight aggregate's unit weight, and sieving time not to exceed 5 min.

Producers of structural lightweight aggregate normally stock materials in several standard size formulations of coarse, intermediate, and fine aggregate. By combining size fractions or by replacing some or all of the fine fraction with a normal-weight sand, a wide range of concrete unit weights may be obtained. The aggregate producer is the best source of information for the proper aggregate combinations to meet fresh unit weight specifications and equilibrium unit weights for dead load design considerations.

Normal-weight sand replacement will typically increase unit weight from about 80 to more than 160 kg/m³ (5 to 10 lb/ft³). Using increasing amounts of cement to obtain high strengths above 35 MPa (5000 psi) concrete will increase air dry density from 32 to 96 kg/m³ (2 to 6 lb/ft³).

ABSORPTION CHARACTERISTICS

Due to their cellular structure, lightweight aggregates absorb more water than their normal-weight aggregate counterparts. Based upon a 24-h absorption test, conducted in accordance with the procedures of ASTM Test Method for Specific Gravity and Absorption of Coarse Aggregate (C 127) and ASTM Test Method for Specific Gravity and Absorption of Fine Aggregate (C 128), structural-grade lightweight aggregates will absorb from 5 to more than 25% by weight of dry aggregate. By contrast, normal-weight aggregates generally absorb less than 2% of moisture. The important difference in measurements of stockpile moisture contents is that with lightweight aggregates the moisture is largely absorbed into the interior of the particles whereas in normal-weight aggregates it is primarily surface adsorption. Recognition of this essential difference is important in mix proportioning, batching, and control. Rate of absorption of lightweight aggregates is dependent on the characteristics of pore size, connection, and distribution, particularly those close to the surface. Internally absorbed water within the particle is not immediately available for chemical interaction with cement as mixing water, but extremely beneficial in maintaining longer periods of curing essential to improvements in the aggregate/matrix contact zone. Internal curing will also bring about reduction of permeability by extending the period in which additional products of hydration are formed in the pores and capillaries of the binder.

MODULUS OF ELASTICITY OF LIGHTWEIGHT AGGREGATE PARTICLES

The modulus of elasticity of concrete is a function of the moduli of its constituents. Concrete may be considered as a two-phase material consisting of coarse aggregate inclusions within a continuous "mortar" fraction that includes cement, water, entrained air, and fine aggregate. Dynamic measurements made on aggregates alone have shown a relationship corresponding to the function: $E = 0.008 \rho^2$, where E is the dynamic modulus of elasticity of the particle in MPa and ρ is the dry mean specific gravity in kg/m³ [4] (Fig. 2). Dynamic moduli for usual expanded aggregates have a range of 10 to 16 GPa (1.45 to 2.3 × 10⁶ psi), whereas the range for strong ordinary aggregates

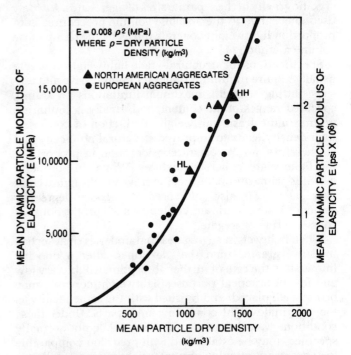

FIG. 2—Relationship between mean particle density and the mean dynamic modulus of elasticity for the particles of lightweight aggregate [12].

is approximately 30 GPa (4.35×10^6 psi) to 100 GPa (14.5×10^6 psi).

PROPERTIES AND PRODUCTION OF LIGHTWEIGHT AGGREGATE CONCRETE

Comprehensive reports detailing the properties of lightweight concretes and lightweight aggregates have been published by Shideler [5], Reichard [6], Holm [7], Carlson [8], and Valore [9,10]. The first three deal with structural-grade concretes, Carlson reported on lightweight aggregate for concrete masonry units, and Valore covered both structural and insulating concretes. In most instances, test procedures for measuring properties of lightweight concretes were the same as commonly used for normal-weight concretes. In limited cases, special test procedures particularly suited to measure lightweight concrete characteristics were developed.

PROPORTIONING

In general, proportioning rules and techniques used for ordinary concrete mixes apply to lightweight concrete with added attention given to concrete unit weight and the influence of the water absorption characteristics of the lightweight aggregate [11]. Most structural-grade lightweight concretes are proportioned by absolute volume methods in which the fresh concrete produced is considered equal to the sum of the absolute volumes of cement, aggregates, net water, and entrained air. Proportioning by this method requires the determination of absorbed and adsorbed moisture contents and the as-used specific gravity of the separate sizes of aggregates. A widely used alternative to the absolute volume procedures is to proportion lightweight concrete mixes by the damp loose volume method [11].

Specifications for structural-grade lightweight concrete usually require minimum values for compressive and tensile splitting strength, maximum limitations on slump, specified ranges of air content, and, finally, a limitation on maximum fresh unit weight. Reduction of concretes' high density leads to improved structural efficiency and is, therefore, an important consideration in proportioning lightweight concrete mixtures. While this property depends primarily on the specific gravity of the lightweight aggregates, it is also influenced to a lesser degree by cement, water, and air contents, and proportions of coarse-to-fine aggregate.

When lightweight aggregates contain levels of absorbed moisture greater than that developed after a one-day immersion, the rate of further absorption will be very low and for all practical purposes lightweight concrete may be batched, placed, and finished with the same facility as their normal-weight concrete counterparts. Under these conditions water/cement (w/c) ratios, while not normally specified, may be established with precision comparable to concretes containing normal-weight aggregates. Water absorbed within the lightweight aggregate prior to mixing is not available for calculating the volume of cement paste at the time of setting. This absorbed water is available, however, for continued cement hydration after external curing has ended. The general practice is to proportion the mix for a particular lightweight aggregate on the basis of a cement content at a given slump.

As with normal-weight concrete, air entrainment in lightweight concrete significantly improves durability and resistance to scaling. In concretes made with angular lightweight aggregates, it is also an effective means of improving workability of otherwise harsh mixtures. With moderate air contents, bleeding and segregation are reduced and mixing water requirements lowered while maintaining optimum workability. Because of the elastic compatibility of the lightweight aggregate and cementitious binder phases, strength reduction penalties due to high air contents will be lower for structural lightweight concrete than for normal-weight concretes [12]. Recommended ranges of total air content of usual structural lightweight concretes are shown in Table 3.

Air content of lightweight aggregate concretes is determined in accordance with the procedures of ASTM Test Method for Air Content of Freshly Mixed Concrete by the Volumetric Method (C 173). Volumetric measurements assure reliable results while pressure meters will provide erratic data due to the influence of aggregate porosity.

Air contents higher than are required for durability considerations are frequently developed for high thermal resistance, or for lowering unit weight of semi-structural "fill" concrete, with reduced compressive strength as a natural consequence. Use of water reducers, retarders, and superplasticizers will result in improved lightweight concrete characteristics in a manner similar to that of normal-weight concretes, however, superplasticizers, while effective, will increase the density of lightweight as well as other concretes.

MIXING, PLACING, FINISHING, AND CURING

When properly proportioned, structural lightweight concrete can be delivered and placed with the same facility as ordinary concretes. The most important consideration in handling any type of concrete is to avoid separation of coarse aggregate from the mortar fraction. Basic principles required to secure a well-placed lightweight concrete include:

(a) well-proportioned, workable mixes that use a minimum amount of free water;
(b) equipment capable of expeditiously moving the concrete;
(c) proper consolidation in the forms; and
(d) quality workmanship in finishing.

TABLE 3—Total Air Content for Lightweight Concretes.

Maximum Size of Aggregate	Air Content, % by volume
(20 mm) (3/4 in.)	4 to 8
(10 mm) (3/8 in.)	5 to 9

Well-proportioned structural lightweight concretes can be placed and screeded with less physical effort than that required for ordinary concrete. Excessive vibratation should be avoided, as this practice serves to drive the heavier mortar fraction down from the surface where it is required for finishing. On completion of final finishing, curing operations similar to ordinary concrete should begin as soon as possible. Lightweight concretes batched with aggregates having high absorptions carry their own internal water supply for curing within the aggregate and as a result are more forgiving to poor curing practices or unfavorable ambient conditions. This "internal curing" water is transferred from the lightweight aggregate to the mortar phase as evaporation takes place on the concrete surface, thus maintaining continuous moisture balance by replacing moisture essential for an extended continuous hydration period determined by ambient conditions and the as-batched lightweight aggregate moisture content.

Lightweight aggregates may absorb part of the mixing water when exposed to increased pumping pressures. To avoid loss of workability, it is essential to raise the presoak absorption level of lightweight aggregates prior to pumping. Presoaking is best accomplished at the aggregate production plant where uniform moisture content is achieved by applying water from spray bars directly to the aggregate moving on belts. This moisture content can be maintained and supplemented at the concrete plant by stockpile hose and sprinkler systems.

Presoaking will significantly reduce the lightweight aggregates' rate of absorption, minimizing water transfer from the mortar fraction that, in turn, causes slump loss during pumping. Higher moisture contents developed during presoaking will result in increased specific gravity that, in turn, develops higher fresh concrete unit weight. Higher water content due to presoaking will eventually diffuse out of the concrete, developing a longer period of internal curing as well as a larger differential between fresh and equilibrium unit weight than that associated with lightweight concretes placed with lower moisture contents. Aggregate suppliers should be consulted for mix design recommendations necessary for consistent pumpability.

LABORATORY AND FIELD CONTROL

Changes in lightweight aggregate moisture content, gradation, or specific gravity as well as usual job site variation in entrained air suggest frequent checks of the fresh concrete to facilitate adjustments necessary for consistent concrete characteristics. Standardized field tests for consistency, fresh unit weight, and entrained-air content should be employed to verify conformance of field concretes with design mixes and the project specification. Sampling should be conducted in accordance with ASTM Practice for Sampling Freshly Mixed Concrete (C 172) and ASTM Test Method for Air Content of Freshly Mixed Concrete by the Volumetric Method (C 173). The ASTM Test Method for Unit Weight of Structural Lightweight Concrete (C 567) describes methods for calculating the in-service, equilibrium unit weight of structural lightweight concrete. In general, when variations in fresh density exceed ±2%, an adjustment in batch weights may be required to restore specified concrete properties. To avoid adverse effects on durability, strength, and workability, air content should not vary more than ±1.5% from specified values.

DENSITY OF STRUCTURAL LIGHTWEIGHT CONCRETE

Although there are numerous structural applications of all lightweight concretes (coarse and fine lightweight aggregate), usual commercial practice in North America is to design sanded lightweight concretes where part or all of the fine aggregates used is natural sand. Long-span bridges using concretes with three-way blends (coarse and fine lightweight aggregates and small supplemental natural sand volumes) have provided long-term durability and structural efficiency (density/strength ratios) [14]. Earliest research reports [5,6,15,16] compared all lightweight concretes with "reference" normal-weight concrete while later studies reported in Refs 13,17–19 supplemented the early findings with data based upon sanded lightweight concretes.

The fresh unit weight of lightweight aggregate concretes is a function of mix proportions, air contents, water demand, and the specific gravity and moisture content of the lightweight aggregate. Decrease in density of exposed concrete is due to moisture loss that, in turn, is a function of ambient conditions and surface area/volume ratio of the member. Design professionals should specify a maximum fresh density for lightweight concrete, as limits of acceptability should be controlled at time of placement.

Dead loads used for design should be based upon equilibrium density that, for most conditions and members, may be assumed to be reached after 90 days. Extensive tests conducted during North American durability studies demonstrated that despite wide initial variations of aggregate moisture content, equilibrium density was found to be 50 kg/m^3 (3.1 lb/ft^3) above oven-dry density (Fig. 3). European recommendations for in-service density are similar [4].

When weights and moisture contents of all the constituents of the batch of concrete are known, an approximate calculated equilibrium density may be determined according to ASTM C 567 from the following equation

$$E = O + 50 \text{ kg/m}^3 \ (E = O + 3 \text{ lb/ft}^3) \qquad (1)$$

where

$O = A + 1.2W$,
E = calculated equilibrium unit weight, kg/m^3 (lb/ft^3);
O = approximate oven-dry weight, kg/m^3 (lb/ft^3);
A = weight of dry aggregates in batch, kg (lb);
W = weight of cement in batch, kg (lb); and
1.2 = weight of hydrated water of hydration (estimated at 20% by weight of cement).

COMPRESSIVE STRENGTH

Compressive strength test procedures for structural lightweight aggregate concretes are similar to those for

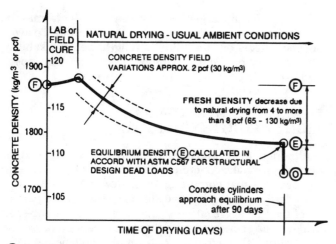

FIG. 3—Concrete density versus time of drying for structural lightweight concrete.

normal-weight concretes with the exception of the 21-day laboratory air 23°C (73.4°F) and 50% relative humidity drying period required by the procedures of ASTM Test Method for Splitting Tensile Strength of Cylindrical Concrete Specimens (C 496) and ASTM C 567. While most structural-grade lightweight aggregates are capable of producing concretes with compressive strengths in excess of 35 MPa (5000 psi), a limited number of lightweight aggregates can be used in concretes that develop cylinder strengths from 48–69 MPa (7000 to 10 000 psi).

While compressive strengths of 21 to 35 MPa (3000 to 5000 psi) are common for cast-in-place structural lightweight concretes, 41 MPa (7000 psi) strengths are presently being specified for offshore applications. Light weight aggregate concrete will demonstrate a strength "ceiling" where further additions of cementitious materials will not significantly raise the maximum attainable strength. Strength ceilings that differ for each lightweight aggregate source are the result of pore size and distribution as well as the strength characteristics of the pore-free vitreous material surrounding the pores. The strength ceiling of a particular lightweight aggregate may be considerably increased by reduction of the top size in a particular grading formulation.

Compressive strength tests of lightweight insulating concrete having oven-dry unit weights not exceeding 800 kg/m³ (50 lb/ft³) are conducted in accordance with ASTM Test Method for Compressive Strength of Lightweight Insulating Concrete (C 495) on 75 × 150 mm (3 × 6 in.) cylinders. Twenty-five days after molding, the specimens are oven-dried at 60 ± 2.8°C (140 ± 5°F) for three days, cooled to room temperature, and tested for compressive strength at 28 days.

ASTM Methods of Securing, Preparing, and Testing Specimens from Hardened Lightweight Insulating Concrete for Compressive Strength (C 513) provides procedures for the determination of the compressive strength of cube specimens from hardened, field lightweight insulating concretes.

TENSILE STRENGTH

Shear, torsion, anchorage, bond strengths, and crack resistance are related to tensile strength that is, in turn, dependent upon tensile strength of the coarse aggregate and mortar phases and the degree to which the two phases are securely bonded. Traditionally, tensile strength has been defined as a function of compressive strength, but this is known to be only a first approximation that does not reflect aggregate particle strength, surface characteristics, nor the concrete's moisture content and distribution. The splitting tensile strength, as determined by ASTM C 496, is used throughout North America as a simple, practical design criteria that is known to be a more reliable indicator of tensile-related properties than beam flexural tests. Splitting tests are conducted by applying diametrically opposite compressive line loads to a concrete cylinder laid horizontally in a testing machine. A minimum tensile splitting strength of 2.0 MPa (290 psi) is a requirement for structural-grade lightweight aggregates conforming to the requirements of ASTM C 330.

Tests have shown that diagonal tensile strengths of beams and slabs correlate closely with the concrete splitting strengths [20,21]. As tensile splitting results vary for different combinations of materials, the specifier should consult with the aggregate suppliers for laboratory-developed splitting strength data. Special tensile strength test data should be developed prior to the start of unusual projects where development of early-age tensile-related handling forces occur as in precast or tilt-up members.

Tensile strength tests on structural lightweight concrete specimens that undergo some drying correlate better with the behavior of concrete in actual structures. Moisture loss progressing slowly into the interior of concrete members will result in the development of outer envelope tensile stresses that balance the compressive stresses in the still-moist interior zones. ASTM C 496 requires a seven-day moist and 21-day laboratory air drying at 23°C (73.4°F) and 50% relative humidity prior to conducting splitting tests. Structural-lightweight-concrete splitting tensile strengths vary from approximately 75 to 100% of normal-weight concretes of equal compressive strength. Replacing lightweight fine aggregate with normal-weight fines will normally increase tensile strength.

ELASTIC PROPERTIES

The modulus of elasticity of concrete is a function of the modulus of each constituent (binder matrix, expanded and normal density aggregates) and their relative mix proportion. The elastic modulus of normal-density concretes is higher because the moduli of the natural aggregate particles (and parent rock formations) are greater than the moduli of lightweight aggregate particles. For practical design conditions, the modulus of elasticity of concretes with densities between 1400 to 2500 kg/m³ (90 to 155 lb/

ft³) and within normal strength ranges may be assumed to follow the formula [1,4]

$$E = 33 p^{1.5} \sqrt{f_c} \qquad E = 0.04 \sqrt{p^3 f_c} \qquad (2)$$

where

E = denotes the secant modulus in psi (MPa),
p = the density in kg/m³ (lb/ft³), and
f_c = the compressive strength in MPa (psi) of a 150 by 300 mm (6 by 12 in.) cylinder (100 mm cube).

This or any other formula should be considered as only a first approximation, as the modulus is significantly affected (±25%) by moisture, aggregate type, and other variables. The formula clearly overestimates the modulus for high-strength lightweight concretes where limiting values are determined by the modulus of the lightweight aggregate. When design conditions require accurate elastic modulus data, laboratory tests should be conducted on specific concretes proposed for the project according to procedures of ASTM Test Method for Static Modulus of Elasticity and Poisson's Ratio of Concrete in Compression (C 469).

Tests to determine Poisson's ratio by the static method for lightweight and sand-and-gravel concrete gave values that varied between 0.15 and 0.25 and averaged 0.20. Dynamic tests yielded slightly higher values [6]. A value of 0.20 may be assumed for design purposes for both types of concretes.

SHRINKAGE

As with ordinary concretes, shrinkage of structural lightweight concretes is principally determined by

(a) shrinkage characteristics of the cement paste fraction,
(b) internal restraint provided by the aggregate fraction,
(c) the relative absolute volume fractions occupied by the shrinkage medium (cement paste fraction) and the restraining skeletal structure (aggregate fraction), and
(d) humidity and temperature environments.

Aggregate characteristics influence cementitious binder quantities (the shrinking fraction) necessary to produce a required strength at a given slump. Particle strength, shape, and gradation influence water demand and directly determine the fractional volume and quality of the cement paste necessary to meet specified strength levels. Once that interaction has been established, it is the rigidity of the aggregate fraction that restrains shrinkage of the cement paste. When structural lightweight aggregate concretes are proportioned with cementitious binder amounts similar to that required for normal aggregate concretes, the shrinkage of lightweight concrete is generally, but not always, slightly greater than that of ordinary concrete due to the lower aggregate stiffness. The time rate of shrinkage strain development in structural lightweight concrete is slower, and the time required to reach a plateau of equilibrium is longer when the as-batched, lightweight-aggregate absorbed moisture is high. Maximum shrinkage strains of high-strength lightweight concretes are slightly greater than high-strength normal-weight concretes containing similar binder content [22].

ASTM C 330 limits shrinkage of structural lightweight concretes to less than 0.07% after 28 days of drying in a curing cabinet maintained at 37.8°C (100°F) at a relative humidity of 32%. Concrete mixtures used in the specimen prisms are prepared with a cement content of 335 kg/m³ (564 lb/yd³) with water contents necessary to produce a slump of 50 to 100 mm (2 to 4 in.) and air content of 6 ± 1%. Specimens are removed from the molds at one day, and moisture cured until seven days at which time the accelerated drying is initiated.

Shrinkage of block concrete is limited to 0.10% when determined in accordance with procedures outlined in ASTM C 331. The ASTM Test Method for Length Change of Hardened Hydraulic Cement Mortar and Concrete (C 157) is followed using fixed proportions of one part cement to six parts aggregate by dry loose volumes, with sufficient water to produce a slump of 50 to 76 mm (2 to 3 in.). Initial length measurements are made after seven days moist storage with final shrinkage measurements at the age of 100 days after storage in laboratory air at 23°C (73.4°F) and 50% relative humidity.

CREEP

Time-related increases in concrete strain due to sustained stress can be measured according to procedures of ASTM Test Method for Creep of Concrete in Compression (C 512). Creep and shrinkage characteristics on any concrete type are principally influenced by aggregate characteristics, water and cement content (paste volume fraction), age at time of loading, type of curing, and applied stress-to-strength ratio. Other second-level variables also influence creep and shrinkage but to a lesser degree. As creep and shrinkage strains will cause increase in long-time deflections, loss of prestress, reduction in stress concentration, and changes in camber, it is essential for design engineers to have an accurate assessment of these time-related characteristics as a necessary design input. ACI Committee 213 [11] demonstrates wide envelopes of one-year specific creep values for low-strength all-lightweight, normally cured concretes. Test results for higher-strength, steam-cured sanded-lightweight concretes have a range of values that narrows significantly and closely envelopes the performance of the normal-weight "reference" concrete. These values are principally based upon the results of the comprehensive testing program of Shideler [5]. Long-term investigations by Troxell [23] on normal-weight concretes report similar wide envelopes of results for differing natural aggregate types so comparisons with "reference" concretes should be based upon data specific to the concretes considered.

Additional large-scale creep testing programs are reported in Refs 7 and 19, and Valore [10] has provided a comprehensive report that also includes European data on structural as well as insulating-grade lightweight concretes.

DURABILITY

Numerous accelerated freeze/thaw testing programs conducted on structural lightweight concrete in North

America [24,25] and in Europe [4] researching the influence of entrained-air volume, cement content, aggregate moisture content, specimen drying times, and testing environment have arrived at similar conclusions: air-entrained lightweight concretes proportioned with high-quality binder provide satisfactory durability results when tested under usual laboratory freeze/thaw programs. Observations of the resistance to deterioration in the presence of deicing salts on mature bridges indicate similar performance between structural lightweight and normal-weight concretes [3]. Comprehensive investigations into the long-term weathering performance of bridge decks [26] and marine structures [27] exposed for many years to severe environments support the findings of laboratory investigations and suggest that properly proportioned and placed lightweight concretes perform equal to or better than normal-weight concretes.

Core samples taken from hulls of 70-year-old lightweight concrete ships as well as 30- to 40-year-old lightweight concrete bridges have demonstrated concretes with high integrity contact zone between aggregate and the matrix with low levels of microcracking. Explanation of this proven record of high resistance to weathering and corrosion is due to several physical and chemical mechanisms including superior contact zone resistance to microcracking developed by significantly higher aggregate/matrix adhesion as well as internal stress reduction due to elastic matching of coarse aggregate and matrix phases. High ultimate strain capacity is also provided by concretes with a high strength/modulus ratios. The ratio at which the disruptive dilation of concrete starts is higher for lightweight concrete than for equal strength normal-weight concrete. A well-dispersed void system provided by the lightweight fine aggregates may also assist the air entrainment pore system and serve an absorption function by reducing salt concentration levels in the matrix phase [27].

Long-term pozzolanic action is provided when the silica-rich expanded aggregate combines with calcium hydroxide liberated during cement hydration. This will minimize leaching of soluble compounds and may also reduce the possibility of sulphate salt disruptive behavior [28].

It is widely recognized that while ASTM Test Method for Resistance of Concrete to Rapid Freezing and Thawing (C 666) provides a useful comparative testing procedure, there remains an inadequate correlation between accelerated laboratory test results and the observed behavior of mature concretes exposed to natural freezing and thawing. The inadequate laboratory/field correlation observed for normal-weight concrete is compounded when interpreting results from laboratory tests on structural lightweight concretes prepared with aggregate moisture contents typical of commercial operations. A proposed modification to ASTM C 666 [29] suggests that a 14-day air-drying period prior to the first freezing cycle will improve correlation between laboratory test data and observed field performance. Durability characteristics of any concrete, both normal weight and lightweight, are primarily determined by the protective qualities of the cement paste matrix. It is imperative that permeability characteristics of the concrete matrix be of high quality in order to protect steel reinforcing from corrosion, which is clearly the dominant form of structural deterioration observed in current construction. The matrix protective quality of concretes proportioned for thermal resistance by using high-air and low-cement contents will be significantly reduced. Very low density, non-structural concretes will not provide resistance to the intrusion of chlorides, carbonation, etc., comparable to the long-term satisfactory performance demonstrated with high-quality, structural-grade lightweight concretes [29].

For a number of years, field exposure testing programs have been conducted by the Canadian Department of Minerals, Energy and Technology (CANMET) on various types of concretes exposed to a cold marine environment at the Treat Island Severe Weather Exposure Station maintained by the U.S. Army Corps of Engineers at Eastport, Maine [29]. Concrete specimens placed on a mid-tide wharf experience alternating conditions of seawater immersion followed by cold air exposure at low tide. In typical winters, the specimens experience about 100 cycles of freezing and thawing. In 1978, a series of prisms were cast using commercial normal-weight aggregates with various cement types and including supplementary cementitious materials. Water-to-cement ratios of 0.40, 0.50, and 0.60 were used to produce 28-day compressive strengths of 30, 26 and 24 MPa (4350, 3770, and 3480 psi), respectively. In 1980, these mixes were essentially repeated with the exception being that the 40 mm (1½ in.) gravel aggregate was replaced with a 25 mm (1 in.) expanded-shale lightweight aggregate. Fine aggregates used in both 1978 and 1980 were commercially available natural sands. Cement contents for the semi-lightweight concrete mixtures were approximately 480, 360 and 240 kg/m^3 (800, 600, and 400 lb/yd^3) that produced compressive strengths of 36, 30 and 19 MPa (5220, 4350, and 2755 psi), respectively. All specimens continue to be evaluated annually for ultrasonic pulse velocity and resonant frequence as well as being rated visually. Ultrasonic pulse velocities are measured centrally along the long axis of the prisms. There were no significant differences between the structural lightweight concrete (eight years) and normal-weight concrete (ten years) after exposure to twice-daily seawater submersion and approximately 1000 cycles of freezing and thawing [29].

CHEMICAL REACTION

ACI Committee 201 on Durability of Concrete reports no documented instance of in-service distress caused by alkali reactions with lightweight aggregate. Mielenz [30] indicates that although the potential exists for alkali aggregate reaction with some natural lightweight aggregates and expanded perlite, the volume change may be accommodated without necessarily causing structural distress. Granulated blast furnace slag has been shown to be an effective inhibitor of such reactions [31], and the fine aggregate fractions of expanded shales, clays, and slates are known to be pozzolanic and may also serve to inhibit disruptive expansion. Bremner [32] reports that no evidence of alkali lightweight aggregate reaction was observed in tests conducted on 70-year-old marine and more than 30-year-old lightweight concrete bridge decks.

POPOUTS

Popouts may result from delayed disruptive expansions caused by the slow hydration of particles of hard-burned lime or magnesia, calcium sulfate, or unstable iron compounds. To test for the presence of these materials, concrete bars prepared by methods similar to those used for the shrinkage tests are cured and tested according to the procedures of ASTM Test Method for Autoclave Expansion of Portland Cement (C 151). No popouts are permitted by ASTM C 331 and C 330 since this disruptive expansion would cause unacceptable aesthetic blemishes on exposed concrete and masonry.

ABRASION RESISTANCE

Abrasion resistance of concrete depends on strength, hardness, and toughness characteristics of the cement paste and the aggregates, as well as on the bond between these two phases. Most lightweight aggregates suitable for structural concretes are composed of solidified glassy material comparable to quartz on the Moh scale of hardness. However, due to its porous system, the net resistance to wearing forces may be less than that of a solid particle of most natural aggregates. Structural-lightweight-concrete bridge decks that have been subjected to more than 100 million vehicle crossings including truck traffic show wearing performance similar to that of normal-weight concretes [33]. Limitations are necessary in certain commercial applications where steel-wheeled industrial vehicles are used, but such surfaces generally receive specially prepared surface treatments. Hoff [34] reports that specially developed testing procedures that measured ice abrasion of concrete exposed to arctic conditions demonstrated essentially similar performance for lightweight and normal-weight concretes.

BOND STRENGTH AND DEVELOPMENT LENGTH

Field performance has demonstrated satisfactory performance for lightweight concrete with respect to bond and development length. Because of the lower particle strength, lightweight concretes have somewhat lower bond splitting capacities than normal-weight concrete. Usual North American design practice (ACI 318 Standard Building Code for Reinforced Concrete) is to dispense with concepts of bond and require slightly longer embedment lengths for reinforcement in lightweight concretes than that required for normal-weight concrete.

FIRE RESISTANCE

When tested according to the procedures of ASTM Method for Fire Tests of Building Construction and Mate-

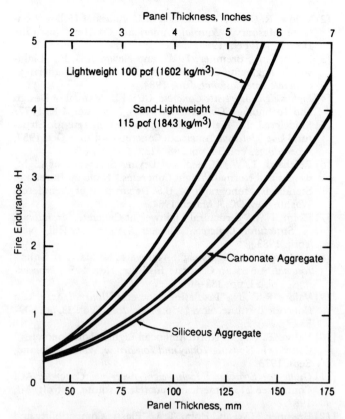

FIG. 4—Fire endurance (heat transmission) of concrete slabs as a function of thickness for naturally dried specimens [11].

rials (E 119), structural lightweight aggregate concrete slabs, walls, and beams have demonstrated greater fire endurance periods than equivalent thickness members made with normal-weight aggregates (Fig. 4). Superior performance is due to a combination of lower thermal conductivity (lower temperature rise on unexposed surfaces), lower coefficient of thermal expansion (lower forces developed under restraint), and the inherent thermal stability developed by aggregates that have been exposed to temperatures greater than 1093°C (2000°F) during pyroprocessing.

Acknowledgments

The principal sources of information for this chapter include the *Guide for Structural Lightweight Aggregate Concrete* (ACI 213) [11], ACI 318 Building Code Requirements for Reinforced Concrete [1], the CEB-FIP Manual of Design and Technology: *Lightweight Aggregate Concrete* [4], and the *Handbook of Structural Concrete* [7]. References to specific building codes and national standards are incorporated because these recommendations have resulted in the satisfactory field service of structural lightweight concrete structures.

REFERENCES

[1] "Standard Building Code for Reinforced Concrete," ACI Committee 318, American Concrete Institute, Detroit, MI, 1989.

[2] Valore, R. C., Jr., "Calculation of U-Values of Hollow Concrete Masonry," *Journal*, American Concrete Institute, Detroit, MI, Feb. 1980.

[3] Holm, T. A., Bremner, T. W., and Newman, J. B., "Lightweight Aggregate Concrete Subject to Severe Weathering," *Concrete International*, June 1984.

[4] *Lightweight Aggregate Concrete*, CEB-FIP Manual of Design and Technology, Construction Press, Lancaster, UK, 1977.

[5] Shideler, J. J., "Lightweight Aggregate Concrete for Structural Use," *Journal*, American Concrete Institute, Oct. 1957; *Proceedings*, Vol. 54, pp. 298–328.

[6] Reichard, T. W., "Creep and Drying Shrinkage of Lightweight and Normal Weight Concretes," National Bureau of Standards, Monograph 74, U.S. Department of Commerce, Washington, DC, 4 March 1964.

[7] Holm, T. A., "Structural Lightweight Concrete," *Handbook of Structural Concrete*, Chapter 7, McGraw-Hill, New York, 1983.

[8] Carlson, C. C., "Lightweight Aggregates for Masonry Units," *Journal*, American Concrete Institute, Nov. 1956; *Proceedings*, Vol. 53, pp. 383–402.

[9] Valore, R. C., Jr., "Insulating Concretes," *Journal*, American Concrete Institute, Nov. 1956; *Proceedings*, Vol. 53, pp. 509–532.

[10] Valore, R. C., Jr., "North American Lightweight Concretes," *Concrete in Housing-Today and Tomorrow*, Warsaw, Poland, Sept. 1973.

[11] *Guide for Structural Lightweight Aggregate Concrete*, ACI Committee 213, American Concrete Institute, Detroit, MI, 1987.

[12] Bremner, T. W. and Holm, T. A., "Elastic Compatibility and the Behavior of Concrete," *Journal*, American Concrete Institute, March/April 1986.

[13] Hansen, J. A., "Replacement of Lightweight Aggregate Fines with Natural Sand in Structural Concrete," *ACI Journal Proceedings*, American Concrete Institute, Vol. 61, No. 7, July 1964.

[14] Holm, T. A. and Bremner, T. W., "70 Year Performance Record for High Strength Structural Lightweight Concrete," *Proceedings*, First Materials Engineering Congress, Materials Engineering Division, American Society of Civil Engineers, Denver, CO, Aug. 1990.

[15] Price, W. H. and Cordon, W. A., "Tests of Lightweight Aggregate Concrete Designed for Monolithic Construction," *Journal*, American Concrete Institute, April 1949; *Proceedings*, Vol. 45, pp. 581–600.

[16] Kluge, R. W., Spanks, M. M., and Tuma, E. C., "Lightweight Aggregate Concrete," *ACI Journal Proceedings*, American Concrete Institute, Vol. 45, No. 9, May 1949.

[17] Pfeifer, D. W., "Sand Replacement in Structural Lightweight Concrete-Splitting Tensile Strength," *ACI Journal Proceedings*, American Concrete Institute, Vol. 64, No. 7, July 1967.

[18] Pfeifer, D. W., "Sand Replacement in Structural Lightweight Concrete-Freezing and Thawing Tests," *ACI Journal Proceedings*, American Concrete Institute, Vol. 64, No. 11, Nov. 1967.

[19] Pfeifer, D. W., "Sand Replacement in Structural Lightweight Concrete-Creep and Shrinkage Studies," *ACI Journal Proceedings*, American Concrete Institute, Vol. 65, No. 2, Feb. 1968.

[20] Hansen, J. A., "Shear Strength of Reinforced Lightweight Concrete Beams," *ACI Journal Proceedings*, American Concrete Institute, Vol. 55, No. 3, Sept. 1958.

[21] Hansen, J. A., "Tensile Strength and Diagonal Tension Resistance of Structural Lightweight Concrete," *Journal*, American Concrete Institute, July 1961; *Proceedings*, Vol. 58, pp. 1–37.

[22] Holm, T. A., "Physical Properties of High Strength Lightweight Aggregate Concretes," *Proceedings*, Second International Congress of Lightweight Concrete, London, April 1980.

[23] Troxell, G. E., Raphael, J. M., and Davis, R. E., "Long Time Creep and Shrinkage Tests of Plain and Reinforced Concrete," *Proceedings*, Vol. 58, American Society for Testing and Materials, Philadelphia, 1958.

[24] Klieger, P. and Hansen, J. A., "Freezing and Thawing Tests of Lightweight Aggregate Concrete," *Journal*, American Concrete Institute, Jan. 1961.

[25] "Freeze-Thaw Durability of Structural Lightweight Concrete," Lightweight Concrete Information Sheet No. 13, Expanded Shale Clay & Slate Institute, Salt Lake City, 1970.

[26] Walsh, R. J., "Restoring Salt Damaged Bridges," *Civil Engineering*, May 1967.

[27] Holm, T. A., "Performance of Structural Lightweight Concrete in a Marine Environment," *Performance of Concrete in a Marine Environment*, ACI SP-65, American Concrete Institute, International Symposium, St. Andrews-By-The-Sea, Canada, Aug. 1980.

[28] Bremner, T. W., Holm, T. A., and deSouza, H., "Aggregate-Matrix Interaction in Concrete Subject to Severe Exposure," FIP-CPC International Symposium on Concrete Sea Structures in Arctic Regions, Calgary, Canada, 29 Aug. 1984.

[29] Holm, T. A. and Bremner, T. W., "The Durability of Structural Lightweight Concrete," *Durability of Concrete*, ACI SP-126, American Concrete Institute, Second International Conference, Montreal, Canada, Aug. 1991.

[30] Mielenz, R. C., "Chapter 33—Petrographic Examination," *Significance of Tests and Properties of Concrete and Concrete-Making Materials*, ASTM STP 169B, American Society for Testing and Materials, Philadelphia, 1978.

[31] Pepper, L. and Mather, B., "Effectiveness of Mineral Admixtures in Preventing Excessive Expansion of Concrete Due to Alkali-Aggregate Reaction," *Proceedings*, American Society for Testing and Materials, Philadelphia, Vol. 59, 1959, pp. 1178–1202.

[32] Bremner, T. W., "Alkali-Aggregate Tests on Structural Lightweight Aggregate Concrete," unpublished private communication, Nov. 1991.

[33] "Criteria for Designing Lightweight Concrete Bridges," Federal Highway Administration Report No. FHWA/RD-85/045, Final Report, McLean, VA, Aug. 1985.

[34] Hoff, G. C., "High Strength Lightweight Aggregate Concrete for Arctic Applications," *Proceedings*, symposium on the Performance of Structural Lightweight Concrete, American Concrete Institute, Detroit, MI, Nov. 1991.

Cellular Concrete

Leo A. Legatski[1]

PREFACE

The chapter on Cellular Concrete was first presented in *ASTM STP 169B* in 1978 and was authored by my father, Professor Leo M. Legatski (Civil Engineering) of the University of Michigan, Ann Arbor. Professor Legatski's special interest in cellular concrete began in the late 1940s continuing until his death in 1986. He was active in this field with contributions to both the American Society for Testing and Materials and the American Concrete Institute. Although I have tried to follow the general presentation of the previous chapter, I have updated the sections and stressed new applications of the material.

INTRODUCTION

Cellular concrete is usually cast in densities ranging from 320 to 1920 kg/m³ (20 to 120 lb/ft³). Density control is achieved by adding a calculated amount of air as a preformed foam to a cementitious slurry with or without the addition of sand or other materials.

Reducing the mixture density by the addition of preformed foam is accompanied by a reduction in strength properties and thermal conductivity. Normally, it is possible to select a concrete density to satisfy strength requirements and provide increased insulating value at a reduced density.

As a result, typical applications for cellular concrete are usually concrete applications that are required to resist modest loads over relatively small spans. Examples of nonstructural cellular concrete applications include: floor fills, sloping roof screeds, filling voids to support slabs and roadways, and reducing lateral loads on wall structures.

The purpose of this chapter is to discuss cellular concrete applications, physical properties, and mixture proportioning.

BACKGROUND

Cellular concretes are lightweight concretes. They consist of a system of macroscopic air cells uniformly distributed in either a cement slurry or a cement grout (with aggregate). The cell size varies approximately from 0.10 to 1 mm (0.004 to 0.04 in.). The air cells must be tough and persistent in order to withstand the rigors of mixing and placing as the air cells are separated, coated with cement paste, and the concrete is pumped or otherwise transported to the point of placement.

This chapter deals only with air cells preformed as foam and added to the cementitious mixture either in the mixer or in the pumping hose. It is also possible to form air cells in a slurry by chemical reaction or by vigorous mixing of the slurry with a foaming additive in a high-speed mixer. In the high-speed mixing method, the volume of air cells produced depends on the amount and properties of the foam concentrate, the mixing time, and the temperature of the water and other materials. Batch density control of this process is difficult at best.

Another method for forming air cells is a chemical reaction forming gas concrete. This European process is well documented. It is suited for precasting operations and requires a large capital investment in plant [1].

Classification of Cellular Concretes

Neat-Cement Cellular Concrete

Neat-cement cellular concrete consists of portland cement, water, and preformed foam. It contains no solid aggregates. Neat-cement cellular concrete is usually limited for practical purposes by the cement content of the mixture. An upper limit is a cast density of about 800 kg/m³ (50 lb/ft³). Substitution of pozzolanic materials for a portion of the cement lowers the actual cement content and permits higher densities of these neat-cement mixtures without producing excessive heat of hydration.

Sanded Cellular Concrete

Sanded cellular concrete is cellular concrete that contains fine aggregate (sand) in addition to cement, water, and preformed foam. These concretes are usually produced in cast densities of 800 to 2080 kg/m³ (50 to 130 lb/ft³). The mixture properties are primarily dependent on the cement content, the water/cement ratio, and the specific characteristics of the sand.

Lightweight Aggregate Cellular Concrete

Lightweight aggregate cellular concrete is similar to sanded cellular concrete but with lightweight aggregate replacing all or part of the sand. The lightweight aggregates are usually a structural grade in order to increase the strength/density ratio. This chapter will not discuss

---

[1] President, Elastizell Corporation of America, Ann Arbor, MI 48106.

expanded lightweight aggregates such as vermiculite or perlite.

Cellular Concrete Modified with Admixtures

Cellular concrete modified with admixtures can improve the properties of cellular concrete mixtures. The following admixtures have been used successfully in cellular concrete.

Cement dispersing agents

Cement dispersing agents disperse cement particles, improve their efficiency in the batch, and increase the compressive strength of the mixture. These high-range water-reducing admixtures (superplasticizers) reduce the water/cement ratio to produce higher strength concrete at a given density than if they were not used.

Chopped fiber

Chopped fiber increases the tensile strength and helps control shrinkage cracking. Various fibers used include steel, glass, polypropylene, polyester, nylon, or other special composites. The fiber quantity is a compromise based on the required workability of the concrete, the fiber's efficiency, and its cost.

A volume of fiber equal to about 0.5% of the absolute volume of cement is often the starting point for research. The upper limit depends upon the projected use of the concrete, its required workability, and cost. The proposed fiber length ranges from about 13 to 38 mm (0.50 to 1.5 in.). A lower fiber length limit is based on the fiber's ability to bond with the cement paste. The upper length limit depends on the increasing difficulty of dispersing fibers in the mixture as the length is increased. A length of 19 mm (0.75 in.) has been found to be very efficient for cellular concretes.

Fly ash

Fly ash is no longer considered a waste material. Its production and properties can be controlled within ASTM Specification for Fly Ash and Raw or Calcined Natural Pozzolan for Use as a Mineral Admixture in Portland Cement Concrete (C 618) resulting in a commercially viable product. Substituting fly ash for a portion of the cement usually has reasonable cost savings without adversely affecting significant properties of the cellular concrete. Its pozzolonic properties improve flowability, increase compressive strength, reduce the heat of hydration, and reduce water permeability.

Latex

Latex, if compatible, may improve certain strength characteristics. Depending on its chemical structure, latex can set up an additional matrix within the normal cement structure and impart specialized properties to the material.

Waste materials

Waste materials may be incorporated into cellular concrete mixes to control strength, density, heat of hydration, and other physical properties. Common waste materials include recycled plastics and rubber. Some applications incorporate difficult to handle waste materials into cellular concrete for ultimate disposal.

Density

In referring to the density of cellular concrete, confusion may be avoided by stating the moisture condition of the material at a specific density. Significant moisture conditions include cast density (plastic concrete density), air-dry density (at a stated age and curing condition), and the oven-dry density.

The air-dry density of cellular concrete usually represents the condition of the in-place material. The change in density due to air drying is a function of temperature, duration of the drying period, humidity, the cast density of the concrete, the water/cement (w/c) ratio, and the surface/area ratio of the element. Although the relationship between air-dry density and cast density seems complicated, the air-dry density of cellular concrete is usually about 80 kg/m^3 (5 lb/ft^3) less than its cast density. Cellular concrete cast, cured, and air dried under job conditions in low-humidity environments may have density losses approaching 160 kg/m^3 (10 lb/ft^3).

Oven-dry density of cellular concrete usually is used only for the determination of thermal conductivity by the guarded hot plate method, that is, ASTM Test Method for Steady-State Heat Flux Measurements and Thermal Transmission Properties by Means of the Guarded-Hot-Plate Apparatus (C 177). For this purpose, the oven-dry density may be calculated with sufficient accuracy from the mixture data by assuming that the water required for hydration of the cement is 20% of the weight of the cement. The oven-dry density (D) is calculated as follows

$$D = [1.2C + A] \text{ kg/m}^3 \quad \text{or} \quad [(1.2C + A)/27] \text{ lb/ft}^3$$

where

C = weight of cement, kg/m^3 (lb/yd^3) of concrete; and
A = weight of aggregate, kg/m^3 (lb/yd^3) of concrete.

There are no defined upper and lower limits for the cast density of cellular concrete mixtures. In practice, the approximate cast density range is considered to be from 320 to 1920 kg/m^3 (20 to 120 lb/ft^3) although concretes at densities from 160 to 2080 kg/m^3 (10 to 130 lb/ft^3) have been produced [2].

APPLICATIONS

Floor Fill

Floor fill mixtures are job-site produced from a sand/cement grout at 1600 to 1760 kg/m^3 (100 to 110 lb/ft^3) having a cement content of 256 to 298 kg/m^3 (564 to 658 lb/yd^3). Preformed foam is added to the sand/cement grout. This grout is pumped into place over wood-frame construction that is typically wood joists and a plywood subfloor. The floor fill is cast 3.8 cm (1.5 in.) thick, the dimension of a standard wood bottom plate.

The purpose of floor fill is to provide fire resistance and sound attenuation characteristics to floor/ceiling systems. The noncombustible floor fill reduces flame penetration through the subfloor. The mass of the cellular concrete, the ceiling board attachment (direct nailed or resilient channels), and the floor covering (carpet and pad) provide airborne (sound transmission class (STC)) and impact (impact insulation class (IIC)) sound attenuation for these constructions.

Floor fill mixtures may also top precast concrete units for leveling the camber of these members to provide a flatter floor. Floor fills are occasionally cast over corrugated steel deck 38 to 51 mm (1.5 to 2 in.) thick above the top of the flutes. Fire ratings exist for many of these constructions.

Cellular concrete floor fills are very effective in floor rehabilitation for the renovation of older buildings. It levels the floors of these structures while minimizing added dead load. For thick fills, lighter materials can occupy much of the volume and then be topped with the cellular concrete floor fill. In floor rehabilitation, cellular concrete floor fills are cast over a variety of floor substrates such as concrete, wood, terrazzo, wood sleepers and cinder fill, tile, etc. Each application must be analyzed on its individual requirements.

Roof Decks

Roof deck applications for new construction are cast over galvanized steel decking (corrugated or fluted) to provide fire ratings, additional seismic resistance, thermal insulation, and a slope-to-drain roof deck. The finished deck is then covered with a waterproofing membrane. Substrates such as concrete (precast and cast-in-place) and wood decking are also common for this application.

The typical mixture utilizes a predetermined quantity of foam added to a cement/water slurry resulting in a density of 480 to 640 kg/m^3 (30 to 40 lb/ft^3). Expanded polystyrene insulation board (EPS) is often included in a sandwich construction over the structural deck. This involves filling the steel deck flutes or otherwise slurrying to a deck and laying the EPS board in the slurry so that it is "cemented" or bonded to the structural deck. It is then topped with the final cellular concrete roof deck fill resulting in a slope-to-drain surface ready for the membrane roofing.

Cellular concrete is an excellent material for reroofing applications. If the existing roofing is not removed, the existing structure must be checked for its ability to support additional load. The sloping roof deck usually incorporates EPS board sandwiched within the fill. If the existing roofing system is wet or damaged, those portions must be replaced or repaired prior to casting the cellular concrete. This solution provides positive drainage for "ponded roof decks" and protects and preserves the insulation system for future reroofing.

Standard roofing membranes are used over cellular concrete roof decks. These include built-up systems as well as single-ply membranes such as ethylene propylene diene monomer (EPDM), polyvinyl chloride (PVC), polyisobutylene (PIB), and modified bitumens. It is not the purpose of this chapter to further discuss these roofing membrane systems.

Engineered Fills

Cellular concretes provide alternate solutions to standard geotechnical procedures. In many cases, engineered fills can economically replace common geotechnical solutions such as piling, removal and replacement of poor soils, surcharging, and bridging over poor soils to name a few. In most applications load balancing or weight credit techniques are combined within the engineered fill solution. For example, to reduce the load over poor soils, it is advantageous to remove part of the existing soil, typically weighing 1920 kg/m^3 (120 lb/ft^3), and replace it with 480 kg/m^3 (30 lb/ft^3) cellular concrete. Thus, for every unit of soil removed, four units of cellular concrete can raise the elevation without additional load.

The excellent flow characteristics of cellular concrete and the fact that cellular concrete does not require the compaction granular fills require are major advantages in these applications. Each application is unique and must be considered on its individual merits. Common applications include roadway rehabilitation, bridge approaches, fills behind retaining walls, load-reducing fills over structures, and various pipeline and culvert applications.

Precast Elements

Precast cellular concrete elements may be produced from specially designed mixtures. Although these are not commercially available in North America, they are more common in Europe.

CONSTRUCTION AND APPLICATION TECHNIQUES

Cellular concrete mixtures are typically job-site produced and placed. If transit mix trucks are used for sanded mixtures (floor-fill applications), the grout is delivered to the job site and the preformed foam is added just prior to placement. This maintains the quality and freshness of the material. The rotary drum action of a transit mix truck is acceptable for sanded mixtures with densities greater than 800 kg/m^3 (50 lb/ft^3).

For low-density applications (roof deck and engineered fill) with neat cement slurries at densities less than 800 kg/m^3 (50 lb/ft^3), rotary drum mixing action is not deal. Instead, paddle type or shear mixers are common methods for both batch and continuous mixing procedures. After the cement/water slurry is produced in these mixers, the preformed foam is added and blended prior to or during placement with a positive displacement pump. As the mixture is pumped, density is measured at the point of placement for quality control. Mix adjustments can then be made to account for pumping distances and other special application conditions.

PHYSICAL PROPERTIES

Because the density of cellular concrete may be varied over such a wide range, 320 to 1920 kg/m³ (20 to 120 lb/ft³), it is an additional variable for selecting the physical properties and a suitable mixture design. For example, low-density cellular concrete has lower thermal conductivity (higher insulation) accompanying its lighter weight. Although strength properties are reduced as the density decreases, they are predictable and reproducible within given parameters.

Compressive Strength

The principal factors affecting the compressive strength of cellular concrete include cast density, cement content, water/cement ratio, aggregate type and amount, special admixtures, and curing conditions. Figure 1 is a typical curve of compressive strength versus cast density (unit weight) for cellular concrete. Similar curves can be developed for different cement factors, water/cement ratios, and various ingredients or admixtures.

Tensile Strength

The tensile strength of cellular concrete bears a similar relationship to the compressive strength as with stone aggregate concrete. Tensile strength is typically 10 to 15% of the compressive strength. Since low-density cellular concretes have very low tensile strengths, adding fiber to increase the tensile strength is beneficial and usually cost effective for specific applications. The addition of fiber often doubles the tensile strength of cellular concrete.

Shear Strength

The shear strength of cellular concrete may be calculated from the formulas for lightweight concrete per ACI 213R-87.

Bond Strength

Because steel reinforcing is not usually used with cellular concrete, current data are not available. Any testing would be more applicable for higher densities since that is where the reinforcing steel is most effective. Some data are available in ACI 523.2R.

Modulus of Elasticity

The modulus of elasticity of cellular concrete is a function of its density and compressive strength. It is reasonable that cellular concrete has a lower modulus of elasticity than concrete of the same density but made with lightweight aggregate, because the cellular concrete has a lower compressive strength.

In a laboratory study of modulus of elasticity of cellular concretes whose cast density varied from 1280 to 1872 kg/m³ (80 to 117 lb/ft³), the following equation was selected to best represent the test results

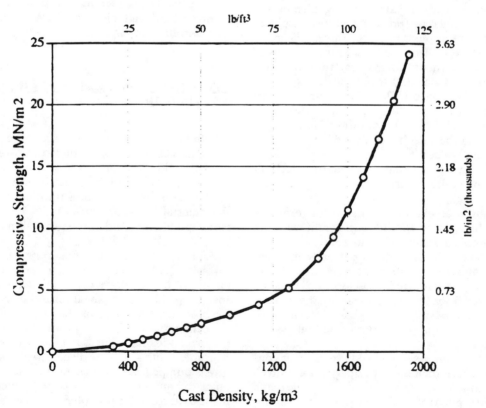

FIG. 1—Compressive strength versus unit weight.

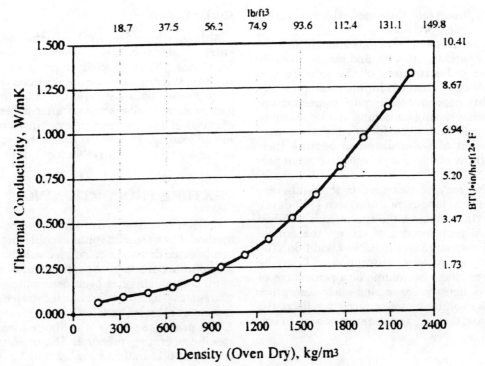

FIG. 2—Thermal conductivity versus density.

$$E_c = W^{1.5} \cdot (37.04) \cdot \sqrt{f'_c} \, (N/m^2)$$

or

$$E_c = (W^{1.5}) \cdot (28.6) \cdot \sqrt{f'_c} \, (lb/in.^2)$$

where

W = (cast density − 80) kg/m³ or (cast density − 5) lb/ft³, and
f'_c = 28-day compressive strength, N/m² (lb/in.²).

Tests have confirmed that the same equation is applicable for low density cellular concretes as well.[2]

Thermal Conductivity, k

The thermal conductivity of a material is the time rate of transfer of heat by conduction, through a unit thickness, across a unit area for a unit difference of temperature. The units of k are watts/metre-Kelvin or W/mK (BTU · in./h · ft² · °F).

Information from several sources was used to develop Fig. 2. These test results represent determinations of thermal conductivity of cellular concretes for a limited range of low densities and from the National Institute of Standards and Technology (NIST) curve for concrete made using no-fines gravel and lightweight manufactured aggregate. This curve reasonably demonstrates the relationship of thermal conductivity to oven-dry density for cellular concretes over the entire density range [3].

Permeability

Data are available on the permeability of lower densities of certain cellular concretes. For cellular concretes with a discrete cell system, the permeability values range from 1×10^{-4} to 1×10^{-8} cm/s (2×10^{-4} to 2×10^{-8} ft/min). Although they are somewhat density dependent, these values are more dependent on the confining head pressure of the test rather than on the density of the cellular concrete.[3]

The permeability test for cellular concrete is different from ASTM Test Method for Permeability of Granular Soils (Constant Head) (D 2434) that is a soil testing procedure for clean sand. Cellular concrete permeability is most commonly tested by the flexible-wall method. This procedure varies the confining head pressure and measures the permeability during steady-state flow.

Water Absorption

Since cellular concrete has a lower density than normal-weight concrete, it may, under certain conditions, absorb and retain water. The consequence of water absorption is a unit weight greater than expected. The amount of water absorbed by cellular concrete varies not only with the density of the material but also with the quality of the mixture ingredients—specifically, the cement, fly ash (if used), and preformed foam concentrate. Some concentrates do not produce a discrete cell structure within the cement matrix. This may be related to the chemistry of the

---

[2] L. M. Legatski, personal communication.

[3] L. A. Legatski, Elastizell Corporation of America, personal correspondence.

ingredients, their dilution rate, their method of generating foam, or their expansion pressure.

If a preformed foam is not tenacious enough to withstand the rigors of batching, mixing, and placing (usually by pumping), the cell structures of the mixture will interconnect resulting in channels for high water absorption. Lower quality concentrates require more mixture water that evaporates during hardening and leaves more pore spaces. Fly ash and silica fume admixtures may decrease the amount of water absorbed because their smaller particles fill available spaces between cement particles.

Water absorption may be measured in three different ways: 24-h immersion, long-term immersion (120 days), and by tide cycle. In each case, the specimen is weighed prior to testing. At each interval of testing, the sample is reweighed and compared to the initial condition. The increase in density is the water absorption and may be expressed as a percentage by volume or a percentage of the initial weight or density. Expressing water absorption as a percentage by volume more accurately reflects the effect of sample size, density, and the surface area to volume relationship.

Frost-Heave Resistance

Since fly ash mixtures are used in highway construction, a procedure has been developed for testing their resistance to frost heave. Low-density cellular concrete tested in this manner exhibits excellent results even with high fly ash substitution. This is because of the high air content of the cellular concrete. However, a poor quality fly ash can reduce otherwise excellent performance for a specific density and replacement percentage.

Freeze-Thaw Resistance

Cellular concrete has excellent freeze-thaw resistance because of its high air content and its relatively high-cement content. ASTM Test Method for Resistance of Concrete to Rapid Freezing and Thawing (C 666) for standard-weight concrete must be modified because the insulating properties of cellular concrete prevent the samples from reaching the specified temperatures in the designated period of the freeze and thaw cycles. Although the samples do not completely freeze or thaw in their centers during the specified time cycle, freezing and thawing occur at less than full depth so that this test can still be considered a measure of the specimen's durability. As a result, the test is often referred to as a modified freeze-thaw test.

Deformation Testing

Low-density cellular concretes possess excellent deformation resistance because of its cellular structure and its ability to absorb loads and crush in a controlled manner. The absorbed energy is greater at lower densities that can withstand more deformation before "bridging and locking" of the material occurs.

Cyclic Loading

Although limited testing has been conducted, some densities of cellular concrete exhibit excellent cyclic loading resistance. This is especially important in highway, airport runway, and railroad base applications when cellular concrete is used as a subbase. Cyclic testing measures the loss in compressive strength after a designated number of cycles at a given frequency varying between a predetermined loading. The results are expressed as a percentage of the actual compressive strength of the sample.

MIXTURE PROPORTIONING

Cellular concretes are proportioned by the solid volume method. For a specific volume (cubic meter or cubic yard) at a selected density, a calculated volume of air completes the volume portion of the mixture after the solid volume of all components has been determined—water, cement, aggregates—based on a selected water/cement ratio or cement content or both.

The preformed foam is calibrated from a nozzle at a specific operating pressure. The preformed foam has a density of 0.16 to 0.22 kg/m^3 (2.5 to 3.5 lb/ft^3), a foam/air volume ratio for the pressure at which it is generated, and a volume output per unit of time at this generated pressure. All three calibration factors are included in mixture proportioning.

Slurry Mixtures

Slurry mixtures are low-density mixtures without aggregates. Proportioning a slurry mixture involves selecting a density and then selecting a water/cement ratio that is compatible for both that density and the specific application. The water/cement ratio can be within a reasonable range for a selected density.

If the water/cement ratio is too low, the cement may be starved for water of convenience and it will seek water from the preformed foam resulting in a loss of volume (yield) and an increase in density. Too high a water/cement ratio results in excessive water bleeding and a weaker concrete mixture.

Grout Mixtures

Grout mixtures are slurry mixtures with aggregates. Their proportioning involves selecting a cement content and a water/cement ratio for a specific density. The selection of these two variables will result in a cellular concrete of a given strength based on the characteristics of the ingredients (cement and aggregate). If different properties such as higher strength are required, the density, cement content, and water/cement ratio may be varied according to the designer's experience.

APPLICATION TECHNIQUES

Accurate batching of ingredients is more critical with cellular concrete mixtures than with regular concrete.

Because a significant portion of the material volume (one fourth to three fourths) is air, the other materials must be batched accurately. For these mixtures, the cement and sand (or other aggregate) are usually weighed into the batch. The mixture water is metered while the preformed foam is time injected into the mixture through a calibrated nozzle.

The mixing technique should be compatible with the type of mixture, its ingredients, the job requirements, and the method of placement. Mixing should efficiently mix the cement and water and properly blend them with the other ingredients including the preformed foam. Paddle, high shear, continuous, and rotary drum mixers may all be acceptable for specific applications depending on the quality requirements of the final product.

Positive displacement pumps such as moyno or peristaltic pumps are used for low-density mixtures. Although piston pumps are efficient for grout mixtures at densities greater than 1440 kg/m^3 (90 lb/ft^3), they do not efficiently pump low-density mixtures.

Casting techniques are different for each type of cellular concrete application. The thinner floor fill mixtures utilize a rolling screed to provide a constant thickness. Since roof deck applications are cast slope-to-drain, string lines provide guides for casting and darby finishing the material by experienced tradesmen. Screed rails may also be used.

Geotechnical fill applications have the greatest variation in casting techniques. Generally, these fills are many meters (feet) thick so that they are cast in lifts of up to 1.0 m (3.3 ft) thick based on the available area to be cast. The succeeding lifts are cast on a daily basis until the final fill profile is reached.

QUALITY CONTROL AND TESTING

Field quality control procedures for cellular concrete involve measuring the cast density of the material at the point of placement and comparing this density with the design density of the mixture. Slight adjustments in the mixture may be necessary to bring the cast density within the specified range. If batching, mixing, and placing procedures are consistent, the material should have only minor deviations in density.

The physical properties of cellular concrete are closely related to the mixture design parameters such as cement content, water/cement ratio, and the quality of other ingredients such as sand gradation and fly ash quality. The physical properties of the mixture are reproducible within acceptable limits.

The actual quality control procedure involves securing cylindrical test specimens from the measured cast density at the point of placement. After a few days, these samples are moved from protection at the casting location to the laboratory for subsequent curing and compressive strength testing. Compressive strength and density are the primary indicators of mixture quality. Properties such as permeability, freeze-thaw resistance, and water absorption correspond to these measured parameters—strength and density.

SUMMARY

Cellular concrete has many potential applications. There are significant opportunities for expanding existing applications as well as continuing research into new areas such as specialized material performance properties. Whenever a lightweight, inert, insulative, and cost-effective material is desired, it is possible that a specific density and mixture of cellular concrete could be designed for the application.

REFERENCES

[1] "Autoclaved Aerated Concrete," RILEM Committees on Autoclaved Aerated Concrete, International Union of Research and Testing Laboratories for Materials and Structures, RILEM, Paris, 1993.

[2] Legatski, L. M. and Mansour, A. E., Jr., "Properties and Proportioning of Elastizell Cellular Concrete," University of Michigan Research Institute, Report 2326-32-F, Aug. 1958.

[3] Valore, R. C., Jr., "A Study of Cellular Concretes," *NBS Technical News Bulletin*, National Bureau of Standards, March 1955, pp. 41–43.

BIBLIOGRAPHY

American Concrete Institute, *Guide for Structural Lightweight Aggregate Concrete*, ACI 213R-87, Detroit, MI.

American Concrete Institute, *Guide for Cast-In-Place Cellular Concrete*, ACI 523.1R-92, Detroit, MI.

American Concrete Institute, *Guide for Low Density Precast Cellular Concrete*, ACI 523.2R-93, Detroit, MI.

American Concrete Institute, *Guide for Cellular Concretes above 50 pcf*, ACI 523.3R-93, Detroit, MI.

American Concrete Institute, *Lightweight Concrete*, SP-29, Detroit, 1971.

American Society for Testing and Materials, Test Method for Steady-State Heat Flux Measurements and Thermal Transmission Properties by Means of the Guarded-Hot-Plate Apparatus (C 177), *Annual Book of ASTM Standards*, Philadelphia, 1993.

American Society for Testing and Materials, Test Method for Compressive Strength of Lightweight Insulating Concrete (C 495), *Annual Book of ASTM Standards*, Philadelphia, 1993.

American Society for Testing and Materials, Test Method for Steady-State Heat Flux Measurements and Properties by Means of the Heat Flow Meter Apparatus (C 518), *Annual Book of ASTM Standards*, Philadelphia, 1993.

American Society for Testing and Materials, Test Method for Resistance of Concrete to Rapid Freezing and Thawing (C 666), *Annual Book of ASTM Standards*, Philadelphia, 1993.

American Society for Testing and Materials, Method of Testing Foaming Agents for Use in Producing Cellular Concrete Using Preformed Foam (C 796), *Annual Book of ASTM Standards*, Philadelphia, 1993.

American Society for Testing and Materials, Specification for Foaming Agents Used in Making Preformed Foam for Cellular Concrete (C 869), *Annual Book of ASTM Standards*, Philadelphia, 1993.

American Society for Testing and Materials, Test Method for Permeability of Granular Soils (Constant Head) (D 2434), *Annual Book of ASTM Standards*, Philadelphia, 1993.

Concrete for Radiation Shielding

Douglas E. Volkman[1]

PREFACE

Milos Polivka and Harold Davis were pioneers in the field of nuclear shielding. Mr. Polivka was Professor of Civil Engineering at the University of California, Berkeley, for many years, while Mr. Davis was affiliated with the Department of Energy's Hanford Site, Richland, Washington, during his long career. Both individuals had a deep desire to advance radiation shielding concepts and standards. M. Dean Keller, a member of ASTM, who knew both men, described Mr. Polivka as an expert in expansive cement as well as concrete for radiation shielding. He added that Mr. Davis was noted as the first person to add boron to concrete to enhance shielding and was an expert in heavy-weight concrete. Both individuals were long-time members of ASTM and both shared authorship of this chapter in the previous edition of this publication, *ASTM STP 169B*. Their leadership in this field paved the way for this newly revised chapter.

SUBCOMMITTEE C09.41 ON CONCRETE FOR RADIATION SHIELDING

History and Responsibilities

Health, safety, environmental issues, and perceptions have been difficult for the Nuclear Power Industry to surmount. In the entire decade of the 1980s, not one nuclear power plant was ordered for construction in the United States. This trend seemingly will hold for the 1990s as well. With the lack of interest in nuclear power, human and monetary resources have been drained from all aspects of the field. Subcommittee C09.41 on Concrete for Radiation Shielding is a concrete systems subcommittee assigned to establishing and maintaining standards in the general discipline of concrete for radiation shielding.

A grand symposium, jointly hosted by ASTM and the American Nuclear Society, was held in 1965. This symposium was the catalyst of Subcommittee C09.02.08, the predecessor of C09.41. Results of three years of study by this subcommittee were presented and led to the development of ASTM Specification for Aggregates for Radiation-Shielding Concrete (C 637), and ASTM Descriptive Nomenclature of Constituents of Aggregates for Radiation-Shielding Concrete (C 638) [1].

Preplaced-aggregate concrete is a method used to make very dense, uniform radiation shields. Subcommittee C09.41 is responsible for the preplaced-aggregate concrete standards in addition to ASTM C 637 and ASTM C 638. The designation and titles for these standards are as follows:

1. ASTM Specification for Grout Fluidifier for Preplaced-Aggregate Concrete (C 937),
2. ASTM Practice for Proportioning of Grout Mixtures for Preplaced-Aggregate Concrete (C 938),
3. ASTM Test Method for Flow of Grout for Preplaced-Aggregate Concrete (C 939),
4. ASTM Test Method for Expansion and Bleeding of Freshly Mixed Grouts for Preplaced-Aggregate Concrete in the Laboratory (C 940),
5. ASTM Test Method for Water Retentivity of Grout Mixtures for Preplaced-Aggregate Concrete in the Laboratory (C 941),
6. ASTM Test Method for Compressive Strength of Grouts for Preplaced-Aggregate Concrete in the Laboratory (C 942),
7. ASTM Practice for Making Test Cylinders and Prisms for Determining Strength and Density of Preplaced-Aggregate Concrete in the Laboratory (C 943), and
8. ASTM Test Method for Time of Setting of Grouts for Preplaced-Aggregate Concrete in the Laboratory (C 953).

Recent Subcommittee Activities

A recent symposium was hosted by Subcommittee C09.41 in June of 1991 at Atlantic City, New Jersey. The symposium was widely advertised to be general and not limited to concrete. The response was disappointing with a total of five papers presented. Only two of these presented new data in the use of concrete as a shielding material. Proceedings from this symposium are available [2], but relevant information pertaining to concrete is outlined later in this chapter.

RADIATION

From the early days of the nuclear industry, concrete has been used for shielding due to its strength, chemical

---

[1] Technical coordinator, Civil/Structural Engineering Section, Los Alamos National Laboratory, Los Alamos, NM 87545.

composition, and ability to be formed into desired shapes. Different additives and aggregates have been used in concrete to enhance shielding for various types of radiation. In order to establish a firm basis for using different concrete mixes for radiation shielding, a knowledge of radiation is essential for those working in the field. Beginning with the basics, simple models of nuclear activity and interaction with materials are developed. These are used to define the properties of concrete that are valuable in radiation shielding.

Simple Nuclear Models

All elements are composed of unique atoms. These atoms contain a certain number of negatively charged particles called electrons as well as an equal number of positively charged particles called protons. The equal numbers of protons and electrons create an electrically balanced atom. In addition to protons and electrons, the atom has some number of neutrally charged particles, neutrons, associated with a particular element. An element consists of an atom uniquely comprising the same number of protons, but possibly a variable number of neutrons attributed to it. Like elements with differing neutron counts are called the isotopes of the element. The nucleus or center of the atom is arranged to contain essentially all the mass. That is, all the protons and neutrons comprise the nucleus, while the electrons orbit the center.

Light elements, so called since these are attributed with the fewest number of protons, have very stable nuclei. The nuclei of the heavy elements are somewhat unstable due to the large ratio of neutrons to protons. This inequity creates a potential for instability. To achieve a more stable condition, particles or energy, or both, are released from the atom causing the atom to be radioactive [3]. However, only ionizing radiation, which interacts with atoms to cause unbalanced electrical charges, needs shielding. There are two categories of ionizing radiation:

(*a*) electromagnetic waves, and
(*b*) nuclear particles [1].

Electromagnetic waves consist essentially of gamma rays, which are similar to X-rays, but have shorter wavelengths. Traditional nuclear particles are alpha, beta, and neutron. Other nuclear particles exist, but the theme of positive, negative, and neutral charges of the particles are maintained [4].

Alpha particles consist of positively charged helium atoms. During atom decay into more stable states, emission of pairs of protons and neutrons (He^{++}) can occur. This type of radioactive emission defines alpha particles.

Beta particles consist of negatively charged electrons. As an example, consider an atom being split during fission. A neutron interacts with a radioactive element, creating two new elements. Binding energy is released. Neutrons split into protons and electrons to stabilize the atomic matter. These free electrons are ejected from the nucleus as beta particles.

Neutron particles are released during fission also. These particles are important in maintaining the fission process by striking other potentially unstable atoms and creating a chain of fission activity. In addition to fission, scientific accelerators produce beams of neutron particles for a variety of experimentation purposes. These processes must be shielded to address health, safety, and environment issues.

Generally, there are three types of neutron particles that are grouped according to the level of kinetic energy. Slow neutrons are referred to as thermal neutrons and have associated energy levels less than 0.5 electron volts (eV). Intermediate or epithermal neutrons have kinetic energy in the range of 5000 eV. Fast neutrons have kinetic energy levels exceeding 500 000 eV [3].

PHYSICAL AND BIOLOGICAL EXPLANATION

Any material of sufficient thickness can be used as shielding against alpha, beta, and gamma radiation. Alpha and beta particles are deflected and absorbed by interacting with the electrical charges of the atoms comprising the shield. In contrast, gamma rays are absorbed by the mass of the shield. Dense materials comprise the best shielding for gamma radiation. Obviously, neutrons are not affected by the electrical fields associated with atoms in a shielding material. Instead, these particles collide elastically or inelastically with the nucleus of light or heavy atoms, respectively.

Shielding is required to protect people and the environment from the biological effects of radiation. Alpha particles cannot penetrate human skin, but are harmful if ingested. Beta particles can penetrate human skin, but not reach the body organs. These also are harmful if ingested. Radon gas, situated in certain geological formations, is an example of naturally occurring alpha and beta particles [5], which unfortunately, is easily inhaled to the lungs. Alpha and beta particles are considered non-penetrating. Penetrating radiation consists of gamma rays and neutrons. This type of radiation can penetrate the body and damage the atoms of the cells composing the various organs [4].

Both non-penetrating and penetrating radiation can ionize or break the bonds of atoms in human body cells, however, non-penetrating radiation is only a biological threat if the particles are introduced internally, while penetrating radiation comprises a biological threat from sources external to the being. Radiation exposure effects on humans consist of radiation sickness, cataracts, sterility, genetic mutations, coma, cancer, and death—depending on the amount and length of time of the exposure [4].

TERMS FOR SHIELD DESIGN CALCULATIONS

Of the numerous material properties and constants associated with shield design, a few are fundamentally important to allow communication between people in the concrete industry with clients or designers in the nuclear industry; four such terms are explained in the following paragraphs.

Absorption Cross Section

This is the material property representing the probability that a neutron will be absorbed by an elemental nucleus. Absorption is an ion exchange in which the target nucleus captures a neutron particle, splits into new elements, and releases gamma-ray energy in the process. An elemental isotope, such as boron-10, that has a very large absorption cross section and releases low energy gamma rays is very advantageous as a shielding material.

Scattering Cross Section

This property is an expression of the probability that a neutron will be scattered by an elemental nucleus. Scattering is accomplished through collisions of a neutron with a nucleus in which the neutron's direction of travel is altered and its energy is lost in the process. An element, such as hydrogen, which has a large scattering cross section is very advantageous for use as an elastic scattering material. Large, heavy elements, such as iron, are important for inelastic scattering of neutrons.

Relaxation Length

This is the thickness of material required to reduce radiation flux to a factor of $1/e$, the inverse of the natural logarithm base number, or 0.368 of the initial radiation [6]. Minimizing the relaxation length of a shield is essential for optimization of the shield.

Absorption Coefficient

This is a term for gamma-ray shielding that is a prediction of the amount of energy absorbed by a shield. The coefficient for any element will vary nonlinearly based on the energy of the gamma rays. The larger the coefficient, the better the shield is stopping gamma penetration [6]. Dense materials consisting of heavy elements are the best materials to control gamma-ray energy. For concrete mixes, steel, iron, and iron ores (such as magnetite, barytes, ilmenite, and hematite) are used to increase mix density for shielding. The final mix density, rather than the particular heavy aggregate used, is the most important aspect of gamma shield design since shield thickness varies proportionately with density [1].

CONCRETE

Normal-weight concrete consists of cement, water, sand, and coarse aggregate. The addition of natural or manufactured minerals, in the form of sand and coarse aggregate, can alter the composition of normal-weight concrete into an efficient radiation shielding material. This conglomeration of materials results in a mix of light and heavy elements. The light elements moderate and absorb neutrons, while the heavy elements are required to stop gamma rays. Proper mixing and placing of the concrete is essential to make as homogeneous a shield as possible.

Natural Minerals

Natural minerals, added to concrete to increase density for gamma-ray attenuation, consist of magnetite ($Fe_2O_3 \cdot FeO$), ilmenite ($FeO \cdot TiO_2$), barite ($BaSO_4$) and hematite (Fe_2O_3) [7].

Serpentine ($3MgO \cdot 2SiO_2 \cdot 2H_2O$), goethite ($Fe_2O_3 \cdot H_2O$), limonite ($Fe_2O_3 \cdot XH_2O$ and referred to as impure goethite), as well as colemanite ($2CaO \cdot 3B_2O_3 \cdot 5H_2O$) [7] are natural minerals commonly used to increase shielding characteristics for neutrons. Caution needs to be exercised with the use of serpentine, however, because it contains asbestos, a known carcinogen.

Manufactured Aggregates

Manufactured products used for gamma shielding typically consists of ferrophosphorous (FeP or Fe_2P or Fe_3P), iron or steel shot, and steel punchings [7].

Boron frit, which consists of the fritted product of silica (SiO_2) with borax ($Na_2B_4O_7 \cdot 10H_2O$), as well as ferro boron and boron carbide (B_4C) [7] are manufactured products that are efficient for neutron shielding. Table 1 lists these aggregates along with the density of each material.

Densities

Standard heavy-weight concrete, using natural iron ore aggregate, can achieve densities up to 3850 kg/m³ (240 lb/ft³) [1]. Barite, magnetite, and ilmanite are the traditional materials used for this endeavor. Table 2 lists concrete densities attainable using heavy aggregate in the mix. Serpentine, goethite and limonite are commonly used for

TABLE 1—Aggregates Used in Shielding Concrete [1].

Natural Mineral		Manufactured	
Local sand and gravel		Crushed aggregates	
calcareous	(2.5 to 2.7)[a]	heavy slags	(~5.0)
siliceous		ferrophosphorus	(5.8 to 6.3)
basaltic	(2.7 to 3.1)	ferrosilicon	(6.5 to 7.0)
Hydrous ore		Metallic iron products	
bauxite	(1.8 to 2.3) [15 to 25%][b]	sheared bars steel punching	(7.7 to 7.8)
serpentine	(2.4 to 2.6) [10 to 13%]	iron or steel shot	(7.5 to 7.6)
goethite	(3.4 to 3.8)		
limonite	[8 to 12%]		
Heavy ore		Boron additives	
barite	(4.0 to 4.4)	boron frit	(2.4 to 2.6)
magnetite		ferroboron	(5.0)
ilmenite	(4.2 to 4.8)	borated diatomaceous earth	(~1.0)
hematite		boron carbide	(2.5 to 2.6)
Boron additives calcium borates borocalcite colemanite gerstley borate	(2.3 to 2.4) (2.0)		

[a] Specific gravity is shown in ().
[b] Water of hydration is indicated in [].

TABLE 2—Densities of Shielding Concretes Commonly Specified for Given Heavy Aggregates [1].

Type of Aggregate	Density, lb/ft³	Density, kg/m³
Barite	3450 to 3600	215 to 225
Magnetite	3500 to 3750	220 to 235
Ilmenite	3500 to 3850	220 to 240
Ferrophosphorus	4550 to 4800	285 to 300
Barite and boron additive	3200 to 3450	200 to 215
Magnetite and boron additive	3350 to 3600	210 to 225

neutron attenuation, due to the fixed water (H_2O) content of these ores. Manufactured iron and steel aggregates are used to achieve concrete densities greater than 4000 kg/m³ (250 lb/ft³) [1].

ABSORPTION AND SCATTERING INTERACTION

The absorption and scattering cross sections of elements commonly found in concrete shields are listed in Table 3. The element listed with the most significant absorption cross section is boron (isotope ^{10}B), while the element with the largest scattering cross section is hydrogen. An abundance of these two elements is very desirable in neutron shields. Boron must be added to the concrete mix. However, hydrogen is present in concrete as mixing water to hydrate the cement or as fixed water content in aggregate. Mixing water is represented in two components, as free water content and fixed water, to hydrate the cement. Early in the life of the concrete, the free water is completely evaporated. Over the useful life of the concrete shield, as much as two thirds of the total original water content may be lost [9].

Irradiation Effects

There are damaging effects to concrete used as shielding from the scattering and absorption processes. Irradiation of concrete causes dissociation of water into its hydrogen and oxygen components. Compressive and rupture strengths both decrease with time due to radiation exposure and high temperature, although the effect is tolerable under normal reactor conditions [10].

TABLE 3—Absorption and Scattering Cross Sections for Elements of Shielding Concrete [8].

Element	Absorption Cross Section (barns)	Scattering Cross Section (barns)
Hydrogen (H)	0.3	99.0
Oxygen (O)	0.0	4.2
Boron (B)	755.0	4.0
Silicon (Si)	0.2	1.7
Iron (Fe)	2.5	11.0
Barium (Ba)	1.2	8.0
Aluminum (Al)	0.2	1.4
Calcium (Ca)	0.4	3.2
Sodium (Na)	0.5	4.0
Sulfur (S)	0.5	1.1

Temperature Effects

Although concrete used in a nuclear reactor shield are cured and aged before being placed in service, compressive strength loss is as high as 50% when it is exposed to temperatures of 430°C (800°F) over a long period of time. Between 430°C (800°F) and 760°C (1400°F), concrete should be made with heat-resistant aggregates to maintain compressive strength losses to 50% and to avoid dehydration. If shielding requires aggregates high in fixed water, the concrete must be subjected to lower temperatures. Limonite dehydrates above 200°C (400°F), while serpentine can withstand temperatures up to 370°C (700°F) [1].

SIZE AND ECONOMICS

Normal-weight concrete, with no special aggregates or additives is satisfactory to shield any nuclear radioactive source. However, high-energy gamma rays and neutrons from a reactor or accelerator require concrete masses that are prohibitive in size. Shielding usually has size constraints. These constraints necessitate the use of special shielding materials to increase effectiveness. However, these materials also increase the cost of shields through higher material costs and the expense associated with construction. The mixing and placing operations are more difficult because very heavy aggregates or special fine aggregates may be used making uniformity hard to achieve by normal concrete construction practice. Therefore, specifications for the material components and construction of concrete shields must be written for specific jobs. The specifier must have a knowledge of cost of the various materials available for shielding and the difficulties involved in constructing the shield with these materials. Table 4 shows comparative costs for general categories of concrete shields [11]. These values must be considered "ball park" values because material and labor costs fluctuate, competitiveness of contractors vary, and other factors such as material availability may affect cost.

ASTM STANDARDS

Cast-in-Place or Precast Concrete

Cast-in-place concrete shielding or precast concrete shield blocks should comply with normal-weight ASTM concrete standards, as amended by ASTM C 637 for aggregates. ASTM C 638 is useful as a supplement to ASTM C 637, because it provides specific descriptions of special aggregates used in concrete for radiation shielding. The following paragraphs briefly summarize the two standards.

TABLE 4—Cost Comparison of Concrete Shields.

General Category of Concrete Shield	Cost Index
Normal weight (reference)	1.0
High density/natural mineral	3.0
High density/manufactured aggregate	4.0 to 9.0
High hydrogen content	1.3

ASTM C 637

This standard is a specification of fine and coarse aggregates specifically used in concrete for radiation shielding. Composition, specific gravity, grading, fixed water content, deleterious substances, and abrasion resistance test requirements are listed, as well as requirement exceptions for the various referenced ASTM Standards.

ASTM C 638

Descriptions of the various fine and coarse aggregates specifically used in concrete for radiation shielding are presented. This standard is useful to obtain an understanding of the aggregate components used in concrete shielding. Specifically covered are iron minerals and ores, barium minerals, ferrophosphorus, boron minerals, and boron frit glasses.

Preplaced Aggregate Concrete

When very heavy-weight concrete, greater than 4000 kg/m^3 (250 lb/ft^3) is required for shielding radiation, or when an especially high level of uniformity of mix is required, preplaced aggregate concrete is one method of construction [12]. Special forms are required in which reinforcing steel and mix aggregates are placed through the aggregate in a manner that fills the voids to make a dense concrete shield first. The precise location of these components and inspection necessary for the work provides for a shield of uniform matrix, unmatched by normal-weight concrete placement techniques. The following ASTM standards deal specifically with making preplaced concrete.

ASTM C 937

This is a specification for fluidifier used in grout to enable the grout to maintain fluid characteristics, while reducing water. Fluidifier is used in preplaced-aggregate concrete where a very dense concrete is required. This standard covers physical requirements, composition, sampling, and test methods for the fluidifier.

ASTM C 938

This practice describes the method of testing required to select correct mix proportions for grout used in preplaced-aggregate concrete. Materials, sampling, and procedure are covered.

ASTM C 939

This test method establishes the time of efflux for grout through a standard flow cone in order to determine the fluidity of the grout for preplaced-aggregate concrete. It may be performed in the laboratory or the field. Calibration of apparatus and procedures are presented.

ASTM C 940

This test determines the amount of expansion exhibited by freshly mixed hydraulic-cement grout over a period of time. Also, the amount of bleed water, or the free water that rises to the top of grout, is determined. Both expansion and bleeding characteristics are necessary to make satisfactory preplaced-aggregate concrete.

ASTM C 941

This test method measures the property of freshly mixed hydraulic-cement grout to retain mixing water. This is used to qualify fluidifiers or determine the effects of admixtures in grout for preplaced-aggregate concrete.

ASTM C 942

This test method provides a measure of the compressive strength of hydraulic-cement grout, which has expansive capability, but hardens in conditions that tend to restrain expansion. This test also is a measure of the effect the fluidifier has on the compressive strength of the grout for preplaced-aggregate concrete. Temperature, humidity, sampling, preparation of specimen, and procedure standards are covered.

ASTM C 943

This practice covers procedures for making standard test cylinders and prisms for preplaced-aggregate concrete. It follows the process used in the field to construct preplaced-aggregate concrete. The cylinders are used to determine compressive strength, while the density is calculated from prism measurements.

ASTM C 953

This test is essential to determine the setting time of grout, as well as the acceptability of components of grout mixed to fluid consistency for use in preplaced-aggregate concrete. Apparatus, sampling, and procedures are presented.

RESEARCH

The following four topics of research are from papers presented at the June, 1991, Symposium on Radiation Shielding for the 21st Century, sponsored by ASTM Subcommittee C09.02.08.

Radiation Exposure Effects of Concrete

P. K. Mukherjee, affiliated with the Ontario Hydro Research Division, presented an interesting paper regarding the effects of radiation exposure on concrete [13]. Mr. Mukherjee outlined research efforts conducted at an Ontario hydro nuclear power plant, which had been operating for over 20 years. Concrete core samples from the shielding of this plant were obtained and tested for physical and thermal properties. The results from this testing were compared with the test results of samples from the original construction, as well as laboratory test specimens.

A summary of some of the findings of this research follow, however, conclusions regarding these results have not been drawn:

1. The average compressive strength of the 20-year-old heavy-weight concrete in the reactor shield increased within the thickness of the shield. The samples taken closest to the radiation source showed compressive

strengths 10% to 20% less than the samples on the farthest face from the radiation source.
2. The modulus of elasticity of the concrete shield also displayed decreased values (approximately 10%) between samples on the farthest face compared with samples closest to the radiation source.
3. The thermal diffusivity, which helps explain the rate a material undergoes temperature change, and the thermal conductivity, which is a measure of heat transfer through a material, were roughly 30% less for the 20-year-old heavy-weight concrete shield samples compared with the laboratory-tested heavy-weight concrete specimen.

Shield Construction Verification

A. J. Oswald and J. F. Shaffer wrote a paper [14] on shield verification practice for nuclear facilities at the Idaho National Engineering Laboratory. These authors outlined the shield verification process of a constructed concrete shield using a test radiation source.

A visual inspection of the shield is the initial phase of the work. All shield penetrations and other deficiencies are recorded in a log, so further testing with a radiation source will be accomplished. The test radiation source is sufficiently large to emulate actual operating conditions of the facility. Radiation flux through the shield is measured, recorded, and compared with calculated values. In this way shielding deficiencies can be corrected during facility construction without impacting construction schedules.

Concrete Properties with Fine-Particle Boron Products

A paper authored by D. E. Volkman and P. L. Bussolini [3], both with the Los Alamos National Laboratory, was a study to determine the effect of fine-particle boron products on concrete. Fine-particle boron additives in concrete are known to retard concrete setting and compressive strength properties. Usually, boron additives above the No. 30 sieve is considered concrete mix quality material. The finer particles are more soluble, resulting in interference with concrete hardening properties. However, finer particles allow better distribution of the boron within the concrete matrix making a better neutron absorbing mix.

The purpose of this testing was to establish the amount of fine-particle colemanite and boron frit that could be added to a concrete mix without substantial adverse reactions. Although additional testing is necessary to confirm these results, the following are offered for information:

1. Fine-particle colemanite (80% passing No. 200 sieve), with 41.0% B_2O_3 will attain 99% compressive strength at 28 days curing at a dosage of 428.4 N/m^3 (73.6 lb/yd^3).
2. Sand-size boron frit (100% passing No. 30, but only 6% passing No. 200) with 51.0% B_2O_3 will attain 83% compressive strength at 28 days curing at a dosage of 428.4 N/m^3 (73.6 lb/yd^3).
3. Fine-particle boron frit (100% passing No. 200 sieve) with 60.0% B_2O_3 will attain 80% compressive strength at 28 days curing at a dosage of 213.6 N/m^3 (36.7 lb/yd^3).

Shield Optimization

Dr. Ehud Greenspan, a visiting scholar at the University of California, Berkeley, presented a paper [15] on optimization of shielding materials. His work dealt primarily with hydrocarbons and certain metal hydrides, rather than concrete materials; however, the optimization code he developed, SWAN, may have application to concrete shielding. SWAN is a program that optimizes in a four-step, iterative process.

CONCLUSIONS

ASTM has been active in preparing and maintaining standards for concrete used in radiation shielding. Subcommittee C09.41, and its predecessor C09.02.08, have been given the task to perform this service.

Normal-weight concrete provides good radiation shielding characteristics, but the shielding can be optimized by altering the mix. The use of iron, steel, or iron ore for aggregates produces a heavy-weight concrete that is most suitable to shield electromagnetic radiation. Aggregates with high fixed-water content are useful in concrete to shield for neutron particles. A boron component is added to concrete for better absorption of thermal neutrons.

Concrete shields are constructed using standard cast-in-place concrete techniques or by using preplaced-aggregate construction methods. The preplaced-aggregate concrete is normally used to ensure density and uniformity requirements of a shield.

Recent research activity has focused on radiation effects on concrete shielding. Besides loss of water, compressive strength and modulus of elasticity is reduced, although none of these is significant enough to detract from the effectiveness of the shield. Other research has been directed toward optimization of shielding materials and verification of shielding effectiveness. Additional work in these areas is necessary to provide refinements in concrete shielding and to determine the effective life span of such shielding.

Acknowledgments

The author wishes to note the importance of the previous authors' work for this chapter. Also, the work and dedication of all members of Subcommittee C09.41 on Concrete for Radiation Shielding is appreciated. Finally, I gratefully acknowledge the assistance of Cheryl L. Blizzard and Lindy Wynn for the presentation of this material.

REFERENCES

[1] Polivka, M. and Davis, H. S., "Radiation Effects and Shielding," *Significance of Tests and Properties of Concrete and Concrete-Making Materials, ASTM STP 169B,* American Society for Testing and Materials, Philadelphia, 1978, pp. 420–434.

[2] *Journal of Testing and Evaluation*, Vol. 20, No. 1, Jan. 1992.
[3] Volkman, D. E. and Bussolini, P. L., "Comparison of Fine Particle Colemanite and Boron Frit in Concrete for Time-Strength Relationship," *"Journal of Testing and Evaluation,"* Vol. 20, No. 1, Jan. 1992.
[4] "Basic Radiation Worker Training," HS Division of Los Alamos National Laboratory, Los Alamos, NM, Aug. 1991.
[5] Glen, H. M., "What is Nuclear Shielding?," *Civil Engineering*, American Society of Civil Engineers, New York, Aug. 1951, pp. 31–33.
[6] *Engineering Compendium on Radiation Shielding; Vol. II—Shielding Materials*, R. G. Jaeger, Ed., Springer-Verlag, Berlin/Heidelberg, 1975, pp. 176–181.
[7] *Engineering Compendium on Radiation Shielding; Vol. II—Shielding Materials*, R. G. Jaeger, Ed., Springer-Verlag, Berlin/Heidelburg, 1975, pp. 87, 136–167.
[8] *Engineering Compendium on Radiation Shielding; Vol. II—Shielding Materials*, R. G. Jaeger, Ed., Springer-Verlag, Berlin/Heidelburg, 1975, p. 237.
[9] *Engineering Compendium on Radiation Shielding; Vol. II—Shielding Materials*, R. G. Jaeger, Ed., Springer-Verlag, Berlin/Heidelburg, 1975, p. 94.
[10] *Engineering Compendium on Radiation Shielding; Vol. II—Shielding Materials*, R. G. Jaeger, Ed., Springer-Verlag, Berlin/Heidelburg, 1975, p. 116.
[11] *Engineering Compendium on Radiation Shielding; Vol. II—Shielding Materials*, R. G. Jaeger, Ed., Springer-Verlag, Berlin/Heidelburg, 1975, pp. 247–249.
[12] *Engineering Compendium on Radiation Shielding; Vol. II—Shielding Materials*, R. G. Jaeger, Ed., Springer-Verlag, Berlin/Heidelburg, 1975, p. 169.
[13] Mukherjee, P. K., "Properties of High-Density Concrete," *Journal of Testing and Evaluation*, Vol. 20, No. 1, Jan. 1992.
[14] Oswald, A. J. and Shaffer, J. F., "Shield Verification Testing at the Idaho Chemical Processing Plant-Fuel Processing Facility," *Journal of Testing and Evaluation*, Vol. 20, No. 1, Jan. 1992.
[15] Greenspan, E., "High Effectiveness Shielding Materials and Optimal Shield Design," *Journal of Testing and Evaluation*, Vol. 20, No. 1, Jan. 1992.

Fiber-Reinforced Concrete

Colin D. Johnston[1]

PREFACE

The subject of fiber-reinforced concrete has not been covered in previous editions of *ASTM STP 169* mainly because ASTM standard specifications and test methods relating to it have only been developed since the formation in 1980 of Subcommittee C09.03.04 (now C09.42) on Fiber-Reinforced Concrete.

INTRODUCTION

Fiber-reinforced concrete (FRC) in the context of this chapter is conventionally mixed concrete containing discontinuous fibers that initially are randomly oriented in three dimensions in the mixture, but may subsequently become partially aligned by vibratory consolidation, surface finishing, or shotcreting and by geometrical constraints at mold surfaces, formwork, or interfaces with existing concrete, rock, or subgrade. The standards applicable specifically to it are under the jurisdiction of ASTM Subcommittee C09.42 on Fiber-Reinforced Concrete, but many of the standards of ASTM Committee C9 on Concrete and Aggregates may apply to FRC with or without modification.

Thin-section, glass fiber-reinforced, cementitious mixtures that generally do not contain coarse aggregate and are prepared by the spray-up or other special processes are not discussed. They are under the jurisdiction of ASTM Committee C27 on Precast Concrete Products and ASTM Subcommittee C27.40 on Glass Fiber Reinforced Concrete Made by the Spray-Up Process. Also excluded are glass fiber-reinforced mortars for surface-bonded masonry that are under the jurisdiction of ASTM Committee C12 on Mortars for Unit Masonry and ASTM Subcommittee C12.06 on Surface Bonding.

NATURE OF FRC

A brief explanation of the nature of FRC insofar as it differs from concrete without fibers is necessary to understand the rationale governing development of tests and the significance of the properties determined in the tests.

Whether the fibers are steel, polypropylene, polyester, nylon, glass, carbon, or other materials yet to be evaluated, they act like very long slender needlelike particles in freshly mixed FRC. During mixing, they are subject to bending, impact, and abrasion by the action of conventional mixers, and must resist both damage caused by breakage that produces shortening and, in the case of multifilament strands, separation that causes greatly increased specific surface. The high specific surface of fibers, either monofilaments or multifilaments, and the needlelike shape, tend to impart considerable cohesion to FRC mixtures. This makes the reduction in workability associated with adding fibers to concrete appear very severe when judged under the static conditions of the standard slump test. However, many FRC mixtures with low slump flow quite easily when vibrated, so workability tests that employ vibration to produce dynamic conditions are often more appropriate than the standard slump test because placement in practice normally employs vibration.

The workability-reducing effect of fibers depends largely on fiber content and fiber aspect ratio. Aspect ratio is the ratio of fiber length to diameter, or equivalent diameter for fibers of noncircular cross section, and is defined for fiber monofilaments like steel in the ASTM Specification for Steel Fibers for Fiber-Reinforced Concrete (A 820). Fiber content is defined most conveniently in practice in kilograms per cubic metre (kg/m^3) of concrete, but fundamentally it is the volume fraction of fibers regardless of the density of the fiber material that influences the effect of fibers in concrete. Both fiber content and aspect ratio are often limited by workability considerations, although the advent of high-range water reducers has helped to address this problem.

Aspect ratio and fiber content also figure prominently in determining property improvements in hardened FRC. Ideally, the aspect ratio should be as small as possible to minimize loss of workability and as large as possible to maximize the resistance of fibers to pullout from the matrix and thus their reinforcing effectiveness. The fiber content should be as small as possible to minimize loss of workability and as large as possible to maximize reinforcing effectiveness.

What is possible in terms of miscibility, placeability, and property improvements, is therefore a compromise

---

[1] Professor of Civil Engineering, The University of Calgary, Calgary, Alberta, Canada T2N 1N4; chairman of ASTM Subcommittee on Fiber-Reinforced Concrete from 1980 to 1990.

that varies widely with the nature of the proposed application. More than 1% fibers by volume of concrete (80 kg/m^3 of steel, 9.1 kg/m^3 of polypropylene) is uncommon in conventionally mixed FRC or in fiber-reinforced shotcrete. Steel fibers are quite widely used in floors at as little as 0.25% by volume, and polypropylene fibers in slab-on-grade at as little as 0.1% by volume. Aspect ratios for steel fibers are generally 50 to 80, and usually less than 100. Aspect ratio is not easily determinable for multifilament or fibrillated fiber types such as polypropylene or polyester. Nominal length is generally 15 to 50 mm, and rarely more than 65 mm.

Innovations in fiber development that combine reductions in fiber length or aspect ratio, thus minimizing the workability-reducing effect, with improvements in the resistance of the fiber to pullout from the matrix, thus maximizing reinforcing effectiveness in the hardened state, have taken place over the years. Compared with fibers produced 20 years ago, straight smooth monofilaments of uniform cross section are now quite rare. Examples of innovative techniques adopted specifically to optimize fiber performance in both the freshly mixed and hardened states are surface texturing, surface deformation (similar to deformed rebar), hooked or enlarged ends, crimping to a wavy rather than a straight profile, partial splitting (fibrillation) to produce multifilament strands with cross links that separate into branched monofilaments during mixing, and bundling of fibers with a water-soluble glue prior to separation in the mixer.

Fiber size determines the number of fibers per kilogram or pound of their batch weight and their number per cubic metre or cubic yard of concrete. For example, in five commercially available types of steel fibers, the number varied from 4000 to 100 000/kg [1]. For fibrillated polypropylene or bundled polyester fibers, the real number dispersed as monofilaments in freshly mixed concrete is difficult to determine because it depends on how completely the bundles or fibrils separate during mixing and what is their average cross-sectional size. Obviously, the greater the number of fibers per unit volume of concrete, the lesser the average spacing between them and possibly the lower the probability of zones of weakness due to fiber deficiency arising from nonuniform distribution. However, the true influence of fiber size and number per unit volume on the performance of hardened FRC is uncertain. Perhaps the only point of fairly universal agreement regarding fiber size is that the fiber length should be at least as large as the nominal maximum aggregate size.

Readers interested in more details regarding the nature of FRC and the range of FRCs possible with different amounts and types of fibers, together with their proven and potential applications, should consult appropriate references [2–4].

FRESHLY MIXED FRC

Sampling and Consolidating Test Specimens

Sampling FRC mixtures for testing should be in accordance with the ASTM Practice for Sampling Freshly Mixed Concrete (C 172). Other sampling requirements specific to FRC are identified in the ASTM Test Method for Time of Flow of Fiber-Reinforced Concrete Through Inverted Slump Cone (C 995), the ASTM Test Method for Flexural Toughness and First-Crack Strength of Fiber-Reinforced Concrete (Using Beam with Third-Point Loading) (C 1018), and in the ASTM Specification for Fiber-Reinforced Concrete and Shotcrete (C 1116). Most important is the requirement that prohibits wet-sieving to remove large aggregate because of possible adverse effects on fiber content and uniformity of distribution.

For consolidation of test specimens, vibration should be chosen for FRC mixtures when ASTM standards permit a choice between rodding and vibration. When ASTM standards for conventional concrete do not provide for vibration or require rodding, external vibration is preferable for FRC when disturbing the natural fiber distribution may affect the test result. For example, uniform fiber distribution is important in beams prepared according to ASTM C 1018, so external vibration is preferable to internal vibration or rodding that may disrupt the fiber distribution. In contrast, disturbing the fiber distribution is much less important in, for example, air content or density tests, so internal vibration is acceptable.

Workability

Adding fibers to freshly mixed concrete significantly alters its behavior in terms of the three widely recognized rheological parameters that determine workability; namely, stability, mobility, and compactability [5]. Fibers impart considerable stability or cohesion to FRC mixtures that may cause them to appear unworkable when judged only in terms of slump, which is a measure primarily of stability under static conditions. However, when properly proportioned, FRC mixtures flow readily under the dynamic conditions produced by the internal or external vibrators normally used for placement in practice. Accordingly, they are satisfactory from the view point of mobility. Consolidation is accomplished quite easily under vibration, indicating satisfactory compactability.

One test that assesses the workability of FRC under vibration is ASTM C 995. It determines the time required for a sample of FRC to exit the narrow end of an inverted slump cone after an internal vibrator is inserted (Fig. 1). It was first recommended [6] on the basis of using readily available equipment (that is, a slump cone, internal vibrator, and unit weight bucket) to develop a test more appropriate for FRC than the slump test because the slump of FRC mixtures is lower than that of conventional concrete mixtures with the same time of flow [6] (Fig. 1). The test works well with steel or other rigid fibers, demonstrating the effects of fiber content and aspect ratio quite clearly [7] (Fig. 2). Inverted cone (I.C.) flow times in the range of 8 to 30 s are appropriate for placement and consolidation by vibration [7]. Times more than 30 s mean that the FRC will be very difficult to place and consolidate fully. Some types of long (greater than 50 mm) flexible fibers like polypropylene may tend to wrap around the specified 25-mm-diameter internal vibrator, thus nullifying the validity of the test result. For some mixtures rendered

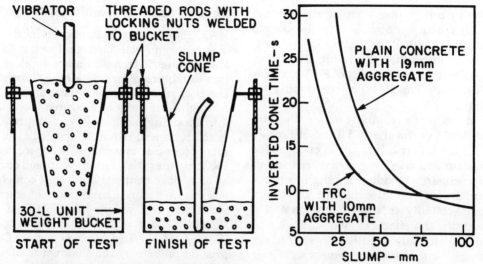

FIG. 1—Inverted slump cone test with relationships between inverted cone time and slump for FRC and plain concrete [6].

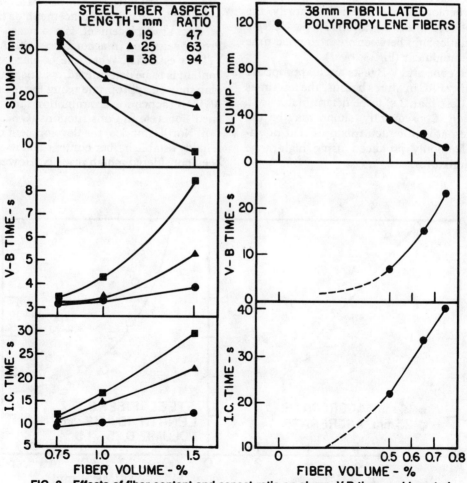

FIG. 2—Effects of fiber content and aspect ratio on slump, V-B time, and inverted cone time [7].

extremely cohesive by a high fiber content, the internal vibrator may simply create a central hole in the sample with the outer portions retained indefinitely in the cone, again nullifying the validity of the test result. Otherwise, the test is a good indicator of the mobility of FRC mixtures under internal vibratory conditions representative of placement in practice.

Another test that assesses workability under vibration is the British Standard Test for the V-B Time of Freshly Mixed Concrete, BS 1881, Part 2. It is just as effective for FRC as for low-slump and no-slump concretes without fibers. It measures primarily mobility, but slight consolidation may occur during the test implying secondary assessment of compactability. It is as effective as ASTM C 995 in demonstrating the effects of fiber content and aspect ratio on the workability of FRC with steel fibers (Fig. 2). It is probably better for FRC mixtures with long flexible fibers or mixtures that are very cohesive, since the test end-point can nearly always be reached even though the vibration time may be abnormally large. V-B times in the range 3 to 10 s represent adequate workability for placement by vibration [7].

Both the inverted slump cone and V-B test results correlate closely, and the essentially linear relationship passing through the origin suggests that both are measuring primarily the same rheological characteristic; namely, the mobility or flow of FRC under vibration [7] (Fig. 3, *left*). In contrast, the relationship between inverted cone time and slump is quite nonlinear (Fig. 3, *right*).

Both the inverted cone and V-B tests are inappropriate for higher workability FRC mixtures because the test times become too short (less than 3 s) to be determinable with reasonable precision. Consequently, slump may be the most practical alternative for such mixtures that nowadays are quite commonly produced using high-range water-reducing admixtures.

Air Content, Yield, and Unit Weight

Air content is just as important for the durability of FRC as for conventional concrete. The standard pressure meter and volumetric techniques are applicable with modification to require consolidation by vibration when slump is less than about 75 mm. Vibration may also be desirable for high fiber content mixtures when slump exceeds 75 mm if rodding becomes difficult or produces visibly unsatisfactory consolidation. The potential for excessive entrapment of air due to incomplete consolidation is greater for FRC mixtures than for conventional concretes, especially at high fiber contents, and only complete consolidation will ensure accurate assessment of entrained air content, yield, and unit weight in ASTM Test Method for Unit Weight, Yield, and Air Content (Gravimetric) of Concrete (C 138), ASTM Test Method for Air Content of Freshly Mixed Concrete by the Volumetric Method (C 173), and ASTM Test Method for Air Content of Freshly Mixed Concrete by the Pressure Method (C 231). ASTM C 173 may prove as difficult for stiff highly cohesive FRC mixtures as for similar mixtures without fibers.

Fiber Content

Verification of the fiber content in freshly mixed FRC and its variation from uniformity throughout a truck load or mixer batch prior to placement, or throughout the end product after placement, is a concern in quality control. Proper sampling in accordance with ASTM C 172 and C 1116 is essential. When the in-place uniformity of fiber content is to be determined, a statistically based sampling plan may be appropriate using the principles given in the ASTM Practice for Examination and Sampling of Hardened Concrete in Constructions (C 823).

In North America, the development of standard tests to reliably establish fiber content and its uniformity has not been considered a high priority. However, the practice of

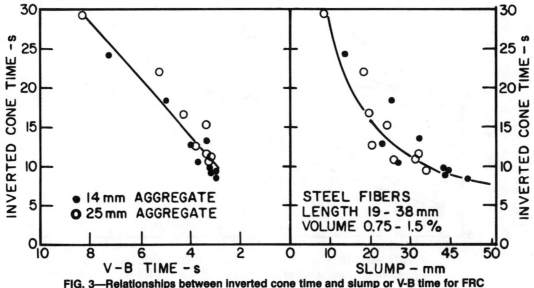

FIG. 3—Relationships between inverted cone time and slump or V-B time for FRC with steel fibers [7].

using relatively low fiber contents of steel fibers (0.25 to 0.38% by volume, 20 to 30 kg/m³) in industrial floors that originated in Europe and is now common in North America has made this issue more important, particularly when large fibers are used and the resulting number of fibers per unit volume of concrete is small. The practice of using very small fiber contents of synthetic fibers primarily to control plastic shrinkage cracking in slabs on grade that is now quite common in North America also highlights the issue, particularly when fiber contents are as low as 0.1% by volume, (about 1 kg/m³ of polypropylene or polyester).

For freshly mixed FRC, washout appears to be the simplest, most direct, and therefore potentially the most accurate method of determining fiber content and its variation within a load or placement unit of FRC. The Japan Society of Civil Engineers standards [8] for testing steel FRC include a Standard Method of Test for Fiber Content of Steel Fiber Concrete, JSCE-SF7, that employs a washout procedure using a container not less than 6 L in volume. The fibers are collected after washout using a magnet. They are then weighed to determine the weight per unit volume of concrete. An alternative procedure that indirectly determines the fiber content of a 100 by 200 mm cylindrical sample in a nonmagnetic (paper or plastic) mold employs an electromagnetic induction coil that surrounds the mold. Fiber content is proportional to the induced current, and the standard requires calibration of the apparatus with data obtained using the washout procedure, as shown in Fig. 4 [9]. This procedure is obviously inapplicable for stainless or other alloyed steel fibers that are not magnetic.

Other types of fiber may in principle be isolated by washout and collected by appropriate means, such as floatation in water in the case of polypropylene or other fibers that float. However, there is remarkably little information on this subject.

For routine quality control, it is arguable that if fiber inventory and concrete production are carefully monitored tests for determining fiber content are no more necessary than tests for determining the water and cement contents of conventional concrete. However, when uniformity of distribution is the concern rather than just the average fiber content, such tests become necessary.

HARDENED FRC

Fiber Content and Orientation

Verification of fiber content and its uniformity in hardened FRC is possible for some fibers. The electromagnetic technique standardized in JSCE-SF7 [8] is applicable to 100 by 200-mm cylindrical samples of hardened FRC, either molded specimens or cores, but is of course limited to the magnetic types of steel fiber. Nevertheless, the results are reported to depend somewhat on fiber orientation [8,9] and on proximity of the surrounding coil to the ends of the specimen [9].

The electromagnetic technique can also be employed using a conventional covermeter placed on a specimen surface [10]. Precast concrete products are monitored in this way in Sweden. However, because the electromagnetic coil no longer surrounds the specimen, and the magnetic field reduces in proportion to the square of the distance from the source, steel fibers close to the coil have more influence on the instrument reading than fibers at greater depth. To reduce the influence of fibers close to the surface, non-metallic interlayer materials (wood, plastic etc.) are placed between the coil and the concrete (Fig. 5). Also, the strength of the magnetic field influences the depth to which it can detect fibers, and specimen boundaries may intrude into the normally spherical shape of the field. This means that separate calibration charts have to be developed for each instrument and interlayer thickness, so steel fiber content is characterized in terms of instrument readings that depend on specimen thickness [10] (Fig. 5). Fiber orientation is again reported to influence the result, reducing it by about 30% for fibers perpendicular to the coil compared with fibers parallel to the coil [10].

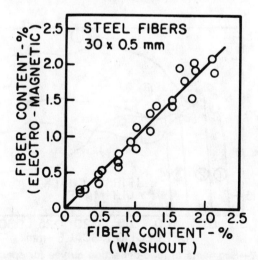

FIG. 4—Steel fiber contents of freshly mixed FRC by electromagnetic and washout techniques [8,9].

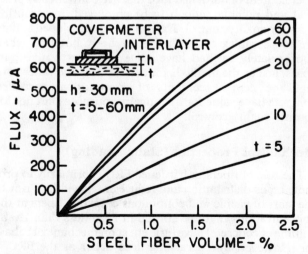

FIG. 5—Steel fiber content of hardened FRC using a covermeter [10].

Fiber orientation and uniformity of distribution have been examined using X-ray radiography to produce visual images of the arrangement of steel fibers in a planar cross section of FRC [11,12]. Although apparently only employed with steel fibers, this technique is applicable in principle to any composite where the fiber and matrix densities are significantly different, as between bone and flesh in the human body. It can be useful in relating crack development under load to local deficiencies in fiber content or to fiber orientation predominantly parallel to the crack [12]. Determination of fiber content from the cross-sectional radiographs requires manual counting, automatic counting by image analyzer, or mathematically derived equations that relate number of fibers per unit area and their resultant spacing to their volume concentration for one, two, or three-dimensional fiber orientations [11,12]. When the results are supported by independent verification, such as crushing of the test samples to extract the fibers, the X-ray method seems feasible although perhaps impractically cumbersome for routinely determining fiber content and detecting anomalies in distribution and orientation. Whether the X-ray technique can be effective for other types of fiber of density significantly different from that of the matrix remains to be established. Polypropylene fibers seem an obvious case because matrix density is about 2.6 times fiber density (steel fibers have density about 3.2 times matrix density).

X-ray fluorescence spectrometry widely used for elemental analysis of materials has been applied to determining the fiber content of zirconia-based glass fibers in glass fiber-reinforced cement [13]. The measurement was based on detection of zirconia present in a known amount in the parent glass, but there were calibration problems attributed to the matrix and varying water contents in the samples. Interferences from other trace elements such as strontium in the cement are also possible. Nevertheless, it appears that fibers containing known amounts of elements having distinct radiation intensity bands not characteristic of anything in the matrix may in principle be detected and quantified in concentration using this technique.

Clearly, there are no techniques for determining the fiber content of hardened FRC that are conveniently practical and reliable enough to be employed on a routine basis for quality control. The techniques developed to date are too complex or unreliable, or both, but may when appropriately verified have value for investigative purposes, just as the methods for determining the water content and cement contents of conventional hardened concrete have value for investigative purposes but not for routine quality control.

Mechanical Properties (Static Loading)

The role of fibers in hardened FRC is primarily to promote crack distribution and reduce crack widths. Prior to the start of visible and continuous cracking, fibers at the concentrations that are normal in FRC (less than 1% by volume of concrete) have little effect on mechanical behavior. However, microcracking does occur as the FRC is loaded, and there are characteristic levels of load and deformation at which the FRC eventually starts to exhibit

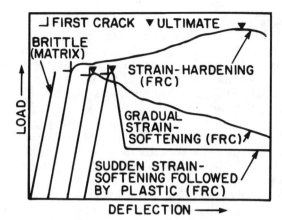

FIG. 6—Load-deflection schematics for matrix and typical FRCs in flexure.

cracks that are significant in continuity, visibility, total length and width. At this stage, the microcracks have become macrocracks, and acoustic emission measurements [14] confirm that the maximum acoustic event rate corresponds closely to these characteristic levels of load and deformation. The condition is termed "first crack" and is clearly identifiable for direct tension and flexure by a sharp reduction in stiffness (Fig. 6). For compression, where the behavior of both matrix and FRC is often quite curvilinear, first crack is not easily identifiable, but the use of fibers only in zones of compressive stress is usually of less interest than in flexure, tension, or shear.

Strengthening

Below first crack and the associated values of load and deformation, matrix strength contributes substantially to composite FRC strength and stiffness. For FRC in tension and flexure, first-crack strength is a more important and meaningful parameter than the ultimate strength based on maximum load because it corresponds to a specific serviceability condition, that is, the onset of macrocracking. In contrast, the maximum load and any strength based upon it can correspond to widely different serviceability

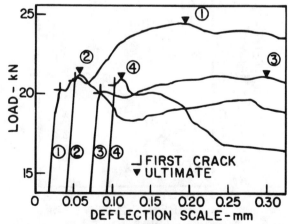

FIG. 7—Segments of load-deflection curves for specimens from the same batch of FRC (1.23% by volume of steel fibers) [15].

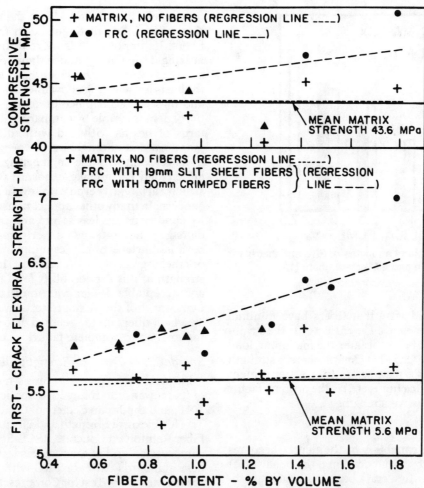

FIG. 8—Strengthening effects of steel fibers in concrete for compression and flexure [15].

conditions in terms of deflection and cracking, depending on whether the FRC exhibits strain-hardening or strain-softening (Fig. 6). It is even possible for such differences to occur within the same batch of specimens when their performance approximates elastic-plastic, which clearly shows that the ultimate or maximum load bears no consistent relationship to serviceability (Fig. 7).

Increases in first-crack strength attributable to fibers are quite small for both flexure and direct tension, even at high fiber contents well above the 1% maximum normally possible in practice. For example, although the relationship in Fig. 8 between first-crack flexural strength and fiber content is statistically significant at the 1% level, the increase in first-crack strength attributable to steel fibers of medium aspect ratio in concrete is only about 15% for a 1.5% fiber content of two types of steel fiber [15]. For direct tension, strength increases are also modest, only about 25% for the same (1.5%) steel fiber content even when the fiber aspect ratio is as high as 100 that is only possible using a mortar matrix [16,17] (Fig. 9). Generally, increasing the fiber content or aspect ratio are not nearly as effective as measures to improve matrix strength, for example the addition of silica fume [18] (Fig. 10). Clearly, the first-crack strength of FRC in tension or flexure is governed primarily by matrix parameters rather than fiber parameters at the fiber contents possible in concrete in practice.

For compression, the relationship between strength and fiber content is not statistically significant (Fig. 8), con-

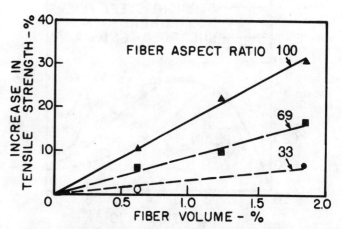

FIG. 9—Strengthening effect of steel fibers in mortar for direct tension [16,17].

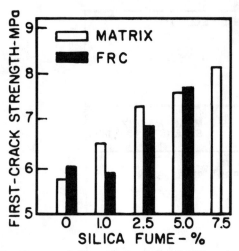

FIG. 10—First-crack flexural strengths of FRC and matrix for concretes with silica fume and 0.5% steel fibers [18].

firming the widely held view that fibers have minimal effect on compressive strength. They affect only the strain-softening phase of FRC behavior after the maximum load has been reached [19] (Fig. 11). Again, matrix strength as determined by matrix parameters (like water-cement ratio, the presence of silica fume, etc.) is the primary factor governing FRC compressive strength.

Toughening

As the load-carrying capability of the matrix becomes progressively reduced by macrocracking, the behavior of the FRC composite after first crack becomes largely attributable to load carried by the fibers. Depending on fiber type, amount, and aspect ratio, it can vary over a wide range in terms of the degree of strain-hardening or strain-softening that follows first crack, and in terms of the strength retained over any selected interval of deformation (Figs. 6 and 11).

Toughness, which is determined largely from the area under the stress-strain or load-deformation curve after first crack (Figs. 6 and 11), is the characteristic of FRC that most clearly distinguishes it from concrete without fibers. Its importance is more difficult to relate to engineering design and practice than, for example, flexural or compressive strength, which have long been accepted as the basis for selecting allowable concrete stresses in design and material acceptance in quality control. It is qualitatively demonstrable by comparing the flexural failure patterns of beams with and without fibers (Fig. 12). This manifestation of toughness or energy absorption capability is relevant in some engineering applications, such as anticipated earthquake exposure where preservation of structural integrity even with severe damage is the primary concern. In many other applications where the serviceability conditions are less severe and catastrophic structural damage is not anticipated, performance up to a specified permissible level of serviceability in terms of deformation or cracking may be more relevant. In such cases, retained strength at this serviceability level may be a more useful and acceptable design criterion than toughness. Accordingly, most of the standard tests evaluate the toughening effects of fibers up to specified deformation or load limits rather than to complete failure of the test specimen.

Standardized Tests for Strength and Toughness of FRC

Flexural performance receives most attention because of its relevance to many FRC applications, as in ASTM C 1018, and the Japan Society of Civil Engineers Method of Test for Flexural Strength and Flexural Toughness of Steel Fiber Reinforced Concrete, JSCE-SF4 [8]. Performance in compression and shear receives less attention, but Japanese standards exist for FRC with steel fibers [8]. They are the Method of Test for Compressive Strength and Compressive Toughness of Steel Fiber Reinforced Concrete, JSCE-SF5, and the Method of Test for Shear Strength of Steel Fiber Reinforced Concrete, JSCE-SF6.

With regard to evaluation of strength, these standards differ considerably in that the test result may be expressed as the strength at first crack, or at the maximum load, or as an average over a prescribed strain or deformation interval. In some of them, the rationale for determining strength after first crack is questionable because load is

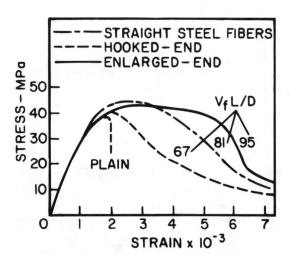

FIG. 11—Typical stress-strain curves for FRC with steel fibers in compression [19].

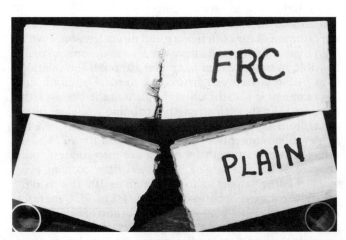

FIG. 12—Toughness of FRC in flexure compared with plain concrete.

converted to stress assuming the material behavior is linear elastic, notably in flexure when the maximum load is not reached until well after first crack (Fig. 7) and material behavior is no longer linear elastic.

With regard to toughness, the standards differ in that the test result may be expressed either simply as the area under the load-deformation or stress-strain curve or as a toughness factor or toughness index derived from portions of the area under these curves. The problem with tests that determine only the area under these curves is that the result is specimen-specific, that is highly dependent on specimen size, shape, and loading configuration. Consequently, the test result fails to characterize material behavior in a manner conceptually independent of specimen and testing variables, unlike, for example, compressive strength that has the same significance (but not necessarily the same value) whether determined on a 100-mm cube or a 150 by 300-mm cylinder.

In most existing standards for toughness testing, the choice of test end-point in terms of deformation or deflection is quite arbitrary and therefore unrelatable to anticipated serviceability conditions. However, in principle, the test end-point may be specified to reflect one or more specific aspects of serviceability. In general, these may be stated qualitatively as somewhere between the two extremes of "very small deformation/deflection and very fine cracks permissible," for example, in bridge decks with deicer exposure, or "very large deformation/deflection and very wide cracks permissible," for example, in earthquake loading of a structure or in shotcretes for tunnel or rock slope applications where short-term structural integrity is of paramount importance. Naturally, there is the option to quantify them by specifying numerical limits on deflection or crack width, or both, anywhere within these qualitative extremes.

In the Japanese standard for flexural testing, JSCE-SF4 [8], the test end-point deflection, δ_{tb} (Fig. 13), is specified arbitrarily as 1/150 of the span. Therefore, there is no possibility of selecting the test end-point to reflect anticipated serviceability conditions. The specimen size, shape, and span limits (only two square cross-sectional sizes 100 or 150 mm and spans, 300 or 450 mm, permitted depending on fiber length) restrict the span-depth ratio to 3.0. Since the formula [20] for determining first-crack deflection shows that it is proportional to L^2/D where L and D are the beam span and depth, respectively, it follows that the test end-point deflection represented by $L/150$ is a fixed multiple of the first-crack deflection (approximately 50 depending on the modulus of elasticity and flexural strength of the matrix) so long as the span-depth ratio, L/D, is fixed at 3.0. If L/D changes, the multiple changes in inverse proportion to it, and the area, T_b (Fig. 13), and parameters derived from it do not have the same meaning in terms of end-point serviceability from one specimen shape to another and from one span to another. Consequently, the standard is not adaptable in principle to circumstances where the minimum standard thickness of 100 mm or an L/D of 3.0 are not possible for practical reasons, as in thin specimens representative of shotcrete or bridge deck overlays, or thick specimens sampled from thick FRC placements like pavements. The main parameter derived from T_b, the flexural toughness factor

$$\overline{\sigma}_b = \frac{T_b L}{\delta_{tb} bD^2}$$

is subject to the same limitations. It represents an average load (T_b/δ_{tb}) for the whole deflection interval, comprising the segments both before and after first crack, converted to a stress using the linear elastic formula for flexural stress. While the load retained at or near a test end-point selected on the basis of serviceability may be useful for design, it is difficult to rationalize how the overall average load can be of direct use in design. Moreover, converting it to a stress using a formula that is clearly invalid after first crack is analytically questionable. Clearly, both T_b and $\overline{\sigma}_b$ are specimen-specific parameters that cannot reflect FRC material behavior independent of specimen size, shape, and span.

The corresponding standard for compression, JSCE-SF5 [8], is much the same in principle with T_c measured to a limiting deformation, δ_{tc}, corresponding to a strain of 0.75% (Fig. 14). A compressive toughness factor

$$\overline{\sigma}_c = \frac{4T_c}{\pi d^2 \delta_{tc}}$$

is derived as the average load divided by the cylinder cross-sectional area to convert it to a stress. Again, it is difficult to rationalize how the overall average rather than the strength retained at or near the test end-point can be useful in design, but the conversion of load to strength is simpler and more justifiable than for flexure.

The standard for shear, JSCE-SF6 [8], provides for double shear of a prismatic specimen (essentially the same

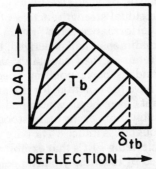

FIG. 13—Toughness in flexure, JSCE-SF4 [8].

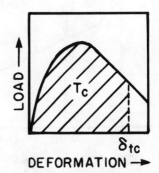

FIG. 14—Toughness in compression, JSCE-SF5 [8].

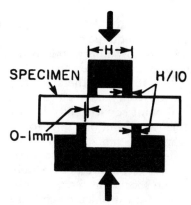

FIG. 15—Shear strength test, JSCE-SF6 [8].

as the beams of JSCE-SF4) using an appropriate specified apparatus (Fig. 15). Shear strength is determined simply as load divided by cross-sectional area. No measure of toughness is established in this test.

In the ASTM C 1018 standard for flexural testing of FRC, material performance is characterized in terms of three parameters or sets of parameters defined to ensure in principle their independence of specimen size, shape, span-depth ratio, and loading configuration [20–22]. They are therefore equally adaptable to test specimens representing thin- or thick-section FRC construction. The rationale for defining these parameters provides for selection of the test end-point deflection to reflect a wide range of anticipated serviceability conditions (Fig. 16). It also ensures that each parameter has a readily understandable meaning and significance relative to an established reference level of material performance, elastic-plastic (Fig. 16), typified by mild steel in tension. This is a level of performance to which practicing engineers can easily relate, and one that is readily achievable in FRCs with amounts and types of fibers that produce superior reinforcing effectiveness, for example, 80 kg/m^3 of HE 60 fibers in Fig. 17.

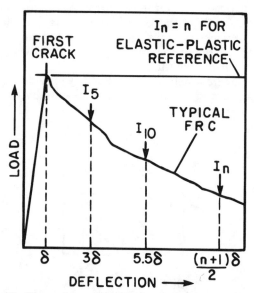

FIG. 16—Flexural toughness indices, ASTM C 1018 [22].

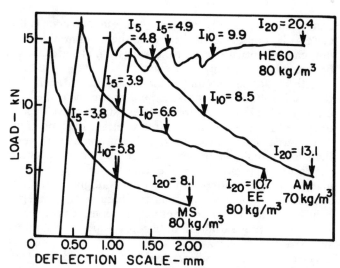

FIG. 17—Influence of steel fiber type on ASTM C 1018 toughness indices for 750 by 150 by 100-mm beams [1].

The first parameter of importance is the first-crack flexural strength (Fig. 16) that is analogous to the yield point for steel. It represents the maximum stress that can be sustained without serious macrocracking and is largely dependent on the matrix, as previously discussed (Fig. 10).

The second important set of parameters are toughness indices, I_5, I_{10}, I_{20}, etc. (Fig. 16), derived as the area under the curve up to the specified end-point deflection divided by the area up to first crack. The index subscript, the corresponding end-point deflection, and the index value are related in such a way that index subscripts and actual values correspond for the elastic-plastic reference level of material performance, that is $I_n = n$ (Fig. 16). Index values more or less than the subscript value indicate the average slope of the curve and degree of strain-hardening or strain-softening behavior observed after first crack (Fig. 17). Performance differences attributable to fiber type (Fig. 17) or amount [1] are clearly identifiable, for example in Fig. 17 between the HE 60 fibers with approximately elastic-plastic behavior ($I_5 = 4.9$, $I_{20} = 20.4$) and the MS fibers with marked strain-softening ($I_5 = 3.8$, $I_{20} = 8.1$).

The third important set of parameters are residual strength factors, $R_{5,10}$, $R_{10,20}$, etc., derived directly from toughness indices. They represent the average strength retained over a specified deflection interval expressed as a percentage of the first-crack strength and are derived so that elastic-plastic or yield-like material performance corresponds to residual strength factors of 100 and fully brittle material performance after first crack to a factor of 0. They are nondimensional as derived, but are readily converted to a retained load capability using the first-crack strength and its corresponding specimen-specific load. Thus, they have the potential for use in design where serviceability limits have been established, and their significance is probably more easily understood by practicing engineers than toughness indices. They are probably most useful for characterizing FRCs that exhibit sudden strain-softening followed by essentially plastic behavior (Fig. 6), where successive factors, $R_{5,10}$, $R_{10,30}$, etc., remain approxi-

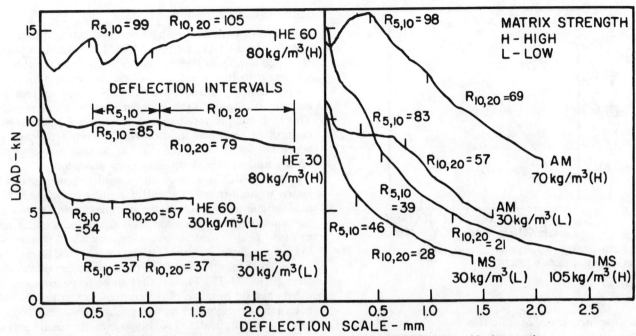

FIG. 18—Influence of steel fiber type and amount on ASTM C 1018 residual strength factors [1].

mately constant and performance differences by fiber type or amount are clearly distinguishable, for example in Fig. 18 (left) between equal amounts of HE 60 and HE 30 fibers and between 30 kg/m³ and 80 kg/m³ of the same fiber. For FRCs characterized by gradual strain-softening (Fig. 6), residual strength factors decrease with increase in deflection (Fig. 18, right), and are probably only useful for design when deflection serviceability limits are known.

ASTM C 1018 requires reporting of first-crack strength; toughness indices, I_5 and I_{10}; and residual strength factor, $R_{5,10}$ as a minimum. Prior to 1992 determination of I_{20} and $R_{10,20}$ was optional but is now mandatory. Testing to larger deflections as appropriate to anticipated serviceability conditions is also recommended, and an appendix provides the rationale for establishing appropriate toughness indices and residual strength factors.

In Sweden, task groups dealing with industrial floor design and shotcrete tunnel linings have prepared recommendations giving values of residual strength calculated as the product of first-crack strength and residual strength factors determined as in ASTM C 1018 but on different specimen sizes. In Belgium, the design of FRC slabs based on residual strength factors is also being discussed [23]. In Canada, the ASTM C 1018 toughness indices have been used in design in many FRC shotcrete projects [24]. In Japan, the design of FRC with steel fibers [8] is recommended on the basis of minima for conventional flexural strength and the flexural toughness factor, $\bar{\sigma}_b$.

When comparing the ASTM C 1018 and JSCE-SF4 standards, it is important to recognize that, in principle and in experimental reality (Fig. 19), it is possible to have two different FRCs with the same T_b and $\bar{\sigma}_b$ at the end-point deflection stipulated in JSCE-SF4. Yet, they can have quite different performance characteristics that are clearly distinguished by all ASTM C 1018 toughness parameters except perhaps I_5, as shown in Fig. 19 where the end-point deflection is 1/300 of the span instead of 1/150 of the span specified in JSCE-SF4.

Mechanical Properties (Dynamic Loading)

Single-Cycle, High-Rate Loading

Uses of FRC that involve resistance to explosives or earthquake loading, or impacts on pavements and bridges due to vehicles or aircraft, may justify impact testing. Weighted pendulum rigs based on modified Charpy impact testers and instrumented drop-weight arrangements with fast-response load and deformation-measuring equipment coupled to appropriate data-acquisition systems have been used in research to evaluate flexural

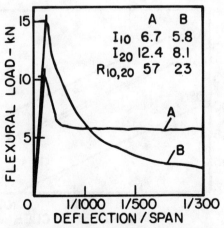

FIG. 19—Equal toughness based on JSCE-SF4 (8) but not in terms of ASTM C 1018 parameters [22].

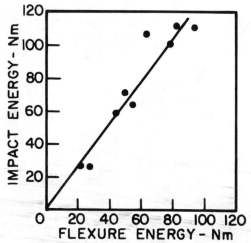

FIG. 20—Energy absorption or toughness compared for impact and slow flexure [20,25].

impact performance [25–27]. However, the experimental complexity and expense of the instrumentation needed to eliminate the influence of extraneous energy losses mitigates against widespread acceptance and routine use of such tests in quality control of FRC. Furthermore, comparison of areas under the load-deflection curves for impact and slow flexure shows that the slow (static) flexure test provides a conservative estimate of toughness determined in the more complex and expensive instrumented impact test [3,20,25–27] (Fig. 20). Impact values can exceed static test values by as little as 20 to 70% [25] up to as much as several hundred percent [27], reflecting the widely different strain rates actually imposed by a diversity of impact testing rigs, each with its own equipment-specific strain rate. Strain-rate sensitivity of data may also be affected by the rate sensitivity of the fibers, which probably is greater for polymers like polypropylene than for metals like steel. These factors make it unlikely that a fully instrumented impact test will be standardized for FRC in the near future.

Multiple-Cycle High-Rate Loading

A form of empirical impact drop-weight test is recommended by ACI Committee 544 [6] in which impact is recorded simply as the number of blows sustained by the specimen to first crack or failure. Determination of the energy actually absorbed by the specimen exclusive of extraneous losses to the support system is not possible. This test was proposed as a standard by the ASTM Subcommittee on Fiber-Reinforced Concrete, but was not accepted for several reasons, including its empirical nature and its reportedly very poor precision.

Uses of FRC in airport and highway pavements subjected to high-frequency traffic loading may justify flexural fatigue testing. Again, the complexity and expense of the equipment and the time needed to conduct such tests mitigate against widespread acceptance and routine use in quality control. Flexural fatigue tests in FRC have shown some benefits attributable to fibers in terms of stress or strength [6, 28–30] (Fig. 21) that are to some extent equipment-specific because of differences in testing variables such as reversing/nonreversing stress, stress range, cycling rate, etc. Incorporating toughness measurements into the test has been attempted [31], but adds further to its complexity. The combined influence of all these factors makes it unlikely that fatigue testing will be standardized for FRC in the near future.

Resistance to Cracking

Cracking can develop in FRC either directly due to application of load, or as a result of loss of moisture and associated differential shrinkage of the exterior of the FRC against restraint provided by the relatively more moist interior.

Cracking Under Restrained Shrinkage

Most investigations [32–34] have employed a steel ring or cylindrical core around which the FRC is cast and against which it subsequently undergoes drying shrinkage. Drying may take place through all three exposed surfaces or may be restricted to the outer one by applying sealant

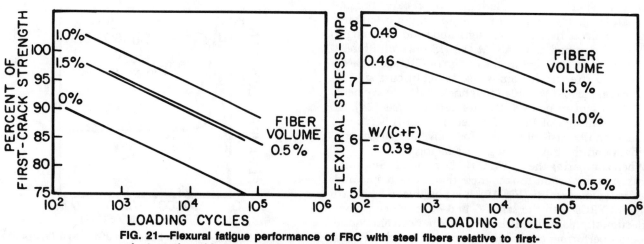

FIG. 21—Flexural fatigue performance of FRC with steel fibers relative to first-crack strength and in terms of actual stress [30].

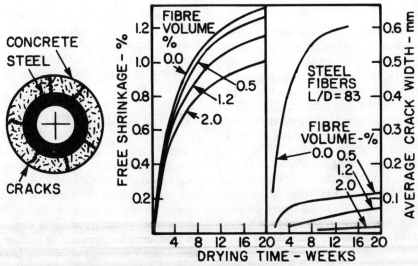

FIG. 22—Free shrinkage and crack development under restrained shrinkage in ring test for matrix and FRC with steel fibers [32].

to the others. In all cases, the number, width, location, and spacing of cracks are monitored as drying progresses. Free (unrestrained) shrinkage measurements on companion prisms tested according ASTM Test Method for Length Change of Hardened Hydraulic Cement Mortar and Concrete (C 157), are also possible. Typical results [32] (Fig. 22) show that fibers have much more effect on crack development in the ring test than on free shrinkage. Both effects increase with increase in fiber content and aspect ratio [32,35] (Fig. 22). Since the detection and measurement of very fine cracks depends on the means of detection and magnification available, the development of refinements incorporating laser holography for detection and high-power microscopes for crack width measurement [34] have facilitated more comprehensive and reliable testing. This has helped to quantify the influence of fiber parameters and fiber properties such as modulus of elasticity, E, and frictional bond strength, τ_f, on cracking (Fig. 23).

While some form of ring test may eventually be standardized for comparing the performance of different FRCs under a specified drying regime, applying the results to specification of joint spacings is difficult in field applications because specimen size and drying conditions are very different. A further problem is that providing restraint sufficient to ensure cracking of the test specimen is only practical when the specimens are quite thin, 35 to 75 mm [32–34], so preferential fiber alignment is more pronounced than in thick sections. This means that the relatively thin ring specimens can be expected to perform better than thick-section slabs or other structural elements.

In principle, the ring test can also be used to monitor the plastic shrinkage and associated cracking that sometimes occur within a few hours of placement under adverse conditions of temperature, humidity, and wind speed causing high surface evaporation. Claims by manufacturers that fibers, particularly small volumes of polypropylene, nylon, or polyester fibers, are effective in preventing this problem are questionable, and there is a need for some form of standard test to confirm such claims [36]. In addition to the ring test, alternatives that require further study are slabs restrained at the edges [37] or fitted with a centrally placed cylindrical insert [38]. A standard to assess the plastic shrinkage potential of restrained FRC slabs is being developed by the ASTM Subcommittee for Fiber-Reinforced Concrete. It compares the area of cracking, determined as the product of crack length times width, for fiber-reinforced and unreinforced concrete slabs. It involves fan-forced air flow over the surface of newly molded slabs under controlled conditions of temperature and humidity sufficiently severe to induce plastic shrinkage cracking. Recent results [39] obtained with this method using fibrillated polypropylene fibers of 13-, 19-, and 51-mm lengths in amounts of 0.05, 0.1, and 0.2% by volume indicate that plastic shrinkage cracking can be reduced from 20 to 90%, but was not entirely eliminated under the conditions of the test (water-cement ratio = 0.48, relative humidity = 40%, and temperature = 35°C). The best results were achieved using 0.2% of 19- and 51-mm fibers.

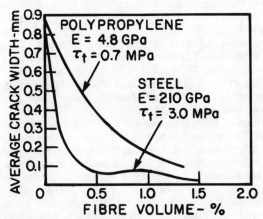

FIG. 23—Crack development compared for 19-mm polypropylene and 25-mm steel fibers in ring test [34].

Cracking Under Load

Crack widths and spacings between cracks can be measured quite easily with appropriate visual magnification of the cracked surface; for example, on the tension face of a test specimen loaded in slow flexure. Monitoring crack development in this way makes it possible to quantify the ability of fibers to control cracking in concrete, the third benefit imparted by them in addition to strengthening and toughening. It also allows serviceability in terms of crack development to be related to the end-point deflection selected in toughness tests according to ASTM C 1018. However, monitoring crack development is not part of the standard, and no comprehensive and conclusive data on the effects of fiber content or type on crack development have yet been obtained, although it seems reasonable to expect such effects from the example shown in Fig. 24 [22].

Durability

No durability tests have been standardized specifically for FRC. The durability of the matrix is assured if it meets the relevant ASTM standards for conventional concrete subjected to freezing and thawing, salt scaling, sulfate resistance, and aggregate reactivity tests. Fiber-specific problems can arise in uncracked FRC if the fibers are chemically incompatible with cement paste or admixtures, and fiber compatibility tests for fibers that may be suspect are being considered by the responsible ASTM subcommittee. In cracked FRC, the fibers may deteriorate in certain exposure environments.

The fibers commonly used in FRC, carbon-steel and polypropylene, pose no problem in uncracked FRC, but, in cracked FRC steel, fibers may rust if exposed to moisture and air and polypropylene fibers may deteriorate if exposed to sufficient ultraviolet radiation. Alloyed steels address the rusting problem, and stabilizing additives may inhibit any ultraviolet attack for polypropylene. As new fibers emerge, there is a need for vigilance and appropriate tests to avoid possible fiber-specific durability problems, such as those that are now well known for glass fibers and many natural cellulose-based fibers [2–4] and those that remain to be fully investigated with aramid, polyester, nylon, and acrylic fibers [40]. In the absence of an established satisfactory performance record, ASTM C 1116 requires credible evidence that unfamiliar fiber types will not react adversely with the cementitious matrix or any chemical or mineral admixtures it contains.

THE FUTURE OF FRC

More and more new fiber types or modifications of existing types may be proposed for use in FRC. Hindsight suggests that compatibility of fibers with the moist alkaline environment of cement paste and tests to confirm it should be a high priority. Glass fibers are an example of where vast sums have been expended trying to develop mechanisms for protecting a fiber that could reasonably have been rejected initially in favor of alternative fibers more chemically compatible with cement paste. Compatibility with conventional mixing and placement techniques was also neglected by the producers of the early forms of steel and polypropylene fibers with smooth surfaces, uniform cross sections, and aspect ratios too high to avoid fiber balling during batching and mixing. The fibers of the future must be engineered to minimize mixing and placement difficulties, maximize short-term property improvements, and ensure that these improvements are sustained over the long term. Much remains to be done in developing standard test methods and performance-based specifications for FRC to meet all of these objectives.

REFERENCES

[1] Johnston, C. D. and Skarendahl, A., "Comparative Performance Evaluation of Steel Fibre-Reinforced Concretes According to ASTM C 1018 Shows Importance of Fibre Parameters," *Materials and Structures* (RILEM), Vol. 25, No. 148, May 1992, pp. 191–200.

[2] "State-of-the-Art Report on Fiber Reinforced Concrete," *Concrete International*, Vol. 4, No. 5, May 1982, pp. 9–30.

[3] Johnston, C. D., "Fibre-Reinforced Cement and Concrete," *Advances in Concrete Technology*, Canada Centre for Mining and Energy Technology, V. M. Malhotra, Ed., CANMET, National Resources Canada, Ottawa, Cat. No. M39–6/94E, Jan. 1994, pp. 603–673.

[4] Hannant, D. J., *Fibre Cements and Fibre Concretes*, Wiley, New York, 1978.

[5] "Behavior of Fresh Concrete During Vibration," Report ACI 309.1R-81, *ACI Journal*, American Concrete Institute; *Proceedings*, Vol. 78, No. 1, Jan.–Feb. 1981, pp. 36–53.

[6] "Measurement of Properties of Fiber Reinforced Concrete," Report ACI 544.2R, *ACI Materials Journal*, Vol. 85, No. 6, Nov.–Dec. 1988, pp. 583–593.

[7] Johnston, C. D., "Measures of the Workability of Steel Fiber Reinforced Concrete and Their Precision," *Cement, Concrete, and Aggregates*, Vol. 6, No. 2, Winter 1984, pp. 74–83.

[8] "Recommendation for Design and Construction of Steel Fiber Reinforced Concrete," *Concrete Library International*, Japan Society of Civil Engineers, No. 3, 1984, pp. 41–69.

[9] Uomoto, T. and Kobayashi, K., "Measurement of Fiber Content of Steel Fiber Reinforced Concrete by Electro-Magnetic Method," *SP-81*, American Concrete Institute, Detroit, MI, 1984, pp. 233–246.

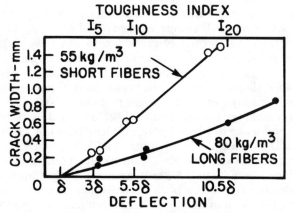

FIG. 24—Crack development in the ASTM C 1018 flexure test for two types of steel fiber [22].

[10] Malmberg, B. and Skarendahl, A., "Determination of Fibre Content, Distribution and Orientation in Steel Fibre Concrete by Electromagnetic Technique," *Proceedings*, RILEM Symposium on Testing and Test Methods of Fibre Cement Composites, Construction Press Ltd., 1978, pp. 289–295.

[11] Kasperkiewicz, J., Malmberg, B., and Skarendahl, A., "Determination of Fibre Content, Distribution and Orientation in Steel Fibre Concrete by X-ray Technique," *Proceedings*, RILEM Symposium on Testing and Test Methods of Fibre Cement Composites, Construction Press Ltd., 1978, pp. 297–305.

[12] Stroeven, P. and Shah, S. P., "Use of Radiography-Image Analysis for Steel Fibre Reinforced Concrete," *Proceedings*, RILEM Symposium on Testing and Test Methods of Fibre Cement Composites, Construction Press Ltd., 1978, pp. 275–288.

[13] Ashley, D. G., "Measurement of Glass Content in Fibre Cement Composites by X-Ray Fluorescence Analysis," *Proceedings*, RILEM Symposium on Testing and Test Methods of Fibre Cement Composites, Construction Press Ltd., 1978, pp. 265–274.

[14] Mobasher, B., Stang, H., and Shah, S. P., "Microcracking in Fiber Reinforced Concrete," *Cement and Concrete Research*, Vol. 20, Pergamon, Elmsford, N.Y., 1990, pp. 665–676.

[15] Johnston, C. D., unpublished data for paper in preparation.

[16] Johnston, C. D. and Coleman, R. A., "Strength and Deformation of Steel Fiber Reinforced Mortar in Uniaxial Tension," *Fiber Reinforced Concrete, SP-44*, American Concrete Institute, Detroit, MI, 1974, pp. 177–193.

[17] Johnston, C. D., "Properties of Steel Fibre Reinforced Mortar and Concrete," *Fibrous Concrete, Concrete International*, CI80, Concrete Society (UK), Construction Press Ltd., 1980, pp. 29–47.

[18] Johnston, C. D. and Gray, R. J., "Flexural Toughness and First-Crack Strength of Fibre Reinforced Concrete Using ASTM C 1018," *Proceedings*, Symposium on Developments in Fibre Reinforced Cement and Concrete, RILEM Committee 49–TFR, 1986, Paper 5.1.

[19] Shah, S. P. et al., "Complete Stress-Strain Curves for Steel Fibre Reinforced Concrete in Uniaxial Tension and Compression," *Proceedings*, RILEM Symposium on Testing and Test Methods of Fibre Cement Composites, Construction Press Ltd., 1978, pp. 399–408.

[20] Johnston, C. D., "Toughness of Steel Fiber Concrete," *Steel Fiber Concrete*, US-Sweden Joint Seminar, Elsevier Science Publishing Co., New York, 1986, pp. 333–360.

[21] Johnston, C. D., "Definition and Measurement of Flexural Toughness Parameters for Fiber Reinforced Concrete," *Cement, Concrete, and Aggregates*, Vol. 4, No. 2, Winter 1982, pp. 53–60.

[22] Johnston, C. D., "Methods of Evaluating the Performance of Fiber-Reinforced Concrete," *Symposium Proceedings*, Vol. 211, Materials Research Society, Pittsburgh, 1991, pp. 15–24.

[23] Moens, J. and Nemegeer, D. E., "Designing Fiber Reinforced Concrete Based on Toughness Characteristics," *Concrete International*, Vol. 13, No. 11, Nov. 1991, pp. 38–43.

[24] Morgan, D. R., "Steel Fiber Reinforced Shotcrete for Support of Underground Openings in Canada," *Concrete International*, Vol. 13, No. 11, Nov. 1991, pp. 56–64.

[25] Hibbert, A. P. and Hannant, D. J., "Impact Resistance of Fibre Concrete," Report SR 654, U.K. Transport and Road Research Laboratory, 1981, p. 25.

[26] Gopalaratnam, V. S. and Shah, S. P., "Properties of Steel Fiber Reinforced Concrete Subject to Impact Loading," *Journal*, American Concrete Institute; *Proceedings*, Vol. 83, No. 1, Jan.–Feb. 1986, pp. 117–126.

[27] Banthia, N., Mindess, S., and Bentur, A., "Impact Behavior of Concrete Beams," *Materials and Structures*, RILEM, Vol. 20, No. 118, July–Aug. 1987, pp. 293–302.

[28] Batson, G. B. et al., "Flexural Fatigue Strength of Steel Fibre Reinforced Concrete Beams," *Journal*, American Concrete Institute; *Proceedings*, Vol. 69, No. 11, Nov. 1972, pp. 673–677.

[29] Nagabhushanam, M., Ramakrishnan, V., and Vandran, G., "Fatigue of Fibrillated Polypopylene Fiber Reinforced Concretes," *International Symposium on Recent Development in Concrete Fiber Composites*, Transportation Research Record 1226, National Research Council, Washington, DC, 1989, pp. 36–47.

[30] Johnston, C. D. and Zemp, R. W., "Flexural Fatigue Behavior of Steel Fiber-Reinforced Concrete—Influence of Fiber Content, Aspect Ratio, and Type," *ACI Materials Journal*, Vol. 88, No. 4, July–Aug. 1991, pp. 374–383.

[31] Otter, D. E. and Naaman, A. E., "Properties of Steel Fiber Reinforced Concrete Under Cyclic Loading," *ACI Materials Journal*, Vol. 85, No. 4, July–Aug. 1988, pp. 254–261.

[32] Malmberg, B. and Skarendahl, A., "Method of Studying the Cracking of Fibre Concrete Under Restrained Shrinkage," *Proceedings*, RILEM Symposium on Testing and Test Methods of Fibre Cement Composites, Construction Press Ltd., 1978, pp. 173–179.

[33] Swamy, R. N. and Stavrides, H., "Influence of Fiber Reinforcement on Restrained Shrinkage and Cracking," *Journal*, American Concrete Institute; *Proceedings*, Vol. 76, No. 3, March 1979, pp. 443–460.

[34] Grzybowski, M. and Shah, S. P., "Shrinkage Cracking of Fiber Reinforced Concrete," *ACI Materials Journal*, Vol. 87, No. 2, March–April 1990, pp. 138–148.

[35] Chern, J. and Young, C., "Study of Factors Influencing Drying Shrinkage of Steel Fiber Reinforced Concrete," *ACI Materials Journal*, Vol. 87, No. 2, March–April 1990, pp. 123–129.

[36] Schupack, M. and Stanley, W. R., "Seven Case Studies of Synthetic Fiber Reinforced Slabs," *Concrete International*, Vol. 14, No. 2, Feb. 1992, pp. 50–56.

[37] Kraai, P. O., "Proposed Test to Determine the Cracking Potential Due to Drying Shrinkage of Concrete," *Concrete Construction*, Vol. 30, No. 9, Sept. 1985, pp. 775–778.

[38] Padron, I. and Zollo, R. F., "Effect of Synthetic Fibers on Volume Stability and Cracking of Portland Cement Concrete and Mortar," *ACI Materials Journal*, Vol. 87, No. 4, July–Aug. 1990, pp. 327–332.

[39] Berke, N. S. and Dallaire, M. P., "The Effect of Low Addition Rates of Polypropylene Fibers on Plastic Shrinkage Cracking and Mechanical Properties," *SP-142*, American Concrete Institute, Detroit, MI, 1992, 1994, pp. 19–42.

[40] Wang, Y., Backer, S., and Li, V. C., "An Experimental Study of Synthetic Fiber Reinforced Cementitious Composites," *Journal of Materials Science*, Vol. 22, 1987, pp. 4281–4291.

Preplaced Aggregate Concrete

Raymond E. Davis, Jr.[1]

PREFACE

The original chapter on preplaced aggregate (PA) concrete was authored by B. A. Lamberton (deceased) former Vice President, Engineering, Intrusion-Prepakt, Inc. At the time of its publication, the only recognized standard test methods for PA concrete were those of the U.S. Army Corps of Engineers that were based on the research and development work conducted at the University of California, Berkeley, under the direction of Professor Raymond E. Davis. Subsequent to Mr. Lamberton's original text, all of the Corps of Engineer test methods for PA concrete were adopted by ASTM Committe C9. These changes are reflected in the revisions to the chapter along with minor additions and corrections to the text, references, and bibliography.

INTRODUCTION

The term "preplaced aggregate" (PA) concrete describes a method of placing gap-graded concrete. As an alternative to blending the fine and coarse aggregate, cement, and water and placing the mixture in a form, coarse aggregate is preplaced in the form and a structural grout is injected into the coarse aggregate mass in such a way as to fill void spaces where it hardens to form dense homogeneous concrete. The process was developed by Wertz to whom a U.S. patent [1] was issued in 1943.

The drying shrinkage of PA concrete is less than that of conventionally placed concrete with similar material proportions [2]. This is due apparently to the point-to-point contact of coarse aggregate particles. Because of this low shrinkage characteristic, it was first used as a method of repairing concrete and masonry structures, particularly railroad bridge piers and tunnel linings [3,4]. The method is now widely accepted for other specialized applications such as placement of mass concrete under water [5], in heavily reinforced structures [6], and for high density biological shielding [7].

While the two-step placement procedure would increase costs over that of conventional methods in a great majority of routine concrete work, it has the advantage of permitting placement of coarse aggregate, representing roughly 60% of the total concrete volume, in a timed sequence independent of mixing and placing cementitious constituents. The coarse aggregate fraction can be preplaced either very rapidly with bulk handling equipment or slowly and carefully by hand labor, depending on the application, followed by grout injection at a time convenient to the overall construction schedule. The method is particularly applicable to placement of concrete in structures containing a profusion of inserted fixtures [8].

MATERIALS

The slurry containing a mixture of portland cement, sand, admixtures, and water that is injected into the coarse aggregate mass is referred to as "grout." Actually a highly fluid structural mortar, it bears no resemblance to the low-strength, generally high water/cement (w/c) ratio grouts that are pressure injected into soil or rock formations to increase strength or reduce permeability. Grout used in PA concrete work is a structural material. Injected into coarse aggregate, it produces concrete comparable in strength and other physical characteristics to that mixed and placed by conventional methods. The same working stresses used in conventional concrete design are applicable to PA concrete.

Cementing Materials

Cementing materials suitable for use in conventionally placed concrete may be used with equal confidence for placement by the PA concrete method. It is customary practice to substitute pozzolanic quality fly ash for part of the portland cement. Substitution in the range of 15 to 35% by weight of portland cement is common. In addition to the beneficial properties normally imparted by any pozzolan to portland cement concrete such as lower heat generation, reduced permeability, and improved resistance to chemical attack, fly ash is particularly beneficial in the PA concrete process in that it improves pumpability and retards initial set. Fly ash should meet the Class F requirements of the ASTM Specification for Fly Ash and Raw or Calcined Natural Pozzolan for Use as a Mineral Admixture in Portland Cement Concrete (C 618).

Fly ash is often omitted when concrete is to be used for high density biological shielding. Since the specific gravity of this material is appreciably less than that of cement,

---

[1] Consulting engineer, South Pasadena, CA 91030, formerly of Intrusion-Prepakt, Inc., Cleveland, OH.

partial replacement of cement with fly ash causes a slight reduction in unit weight of the concrete. Fly ash is also less effective than cement in chemically fixing water and so is undesirable in shielding subject to high neutron flux that is more effectively attenuated by hydrogen atoms.

Fine Aggregate

Fine aggregate, properly graded, is essential to successful execution of the work. Oversize material will cause obstruction of the void channels in the coarse aggregate mass. Excess fines will increase water requirement and so reduce compressive strength and increase drying shrinkage. Gap-graded fine aggregate may cause excessive bleeding. A typical gradation appears in Table 2, Grading 1 of ASTM Specification for Aggregates for Radiation-Shielding Concrete (C 637). Although described in this specification in connection with high-density aggregate for biological shielding, the grading is equally suitable for use with normal-density aggregates. Either natural or manufactured sand may be used, although the former is usually preferable from the standpoint of pumpability. As indicated in Table 2, Grading 2 of ASTM C 637, somewhat finer grading is required where high-density sand is used or where coarse aggregate grading is finer than normal.

Coarse Aggregate

Coarse aggregate grading is far less critical than is fine aggregate grading. The only absolute requirements are that it be (a) free of surface dust that would prevent bond of grout to the aggregate particles, (b) sufficiently saturated that it will not absorb water from the grout and so cause premature thickening, and (c) of such a grading that grout will flow readily by gravity alone through the void system. The particles should be of such toughness and hardness that they do not fracture or degrade during transport and placement into the forms. Extra caution must be exercised during placement to avoid degrading friable aggregate.

Normal aggregate grading limits for most structural applications are shown in Table 2, Grading 1 of ASTM C 637. Aggregate graded within these limits will exhibit a void content in the range of 43 to 48%. In the absence of very closely spaced reinforcing or restricted form configuration, it is common practice to scalp coarse aggregate on a 20-mm (¾-in.) wash screen. A trommel-type screen is generally more effective than a deck screen in washing the aggregate particles to the required surface cleanliness.

There is no practical limit for maximum aggregate size other than that imposed by handling equipment, section thickness, and spacing of reinforcing bars and embedments. Aggregate for mass concrete work such as bridge piers may be graded from 25 to 150 mm (1 to 6 in.), for example, while man-sized pieces of waste rock may be used for deep mine concreting. Where minimum grout consumption is desired, consideration should be given to extreme gap grading of the coarse aggregate such as a blend of 20 to 40-mm (¾ to 1½-in.) aggregate with 200 to 300-m (8 to 12-in.) aggregate. Although the economies of aggregate manufacture often preclude such a blend, void contents as low as 32% have been reported for such a grading.

A wide variety of high-density aggregate, both fine and coarse as described in ASTM C 637, have been placed by the PA concrete method that is suited particularly for construction of high-density biological shielding. The advantage of the method for this type of work is two-fold. The heavy particles of mineral ore or steel punchings or a blend of these coarse aggregates is placed in the form with no possibility of segregation such as may occur when coarse aggregate and grout is combined in a plastic mass and placed by conventional methods. Second, the method is well suited to concrete placement around multiple embedments, closely spaced reinforcing, and within the tight confines of complex formwork characteristic of most shielding installations.

Grout Fluidifier

Grout fluidifier is required on all PA concrete work to promote fluidity of the grout mixture and, most importantly, to cause expansion of the grout prior to initial set in an amount sufficient to more than offset setting shrinkage of the grout that would otherwise take place beneath coarse aggregate particles. Grout fluidifier is customarily used as a commercially preblended material conforming to ASTM Specification for Grout Fluidifier for Preplaced-Aggregate Concrete (C 937). It contains a water-reducing agent, a suspending agent, aluminum powder, and a chemical buffer to assure properly timed reaction of the aluminum powder with the alkalies of the cement. This reaction results in the formation of hydrogen gas to expand the fluid grout, usually in the range of 5 to 8%. This gas-forming reaction should be so timed that, at 21°C (70°F), it is approximately 50% complete in 1 h and 95% complete in 3 h. At a slow expansion rate, bleeding tends to increase. If the expansion time is more rapid, grout mixing and pumping time are shortened unnecessarily.

Since the expansive characteristic of the grout is dependent on the alkali content of the cement, it may be depressed when the standard grout fluidifier formulation is used with low alkali cements. Performance of grout fluidifier in combination with job cement should be determined, therefore, and the product custom formulated by the manufacturer if necessary when use of cement is anticipated, the alkali content of which is less than 0.40. The release of hydrogen bubbles and the expansive reaction have an effect similar to that of an air-entraining agent with respect to freeze-thaw durability. An air-entraining agent or air-entrained cement should not be used routinely in the PA concrete process. However, where freeze-thaw conditions are severe and a specific minimum air content is required, an air content determination should be made using grout fluidifier and job cement and an air-entraining agent incorporated in the grout mix if required.

Grout Mix Proportions

Grout mix proportions may be selected in accordance with ASTM Practice for Proportioning Grout Mixtures for Preplaced-Aggregate Concrete (C 938). For typical structural applications, proportions are normally in the range of 2:1:3 to 4:1:5 by weight in which the ratio represents, respectively, the weights of cement, fly ash, and sand. The ratio of water to cementing materials (cement plus fly ash) is in the range of 0.43 to 0.47. Grout fluidifier is added in an amount equal to 1% by weight of the cementing materials. For mass concrete applications, the ratio of cementing materials to sand may be reduced to as low as 1:2. Further decrease in this ratio is usually precluded by pumpability limitations.

SPECIFICATIONS

Specifications for PA concrete placement vary widely in detail depending on job complexity. For routine structural repair of concrete or masonry, performance type specifications normally are employed, requiring only that a minimum stipulated compressive strength be exhibited by test cylinders and that good construction practice be followed in accordance with such accepted publications as ACI Standard 304 [9]. For difficult projects such as large heavily reinforced structures or biological shielding, specifications may be entirely of the prescription type, setting forth not only compressive strength required but aggregate composition and grading, grout insert spacing, grout surface monitoring procedures, placement temperature, and in-place density.

High-Density Aggregates

High-density aggregates typically used in PA concrete construction are described in the ASTM Descriptive Nomenclature of Constituents of Aggregates for Radiation-Shielding Concrete (C 638) and typical specifications for these aggregates are presented in ASTM C 637. All aggregates described in ASTM C 638 are suitable for placement by the PA concrete method, although particular care should be exercised with the softer natural minerals such as barite, geothite, and limonite. Ferrophosphorous varies widely in hardness and some of it may be so soft and friable as to be unsuitable for placement by the PA concrete method. All of these materials are obviously far more expensive than normal density aggregate and purchase specifications for these materials should be written with particular care. They should not only call for specific gravity limitations but also carefully describe sampling procedure with respect to quantity represented by each sample, frequency of sampling, and the location where tests are to be performed. Aggregates should always be tested prior to shipment, with conformance sampling and testing performed at the job site. If the aggregate is relatively friable and breakage during transit is high, consideration should be given to reclaiming undersize material and incorporating it in concrete to be placed by conventional methods.

Where blended aggregates are to be used, such as a combination of iron ore and steel punchings, the two materials must be inspected separately and the final blend again inspected for particle size gradation and homogeneity of the blend. Aggregates may be blended continuously by streaming through metering gates onto a belt ahead of a washing screen that completes the blending process, or they may be batch blended by a few rotations in a concrete mixer followed by washing and screening. Although the latter method may appear to be somewhat more accurate, in practice either type of blending operation causes some size degradation of the more friable of the two aggregate fractions, and allowance must be made for a loss of some of this friable material in the blending operation.

All high-density aggregate should be weighed before placement and sampled frequently to determine unit density and void content. Based on these determinations, calculations can be made to determine required grout density. Where watertight steel forms are used, void content of the entire coarse aggregate mass may be determined in-place by metering water into and out of the aggregate-filled forms. Using careful testing and sampling procedures, it is possible to specify in-place density of concrete placed by the PA concrete method as close as ±1%, although a tolerance of ±2% is a more common limitation.

Temperatures

Temperatures at the time of grout injection are often of great importance where minimum thermal shrinkage is an important consideration. The aggregate may be precooled in the form by flooding it with chilled water or covering the coarse aggregate with flaked ice and allowing the ice to melt and drip down through the aggregate mass. Using this technique, in combination with precooling the grout ingredients and using ice in the mixing water, permits placement temperatures to be achieved within the range of 5 ± 3°C (40 ± 5°F).

Grout Surface Monitoring

Grout surface monitoring is of particular importance in ensuring complete penetration of all voids within the coarse aggregate mass and complete contact with embedments and penetrations. For routine repair work and for many structural concrete applications, location of grout surface can be determined with sufficient accuracy by simply observing grout or moisture seepage through the forms or through insert or inspection holes drilled in the side of the forms. Slotted sounding wells may also be used, particularly for deep mass concrete placements, in which a float on the end of a line is dropped into the sounding well and the depth to the grout surface physically measured. Using such methods, location to the grout surface can be determined with an accuracy of about ±300 mm (12 in.). For critical applications, location and frequency of sounding wells may be specified, as well as location and spacing of grout insert pipes and vent pipes. The Time Domain Reflectometer provides a more accurate means of determining grout surface location in a non-flooded aggregate mass. Short voltage pulses are transmitted on electrically calibrated detector wires located within the aggregate mass and the pulse reflections assembled and displayed on an oscilloscope. Grout or water around the sensor wires alters the impedance permitting an experienced operator to locate the grout surface with an accuracy of about ±80 mm (±3 in.) [9].

Fluid Grout Characteristics

Fluid grout characteristics are always routinely tested during any placement by the PA concrete method. Measurements of grout viscosity or flow, measuring the time in seconds for discharge of 1725 cm^3 of grout through a 12.7-mm (½-in.) orifice in accordance with ASTM Test Method for Flow of Grout for Preplaced-Aggregate Concrete (C 939) should be made routinely at about 2-h intervals. This test is analogous to the slump test for conventionally placed concrete. Typical consistencies range from 21 to 30 s with the upper end of the range suitable for use with plus 20 mm (¾ in.) coarse aggregate and the lower end of the range required for plus 12 mm (½ in.) coarse aggregate. Specifications usually require grout consistency to be controlled within a range of ±2 of a stipulated value.

The test for expansion in accordance with ASTM Test Method for Expansion and Bleeding of Freshly Mixed Grouts for Preplaced-Aggregate Concrete in the Laboratory (C 940) is performed less frequently, often no more than once a day. This test requires only observation of total grout expansion 3 h after mixing. The utility and significance of the test procedure is greatly enhanced if observations are made of both total grout expansion and accumulated bleed water at four 15-min intervals following mixing and at 30-min intervals thereafter until all expansion has ceased. Although with normal cement and at normal temperature, expansion is virtually complete at 3 h, expansion may continue as long as 4 h at lower temperatures and with some low alkali cements.

Following the procedures of ASTM C 940, approximately 800 cm^3 of fresh grout are placed in a 1000-cm^3 graduate. The observer should then note and report not only the total volume of the specimen, including grout and accumulated bleed water, at the aforementioned intervals, but the total volume of grout and the total volume of bleed water. Expansion, expressed in percent, is calculated as total volume minus initial volume divided by initial volume. Bleeding, expressed in percent, is reported as the volume of bleed water (total volume minus grout volume) divided by initial volume.

Water retentivity of freshly mixed grouts should be determined in accordance with ASTM Test Method for Water Retentivity of Grout Mixtures for Preplaced-Aggregate Concrete in the Laboratory (C 941).

Grout bleeding, grout expansion, and grout consistency are all strongly influenced by the effectiveness of mixing equipment and procedures. Before any adjustment is made to the mix design, adequacy of the mixing procedure should be evaluated carefully.

Compressive Strength

Compressive strength of hardened grouts should be determined in accordance with ASTM Test Method for Compressive Strength of Grouts for Preplaced-Aggregate Concrete in the Laboratory (C 942).

Time of Setting

Time of setting of grouts should be determined in accordance with ASTM Test Method for Time of Setting of Grouts for Preplaced-Aggregate Concrete in the Laboratory (C 953).

Density

Density of hardened concrete is normally determined by casting test cylinders in accordance with ASTM Practice for Making Test Cylinders and Prisms for Determining Strength and Density of Preplaced-Aggregate Concrete in the Laboratory (C 943). From these test cylinders, 100 by 100 by 150-mm (4 by 4 by 6-in.) prisms are cut on which final acceptance tests are performed. Density determinations on the standard 150 by 300-mm (6 by 12-in.) test cylinder will result in an erroneously low figure. This procedure is particularly important in counter-weight construction where excessive density may be as objectionable as deficient density.

Hardened Concrete

Hardened concrete placed by the PA method is evaluated by the same methods employed with conventionally placed concrete. However, two notes of caution are in order. In accordance with good practice, cores extracted from conventionally placed concrete should be of a diameter equal to at least four times that of the maximum coarse aggregate particle size. This limitation on minimum core diameter is even more important with PA concrete by reason of the fact that there is a higher percentage of large aggregate particles with a correspondingly greater possibility that these particles will be torn loose from the grout matrix. Secondly, nondestructive testing procedures such as the Schmidt hammer and the Windsor probe should be used and interpreted with particular caution. Because of the fact that densely packed large coarse aggregate particles are in close proximity to the formed surface, covered only with a thin layer of structural mortar, such tests may show a substantial variation in strength between what is essentially a test on an aggregate particle and another test, fractions of an inch away, performed on the structural mortar. Discrepancies in test results might be expected, for example, on attempting to determine 28-day strength on concrete placed by the PA method using very dense aggregate and grout with a relatively high percentage of fly ash replacement of the cement.

CLOSURE

The PA concrete process is an internationally accepted method of concrete placement appropriate to certain specialized construction problems. Test methods for evaluating the properties of the fresh grout and coarse aggregate fractions employed in this method are well established.

REFERENCES

[1] Wertz, L. S., "Process for Filling Cavities," Patent No. 2,313,110, 9 March 1943.
[2] "Investigation of the Suitability of Prepakt Concrete for Mass and Reinforced Concrete Structures," Technical Memorandum No. 6-330, U.S. Army Corps of Engineers, Vicksburg, MS, Oct. 1951, p. 34.
[3] Davis, R. E., "Prepakt Method of Concrete Repair," *Journal*, American Concrete Institute; *Proceedings*, Vol. 57, No. 2, Aug. 1960, pp. 155–172.
[4] Keats, B. D., "The Maintenance and Reconstruction of Concrete Tunnel Linings with Treated Mortar and Special Concrete," *Journal*, American Concrete Institute; *Proceedings*, Vol. 43, No. 7, March 1947, pp. 813–826.
[5] Davis, R. E., Jr., and Haltenhoff, C. E., "Mackinac Bridge Pier Construction," *Journal*, American Concrete Institute; *Proceedings*, Vol. 28, No. 6, Dec. 1956, pp. 581–595.
[6] Klein, A. M. and Crockett, J. A. J., "Design and Construction of a Fully Vibration-Controlled Forging Hammer Foundation," *Journal*, American Concrete Institute; *Proceedings*, Vol. 24, No. 5, Jan. 1953, pp. 421–444.
[7] Davis, S., "High-Density Concrete for Shielding Atomic Energy Plants," *Journal*, American Concrete Institute; *Proceedings*, Vol. 29, No. 11, May 1958, pp. 965–977.
[8] Tirpak, E. G., "Report on Design and Placement Techniques of Barytes Concrete for Reactor Biological Shielding," Report No. 1739, Oak Ridge National Laboratory, Oak Ridge, TN, May 1954.
[9] "Preplaced Aggregate Concrete for Structural and Mass Concrete," ACI Committee 304, *Journal*, American Concrete Institute; *Proceedings*, Vol. 66, No. 10, Oct. 1969, pp. 785–797.

BIBLIOGRAPHY

Akatsuka, Y. and Moriguchi, H., "Strengths of Prepacked Concrete and Reinforced Concrete Beams," *Journal*, American Concrete Institute; *Proceedings*, Vol. 64, No. 4, April 1967, pp. 204–212.
Davis, H. S., "How to Choose and Place Mixes for High Density Concrete," *Nucleonics*, June 1955, pp. 60–65.
Davis, R. E., Jansen, E. C., and Neelands, W. T., "Restoration of Barker Dam," *Journal*, American Concrete Institute; *Proceedings*, Vol. 44, No. 8, April 1948, pp. 633–667.
Davis, R. E., Jr., Johnson, G. D., and Wendell, G. E., "Kemano Penstock Tunnel Liner Backfilled with Prepacked Concrete," *Journal*, American Concrete Institute; *Proceedings*, Vol. 52, No. 3, Nov. 1955, pp. 287–308.
Downs, L. V., "Floating Caisson Facilitates Repair of Grand Coulee Spillway Bucket," *Civil Engineering*, April 1950, pp. 35–39.
King, J. C., "Special Concretes and Mortars," *Handbook of Heavy Construction*, 2nd ed., McGraw-Hill, New York, 1971, pp. 22–1 to 22–17.
Lamberton, B. A., "Placing Concrete in Deep Mines," *Civil Engineering*, 1956, pp. 37–39.
Mass, G. R. and Meier, J. G., "Investigation and Repair of Hoist Dam," *Concrete International*, Vol. 2, 1980, p. 49.
Olds, F. C., "QA Put to Work on the PCRV," *Power Engineering*, March 1970, pp. 45–48.
Steinman, D. B., *Miracle Bridge at Mackinac*, Wm. B. Eerdmans Publishing Co., Grand Rapids, MI, 1957, pp. 93–103.
U.S. Department of the Interior, *Concrete Manual*, 8th ed., 1975, pp. 446–449.

Roller-Compacted Concrete (RCC)

Kenneth L. Saucier[1]

PREFACE

Roller compacted concrete (RCC) is discussed in this special technical publication for the first time. The widespread use of RCC has occurred in the period since publication of the last version of *ASTM STP 169B* in 1978. Initial development, in the 1970s, was focused primarily on mass concrete for construction of locks and dams. In the 1980s, RCC was adapted for use in construction of pavements and other types of horizontal structures. Recently, in the 1990s, it has been applied as a repair technique to several types of structures.

DEFINITION

The American Concrete Institute (ACI) in *Cement and Concrete Terminology* (ACI 116R-90) defines RCC as, "concrete compacted by roller compaction; concrete that, in its unhardened state, will support a roller while being compacted." The term "roller compaction" is defined as follows by the ACI: "a process for compacting concrete using a roller, often a vibrating roller." The terms "rollcrete" and "rolled concrete" are no longer appropriate. Properties of hardened RCC are similar to those of conventionally placed concrete.

HISTORY

Dams

RCC developed as a result of efforts to design and build commercial concrete dams that could be constructed rapidly [1]. At the Rapid Construction of Concrete Dams Conference in 1970, Raphael [2] presented a paper in which he extrapolated from soil-cement applications the concept of placement and compaction of an embankment with cement-enriched granular bank or pit-run material using high-capacity earth-moving and compaction equipment. He noted that the increase in shear strength of cement-stabilized material would result in a significant reduction of the cross section when compared with a typical embankment dam and that use of continuous placement methods, similar to those used in earth dams, would generate savings in time and money as compared with traditional concrete gravity dam construction.

In 1972, Cannon [3] presented results of tests on a lean concrete using 75-mm maximum size aggregates transported by truck, spread by a front-end loader, and compacted by a vibratory roller at a Tennessee Valley Authority (TVA) project. The U.S. Corps of Engineers (USCE) soon thereafter constructed RCC field test sections at Jackson, Mississippi, [4] and Lost Creek Dam in Oregon [5] in 1972 and 1973.

Starting in 1974 and continuing through 1982, more that 2.5 million m³ of RCC were place at Tarbela Dam, Pakistan [6]. The initial application was to replace rock and embankment after one of four 14-m diameter outlet tunnels collapsed during initial filling of the reservoir. This repair was followed by rehabilitation work on both the auxiliary and service spillways consisting of massive groins, stilling basins, and cofferdams. In the tunnel repairs, 350 000 m³ of RCC were placed in 42 working days.

One of the first effective uses of RCC in the United States was in 1976 at TVAs Bellefonte Nuclear Plant [7], where 6000 m³ were used to raise the supporting base under the turbine building approximately 3 m.

In Japan, research into RCC for dams was initiated in 1974 under the guidance of its committee on Rationalized Construction of Concrete Dams [8]. The results led to the use of RCC (referred to as the "rolled concrete dam," RCD, method in Japan) in the main body of Shimajigawa Dam starting in 1978 and for the foundation slab in Ohkawa Dam starting in 1979.

The 50-m-high Willow Creek Dam confirmed the economy and rapid construction possible with RCC. More than 330 000 m³ of RCC were placed in less than five months at an approximate cost of $17 per m³, including the precast concrete panels that formed the vertical upstream face and all incidental costs for the RCC mass [9]. Figure 1 shows the construction process of Willow Creek Dam in 1982.

The U.S. Bureau of Reclamation's (USBR) 90-m-high Upper Stillwater Dam, completed in 1987, contains 1.12 million m³ of RCC placed within horizontally slipformed, air-entrained concrete facing elements [10].

Since these first projects, RCC has rapidly gained popularity and it has been used in a number of completed

---

[1] Chief, Concrete Technology Division, Structures Laboratory, U.S. Army Engineer Waterways Experiment Station, Vicksburg, MS 39180-8199.

FIG. 1—Construction of Willow Creek Dam.

structures in Brazil, Venezuela, France, Australia, and South Africa as well as in the United States.

Pavements

The first known RCC test pavement in the United States was installed in 1975 by the U.S. Army Corps of Engineers (USACE) at Vicksburg, Mississippi. This 4-m by 80-m service road proved the feasibility of RCC for use in pavement construction [11]. Figure 2 shows a typical RCC paving project.

Meanwhile, in Canada the forest-products industry was experimenting with RCC paving for log sorting yards. The log-sort yard at Caycuse, British Columbia, built in the fall of 1976, included 16 350 m² of 0.35-m-thick RCC pavement placed in a two-lift operation on a crushed-rock base. The yard size was doubled in 1979 with a second RCC application. When inspected in 1984, these pavements were in excellent condition [12].

In 1983, a small area of RCC pavement was placed on a tank road at Fort Stewart, Georgia, using troop labor from the post. Although the road was placed in an area of poor subgrade using rather primitive paving methods, the users have been most pleased with the finished product [12].

The first significant RCC pavement in the United States was constructed at Fort Hood, Texas, by the Corps of Engineers in August 1984. This was a large parking area for tanks and other tracked vehicles surrounding a maintenance shop. An 15 000 m² area of 0.25-m pavement was placed in one lift, at a cost of about $75 per m³ in place [12].

During 1988 and 1989, approximately 162 000 m² of 0.25-m-thick RCC was placed for tank hardstands and service roads at Fort Drum, New York. This placement marked the first known placement of air-entrained RCC [13].

Other significant applications in the private sector include the intermodal freight yards in Tacoma, Washington, and Denver, Colorado; a city street in Portland, Oregon; docks for the Massachusetts Port Authority in Boston; a coal yard in Hugo, Oklahoma; a large truck-stop parking area in Austin, Texas; and the Burlington Northern Railroad freight yard in Omaha, Nebraska [14]. The largest application in the private sector to date was at the Saturn automobile plant in Tennessee in 1988–1989. Approximately 500 000 m² of a 180-mm-thick pavement was placed in parking areas and roads [15].

APPLICATIONS

RCC may be considered for application where no-slump concrete can be transported, placed, and compacted using earth- and rock-fill construction equipment or, in the case of pavements, asphalt laydown equipment. Ideal RCC projects will involve large placement areas, little or no reinforcement, little or no embedded metal work, or other discontinuities. Application of RCC is often considered when it is economically competitive with other construction methods. It may be considered for large work pads, aprons, massive open foundations, base slabs, cofferdams, massive backfill, riprap for bank protection, as a repair material, and for pavement construction [16].

Use of RCC for pavements evolved from the use of soil cement and cement-treated base (CTB) courses. Although equipment for batching or feeding and mixing has developed from that used for the base courses, RCC for pavements requires better controls on proportioning. Also, a

FIG. 2—Typical RCC paving project.

true paver or laydown machine is normally used for placing and finishing the RCC pavement. The RCC pavement mixture has considerably more cementitious material than CTB, and differs from most soil cement in that it contains coarse aggregate. The most important difference between RCC pavement and CTB or soil cement is that the RCC is designed to be a true portland-cement concrete pavement with structural strength at least comparable to that of conventional portland-cement concrete, and often higher [12]. The RCC pavement is also designed to have resistance to the abrasion of traffic, durability when exposed to severe weather, and a surface finish and straightedge tolerance satisfactory for the usual requirements of the traffic involved.

RCC for pavements differs significantly from RCC for dams, which is simply a form of low-cement-content mass concrete. The RCC pavement mixtures have a much higher cementitious paste content and much smaller nominal maximum-size of coarse aggregate. These factors along with a different approach to mixture-proportioning design produce a more workable mixture than that used for dams, although it is still a no-slump concrete, stiff enough to support vibratory rollers.

ADVANTAGES

The primary advantage of RCC over conventional construction is in cost savings. Construction cost histories of RCC and conventional concrete show that the unit cost per cubic metre of RCC is considerably less than conventionally placed concrete. Approximate costs of RCC range from 20 to 30% less than conventionally placed concrete [12,16]. The difference in percentage savings usually depends on complexity of placement and on total quantities of concrete placed. Savings associated with RCC are primarily due to reduced forming, placement, and compaction costs, as well as reduced construction times. To achieve the highest measure of cost effectiveness and achieve a high-quality product similar to what is expected of conventional concrete structures, the following design and construction objectives are desired: RCC should be placed as quickly as practical after mixing; operations should include as few laborers as possible; design should avoid, as much as possible, multiple mixtures and other construction or forming requirements that tend to interfere with production; and the design should not require complex construction procedures.

Maximum placement rates of 9500 m^3 per day have been achieved in dam construction [16]. These production rates make dam construction in one season readily achievable for even large structures. When compared to embankment or conventional dams, construction time for large projects can be reduced by one to two years. Other benefits from rapid construction include reduced administration costs, earlier project benefits, and possible use at sites with limited construction seasons. Basically, RCC construction offers economic advantages in all aspects of construction that are related to time.

MATERIALS

Cementitious Materials

RCC can be made with any of the basic types of hydraulic cement or a combination of hydraulic cement and pozzolan. Selection of materials for chemical resistance to sulfate attack, potential alkali reactivity, and resistance to abrasion with certain aggregates should follow procedures used for conventional concrete construction.

The strength of RCC is primarily dependent upon the quality of the aggregate; degree of compaction; and the proportions of cement, pozzolan, and water. The type of cementitious material has a significant effect on the rate of hydration and the rate of strength development and, therefore, significantly affects strengths at early ages.

Cement

RCC can be made using any of the basic types of portland cement given in ASTM Specification for Portland Cement (C 150) or blends of these with ground granulated blast-furnace slag as specified in ASTM Specification for Ground Granulated Blast-Furnace Slag for Use in Concrete and Mortar (C 989). To minimize thermal cracking in mass applications, portland cements with lower heat-generation characteristics than Type I are often specified. They include Type II (moderate heat), Type IP (portland-pozzolan cement), and Type IS (portland blast-furnace slag cement). Strength development for these lower-heat cements is usually slower than for Type I.

Pozzolans

The selection of a pozzolan suitable for RCC should be based on its conformance with ASTM Specification for Fly Ash and Raw or Calcined Natural Pozzolan for Use as a Mineral Admixture in Portland Cement Concrete (C 618) or other applicable standard and its cost and availability. Nearly all RCC projects using pozzolans have used Class F fly ash, due primarily to the effect of its spherical particles on workability. Use of Class F fly ash in RCC serves three purposes: (1) as a partial replacement for portland cement to reduce heat generation, (2) to reduce cost, and (3) as a mineral addition to the mixture to provide fines to improve workability. In general, performance of fly ash in RCC has differed from the results expected in conventional concrete only in those instances involving aggregates containing large quantities of natural fines [1].

Aggregates

As with conventional concrete, aggregates for RCC should be evaluated for quality and grading. Aggregate for RCC should meet the same standards for quality and grading as required for conventional concrete construction. Only where unusual circumstances exist, such as construction during an emergency or when the use of a poorer-quality aggregate meets quality requirements of the concrete, could aggregate of lesser quality be justified. Changes from the grading or quality requirements of ASTM Specification for Concrete Aggregates (C 33) should be supported by laboratory or field test results that show that the concrete produced from the proposed materials fulfills the requirements of the project as is provided for in ASTM C 33. The nominal maximum size of aggregate (NMSA) particle that has been handled and compacted in most roller compacted mass concrete construction is 75 mm. While larger sizes have been successfully used in Japan and at Tarbela Dam, the use of NMSA larger than 75 mm will often not be technically justified or economically viable [16].

Almost all RCC pavements have used a nominal maximum size aggregate of either 19.0-mm or 9.5-mm conforming to the quality requirements of ASTM C 33 [14,17]. The required amount of material passing the 75-μm may be greater for RCC than acceptable for conventional concrete. The larger percentage of fines is used to increase the paste content in the mixture to fill voids and contribute to compactibility. The additional fines are usually made up of naturally occurring nonplastic silt and fine sand, manufactured fines, or extra pozzolan.

Grading

The largest practical nominal maximum size aggregate is usually used in construction of RCC dams. However, RCC-containing aggregate having a nominal maximum size greater than 75 mm will often experience excessive segregation during the spreading and compaction operations. The grading limits of individual coarse aggregate size fractions should comply with those used in conventional concrete [16]. Individual size groups are normally combined to produce gradings approaching ideal gradings, as given in Table 1. Fine aggregate gradings are also specified as shown in Table 2. Approximate fine aggregate contents, expressed as a percentage of the total aggregate volume, for mass RCC are given in Table 3.

Most RCC paving projects now require two sizes of aggregates, 19.0-mm maximum, split on the 4.75-mm sieve such that they can be combined to produce a total grading within a band ranging from 83 to 100% by weight passing the 19.0-mm sieve to 2 to 8% passing the 75-μm sieve [17].

TABLE 1—*Ideal Coarse Aggregate Grading.*

Sieve Size (mm)	Cumulative Percent Passing		
	4.75 to 75 mm	4.75 to 37.5 mm	4.75 to 19.0 mm
75	100		
63	88		
50	76		
37.5	61	100	
25.0	44	72	
19.0	33	55	100
12.5	21	35	63
9.5	14	23	41
4.75	...	...	...

TABLE 2—*Fine Aggregate Grading Limits.*

Sieve Size	Cumulative Percent Passing Ref 16	Cumulative Percent Passing ASTM C 33
9.5 mm	100	100
4.75 mm	95 to 100	95 to 100
2.36 mm	75 to 95	80 to 100
1.18 mm	55 to 80	50 to 85
600 μm	35 to 60	25 to 60
300 μm	24 to 40	10 to 30
150 μm	12 to 28	2 to 10
75 μm	8 to 18	...
Fineness modulus	2.10 to 2.75	2.3 to 3.1

TABLE 3—Approximate Ratio of Fine to Total Aggregate Volume.

Nominal Maximum Size and Type of Coarse Aggregate	Fine Aggregate Ratio, percent of total aggregate volume
75 mm, crushed	29 to 36
75 mm, rounded	27 to 34
37.5 mm, crushed	39 to 47
37.5 mm, rounded	35 to 45
19.0 mm, crushed	48 to 59
19.0 mm, rounded	41 to 45

Admixtures

Water-Reducing and Retarding Admixtures

The use of a water-reducing and retarding admixture or a retarding admixture as specified in the ASTM Specification for Chemical Admixtures for Concrete (C 494) may be considered for any RCC placement. The use of a water-reducing and retarding admixture has been proven beneficial for extending placeability of mass RCC for at least 1 h and maintaining lift surfaces in an unhardened state until the next layer of RCC was placed, thereby creating a better bond and increased likelihood of a watertight joint [16]. The extended workability was especially beneficial during warmer weather, during RCC startup activities, and for placement of thick lifts. Required dosages of water-reducing and retarding admixtures are normally several times as much as recommended for conventionally placed concrete.

Air-Entraining Admixtures

Air-entraining admixtures have been added to RCC mixtures in an attempt to entrain air in a proper air-void size and spacing to resist damage to the concrete when it is subjected to repeated cycles of freezing and thawing while in a critically saturated state. Research continues to determine if and how a proper air-void system can be achieved in RCC. From what has been learned, it appears that the dosages of air-entraining admixtures required for RCC may be considerably greater than required for conventionally placed concrete [16].

MIXTURE PROPORTIONING

As with conventional concrete construction, the primary considerations for mixture proportioning are normally durability, strength, workability, and, in the case of RCC, compactibility. Another important consideration for mass RCC is the minimization of heat rise due to the chemical reactions of the cementitious ingredients. Again, as with conventional mass concrete, factors such as use of (1) the largest nominal maximum-size of aggregate, (2) minimum amount of cementitious material, (3) pozzolans or blended cements, and (4) cooling procedures for the materials are evaluated on a job-specific basis.

A number of mixture-proportioning methods have been successfully used for RCC structures throughout the world, making it difficult to generalize any one procedure as being standard. These methods have differed significantly due to the location and structural requirements of the structure, availability of materials, the mixing and placing equipment used, and time constraints. Approaches to mixture proportioning also differ significantly depending on the philosophy of the treatment of aggregates as either conventional concrete aggregates or as aggregates used in the placement of stabilized embankments. ACI 207.5R [1] discusses the three predominant mixture proportioning methods. They are: (1) proportioning RCC to meet specified limits of workability, (2) relying on trial mixture tests to select the most economical aggregate-cementitious materials combination, and (3) proportioning RCC using soils compaction concepts.

Proportioning to Meet Specified Limits of Consistency

Proportioning for optimum workability suitable for compaction uses the modified Vebe compactibility test as the basis for determining workability and optimizing aggregate proportions. The modified Vebe apparatus consists of a vibrating table of fixed frequency and amplitude supporting an attached 0.01 m^3 container. A loose RCC sample is placed in the container under a surcharge and the sample is vibrated until consolidated. The vibration time for full consolidation is measured and compared with onsite compaction tests with vibratory rollers. The desired time is determined based on density tests and evaluation of core samples. This vibration time is influenced by mixture proportions, particularly water content, overall aggregate grading, NMSA, fine aggregate content, and fines content.

Proportioning for the Most Economical Materials Combination

Mixtures for a number of RCC structures have been proportioned based on the results of physical tests of samples made from trials using a fixed aggregate grading while varying the amount of cementitious material and comparing results. Based on these results, supplemental tests may be appropriate at a constant amount of cementitious material while adjusting the aggregate proportions. The most economical combination of cementitious materials and aggregate that provides the required strength and a field-usable mixture is then selected for the project. The proportioning of these mixtures has resulted in cementitious material paste contents that essentially fill the voids between aggregates.

Proportioning Using Soils Compaction Concepts

The soils-compaction procedure is more suited to smaller aggregate mixtures with higher cementitious material contents. It involves determining the maximum dry density of materials using modified Proctor compaction procedures and can be considered an extension of soil-cement technology. Optimum water contents are established using the same procedures to establish the optimum water content of embankment material and soil cement. Compaction is dependent upon the energy imparted to the specimen. The compactive effort of the modified Proctor test has been found to correspond closely to in-place density measurements of the smaller NMSA mixtures with which it is used.

Trial Mixtures

All of the methods include the preparation of trial mixtures to confirm that the workability and compactibility are suitable for roller compaction. This is usually confirmed in a test section using the placing methods and equipment that are planned for use on the job. If the laboratory-proportioned mixture proves unsuitable for construction, the mixture is adjusted accordingly. A compilation of many of the mixtures used in mass RCC has been made by Hansen [18], and for pavement mixtures by Ragan [19]. Tables 4 and 5 give typical proportions for a mass mixture and a pavement mixture, respectively.

BATCHING AND MIXING

The batching and mixing plant requirements for a project to be constructed using RCC are essentially the same as for a project built with conventional concrete [16]. The production, stockpiling, and reclamation of aggregate from the stockpiles are done in the same way and with the same equipment as for conventional concrete. Likewise, the batching and mixing equipment used for RCC is the same as would be used if the concrete were to be conventionally placed. RCC has been mixed both in tilting-drum mixers and in pugmill mixers. Experience indicates that pugmills produce faster and more effective mixing due to their intense scrubbing action, a feature that has made them almost a necessity for RCC paving projects [17]. Continuous batching and mixing plants have also been used. Truck mixers have been used to mix RCC having a NMSA of 25 mm, however, their use is not recommended due to problems with discharge of the concrete.

TRANSPORTING

The entire process of mixing, transporting, placing, spreading, and compacting is accomplished as rapidly as possible with as little rehandling as possible. Local environment and placing conditions at different jobsites will affect the reasonable time limits for these operations. The time lapse between the start of mixing and completion of compaction must be less than the time of initial setting of the mixture under the conditions in which it is being used. A general rule to follow with nonretarded mixtures is that placing, spreading, and compacting should be accomplished within 40 min of mixing [1].

Mass Concrete

The two primary methods of transporting mass RCC are by conveyor and vehicle. Conveyor systems have proven to be an efficient and safe way to transport concrete from the mixer to the placement area. Conveyors are designed to meet the production requirements, avoid segregating the mixture, and be covered to protect the concrete from drying and wetting by rain.

Hauling and dumping of RCC with end-dump trucks combined with remixing and spreading of RCC by bulldozers has proven to be an economical and effective method of placing mass RCC. If 75-mm NMSA is used, the bulldozer spreading and remixing procedures are closely controlled to reduce or eliminate segregation. Scrapers and bottom-dump trucks place RCC while moving. Except at the margin of spread lanes, segregation is minimal. At the edge of spread lanes, RCC is susceptible to segregation which is controlled by blending of the concrete at the interface of the lanes.

RCC has been successfully placed in lift thicknesses ranging from a minimum of 0.15 m (compacted thickness), to over 1 m although no general production in the United States has exceeded 0.6 m. The design of dams where lift thicknesses greater than 0.3 m have been used has been based on the realization that the spreading of the RCC with heavy dozers not only remixes and redistributes the concrete to overcome segregation but also provides compaction. These procedures have been established and proven by large-scale, well-controlled test section construction and testing, as well as in full-scale production of RCC for dams in Japan and at Elk Creek Dam [16].

Pavements

Transporting of RCC for paving has been accomplished previously with dump trucks. There are therefore two critical operations in the process: (1) timely scheduling of the trucks and (2) communication with the mixing plant. Trucks are scheduled to provide a continuous supply of concrete, but spaced so that they will not be delayed at the paving machine and thus permit the mixed concrete to dry out and lose workability. Communication between

TABLE 4—Typical Mass RCC Mixture.

Item	
Cementitious material,[a] kg/m^3	175[c]
Coarse aggregate,[b] kg/m^3	1300
Fine aggregate, kg/m^3	800
Water, kg/m^3	122
W/C + P,[d] by mass	0.70
Admixture	none

[a] Includes 20 to 50% pozzolan.
[b] 75-mm NMSA.
[c] Wide variation between applications; may be as low as 100 kg/m^3.
[d] W/C + P = water to cement plus pozzolan ratio.

TABLE 5—Typical RCC Pavement Mixture.

Item	
Cementitious material,[a] kg/m^3	350
Coarse aggregate,[b] kg/m^3	1100
Fine aggregate, kg/m^3	900
Water, kg/m^3	122
W/C,[c] by mass	0.35
Admixture	water reducer

[a] Includes 20 to 30% pozzolan.
[b] 19-mm NMSA.
[c] W/C = water to cement ratio.

the placing site and the mixing plant is required so that mixture adjustments can be made quickly. Adjustments in water will be necessary because of changing conditions, especially variable moisture content of the aggregates in the stockpiles and changing ambient conditions throughout the day. Moisture content is critical because even a 0.1 or 0.2% change can have a significant effect on workability (compactibility) of the mixture. Normally, it is also required that the plant have a transfer hopper at the end of the discharge belt or discharge from the pugmill to contain the concrete between trucks so that the least amount of stoppage of the plant is required. Each time the plant is stopped and restarted, the mixture changes in consistency, which, in turn, affects the compaction process [14].

PLACEMENT

Mass RCC

Compaction

The key to a successful RCC job is attainment of good compaction. Each lift of a mass placement is compacted using a vibrating steel-wheel roller. It has been determined from various test sections and actual construction projects that RCC can be adequately compacted using a variety of vibratory compactors. For most applications, it is recommended that a double-drum, self-propelled, mid-size asphalt roller be used. Rollers of this type should have a high frequency, low amplitude, and a dynamic force of between 40 and 60 N/m of drum length. Rollers with these characteristics have been used at numerous U.S. RCC dams [16].

Compaction of RCC should be accomplished as soon as possible after it is spread, especially in hot weather. Typically, compaction is specified to be within 10 min of spreading and 40 min from the time of initial mixing. These times can be increased for RCC mixtures with extended times of setting [1]. The fresh mixture surface is smoothly spread so that the roller drum produces a consistent compactive pressure under the entire width of the drum. If the uncompacted lift surface of less workable RCC is not smooth, the drum may overcompact high spots and undercompact adjacent to them. Each RCC mixture will have its own characteristic behavior for compaction depending on temperature, humidity, wind, plasticity of the aggregate fines, overall grading, and the NMSA. Mass RCC mixtures should compact to a close-textured, relatively smooth surface. Experience indicates that four to six roller passes (a round trip with a double-drum roller across the same area constitutes two passes) are adequate for most RCC applications. In general, the material should not pick up onto the roller drum, nor should there be free surface moisture or pumping of water from the mixture. These conditions can be observed and adjustments made in the water content if they occur.

Lift Surface Treatment

At the completion of rolling, lift surfaces are moistened and kept damp until the next lift is placed or until the required curing period has ended. Fog spray nozzles that provide an extremely fine spray are recommended. If coarse sprays are used, paste and fine aggregates sometimes erode away from the surface. The surface is maintained in a moist condition commencing immediately following compaction; and, as necessary, the lift surface cleaned prior to placement of the next lift. The cleanup includes the removal of all loose material, debris, standing or running water, snow, ice, oil, and grease. Research has indicated that excellent bond is obtained when the top lift is placed before the lower lift has achieved its time of final setting and good bond is obtained when the surfaces are clean even after time of final setting of the lower lift [20]. The design of RCC structures where watertightness and bonding are required between lifts may require placing a bedding (bonding) mortar over the entire surface area between all lift placements. A bedding mortar is a fluid, cementitious material that is used to increase bond between RCC lifts and to improve watertightness by filling any voids that may occur at the bottom of an RCC lift during placement and compaction. Typical bedding mortar uses 4.75-mm sieve NMSA, is retarded, has a high slump, and contains a high portland-cement and fly ash content. Application of this type of bedding mortar has been used for lift joint preparation for nearly all RCC dams in Japan and many dams in the United States [16].

Pavements

Placement of RCC pavements is accomplished by the use of heavy-duty track-equipped paving machines of the self-propelled type. Prior to mid-1985, American-made asphalt paving machines were used to lay the RCC material. These pavers were equipped only with vibrating screeds, which meant that almost all the compaction had to be provided by the vibratory rollers.

Since then, heavy-duty pavers with tamping screeds and a vibrating screed have been made specifically for RCC pavement. One model introduced into the United States from Germany compacts the RCC course to 95% modified Proctor density, leaving less compaction to be performed by the vibratory rollers [14]. An American-made paver has been developed that has a tamping screed in addition to a vibrating screed. It can compact the RCC to about 90% of modified Proctor density. Specifications usually require that grade be controlled on one side of the paver with electronic controls operating from a stringline. The other side must also be electronically controlled in the same manner or from a ski riding on an adjacent paved lane.

Primary compaction equipment is the heavy double-drum steel wheel vibratory roller weighing at least 2650 kg/m of drum length and vibrating at least 25 Hz [21]. Many specifications require at least four complete passes of the double-drum vibratory roller to produce a density of at least 98.5%. After required density is attained, the surface may be rolled with rubber-tired roller to close up the surface. One final pass of a non-vibratory steel wheel roller may be required. It is extremely important that rollers follow immediately behind the paver and that rolling be completed as rapidly as possible, before drying occurs.

A rolling pattern should be established during construction of the test section and strictly adhered to. Normally, each edge is rolled first, to within 0.3 to 0.5 m of the edge. Next comes the lane center; if it abuts an existing lane, then that joint is rolled.

CURING

As with conventional concrete, RCC is cured for a prescribed curing period to develop the required strength and prevent cracking. However, due to the large areas of exposed surfaces, procedures to achieve the necessary curing vary with the job. Because RCC is dryer than conventional concrete, surfaces tend to dry more rapidly during warm weather. During such conditions, considerable effort is required to maintain a uniformly moist surface. Good specifications address the proper procedures and equipment required for curing of RCC.

TEST SECTION

As an aid in training personnel and discovering problems in techniques and methods, construction of a project test section prior to the start of production operations is essential in almost every case where RCC is used. The experience gained on a test section will provide a common basis of knowledge for all concerned. A mating of the mixture to the equipment to be used is a prime objective of the test section exercise. It thus provides an opportunity to adjust the RCC mixture proportions if necessary. The test section should be planned to demonstrate the capability to produce the quality and quantity of RCC required by contract specifications. The test section should be constructed sufficiently early in the contract to allow the contractor time to increase the size of his batching, mixing, or transporting system; or to modify placing, spreading, and compaction techniques; or to modify any other operation that is considered essential to the success of the job.

PROPERTIES OF HARDENED RCC

The significant material properties of hardened RCC include compressive strength, tensile strength, modulus of elasticity, tensile strain capacity, Poisson's ratio, volume change (thermal, drying, and autogenous), thermal coefficient of expansion, specific heat, creep, permeability, and durability. The hardened properties of RCC and conventional concrete are quite similar and differences are primarily due to differences in mixture proportions, gradings, and voids content [1]. A wide range of RCC mixtures can be proportioned, just as there is a wide range of mixtures for conventionally placed concrete. It is difficult to quantify typical values in either case. In general, RCC will have lower cement, paste, and water contents and may contain nonplastic fines to fill aggregate voids. Aggregate quality, grading, and physical properties have a major influence on the physical properties of RCC.

Volume Change

The two significant changes in volume experienced with RCC are due to drying shrinkage (primarily in pavements) and thermal expansion and contraction in mass concrete. Volume change associated with drying shrinkage is normally less than that in comparable conventional concrete mixtures due to the lower water content. This lower shrinkage has resulted in less cracking and revised design considerations for RCC pavements [22]. With respect to thermal considerations, heat rise that causes expansion of a massive concrete structure is due almost entirely to the chemical reactions of the cementitious material. Therefore, the use of lesser amounts of cementitious material in mass RCC construction lowers the potential for thermal cracking. Indeed, Hansen [18] has reported on several dams where this has appeared to be the result.

Durability

Durability of RCC is usually considered in terms of abrasion, erosion, or damage from freezing and thawing, but not often in terms of all three together. Pavements at heavy-duty facilities such as log-storage yards and coal-storage areas have shown no appreciable wear from traffic and industrial abrasion under severe conditions [1,12]. Erosion tests in test flumes have indicated the excellent erosion resistance for RCC [1,20]. However, there have been conflicting results with respect to resistance of RCC to freezing and thawing. Examples do exist of good performance of nonair-entrained concrete in the field [1,12,23,24], apparently due to the fact that the concrete was not critically saturated at the time of freezing. Laboratory specimens of nonair-entrained RCC tested according to ASTM Test Method for Resistance of Concrete to Rapid Freezing and Thawing (C 666), Procedure A (in water) [20] and large blocks of mass concrete material exposed to natural weathering of Treat Island, Maine [1], have typically performed very poorly. The basis of the problem is the difficulty in securing a proper air-void system in the relatively dry matrix of the RCC. Recently, however, research has shown that properly air-entrained RCC can be achieved and will, in fact, have good resistance to freezing and thawing [13,23].

QUALITY CONTROL AND QUALITY ASSURANCE PROCEDURES

Quality control and quality assurance measures for RCC have evolved with the development of RCC for both mass and pavement applications during the 1980s. Visual observations were used almost totally for quality control of placement during the early years of RCC development. Although still an important tool, observations are now supplemented by several new ASTM standard tests and the adaptation of the nuclear density test to RCC construction.

Monitoring Consistency and Compactibility

A modified Vebe test, conducted according to either Method A or B of ASTM Test Methods for Consistency and

Density of Roller-Compacted Concrete Using a Vibrating Table (C 1170) is used to determine consistency and compactibility of the freshly mixed RCC. Method A, which specifies a surcharge on the test material, is normally used due to the stiffness of most RCC mixtures. The Vebe time used during construction is determined initially during the mixture proportioning studies. The time is then adjusted as necessary when the test section is constructed. Still further adjustments, as necessary, may be made to the Vebe time during construction. Once a Vebe time is established, the normal procedure is to maintain a consistent Vebe time for the RCC being produced by making batch water adjustments as necessary to compensate for changes in aggregate moisture, changes in humidity, wind, and temperature. Batch water adjustments are usually made if two consecutive Vebe readings vary from a target Vebe time by five or more seconds.

Preparation of Test Specimens

Cylindrical test specimens for determination of compressive strength of RCC cannot be fabricated using the standard procedures used for conventional concrete; those being, ASTM Test Methods of Making and Curing Concrete Test Specimens in the Field (C 31) and ASTM Test Method of Making and Curing Concrete Test Specimens in the Laboratory (C 192). Test specimens of RCC are fabricated according to ASTM Practice for Making Roller-Compacted Concrete in Cylinder Molds Using a Vibrating Table (C 1176). It is recognized that the consolidation of the concrete in the cylinders is not identical to that achieved under a vibrating roller, however, the method has been accepted on many jobs. The concrete in the molds is consolidated in lifts by external vibration that equates generally to compaction by a vibrating roller. Each of these lifts is consolidated in the cylinder by observing mortar filling in the voids as the concrete is vibrated. If a satisfactory degree of the void filling does not occur, the concrete probably has insufficient mortar due to either improper sampling, segregation, or improper mixture proportioning or the concrete has lost workability due to chemical reaction of the cementitious material.

Monitoring In-Place Density

Density measurements are taken during placing of RCC using a nuclear density gage and following compaction according to one or both of the procedures covered by ASTM Test Method for Density of Unhardened and Hardened Concrete in Place by Nuclear Methods (C 1040). A single-probe or double-probe nuclear gage provides reliable information when large numbers of readings are taken. However, the two-probe gage provides the capability of monitoring RCC densities at all depths within the limit of fresh RCC. Data from the nuclear-gage readings can be used during the compaction process to confirm that the mixture proportions are correct, for determining if densities are uniform throughout the lift, and for identifying rock pockets or segregation. Field nuclear-gage readings are compared to values obtained for density of RCC in the project laboratory. To ensure the accuracy of the nuclear gages being used, a test block is made during the early stages of the project and kept available.

Making Visual Observations

An inspector should be present at all times that RCC is being placed. As RCC is delivered, spread, and compacted, visual features are observed to determine that it is of the correct workability. Usually, if the RCC is too dry for proper compaction, the obvious signs are: (1) increased segregation of the mixture, (2) aggregate particles on the surface that are cracked by the roller, and (3) little reworking of the mixture adjacent to the bulldozer as the RCC is spread. In addition, concrete that is too dry will not show the development of paste at the surface after three or four roller passes. If closely spaced surface cracking is observed as the roller moves over the surface, the mixture is probably slightly dry. The RCC is too wet if the equipment sinks into the mass. It may be somewhat wet if rutting is produced by the heavy equipment.

CLOSURE

Roller-compacted concrete has advanced significantly as a viable construction technique from a slow start in the early 1970s to the present. The early development, primarily on dams, was nurtured by research, trial placements, and field adjustments at the jobsite. During the 1980s, RCC technology accelerated by application to pavements and in repair of mass concrete structures. Additional research and development is needed in the areas of: (1) lift joint treatment of mass structures, (2) pavement design procedures to include spacing of expansion joints, (3) placement techniques to eliminate segregation problems in mass concrete construction and to secure smooth surfaces in pavement applications, (4) reliable quality control methods for use during the placement of RCC, and (5) improved consolidation (compaction) equipment and techniques.

REFERENCES

[1] "Roller Compacted Mass Concrete," ACI 207.5R-89, *Manual of Concrete Practice*, Part 1, American Concrete Institute, Detroit, MI, 1990.
[2] Raphael, J. M., "The Optimum Gravity Dam," *Rapid Construction of Concrete Dams*, American Society of Civil Engineers, New York, 1971, pp. 221–247.
[3] Cannon, R. W., "Concrete Dam Construction Using Earth Compaction Methods," *Economical Construction of Concrete Dams*, American Society of Civil Engineers, New York, 1972, pp. 143–152.
[4] Tynes, W. O., "Feasibility Study of No-Slump Concrete for Mass Concrete Construction," Miscellaneous Paper No. C-73-10, U.S. Army Engineer Waterways Experiment Station, Vicksburg, MS, Oct. 1973.
[5] Hall, D. J. and Houghton, D. L., "Roller Compacted Concrete Studies at Lost Creek Dam," U.S. Army Engineer District, Portland, OR, June 1974.

[6] Johnson, H. A. and Chao, P. C., "Rollcrete Usage at Tarbela Dam," *Concrete International: Design and Construction*, Vol. 1, No. 11, Nov. 1979, pp. 20–33.

[7] Cannon, R. W., "Bellefonte Nuclear Plant—Roller Compacted Concrete—Summary of Concrete Placement and Evaluation of Core Recovery," Report No. CEB-76-38, Tennessee Valley Authority, Knoxville, TN, 1977.

[8] Hirose, T. and Yanagida, T., "Dam Construction in Japan: Burst of Growth Demands Speed, Economy," *Concrete International: Design and Construction*, Vol. 6, No. 5, May 1984, pp. 14–19.

[9] Schrader, E. and McKinnon, R., "Construction of Willow Creek Dam," *Concrete International: Design and Construction*, Vol. 6, No. 5, May 1984, pp. 38–45.

[10] Oliverson, J. E. and Richardson, A. T., "Upper Stillwater Dam—Design and Construction Concepts," *Concrete International: Design and Construction*, Vol. 6, No. 5, May 1984, pp. 20–28.

[11] Burns, C. D., "Compaction Study of Zero-Slump Concrete," Miscellaneous Paper No. S-76-16, U.S. Army Engineer Waterways Experiment Station, Vicksburg, MS, Aug. 1978.

[12] Keifer, O., Jr., "Paving with Roller Compacted Concrete," *Concrete Construction*, March 1986, pp. 287–297.

[13] Ragan, S. A., Pittman, D. W., and Grogan, W. P., "An Investigation of the Frost Resistance of Air-Entrained and Nonair-Entrained Roller-Compacted Concrete (RCC) Mixtures for Pavement Applications," Technical Report GL-90-18, U.S. Army Waterways Experiment Station, Vicksburg, MS, 1990.

[14] Keifer, O., Jr., "Corps of Engineers Experience with RCC Pavements," *Roller Compacted Concrete II*, Conference Proceedings, American Society of Civil Engineers, San Diego, CA, March 1988, pp. 429–437.

[15] Munn, W. D., "Roller Compacted Concrete Paves Factory Roads," *Highway and Heavy Construction*, Sept. 1989.

[16] "Roller Compacted Concrete," Engineer Manual No. 1110-2-2006, U.S. Department of the Army, Corps of Engineers, Washington, DC, 1 Feb. 1992.

[17] Keifer, O., Jr., "Paving with RCC," *Civil Engineering*, Oct. 1987, pp. 65–68.

[18] Hansen, K. D. and Reinhardt, W. G., *Roller-Compacted Concrete Dams*, McGraw-Hill, New York, 1991.

[19] Ragan, S. A., "Proportioning RCC Pavement Mixtures," *Roller Compacted Concrete II*, Conference Proceedings, American Society of Civil Engineers, San Diego, CA, March 1988, pp. 380–393.

[20] Saucier, K. L., "No-Slurry Roller Compacted Concrete (RCC) for Use in Mass Concrete Construction," Technical Report SL-84-17, U. S. Army Corps of Engineers, Waterways Experiment Station, Vicksburg, MS, Oct. 1984.

[21] "Roller Compacted Concrete (RCC) Pavement for Airfields, Roads, Streets, and Parking Lots," Guide Specifications CEGS-02520, U.S. Army Corps of Engineers, Washington, DC, Jan. 1988.

[22] Rollings, R. S., "Design of Roller Compacted Concrete Pavements," *Roller Compacted Concrete II*, Conference Proceedings, American Society of Civil Engineers, San Diego, CA, March 1988, pp. 454–466.

[23] Ragan, S. A., "Evaluation of the Frost Resistance of Roller-Compacted Concrete Pavements," Miscellaneous Paper SL-86-16, U.S. Army Corps of Engineers, Waterways Experiment, Vicksburg, MS, Oct. 1986.

[24] Waddell, J. J. and Dobrowolski, J. A., "Special Concretes and Techniques," *Concrete Construction Handbook*, 3rd ed., McGraw-Hill, New York, 1993, p. 30.12.

Polymer-Modified Concrete and Mortar

Lou A. Kuhlmann[1] and Michael O'Brien[2]

PREFACE

This chapter on Polymer Modified Concrete and Mortar is a new contribution to *ASTM STP 169*. In January 1987, a Task Group was first formed in Subcommittee C09.03.18 (C09.25) due to the need identified for standards development in the area of Polymer Modified Cementitious materials. This Task Group started the process, and then in June of 1989 this activity was organized under a new subcommittee, C09.03.19 (C09.44) where the work continues. Membership in this subcommittee has grown to 47. This group is working on developing specifications and test methods pertaining to the use of latexes and redispersible dry polymers as modifiers for hydraulic cement concrete and mortars. These polymer modifiers mainly contribute to adhesion, water resistance, reduced permeability, and increased durability. Polymer modifiers are used in a variety of applications, such as, patching compounds, stucco, ceramic tile thin sets and grouts, and bridge deck overlays. The first ASTM sponsored Symposium on Polymer Modified Concrete and Mortar was held in Louisville in June of 1992. A second symposium is in the planning stages for 1995.

INTRODUCTION

Latex is a dispersion of small organic polymer particles in water. When latex is used as an additive to portland cement mixes, the resultant mixture is called polymer-modified mortar or concrete. The polymer particles in a latex are spheres and typically between 0.05 and 0.50 µm in diameter. (A cubic centimetre of a latex containing 50% polymer solids of 0.20 µm-diameter particles would contain approximately 10×10^{12} particles). Surfactants are added to the latex formulation in the manufacturing process (emulsion polymerization) to prevent coagulation of the particles from the mechanical stress of the process, as well as premature chemical reactions with the portland cement. These surfactants also function as water reducers, thus contributing to the improved properties that the latex adds to the mortar and concrete. However, it is the nature of these surfactants to foam when agitated. It is therefore necessary to incorporate an antifoam agent in the latex prior to use in order to control the air content of the portland cement mix.

Latex was originally produced from the sap of the rubber tree and, prior to World War II, this "natural" latex from Southeast Asia was the raw material that established the rubber industry. The war stimulated research into synthetic latex processes resulting in the production of latex by the emulsion process. This original process used a styrene-butadiene (SB) polymer, and the product therefore became known as styrene-butadiene rubber (SBR), a designation that has frequently been used incorrectly today when referring to all latexes, no matter what polymer is dispersed.

One of the first widespread uses for latex-modified portland cement was as mortar for bridge deck overlays. In 1957, 12.7-mm (½-in.) thick mortar modified with SB latex was installed as an experimental coating on a bridge deck in Cheboygan, Michigan, to determine if this would provide a long-lasting wearing surface. Since then, thousands of bridges have been overlaid with latex-modified mixes, initially with 20-mm (¾-in.) mortar, and now with 38-mm (1½-in.) concrete, all of them relying on the adhesion properties of the latex to permanently bond the overlay to the deck concrete [1].

Acrylic polymers have also been used widely for more than 30 years to increase the strength properties and durability of mortar in thin sections [2]. Latex-modified mortars are used in a variety of functional and decorative coating applications such as ceramic tile thinsets and grouts, water-resistant coatings for basement masonry walls, overlayments, and self-leveling flooring applications. These latex-modified mortars are also used in many applications where aesthetics are important. Acrylic latexes display resistance to ultraviolet radiation so they resist yellowing and chalking.

LATEX TYPES

Although there are many types and formulations of latexes manufactured, only those formulated specifically for use in portland cement are suitable in mortar and concrete applications. Descriptions of the formulations and manufacturing processes are available in the literature [3–5]. The latexes most commonly used are copolymers of styrene-butadiene (SB), acrylates (PAE), styrene-

---

[1] Development associate, Dow Chemical Co., Larkin Laboratory, Midland, MI 48640.
[2] Senior research scientist, Rohm and Haas Research Laboratories, Spring House, PA 19477.

acrylate (SA), polyvinyl acetate (PVA), and vinyl acetate-ethylene (VAE). Typical properties of these latexes are given in Table 1. As the names indicate, these latexes are composed of organic polymers that are combinations of various monomers, that is, styrene, acrylate, butadiene, vinyl acetate, etc.

Styrene-butadiene is the most commonly used latex for concrete overlays on bridge decks and is the only latex that has been evaluated for overlays by the Federal Highway Administration [6]. The other latexes, plus SB, are used in mortar applications, such as patching, floor leveling, tile grout, and stucco.

Advancements in processing technology allow producers to now convert some latexes into a dry, powder form, reducing shipping cost and simplifying the mixing process [7,8]. As a powder, the product is used by mixing with water and the other dry ingredients (cement and fine aggregate). These powders have typically been made from polyvinyl acetate and were only applicable to areas where there would be no exposure to moisture. Recently, however, powders from other polymers have been produced that allow the use to be extended to applications where moisture is present. In all cases, manufacturers' recommendations should be followed to assure proper application of these products.

MECHANISM OF LATEX MODIFICATION

The mechanism of latex modification of portland cement involves two processes: the hydration of the cement and the coalescence of the latex. The chemistry and reaction processes of cement hydration occur the same as in conventional mortar and concrete. However, while this is taking place, water is being consumed and removed from the latex, concentrating the latex particles and bringing them closer together. With continual water removal, both by cement hydration and evaporation, the latex particles eventually coalesce into a film that is interwoven throughout the hydrated cement particles, coating these particles and the aggregate surfaces with a semicontinuous plastic film. This coalescence results in partially filled void spaces (Fig. 1), as well as adhesion between the aggregate and cement hydrates.

The success of this process is dependent on several factors:

1. physical properties of the latex, that is, particle size and quality of dispersion;
2. composition of the latex, that is, chemical and physical structure of the polymer; and
3. environment, that is, time and temperature of application.

FIG. 1—Electron microphotograph of latex-modified concrete (11 600×).

A compact configuration of the latex particles is a necessary stage in the process of film formation, and this will occur only if the particles are dispersed sufficiently to allow them to easily move past each other in the water phase. With a poor dispersion, in which the particles are flocculated, the particles will not become closely packed but will form intermittent voids, and thus a poor quality (spongy) film. Figure 2 is a schematic of the proper film formation process.

Isenburg et al. [9] proposed the following hypothesis for the mechanism of latex reinforcement of portland cement:

1. Latex provides equal workability at significantly lower water/cement ratios.
2. Latex particles coat the cement grains and aggregate forming a continuous polymer matrix throughout the structure.
3. Microcracks form to relieve shrinkage strain due to drying at less than 80 to 100% relative humidity.
4. Microcrack propagations are restrained and held together by the polymer network.

Laboratory studies have been conducted that confirm most of the features of Vanderhoff's mechanism [9].

Proper film formation of the latex is required to retard water loss. This film ensures an adequate supply of water to allow hydration of the cement as well as the development of adhesion and impermeability properties. This film formation is governed by the latexes' minimum film formation temperature (MFFT), the temperature below which the polymer spheres will not coalesce to form a

TABLE 1—Typical Physical Properties of Latexes.

Latex Type	Acronym	% Solids	Viscosity, Hz	MFFT[a], °C	pH
Styrene-butadiene	SB	47	20 to 50	5 to 15	9 to 11
Acrylic copolymers	PAE	47	20 to 100	10 to 12	9 to 10
Styrene-acrylic copolymers	SA	48	75 to 5000	10 to 18	6 to 9
Polyvinyl acetate	PVA	55	1000 to 2500	15 to 30	4 to 5
Vinyl acetate-ethylene	VAE	55	500 to 2500	10 to 15	5 to 6

[a] Minimum film forming temperature.

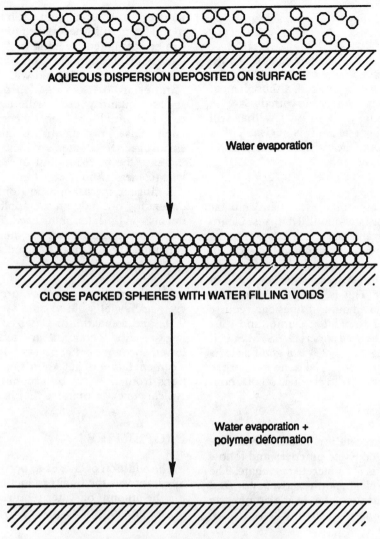

FIG. 2—Process of latex film formation.

film. The MFFT of latexes used in cement will vary with latex composition but typically ranges between 4 and 10°C (40 and 50°F). Application temperature, therefore, should be maintained above the MFFT so that film formation can occur. This is typically 24 h for thin sections, ≤13 mm (≤½ in.) thick, and three to seven days for thicker sections.

FORMULATING WITH LATEXES

The level of latex modification is usually measured as the weight ratio of polymer solids to cement. The level of latex selected is dependent on the level of performance required. Most performance properties plateau between 15 and 20% solids. The 10 to 20% level represents the range for optimum, cost effective performance.

Some other formulation considerations that are important are: the type and amount of antifoam added, water content, aggregate/cement ratio, and the particular choice of aggregate.

When latex-modified mortar and concrete are mixed, air will be entrained within the mix due to the surfactants in the latex. The amount of air generated can be controlled by the inclusion of an antifoam in the latex. The requirements of the end use determine how much air is required and therefore how much antifoam to use. The latex producers can make recommendations for the proper antifoam to be used with their product. And since the cement and aggregate can affect the air content, a trial mix should also be made with all of the recommended ingredients. The kind of air that is entrained in latex-modified mixes has been shown to be small, discrete bubbles [10].

The amount of water used in the mix should be kept to the minimum required to obtain the desired workability. The water level of a latex-modified mix should be lower than a corresponding unmodified mix at equal workability.

The choice of the cement, fine aggregate, and stone used in the appropriate mix is based on the requirements of the application.

Materials

Cement

For most applications, Type I/II portland cement is suitable with latex-modified mixes. Where a faster setting time is required, Type III cement can be utilized. Although most brands of portland cement are compatible with the latexes used in mortar and concrete, there are a few that will entrain excessive air. For this reason, it is advisable that the cement be evaluated prior to the work for compatibility with the particular latex being used.

Fine Aggregate

Most fine aggregates that are suitable for quality mortar and concrete are suitable for latex-modified mixes. Cleanliness is particularly important, since fine contaminants such as dirt or clay can increase the demand for water in the mix and adversely affect workability. Good gradation is important since there is little excess water available in these mixes for workability. This is particularly true for concrete sand, where a large amount of fines can require excessive amounts of water to achieve slump, and thus degrade the properties of the cured concrete. ASTM Specification for Concrete Aggregates (C 33) is a good criteria for gradation, but trial mixes should be done to confirm the suitability of the fine aggregate to produce quality concrete.

Coarse Aggregate

The choice of coarse aggregate for modified concrete should be based on quality concrete practices and follow appropriate ASTM standards for concrete aggregate. The coarse aggregate should be clean, non-reactive, and sized for the thickness of the application, that is, maximum size should be no greater than one-third the thickness.

Water

Water should be clean, potable, and meet the minimum chloride requirements of the application.

Admixtures

No admixtures should be used in conjunction with a latex mix, without prior knowledge and approval of the latex manufacturer.

Mixture Proportions

Mortar

Mortar incorporating latex modification consists of portland cement, fine aggregate, latex, and water. The selection of the aggregate size will depend on the thickness of the application, just as in non-modified mortar, that is, maximum size should be no greater than one-half the thickness of the application.

It is important to know the solids content of each particular latex product before designing the mix. From this, the amount of latex required can be determined, based on a specified latex solids-cement ratio. The next step would be to select an appropriate water-cement ratio. (Because most latex formulations function as water reducers, a water-cement ratio of less than 0.40 is common.) From this, the amount of additional water required can be calculated, keeping in mind that the water in the latex is included in the total water calculation.

Portland cement is used with fine aggregate at a range of ratios, from 1:1 to 1:4, depending on the end use. The specific quantity used will depend on a the strength required, with higher cement contents generally providing an increase in strength, but also a tendency for more shrinkage. Of course, an increase in cement will also increase the total amount of latex in the mixture if the latex-cement ratio is constant.

A range of mixture proportions are possible for mortar, depending not only on the application being considered, but also on the latex being used. Table 2 gives a representative mixture proportion for mortar to be used to repair concrete. Table 3 gives a formulation for a metal coating.

Concrete

Mixture proportion procedures for latex-modified concrete are usually based on a latex solids/cement ratio of 0.15, and a maximum water-cement ratio of 0.40. Most latex-modified concrete mixture proportions have followed conventional concrete criteria for quality and used a cement factor of 300 kg/m^3 (7 sacks/yd^3), although deviations from this can be accommodated. Table 4 gives a typical concrete mixture that is used for overlays.

PROPERTIES

The addition of latexes to portland cement mixes generally improves the final product in two ways: (1) by reducing the amount of water required in the plastic state and (2) by providing dispersed polymer in the matrix of the

TABLE 2—Representative Mixture Proportions for Latex-Modified Mortar Suitable for Concrete Patching.

	Parts, by weight
Portland cement	1.00
Fine aggregate	3.50
Latex (48% solids)	0.31
Water, approximately	0.24

Formulation constants: latex solids/cement = 0.15; water/cement = 0.40

TABLE 3—Sprayable Formulation for an Acrylic Latex Cementitious Metal Primer.

Materials	Parts, by weight
Silica flour 120	200
Portland cement Type I	100
Acrylic latex (48% solids)	62.5
Defoamer	0.3
Water	90
Formulation Constants	
Solids content, %	72.7
Weight, kg/L (lb/gal)	1.8 (15.2)
Polymer solids/cement	0.30
Filler/cement	2

TABLE 4—*Representative Mixture Proportions for Latex-Modified Concrete for an Overlay.*

Material	Quantity	
Portland cement	300 kg	(658 lb)
Fine aggregate	785 kg	(1725 lb)
Coarse aggregate	520 kg	(1150 lb)
Latex (48% solids)	93 L	(24.5 gal)
Water	72 L, max	(19 gal)

NOTE—This mix provides a latex solids/cement of 0.15 and a water/cement of 0.37.

hardened state. Combined, these result in the following improved properties: bond strength, flexural strength, freeze/thaw durability, and permeability. The degree of improvement of any particular property will vary somewhat with the particular latex used. Complete and detailed information on these latexes can be found in Ref *11*.

Plastic Properties

The water-reducing property of the latexes is evidenced by good workability at a low water-cement ratio. Mortar with a flow of 120 cm and concrete with a slump of 15 to 20 cm (6 to 8 in.) is typical for mixes made with a water-cement ratio of 0.37.

The air content of the mixes can vary widely, depending on the requirements of the end use. For some thin-layer applications, air contents above 10% are desired to improve trowelability. For mortar and concrete repair work, air contents in the range of 4 to 6% are desired. All of these can be achieved by the proper selection of the amount of latex and antifoam used.

The setting time of latex-modified mixes is controlled primarily by the reaction of portland cement with water and is similar to that of unmodified mixtures [*12,13*] (see Fig. 3). There are some differences, however, due to the influence of each particular polymer type. The available working time for finishing, however, can be considerably shorter than unmodified mixes due to evaporation of water from the surface of the mix. This evaporation causes the formation of a film, or crust, on the concrete or mortar surface that can tear if over-finished. Producers usually recommend that trial mixes be made to evaluate the characteristics of the components of each particular mix to ensure that the mix fully meets the customer's requirements.

Hardened Properties

Air Void System

Air voids in latex-modified concrete do not always conform to the size criteria established for conventional concrete [*10,14*]. The same void spacing factor required for resistance to freezing and thawing in conventional concrete is not necessary in latex-modified concrete containing 10 to 20% polymer solids/cement, since the latex polymer tends to seal the matrix and limit the amount of water permeating the concrete. The voids are, however,

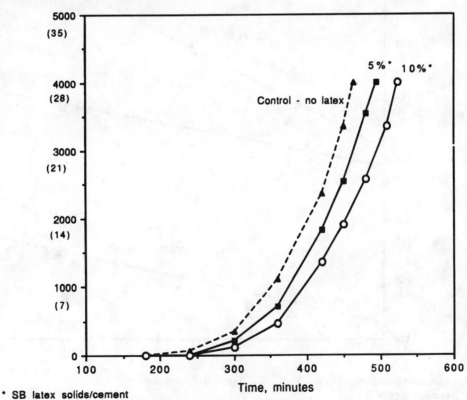

* SB latex solids/cement

FIG. 3—Setting time of latex-modified concrete.

Bond

The adhesion property of latex-modified mixes has been demonstrated both in the laboratory and in the field. A study of SB latex-modified mortar [9] indicated that its bond to the concrete surface is chemical in nature, thus creating a somewhat homogeneous combination of the two materials. Bond tests of this mortar [15] have demonstrated that ultimate failure should be in the parent concrete, that is, the bond strength of a modified mortar to a properly cleaned concrete surface usually exceeds the tensile strength of the substrate concrete. Bond studies of other latexes have also been reported. In a five-year outdoor exposure study, where samples experienced over 70 freeze/thaw cycles and 130 cm (50 in.) of rain per year, shear bond values of acrylic modified mortar [2] more than doubled those of the unmodified control (Fig. 4). Additionally, when the bonds were tested, the latex-modified samples failed cohesively, whereas the controls failed at the bond line. In adhesion tests of samples cured normally in air, Nihon University [12] reported that mortars modified with SB, PAE, and EVA latexes exceeded the bond strength of the unmodified control by factors ranging from 2 to 3. In this study, samples were also submerged in water before testing. All showed a decrease in strength, although the modified mortars always exceeded the control. Another laboratory study [16] compared the bond strengths of five different latexes. The results, shown in Fig. 5, indicate that SB and VAE had the highest values and that all but the PVA failed in the parent (substrate) material. The PVA failed at the bond line at a strength of 1.28 MPa (185 psi). Adhesion performance in the field was reported in a study of 20-year old bridge deck overlays [17], two of which had SB latex mortar overlays. Cores were tested for overlay adhesion and resulted in failure of the base concrete. The bond properties of concrete modified with SB latex have been reported frequently [18–21] since this material has been used extensively for thin-bonded overlays on concrete bridge and parking garage decks. Values of 2.07 to 2.76 MPa (300 to 400 psi) at 28 days cure for direct tensile bond strength are typical, with failure occurring in the substrate if surface preparation is proper.

Durability

The durability of concrete modified with SB latex has been demonstrated by superior resistance to freezing and thawing, and wearability. The low permeability to water results in improved resistance to freezing and thawing [2,6] as measured by ASTM Test Method for Resistance of Concrete to Rapid Freezing and Thawing (C 666) and ASTM Test Method for Scaling Resistance of Concrete

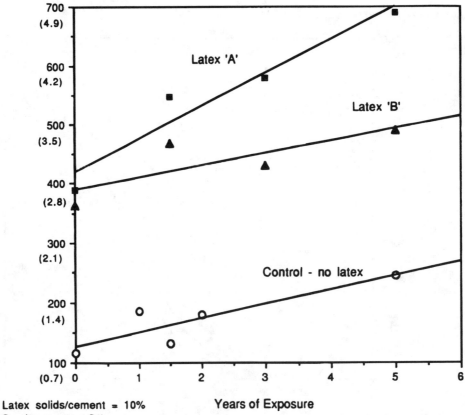

FIG. 4—Effect of weathering on adhesion of acrylic-modified mortar.

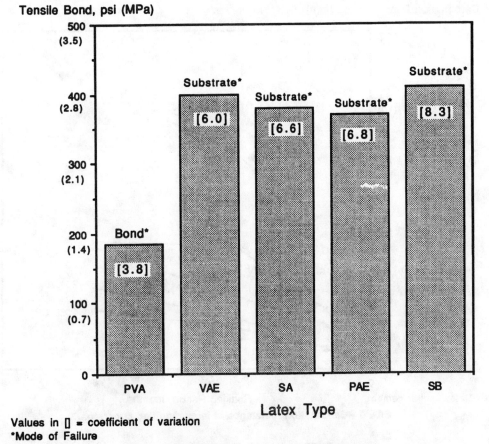

FIG. 5—Bond strength of latex-modified mortars.

Surfaces Exposed to Deicing Chemicals (C 672). Wear resistance, as evidenced by the thousands of bridge decks in service, was documented by the Oregon Department of Transportation [22]. Their study of the wear characteristics of latex-modified concrete bridge deck overlays indicated a life expectancy of 23 to 45 years for a lane having an average daily traffic of over 20 000 vehicles.

Permeability

The pore-sealing effect of latex in a concrete mix results in a major reduction of its permeability to both gases and liquids. For instance, carbonation studies [23] have shown that inclusion of latexes in concrete significantly reduces the carbonation depth of the concrete (Fig. 6). Chloride permeability is another property that has been measured frequently [2,13,24] on latex-modified concrete, primarily on SB-modified concrete, since this is of major interest for bridge and parking deck applications. Figure 7 gives the results of a study [25] of SB-modified compared to conventional concretes, using the ASTM Test Method for Electrical Indication of Concrete's Ability to Resist Chloride Ion Penetration (C 1202). Figure 8 shows chloride penetration from a ponding study [2] of PAE-modified concrete. In all cases, the resistance to permeability performance of latex-modified concrete is evident.

Strain Capacity

The ability to absorb movement (strain capacity) is an important feature of latex-modified mixes. Figure 9 [12] clearly indicates that the strain capacity of mortar that is modified with an acrylate copolymer increases as latex content increases.

APPLICATIONS

Portland cement mixes containing latex are typically used where the following properties are desired: adhesion, durability, low permeability, weatherability, and flexibility.

As with conventional portland cement systems, the choice of whether to use mortar or concrete is based on the thickness of the application, that is, for thin layers, 20 mm (¾ in.) or less, mortar is used, and for cross sections greater than 20 mm (¾ in.), concrete is appropriate.

Mortar

Because of its adhesion characteristics, latex-modified mortar has no minimum thickness requirement. Mortar coatings, as thin as featheredge, are possible as long as the appropriate fine aggregate is selected and an adequate amount of latex is used. A wide variety of applications have been developed since the material was introduced in the 1950s. These include:

1. concrete repair for parking garage floors, bridge decks, swimming pools, and industrial floors; and

584 TESTS AND PROPERTIES OF CONCRETE

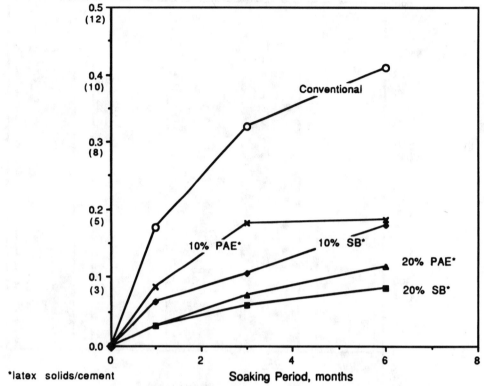

*latex solids/cement

FIG. 6—Carbonation resistance of latex-modified concretes.

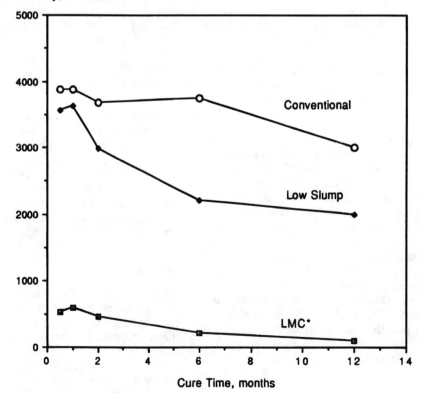

*15% SB latex solids/cement

FIG. 7—Permeability of concretes versus cure time.

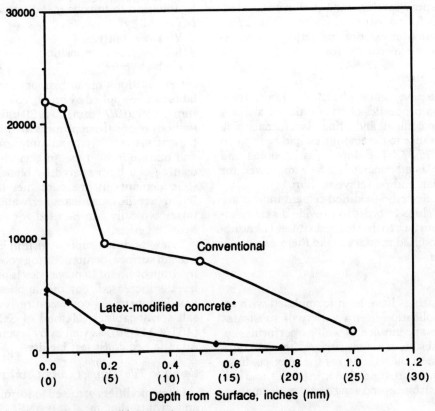

*15% PAE latex solids/cement

FIG. 8—Chloride ion resistance of latex-modified concrete.

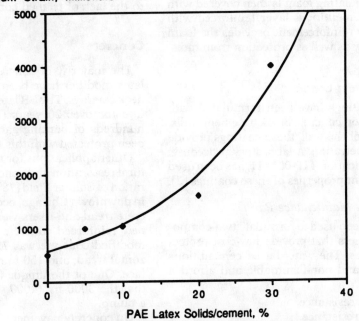

FIG. 9—Effect of latex content on strain capacity of mortar.

2. functional coatings for water-resistant basement coatings, decorative spray coatings for exterior walls, skid-resistant coatings for pavement and ship decks, adhesives and base coats for exterior insulation finish systems, and maintenance coatings for metal.

Skid-Resistant Coatings

A unique concrete pavement application, taking advantage of the adhesion properties of latex, utilized a slurry of SB latex, portland cement, and a fine blast furnace slag to restore skid resistance to pavement that had been worn smooth by traffic [26,27]. The slurry was broomed and screeded onto the blasted concrete surface and cured for one day under burlap and polyethylene film.

Thin coatings of an acrylic-modified cement mortar are applied to decks of ships in order to provide a skid-resistant and protective surface to the steel substrate. The adhesion of the latex-modified mortars make them especially suitable for this use.

Decorative Coatings

Cement-based coatings have been formulated with an acrylic latex for application over a variety of substrates in order to improve appearance as well as performance. Wood, concrete, and steel are some surfaces that are typically coated. Because of the adhesion of the latex, coatings can be relatively thin, approximately 3 mm (⅛ in.), and still provide weather resistance and long-term performance.

Adhesives and Base Coats for Exterior Insulation Finish Systems

Exterior insulation finish systems (EIFS) are another unique application for acrylic latex-modified cement mortar. In this application, an insulating material such as expanded polystyrene foam is attached to the outside surfaces of walls of buildings. The insulating foam is typically attached to the substrate with an acrylate latex-modified cement mortar. The insulating foam is then covered with an acrylate-modified cementitious layer reinforced with a fiberglass scrim. This reinforcement provides the foam with structural integrity as well as protection from moisture and sunlight.

Water-Resistant Basement Coatings

Portland cement coatings have been formulated with acrylic latex to be used on concrete block basement walls. Applied to the outside of the wall, these coatings provide resistance to water penetration. A laboratory procedure, based on Federal Specification TTP-001411, has been used to test the water-resistant properties of these coatings [2].

Latex-Modified Cement Maintenance Paint

Acrylic latexes have been used to formulate two-component cementitious primers that provide low-cost protection of metal substrates. The water-based cementitious mortars are low odor and non-flammable and afford a number of benefits:

1. Corrosion and water resistance.
2. Flash and early rust resistance.
3. Protection of rusty as well as clean metal surfaces.
4. Capability of curing in damp enclosed areas.
5. Adhesion to ferrous metal surfaces.
6. Flexibility.
7. Weatherability.
8. Solvent-free composition.
9. Ambient cure.

Cementitious metal primers modified with an acrylic latex can be applied by brush, roll, and airless spray. Thick films, 0.50 to 0.75 mm (20 to 30 mils), can be easily applied without sag as these primers are high in solids. Two-component systems are commonly used in the maintenance and marine industries. An acrylic-modified cementitious coating can be prepared by blending the dry ingredient (filler/cement) in one container that is mixed with the liquid (acrylic latex/defoamer/water) in the other. A power mixer typically can be used for mixing the two components together.

Latex-modified mortar in thin coatings will adhere to a metal surface, demonstrating good adhesion and flexibility. Improvement in properties is observed as the polymer level is increased. For this application, 30% latex solids has demonstrated corrosion resistance as evaluated in a salt spray cabinet (Method of Salt Spray (Fog) Testing (ASTM 117-90)) as well as by several years exposure at an Atlantic Sea coast test facility.

Ceramic Tile Thinsets and Grouts

Latex modifiers are used to formulate thin-set adhesives and grouts that meet the ANSI Specification for Latex Portland Cement Mortar A118.4 and A118.6 for thin-sets and grouts, respectively. Latexes provide high adhesion, improved water resistance, flexibility, and impact strength.

Shotcrete

Latex-modified shotcrete has been used for vertical and overhead concrete repair, but only to a limited degree due to the lack of trained contractors.

Concrete

The majority of concrete applications incorporating a latex modifier have been as a bridge and parking garage deck overlays. The SB latex has been used for this application for over 20 years [17]. Thousands of bridges and hundreds of parking garages in the United States have been protected with this type of overlay during this time.

Other applications for this modified concrete are structural restoration and concrete pavement repair. The restoration of Soldier Field [18], a 58-year-old concrete stadium in downtown Chicago, occurred in 1981. The existing concrete tread-and-risers were used as the form for pouring new reinforced concrete containing SB latex. This new modified concrete was 75 mm (3 in.) thick on the horizontal tread, and 150 mm (6 in.) against the vertical surface. One of the unique features of this application was that the 3000 m³ (4000 yd³) of concrete were placed by a pump.

On concrete pavement, polymer-modified concrete has been used to repair deterioration at the centerline of 56 km (35 miles) of interstate highways in Pennsylvania [28]. A 50-cm (20-in.) wide strip, 25 mm (1 in.) deep, was removed by a scarifier and replaced with concrete containing SB latex. The construction techniques were similar to those of an overlay; that is, blast clean the old surface, scrub in the paste to assure bond, and cure for one day under wet burlap and polyethylene. Because of the adhesion properties of the modified concrete, this thin section is well bonded and serves as an expedient repair.

EQUIPMENT CONSIDERATIONS

Most equipment and tools used for conventional mortar and concrete have been used successfully with latex-modified systems. This includes pumps, buggies, trowels, and screeds, to name a few. Whatever equipment is being used, it is important to minimize the time that the surface of the freshly mixed material is exposed to the atmosphere so that crusting does not occur. This would be of more concern, for instance, if a buggy were used rather than a pump to transport latex-modified concrete. Whereas no drying would occur while the modified material is in the pump hose, the buggy might have to be covered to prevent drying if the transport time were long.

Because latex-modified mortars and concretes bond so well, a major concern with any equipment that is used with these systems is cleaning. It is important, therefore, that equipment be cleaned thoroughly with water immediately after use and before drying occurs.

LIMITATIONS

For exterior applications, or where exposure to moisture is possible, PVA latex should not be used. For outdoor applications where color is important, such as stucco that uses white portland cement, the SB is not the preferred choice due to a discoloration from the effect of sunlight on the butadiene. Those latexes/polymers not containing butadiene, that is, SA, PAE, and VAE, are recommended.

As with conventional portland cement, these materials should be used with caution during extreme weather conditions, that is, between 4 and 30°C (40 and 85°F). During conditions of rapid drying, these latex-modified systems are more sensitive than conventional mortars and concrete with low water contents because of the film-forming characteristics of the latex. Rapid drying will cause a skin (or crust) to form on the surface if it is allowed to dry, making the finishing operation difficult. Care should be exercised when ambient conditions, that is, relative humidity, wind, and temperature, create an environment for rapid evaporation of water.

Since latex-modified mortars and concretes require air drying to achieve their optimum properties, these systems are not suitable for underwater applications unless sufficient cure time is allowed for latex coalescence. Typically, 28 days is recommended. This cure time can be monitored by samples cured under identical conditions at the field installation and tested periodically until the design properties are achieved.

SPECIFICATIONS

A specification for latexes to be used with portland cement mixes is currently being prepared at ASTM. A specification for using SB latex-modified concrete for bridge and parking garage overlays has been published by the American Concrete Institute [29].

REFERENCES

[1] Kuhlmann, L. A., "Performance History of Latex-Modified Concrete Overlays," *Applications of Polymer Concrete, SP-69*, American Concrete Institute, Detroit, MI, 1981, pp. 123–144.

[2] Lavelle, J. A., "Acrylic Latex-Modified Portland Cement," *ACI Materials Journal*, Jan.–Feb. 1988.

[3] Thompson, S. J., *The S/B Latex Story*, Pendell Publishing Co., Midland, MI, 1989.

[4] Walters, D. G., "What Are Latexes?" *Concrete International*, Vol. 9, No. 12, Dec. 1987.

[5] Dennis, R., "Latex in the Construction Industry," *Chemistry & Industry*, 5 Aug. 1985.

[6] Clear, K. C. and Chollar, B., "Styrene-Butadiene Latex Modifiers for Bridge Deck Overlay Concrete," Report No. FHWA-RD-78-35, Federal Highway Administration, Washington, DC, 1978.

[7] Walters, D. G., "Vinyl Acetate-Ethylene Copolymer Redispersible Powder Hydraulic Cement Admixtures," presented at American Concrete Institute, Fall Convention, Philadelphia, 1990.

[8] Tsai, M. C., Burch, M., and Lavelle, J., "Solid Grade Acrylic Cement Modifiers," *Polymer-Modified Hydraulic Cement Mixtures, ASTM STP 1176*, L. A. Kuhlmann and D. G. Walters, Eds., American Society for Testing Materials, Philadelphia, 1993.

[9] Isenburg, J. E., Rapp, D. E., Sutton, E. J., and Vanderhoff, J. W., "Microstructure and Strength of the Bond Between Concrete and Styrene-Butadiene Latex-Modified Mortar," *Highway Research Record 370*, 1971, pp. 75–90.

[10] Kuhlmann, L. A. and Foor, N. C., "Chloride Permeability versus Air Content of Latex Modified Concrete," *Cement, Concrete and Aggregates*, Vol. 6, No. 1, 1984, pp. 11–16.

[11] *State-of-the-Art Report on Polymer Modified Concrete*, ACI Subcommittee 548A, American Concrete Institute, Detroit, MI, 1991, Chapters 1–3.

[12] Ohama, Y., "Polymer-Modified Mortars and Concretes," *Concrete Admixtures Handbook*, V. S. Ramachandran, Ed., Noyes Publication, Park Ridge, NJ, 1984, pp. 337–429.

[13] Smutzer, R. K. and Hockett, R. B., "Latex Modified Portland Cement Concrete—A Laboratory Investigation of Plastic and Hardened Properties of Concrete Mixtures Containing Three Formulations Used in Bridge Deck Overlays," Indiana State Highway Commission, Indianapolis, IN, Feb. 1981.

[14] Marusin, S. L. "Microstructure, Pore Characteristics and Chloride Ion Penetration in Conventional Concrete and Concrete Containing Polymers," *SP-99*, American Concrete Institute, Detroit, MI, 1987, pp. 135–150.

[15] Kuhlmann, L. A., "A Test Method for Measuring the Bond Strength of Latex Modified Concrete and Mortar," *ACI Materials Journal*, Vol. 87, No. 4, July–Aug. 1990, pp. 387–394.

[16] Walters, D. G., "A Comparison of Latex-Modified Portland Cement Mortars," *ACI Materials Journal*, Vol. 83, No. 4, July–Aug. 1990, pp. 597–605.

[17] Sprinkel, M. J., "Twenty-Year Performance of Latex-Modified Concrete Overlays," Transportation Research Board, Washington, DC, 1992.

[18] Pfeifer, D. W., "Utilization of Latex-Modified Concrete for Restoration of Seats in Soldier Field for Chicago Park District," Wiss, Janney, Elstner and Assoc., Inc., Chicago, Sept. 1978.

[19] Knab, L. I. and Spring, C. B., "Evaluation of Test Methods for Measuring the Bond Strength of Portland Cement Based Repair Materials to Concrete," *Cement, Concrete, and Aggregates*, Vol. 11, No. 1, 1989, pp. 3–14.

[20] Knab, L. I., Spring, C. B., Lane, O. J., and Sprinkel, M. M., "Preliminary Minimum Strength Levels for Portland Cement and Latex Modified Concrete Repair Materials for Concrete Pavement Overlays Using Direct Shear, Uniaxial Tension, and Slant Shear Bond Test Methods," National Institute of Standards and Technology, Gaithersburg, MD, 1989.

[21] Sprinkel, M., "Overview of Latex Modified Concrete Overlays," Virginia Highway & Transportation Research Council, Charlottesville, VA, July 1984.

[22] Howard, J. D., "Marquam Bridge Repair, Latex Modified Concrete Overlay and Joint Replacement," Transportation Research Board, Washington, DC, 25 Jan. 1988.

[23] Ohama, Y. and Miyake, T., "Accelerated Carbonation of Polymer-Modified Concrete," *Transactions*, Vol. 2, Japan Concrete Institute, 1980.

[24] Clear, K. C., "Time-to-Corrosion of Reinforcing Steel in Concrete Slabs, Volume 3: Performance After 830 Daily Salting Applications," Report No. FHWA-RD-76-70, Federal Highway Administration, Washington, DC, 1976.

[25] Whiting, D. and Kuhlmann, L. A., "Curing and Chloride Permeability," *Concrete International*, April 1987.

[26] Scholer, C. F., "Thin Applied Surfacing for Improving Skid Resistance of Concrete Pavement," Report No. FHWA/RD/JHRP-80/16, Federal Highway Administration, Washington, DC, 1980.

[27] Sprinkel, M. and Milliron, M., "Restoring Skid Resistance to Concrete Pavements and Bridge Decks Using a Latex-Modified Portland Cement Slag Slurry," Transportation Research Board, Washington, DC, Jan. 1991.

[28] "LMC Specified for Centerline Repairs," *Roads & Bridges*, April 1990, pp. 48–49.

[29] "Standard Specification for Latex-Modified Concrete (LMC) Overlays (ACI 548.4)," *ACI Materials Journal*, Vol. 89, No. 5, Sept.–Oct. 1992, pp. 521–526.

Shotcrete

I. Leon Glassgold[1]

PREFACE

This chapter on shotcrete has not appeared in prior editions of *ASTM STP 169* and makes its debut in this version, *ASTM STP 169C*. Shotcrete has been a viable and important process in concrete construction for almost 80 years, yet it has only been universally accepted worldwide in the past 30 years or so. Its versatility and applicability to a large variety of singular uses provides an economical alternative and complement to conventional concrete construction. The upsurge in its use has led to greater interest in shotcrete, its capabilities, liabilities, and the nature of its distinct characteristics. It is said that the design and application of shotcrete is more art than science and, consequently, the process is less susceptible to standardization than concrete. Nevertheless, it has become apparent in recent years that there is a need to improve and assure the quality of shotcrete installations by the establishment of appropriate standards.

In the United States, responsibility for creating and maintaining standards for concrete and its associated technologies, including shotcrete, has been shared by the American Concrete Institute (ACI) and ASTM. In a memorandum of understanding in 1936, ASTM agreed to address material specifications and test methods while ACI would cover concrete design and construction practice.

ACI has been publishing shotcrete articles and reports since 1911 and standards on shotcrete practice since 1951. In addition, these ACI documents have been referencing ASTM standards for material specifications and test methods. As early as 1920, Professor M. O. Fuller of Lehigh University performed experimental tests on gunite (shotcrete) slabs using standard specification ASTM C 9-17 for cement [1] that was replaced by ASTM Specification for Portland Cement (C 150). However, until 1988, the primary ASTM standards in use in shotcrete construction were based on concrete technology.

Since 1988, only five shotcrete standards have been published by various subcommittees of ASTM Committee C9. Both the number of standards and rate at which they are being generated needed to be increased to adequately cover the technology. In 1990, ASTM Committee C9 decided to concentrate its efforts through a subcommittee on shotcrete, C09.46. Its mission would be to review and codify ASTM standards currently being used for shotcrete and develop new standards that would promote and enhance its overall quality. Subcommittee C09.46 is currently in place and has begun the process of upgrading ASTM shotcrete standards as described later in this chapter in the section on ASTM and Shotcrete.

INTRODUCTION

Shotcrete Defined

The shotcrete process is a construction method utilizing compressed air as the primary source of motive power to project mortar and concrete against a surface with high velocity of impact. It is currently defined by ACI Committee 506 (ACI 506) on Shotcrete as "Mortar or concrete pneumatically projected at high velocity onto a surface" [2]. Based on this definition, only those processes and equipment using compressed air to impart high velocity to a moving body of material passing through a hose can be considered shotcrete. The term shotcrete also describes two distinct processes, namely, dry-mix and wet-mix. Both utilize either fine aggregate producing a mortar, or fine plus coarse aggregate that results in concrete. A more comprehensive description of each process, the plastic and hardened physical properties of the in-place shotcrete, variety of applications, and relative merits are found in the *Guide to Shotcrete* [2] and other ACI 506 documents.

Shotcrete Terms

The term shotcrete is a coined word introduced to the construction industry by the American Railway Engineering Association (AREA) around 1930 to provide a generic term to describe the dry-mix process, then known as Gunite, and such other proprietary terms as Blastcrete, Blocrete, Jetcrete, Guncrete, Spraycrete, Nucrete, etc. [3]. ACI initially used the term "pneumatically placed mortar" to describe the dry-mix process in its Recommended Practice for the Application of Mortar by Pneumatic Pressure (ACI 805-51). However, in the interest of brevity the term "shotcrete" was used in the text of this first ACI shotcrete standard [4]. This normalized the use of the term in the United States, Canada, and all those countries basing their shotcrete technology on ACI documents. It should be noted that currently, in some localized areas of the United

---

[1] President, Masonry Resurfacing & Construction Co., Inc., Baltimore, MD 21226.

590 TESTS AND PROPERTIES OF CONCRETE

States, the term gunite describes the dry-mix method while shotcrete is used for wet-mix. The European Community favors the use of the term "sprayed concrete." In the interest of good practice and reducing confusion, it would be beneficial to all parties concerned if one term would be universally used to encompass a technology whose importance in the construction industry is increasing at a rapid pace.

Shotcrete Beginnings

In 1911, ACI, then known as the National Association of Cement Users (NACU), published one of the earliest, if not the earliest article on the Gunite (dry-mix) process [5]. An article published in Cement World magazine in 1916 described what would later be called the wet-mix process [6]. However, the wet-mix process did not become a fully accepted technology until the early 1950s when the True Gun, a dual tank pneumatic device, was introduced. It was not long thereafter that various types of concrete pumps were adapted to the shotcrete process that resulted in the acceptance of the wet-mix process as a viable and economical method for many shotcrete applications.

Shotcrete Properties

Shotcrete differs from concrete in manufacture, method of placement, and physical properties, though the final hardened product is very similar in appearance. Methods and techniques of shotcrete manufacture and placement are important factors in the shotcrete process and do affect its ultimate quality and properties. During the Gunite period, emphasis was primarily placed on such properties as compressive, tensile and flexural strength in addition to bond, permeability, shrinkage, and soundness. In most cases, when shotcrete was compared to concrete, the comparison was most favorable. However, with the advent of the concept of entrained air and the maturing of the shotcrete process, freeze-thaw testing and durability evaluations became important factors in comparing shotcrete with concrete. Because dry-mix shotcrete does not entrain air and some failures have been reported in the literature, questions have surfaced about its durability and other properties [7]. The actual causes or problems that have generated this controversy are difficult to pinpoint or document; however, significant studies, tests, and evaluations by responsible researchers have indicated that good quality shotcrete is a durable material [7–14]. Some of the standards utilized in the durability investigations include:

1. ASTM Test Method for Specific Gravity, Absorption, and Voids in Hardened Concrete (C 642)
2. ASTM Test Method for Resistance of Concrete to Rapid Freezing and Thawing (C 666)
3. ASTM Test Method for Critical Dilation of Concrete Specimens Subjected to Freezing (C 671)
4. ASTM Test Method for Scaling Resistance of Concrete Surfaces Exposed to Deicing Chemicals (C 672)
5. ASTM Practice for Evaluation of Frost Resistance of Coarse Aggregates in Air-Entrained Concrete by Critical Dilation Procedures (C 682)
6. ASTM Test Method for Electrical Indication of Concrete's Ability to Resist Chloride Ion Penetration (C 1202)

In recent years, silica fume has been used as an additive to improve the strength and durability of both wet- and dry-mix shotcrete [7]. The overwhelming evidence favors the thesis that shotcrete, properly applied with sound materials, can be a very durable material.

ACI and Shotcrete

As noted earlier, ACI has been reporting on shotcrete almost from its inception in 1911, as a concrete technology. In 1942, ACI formed its Committee 805 on Pneumatically Placed Mortar that was retired after completion of the standard (ACI 805-51). In 1957, ACI Committee 506 (ACI 506) on Shotcrete was reactivated to revise and update the aforementioned recommended practice for shotcreting. Since that time, ACI 506 has sponsored many ACI symposia and seminars in addition to producing a number of reports and standards. One of its best known, most popular documents, is the special publication, *Shotcreting SP-14*, [15], that contains many papers describing the state of the art of shotcreting in the early 1960s. ACI 506 has also produced the 1966 Standard Recommended Practice for Shotcreting (ACI 506-66) [16]; the 1977 Specifiction for Materials, Proportioning and Application of Shotcrete (ACI 506.2-77) [17]; the 1982 Guide to the Certification of Shotcrete Nozzlemen (506.3R-82) [18], and the 1984 State of the Art Report on Fiber Reinforced Shotcrete (ACI 506.2-77) [19]. The Guide to Shotcrete (ACI 506R-85) [2] was published in 1985 to replace the 1966 recommended practice (ACI 506-66) [16] that was out of date.

ASTM and Shotcrete

ASTM Committee C9 has produced five shotcrete standards since 1988 [20]. Its Chemical Admixtures subcommittee published two shotcrete standards in January 1990: the ASTM Practice for Preparing and Testing Specimens from Shotcrete Test Panels (C 1140) and the ASTM Specification for Admixtures for Shotcrete (C 1141). In July of 1989, the Fiber Reinforced Concrete subcommittee published ASTM Specification of Fiber-Reinforced Concrete and Shotcrete (C 1116). In June of 1989, the ASTM Test Method for Time of Setting of Shotcrete Mixtures by Penetration Resistance (C 1117) was published. Still earlier, in November of 1988, ASTM published the ASTM Test Method for Time of Setting of Portland-Cement Pastes Containing Accelerating Admixtures for Shotcrete by Use of Gillmore Needles (C 1102).

These five ASTM standards, four of which focus directly on shotcrete and one that divides its attention between concrete and shotcrete, do not cover the entire shotcrete process or technology. At present, the majority of the ASTM standards referenced in ACI shotcrete documents refer to ASTM Committee C9 concrete standards for lack of comparable shotcrete standards. They may or may not be entirely adaptable or suitable for use in shotcreting, since shotcrete and conventional concrete employ different placement procedures. The application of these concrete standards to shotcrete placement and to shotcrete in the plastic state is generally acceptable, however, they should be checked to ensure compatibility with the nature of the project. In most cases, ASTM standards usually are adequate and adaptable when applied to in-place hardened shotcrete.

Current shotcrete standards were generated by various ASTM C9 subcommittees whose missions are concrete oriented. However, ACI 506 has seen fit to adopt many of them and they can be found in current ACI 506 documents. In addition to concrete standards, ACI 506 references ASTM standards for reinforcing bars, welded wire fabric, and prestressed wire. This group of standards comes under the jurisdiction of ASTM Committee A1 on Ferrous Metals and cannot be revised or rewritten for shotcrete use by a subcommittee of Committee C9. Therefore, when they are applied to a shotcrete project, they should be reviewed for suitability. For example, the use of epoxy-coated reinforcing steel bars in shotcrete applications is questionable since the epoxy coating could be damaged during gunning operations. Therefore, the use of ASTM Specification for Epoxy-Coated Reinforcing Steel Bars (A 775) would only be suitable under very special circumstances.

The newly established shotcrete subcommittee, C09.46, will review all Committee C9 standards to determine which of them are applicable to current shotcrete practice. Where meaningful changes are required, Subcommittee C09.46 will rewrite the standards for shotcrete application. In addition, it is also apparent that the small number of current ASTM shotcrete standards do not address all the needs of shotcrete construction. It will be the task of the shotcrete subcommittee to identify and develop new, pertinent specifications and test methods that can better define and describe the physical properties of shotcrete in its plastic and hardened states.

Refractory Shotcrete

Standards for shotcreting are also being produced by ASTM Committee C8 on Refractories. Subcommittee C08.09 on Monolithic Refractories describes their shotcrete product as "cold nozzle-mix gunning" and has produced ASTM Practice for Preparing Refractory Concrete Specimens by Cold Gunning (C 903). In addition, Subcommittee C08.08 on Basic Refractories has also produced ASTM Practice for Preparing Test Specimens from Basic Refractory Gunning Products by Pressing (C 973). A recent article on the gunning of prepackaged refractory castables indicates that these two standards are inadequate and that there is considerable need for various new and reliable standards in the field of refractory shotcrete [21]. Subcommittee C09.46 has a full agenda for the next few years, but it appears a joint committee of Committees C8 and C9 should eventually develop refractory shotcrete standards utilizing a common terminology and nomenclature.

HISTORY OF SHOTCRETE

Early Origins

The early 1900s saw the use of a concrete made from the relatively new binder, portland cement, become an important factor in many types of building construction. Until 1871 when the new Coplay cement plant in Pennsylvania began producing portland cement, cement was imported from Europe to the United States. The development of the new concept of concrete in construction, especially when it is reinforced with steel bars, gave great impetus to the use of concrete in all types of applications in the years that followed. This period saw the beginning of a search for new applications and uses and for effective and alternative means for placing concrete. Therefore, the appearance of the Gunite process in those early years was not a surprising development. The shotcrete process actually had its origins around 1911 with the invention of a device called the cement gun, which provided the mechanism for projecting cement mortar and plaster onto a surface [22–24]. Carl E. Akeley, a well-known explorer and naturalist, attempted unsuccessfully to move wet plaster mortar from a pressure tank through a hose using air pressure. He finally conceived the concept of forcing dry plaster from a pressure tank through a hose and adding water to the mixture on exit from the hose. His experiments, which began around 1908, culminated in the issuance of U.S. patents in February and May of 1911, covering the comprehensive details of a double-chambered Gunite device known as the cement gun.

For whatever reason, Akeley's genius could not make a commercial success of the gun or the process and he eventually sold his patent rights to the cement gun several years after the issuance of the patents. His successors, the Cement Gun Company, immediately set about improving the gun and promoting its use. They established a contracting operation and undertook considerable experimentation and research to illustrate and define the Gunite process, its properties, applications, and qualities. The years prior to World War II saw the development of many uses and applications for this new technology by this company and other applicators. While similar equipment and processes appeared on the scene during the 1920s and 30s, Gunite was the best known term for what was to become the shotcrete process as we know it today. During these early years, the majority of Gunite that was placed was of low volume using relatively small mixers or hand mixing. Also, dry-mix shotcrete in the United States originally utilized only graded fine aggregate, eventually as specified by ASTM Specification for Concrete Aggregates (C 33). Coarse aggregate dry-mix shotcrete never took hold as a more economical shotcrete material because of its high rebound factor.

Later Developments

The post World War II period saw a great upsurge in the use of shotcrete for many old and new applications and the introduction of new varieties of delivery equipment, along with improvement in methods for gunning shotcrete. The wet-mix shotcrete process was reintroduced using a pneumatic feed gun and soon was followed by the adaption of postive displacement pumps that were rapidly becoming standard equipment for placing concrete on many large projects. The wet process would soon fill a need for extra-high-volume production that would increase the economy of certain shotcrete applications. The dry-mix process, which had been monopolized by single- and double-chamber delivery equipment, now saw the introduction and increased use of the continuous feed rotary machines of both the feedbowl and barrel types. More sophisticated mixing equipment was also developed for increased production, as were continuous proportioning elevating screw mixers with the capacity for blending two or three separate materials. The wet-mix process also utilized continuous proportioning mobile mixers as well as ready-mix trucks to supply mortar and concrete for gunning. Special nozzles and water/air rings were also introduced to improve the gunning and plastic state properties of both wet- and dry-mix shotcrete.

SHOTCRETE APPLICATIONS

The very first literature on Gunite dating back to 1911, and that which followed, describes the early applications and process development. The first practical application by Akeley was the gunning of a stucco-like mortar coating to a building at the Field Museum of Natural History in Chicago. The early applications, as described in the engineering literature from 1911 to 1918, were of great variety and included: exterior applications of thin layers of a stucco-like mortar over wood frames; fireproofing of steel structural members; rock surfacing and stabilization; repair of concrete bridges, dams, and sea walls; rust preventative linings for large steel water pipelines; coating of brick, block, stone, pile, and wood structures; partition walls; boat building; relining of reservoirs, canals and aqueducts; tunnel lining and relining; abrasive resistant linings; refractory linings in the manufacture of steel; coke ovens and furnaces; coal bunkers; and, in general, a replacement for hand-placed mortars and cement plaster. It appears that the majority of the Gunite applications developed in the first ten years of its existence are still in use today [24,25].

ACI classifies shotcrete as conventional, refractory, or special [2]. Conventional shotcrete uses portland cement and normal aggregates. Refractory shotcrete uses high-temperature binders such as calcium aluminate cements and refractory aggregates. Special shotcretes are proprietary mixtures of binder and aggregates or conventional shotcrete with special additives that alter the physical properties of the shotcrete for special purposes. Conventional shotcretes are gunned in almost any thickness in structural and non-structural applications; linings and coatings; concrete, masonry, steel repair; and strengthening and reinforcing all types of structures. Refractory shotcrete is used as a structural, insulating, or lining material in high-temperature applications. Special shotcretes using silicates, phosphates, and polymers as a binder are utilized in corrosive and other highly aggressive environments such as acid and caustic storage tanks, chimneys, and stacks. Other special applications include abrasion-resistant shotcrete linings incorporating special aggregates for flumes, coal bunkers, and similar applications.

SHOTCRETE EQUIPMENT

Dry-Mix Delivery Equipment

Carl Akeley received his first patent for a dry-mix delivery device in February 1911 [*12*]. This "cement gun" consisted of a single chamber with a vertical feedwheel to provide more uniform and consistent material flow. This was an intermittent feed and flow device, cumbersome from the standpoint of a continuous operation. In May of 1911, Akeley obtained a patent for a second device, an outgrowth of the first patent. It had two chambers and placed the feed wheel in a horizontal position, as it is in most of today's shotcrete devices. This allowed for intermittent feeding of the top chamber while the bottom chamber maintained continuous flow. This machine was the forerunner for many machines, some without feed wheels, introduced after the original (1911) cement gun patents ran out. A second group of machines that have become popular over the last 30 years are the continuous feed and flow rotary guns, the barrel, and feed-bowl types. These guns are primarily for dry-mix shotcrete but can be adapted to wet-mix use. The continuous feeding action of these guns utilize a rotating airlock principle. The barrel-type rotary gun has seal plates at the top and bottom of the rotating chamber while the feed-bowl type rotary uses a single seal on top of the rotating feed bowls.

Wet-Mix Delivery Equipment

Early wet-mix delivery equipment, as reported in an 1916 article, consisted of a pressure vessel utilizing either compressed air or steam for motive power to transport concrete in large pipelines fairly long distances at the approximate rate of 15.2 m/s (50 ft/s) [5]. However, the True Gun of the early 1950s appears to be the first commercially available and viable pneumatic feed gun utilizing a dual chamber tank that also mixed the materials and permitted continuous flow operation. Up to this time, displacement (piston) pumps with 150 to 200 mm (6 to 8 in.) pipe lines had been used for pumping concrete. As smaller pumps with smaller hose sizes of 50 to 100 mm (2 to 4 in.) became available in the 1960s and 70s, they were adapted to shotcreting by introducing compressed air at the nozzle. The peristaltic or squeeze pump was also adapted to shotcrete use at this time. Today, almost any type of concrete pump can be used for shotcreting, provided it can handle very low slump concrete of 0 to 50 mm (0 to 2 in.), and a large volume of compressed air can be added at the nozzle.

Air Compressors

In order to impart the properties that are characteristic of good, high-quality dry- or wet-mix shotcrete, it is essential that sufficient quantities of air at proper pressure be available to guarantee high nozzle velocity that results in high velocity of impact [25]. It must be understood that the properties of high-velocity shotcrete differentiate it from ordinary low-velocity pumped concrete. The larger the delivery equipment and the larger the delivery hose, the greater the shotcrete production. Large volumes of shotcrete gunned at a rate of 6 to 9 m^3/h (8 to 12 yd^3/h) require large air compressor capacities in the range of 26 to 28 m^3/min. (750 to 1000 ft^3/min) or more. To provide guidance in determining the minimum air requirements for a particular hose size, refer to Table 3.1 in ACI 506R-90.

Mixing Equipment

Applying quality shotcrete that will perform as designed requires that batching and mixing conform to ASTM Specification for Concrete Made by Volumetric Batching and Continuous Mixing (C 685). Dry-mix shotcrete is usually proportioned and mixed in the field by utilizing portable proportioning/mixing equipment. As an alternative, it may be delivered to the job site by truck from a ready-mix plant. In some cases, because access is poor or bulk materials are unavailable, blended prepackaged materials may be used. It is recommended that prepackaged materials be remixed at the job site adding the required predampening water. Mixing of dry-mix materials can be done either by batch or continuous mixer. When powdered admixtures are used, batch-type mixing provides more uniform introduction to and blending of the mixture. Continuous mixers have separate hoppers for both cement and aggregates with the individual ingredients being fed to the mixing device by proportioning augers, belt systems, or a combination thereof.

Wet-mix shotcrete mixtures can be produced at the job site utilizing concrete mixers, proportioning devices, and admixture dispensing equipment. This approach is probably most economical for small-volume projects; however, for larger projects, the most convenient approach is to use shotcrete material supplied from a nearby batch plant by truck mixer.

Ancillary Equipment

The delivery hose and fittings used for the shotcrete installation must be matched to the pressure and volume requirements of the delivery equipment. Nozzles should be sized to fit the work being done. Specially designed nozzles are available that claim to improve gunning and the in-place properties of the shotcrete. However, there is insufficient test data available in the literature to confirm specific improvements. The nozzle body of the dry-mix process is used to introduce water and liquid admixtures to the mix. In the wet-mix process, the nozzle body is used primarily to introduce compressed air at the nozzle to increase velocity, though it also may be used to introduce liquid quickset accelerators to the shotcrete. In the dry-mix process, the use of a hydro-nozzle or a prewetting "long" nozzle may improve in-hose mixing of solids and water with a resultant lower rebound loss and improvement of physical properties.

Other equipment occasionally needed or used in shotcrete applications are water booster pumps, water heaters, air movers, space heaters, communication devices, lighting, blow pipes, fiber feeders, and admixture dispensers. Additional systems and equipment in use today include remote-controlled and semi-automatic nozzle booms that are operated at a distance from the area being gunned. Dry-mixed equipment can be used for sandblasting purposes and also gun casting, a process that reduces the exit velocity drastically allowing mortar or concrete to be placed in almost inaccessible and distant locations with little or no rebound. In addition, some dry- and wet-mix equipment can be readily adapted to pressure-grouting applications.

MATERIALS

When Gunite appeared in 1911, the basic materials for shotcrete were portland cement and well-graded fine concrete sand, with an occasional addition of lime. However, it was apparent from the beginning that the dry-mix process could be used with any type of binder, filler, and liquid provided the materials in the mixture did not react with themselves before the introduction of the liquid at the nozzle. This principle was introduced to the glass, steel, and similar heavy industry around 1916 where heat containing devices utilizing refractory ingredients were repaired.

Binders

The primary binder used in the shotcrete process has been portland cement that conforms to ASTM C 150, however, blended cements meeting ASTM Specification for Blended Hydraulic Cements (C 595) and ASTM Specification for Ground Iron Blast Furnace Slag for Use in Concrete and Mortars (C 989) can also be used. For some refractory applications, portland cement can be used, however, the predominant binders are calcium aluminate cements with varying purities and alumina contents. There are no ASTM specifications for the calcium aluminate cements, however, information on a variety of refractory cements can be found in ACI 547R, *Refractory Concrete, SP-57*.

Aggregates

Aggregates for shotcrete may be normal weight, lightweight, or heavy weight depending on the application. Normal-weight aggregates should conform to ASTM C 33; lightweight aggregates should conform to ASTM Specification for Lightweight Aggregates for Structural Concrete (C 330); and heavy-weight aggregates should conform to ASTM Specification for Aggregates for Radiation-Shielding Concrete (C 637). Gradation should meet the requirement of Table 2.1 of ACI 506R-90.

Water

Water used for mixing and curing shotcrete in both wet and dry processes should be potable, that is, free from substances that could be injurious to shotcrete. Mixing water should not increase the chloride ion content of the shotcrete. If it does, the total chloride ion content of the water, cement, aggregates, and admixtures should not exceed the values as outlined in Table 4.5.4 of ACI 318, Building Code Requirements for Reinforced Concrete.

Admixtures

Admixtures for both wet- and dry-mix shotcrete should meet the requirements of ASTM (C 1141). Quick-setting accelerators (quicksets) that induce both early initial and final set in dry- and wet-mix shotcrete are commonly used to apply relatively thick, overhead layers of material to tunnel linings and similar applications. Presently, there is no standard ASTM specification that properly defines and differentiates quicksets from the commonly available chloride and non-chloride accelerators. However, in shotcrete practice, it is customary to consider quicksets as accelerators that generate an initial set of 3 min or less and a final set of 12 min or less. Presently, ASTM C 1141 does not clearly or adequately address this situation but is being revised to do so. ASTM C 1102 should be used to determine the time of set for a portland cement-accelerator combination. ASTM C 1102 also provides information on the compatability of the combination being tested. Since some of the quickset accelerators cause large reductions in ultimate shotcrete strength, decreased freeze-thaw resistance, and possible reduction in other durability properties, they should be evaluated using ASTM C 1140 [17]. According to ACI 506R, calcium chloride meeting the requirements of ASTM Specification for Calcium Chloride (D 98) may be used in shotcrete in amounts not greater than 2%/cwt of cement. Also, the amount added should be restricted by the limits for chloride ion content set in Table 4.5.4. of ACI 318. Air-entraining admixtures are not used in the dry-mix shotcrete process; however, in wet-mix shotcrete, they should meet requirements of ASTM Specification for Air-Entraining Admixtures for Concrete (C 260).

Pozzolans

Pozzolans in the form of fly ash, both Class C and Class F, and silica fume with silicon dioxide contents of 85% or greater may be used as a cement replacement or additive in shotcrete. Silica fume in dosages of 3-12%/cwt cement when added to a shotcrete mixture will improve strength, reduce permeability, and increase electrical resistivity. Its use in marine or corrosive environments should increase the service life of the concrete. Fly ash for shotcrete should conform to ASTM Specification for Fly Ash and Raw or Calcined Natural Pozzolan for Use as a Mineral Admixture in Portland Cement Concrete (C 618). ASTM Subcommittee C09.24 on Fly Ash, Slag, Mineral Admixtures, and Supplementary Cementitious Materials is currently preparing an ASTM Specification for Silica Fume for Use in Hydraulic Cement Concrete and Mortar (C 1240) that should be available in the immediate future.

Polymers

Polymers such as latexes and acrylics may be used as additives to produce a polymer-modified shotcrete. This material is reported to increase flexural and tensile strengths in addition to improving bond and reducing permeability and absorptivity. Currently, ASTM C 1141 references ASTM Specification for Latex Agents for Bonding Fresh to Hardened Concrete (C 1059) for Grade 6, latex even though the document is for bonding compounds. It appears there is a need for a standard for latexes or acrylics as admixtures in shotcrete.

Reinforcing

Reinforcing bars used in shotcrete should conform to ASTM Specification for Deformed and Plain Billet-Steel Bars for Concrete Reinforcement (A 615), ASTM for Rail-Steel Deformed and Plain Bars for Concrete Reinforcement (A 616), ASTM Specification for Axle-Steel Deformed and Plain Bars for Concrete Reinforcement (A 617), ASTM Specification for Low-Alloy Steel Deformed Bars for Concrete Reinforcement (A 706), ASTM Specification for Zinc-Coated (Galvanized) Bars for Concrete Reinforcement (A 767), or ASTM A 775. Welded wire fabric for shotcrete should conform to ASTM Specification for Steel Welded Wire, Fabric, Plain, for Concrete Reinforcement (A 185) or ASTM Specification for Welded Deformed Steel Wire Fabric for Concrete Reinforcement (A 497) and may be uncoated or coated with zinc or epoxy. If galvanized, it should conform to ASTM Practice for Providing High-Quality Zinc Coatings (Hot-Dip) (A 385) and ASTM Specification for Zinc-Coated (Galvanized) Carbon Steel Wire (A 641). Prestressing steel, when used, should conform to ASTM Specification for Steel Strand, Uncoated Seven-Wire for Prestressed Concrete (A 416), ASTM Specification for Uncoated Stress-Relieved Steel Wire for Prestressed Concrete (A 421), or ASTM Specification for Uncoated High-Strength Steel Bar for Prestressing Concrete (A 722). When fibers of any type are used, they should conform to the requirement of ASTM C 1116. An ASTM standard for anchors used in shotcrete has not yet been developed.

Miscellaneous

Bonding agents are usually not needed in normal shotcrete application. However, if epoxies are used as bonding agents, they should comply with ASTM Specification for Epoxy-Resin-Base Bonding Systems for Concrete (C 881). It should be noted that bonding compounds not properly applied or allowed to harden before gunning begins can act as bond breakers. They should be tested for compatability with project construction procedures prior to use. Curing materials for shotcrete should conform to ASTM Specification for Sheet Materials for Curing Concrete (C 171) and ASTM Specification for Liquid Membrane-Forming Compounds for Curing Concrete (C 309).

PROPORTIONING SHOTCRETE MIXTURES

Prescription versus Performance Specifications [2]

The design and proportioning of shotcrete mixtures is primarily dependent on choice of shotcrete process, type of delivery equipment, ultimate use, location and environmental considerations, project size, and thickness and volume of shotcrete to be placed in a given time frame. An analysis of a project considering these factors will usually dictate which of the two processes, wet-mix or dry-mix, is most feasible and what type of specification, performance, or prescription fits the work at hand. Usually, performance specifications are preferred giving the applicator, within limits, the choice of materials, proportions, equipment, and methods to achieve the desired result. Prescription specifications are more detailed and restrictive, limiting an innovative approach to construction. Performance and prescription specifications should not be interwoven with each other so as to minimize conflicting and confusing requirements and results.

Some performance requirements that should be considered are:

1. cement/binder type,
2. aggregate gradation,
3. compressive strength (minimum at specified age),
4. bond strength,
5. permeability or absorptivity,
6. air content (wet-mix only),
7. initial/final setting time,
8. chloride ion content, and
9. use of admixtures.

In most applications, the first three requirements just described may be required with the inclusion of the remainder depending on the peculiarities and needs of the application.

Some prescription requirements are:

1. cement type and amount,
2. aggregate type and gradation,
3. permitted admixtures and dosage,
4. pozzolan type and content,
5. mix proportions or limitations,
6. slump (wet-mix only),
7. air content (wet-mix only),
8. water-cement ratio (wet-mix only),
9. chloride ion content,
10. initial/final setting time, and
11. compressive strength (minimum).

In this case, the designer may also elect to specify the process type and the mixing and shotcreting equipment. In addition, under special circumstances, the Toughness Index may be specified, as described earlier in ASTM Test Method for Flexural Toughness and First-Crack Strength of Fiber-Reinforced Concrete (Using Beam with Third-Point Loading) (C 1018). It should be understood that this is not a shotcrete standard.

Proportioning—General

The basic approach to proportioning shotcrete mixes is similar to regular or normal concrete practice, especially in the wet-mix process. However, the nature of the shotcrete process may create some differences in approach. Rebound of larger aggregate particles usually results in changed in-place physical properties, including higher cement factors, lower coarse-to-fine aggregate ratios, lower water-cement ratios and a higher specific surface for the aggregate remaining in-place. Attention must be paid to these factors because of the greater shrinkage potential and the possible development of fine surface cracks in the shotcrete. The best way to quantify and qualify a mixture design is by field testing and pre-construction testing as outlined in Section 6.4 of ACI 506R-90. Also, in shotcrete work, strength is not solely derived from the material proportions and water-cement ratio. Important contributing factors are high velocity of impact, nozzle technique and the type and capacity of the equipment used. When properties such as durability and very high strength are a consideration, the shotcrete can be enhanced by the use of additives such as silica fume and fly ash.

Wet-Mix Proportioning

Proportioning for the wet-mix process follows the procedures of ACI 211.1 and ACI 211.2. However, the maximum allowable size of the coarse aggregate is dependent on the material hose size, and the quantity of coarse aggregate used decreases as the size of the aggregate increases. In many wet-mix applications, the sand-aggregate ratio may be as high as 60 to 65%. Meeting the air content requirements of the hardened shotcrete can be a problem since entrained air is lost through the pump and significant amounts of air are lost on exit from the nozzle. If 4% air entrainment is required in the shotcrete, somewhere between 8% and 10% may be required at the pump hopper. Maintaining optimum slump is an important factor in wet-mix shotcrete with 50 to 75 mm (2 to 3 in.) or less normally being required for consistently uniform placement. The use of water reducers is usually beneficial in helping to achieve the specified properties of wet-mix shotcrete.

Dry-Mix Proportioning

Proportioning for dry-mix shotcrete is somewhat less complicated because in current practice, most mix designs use very little coarse aggregate, the maximum aggregate size being between 6 mm (¼ in.) and 9 mm (⅜ in.). ASTM C 33 contains fine aggregate gradations that can be adapted readily to dry-mix shotcrete application with minor changes. In most performance specifications, the main design requirement is a minimum compressive strength, usually 27.6 MPa (4000 psi), though strengths as low as 20.7 MPa (3000 psi) have been specified in certain areas of the United States.

The basic approach to proportioning dry-mix shotcrete is to establish a ratio of cement to fine aggregate that will achieve the strength required plus a factor of safety. In most areas, a history of prior experience will usually approximate this proportion that usually ranges from 1:3 to 1:4 ½ by weight or volume though special applications could require leaner or richer mixtures. Preconstruction testing to determine the mixture proportions is a suitable alternative, but only if it is cost effective or the nature of the application dictates. Generally, preconstruction testing is most suitable for very large projects where its cost is a very small proportion of the overall project. It cannot be overemphasized that preconstruction testing is usually not a viable alternative in most small projects—prior experience is.

BATCHING AND MIXING

General

A considerable quantity, if not the majority of shotcrete that is placed is batched and mixed in the field by portable equipment using bulk or bagged materials. The remainder, and its portion is increasing rapidly, uses ready-mix trucks to deliver the shotcrete material, wet or dry, from a central ready-mix concrete plant. For those applications where supplying ingredients in bulk is inconvenient or impractical, premixed prepackaged dry materials may be utilized. These packages may contain the cement, aggregate, dry admixtures, and other additives such as pozzolans and fibers. They are usually available in small bags of 22.7 to 45.4 kg (50 to 100 lb) or bulk bags of 454 to 1820 kg (1000 to 4000 lb) or other large containers. The individual ingredients of the mixture should meet the requisite ASTM and ACI requirements for shotcrete materials. For proper dry-mix application, these prepackaged materials should be predampened and remixed in the field just prior to use.

Batching

Batching of shotcrete materials is done by weight or volume with weight batching the preferred choice. On-site mixers use volumetric or manual batching while ready-mix plants use weight batching for cement and aggregates and volumetric batching for water and admixtures. Pozzolans are introduced either by weight or volume depending on the capabilities of the plant. In the dry-mix process, water is added at the nozzle by the nozzleman at a rate that depends on the appearance of the shotcrete. On larger projects using volumetric batching, sand bulking can cause errors in proportioning. Bulk density should be checked on a daily basis, if sand bulking becomes a problem.

Specifications for weight batching are found in ASTM Specification for Ready-Mix Concrete (C 94) and volume batching in ASTM C 685. When using fibers, additional batching information can be found in ASTM C 1116. The tolerances found in these specifications are more restrictive than is needed to produce quality shotcrete. ACI 506 recommends the following tolerances be followed:

1. cement: 2% +/- of weight,
2. aggregate: 4% +/- of weight, and
3. admixtures: 6% +/- of weight.

As mentioned earlier, proportioning for batch mixing can be handled manually by using cement in 42.6-kg (94-lb) bags and weight-calibrated containers for aggregates and pozzolans. In the dry-mix process, standard or quickset accelerators can be added in non-liquid (dry) form during mixing or at the nozzle in liquid form. In the wet-mix process, all admixtures are added during mixing except for conventional accelerators and quicksetting accelerators that should be added at the nozzle in liquid form.

Mixing

Currently, ASTM has only one specification, ASTM C 685 that can be applied to dry-mix batching and mixing procedures. However, ACI 506R is an excellent source of supplementary information on good mixing practice. Proper and thorough mixing is absolutely essential to produce uniform quality dry-mix shotcrete, minimizing sand pockets and laminations.

In the wet-mix process, mixing requirements should conform to either ASTM C 685 or ASTM (C 94) depending on the mode of mixing. Additional information on the mixing of pumped concretes is available in ACI 304.2 and ACI 304. Here also, uniformity and consistency of the mixture and its slump from batch to batch helps to provide a quality shotcrete installation.

MISCELLANEOUS

Curing

Curing of in-place shotcrete requires the same attention to detail as does ordinary concrete so as to develop the shotcrete's potential strength and durability properties. Since much shotcrete is applied in relatively thin sections of 50 to 100 mm (2 to 4 in.), it is especially important that proper curing procedures be used as specified in ACI 506.2 Ponding or continuous sprinkling provide the best cures, however, where these procedures are not convenient or feasible, the use of curing compounds that comply with ASTM C 309 or sheet materials meeting ASTM C 171 can be used.

Quality Assurance

Quality assurance requirements vary with the size of a project. With an applicator of known reputation and consistent performance, quality assurance procedures may not be required on small projects where its benefits would be outweighed by cost. However, on large projects, quality assurance takes the form of preconstruction testing to qualify materials and proportions, in addition to prequalifying the nozzlemen and the contractor's setup. Preconstruction testing is described in ACI 506.2 and uses ASTM C 1140 and ASTM Test Method for Obtaining and Testing Drilled Cores and Sawed Beams of Concrete (C 42) for strength determination. These references are also used for quality control during the course of the project.

Acceptance of shotcrete is usually based on cores or cubes from field-test panels or cores from the structure tested in accordance with ASTM C 42. Nondestructive testing devices such as impact hammers (ASTM Test Method for Rebound Number of Hardened Concrete (C 805)); probes (ASTM Test Method for Penetration Resistance of Hardened Concrete (C 803)); ultrasonic and pulse velocity (ASTM Test Method for Pulse Velocity Through Concrete (C 597)) and where applicable, pullout devices (ASTM Test Method for Pullout Strength of Hardened Concrete (C 900)) can be used to determine the uniformity of the in-place shotcrete but should not be used as the basis for acceptance of the shotcrete.

TESTING

Testing requirements for shotcrete are, in general, similar to those established for ordinary portland-cement concrete. Differences between concrete and shotcrete arise primarily in the area of placement and to some degree in manufacture. The methods for placing ordinary concrete utilize procedures that place concrete in its final position at close to zero velocity with slump varying between zero and 200 mm (8 in.). With qualifying shotcrete in both the dry-mix and wet-mix processes, the velocity of placement and impact should be high, sufficient to increase in-place compaction and thereby improve the properties of the shotcrete. Wet-mix shotcrete may have a slump at the pump hopper that varies between 38 and 75 mm (1½ and 3 in.) with the in-place material not sagging or sloughing. Properly placed dry-mix shotcrete normally has zero in-place slump also without any subsidience.

Current ASTM specifications and test methods are utilized to qualify the individual materials used in the shotcrete process except for admixtures where ASTM C 1141 replaces ASTM Specification for Chemical Admixtures for Concrete (C 494).

The characteristics that differentiate concrete from shotcrete may dictate an alternate approach to test methods for shotcrete and shotcrete materials. Most of the current ASTM test methods, specifications, and recommended practices for concrete and concrete-related materials include four general divisions of concrete technology: *Materials, Manufacture, Plastic-State Properties*, and *Hardened-State Properties*. Except for ASTM C 1102, ASTM C 1116, ASTM C 1117, ASTM C 1140, and ASTM C 1141, applicable concrete standards have been adapted to shotcrete practice. For all practical purposes, most of these standards are probably adequate, but there may be areas where the concrete standard may not be readily adaptable to the shotcrete process. ASTM Subcommittee C09.46 on Shotcrete is presently reviewing this situation so as to bring the standards in line with field practice.

CLOSURE

At this point in time, it appears that of the four divisions mentioned in the sections on *Testing, Materials*, and *Manufacture* utilize existing ASTM standards most effectively, with materials requiring the least amount of change. In the area of manufacture for both dry- and wet-mix shotcrete, there are some unanswered questions about the efficiency of different types of equipment and the effect of velocity of impact on the properties of the shotcrete [26]. It is conceivable that standards may have to be developed to qualify shotcrete delivery, mixing, and ancillary equipment or combinations thereof. In addition, a shotcrete version of ASTM C 685 might also be advisable [9]. Shotcrete in the *Plastic State* followed by the *Hardened State* appears to present the biggest challenge for the development of new standards. Both areas may require new and creative innovations, especially in quality assurance and in-place evaluation.

The shotcrete process has a long history as an effective and important construction tool for many applications in the construction field and, currently, its use is increasing by leaps and bounds with the influx into the shotcrete field of many new, and, in some cases, inexperienced applicators and designers. It is an absolute necessity that innovative and more efficient guidelines and standards be established for quicker and more effective testing leading to better quality control of the in-place product. Hopefully, in the not too distant future, means and methods for balanced and objective evaluation of shotcrete will be available to provide quality assurance criteria acceptable to both owner and contractor.

REFERENCES

[1] Fuller, M. O. and Durham, G. E., "Report Showing Results of Tests Made on Gunite Slabs," Cement Gun Co., Inc. Allentown, PA, 1926.

[2] *Guide to Shotcrete*, (ACI 506-85), ACI Committee 506, American Concrete Institute, Detroit, MI, 1985.

[3] Lorman, W. R., *Engineering Properties of Shotcrete, SP-14A*, American Concrete Institute, Detroit, MI, 1968.

[4] "Recommended Practice for the Application of Mortar by Pneumatic Pressure," (ACI 805-51), ACI Committee 805, American Concrete Institute, Detroit, MI, 1951.

[5] Prentiss, G. L., "The Use of Compressed Air in Handling Mortars and Concrete," National Association of Cement Users, 1911, pp. 504–521.

[6] Springer, J. L., "Compressed Air and Live Steam for Placing Concrete," *Cement World*, Aug. 1916, pp. 13–20.

[7] Glassgold, I. L., "Shotcrete Durability: An Evaluation," *Concrete International*, Vol. 11, No. 8, 1989, pp. 78–85.

[8] Litvin, A. and Shideler, J. J., "Laboratory Study of Shotcrete," *Shotcreting, SP-14*, American Concrete Institute, Detroit, MI, 1966, pp. 165–184.

[9] Reading, T. J., "Durability of Shotcrete," *Concrete International*, Vol. 3, No. 1, 1981, pp. 27–33.

[10] Morgan, D. R., "Freeze Thaw Durability of Shotcrete," *Concrete International*, Vol. 11, No. 8, 1989, pp. 86–93.

[11] Seegebrecht, G. W., Litvin, A., and Gebler, S. H., "Durability of Dry-Mix Shotcrete," *Concrete International*, Vol. 11, No. 10, 1989, pp. 47–50.

[12] Gebler, S. H., "Durability of Dry-Mix Shotcrete Containing Regulated-Set Cement," *Concrete International*, Vol. 11, No. 10, 1989, pp. 56–58

[13] Glassgold, I. L., *Shotcrete in the United States: A Brief History, SP-128*, Vol. 1, American Concrete Institute, Detroit, MI, pp. 289–305.

[14] Gebler, S. H., Litvin, A. W. J., and Schutz, R., "Durability of Dry-Mix Shotcrete Containing Rapid Set Accelerators," *ACI Materials Journal*, American Concrete Institute, Vol. 89, No. 3, May–June 1992, pp. 259–262.

[15] *Shotcreting, SP-14*, American Concrete Institute, Detroit, MI, 1986.

[16] "Recommended Practice for Shotcreting," (ACI 506-66), ACI Committee 506, American Concrete Institute, Detroit, MI, 1966, (withdrawn as an ACI standard in 1983).

[17] "Specification for Materials, Proportioning and Application of Shotcrete," (ACI 506.2-77), ACI Committee 506, American Concrete Institute, Detroit, MI, 1977.

[18] "Guide to the Certification of Shotcrete Nozzlemen," (ACI 506.3R-91), ACI Committee 506, American Concrete Institute, Detroit, MI, 1984.

[19] "State of the Art on Fiber Reinforced Shotcrete," (ACI 506.1R-84), ACI Committee 506, American Concrete Institute, Detroit, MI, 1984.

[20] *Annual Book of ASTM Standards*, Vol. 4.01 and 4.02, American Society for Testing and Materials, Philadelphia, 1992.

[21] Fisher, R., "Installation Properties of Gunning Mixes," presented at the 28th Annual Symposium on Refractories, American Ceramic Society, St. Louis, MO, 27 March 1992 (unpublished).

[22] Rodriguez, L., "Samuel W. Traylor, The Cement Gun and the Process of Innovation," M.A. thesis, Lehigh University, Bethlehem, PA, 1989.

[23] Jordan, W. A., "The Cement Gun," Paper No. 33, *Journal*, American Society of Engineering Contractors, New York, December 1911, pp. 407–434.

[24] Weber, C., "The Cement Gun, and It's Work," *Journal*, Western Society of Engineers, Vol. XIX, No. 3, March 1914, pp. 272–315.

[25] Collier, B. C., "The Cement Gun, It's Application and Uses," *Bulletin 22*, The Cement Gun Co., Inc., Allentown, PA, Dec. 1918.

[26] Stewart, E. P., "New Test Data Aid Quality Control of Gunite," *Engineering News-Record*, reprint, 9 Nov. 1933.

BIBLIOGRAPHY

American Concrete Institute, "Application and Use of Shotcrete," American Compilation No. 6, Detroit, MI, 1981.

American Concrete Institute, *Refractory Concrete, SP-57*, Detroit, MI, 1978.

American Concrete Institute, "Refractory Concrete: State-of-the-Art Report," ACI 547-79, ACI Committee 547, Detroit, MI, 1979.

American Concrete Institute/American Society of Civil Engineers, *Shotcrete for Ground Support, SP-54*, Detroit, MI, 1976.

American Concrete Institute/American Society for Civil Engineers, *Use of Shotcrete for Underground Structural Support, SP-43*, Detroit, 1973.

Engineering Foundation, *Shotcrete for Underground Support III*, New York, 1978.

Engineering Foundation, *Shotcrete for Underground Support IV*, New York, 1982.

Litvin, A. and Shideler, J. J., *Structural Applications of Pumped and Sprayed Concrete*, Bulletin No. D 72, Development Department, Portland Cement Association, Skokie, IL, 1964.

Mahar, J. W., Parker, H. W., and Wuellner, W. W., "Shotcrete Practice in Underground Construction," Report No. UILU-ENGf-75-2018, University of Illinois, Urbana, IL, Aug. 1975.

Parker, H. W., Fernandez-Delgado, G., and Lorig, L. J., "Field-Oriented Investigation of Conventional and Experimental Shotcrete for Tunnels," Report No. FRA-OR&D 76-06, U.S. Department of Transportation, Washington, DC, Aug. 1975.

Ryan, T. F., *Gunite, A Handbook for Engineers*, Cement and Concrete Association, London, UK, 1973.

Ward, W. H. and Hills, D. L., *Sprayed Concrete: Tunnel Support Requirements and the Dry-Mix Process*, Building Research Station, Watford, UK, 1977.

Organic Materials for Bonding, Patching, and Sealing Concrete

Raymond J. Schutz[1]

PREFACE

Organic Materials for Bonding, Patching, and Sealing Concrete were not covered in the original *ASTM STP 169* published in 1956. The subject was discussed by M. R. Smith in *ASTM STP 169A* published in 1966. A much expanded chapter covering the then newer materials such as epoxy resins, latices and sealers authored by R. J. Schutz, appeared in *ASTM STP 169B* published in 1978.

INTRODUCTION

Concrete is one of the most durable materials of construction. However, because concrete is inherently porous and alkaline, it can be attacked both physically and chemically by certain injurious solutions including acids, salts, and water.

Attack of concrete by water or injurious solutions is especially severe where the concrete is exposed to alternate cycles of wetting and drying or freezing and thawing, or both.

Since such attack can only occur in a wet environment, coatings quite often are applied to concrete to prevent ingress of the solution. There is no frost damage to concrete if it is dry [1], and all chemical attack requires the presence of water [2].

Absorption and subsequent chemical or physical attack can also be reduced by polymer impregnation [3] or by inclusion of the polymer in the plastic-concrete mixture in the form of a latex or unreacted premixed epoxy resin and curing agent [4].

Organic materials that have been used for bonding, patching, and sealing of concrete include epoxy resins, silicones, bitumens, linseed oil, oil-based paints, acrylics, urethanes, polyvinyls, styrene butadiene, chlorinated rubber, polyesters, and vinylesters.

BONDING AND PATCHING MATERIALS

Materials in common use for bonding and patching of concrete fall into three general groups: (1) bonding admixtures, (2) adhesives, and (3) resinous mortars.

Admixtures

Admixtures that are used to improve the adhesion of plastic mortar or concrete to hardened concrete are classified by the American Concrete Institute (ACI) as bonding admixtures. These admixtures are latices of either acrylic, styrene butadiene, or polyvinyl acetate. They are generally used in one of three methods: (1) applied directly as an adhesive, (2) mixed with portland cement or portland cement and sand and applied as a bonding grout for conventional mortar or concrete, or (3) used in a similar manner with concrete or mortar containing the latex as an admixture.

The inclusion of the latex as an admixture in the mortar or concrete will improve both freeze/thaw and chemical resistance. However, the working time of such mortars and concretes is extremely short—the latex will tend to coalesce in 10 to 20 min depending on the type, temperature, and other factors. Handling beyond that time generally results in profuse cracking on drying. In order to be effective as an admixture, these latices must be used at a concentration of at least 12% polymer solids based on the cement content.

ASTM Specification for Latex Agents for Bonding Fresh to Hardened Concrete (C 1059) provides for the classification of latex bonding agents according to use: Type 1 (redispersable) for use in interior work not subject to immersion in water or high humidity, and Type II (non-redispersable) for use in areas subject to high humidity or immersion in water and is suitable for use in other areas.

Type I latices are generally based on polyvinyl acetate, and Type II latices are generally based on styrene butadiene or acrylic polymers.

ASTM Test Method for Bond Strength of Latex Systems Used With Concrete By Slant Shear (C 1042) covers the determination of the bond strength of these systems.

Type I systems are applied neat either by brush, spray, or roller. The fresh mortar or concrete can be applied at any time after application of the Type 1 latex. Since Type I systems are redispersable, the latex may be allowed to dry before application of the fresh mortar or concrete.

Type II latices are mixed with cement or cement and sand and are applied as a slurry or grout. Since Type II latices are non-redispersable, application of the fresh mortar or concrete must be accomplished before the slurry drys.

Mortar and concrete containing Type II latices, due to their excellent resistance to freeze/thaw cycles and deicing

---
[1] Construction materials consultant, Waupun, WI 53963.

salts, have been used to repair and overlay bridge decks with an excellent service record [5].

Many prepackaged latex mortars are available. These may be two-component systems (a container of latex and a container of portland-cement mortar) or a single component system (the latex in powder form is packaged with the cement and aggregate).

ADHESIVES, PATCHING, AND OVERLAYING MATERIALS

Epoxy resins are the most widely used organic systems for bonding, patching, and overlaying concrete. Suitable systems are available for use as adhesives (for bonding new plastic mortar or concrete to existing hardened concrete) and also as binders (for the manufacture of resinous mortar and concrete patches or overlays). For these purposes, low molecular-weight liquid epoxy resins of the bisphenol A-type (reaction products of bisphenol A and epichlorohydrin) are generally used. These resins are mixed with a curing agent immediately prior to use. These curing agents are generally of the chemical groups such as amines, polyamines, or polyamides.

The epoxy resins cure by cross-linking; therefore, curing shrinkage is much less with these than with the polyester or vinyl ester systems that cure by polymerization. The cured-resin systems will have tensile strengths and compressive strengths between 10 and 20 times that of good concrete. They will be resistant to alkalies, mild acids, solvents, and oils.

A specification is available for epoxy resin systems for use with concrete; namely, ASTM Specification for Epoxy-Resin-Base Bonding Systems for Concrete (C 881). This specification classifies epoxy resin systems by type, class, grade, and color.

Seven types are described by this specification according to their proposed use:

1. Type I—For use in non-load-bearing applications for bonding hardened concrete to hardened concrete and other materials, and as a binder in epoxy mortars or epoxy concretes.
2. Type II—For use in non-load-bearing applications for bonding freshly mixed concrete to hardened concrete.
3. Type III—For use in bonding skid-resistant materials to hardened concrete, and as a binder in epoxy mortars or epoxy concretes used on traffic-bearing surfaces (or surfaces subject to thermal or mechanical movements).
4. Type IV—For use in load-bearing applications for bonding hardened concrete to hardened concrete and other materials and as a binder for epoxy mortars and concretes.
5. Type V—For use in load-bearing applications for bonding freshly mixed concrete to hardened concrete.
6. Type VI—For bonding and sealing segmental precast elements with internal tendons and for span-by-span erection when temporary post tensioning is applied.
7. Type VII—For use as a non-stress-carrying sealer for segmental precast elements when temporary post tensioning is not applied as in span-by-span erection.

Three grades of systems are defined according to their flow characteristics and are distinguished by the viscosity and consistency requirements:

Grade 1—low viscosity,
Grade 2—medium viscosity, and
Grade 3—non-sagging consistency.

Classes A, B, and C are defined by Types 1 through 5, and Classes D, E, and F are defined for Types 6 and 7, according to the range of temperatures for which they are suitable.

1. Class A—For use below 4.5°C (40°F), the lowest allowable temperature to be defined by the manufacturer of the product.
2. Class B—For use between 4.5 and 15.5°C (40 and 60°F).
3. Class C—For use above 15.5°C (60°F), the highest allowable temperature to be defined by the manufacturer of the product.
4. Class D—For use between 4.5 and 18.0°C (40 and 65°F).
5. Class E—For use between 15.5 and 26.5°C (60 and 80°F).
6. Class F—For use between 24.0 and 32.0°C (75 and 90°F).

The temperature in question is usually that of the surface of the hardened concrete to which the bonding system is to be applied. This temperature may be considerably different from that of the air. Where unusual curing rates are desired, it is possible to use a class of bonding agent at a temperature other than that for which it is normally intended. For example, a Class A system will cure rapidly at room temperature.

Three ASTM test methods are available to determine the suitability of epoxy resins for use with concrete:

1. ASTM Test Method for Bond Strength of Epoxy-Resin Systems Used With Concrete (C 882) is a test method that determines the bond strength by using the epoxy system to bond together two equal sections of a 76.2 by 152.4-mm (3 by 6-in.) portland-cement mortar cylinder, each section of which has a diagonally cast bonding area at a 30° angle from vertical. After suitable curing of the bonding agent, the test is performed by determining the compressive strength of the composite cylinder.
2. ASTM Test Method for Effective Shrinkage of Epoxy-Resin Systems Used With Concrete (C 883). In this test method, a laminate is constructed of the epoxy resin applied to a glass plate. As the epoxy cures, any shrinkage will cause a bowing of the glass plate. Failure in this test consists of shrinkage sufficient to fracture the glass.
3. ASTM Test Method for Thermal Compatibility Between Concrete and an Epoxy-Resin Overlay (C 884). Although developed for epoxy-resin systems, this test method may be used to determine the thermal compatibility of any resinous mortar and concrete. In this test method, a layer of epoxy-sand mortar is applied to a slab of cured and dried concrete. After the epoxy has cured, the sample is subjected to five cycles of temperature change between 25°C (77°F) and −21.1°C (−6°F). Crack lines between the concrete and the epoxy mortar constitute failure of the test.

Epoxy-resin systems conforming to ASTM C 881 Types I, II, IV, and V have relatively high elastic moduli, since a structural bond is required in most adhesive applications. In the thin glue lines usually used, the differences in linear coefficient of expansion between the epoxy resin and the concrete is inconsequential because the total tensile strength of the thin glue line (1 mm or less) is less than the tensile strength of the concrete. In thin glue lines, creep of the epoxy resin is not a problem because the effective modulus of elasticity as a material in a thin glue line will be extremely high [6].

For applications as an adhesive, the epoxy resin should contain no solvent, since solvent may be entrapped during cure and cause both a rubbery cure and later shrinkage of the adhesive. The epoxy-resin systems may be applied by brush, roller, or spray to properly prepared concrete surfaces. Where reinforcing steel is exposed, the steel should be cleaned of rust, oil, and foreign material and coated with the epoxy-resin adhesive to prevent further corrosion.

ACI 503.1, Standard Specification for Bonding Hardened Concrete, Steel, Wood, Brick and Other Materials to Hardened Concrete with a Multi-Component Epoxy Adhesive, and ACI 503.2, Standard Specification for Bonding Plastic Concrete to Hardened Concrete with a Multi-Component Epoxy Adhesive, cover bonding techniques in detail [7].

Epoxy-resin systems are used quite commonly as the sole binder for concrete patches and overlays. Since such patches and overlays are generally thick, a resin with a low modulus of elasticity, such as those conforming to ASTM C 881, Type III, must be used.

If high-modulus binders are used, the difference in thermal coefficient of expansion between the thick resin mortar and the base concrete will result in failure of the overlay or patch by shearing of the concrete adjacent to the resinous mortar or concrete. When applied in layers 40-mm (1½-in.) or greater in thickness, it is desirable to include coarse aggregate in the system, both for reasons of economy and thermal compatibility. Figure 1 illustrates the effect of aggregate loading on linear coefficient of expansion of several epoxy-mortar systems [7].

The use of epoxy resins as a binder for mortars is covered by ACI 503.4, Standard Specification for Repairing Concrete with Epoxy Mortars [3].

Thin resinous mortars (also referred to as skid-resistant surfacings, or resinous overlays) may be economically produced by applying the resin by roller, squeegee, or spray to a horizontal concrete surface and broadcasting fine aggregate in excess onto the uncured resin. After cure, the excess aggregate may be swept off and recovered.

Several applications may be applied if desired. This method eliminates the need for mortar mixers, troweling, and transporting of resinous mortars. These thin overlays have been used for sealing bridge decks and other flatwork for over 20 years and have a good service record.

ACI 503.3, Standard Specification for Producing a Skid-Resistant Surface on Concrete by the Use of a Multi-Component Epoxy System, details this method [7].

Some epoxy-resin systems are water sensitive before cure. Therefore, they can only be used under dry conditions. Systems conforming to ASTM C 881 are able to cure under humid conditions and bond to damp surfaces.

Where large cavities are to be repaired by use of an epoxy-resin concrete, the most economical method is to prepack aggregates into the void and then inject the epoxy resin as a grout from the bottom of the void. This approach displaces air or displaces water in the event the repair is made under water.

Proportioning epoxy-resin mortar or concrete follows the same guidelines as proportioning portland-cement concrete mixtures. However, since the resin systems are sticky and viscous, the sand/aggregate ratio is usually the reverse of that used for portland-cement concrete.

The total of voids in the system should be less than 12% by volume. Voids in excess of 12% will result in a permeable mass with resultant poor chemical and freeze/thaw resistance. When using forms for epoxy-resin mortar or concrete, the forms should be coated with polyethylene rather than oil. Oil may be absorbed into the resin impairing the cure of the system.

ACI Committee 503 has issued a guide covering the use of epoxy resin in concrete construction [7]. This report should be studied by those planning to use these materials. It stresses the necessity for preparing the concrete surface to be treated before application of the resin system. Three surface conditions must be met if an application is to be successful: (1) the surface must be strong and sound, (2) the surface must be clean (that is, free of oil, grease, or other contaminants such as residues of curing compounds, waxes, or polishes that may have been applied to the surface of the concrete) and (3) that water vapor is not being expelled from the concrete (out-gassing) during application and curing of the system.

It is often desirable to use resins other than epoxies for mortars and overlays. Acrylic, polyester, and vinylester resins have been used successfully. The choice may be dictated by economy, viscosity, cure rate, and specific chemical resistance.

The thermal coefficient of expansion of these resins is similar to that of the epoxy resins, and it is recommended that a compatibility test, ASTM C 884, be performed prior to use.

In general, the vinylester and the polyester resins have better acid resistance than the epoxies, but, when used for chemical protection, each system should be evaluated or pretested for resistance to that particular chemical exposure and temperature.

Epoxy resins have been used extensively for pressure grouting cracks to establish the structural integrity of a concrete member [7]. For gravity-flow sealing of cracks in concrete decks, the high-molecular-weight methocrylate resins are appropriate because of their low viscosity. These resins are available with viscosities lower than 25 cps whereas the lowest viscosity epoxy resins will be in the range of 200 to 400 cps [8].

COATINGS

Bituminous materials include both asphaltic and coal-tar-based materials.

There is considerable difference between asphalts and coal tars. Asphalts are resistant to many mildly corrosive materials, but are attacked readily by solvents, oils, and gasoline. Further, unless they are specially formulated, asphaltic coatings may be permeable to water. Asphaltic coatings generally are used to protect foundations, basements, and similar structures from water.

Coal-tar coatings possess a high degree of water resistance and should be considered wherever continuous immersion is encountered. Coal-tar coatings have a tendency to crack and craze when exposed to ultraviolet light or high temperature. The most common use for coal-tar coatings is the protection of underground pipelines. When formulated with epoxy-resin systems, excellent long-term performance has been recorded [9].

Bituminous coatings may be used with reinforcing fabric and tape to increase their resistance to impact on backfilling and handling. Such reinforcing also increases their thickness as well as their impermeability and service life.

Synthetic-Resin Coatings

Synthetic-resin coatings other than latex-based coatings that are successfully used for concrete are generally applied as solvent solutions and include: coumarone-indene, styrene-acrylic, chlorinated rubber, chlorosulfonated polyethylene, polyurethane, neoprene, and epoxy resins.

The most commonly used materials in this group are the chlorinated-rubber, the styrene-acrylate, the polyurethane, and the epoxy-resin-based coatings. For protection for interior use, where aggressive solutions such as mild concentrations of inorganic acids and low concentrations of organic acids and caustics are encountered, the epoxy-resin-based coatings have proven very successful. These may be either polyamine, polyamide, amine-adduct, or acrylic-amine systems [9]. They may be applied as solvent solution, water-based, or 100% solids systems.

Because these epoxy-resin systems will tend to chalk and yellow, they are not suitable as exterior architectural coatings. The acrylic-amine epoxy-resin systems exhibit good resistance to ultraviolet radiation and may be used in both interior and exterior applications. However, they are not quite as chemically resistant as the epoxy-resin systems.

The polyurethane-based coatings exhibit extremely good scuff and abrasion resistance and form the basis for many coatings for concrete floors. They are often used in conjunction with an epoxy-resin-based prime coat.

The aromatic polyurethanes exhibit good chemical resistance but tend to yellow on outdoor exposure. The aliphatic polyurethanes have excellent ultraviolet resistance and are the choice for protection of exterior concrete.

Chlorinated rubber and styrene-acrylate-based clear coatings are used as both curing compounds and so-called floor hardeners. Applied after the concrete has hardened, usually 24 h after placing the concrete, these coatings retain moisture efficiently and aid in curing the concrete.

When used as curing compounds, these coating should conform to the water retention requirements of ASTM Specification for Liquid Membrane-Forming Compounds for Curing Concrete (C 309). For this purpose, the application rate is usually 5 m²/L (200 ft²/gal) Although these systems do not actually harden the concrete surface, they do protect the concrete from abrasion and subsequent dusting.

Both the chlorinated rubber and syrene-acrylate coatings, when subject to forklift or other vehicular traffic may exhibit the phenomenon of "rubber-burn," that is, permanent markings or stains from rubber tires. All organic coatings may exhibit this phenomenon to some

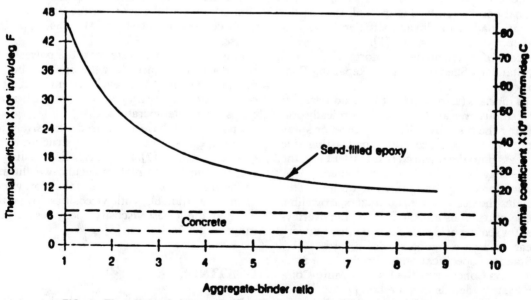

FIG. 1—The effect of changes in the sand aggregate-binder ratio on the thermal coefficient of an epoxy system.

degree. The two-component polyurethane or epoxy-resin-based coatings have better resistance to rubber-burn than single-component coatings. They may, however, depending on the composition of the tires, also exhibit staining.

Latex Coatings

Latex coatings are relatively insensitive to dampness and alkalinity and have become the most widely used coatings for interior and exterior concrete surfaces. Two types of latices in common use are acrylic and polyvinylacetate latices. While both are suitable for interior concrete coatings, the acrylic gives the best results in exterior exposure due to its superior ultraviolet resistance. Since these latices are water based, surface dampness is not a problem. When heavily filled or pigmented, they form permeable coatings, and the presence of vapor pressure does not result in failure.

SEALERS

Materials commonly termed "sealers" or "penetrating sealers" are used to reduce the water absorption and chloride ion penetration of concrete. Most of these materials do not change the appearance of the concrete significantly. Sealers do not actually seal the concrete but reduce water inflow. Penetration is also minimal, 3 mm (⅛ in.) being reported as a general maximum [10]. Many act as effective chloride ion shields.

Linseed Oil

Numerous tests carried out by state highway departments and private laboratories indicate that one of the most economical sealers is linseed oil. Linseed oil is available as a solution or emulsion [11,12].

The solution type is a mixture of raw or polymerized (boiled) linseed oil and a solvent such as mineral spirits, usually in a 50% concentration. The solution types are more economical and more widely used.

The emulsion type contains polymerized linseed oil, kerosene, detergent, and trisodium phosphate in water.

The protective effect is obtained from penetration of the linseed oil into the concrete to a depth of approximately 3 mm (⅛ in.). Upon evaporation of the solvent or water, the linseed oil cures by the process of auto-oxidative polymerization forming an effective seal. Linseed oil has the disadvantage of rendering the concrete blotchy and unsightly and, if applied in excess, will produce a slippery surface.

Silanes

Many materials are marketed for this application. The alkoxysilanes, oligomeric alkoxysiloxs, and the polysilox silanes do not form films but react with the concrete to form a hydrophobic layer.

The sealers based on urethanes, epoxies, and acrylics form semipermeable films.

One extensive test program [12] indicated that an alklyl-alkoxy silane, a methyl methacrylate, an amine-cured epoxy, a moisture-cured polyurethane, and a linseed oil-based sealer performed well in simulated exposure tests.

Other tests [10] indicate that the performance of these sealers is highly dependent on the characteristics of the concrete treated. These tests also indicate that although there are trends in the performance of a generic type, being of a particular generic type does not ensure the effectiveness of a sealer. Therefore, it is recommended that pretesting should be carried out with the candidate sealers and the concrete in question. Such a test method is described by Marusin [13]. The test consists of determining the depth of sealer penetration, and weight gain and chloride ion profile after 21 days of ponding with a salt solution.

REFERENCES

[1] Woods, H., "Durability of Concrete Construction," ACI Monograph 4, American Concrete Institute, Detroit, MI, 1968, p. 15.
[2] Kleinlogel, A., *Influences on Concrete*, Frederick Ungar Publishing Company, New York, 1950, p. 10.
[3] "Concrete-Polymer Materials," Third Topical Report, Bureau of Reclamation, Brookhaven National Laboratory, Denver, CO, Jan. 1971.
[4] Nawy, E. G., Sauer, J. A., and Sun, P. F., "Polymer Cement Concrete," *Journal*, American Concrete Institute, Nov. 1975.
[5] Kuhlmann, L. A., "Peace Bridge Rehabilitation," *Concrete International*, Nov. 1989, pp. 30–32.
[6] Schutz, R. J., "Epoxy Adhesives in Pretressed and Precast Bridge Construction," *Journal*, American Concrete Institute, Title No. 73-14, March 1976, pp. 155–159.
[7] *Manual of Concrete Practice*, Part 5, American Concrete Institute, Detroit, MI, pp. 503R, 503.1–503.4.
[8] "High Molecular Weight Methyacrylate Concrete Crack Bonder," Technology Transfer, Demonstration Project No. 51, Federal Highway Administration, Washington, DC, April 1988, p. 2.
[9] Roberts, A. D., "Organic Coatings: Properties, Selection, and Use," *Building Science Series* 7, U.S. Department of Commerce, Washington, DC, Feb. 1968, p. 56.
[10] Aitken, C. T. and Litvan, G. G., "Laboratory Investigation of Concrete Sealers," *Concrete International*, Nov. 1989, pp. 37–42.
[11] "Linseed Oil in Roads," *Chemical Week*, 21 Nov. 1964.
[12] "Concrete Sealers for Protection of Bridge Structures," National Cooperative Highway Research Program Report 244, Transportation Research Board, Washington, DC, Dec. 1981, pp. 9–10.
[13] Marusin, S. L., "Evaluating Sealers," *Concrete International*, Nov. 1989, pp. 79–81.

Packaged, Dry, Cementitious Mixtures

Owen Brown[1]

PREFACE

Packaged, Dry, Combined Materials for Mortar and Concrete was covered in the two previous editions of ASTM special technical publications, *ASTM STP 169A* and *ASTM STP 169B*. Professor A. W. Brust, the Department of Civil Engineering, Washington University, St. Louis, Missouri, was the original author of this chapter for *STP 169A*. A. C. Carter, Manager of Quality Assurance, Texas Industries, Inc., Arlington, Texas, updated the chapter in *ASTM STP 169B*. Additional prepackaged cementitious products now covered by ASTM specifications required a change in title for this chapter to adequately reflect the content thereof. This edition covers these additional products and suggests where additional specification activity may be required as even more packaged, dry, cementitious products become available to the construction industry.

EARLY HISTORY

Prior to the development of prepackaged concrete, sand, and mortar mixes, the homeowner had to purchase bagged cement, bulk sand, and coarse aggregate, and had to hand mix from scratch the concrete or mortar he needed. About 55 years ago, the packaged concrete mixes began what was to become an ever-growing variety of packaged cementitious products, for the homeowner and the contractor.

The very first producers of packaged, dry, combined materials for mortar and concrete, either introduced the ingredients separately into the package, or mixed them in small increments and placed them in the bag. Processes used weigh or volume measurements, or a combination of both to proportion the ingredients of the mix. Today, high-capacity production lines still use either mode of measure, but incorporate new technology for the control of proportions. Automated process lines help keep uniformity and quality at a high level of control.

The first publication of ASTM tentative specification for packaged, dry, combined materials for mortar and concrete, ASTM C 387, was as information only, published in the *ASTM Bulletin* in April 1955. The tentative specification was accepted by the Administrative Committee on Standards, 5 Sept. 1956. The compressive strength minimums for mortar for unit masonry and for normal-strength concrete were the same as those required by the current ASTM C 387. Compressive strength minimums for concrete mortar (sand topping mix) and other concretes were slightly lower than for the same mixes today (see Table 1 from ASTM C 387-56T, Table I).

In the beginning, package sizes were generally larger than demanded by the market today. Products were available from hardware stores, building supply yards, lumber yards, and from the producing locations. These locations were accustomed to handling 42.6 kg (94 lb) bags of portland cement.

HAZARDOUS CONSIDERATIONS

ASTM addresses the hazards of a standard or method with a somewhat generic caveat, while it is the responsibility of the producer to provide the customer with a detailed hazardous warning covering the ingredients of the product. This is accomplished with a Material Safety Data Sheet (MSDS) that must comply with Federal Regulation 29 CFR 1910.1200.

Portland cement is known to produce a high pH when in contact or mixed with water. It therefore has been listed by the Occupational Safety and Health Administration (OSHA) and the American Conference of Governmental Industrial Hygienists (ACGIH) as a hazardous material with specific permissible exposure limits listed for workers in contact with this material. Portland cement is classified as a nuisance dust, and exposure can affect the skin, the eyes, and the mucous membrane. The MSDS for products containing portland cement must address these issues, classify them by acute and chronic exposure, and provide other information including, but not limited to, emergency first aid procedures, fire and explosion hazard data, precautions for safe handling and use, control measures, and disposal of spills.

Silica sand was concluded to be, in 1987, a potential carcinogen by The International Agency for Research on Cancer (IARC). Although neither OHSA nor the National Institute of Safety and Health (NIOSH) have officially accepted this position, those using silica sand in the blending of cement products must notify users of this potential danger. The MSDS for products containing silica sand, therefore, contains a hazardous statement warning not only of the danger of the inhalation of dust from silica sand and its effect in respiratory disease, including silicosis,

---

[1] Consultant, Savannah, GA 31411.

TABLE 1—ASTM C 387-56T (Table I).

Kind of Material	Water Retention, minimum, %	Compressive Strength, minimum, psi (MPa)			
		1 day	3 days	7 days	28 days
Mortar for unit masonry		...	...	...	...
Type M	70				2500 (17.2)
Type S	70				1800 (12.4)
Type N	70				750 (5.2)
Type O	70				350 (2.4)
Concrete Mortar	...	...	...	3000 (20.7)	4000 (27.6)
Normal strength concrete	...	...	...	2500 (17.2)	3500 (24.1)
High early strength concrete	...	1500 (10.3)	2500 (17.2)	...	...
Lightweight concrete (using natural sand)	...	...	...	2000 (13.8)	3000 (20.7)
Lightweight concrete	...	...	...	1000 (6.9)	1500 (10.3)

pneumoconiosis and pulmonary fibrosis, but now must warn of the possibility of cancer.

Such warning statements are not only provided in the form of an MSDS, but are required on the package as well.

CURRENT SPECIFICATIONS

The Products

Although the packaged industry began with mortar and concrete, the list of specifications covering packaged materials continues to grow. The most important step in standardization of these materials is the requirement that the ingredients used to produce the packaged materials, such as the cements, aggregates, admixtures, etc., must conform with the applicable ASTM standard for that ingredient.

Furthermore, the testing required to verify the performance of the packaged materials utilizes ASTM standard test methods. Therefore, while the packaged material may be used by a homeowner, he is afforded the same protection of performance as the contractor or highway engineer who may also use prepackaged products or may proportion his materials at the job site in accordance with other applicable ASTM specifications.

ASTM C 387

This specification currently covers a number of different kinds of concrete, incorporating both normal-weight and lightweight aggregates. Regardless of the type of aggregate, the required compressive strength of normal-strength concrete is the same. High early-strength concrete is included in the specification, as are a high-strength mortar and three kinds of mortar for unit masonry. The identification of the mortars for unit masonry are the same as those found in ASTM Specification for Mortar for Unit Masonry (C 270) that is under the jurisdiction of ASTM Committee C12 and specifies the same performance levels for the Types N, S, and M mortars as ASTM C 270.

A materials list of ingredients to be used for producing these mortars and concretes contains the appropriate ASTM specification for ingredient compliance. Since it is essential that all ingredients be dry before packaging, a section on aggregate preparation limits the moisture content of dried aggregates to 0.1% by weight. The proportioning of materials in each package must produce at least the minimum physical requirements specified (see Table 2 from ASTM C 387, Table I). Sampling and testing procedures require that standard ASTM methods be utilized.

Packaging and marking specify the proper identification. Container construction is established to ensure adequate protection against breakage and water absorption by the ingredients in the package. Package weights are addressed to ensure a fair measure of materials are received. Rejection of products can occur for reasons of lightweight packages or failure to meet the physical requirements of the intended material.

ASTM C 928

This specification was first issued in 1980, in response to State Highway Departments and other requests, to standardize the performance of concrete repair materials being offered in the marketplace. The responsibility for developing this specification was presented to Subcommittee C09.03.17, because the materials would be packaged for distribution.

There were already a number of materials being offered, stating a variety of performances at varying test ages. The

TABLE 2—ASTM C 387-87 (Table I).

Kind of Material	Water Retention, minimum, %	Compressive Strength, minimum, psi (MPa)		
		3 days	7 days	28 days
Concrete				
High early strength	...	2500 (17.2)	3500 (24.1)	
Normal strength, normal weight	...	...	2500 (17.2)	3500 (24.1)
Lightweight using normal weight sand	...	...	2500 (17.2)	3500 (24.1)
Lightweight	...	...	2500 (17.2)	3500 (24.1)
High strength mortar	...	...	3000 (20.7)	5000 (34.5)
Mortar for unit masonry				
Type M	70	...	...	2500 (17.2)
Type S	70	...	...	1800 (12.4)
Type N	70	...	...	750 (5.2)

ASTM specification development procedure provides a forum to obtain consensus product requirements. In the process, it was determined that the market needed two levels of performance for concrete repair materials. A rapid hardening product that would provide a reasonable level of compressive strength gain in 3 h, and a very rapid hardening product that would yield high levels of compressive strength in 3 h. The minimum physical requirements also include bond strength, shrinkage, and workability requirements (see Table 3 from ASTM C 928, Table I).

Ingredients used in producing these repair materials are not referenced to ASTM specifications, since many of the cements used for these products are not currently addressed in ASTM documents. This specification therefore uses performance criteria to accept compliance rather than address the constituents of the product by prescription. Testing to verify performance and compliance to the specifications are methods of test that are ASTM standards that have been examined and approved by the consensus process. Packaging and marking instructions are part of the specification, and reasons for rejection of products are clearly defined.

A recent revision to this specification, incorporates a minimum requirement for bond strength when tested in accordance with ASTM Test Method for Bond Strength of Epoxy-Resin Systems Used with Concrete (C 882). Both rapid hardening and very rapid hardening materials for concrete repair must exhibit a minimum bond strength of 6.8 MPa (1000 psi) at one day and 10.2 MPa (1500 psi) at seven days when tested by this method. This is another example of the continuing review and improvement of ASTM specifications by the subcommittee and committee structure.

ASTM C 1107

This specification was issued in 1989, and was subsequently adopted by the Corps of Engineers under CRD-C 621-89a replacing their own specification that was the industry standard before ASTM adopted ASTM C 1107. The specification currently covers three grades of nonshrink grout, as follows:

Grade A pre-hardening volume-adjusting,
Grade B post-hardening volume-adjusting, and
Grade C combination volume-adjusting.

In addition to the requirement that the various grades of nonshrink grout must comply with specific minimum and maximum height changes, all grades are required to attain minimum compressive strengths at specified ages.

The test methods required are standard ASTM methods. The pre-hardening volume change is determined by ASTM Test Method for Early Volume Change of Cementitious Mixtures (C 827). Volume control of the nonshrink grout, meeting the Grade A requirement, is caused by expansion before hardening. The post-hardening volume change is determined by ASTM Test Method for Measuring Change in Height of Cylindrical Specimens from Hydraulic Cement Grout (C 1090). Volume control for Grade B nonshrink grout is caused by expansion after hardening. The Grade C grout exhibits expansion in both pre- and post-hardening conditions. The minimum physical requirements also include compressive strength limits (see Table 4 from ASTM C 1107, Table I). Nonshrink grouts may be placed at differing consistencies but are required to meet the same minimum performance criteria. This also applies to temperature of placement and maximum working time, thereby making this a somewhat cumbersome specification to qualify products on a batch-to-batch basis.

ASTM C 887

This specification is under the jurisdiction of ASTM Committee C12 on Mortars for Unit Masonry and the direct responsibility of Subcommittee C12.06 on Surface Bonding. This specification was first issued in 1979 and provides for a new innovation in masonry construction, that is, dry-stacked block with surface bonding on both sides. The standard practice for construction of surface-bonded walls can be found in ASTM Practice for Construction of Dry-Stacked, Surface-Bonded Walls (C 946).

A limited number of standard laboratory tests are required to show compliance with ASTM C 887. The physical properties include flexural and compressive strength (see Table 5 from ASTM C 887, Table I). The Appendix of the specification contains other recommended tests to be performed on surface-bonded masonry assemblages in accordance with test procedures of other ASTM methods,

TABLE 3—ASTM C 928-91 (Table I).[a]

Compressive Strength, minimum, psi (MPa)	3 h	1 day	7 days	28 days
Rapid hardening	500 (3.4)	2000 (13.8)	4000 (27.6)	[b]
Very rapid hardening	1000 (6.9)	3000 (20.7)	4000 (27.6)	[b]

Length change, maximum	
Allowable increase, after 28 days in water, based on length at 3 h, %	+0.15
Allowable decrease, after 28 days in air, based on length at 3 h, %	−0.15

Consistency of concrete and mortar:	Concrete slump min, in. (mm)	Flow of mortar min, %
Rapid hardening, consistency at 15 min after addition of the mixing liquid	3 (76)	100
Very rapid hardening, consistency at 5 min after addition of the mixing liquid	3 (76)	100

Scaling resistance to deicing chemicals after 25 cycles:
Concrete specimens, max visual rating 2.5
Mortar specimens, max 1.0 lb/ft² (5.0 kg/m²) of scaled material[c]

[a] It is recognized that the characteristics and qualities of hardened repair material other than those mentioned in Table 3 might need consideration when certain kinds of concrete repairs are to be made. For the more severe use and exposures that require a higher level of performance, the user is advised to consult with individuals skilled in dealing with such matters.

[b] The strength at 28 days shall be not less than the strength at seven days.

[c] A 10-in. (254-mm) square spalled to an average depth of ⅛ in. (3.17 mm) for 100% of its surface would have about 2.0 lb/ft² (10 kg/m²) of scaled material.

TABLE 4—ASTM C 1107-91 (Table I).

Compressive Strength, minimum	psi	(MPa)
1 day[a]	1000	(6.9)
3 day	2500	(17.2)
7 day	3500	(24.1)
28 day	5000	(34.5)

	Grade Classification		
	A Prehardening Volume Controlled Type	B Post-Hardening Volume Controlled Type	C Combination Volume Controlled Type
Early-age height change			
Maximum % at final set	+4.0	NA[b]	+4.0
Minimum % at final set	0.0	NA	0.0
Height change of moist cured hardened grout at 1, 3, 14, and 28 Days			
Maximum, %	NA	+0.3	+0.3
Minimum, %	0.0	0.0	0.0

[a] When required, the purchaser must so specify in the purchase contract.
[b] NA = not available.

primarily those of Committee E6 on Building Systems. These tests are required by the major national building codes to verify the true performance of a surface-bonding mortar and are used to establish design loads for the composite wall. This is further evidence of the way ASTM coordinates its efforts between committees.

Surface bonding mortar is a pre-mixed, pre-packaged combination of materials that incorporate glass fibers to eliminate the shrinkage cracking generally associated with cement parget coats. These glass fibers must exhibit resistance to attack by the alkaline components of the matrix, so as not to be affected on long-term performance. AR glass fibers, because of their high strontium composition, are used not only in surface-bonding mortars, but in other cementitious systems where the improved flexural strength performance offered by these fibers is desired. Such universally accepted applications as glass fiber reinforced cement (GFRC) attests to the acceptability of AR glass fibers for cement systems.

USE OF THE PRODUCTS

By Whom

One of the major decisions for the original production of mortar and concrete in packaged units was the desire on the part of the manufacturers to provide these materials in smaller quantities to the do-it-yourself (DIY), homeowner market.

Prior to the availability of packaged mixes, the homeowner would have to buy a minimum of one 42.6-kg (94-lb) bag of cement, and enough bulk sand or sand and gravel (in his own containers or directly into the trunk of his car) for his project. This was not a very convenient method for those who lived in city or urban areas. Ready-mixed concrete then as today served those who purchased a minimum of one or more cubic metres (one or more cubic yards).

Today the DIY market is served with a variety of package sizes of mortar and concrete from ~4.5- to 40.8-kg (10- to 90-lb) bags. The sizing of packages permits the male or female of the family to purchase and carry mortar and concrete home for those DIY projects.

Although the DIY market was the primary purpose for the packaging of mortar and concrete, other user purposes have been served. State highway departments are the largest users of ASTM C 928 rapid and very rapid hardening concrete repair materials. These packaged mortars

TABLE 5—ASTM C 887-79a (Table I).

FLEXURAL STRENGTH[a] (average of three prisms)

The flexural strength of prisms of surface bonding mortar prepared and tested in accordance with this specification shall be equal to or higher than the values specified for the ages indicated as follows:

	psi	(MPa)
1 day	450	(3.1)
7 days	700	(4.8)
28 days	800	(5.5)

COMPRESSIVE STRENGTH (average of six modified cubes)

The compressive strength of modified cubes of surface-bonding mortar prepared and tested in accordance with this specification shall be equal to or higher than the values specified for the ages indicated as follows:

	psi	(MPa)
1 day	1600	(11.0)
28 days	3500	(24.1)

Time of setting, Vicat needle, initial set,
  Minimum, min — 45
  Final set, maximum, h — 8
Water retention flow after suction, minimum, %
  of original flow, minimum — 75

and concretes are particularly effective in highway repair work where small patches can be made and highways reopened in a matter of a few hours.

Various service companies, such as telephone, natural gas, water, and others, are finding that packaged mortars and concrete are convenient for their repair jobs as they install their much-needed utilities in homes and industry.

The contractors, both small and large, are also learning that the packaged materials are essential to their use in both repair and in construction. This is particularly true in the larger metropolitan areas, where space limitations make it more difficult to deposit bulk materials for job-mixed concretes and mortars. Environmental considerations and disposal of excess or waste materials add to the contractors problem and lead him toward the prepackaged concept.

Applications

The applications for prepackaged materials are many, and the list continues to grow. The concretes covered by ASTM C 387 are used by both DIY homeowners and contractors for constructing slabs on grade for such uses as driveways, sidewalks, and patios. Prepackaged concrete also finds uses for repair of existing slabs, either as partial replacement or refurbishment. Concrete mix is also used for other construction and repair needs.

High-strength mortar or sand mix (sand topping) are used where thin layers (less than 50.8 mm (2 in.)) are required. Sand mix is frequently employed for overlays of deteriorated slabs. These applications require the use of a bonding media to ensure that the new topping will not separate from the base slab. In many cases, the overlay may be as thin as 6.35 mm (¼ in.) or even less. High-strength mortar also serves as a mortar to lay brick, especially when the brick is to serve as a walkway, patio surface, or driveway. In these applications, mortar for unit masonry provides inadequate strength.

Mortar for unit masonry is used for the typical laying of brick and block for various DIY projects, such as the construction of party walls, retaining walls, and barbecue pits. While most prepackaged mortar for unit masonry is not used on major construction sites, small contractors do use this product for small building projects. Prepackaged mortar for unit masonry is applicable for tuckpointing, where mortar deterioration is evident on existing brick and block buildings.

Rapid hardening and very rapid hardening repair products, covered by ASTM C 928, have their largest application with the highway engineer, where the urgency to patch and reopen highways quickly justifies the use of more costly materials. The potential for accidents resulting from closed highway lanes can be sufficient to make the slight additional cost of materials seem trivial.

Because many of these repair products have demonstrated their capabilities in highway repair, they are becoming increasingly popular for use in other concrete repair projects with contractors active in industrial and commercial operations where floors of garages, warehouses, and manufacturing plants need instant repair so that the areas can be quickly reusable for vehicular traffic.

Nonshrink grout covered by ASTM C 1107 is the least likely of the prepackaged products to find application by the homeowner or do-it-yourselfer. This material in its many consistencies is used by contractors to set building columns and heavy machinery. Its uses are established by the need to ensure positive and complete contact by a supporting member.

The development of surface bonding mortar as covered by ASTM C 887 was for the construction of concrete masonry walls without the use of mortar.

Surface-bonded block walls are stronger than block and mortar walls in flexure and provide additional benefits of a finished textured look, waterproofness for below-grade walls, and a level of insulation for above-grade walls due to its unique capability to completely stop the penetration of air. This latter characteristic also applies to the stoppage of methane and radon gases.

Availability

The concrete and mortars covered by ASTM C 387 and C 887 are available in most if not all retail home center, lumberyard, hardware stores. These products together with the repair products covered by ASTM C 928 and the nonshrink grouts covered by ASTM C 1107 can be found in most construction supply houses. Most producers of these products do not offer the products or sale from their manufacturing facilities, although arrangements may be made with construction supply houses to make deliveries of full truckload quantities directly to large construction projects.

QUALITY CONTROL

The term quality control has many meanings. For the producer, quality control is the maintenance of the performance of a product that begins with the raw materials used to produce that product. All manufacturers maintain some form of quality control program in order to ensure that the product they make meets the requirements of the user. The producer is interested in maintaining a uniform production, since this makes his process run more efficiently. Therefore, the uniformity of the raw materials he uses is vital to his efficiency.

For the user, quality control provides the assurance that the product meets the applicable specification. In addition, the user also hopes that uniformity of performance is maintained by the producer.

It is obvious, therefore, that both producer and user have the same goals that can only be realized through adequate specifications of the materials used to produce the end product. In the specifications covered in this chapter, this is attempted by referenced ASTM documents covering the individual ingredients. Further revision of the documents may be required to gain improved uniformity of finished products. Such revisions are essential to improving the quality control of prepackaged mortars and concretes.

The remainder of this section will deal with quality control as it relates to material specifications.

Materials

The cementitious components, with few exceptions, are listed within the specifications covered in this chapter by reference to existing ASTM documents. These components are essentially cements covered either by ASTM Specification for Portland Cement (C 150) or ASTM Specification for Blended Hydraulic Cements (C 595). Mortars for unit masonry produced in accordance with ASTM C 387, utilize masonry cements under ASTM Specification for Masonry Cement (C 91) as well as portland-lime varieties proportioned in accordance with ASTM C 270 where the hydrated lime meets ASTM Specification for Hydrated Lime for Masonry Purposes (C 207) requirements.

It can be stated that the cementitious components permitted in the production of products to meet both ASTM C 387 and C 887 prepackaged products are required to meet existing ASTM specifications. This results from the fact that these specifications are basically prescription specifications. This is not the case with ASTM C 928 and C 1107, prepackaged mortars, where the specifications are written to performance criteria.

The proprietary components used in ASTM C 928 and C 1107 are not covered by ASTM specifications. In many instances, the cementitious components represent special formulations or combination of ingredients that have limited potential in the area of mortars and concrete, and therefore have not been considered for coverage by ASTM specifications. Some of these special cements are being studied in a new Committee C1 subcommittee, where a specification is eminent.

Aggregate quality is specified in ASTM C 387 by reference to ASTM Specification for Concrete Aggregates (C 33) and ASTM Specification for Aggregate for Masonry Mortar (C 144). These specifications provide for grading and for other aggregate qualities. Aggregates for ASTM C 887, C 928, and C 1107 are not specifically referenced but are left to the discretion of the manufacturer through the use of performance criteria.

The dryness of aggregates is covered in ASTM C 387 under the section, Preparation of Aggregates. In this section, aggregates shall have a moisture content of less than 0.1% by weight. There are no similar provisions in the other specifications.

Packaging

Package sizes are not a requirement within these ASTM specifications. Package sizes are actually user driven in most instances. Because of increased workmen's compensation claims for back injuries, even packages for industrial and commercial uses are decreasing in size.

The construction of packages are included in the specifications, in order to assure that the ingredients have a reasonable useful life. This is frequently referred to as the "shelf life" of the material. Container construction is specified in terms of water-vapor transmission of the package, determined in accordance with ASTM Test Methods for Water Vapor Transmission of Materials (E 96).

Physical Requirements

Each of the prepackaged mortar and concrete specifications have a table of physical requirements. These tables have been included in this chapter for reference. In general, the physical requirements involve strength minimums to be met at various test ages. Other characteristics are included, and vary by product.

SIGNIFICANCE OF SPECIFICATIONS

Specifications are written because of a need to standardize the performance of products between what the manufacturer visualizes and what the user demands. For this reason, it is frequently difficult to develop a specification with a minimum of options.

The prepackaged mortar and concrete specifications contain a list of physical requirements, usually with minimum values. There are no other limits that the manufacturer must adhere to. Occasionally, the producer is required to produce a mortar or concrete to some higher guaranteed performance.

Rejection of Product

The prepackaged mortar and concrete specifications incorporate provisions for the user to reject the product failing to meet requirements of the specification.

The most obvious reason for product rejection is that the product fails to meet some or all of the physical requirements of the specification.

In addition, the user may also reject a shipment if the package weights vary from the printed weight by more than a specified amount, or if the average weight of a given number of packages is less than that printed on the bag. All broken packages may also be rejected.

Standards for Producer and Consumer

The entire standardization process in ASTM assures that the development of specifications is unbiased and, therefore, the standards for prepackaged mortars and concrete provide protection for both producer and consumer.

FUTURE NEEDS OF PREPACKAGED PRODUCT SPECIFICATIONS

Concrete Products

There are a number of additional prepackaged concrete products available on the market that should be applicants for coverage by ASTM specifications.

Fast-setting concrete mix provides both the DIY and contractor user with a prepackaged concrete with fast-setting characteristics. This type of concrete permits rapid utilization of the finished slab or footer and is particularly adaptable for that week-end project or for the small-contractor one-day project.

A fiber-reinforced concrete mix aids the homeowner in producing a finished slab with a minimum of drying shrinkage cracks. Because of the general tendency for the homeowner to use too much water in placing his concrete, this product will help the average user produce a better crack-resistant slab.

Mortar Products

There are a number of prepackaged products that are designed to provide specific performance characteristics for the user. The user may be the typical homeowner or the large contractor. These products are essentially mortar products, since they contain cementitious components and a grading of fine aggregate. The fine aggregate grading requirements vary appreciably, depending on their thickness of use.

Shotcrete

Prepackaged shotcrete, frequently called Gunite when the material is to be dry-applied, has been available to users for many years. Both fiber and non-fiber reinforced, modified and unmodified mixes have been offered to meet specific project requirements. These materials generally utilize fine aggregate conforming to ASTM C 33 and can utilize coarse aggregate meeting the same specification, if desired.

Shotcrete materials are frequently used in underground coal mines to cover and reinforce the ribs and roof to prevent sloughing. They also are employed in major tunnel projects as temporary support during construction of tunnel linings. Shotcrete is also a reinforcing technique used to repair deteriorated poured-in-place concrete structures. Subcommittee C09.03.20 on Shotcrete is preparing a specification for this class of materials. It is anticipated that a prepackaged version, under the jurisdiction of Subcommmittee C09.03.17, will follow.

Floor Resurfacers

A new class of floor resurfacing products has emerged for use both in the construction and reconstruction markets. Poured at high water contents, many of the products need no troweling or secondary placement operations. Many of these materials are pumped in-place and self-level much as the spreading of water. These products are offered in normal or fast-setting varieties, depending on the need to place the finished floor in service.

These self-leveling floor resurfacing materials are used in new construction when the surface of a newly installed floor is damaged by rain or other actions before obtaining final set. They are also employed when the floor is improperly leveled or placed at the wrong elevation and a concrete topping is not practical. Usually these surfaces are to be covered with floor coverings, so that the wear surface of the material is not a factor.

The use of self-leveling floor resurfacers in reconstruction is particularly advantageous, since these materials can be installed at thicknesses from 1.58 mm (¹⁄₁₆ in.). The disadvantage when using these materials is the need to take existing floor surfaces to bare concrete. This can frequently require scarifying or shotblasting the concrete surface to remove such materials as mastics from previous floor coverings.

Latex-Modified Patching Products

A new specification for Latex & Redispersible Powder Polymer Modifiers for Hydraulic Cement Concrete and Mortar is being developed by Subcommittee C09.03.19. Once this specification becomes available, it would be appropriate for Subcommittee C09.03.17 on Packaged, Dry, Combined Materials to develop a specification for packaged latex-modified concrete patching products. Many products are currently offered to both DIY and contractor markets, but their performance is not optimized because they are not required to meet a given level of performance. An ASTM specification would be of major benefit to both user and producer.

Stucco Products

Committee C11 on Gypsum and Related Building Materials and Systems is responsible for ASTM Specification for Application of Portland Cement-Based Plaster (C 926). This specification covers the minimum requirements for the application of full-thickness portland cement-based plaster for exterior (stucco) and interior work. The specification also sets forth tables necessary for the proportions of various plaster mixes. The specification does not address the prepackaged stucco mixes that are available today in the marketplace.

Base-coat stuccos and finish-coat stuccos are offered to both consumer and contractor for new construction and repair. These products do not reference an ASTM specification nor are they bound by the requirements of ASTM C 926.

A new generation of fiber-reinforced stuccos, referred to as one-coat stuccos, are also available and not governed by ASTM specifications. These products are prime candidates for standards development in the prepackaged area.

Acknowledgments

The author wishes to acknowledge the work of the original author of this chapter in *ASTM STP 169A*, A. W. Brust, Department of Civil Engineering, Washington University, St. Louis, Missouri; and the work of the second author of this chapter in *ASTM STP 169B*, A. C. Carter, manager of Quality Assurance, Texas Industries, Inc., Arlington, Texas.

Subject Index

A

AASHTO Accreditation Program, 40
AASHTO Materials Reference Laboratory, 40–41
AASHTO T 259, 167–168
AASHTO T 260, 171
AASHTO T 277–83, 255
Abrasion
  definitions, 182
  test methods
    application, 190–191
    ASTM C 418, 186
    ASTM C 779, 186–188
    ASTM C 944, 188–189
    ASTM C 1138, 189–190
  types, 182
Abrasion resistance, 182–190
  aggregate quality and, 183
  compressive strength and, 183–184
  concrete types and, 184
  curing and, 185
  factors affecting, 182–185
  finishing procedures and, 184–185
  lightweight aggregate concrete, 531
  mixture proportioning and, 183–184
  surface treatment and, 185
Absorption (*see also* Water, absorption)
  characteristics, lightweight aggregates, 525
  method, cement content analysis, 113–114
  soundness tests, 414
Absorption coefficient, 542
Absorption cross section, 542–543
Accelerating admixtures, 494–495
ACI 116, 182
ACI 116R, 49
ACI 116R-90, 491, 567
ACI 117, 512
ACI 207.5R, 571
ACI 208-58, 202
ACI 211, 59, 308
ACI 211.1, 54, 75, 404, 595
ACI 211.2, 75, 595
ACI 212.3R, 491–492
ACI 213, 531
ACI 213R-87, 536
ACI 214, 71, 73
ACI 214-77, 53
ACI 226.1R-87, 500, 506
ACI 226.3R-87, 500
ACI 228, 73
ACI 304, 50, 596
ACI 304.2, 596
ACI 305R-89, 480
ACI 306, 238
ACI 306R-88, 480
ACI 308, 74
ACI 308-81, 481
ACI 309.1R, 49
ACI 310, 513
ACI 318, 38, 272, 513, 531, 594
ACI 318-63, 204, 206
ACI 318-71, 204–205, 207
ACI 318-89, 205–206
ACI 503.1, 601
ACI 503.2, 601
ACI 503.3, 601
ACI 503.4, 601
ACI 506, 589–591
ACI 506-66, 590
ACI 506R, 594, 596
ACI 506R-90, 593
ACI 506.2, 596
ACI 506.2-77, 590
ACI 506.3R-82, 590
ACI 523.2R, 536
ACI 547R, 593
ACI 805-51, 589–590
Acid attack, 272–274
Acoustic properties, unit weight and, 309–310
Activity index, hydraulic cements, 451
Additives, content analysis, 318
Adhesives, 600–602
Admixtures (*see also* Chemical admixtures; Mineral admixtures; specific materials)
  air-entraining, 299
  alkali–silica reactivity and, 370–371
  bonding and patching, 599–600
  content analysis, 318
  definitions, 491
  effects on permeability, 256
  proportioning, abrasion resistance and, 183–184
  ready mixed concrete, 515
  in recycled concrete, 359
  roller–compacted concrete, 571
  sampling, 19–20
  shotcrete, 594
Adsorption, 247–248
Age
  of cement paste, permeability coefficient and, 245
  of concrete, petrographic examination and, 214
Aggregates (*see also* Lightweight aggregates; Petrographic evaluation, aggregates)
  air voids, 422–423
  alkali–silica reactivity, 366–369
  as a whole, 388
  coarse
    degradation, 390–391
    grading, 402–404
    latex–modified mixes, 580
    particle size, workability and, 54
    pore system, 429–437
    preplaced aggregate concrete, 563
    size, effect on flexural strength, 132
  coatings, 407–408, 418–419
  concrete permeability and, 168–169
  configuration, 421–422
  content analysis, 318
  definition, 5
  degradation, impact on specifications, 398
  deleterious substances, 415–418
  drying shrinkage and, 222–223
  effect on
    bleeding, 104
    compressive strength, 287
    unit weight, 311
  elastic properties, 395
  fine
    entrained air and, 487
    grading, 404
    latex–modified mixes, 580
    preplaced aggregate concrete, 563
  fineness modulus, 401–402
  fire resistance, 283, 285
  frictional properties, 397–398
  frost resistance, 298
  grading, 401–405, 422
  hardness, 395–397
  heavyweight, 293
  high–density, preplaced aggregate concrete specification, 564
  lightweight, 293
  mobility, 421–422
  permeability, 246
  petrographic examination, 211
  properties, 11
  Poisson's ratio, 395
  porosity and permeability, 240–241
  quality, abrasion resistance and, 183
  in radiation shielding, 542
  ready mixed concrete, 515
  roller–compacted concrete, 570–571
  sampling, 19
  shape, 405–406, 421–422
  shotcrete, 593

soundness, 412–415
specific gravity, 424–425
specific heat, 442
standards, 389
strength, 388–389, 394–395
surface moisture, 425–426
texture, 406–408, 421–422
thermal conductivity, 230, 441–442
thermal diffusivity, 442
thermal expansion, 233–234, 440–441
thermal properties, 438–443
water absorption, 423–424, 426–428
wear, 396–398
wet degradation and attrition tests, 391–394
wood, 178–179
Aggregate-to-mortar bond, 395
Air compressors, shotcrete, 592–593
Air content (*see* Air voids)
cohesion and, 297
consistency and, 487
effects of algae in mixing water, 475–476
fresh fiber-reinforced concrete, 550
frost resistance and, 297
hardened concrete, 296–300, 308
hardness effect on mixing water, 476
hydraulic cements, 458–459
measurement
hardened concrete, 300
fresh concrete, 65–66
methods, fresh concrete, 66–69
optimum, air-void parameters, 485
ready mixed concrete, 514
strength reduction and, 297
uniformity test, 51–52
unit weight and yield, 297, 311
water/cement ratio and, 487
workability and, 297
Air-entraining admixtures, 484–489
chemical, 487
current specifications and test results, 489
definitions, 484
determining air void characteristics, 488–489
entrained air, factors affecting amount and character, 486–488
fine aggregate, 487
hydraulic cements, 458
materials, 484–485
mixing action, 487–488
potential, cement content and, 487
roller-compacted concrete, 571
supplementary cementing materials, 487
temperature, 488
Air entrainment
grading effect on, 404–405
workability and, 54
Air-free unit weight test, 50–51
Air voids
aggregates, 422–423
characteristics, 486, 488–489
composite gradation, 299–300
content analysis, 318
dispersion and spacing, 398, 301–302, 486
expulsion with vibration, 488
freeze-thaw damage and, 297–298, 486
frost resistance and, 298, 308
lightweight aggregates, 524–525
shape, 300
size, 486
Air-void system
geometry evaluation, 302–308
ASTM C 457 microscopic analysis, 302–303
comparing parameters in fresh and hardened concrete, 307

image analysis techniques, 307
linear traverse method, 303–304
modified point count method, 304
precision and bias, 304
sources of variability and uncertainty, 304–307
test result interpretation, 307–308
latex-modified portland cement, 581–582
specific surface, 300–301
with and without air-entraining admixtures, 298–299
Akeley, Carl E., 591
Algae, in mixing water, effects on air content and strength, 475–476
Alite, 464
Alkali-aggregate reaction, research needs, 45
Alkali-carbonate rock reactivity, 372–386
ASTM standards, 372–373
chemical and mineralogical composition, 378–379
compared with alkali-silica reaction, 375
distress manifestations, 374–375
evaluating potential for
autoclaved concrete microbars, 385
concrete prism expansion test, 382–383
determination by chemical composition, 384–385
field service record, 381–382
freeze-thaw in salt solution, 385
miniature rock prisms, 385
petrographic examination, 384
rock cylinder expansion test, 383–384
expansive dedolomization reaction, 373–374
mechanism of reaction and expansion, 381
petrographic characteristics, 375–378
quarry sampling, 385
types of reactions, 373
Alkali reactivity, aggregate constituents of hardened concrete, 348
Alkali-silica reactivity, 365–371
admixed materials, 370–371
avoiding expansive reactions, 369–371
cement alkali level, 369–370
compared with alkali-carbonate rock reactivity, 375
fly ash, 504
hydraulic cements, 459
identifying potentially reactive aggregate, 367–369
mechanism of reactions and distress, 366–367
moisture availability and environmental effects, 367
mortar bar tests, 367–368
petrographic examination, 368
Quick Chemical Test, 368
rapid mortar-bar expansion test, 368–369
symptoms, 365–366
types of reactive aggregate, 366–367
Alkali sulfates, hydraulic cements, 471
Aluminate, 464–465
Aluminum, corrosion, 174–176
American Association of Laboratory Accreditation, 40
Amino alcohol derivatives, 221
Anti-corrosion agent, 169
AS1342, 253
Asbestos, corrosion, 180
Aspect ratio, 547, 553
effect on setting time, 128
ASTM 117-90, 586
ASTM A 185, 594
ASTM A 305, 202, 204, 206
ASTM A 305-56T, 202
ASTM A 385, 594

ASTM A 408, 204
ASTM A 416, 594, 594
ASTM A 421, 594
ASTM A 497, 594
ASTM A 615, 202, 204, 594
ASTM A 616, 202, 594
ASTM A 617, 202, 594
ASTM A 641, 594
ASTM A 706, 594
ASTM A 722, 594
ASTM A 767, 594
ASTM A 775, 206, 594
ASTM A 820, 547
ASTM C 29, 423, 430, 523
ASTM C 31, 71–75, 124–126, 137, 149, 151, 519, 575
ASTM C 33, 54, 178, 180, 183, 299, 341, 359–360, 368, 372–373, 389, 396, 398, 401–402, 404–406, 408, 411, 414–417, 423, 459, 580, 591, 593, 595, 608–609
ASTM C 39, 71, 125–126, 128–131, 137, 286, 519
ASTM C 40, 26, 417
ASTM C 42, 20, 125–126, 128, 330–331, 596–597
ASTM C 70, 426
ASTM C 71, 481
ASTM C 78, 71, 130–132, 287
ASTM C 85, 315–317
ASTM C 87, 417
ASTM C 88, 267, 353, 411–413, 415, 419
ASTM C 91, 40, 609
ASTM C 94, 20, 50–51, 53, 71–72, 473, 511–520, 596
ASTM C 109, 33, 450, 452, 457–458, 506
ASTM C 114, 468–469, 502
ASTM C 115, 300, 450
ASTM C 116, 125–126, 128, 130, 548, 550, 560
ASTM C 117, 344, 408, 416–417
ASTM C 123, 415, 417
ASTM C 125, 5, 11, 53–54, 300, 347, 353–354, 403, 484, 500, 505
ASTM C 127, 75, 414, 423–428, 432, 525
ASTM C 128, 75, 414, 423–428, 525
ASTM C 131, 183, 389–391, 415, 417
ASTM C 136, 344, 401, 525
ASTM C 138, 51, 69, 75, 297, 307–308, 488, 512, 550
ASTM C 142, 342, 415–416
ASTM C 143, 52, 55, 66, 72–73, 75
ASTM C 144, 609
ASTM C 150, 40, 107, 223–224, 226, 236, 270, 272, 317, 449, 462–463, 465–467, 471, 570, 589, 593, 609
ASTM C 151, 224–226, 450, 456, 531
ASTM C 156, 478, 481–482
ASTM C 157, 225–227, 450, 452, 455, 529, 559
ASTM C 171, 594, 596
ASTM C 172, 20, 66, 73, 527, 548, 550
ASTM C 173, 51–52, 68, 73, 75, 307, 486, 526, 527, 550
ASTM C 177, 231, 288, 443, 523, 534
ASTM C 183, 20, 26
ASTM C 184, 450
ASTM C 185, 450, 452, 459
ASTM C 186, 235, 450, 455
ASTM C 187, 453, 450, 452
ASTM C 188, 432, 450–451, 524
ASTM C 190, 457
ASTM C 191, 450, 453–454
ASTM C 192, 71–73, 75–76, 124–126, 159, 575
ASTM C 200, 344
ASTM C 204, 300, 450–451
ASTM C 207, 609

INDEX 613

ASTM C 215, 154, 156, 321–322
ASTM C 219, 5, 462, 484
ASTM C 227, 178, 216, 358–359, 367–369, 450, 452, 456, 459
ASTM C 231, 51, 66, 69, 73, 75, 307, 486, 488, 514, 550
ASTM C 232, 105, 108
ASTM C 233, 489, 492
ASTM C 235, 417
ASTM C 236, 288, 523
ASTM C 243, 108–109, 496
ASTM C 260, 157, 159, 492, 594
ASTM C 266, 450, 453–454
ASTM C 267, 274
ASTM C 270, 605, 609
ASTM C 289, 178, 367–368
ASTM C 290, 154–156
ASTM C 291, 154, 156
ASTM C 292, 154
ASTM C 293, 71, 130–132
ASTM C 295, 211, 341, 343–345, 348, 355, 361, 367–368, 372–373, 415–416, 418–419, 439
ASTM C 298, 605–609
ASTM C 309, 20, 478, 481–482, 594, 596, 602
ASTM C 310, 154–155
ASTM C 311, 20, 370, 450–451, 502–504, 507
ASTM C 311–53T, 501
ASTM C 325, 450
ASTM C 330, 157, 398, 423, 522–524, 528–529, 593
ASTM C 331, 398, 423, 524, 529
ASTM C 332, 423, 523–524
ASTM C 342, 452
ASTM C 348, 457
ASTM C 349, 457
ASTM C 350, 501, 505
ASTM C 350–54T, 501
ASTM C 359, 454
ASTM C 360, 58
ASTM C 387, 604–605, 608–609
ASTM C 387–56T, 604–605
ASTM C 402, 501, 505
ASTM C 402–57T, 501
ASTM C 403, 79–80, 82, 84–86, 141
ASTM C 403–57T, 77, 79
ASTM C 403–90, 86
ASTM C 418, 186, 188, 191, 389
ASTM C 430, 450, 503
ASTM C 441, 370, 459
ASTM C 451, 450, 454
ASTM C 452, 270, 272, 450, 452, 460
ASTM C 457, 67–68, 247, 296, 300–307, 316, 344, 347, 351, 353, 361–362, 433, 486, 489
ASTM C 469, 193–195, 529
ASTM C 470, 72
ASTM C 490, 154, 234
ASTM C 493, 432
ASTM C 494, 84–85, 157, 159, 169, 491–493, 495–497, 571, 597
ASTM C 495, 528
ASTM C 496, 130–131, 133, 528
ASTM C 511, 74–75
ASTM C 512, 72, 196–198
ASTM C 513, 528
ASTM C 518, 231
ASTM C 535, 389–390, 415, 417
ASTM C 563, 450, 458, 466
ASTM C 566, 426
ASTM C 567, 309–311, 527–528
ASTM C 586, 373, 377, 379–380, 382–385
ASTM C 595, 40, 236, 315, 449, 451, 456, 459, 467–468, 593, 609

ASTM C 596, 225, 227, 452
ASTM C 597, 325, 597
ASTM C 617, 125–126
ASTM C 618, 20, 235, 266, 270, 370, 456, 468, 501–505, 534, 562, 570, 594
ASTM C 637, 398, 540, 543–544, 563–564, 593
ASTM C 638, 540, 543–544, 564
ASTM C 642, 253, 310–311, 590
ASTM C 666, 154–159, 161, 216, 414, 530, 574, 582, 590
ASTM C 670, 26–27, 144, 157, 188, 413, 437
ASTM C 671, 154–155, 160–161, 414, 590
ASTM C 672, 26, 154–155, 161, 276, 583, 590
ASTM C 682, 154–155, 160–161, 414, 590
ASTM C 684, 140, 142, 144–145, 150
ASTM C 685, 511–513, 518–520, 596–597
ASTM C 702, 373
ASTM C 778, 457
ASTM C 779, 186–188, 191, 389
ASTM C 786, 450
ASTM C 796, 492
ASTM C 801, 134, 192
ASTM C 802, 26
ASTM C 803, 328, 331, 597
ASTM C 805, 327, 597
ASTM C 806, 225, 227, 450
ASTM C 807, 450, 453–454
ASTM C 823, 20, 211, 342, 373, 382, 550
ASTM C 827, 452
ASTM C 830, 432
ASTM C 845, 224, 227, 449, 456
ASTM C 851, 417
ASTM C 856, 171, 211–212, 316, 342, 361, 373
ASTM C 869, 492
ASTM C 873, 124–126
ASTM C 876, 170
ASTM C 878, 225, 227
ASTM C 881, 594, 600–601
ASTM C 882, 600, 606
ASTM C 883, 600
ASTM C 884, 600, 602
ASTM C 887, 606–608
ASTM C 887–79a, 607
ASTM C 900, 333, 597
ASTM C 903, 591
ASTM C 917, 26, 33, 35, 457, 589
ASTM C 918, 140, 147–150, 335
ASTM C 926, 610
ASTM C 937, 540, 544, 563
ASTM C 938, 540, 544, 563
ASTM C 939, 540, 544, 565
ASTM C 940, 108, 540, 544, 565
ASTM C 941, 108, 110, 540, 544, 565
ASTM C 942, 540, 544, 565
ASTM C 943, 540, 544, 565
ASTM C 944, 188–189, 389
ASTM C 946, 606
ASTM C 953, 540, 544, 565
ASTM C 973, 591
ASTM C 974, 492
ASTM C 989, 370, 505–506, 570, 593
ASTM C 995, 548, 550
ASTM C 1012, 270, 272, 450, 452, 460
ASTM C 1017, 157, 492–493
ASTM C 1018, 548, 554–557, 560
ASTM C 1038, 466, 450, 456
ASTM C 1040, 310, 575
ASTM C 1042, 599
ASTM C 1059, 594, 599
ASTM C 1064, 73, 75
ASTM C 1073, 506
ASTM C 1074, 148, 335
ASTM C 1077, 518
ASTM C 1076, 27

ASTM C 1077, 39–40
ASTM C 1078, 52, 112–113
ASTM C 1079, 52, 112, 117
ASTM C 1084, 272, 315–317
ASTM C 1102, 492, 590, 594, 597
ASTM C 1105, 373, 382–386
ASTM C 1107, 606–609
ASTM C 1116, 590, 594, 596–597
ASTM C 1117, 85, 590, 597
ASTM C 1137, 389, 393
ASTM C 1138, 184, 189–191, 389, 391
ASTM C 1140, 590, 594, 596–597
ASTM C 1141, 491, 590, 594, 597
ASTM C 1150, 333
ASTM C 1151, 253, 482
ASTM C 1170, 575
ASTM C 1176, 575
ASTM C 1202, 73, 168, 255, 583, 590
ASTM C 1218, 171
ASTM C 1222, 40
ASTM C 1240, 507, 594
ASTM D 75, 19, 343, 373
ASTM D 98, 492
ASTM D 448, 401
ASTM D 854, 432
ASTM D 2419, 417
ASTM D 2434, 537
ASTM D 2766, 232
ASTM D 2936, 130–131
ASTM D 3042, 389, 397
ASTM D 3319, 389, 397–398
ASTM D 3398, 405
ASTM D 3663, 434
ASTM D 3665, 19–20, 25
ASTM D 3744, 389, 391–392
ASTM D 4222, 433
ASTM D 4404, 434
ASTM D 4460, 27
ASTM D 4525, 435
ASTM D 4567, 434
ASTM D 4580, 99
ASTM D 4641, 433
ASTM D 4780, 434
ASTM D 4791, 405
ASTM E 6, 192–193, 195
ASTM E 11, 401
ASTM E 105, 19
ASTM E 119, 282, 291, 531
ASTM E 122, 19, 25
ASTM E 141, 19
ASTM E 274, 389
ASTM E 303, 397
ASTM E 329, 39, 518
ASTM E 548, 39
ASTM E 660, 389, 397
ASTM E 707, 389
ASTM E 867, 397
ASTM E 994, 39
ASTM E 1085, 506
ASTM E 1187, 39
ASTM E 1224, 39
ASTM E 1301, 39
ASTM E 1322, 39
ASTM E 1323, 39
ASTM G 40, 182
ASTM P 214, 368–369
Attrition, tests, aggregates, 391–394

B

Ball penetration test, 58–59
Batching
　ready mixed concrete, 514–515
　roller–compacted concrete, 572
　shotcrete, 596

614 TESTS AND PROPERTIES OF CONCRETE

Batching plant, ready mixed concrete, 515–516
Beam
   reinforcement bond, 202–204
   stub–cantilever, 203
Bearing strips, effect on splitting tensile strength, 133
Belite, 464
Binders, shotcrete, 593
Bituminous materials, as contaminant, recycled concrete, 360
Blank test, 117
Bleeding, 88–111
   characteristics and setting time, 78
   consolidation and revibration effects, 107
   controlling, 107–108
   duration, 91–92
   effect on hardened concrete, 95–102
      blisters, 99, 101
      density, 96
      durability, 97–98
      mortar flaking, 98–99
      paste–aggregate bond, 96
      paste–steel bond, 96–97
      scaling, 97–98
      strength, 95–96
      surface delamination, 99–100
      surface appearance, 100, 102
   effects on plastic concrete
      placing and finishing, 95
      plastic shrinkage, 93–95
      postbleeding expansion, 93
      thixotropic mixtures, 94
      volume change, 92–93
      water/cement ratio, 94
   fundamentals, 88, 90–91
   impermeable subbase effect, 106
   increasing, 108
   ingredient effects, 101–105
      aggregate, 104
      cement, 101–103
      chemical admixtures, 104–106
      supplementary materials, 102–104
      water content and water/cement ratio, 101–102
   mathematical models, 110
   placement size and height effects, 105–106
   planes of weakness due to, 129
   reducing, 108
   remixing effect, 107
   significance, 88
   special applications, 110
   test methods, 108–110
   uniform seepage, 90
   water/cement ratio and, 90–92
   weather conditions effects, 106–107
   zones, 90–91
Bleed water, 88–89
   removal, 108
Blisters, bleeding effect, hardened concrete, 99–101
BNQ 2560–070182, 393
Bogue calculations, portland cement, 463
Bond
   latex–modified portland cement, 582–583
   with reinforcing steel, 202–207
   strength
      cellular concrete, 536
      lightweight aggregate concrete, 531
Bonding agents, 599–600
   shotcrete, 594
Bond pullout pin test, setting time determination, 79–80
Boron, fine-particle products, in radiation shielding, 545

Break–off method, nondestructive tests, 333–334
Brickwork, as contaminant, recycled concrete, 360
BS 812, 393, 396
BS 1881, 253, 550
BS 5075, 79
Bulk flow, 250–251
Bulk modulus, temperature effect, 287

C

Calcium chloride, as accelerating admixture, 494
Calcium hydroxide, 270
   chemical resistance and, 264
   dehydration, 284
   leaching and deposition, 265
Calcium nitrate, 169
Calcium oxide, hydration, expansion due to, 224, 226
Calcium sulfoaluminate, 267
Calcium sulfate, reaction, 267–268
Calorimeter, 235
Capacitancy methods, water content analysis, 118
Capillary absorption tests, 253
Carbonate, addition to hydraulic cements, 471
Carbonate aggregates, fire resistance, 285
Carbonate rock (see also Alkali–carbonate rock reactivity)
   chemical composition and alkali–carbonate rock reactivity, 384–385
   expansive, composition, 377
   petrographic examination, 384
   reactive
      factors affecting expansion, 379–381
Carbonation, 275
   depth of, 252
   portland–cement paste, 173
   shrinkage, 220–222
Casting direction, effect on setting time, 129
Cast–in–place concrete shielding, 543–544
Cathodic protection, corrosion, 171
Cellular concrete, 533–539
   applications, 534–535
   classification, 533–534
   construction and application techniques, 535, 538–539
   density, 534
   mixture proportioning, 538
   physical properties, 536–538
   quality control, 539
Cement
   alkali level, alkali–silica reactivity and, 369–370
   brand, setting time and, 82
   carbonation behavior and, 221–222
   chemistry, accelerated curing and, 144–145
   effect on bleeding, 101–103
   fineness, heat of hydration and, 235
   latex–modified mixes, 580
   ready mixed concrete, 515
   roller–compacted concrete, 570
   type, setting time and, 82
   workability and, 53–54
Cement and Concrete Laboratory, 40
Cement content
   air-entraining admixture potential and, 487
   analysis, 112–117, 315–316
      absorption method, 113–114
      ASTM C 1078, 112–113
      ASTM C 1084 analysis, 317
      calculations and report, 317
      cement-type analysis, 317
      concrete consistency method, 117
      conductimetric method, 117
      constant neutralization method, 113
      Dunagan Bouyancy Method, 116
      flotation method, 115
      hydrometer analysis method, 116–117
      maleic acid analyses, 317
      sampling, 316–317
      nuclear cement gage, 116
      rapid analysis machine, 114–115
      thermal neutron activation, 116
      Willis and Hime Separation Method, 115–116
   dispersing agents, cellular concrete, 534
   fresh concrete, 52–53
   setting time and, 83, 86
Cement gel, 7–8
Cementing materials (see also Packaged, dry, cementitious mixtures)
   air-entraining admixtures, 487
   preplaced aggregate concrete, 562–563
   roller–compacted concrete, 569–570
Cement paste
   composition
      environmental influences, 215–216
      high temperature properties and, 284–285
   creep, 242
   drying shrinkage and, 222–223
   fresh
      rheology, 9
      structure, 6–7
   structure, 12
   thermal conductivity, 230–231
Central mixing, ready mixed concrete, 516–517
Centrifuge test, 51
Ceramic tile, thinsets and grouts, latex-modified, 586
Chemical admixtures, 487, 491–497
   accelerating, 494
   applications, 494–495
   effect on bleeding, 104–106
   freeze–protection admixture, 497
   high–range water-reducing, 493–494
   pumping aids, 496–497
   sampling, 19–20
   set-retarding and water-reducing, 492–493
   specifications, 495–496
   standards, 491–492
   workability and, 54–55
Chemical analysis, cement content, 315–316
Chemical contamination, recycled concrete, 360
Chemical degradation, moisture and, 173
Chemical reaction, lightweight aggregate concrete, 530–531
Chemical resistance (see also Corrosion, reinforcing steel; Sulfate resistance), 263–276
   acid attack, 272–274
   attack by other chemicals, 275–276
   calcium hydroxide and, 264
   carbonation, 275
   scaling, efflorescence, and leaching, 264–266
   seawater and brines, 274–275
   types of mechanisms, 263
   weathering processes, 264
Chert, 415, 417
Chloride
   induced corrosion, 164–167, 171
   in seawater, 274
   surface concentration, 255
Clay
   expanded, petrographic examination, 357
   lumps, 415–416
Clinker, impurities, 464
Coal, 416
Coatings, 602–603
   aggregate, 407–408, 418–419

artificially generated, 418
decorative, latex–modified, 586
definition, 411–412
effect on concrete, 418–419
latex, 603
naturally occurring, 418
skid–resistant, latex–modified, 586
synthetic–resin, 602–603
water–resistant basement, latex–modified, 586
Coefficient of thermal expansion, 439–440
Cohesion, air content and, 297
Compactibility, monitoring, roller–compacted concrete, 574–575
Compacting factor test, 59–60
Compaction, roller–compacted concrete, 573
Compressibility, changes and setting time, 78–79
Compressive strength, 10
abrasion resistance and, 183–184
aspect ratio, 128
cellular concrete, 536
diameter–to–aggregate ratio and, 128
hardened grout, 565
length of broken beam ends and, 128
loading direction versus casting direction and, 129
loading rate and, 130
lightweight aggregate concrete, 527–528
ready mixed concrete, 519–520
recycled concrete, 359
specimens
end conditions and, 126–127
geometry and, 130
moisture condition and temperature, 129
size and, 127–128
temperature effect, 286–287, 290
testing, 125–126, 130
testing machine properties and, 129–130
uniformity measurement, 53
Compressive stresses, modulus of elasticity, 193–194
Compressive toughness factor, 555
Compressometer, 193–194
Concrete
definition, 5
types, abrasion resistance and, 184
Concrete consistency method
cement content analysis, 117
water content analysis, 118
Concrete cores, corrosion, 171
Concrete–making materials, 31–37
definitions, 5
properties affecting concrete performance, 31–33
uniformity and, 31–33, 35–37
Concrete Materials Engineering Council, 40
Concrete prism expansion test, 382–383
Concrete products, prepackaged, 609–610
Conductimetric method, cement content analysis, 117
Consistency, 11–12
air content and, 487
concrete consistency method, 117–118
hydraulic cements, 451–452
measurement, setting time determination, 77–78
roller–compacted concrete, 571, 574–575
workability and, 54
Consolidation
degree of, unit weight, 309
effect on bleeding, 107
fresh fiber–reinforced concrete, 548
Constant neutralization method, cement content analysis, 113
Contaminants

detection, 343
recycled concrete, 360–361
Continuous mixing, ready mixed concrete, 518
Copper and copper alloys, corrosion, 176–177
Core and pullout test, 332–333
Corrosion, 164–171, 173–174
aluminum, 174–176
anti–corrosion agent, 169
asbestos, 180
assessing severity of, 170–171
cathodic protection, 171
chloride–induced, 164–167
chloride ion effect, 167–168
chloride samples, 171
concrete cores, 171
copper and copper alloys, 176–177
damage caused by, 166–167
half–cell potential surveys, 170
lead, 176
mechanisms, 164–166
precautionary steps against, 168–169
prestressed concrete, 169–170
products, 167
repairs to deteriorated structures, 171
steel–fiber–reinforced concrete, 179–180
zinc, 169, 177
Cracking
alkali–carbonate rock reactivity, 374–375
hardened fiber–reinforced concrete, 558–560
high temperatures and, 291–292
reinforcing steel bond effects, 207
research, 44–45
Cracking resistance, hardened fiber–reinforced concrete, 558–560
CRD–C 36, 232
CRD–C 37, 232
CRD–C 38, 235
CRD–C 44, 231
CRD–C 45, 231
CRD–C 124, 232
CRD–C 141, 394
CRD–C 242, 232
CRD–C 401, 476
Creep, 9–10, 195–198
cement paste, 242
equations, 196–197
gel water, 196
at high temperature, 287–288
lightweight aggregate concrete, 529
measurement, 196
mechanism, 243–244
paste content effect, 198
prediction equations, 198
principle of superposition, 197–198
rheological models, 197
significance of, 198–199
specifications and, 199–200
specimen size effect, 198
Crushed stone, petrographic examination, 352–354
CSA A23.2–14A, 382
Curing
abrasion resistance and, 185
accelerated, 140–147
apparatus, 141–142
cement chemistry effect, 144–145
experimental program, 140–141
high temperature and pressure method, 145–146
prediction of later–age strength, 146–147
results, 142–143
significance of test procedures, 143–144
test precision, 144
atmospheric conditions effect, 479–480

autogenous, 143
carbonation behavior and, 221
compounds, 481–482
effect on
concrete properties, 478–479
permeability, 256–258
future work, 482–483
lightweight aggregate concrete, 526–527
methods and materials for moisture retention, 480–481
new developments, 482
roller–compacted concrete, 574
shotcrete, 596
standard, 74
test methods and specifications, 481–482
test specimens
in field, 74
in laboratory, 75
water, 476–477
Cyclic loading, cellular concrete, 538

D

Damping properties, 322
Dams, roller–compacted concrete, 567–568
Darcy's law, 431
Darcy's permeability, 251
Deflection
reinforcing steel bond effects, 207
formula, 194–195
Deformation testing, cellular concrete, 538
Degradation
aggregates
coarse, 390–391
impact on specifications, 398
wet, 391–394
chemical, moisture and, 173
Dehydration, calcium hydroxide, 284
Deleterious substances, 415–418
chert, 415, 417
clay lumps, 415–416
coal and lignite, 416
definition, 411
effect on concrete, 416
friable particles, 415–416
material finer than 75 &mgr;m, 416–417
organic impurities, 417–418
soft particles, 417
tests, 416–418
Density
bleeding effect, hardened concrete, 96
cellular concrete, 534
hydraulic cements, 451
in–place, roller–compacted concrete, 575
preplaced aggregate concrete, 565
radiation shielding, 542–542
structural lightweight aggregate concrete, 527–528
Deposition, 264–266
Development length
lightweight aggregate concrete, 531
properties influencing, bond with reinforcing steel, 207
Diameter–to–aggregate ratio, effect on setting time, 128
Diatomite, petrographic examination, 358–359
Dilation methods, freezing and thawing, 159–161
DIN 1048, 253–254
Direct tension test, setting time, 131
Distress
development of, 284
manifestations, alkali–carbonate rock reactivity, 374–375
Dolomite, floating, 375, 377
Double–punch test, 131, 133

Dressing–wheel abrasion test machine, 187–188
Drying
  effects, 8–9
  rate, 173–174
  shrinkage, 222–224, 275
Dry–mix delivery equipment, shotcrete, 592
Dunagan Bouyancy Method, 50, 116, 118
Durability
  bleeding effect, hardened concrete, 97–98
  effect of cement paste capillary porosity, 478–479
  exposure to seawater and brines, 275
  hardened fiber–reinforced concrete, 560
  hydraulic cements, 458–460
  intrinsic, improvement, 264
  latex–modified portland cement, 582–583
  lightweight aggregate concrete, 529–530
  research needs, 44–45
  roller–compacted concrete, 574
Durability factor, 156–158, 436
Dynamic loading, hardened fiber–reinforced concrete, 557–558

E

Echo method, nondestructive tests, 325
Economics, significance of, sampling, 17
Efflorescence, 265
Elasticity, 9–10
  relative dynamic modulus, 156
Elastic properties, 192–195
  aggregates, 395
  lightweight aggregate concrete, 528–529
  modulus of elasticity, 193–195
  Poisson's ratio, 193, 195
  significance of, 198–199
  specifications and, 199–200
Electrical resistance
  measurement, 77
  water content analysis, 118–119
Embedded metals and materials, 173–180
  aluminum, 174–176
  asbestos, 180
  copper and copper alloys, 176–177
  fibers, 179
  glass, 178
  glass fibers, 180
  lead, 176
  organic materials, 180
  other metals, 178
  plastics, 179
  recycled concrete, 180
  steel fibers, 179–180
  wood, 178–179
  zinc, 169, 177
Engineered fills, cellular concrete, 535
ENV 197, 226
Environment
  alkali–silica reactivity and, 367
  effect on
    bleeding, 106–107
    cement paste composition, 215–216
    curing, 479–480
  volume change and, 219
Epoxy
  fusion–bonded, 169
  coating, effects on bond of bars, 205–206
  resin systems, 600–602
Ettringite, 267–268, 224
Evaluation, sampling, 26
Evaporation retardant, spray–applied, 480
Expansion
  reactive carbonate rock, 379–381
    mechanism, 381
  thermal, 284–285, 289

Expansive cement, volume change, 224–225, 227
Expansive dedolomization reaction, 373–374
Exterior insulation finish systems, latex–modified, 586

F

Fatigue strength, 135–137
Feldman and Sereda's model, porosity, 242–243
Fiber
  applications, 179
  chopped, cellular concrete, 534
  content, fresh fiber–reinforced concrete, 550–551
  glass, 180
Fiber–reinforced concrete, 547–560
  fresh, 548–551
  future, 560
  hardened, 551–560
    cracking resistance, 558–560
    durability, 560
    dynamic loading, 557–558
    fiber content and orientation, 551–552
    static loading, 552–557
    strengthening, 552–554
    tests for strength and toughness, 554–557
    toughening, 554
  nature of, 547–548
Fick's law, 249
Field concrete, deteriorated, 215–216
Finely divided material, workability and, 54
Fineness
  cement, setting time and, 82–83
  fly ash, 503–504
  hydraulic cements, 449–451
Fineness modulus, aggregates, 401–402
Finishing
  abrasion resistance and, 184–185
  bleeding effect, fresh concrete, 95
  lightweight aggregate concrete, 526–527
Fire
  damage, investigation and repair, 292–293
  endurance ratings, 283
  protection requirements, 282
  testing, 282–284
Fire resistance
  aggregate component effect, 285
  effect of embedded steel and structural systems, 285–286
  moisture effects, 292
Flexural bond stresses, 206
Flexural members, deflection, 198
Flexural strength
  factors affecting, 131–132
  relationship to splitting tensile strength, 133
  temperature effect, 287
  testing, 130–131
Flexural toughness factor, 555
Flexure, modulus of elasticity, 194–195
Floor fill, cellular concrete, 534–535
Floor resurfacers, prepackaged, 610
Flotation method, 115, 118
Flow cone, workability measurement, 60–61
Fluid penetration coefficient, 254
Fly ash
  alkali–silica reactivity, 370, 504
  benefits, 501
  carbonation behavior and, 221–222
  cellular concrete, 534
  cement paste thermal properties and, 284
  chemical composition, 468
  chemical requirements, 501–502
  classification, 501–505
  effect on bleeding, 102–103
  heat of hydration, 235

  physical requirements, 503–504
  preplaced aggregate concrete, 562–563
  quality assurance, 504–505
  sulfate resistance, 270–271
  supplementary optional requirements
    chemical, 502–503
    physical, 504
  test requirements not in specifications, 504
Freeze–protection admixture, 497
Freeze–thaw, 153–162
  carbonation behavior and, 221
  damage
    air voids and, 297–298
    mechanism, 297
    slow–freeze conditions, 298
  dilation methods, 159–161
  durability problems, 436
  historical evolution, 154–155
  lightweight aggregate concrete, 530
  petrographic examination, 216
  rapid tests, 156–159
  research needs, 45
  in salt solution, alkali–carbonate rock reactivity potential, 385
  salt water use, 159
  scaling resistance, 161
  standardized test methods, 154
  testing
    concrete, 414–415
    unconfined aggregates, 414
  theoretical considerations, 155–156
Freeze–thaw resistance
  aggregates, 298
  air content and, 297
  air voids and, 298, 486
  cellular concrete, 538
  recycled concrete, 359
Fresh concrete
  air content measurement, 65–68
  bleeding effects
    placing and finishing, 95
    plastic shrinkage, 93–95
    postbleeding expansion, 93
    thixotropic mixtures, 94
    volume change, 92–93
    water/cement ratio, 94
  cement content (see Cement content, analysis)
  determining air void characteristics, 488
  entrained air
    factors affecting amount and character, 486–488
    function, 485–486
  sampling, 20, 66
  temperature measurement, 65–66
  unit weight measurement, 65
  water content analysis, 52, 117–119
Friable particles, 415–416
  tests for, 416
Frictional properties, aggregates, 397–398
Frost (see Freeze–thaw)

G

Gas, permeability testing and standards, 252
Gas diffusion, 252
Glass
  as contaminant, recycled concrete, 360
  content, blast–furnace slag, 506
  fibers, reactivity, 180
  reactivity, 178
Glycol ether derivatives, 221
Goodman diagram, 136
Grading, aggregates, 401–405, 422
  definition, 401
  lightweight, 525

roller–compacted concrete, 570–571
significance, 402
specification, 405
test method, 401–402
Gravimetric method, unit weight of fresh concrete, 69
Grout
analysis, 317
ceramic tile, latex–modified, 586
fluid characteristics, preplaced aggregate concrete, 565
fluidifier, preplaced aggregate concrete, 563
mix proportions, preplaced aggregate concrete, 563–564
mixtures, cellular concrete, 538
surface monitoring, preplaced aggregate concrete, 564–565
Gypsum, 267

H

Half–cell potential surveys, corrosion, 170
Hardened cement paste
compressive strength, 10
drying effects, 8–9
elasticity and creep, 9–10
microstructure, 241–242
permeability, 10
porosity, 240–241, 244–245
properties, 13
thermal expansion, 10
Hardened concrete
air content, 296–300
bleeding effect, 95–102
blisters, 99, 101
density, 96
durability, 97–98
mortar flaking, 98–99
paste–aggregate bond, 96
paste–steel bond, 96–97
scaling, 97–98
strength, 95–96
surface appearance, 100, 102
surface delamination, 99–100
determining air void characteristics, 489
entrained air function, 485–486
petrographic examination, 211–212
aggregates, 347–348
placed by preplaced aggregate method, 565
porosity, 246
sampling, 20
specimens cut from existing structures, 125
unit weight, determining, 310–311
Hardening reactions, microstructure and, 7–8
Hardness
aggregates, 395–397
mixing water, effect on air content, 476
Heat flow, 236–237
Heat generation, importance of, 236
Heat of hydration, 234–236
hydraulic cements, 455
setting time and, 78
High–range water–reducing admixtures, 169, 493–494
High temperature and pressure method, accelerated curing, 145–146
History, of field concrete, reconstruction, 215
Hydration, 77
cement paste permeability and, 245
early reactions, 7
effects on permeability, 256–257
free CaO and MgO, expansion due to, 224
heat of, 234–236
portland cement, 464

Hydraulic cements (*see also* Portland cement), 449–461
activity index, 451
air content, 458–459
alkali–silica reactivity, 459
alkali sulfates, 471
blended, 467–468
carbonate additions, 471
chemical properties, 462–472
composition test methods, 468–470
consistency, 451–452
definition, 5
density, 451
durability, 458–460
fineness, 449–451
heat of hydration, 455
optical microscopy, 470
performance versus prescription standards, 471–472
physical properties test methods, 450
quantitative phase analysis, 470–471
quantitative X–ray diffraction, 470
sampling, 20
scanning electron microscopy, 470
setting time, 452–455
SO_3 content, 471
optimum, 458
strength, 456–458
sulfate attack, 459
trace elements, 471
user's concerns, 471
volume change, 455–456
Hydraulic pressure theory, 155
Hydrochloric acid, 273
Hydrometer analysis method, cement content analysis, 116–117

I

Image analysis techniques, air–void systems, 307
Impermeable subbase, effect on bleeding, 106
Impulse response method, nondestructive tests, 325
Impurities, effect on mixing water, 474–475
Industrial cinders, petrographic examination, 357
Insoluble residue, limits in portland cement, 465
Inspector, sampling as duty of, 17–18
Insulation, winter concreting, 238
Ionic diffusion, 254–255
Irradiation, effects on radiation shielding, 543
ISO Guide 49, 39
ISO Guide 25, 39

J

JSCE-SF4, 554–555, 557
JSCE-SF5, 554–555
JSCE-SF6, 554–555
JSCE-SF7, 551

K

Kelly ball test, 58–59
Kozeny equation, 431
K–slump tester, 55–57

L

Langlier Index, 265
Lapped splice, reinforcing steel bond, 205
Latex (*see also* Portland cement, latex–modified), 577
carbonation resistance, 221
cellular concrete, 534
coatings, 603
formulation, 579–580
modification mechanism, 578–579
permeability and, 169
prepackaged patching products, 610
shotcrete, 594
types, 577–578
Leaching, 264–266, 273
Lead, corrosion, 176
Lift surface, treatment, roller–compacted concrete, 573
Lightweight aggregate concrete, 522, 526–531
abrasion resistance, 531
bond strength and development length, 531
cellular, 533–534
chemical reaction, 530–531
classification, 522
compressive strength, 527–528
creep, 529
density, 527–528
durability, 529–530
elastic properties, 528–529
insulating–grade, 523
mixing, placing, finishing, and curing, 526–527
popouts, 531
proportioning, 526
shrinkage, 529
structural–grade, 522–523
structural/insulating, 523
tensile strength, 528
Lightweight aggregates, 522–526
absorption characteristics, 525
bulk unit weight, 524
classification, 522
grading, 525
insulating–grade, 523
internal structure, 523–524
modulus of elasticity, 525–526
petrographic examination, 357–359
shape and texture, 523–524
specific gravity, 524
structural–grade, 522–523
total porosity, 524–525
Lignite, 416
Lignosulfonate water–reducing retarders, 493
Lime, in fly ash, 503
Lime–saturated water, 265
Linear traverse method, air–void systems, 303–304
Linseed oil, 603
Liquid displacement techniques, porosity, 246–247
Lithium salts, alkali–silica reactivity and, 371
Loading
center–point versus third–point, effect on flexural strength, 132
direction, effect on setting time, 129
rate, effect on
flexural strength, 132
setting time, 130
splitting tensile strength, 133
Lorman equation, 197
Los Angeles abrasion, 390–391
Loss on ignition
fly ash, 502
portland cement, 465

M

Magnesium, sulfate reaction, 268–269
Magnesium oxide
in fly ash, 503
hydration, expansion due to, 224, 226
limits in portland cement, 465
Magnetite, 167
Maleic acid, cement content determination, 317

Mass concrete, roller-compacted concrete, transporting, 572
Mass transfer, 247–252
　adsorption, 247–248
　ionic diffusion, 254–255
Mathematical models, bleeding, 110
Maturity functions, 334–335
Maturity index, 148
Maturity method, 147–150
　application, 148–149
　interpretation of results, 144–150
　nondestructive tests, 334–335
　precautions, 150
　strength–maturity relationship, 148
Mechanical properties
　hardened fiber-reinforced concrete, 552–557
　lightweight aggregate concrete, 293
　moisture content effect, 290–291
　permeability and porosity and, 255–259
　temperature effects, 286
　very high strength concrete, 293
Mercury intrusion porosimetry, 247
Metallic contaminants, recycled concrete, 360
Microbars, autoclaved, 385
Micro-Deval test, 393
Microsilica, chemical composition, 468
Microstructure, hardening reactions and, 7–8
Microwave-absorption method, water content analysis, 119
Microwave oven, concrete separation, 118
Mineral admixtures, 500–507
　blast-furnace slag, 505–506
　definitions, 500
　pozzolans, 500–505
　sampling, 20
　silica fume, 507
Minerals, in radiation shielding, 542
Mixing
　action and air entrainment, 487–488
　equipment, shotcrete, 593
　lightweight aggregate concrete, 526–527
　proportioning (see Proportioning)
　ready mixed concrete, 516–517
　roller-compacted concrete, 572
　shotcrete, 596
Mixing water, 473–476
　algae effects, 475–476
　hardness and air content, 476
　impurity effect, 474–475
　ready mixed concrete, 515
　seawater, 475
Modeling, fire resistance, 283
Modified point count method, air-void systems, 304
Modulus of elasticity
　aggregates, 395
　cellular concrete, 536–537
　in compression, 193–194
　dynamic, 320–321
　　calculation, 321–322
　　factors affecting, 322
　　from pulse velocity, 324
　in flexure, 194–195
　lightweight aggregates, 525–526
　in shear, 195
　specifications and, 199
　structural design and, 198
　temperature effect, 287
　in tension, 194
Modulus of rigidity, dynamic, 322
Moisture
　availability, alkali-silica reactivity and, 367
　chemical degradation and, 173
　fire resistance and, 283, 292

flexural strength and, 132
mechanical and thermal performance and, 290–291
retention, curing methods and materials, 480–481
surface, aggregates, 425–426
test specimen, effect on setting time, 129
Moisture clog spalling, 292
Mortar
　analysis, 317
　flaking, bleeding effect, hardened concrete, 98–99
　latex-modified
　　mixes, 580
　　portland cement, 583, 586
　prepackaged products, 610
Mortar bar method, alkali-silica reactivity, 367–368
MTO LS-614, 393
MTC LS-615, 385
Munich model, porosity, 243–244

N

National Voluntary Laboratory Accreditation Program, 40–41
Neat-cement cellular concrete, 533
Neutron-scattering methods, water content analysis, 119
Nondestructive tests, 320–336
　break-off method, 333–334
　combined methods, 335–336
　maturity method, 334–335
　pin penetration test, 331
　probe penetration test, 327–331
　pulloff tests, 335
　pullout test, 332–333
　pulse velocity method, 323–325
　rebound method, 325–327
　resonant frequency methods, 320–323
　stress wave propagation methods, 325
Nuclear cement gage, cement content analysis, 116
Nuclear models, 541
Nuclear shielding (see Radiation shielding)
Nurse-Saul maturity function, 334

O

Optical microscopy, hydraulic cements, 470
Organic impurities, tests for, 417–418
Organic materials, 180
　adhesives, patching, and overlaying materials, 600–602
　bonding and patching materials, 599–600
　coatings, 602–603
　sealers, 603
Otto Graf viscosimeter, 61–62
Overlaying materials, 600–602

P

Packaged, dry, cementitious mixtures, 604–610
　applications, 608
　availability, 608
　future needs, 609–610
　hazardous considerations, 604–605
　history, 604
　materials, 609
　packaging, 609
　quality control, 608–609
　specifications, 605–607, 609
　use, 607–608
Paint, latex-modified cement maintenance, 586
Particles, strength, concrete aggregates, 11
Paste-aggregate bond
　bleeding effect, hardened concrete, 96

permeability, 246
Paste-steel bond, bleeding effect, hardened concrete, 96–97
Patching materials, 599–602
　latex-modified, prepackaged, 610
Pavements
　polymer-modified concrete, 586–587
　roller-compacted concrete, 568–569
　　placement, 573–574
　　transporting, 572–573
　wear, 397–398
Penetration resistance method, setting time determination, 78–79
Perlite, petrographic examination, 358
Permeability, 10, 435–436
　admixtures and, 256
　aggregate, 246
　cellular concrete, 537
　cement paste, hydration and, 245
　concrete aggregates, 11
　correlation with strength, 259
　corrosion and, 168–169
　effect of curing and conditioning, 256–258
　factor in seawater attack, 274
　hydration and, 256–257
　latex-modified portland cement, 583–585
　paste-aggregate interface, 246
　pore structure and, 240–241
　porosity and, 255–259
　testing and standards, 252–255
　　gas, 252
　　ionic diffusion, 254–255
　　water, 253–254
　unit weight and, 309
　water/cement ratio and, hardened cement paste, 245, 255–257
Permeability coefficient, intrinsic, 252
Petrographic examination, 210–217
　age of concrete, 214
　aggregates, 341–362
　　alkali reactivity, 348
　　contamination detection, 343
　　correlation of samples with aggregates previously tested or used, 342
　　determining processing effects, 343
　　establishing properties and probable performance, 342
　　examination in field, 343, 348–349
　　examination in laboratory, 343–345, 349–352
　　hardened concrete, 347–348
　　lightweight, 357–359
　　natural, 348–352
　　observations included, 345–346
　　particle condition, 346–347
　　potentially reactive, identifying, 368
　　preliminary determination of quality, 341–342
　　proportions of coarse and fine aggregates, 347–348
　　purpose, 341–343
　　samples, 343
　　selecting and interpreting tests, 342
　approach, 213
　blast-furnace slag, 354–357
　cement content, 316
　communications, 211
　comparisons, 214
　composition and, 215
　crushed stone, 352–354
　dolomitic carbonate rocks, 384
　fabric and composition, 213–214
　freezing and thawing, 216
　interpretation of observations, 214

methods, 211–212
normal and unusual concrete, 215
purpose, 213
reconstruction of history of field concrete, 215
recycled concrete, 359–362
for soundness, 415
sources of concrete, 214–215
Petrography, 210–211
Pin penetration test, nondestructive tests, 331
Placeability, workability, measurement, 57
Placement
bleeding effect, fresh concrete, 95
lightweight aggregate concrete, 526–527
at low temperature, 238
roller–compacted concrete, 573–574
size and height, effect on bleeding, 105–106
Plastic, resistance to strong alkalies, 179
Plasticity, aggregate structure and, 12
Plastic properties
latex–modified portland cement, 581
shrinkage, fresh concrete, 93–95
Poisson's ratio, 193, 195
aggregates, 395
calculation, 321–322
temperature effect, 287
Polishing, 396
Polymers, 180
abrasion resistance and, 184
carbonation behavior and, 221
shotcrete, 594
Popouts, lightweight aggregate concrete, 531
Pore size
measurement, 433–434
parameters, 430–431
Pore space, quantitative assessments, 429
Pore structure, permeability and, 240–241
Pore system, coarse aggregates, 429–437
concepts and definitions, 429–431
measurement of other parameters, 435
pore size, 430–431, 433–434
pore volume, 430, 432–433
significance, 435–437
Pore volume
measurement, 432–433
parameters, 430
Porosity, 240–247, 433
aggregates, 246
bulk flow, 250–251
concrete aggregates, 11
Feldman and Sereda's model, 242–243
hardened cement paste, 244–245
microstructure, 241–242
in hardened concrete, 246
lightweight aggregates, 524–525
measurement methods, 246–247
mixing–and–placing, 246
Munich model, 243–244
paste–aggregate interface, 246
permeability and, 255–259
Powers' model, 241–242
surface diffusion, 248–249
vapor diffusion, 249–250
water/cement ratio, 297
Portland cement (see also Latex), 462–465
ASTM C 150 chemical requirements, 465–467
autogenous shrinkage, 219–220
Bogue calculations, 463
chemical composition, 462–463
definition, 5
hydration reactions, 464
impurities in clinker phases, 464
insoluble residue limits, 465
latex–modified, 577–587
air void system, 581–582
applications, 583, 586–587
bond, 582–583
durability, 582–583, 583
equipment, 587
latex types, 577–578
limitations, 587
mechanism, 578–579
plastic properties, 581
properties, 580–583
specifications, 587
strain capacity, 583
loss on ignition, 465
magnesium oxide limits, 465
major phases and cement properties, 464–465
microstructure, 241–242
optional chemical requirements, 467
paste
carbonation, 173
microstructure, 7–8
phase composition, 463
slag activity test, 506
SO_3, 466
sulfate resistance, 269–270
Type I, ASTM C 150 chemical requirements, 466
Type II, ASTM C 150 chemical requirements, 466–467
Type III, ASTM C 150 chemical requirements, 467
Type IV, ASTM C 150 chemical requirements, 467
Type V, ASTM C 150 chemical requirements, 467
volume change, tests, 226, 226–227
Postbleeding expansion, fresh concrete, 93
Powers' model, 241–242
Power's spacing factor, 301–302
Pozzolans (see also Fly ash), 500–505
alkali–silica reactivity and, 370
chemical composition, 468
effect on bleeding, 104
heat of hydration, 235
history and use, 500–501
natural, 505
ready mixed concrete, 515
roller–compacted concrete, 570
shotcrete, 594
sulfate resistance, 270–271
Precast cellular concrete elements, 535
Precast concrete shield blocks, 543–544
Precision, ASTM C 403, 82
Preplaced aggregate concrete, 562–565
aggregates, 563
cementing materials, 562–563
grout fluidifier, 563
grout mix proportions, 563–564
radiation shielding, 544
specifications, 564–565
Pressure air measurement, 66–68
Prestress, loss of, 198
Prestressed concrete
corrosion, 169–170
fire tests, 292
reinforcing steel bond, 206–207
Principle of superposition, creep, 197–198
Probability sampling, 21
Probe penetration test
advantages and disadvantages, 331
nondestructive tests, 327–331
Production, uniform concrete, 50–53
Proportioning, 11–12
abrasion resistance and, 183–184
cellular concrete, 538
grout, preplaced aggregate concrete, 563–564

lightweight aggregate concrete, 526
roller–compacted concrete, 571–572
shotcrete, 595–596
workability and, 55
Pulloff tests, 335
Pullout test, 332–333
Pulse velocity method, 323–325
Pumice, petrographic examination, 357–358
Pumping aids, 496–497

Q

Quality assurance
fly ash, 504–505
roller–compacted concrete, 574–575
shotcrete, 596
Quality control
cellular concrete, 539
packaged, dry, cementitious mixtures, 608–609
roller–compacted concrete, 574–575
sampling and, 18
slag, 506
Quantitative phase analysis, hydraulic cements, 470–471
Quantitative X–ray diffraction, hydraulic cements, 470
Quick Chemical Test, 368

R

Radiation, 540–541
exposure effects on concrete, 544–545
Radiation shielding, 293, 540–545
absorption and scattering interaction, 543
concrete, 542–543
construction verification, 545
fine–particle boron products, 545
optimization, 545
physical and biological explanation, 541
research, 544–545
size and economics, 543
standards, 543–544
Subcommittee C09.41, 540
terminology, 541–542
unit weight and, 310
Radioactivity, recycled concrete, 360
Rapid analysis machine, cement content analysis, 114–115
Rapid mortar–bar expansion test, alkali–silica reactivity, 368–369
Ready mixed concrete, 511–520
admixtures, 515
aggregates, 515
basis of purchase, 512
batching and measuring materials, 514–515
batching plant, 515–516
cement and pozzolans, 515
compressive strength, 519–520
history, 511
materials, 513–514
mixing, 516–517
mixing water, 515
ordering information, 512–513
recorders, 516
sampling and testing, 518–520
slump and air content, 514
specifications, 511–512
volume in–place, 512
volumetric batching and continuous mixing, 518
water addition control, 517–518
water/cement ratio, 513
water quality, 514
yield, 512
Rebound hammer test, 325–327, 392

Recycled concrete, 180
  admixtures, 359
  compressive strength, 359
  field examination, 361
  freeze–thaw resistance, 359
  petrographic examination, 359–362
  unsound, 359–360
Refractory concrete, thermal properties, 293–294
Refractory shotcrete, 591
Regression lines, 24–25
Reinforcing, shotcrete, 594
Reinforcing steel (*see also* Corrosion)
  bond with, 202–207
    epoxy coating effects, 205–206
    influence in control of cracking and deflections, 207
    lapped splices, 205
    prestressed concrete, 206–207
    properties influencing development length, 204–205
    splitting failures, 205
    tests, 202–204
  coating, 169
  fire resistance and, 285–286
Relaxation length, 542
Remixing, effect on bleeding, 107
Repairs, deteriorated structures, 171
Research needs
  basic research, 42–43
  reappraisal of past research, 43
  training of researchers, 43
  use of standard tests, 44–45
Resonant frequency methods, nondestructive tests, 320–323
Retarding admixtures (*see* Set-retarding admixtures)
Revibration, effect on bleeding, 107
Revolving–disk abrasion test machine, 186–187
Rheological models, creep, 197
Rheology, fresh cement paste, 9
Rice husk ash, effect on bleeding, 104
Ring anode effect, 171
Rock cylinder expansion test, 383–384
Rock prisms, 385
Roller–compacted concrete, 567–575
  admixtures, 571
  advantages, 569
  aggregates, 570–571
  applications, 568–569
  batching and mixing, 572
  cement, 570
  cementitious materials, 569–570
  curing, 574
  definition, 567
  hardened, properties, 574
  history, 567–568
  mixture proportioning, 571–572
  placement, 573–574
  pozzolans, 570
  quality control and assurance, 574–575
  transporting, 572–573
Roof decks, cellular concrete, 535
Rotating–cutter drill press, 188–189
Rust, 167

S

Salt attack, 276
Salt scaling, 276
Samples (*see also* Test specimens)
  aggregates for petrographic examination, 343
  compositing, 18
  definition, 15
  for field, 73
  obtaining, 16–17
  preparation, cement analysis, 316–317
  protection, 18
  selection, cement analysis, 316
  variability, recognition of, 21
Sampling, 15–22
  admixtures
    chemical, 19–20
    mineral, 20
  aggregates, 19
  for alkali–carbonate rock reactivity, 385
  ASTM standards, 16, 21
  definition, 15–16
  as duty of inspector, 17–18
  economic significance, 17
  evaluation, 26
  fresh concrete, 20, 66
  fresh fiber-reinforced concrete, 548
  hardened concrete, 20
  hydraulic cement, 20
  plan, 16
  probability, 21
  procedural sources of uncertainly and variability, 306
  quality control and, 18
  ready mixed concrete, 519
  statistics, 24–26
  stratified random, 25–26
  test specimens (*see* Test specimens)
  trained personnel, 21
Sand, workability and, 54
Sandblasting, abrasion testing, 186
Sanded cellular concrete, 533
Saturated flow, water, 254
Saturation, degree of, freezing and thawing tests, 159
Scaling, 156, 265
  bleeding effect, hardened concrete, 97–98
  resistance, 161
  salt, 276
Scanning electron microscopy, hydraulic cements, 470
Scattering cross section, 542–543
Schmidt rebound hammer, 325–327
Scoria, petrographic examination, 357–358
Sealers, 169, 171, 603
  abrasion resistance and, 185
  for curing, 481
Seawater, 274–275
  as mixing water, 475
Selective dissolution, hydraulic cements, 469
Set-accelerating admixtures, setting time and, 84
Set-retarding admixtures, 492–493
  applications, 495
  roller–compacted concrete, 571
  setting time and, 84
Setting shrinkage, fresh concrete, 93–95
Setting time, 77–87
  aspect ratio and, 128
  bleeding characteristics and, 78
  bond pullout pin test, 79–80
  casting direction effect, 129
  cement brand and type, 82
  cement content, 83, 86
  compressabilty changes and, 78–79
  compressive, 125–126
    test result significance, 130
  concrete mixtures and, 81–82
  concrete under combined states of stress, 134–135
  consistency and, 77–78
  diameter–to–aggregate ratio effect, 128
  direct tension test, 131
  double–punch test, 131, 133
  fatigue strength, 135–137
  fineness of cement and, 82–83
  flexural, 130–132
  future considerations, 86–87
  heat of hydration, 78
  history, 77–78
  hydraulic cements, 452–455
  loading direction effect, 129
  loading rate effect, 130
  moisture effect, 129
  obtaining test specimens, 124–125
  penetration resistance method, 78–79
  preplaced aggregate concrete, 565
  purposes, 123
  set–retarding and –accelerating admixtures, 84
  shear and torsional strength, 134
  sound velocity and frequency, 78
  splitting tensile strength, 131–133
  strength relationships, 78, 133
  temperature and, 83–84, 129
  tensile strength, 130–131, 133
  tests
    acceptance specifications, 84–85
    field concreting, 85
  test specimens, 124–125
  treatment of data, 80–81
  volume change and, 78
  water/cement ratio, 83, 85
Sewer lines, acid attack, 273
Shale, expanded, petrographic examination, 357
Shape, 405–406
  aggregates, 421–422
  lightweight aggregates, 523–524
Shear, modulus of elasticity, 195
Shear strength, 134, 536
Shotcrete, 589–597
  ACI recommended practices, 590
  admixtures, 594
  aggregates, 593
  air compressors, 592–593
  ancillary equipment, 593
  applications, 592
  ASTM standards, 590–591
  batching and mixing, 596
  binders, 593
  bonding agents, 594
  curing, 596
  definition, 589
  dry–mix delivery equipment, 592
  history, 590, 591–592
  latex–modified, 586
  mixing equipment, 593
  mixture proportioning, 595–596
  polymers, 594
  pozzolans, 594
  prepackaged, 610
  properties, 590
  quality assurance, 596
  refractory, 591
  reinforcing, 594
  terminology, 589–590
  testing, 597
  water, 593–594
  wet–mix delivery equipment, 592
Shrinkage
  autogenous, portland cement, 219–220
  carbonation, 220–222
  drying, 222–224, 275
  lightweight aggregate concrete, 529
  mechanism, 243–244
  plastic, 93–95
Shrink mixing, ready mixed concrete, 517
Silanes, 169, 171, 603
Silica, acid-soluble, 316
Silica fume, 169, 507

abrasion resistance and, 184
alkali–silica reactivity and, 371
cement paste thermal properties and, 284
chemical requirements, 507
drying shrinkage and, 223–224
effect on bleeding, 104
history and use, 507
physical requirements, 507
sulfate resistance, 270–271
Siliceous aggregates, fire resistance, 285
Slag, blast-furnace, 505–506
   activity test, portland cement, 506
   advantages, 505
   blast-furnace, 354–355
     alkali–silica reactivity, 370–371
     chemical composition, 468
   carbonation behavior and, 222
   cement paste thermal properties and, 284
   classification, 505–506
   effect on bleeding, 103
   expanded, petrographic examination, 357
   glass content, 506
   history and use, 505
   petrographic examination, 354–357
   sulfate resistance, 270–271
Slate, expanded, petrographic examination, 357
Slump, ready mixed concrete, 514
Slump test, 52, 55
Slurry mixtures, cellular concrete, 538
SO_3
   in blast-furnace slag, 506
   in fly ash, 502
   hydraulic cement content, 471
   limit in portland cement, 466
   optimum
     hydraulic cements, 458
     portland cement, 466
Sodium sulfate, reaction, 268
Soft particles, tests for, 417
Soils compaction, roller-compacted concrete, 571–572
Soniscope, 323–324
Sound, velocity and frequency measurements, setting time, 78
Soundness, 412–415
   absorption tests, 414
   definition, 411
   fly ash, 503
   freeze–thaw testing
     concrete, 414–415
     unconfined aggregate, 414
   petrographic examination, 415
   recycled concrete, 359–360
   sulfate soundness test, 412–414
Spalling, high temperatures and, 291–292
Specific gravity, aggregates, 11, 424–425
   lightweight, 524
Specific heat, 231–232
   aggregates, 442
   determining, 288–289
Specific surface, air-void system, 300–301
Splitting tensile strength
   factors affecting, 132–133
   relationship to flexural strength, 133
   setting time, 131
Standardization
   break-off test, 333–334
   maturity method, 335
   penetration resistance techniques, 331
   pullout tests, 333
   pulse velocity method, 324–325
   rebound test method, 327
   resonant frequency methods, 322

Standards
   aggregates, 389
   alkali–carbonate rock reactivity, 372–373
   chemical admixtures, 491–492
   hydraulic cement, 471–472
   permeability, 252–255
   radiation shielding, 543–544
   sampling, 16, 21
   shotcrete, 590–591
   uniformity, guidelines for future, 35–37
Static loading, hardened fiber-reinforced concrete, 552–557 Statistics
   parameters, 23–24
   precision and bias statements, 26–27
   regression lines, 24–25
   testing, 26
Steel-fiber-reinforced concrete, corrosion, 179–180
Steel fibers, in fiber-reinforced concrete, 551
Stiffness, determination and setting time, 78
Strain capacity, latex-modified portland cement, 583, 585
Strength (*see also* Compressive strength; Tensile strength)
   aggregates, 388–389, 394–395
   bleeding effect, hardened concrete, 95–96
   cement paste capillary porosity and, 478–479
   correlation with permeability, 259
   determination and setting time, 78
   effects of algae in mixing water, 475–476
   factors affecting, 71
   fracture mechanics, 44
   hydraulic cements, 456–458
   nature of, 123–124
   predicting at later ages, 140–151
     accelerated curing methods, 140–147
     maturity method, 147–150
     rebound hammer test, 326–327
   reduction, air content and, 297
   relationships, 133
   testing, 123–137
   thermal cycling effect, 290
Strengthening, hardened fiber-reinforced concrete, 552–554
Strength–maturity relationship, 148
Stress
   calculations, rheological behavior and, 198–199
   combined states, 134–135
   levels and fire resistance, 283
Stress–strain curve, 192–193
Stress wave propagation methods, nondestructive tests, 325
Structural design, elastic properties, 198
Structure, cement paste, 12
Stucco products, prepackaged, 610
Sulfate attack, 266–271
   calcium sulfate, 267–268
   control, 269–271
   factors governing reactions, 266
   hydraulic cements, 459
   magnesium sulfate, 268–269
   sodium sulfate, 268
Sulfate resistance, 266–272
   sulfate attack, 266–269
   control of, 269–271
   tests, 271–272
Sulfate soundness test, 412–414
Sulfuric acid, 273
Supplementary cementing materials, effect on bleeding, 102–104
Surface
   appearance, bleeding effect, hardened concrete, 100, 102

   delamination, bleeding effect, hardened concrete, 99–100
   diffusion, 248–249
   hardness
     pin penetration test, 331
     probe penetration test, 327–331
     rebound method, 325–327
   treatment, abrasion resistance and, 185
Sweating (*see* Bleeding)
Swelling, mechanism, 243–244

T

Temperature
   air-entraining admixtures, 488
   effect on
     compressive strength, 286–287, 290
     creep, 287–288
     flexural strength, 132, 287
     mechanical properties, 286
     modulus of elasticity, Poisson's ratio and bulk modulus, 287
     radiation shielding, 543
   high
     coupled with blast, 294
     spalling and cracking, 291–292
   measurement, fresh concrete, 65–66
   setting time and, 83–84, 129
   at time of grout injection, preplaced aggregate concrete, 564
Tensile strength
   cellular concrete, 536
   lightweight aggregate concrete, 528
   splitting, setting time, 131
   test
     procedures, 130–131
     result significance, 133
Tension, modulus of elasticity, 194
Testing (*see also* specific materials)
   concerns, 38
   statistics, 26
   trends in promoting quality, 38–40
Testing laboratories
   evaluating authorities, 40–41
   quality promotion trends, 38–40
   technician competency, 41
   testing concerns, 38
Testing personnel, test specimens, 71–72
Test specimens (*see also* Samples), 71–76
   air-void systems, 302
   applications, 71–72
   beam-end, 204
   compressive strength
     aspect ratio, 128
     diameter-to-aggregate ratio, 128
     end condition effect, 126–127
     geometry effect, 130
     length of broken beam ends effect, 128
     loading rate effect, 130
     loading versus casting direction, 129
     moisture condition and temperature effect, 129
     size effect, 127–128
   consolidation, 75
   cut from existing structures, 125
   dimension
     effect on flexural strength, 131
     effect on splitting tensile strength, 132–133
   making and curing
     in field, 72–74
     in laboratory, 75
   materials conditioning and testing, 75
   mixing and testing, 75
   moisture
     effect on splitting tensile strength, 133

and temperature effect on flexural strength, 132
molded, 124–125
preparation, roller–compacted concrete, 575
sampling, 72
setting time, 124–125
size, 73
   creep effect, 198
   effect on flexural strength, 131–132
   effect on splitting tensile strength, 132–133
specifications, for molds, 72
test data, 73
testing personnel, 71–72
Texture, aggregates, 406–408, 421–422
   lightweight, 523–524
Thaulow concrete tester, 59
Thaumasite, 267
Thawing (see Freezing–thawing)
Thermal coefficient, epoxy system, sand aggregate–binder ratio, 601
Thermal conductivity, 229–231
   aggregates, 441–442
   cellular concrete, 537
   determining, 288–289
Thermal conductivity methods, water content analysis, 118
Thermal cycling, 289–290
Thermal diffusivity, 232–233
   aggregates, 442
   determining, 288–289
Thermal expansion, 10, 233–234, 284–285, 289
   aggregates
      determination, 440–441
      coefficient, 439–440
Thermal incompatibilities, 289
Thermal neutron activation, cement content analysis, 116
Thermal properties, 229–238
   aggregates, 438–443
   cement paste component effects, 284–285
   concepts, 438–439
   conductivity, 229–231
   determining, 288–289
   diffusivity, 232–233
   expansion, 233–234
   heat flow, 236–237
   heat generation and temperature rise, 236
   heat of hydration, 234–236
   heavyweight aggregate concrete, 293
   lightweight aggregate concrete, 293
   moisture content effect, 290–291
   refractory concrete, 293–294
   specific heat, 231–232
   test methods, 231
   unit weight and, 309–310
   very high strength concrete, 293
   volume and length changes, 237–238
   winter concreting and insulation, 238
Thermal stresses, 292
Thermal volume change, 289
Thinsets, ceramic tile, latex–modified, 586
Thixotropic mixtures, bleeding effects, fresh concrete, 94
Torsional strength, 134
Toughening, hardened fiber–reinforced concrete, 554
Trace elements, hydraulic cements, 471
Transporting
   roller–compacted concrete, 572–573
   test specimens, 74
Truck mixing, ready mixed concrete, 517
Tuff, petrographic examination, 357–358
Tuthill–Cordon test procedure, 78–79
Two–point workability test, 61–63

Type K cement, drying shrinkage, 224–225

U

Uniformity, 31–33, 35–37
   ASTM C 917, 33, 35
   batch–to–batch variations, 50
   guidelines for future standards, 35–37
   measurement, 50–53
   production, 50–53
   unit weight and, 309
   within–batch variations, 50
Uniform seepage, 90
Unit weight, 308–311, 422–423
   aggregate density effect, 311
   air content and, 297, 311
   definition, 422–423
   degree of consolidation, 309
   fresh fiber–reinforced concrete, 550
   hardened concrete, determining, 310–311
   inferring batch weights and composition, 310
   lightweight aggregates, 524
   measurement
      fresh concrete, 65
      gravimetric method, 69
   paste content effect, 311
   permeability, 309
   significance, 308–311
   thermal, acoustic, and nuclear shielding properties, 309–310
   typical values, 310
   uniformity and, 309
   voids content, 309
Unit weight test, 51
U.S. Army Corp of Engineers, method of test for concrete mixer performance, 51
USBR 4907, 232
USBR 4908, 270, 272
USBR 4909, 232
USBR 4910, 234
USBR 4911, 236
U.S. Bureau of Reclamation, test of mixer performance, 50

V

Vapor diffusion, 249–250
Vebe apparatus, 59
Vermiculite, exfoliated, petrographic examination, 358
Vibration, air void expulsion, 488
Volcanic cinders, petrographic examination, 357–358
Volume change (see also Expansion), 219–227
   autogenous, 219–220
   bleeding, fresh concrete, 92–93
   carbonation shrinkage, 220–222
   drying shrinkage, 222–224
   expansion due to hydration of free CaO and MgO, 224, 226
   expansive cements, 224–225
   hydraulic cements, 455–456
   roller–compacted concrete, 574
   setting time and, 78
   test methods, 225–227
   thermal, 289
   thermal properties and, 237–238
Volumetric method, air content of fresh concrete, 68–69

W

Washington degradation test, 392
Waste materials, cellular concrete, 534
Water
   absorption, 253
   aggregates, 423–424
   cellular concrete, 537–538
      measurement validity, 426–428
   addition to ready mixed concrete, 517–518
   classification, in gel, 243
   in concrete, 5–6
   content
      analysis, fresh concrete, 52
      effect on bleeding, 101–102
   curing, 476–477
   mixing (see Mixing water)
   municipal, analysis, 474
   penetration, 253–254
   permeability testing and standards, 253–254
   quality, ready mixed concrete, 514
   in shotcrete, 593–594
   thermal conductivity, 230
Water/cement ratio
   air content and, 487
   bleeding effect, 90–92, 94, 101–102
   drying shrinkage and, 223
   permeability and, 168–169
      hardened cement paste, 245, 255–257
   porosity, 297
   ready mixed concrete, 513
   setting time and, 83, 85
Water content
   analysis, 117–119
      ASTM C 1079, 117
      capacitance methods, 118
      concrete consistency method, 118
      Dunagan Bouyancy Method, 118
      electrical resistance, 118–119
      flotation method, 118
      microwave–absorption method, 119
      microwave oven separation method, 118
      neutron–scattering methods, 119
      thermal conductivity methods, 118
   determination, 318
Water gain (see Bleeding)
Water–reducing admixtures, 492–493
   applications, 494–495
   high–range, 493–494
   roller–compacted concrete, 571
Wear, aggregates, 396–398
Weather (see Environment)
Weathering processes (see also Freezing–thawing), 161–162
   effect on acrylic–modified mortar adhesion, 582
   salt, 276
Weeping (see Bleeding), 88
Weight loss, 158
Wet abrasion tests, aggregates, 391–394
Wet–mix delivery equipment, shotcrete, 592
Wigmore consistometer, 59–60
Willis and Hime Separation Method, cement content analysis, 51, 115
Windsor probe test system, 328–329
Wood, problems incidental to use, 178
Workability, 49–63
   air content and, 51, 297
   air entrainment and, 54
   cement and, 53–54
   chemical admixtures and, 54–55
   coarse aggregate and, 54
   consistency and, 54
   definition, 53
   fiber effect, 547
   finely divided material and, 54
   fresh fiber–reinforced concrete, 548–550
   measurement, 55–63
      ball penetration test, 58–59
      compacting factor, 59–60
      flow cone, 60–61

grout consistency meter, 60–62
K-slump tester, 55–57
Otto Graf viscosimeter, 61–62
placeability, 57
rational measure, 44
remolding test, 58
slump test, 55
Thaulow concrete tester, 59
two-point workability test, 61–63
Vebe apparatus, 59
Wigmore consistometer, 59–60
mixing-and-placing porosity, 246
mixture proportions and, 55
recycled concrete aggregates, 180
sand and, 54
terminology, 49

X

X-ray fluorescence, hydraulic cements, 469

Y

Yield
air content and, 297, 311
fresh fiber-reinforced concrete, 550
ready mixed concrete, 512

Z

Zinc, corrosion, 169, 177